SECOND
EDITION

Discrete and Combinatorial Mathematics

AN APPLIED INTRODUCTION

RALPH P. GRIMALDI

ROSE–HULMAN INSTITUTE OF TECHNOLOGY

 ADDISON-WESLEY PUBLISHING COMPANY

Reading, Massachusetts · Menlo Park, California · New York
Don Mills, Ontario · Wokingham, England · Amsterdam · Bonn
Sydney · Singapore · Tokyo · Madrid · San Juan

Sponsoring Editor: Charles B. Glaser
Production Supervisor: Jack Casteel
Text Designer: Marie McAdam
Copy Editor: Connie Day
Illustrator: Illustrated Arts
Art Consultant: Loretta Bailey
Manufacturing Supervisor: Roy Logan
Cover Design: Marshall Henrichs

Library of Congress Cataloging-in-Publication Data
Grimaldi, Ralph P.
  Discrete and combinatorial mathematics : an
applied introduction / by Ralph P. Grimaldi.—2nd ed.
    p. cm.
  Includes index.

    1. Mathematics—1961– 2. Electronic data processing—
Mathematics. 3. Algebra. I. Title
QA39.2.G7478 1989
510—dc19   88-15398
                                                 CIP
ISBN 0-201-11954-4

Reprinted with corrections, June 1989

6 7 8 9 10 DO 9594939291

NOTATION

RELATIONS	$A \times B$	the cross, or Cartesian, product of sets A, B
	$\mathcal{R} \subseteq A \times B$	$\mathcal{R}$ is a relation from A to B
	$a \, \mathcal{R} \, b; \; (a, b) \in \mathcal{R}$	a is related to b
	$a \, \not\mathcal{R} \, b; \; (a, b) \notin \mathcal{R}$	a is not related to b
	$\mathcal{R}^c$	the converse of relation $\mathcal{R}$: $(a, b) \in \mathcal{R}$ iff $(b, a) \in \mathcal{R}^c$
	$\mathcal{R} \circ \mathcal{S}$	the composite relation for $\mathcal{R} \subseteq A \times B$, $\mathcal{S} \subseteq B \times C$: $(a, c) \in \mathcal{R} \circ \mathcal{S}$ if $(a, b) \in \mathcal{R}$, $(b, c) \in \mathcal{S}$ for some $b \in B$
	$\mathrm{lub}\,\{a, b\}$	the least upper bound of a and b
	$\mathrm{glb}\,\{a, b\}$	the greatest lower bound of a and b
	$[a]$	the equivalence class of element a (relative to an equivalence relation $\mathcal{R}$ on a set A): $\{x \in A \mid x \, \mathcal{R} \, a\}$
FUNCTIONS	$f : A \rightarrow B$	f is a function from A to B
	$f(A_1)$	for $f : A \rightarrow B$ and $A_1 \subseteq A$, $f(A_1)$ is the image of A_1 under f, i.e., $\{f(a) \mid a \in A_1\}$
	$f(A)$	for $f : A \rightarrow B$, $f(A)$ is the range of f.
	$f : A \times A \rightarrow A$	f is a binary operation on A
	$1_A : A \rightarrow A$	the identity function on A: $1_A(a) = a$ for each $a \in A$
	$f|_{A_1}$	the restriction of $f : A \rightarrow B$ to $A_1 \subseteq A$
	$g \circ f$	the composite function for $f : A \rightarrow B$, $g : B \rightarrow C$
	f^{-1}	the inverse of function f
	$f^{-1}(B_1)$	the preimage of $B_1 \subseteq B$ for $f : A \rightarrow B$
	$f \in O(g)$	f is "big-Oh" of g; f is of order g
THE ALGEBRA OF STRINGS	Σ	a finite set of symbols called an alphabet
	λ	the empty string
	$\|x\|$	the length of string x
	Σ^n	$\{x_1 x_2 \ldots x_n \mid x_i \in \Sigma\}$, $n \in \mathbf{Z}^+$
	Σ^0	$\{\lambda\}$
	Σ^+	$\bigcup_{n \in \mathbf{Z}^+} \Sigma^n$: the set of all strings of positive length
	Σ^*	$\bigcup_{n \geq 0} \Sigma^n$: the set of all finite strings
	$A \subseteq \Sigma^*$	A is a language
	AB	the concatenation of languages $A, B \subseteq \Sigma^*$: $\{ab \mid a \in A, b \in B\}$
	A^n	$\{a_1 a_2 \ldots a_n \mid a_i \in A \subseteq \Sigma^*\}$, $n \in \mathbf{Z}^+$
	A^0	$\{\lambda\}$
	A^+	$\bigcup_{n \in \mathbf{Z}^+} A^n$
	A^*	$\bigcup_{n \geq 0} A^n$: the Kleene closure of language A
	$M = (S, \mathcal{I}, \mathbb{O}, \nu, \omega)$	A finite state machine M with internal states S, input alphabet $\mathcal{I}$, output alphabet $\mathbb{O}$, next state function $\nu : S \times \mathcal{I} \rightarrow S$ and output function $\omega : S \times \mathcal{I} \rightarrow \mathbb{O}$

DEDICATED TO
ANDONETTA AND PAUL

Preface

The technological advances of the last two decades have resulted in several changes in the undergraduate curriculum. These changes include the introduction of material, if not entire courses, in

1. discrete methods that stress the finite nature inherent in certain problems and structures;
2. combinatorics—the algebra of enumeration, or counting;
3. graph theory with its applications and interrelations with areas such as data structures and methods of optimization; and,
4. finite algebraic structures that arise in conjunction with disciplines such as coding theory, methods of enumeration, gating networks, and combinatorial designs.

A primary reason for studying the material in any or all of these four major topics is the abundance of applications one finds in computer science—especially in the areas of data structures, the theory of computer languages, and the analysis of algorithms. In addition, there are also applications in engineering and the physical and life sciences, as well as in statistics and the social sciences. Consequently, the subject matter of discrete and combinatorial mathematics provides valuable material for students in many majors, not just for those in mathematics and computer science.

The purpose of this book is to provide an introductory survey in both discrete and combinatorial mathematics. The coverage is intended for the beginning student, so there are a great number of examples with detailed explanations. (The examples are numbered separately, and an open square is used to denote the end of each example.) In addition, wherever proofs are given, they too are presented with sufficient detail.

The text strives to accomplish the following objectives:
1. To introduce the student at the sophomore-junior level, if not earlier, to the topics and techniques of discrete methods and combinatorial reasoning. Problems in counting, or enumeration, require a careful analysis of structure (for example, whether or not order is important) and logical possibilities. There may even be a question of existence for some situations. Following such a careful analysis, we often find that the solution of a problem requires simple techniques for counting the possible outcomes that evolve from the breakdown of the given problem into smaller subproblems.
2. To introduce a wide variety of applications. In this regard, where structures from abstract algebra are required, only the basic theory needed for the application is developed. Furthermore, the solutions of some applications lend themselves to iterative procedures that lead to specific algorithms. The algorithmic approach to the solution of problems is fundamental in discrete mathematics,

and this approach reinforces the close ties between this discipline and the area of computer science.

3. To develop the mathematical maturity of the student through the study of an area that is so different from the calculus. Here, for example, there is the opportunity to establish results by counting a certain collection of objects in more than one way. This provides what are called combinatorial identities; it also introduces a novel proof technique. In this edition the nature of proof, along with what constitutes a valid argument, is developed in Chapter 2, in conjunction with the laws of logic and rules of inference. Proofs by mathematical induction are introduced in Chapter 4 and then used throughout the subsequent chapters of the text.

With regard to proofs in general, in many instances an attempt has been made to motivate theorems from observations on specific examples. In addition, whenever a finite situation provides a result that is not true for the infinite case, this situation is singled out for attention. Closed squares are used in the text to indicate the end of a proof. Proofs that are extremely long and/or rather special in nature are omitted. However, for the small number of proofs that are omitted, references are supplied for the reader interested in seeing the validation of these results. (The amount of emphasis placed on proofs will depend on the goals of the individual instructor and on those of his or her student audience.)

4. To present an adequate survey of topics for the computer science student who will be taking more advanced courses in data structures, the theory of computer languages, and the analysis of algorithms. The coverage here on groups, rings, fields, and Boolean algebras will also provide an applied introduction for mathematics majors who wish to continue their study of abstract algebra.

The prerequisites needed for this book are primarily a sound background in high school mathematics and an interest in attacking and solving a variety of problems. No particular programming ability is assumed. However, there are several program segments (presented primarily in Pascal) in the text and these are designed and explained in order to reinforce particular examples. With regard to calculus, we shall mention later in this preface its extent in Chapters 9 and 10.

My major motivation for writing the first edition of this book was the encouragement I'd received over the years from my students and colleagues. That text reflected both my interests and those of my students, as well as the recommendations of the Committee on the Undergraduate Program in Mathematics and the Association of Computing Machinery. This second edition continues in the same vein and now reflects the recommendations of the students and instructors who have used or are presently using the first edition.

FEATURES
The following briefly describes some of the major features of the text. These are designed to assist the student in learning the fundamentals of discrete and combinatorial mathematics.

Emphasis on algorithms and applications.   Algorithms and applications in many areas are presented throughout the text. For example,

1. Chapter 1 includes several instances where the introductory topics on enumeration are needed;

2. Section 7 of Chapter 5 provides an introduction to computational complexity. This material is then used in Section 8 of this chapter in order to analyze the running times of some elementary program segments.

3. The material in Chapter 6 deals with languages and finite state machines. This introduces the reader to an important area in computer science—the theory of computer languages.

4. Chapters 7 and 12 include discussions on the applications and algorithms dealing with topological sorting and the searching techniques known as the depth-first search and the breadth-first search.

5. In Chapter 10 we find the topic of recurrence relations. The coverage here includes applications on (a) the bubble sort, (b) binary search, (c) the Fibonacci numbers, (d) linear resistance networks, (d) the data structure called the stack, and (f) binary trees.

6. Chapter 16 introduces the fundamental properties of the algebraic structure called the group. The coverage here shows how this structure is used in the study of algebraic coding theory and Polya's method of enumeration.

Detailed explanations.   Whether it is an example or the proof of a theorem, explanations are designed to be careful and thorough. The presentation is primarily concerned with better understanding on the part of the reader who is seeing this material for the first time.

Exercises.   The role of the exercises in any mathematics text is a dominant one. The amount of time spent on the exercises greatly influences the pace of the course. Depending on the interest and mathematical background of the student audience, an instructor should find that the class time spent on discussing exercises will vary.

There are slightly over 1400 exercises in the text. Those that appear at the end of each section generally follow the order in which the section material is developed. These exercises are designed to: (a) review the basic concepts in the section; (b) tie together ideas presented in earlier sections of the chapter; and, (c) introduce additional concepts that are related to the material in the section. Some exercises call for the development of an algorithm, or the writing of a computer program, to solve a certain instance of a general problem. These usually require only a minimal amount of programming experience.

Each chapter concludes with a set of miscellaneous exercises. These provide further review of the ideas presented in the chapter, and also use material developed in earlier chapters of the text.

Answers are provided at the back of the text for almost all of the odd-numbered exercises.

Chapter summaries.   The last numbered section in each chapter provides a summary and historical review of the major ideas covered in that chapter. This should give the reader an overview of the contents of the chapter and provide information for further study and applications. Such further study can be readily assisted by the list of references that is supplied.

In particular, the summaries at the ends of Chapters 1, 5, and 9 include tables on the enumeration formulas developed within each of these chapters. Sometimes, these tables include results from earlier chapters in order to make comparisons and to show how the new results extend the prior ones.

ORGANIZATION

The areas of discrete and combinatorial mathematics are somewhat new to the undergraduate curriculum, so there are several options as to which topics should be covered in these courses. Each instructor and each student may have different interests. Consequently, the coverage here is fairly broad, as a survey course mandates. Yet there are further topics that some readers may still feel should be included. Furthermore, there will also be some differences of opinion with regard to the order in which some topics are presented in this text.

The nature and importance of the algorithmic approach to problem solving is stressed throughout the text. Ideas and approaches on problem solving are further strengthened by the interrelations between enumeration and structure, two other major topics in the book.

The material in the text is subdivided into four major areas. The first seven chapters form the underlying core of the book and present the fundamentals of discrete mathematics. The coverage here provides enough material for a one-quarter or one-semester course in discrete mathematics. A second course—one that emphasizes combinatorics—should include Chapters 8, 9, and 10 (and, time permitting, sections 1, 2, 3, 9, 10, and 11 of Chapter 16). In Chapter 9 some results from calculus are used; namely, fundamentals on differentiation and partial fraction decompositions. Despite this, for those who wish to skip this chapter, sections 1, 2, 3, 6, and 7 of Chapter 10 can still be covered.

A course that emphasizes the theory and applications of finite graphs can be developed from Chapters 11, 12, and 13. These chapters form the third major subdivision of the text. For a course in applied algebra, Chapters 14, 15, 16, and 17 (the fourth, and final, subdivision) deal with the algebraic structures—group, ring, Boolean algebra, and field—and include applications on switching functions, algebraic coding theory, and combinatorial designs. Finally, a course on the role of discrete structures in computer science can be developed from the material in Chapters 11, 12, 13, 15, and sections 1–8 in Chapter 16. For here we find applications on switching functions and algebraic coding theory, as well as an introduction to graph theory and trees, and their role in optimization. Other possible courses can be developed by considering the following chapter dependencies.

Chapter	Dependence on Prior Chapters
1	No dependence
2	No dependence (Hence an instructor can start a course in discrete mathematics with either the study of logic or an introduction to enumeration.)
3	1, 2
4	1, 2, 3

5	1, 2, 3
6	1, 2, 3, 5 (Minor dependence on Section 4.1 in Section 6.1)
7	1, 2, 3, 5, 6 (Minor dependence on Section 4.1 in Section 7.2.)
8	1, 3 (Minor dependence on Section 5.3 in Example 8.3 of Section 8.1.)
9	1, 3
10	1, 3, 4, 5, 9
11	1, 2, 3, 4, 5 (Although some graph theoretic ideas are mentioned in Chapters 5, 6, 7, 8, and 10, the material in this chapter is developed with no dependence on these earlier results.)
12	1, 2, 3, 4, 5, 11
13	3, 5, 11, 12
14	2, 3, 4, 5, 7 (The Euler phi function (ϕ) is used in Section 14.3. This function is derived in Section 8.1 but the result can be used here in Chapter 14 without covering Chapter 8.)
15	2, 3, 5, 7
16	1, 2, 3, 4, 5, 7
17	2, 3, 4, 5, 7, 14

In addition the index has been carefully developed in order to make the text even more flexible. Terms are presented with primary listings and several secondary listings. Also there is a great deal of cross referencing. This is designed to help the instructor who may want to change the order of presentation and deviate from the straight-and-narrow.

CHANGES IN THE SECOND EDITION

The changes made in the second edition of *Discrete and Combinatorial Mathematics* vary from moderate to extensive. However, the tone and purpose of the text have been retained. It is still the author's goal to provide within these pages a sound, readable, and understandable introduction to discrete and combinatorial mathematics—for the beginning student. Based on the comments and criticisms of the instructors and students who used the first edition, here are some of the changes one will find in the second edition:

• The order of the chapters has been changed to improve the structure of the presentation. As we mentioned earlier in the section on organization, the seventeen chapters of this text now fall rather naturally into the four subdivisions:
  (i) Fundamentals of Discrete Mathematics (Chapters 1–7)
  (ii) Further Topics in Enumeration (Chapters 8–10)
  (iii) Graph Theory and Applications (Chapters 11–13)
  (iv) Modern Applied Algebra (Chapters 14–17)

• In some instances—such as in Chapters 5, 8, and 12—sections and/or exercise sets have been subdivided to improve classroom presentation and reader understanding.

• In the first edition there was one section on logic (in the chapter dealing with set theory). Now there is an entire chapter, (Chapter 2) that deals with the fundamentals of logic, and this includes an introduction to rules of inference and their role in developing proofs. In addition, quantifiers and quantified statements are examined in this new chapter.

• The material on mathematical induction is now presented earlier in the text (in Chapter 4). This chapter also has more on the properties of algorithms in general—along with its coverage on the Division algorithm and Euclidean algorithm for integers.

• The very important topics of computational complexity and the analysis of algorithms are given much more coverage in this edition. The topics are introduced at the end of Chapter 5 (on functions). (However, the coverage may be delayed until Chapter 10 if one so desires.) There is also an optional section in Chapter 10 on divide-and-conquer algorithms.

• Chapter 13 (on optimization) now includes a section on Dijkstra's shortest path algorithm. In addition, the computational complexity of this algorithm (as well as that of Kruskal's algorithm and Prim's algorithm) is examined in this presentation.

• There are slightly more than 1400 exercises in the second edition. This is an increase of more than 350 over the first edition.

• Throughout the entire text more examples have been added. These additions are a result of the comments made by my students in using the first edition of this text.

ACKNOWLEDGMENTS

If space permitted, I should like to mention each of the students who provided help and encouragement when I was writing both the first and second editions of this book. Their suggestions helped to remove many mistakes and ambiguities, thus improving the exposition. Most helpful in this category were Paul Griffith, Meredith Vannauker, Paul Barloon, Byron Bishop, Lee Beckham, Brett Hunsaker, Tom Vanderlaan, Michael Bryan, John Breitenbach, Dan Johnson, Brian Wilson, Allen Schneider, Scott Krutsch, John Dowell, Charles Wilson, and Richard Nichols.

I thank Larry Alldredge and Martin Rivers for their comments on the computer science material, and Barry Farbrother, Paul Hogan, Dennis Lewis, Charles Kyker, and Jerome Wagner for their enlightening remarks on some of the applications given in the text.

I gratefully acknowledge the persistent enthusiasm and encouragement of the staff at Addison-Wesley (both past and present), especially Wayne Yuhasz, Thomas Taylor, Jeffrey Pepper, Charles Glaser, Mary Crittendon, Herb Merritt, Maria Szmauz, Adeline Ruggles, Stephanie Botvin, and Jack Casteel. Deborah Schneider deserves special recognition for her outstanding contributions to this second edition. I am also indebted to my colleagues, John Kinney, Gary Sherman, and especially Alfred Schmidt, for their interest and encouragement throughout the writing of both editions of this text.

Thanks and appreciation are due the following reviewers of the first and/or second editions of this book.

Larry Alldredge, *Telesoft Corporation of San Diego;* V. K. Balakrishnan, *University of Maine at Orono;* Robert Barnhill, *University of Utah;* Dale Bedgood, *East Texas State*

University; Jerry Beehler, *Tri-State University;* Allan Bishop, *Western Illinois University;* Monte Boisen, *Virginia Polytechnic Institute;* Samuel Councilman, *California State University at Long Beach;* Robert Crawford, *Western Kentucky University;* Carl DeVito, *Naval Postgradaute School;* Thomas Dowling, *Ohio State University;* Carl Eckberg, *San Diego State University;* Robert Geitz, *Oberlin College;* Maryann Hastings, *Marymount College;* Richard Iltis, *Willamette University;* Akihiro Kanamori, *Boston Univeristy;* James T. Lewis, *University of Rhode Island;* Hugh Montgomery, *University of Michigan;* John Rausen, *New Jersey Institute of Technology;* Martin Rivers, *Applied Computing Devices, Inc.;* Debra Diny Scott, *University of Wisconsin at Green Bay;* Dalton Tarwater, *Texas Tech University;* Jeff Tecosky-Feldman, *Harvard University;* W. L. Terwilliger, *Bowling Green State University;* Donald Thompson, *Pepperdine University;* Thomas Upson, *Rochester Institute of Technology;* W. D. Wallis, *Southern Illinois University;* Yixin Zhang, *University of Nebraska at Omaha.*

Special thanks is due to Douglas Shier, of the College of William and Mary, for the outstanding work he did in reviewing the manuscripts of both the first and second editions.

The last, and surely the most important, note of thanks belongs once again to the ever-patient and encouraging secretary of the Rose-Hulman mathematics department—Mrs. Mary Lou McCullough. Thank you once again, Mary Lou, for all of your work!

Alas, the remaining errors, ambiguities, and misleading comments are the sole responsibility of the author.

R.P.G.
Terre Haute, Indiana

Contents

Fundamental Principles of Counting

Enumeration, or counting, may strike one as an obvious process that a student learns when first studying arithmetic. But then, it seems, very little attention is paid to further development in counting as the student turns to "more difficult" areas in mathematics, such as algebra, geometry, trigonometry, and calculus. Consequently, this first chapter should provide some warning as to the seriousness and difficulty of "mere" counting.

Enumeration does not end with arithmetic. It also has applications in such areas as coding theory, probability, and statistics (in mathematics) and in the analysis of algorithms (in computer science). Later chapters will offer some specific examples of these applications.

As we enter this fascinating field of mathematics, we shall come upon many problems that are very simple to state but somewhat "sticky" to solve. Beware of formulas! Without an analysis of each problem, a mere knowledge of formulas is next to useless. Instead, welcome the challenge to solve unusual problems or those that are different from problems you have encountered in the past. Seek solutions based on your own scrutiny, regardless of whether it reproduces what the author provides. There are often several ways to solve a given problem.

1.1

THE RULES OF SUM AND PRODUCT

Our study of discrete and combinatorial mathematics begins with two basic principles of counting: the rules of sum and product. The statements and initial applications of these rules appear quite simple. In analyzing more complicated problems, one often is able to break such problems down into parts that one can handle by using these basic principles. We want to develop the ability to "decompose" such problems and piece together our partial solutions to arrive at the final answer. A good way to do this is to analyze and solve many diverse enumeration problems, taking note, all the while, of the principles being used in the solutions. This is the approach we will follow here. Our first principle of counting can be stated as follows:

The Rule of Sum: If a first task can be performed in m ways, while a second task can be performed in n ways, and the two tasks cannot be performed simultaneously, then performing either task can be accomplished in any one of $m + n$ ways.

Note that when we say that a particular occurrence, such as a first task, can come about in m ways, these m ways are assumed to be distinct, unless a statement is made to the contrary. This will be true throughout the entire text.

Example 1.1   A college library has 40 textbooks on sociology and 50 textbooks dealing with anthropology. By the rule of sum, a student at this college can select among $40 + 50 = 90$ textbooks in order to learn more about one or the other of these two subjects.   □

Example 1.2   The rule can be extended beyond two tasks as long as no pair of the tasks can occur simultaneously. For instance, a computer science instructor who has, say, five introductory books each on APL, BASIC, FORTRAN, and Pascal can recommend any one of these 20 books to a student who is interested in learning a first programming language.   □

Example 1.3   The computer science instructor of Example 1.2 has two colleagues. One of these colleagues has three textbooks on algorithm analysis, and the other has five such textbooks. If n denotes the number of books on this topic that this instructor can borrow from them, then $5 \le n \le 8$, for here both colleagues *may* own copies of the same textbook(s).   □

The following example introduces the rule of product.

Example 1.4   In trying to reach a decision on plant expansion, an administrator assigns 12 of her employees to two committees. Committee A consists of five members and is to investigate possible favorable results from such an expansion. The other seven employees, committee B, will scrutinize possible unfavorable repercussions. Should the administrator decide to speak to just one committee member before making her decision, then by the rule of sum there are 12 employees she can call upon for input. However, in order to be a bit more unbiased, she decides to speak with a member of committee A on Monday, and then with a member of committee B on Tuesday, before reaching a decision. Using the following principle, we find that she can speak with two such employees in $5 \times 7 = 35$ ways.   □

> **The Rule of Product:** If a procedure can be broken down into first and second stages, and if there are m possible outcomes for the first stage and n for the second stage, then the total procedure can be carried out, in the designated order, in mn ways.

This rule is sometimes referred to as the *principle of choice*.

Example 1.5 The drama club of Central University is holding tryouts for a spring play. With six men and eight women auditioning for the leading male and female roles, by the rule of product the director can cast his leading couple in $6 \times 8 = 48$ ways. □

Example 1.6 Here various extensions of the rule are illustrated by considering the manufacture of license plates consisting of two letters followed by four digits.

a) If no letter or digit can be repeated, there are $26 \times 25 \times 10 \times 9 \times 8 \times 7 = 3{,}276{,}000$ different possible plates.

b) With repetitions of letters and digits allowed, $26 \times 26 \times 10 \times 10 \times 10 \times 10 = 6{,}760{,}000$ different license plates are possible.

c) If repetitions are allowed, on how many of the plates in part (b) are both letters vowels (a, e, i, o, u) and all digits even? (0 is an even integer.) □

Example 1.7 In the main memory of a computer, information is stored in memory cells. To identify the cells in a computer's main memory, each cell is assigned a unique name called its *address*. For some computers an address is represented as an ordered list of eight symbols, where each symbol is one of the *bits* 0 or 1. This list of eight bits is called a *byte*. Using the rule of product, we find that there are $2 \times 2 \times 2 \times 2 \times 2 \times 2 \times 2 \times 2 = 2^8 = 256$ such bytes. Hence we have 256 addresses for memory cells where certain information can be stored.

Some machines (including the PDP 11 family†) use two-byte addresses. Such an address is made up of two consecutive bytes, or 16 consecutive bits, so it is possible to store as many as $256 \times 256 = 2^8 \times 2^8 = 2^{16} = 65{,}536$ pieces of information. Other computers (including the IBM PC/RT‡) use addressing systems of four bytes. In this case, as many as $2^8 \times 2^8 \times 2^8 \times 2^8 = 2^{32} = 4{,}294{,}967{,}296$ addresses are available for the storage of information in the machine's memory cells. □

Example 1.8 At times it is necessary to combine several different counting principles in the solution of one problem. Here we find that both the rules of sum and product are needed to attain the answer.

In certain versions of the programming language BASIC, a variable name consists of a single letter (A, B, C, . . .) or a single letter followed by a single digit. Because the computer does not distinguish between capitals and lowercase letters, a and A are considered the same variable name, as are E7 and e7. By the

---

† The PDP 11 family of computers is a product of the Digital Equipment Corporation.
‡ The IBM PC/RT processor is manufactured by the International Business Machines Corporation.

rule of product there are $26 \times 10 = 260$ variable names consisting of a letter followed by a digit; and because there are 26 variable names consisting of a single letter, by the rule of sum there are a total of $26 + 260 = 286$ variable names in this programming language.   □

1.2
PERMUTATIONS

Continuing to examine applications of the rule of product, we turn now to counting arrangements of distinct objects in a specified order. Generally we consider the objects to be placed in a row, or straight line, and this results in a certain linear arrangement. These arrangements are often called *permutations*, and we shall develop some systematic methods for dealing with them. We start with a typical example.

Example 1.9   In a class of 10 students, five are to be chosen and seated in a row for a picture. How many such linear arrangements are possible?

The key word here is *arrangement*, which designates the importance of order. If A, B, C, . . . , I, J denote the ten students, then BCEFI, CEFIB, and ABCFG are three such different arrangements, even though the first two involve the same five students.

To answer the question, we consider the positions and possible numbers of students we can choose from in order to fill each position. The filling of a position is a stage of our procedure.

10	×	9	×	8	×	7	×	6
1st position		2nd position		3rd position		4th position		5th position

Any of the 10 students can occupy the first position in the row. Because repetitions are not possible here, we can select only one of the nine remaining students to fill the second position. Continuing in this way, we find only six students to select from to fill the fifth and final position. This yields a total of 30,240 possible arrangements of five students selected from the class of 10.

Exactly the same answer is obtained if the positions are filled in the opposite order ($6 \times 7 \times 8 \times 9 \times 10$). If the order followed is 3rd, 1st, 4th, 5th, and 2nd (that is, the 3rd position is filled first, the 1st second, the 4th third, the 5th fourth, and the 2nd fifth), then the answer is $9 \times 6 \times 10 \times 8 \times 7$.   □

DEFINITION 1.1   For an integer $n \geq 0$, *n factorial* (denoted $n!$) is defined by

$$0! = 1,$$
$$n! = (n)(n-1)(n-2)\cdots(3)(2)(1), \quad \text{for} \quad n \geq 1.$$

One finds that $1! = 1$, $2! = 2$, $3! = 6$, $4! = 24$, and $5! = 120$. In addition, for any $n \geq 0$, $(n+1)! = (n+1)(n)!$.

Utilizing the factorial notation, we find that the answer to Example 1.9 can be expressed in the following, more compact form:

$$10 \times 9 \times 8 \times 7 \times 6 = 10 \times 9 \times 8 \times 7 \times 6 \times \frac{5 \times 4 \times 3 \times 2 \times 1}{5 \times 4 \times 3 \times 2 \times 1} = \frac{10!}{5!}$$

DEFINITION 1.2　Given a collection of n distinct objects, any (linear) arrangement of these objects is called a *permutation* of the collection.

Starting with the letters a, b, c, we find that there are six ways to arrange, or permute, all of the letters: abc, acb, bac, bca, cab, cba. If we are interested in arranging only two of the letters at a time, there are six such size-two permutations from the collection: ab, ba, ac, ca, bc, cb.

In general, if there are n distinct objects, denoted $a_1, a_2, \ldots, a_n$, and r is an integer, with $1 \leq r \leq n$, then by the rule of product, the number of permutations of size r for the n objects is

$$n \times (n-1) \times (n-2) \times \cdots \times (n-r+1) =$$

1st　　　2nd　　　3rd　　　　　　　rth
position　position　position　　　　　position

$$(n)(n-1)(n-2)\cdots(n-r+1) \times \frac{(n-r)(n-r-1)\cdots(3)(2)(1)}{(n-r)(n-r-1)\cdots(3)(2)(1)} = \frac{n!}{(n-r)!}.$$

We denote this number by $P(n,r)$. For $r = 0$, $P(n,0) = 1 = n!/(n-0)!$, so $P(n,r) = n!/(n-r)!$, where $0 \leq r \leq n$. A special case of this result is Example 1.9, where $n = 10$, $r = 5$, and $P(10,5) = 30,240$. When permuting all of the n objects in the collection, we have $r = n$ and find that $P(n,n) = n!/0! = n!$.

The number of permutations of size r, where $0 \leq r \leq n$, from a collection of n objects, is $P(n,r) = n!/(n-r)!$. However, if repetitions are allowed, then by the rule of product there are n^r possible arrangements, with $r \geq 0$.

Example 1.10　The number of permutations of the letters in the word COMPUTER is 8!. If only four of the letters are used, the number of permutations (of size four) is $P(8,4) = 8!/(8-4)! = 8!/4! = 1680$. If repetitions of letters are allowed, the number of possible 12-letter sequences is $8^{12} \doteq 6.872 \times 10^{10}$.　□

Example 1.11　Unlike Example 1.10, the number of arrangements of the letters in the word BALL is 12, not 4!, or 24. The reason is that we do not have four distinct letters to arrange. To get the 12 arrangements, we can list them as shown in Table 1.1(a).
　　　If the two L's are distinguished as L_1, L_2, then we can use our previous ideas on permutations of distinct objects; with the four distinct symbols B, A, L_1, L_2, we have 4! = 24 permutations. These are listed in Table 1.1(b). Table 1.1 reveals that to each arrangement for which the L's are indistinguishable there corre-

sponds a *pair* of permutations with distinct L's. Consequently,

$2 \times$ (Number of arrangements of the symbols B, A, L, L)

$= $ (Number of permutations of the symbols B, A, L_1, L_2),

and the answer to the original problem of finding all arrangements of the letters in BALL is $4!/2 = 12$.   □

Table 1.1

A B L L	A B L_1 L_2	A B L_2 L_1
A L B L	A L_1 B L_2	A L_2 B L_1
A L L B	A L_1 L_2 B	A L_2 L_1 B
B A L L	B A L_1 L_2	B A L_2 L_1
B L A L	B L_1 A L_2	B L_2 A L_1
B L L A	B L_1 L_2 A	B L_2 L_1 A
L A B L	L_1 A B L_2	L_2 A B L_1
L A L B	L_1 A L_2 B	L_2 A L_1 B
L B A L	L_1 B A L_2	L_2 B A L_1
L B L A	L_1 B L_2 A	L_2 B L_1 A
L L A B	L_1 L_2 A B	L_2 L_1 A B
L L B A	L_1 L_2 B A	L_2 L_1 B A

(a)   (b)

Example 1.12   Using the idea developed in Example 1.11, we now consider the arrangements of all letters in PEPPER.

There are $3! = 6$ arrangements with the P's distinguished for each arrangement in which the P's are not distinguished. For example, $P_1 E P_2 P_3 E R$, $P_1 E P_3 P_2 E R$, $P_2 E P_1 P_3 E R$, $P_2 E P_3 P_1 E R$, $P_3 E P_1 P_2 E R$, and $P_3 E P_2 P_1 E R$ all correspond to PEPPER when we remove the subscripts on the P's. In addition, to the arrangement $P_1 E P_2 P_3 E R$ there corresponds the pair of permutations $P_1 E_1 P_2 P_3 E_2 R$, $P_1 E_2 P_2 P_3 E_1 R$, when the E's are distinguished. Consequently,

$(2!)(3!)$(Number of arrangements of the letters in PEPPER)

$= $ (Number of permutations of the symbols P_1, E_1, P_2, P_3, E_2, R),

so the number of arrangements of the letters in PEPPER is $6!/(2!\,3!) = 60$.   □

Before stating a general principle for arrangements with repeated symbols, we want to point out that in our prior two examples, we solved a new type of problem by relating it to previous enumeration principles. This is a common practice in mathematics in general, and it often occurs in the derivations of discrete and combinatorial formulas.

In general, if there are n objects with n_1 of a first type, n_2 of a second type, ..., and n_r of an rth type, where $n_1 + n_2 + \cdots + n_r = n$, then there are $\dfrac{n!}{n_1!\,n_2!\cdots n_r!}$ arrangements of the given n objects. (Objects of the same type are indistinguishable.)

Example 1.13　The MASSASAUGA is a brown and white venomous snake indigenous to North America. Arranging all of the letters in MASSASAUGA, we find that there are

$$\frac{10!}{4!\,3!\,1!\,1!\,1!} = 25{,}200$$

possible arrangements. Among these there are

$$\frac{7!}{3!\,1!\,1!\,1!\,1!} = 840$$

in which all four A's are together. To get this last result we considered all arrangements of the seven symbols AAAA (one symbol), S, S, S, M, U, G. □

Example 1.14　Determine the number of (staircase) paths in the xy plane from $(2,1)$ to $(7,4)$ where each such path is made up of individual steps going one unit to the right (R) or one unit upward (U). The darkened lines in Fig. 1.1 show two of these paths.

　　Beneath each path in Figure 1.1 we have listed the individual steps. For example, in part (a) of the figure, the list R, U, R, R, U, R, R, U indicates that starting at the point $(2,1)$, we first move one unit to the right [to $(3,1)$], then one unit upward [to $(3,2)$], followed by two units to the right [to $(5,2)$], and so on, until we reach the point $(7,4)$. The path consists of 5 R's for moves to the right and 3 U's for moves upward.

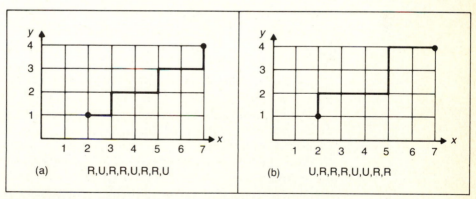

Figure 1.1

　　The path in part (b) of the figure is also made up of 5 R's and 3 U's. In general, the overall trip from $(2,1)$ to $(7,4)$ requires $7 - 2 = 5$ horizontal moves to the right and $4 - 1 = 3$ vertical moves upward. Consequently, each path corresponds with a list of 5 R's and 3 U's, and the solution for the number of paths emerges as the number of arrangements of the 5 R's and 3 U's, which is $\frac{8!}{5!\,3!} = 56.$ □

Example 1.15   We now do something a bit more abstract and prove that if n and k are positive integers with $n = 2k$, then $n!/2^k$ is an integer. Because our argument relies on counting, it is an example of a combinatorial proof.

Consider the n symbols $x_1, x_1, x_2, x_2, \ldots, x_k, x_k$. The number of ways in which we can arrange all of these $n = 2k$ symbols is an integer that is

$$\underbrace{\frac{n!}{2!\,2!\cdots 2!}}_{k \text{ factors of } 2!} = \frac{n!}{2^k}. \quad \square$$

Finally, we will apply what has been developed so far to a situation in which the arrangements are no longer linear.

Example 1.16   If six people, designated as A, B, . . . , F, are seated about a round table, how many different circular arrangements are possible, if arrangements are considered the same when one can be obtained from the other by rotation? (In Fig. 1.2, arrangements (a) and (b) are considered identical, whereas (b), (c), and (d) are three distinct arrangements.)

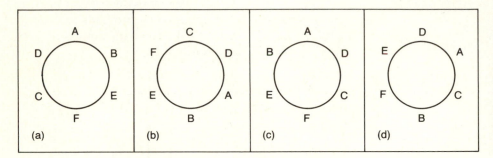

Figure 1.2

As in so many new circumstances we try to relate to past ideas. Considering Figs. 1.2(a) and 1.2(b), and starting at the top of the circle and moving clockwise, we list the distinct linear arrangements ABEFCD and CDABEF, which correspond to the same circular arrangement. In addition to these two, four other linear arrangements—BEFCDA, DABEFC, EFCDAB, and FCDABE—are found to correspond to the same circular arrangement as in (a) or (b). So inasmuch as each circular arrangement corresponds with six linear arrangements, we have $6 \times$ (Number of circular arrangements of A, B, . . . , F) = (Number of linear arrangements of A, B, . . . , F) = 6!.

Consequently there are $6!/6 = 5! = 120$ arrangements of A, B, . . . , F around the circular table. $\square$

Example 1.17   Suppose now that the six people of Example 1.16 are three married couples and that A, B, and C are the females. We want to arrange the six people around the table so that the sexes alternate. (Once again, arrangements are considered identical if one can be obtained from the other by rotation.)

Before we solve this problem, let us solve Example 1.16 by an alternative method, which will assist us in solving our present problem.

If we place A at the table as shown in Fig. 1.3(a), five locations (clockwise from A) remain to be filled. The filling of these locations with B, C, ... , F is the problem of permuting B, C, ... , F in a linear manner, and this can be done in $5! = 120$ ways.

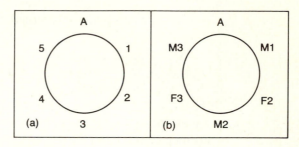

Figure 1.3

To solve the new problem of alternating the sexes, consider the method shown in Fig. 1.3(b). A (a female) is placed as before. The next position, clockwise from A, is marked M1 (Male 1) and can be filled in three ways. Continuing clockwise from A, position F2 (Female 2) can be filled in two ways. Proceeding in this manner, by the rule of product, there are $3 \times 2 \times 2 \times 1 \times 1 = 12$ ways in which these six people can be arranged with no two men or women seated next to each other.   □

EXERCISES 1.1 AND 1.2

1. During a local campaign, eight Republican and five Democratic candidates are nominated for president of the school board.

   a) If the president is to be one of these candidates, how many possibilities are there for the eventual winner?

   b) How many possibilities exist for a pair of candidates (one from each party) to oppose one another for the eventual election?

   c) Which counting principle is used in part (a)? in part (b)?

2. Answer part (c) of Example 1.6.

3. Buick automobiles come in 4 models, 12 colors, 3 engine sizes, and 2 transmission types.

   a) How many distinct Buicks can be manufactured?

   b) If one of the available colors is blue, how many different blue Buicks can be manufactured?

c) If one engine size is V-8, how many distinct blue Buicks have a V-8 engine?

4. In order to raise money for a new municipal pool, the chamber of commerce in a certain city sponsors a race. Each participant pays a $5 entrance fee and has a chance to win one of the different-sized trophies that are to be awarded to the first eight runners who finish.

   a) If 30 people enter the race, in how many ways will it be possible to award the trophies?

   b) If Roberta and Candice are two participants in the race, in how many ways can the trophies be awarded with these two runners among the top three?

5. A certain "Burger Joint" advertises that a customer can have his or her hamburger with or without any or all of the following: catsup, mustard, mayonnaise, lettuce, tomato, onion, pickle, cheese, or mushrooms. How many different kinds of hamburger orders are possible?

6. Matthew works as a computer operator at a small university. One evening he finds that 12 computer programs have been submitted earlier that day for batch processing. In how many ways can Matthew order the processing of these programs if (a) there are no restrictions? (b) he considers four of the programs higher in priority than the other eight and wants to process those four first? (c) he first separates the programs into four of top priority, five of lesser priority, and three of least priority, and he wishes to process the 12 programs in such a way that the top-priority programs are processed first and the three programs of least priority are processed last?

7. In how many ways can the symbols a, b, c, d, e, e, e, e, e be arranged so that no e is adjacent to another e? $4! = 24$

8. An alphabet of 40 symbols is used for transmitting messages in a communication system. How many distinct messages (lists of symbols) of 25 symbols can the transmitter generate if symbols can be repeated in the message? How many if 10 of the 40 symbols can appear only as the first and/or last symbols of the message, the other 30 symbols can appear anywhere, and repetitions of all symbols are allowed?

9. In a certain implementation of the programming language Pascal, an identifier consists of a single letter or a letter followed by up to seven symbols, which may be letters or digits. (We assume that the computer does not distinguish between capital and lowercase letters; there are 26 letters and 10 digits.) How many distinct identifiers are possible for this version of Pascal?

10. The production of a machine part consists of four stages. There are six assembly lines for the first stage, four assembly lines for the second stage, five for the third stage, and five for the last. Determine the number of different ways in which a machine part can be totally assembled in this production process.

11. A computer science professor has seven different programming books on a bookshelf. Three of the books deal with FORTRAN; the other four are

concerned with BASIC. In how many ways can the professor arrange these books on the shelf (a) if there are no restrictions? (b) if the languages should alternate? (c) if all the FORTRAN books must be next to each other? (d) if all FORTRAN books must be next to each other and all BASIC books must be next to each other? (e) if the three FORTRAN books are placed on the shelf with two BASIC books on either side?

12. **a)** In how many ways can the letters in VISITING be arranged?

    b) For the arrangements of part (a), how many have all three I's together?

13. **a)** How many arrangements are there of all the letters in SOCIOLOG-ICAL?

    b) In how many of these arrangements are A and G adjacent?

    c) In how many of the arrangements in part (a) are all the vowels adjacent?

    d) How many of the arrangements in part (c) have the vowels in alphabetical order? How many in part (a)?

14. Show that for integers $n, r \geq 0$, with $n + 1 > r$,

$$P(n + 1, r) = \left(\frac{n + 1}{n + 1 - r} \right) P(n, r).$$

15. Find the value(s) of n in each of the following: (a) $P(n, 2) = 90$, (b) $P(n, 3) = 3P(n, 2)$, and (c) $2P(n, 2) + 50 = P(2n, 2)$.

16. How many different paths in the xy plane are there from $(0, 0)$ to $(7, 7)$ if a path proceeds one step at a time by going either one space to the right (R) or one space upward (U)? How many such paths are there from $(2, 7)$ to $(9, 14)$? Can any general statement be made that incorporates these two results?

17. **a)** How many distinct paths are there from $(-1, 2, 0)$ to $(1, 3, 7)$ in Euclidean three-space if each move is one of the following types?

    (H): $(x, y, z) \rightarrow (x + 1, y, z)$;   (V): $(x, y, z) \rightarrow (x, y + 1, z)$;
    (A): $(x, y, z) \rightarrow (x, y, z + 1)$.

    b) How many such paths are there from $(1, 0, 5)$ to $(8, 1, 7)$?

    c) Generalize the results in parts (a) and (b).

    d) How many of the paths in part (a) start with an even number of A's?

18. Consider the following Pascal program segment where i, j, and k are integer variables.

```
For i := 1 to 12 do
    For j := 5 to 10 do
        For k := 15 downto 8 do
            Writeln ((i - j)*k);
```

    a) How many times is the *Writeln* statement executed?

    b) Which counting principle is used in part (a)?

19. Determine the number of six-digit integers (no leading zeros) in which (a) no digit may be repeated; (b) digits may be repeated. Answer parts (a) and (b)

with the extra condition that the six-digit integer is (i) even; (ii) divisible by 5; (iii) divisible by 4.

20. A sequence of letters of the form abcba, where the expression is unchanged upon reversing order, is an example of a *palindrome* (of five letters).

  a) If a letter may appear more than twice, how many palindromes of five letters are there? of six letters?

  b) Repeat part (a) under the condition that no letter appears more than twice.

21. a) In how many possible ways could a student answer a 10-question True–False examination?

  b) In how many ways can the student answer the test in part (a) if it is possible to leave a question unanswered in order to avoid an extra penalty for a wrong answer?

  c) Answer parts (a) and (b) if the examination consists of 10 multiple-choice questions, each question followed by four choices.

22. a) Provide a combinatorial argument to show that if n and k are positive integers with $n = 3k$, then $n!/(3!)^k$ is an integer.

  b) Generalize the result of part (a).

23. a) In how many ways can seven people be arranged about a circular table?

  b) If two of the people insist on sitting next to each other, how many arrangements are possible?

24. a) In how many ways can eight people, denoted A, B, . . . , H, be seated about the square table shown in Fig. 1.4, where Figs. 1.4(a) and 1.4(b) are considered the same but are distinct from Fig. 1.4(c)?

  b) If two of the eight people, say A and B, do not get along well, how many different seatings are possible with A and B not sitting next to each other?

  c) How many of the seatings in part (b) avoid having A and B sitting across the table from each other?

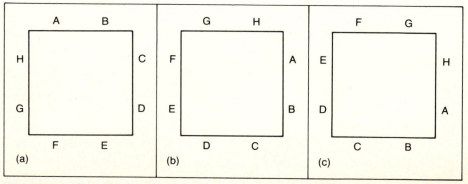

Figure 1.4

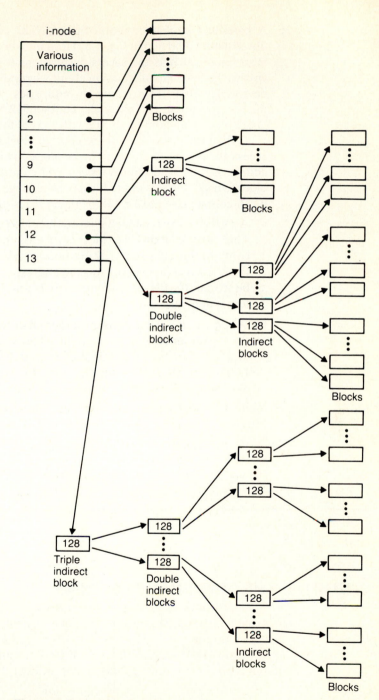

Figure 1.5

25. A particular implementation of the UNIX† operating system provides the file structure modeled in Fig. 1.5. Here the *i-node* for the file contains certain information such as the access permissions for the file. This is followed by entries that contain information about where the file is located on the storage device. The first 10 entries are the addresses of blocks where actual data for the file are stored. If a block can contain 512 bytes of information, these 10 *direct* blocks can hold up to $512 \times 10 = 5120$ bytes of data.

When the size of the file exceeds 10 blocks, the eleventh entry provides access to, or *points* to, an *indirect block*. This indirect block contains the addresses for 128 additional blocks where data can be stored. If this indirect block is used, the size of the file can be as large as $10 + 128 = 138$ blocks, and it can contain up to $512 \times 138 = 70,656$ bytes of information.

   a) A twelfth entry is needed if more than 138 blocks are required for the file. This entry contains the address of a double indirect block that in turn contains the addresses of 128 indirect blocks. As we noted earlier, each of these indirect blocks contains the addresses for 128 blocks where data can be stored. If a file uses 12 entries in its i-node, what is its maximum size in terms of blocks and bytes?

   b) When the file must be even larger than that provided by 12 entries, a thirteenth entry is needed. It provides the address to a triple indirect block that points to 128 double indirect blocks, as shown in the figure. Determine the maximum file size, for implementation of this UNIX operating system, under these conditions, in terms of blocks and bytes.

26. Write a computer program (or develop an algorithm) to compute $n!$ for any integer $n \geq 0$.

27. Write a computer program (or develop an algorithm) to compute $P(n, r)$ for any integers $n, r \geq 0$.

† UNIX is a trademark of the Bell Laboratory System.

1.3
COMBINATIONS: THE BINOMIAL THEOREM

The standard deck of playing cards consists of 52 cards comprising four suits: clubs, diamonds, hearts, and spades. Each suit has 13 cards: ace, 2, 3, . . . , 9, 10, jack, queen, king.

If we are asked to draw three cards from a standard deck, in succession and without replacement, then by the rule of product there are

$$52 \times 51 \times 50 = \frac{52!}{49!} = P(52, 3)$$

possibilities, one of which is AH (ace of hearts), 9C (nine of clubs), KD (king of diamonds). If instead we simply select three cards at one time from the deck so that the order of selection of the cards is no longer important, then the six permutations AH–9C–KD, AH–KD–9C, 9C–AH–KD, 9C–KD–AH, KD–9C–AH, and KD–AH–9C all correspond to just one (unordered) selection. Consequently, each selection, or combination, of three cards, *with no reference to order*, corresponds with 3! permutations of three cards. In equation form this translates into

(3!) × (Number of selections of size three from a deck of 52)

$$= \text{Number of permutations of the 52 cards taken three at a time}$$

$$= P(52, 3) = \frac{52!}{49!}.$$

Consequently three cards can be drawn, without replacement, from a standard deck in $52!/(3!\,49!) = 22{,}100$ ways.

In general, if we start with n distinct objects, each *selection* or *combination* of r of these objects, with no reference to order, corresponds to $r!$ permutations of size r from the n objects. Thus the number of combinations of size r from a collection of size n, denoted $C(n, r)$, where $0 \le r \le n$, satisfies $(r!) \times C(n, r) = P(n, r)$ and

$$C(n, r) = \frac{P(n, r)}{r!} = \frac{n!}{r!(n - r)!}, \qquad 0 \le r \le n.$$

In addition to $C(n, r)$ the symbol $\binom{n}{r}$ is also frequently used. Note that $C(n, 0) = 1$.

A word to the wise! When dealing with any counting problem, we should ask ourselves about the importance of order in the problem. When order is relevant, we think in terms of permutations and arrangements and the rule of product. When order is not relevant, combinations could play a key role in solving the problem.

Example 1.18   A hostess is having a dinner party for some members of her charity committee. Because of the size of her home, she can invite only 11 of the 20 committee members. Order is not important, so she can invite "the lucky 11" in $\binom{20}{11} = 20!/(11!9!) = 167{,}960$ ways. However, once the 11 arrive, how she arranges them around her rectangular dining table is an arrangement problem. Unfortunately, no part of the theory of combinations and permutations can help our hostess deal with "the offended nine" who were not invited.   □

Example 1.19   **a)** A student taking a history examination is directed to answer any seven of 10 essay questions. There is no concern about order here, so the student can answer the examination in

$$\binom{10}{7} = \frac{10!}{7!\,3!} = \frac{10 \times 9 \times 8}{3 \times 2 \times 1} = 120 \text{ ways.}$$

b) If the student must answer three questions from the first five and four questions from the last five, three questions can be selected from the first five in $\binom{5}{3} = 10$ ways, and the other four questions can be selected in $\binom{5}{4} = 5$ ways. Hence, by the rule of product, the student can complete the examination in $\binom{5}{3}\binom{5}{4} = 10 \times 5 = 50$ ways. $\square$

Example 1.20

a) At Rydell High School, the gym teacher must select nine girls from the junior and senior classes for a volleyball team. If there are 28 juniors and 25 seniors, she can make the selection in $\binom{53}{9} = 4,431,613,550$ ways.

b) If two juniors and one senior are the best spikers and must be on the team, then the rest of the team can be chosen in $\binom{50}{6} = 15,890,700$ ways.

c) For a certain tournament the team must comprise four juniors and five seniors. The teacher can select the four juniors in $\binom{28}{4}$ ways. For each of these selections she has $\binom{25}{5}$ ways to choose the five seniors. Consequently, by the rule of product, she can select her team in $\binom{28}{4}\binom{25}{5} = 1,087,836,750$ ways for this particular tournament. $\square$

Some problems can be treated from the viewpoint of either arrangements or combinations, depending on how one analyzes the situation. The following example demonstrates this.

Example 1.21

The gym teacher of Example 1.20 must make up four volleyball teams of nine girls each from the 36 freshman girls in her P.E. class. In how many ways can she select these four teams? Call the teams A, B, C, and D.

a) To form team A, she can select any nine girls from the 36 enrolled in $\binom{36}{9}$ ways. For team B the selection process yields $\binom{27}{9}$ possibilities. This leaves $\binom{18}{9}$ and $\binom{9}{9}$ possible ways to select teams C and D, respectively. So by the rule of product, the four teams can be chosen in

$$\binom{36}{9}\binom{27}{9}\binom{18}{9}\binom{9}{9} = \left(\frac{36!}{27!\,9!}\right)\left(\frac{27!}{18!\,9!}\right)\left(\frac{18!}{9!\,9!}\right)\left(\frac{9!}{9!\,0!}\right)$$

$$= \frac{36!}{9!\,9!\,9!\,9!} \doteq 2.145 \times 10^{19} \text{ ways.}$$

b) For an alternative solution, consider the 36 students lined up as follows:

1st	2nd	3rd		35th	36th
student	student	student	. . .	student	student

In order to select the four teams, we must distribute nine A's, nine B's, nine C's, and nine D's in the 36 spaces. The number of ways in which this can be done is the number of arrangements of 36 letters comprising nine each of A, B, C, and D. This is now the familiar problem of arrangements of nondistinct objects, and the answer is

$$\frac{36!}{9!\,9!\,9!\,9!}, \quad \text{as in part (a).} \quad \square$$

Our next two examples point out how some problems require the concepts of both arrangements and combinations for their solutions.

Example 1.22   The number of arrangements of the letters in TALLAHASSEE is

$$\frac{11!}{3!\,2!\,2!\,2!\,1!\,1!} = 831{,}600.$$

How many of these arrangements have no adjacent A's?

When we disregard the A's, there are

$$\frac{8!}{2!\,2!\,2!\,1!\,1!} = 5040$$

ways to arrange the remaining letters. One of these 5040 ways is shown in the following figure, where the upward arrows indicate nine possible locations for the three A's.

Three of these locations can be selected in $\binom{9}{3} = 84$ ways; and because this is also possible for all the other 5039 arrangements of E, E, S, T, L, L, S, H, by the rule of product there are $5040 \times 84 = 423{,}360$ arrangements of the letters in TALLAHASSEE with no consecutive A's.   □

Example 1.23   In the studies of algebraic coding theory and the theory of computer languages, we consider certain arrangements, called *strings*, made up from a prescribed *alphabet* of symbols. If the prescribed alphabet consists of the symbols 0, 1, and 2, for example, then 01, 11, 21, 12, and 20 are five of the nine strings of *length* two. Among the 27 strings of length three are 000, 012, 202, and 110.

In general, if n is any positive integer, then by the rule of product there are 3^n strings of length n for the alphabet 0, 1, and 2. If $x = x_1 x_2 x_3 \ldots x_n$ is one of these strings, we define the *weight* of x, denoted $\text{wt}(x)$, by $\text{wt}(x) = x_1 + x_2 + x_3 + \cdots + x_n$. For example, $\text{wt}(12) = 3$ and $\text{wt}(22) = 4$ for the case where $n = 2$; $\text{wt}(101) = 2$, $\text{wt}(210) = 3$, and $\text{wt}(222) = 6$ for $n = 3$.

Among the 3^{10} strings of length 10, we wish to determine how many have even weight. Such a string has even weight when the number of 1's in the string is even.

There are six different cases to consider. If the string x contains no 1's, then each of the 10 locations in x can be filled with either 0 or 2, and by the rule of product there are 2^{10} such strings. When the string contains two 1's, the locations for these two 1's can be selected in $\binom{10}{2}$ ways. Once these two locations have been specified, there are 2^8 ways to place either 0 or 2 in the other eight positions. Hence there are $\binom{10}{2}2^8$ strings of even weight that contain two 1's. The numbers of strings for the other four cases are given in Table 1.2.

Table 1.2

Number of 1's	Number of strings	Number of 1's	Number of strings
4	$\binom{10}{4}2^6$	8	$\binom{10}{8}2^2$
6	$\binom{10}{6}2^4$	10	$\binom{10}{10}$

Consequently, by the rule of sum, the number of strings of length 10 that have even weight is $2^{10} + \binom{10}{2}2^8 + \binom{10}{4}2^6 + \binom{10}{6}2^4 + \binom{10}{8}2^2 + \binom{10}{10} = \sum_{n=0}^{5}\binom{10}{2n}2^{10-2n}$.   □

We shall close this section with three results related to the concept of combinations.

First we note that for integers n, r, with $n \geq r \geq 0$, $\binom{n}{r} = \binom{n}{n-r}$. This can be established algebraically from the formula for $\binom{n}{r}$, but we prefer to observe that when dealing with a selection of size r from a collection of n distinct objects, the selection process leaves behind $n - r$ objects. Consequently, $\binom{n}{r} = \binom{n}{n-r}$ affirms the existence of a correspondence between the selections of size r (objects chosen) and the selections of size $n - r$ (objects left behind). An example of this correspondence is shown in Table 1.3, where $n = 5$, $r = 2$, and the distinct objects are 1, 2, 3, 4, and 5. This type of correspondence will be more formally defined in Chapter 5 and used in other counting situations.

Table 1.3

Selections of size $r = 2$ (objects chosen)		Selections of size $n - r = 3$ (objects left behind)	
1. 1,2	6. 2,4	1. 3,4,5	6. 1,3,5
2. 1,3	7. 2,5	2. 2,4,5	7. 1,3,4
3. 1,4	8. 3,4	3. 2,3,5	8. 1,2,5
4. 1,5	9. 3,5	4. 2,3,4	9. 1,2,4
5. 2,3	10. 4,5	5. 1,4,5	10. 1,2,3

Our second result is a theorem from our past experience in algebra.

THEOREM 1.1   (*The Binomial Theorem*) If x and y are variables and n is a positive integer, then

$$(x + y)^n = \binom{n}{0}x^0 y^n + \binom{n}{1}x^1 y^{n-1} + \binom{n}{2}x^2 y^{n-2} + \cdots$$

$$+ \binom{n}{n-1}x^{n-1}y^1 + \binom{n}{n}x^n y^0 = \sum_{k=0}^{n}\binom{n}{k}x^k y^{n-k}.$$

Before considering the general proof, we examine a special case. If $n = 4$, the coefficient of $x^2 y^2$ in the product

$$(x + y)(x + y)(x + y)(x + y)$$

1st   2nd   3rd   4th

factor   factor   factor   factor

is the number of ways in which we can select two x's (leaving two y's) from the four x's, one of which is available in each factor. (Although the x's are the same in appearance, we distinguish them as the x in the first factor, the x in the second factor, . . . , and the x in the fourth factor.) For example, among the possibilities, we can select (1) x from the first two factors and y from the last two; or (2) x from the first and third factors and y from the second and fourth. Table 1.4 summarizes the six possible selections.

Table 1.4

Factors selected for x		Factors selected for y	
(1)	1, 2	(1)	3, 4
(2)	1, 3	(2)	2, 4
(3)	1, 4	(3)	2, 3
(4)	2, 3	(4)	1, 4
(5)	2, 4	(5)	1, 3
(6)	3, 4	(6)	1, 2

Consequently the coefficient of $x^2 y^2$ in $(x + y)^4$ is $\binom{4}{2} = 6$, the number of ways to select two distinct objects from a collection of four distinct objects.

Now we turn to the proof of the general case.

Proof   In the product

$$(x + y)(x + y)(x + y)\cdots(x + y)$$

$$\underset{\substack{\text{1st}\\\text{factor}}}{} \quad \underset{\substack{\text{2nd}\\\text{factor}}}{} \quad \underset{\substack{\text{3rd}\\\text{factor}}}{} \quad \underset{\substack{n\text{th}\\\text{factor}}}{}$$

the coefficient of $x^k y^{n-k}$, where $0 \le k \le n$, is the number of different ways in which we can select k x's (and consequently $(n - k)$ y's) from the n available factors. (One way, for example, is to choose x from the first k factors and y from the last $n - k$ factors.) The total number of such selections of size k from a collection of size n is $C(n, k) = \binom{n}{k}$, and from this the binomial theorem follows. ∎

From this result we note that

$$(x + y)^n = \sum_{k=0}^{n} \binom{n}{n - k} x^k y^{n-k}.$$

Also, $\binom{n}{k}$ is often referred to as a *binomial coefficient* because of this theorem.

Example 1.24
a) From the binomial theorem it follows that the coefficient of $x^5 y^2$ in $(x + y)^7$ is $\binom{7}{5} = \binom{7}{2} = 21$.

b) The coefficient of $a^5 b^2$ in $(2a - 3b)^7$ is $\binom{7}{5}(2)^5(-3)^2$. We obtain this from the binomial theorem by setting $x = 2a$ and $y = -3b$. □

COROLLARY 1.1   For any integer $n > 0$,

a) $\binom{n}{0} + \binom{n}{1} + \binom{n}{2} + \cdots + \binom{n}{n} = 2^n$, and

b) $\binom{n}{0} - \binom{n}{1} + \binom{n}{2} - \cdots + (-1)^n \binom{n}{n} = 0$.

Proof   Part (a) follows from the binomial theorem when we set $x = y = 1$. When $x = -1$ and $y = 1$, part (b) results. ∎

Our third and final result generalizes the binomial theorem and is called the *multinomial theorem*.

THEOREM 1.2   For positive integers n, t, the coefficient of $x_1^{n_1} x_2^{n_2} x_3^{n_3} \cdots x_t^{n_t}$ in $(x_1 + x_2 + x_3 + \cdots + x_t)^n$ is

$$\frac{n!}{n_1! \, n_2! \, n_3! \cdots n_t!},$$

where each n_i is an integer with $0 \le n_i \le n$, for all $1 \le i \le t$, and $n_1 + n_2 + n_3 + \cdots + n_t = n$.

Proof   As in the proof of the binomial theorem, the coefficient of $x_1^{n_1} x_2^{n_2} x_3^{n_3} \cdots x_t^{n_t}$ is the number of ways we can select x_1 from n_1 of the n factors, x_2 from n_2 of the $n - n_1$ remaining factors, x_3 from n_3 of the $n - n_1 - n_2$ now remaining factors, . . . , and x_t from n_t of the last $n - n_1 - n_2 - n_3 - \cdots - n_{t-1} = n_t$ remaining factors. This can be carried out, as in part (a) of Example 1.21, in

$$\binom{n}{n_1}\binom{n - n_1}{n_2}\binom{n - n_1 - n_2}{n_3} \cdots \binom{n - n_1 - n_2 - \cdots - n_{t-1}}{n_t}$$

ways. We leave to the reader the details of showing that this is equal to

$$\frac{n!}{n_1! \, n_2! \, n_3! \cdots n_t!},$$

which is also written as

$$\binom{n}{n_1, n_2, n_3, \ldots, n_t}$$

and is called a *multinomial coefficient*. ∎

EXERCISES 1.3

1. Calculate $\binom{6}{2}$ and check your answer by listing all the selections of size two that can be made from the objects a, b, c, d, e, and f.

2. Facing a four-hour bus trip back to college, Diane decides to take along five magazines from the 12 that her sister Ann Marie has recently acquired. In how many ways can Diane make her selection?

3. A committee of 12 is to be selected from 10 men and 10 women. In how many ways can the selection be carried out if (a) there are no restrictions? (b) there must be six men and six women? (c) there must be an even number of women? (d) there must be more women than men? (e) there must be at least eight men?

4. In how many ways can a gambler draw five cards from a standard deck and get (a) a flush (five cards of the same suit)? (b) four aces? (c) four of a kind?

(d) three aces and two jacks? (e) a full house (three of a kind and a pair)? (f) three of a kind? (g) two pairs?

5. How many bytes contain (a) exactly two 1's; (b) exactly four 1's; (c) exactly six 1's; (d) at least six 1's?

6. How many ways are there to pick a five-person basketball team from 12 possible players? How many selections include the weakest and the strongest players?

7. A student is to answer seven out of 10 questions on an examination. In how many ways can he make his selection if (a) there are no restrictions? (b) he must answer the first two questions? (c) he must answer at least three of the first five questions?

8. In ordering the daily special at a diner, a customer has a choice of three entrees and may select any two of six available vegetables.

   a) How many different dinners can she select if (i) she must select two different vegetables? (ii) she is permitted to have two helpings of the same vegetable?

   b) Answer parts (i) and (ii) of (a) if she also has a choice of tomato juice, orange juice, or bean soup as an appetizer.

9. In how many ways can 12 different books be distributed among four children so that (a) each child gets three books? (b) the two oldest children get four books each while the two youngest get two books each?

10. At Pete's Pizza Parlor, pizzas come in four sizes: small, medium, large, and colossal. A customer can order a plain cheese pizza or request any combination of the following additional seven ingredients: anchovies, green peppers, mushrooms, olives, onions, pepperoni, and sausage. Determine the number of different pizzas that (a) are medium in size and have exactly two additional ingredients; (b) have exactly two additional ingredients; (c) are large or colossal and have exactly three additional ingredients; (d) are large or colossal and have at most three additional ingredients; (e) have at least two but no more than five additional ingredients.

11. How many arrangements of the letters in MISSISSIPPI have no consecutive S's?

12. A gym coach must select 11 seniors to play on a football team. If he can make his selection in 12,376 ways, how many seniors are eligible to play?

13. a) Fifteen points, no three of which are collinear, are given on a plane. How many lines do they determine?

    b) Twenty-five points, no four of which are coplanar, are given in space. How many triangles do they determine? How many planes? How many tetrahedra (pyramidlike solids with four triangular faces)?

14. For the strings of length 10 in Example 1.23, how many have (a) four 0's, three 1's, and three 2's; (b) at least eight 1's; (c) weight 4?

15. Consider the collection of all strings of length 10 made up from the alphabet

0, 1, 2, and 3. How many of these strings have weight 3? How many have weight 4? How many have even weight?

16. How many triangles are determined by the vertices of a regular polygon of n sides? How many if no side of the polygon is to be a side of any triangle?

17. Determine the coefficient of $x^9 y^3$ in (a) $(x + y)^{12}$, (b) $(x + 2y)^{12}$, and (c) $(2x - 3y)^{12}$.

18. Complete the details in the proof of the multinomial theorem.

19. Determine the coefficient of

   a) xyz^2 in $(x + y + z)^4$ b) xyz^2 in $(w + x + y + z)^4$

   c) xyz^2 in $(2x - y - z)^4$ d) xyz^{-2} in $(x - 2y + 3z^{-1})^4$

   e) $w^3 x^2 yz^2$ in $(2w - x + 3y - 2z)^8$.

20. Determine the sum of all the coefficients in

   a) $(x + y)^3$

   b) $(x + y)^{10}$

   c) $(x + y + z)^{10}$

   d) $(w + x + y + z)^5$

   e) $(2s - 3t + 5u + 6v - 11w + 3x + 2y)^{10}$.

21. Show that if n is an integer, with $n \geq 1$, then

$$\binom{2n}{n} + \binom{2n}{n-1} = \frac{1}{2}\binom{2n+2}{n+1}.$$

22. Show that for all $n \geq 2$, $\binom{n+1}{2} = \binom{n}{2} + n$.

23. Show that for all $m, n \geq 1$, $n\binom{m+n}{m} = (m+1)\binom{m+n}{m+1}$.

24. With n a positive integer, evaluate the sum

$$\binom{n}{0} + 2\binom{n}{1} + 2^2\binom{n}{2} + \cdots + 2^k\binom{n}{k} + \cdots + 2^n\binom{n}{n}.$$

25. For x a real number and n a positive integer, show that

   a) $1 = (1 + x)^n - \binom{n}{1}x^1(1 + x)^{n-1} + \binom{n}{2}x^2(1 + x)^{n-2} - \cdots + (-1)^n\binom{n}{n}x^n.$

   b) $1 = (2 + x)^n - \binom{n}{1}(x + 1)(2 + x)^{n-1} + \binom{n}{2}(x + 1)^2(2 + x)^{n-2} - \cdots$

   $\qquad + (-1)^n\binom{n}{n}(x + 1)^n.$

   c) $2^n = (2 + x)^n - \binom{n}{1}x^1(2 + x)^{n-1} + \binom{n}{2}x^2(2 + x)^{n-2} - \cdots + (-1)^n\binom{n}{n}x^n.$

26. Determine x if $\sum_{i=0}^{50} \binom{50}{i}8^i = x^{100}$.

27. Write a computer program (or develop an algorithm) to compute $C(n, r)$ for any integers $n, r \geq 0$.

28. a) Write a computer program (or develop an algorithm) that lists all selections of size 2 from the objects 1, 2, 3, 4, 5, 6.

b) Repeat part (a) for selections of size 3.

1.4

COMBINATIONS WITH REPETITION: DISTRIBUTIONS

When repetitions are allowed, we have seen that for n distinct objects an arrangement of size r of these objects can be obtained in n^r ways, for an integer $r \geq 0$. We now turn to the comparable problem for combinations and once again obtain a related problem whose solution follows from our previous enumeration principles.

Example 1.25   On their way home from track practice, seven high-school freshmen stop at a fast-food restaurant, where each of them has one of the following: a cheeseburger, a hot dog, a taco, or a fish sandwich. How many different purchases are possible?

Let c, h, t, and f represent cheeseburger, hot dog, taco, and fish sandwich, respectively. Here we are concerned with how many of each item are purchased, and not with the order in which they are purchased, so the problem is one of selections or combinations with repetition.

In Table 1.5 we have listed some possible purchases in column (a) and another means of representing each purchase in column (b).

Table 1.5

(a)	(b)
1. c, c, h, h, t, t, f	1. xx|xx|xx|x
2. c, c, c, c, h, t, f	2. xxxx|x|x|x
3. c, c, c, c, c, c, f	3. xxxxxx|||x
4. h, t, t, f, f, f, f	4. |x|xx|xxxx
5. t, t, t, t, t, f, f	5. ||xxxxx|xx
6. t, t, t, t, t, t, t	6. ||xxxxxxx|
7. f, f, f, f, f, f, f	7. |||xxxxxxx

For a purchase in column (b) of Table 1.5 we realize that each x to the left of the first bar (|) represents a c, each x between the first and second bars represents an h, the x's between the second and third bars stand for t's, and any x to the right of the third bar stands for an f. The third purchase, for example, has three consecutive bars because no one bought a hot dog or taco; the bar at the start of the fourth purchase indicates that there were no cheeseburgers in that purchase.

Once again a correspondence has been established between two collections of objects, where we know how to count the number in one collection. For column (b) of Table 1.5, we are enumerating all arrangements of 10 symbols consisting of seven x's and three |'s, so by our correspondence the number of different orders for column (a) is

$$\frac{10!}{7!\,3!} = \binom{10}{7}.$$

In this example we note that the seven x's (one for each freshman) correspond to the size of the selection and that the three bars are needed to separate the $3 + 1 = 4$ possible food items that can be chosen.   □

In general, when we wish to select, *with repetition*, r of n distinct objects, we find (as in Table 1.5) that we are considering all arrangements of r x's and $n - 1$ |'s and their number is

$$\frac{(n + r - 1)!}{r!(n - 1)!} = \binom{n + r - 1}{r}.$$

Consequently the number of combinations of n objects taken r at a time, *with repetition*, is $C(n + r - 1, r)$.

(In Example 1.25, $n = 4, r = 7$, so it is possible for r to exceed n when repetitions are allowed.)

Example 1.26   A donut shop offers 20 different kinds of donuts. Assuming that there are at least a dozen of each kind when we enter the shop, we can select a dozen donuts in $C(20 + 12 - 1, 12) = C(31, 12) = 141,120,525$ ways. (Here $n = 20, r = 12$.)   □

Example 1.27   President Helen has four secretaries: (1) Betty, (2) Goldie, (3) Mary Lou, and (4) Mona, and she wishes to distribute to them $100, in ten-dollar bills, as a Christmas bonus.

a) Allowing the situation in which one or more of the secretaries get nothing, President Helen is making a selection of size 10 (one for each ten-dollar bill) from a collection of size 4 (four secretaries), with repetition. This can be done in $C(4 + 10 - 1, 10) = C(13, 10) = 286$ ways.

b) If there are to be no hard feelings, each secretary must get at least $10. With this restriction, President Helen is now faced with making a selection of size 6 (the remaining six ten-dollar bills) from the same collection of size 4, and the choices now number $C(4 + 6 - 1, 6) = C(9, 6) = 84$. (For example, here the selection 2, 3, 3, 4, 4, 4 means that Betty does not get anything extra, whereas Goldie gets an additional $10, Mary Lou receives an additional $20, and Mona ends up with $40 total.)

c) If each secretary must get at least $10 and Mona, as head secretary, gets at least $50, the number of ways President Helen can distribute the bonus money is

$$C(3 + 2 - 1, 2) + C(3 + 1 - 1, 1)$$

Mona gets
exactly \$50

Mona gets
exactly \$60

$$+ C(3 + 0 - 1, 0) = 10 = C(4 + 2 - 1, 2) \quad \square$$

Mona gets
exactly \$70

Using the
technique in part (b)

Having worked examples utilizing combinations with repetition, we now consider two examples involving other counting principles as well.

Example 1.28 In how many ways can we distribute seven apples and six oranges among four children so that each child receives at least one apple?

Giving each child an apple, we have $C(4 + 3 - 1, 3) = 20$ ways to distribute the other three apples and $C(4 + 6 - 1, 6) = 84$ ways to distribute the six oranges. So by the rule of product, there are $20 \times 84 = 1680$ ways to distribute the fruit under the stated conditions. $\square$

Example 1.29 A message made up of 12 different symbols is to be transmitted through a communication channel. In addition to the 12 symbols, the transmitter will also send a total of 45 (blank) spaces between the symbols, with at least three spaces between each pair of consecutive symbols. In how many ways can the transmitter send the message?

There are 12! ways to arrange the 12 different symbols, and for any of these arrangements there are 11 positions between the 12 symbols. Because there must be at least three spaces between successive symbols, we use up 33 of the 45 spaces and must now locate the remaining 12 spaces. This is now a selection, with repetition, of size 12 (the spaces) from a collection of size 11 (the locations), and this can be accomplished in $C(11 + 12 - 1, 12) = 646{,}646$ ways.

Consequently, by the rule of product the transmitter can send the message with the required spacing in $(12!)\binom{22}{12} \doteq 3.097 \times 10^{14}$ ways. $\square$

In the next example an idea is introduced that appears to have more to do with number theory than with combinations or arrangements. Nonetheless, the solution of this example will turn out to be equivalent to counting combinations with repetitions.

Example 1.30 Determine all integer solutions to the equation

$$x_1 + x_2 + x_3 + x_4 = 7, \quad \text{where } x_i \geq 0 \quad \text{for all } 1 \leq i \leq 4.$$

One solution of the equation is $x_1 = 3, x_2 = 3, x_3 = 0, x_4 = 1$. (This is different from a solution such as $x_1 = 1, x_2 = 0, x_3 = 3, x_4 = 3$, even though the same four integers are being used.) A possible interpretation for the solution $x_1 = 3, x_2 = 3$,

$x_3 = 0$, $x_4 = 1$ is that we are distributing seven pennies (identical objects) among four children (distinct containers), and here we have given three pennies each to the first two children, nothing to the third child, and the last penny to the fourth child. Continuing with this interpretation, we see that each nonnegative integer solution of the equation corresponds to a selection, with repetition, of size 7 (the *identical* pennies) from a collection of size 4 (the *distinct* children), so there are $C(4 + 7 - 1, 7) = 120$ solutions. □

At this point it is crucial that we recognize the equivalence of the following:

a) The number of integer solutions of the equation

$$x_1 + x_2 + \cdots + x_n = r, \qquad x_i \geq 0, \qquad 1 \leq i \leq n.$$

b) The number of selections, with repetition, of size r from a collection of size n.

c) The number of ways r identical objects can be distributed among n distinct containers.

In terms of distributions, (c) is valid only when the r objects being distributed are identical and the n containers are distinct. When both the r objects and the n containers are distinct, we can select any of the n containers for each one of the objects and get n^r distributions by the rule of product.

When the objects are distinct but the containers are identical, we shall solve the problem using the Stirling numbers of the second kind (Chapter 5). For the final case, in which both objects and containers are identical, the theory of partitions of integers (Chapter 9) will provide some necessary results.

Example 1.31   In how many ways can one distribute 10 (identical) white marbles among six distinct containers?

Solving this problem is equivalent to finding the number of nonnegative integer solutions to the equation $x_1 + x_2 + \cdots + x_6 = 10$. That number is the number of selections of size 10, with repetition, from a collection of size 6. Hence the answer is $C(6 + 10 - 1, 10) = 3003$. □

We now examine two other examples related to the theme of the section.

Example 1.32   From Example 1.31 we know that there are 3003 nonnegative integer solutions to the equation $x_1 + x_2 + \cdots + x_6 = 10$. How many such solutions are there to the inequality $x_1 + x_2 + \cdots + x_6 < 10$?

One approach that may seem feasible in solving this inequality is to determine the number of such solutions to $x_1 + x_2 + \cdots + x_6 = k$, where k is an integer and $0 \leq k \leq 9$. Although feasible now, the technique becomes unrealistic if 10 is replaced by a somewhat larger number, say 100. In Chapter 3, however, we shall establish a combinatorial identity that will help us obtain an alternative solution to the problem by using this approach.

For the present we transform the problem by noting the correspondence between the nonnegative integer solutions of

$$x_1 + x_2 + \cdots + x_6 < 10 \tag{1}$$

and the integer solutions of

$$x_1 + x_2 + \cdots + x_6 + x_7 = 10, \qquad 0 \leq x_i, \qquad 1 \leq i \leq 6, \qquad 0 < x_7. \tag{2}$$

The number of solutions of (2) is the same as the number of nonnegative integer solutions of $y_1 + y_2 + \cdots + y_6 + y_7 = 9$, where $y_i = x_i$ for $1 \leq i \leq 6$, and $y_7 = x_7 - 1$. This is $C(7 + 9 - 1, 9) = 5005$. □

Our next result takes us back to the binomial and multinomial expansions.

Example 1.33   In the binomial expansion for $(x + y)^n$, each term is of the form $\binom{n}{k} x^k y^{n-k}$, so the total number of terms in the expansion is the number of nonnegative integer solutions of $n_1 + n_2 = n$ (n_1 is the exponent for x, n_2 the exponent for y). This number is $C(2 + n - 1, n) = n + 1$.

Perhaps it seems that we have used a rather long-winded argument to get this result. Many of us would probably be willing to believe the result on the basis of our experiences in computing $(x + y)^n$ for various small values of n.

Although experience is worthwhile in pattern recognition, it is not always enough to find a general principle. Here it would prove of little value if we wanted to know how many terms there are in the expansion of $(w + x + y + z)^{10}$.

Each distinct term here is of the form $\binom{10}{n_1, n_2, n_3, n_4} w^{n_1} x^{n_2} y^{n_3} z^{n_4}$, where $0 \leq n_i$ for $1 \leq i \leq 4$, and $n_1 + n_2 + n_3 + n_4 = 10$. This last equation can be solved in $C(4 + 10 - 1, 10) = 286$ ways, so there are 286 terms in the expansion of $(w + x + y + z)^{10}$. □

Our last two examples for this section provide applications from the area of computer science. Furthermore, the final example will lead to an important summation formula that we shall use in many later chapters.

Example 1.34   Consider the following Pascal program segment, where i, j, and k are integer variables.

```
For i := 1 to 20 do
   For j := 1 to i do
      For k := 1 to j do
         Writeln (i*j + k);
```

How many times is the *Writeln* statement executed in this program segment?

Among the possible choices for i, j, and k (in the order i − first, j − second, k − third) that will lead to execution of the *Writeln* statement, we list (1) 1, 1, 1; (2) 2, 1, 1; (3) 15, 10, 1; and (4) 15, 10, 7. We note that $i = 10$, $j = 12$, $k = 5$ is not one of the selections to be considered, because $j = 12 > 10 = i$; this violates the condition set forth in the second *For* loop. Each of the above four selections where the *Writeln* statement is executed satisfies the condition $1 \leq k \leq j \leq i \leq 20$.

In fact, any selection a, b, c ($a \leq b \leq c$) of size 3, with repetitions allowed, from the list $1, 2, 3, \ldots, 20$ results in one of the correct selections: here, $k = a$, $j = b$, $i = c$. Consequently the *Writeln* statement is executed

$$\binom{20 + 3 - 1}{3} = \binom{22}{3} = 1540 \text{ times.}$$

If there had been r (≥ 1) *For* loops instead of three, the *Writeln* statement would have been executed $\binom{20 + r - 1}{r}$ times. □

Example 1.35   Here we shall use the following Pascal program segment in order to derive a summation formula. In this given program segment, the variables counter, i, j, and n are integer variables. We assume that earlier in the program, the user is prompted for a positive integer; this input establishes the value of n.

```
counter := 0;
For i := 1 to n do
   For j := 1 to i do
      counter := counter + 1;
```

From the results in Example 1.34, after this segment is executed the value of (the variable) counter will be $\binom{n + 2 - 1}{2} = \binom{n + 1}{2}$. (This is also the number of times that the statement

(*)   counter := counter + 1

is executed.)

This result can also be obtained as follows: When $i := 1$, then j varies from 1 to 1 and (*) is executed once; when i is assigned the value 2, then j varies from 1 to 2 and (*) is executed twice; j varies from 1 to 3 when i is assigned the value 3, and (*) is executed three times; in general, for $1 \leq k \leq n$, when $i := k$, then j varies from 1 to k and (*) is executed k times. In total, the variable counter is incremented [and the statement (*) is executed] $1 + 2 + 3 + \cdots + n$ times. Consequently,

$$\sum_{i=1}^{n} i = 1 + 2 + 3 + \cdots + n = \binom{n + 1}{2} = \frac{n(n + 1)}{2}.$$

The derivation of this summation formula, obtained by counting the same result in two different ways, constitutes a combinatorial proof. □

EXERCISES 1.4

1. In how many ways can 10 (identical) dimes be distributed among five children if (a) there are no restrictions? (b) each child gets at least one dime? (c) the oldest child gets at least two dimes?

2. In how many ways can 15 (identical) candy bars be distributed among five children so that the youngest gets only one or two of them?

3. Determine how many ways 20 coins can be selected from four large containers filled with pennies, nickels, dimes, and quarters. (Each container is filled with only one type of coin.)

4. A certain ice cream store has 31 flavors of ice cream available. In how many ways can we order a dozen ice cream cones if (a) we do not want the same flavor more than once? (b) a flavor may be ordered as many as 12 times? (c) a flavor may be ordered no more than 11 times? (d) more than half of them must be chocolate?

5. a) In how many ways can we select five coins from a collection of 10 consisting of one penny, one nickel, one dime, one quarter, one half-dollar, and five (identical) Susan B. Anthony dollars?

   b) In how many ways can we select n objects from a collection of size $2n$ that consists of n distinct and n identical objects?

6. Answer Example 1.29 where the 12 symbols being transmitted are four A's, four B's, and four C's.

7. Determine the number of integer solutions of $x_1 + x_2 + x_3 + x_4 = 32$, where

   a) $x_i \geq 0, \quad 1 \leq i \leq 4$.
   b) $x_i > 0, \quad 1 \leq i \leq 4$.
   c) $x_1, x_2 \geq 5, \quad x_3, x_4 \geq 7$.
   d) $x_i \geq 8, \quad 1 \leq i \leq 4$.
   e) $x_i \geq -2, \quad 1 \leq i \leq 4$.
   f) $x_1, x_2, x_3 > 0, \quad 0 < x_4 \leq 25$.

8. In how many ways can a teacher distribute eight chocolate donuts and seven jelly donuts among three student helpers if each helper wants at least one donut of each kind?

9. Two n-digit integers (leading zeros allowed) are considered equivalent if one is a rearrangement of the other. (For example, 12033, 20331, and 01332 are considered equivalent five-digit integers.)

   a) How many five-digit integers are not equivalent?

   b) If the digits 1, 3, and 7 can appear at most once, how many nonequivalent five-digit integers are there?

10. Determine the number of integer solutions for $x_1 + x_2 + x_3 + x_4 + x_5 < 40$, where

    a) $x_i \geq 0, \quad 1 \leq i \leq 5$.
    b) $x_i \geq -3, \quad 1 \leq i \leq 5$.

11. In how many ways can we distribute eight identical white balls into four distinct containers so that (a) no container is left empty? (b) the fourth container has an odd number of balls in it?

12. a) Find the coefficient of $v^2 w^4 xz$ in $(3v + 2w + x + y + z)^8$.

    b) How many distinct terms arise in the expansion in part (a)?

13. How many ways are there to place 12 marbles of the same size in five distinct jars if (a) the marbles are all black? (b) each marble is a different color?

14. Find the number of nonnegative integer solutions for

    a) $2x_1 + x_2 + x_3 + x_4 = 10$;
    b) $2x_1 + 2x_2 + x_3 + x_4 = 10$.

15. In how many ways can one quarter, one dime, one nickel, and 25 1982 pennies be distributed among five children (a) with no restrictions? (b) so that the oldest child gets either 20¢ or 25¢?

16. A chemistry teacher has seven cartons each containing 36 test tubes of "unknowns" for a laboratory experiment. The first carton's 36 unknowns comprise four different compounds occurring 5, 12, 7, 12 times, respectively. In how many ways can the contents of this carton be distributed among five different chemistry labs?

17. a) How many nonnegative integer solutions are there to the pair of equations $x_1 + x_2 + x_3 + \cdots + x_7 = 37$, $x_1 + x_2 + x_3 = 6$?

   b) How many solutions in part (a) have $x_1, x_2, x_3 > 0$?

18. In how many ways can 24 pieces of chalk be distributed among four classrooms so that the largest classroom has at least as many pieces as the other three classrooms combined?

19. How many times is the *Writeln* statement executed during the following Pascal program segment? (Here i, j, k, and m are integer variables.)

```
For i := 1 to 20 do
    For j := 1 to i do
        For k := 1 to j do
            For m := 1 to k do
                Writeln ((i*j) + (k*m));
```

20. In the following Pascal program segment, i, j, k, and counter are integer variables. Determine the value that counter will have after the program segment is executed.

```
counter := 10;
For i := 1 to 15 do
    For j := i to 15 do
        For k := j to 15 do
            counter := counter + 1;
```

21. Find the value of sum after the given Pascal program segment is executed. (Here i, j, k, increment, and sum are integer variables.)

```
increment := 0;
sum := 0;
For i := 1 to 10 do
    For j := 1 to i do
        For k := 1 to j do
            Begin
                increment := increment + 1;
                sum := sum + increment
            End;
```

22. Consider the following Pascal program segment, where i, j, k, n, and counter are integer variables. Earlier in the program, the user is prompted for a

positive integer establishing the value of n for that particular run of the program.

```
counter := 0;
For i := 1 to n do
    For j := 1 to i do
        For k := 1 to j do
            counter := counter + 1;
```

We shall determine, in two different ways, the number of times the statement

```
counter := counter + 1
```

is executed. (This is also the value of counter after execution of the program segment.) From the result in Example 1.34, we know that the statement is executed $\binom{n+3-1}{3} = \binom{n+2}{3}$ times. For a fixed value of i, the *For* loops involving j and k result in $\binom{i+1}{2}$ executions of the counter increment statement. Consequently, $\binom{n+2}{3} = \sum_{i=1}^{n} \binom{i+1}{2}$. Use this result to obtain a summation formula for $1^2 + 2^2 + 3^2 + \cdots + n^2 = \sum_{i=1}^{n} i^2$.

23. **a)** Use the ideas in Example 1.35 and Exercise 22 to explain why $\binom{n+3}{4} = \sum_{i=1}^{n} \binom{i+2}{3}$.

  b) Use the result in part (a) to obtain a summation formula for

$$1^3 + 2^3 + 3^3 + \cdots + n^3 = \sum_{i=1}^{n} i^3.$$

24. Show that the number of ways to place n distinct objects into r different containers, with the objects in each container ordered, is $P(r + n - 1, r - 1)$.

25. **a)** Given positive integers m, n with $m \geq n$, show that the number of ways to distribute m identical objects into n distinct containers with no container left empty is $C(m - 1, m - n) = C(m - 1, n - 1)$.

  b) Show that the number of distributions in part (a) where each container holds at least r objects ($m \geq nr$) is $C(m - 1 + (1 - r)n, n - 1)$.

26. Write a computer program (or develop an algorithm) to compute the integer solutions for

$$x_1 + x_2 + x_3 = 10, \quad 0 \leq x_i, \quad 1 \leq i \leq 3.$$

27. Write a computer program (or develop an algorithm) to compute the integer solutions for

$$x_1 + x_2 + x_3 + x_4 = 15, \quad 0 \leq x_i, \quad 1 \leq i \leq 4.$$

28. Write a computer program (or develop an algorithm) to compute the integer solutions for

$$x_1 + x_2 + x_3 + x_4 = 4, \quad -2 \leq x_i, \quad 1 \leq i \leq 4.$$

1.5
AN APPLICATION IN THE PHYSICAL SCIENCES (OPTIONAL)

In this section we discuss an application of the counting techniques developed earlier in this chapter. This application arises in statistical mechanics and statistical thermodynamics. In these areas, one is interested in the number of ways r subatomic particles can be distributed among n distinct energy states.

For the Maxwell–Boltzmann model the particles are assumed to be distinct, and any number of them can be in any energy state. Here we get n^r possible distributions, because there are n possible energy states for each of the r particles. (Modern theory of quantum mechanics has shown that this model is inappropriate for those subatomic particles known at present.)

Two more successful models are the Bose–Einstein and the Fermi–Dirac models, which were originally based on the intrinsic angular momenta of the particles.

The Bose–Einstein model requires the r particles to be identical, with any number of them allowed in any energy state. The number of different distributions here is the number of nonnegative integer solutions to $x_1 + x_2 + \cdots + x_n = r$, which is $C(n + r - 1, r)$. Particles with integer spin (in units of $(h/2\pi)$, where $h = $ Planck's constant) follow this model. Such particles, called *bosons*, include photons, pi mesons, and the conduction electron pairs in superconductive material, such as lead or tin.

For particles with half-integer spin, such as protons, electrons, and neutrons, the Fermi–Dirac model is more appropriate, and the particles are called *fermions*. In this model the r particles are again identical, but an energy state can contain at most one particle. Consequently, here $r \leq n$, and the number of possible distributions is $\binom{n}{r}$. (This model is quite useful in the study of semiconductor band theory.)

1.6
SUMMARY AND HISTORICAL REVIEW

In this first chapter we introduced the fundamentals for counting combinations, permutations, and arrangements in a large variety of problems. The breakdown of problems into components requiring the same or different formulas for their solutions provided a key insight into the areas of discrete and combinatorial mathematics. This is somewhat similar to the *top-down approach* for developing algorithms in a structured language such as Pascal. Here one develops the algorithm for the solution of a difficult problem by first considering certain major subproblems that need to be solved. These subproblems are then further *refined*—subdivided into more easily workable programming tasks. Each level of refinement improves on the clarity, precision, and thoroughness of the algorithm until it is readily translatable into the code of the programming language. (We shall have more to say about algorithms in later chapters.)

The following chart (Table 1.6) summarizes the major counting formulas we have developed so far. Here we are dealing with a collection of n distinct objects. The formulas count the number of ways to select, or order, with or without repetitions, r of these n objects. The summaries of Chapters 5 and 9 include other such charts that evolve as we extend our investigations into other counting methods.

Table 1.6

Order is relevant	Repetitions are allowed	Type of result	Formula	Location in text
Yes	No	Permutation	$P(n,r) = n!/(n-r)!,$ $\quad 0 \le r \le n$	Page 5
Yes	Yes	Arrangement	$n^r, \quad n, r \ge 0$	Page 5
No	No	Combination	$C(n,r) = n!/[(n-r)!r!] = \binom{n}{r},$ $\quad 0 \le r \le n$	Page 15
No	Yes	Combination with repetition	$\binom{n+r-1}{r}, \quad n, r \ge 0$	Page 24

As we continue to investigate further principles of enumeration, as well as discrete mathematical structures for applications in coding theory, enumeration, optimization, and sorting schemes in computer science, we shall rely on the fundamental ideas introduced in this chapter.

The notion of permutation can be found in the Hebrew work *Sefer Yetzirah* (*The Book of Creation*), a manuscript written by a mystic sometime between 200 and 600. However, even earlier, it is of interest to note that a result by Xenocrates of Chalcedon (396–314 B.C.) may possibly contain "the first attempt on record to solve a difficult problem in permutations and combinations." For further details consult page 319 of the text by T. L. Heath [3], as well as page 113 of the article by N. L. Biggs [1], a valuable source on the history of enumeration. The first textbook dealing with some of the material we discussed in this chapter was *Ars Conjectandi* by Jakob Bernoulli (1654–1705). The text was published posthumously in 1731 and contained a reprint of the first formal treatise on probability, written by Christiaan Huygens in 1657.

The binomial theorem for $n = 2$ appears in the work of Euclid (300 B.C.), but it was not until the sixteenth century that the term "binomial coefficient" was actually introduced by Michel Stifel (1486–1567). In his *Arithmetica Integra* (1544) he gives the binomial coefficients up to the order of $n = 17$. Blaise Pascal (1623–1662), in his research on probability, published in the 1650s a treatise dealing with the relationships among binomial coefficients, combinations, and polynomials. These results were used by Jakob Bernoulli in proving the general form of the binomial theorem in a manner analogous to that presented in this chapter. Actual use of the symbol $\binom{n}{r}$ did not begin until the nineteenth century, when it was used by Andreas von Ettinghausen (1796–1878).

It was not until the twentieth century, however, that the advent of the computer made possible the systematic analysis of processes and algorithms used to generate permutations and combinations. We shall examine one such algorithm in Section 10.1.

The first comprehensive textbook dealing with topics in combinations and permutations was written by W. Whitworth [9]. Chapter 2 of D. Cohen [2], Chapter 1 of C. L. Liu [4], Chapter 2 of F. S. Roberts [6], Chapter 1 of H. Ryser [7], and Chapter 2 of A. Tucker [8] also deal with the material of this chapter. For more on the ideas of statistical mechanics and statistical thermodynamics, refer to the textbook by R. Reed and R. Roy [5].

REFERENCES

1. Biggs, Norman L. "The Roots of Combinatorics." *Historia Mathematica* 6 (1979): pp. 109–136.
2. Cohen, Daniel I. A. *Basic Techniques of Combinatorial Theory*. New York: Wiley, 1978.
3. Heath, Thomas Little. *A History of Greek Mathematics,* vol. 1. Reprint of the 1921 edition. New York: Dover Publications, 1981.
4. Liu, C. L. *Introduction to Combinatorial Mathematics*. New York: McGraw-Hill, 1968.
5. Reed, Robert D., and Roy, R. R. *Statistical Physics for Students of Science and Engineering*. Scranton, Penn.: Intext Educational Publishers, 1971.
6. Roberts, Fred S. *Applied Combinatorics*. Englewood Cliffs, N.J.: Prentice-Hall, 1984.
7. Ryser, H. J. *Combinatorial Mathematics*. Published by the Mathematical Association of America. New York: Wiley, 1963.
8. Tucker, Alan. *Applied Combinatorics*. New York: Wiley, 1980.
9. Whitworth, W. A. *Choice and Chance*. Reprint of the 1901 edition. New York: Hafner, 1965.

MISCELLANEOUS EXERCISES

1. In the manufacture of a certain type of automobile, four kinds of major defects and seven kinds of minor defects can occur. For those situations in which defects do occur, in how many ways can there be twice as many minor defects as there are major ones?

2. There are nine different dials on a machine, and each dial has five settings labeled 0, 1, 2, 3, and 4.

   a) In how many ways can all the dials on the machine be set?

   b) If the nine dials are arranged in a line at the top of the machine, how many of the machine settings have no two adjacent dials with the same setting?

   c) How many machine settings in part (b) use only 0, 2, and 4 as dial settings?

   d) How many machine settings in part (a) use only three different dial settings, where each dial setting appears three times?

3. The word processing software for a certain microcomputer allows filenames of from one to eight characters. Each character may be any of the 36 alphanumeric characters (26 letters and 10 digits) or any of 15 other designated symbols. (For filenames the

computer reads all letters as upper case, regardless of whether they are typed as upper case or lower case.) It is also possible to follow the filename with an optional filename extension. This is provided by writing a period and from one to three additional alphanumeric characters. Six examples of filenames that may therefore be used are (1) TAXES, (2) BUDGET, (3) A13B4781, (4) INCOME.87, (5) SCORES.JUN, and (6) Z35.P72. Here the last three examples use the optional filename extension.

   a) How many filenames use only the 36 alphanumeric characters and no extensions?

   b) How many of the filenames in part (a) start with two A's?

   c) How many filenames use extensions of exactly three alphanumeric characters?

   d) For the filenames in part (c), how many have no repeated character in the filename extension?

  4. A choir director must select six hymns for a Sunday church service. She has three hymn books and each contains 25 hymns (there are 75 different hymns in all). In how many ways can she select the hymns if she wishes to select (a) two hymns from each book? (b) at least one hymn from each book?

  5. How many ways are there to place 25 different flags on 10 numbered flagpoles if the order of the flags on a flagpole is (a) not relevant? (b) relevant? (c) relevant and every flagpole flies at least one flag?

  6. How many distinct four-digit integers can one make from the digits 1, 3, 3, 7, 7, and 8?

  7. There are 12 men at a dance. (a) In how many ways can eight of them be selected to form a cleanup crew? (b) How many ways are there to pair off eight women at the dance with eight of these 12 men?

  8. How many n-digit quaternary $(0, 1, 2, 3)$ sequences have exactly r 1's?

  9. In how many ways can the letters in WONDERING be arranged with exactly two consecutive vowels?

10. An organic solvent is made by mixing six different liquid compounds. After a first compound is poured into a vat, the other compounds are added in a prescribed order. All possible orders are tested to determine which produces the best yield. How many tests are needed?

11. In how many ways can the 10 identical horses on a carousel be painted so that three are brown, three are white, and four are black?

12. In how many ways can a teacher distribute 12 different science books among 16 students if (a) no student gets more than one book? (b) the oldest student gets two books but no other student gets more than one book?

13. Four numbers are selected from the following list of numbers: $-5, -4, -3, -2, -1,$ $1, 2, 3, 4$. (a) In how many ways can the selections be made so that the product of the four numbers is positive and (i) the numbers are distinct? (ii) each number may be selected as many as four times? (iii) each number may be selected at most three times? (b) Answer part (a) with the product of the four numbers negative.

14. **a)** Find the coefficient of $x^2 yz^2$ in $[(x/2) + y - 3z]^5$.

   b) How many distinct terms are there in the complete expansion of

$$\left(\frac{x}{2} + y - 3z \right)^5 ?$$

   c) What is the sum of all coefficients in the complete expansion?

15. a) In how many ways can 10 people, denoted $A, B, \ldots, I, J$, be seated about the rectangular table shown in Fig. 1.6, where Figs. 1.6(a) and 1.6(b) are considered the same but are considered different from Fig. 1.6(c)?

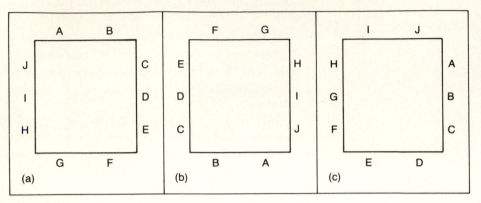

Figure 1.6

b) In how many of the arrangements of part (a) are A and B seated on longer sides of the table across from each other?

16. a) Determine the number of nonnegative integer solutions to the pair of equations

$$x_1 + x_2 + x_3 = 6, \quad x_1 + x_2 + \cdots + x_5 = 15, \quad x_i \geq 0, 1 \leq i \leq 5.$$

b) Answer part (a) with the pair of equations replaced by the pair of inequalities

$$x_1 + x_2 + x_3 \leq 6, \quad x_1 + x_2 + \cdots + x_5 \leq 15, \quad x_i \geq 0, 1 \leq i \leq 5.$$

17. A closed system consists of four photons and satisfies the Bose–Einstein model for statistical mechanics. The total energy of the system is $4E$, where each photon can have an energy level of kE, with k an integer, $0 \leq k \leq 4$, and E a positive constant. A photon of energy kE can occupy any one of $k^2 + 1$ distinct energy states at that energy level. In terms of energy states occupied by the photons, how many different configurations can this closed system assume?

18. A closed system consists of four electrons and satisfies the Fermi–Dirac model for statistical mechanics. The total energy of the system is $4E$, where each electron can have an energy level of kE, with k an integer, $0 \leq k \leq 4$, and E a positive constant. An electron of energy kE can occupy any one of $2(k^2 + 1)$ distinct energy states at that energy level. In terms of energy states occupied by electrons, how many different configurations can this closed system assume?

19. In how many ways can a dozen apples be distributed among five children so that no child gets more than seven apples?

20. For any given set in a tennis tournament, opponent A can beat opponent B in seven different ways. (At 6–6 they play a tie breaker.) The first opponent to win three sets wins the tournament. (a) In how many ways can scores be recorded with A winning in five sets? (b) In how many ways can scores be recorded with the tournament requiring at least four sets?

21. Let n be odd. In how many ways can we arrange n 1's and r 0's with a run (list of consecutive identical symbols) of exactly k 1's with $k \leq n < 2k$?

22. Given n distinct objects, determine in how many ways r of these objects can be arranged in a circle, where arrangements are considered the same if one can be obtained from the other by rotation.

23. For any positive integer n, show that

$$\binom{n}{0} + \binom{n}{2} + \binom{n}{4} + \cdots = \binom{n}{1} + \binom{n}{3} + \binom{n}{5} + \cdots .$$

24. a) In how many ways can the letters in UNUSUAL be arranged?

   b) For the arrangements in part (a), how many have all three U's together?

   c) How many of the arrangements in part (a) have no consecutive U's?

   d) How many of the arrangements in part (a) have all four vowels together?

25. a) Find the number of ways to write 17 as a sum of 1's and 2's if order is relevant.

   b) Answer part (a) for 18 in place of 17.

   c) Generalize the results in parts (a) and (b) for n odd and for n even.

26. a) In how many ways can 17 be written as a sum of 2's and 3's if the order of the summands is (i) not relevant? (ii) relevant?

   b) Answer part (a) for 18 in place of 17.

27. a) If n and r are positive integers with $n \geq r$, how many solutions are there to

$$x_1 + x_2 + \cdots + x_r = n,$$

   where each x_i is a positive integer, for $1 \leq i \leq r$?

   b) In how many ways can a positive integer n be written as a sum of r positive integer summands ($1 \leq r \leq n$) if the order of the summands is relevant?

28. a) In how many ways can one travel in the xy plane from $(1, 2)$ to $(5, 9)$ if each move is one of the following types:

$$\text{(H): } (x, y) \rightarrow (x + 1, y); \qquad \text{(V): } (x, y) \rightarrow (x, y + 1)?$$

   b) Answer part (a) if a third (diagonal) move (D): $(x, y) \rightarrow (x + 1, y + 1)$ is also possible.

29. a) In how many ways can a particle move in the xy plane from the origin to the point $(7, 4)$ if the moves that are allowed are of the form:

$$\text{(H): } (x, y) \rightarrow (x + 1, y); \qquad \text{(V): } (x, y) \rightarrow (x, y + 1)?$$

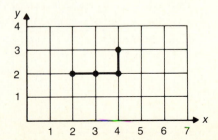

Figure 1.7

b) How many of the paths in part (a) do not use the path from $(2, 2)$ to $(3, 2)$ to $(4, 2)$ to $(4, 3)$ shown in Fig. 1.7?

c) Answer parts (a) and (b) if a third type of move (D): $(x, y) \rightarrow (x + 1, y + 1)$ is also allowed.

30. The following exercise illustrates an important counting method known as the *reflection principle*. Here a particle moves in the xy plane according to the following moves.

$$U: (m, n) \rightarrow (m + 1, n + 1); \qquad L: (m, n) \rightarrow (m + 1, n - 1);$$

where m and n are integers, $m, n \geq 0$.

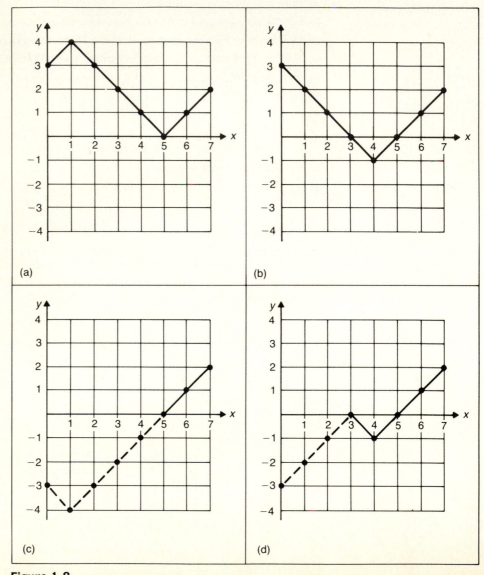

Figure 1.8

In Figs. 1.8(a) and 1.8(b), we have two such paths from $(0,3)$ to $(7,2)$.

a) How many such paths are there from $(0,3)$ to $(7,2)$ under these restrictions?

b) Figures 1.8(c) and 1.8(d) demonstrate the following idea for the paths in parts (a) and (b), respectively, of the figure. When a path from $(0,3)$ to $(7,2)$ touches or crosses the x axis, there is a corresponding path from $(0,-3)$ to $(7,2)$ obtained by *reflecting* the initial segment of the path before it first touches or crosses the x axis. Use this observation to find the number of paths from $(0,3)$ to $(7,2)$ that touch or cross the x axis at least once.

c) How many paths from $(0,3)$ to $(7,2)$ never touch or cross the x axis?

d) The cost of admission at a neighborhood movie theatre is $5. When Donna the cashier opens the ticket booth, she has one $10 bill and two $5 bills. The line for early admission contains 11 patrons, five of whom have a $5 bill each (ready to purchase a ticket) and six others, each of whom has a $10 bill. (Each time that Donna is unable to give the correct change she summons the manager for a $5 bill.) In how many different ways can these 11 people be lined up so that Donna always has at least one $5 bill and never has to summon the manager?

CHAPTER 2

Fundamentals of Logic

In the first chapter we derived a summation formula in Example 1.35 (Section 1.4). We obtained this formula by counting the same collection of objects (the number of times a statement was executed in a certain program segment) in two different ways and then equating the results. Consequently we say that the result was established by a *combinatorial proof*. This is one of many different techniques for arriving at a proof that we shall work with throughout the text.

In this chapter we take a close look at what constitutes a more conventional proof or valid argument. When a mathematician wishes to provide a proof for a given situation, he or she must use a system of logic. This is also true when a computer scientist develops the algorithms needed for a program or system of programs. The logic of mathematics is applied to decide whether one statement follows from, or is a logical consequence of, one or more other statements.

Some of the rules that govern this process are described in this chapter. We shall then use these rules in proofs (provided in the text and required in the exercises) throughout subsequent chapters. However, at no time can we hope to arrive at a point where we can apply the rules in an automatic fashion. As in applying the counting ideas discussed in Chapter 1, we should always analyze and seek to understand the situation given. This often calls for attributes we cannot learn in a book, such as insight and creativity. Merely trying to apply formulas or invoke rules will not get us very far either in proving results (such as theorems) or in doing enumeration problems.

2.1
BASIC CONNECTIVES AND TRUTH TABLES

In the development of any mathematical theory, assertions are made in the form of sentences. Such verbal or written assertions, called *statements* or *propositions*, are declarative sentences that are either true or false. For example, the following are propositions:

a) p: Combinatorics is a required course.

b) q: I am a computer science major.

c) r: The plotter is out of order today.

These propositions can be thought of as *primitive* statements, for there is really no way to break them down into anything simpler. Primitive statements are used with *logical connectives* to form *compound* statements.

We can negate a proposition or combine two propositions as follows.

1. Negation: The *negation* of a statement p is denoted by $\bar{p}$, which is read "Not p." For p as above, $\bar{p}$ is the statement "Combinatorics is not a required course."

2. Conjunction: The *conjunction* of p, q is denoted by $p \wedge q$, which is read "p and q." In our example the compound statement $p \wedge q$ is read "Combinatorics is a required course, **and** I am a computer science major."

3. Disjunction: The expression $p \vee q$ denotes the *disjunction* of p, q, which is read "p or q." Hence "Combinatorics is a required course, **or** I am a computer science major" is the verbal translation for $p \vee q$. We use the word "or" in the *inclusive* sense here. Consequently, $p \vee q$ is true if one or the other *or both* of the statements p, q are true. In English we sometimes write "and/or" to point this out. The *exclusive* "or" is denoted by $p \veebar q$. The compound statement $p \veebar q$ is true if one or the other but *not both* of the statements p, q is true. One way to express $p \veebar q$ for the example here is "Combinatorics is a required course, or I am a computer science major, but not both."

4. Implication: We say that "p implies q" and write $p \rightarrow q$ to designate the *implication* of q by p. Alternatively, we can say (a) If p, then q; (b) p is sufficient for q; (c) p only if q; and (d) q is necessary for p. A verbal translation of $p \rightarrow q$ for our example is "If combinatorics is a required course, then I am a computer science major." The statement p is called the *hypothesis* of the implication; q is called the *conclusion*. When statements are combined in this manner, there need not be any causal relationship between the statements. We simply write $p \rightarrow q$ with the foregoing definition for implication, whether p, q are related in some way or not.

5. Equivalence: Last, the *equivalence* of two statements p, q, is denoted by $p \leftrightarrow q$, which is read "p is equivalent to q," "p if and only if q," or "p is necessary and sufficient for q." For our p, q, "Combinatorics is a required course if and only if I am majoring in computer science" conveys the meaning of $p \leftrightarrow q$. We sometimes abbreviate "p if and only if q" as "p iff q."

Throughout our discussion on logic, we shall refuse to recognize as statements certain types of declarative sentences, such as those that require an individual's opinion or depend on the current time, and hence may not be either true or false. In addition, we must also realize that a sentence such as

The number x is an integer.

is *not* a statement because its *truth value* (true or false) cannot be determined until a numerical value is assigned for x. If x were assigned the value 7, the result would be a true statement. Assigning x a value such as ½, $\sqrt{2}$, or π, however, would make the statement false.

The presence of a variable, or unknown quantity, in a sentence does not automatically mean that it is not a statement. For example, the sentence

$$\text{If } x = 3, \text{ then } x^2 = 9.$$

is a true statement, whereas the statement

$$\text{If } x = 3, \text{ then } x + 2 = 6.$$

is false.

In the foregoing discussion, we mentioned the circumstances under which the statements $p \vee q$, $p \veebar q$ are considered true, on the basis of the truth of their primitive components p, q. This idea of the truth or falsity of a compound statement being a function of the truth values of its primitive components is worth further investigation. Tables 2.1 and 2.2 summarize the truth and falsity of the negation and other compound statements on the basis of the truth values of their primitive components. In constructing such *truth tables*, we write "0" for false and "1" for true.

Table 2.1

p	$\bar{p}$
0	1
1	0

Table 2.2

p	q	$p \wedge q$	$p \vee q$	$p \veebar q$	$p \to q$	$p \leftrightarrow q$
0	0	0	0	0	1	1
0	1	0	1	1	1	0
1	0	0	1	1	0	0
1	1	1	1	0	1	1

The four possible truth assignments for p, q can be listed in any order. For later work, the listing presented here will prove useful.

We see that the columns of truth values for p, $\bar{p}$ are the opposite of each other. The statement $p \wedge q$ is true only when both p, q are true, whereas $p \vee q$ is false only when both primitives are false. As we noted before, $p \veebar q$ is true when exactly one of p, q is true.

For the implication $p \to q$, the result is true in all cases except where p is true and q is false. We do not want a true statement to lead us into believing something that is false. However, we regard as true a statement such as "If $2 + 3 = 6$, then $2 + 4 = 7$," even though the statements "$2 + 3 = 6$" and "$2 + 4 = 7$" are both false.

Finally, the statement $p \leftrightarrow q$ is true when the primitive statements have the same truth value.

One's first encounter with the truth table for implication ($p \to q$), as shown in Table 2.2, is usually difficult to accept, especially the results in the first two rows (where p has the truth value 0). The following example should help make these truth value assignments easier to grasp.

Example 2.1   Consider the following scenario. It is almost the week before Christmas and Penny will be attending several parties that week. Ever conscious of her weight, she plans not to weigh herself until the day after Christmas. Considering what those parties may do to her waistline by then, she makes the following resolution for the December 26 outcome: "If I weigh more than 120 pounds, then I shall enroll in an exercise class."

Here we let p and q denote the (primitive) statements.

p: I weigh more than 120 pounds.
q: I shall enroll in an exercise class.

Then Penny's statement (implication) is given by $p \rightarrow q$.

We shall consider the truth values of this particular example of $p \rightarrow q$ for the rows of Table 2.2. Consider first the easier cases in rows 4 and 3.

- Row 4: p and q both have the truth value 1. On December 26 Penny finds that she weighs more than 120 pounds and promptly enrolls in an exercise class, just as she said she would. Here we consider $p \rightarrow q$ to be true and assign it the truth value 1.

- Row 3: p has the truth value 1, q has the truth value 0. Now that December 26 has arrived, Penny finds her weight to be over 120 pounds, but she makes no attempt to enroll in an exercise class. In this case we feel that Penny has broken her resolution—in other words, the implication $p \rightarrow q$ is false (and has the truth value 0).

The cases in rows 1 and 2 may not immediately agree with our intuition, but the example should make these results easier to accept.

- Row 1: p and q both have the truth value 0. Here Penny finds that on December 26 her weight is 120 pounds or less and she does not enroll in an exercise class. She has not violated her resolution; we take her statement $p \rightarrow q$ to be true and assign it the truth value 1.

- Row 2: p has the truth value 0, q has the truth value 1. This last case finds Penny weighing 120 pounds or less on December 26 but still enrolling in an exercise class. Perhaps her weight is 119 or 120 pounds and she feels this is still too high. Or maybe she wants to join an exercise class because she thinks it will be good for her health. No matter what the reason, she has not contradicted her resolution $p \rightarrow q$. Once again, we accept this compound statement as true, assigning it the truth value 1.   □

Our next example discusses a related notion: the *decision* (or *selection*) structure in computer programming.

Example 2.2   In computer science the If-Then and If-Then-Else decision structures arise in such languages as BASIC and Pascal. The hypothesis p is often a relational expression such as $x > 2$. This expression then becomes a (logical) statement that has the truth value 0 or 1, depending on the value of the variable x at that point in the program. The conclusion q may be an executable statement directing the program to another line or causing some result to be printed. (Consequently, as an "executable statement," q is not one of the statements or propositions, which we have been discussing.) When dealing with "If p Then q," in this context, the computer executes q only on the condition that p is true. For p false, the computer goes to the next (perhaps numbered) line in the program sequence. For the decision structure "If p Then q Else r," q is executed when p is true and r is executed when p is false. ☐

Before continuing, a word of caution: Be careful when using the symbols → and ↔. Implication and equivalence are not the same, as evidenced by the last two columns of Table 2.2.

In *definitions*, however, the two ideas do come together. For example, in defining an isosceles triangle we may write, "If two sides of a triangle are of equal length, then the triangle is called isosceles." But we must also realize that if a triangle is described as isosceles, then it must have two sides of equal length. Consequently, when a definition is given in the form $p \rightarrow q$, we understand that what is really meant is $p \leftrightarrow q$.

What happens in the case of definitions does not carry over to the implication statement in general. We shall examine this point further in Example 2.13.

Although the results of Tables 2.1 and 2.2 were given for primitive statements p, q, these ideas also apply when compound statements are substituted for the symbols p, q. Examples 2.3 through 2.5 demonstrate this.

Example 2.3 Let us examine the truth table for the compound statement "I am a computer science major, and if the plotter is not out of order today, then combinatorics is a required course." In symbolic notation this is $q \wedge (\bar{r} \rightarrow p)$, where p, q, and r represent the propositions introduced at the start of this section. The last column of Table 2.3 contains the truth values for this result. The preceding columns show how we build the truth table up by considering smaller parts of the compound statement and using the results from Tables 2.1 and 2.2. ☐

Table 2.3

p	q	r	$\bar{r}$	$\bar{r} \rightarrow p$	$q \wedge (\bar{r} \rightarrow p)$
0	0	0	1	0	0
0	0	1	0	1	0
0	1	0	1	0	0
0	1	1	0	1	1
1	0	0	1	1	0
1	0	1	0	1	0
1	1	0	1	1	1
1	1	1	0	1	1

Example 2.4 In Table 2.4 we develop the truth tables for the compound statements $p \vee (q \wedge r)$ (column 5) and $(p \vee q) \wedge r$ (column 7).

Table 2.4

p	q	r	$q \wedge r$	$p \vee (q \wedge r)$	$p \vee q$	$(p \vee q) \wedge r$
0	0	0	0	0	0	0
0	0	1	0	0	0	0
0	1	0	0	0	1	0
0	1	1	1	1	1	1
1	0	0	0	1	1	0
1	0	1	0	1	1	1
1	1	0	0	1	1	0
1	1	1	1	1	1	1

Because the truth values for columns 5 and 7 differ (in rows 5 and 7), we must avoid writing a compound statement such as $p \vee q \wedge r$. Without parentheses to indicate which of the connectives $\wedge$ and $\vee$ should be applied first, we have no idea whether we are dealing with $p \vee (q \wedge r)$ or $(p \vee q) \wedge r$. □

Our last example for this section illustrates two special types of propositions.

Example 2.5 The results in columns 4 and 6 of Table 2.5 reveal that the proposition $p \rightarrow (p \vee q)$ is always true, whereas the proposition $p \wedge (\bar{p} \wedge q)$ is always false. □

Table 2.5

p	q	$p \vee q$	$p \rightarrow (p \vee q)$	$(\bar{p} \wedge q)$	$p \wedge (\bar{p} \wedge q)$
0	0	0	1	0	0
0	1	1	1	1	0
1	0	1	1	0	0
1	1	1	1	0	0

DEFINITION 2.1 A proposition that is always true is called a *tautology*; one that is always false is labeled a *contradiction*.

Throughout this chapter we shall use the symbol T_0 to denote any tautology and the symbol F_0 to denote any contradiction.

We can use the ideas of tautology and implication to describe what we mean by a theorem. In general, a *theorem* is a mathematical statement that can be shown (by means of a proof) to be true. You undoubtedly encountered theorems and their proofs in high school algebra and geometry.

Many theorems can be given in the following form. If the statements $p_1, p_2, p_3, \ldots, p_n$ denote the hypotheses of an implication and q is its conclusion, then we say that the implication $(p_1 \wedge p_2 \wedge p_3 \wedge \cdots \wedge p_n)^{\dagger} \rightarrow q$ is a theorem when the truth of $p_1, p_2, p_3, \ldots, p_n$ leads to the truth of q. (So if each of $p_1, p_2, p_3, \ldots, p_n$ has the truth value 1, so does q.) If any one of $p_1, p_2, p_3, \ldots, p_n$ is false, then

† At this point we have dealt only with the conjunction of two statements, so we must point out that the conjunction $p_1 \wedge p_2 \wedge p_3 \wedge \cdots \wedge p_n$ of n statements is true only when each p_i, $1 \le i \le n$, is true. We shall deal with this generalized conjunction in detail in Example 4.8 of Section 4.1.

$p_1 \wedge p_2 \wedge p_3 \wedge \cdots \wedge p_n$ is false, and no matter what truth value q has, the implication $(p_1 \wedge p_2 \wedge p_3 \wedge \cdots \wedge p_n) \to q$ is true. Consequently, for the circumstances we have described here, the implication $(p_1 \wedge p_2 \wedge p_3 \wedge \cdots \wedge p_n) \to q$ is a theorem (and a tautology), as long as the truth value of q is 1 when each p_i, $1 \le i \le n$, has the truth value 1.

We will have more to say about theorems of this form and how we can prove them in Section 2.3.

EXERCISES 2.1

1. Determine whether each of the following sentences is a statement.
   a) In 1984 Ronald Reagan was the president of the United States.
   b) $x + 3$ is a positive integer. *no*
   c) If $x + 3$ is a positive integer, then $x + 5$ is a positive integer.
   d) If only every morning could be as sunny and clear as this one!
   e) For every real number x, $\sin^2 x + \cos^2 x = 1$.
   f) When $x = 3$, $4 - x > 0$ and $x - 5 < 0$.
   g) Seventeen is an even number. *yes*
   h) If Jennifer is late for the party, then her cousin Zachary will be quite angry.
   i) From the halls of Montezuma to the shores of Tripoli.
   j) For any real number x, $|x| < 2$ if and only if $-2 < x < 2$.
   k) What time is it?

2. Let p, q, r, s denote the following statements: p: I finish writing my computer program before lunch; q: I play tennis in the afternoon; r: The sun is shining; s: The humidity is low. Write the following in symbolic form.
   a) If the sun is shining, I shall play tennis this afternoon.
   b) Finishing the writing of my computer program before lunch is necessary for my playing tennis this afternoon.   *q → p*
   c) The sun is shining and I shall play tennis this afternoon.
   d) Low humidity and sunshine are sufficient for me to play tennis this afternoon.   *r ∧ s → q*

3. Let p, q, r denote the following statements: p: Triangle ABC is isosceles; q: Triangle ABC is equilateral; r: Triangle ABC is equiangular. Translate each of the following into an English sentence.
   a) $q \to p$ b) $\bar{p} \to \bar{q}$ c) $q \leftrightarrow r$
   d) $p \wedge \bar{q}$ e) $r \to p$

4. Construct a truth table for each of the following propositions.
   a) $p \to (p \vee q)$ b) $p \to (q \to r)$ c) $(p \to q) \to r$
   d) $(p \to q) \to (q \to p)$ e) $[p \wedge (p \to q)] \to q$ f) $(p \wedge q) \to p$
   g) $q \leftrightarrow (\bar{p} \vee \bar{q})$ h) $[(p \to q) \wedge (q \to r)] \to (p \to r)$

5. Are any of the propositions in Exercise 4 tautologies?

6. Verify that $[p \rightarrow (q \rightarrow r)] \rightarrow [(p \rightarrow q) \rightarrow (p \rightarrow r)]$ is a tautology.

7. a) How many rows are needed for the truth table of the proposition $(p \vee \bar{q}) \leftrightarrow [(\bar{r} \wedge s) \rightarrow t]$?

   b) If $p_1, p_2, \ldots, p_n$ are primitive propositions and the compound statement p contains at least one occurrence of each p_i, where $1 \leq i \leq n$, how many rows are needed to construct the truth table for p?

8. Determine all truth value assignments, if any, for p, q, r, s, t that make each of the following propositions false.

   a) $(p \wedge q \wedge r) \rightarrow (s \vee t)$

   b) $(p \wedge q \wedge r) \rightarrow (s \veebar t)$

9. a) If statement q has the truth value 1, determine all truth value assignments for the statements $p, r,$ and s for which the truth value of the statement

$$(q \rightarrow [(\bar{p} \vee r) \wedge \bar{s}]) \wedge [\bar{s} \rightarrow (\bar{r} \wedge q)]$$

   is 1.

   b) Answer part (a) if q has the truth value 0.

10. In the following Pascal program segment $i, j, m,$ and n are integer variables. The values of m and n are supplied by the user earlier in the execution of the (total) program.

```
For i := 1 to m do
  For j := 1 to n do
    If i <> j then
      Writeln ('The sum of i and j is ', i + j);
```

How many times is the Writeln statement in the segment executed when (a) $m = 10, n = 10$; (b) $m = 20, n = 20$; (c) $m = 10, n = 20$; (d) $m = 20, n = 10$?

11. For the following BASIC program, how many times is the PRINT statement of line 40 executed?

```
10 X = 10
20 FOR I = 1 TO 7
30    FOR J = 1 TO I + 3
40        IF ((X > 8) OR (I > 5 AND J < 10)) THEN PRINT X
50    NEXT J
60    X = X - 1
70 NEXT I
80 END
```

12. A segment of a Pascal program contains a Repeat-Until loop structured as follows

```
Repeat
. . . . . . . .
. . . . . . . .
Until ((x <> 0) and (y > 0)) or (not ((w > 0) and (t = 3)));
```

For each of the following assignments for the variables x, y, w, and t, determine whether or not the loop terminates.

a) $x = 7, y = 2, w = 5, t = 3$.   **b)** $x = 0, y = 2, w = -3, t = 3$.

c) $x = 0, y = -1, w = 1, t = 3$.   **d)** $x = 1, y = -1, w = 1, t = 3$.

e) $x = 0, y = 3, w = 7, t = 4$.

2.2
LOGICAL EQUIVALENCE: THE LAWS OF LOGIC

In all areas of mathematics we need to know when the entities we are studying are equal or essentially the same. For example, in arithmetic and algebra we know that two nonzero real numbers are equal when they have the same magnitude and algebraic sign. Hence, for two nonzero real numbers x, y, we have $x = y$ if $|x| = |y|$ and $xy > 0$, and conversely (that is, if $x = y$, then $|x| = |y|$ and $xy > 0$). When we deal with triangles in geometry, the notion of congruence arises. Here triangle ABC and triangle DEF are congruent if, for instance, they have equal corresponding sides—that is, the length of side AB = the length of side DE, the length of side BC = the length of side EF, and the length of side CA = the length of side FD.

Our study of logic is often referred to as the *algebra of propositions* (as opposed to the algebra of real numbers). In this algebra we shall use the truth tables of the statements, or propositions, to develop an idea of when two such entities are essentially the same. We begin with an example.

Example 2.6   In Table 2.6 we have the truth tables for the statements $\bar{p} \vee q$ and $p \rightarrow q$. Here we

Table 2.6

p	q	$\bar{p}$	$\bar{p} \vee q$	$p \rightarrow q$
0	0	1	1	1
0	1	1	1	1
1	0	0	0	0
1	1	0	1	1

see that the corresponding truth values for $\bar{p} \vee q$ and $p \rightarrow q$ are exactly the same. $\square$

This situation leads us to the following idea.

DEFINITION 2.2   Two statements s_1, s_2 are said to be *logically equivalent*, and we write $s_1 \Leftrightarrow s_2$, when the truth tables for s_1, s_2 are exactly the same.   ▄

As a result of this concept, we see that we can express the connective for implication in terms of negation and disjunction. In the same manner, from the result in Table 2.7 we have $(p \leftrightarrow q) \Leftrightarrow (p \rightarrow q) \wedge (q \rightarrow p)$. Using the logical equivalence from Table 2.6, we can also write $(p \leftrightarrow q) \Leftrightarrow (\bar{p} \vee q) \wedge (\bar{q} \vee p)$. Consequently, if we so choose, we can eliminate the connectives $\rightarrow$ and $\leftrightarrow$ from compound statements.

Examining Table 2.8, we find that negation, along with the connectives $\wedge$ and $\vee$, are all we need to replace the *exclusive or* connective, $\veebar$. In fact, we may even eliminate either $\wedge$ or $\vee$. However, for the applications we want to study, we need both $\wedge$ and $\vee$ as well as negation.

Table 2.7

p	q	$p \rightarrow q$	$q \rightarrow p$	$(p \rightarrow q) \wedge (q \rightarrow p)$	$p \leftrightarrow q$
0	0	1	1	1	1
0	1	1	0	0	0
1	0	0	1	0	0
1	1	1	1	1	1

Table 2.8

p	q	$p \veebar q$	$(p \vee q)$	$\overline{(p \wedge q)}$	$(p \vee q) \wedge \overline{(p \wedge q)}$
0	0	0	0	1	0
0	1	1	1	1	1
1	0	1	1	1	1
1	1	0	1	0	0

We now use the idea of logical equivalence to examine some of the important properties that hold for the algebra of propositions.

For any real numbers a, b, we know that $-(a + b) = (-a) + (-b)$. Is there a comparable result for propositions p, q?

Example 2.7   In Table 2.9 we have constructed the truth tables for the statements $\overline{p \wedge q}, \bar{p} \vee \bar{q}$, $\overline{p \vee q}$, and $\bar{p} \wedge \bar{q}$.

Table 2.9

p	q	$p \wedge q$	$\overline{p \wedge q}$	$\bar{p} \vee \bar{q}$	$p \vee q$	$\overline{p \vee q}$	$\bar{p} \wedge \bar{q}$
0	0	0	1	1	0	1	1
0	1	0	1	1	1	0	0
1	0	0	1	1	1	0	0
1	1	1	0	0	1	0	0

Columns 4 and 5 of the table reveal that $\overline{p \wedge q} \Leftrightarrow \bar{p} \vee \bar{q}$; columns 7 and 8 reveal that $\overline{p \vee q} \Leftrightarrow \bar{p} \wedge \bar{q}$. These results are known as *DeMorgan's Laws*. They are similar to the familiar law for real numbers,

$$-(a + b) = (-a) + (-b),$$

which shows the negative of a sum to be equal to the sum of the negatives. Here, however, a crucial difference emerges: The negation of the *conjunction* of two statements p, q results in the *disjunction* of their negations $\bar{p}$, $\bar{q}$, whereas the negation of the *disjunction* of p, q is logically equivalent to the *conjunction* of the negations $\bar{p}$, $\bar{q}$. □

In the arithmetic of real numbers, the operations of addition and multiplication are both involved in the principle called the Distributive Law of Multiplication over Addition: For all real numbers a, b, c,

$$a \times (b + c) = (a \times b) + (a \times c).$$

The next example shows that there is a similar law for propositions. There is also a second, related law for propositions that has no counterpart in the arithmetic of real numbers.

Example 2.8   Table 2.10 contains the truth tables for the statements $p \wedge (q \vee r)$, $(p \wedge q) \vee (p \wedge r)$, $p \vee (q \wedge r)$, and $(p \vee q) \wedge (p \vee r)$. From the table it follows that for any propositions p, q, and r,

Table 2.10

p	q	r	$p \wedge (q \vee r)$	$(p \wedge q) \vee (p \wedge r)$	$p \vee (q \wedge r)$	$(p \vee q) \wedge (p \vee r)$
0	0	0	0	0	0	0
0	0	1	0	0	0	0
0	1	0	0	0	0	0
0	1	1	0	0	1	1
1	0	0	0	0	1	1
1	0	1	1	1	1	1
1	1	0	1	1	1	1
1	1	1	1	1	1	1

$$p \wedge (q \vee r) \Leftrightarrow (p \wedge q) \vee (p \wedge r) \qquad \text{The Distributive Law of } \wedge \text{ over } \vee$$
$$p \vee (q \wedge r) \Leftrightarrow (p \vee q) \wedge (p \vee r) \qquad \text{The Distributive Law of } \vee \text{ over } \wedge \quad □$$

The second distributive law has no counterpart in the arithmetic of real numbers. That is, it is not true that for all real numbers a, b, and c, $a + (b \times c) = (a + b) \times (a + c)$. For $a = 2$, $b = 3$, and $c = 5$, for instance, $a + (b \times c) = 17$ but $(a + b) \times (a + c) = 35$.

Before going any further, we note that, in general, if s_1, s_2 are statements and $s_1 \leftrightarrow s_2$ is a tautology, then s_1, s_2 must have the same truth tables and $s_1 \Leftrightarrow s_2$. When s_1 and s_2 are logically equivalent statements, (that is, $s_1 \Leftrightarrow s_2$), then the compound statement $s_1 \leftrightarrow s_2$ is a tautology. Under these circumstances it is also true that $\bar{s}_1 \Leftrightarrow \bar{s}_2$, and $\bar{s}_1 \leftrightarrow \bar{s}_2$ is a tautology.

If s_1, s_2, and s_3 are statements where $s_1 \Leftrightarrow s_2$ and $s_2 \Leftrightarrow s_3$, then $s_1 \Leftrightarrow s_3$. When two statements s_1 and s_2 are not logically equivalent, we may write $s_1 \not\Leftrightarrow s_2$ to designate this situation.

Using the concepts of logical equivalence, tautology, and contradiction, we state the following list of laws for the algebra of propositions.

The Laws of Logic

For any propositions p, q, r,

1. $\overline{\overline{p}} \Leftrightarrow p$ Law of *Double Negation*

2. $\overline{p \vee q} \Leftrightarrow \overline{p} \wedge \overline{q}$ *DeMorgan's* Laws
   $\overline{p \wedge q} \Leftrightarrow \overline{p} \vee \overline{q}$

3. $p \vee q \Leftrightarrow q \vee p$ *Commutative* Laws
   $p \wedge q \Leftrightarrow q \wedge p$

4. $p \vee (q \vee r) \Leftrightarrow (p \vee q) \vee r$†   *Associative* Laws
   $p \wedge (q \wedge r) \Leftrightarrow (p \wedge q) \wedge r$

5. $p \vee (q \wedge r) \Leftrightarrow (p \vee q) \wedge (p \vee r)$ *Distributive* Laws
   $p \wedge (q \vee r) \Leftrightarrow (p \wedge q) \vee (p \wedge r)$

6. $p \vee p \Leftrightarrow p$ *Idempotent* Laws
   $p \wedge p \Leftrightarrow p$

7. $p \vee F_0 \Leftrightarrow p$ *Identity* Laws
   $p \wedge T_0 \Leftrightarrow p$

8. $p \vee \overline{p} \Leftrightarrow T_0$ *Inverse* Laws
   $p \wedge \overline{p} \Leftrightarrow F_0$

9. $p \vee T_0 \Leftrightarrow T_0$ *Domination* Laws
   $p \wedge F_0 \Leftrightarrow F_0$

10. $p \vee (p \wedge q) \Leftrightarrow p$ *Absorption* Laws
    $p \wedge (p \vee q) \Leftrightarrow p$

   To prove all these results, we could write down the truth tables and compare the corresponding truth values in each case, as we did in Examples 2.7 and 2.8. However, aside from the Law of Double Negation, these laws fall naturally into pairs. This leads us to the following concept.

DEFINITION 2.3   If s is a proposition (involving only $-$, $\wedge$, and $\vee$), the *dual* of s, denoted s^d, is the proposition obtained by replacing each occurrence of $\wedge (\vee)$ in s by $\vee (\wedge)$ and replacing each occurrence of $T_0 (F_0)$ by $F_0 (T_0)$.

   The statements $p \vee \overline{p}$ and $p \wedge \overline{p}$ are duals of each other, as are the statements $p \vee T_0$ and $p \wedge F_0$.

   We now state and use a theorem without proving it. However, in Chapter 15 we shall justify the result that appears here.

THEOREM 2.1   *(The Principle of Duality)* Let s, t be propositions as described in Definition 2.3. If $s \Leftrightarrow t$, then $s^d \Leftrightarrow t^d$.

   As a result, laws 2 through 10 in our list can be established by proving one of the laws in each pair and then invoking this principle.

† We note that because of the Associative Laws, there is no ambiguity in statements of the form $p \vee q \vee r$ or $p \wedge q \wedge r$.

We may derive many other logical equivalences by using the following *Substitution Rules*:

1. Let P denote a compound statement that is a tautology. If p is a statement that appears in P and we replace each occurrence of p by the *same* statement q, then the resulting compound statement P_1 is also a tautology.

2. Let P be a compound statement where p is a statement that appears in P, and let q be a statement such that $q \Leftrightarrow p$. Suppose that in P we replace p by q. Then this replacement yields the compound statement P_1. Under these circumstances $P_1 \Leftrightarrow P$.

These rules are illustrated in the following examples.

Example 2.9
a) From the first of DeMorgan's Laws we know that for any propositions p, q, the compound statement

$$P: \quad \overline{p \vee q} \leftrightarrow (\overline{p} \wedge \overline{q})$$

is a tautology. When we replace each occurrence of p by $r \wedge s$, it follows from the first substitution rule that

$$P_1: \quad \overline{(r \wedge s) \vee q} \leftrightarrow [\overline{(r \wedge s)} \wedge \overline{q}]$$

is also a tautology. Extending this result one step further, we may replace each occurrence of q by $t \rightarrow u$. The same substitution rule now yields the tautology

$$P_2: \quad \overline{(r \wedge s) \vee (t \rightarrow u)} \leftrightarrow [\overline{(r \wedge s)} \wedge \overline{(t \rightarrow u)}],$$

and the logical equivalence

$$P_2': \quad \overline{(r \wedge s) \vee (t \rightarrow u)} \Leftrightarrow \overline{(r \wedge s)} \wedge \overline{(t \rightarrow u)}.$$

b) The second Domination Law tells us that for any statement p, the compound statement

$$P: \quad (p \wedge F_0) \leftrightarrow F_0$$

is a tautology. If we replace p by the statement $[(q \vee r) \rightarrow s]$, then the same first substitution rule results in the new tautology

$$P_1: \quad ([(q \vee r) \rightarrow s] \wedge F_0) \leftrightarrow F_0,$$

and the logical equivalence

$$P_1': \quad [(q \vee r) \rightarrow s] \wedge F_0 \Leftrightarrow F_0. \quad \square$$

Example 2.10
a) For an application of the second substitution rule, let P denote the compound statement $(p \rightarrow q) \rightarrow r$. Because $(p \rightarrow q) \Leftrightarrow \overline{p} \vee q$ (as shown in Example 2.6 and Table 2.6), if P_1 denotes the compound statement $(\overline{p} \vee q) \rightarrow r$, then $P_1 \Leftrightarrow P$.

b) Now let P represent the compound statement $p \rightarrow (p \vee q)$. Because $\overline{\overline{p}} \Leftrightarrow p$, the compound statement $P_1: p \rightarrow (\overline{\overline{p}} \vee q)$ is derived from P by replacing the second occurrence (but not the first occurrence) of p by $\overline{\overline{p}}$. The second substitution rule still implies that $P_1 \Leftrightarrow P$. (Note that $P_2: \overline{\overline{p}} \to (\overline{\overline{p}} \vee q)$, derived by replacing both occurrences of p by $\overline{\overline{p}}$, is also logically equivalent to P.) ☐

Our next example demonstrates how we can use the idea of logical equivalence together with the laws of logic and the substitution rules.

Example 2.11   Negate and simplify the compound statement $(p \vee q) \to r$.
We organize our explanation as follows:

1. $(p \vee q) \to r \Leftrightarrow \overline{(p \vee q)} \vee r$ (by the first substitution rule, because $(s \to t) \leftrightarrow (\overline{s} \vee t)$ is a tautology for any statements s, t).

2. Negating the results in step 1, we have $\overline{(p \vee q) \to r} \Leftrightarrow \overline{\overline{(p \vee q)} \vee r}$.

3. From the first of DeMorgan's Laws and the first substitution rule, $\overline{\overline{(p \vee q)} \vee r} \Leftrightarrow \overline{\overline{(p \vee q)}} \wedge \overline{r}$.

4. The Law of Double Negation and the first substitution rule now give us $\overline{\overline{(p \vee q)}} \wedge \overline{r} \Leftrightarrow (p \vee q) \wedge \overline{r}$.

From steps 1 through 4 we have $\overline{(p \vee q) \to r} \Leftrightarrow (p \vee q) \wedge \overline{r}$. ☐

Example 2.12   In Definition 2.3 the dual s^d of a statement s was defined only for statements involving negation and the basic connectives $\wedge$ and $\vee$. How does one determine the dual of a statement such as $s: p \to q$?
Because $(p \to q) \Leftrightarrow \overline{p} \vee q$, s^d is logically equivalent to the statement $(\overline{p} \vee q)^d$, which is $\overline{p} \wedge q$. ☐

The implication $p \to q$ and certain statements related to it are examined in the following example.

Example 2.13   Table 2.11 gives the truth tables for the statements $p \to q$, $\overline{q} \to \overline{p}$, $q \to p$, and $\overline{p} \to \overline{q}$.

Table 2.11

p	q	$p \to q$	$\overline{q} \to \overline{p}$	$q \to p$	$\overline{p} \to \overline{q}$
0	0	1	1	1	1
0	1	1	1	0	0
1	0	0	0	1	1
1	1	1	1	1	1

The third and fourth columns of the table reveal that

$$(p \to q) \Leftrightarrow (\overline{q} \to \overline{p}).$$

The statement $\bar{q} \rightarrow \bar{p}$ is called the *contrapositive* of the implication $p \rightarrow q$. Columns 5 and 6 of the table show that

$$(q \rightarrow p) \Leftrightarrow (\bar{p} \rightarrow \bar{q}).$$

The statement $q \rightarrow p$ is called the *converse* of $p \rightarrow q$; $\bar{p} \rightarrow \bar{q}$ is termed the *inverse* of $p \rightarrow q$.

In the case where p, q represent the statements

p: Quadrilateral $ABCD$ is a square.
q: Quadrilateral $ABCD$ is equilateral.

we obtain

- (The implication: $p \rightarrow q$) If quadrilateral $ABCD$ is a square, then quadrilateral $ABCD$ is equilateral. (TRUE)
- (The contrapositive: $\bar{q} \rightarrow \bar{p}$) If quadrilateral $ABCD$ is not equilateral, then it is not a square. (Likewise TRUE)
- (The converse: $q \rightarrow p$) If quadrilateral $ABCD$ is equilateral, then quadrilateral $ABCD$ is a square. (FALSE; any rhombus that is not equiangular is still equilateral.)
- (The inverse: $\bar{p} \rightarrow \bar{q}$) If quadrilateral $ABCD$ is not a square, then it is not equilateral. (Likewise FALSE)

As this particular example demonstrates, we must avoid "arguing by the converse." The fact that a certain implication (or theorem) $p \rightarrow q$ is true does *not* require that the converse $q \rightarrow p$ must also be true. However, it does necessitate the truth of the contrapositive $\bar{q} \rightarrow \bar{p}$. □

The next example shows that logically equivalent statements may lead to different situations in a computer science application.

Example 2.14 Table 2.12 reveals that the compound statements $(p \wedge q) \rightarrow r$ and $p \rightarrow (q \rightarrow r)$ are logically equivalent.

Table 2.12

p	q	r	$p \wedge q$	$(p \wedge q) \rightarrow r$	$q \rightarrow r$	$p \rightarrow (q \rightarrow r)$
0	0	0	0	1	1	1
0	0	1	0	1	1	1
0	1	0	0	1	0	1
0	1	1	0	1	1	1
1	0	0	0	1	1	1
1	0	1	0	1	1	1
1	1	0	1	0	0	0
1	1	1	1	1	1	1

In the Pascal program segments shown in Fig. 2.1, x, y, z, and i are integer variables. Part (a) of the figure uses a decision structure *comparable to* a statement of the form $(p \land q) \to r$. Here, as well as in part (b), we have p: $x > 0$, q: $y > 0$, which become statements when the variables x, y are assigned the values $4 - i$ (for x) and $4 + 3*i$ (for y). Now be careful! The letter r denotes the *Writeln* statement, an 'executable statement' which is actually *not* a statement in the usual sense of a declarative sentence that can be labeled true or false.

In the segment shown in part (a), the total number of comparisons, $(x > 0)$ and $(y > 0)$, that are carried out during program execution is 10 (for $x > 0$) + 10 (for $y > 0$) = 20. The segment shown in part (b), on the other hand, uses a statement form *comparable to* the nested implications $p \to (q \to r)$. In this case the comparison $(y > 0)$ is not executed unless the comparison $(x > 0)$ is executed and evaluated as true. Consequently, the total number of comparisons here is 10 (for $x > 0$) + 3 (for $y > 0$, when i is assigned 1, 2, 3) = 13. Hence, in terms of the total number of comparisons made in each of these two cases, the program segment shown in part (b) is more efficient than the program segment shown in part (a). □

```
    z := 4;
    For i := 1 to 10 do
      Begin
        x := z - i;
        y := z + 3*i;
        If (x > 0) and (y > 0) then
            Writeln ('The value of the sum x + y is ', x + y)
      End;

(a)
```

```
    z := 4;
    For i := 1 to 10 do
      Begin
        x := z - i;
        y := z + 3*i;
        If x > 0 then
          If y > 0 then
              Writeln ('The value of the sum x + y is ', x + y)
      End;

(b)
```

Figure 2.1

We close this section with an application in simplifying switching networks. Here the laws of logic prove quite helpful.

A switching network is made up of wires and switches connecting two terminals T_1 and T_2. In such a network, any switch is either open (0), so that no current flows through it, or closed (1), so that current does flow through it.

In Fig. 2.2 we have in part (a) a network with one switch. Each of parts (b) and (c) contains two (independent) switches.

For the network in part (b), current flows from T_1 to T_2 if either of the switches p, q is closed. We call this a *parallel* network and represent it by $p \vee q$. The network in part (c) requires that each of the switches p, q be closed in order for current to flow from T_1 to T_2. Here the switches are in *series*, and this network is represented by $p \wedge q$.

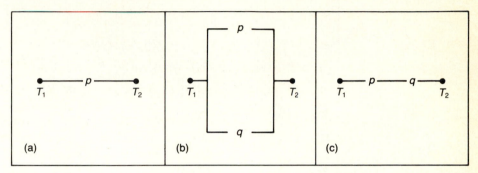

Figure 2.2

Example 2.15   The switches in a network need not act independently of each other. Consider the network shown Fig. 2.3(a). Here the switches labeled t and $\bar{t}$ are not independent. We have coupled these two switches so that t is open (closed) if and only if $\bar{t}$ is simultaneously closed (open). The same is true of the switches at q, $\bar{q}$. (Also, for example, the three switches labeled p are not independent.)

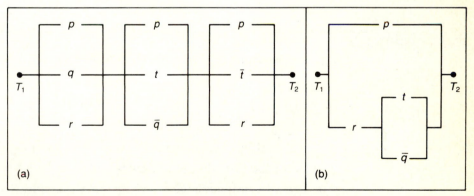

Figure 2.3

This network is represented by $(p \vee q \vee r) \wedge (p \vee t \vee \bar{q}) \wedge (p \vee \bar{t} \vee r)$. Using the laws of logic, we simplify this proposition, which represents the network, as follows. (Here, for simplicity, we shall not mention applications of the substitution rules, but we shall list the major laws of logic being used.)

$(p \vee q \vee r) \wedge (p \vee t \vee \bar{q}) \wedge (p \vee \bar{t} \vee r)$	**Reason**
$\Leftrightarrow p \vee [(q \vee r) \wedge (t \vee \bar{q}) \wedge (\bar{t} \vee r)]$	Distributive Law of $\vee$ over $\wedge$
$\Leftrightarrow p \vee [([(q \vee r) \wedge t] \vee [(q \vee r) \wedge \bar{q}]) \wedge (\bar{t} \vee r)]$	Distributive Law of $\wedge$ over $\vee$
$\Leftrightarrow p \vee [([(q \vee r) \wedge t] \vee [(q \wedge \bar{q}) \vee (r \wedge \bar{q})]) \wedge (\bar{t} \vee r)]$	Distributive Law of $\wedge$ over $\vee$
$\Leftrightarrow p \vee [([(q \vee r) \wedge t] \vee (r \wedge \bar{q})) \wedge (\bar{t} \vee r)]$	$q \wedge \bar{q} \Leftrightarrow F_0$, the identity for $\vee$
$\Leftrightarrow p \vee [([(q \vee r) \wedge t] \wedge (\bar{t} \vee r)) \vee ((r \wedge \bar{q}) \wedge (\bar{t} \vee r))]$	Why?
$\Leftrightarrow p \vee [[(q \vee r) \wedge (t \wedge (\bar{t} \vee r))] \vee [(r \wedge \bar{q} \wedge \bar{t}) \vee (r \wedge \bar{q} \wedge r)]]$	Why?
$\Leftrightarrow p \vee [[(q \vee r) \wedge (t \wedge r)] \vee [(r \wedge \bar{q} \wedge \bar{t}) \vee (r \wedge \bar{q})]]$	Why?
$\Leftrightarrow p \vee [[(q \vee r) \wedge (t \wedge r)] \vee (r \wedge \bar{q})]$	Absorption Law
$\Leftrightarrow p \vee [(q \wedge t \wedge r) \vee (r \wedge t \wedge r) \vee (r \wedge \bar{q})]$	Distributive Law of $\wedge$ over $\vee$
$\Leftrightarrow p \vee [((q \wedge t \wedge r) \vee (r \wedge t)) \vee (r \wedge \bar{q})]$	Idempotent Law
$\Leftrightarrow p \vee [(r \wedge t) \vee (r \wedge \bar{q})]$	Absorption Law
$\Leftrightarrow p \vee [r \wedge (t \vee \bar{q})]$	Distributive Law of $\wedge$ over $\vee$

Hence $(p \vee q \vee r) \wedge (p \vee t \vee \bar{q}) \wedge (p \vee \bar{t} \vee r) \Leftrightarrow p \vee [r \wedge (t \vee \bar{q})]$, and the network shown in Fig. 2.3(b) is equivalent to the original network in the sense that current flows from T_1 to T_2 in network (a) exactly when it does so in network (b). But in (b) the network has only four switches, five fewer than in network (a). □

EXERCISES 2.2

1. **a)** Use truth tables to verify the following logical equivalences.

   (i) $p \rightarrow (q \wedge r) \Leftrightarrow (p \rightarrow q) \wedge (p \rightarrow r)$

   (ii) $[(p \vee q) \rightarrow r] \Leftrightarrow [(p \rightarrow r) \wedge (q \rightarrow r)]$

   (iii) $[p \rightarrow (q \vee r)] \Leftrightarrow [\bar{r} \rightarrow (p \rightarrow q)]$

   b) Use the substitution rules to show that $[p \rightarrow (q \vee r)] \Leftrightarrow [(p \wedge \bar{q}) \rightarrow r]$.

2. Verify the first Absorption Law by means of a truth table.

3. Use the substitution rules to verify that each of the following are tautologies.

   a) $[p \vee (q \wedge r)] \vee \overline{[p \vee (q \wedge r)]}$

   b) $[(p \vee q) \rightarrow r] \leftrightarrow [\bar{r} \rightarrow \overline{(p \vee q)}]$

   c) $[[(p \vee q) \rightarrow r] \vee (s \wedge t)] \leftrightarrow [[(p \vee q) \rightarrow r] \vee s] \wedge [[(p \vee q) \rightarrow r] \vee t]$

4. Simplify the statement $[[[(p \wedge q) \wedge r] \vee [(p \wedge q) \wedge \bar{r}]] \vee \bar{q}] \rightarrow s$.

5. Negate each of the following statements.

a) Kelsey will get a good education if she puts her studies before her interest in cheerleading.

b) Norma is doing her mathematics homework and Karen is practicing her piano lessons.

c) If Logan goes on vacation, then he will enjoy himself if he doesn't worry about traveling by airplane.

d) If Harold passes his Pascal course and finishes his data structures project, then he will graduate at the end of the semester.

6. Negate each of the following, and simplify the resulting proposition.

   a) $p \wedge (q \vee r) \wedge (\bar{p} \vee \bar{q} \vee r)$ **b)** $(p \wedge q) \rightarrow r$

   c) $p \rightarrow (\bar{q} \wedge r)$ **d)** $p \vee q \vee (\bar{p} \wedge \bar{q} \wedge r)$

   e) $p \leftrightarrow q$ **f)** $p \wedge (p \rightarrow q) \wedge (q \rightarrow r)$

7. **a)** Prove that $(\bar{p} \vee q) \wedge (p \wedge (p \wedge q)) \Leftrightarrow (p \wedge q)$.

   b) Write the dual of the result in part (a).

8. Write the dual for (a) $q \rightarrow p$, (b) $p \rightarrow (q \wedge r)$, (c) $p \leftrightarrow q$, and (d) $p \veebar q$.

9. Write the converse, inverse, and contrapositive of each of the following statements.

   a) If quadrilateral $ABCD$ is a square, then quadrilateral $ABCD$ is a rectangle.

   b) If n is an even integer, then n^2 is an even integer.

   c) If two distinct lines in the xy plane are parallel, then they have the same slope.

10. Determine whether each of the following statements is true or false. Here p, q, and r are statements.

   a) One form of the converse of "p is sufficient for q" is "p is necessary for q."

   b) One form of the inverse of "p is necessary for q" is "$\bar{q}$ is sufficient for $\bar{p}$."

   c) One form of the contrapositive of "p is necessary for q" is "$\bar{q}$ is necessary for $\bar{p}$."

   d) One form of the contrapositive of "p only if q" is "$\bar{q}$ is necessary for $\bar{p}$."

   e) One form of the converse of $p \rightarrow (q \rightarrow r)$ is $(\bar{q} \vee r) \rightarrow p$.

11. Find a form of the contrapositive of $p \rightarrow (q \rightarrow r)$ with (a) only one occurrence of the connective $\rightarrow$; (b) no occurrences of the connective $\rightarrow$.

12. Write the converse, inverse, contrapositive, and negation of the following statement: "If Sandra finishes her work, she will go to the basketball game unless it snows."

13. Consider the two Pascal program segments shown in Fig. 2.1 (Example 2.14). Find the total number of comparisons executed in each segment if, instead of being assigned the value 4, z were assigned the value (a) 2; (b) 6; (c) 9; (d) 10; (e) 15; (f) n, an integer greater than 10.

14. Show that $p \veebar q \Leftrightarrow [(p \wedge \bar{q}) \vee (\bar{p} \wedge q)] \Leftrightarrow \overline{(p \leftrightarrow q)}$.

15. Verify that $[(p \leftrightarrow q) \wedge (q \leftrightarrow r) \wedge (r \leftrightarrow p)] \Leftrightarrow [(p \rightarrow q) \wedge (q \rightarrow r) \wedge (r \rightarrow p)]$.

16. **a)** Verify that $p \rightarrow [q \rightarrow (p \wedge q)]$ is a tautology.

    b) Verify that $(p \vee q) \rightarrow [q \rightarrow q]$ is a tautology by using the result from part (a) along with the substitution rules and the laws of logic.

    c) Is $(p \vee q) \rightarrow [q \rightarrow (p \wedge q)]$ a tautology?

17. Define the connective "Nand," or "Not . . . and," by $(p \uparrow q) \Leftrightarrow \overline{(p \wedge q)}$ for propositions p, q. Represent the following using only this connective.

    a) $\bar{p}$ **b)** $p \vee q$ **c)** $p \wedge q$

    d) $p \rightarrow q$ **e)** $p \leftrightarrow q$

18. The connective "Nor," or "Not . . . or," is defined by $(p \downarrow q) \Leftrightarrow \overline{(p \vee q)}$ for propositions p, q. Represent the propositions in parts (a) through (e) of Exercise 17, using only this connective.

19. If p, q are propositions, prove that

    a) $\overline{(p \downarrow q)} \Leftrightarrow (\bar{p} \uparrow \bar{q})$ **b)** $\overline{(p \uparrow q)} \Leftrightarrow (\bar{p} \downarrow \bar{q})$

20. Simplify each of the networks shown in Fig. 2.4.

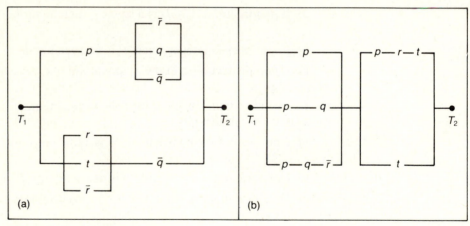

(a)   (b)

Figure 2.4

2.3
LOGICAL IMPLICATION: RULES OF INFERENCE

In Section 2.1 we mentioned the notion of a *theorem*: a statement (generally mathematical) that we can prove to be true. Now it is time for us to begin a formal study of the notion of *proof*, the way in which we argue the truth of a theorem.

There is no list of rules one may apply to decide whether a given statement is a theorem. The construction of a proof is a skill that one learns by carefully examining the forms of correct proofs and practicing the developing and writing of proofs of statements, some of which are already known to be theorems. As in all types of mathematical problems, we cannot be mere spectators. To acquire the ability to do proofs, we must study and practice.

Let us first consider the general form of a statement that we want to show is a theorem:

$$(p_1 \wedge p_2 \wedge p_3 \wedge \cdots \wedge p_n) \to q.$$

Here the statements $p_1, p_2, p_3, \ldots, p_n$ denote the hypotheses of the implication. These often come from (1) definitions in the subject area being studied, (2) axioms and postulates—statements in the subject area that are accepted without proof, and (3) theorems that we have already proved. In the attempt to verify that the foregoing statement is a theorem, we want to show that when all the hypotheses $p_1, p_2, p_3, \ldots, p_n$ are true, then q is also true. (Note that if any one of $p_1, p_2, p_3, \ldots, p_n$ is false, then the implication $(p_1 \wedge p_2 \wedge p_3 \wedge \cdots \wedge p_n) \to q$ is automatically true.) Consequently, one way to establish the given statement as a theorem is to show that $(p_1 \wedge p_2 \wedge p_3 \wedge \cdots \wedge p_n) \to q$ is a tautology.

The following example illustrates this technique.

Example 2.16 Let p_1, p_2, p_3, and q be the following statements:

p_1: If Roger studies, then he will pass discrete mathematics.
p_2: If Roger doesn't play tennis, then he'll study.
p_3: Roger failed discrete mathematics.
q: Therefore, Roger played tennis.

We want to know whether the truth of the hypotheses p_1, p_2, and p_3 leads to the truth of the conclusion q. To determine this, let p, r, and s be given as

p: Roger studies.
r: Roger passes discrete mathematics.
s: Roger plays tennis.

We then rewrite p_1, p_2, p_3, and q as

p_1: $p \to r$ p_2: $\bar{s} \to p$ p_3: $\bar{r}$ q: s.

We can now determine whether or not $(p_1 \wedge p_2 \wedge p_3) \to q$ is a theorem by considering the truth table (Table 2.13) for

$$[(p \to r) \wedge (\bar{s} \to p) \wedge \bar{r}] \to s.$$

Table 2.13

			p_1	p_2	p_3	$(p_1 \wedge p_2 \wedge p_3) \to q$
p	r	s	$p \to r$	$\bar{s} \to p$	$\bar{r}$	$[(p \to r) \wedge (\bar{s} \to p) \wedge \bar{r}] \to s$
0	0	0	1	0	1	1
0	0	1	1	1	1	1
0	1	0	1	0	0	1
0	1	1	1	1	0	1
1	0	0	0	1	1	1
1	0	1	0	1	1	1
1	1	0	1	1	0	1
1	1	1	1	1	0	1

Because the final column in Table 2.13 contains all 1's, the result is a tautology. Hence we can say that $(p_1 \wedge p_2 \wedge p_3) \to q$ is a theorem.   □

This idea leads to the following definition.

DEFINITION 2.4   If p, q are statements such that $p \to q$ is a tautology, then we say that p *logically implies* q, and we write $p \Rightarrow q$ to denote this.   ▬

In the case where p, q are statements and $p \Rightarrow q$, the implication $p \to q$ is a tautology and is referred to as a *logical implication*.

In Example 2.5 we found that for any statements p, q, the proposition $p \to (p \vee q)$ is always true (it is a tautology). Therefore we can say that p logically implies $p \vee q$ and write $p \Rightarrow (p \vee q)$.

Let p, q be any statements.

1.   If $p \Leftrightarrow q$, then the statement $p \leftrightarrow q$ is a tautology, so the statements p, q always have the same truth values. Under these conditions, the statements $p \to q$, $q \to p$ will always be true, so we have $p \Rightarrow q$ and $q \Rightarrow p$.

2.   Conversely, suppose that $p \Rightarrow q$ and $q \Rightarrow p$. The logical implication $p \Rightarrow q$ tells us that we never have statement p with the truth value 1 and statement q with the truth value 0. But could we have q with the truth value 1 and p with the truth value 0? If this occurred, we could not have the logical implication $q \Rightarrow p$. Therefore, when $p \Rightarrow q$ and $q \Rightarrow p$, the statements p, q have the same truth values and $p \Leftrightarrow q$.

Finally, the notation $p \not\Rightarrow q$ is used to indicate that $p \to q$ is *not* a logical implication.

Example 2.17   From the results in Example 2.7 (see Table 2.9), we know that for any statements p, q,

$$\overline{p \wedge q} \Leftrightarrow \overline{p} \vee \overline{q}.$$

Consequently,

$$\overline{p \wedge q} \Rightarrow \overline{p} \vee \overline{q} \quad \text{and} \quad \overline{p} \vee \overline{q} \Rightarrow \overline{p \wedge q}$$

for any statements p, q. We can also denote this by realizing that because each of the statements

$$\overline{p \wedge q} \to \overline{p} \vee \overline{q} \quad \text{and} \quad \overline{p} \vee \overline{q} \to \overline{p \wedge q}$$

is a tautology, we may write

$$[\overline{p \wedge q} \to \overline{p} \vee \overline{q}] \Leftrightarrow T_0 \quad \text{and} \quad [\overline{p} \vee \overline{q} \to \overline{p \wedge q}] \Leftrightarrow T_0.   □$$

Returning now to our study of techniques for proving theorems (or logical implications), we must take a careful look at the size of Table 2.13. The table has eight rows because we were able to represent the three hypotheses p_1, p_2, and p_3

and the conclusion q in terms of the three statements p, r, and s. If we were confronted, for example, with establishing whether or not

$$[(p \to r) \wedge (r \to s) \wedge (t \vee \bar{s}) \wedge (\bar{t} \vee u) \wedge \bar{u}] \to \bar{p}$$

is a logical implication (or theorem), the needed table would require $2^5 = 32$ rows. As the number of hypotheses gets larger and our truth tables grow to 64, 128, 256 or more rows, this approach rapidly loses its appeal.

Furthermore, looking at Table 2.13 once again, we realize that in establishing whether or not

$$[(p \to r) \wedge (\bar{s} \to p) \wedge \bar{r}] \to s$$

is a logical implication we only need to consider those rows of the table where each of the three hypotheses $p \to r$, $\bar{s} \to p$, and $\bar{r}$ has the truth value 1. (Remember that if the left-hand statement is false, we do not care about the truth value of the right-hand statement.) This happens only in the second row, so a good deal of Table 2.13 is not really necessary. (It is not always the case that only one row has all of the hypotheses true. This is demonstrated in the first of the Section 2.3 exercises.)

Consequently, what is needed here is a technique or list of techniques that somehow bypass the need for constructing any truth tables, especially large ones. These techniques are called *rules of inference*. They help us as follows:

1. Using these techniques enables us to consider only the cases wherein all the hypotheses are true. Hence we consider only the rows of a truth table wherein each hypothesis has the truth value 1 and we do *not* construct the truth table.

2. The rules of inference are fundamental in the development of a step-by-step *proof* showing how the conclusion q logically follows from the hypotheses $p_1, p_2, p_3, \ldots, p_n$ in an implication

$$(p_1 \wedge p_2 \wedge p_3 \wedge \cdots \wedge p_n) \to q.$$

Such a development establishes the validity of our proof (or *argument*).

Each rule of inference arises from a logical implication or logical equivalence. In each case, the logical implication or logical equivalence is stated without proof. (Several of these proofs will be dealt with in the section exercises.)

Many rules of inference arise in the study of logic. We concentrate on those that we need to help us establish the theorems discussed in the remaining chapters of this textbook. Table 2.14 (on p. 68) summarizes the rules we shall now start to investigate.

Example 2.18   As a first example we consider the rule of inference called *Modus Ponens*, or the *Rule of Detachment*. In symbolic form this rule is expressed by the logical implication

$$[p \wedge (p \to q)] \Rightarrow q.$$

The actual rule will be written in the tabular form

$$
\begin{array}{l}
p \\
\underline{p \rightarrow q} \\
\therefore\ q
\end{array}
$$

where the three dots ($\therefore$) stand for the word "Therefore," indicating that q is a logical consequence of the hypotheses p and $p \rightarrow q$, which appear above the horizontal line. We also say that the hypotheses (statements above the horizontal line) present a *valid argument* for, or *proof* of, the conclusion q.

This rule arises in situations where we are arguing that *if* (1) p is true, *and* (2) $p \rightarrow q$ is true (or $p \Rightarrow q$), *then* the conclusion q must also be true. (After all, if q were false and p were true, then we could not have $p \rightarrow q$ true.)

The following proof illustrates how you may have applied the Rule of Detachment in high school geometry.

(1) Triangle ABC is equilateral. p

(2) If triangle ABC is equilateral, then triangle ABC is isosceles. $\underline{p \rightarrow q}$

(3) Therefore triangle ABC is isosceles. $\therefore\ q$

                                                      □

Example 2.19   A second rule of inference is given by the logical implication

$$[(p \rightarrow q) \wedge (q \rightarrow r)] \Rightarrow p \rightarrow r,$$

where p, q, and r are any statements. In tabular form it is written

$$
\begin{array}{l}
p \rightarrow q \\
\underline{q \rightarrow r} \\
\therefore\ p \rightarrow r
\end{array}
$$

This rule, which is referred to as the *Law of the Syllogism*, arises in many proofs and other mathematical arguments. In fact, we use it (perhaps unknowingly) many times in solving simple algebra equations, as the following solution (proof) demonstrates.

(1) If $3x - 7 = 20$, then $3x = 27$. $p \rightarrow q$

(2) If $3x = 27$, then $x = 9$. $\underline{q \rightarrow r}$

(3) Therefore, if $3x - 7 = 20$, then $x = 9$. $\therefore\ p \rightarrow r$ □

The next example involves a slightly longer proof that uses the rules of inference developed in Examples 2.18 and 2.19. In fact, we find here that there may be more than one way to establish the validity of an argument.

Example 2.20   Consider the following argument.

(1) In triangle ABC, the lengths of sides AB and AC are equal. p

(2) If a triangle ABC has two equal sides, then the triangle is
     isosceles. $p \rightarrow q$

(3) If triangle ABC is isosceles, then the angles opposite the sides of equal length are equal.

$$q \to r$$

(4) Therefore, the angles at B and C in triangle ABC are equal. $\quad \therefore \; r$

Concentrating on the forms of the statements in the preceding argument, we may write the argument more compactly as

(*)
$$\begin{array}{c} p \\ p \to q \\ \underline{q \to r} \\ \therefore \; r \end{array}$$

Now we need no longer worry about what the statements actually stand for. Our objective is to use the two rules of inference that we have studied so far in order to deduce the truth of statement r from the truth of the three hypotheses p, $p \to q$, and $q \to r$.

Our proof is demonstrated as follows:

Steps	Reasons
(1) $p \to q$	Hypothesis
(2) $q \to r$	Hypothesis
(3) $p \to r$	This follows from steps (1) and (2) and the Law of the Syllogism.
(4) p	Hypothesis
(5) $\therefore \; r$	This follows from steps (4) and (3) and the Rule of Detachment.

Before continuing with a third rule of inference, we shall show that a second proof can be given for the argument presented at (*). Here our "reasons" will be shortened to the form we shall use for the rest of the section. However, we shall always list whatever is needed to demonstrate how each step in a proof comes about, or follows, from prior steps.

A second proof for the argument is

Steps	Reasons
(1) p	Hypothesis
(2) $p \to q$	Hypothesis
(3) q	(1), (2), and the Rule of Detachment
(4) $q \to r$	Hypothesis
(5) $\therefore \; r$	(3), (4), and the Rule of Detachment $\quad \square$

Example 2.21   The rule of inference called *Modus Tollens* is given by

$$\begin{array}{c} p \to q \\ \underline{\bar{q}} \\ \therefore \; \bar{p} \end{array}$$

This follows from the logical implication $[(p \rightarrow q) \wedge \bar{q}] \Rightarrow \bar{p}$.

We will use this rule in developing a proof for the argument

$$
\begin{array}{c}
p \rightarrow r \\
r \rightarrow s \\
t \vee \bar{s} \\
\bar{t} \vee u \\
\underline{\bar{u}} \\
\therefore \bar{p}
\end{array}
$$

Both Modus Tollens and the Law of the Syllogism come into play, along with the logical equivalence we developed in Example 2.6.

Steps	**Reasons**
(1) $p \rightarrow r, r \rightarrow s$	Hypotheses
(2) $p \rightarrow s$	(1) and the Law of the Syllogism
(3) $t \vee \bar{s}$	Hypothesis
(4) $\bar{s} \vee t$	(3) and the Commutative Law of $\vee$
(5) $s \rightarrow t$	(4) and the logical equivalence $\bar{s} \vee t \Leftrightarrow s \rightarrow t$
(6) $p \rightarrow t$	(2), (5) and the Law of the Syllogism
(7) $\bar{t} \vee u$	Hypothesis
(8) $t \rightarrow u$	(7) and the logical equivalence $\bar{t} \vee u \Leftrightarrow t \rightarrow u$
(9) $p \rightarrow u$	(6), (8) and the Law of the Syllogism
(10) $\bar{u}$	Hypothesis
(11) $\therefore \bar{p}$	(9), (10) and Modus Tollens □

Before proceeding further, we should mention a fairly simple but important rule of inference.

Example 2.22   The following rule of inference arises from the observation that if p, q are statements where $p \Leftrightarrow T_0$ and $q \Leftrightarrow T_0$, then $p \wedge q \Leftrightarrow T_0$.

Now suppose that statements p, q occur in the development of a proof. These statements may be (given) hypotheses or results that are derived from hypotheses and/or from results developed earlier in the proof. Then under these circumstances the two statements p, q can be combined into their conjunction $p \wedge q$, and this new statement can be used in later steps as the proof continues.

We call this rule the *Rule of Conjunction* and write it in tabular form as

$$
\begin{array}{c}
p \\
\underline{q} \\
\therefore p \wedge q
\end{array}
$$

Its use is demonstrated in Example 2.23.   □

Our next rule of inference is sometimes confused with the contrapositive method of proof used in Modus Tollens, because both rules use the negation of a statement. However, they are two distinct rules.

Example 2.23 Let us first recall that if s is a statement and $\bar{s} \Rightarrow F_0$, then the statement $\bar{s} \rightarrow F_0$ is always true. Therefore $\bar{s}$ is false and, consequently, s is true.

From the logical implication $\bar{s} \Rightarrow F_0$ we obtain the rule of inference called the *Proof by Contradiction*, or *Reductio ad Absurdum*. In tabular form this is written as

$$\frac{\bar{s} \Rightarrow F_0}{\therefore \ s}$$

We demonstrate the use of this rule in the following proof. First let us recall that an integer n is called *even* if there exists an integer k such that $n = 2k$. An integer m is called *odd* if there exists an integer k such that $m = 2k + 1$. A real number x is called *rational* if $x = a/b$, where a and b are integers and $b \neq 0$. If a real number is not rational it is termed *irrational*.

Now suppose that we wish to prove the result

s: The real number $\sqrt{2}$ is an irrational number.

Instead of trying to prove the statement (directly), we shall demonstrate (indirectly) how $\bar{s} \Rightarrow (r \wedge \bar{r})$ for some statement r. (Hence the Proof by Contradiction is sometimes referred to as an *indirect proof*).

Now suppose that $\sqrt{2}$ is rational (that is, assume $\bar{s}$ is true). Then $\sqrt{2} = a/b$, where a and b are integers and $b \neq 0$. We further assume that a/b is written in lowest terms. Therefore we have the statement

r: The only positive integer that divides both a and b is 1.

Because $(\sqrt{2})^2 = 2$, it follows that $(a/b)^2 = 2$, or $a^2 = 2b^2$. With $a^2 = 2b^2$ we find that a^2 is even; if a^2 is even, it follows that a is also even. So $a = 2c$ for some integer c.

This leads to $2b^2 = a^2 = (2c)^2 = 4c^2$, and now $b^2 = 2c^2$. But with b^2 even, b must also be even. Consequently, the integers a and b have 2 as a second positive divisor. So 1 is *not* the only positive integer that divides both a and b; this gives us the statement $\bar{r}$.

Now how did we end up with both r and $\bar{r}$ (and the contradiction $r \wedge \bar{r}$, by the Rule of Conjunction)? What caused this to happen is that we assumed $\bar{s}$ to be true when we wrote "Suppose that $\sqrt{2}$ is rational." Consequently, this supposition is actually false, so its negation, namely s, must be true.

Therefore, because $\bar{s} \Rightarrow (r \wedge \bar{r})$, we have $\bar{s} \Leftrightarrow F_0$. As a result, $s \Leftrightarrow T_0$, and this means that $\sqrt{2}$ is irrational. □

A special case of the Proof by Contradiction,

$$[\bar{s} \Rightarrow (r \wedge \bar{r})] \Rightarrow (s \Leftrightarrow T_0),$$

is worth noting here.

Suppose we wish to establish the theorem $p \Rightarrow q$. Replacing statement s above by the statement $p \rightarrow q$, we see that it may be possible to verify that $\overline{(p \rightarrow q)} \Rightarrow (r \wedge \bar{r})$ for some statement r. It will then follow that $(p \rightarrow q) \Leftrightarrow T_0$, or $p \Rightarrow q$.

Because $\overline{(p \rightarrow q)} \Leftrightarrow \overline{(\bar{p} \vee q)} \Leftrightarrow p \wedge \bar{q}$, this special case of the Proof by Contradiction may also be expressed as

$$[(p \wedge \bar{q}) \Rightarrow (r \wedge \bar{r})] \Rightarrow (p \Rightarrow q).$$

It will arise in many situations throughout later chapters of the text.

   Now that we have examined five rules of inference, we shall summarize these rules and introduce several others in Table 2.14.

Table 2.14

Rule of inference	Related logical implication or logical equivalence	Name of rule
1. p $p \rightarrow q$ $\therefore\ q$	$[p \wedge (p \rightarrow q)] \Rightarrow q$	Rule of Detachment (Modus Ponens)
2. $p \rightarrow q$ $q \rightarrow r$ $\therefore\ p \rightarrow r$	$[(p \rightarrow q) \wedge (q \rightarrow r)] \Rightarrow (p \rightarrow r)$	Law of the Syllogism
3. $p \rightarrow q$ $\bar{q}$ $\therefore\ \bar{p}$	$[(p \rightarrow q) \wedge \bar{q}] \Rightarrow \bar{p}$	Modus Tollens
4. p q $\therefore\ p \wedge q$		Rule of Conjunction
5. $\bar{s} \Rightarrow F_0$ $\therefore\ s$	$(\bar{s} \Rightarrow F_0) \Rightarrow (s \Leftrightarrow T_0)$	Proof by Contradiction (Reductio ad Absurdum)
5'. $(p \wedge \bar{q}) \Rightarrow F_0$ $\therefore\ p \Rightarrow q$	$[(p \wedge \bar{q}) \Rightarrow F_0] \Rightarrow (p \Rightarrow q)$	A special case of the Proof by Contradiction
6. $p \wedge q$ $\therefore\ p$	$(p \wedge q) \Rightarrow p$	Rule of Conjunctive Simplification
7. p $\therefore\ p \vee q$	$p \Rightarrow p \vee q$	Rule of Disjunctive Amplification
8. $p \vee q$ $\bar{p}$ $\therefore\ q$	$[(p \vee q) \wedge \bar{p}] \Rightarrow q$	Rule of Disjunctive Syllogism
9. $p \wedge q$ $p \rightarrow (q \rightarrow r)$ $\therefore\ r$	$[p \rightarrow (q \rightarrow r)] \Leftrightarrow [(p \wedge q) \rightarrow r]$	Rule of Conditional Proof
10. $p \rightarrow r$ $q \rightarrow r$ $\therefore\ (p \vee q) \rightarrow r$	$[(p \rightarrow r) \wedge (q \rightarrow r)] \Rightarrow [(p \vee q) \rightarrow r]$	Rule for Proof by Cases
11. $p \rightarrow q$ $r \rightarrow s$ $p \vee r$ $\therefore\ q \vee s$	$[(p \rightarrow q) \wedge (r \rightarrow s) \wedge (p \vee r)] \Rightarrow (q \vee s)$	Rule of the Constructive Dilemma
12. $p \rightarrow q$ $r \rightarrow s$ $\bar{q} \vee \bar{s}$ $\therefore\ \bar{p} \vee \bar{r}$	$[(p \rightarrow q) \wedge (r \rightarrow s) \wedge (\bar{q} \vee \bar{s})] \Rightarrow (\bar{p} \vee \bar{r})$	Rule of the Destructive Dilemma

The next four examples demonstrate how the rules listed in Table 2.14 and other results, such as the laws of logic, are used in constructing proofs.

Example 2.24   Our first example demonstrates the validity of the argument

$$\begin{array}{r} p \rightarrow r \\ \bar{p} \rightarrow q \\ q \rightarrow s \\ \hline \therefore \ \bar{r} \rightarrow s \end{array}$$

Steps	Reasons
(1) $p \rightarrow r$	Hypothesis
(2) $q \rightarrow s$	Hypothesis
(3) $\bar{p} \rightarrow q$	Hypothesis
(4) $p \vee q$	(3) and $(\bar{p} \rightarrow q) \Leftrightarrow (\bar{\bar{p}} \vee q) \Leftrightarrow (p \vee q)$, where the second logical equivalence follows by the Law of Double Negation
(5) $r \vee s$	(1), (2), (4), and the Rule of the Constructive Dilemma
(6) $\therefore \ \bar{r} \rightarrow s$	(5) and $(r \vee s) \Leftrightarrow (\bar{\bar{r}} \vee s) \Leftrightarrow (\bar{r} \rightarrow s)$, where the Law of Double Negation is used in the first logical equivalence □

The next example is somewhat more involved.

Example 2.25   Establish the validity of the argument

$$\begin{array}{r} p \rightarrow q \\ q \rightarrow (r \wedge s) \\ \bar{r} \vee (\bar{t} \vee u) \\ p \wedge t \\ \hline \therefore \ u \end{array}$$

Steps	Reasons
(1) $p \rightarrow q$	Hypothesis
(2) $q \rightarrow (r \wedge s)$	Hypothesis
(3) $p \rightarrow (r \wedge s)$	(1), (2), and the Law of the Syllogism
(4) $p \wedge t$	Hypothesis
(5) p	(4) and the Rule of Conjunctive Simplification
(6) $r \wedge s$	(5), (3), and the Rule of Detachment
(7) r	(6) and the Rule of Conjunctive Simplification
(8) $\bar{r} \vee (\bar{t} \vee u)$	Hypothesis
(9) $\overline{(r \wedge t)} \vee u$	(8), the Associative Law of $\vee$, and DeMorgan's Laws
(10) t	(4) and the Rule of Conjunctive Simplification
(11) $r \wedge t$	(7), (10), and the Rule of Conjunction
(12) $\therefore \ u$	(9), (11), and the Rule of Disjunctive Syllogism □

Example 2.26   This example will provide a proof for the following argument:

> If the band could not play rock music or the refreshments were not delivered on time, then the New Year's party would have been cancelled and Alicia would have been angry. If the party were cancelled, then refunds would have had to be made. No refunds were made.
>
> Therefore the band could play rock music.

First we convert the given argument into symbolic form by using the following statement assignments:

p: The band could play rock music.
q: The refreshments were delivered on time.
r: The New Year's party was cancelled.
s: Alicia was angry.
t: Refunds had to be made.

The argument above now becomes

$$(\bar{p} \vee \bar{q}) \to (r \wedge s)$$
$$r \to t$$
$$\underline{\bar{t}}$$
$$\therefore p$$

And now we prove the validity of this argument in the following:

Steps	**Reasons**
(1) $r \to t$	Hypothesis
(2) $\bar{t}$	Hypothesis
(3) $\bar{r}$	(1), (2) and Modus Tollens
(4) $\bar{r} \vee \bar{s}$	(3) and the Rule of Disjunctive Amplification
(5) $\overline{r \wedge s}$	(4) and DeMorgan's Laws
(6) $(\bar{p} \vee \bar{q}) \to (r \wedge s)$	Hypothesis
(7) $\overline{(\bar{p} \vee \bar{q})}$	(5), (6) and Modus Tollens
(8) $p \wedge q$	(7), DeMorgan's Laws, and the Law of Double Negation
(9) $\therefore p$	(8) and the Rule of Conjunctive Simplification

□

Example 2.27   In this example we shall use the Proof by Contradiction.
Consider the argument

$$\bar{p} \leftrightarrow q$$
$$q \to r$$
$$\underline{\bar{r}}$$
$$\therefore p$$

In proving the validity for this argument, we assume the negation $\bar{p}$ of the conclusion p as another hypothesis. The objective now is to use these four hypotheses to derive a contradiction F_0. Our derivation (proof) follows.

Steps	**Reasons**
(1) $\bar{p} \leftrightarrow q$	Hypothesis
(2) $(\bar{p} \rightarrow q) \wedge (q \rightarrow \bar{p})$	(1) and $[\bar{p} \leftrightarrow q] \Leftrightarrow [(\bar{p} \rightarrow q) \wedge (q \rightarrow \bar{p})]$
(3) $\bar{p} \rightarrow q$	(2) and the Rule of Conjunctive Simplification
(4) $q \rightarrow r$	Hypothesis
(5) $\bar{p} \rightarrow r$	(3), (4), and the Law of the Syllogism
(6) $\bar{p}$	Hypothesis (the one assumed)
(7) r	(5), (6), and the Rule of Detachment
(8) $\bar{r}$	Hypothesis
(9) $r \wedge \bar{r}$ ($\Leftrightarrow F_0$)	(7), (8), and the Rule of Conjunction
(10) $\therefore p$	(6), (9), and the Proof by Contradiction

If we examine further what has happened here, we find that

$$[(\bar{p} \leftrightarrow q) \wedge (q \rightarrow r) \wedge \bar{r} \wedge \bar{p}] \Rightarrow F_0.$$

This logical implication requires the truth value of $[(\bar{p} \leftrightarrow q) \wedge (q \rightarrow r) \wedge \bar{r} \wedge \bar{p}]$ to be 0. Because $\bar{p} \leftrightarrow q$, $q \rightarrow r$, and $\bar{r}$ are the given hypotheses, each of these statements has the truth value 1. Consequently, for $[(\bar{p} \leftrightarrow q) \wedge (q \rightarrow r) \wedge \bar{r} \wedge \bar{p}]$ to have the truth value 0, the statement $\bar{p}$ must have the truth value 0. Therefore p has the truth value 1, and the conclusion p of the argument is true. □

Examples 2.24–2.27 have given us some idea how to establish the validity of an argument. We must now learn how to determine when an argument is *invalid*.

This occurs in an argument

$$p_1$$
$$p_2$$
$$p_3$$
$$\vdots$$
$$\underline{p_n}$$
$$\therefore q$$

where each of the hypotheses $p_1, p_2, p_3, \ldots, p_n$ is true (has the truth value 1), yet the conclusion q is false (has the truth value 0).

The next example illustrates an indirect method whereby we may be able to show that an argument we *feel* is invalid (perhaps because we cannot find a way to show that it is valid) actually *is* invalid.

Example 2.28 Consider the argument

$$p$$
$$p \lor q$$
$$q \to (r \to s)$$
$$t \to r$$
$$\overline{\therefore \ \bar{s} \to \bar{t}}$$

To show that this is an invalid argument, we need *one* assignment of truth values for each of the statements p, q, r, s, and t such that the conclusion $\bar{s} \to \bar{t}$ is false (has the truth value 0) while the four hypotheses are all true (have the truth value 1).

The only time the conclusion $\bar{s} \to \bar{t}$ is false is when $\bar{s}$ is true and $\bar{t}$ is false. This implies that the truth value for s is 0 and that the truth value for t is 1.

Because p is one of the hypotheses, its truth value must be 1. For the hypothesis $p \lor q$ to have the truth value 1, q may be either true (1) or false (0). So let us consider the hypothesis $t \to r$ where we know that t is true. If $t \to r$ is to be true, then r must be true (have the truth value 1). Now with r true (1) and s false (0), the truth value of the hypothesis $q \to (r \to s)$ will be 1 only when q is false (0).

Consequently, under the truth value assignments

$$p: \ 1 \qquad q: \ 0 \qquad r: \ 1 \qquad s: \ 0 \qquad t: \ 1,$$

the four hypotheses

$$p \qquad\qquad p \lor q \qquad\qquad q \to (r \to s) \qquad\qquad t \to r$$

all have the truth value 1, while the conclusion

$$\bar{s} \to \bar{t}$$

has the truth value 0. In this case we have shown the given argument to be invalid. □

The truth values p: 1, q: 0, r: 1, s: 0, and t: 1 of Example 2.28 provide a case that disproves what we thought might have been a theorem. We should now start to realize that in trying to prove that a statement of the form

$$(p_1 \land p_2 \land p_3 \land \cdots \land p_n) \to q$$

is a theorem, we must present a general argument for all possible cases where the hypotheses $p_1, p_2, p_3, \ldots, p_n$ are true. This argument is based primarily on the rules of inference and the laws of logic. We cannot use one specific example as a means of establishing a general result. However, whenever we wish to *disprove* a certain statement, all we need to find is one instance for which the statement is false. This instance we call a *counterexample*.

Let us consider a second example wherein we try the indirect approach of Example 2.28.

Example 2.29   What can we say about the validity or invalidity of the following argument?

$$p \to q$$
$$q \to s$$
$$r \to \bar{s}$$
$$\underline{\bar{p} \veebar r}$$
$$\therefore \bar{p}$$

Can the conclusion $\bar{p}$ be false while the four hypotheses are all true? The conclusion $\bar{p}$ is false when p has the truth value 1. So for the hypothesis $p \to q$ to be true, the truth value of q must be 1. From the truth of the hypothesis $q \to s$, the truth of q forces the truth of s. At this point we have statements p, q, and s all with the truth value 1. Continuing with the hypothesis $r \to \bar{s}$, we find that because s has the truth value 1, the truth value of r must be 0. Hence r is false. But with $\bar{p}$ false and the hypothesis $\bar{p} \veebar r$ true, we also have r true. Therefore we find that $p \Rightarrow (\bar{r} \wedge r)$.

We have failed in our attempt to find a counterexample to the validity of the given argument. However, this failure has resulted in a proof of the validity of the argument—a proof based on the Proof by Contradiction.   □

This introduction to the rules of inference and methods of proof has been far from exhaustive. Several of the books cited among the references listed near the end of this chapter offer additional material for the reader who wishes to pursue this topic further. In Chapter 4 another very important proof technique called *mathematical induction* will be added to our arsenal of weapons for attacking the proofs of mathematical theorems. First, however, the reader should carefully complete the exercises for this section.

EXERCISES 2.3

1. Establish each of the following arguments by means of a truth table. In each case, determine which rows of the table are crucial for assessing the validity of the argument and which rows can be ignored.

   a) $[p \wedge (p \to q) \wedge r] \Rightarrow [(p \vee q) \to r]$
   b) $[[(p \wedge q) \to r] \wedge \bar{q} \wedge (p \to \bar{r})] \Rightarrow (\bar{p} \vee \bar{q})$
   c) $[p \wedge (p \to q) \wedge q] \Rightarrow [(p \vee r) \to q]$

2. Use truth tables to verify that each of the following is a logical implication.

   a) $[p \wedge (p \to q)] \to q$
   b) $[(p \to q) \wedge (q \to r)] \to (p \to r)$
   c) $[(p \to q) \wedge \bar{q}] \to \bar{p}$
   d) $[(p \to r) \wedge (q \to r)] \to [(p \vee q) \to r]$
   e) $(p \to q) \to [(p \wedge r) \to (q \wedge r)]$

3. Verify that each of the following is a logical implication by showing that it is impossible for the conclusion to have the truth value 0 while the hypotheses each have the truth value 1.

   a) $(p \wedge q) \rightarrow p$

   b) $p \rightarrow (p \vee q)$

   c) $[(p \vee q) \wedge \bar{p}] \rightarrow q$

   d) $[(p \rightarrow q) \wedge (r \rightarrow s) \wedge (p \vee r)] \rightarrow (q \vee s)$

   e) $[(p \rightarrow q) \wedge (r \rightarrow s) \wedge (\bar{q} \vee \bar{s})] \rightarrow (\bar{p} \vee \bar{r})$

   f) $(p \rightarrow q) \rightarrow [(p \vee r) \rightarrow (q \vee r)]$

   g) $(p \rightarrow q) \rightarrow [(q \rightarrow r) \rightarrow (p \rightarrow r)]$

4. For each pair of statements, s_1, s_2, determine which of the following is true.

   (i) $s_1 \Leftrightarrow s_2$ **(ii)** $s_1 \Rightarrow s_2, s_2 \not\Rightarrow s_1$

   (iii) $s_2 \Rightarrow s_1, s_1 \not\Rightarrow s_2$ **(iv)** $s_1 \not\Rightarrow s_2, s_2 \not\Rightarrow s_1$

   a) s_1:   Triangle ABC is equilateral.
       s_2:   Triangle ABC is equiangular.

   b) s_1:   Triangle ABC is equilateral.
       s_2:   Triangle ABC is isosceles.

   c) s_1:   Quadrilateral $ABCD$ is a square.
       s_2:   Quadrilateral $ABCD$ is a rhombus.

   d) s_1:   Quadrilateral $ABCD$ is equilateral.
       s_2:   Quadrilateral $ABCD$ is equiangular.

   e) s_1:   Quadrilateral $ABCD$ is a rectangle.
       s_2:   Quadrilateral $ABCD$ is equiangular.

5. Give the reason(s) for each step in the following proof that

$$[p \wedge (p \rightarrow q) \wedge (s \vee r) \wedge (r \rightarrow \bar{q})] \Rightarrow (s \vee t).$$

Steps	**Reasons**
(1) p	
(2) $p \rightarrow q$	
(3) q	
(4) $r \rightarrow \bar{q}$	
(5) $q \rightarrow \bar{r}$	
(6) $\bar{r}$	
(7) $s \vee r$	
(8) s	
(9) $\therefore s \vee t$	

6. Give the reasons for the steps verifying the following argument.

$$(\bar{p} \vee q) \rightarrow r$$
$$r \rightarrow (s \vee t)$$
$$\bar{s} \wedge \bar{u}$$
$$\underline{\bar{u} \rightarrow \bar{t}}$$
$$\therefore p$$

Steps   **Reasons**

   (1) $\bar{s} \wedge \bar{u}$

   (2) $\bar{u}$

   (3) $\bar{u} \rightarrow \bar{t}$

   (4) $\bar{t}$

   (5) $\bar{s}$

   (6) $\bar{s} \wedge \bar{t}$

   (7) $r \rightarrow (s \vee t)$

   (8) $\overline{(s \vee t)} \rightarrow \bar{r}$

   (9) $(\bar{s} \wedge \bar{t}) \rightarrow \bar{r}$

  (10) $\bar{r}$

  (11) $(\bar{p} \vee q) \rightarrow r$

  (12) $\bar{r} \rightarrow \overline{(\bar{p} \vee q)}$

  (13) $\bar{r} \rightarrow (p \wedge \bar{q})$

  (14) $p \wedge \bar{q}$

  (15) $\therefore p$

7. a) Give the reasons for the steps in the following Proof by Contradiction that

$$[(p \rightarrow q) \wedge (\bar{r} \vee s) \wedge (p \vee r)] \Rightarrow (\bar{q} \rightarrow s).$$

Steps   **Reasons**

   (1) $\overline{(\bar{q} \rightarrow s)}$

   (2) $\bar{q} \wedge \bar{s}$

   (3) $\bar{s}$

   (4) $\bar{r} \vee s$

   (5) $\bar{r}$

   (6) $p \rightarrow q$

   (7) $\bar{p} \vee q$

   (8) $\bar{q}$

   (9) $\bar{p}$

  (10) $p \vee r$

  (11) r

  (12) $\bar{r} \wedge r$

  (13) $\therefore \bar{q} \rightarrow s$

b) Give a direct proof for the result in part (a).

c) Give a direct proof for the result in Example 2.27.

8. Establish the validity of the following arguments.

a) $(p \wedge q) \Rightarrow (p \vee q)$

b) $[p \wedge (p \rightarrow q) \wedge (\bar{q} \vee r)] \Rightarrow r$

c) $p \rightarrow q$
$\bar{q}$
$\bar{r}$
$\overline{\quad\quad}$
$\therefore \ \overline{p \vee r}$

d) $p \rightarrow q$
$r \rightarrow \bar{q}$
r
$\overline{\quad\quad}$
$\therefore \ \bar{p}$

e) $p \rightarrow (q \rightarrow r)$
$\bar{q} \rightarrow \bar{p}$
p
$\overline{\quad\quad}$
$\therefore \ r$

f) $p \wedge q$
$p \rightarrow (r \wedge q)$
$r \rightarrow (s \vee t)$
$\bar{s}$
$\overline{\quad\quad}$
$\therefore \ t$

g) $p \rightarrow (q \rightarrow r)$
$p \vee s$
$t \rightarrow q$
$\bar{s}$
$\overline{\quad\quad}$
$\therefore \ \bar{r} \rightarrow \bar{t}$

9. Prove that $\sqrt[3]{2}$ is irrational.

10. Show that each of the following arguments is invalid by providing a counterexample—that is, an assignment of truth values for the given propositions such that all hypotheses are true (have the truth value 1) while the conclusion is false (has the truth value 0).

a) $[(p \wedge \bar{q}) \wedge [p \rightarrow (q \rightarrow r)]] \Rightarrow \bar{r}$

b) $[[(p \wedge q) \rightarrow r] \wedge (\bar{q} \vee r)] \Rightarrow p$

c) $p \leftrightarrow q$
$q \rightarrow r$
$r \vee \bar{s}$
$\bar{s} \rightarrow q$
$\overline{\quad\quad}$
$\therefore \ s$

d) p
$p \rightarrow r$
$p \rightarrow (q \vee \bar{r})$
$\bar{q} \vee \bar{s}$
$\overline{\quad\quad}$
$\therefore \ s$

11. Write each of the following arguments in symbolic form. Then provide a proof for the argument or give a counterexample to show that it is invalid.

a) If Rochelle gets the supervisor's position and works hard, then she'll get a raise. If she gets the raise, then she'll buy a new car. She has not purchased a new car. Therefore either Rochelle did not get the supervisor's position or she did not work hard.

b) If Dominic goes to the racetrack, then Helen will be mad. If Ralph plays cards all night, then Carmela will be mad. If either Helen or Carmela gets mad, then Veronica (their attorney) will be notified. Veronica has not heard from either of these two clients. Consequently, Dominic didn't make it to the racetrack and Ralph didn't play cards all night.

c) If Norma is to go to her Tuesday morning meeting, then she must get up early that morning. If she goes to the rock concert on Monday evening, then she will not get home until after 11:00 P.M. If Norma gets home at such an hour and gets up early the next morning, then she'll have to go to work after getting less than seven hours of sleep. Unfortunately, Norma cannot work on less than seven hours of sleep. Consequently, Norma must either not attend the rock concert or miss her Tuesday morning meeting.

d) Margie will teach the Pascal class if and only if she gets a raise. If Margie gets a raise, then she'll vacation in Hawaii this winter. If she doesn't teach the Pascal class, then this winter will find her bird watching in Tennessee. In the winter Margie either misses the opportunity for her trip to Hawaii or goes bird watching in Tennessee. Therefore, this winter Margie doesn't go bird watching in Tennessee.

e) In a given Pascal program, if the integer variable n is not initialized in the program, then $n > 0$. If $n > 0$, then the value of $n^2 + n$ is printed later in the execution of a "Write" statement. If the value of n is supplied by the user in a "Read" statement, then n is not initialized in the program. The value of $n^2 + n$ was not printed during program execution. Therefore, the user did not supply a value for the integer variable n.

f) If there is a chance of rain or her red headband is missing, then Lois will not mow her lawn. Whenever the temperature is over 80°F, there is no chance for rain. Today the temperature is 85°F and Lois is wearing her red headband. Therefore, (sometime today) Lois will mow her lawn.

2.4
THE USE OF QUANTIFIERS

In Section 2.1 we mentioned how sentences that involve a variable, such as x, need not be statements. For example, the sentence "The number $x + 2$ is an even integer" is not necessarily true or false unless we know what value is substituted in for x. If we restrict our choices to integers, then when x is replaced by -5, -1, or 3, the resulting statement is false. In fact, it is false whenever x is replaced by an odd integer. When an even integer is substituted for x, however, the resulting statement is true.

We refer to the sentence "The number $x + 2$ is an even integer" as an *open statement*, which we formally define as follows:

DEFINITION 2.5   A declarative sentence is an *open statement* if

1. it contains one or more variables, and

2. it is not a statement, but

3. it becomes a statement when the variables in it are replaced by certain allowable choices.

When we examine the sentence "The number $x + 2$ is an even integer" in light of this definition, we find it is an open statement that contains the single variable x. With regard to the third element of the definition, in our earlier discussion we restricted the "certain allowable choices" to integers. These allowable choices constitute what is called the *universe* or *universe of discourse* for the open statement. The universe comprises the choices we wish to consider or allow for the variable(s) in the open statement. (The universe is an example of a *set*, a concept we shall examine in some detail in the next chapter.)

In dealing with open statements, we shall use the following notation:

The open statement "The number $x + 2$ is an even integer" is denoted by $p(x)$ [or $q(x)$, $r(x)$, etc.]. Then $\overline{p(x)}$ may be read "The number $x + 2$ is *not* an even integer."

We shall use $q(x,y)$ to represent an open statement that contains two variables, such as

$q(x,y)$:　The numbers $y + 2$, $x - y$, and $x + 2y$ are even integers.

In the case of $q(x,y)$, there is more than one occurrence of each of the variables x, y. It is understood that when we replace one of the x's by a choice from our universe, we replace the other x by the same choice. Likewise, when a substitution (from the universe) is made for one occurrence of y, that same substitution is made for all other occurrences of the variable y.

With $p(x)$ and $q(x,y)$ as above, and the universe still stipulating the integers as our only allowable choices, we get the following results when we make some replacements for the variables x, y.

$p(5)$:　The number 7 (= 5 + 2) is an even integer. (FALSE)

$\overline{p(7)}$:　The number 9 is not an even integer. (TRUE)

$q(4,2)$:　The numbers 4, 2, and 8 are even integers. (TRUE)

We also note, for example, that $q(5,2)$ and $q(4,7)$ are both false statements, whereas $\overline{q(5,2)}$ and $\overline{q(4,7)}$ are true.

Consequently, we see that for both $p(x)$ and $q(x,y)$, some substitutions result in true statements and others in false statements. Therefore we can make the following true statements.

1.　　　　　　　　　　For some x, $p(x)$.

2.　　　　　　　　　　For some x, y, $q(x,y)$.

Note that in this situation, the statements "For some x, $\overline{p(x)}$" and "For some x, y, $\overline{q(x,y)}$" are also true.

The phrases "For some x" and "For some x, y" are said to *quantify* the open statements $p(x)$ and $q(x,y)$, respectively. Many postulates, definitions, and theorems in mathematics involve statements that are quantified open statements. These result from the two types of *quantifiers*, which are called the *existential* and the *universal quantifiers*.

Statement (1) uses the *existential quantifier* "For some x," which can also be expressed as "For at least one x" or "There exists an x such that." This quantifier is written in symbolic form as $\exists x$. Hence the statement "For some x, $p(x)$" becomes $\exists x\, p(x)$, in symbolic form.

Statement (2) becomes $\exists x\, \exists y\, q(x,y)$ in symbolic form. The notation $\exists x,y$ can be used to abbreviate $\exists x\, \exists y\, q(x,y)$ to $\exists x,y\, q(x,y)$.

The *universal quantifier* is denoted by $\forall x$ and is read "For all x," "For each x," or "For every x."

Taking $p(x)$ as defined earlier and using the universal quantifier, we can change the open statement $p(x)$ into the (quantified) statement $\forall x\, p(x)$, a false statement.

If we consider the open statement $r(x)$: "$2x$ is an even integer" with the same universe (of all integers), then the (quantified) statement $\forall x\ r(x)$ is a true statement. When we say that $\forall x\ r(x)$ is true, we mean that no matter which integer (from our universe) is substituted for x, the resulting statement is true. Also note that the statement $\exists x\ r(x)$ is a true statement, whereas $\forall x\ \overline{r(x)}$ and $\exists x\ \overline{r(x)}$ are both false.

The following example shows how these new ideas about quanitifiers can be used with logical connectives.

Example 2.30 Here the universe comprises all real numbers. The open statements $p(x)$, $q(x)$, $r(x)$, and $s(x)$ are given by

$$
\begin{aligned}
p(x): &\quad x \geq 0\\
q(x): &\quad x^2 \geq 0\\
r(x): &\quad x^2 - 3x - 4 = 0\\
s(x): &\quad x^2 - 3 > 0
\end{aligned}
$$

Then the following statements are true.

1.
$$\exists x\, [p(x) \wedge r(x)].$$

This follows because the real number 4, for example, is a member of the universe and is such that both of the statements $p(4)$ and $r(4)$ are true.

2.
$$\forall x\, [p(x) \rightarrow q(x)].$$

If we replace x in $p(x)$ by a negative real number a, then $p(a)$ is false, but $p(a) \rightarrow q(a)$ is true regardless of the truth value of $q(a)$. Replacing x in $p(x)$ by a nonnegative real number b, we find that $p(b)$ and $q(b)$ are both true, as is $p(b) \rightarrow q(b)$. Consequently, $p(x) \rightarrow q(x)$ is true for all replacements x in the universe of all real numbers, and the (quantified) statement $\forall x\, [p(x) \rightarrow q(x)]$ is true.

This statement may be translated into either of the following:

a) For every real number x, if $x \geq 0$, then $x^2 \geq 0$.

b) Every nonnegative real number has a nonnegative square.

Also, the statement $\exists x\, [p(x) \rightarrow q(x)]$ is true.

The next statements we examine are false.

1′.
$$\forall x\, [q(x) \rightarrow s(x)].$$

We want to show that the statement is false, so we need only exhibit one *counterexample*—that is, *one value of* x for which $q(x) \rightarrow s(x)$ is false—rather than proving anything for all x as we did for statement (2). Replacing x by 1, we find that $q(1)$ is true and $s(1)$ is false. Therefore $q(1) \to s(1)$ is false, and consequently the (quantified) statement $\forall x\, [q(x) \to s(x)]$ is false. (Note that $x = 1$ does not produce the only counterexample: Any real number a between $-\sqrt{3}$ and $\sqrt{3}$ will make $q(a)$ true and $s(a)$ false.)

2′. $$\forall x\, [r(x) \vee s(x)].$$

Here there are many counterexamples, such as $1, \frac{1}{2}, -\frac{3}{2}$, and 0. Upon changing quantifiers, however, we find that the statement $\exists x\, [r(x) \vee s(x)]$ is true.

3′. $$\forall x\, [r(x) \to p(x)].$$

The real number -1 is a solution of the equation $x^2 - 3x - 4 = 0$, so $r(-1)$ is true while $p(-1)$ is false. Therefore the choice of -1 provides the unique counterexample we need to disprove this (quantified) statement.

Statement (3′) may be translated into either of the following:

a) For every real number x, if $x^2 - 3x - 4 = 0$, then $x \geq 0$.

b) For every real number x, if x is a solution of the equation $x^2 - 3x - 4 = 0$, then $x \geq 0$. □

From Example 2.30 we make the following observations. Let $p(x)$ denote any open statement (in the variable x) with a prescribed *nonempty* universe (that is, the universe contains at least one entry). Then if $\forall x\, p(x)$ is true, so is $\exists x\, p(x)$, or

$$\forall x\, p(x) \Rightarrow \exists x\, p(x).$$

Also, if $\exists x\, p(x)$ is true, it does not follow the $\forall x\, p(x)$ must be true. Hence $\exists x\, p(x)$ does not logically imply $\forall x\, p(x)$.

The next example demonstrates how the truth value of a quantified statement may depend on the universe prescribed.

Example 2.31   Consider the open statement $p(x)$: $x^2 \geq 0$.

1. If the universe consists of all real numbers, then the quantified statement $\forall x\, p(x)$ is true.

2. For the universe of all complex numbers, however, the same quantified statement $\forall x\, p(x)$ is false. The complex number i provides one of many possible counterexamples.

Yet for either universe, the quantified statement $\exists x\, p(x)$ is true. □

One use of quantifiers in computer science is illustrated in the following example.

Example 2.32   In the following Pascal program segment, n is an integer variable and the variable A is an array $A[1], A[2], \ldots, A[20]$, of 20 integer values.

```
For n := 1 to 20 do
    A[n] := n*n - n;
```

The following statements about the array A can be represented in quantified form, where the universe consists of all integers from 1 to 20, inclusive.

1. Every entry in the array is nonnegative:
$$\forall n \, (A[n] \geq 0).$$

2. The integer $A[20]$ is the largest entry in the array:
$$\forall n \, [(1 \leq n \leq 19) \rightarrow (A[n] < A[20])].$$

3. There exist two consecutive entries in A where the larger entry is twice the smaller:
$$\exists n \, (A[n+1] = 2A[n]).$$

4. The entries in the array are sorted in ascending order:
$$\forall n \, [(1 \leq n \leq 19) \rightarrow (A[n] < A[n+1])].$$

Our last statement requires the use of two variables m, n.

5. The entries in the array are distinct:
$$\forall m \, \forall n \, [(m \neq n) \rightarrow (A[m] \neq A[n])], \qquad \text{or}$$
$$\forall m,n \, [(m < n) \rightarrow (A[m] \neq A[n])]. \quad \square$$

Before continuing, we summarize and somewhat extend, in Table 2.15, what we have learned about quantifiers.

Table 2.15

Statement	When is it true?	When is it false?
$\exists x \, p(x)$	For some (at least one) a in the universe, $p(a)$ is true.	For every a in the universe, $p(a)$ is false.
$\forall x \, p(x)$	For every replacement a in the universe, $p(a)$ is true.	There is at least one replacement a from the universe for which $p(a)$ is false.
$\exists x \, \overline{p(x)}$	For at least one choice a in the universe, $p(a)$ is false, so its negation $\overline{p(a)}$ is true.	For every replacement a in the universe, $p(a)$ is true.
$\forall x \, \overline{p(x)}$	For every replacement a in the universe, $p(a)$ is false, and its negation $\overline{p(a)}$ is true.	There is at least one replacement a from the universe for which $\overline{p(a)}$ is false, and $p(a)$ is true.

Because the results in Table 2.15 involve only one open statement, we now wish to further our study of quantifiers by examining statements wherein more than one open statement appears. The next example starts us on our way.

Example 2.33 Here the universe is all of the integers, and the open statements $r(x)$, $s(x)$ are given by

$$r(x): \quad 2x + 1 = 5$$
$$s(x): \quad x^2 = 9.$$

We see that the statement $\exists x\,[r(x) \wedge s(x)]$ is false, because there is no one integer a such that $2a + 1 = 5$ and $a^2 = 9$. However, there is an integer $b\ (= 2)$ such that $2b + 1 = 5$, and there is a second integer $c\ (= 3$ or $-3)$ such that $c^2 = 9$. Therefore the statement $\exists x\,r(x) \wedge \exists x\,s(x)$ is true. Consequently, the existential quantifier $\exists x$ does not distribute over the logical connective $\wedge$. This one counterexample is enough to show that

$$\exists x\,[r(x) \wedge s(x)] \not\Leftrightarrow [\exists x\,r(x) \wedge \exists x\,s(x)],$$

where $\not\Leftrightarrow$ is read "is *not* logically equivalent to." It also demonstrates that

$$[\exists x\,r(x) \wedge \exists x\,s(x)] \not\Rightarrow \exists x\,[r(x) \wedge s(x)],$$

where $\not\Rightarrow$ is read "does *not* logically imply."

What, however, can we say about the converse of a quantified statement of this form? At this point we present a general argument for *any* open statements $p(x)$, $q(x)$ and *any* prescribed universe.

Examining the statement

$$\exists x\,[p(x) \wedge q(x)] \rightarrow [\exists x\,p(x) \wedge \exists x\,q(x)],$$

we find that when the hypothesis $\exists x\,[p(x) \wedge q(x)]$ is true, there is at least one element c in the universe for which the statement $p(c) \wedge q(c)$ is true. By the Rule of Conjunctive Simplification (see Section 2.3), $[p(c) \wedge q(c)] \Rightarrow p(c)$. From the truth of $p(c)$ we have the true statement $\exists x\,p(x)$. Similarly we obtain $\exists x\,q(x)$, another true statement. The Rule of Conjunction (also introduced in Section 2.3) then provides the true statement $\exists x\,p(x) \wedge \exists x\,q(x)$. Consequently,

$$\exists x\,[p(x) \wedge q(x)] \Rightarrow [\exists x\,p(x) \wedge \exists x\,q(x)]. \quad \square$$

Arguments similar to this one provide the tautologies listed in Table 2.16. Proofs for some of these results, and related counterexamples, will be examined in the exercises that follow this section.

Table 2.16
Logical Relations for Quantified Statements in One Variable

For a prescribed universe and any open statements $p(x)$, $q(x)$ in the variable x:
$\exists x\,[p(x) \wedge q(x)] \Rightarrow [\exists x\,p(x) \wedge \exists x\,q(x)]$
$\exists x\,[p(x) \vee q(x)] \Leftrightarrow [\exists x\,p(x) \vee \exists x\,q(x)]$
$\forall x\,[p(x) \wedge q(x)] \Leftrightarrow [\forall x\,p(x) \wedge \forall x\,q(x)]$
$[\forall x\,p(x) \vee \forall x\,q(x)] \Rightarrow \forall x\,[p(x) \vee q(x)]$

In addition to those listed in Table 2.16, many other tautologies can be derived from the laws of logic and the logical implications and logical equivalences given in Sections 2.2 and 2.3. Our next example lists several of these and demonstrates how two of them are verified.

Example 2.34   Let $p(x)$, $q(x)$, and $r(x)$ denote open statements for a given universe. We find the following tautologies. (Many more are also possible.)

1. $\forall x\,[p(x) \wedge (q(x) \wedge r(x))] \Leftrightarrow \forall x\,[(p(x) \wedge q(x)) \wedge r(x)]$

To establish this logical equivalence, we proceed as follows:
For all a in the universe, consider the *statements* $p(a) \wedge (q(a) \wedge r(a))$ and $(p(a) \wedge q(a)) \wedge r(a)$. By the Associative Law for $\wedge$, we have

$$p(a) \wedge (q(a) \wedge r(a)) \Leftrightarrow (p(a) \wedge q(a)) \wedge r(a).$$

Consequently, for the *open statements* $p(x) \wedge (q(x) \wedge r(x))$ and $(p(x) \wedge q(x)) \wedge r(x)$, it follows that

$$\forall x\, [p(x) \wedge (q(x) \wedge r(x))] \Leftrightarrow \forall x\, [(p(x) \wedge q(x)) \wedge r(x)].$$

2. $\exists x\, [p(x) \rightarrow q(x)] \Leftrightarrow \exists x\, [\overline{p(x)} \vee q(x)]$

For any c in the universe, it follows from Example 2.6 that

$$[p(c) \rightarrow q(c)] \Leftrightarrow [\overline{p(c)} \vee q(c)].$$

Therefore, the statements $\exists x\, [p(x) \rightarrow q(x)]$ and $\exists x\, [\overline{p(x)} \vee q(x)]$ are simultaneously true (or false), so

$$\exists x\, [p(x) \rightarrow q(x)] \Leftrightarrow \exists x\, [\overline{p(x)} \vee q(x)].$$

3. Other tautologies that we shall often find useful include
   a) $\forall x\, \overline{\overline{p(x)}} \Leftrightarrow \forall x\, p(x)$
   b) $\forall x\, [\overline{p(x) \wedge q(x)}] \Leftrightarrow \forall x\, [\overline{p(x)} \vee \overline{q(x)}]$
   c) $\forall x\, [\overline{p(x) \vee q(x)}] \Leftrightarrow \forall x\, [\overline{p(x)} \wedge \overline{q(x)}]$

4. The results for tautologies 3(a), (b), and (c) remain valid when all of the universal quantifiers are replaced by existential quantifiers. $\square$

The results of Tables 2.15 and 2.16, and Examples 2.33 and 2.34 will now help us with a very important concept. How do we negate quantified statements that involve a single variable?

Consider the statement $\forall x\, p(x)$. One form of its negation $\overline{\forall x\, p(x)}$ can be given as "It is not the case that for all x, $p(x)$ holds."

This is not a very useful result. But considering it further causes us to realize that *for some* replacement a from the universe, the resulting statement $p(a)$ does not hold, so $\exists x\, \overline{p(x)}$ is true.

Conversely, when we consider the statement $\exists x\, \overline{p(x)}$, we find that for at least one replacement a from the universe, $\overline{p(a)}$ is true, so $p(a)$ is false. Then the statement $\forall x\, p(x)$ is false, whence $\overline{\forall x\, p(x)}$ is true.

These observations lead to the following rule for negating the statement $\forall x\, p(x)$.

$$\overline{\forall x\, p(x)} \Leftrightarrow \exists x\, \overline{p(x)}$$

In a similar way, Table 2.15 shows that the statement $\exists x\, p(x)$ is true (false) precisely when the statement $\forall x\, \overline{p(x)}$ is false (true). This observation then motivates a rule for negating the statement $\exists x\, p(x)$. It is

$$\overline{\exists x\, p(x)} \Leftrightarrow \forall x\, \overline{p(x)}.$$

These two rules for negation, and two others that follow from them, are given in Table 2.17 for convenient reference.

Table 2.17
Rules for Negating Statements
with One Quantifier

$\overline{\forall x\, p(x)} \Leftrightarrow \exists x\, \overline{p(x)}$
$\overline{\exists x\, p(x)} \Leftrightarrow \forall x\, \overline{p(x)}$
$\overline{\forall x\, \overline{p(x)}} \Leftrightarrow \exists x\, \overline{\overline{p(x)}} \Leftrightarrow \exists x\, p(x)$
$\overline{\exists x\, \overline{p(x)}} \Leftrightarrow \forall x\, \overline{\overline{p(x)}} \Leftrightarrow \forall x\, p(x)$

We use the rules for negating quantified statements in the following example.

Example 2.35   Here we find the negation of two statements, where the universe comprises all of the integers.

1. Let $p(x)$ and $q(x)$ be given by

$$p(x): \quad x \text{ is odd.}$$
$$q(x): \quad x^2 - 1 \text{ is even.}$$

The statement "If x is odd, then $x^2 - 1$ is even" can be symbolized as $\forall x\, [p(x) \rightarrow q(x)]$. (This is a true statement.)
The negation of this statement is determined as follows:

$$\overline{\forall x\, [p(x) \rightarrow q(x)]} \Leftrightarrow \exists x\, \overline{[p(x) \rightarrow q(x)]} \Leftrightarrow \exists x\, \overline{[\overline{p(x)} \vee q(x)]}$$
$$\Leftrightarrow \exists x\, [\overline{\overline{p(x)}} \wedge \overline{q(x)}] \Leftrightarrow \exists x\, [p(x) \wedge \overline{q(x)}].$$

In words, the negation says, "There exists an integer x such that x is odd and $x^2 - 1$ is odd (that is, not even)." (This statement is false.)

2. As in Example 2.33, let $r(x)$ and $s(x)$ be the open statements.

$$r(x): \quad 2x + 1 = 5.$$
$$s(x): \quad x^2 = 9.$$

The quantified statement $\exists x\, [r(x) \wedge s(x)]$ is false because it asserts the existence of at least one integer a such that $2a + 1 = 5$ ($a = 2$) and $a^2 = 9$ ($a = 3$ or -3). Consequently, its negation

$$\overline{\exists x\, [r(x) \wedge s(x)]} \Leftrightarrow \forall x\, \overline{[r(x) \wedge s(x)]} \Leftrightarrow \forall x\, [\overline{r(x)} \vee \overline{s(x)}]$$

is true. This negation may be given in words as "For every integer x, $2x + 1 \neq 5$ or $x^2 \neq 9$."   □

Because a mathematical statement may involve more than one quantifier, we continue this section by offering some examples and making some observations on these types of statements.

Example 2.36   Here we have two variables x, y, both real numbers. (The universe consists of all real numbers.) The commutative law for the addition of real numbers may be expressed by

$$\forall x \ \forall y \ (x + y = y + x).$$

This statement may also be given as

$$\forall y \ \forall x \ (x + y = y + x).$$

Likewise, in the case of the multiplication of real numbers, we may write

$$\forall x \ \forall y \ (xy = yx) \qquad \text{or} \qquad \forall y \ \forall x \ (xy = yx).$$

These two examples suggest the following general result. If $p(x,y)$ is an open statement in the two variables x, y (with either a prescribed universe for both x and y or one prescribed universe for x and a second for y), then the statements $\forall x \ \forall y \ p(x,y)$ and $\forall y \ \forall x \ p(x,y)$ are logically equivalent, that is,

$$\forall x \ \forall y \ p(x,y) \Leftrightarrow \forall y \ \forall x \ p(x,y). \quad \square$$

Example 2.37   When dealing with the associative law for the addition of real numbers, we find that for any real numbers x, y, and z,

$$x + (y + z) = (x + y) + z.$$

Using universal quantifiers (with the universe of all real numbers), we may express this by

$$\forall x \ \forall y \ \forall z \ [x + (y + z) = (x + y) + z], \text{ or}$$
$$\forall y \ \forall x \ \forall z \ [x + (y + z) = (x + y) + z].$$

In fact, there are $3! = 6$ ways to order these three universal quantifiers, and all six of these quantified statements are logically equivalent to one another.
     This is actually true for all open statements $p(x,y,z)$, and to shorten the notation, one may write, for example,

$$\forall x,y,z \ p(x,y,z) \Leftrightarrow \forall y,x,z \ p(x,y,z) \Leftrightarrow \forall x,z,y \ p(x,y,z),$$

describing the logical equivalence for three of the six statements. $\square$

Similar results hold for statements with more than one existential quantifier. This is demonstrated in the following example.

Example 2.38   For the universe of all integers, consider the true statement "There exist integers x, y such that $x + y = 6$." We may represent this in symbolic form by

$$\exists x \ \exists y \ x + y = 6.$$

Here $p(x,y)$ denotes the open statement "$x + y = 6$." The same result can also be given by $\exists y \ \exists x \ p(x,y)$.

In general, for any open statement $p(x,y)$ and universe(s) prescribed for the variables x, y,

$$\exists x\ \exists y\ p(x,y) \Leftrightarrow \exists y\ \exists x\ p(x,y).$$

Similar results follow for statements involving three or more such quantifiers. □

When a statement involves both existential and universal quantifiers, however, we must be careful about the order in which the quantifiers are written. Example 2.39 illustrates this.

Example 2.39 We restrict ourselves here to the universe of all integers and let $p(x,y)$ denote the open statement "$x + y = 17$."

1. The statement

$$\forall x\ \exists y\ p(x,y)$$

says that "For every integer x, there exists an integer y such that $x + y = 17$." (We read the quantifiers from left to right.)

This statement is true; once we select *any* x, the integer $y = 17 - x$ does *exist* and $x + y = x + (17 - x) = 17$. But we realize that each value of x gives rise to a different value of y.

2. Now consider the statement

$$\exists y\ \forall x\ p(x,y).$$

This statement is read "There exists an integer y so that for all integers x, $x + y = 17$." This statement is definitely false. Once *an* integer y is selected, the *only* value that x can have (and still satisfy $x + y = 17$) is $17 - y$.

If the statement $\exists y\ \forall x\ p(x,y)$ were true, then every integer (x) would equal $17 - y$ (for some one fixed y). This says, in effect, that all integers are equal!

Consequently, there are times when the statements $\forall x\ \exists y\ p(x,y)$ and $\exists y\ \forall x\ p(x,y)$ are not logically equivalent. □

Translating mathematical statements—be they postulates, definitions, or theorems—into symbolic form can be helpful for two important reasons.

1. Doing so forces us to be very careful and precise about the meanings of statements, the meanings of phrases such as "For all x" and "There exists an x," and the order in which such phrases appear.

2. After we translate a mathematical statement into symbolic form, the rules we have learned should then apply when we want to determine such related statements as the negation or, if appropriate, the contrapositive, converse, or inverse.

Our last two examples illustrate this, and in so doing, extend the results in Table 2.17.

Example 2.40   Let $p(x,y)$, $q(x,y)$, and $r(x,y)$ represent three open statements, with replacements for the variables x, y chosen from some prescribed universe(s). What is the negation of the following statement?

$$\forall x \, \exists y \, [(p(x,y) \land q(x,y)) \to r(x,y)]$$

We find that

$$\overline{\forall x \, \exists y \, [(p(x,y) \land q(x,y)) \to r(x,y)]}$$
$$\Leftrightarrow \exists x \, \overline{\exists y \, [(p(x,y) \land q(x,y)) \to r(x,y)]}$$
$$\Leftrightarrow \exists x \, \forall y \, \overline{[(p(x,y) \land q(x,y)) \to r(x,y)]}$$
$$\Leftrightarrow \exists x \, \forall y \, \overline{[\overline{(p(x,y) \land q(x,y))} \lor r(x,y)]}$$
$$\Leftrightarrow \exists x \, \forall y \, [(p(x,y) \land q(x,y)) \land \overline{r(x,y)}]$$
$$\Leftrightarrow \exists x \, \forall y \, [(p(x,y) \land q(x,y)) \land \overline{r(x,y)}]$$

Now suppose that we are trying to establish a theorem for which

$$\forall x \, \exists y \, [(p(x,y) \land q(x,y)) \to r(x,y)]$$

is the conclusion. Should we want to try to prove the result by the method of Proof by Contradiction, we assume as an additional hypothesis the negation of this conclusion. Consequently, our additional hypothesis is the statement

$$\exists x \, \forall y \, [(p(x,y) \land q(x,y)) \land \overline{r(x,y)}]. \quad \square$$

Last, we consider how to negate the definition of *limit*, a fundamental concept in calculus.

Example 2.41   In calculus, one studies the properties of real-valued functions. (Functions will be examined in Chapter 5 of this text.) Among these properties is the existence of limits, and one finds the following definition: $\lim_{x \to a} f(x) = L$ if (and only if) for every $\epsilon > 0$ there exists a $\delta > 0$ so that, for all x (where $f(x)$ is defined), $(0 < |x - a| < \delta) \to (|f(x) - L| < \epsilon)$.

This can be expressed in symbolic form as

$$\lim_{x \to a} f(x) = L \Leftrightarrow \forall \epsilon > 0 \ \exists \delta > 0 \ \forall x \, [(0 < |x - a| < \delta) \to (|f(x) - L| < \epsilon)].$$

(Here the universe comprises the real numbers and we only consider those real values x for which $f(x)$ is defined. Also, the quantifiers $\forall \epsilon > 0$ and $\exists \delta > 0$ now contain some restrictive information.) Then, to negate this definition, we do the following (we have combined some steps):

$$\overline{\lim_{x \to a} f(x) \neq L}$$

$$\Leftrightarrow \forall \epsilon > 0 \ \exists \delta > 0 \ \ \overline{\forall x \, [(0 < |x - a| < \delta) \to (|f(x) - L| < \epsilon)]}$$
$$\Leftrightarrow \exists \epsilon > 0 \ \forall \delta > 0 \ \ \exists x \, \overline{[(0 < |x - a| < \delta) \lor (|f(x) - L| < \epsilon)]}$$
$$\Leftrightarrow \exists \epsilon > 0 \ \forall \delta > 0 \ \ \exists x \, [(0 < |x - a| < \delta) \land (|f(x) - L| < \epsilon)]$$
$$\Leftrightarrow \exists \epsilon > 0 \ \forall \delta > 0 \ \ \exists x \, [(0 < |x - a| < \delta) \land (|f(x) - L| \geq \epsilon)]$$

Translating into words, we find that $\lim_{x \to a} f(x) \neq L$ if and only if there exists a positive (real) number ϵ such that for every positive (real) number δ, there is an x value (where $f(x)$ is defined) such that $0 < |x - a| < \delta$ (that is, $x \neq a$ and its distance from a is less than δ) but $|f(x) - L| \geq \epsilon$ (that is, the value of $f(x)$ differs from L by at least ϵ).   □

EXERCISES 2.4

1. For the universe of all integers, let $p(x)$, $q(x)$, $r(x)$, $s(x)$, and $t(x)$ be the following open statements.

$$
\begin{aligned}
p(x) &: \quad x > 0 \\
q(x) &: \quad x \text{ is even} \\
r(x) &: \quad x \text{ is a perfect square} \\
s(x) &: \quad x \text{ is (exactly) divisible by 4} \\
t(x) &: \quad x \text{ is (exactly) divisible by 5}
\end{aligned}
$$

a) Write the following statements in symbolic form.

   (i) There exists a positive integer that is even.

   (ii) If x is even, then x is not divisible by 5.

   (iii) Every even integer is not divisible by 5.

   (iv) There exists an even integer divisible by 5.

   (v) If x is even and x is a perfect square, then x is divisible by 4.

b) Determine whether each of the five statements in part (a) is true or false. For each that is false, provide a counterexample.

c) Express each of the following symbolic representations in words.

   (i) $\forall x \, [r(x) \rightarrow p(x)]$

   (ii) $\forall x \, [s(x) \rightarrow q(x)]$

   (iii) $\forall x \, [s(x) \rightarrow \overline{t(x)}]$

   (iv) $\exists x \, [s(x) \wedge \overline{r(x)}]$

   (v) $\forall x \, [\overline{r(x)} \vee \overline{q(x)} \vee s(x)]$

d) Provide a counterexample for each of the five statements in part (c) that is false.

2. Let $p(x)$, $q(x)$, and $r(x)$ denote the following open statements.

$$
\begin{aligned}
p(x) &: \quad x^2 - 8x + 15 = 0 \\
q(x) &: \quad x \text{ is odd} \\
r(x) &: \quad x > 0
\end{aligned}
$$

For the universe of all integers, determine the truth or falsity of each of the following statements. If a statement is false, give a counterexample.

a) $\forall x \, [p(x) \rightarrow q(x)]$

b) $\forall x \, [q(x) \rightarrow p(x)]$

c) $\exists x \, [p(x) \rightarrow q(x)]$

d) $\exists x \, [q(x) \rightarrow p(x)]$

e) $\exists x \, [r(x) \wedge p(x)]$

f) $\forall x \, [p(x) \rightarrow r(x)]$

g) $\exists x \, [r(x) \rightarrow p(x)]$

h) $\forall x \, [\overline{q(x)} \rightarrow \overline{p(x)}]$

i) $\exists x \, [p(x) \rightarrow (q(x) \wedge r(x))]$

j) $\forall x \, [(p(x) \vee q(x)) \rightarrow r(x)]$

3. Let $p(x)$, $q(x)$, and $r(x)$ be the following open statements.

$$p(x): \quad x^2 - 7x + 10 = 0$$
$$q(x): \quad x^2 - 2x - 3 = 0$$
$$r(x): \quad x < 0$$

a) Determine the truth or falsity of the following statements, where the universe is all integers. If a statement is false, provide a counterexample or explanation.

   (i) $\forall x \, [p(x) \to \overline{r(x)}]$ (iii) $\exists x \, [q(x) \to r(x)]$

   (ii) $\forall x \, [q(x) \to r(x)]$ (iv) $\exists x \, [p(x) \to r(x)]$

b) Find the answers to part (a) when the universe consists of all positive integers.

c) Find the answers to part (a) when the universe contains only the integers 2 and 5.

4. For the following Pascal program segment, m and n are integer variables. The variable A is a two-dimensional array $A[1,1]$, $A[1,2], \ldots$, $A[1,20]$, $\ldots$, $A[10,1], \ldots$, $A[10,20]$, with 10 rows (indexed from 1 to 10) and 20 columns (indexed from 1 to 20).

```
For m := 1 to 10 do
   For n := 1 to 20 do
      A[m,n] := m + 3*n;
```

Write the following statements in symbolic form. (The universe for the variable m contains only the integers from 1 to 10 inclusive; for n the universe consists of the integers from 1 to 20 inclusive.)

a) All entries of A are positive.

b) All entries of A are positive and less than or equal to 70.

c) Some of the entries of A exceed 60.

d) The entries in any row of A are sorted into ascending order.

e) The entries in any column of A are sorted into ascending order.

f) The entries in the first three rows of A are distinct.

g) The entries in any three consecutive rows of A are distinct.

h) For any two consecutive rows of A, the sum of the entries in the later row (the one with the higher row index) is 20 more than the sum of the entries in the prior row.

i) In any two adjacent columns of A there are repeated entries.

j) All entries in the third, sixth, and ninth rows of A are divisible by 3.

5. Negate and simplify each of the following.

a) $\exists x \, [p(x) \vee q(x)]$ b) $\forall x \, [p(x) \wedge \overline{q(x)}]$

c) $\forall x \, [p(x) \to q(x)]$ d) $\exists x \, [(p(x) \vee q(x)) \to r(x)]$

e) $\exists x \, [p(x) \to (q(x) \wedge r(x))]$

6. a) Let $p(x)$, $q(x)$ be open statements in the variable x, with a given universe. Prove that

$$\forall x\, p(x) \lor \forall x\, q(x) \Rightarrow \forall x\, [p(x) \lor q(x)].$$

b) Find a counterexample for the converse in part (a). That is, find open statements $p(x)$, $q(x)$ and a universe such that $\forall x\, [p(x) \lor q(x)]$ is true, while $\forall x\, p(x) \lor \forall x\, q(x)$ is false.

7. For a prescribed universe and any open statements $p(x)$, $q(x)$ in the variable x, prove that

a) $\exists x\, [p(x) \lor q(x)] \Leftrightarrow \exists x\, p(x) \lor \exists x\, q(x)$

b) $\forall x\, [p(x) \land q(x)] \Leftrightarrow \forall x\, p(x) \land \forall x\, q(x)$

8. For the following statements the universe comprises all nonzero integers. Determine the truth value of each statement.

a) $\exists x\, \exists y\, [xy = 1]$

b) $\exists x\, \forall y\, [xy = 1]$

c) $\forall x\, \exists y\, [xy = 1]$

d) $\forall x\, \forall y\, [\sin^2 x + \cos^2 x = \sin^2 y + \cos^2 y]$

e) $\exists x\, \exists y\, [(2x + y = 5) \land (x - 3y = -8)]$

f) $\exists x\, \exists y\, [(3x - y = 7) \land (2x + 4y = 3)]$

9. Answer Exercise 8 for the universe of all nonzero real numbers.

10. In the arithmetic of real numbers, there is a real number, namely 0, called the identity of addition because $a + 0 = 0 + a = a$ for every real number a. This may be expressed in symbolic form by

$$\exists z\, \forall a\, [a + z = z + a = a].$$

(We agree that the universe comprises all real numbers.)

a) In conjunction with the existence of an additive identity is the existence of additive inverses. Write a quantified statement that expresses "Every real number has an additive inverse." (Do not use the minus sign anywhere in your statement.)

b) Write a quantified statement dealing with the existence of a multiplicative identity for the arithmetic of real numbers.

c) Write a quantified statement covering the existence of multiplicative inverses for the nonzero real numbers. (Do not use the exponent -1 anywhere in your statement.)

d) Do the results in parts (b) and (c) change in any way when the universe is restricted to the integers?

11. Consider the quantified statement $\forall x\, \exists y\, [x + y = 17]$. Determine whether this statement is true or false for each of the following universes: (a) the integers; (b) the positive integers; (c) the integers for x, the positive integers for y; (d) the positive integers for x, the integers for y.

12. Let the universe for the variables in the following statements consist of all real numbers. In each case negate and simplify the given statement.

   a) $\forall x \, \forall y \, [(x > y) \rightarrow (x - y > 0)]$

   b) $\forall x \, \forall y \, [[(x > 0) \wedge (y = \log_{10} x)] \rightarrow (x = 10^y)]$

   c) $\forall x \, \forall y \, [(x < y) \rightarrow \exists z \, (x < z < y)]$

   d) $\forall x \, \forall y \, [(|x| = |y|) \rightarrow (y = \pm x)]$

   e) $[\forall x \, \forall y \, ((x > 0) \wedge (y > 0))] \rightarrow [\exists z \, (xz > y)]$

13. In mathematics one often wants to assert not only the existence of an object a (be it a number, a triangle, or whatever) that satisfies an open statement $p(x)$ but also the fact that this object a is the only such object for which $p(x)$ is satisfied (true). Hence the object is *unique*. This is denoted by the quantifier $\exists! x$, which is read "There exists a unique x." This quantifier can be defined in terms of the existential and universal quantifiers:

$$[\exists! x \, p(x)] \Leftrightarrow \{[\exists x \, p(x)] \wedge [\forall x \, \forall y \, [(p(x) \wedge p(y)) \rightarrow (x = y)]]\}$$

This definition indicates that "a proof for unique existence" requires both "an existence proof," which is often accomplished by constructing an example satisfying $p(x)$, and "a uniqueness proof."

   a) Write the following in symbolic form, using this new quantifier. (The universe consists of all real numbers.)

   (i) Every nonzero real number has a unique multiplicative inverse.

   (ii) The sum of any two real numbers is unique.

   (iii) For each x coordinate, the corresponding y coordinate on the line $y = 3x + 7$ is unique.

   b) Let $p(x,y)$ denote the open statement "$y = -2x$." (Here the universe is all integers.) Determine whether each of the following is true or false.

   (i) $[\forall x \, \exists! y \, p(x,y)] \rightarrow [\exists! y \, \forall x \, p(x,y)]$

   (ii) $[\exists! y \, \forall x \, p(x,y)] \rightarrow [\forall x \, \exists! y \, p(x,y)]$

   c) Answer part (b) for the open statement $p(x,y)$: $x + y$ is even.

   d) Consider the statement p: $\exists! x \, (x > 1)$. Give an example of a universe where p is true and an example of one where p is false.

2.5
SUMMARY AND HISTORICAL REVIEW

In this second chapter we have introduced some of the fundamentals of logic—in particular, some of the rules of inference necessary for establishing mathematical theorems.

The first systematic study of logical reasoning is found in the work of the Greek philosopher Aristotle (384–322 B.C.). In a modified form, this type of logic was taught up to and throughout the Middle Ages.

The German mathematician Gottfried Wilhelm Leibniz (1646–1716) is often considered the first scholar who seriously pursued the development of symbolic logic as a universal scientific language. This he professed in his essay *De Arte Combinatoria*, published in 1666. His research in the area of symbolic logic, carried out from 1679 to 1690, gave considerable impetus to the creation of this mathematical discipline.

Following the work by Leibniz, little change took place until the nineteenth century, when the English mathematician George Boole (1815–1864) created a system of mathematical logic that he introduced in 1847 in the pamphlet *The Mathematical Analysis of Logic, Being an Essay towards a Calculus of Deductive Reasoning*. In the same year, Boole's countryman Augustus DeMorgan (1806–1871) published *Formal Logic; or, the Calculus of Inference, Necessary and Probable*. In some ways this treatise extended Boole's work considerably. Then, in 1854, Boole detailed his ideas and further research in the notable work, *An Investigation in the Laws of Thought, on Which Are Founded the Mathematical Theories of Logic and Probability*.

The concepts formulated by Boole were thoroughly examined in the work of another German scholar, Ernst Schröder (1841–1902). These results are known collectively as *Vorlesungen über die Algebra der Logic*; they were published from 1890 to 1895.

Further developments in the area saw an even more modern approach evolve in the work of the German logician Gottlieb Frege (1848–1925) between 1879 and 1903. This work significantly influenced the monumental *Principia Mathematica* (1910–1913) by England's Alfred North Whitehead (1861–1947) and Bertrand Russell (1872–1970). Here what was begun by Boole was finally brought to fruition. Thanks to this remarkable effort and the work of other twentieth-century mathematicians and logicians, in particular the comprehensive *Grundlagen der Mathematik* (1934–1939) of David Hilbert (1862–1943) and Paul Bernays (1888–1977), the more polished techniques of contemporary mathematical logic are now available.

Several sections of this chapter stressed the importance of proof. In mathematics a proof bestows authority on what might otherwise be dismissed as mere opinion. Proof embodies the power and majesty of pure reason. But even more than that, it suggests new mathematical ideas. Our concept of proof goes hand in hand with the notion of a *theorem*—a mathematical statement the truth of which we wish to confirm by means of a logical argument, namely, a *proof*. For those who feel they can ignore the importance of logic and the rules of inference, we submit the following words of wisdom spoken by Achilles in Lewis Carroll's *What the Tortoise Said to Achilles*: "Then Logic would take you by the throat, and force you to do it!"

Comparable coverage of the material presented in this chapter can be found in Chapters 2 and 6 of the text by K. A. Ross and C. R. B. Wright [5] and in Chapter 1 of that by D. F. Stanat and D. F. McAllister [7]. The text by H. Delong [1] provides an historical survey of mathematical logic, together with an examination of the nature of its results and the philosophical consequences of these results. This is also the case with the texts by H. Eves and C. V. Newsom [2],

R. R. Stoll [8], and R. L. Wilder [9], wherein the relationships among logic, proof, and set theory (the topic of our next chapter) are examined in their roles in the foundations of mathematics.

The text by E. Mendelson [3] provides an interesting intermediate introduction for those readers who wish to pursue additional topics in mathematical logic. The objective of the work by R. P. Morash [4] is to prepare the student with a calculus background for the more theoretical mathematics found in abstract algebra and real analysis. It provides an excellent introduction to the basic methods of proof. Finally, the unique text by D. Solow [6] is devoted entirely to introducing the reader who has a background in high school mathematics to the primary techniques used in writing mathematical proofs.

REFERENCES

1. Delong, Howard. *A Profile of Mathematical Logic*. Reading, Mass.: Addison-Wesley, 1970.
2. Eves, Howard, and Newsom, Carroll V. *An Introduction to the Foundations and Fundamental Concepts of Mathematics*, rev. ed. New York: Holt, 1965.
3. Mendelson, Elliott. *Introduction to Mathematical Logic*, 3rd ed. Monterey, Calif.: Wadsworth and Brooks/Cole, 1987.
4. Morash, Ronald P. *Bridge to Abstract Mathematics*: *Mathematical Proof and Structures*. New York: Random House/Birkhäuser, 1987.
5. Ross, Kenneth A., and Wright, Charles R. B. *Discrete Mathematics*. Englewood Cliffs, N.J.: Prentice-Hall, 1985.
6. Solow, Daniel. *How to Read and Do Proofs*. New York: Wiley, 1982.
7. Stanat, Donald F., and McAllister, David F. *Discrete Mathematics in Computer Science*. Englewood-Cliffs, N.J.: Prentice-Hall, 1977.
8. Stoll, Robert R. *Set Theory and Logic*. San Francisco: Freeman, 1963.
9. Wilder, Raymond L. *Introduction to the Foundations of Mathematics*, 2nd ed. New York: Wiley, 1965.

MISCELLANEOUS EXERCISES

1. Construct the truth table for $p \leftrightarrow [(q \wedge r) \rightarrow \overline{(s \vee r)}]$.

2. **a)** Construct the truth table for $(p \rightarrow q) \wedge (\bar{p} \rightarrow r)$.

   b) Translate the statement in part (a) into words such that the word "not" does not appear in the translation.

3. Let p, q, and r denote propositions. Prove or disprove (provide a counterexample for) each of the following:

   a) $[p \leftrightarrow (q \leftrightarrow r)] \Leftrightarrow [(p \leftrightarrow q) \leftrightarrow r]$

   b) $[p \rightarrow (q \rightarrow r)] \Leftrightarrow [(p \rightarrow q) \rightarrow r]$

4. Express the negation of the statement $p \leftrightarrow q$ in terms of the connectives $\wedge$ and $\vee$.

5. **a)** Find the dual of the statement $(\bar{p} \wedge \bar{q}) \vee (T_0 \wedge p) \vee p$.

   b) Use the laws of logic to show that your result from part (a) is logically equivalent to $p \wedge \bar{q}$.

6. Find the converse, inverse, contrapositive, and negation of the statement $p \lor q$.

7. For each of the following, fill in the blank with the word *converse*, *inverse*, or *contrapositive* so that the result is a true statement.

   a) The converse of the inverse of $p \to q$ is the _____ of $p \to q$.

   b) The converse of the inverse of $p \to q$ is the _____ of $q \to p$.

   c) The inverse of the converse of $p \to q$ is the _____ of $p \to q$.

   d) The inverse of the converse of $p \to q$ is the _____ of $q \to p$.

   e) The converse of the contrapositive of $p \to q$ is the _____ of $p \to q$.

   f) The converse of the contrapositive of $p \to q$ is the _____ of $q \to p$.

   g) The inverse of the contrapositive of $p \to q$ is the _____ of $p \to q$.

   h) The contrapositive of the inverse of $p \to q$ is the _____ of $p \to q$.

   i) The converse of the inverse of the contrapositive of $p \to q$ is the _____ of $q \to p$.

   j) The inverse of the contrapositive of the converse of $p \to q$ is the _____ of $q \to p$.

8. Establish the validity of the argument
$$[(p \to q) \land [(q \land r) \to s] \land r] \Rightarrow (p \to s).$$

9. Prove or disprove each of the following, where p, q, and r are any statements.

   a) $[(p \veebar q) \veebar r] \Leftrightarrow [p \veebar (q \veebar r)]$

   b) $[p \veebar (q \to r)] \Leftrightarrow [(p \veebar q) \to (p \veebar r)]$

   c) $[p \to (q \veebar r)] \Rightarrow [(p \to q) \veebar (p \to r)]$

   d) $[(p \to q) \veebar (p \to r)] \Rightarrow [p \to (q \veebar r)]$

10. Write each of the following arguments in symbolic form. Then either provide a proof for the argument or give a counterexample to show that it is invalid.

    a) If it is cool this Friday, then Craig will wear his suede jacket if the pockets are mended. The forecast for Friday calls for cool weather, but the pockets have not been mended. Therefore, Craig won't be wearing his suede jacket this Friday.

    b) If Paul retires this year, then Betty will be happy. Paul's enjoyment of life is sufficient to guarantee Betty's happiness. Consequently, if Paul retires this year, then Betty's happiness is a necessary condition for Paul to enjoy his life.

    c) The contract will be fulfilled if and only if the new windows are installed in the house in June. If the new windows are installed in June, then Evelyn can move into her new house on the first of July. If she can't move in on July 1, then Evelyn must pay the July rent on her apartment. The windows have been installed in June or Evelyn must pay the July rent on her apartment. Therefore, Evelyn won't have to pay rent on her apartment for July.

11. Prove the following via Proof by Contradiction, where the universe consists of the integers.
$$\forall x\, \forall y\, [(3x + \sqrt{2}y = 0) \to [(x = 0) \land (y = 0)]]$$

12. If the universe is all of the real numbers, prove the following by contradiction.

$$\forall x\ \forall y\ [((x \text{ rational}) \wedge (y \text{ irrational})) \rightarrow (x + y \text{ irrational})]$$

13. Let the universe consist of all even integers. Prove or disprove (by finding a counter-example) that $\forall x\ \exists y\ \exists z\ [x = yz]$.

14. Express each of the following in symbolic form. The universe consists of all positive real numbers.

    a) There is no smallest positive real number.

    b) There exists a unique positive real number that equals its square.

    c) Every positive real number has a unique multiplicative inverse.

CHAPTER
3
Set Theory

Underlying the mathematics we study in algebra, geometry, combinatorics, and almost every other area of contemporary mathematics is the notion of a set. Very often this concept provides an underlying structure for a concise formulation of the mathematical topic being investigated. Consequently, many books on mathematics have an introductory chapter on set theory or mention in an appendix those parts of the theory that are needed in the text. Here it may appear that, in opening the book with a chapter on fundamentals of counting, we have neglected set theory. Actually we have relied on intuition; each time the word "collection" was mentioned in Chapter 1, we were dealing with a set. Also, in Section 2.4, the notion of a set (if not the term itself) was invoked when we dealt with the universe (of discourse) for an open statement.

Trying to define a set is rather difficult and often results in the circular use of such synonyms as "class," "collection," and "aggregate." When we first began the study of geometry, we used our intuition to grasp the ideas of point, line, and incidence. Then we started to define new terms and prove theorems, relying on these intuitive notions along with certain axioms and postulates. In our study of set theory, intuition is invoked once again, this time for the comparable ideas of element, set, and membership.

We shall find that the ideas we developed in Chapter 2 on logic are closely tied to set theory. Furthermore, many of the proofs we shall study in this chapter draw on the methods developed in Chapter 2. There will even be a few instances where the combinatorial type of proof (from Chapter 1) will be applied.

3.1
SETS AND SUBSETS

We have a "gut feeling" that a set should be a well-defined collection of objects. These objects are called *elements* and are said to be *members* of the set.

The adjective "well-defined" implies that for any element we care to consider, we are able to determine whether it is in the set under scrutiny. Consequently, we avoid dealing with sets that depend on opinion, such as the set of outstanding major league pitchers for the 1970s.

We use capital letters, such as A, B, C, ..., to represent sets and lowercase letters to represent elements. For a set A we write $x \in A$ if x is an element of A; $y \notin A$ indicates that y is not a member of A.

Example 3.1   A set can be designated by listing its elements within *set braces*. For example, if A is the set consisting of the first five positive integers, then we write $A = \{1, 2, 3, 4, 5\}$. Here $2 \in A$ but $6 \notin A$.

Another standard notation for this set provides us with $A = \{x \mid x$ is an integer and $1 \le x \le 5\}$. Here the vertical line $\mid$ within the set braces is read "such that." The symbols $\{x \mid \ldots\}$ are read "the set of all x such that...." The properties following $\mid$ help us determine the elements of the set that is being described.

Beware! $\{x \mid 1 \le x \le 5\}$ is not an adequate description of the set A unless we have agreed in advance that the elements we are considering are integers. When such an agreement is adopted, we say that we are specifying a *universe*, or *universe of discourse*, which is usually denoted by $\mathcal{U}$. We then select only elements from $\mathcal{U}$ to form our sets. In this particular problem, if $\mathcal{U}$ denotes the set of all integers or the set of all positive integers, then $\{x \mid 1 \le x \le 5\}$ adequately describes A. If $\mathcal{U}$ were the set of all real numbers, then $\{x \mid 1 \le x \le 5\}$ would contain all of the real numbers from 1 to 5 inclusive; if $\mathcal{U}$ consisted of only even integers, then the only members of $\{x \mid 1 \le x \le 5\}$ would be 2 and 4.   □

Example 3.2   For $\mathcal{U} = \{1, 2, 3, \ldots\}$, the set of positive integers, let

a) $A = \{1, 4, 9, \ldots, 64, 81\} = \{x^2 \mid x \in \mathcal{U}, x^2 < 100\}$

b) $B = \{1, 4, 9, 16\} = \{y^2 \mid y \in \mathcal{U}, y^2 < 20\} = \{y^2 \mid y \in \mathcal{U}, y^2 < 23\}$

c) $C = \{2, 4, 6, 8, \ldots\} = \{2k \mid k \in \mathcal{U}\}.$

Sets A and B are examples of *finite* sets, whereas C is an *infinite* set. When dealing with sets like A or C, we can either describe the sets in terms of properties the elements must satisfy or list enough elements to indicate what is, we hope, an obvious pattern. For any finite set A, $|A|$ denotes the number of elements in A and is referred to as the *cardinality*, or *size*, of A.

Here the sets B and A are such that every element of B is also an element of A. This important relationship occurs throughout set theory and its applications, and it leads to the following definition.   □

DEFINITION 3.1   If C, D are sets from a universe $\mathcal{U}$, we say that C is a *subset* of D and write $C \subseteq D$ or $D \supseteq C$, if every element of C is an element of D. If, in addition, there is an element of D that is not in C, then C is called a *proper subset* of D, and this is denoted by $C \subset D$ or $D \supset C$.

Note that for any sets C, D from a universe $\mathcal{U}$, if $C \subseteq D$ then $x \in C \Rightarrow x \in D$.

Also,

$$C \subseteq D \Rightarrow |C| \le |D|, \text{ for } C, D \text{ finite, and}$$
$$C \subset D \Rightarrow C \subseteq D, \text{ but}$$
$$C \subseteq D \not\Rightarrow C \subset D.$$

Example 3.3   A variable name in ANSI (American National Standards Institute) FORTRAN consists of a single letter followed by at most five characters (letters or digits). If $\mathcal{U}$ is the set of all such variable names, then by the rules of sum and product, $|\mathcal{U}| = 26 + 26(36) + 26(36)^2 + \cdots + 26(36)^5 = 26 \sum_{i=0}^{5} 36^i = 1,617,038,306$, so $\mathcal{U}$ is a large, but still finite, set. An integer variable in this programming language must start with I, J, K, L, M, N. So if A is the subset of all integer variables in ANSI FORTRAN, then $|A| = 6 + 6(36) + 6(36)^2 + \cdots + 6(36)^5 = 6 \sum_{i=0}^{5} 36^i = 373,162,686.$ □

The subset concept may now be used to develop the idea of set equality. First we consider the following example.

Example 3.4   For the universe $\mathcal{U} = \{1, 2, 3, 4, 5\}$, consider the set $A = \{1, 2\}$. If $B = \{x \mid x^2 \in \mathcal{U}\}$, then the members of B are $1, 2$. Here A and B contain the same elements—and no other element(s). This leads us to feel that the sets A and B are *equal*.

However, it is also true here that $A \subseteq B$ and $B \subseteq A$, and we prefer to formally define the idea of set equality by using these subset relations. This leads us to the following definition.   □

DEFINITION 3.2   For a given universe $\mathcal{U}$, the sets C and D (taken from $\mathcal{U}$) are said to be *equal*, and we write $C = D$, when $C \subseteq D$ and $D \subseteq C$.

From these ideas on set equality, we find that neither order nor repetition is relevant for a general set. Consequently, $\{1, 2, 3\} = \{3, 1, 2\} = \{2, 2, 1, 3\} = \{1, 2, 1, 3, 1\}$.

Now that we have defined the concepts of subset and set equality, we shall use the quantifiers of Section 2.4 to examine the negations of these ideas.

For a given universe $\mathcal{U}$, let A, B be sets taken from $\mathcal{U}$. Then we may write

$$A \subseteq B \Leftrightarrow \forall x [x \in A \Rightarrow x \in B].$$

(Here the universal quantifier $\forall x$ indicates that we consider every element x in the prescribed universe $\mathcal{U}$.)

From this we have

$$A \not\subseteq B \text{ (that is, } A \text{ is not a subset of } B)$$
$$\Leftrightarrow \overline{\forall x [x \in A \Rightarrow x \in B]}$$
$$\Leftrightarrow \exists x \overline{[x \in A \lor x \in B]}$$
$$\Leftrightarrow \exists x [x \in A \land x \notin B].$$

Hence $A \nsubseteq B$ if there is at least one element x in the universe where x is a member of A but x is not a member of B.

In a similar way, because $A = B \Leftrightarrow A \subseteq B \wedge B \subseteq A$, then

$$A \neq B \Leftrightarrow \overline{A \subseteq B \wedge B \subseteq A} \Leftrightarrow \overline{A \subseteq B} \vee \overline{B \subseteq A} \Leftrightarrow A \nsubseteq B \vee B \nsubseteq A.$$

Therefore two sets A and B are not equal if and only if (1) there exists at least one element x in $\mathcal{U}$ where $x \in A$ but $x \notin B$ or (2) there exists at least one element y in $\mathcal{U}$ where $y \in B$ and $y \notin A$ (or perhaps both).

We also note that for any sets $C, D \subseteq \mathcal{U}$ (that is $C \subseteq \mathcal{U}$ and $D \subseteq \mathcal{U}$),

$$C \subset D \Leftrightarrow C \subseteq D \wedge C \neq D.$$

Now that we have introduced the ideas of set equality, subset, and proper subset, the proof of our first theorem for this chapter will be fairly straightforward.

THEOREM 3.1   Let $A, B, C \subseteq \mathcal{U}$.

    a) If $A \subseteq B$ and $B \subseteq C$, then $A \subseteq C$.   **b)** If $A \subset B$ and $B \subseteq C$, then $A \subset C$.

    c) If $A \subseteq B$ and $B \subset C$, then $A \subset C$.   **d)** If $A \subset B$ and $B \subset C$, then $A \subset C$.

Proof   We shall prove parts (a) and (b) and leave the remaining parts for the exercises.

    a) In order to prove that $A \subseteq C$, we need to verify that for all $x \in \mathcal{U}$, if $x \in A$ then $x \in C$. We start with an element x from A. Since $A \subseteq B$, $x \in A$ implies $x \in B$. Then with $B \subseteq C$, $x \in B$ implies $x \in C$. So $x \in A$ implies $x \in C$ (by the Law of the Syllogism—Rule 2 in Table 2.14), and $A \subseteq C$.

    b) Since $A \subset B$, if $x \in A$ then $x \in B$. With $B \subseteq C$, it then follows that $x \in C$, so $A \subseteq C$. However, $A \subset B \Rightarrow$ there exists an element $b \in B$ such that $b \notin A$. Because $B \subseteq C$, $b \in B \Rightarrow b \in C$. Thus $A \subseteq C$, and there exists an element $b \in C$ with $b \notin A$, so $A \subset C$.   ■

Our next example involves several subset relations.

Example 3.5   Let $\mathcal{U} = \{1, 2, 3, 4, 5\}$ with $A = \{1, 2, 3\}$, $B = \{3, 4\}$, and $C = \{1, 2, 3, 4\}$. Then the following subset relations hold:

    a) $A \subseteq C$ **b)** $A \subset C$

    c) $B \subset C$ **d)** $A \subseteq A$

    e) $B \nsubseteq A$ **f)** $A \not\subset A$ (that is, A is not a proper subset of A)

In addition to A, B, we are interested in determining how many subsets C has in total. Before answering this, however, we need to introduce the set with no members.   □

DEFINITION 3.3   The *null set*, or *empty set*, is a set containing no elements. It is denoted by $\emptyset$ or $\{\ \}$.

We note that $|\emptyset| = 0$ but $\{0\} \neq \emptyset$. Also, $\emptyset \neq \{\emptyset\}$ because $\{\emptyset\}$ is a set with one element, namely the null set.

The empty set satisfies the following property. To establish this we use the method of proof called *Reductio ad Absurdum*. (See Rules 5 and 5' of Table 2.14.)

THEOREM 3.2 For any set A, $\emptyset \subseteq A$; $\emptyset \subset A$ if $A \neq \emptyset$.

Proof If the first result is not true, then $\emptyset \not\subseteq A$, so there is an element x from the universe with $x \in \emptyset$ but $x \notin A$. But $x \in \emptyset$ is impossible. So we reject the assumption $\emptyset \not\subseteq A$ and find that $\emptyset \subseteq A$. In addition, if $A \neq \emptyset$, then there is an element $a \in A$ (and $a \notin \emptyset$), so $\emptyset \subset A$. ∎

Example 3.6 Returning now to Example 3.5 we determine the number of subsets of the set $C = \{1, 2, 3, 4\}$. In constructing a subset of C, we have, for each member x of C, two distinct choices: either include it in the subset or exclude it. Consequently there are $2 \times 2 \times 2 \times 2$ choices, resulting in $2^4 = 16$ subsets of C. These include the empty set $\emptyset$ and the set C itself. Should we need the number of subsets of two elements from C, the result is the number of ways two objects can be selected from a set of four objects, namely $C(4, 2)$ or $\binom{4}{2}$. As a result, the total number of subsets of C, 2^4, is also the sum $\binom{4}{0} + \binom{4}{1} + \binom{4}{2} + \binom{4}{3} + \binom{4}{4}$, where the first summand is for the empty set, the second summand for the four *singleton* subsets, the third summand for the six subsets of size two, etc. So $2^4 = \sum_{k=0}^{4} \binom{4}{k}$. □

DEFINITION 3.4 If A is a set from universe $\mathscr{U}$, the *power set* of A, denoted $\mathscr{P}(A)$, is the collection of all subsets of A.

Example 3.7 For the set C of Example 3.6, $\mathscr{P}(C) = \{\emptyset, \{1\}, \{2\}, \{3\}, \{4\}, \{1, 2\}, \{1, 3\}, \{1, 4\}, \{2, 3\}, \{2, 4\}, \{3, 4\}, \{1, 2, 3\}, \{1, 2, 4\}, \{1, 3, 4\}, \{2, 3, 4\}, C\}$. □

> In general, for any finite set A with $|A| = n \geq 0$, A has 2^n subsets, so $|\mathscr{P}(A)| = 2^n$. For any $0 \leq k \leq n$, there are $\binom{n}{k}$ subsets of size k. Counting the subsets of A according to the number, k, of elements in a subset, we have the combinatorial identity $\binom{n}{0} + \binom{n}{1} + \binom{n}{2} + \cdots + \binom{n}{n} = 2^n$, for $n \geq 0$.

This identity was established earlier in Corollary 1.1(a). The presentation here is another example of a combinatorial proof because the identity is established by counting the same collection of objects (subsets of A) in two different ways.

The ability to count certain subsets of a given set provides a second approach to the solution of one of our earlier examples.

Example 3.8 In Example 1.14, we counted the number of (staircase) paths in the xy plane from $(2, 1)$ to $(7, 4)$ where each such path is made up of individual steps going one unit to the right (R) or one unit upward (U). Figure 3.1 is the same as Fig. 1.1, where two of the possible paths are indicated.

The path in Fig. 3.1(a) has its three upward (U) moves located in positions 2, 5, and 8, of the list below the figure. Consequently, this path determines the

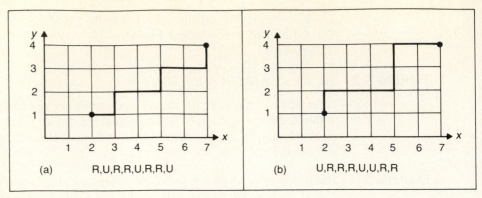

Figure 3.1

three-element subset $\{2, 5, 8\}$ of the set $\{1, 2, 3, \ldots, 8\}$. In Fig. 3.1(b) the path determines the three-element subset $\{1, 5, 6\}$. Conversely, if we start, for example, with the subset $\{1, 3, 7\}$ of $\{1, 2, 3, \ldots, 8\}$, then the path that determines this subset is given by U, R, U, R, R, R, U, R.

Consequently, the number of paths sought here equals the number of subsets A of $\{1, 2, 3, \ldots, 8\}$, where $|A| = 3$. There are $\binom{8}{3} = \dfrac{8!}{3!\,5!} = 56$ such paths (and subsets), as we found in Example 1.14.

If we had considered the moves R to the right, instead of the upward moves U, we would have found the answer to be the number of subsets B of $\{1, 2, 3, \ldots, 8\}$, where $|B| = 5$. There are $\binom{8}{5} = \dfrac{8!}{5!\,3!} = 56$ such subsets. (The idea presented here was examined earlier for the result developed in Table 1.3.)   □

Our next example yields another important combinatorial identity.

Example 3.9   For integers n, r with $n \geq r \geq 1$,

$$\binom{n+1}{r} = \binom{n}{r} + \binom{n}{r-1}.$$

Although this can be established algebraically from the definition of $\binom{n}{r}$ as $n!/(r!(n-r)!)$, we use a combinatorial approach. Let $A = \{x, a_1, a_2, \ldots, a_n\}$ and consider all subsets of A that contain r elements. There are $\binom{n+1}{r}$ such subsets. These fall into exactly one of the following two cases: those that contain the element x and those that do not. To get a subset C of A, where $x \in C$ and $|C| = r$, place x in C and then select $r - 1$ of the elements $a_1, a_2, \ldots, a_n$. This can be done in $\binom{n}{r-1}$ ways. For the other case we want a subset B of A with $|B| = r$ and $x \notin B$. So we select r elements from among $a_1, a_2, \ldots, a_n$, which we can do in $\binom{n}{r}$ ways. It then follows by the rule of sum that $\binom{n+1}{r} = \binom{n}{r} + \binom{n}{r-1}$.   □

Example 3.10   We now investigate how the identity of Example 3.9 can help us solve Example 1.32, where we sought the number of nonnegative integer solutions of the inequality $x_1 + x_2 + \cdots + x_6 < 10$.

For each integer k, $0 \le k \le 9$, the number of solutions to $x_1 + x_2 + \cdots + x_6 = k$ is $\binom{6+k-1}{k} = \binom{5+k}{k}$. So the number of nonnegative integer solutions to $x_1 + x_2 + \cdots + x_6 < 10$ is

$$\binom{5}{0} + \binom{6}{1} + \binom{7}{2} + \binom{8}{3} + \cdots + \binom{14}{9}$$

$$= \left[\binom{6}{0} + \binom{6}{1} \right] + \binom{7}{2} + \binom{8}{3} + \cdots + \binom{14}{9}$$

$$= \left[\binom{7}{1} + \binom{7}{2} \right] + \binom{8}{3} + \cdots + \binom{14}{9} = \left[\binom{8}{2} + \binom{8}{3} \right] + \binom{9}{4} + \cdots + \binom{14}{9}$$

$$= \left[\binom{9}{3} + \binom{9}{4} \right] + \cdots + \binom{14}{9} = \cdots = \binom{14}{8} + \binom{14}{9} = \binom{15}{9} = 5005. \quad \square$$

Example 3.11 In Fig. 3.2 we find a part of the useful and interesting array of numbers called *Pascal's triangle*.

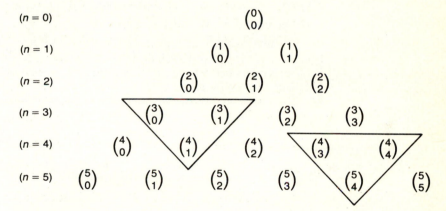

$(n = 0)$ $\binom{0}{0}$

$(n = 1)$ $\binom{1}{0}$ $\binom{1}{1}$

$(n = 2)$ $\binom{2}{0}$ $\binom{2}{1}$ $\binom{2}{2}$

$(n = 3)$ $\binom{3}{0}$ $\binom{3}{1}$ $\binom{3}{2}$ $\binom{3}{3}$

$(n = 4)$ $\binom{4}{0}$ $\binom{4}{1}$ $\binom{4}{2}$ $\binom{4}{3}$ $\binom{4}{4}$

$(n = 5)$ $\binom{5}{0}$ $\binom{5}{1}$ $\binom{5}{2}$ $\binom{5}{3}$ $\binom{5}{4}$ $\binom{5}{5}$

Figure 3.2

Note that in this partial listing the two triangles shown satisfy the condition that the binomial coefficient at the bottom of the inverted triangle is the sum of the other two terms in the triangle. This follows from the identity in Example 3.9.

When we replace each of the binomial coefficients by its numerical value, the Pascal triangle appears as shown in Fig. 3.3. $\square$

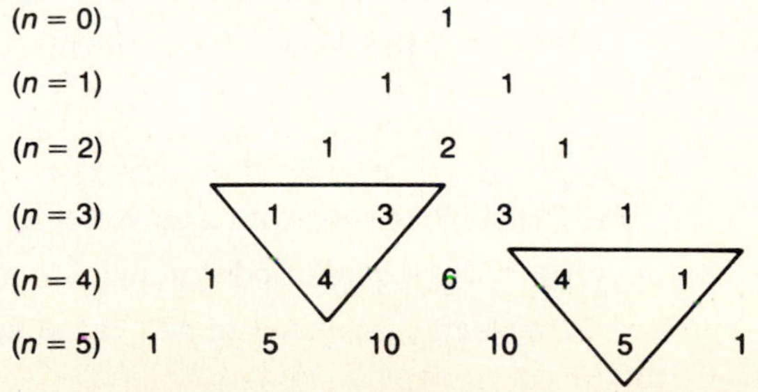

$(n = 0)$ 1

$(n = 1)$ 1 1

$(n = 2)$ 1 2 1

$(n = 3)$ 1 3 3 1

$(n = 4)$ 1 4 6 4 1

$(n = 5)$ 1 5 10 10 5 1

Figure 3.3

There are certain sets of numbers that appear frequently throughout the text. Consequently, we close this section by assigning them the following designations.

---

a) $\mathbf{Z}$ = the set of *integers* = $\{0, 1, -1, 2, -2, 3, -3, \ldots\}$

b) $\mathbf{N}$ = the set of *nonnegative integers* or *natural numbers* = $\{0, 1, 2, 3, \ldots\}$

c) $\mathbf{Z}^+$ = the set of *positive integers* = $\{1, 2, 3, \ldots\}$ = $\{x \in \mathbf{Z} \,|\, x > 0\}$

d) $\mathbf{Q}$ = the set of *rational numbers* = $\{a/b \,|\, a, b \in \mathbf{Z}, b \neq 0\}$

e) $\mathbf{Q}^+$ = the set of *positive rational numbers* = $\{a/b \,|\, a/b \in \mathbf{Q}, a/b > 0\}$

f) $\mathbf{Q}^*$ = the set of *nonzero rational numbers*

g) $\mathbf{R}$ = the set of *real numbers*

h) $\mathbf{R}^+$ = the set of *positive real numbers*

i) $\mathbf{R}^*$ = the set of *nonzero real numbers*

j) $\mathbf{C}$ = the set of *complex numbers* = $\{x + yi \,|\, x, y \in \mathbf{R}, i^2 = -1\}$

k) $\mathbf{C}^*$ = the set of *nonzero complex numbers*

ℓ) For any $n \in \mathbf{Z}^+$, $\mathbf{Z}_n = \{0, 1, 2, \ldots, n-1\}$

m) For real numbers a, b with $a < b$, $[a, b] = \{x \,|\, x \in \mathbf{R}, a \leq x \leq b\}$, $(a, b) = \{x \,|\, x \in \mathbf{R}, a < x < b\}$, $[a, b) = \{x \,|\, x \in \mathbf{R}, a \leq x < b\}$, $(a, b] = \{x \,|\, x \in \mathbf{R}, a < x \leq b\}$. The first set is called a *closed interval*, the second set an *open interval*, and the other two sets *half-open intervals*.

EXERCISES 3.1

1. Which of the following sets are equal?
   a) $\{1, 2, 3\}$
   b) $\{3, 2, 1, 3\}$
   c) $\{3, 1, 2, 3\}$
   d) $\{1, 2, 2, 3\}$

2. Let $A = \{1, \{1\}, \{2\}\}$. Which of the following statements are true?
   a) $1 \in A$ T
   b) $\{1\} \in A$ T
   c) $\{1\} \subseteq A$ T
   d) $\{\{1\}\} \subseteq A$ T
   e) $\{2\} \in A$ F
   f) $\{2\} \subseteq A$ F
   g) $\{\{2\}\} \subseteq A$ T
   h) $\{\{2\}\} \subseteq A$ T

3. For $A = \{1, 2, \{2\}\}$, which of the eight statements in Exercise 2 are true?

4. Which of the following statements are true?
   a) $\emptyset \in \emptyset$ F
   b) $\emptyset \subset \emptyset$ F
   c) $\emptyset \subseteq \emptyset$ T
   d) $\emptyset \in \{\emptyset\}$ T
   e) $\emptyset \subset \{\emptyset\}$
   f) $\emptyset \subseteq \{\emptyset\}$ T

5. Let A, B be sets from a universe $\mathcal{U}$.
   a) Write a quantified statement to express the proper subset relation $A \subset B$.
   b) Negate the result in part (a) to determine when $A \not\subset B$.

6. For $A = \{1, 2, 3, 4, 5, 6, 7\}$, determine the number of
   a) subsets of A.
   b) nonempty subsets of A.
   c) proper subsets of A.
   d) nonempty proper subsets of A.
   e) subsets of A containing three elements.
   f) subsets of A containing 1, 2.
   g) subsets of A containing five elements, including 1, 2.
   h) proper subsets of A containing 1, 2.
   i) subsets of A with an even number of elements.
   j) subsets of A with an odd number of elements.
   k) subsets of A with an odd number of elements, including the element 3.

7. a) If a set A has 63 proper subsets, what is $|A|$?
   b) If a set B has 64 subsets of odd cardinality, what is $|B|$?
   c) Generalize the result of part (b).

8. Which of the following sets are nonempty?
   a) $\{x \mid x \in \mathbf{N}, 2x + 7 = 3\}$ b) $\{x \mid x \in \mathbf{Z}, 3x + 5 = 9\}$
   c) $\{x \mid x \in \mathbf{Q}, x^2 + 4 = 6\}$ d) $\{x \mid x \in \mathbf{R}, x^2 + 4 = 6\}$
   e) $\{x \mid x \in \mathbf{R}, x^2 + 5 = 4\}$ f) $\{x \mid x \in \mathbf{R}, x^2 + 3x + 3 = 0\}$
   g) $\{x \mid x \in \mathbf{C}, x^2 + 3x + 3 = 0\}$

9. About to leave a restaurant counter a man sees that he has one penny, one nickel, one dime, one quarter, and one half-dollar. In how many ways can he leave some (at least one) of his coins for a tip if
   a) there are no restrictions?
   b) he wants to have some change left?
   c) he wants to leave at least 10 cents?

10. a) How many subsets of $\{1, 2, 3, \ldots, 11\}$ contain at least one even integer?
    b) How many subsets of $\{1, 2, 3, \ldots, 12\}$ contain at least one even integer?
    c) Generalize the results of parts (a) and (b).

11. Give an example of three sets W, X, Y such that $W \in X$ and $X \in Y$ but $W \notin Y$.

12. Write the next three rows for the Pascal triangle shown in Fig. 3.3.

13. Complete the proof of Theorem 3.1.

14. For sets $A, B, C \subseteq \mathcal{U}$, prove or disprove (with a counterexample), the statement: If $A \subseteq B$, $B \not\subseteq C$, then $A \not\subseteq C$.

15. One quarter of the five-element subsets of $\{1, 2, 3, \ldots, n\}$ contain the element 7. Determine n.

16. Establish the identity in Example 3.9 algebraically.

17. Give a combinatorial argument to show that for integers n, r with $n \geq r \geq 2$,

$$\binom{n+2}{r} = \binom{n}{r} + 2\binom{n}{r-1} + \binom{n}{r-2}.$$

18. For positive integers n, r show that

$$\binom{n+r+1}{r} = \binom{n+r}{r} + \binom{n+r-1}{r-1} + \cdots + \binom{n+2}{2} + \binom{n+1}{1} + \binom{n}{0}.$$

19. Let $A = \{1, 2, 3, \ldots, 39, 40\}$.

   a) Write a computer program (or develop an algorithm) to generate six-element subsets of A.

   b) For $B = \{2, 3, 5, 7, 11, 13, 17, 19, 23, 29, 31, 37\}$, write a computer program (or develop an algorithm) to generate a six-element subset of A and then determine whether it is a subset of B.

20. Let $A = \{1, 2, 3, \ldots, 7\}$. Write a computer program (or develop an algorithm) that lists all of the subsets B of A, where $|B| = 4$.

21. Write a computer program (or develop an algorithm) that prints out all of the subsets of $\{1, 2, 3, \ldots, n\}$, where $1 \leq n \leq 10$. (The value of n should be supplied during program execution.)

3.2
SET OPERATIONS AND THE LAWS OF SET THEORY

After learning how to count, a student usually faces methods for combining counting numbers. First this is accomplished through addition. Usually the student's world of arithmetic revolves about the set $\mathbf{Z}^+$ (or a subset of $\mathbf{Z}^+$ that can be spoken and written about, as well as punched out on a hand-held calculator) wherein the addition of two elements from $\mathbf{Z}^+$ results in a third element of $\mathbf{Z}^+$, called the sum. Hence the student can concentrate on addition without having to enlarge his or her arithmetic world beyond $\mathbf{Z}^+$. This is also true for the operation of multiplication.

The addition and multiplication of positive integers are said to be *binary* (acting on *two* positive integers) *operations* on $\mathbf{Z}^+$. (The general concept of a binary operation will be covered in Section 5.4.)

In a comparable way, we introduce the following binary operations for sets.

DEFINITION 3.5   For $A, B \subseteq \mathcal{U}$ we define the following:

   a) $A \cup B$ (the *union* of A and B) $= \{x \mid x \in A \vee x \in B\}$.

   b) $A \cap B$ (the *intersection* of A and B) $= \{x \mid x \in A \wedge x \in B\}$.

   c) $A \bigtriangleup B$ (the *symmetric difference* of A and B) $= \{x \mid (x \in A \vee x \in B) \wedge x \notin A \cap B\} = \{x \mid x \in A \cup B \wedge x \notin A \cap B\}$.

Note that if $A, B \subseteq \mathcal{U}$, then $A \cap B$, $A \cup B$, $A \triangle B \subseteq \mathcal{U}$. Consequently, $\cap$, $\cup$, $\triangle$ are binary operations on $\mathcal{P}(\mathcal{U})$, or we can say that $\mathcal{P}(\mathcal{U})$ is *closed* under these operations.

Example 3.12   With $\mathcal{U} = \{1, 2, 3, \ldots, 9, 10\}$, $A = \{1, 2, 3, 4, 5\}$, $B = \{3, 4, 5, 6, 7\}$, $C = \{7, 8, 9\}$, we have:

a) $A \cap B = \{3, 4, 5\}$ b) $A \cup B = \{1, 2, 3, 4, 5, 6, 7\}$

c) $B \cap C = \{7\}$ d) $A \cap C = \emptyset$

e) $A \triangle B = \{1, 2, 6, 7\}$ f) $A \cup C = \{1, 2, 3, 4, 5, 7, 8, 9\}$

g) $A \triangle C = \{1, 2, 3, 4, 5, 7, 8, 9\}$ □

In Example 3.12 we see that $A \cap B \subseteq A \subseteq A \cup B$. This result is not special for just this example but is true in general. This is because

$$x \in A \cap B \Rightarrow x \in A \wedge x \in B \Rightarrow x \in A$$

(by the Rule of Conjunctive Simplification—Rule 6 of Table 2.14), and

$$x \in A \Rightarrow x \in A \vee x \in B \Rightarrow x \in A \cup B$$

(where the first logical implication is a result of the Rule of Disjunctive Amplification—Rule 7 of Table 2.14).

From parts (d), (f), and (g) of Example 3.12, we introduce the following general ideas.

DEFINITION 3.6   Let $S, T \subseteq \mathcal{U}$. The sets S and T are called *disjoint*, or *mutually disjoint*, when $S \cap T = \emptyset$.

THEOREM 3.3   If $S, T \subseteq \mathcal{U}$, then S and T are disjoint if and only if $S \cup T = S \triangle T$.

Proof   We start with S, T disjoint. (To prove that $S \cup T = S \triangle T$ we use Definition 3.2. Since this is the first instance of its use, we shall be somewhat cautious and very detailed.) If $x \in S \cup T$, then $x \in S$ or $x \in T$ (or perhaps both). But with S and T disjoint, $x \notin S \cap T$ so $x \in S \triangle T$. Consequently, because $x \in S \cup T$ implies $x \in S \triangle T$, we have $S \cup T \subseteq S \triangle T$. For the opposite inclusion, if $y \in S \triangle T$, then $y \in S$ or $y \in T$. (But $y \notin S \cap T$; we don't actually use this here.) So $y \in S \cup T$. Therefore $S \triangle T \subseteq S \cup T$ and by Definition 3.2 it follows that $S \triangle T = S \cup T$.

Conversely, if $S \cup T = S \triangle T$ and $S \cap T \neq \emptyset$, let $x \in S \cap T$. Then $x \in S$ and $x \in T$, so $x \in S \cup T$. However, $x \notin S \triangle T$, contradicting the given set equality. Consequently, S and T are disjoint.   ∎

In proving the first part of Theorem 3.3 we showed that if S, T are any sets, then $S \triangle T \subseteq S \cup T$. The disjointedness of S and T was needed only for the opposite inclusion.

After mastering the skill of addition, one usually comes next to subtraction. Here the set **N** causes some difficulty. For example, **N** contains 2 and 5 but

$2 - 5 = -3$, and $-3 \notin \mathbf{N}$. Hence $\mathbf{N}$ is not closed under subtraction, and subtraction is not a binary operation on $\mathbf{N}$. However it is a binary operation on the *superset* $\mathbf{Z}$ of $\mathbf{N}$. So with $\mathbf{Z}$ we can introduce the *unary*, or *monary*, *operation* of negation where we "take the minus" of a number such as 3, getting -3. (The general definition of a unary, or monary, operation is covered in Section 5.4.)

We now introduce a comparable unary operation for sets.

DEFINITION 3.7   For a set $A \subseteq \mathcal{U}$, the *complement* of A, denoted $\mathcal{U} - A$, or $\overline{A}$, is given by $\{x \mid x \in \mathcal{U} \wedge x \notin A\}$.

Example 3.13   For the sets of Example 3.12, $\overline{A} = \{6, 7, 8, 9, 10\}$, $\overline{B} = \{1, 2, 8, 9, 10\}$, and $\overline{C} = \{1, 2, 3, 4, 5, 6, 10\}$.   □

For any universe $\mathcal{U}$ and any set $A \subseteq \mathcal{U}$, we find that $\overline{A} \subseteq \mathcal{U}$. Therefore $\mathcal{P}(\mathcal{U})$ is *closed* under the unary operation defined by the complement.

Related to the concept of the complement we have the following:

DEFINITION 3.8   For $A, B \subseteq \mathcal{U}$, the *(relative) complement* of A in B, denoted $B - A$, is given by $\{x \mid x \in B \wedge x \notin A\}$.

Example 3.14   For the sets of Example 3.12 we have:

a) $B - A = \{6, 7\}$ b) $A - B = \{1, 2\}$ c) $A - C = A$

d) $C - A = C$ e) $A - A = \emptyset$ f) $\mathcal{U} - A = \overline{A}$ □

Our next result will again make use of Definition 3.2. It provides links among the notions of subset, union, intersection, and complement.

THEOREM 3.4   For sets $A, B \subseteq \mathcal{U}$, the following statements are equivalent:

a) $A \subseteq B$ b) $A \cup B = B$

c) $A \cap B = A$ d) $\overline{B} \subseteq \overline{A}$

Proof   In order to prove the theorem, we prove that (a) $\Rightarrow$ (b), (b) $\Rightarrow$ (c), (c) $\Rightarrow$ (d), and (d) $\Rightarrow$ (a). (The reader may wish to refer to Exercise 15 at the end of Section 2.2.)

(i) (a) $\Rightarrow$ (b)   If A, B are any sets, then $B \subseteq A \cup B$. For the opposite inclusion, if $x \in A \cup B$, then $x \in A$ or $x \in B$, but since $A \subseteq B$, in either case we have $x \in B$. So $A \cup B \subseteq B$ and the equality follows.

(ii) (b) $\Rightarrow$ (c)   Given sets A, B, we always have $A \supseteq A \cap B$. For the opposite inclusion, let $y \in A$. With $A \cup B = B$, $y \in A \Rightarrow y \in A \cup B \Rightarrow y \in B \Rightarrow y \in A \cap B$, so $A \subseteq A \cap B$ and we conclude that $A = A \cap B$.

(iii) (c) $\Rightarrow$ (d) $z \in \overline{B} \Rightarrow z \notin B \Rightarrow z \notin A \cap B$ since $A \cap B \subseteq B$. With $A \cap B = A$, $z \notin A \cap B \Rightarrow z \notin A \Rightarrow z \in \overline{A}$, so $\overline{B} \subseteq \overline{A}$.

(iv) (d) $\Rightarrow$ (a)   Last, $w \in A \Rightarrow w \notin \overline{A}$, and since $\overline{B} \subseteq \overline{A}$, $w \notin \overline{A} \Rightarrow w \notin \overline{B}$. Then $w \notin \overline{B} \Rightarrow w \in B$, so $A \subseteq B$.   ∎

With a bit of theorem proving under our belts, we now introduce the major laws that govern set theory. These bear a marked similarity to the laws of logic given in Section 2.2. In many instances these set theoretic principles are similar to the arithmetic properties of the real numbers, where "$\cup$" plays the role of "$+$" and "$\cap$" the role of "$\times$." However, there are several differences.

The Laws of Set Theory

For any sets A, B, and C taken from a universe $\mathcal{U}$

1. $\overline{\overline{A}} = A$ Law of *Double Complement*

2. $\overline{A \cup B} = \overline{A} \cap \overline{B}$ *DeMorgan's* Laws
   $\overline{A \cap B} = \overline{A} \cup \overline{B}$

3. $A \cup B = B \cup A$ *Commutative* Laws
   $A \cap B = B \cap A$

4. $A \cup (B \cup C) = (A \cup B) \cup C$ *Associative* Laws
   $A \cap (B \cap C) = (A \cap B) \cap C$

5. $A \cup (B \cap C) = (A \cup B) \cap (A \cup C)$ *Distributive* Laws
   $A \cap (B \cup C) = (A \cap B) \cup (A \cap C)$

6. $A \cup A = A$ *Idempotent* Laws
   $A \cap A = A$

7. $A \cup \emptyset = A$ *Identity* Laws
   $A \cap \mathcal{U} = A$

8. $A \cup \overline{A} = \mathcal{U}$ *Inverse* Laws
   $A \cap \overline{A} = \emptyset$

9. $A \cup \mathcal{U} = \mathcal{U}$ *Domination* Laws
   $A \cap \emptyset = \emptyset$

10. $A \cup (A \cap B) = A$ *Absorption* Laws
    $A \cap (A \cup B) = A$

All these laws can be established as in the first part of the proof of Theorem 3.3. We demonstrate this by establishing the first of DeMorgan's Laws and the second Distributive Law, that of intersection over union.

Proof   Let $x \in \mathcal{U}$. Then $x \in \overline{A \cup B} \Rightarrow x \notin A \cup B \Rightarrow x \notin A$ and $x \notin B \Rightarrow x \in \overline{A}$ and $x \in \overline{B} \Rightarrow x \in \overline{A} \cap \overline{B}$, so $\overline{A \cup B} \subseteq \overline{A} \cap \overline{B}$. To establish the opposite inclusion, we check to see that the converse of each logical implication is also a logical implication (that is, that each logical implication is, in fact, a logical equivalence). As a result we find that $x \in \overline{A} \cap \overline{B} \Rightarrow x \in \overline{A}$ and $x \in \overline{B} \Rightarrow x \notin A$ and $x \notin B \Rightarrow x \notin A \cup B \Rightarrow x \in \overline{A \cup B}$. Therefore $\overline{A} \cap \overline{B} \subseteq \overline{A \cup B}$. Consequently, with $\overline{A \cup B} \subseteq \overline{A} \cap \overline{B}$ and $\overline{A} \cap \overline{B} \subseteq \overline{A \cup B}$, it follows from Definition 3.2 that $\overline{A \cup B} = \overline{A} \cap \overline{B}$. ∎

In our second proof, we shall establish both subset relations simultaneously by using the logical equivalence ($\Leftrightarrow$) as opposed to logical implications ($\Rightarrow$ and $\Leftarrow$).

Proof   For each $x \in \mathcal{U}$, $x \in A \cap (B \cup C) \Leftrightarrow (x \in A)$ and $(x \in B \cup C) \Leftrightarrow (x \in A)$ and $(x \in B$ or $x \in C) \Leftrightarrow (x \in A$ and $x \in B)$ or $(x \in A$ and $x \in C) \Leftrightarrow (x \in A \cap B)$ or $(x \in A \cap C) \Leftrightarrow x \in (A \cap B) \cup (A \cap C)$. As we have equivalent statements throughout, we have established both subset relations simultaneously, so $A \cap (B \cup C) = (A \cap B) \cup (A \cap C)$. (The equivalence of the third and fourth statements follows from the comparable principle in the laws of logic—namely, the Distributive Law of conjunction over disjunction.) ■

The reader undoubtedly expects the pairing of the laws in items 2 through 10 to have some importance. As with the laws of logic, these pairs of statements are called *duals*. One statement can be obtained from the other by interchanging all occurrences of $\cup$ by $\cap$, and vice versa, and all occurrences of $\mathcal{U}$ by $\emptyset$, and vice versa.

This leads us to the following idea.

DEFINITION 3.9   Let $X \subseteq \mathcal{U}$ denote a set involving a finite number of sets (such as A, $\overline{A}$, B, $\overline{B}$, etc.) and only the set operation symbols $\cap$ and $\cup$. The *dual* of X, denoted X^d, is the subset of $\mathcal{U}$ obtained from X by replacing each occurrence of $\cap$ ($\cup$) in X by $\cup$ ($\cap$), and each occurrence of $\mathcal{U}$ ($\emptyset$) by $\emptyset$ ($\mathcal{U}$).

As in Section 2.2, we shall state and use the following theorem. We shall prove a more general result in Chapter 15.

THEOREM 3.5   (*The Principle of Duality*). Let $X, Y \subseteq \mathcal{U}$, where X and Y are as described in Definition 3.9. If X, Y denote sets (*in general*) and $X = Y$, then (the dual statement) $X^d = Y^d$ is also true.

Using this principle cuts our work down considerably. For each pair of laws in items 2 through 10, one need prove only one of the statements and then invoke this principle to obtain the other statement in the pair.

We must be careful about applying Theorem 3.5. This result cannot be applied to particular situations but only to results (theorems) about sets in general. For example, let us consider the *particular* situation where $\mathcal{U} = \{1, 2, 3, 4, 5\}$ and $A = \{1, 2, 3, 4\}$, $B = \{1, 2, 3, 5\}$, $C = \{1, 2\}$, and $D = \{1, 3\}$. Under these circumstances

$$A \cap B = \{1, 2, 3\} = C \cup D.$$

However, we cannot infer that $A \cap B = C \cup D \Rightarrow (A \cap B)^d = (C \cup D)^d$—that is, that $A \cup B = C \cap D$. For here $A \cup B = \{1, 2, 3, 4, 5\}$, whereas $C \cap D = \{1\}$. The reason why Theorem 3.5 is not applicable here is that although $A \cap B = C \cup D$ in this *particular* example, it is not true in general (that is, for any sets A, B, C, and D taken from a universe $\mathcal{U}$).

Example 3.15   Inasmuch as Definition 3.9 and Theorem 3.5 do not mention anything about subsets, can we find a dual for the statement $A \subseteq B$ (where $A, B \subseteq \mathcal{U}$)?

Here we get an opportunity to use some of the results in Theorem 3.4. In particular, we can deal with the statement $A \subseteq B$ by using the equivalent statement $A \cup B = B$.

The dual of $A \cup B = B$ gives us $A \cap B = B$. But $A \cap B = B \Leftrightarrow B \subseteq A$. Consequently, the dual of the statement $A \subseteq B$ is the statement $B \subseteq A$.

(We could also have obtained this result by using $A \subseteq B \Leftrightarrow A \cap B = A$. The reader will be asked to verify this in the section exercises.) $\square$

When at most four sets are involved in a set-equality or subset statement, we can investigate the problems graphically.

Named in honor of the English logician John Venn (1834–1883), a *Venn diagram* is constructed as follows: $\mathcal{U}$ is depicted as the interior of a rectangle, while subsets of $\mathcal{U}$ are represented by the interiors of circles and other closed curves. Figure 3.4 shows two Venn diagrams. The shaded region in Fig. 3.4(a) represents the set A, whereas $\overline{A}$ is represented by the unshaded area. The shaded region in Fig. 3.4(b) represents $A \cup B$. The set $A \cap B$ is the cross-hatched region in this figure.

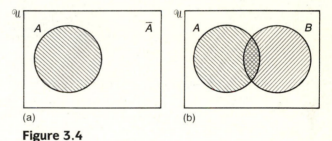

(a)   (b)

Figure 3.4

In Fig. 3.5 Venn diagrams are used to establish one of DeMorgan's Laws. Fig. 3.5(a) has everything except $A \cap B$ shaded. In Fig. 3.5(b), $\overline{A}$ is the region shaded by the lines going from the lower left to the upper right; the lines going from the upper left to the lower right shade the region representing $\overline{B}$. For these Venn diagrams, the shaded area in part (b) is the same as that in part (a). Consequently, $\overline{A \cap B} = \overline{A} \cup \overline{B}$.

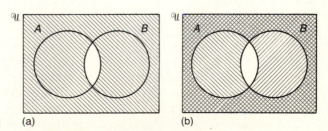

(a)   (b)

Figure 3.5

We further illustrate the use of these diagrams by showing that for any sets $A, B, C \subseteq \mathcal{U}$,

$$\overline{(A \cup B) \cap C} = (\overline{A} \cap \overline{B}) \cup \overline{C}.$$

Instead of shading regions, another approach that also uses Venn diagrams numbers the regions as shown in Fig. 3.6 where, for example, region 3 is $\overline{A} \cap B \cap \overline{C}$ and region 7 is $A \cap \overline{B} \cap C$. Each region is a set of the form $S_1 \cap S_2 \cap S_3$, where S_1 is replaced by A or $\overline{A}$, S_2 by B or $\overline{B}$, and S_3 by C or $\overline{C}$. Consequently, by the rule of product, there are eight possible regions.

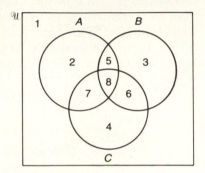

Figure 3.6

Consulting Fig. 3.6, we see that $A \cup B$ comprises regions $2, 3, 5, 6, 7, 8$ and that regions $4, 6, 7, 8$ make up set C. Therefore $(A \cup B) \cap C$ comprises the regions common to $A \cup B$ and C: namely, regions $6, 7, 8$. Consequently, $\overline{(A \cup B) \cap C}$ is made up of regions $1, 2, 3, 4, 5$. The set $\overline{A}$ consists of regions $1, 3, 4, 6$, while regions $1, 2, 4, 7$ make up $\overline{B}$. Consequently, $\overline{A} \cap \overline{B}$ comprises regions 1 and 4. Since regions $4, 6, 7, 8$ comprise C, the set $\overline{C}$ is made up of regions $1, 2, 3, 5$. Taking the union of $\overline{A} \cap \overline{B}$ with $\overline{C}$, we then finish with regions $1, 2, 3, 4, 5$, as we did for $\overline{(A \cup B) \cap C}$.

One more technique for establishing set equalities is the *membership table*. (This method is akin to using the truth table introduced in Section 2.1.)

We observe that for sets $A, B \subseteq \mathcal{U}$, an element $x \in \mathcal{U}$ satisfies exactly one of the following four situations:

a) $x \notin A, x \notin B$ **b)** $x \notin A, x \in B$

c) $x \in A, x \notin B$ **d)** $x \in A, x \in B$.

When x is an element of a given set, we write a 1 in the column representing that set; when x is not in the set, we enter a 0. Table 3.1 gives the membership tables for $A \cap B$, $A \cup B$, $\overline{A}$ in this notation. Here, for example, the third row in part (a) of the table tells us that when an element $x \in \mathcal{U}$ is in set A but not in B, then it is not in $A \cap B$ but it is in $A \cup B$.

Table 3.1

A	B	A∩B	A∪B		A	$\overline{A}$
0	0	0	0		0	1
0	1	0	1		1	0
1	0	0	1			
1	1	1	1			

(a)   (b)

These operations on 0 and 1 are the same as in ordinary arithmetic except that $1 \cup 1 = 1$.

Using membership tables, we can establish the equality of two sets by comparing their respective columns in the table. Table 3.2 demonstrates this for the Distributive Law of union over intersection. We see here how each of the eight rows corresponds with exactly one of the eight regions in the Venn diagram of Fig. 3.6. For example, row 1 corresponds with region 1: $\overline{A} \cap \overline{B} \cap \overline{C}$; and row 6 corresponds with region 7: $A \cap \overline{B} \cap C$.

Table 3.2

A	B	C	B∩C	A∪(B∩C)	A∪B	A∪C	(A∪B)∩(A∪C)
0	0	0	0	0	0	0	0
0	0	1	0	0	0	1	0
0	1	0	0	0	1	0	0
0	1	1	1	1	1	1	1
1	0	0	0	1	1	1	1
1	0	1	0	1	1	1	1
1	1	0	0	1	1	1	1
1	1	1	1	1	1	1	1

Since these columns are identical, we conclude that $A \cup (B \cap C) = (A \cup B) \cap (A \cup C)$.

Now that we have the laws of set theory, what can we do with them? The following examples will demonstrate how the laws are used to simplify a complicated set expression or to derive new set equalities.

Example 3.16   Simplify the expression $\overline{\overline{(A \cup B)} \cap C} \cup \overline{B}$.

	Reasons
$\overline{\overline{(A \cup B)} \cap C} \cup \overline{B}$	
$= \overline{(\overline{(A \cup B)} \cap C)} \cap \overline{\overline{B}}$	DeMorgan's Law
$= ((A \cup B) \cap C) \cap B$	Law of Double Complement
$= (A \cup B) \cap (C \cap B)$	Associative Law of Intersection
$= (A \cup B) \cap (B \cap C)$	Commutative Law of Intersection
$= [(A \cup B) \cap B] \cap C$	Associative Law of Intersection
$= B \cap C$	Absorption Law □

Example 3.17   Express $\overline{A-B}$ in terms of $\cup$ and $^-$.

From the definition of relative complement, $A-B=\{x\,|\,x\in A \wedge x\notin B\}=A\cap\overline{B}$. Therefore,

	Reasons
$\overline{A-B}=\overline{A\cap\overline{B}}$	
$=\overline{A}\cup\overline{\overline{B}}$	DeMorgan's Law
$=\overline{A}\cup B$	Law of Double Complement $\square$

Example 3.18   From the observation made in Example 3.17 we have $A\,\triangle\,B=\{x\,|\,x\in A\cup B\wedge x\notin A\cap B\}=(A\cup B)-(A\cap B)=(A\cup B)\cap\overline{(A\cap B)}$, so

	Reasons
$\overline{A\,\triangle\,B}=\overline{(A\cup B)\cap\overline{(A\cap B)}}$	
$=\overline{(A\cup B)}\cup\overline{\overline{(A\cap B)}}$	DeMorgan's Law
$=\overline{(A\cup B)}\cup(A\cap B)$	Law of Double Complement
$=(A\cap B)\cup\overline{(A\cup B)}$	Commutative Law of $\cup$
$=(A\cap B)\cup(\overline{A}\cap\overline{B})$	DeMorgan's Law
$=[(A\cap B)\cup\overline{A}]\cap[(A\cap B)\cup\overline{B}]$	Distributive Law of $\cup$ over $\cap$
$=[(A\cup\overline{A})\cap(B\cup\overline{A})]\cap$ $[(A\cup\overline{B})\cap(B\cup\overline{B})]$	Distributive Law of $\cup$ over $\cap$
$=[\mathcal{U}\cap(B\cup\overline{A})]\cap[(A\cup\overline{B})\cap\mathcal{U}]$	Inverse Law
$=(B\cup\overline{A})\cap(A\cup\overline{B})$	Identity Law
$=(\overline{A}\cup B)\cap(A\cup\overline{B})$	Commutative Law of $\cup$
$=(\overline{A}\cup B)\cap(\overline{A}\cap B)$	DeMorgan's Law
$=\overline{A}\,\triangle\,B$	
$=(A\cup\overline{B})\cap(\overline{A}\cup B)$	Commutative Law of $\cap$
$=(A\cup\overline{B})\cap\overline{(A\cap\overline{B})}$	DeMorgan's Law
$=A\,\triangle\,\overline{B}$ $\square$	

In closing this section we extend the set operations of $\cup$ and $\cap$ beyond three sets.

DEFINITION 3.10   Let I be a nonempty set and $\mathcal{U}$ a universe. For each $i\in I$ let $A_i\subseteq\mathcal{U}$. Then I is called an *index set* (or *set of indices*), and each $i\in I$ is called an *index*.

Under these conditions,

$$\bigcup_{i\in I}A_i=\{x\,|\,x\in A_i\quad\text{for at least one }i\in I\},\text{ and}$$
$$\bigcap_{i\in I}A_i=\{x\,|\,x\in A_i\quad\text{for every }i\in I\}.$$

We can rephrase the results in Definition 3.10 by using quantifiers:

$$x\in\bigcup_{i\in I}A_i\Leftrightarrow\exists\,i\in I\ (x\in A_i)$$
$$x\in\bigcap_{i\in I}A_i\Leftrightarrow\forall\,i\in I\ (x\in A_i)$$

Then $x \notin \bigcup_{i\in I} A_i \Leftrightarrow \overline{\exists i \in I \ (x \in A_i)} \Leftrightarrow \forall i \in I \ (x \notin A_i)$; that is, $x \notin \bigcup_{i\in I} A_i$ if and only if $x \notin A_i$ for *every* index $i \in I$. Similarly, $x \notin \bigcap_{i\in I} A_i \Leftrightarrow \overline{\forall i \in I \ (x \in A_i)} \Leftrightarrow \exists i \in I \ (x \notin A_i)$; that is, $x \notin \bigcap_{i\in I} A_i$ if and only if $x \notin A_i$ for *at least one* index $i \in I$.

If the index set I is the set $\mathbf{Z}^+$, we can write

$$\bigcup_{i\in\mathbf{Z}^+} A_i = A_1 \cup A_2 \cup \cdots = \bigcup_{i=1}^{\infty} A_i, \qquad \bigcap_{i\in\mathbf{Z}^+} A_i = A_1 \cap A_2 \cap \cdots = \bigcap_{i=1}^{\infty} A_i.$$

Example 3.19 Let $I = \{3, 4, 5, 6, 7\}$, and for $i \in I$ let $A_i = \{1, 2, 3, \ldots, i\} \subseteq \mathcal{U} = \mathbf{Z}^+$. Then $\bigcup_{i\in I} A_i = \bigcup_{i=3}^{7} A_i = \{1, 2, 3, \ldots, 7\} = A_7$, whereas $\bigcap_{i\in I} A_i = \{1, 2, 3\} = A_3$. $\square$

Example 3.20 Let $\mathcal{U} = \mathbf{R}$ and $I = \mathbf{R}^+$. If for each $r \in \mathbf{R}^+$, $A_r = [-r, r]$, then $\bigcup_{r\in I} A_r = \mathbf{R}$ and $\bigcap_{r\in I} A_r = \{0\}$. $\square$

When dealing with generalized unions and intersections, membership tables and Venn diagrams are unfortunately next to useless, but the rigorous approach, as demonstrated in the first part of the proof of Theorem 3.3, is still available.

THEOREM 3.6 (*Generalized DeMorgan Laws*) Let I be an index set where for each $i \in I$, $A_i \subseteq \mathcal{U}$. Then

a) $\overline{\bigcup_{i\in I} A_i} = \bigcap_{i\in I} \overline{A_i}$

b) $\overline{\bigcap_{i\in I} A_i} = \bigcup_{i\in I} \overline{A_i}$

Proof We shall prove Theorem 3.6(a) and leave the proof of part (b) for the reader. For each $x \in \mathcal{U}$, $x \in \overline{\bigcup_{i\in I} A_i} \Leftrightarrow x \notin \bigcup_{i\in I} A_i \Leftrightarrow x \notin A_i$, for all $i \in I \Leftrightarrow x \in \overline{A_i}$, for all $i \in I \Leftrightarrow x \in \bigcap_{i\in I} \overline{A_i}$. ∎

EXERCISES 3.2

1. For $\mathcal{U} = \{1, 2, 3, \ldots, 9, 10\}$ let $A = \{1, 2, 3, 4, 5\}$, $B = \{1, 2, 4, 8\}$, $C = \{1, 2, 3, 5, 7\}$, and $D = \{2, 4, 6, 8\}$. Determine each of the following:

   a) $(A \cup B) \cap C$ **b)** $A \cup (B \cap C)$ **c)** $\overline{C} \cup \overline{D}$

   d) $\overline{C \cap D}$ **e)** $(A \cup B) - C$ **f)** $A \cup (B - C)$

   g) $(B - C) - D$ **h)** $B - (C - D)$ **i)** $(A \cup B) - (C \cap D)$

2. If $A = [0, 3]$, $B = [2, 7)$, with $\mathcal{U} = \mathbf{R}$, determine each of the following:

   a) $A \cap B$ **b)** $A \cup B$ **c)** $\overline{A}$

   d) $A \triangle B$ **e)** $A - B$ **f)** $B - A$

3. Let $\mathcal{U} = \{a, b, c, \ldots, x, y, z\}$, with $A = \{a, b, c\}$ and $C = \{a, b, d, e\}$. If $|A \cap B| = 2$ and $(A \cap B) \subset B \subset C$, determine B.

4. Prove each of the following results without using Venn diagrams or membership tables. (Assume a universe $\mathcal{U}$.)

   a) If $A \subseteq B$ and $C \subseteq D$, then $A \cap C \subseteq B \cap D$ and $A \cup C \subseteq B \cup D$.

b) If $A \subseteq C$ and $B \subseteq C$, then $A \cap B \subseteq C$ and $A \cup B \subseteq C$.

c) If $A \subseteq B$ and $A \subseteq C$, then $A \subseteq B \cap C$ and $A \subseteq B \cup C$.

d) $A \subseteq B$ if and only if $A \cap \overline{B} = \emptyset$

e) $A \subseteq B$ if and only if $\overline{A} \cup B = \mathcal{U}$.

5. Prove or disprove each of the following:

   a) For sets $A, B, C \subseteq \mathcal{U}$, $A \cap C = B \cap C \Rightarrow A = B$.

   b) For sets $A, B, C \subseteq \mathcal{U}$, $A \cup C = B \cup C \Rightarrow A = B$.

   c) For sets $A, B, C \subseteq \mathcal{U}$, $[(A \cap C = B \cap C) \wedge (A \cup C = B \cup C)] \Rightarrow A = B$.

   d) For sets $A, B, C \subseteq \mathcal{U}$, $A \triangle C = B \triangle C \Rightarrow A = B$.

6. Using Venn diagrams, investigate the truth or falsity of each of the following, for sets $A, B, C \subseteq \mathcal{U}$.

   a) $A \triangle (B \cap C) = (A \triangle B) \cap (A \triangle C)$

   b) $A \cap (B \triangle C) = (A \cap B) \triangle (A \cap C)$

   c) $A \triangle (B \cup C) = (A \triangle B) \cup (A \triangle C)$

   d) $A \cup (B \triangle C) = (A \cup B) \triangle (A \cup C)$

   e) $A - (B \cup C) = (A - B) \cap (A - C)$

   f) $A - (B \cap C) = (A - B) \cup (A - C)$

   g) $A \triangle (B \triangle C) = (A \triangle B) \triangle C$

7. If $A = \{a, b, d\}$, $B = \{d, x, y\}$, and $C = \{x, z\}$, how many proper subsets are there for the set $(A \cap B) \cup C$? How many for the set $A \cap (B \cup C)$?

8. For a given universal set $\mathcal{U}$, each subset A of $\mathcal{U}$ satisfies the idempotent laws of union and intersection. (a) Are there any real numbers that satisfy an idempotent property for addition? (That is, can we find any real number(s) x such that $x + x = x$?) (b) Answer part (a) replacing addition by multiplication.

9. Write the dual statement for each of the following set-theoretic results.

   a) $\mathcal{U} = (A \cap B) \cup (A \cap \overline{B}) \cup (\overline{A} \cap B) \cup (\overline{A} \cap \overline{B})$

   b) $A = A \cap (A \cup B)$

   c) $A \cup B = (A \cap B) \cup (A \cap \overline{B}) \cup (\overline{A} \cap B)$

   d) $A = (A \cup B) \cap (A \cup \emptyset)$

10. Use the equivalence $A \subseteq B \Leftrightarrow A \cap B = A$ to show that the dual statement of $A \subseteq B$ is the statement $B \subseteq A$.

11. Prove or disprove each of the following for sets $A, B \subseteq \mathcal{U}$.

   a) $\mathcal{P}(A \cup B) = \mathcal{P}(A) \cup \mathcal{P}(B)$ **b)** $\mathcal{P}(A \cap B) = \mathcal{P}(A) \cap \mathcal{P}(B)$

12. Use membership tables to establish each of the following:

   a) $\overline{A \cap B} = \overline{A} \cup \overline{B}$ **b)** $A \cup A = A$ **c)** $A \cup (A \cap B) = A$

   d) $\overline{(A \cap B) \cup (\overline{A} \cap C)} = (A \cap \overline{B}) \cup (\overline{A} \cap \overline{C})$

13. a) How many rows are needed to construct the membership table for $A \cap (B \cup C) \cap (D \cup \overline{E} \cup F)$?

b) How many rows are needed to construct the membership table for a set made up from the sets $A_1, A_2, \ldots, A_n$, using $\cap$, $\cup$ and $^-$?

c) Given the membership tables for two sets A, B, how can the relation $A \subseteq B$ be recognized?

d) Use membership tables to determine whether or not $(A \cap B) \cup \overline{(B \cap C)} \supseteq A \cup \overline{B}$.

14. Using the laws of set theory, simplify each of the following:

a) $A \cap (B - A)$

b) $(A \cap B) \cup (A \cap B \cap \overline{C} \cap D) \cup (\overline{A} \cap B)$

c) $(A - B) \cup (A \cap B)$

d) $\overline{A} \cup \overline{B} \cup (A \cap B \cap \overline{C})$

e) $\overline{A} \cup (A \cap \overline{B}) \cup (A \cap B \cap \overline{C}) \cup (A \cap B \cap C \cap \overline{D}) \cup \cdots$

15. Let $\mathcal{U} = \mathbf{R}$ and let $I = \mathbf{Z}^+$. For each $n \in \mathbf{Z}^+$ let $A_n = [-2n, 3n]$. Determine each of the following:

a) A_3

b) A_4

c) $A_3 - A_4$

d) $A_3 \triangle A_4$

e) $\bigcup_{n=1}^{7} A_n$

f) $\bigcap_{n=1}^{7} A_n$

g) $\bigcup_{n \in \mathbf{Z}^+} A_n$

h) $\bigcap_{n=1}^{\infty} A_n$

16. Provide the details for the proof of Theorem 3.6(b).

17. Given a universe $\mathcal{U}$ and an index set I, for each $i \in I$ let $B_i \subseteq \mathcal{U}$. Prove that for $A \subseteq \mathcal{U}, A \cap (\bigcup_{i \in I} B_i) = \bigcup_{i \in I} (A \cap B_i)$, and $A \cup (\bigcap_{i \in I} B_i) = \bigcap_{i \in I} (A \cup B_i)$. [Generalized Distributive Laws]

3.3
COUNTING AND VENN DIAGRAMS

With all of the theoretical work and theorem proving we did in the last section, now is a good time to examine some additional counting problems.

For sets A, B from a finite universe $\mathcal{U}$, the following Venn diagrams will help us obtain counting formulas for $|\overline{A}|$ and $|A \cup B|$ in terms of $|A|, |B|$ and $|A \cap B|$.

As Fig. 3.7 demonstrates, $A \cup \overline{A} = \mathcal{U}$ and $A \cap \overline{A} = \emptyset$, so by the rule of sum, $|A| + |\overline{A}| = |\mathcal{U}|$ or $|\overline{A}| = |\mathcal{U}| - |A|$. The sets A, B, in Fig. 3.8, have empty intersection, so here the rule of sum leads us to $|A \cup B| = |A| + |B|$ and necessitates that A, B be finite but does not require any condition on the cardinality of $\mathcal{U}$.

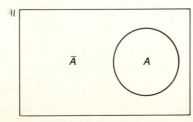

Figure 3.7

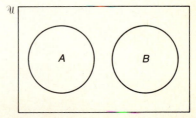

Figure 3.8

Turning to the case where A, B are not disjoint, we motivate the formula for $|A \cup B|$ with the following example.

Example 3.21   In a class of 50 college freshmen, 30 are studying BASIC, 25 are studying Pascal, and 10 are studying both languages. How many freshmen are studying either computer language?

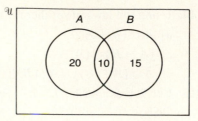

Figure 3.9

We let $\mathcal{U}$ be the class of 50 freshmen, A the subset of those students studying BASIC, and B the subset of those studying Pascal. In order to answer the question, we need $|A \cup B|$. In Fig. 3.9 the numbers in the regions are obtained from the given information: $|A| = 30$, $|B| = 25$, $|A \cap B| = 10$. Consequently, $|A \cup B| = 45 \neq |A| + |B|$, because $|A| + |B|$ counts the students in $A \cap B$ twice. To remedy this overcount, we subtract $|A \cap B|$ from $|A| + |B|$ to obtain the correct formula: $|A \cup B| = |A| + |B| - |A \cap B|$.   □

For the general situation, if A and B are finite sets, then $|A \cup B| = |A| + |B| - |A \cap B|$. Consequently, finite sets A and B are (mutually) disjoint if and only if $|A \cup B| = |A| + |B|$.

This situation extends to three sets, as the following example illustrates.

Example 3.22   A 14-pin logic chip has four AND gates, each with two inputs and one output. (See Fig. 3.10.) The first AND gate (the gate at pins 1, 2, 3) can have any or all of the following defects:

D_1:   the first input (pin 1) is stuck at 0.
D_2:   the second input (pin 2) is stuck at 0.
D_3:   the output (pin 3) is stuck at 1.

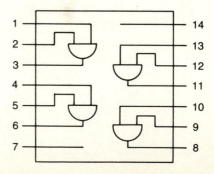

Figure 3.10

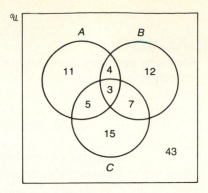

Figure 3.11

For a sample of 100 such logic chips we let A, B, and C be the subsets having defects D_1, D_2, and D_3, respectively. With $|A| = 23$, $|B| = 26$, $|C| = 30$, $|A \cap B| = 7$, $|A \cap C| = 8$, $|B \cap C| = 10$, and $|A \cap B \cap C| = 3$, how many chips in the sample are defective?

Working backwards from $|A \cap B \cap C| = 3$ to $|A| = 23$, we label the regions as shown in Fig. 3.11 and find that $|A \cup B \cup C| = 57 = |A| + |B| + |C| - |A \cap B| - |A \cap C| - |B \cap C| + |A \cap B \cap C|$. $\square$

From the formula for $|A \cup B \cup C|$ and DeMorgan's Law, we find that if A, B, and C are sets from a finite universe $\mathcal{U}$, then $|\overline{A} \cap \overline{B} \cap \overline{C}| = |\overline{A \cup B \cup C}| = |\mathcal{U}| - |A \cup B \cup C| = |\mathcal{U}| - |A| - |B| - |C| + |A \cap B| + |A \cap C| + |B \cap C| - |A \cap B \cap C|$.

We close this section with a problem that uses this result.

Example 3.23 A student visits an arcade each day after school and plays one game of either Laser-Man, Millipede, or Space Conquerors. In how many ways can he play one game each day so that he plays each of the three types at least once during a given school week?

Here there is a slight twist. The set $\mathcal{U}$ consists of all arrangements of size five taken from the set of three games, with repetitions allowed. The set A represents the subset of all sequences of five games played during the week *without playing Laser Man*. The sets B and C are defined similarly, leaving out Millipede and Space Conquerors, respectively. The enumeration techniques of Chapter 1 give $|\mathcal{U}| = 3^5$, $|A| = |B| = |C| = 2^5$, $|A \cap B| = |A \cap C| = |B \cap C| = 1^5 = 1$ and $|A \cap B \cap C| = 0$, so by the formula above there are $|\overline{A} \cap \overline{B} \cap \overline{C}| = 3^5 - 3 \cdot 2^5 + 3 \cdot 1^5 - 0 = 150$ ways the student can select his daily games during a school week and play each type of game at least once.

This example can be expressed in an equivalent distribution form, since we are seeking the number of ways to distribute five distinct objects (Monday, Tuesday, . . . , Friday) among three distinct containers (the computer games) with no container left empty. More will be said about this in Chapter 5. $\square$

3.4
A WORD ON PROBABILITY

When one performs an *experiment* such as tossing a single die or selecting two students from a class of 20 to work on a project, a listing of all possible outcomes for the situation is called a *sample space*. Consequently, $\{1, 2, 3, 4, 5, 6\}$ serves as a sample space for the first experiment mentioned, while $\{\{a_i, a_j\} \,|\, 1 \le i \le 20, 1 \le j \le 20, i \ne j\}$ could be used for the latter experiment, with a_i denoting the *i*th student, $1 \le i \le 20$.

Unfortunately, for a given experiment there can be more than one sample space. If we toss a coin three times and consider the outcomes, a possible sample space for our experiment is $\{0, 1, 2, 3\}$, where the number i, for $0 \le i \le 3$, refers to the number of heads that come up in the three tosses. The outcomes can also be given by the sample space $\mathcal{S} = \{\text{HHH, THH, HTH, HHT, TTH, THT, HTT, TTT}\}$, where we feel that each of the eight possible outcomes has the *same likelihood* of occurrence. This is not the case with our first sample space $\{0, 1, 2, 3\}$, where we feel that there is a better chance of getting one head in the three tosses than there is of getting no heads.

In this text we shall always use a sample space where each element has the same likelihood of occurrence. Under this assumption of equal likelihood, we shall use a definition for probability that was first given by the French mathematician Pierre-Simon de Laplace (1749–1827) in his *Analytic Theory of Probability*.

Under the assumption of equal likelihood, let $\mathcal{S}$ be a sample space for an experiment $\mathcal{E}$. Any subset A of $\mathcal{S}$ is called an *event*. Each element of $\mathcal{S}$ is called an *elementary event*, so if $|\mathcal{S}| = n$ and $a \in \mathcal{S}, A \subseteq \mathcal{S}$, then

$$Pr(a) = \textit{The probability that a occurs} = \frac{1}{n} = \frac{|\{a\}|}{|\mathcal{S}|}, \text{ and}$$

$$Pr(A) = \textit{The probability that A occurs} = \frac{|A|}{|\mathcal{S}|} = \frac{|A|}{n} = \frac{1}{n}\left(\sum_{a \in A} |\{a\}|\right),$$

since A is the disjoint union of its singleton subsets.

We demonstrate these ideas in the following examples.

Example 3.24   In tossing a die once, what is the probability of getting a 5 or a 6?

Here $\mathcal{S} = \{1, 2, 3, 4, 5, 6\}$ and the event A we want to consider is $\{5, 6\}$. Hence

$$Pr(A) = \frac{|A|}{|\mathcal{S}|} = \frac{2}{6} = \frac{1}{3}. \quad \square$$

Example 3.25   If one tosses a coin four times, what is the probability of getting two heads and two tails?

The sample space here consists of all sequences of the form x_1, x_2, x_3, x_4, where each x_i, for $1 \le i \le 4$, is replaced by a T or an H, so $|\mathcal{S}| = 2^4 = 16$. The event A we are concerned about contains all arrangements of the four symbols H, H, T, T, so $|A| = 4!/(2! \, 2!) = 6$. Consequently, $Pr(A) = 6/16 = 3/8$. □

For Example 3.25, each toss is *independent* of the outcome of any previous toss. Such an occurrence is called a *Bernoulli trial*. In a first course in probability, these trials are investigated in conjunction with the Bernoulli distribution, an important example of a discrete probability distribution. We shall come upon this again in Chapter 16 when we study the application of abelian groups in coding theory.

In our last example we use the concepts of Venn diagrams and probability.

Example 3.26 In a survey of 120 passengers, an airline found that 48 preferred wine with their meals, 78 preferred mixed drinks, and 66 iced tea. In addition, 36 enjoyed any given pair of these beverages and 24 passengers enjoyed them all. If two passengers are selected at random from the survey sample of 120, what is the probability that

a) (Event A) they both want only iced tea with their meals?

b) (Event B) they both enjoy exactly two of the three beverage offerings?

From the information provided, we construct the Venn diagram shown in Fig. 3.12. The sample space $\mathscr{S}$ consists of the pairs of passengers we can select from the sample of 120, so $|\mathscr{S}| = \binom{120}{2} = 7140$. The Venn diagram indicates that there are 18 passengers who drink only iced tea, so $|A| = \binom{18}{2}$ and $Pr(A) = 51/2380$. The reader should verify that $Pr(B) = 3/34$. □

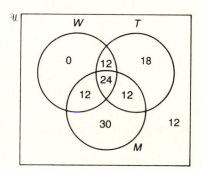

Figure 3.12

EXERCISES 3.3 AND 3.4

1. How many permutations of the 26 different letters of the alphabet contain (a) either the pattern "OUT" or the pattern "DIG"? (b) neither the pattern "MAN" nor the pattern "ANT"?

2. A six-character variable name in ANSI FORTRAN starts with a letter of the alphabet. Each of the other five characters can be either a letter or a digit.

(Repetitions are allowed.) How many six-character variable names contain the pattern "FUN" or the pattern "TIP"?

3. How many permutations of the digits $0, 1, 2, \ldots, 9$ either start with a 3 or end with a 7, or both?

4. A professor has two dozen introductory textbooks on computer science and is concerned about their coverage of the topics (A) compilers, (B) data structures, and (C) interpreters. The following data are the numbers of books that contain material on these topics:

$\lvert A \rvert = 8$	$\lvert B \rvert = 13$	$\lvert C \rvert = 13$
$\lvert A \cap B \rvert = 5$	$\lvert A \cap C \rvert = 3$	$\lvert B \cap C \rvert = 6$
$\lvert A \cap B \cap C \rvert = 2$		

(a) How many of the textbooks include material on exactly one of these topics? (b) How many do not deal with any of the topics? (c) How many have no material on compilers?

5. A letter is selected at random from the word "chemistry." (a) What is the probability that it is either e or i? (b) What is the probability that it does not occur before q in the alphabet?

6. In selecting a new computer for its computing center, a college examines 15 different models, paying attention to the following considerations: (A) magnetic tape drive, (B) terminal for graphics display, (C) semiconductor memory (in addition to core memory). The numbers of computers with any or all of these features are as follows: $\lvert A \rvert = \lvert B \rvert = \lvert C \rvert = 6$, $\lvert A \cap B \rvert = \lvert B \cap C \rvert = 1$, $\lvert A \cap C \rvert = 2$, $\lvert A \cap B \cap C \rvert = 0$. (a) How many of the models have exactly one of the features being considered? (b) How many have none of the features? (c) If a model is selected at random, what is the probability that it has exactly two of these features?

7. a) How many arrangements of the letters in MISCELLANEOUS have no pair of consecutive letters the same?

   b) If an arrangement of these letters is randomly generated, what is the probability that no pair of identical consecutive letters occurs?

8. How many arrangements of the letters in CHEMIST have either H before E, or E before T, or T before M? (Here "before" means anywhere before, not just immediately before.)

9. An integer is selected at random from 3 through 17 inclusive. If A is the event where a number divisible by 3 is chosen and B is the event where the number exceeds 10, determine $Pr(A)$, $Pr(B)$, $Pr(A \cap B)$, and $Pr(A \cup B)$. How is $Pr(A \cup B)$ related to $Pr(A)$, $Pr(B)$, and $Pr(A \cap B)$?

10. The freshman class of a private engineering college has 300 students. It is known that 180 can program in Pascal, 120 in FORTRAN, 30 in APL, 12 in Pascal and APL, 18 in FORTRAN and APL, 12 in Pascal and FORTRAN, and 6 in all three languages.

a) If a student is selected at random, what is the probability that she can program in exactly two languages?

b) If two students are selected at random, what is the probability that they can

   (i) both program in Pascal?

   (ii) both program only in Pascal?

11. Eight different books, three in physics and five in electrical engineering, are placed at random on a library shelf. Find the probability that the three physics books are all together.

12. A shipment of 24 new cars contains 15 in excellent condition, six with minor defects and three with major defects. If two cars are selected from the shipment, what is the probability that (a) both are in excellent condition? (b) both have minor defects? (c) at most one has a minor defect? (d) at least one has a minor defect? (e) exactly one has a minor defect? (f) neither has a minor defect?

How are the results in parts (b), (e), and (f) related?

13. If the letters in the word CORRESPONDENT are arranged in a random order, what is the probability that the arrangement starts and ends with the same letter?

3.5

SUMMARY AND HISTORICAL REVIEW

In this chapter we introduced some of the fundamentals of set theory with consideration to enumeration and elementary probability.

The algebra of set theory evolved during the nineteenth and early twentieth centuries. In England, George Peacock (1791–1858) was a pioneer in mathematical reforms and was among the first, in his *Treatise on Algebra*, to revolutionize the entire conception of algebra and arithmetic. His ideas were further developed by Duncan Gregory (1813–1844), William Rowan Hamilton (1805–1865), and Augustus DeMorgan (1806–1871), who attempted to remove ambiguity from elementary algebra and cast it in the strict postulational form. It was not until 1854, however, when Boole published his *Investigation of the Laws of Thought*, that an algebra dealing with sets and logic was formalized and the work of Peacock and his contemporaries extended.

The presentation here is primarily concerned with finite sets. However, the investigation of infinite sets and their cardinalities has occupied the minds of many mathematicians and philosophers. The intuitive approach to set theory was taken until the time of Georg Cantor (1845–1918) who defined a set, in 1895, in a way comparable to the "gut feeling" we mentioned at the start of Section 3.1. His definition, however, was one of the obstacles Cantor was never able to remove entirely from his theory of sets.

In the 1870s, when Cantor was researching trigonometric series and series of real numbers, he needed a device to compare the sizes of infinite sets of numbers. His treatment of the infinite as an actuality, on the same level as the finite, was quite revolutionary. Some of his work was rejected, but it won wide enough acceptance so that by 1890 the theory of sets, both finite and infinite, was considered a branch of mathematics in its own right.

By the turn of the century the theory was widely accepted, but in 1901 the paradox now known as Russell's paradox showed that set theory, as originally proposed, was internally inconsistent. The difficulty seemed to be in the unrestricted way in which sets could be defined; the idea of a set being a member of itself was considered particularly suspect. In their work *Principia Mathematica*, Bertrand Russell (1872–1970) and Alfred North Whitehead (1861–1947) developed a hierarchy in the theory of sets known as the *theory of types*. This axiomatic set theory, among other twentieth-century formulations, avoided the Russell paradox.

If we wish to summarize the importance of the role of set theory in the development of twentieth-century mathematics, the following quote attributed to the German mathematician David Hilbert (1862–1943) is worth pondering: "No one shall expel us from the paradise which Cantor has created for us."

In Section 3.1 we mentioned the array of numbers known as Pascal's triangle. We could have done this in Chapter 1 with the binomial theorem, but we waited until we had some combinatorial identities that we needed to verify how the triangle is constructed. The array appears in the work of the Chinese algebraist Chu Shi-kie (1303), but its first appearance in Europe was not until the sixteenth century, on the title page of a book by Petrus Apianus (1495–1552). Niccolo Tartaglia (1499–1559) used the triangle in computing powers of $(x + y)$. Because of his work on the properties and applications of this triangle, the array has been named in honor of the French mathematician Blaise Pascal (1623–1662).

Finally, although probability originated with games of chance and enumeration problems, we mention it here because set theory has evolved as the exact medium needed to state and solve problems in this important contemporary area of applied mathematics.

More on the history and development of set theory can be found in Chapter 26 of C. Boyer [1]. Formal developments of set theory, including results on infinite sets, can be found in P. Halmos [3] and P. Suppes [5]. An interesting history of the origins of probability and statistical ideas, up to the Newtonian era, can be found in F. N. David [2]. Chapters 1 and 2 of P. Meyer [4] are an excellent source for those interested in learning more about discrete probability.

REFERENCES

1. Boyer, Carl B. *History of Mathematics*. New York: Wiley, 1968.
2. David, Florence Nightingale. *Games, Gods, and Gambling*. New York: Hafner, 1962.
3. Halmos, Paul R. *Naive Set Theory*. New York: Van Nostrand, 1960.

4. Meyer, Paul L. *Introductory Probability and Statistical Applications,* 2nd ed. Reading, Mass.: Addison-Wesley, 1970.

5. Suppes, Patrick C. *Axiomatic Set Theory.* New York: Van Nostrand, 1960.

MISCELLANEOUS EXERCISES

1. Let $A, B, C \subseteq \mathcal{U}$. Prove that $(A - B) \subseteq C$ if and only if $(A - C) \subseteq B$.

2. If $A, B \subseteq \mathcal{U}$, prove that $A \subseteq B$ if and only if $[\forall C \subseteq \mathcal{U} \ (C \subseteq A \Rightarrow C \subseteq B)]$.

3. Let $A, B, C \subseteq \mathcal{U}$. Prove or disprove (with a counterexample) each of the following:

   a) $A - C = B - C \Rightarrow A = B$

   b) $[(A \cap C = B \cap C) \wedge (A - C = B - C)] \Rightarrow A = B$

   c) $[(A \cup C = B \cup C) \wedge (A - C = B - C)] \Rightarrow A = B$

   d) $\mathcal{P}(A - B) = \mathcal{P}(A) - \mathcal{P}(B)$

4. a) For positive integers m, n, r, with $r \leq \min\{m, n\}$, show that

$$\binom{m + n}{r} = \binom{m}{0}\binom{n}{r} + \binom{m}{1}\binom{n}{r - 1} + \binom{m}{2}\binom{n}{r - 2} + \cdots$$
$$+ \binom{m}{r}\binom{n}{0} = \sum_{k=0}^{r} \binom{m}{k}\binom{n}{r - k}.$$

   b) For n a positive integer, show that

$$\binom{2n}{n} = \sum_{k=0}^{n} \binom{n}{k}^2.$$

5. a) In how many ways can a teacher divide a group of seven students into two teams each containing at least one student? two students?

   b) Answer (a) replacing seven by a positive integer $n \geq 4$.

6. Determine whether each of the following statements is true or false. For each false statement, give a counterexample.

   a) If A and B are infinite sets, then $A \cap B$ is infinite.

   b) If B is infinite and $A \subseteq B$, then A is infinite.

   c) If $A \subseteq B$ with B finite, then A is finite.

   d) If $A \subseteq B$ with A finite, then B is finite.

7. A set A has 128 subsets of even cardinality. (a) How many subsets of A have odd cardinality? (b) What is $|A|$?

8. Let $A = \{1, 2, 3, \ldots, 15\}$.

   a) How many subsets of A contain all of the odd integers in A?

   b) How many subsets of A contain exactly three odd integers?

   c) How many eight-element subsets of A contain exactly three odd integers?

   d) Write a computer program (or develop an algorithm) to generate an eight-element subset of A and have it print out how many of the eight elements are odd.

9. Let $A, B, C \subseteq \mathcal{U}$. Prove that $(A \cap B) \cup C = A \cap (B \cup C)$ if and only if $C \subseteq A$.

10. For any $A, B, C \subseteq \mathcal{U}$, prove that $(A - B) - C = (A - C) - (B - C)$.

11. For any $A, B \subseteq \mathcal{U}$, prove that (a) $A \cup B = \mathcal{U}$ if and only if $\overline{A} \subseteq B$; (b) $A \cap B = \emptyset$ if and only if $\overline{A} \supseteq B$.

12. Let $\mathcal{U}$ be a given universe with $A, B \subseteq \mathcal{U}$, $|A \cap B| = 3$, $|A \cup B| = 8$, and $|\mathcal{U}| = 12$.

   a) How many subsets $C \subseteq \mathcal{U}$ satisfy $A \cap B \subseteq C \subseteq A \cup B$? How many of these subsets C contain an even number of elements?

   b) How many subsets $D \subseteq \mathcal{U}$ satisfy $\overline{A \cup B} \subseteq D \subseteq \overline{A} \cup \overline{B}$? How many of these subsets D contain an even number of elements?

13. Let $\mathcal{U} = \mathbf{R}$ and let the index set $I = \mathbf{Q}^+$. For each $q \in \mathbf{Q}^+$, let $A_q = [0, 2q]$ and $B_q = (0, 3q]$. Determine each of the following:

   a) $A_{7/3}$

   b) $B_{3/5}$

   c) $A_3 - B_4$

   d) $A_3 \triangle B_4$

   e) $\displaystyle\bigcup_{q \in I} A_q$

   f) $\displaystyle\bigcup_{q \in I} B_q$

   g) $\displaystyle\bigcap_{q \in I} A_q$

   h) $\displaystyle\bigcap_{q \in I} B_q$

14. For a universe $\mathcal{U}$ and sets $A, B \subseteq \mathcal{U}$, prove each of the following:

   a) $A \triangle B = B \triangle A$

   b) $A \triangle \overline{A} = \mathcal{U}$

   c) $A \triangle \mathcal{U} = \overline{A}$

   d) $A \triangle \emptyset = A$, so $\emptyset$ is the identity for $\triangle$, as well as for $\cup$

15. Consider the membership table (Table 3.3). If we are given the condition that $A \subseteq B$, then we need consider only those rows of the table for which this is true—rows 1, 2, and 4, as indicated by the arrows. For these rows, the columns for B and $A \cup B$ are exactly the same, so this membership table shows that $A \subseteq B \Rightarrow A \cup B = B$.

Table 3.3

	A	B	$A \cup B$
$\rightarrow$	0	0	0
$\rightarrow$	0	1	1
	1	0	1
$\rightarrow$	1	1	1

Use membership tables to verify each of the following:

   a) $A \subseteq B \Rightarrow A \cap B = A$

   b) $[(A \cap B = A) \wedge (B \cup C = C)] \Rightarrow A \cup B \cup C = C$

   c) $C \subseteq B \subseteq A \Rightarrow (A \cap \overline{B}) \cup (B \cap \overline{C}) = A \cap \overline{C}$

   d) $A \triangle B = C \Rightarrow A \triangle C = B$ and $B \triangle C = A$

16. State the dual of each theorem in Exercise 15. (Here you will want to use the result of Example 3.15 in conjunction with Theorem 3.5.)

17. a) Determine the number of linear arrangements of m 1's and r 0's with no adjacent 1's. (State any needed condition(s) for m, r.)

   b) If $\mathcal{U} = \{1, 2, 3, \ldots, n\}$, how many sets $A \subseteq \mathcal{U}$ are such that $|A| = k$ with A containing no consecutive integers? (State any needed condition(s) for n, k.)

18. At a high school science fair, 34 students received awards for scientific projects. Fourteen awards were given for projects in biology, 13 in chemistry, and 21 in physics. If three students received awards in all three subject areas, how many received awards for exactly (a) one subject area? (b) two subject areas?

19. If eight people are seated around a circular table, what is the probability that a given pair are sitting next to each other?

20. If the letters in the word BOOLEAN are arranged at random, what is the probability that the two O's remain together in the arrangement?

21. If 16 chocolate-chip cookies are distributed among four children, what is the probability that every child gets (a) at least one cookie? (b) at least two cookies? (c) four cookies?

22. Fifty students, each with 75¢, visited the arcade of Example 3.23. Seventeen of the students played each of the three computer games, and 37 of them played at least two of them. No student played any other game at the arcade, nor did any student play a given game more than once. Each game costs 25¢ to play, and the total proceeds from the student visit were $24.25. How many of these students preferred to watch and played none of the games?

23. In how many ways can 15 laboratory assistants be assigned to work on one, two, or three different experiments so that each experiment has at least one person spending some time on it?

24. Professor Diane gave her chemistry class a test consisting of three questions. There are 21 students in her class, and every student answered at least one question. Five students did not answer the first question, seven failed to answer the second question, and six did not answer the third question. If nine students answered all three questions, how many answered exactly one question?

25. Let $\mathcal{U}$ be a given universe with $A, B \subseteq \mathcal{U}$, $A \cap B = \emptyset$, $|A| = 12$, and $|B| = 10$. If seven elements are selected from $A \cup B$, what is the probability the selection contains four elements from A and three from B?

26. For a finite set A of integers, let $\sigma(A)$ denote the sum of the elements of A. Then if $\mathcal{U}$ is a finite universe taken from $\mathbf{Z}^+$, $\displaystyle\sum_{A \in \mathscr{P}(\mathcal{U})} \sigma(A)$ denotes the sum of all elements of all subsets of $\mathcal{U}$. Determine $\displaystyle\sum_{A \in \mathscr{P}(\mathcal{U})} \sigma(A)$ for

   a) $\mathcal{U} = \{1, 2, 3\}$

   b) $\mathcal{U} = \{1, 2, 3, 4\}$

   c) $\mathcal{U} = \{1, 2, 3, 4, 5\}$

   d) $\mathcal{U} = \{1, 2, 3, \ldots, n\}$

   e) $\mathcal{U} = \{a_1, a_2, a_3, \ldots, a_n\}$, where $s = a_1 + a_2 + a_3 + \cdots + a_n$

27. a) In chess, the king can move one position in any direction. Assuming that the king is moved only in a forward manner (one position up, to the right, or diagonally northeast), along how many different paths can a king be moved from the lower-left corner position to the upper-right corner position on the standard 8×8 chessboard?

   b) For the paths in part (a), what is the probability that a path contains (i) exactly two diagonal moves? (ii) exactly two diagonal moves that are consecutive? (iii) an even number of diagonal moves?

CHAPTER 4

Properties of the Integers: Mathematical Induction

Having known about the integers since our first encounters with arithmetic, in this chapter we examine a special property exhibited by the subset of positive integers. This property will enable us to establish certain mathematical formulas and theorems by using a technique called *mathematical induction*. This method of proof will play a key role in many of the results we get in the later chapters of this text.

When $x, y \in \mathbf{Z}$, we know that $x + y, xy, x - y \in \mathbf{Z}$. Thus we say that the set $\mathbf{Z}$ is *closed* under (the binary operations of) addition, multiplication, and subtraction. Turning to division, however, we find, for example, that $2, 3 \in \mathbf{Z}$ but that the rational number $\frac{2}{3}$ is *not* a member of $\mathbf{Z}$. So $\mathbf{Z}$ is not closed under *nonzero* division. To cope with this situation, we shall introduce a somewhat restricted form of division for $\mathbf{Z}$ and shall concentrate on special elements of $\mathbf{Z}^+$ called *primes*. These primes turn out to be the "building blocks" of the integers, and they provide our first example of a representation theorem—in this case the Fundamental Theorem of Arithmetic.

4.1
THE WELL-ORDERING PRINCIPLE: MATHEMATICAL INDUCTION

Given any two distinct integers x, y, we know that we must have either $x < y$ or $y < x$. However, this is also true if, instead of being integers, x and y are rational numbers or real numbers. What makes $\mathbf{Z}$ special in this situation?

Suppose we try to express the subset $\mathbf{Z}^+$ of $\mathbf{Z}$, using the inequality symbols $>$ and $\geq$. We find that we can define the set of positive elements of $\mathbf{Z}$ as

$$\mathbf{Z}^+ = \{x \in \mathbf{Z} \,|\, x > 0\} = \{x \in \mathbf{Z} \,|\, x \geq 1\}.$$

When we try to do likewise for the rational and real numbers, however, we find that

$$\mathbf{Q}^+ = \{x \in \mathbf{Q} \,|\, x > 0\} \qquad \text{and} \qquad \mathbf{R}^+ = \{x \in \mathbf{R} \,|\, x > 0\},$$

but we cannot represent $\mathbf{Q}^+$ or $\mathbf{R}^+$ using $\geq$ as we did for $\mathbf{Z}^+$.

The set $\mathbf{Z}^+$ is different from the sets $\mathbf{Q}^+$ and $\mathbf{R}^+$ in that *every* nonempty subset X of $\mathbf{Z}^+$ contains an integer $a \in X$ such that $a \leq x$, for all $x \in X$ (that is, X contains a *least* element). This is not so for either $\mathbf{Q}^+$ or $\mathbf{R}^+$. The sets themselves do not contain least elements. There is no smallest positive rational number or smallest positive real number. If we could find, for example, a smallest positive rational number q, then since $0 < q/2 < q$, we would have the smaller positive rational number $q/2$.

These observations lead us to the following property of the set $\mathbf{Z}^+ \subset \mathbf{Z}$.

The Well-Ordering Principle. Any *nonempty* subset of $\mathbf{Z}^+$ contains a smallest element. (We often express this by saying that $\mathbf{Z}^+$ is *well-ordered*.)

This principle serves to distinguish $\mathbf{Z}^+$ from $\mathbf{Q}^+$ and $\mathbf{R}^+$. But does it lead anywhere that is mathematically interesting or useful? The answer is a resounding "Yes!" It is the basis of a proof technique known as mathematical induction. This technique helps us prove a general mathematical statement involving a positive integer, when certain instances of that statement suggest a general pattern.

We now establish the basis for this induction technique.

THEOREM 4.1   (*Finite Induction Principle*, or *Principle of Mathematical Induction*) Let $S(n)$ denote a mathematical statement (or set of statements) that involves one or more occurrences of the symbol n, which represents a positive integer.

   a) If $S(1)$ is true; and

   b) If whenever $S(k)$ is true for some $k \in \mathbf{Z}^+$, the truth of $S(k + 1)$ is implied by the truth of $S(k)$;

then $S(n)$ is true for all $n \in \mathbf{Z}^+$.

Proof   Let $S(n)$ be such a statement satisfying conditions (a) and (b), and let $F = \{t \in \mathbf{Z}^+ \,|\, S(t) \text{ is false}\}$. We wish to prove that $F = \emptyset$.

If $F \neq \emptyset$, then by the Well-Ordering Principle, F has a least element s. Since $S(1)$ is true, it follows that $s \neq 1$, so $s > 1$, and consequently $s - 1 \in \mathbf{Z}^+$. With $s - 1 \notin F$, we have $S(s - 1)$ true. So by condition (b) it follows that $S((s - 1) + 1) = S(s)$ is true, contradicting $s \in F$. This contradiction arose from the assumption that $F \neq \emptyset$. Consequently, $F = \emptyset$. ∎

The choice of 1 in the first condition of Theorem 4.1 is not mandatory. All that is needed is for the statement $S(n)$ to be true for some $n_0 \in \mathbf{Z}$ so that the induction process has a starting place. The integer n_0 could be 5 just as well as 1. It could even be zero or negative, because the set $\mathbf{Z}^+$ in union with $\{0\}$ or any *finite* set of negative integers is well-ordered. (And when we do an induction proof and start with $n_0 < 0$, we are considering the set of all *consecutive* negative integers $\geq n_0$ in union with $\{0\}$ and $\mathbf{Z}^+$.)

Under these circumstances, we may then express the Finite Induction Principle, using quantifiers, as

$$[S(n_0) \wedge [\forall k \geq n_0 [S(k) \Rightarrow S(k+1)]]] \Rightarrow \forall n \geq n_0 \; S(n).$$

We may get a somewhat better understanding of why this method of proof is valid by using our intuition in conjunction with the situation presented in Fig. 4.1.

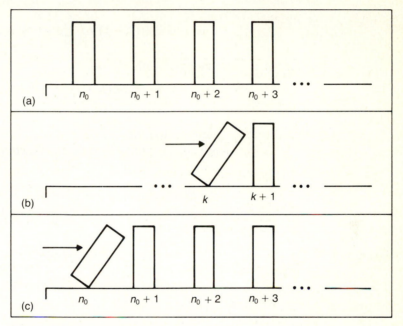

Figure 4.1

In part (a) of the figure we see the first four of an infinite (ordered) arrangement of dominoes, each standing on end. The spacing between any two consecutive dominoes is always the same, and it is such that if any one domino (say the kth) is pushed over to the right, then it will knock over the next $((k+1)$st) domino. This is suggested in Fig. 4.1(b). Our intuition leads us to feel that this process will continue, the $(k+1)$st domino toppling and knocking over (to the right) the $(k+2)$nd domino, and so on. Part (c) of the figure indicates how the truth of $S(n_0)$ provides the push (to the right) to the first domino (at n_0). This sets the process in motion. The truth of $S(k)$ forcing the truth of $S(k+1)$ continues the toppling process. We then infer the fact that $S(n)$ is true for all $n \geq n_0$ as we imagine *all* the successive dominoes toppling (to the right).

We shall now demonstrate several results that call for the use of Theorem 4.1.

Example 4.1   For any $n \in \mathbf{Z}^+$, $\sum_{i=1}^{n} i = 1 + 2 + 3 + \cdots + n = (n)(n+1)/2$.

Proof: For $n = 1$, the statement $S(1)$ is $\sum_{i=1}^{1} i = 1 = (1)(1+1)/2$. So $S(1)$ is true and we have a starting point to begin the induction. Assuming the result true for $n = k$, for some $k \in \mathbf{Z}^+$, we want to show how the truth of $S(k)$ "forces" us to accept the truth of $S(k+1)$. (The assumption of the truth of $S(k)$ is our *induction hypothesis*.) To establish $S(k+1)$, we need to show that

$$\sum_{i=1}^{k+1} i = \frac{(k+1)(k+2)}{2}.$$

We proceed as follows.

$$\sum_{i=1}^{k+1} i = 1 + 2 + \cdots + k + (k+1) = \left(\sum_{i=1}^{k} i\right) + (k+1) = \frac{k(k+1)}{2} + (k+1),$$

for we are assuming the truth of $S(k)$. But

$$\frac{k(k+1)}{2} + (k+1) = \frac{k(k+1)}{2} + \frac{2(k+1)}{2} = \frac{(k+1)(k+2)}{2},$$

establishing condition (b) of the theorem.

Consequently, by the principle of finite induction, $S(n)$ is true for all $n \in \mathbf{Z}^+$. $\square$

Now that we have obtained the summation formula for $\sum_{i=1}^{n} i$ in two ways (see Example 1.35), we shall digress from our main topic to consider an example that uses this summation formula.

Example 4.2   A wheel of fortune has the numbers from 1 to 36 painted on it in a random manner. Show that regardless of how the numbers are situated, there are three in succession which total 55 or more.

Let x_1 be any number on the wheel. Counting clockwise from x_1, label the other numbers $x_2, x_3, \ldots, x_{36}$. For the result to be false, we must have $x_1 + x_2 + x_3 < 55, \ldots, x_{34} + x_{35} + x_{36} < 55, x_{35} + x_{36} + x_1 < 55$, and $x_{36} + x_1 + x_2 < 55$. In these 36 inequalities, each of the terms $x_1, x_2, \ldots, x_{36}$ appears exactly three times, so each of the integers $1, 2, \ldots, 36$ appears three times. Adding all 36 inequalities, we find that $3 \sum_{i=1}^{36} x_i = 3 \sum_{i=1}^{36} i < 36(55) = 1980$. But $\sum_{i=1}^{36} i = (36)(37)/2 = 666$, and this gives us the contradiction that $1998 = 3(666) < 1980$. $\square$

The next summation formula takes us from first powers to squares.

Example 4.3   Prove that for any $n \in \mathbf{Z}^+$, $\sum_{i=1}^{n} i^2 = (n)(n+1)(2n+1)/6$.

Proof: We start with $S(1)$. Here $\sum_{i=1}^{1} i^2 = 1^2 = 1 = (1)(1+1)(2+1)/6$, so $S(1)$ is true. Assuming $S(k)$: $\sum_{i=1}^{k} i^2 = (k)(k+1)(2k+1)/6$, for $k \in \mathbf{Z}^+$, we want to deduce the truth of

$$S(k+1): \sum_{i=1}^{k+1} i^2 = (k+1)((k+1)+1)(2(k+1)+1)/6$$
$$= (k+1)(k+2)(2k+3)/6.$$

Using the induction hypothesis $S(k)$, we find that

$$\sum_{i=1}^{k+1} i^2 = 1^2 + 2^2 + \cdots + k^2 + (k+1)^2 = \sum_{i=1}^{k} i^2 + (k+1)^2$$

$$= [(k)(k+1)(2k+1)/6] + (k+1)^2.$$

From this we have

$$\sum_{i=1}^{k+1} i^2 = (k+1)[(k)(2k+1)/6 + (k+1)] = (k+1)[(2k^2 + 7k + 6)/6]$$

$$= (k+1)(k+2)(2k+3)/6,$$

and the general result follows by mathematical induction.   □

Consider the following Pascal programs. The program in Fig. 4.2 uses a repeat-until loop to accumulate the sum of the squares. The second program (Fig. 4.3) demonstrates how the result of Example 4.3 can be used in place of such a loop. Each program is used to evaluate the sum of the squares of the first 17 positive integers. However, whereas the program in Fig. 4.2 entails a total of $2n \ (= 34)$ additions and $n \ (= 17)$ applications of the squaring process (sqr), the program in Fig. 4.3 requires only two additions, three multiplications, and one (integer) division. And the total number of additions, multiplications, and (integer) divisions is still 6 if the value of n gets larger. Consequently, the program in Fig. 4.3 is considered more efficient. (This idea of a *more efficient* program will be examined further in Sections 5.7 and 5.8.)

```
Program SumOfSquares1 (input,output);
Var
        i,n,s: integer;
Begin
        Writeln('We wish to find the sum of the squares of the first ');
        Write('n positive integers where n= ');
        Read(n);
        s := 0;
        i := 0;
        Repeat
                i := i + 1;
                s := s + sqr(i)
        Until i = n;
        Writeln('The sum of the squares of the first ', n:0);
        Write(' positive integers is ', s:0)
End.

We wish to find the sum of the squares of the first
n positive integers where n = 17
The sum of the squares of the first 17
positive integers is 1785
```

Figure 4.2

```
Program SumOfSquares2 (input,output);
Var
        n,s: integer;
Begin
        Writeln('We wish to find the sum of the squares of the ');
        Write('first n positive integers where n = ');
        Read(n);
        s := (n)*(n + 1)*(2 * n + 1) Div 6;
        Writeln('The sum of the squares of the first ', n:0);
        Write(' positive integers is ', s:0)
End.

We wish to find the sum of the squares of the
first n positive integers where n = 17
The sum of the squares of the first 17
positive integers is 1785
```

Figure 4.3

Having seen two applications of mathematical induction, one may wonder whether this principle applies only to the verification of *known* summation formulas. The next four examples show that mathematical induction is a vital tool in many other circumstances as well.

Example 4.4   Let us consider the sums of consecutive odd positive integers.

1) 1	$= 1$	$(= 1^2)$
2) $1 + 3$	$= 4$	$(= 2^2)$
3) $1 + 3 + 5$	$= 9$	$(= 3^2)$
4) $1 + 3 + 5 + 7$	$= 16$	$(= 4^2)$
5) $1 + 3 + 5 + 7 + 9$	$= 25$	$(= 5^2)$
6) $1 + 3 + 5 + 7 + 9 + 11$	$= 36$	$(= 6^2)$

From these first six cases we *conjecture* the following result: The sum of the first n consecutive odd positive integers is n^2; that is,

$$S(n): \quad \sum_{i=1}^{n} (2i - 1) = n^2.$$

Now that we have developed what we feel is a true summation formula, we use the principle of mathematical induction to verify its truth for *all* $n \geq 1$.

From the above calculations, we see that $S(1)$ is true [as are $S(2)$, $S(3)$, $S(4)$, $S(5)$, and $S(6)$]. Assuming the truth of $S(k)$, we have

$$\sum_{i=1}^{k} (2i - 1) = k^2, \text{ for } k \in \mathbf{Z}^+.$$

We now deduce the truth of $S(k + 1)$: $\quad \sum_{i=1}^{k+1} (2i - 1) = (k + 1)^2$.

Since we have assumed the truth of $S(k)$, our induction hypothesis, we may now write

$$\sum_{i=1}^{k+1} (2i-1) = \sum_{i=1}^{k} (2i-1) + [2(k+1)-1] = k^2 + [2(k+1)-1]$$

$$= k^2 + 2k + 1 = (k+1)^2$$

Consequently, the result $S(n)$ is true for all $n \geq 1$, by the principle of mathematical induction. □

Now it is time to investigate some results that are not summation formulas.

Example 4.5 In Table 4.1, we have listed in adjacent columns the values of $4n$ and $n^2 - 7$ for the positive integers n, where $1 \leq n \leq 8$.

Table 4.1

n	$4n$	$n^2 - 7$	n	$4n$	$n^2 - 7$
1	4	−6	5	20	18
2	8	−3	6	24	29
3	12	2	7	28	42
4	16	9	8	32	57

From the table, we see that $(n^2 - 7) < 4n$ for $n = 1, 2, 3, 4, 5$, but when $n = 6, 7, 8$, we have $4n < (n^2 - 7)$. These last three observations lead us to the conjecture $S(n)$:   For all $n \geq 6$, $4n < (n^2 - 7)$.

Once again, the finite induction principle is the proof technique we need to verify our conjecture.

Table 4.1 confirms that $S(6)$ is true [as are $S(7)$ and $S(8)$]. (At last we have an example wherein the starting point is an integer $n_0 \neq 1$.)

In this example, the induction hypothesis is $S(k)$: $4k < (k^2 - 7)$, where $k \in \mathbf{Z}^+$ and $k \geq 6$. We need to obtain the truth of $S(k+1)$ from that of $S(k)$. That is, from $4k < (k^2 - 7)$ we must conclude that $4(k+1) < [(k+1)^2 - 7]$. Here are the necessary steps:

$$4k < (k^2 - 7) \Rightarrow 4k + 4 < (k^2 - 7) + 4 < (k^2 - 7) + (2k + 1)$$

(because for $k \geq 6$, we find $2k + 1 \geq 13 > 4$), and

$$4k + 4 < (k^2 - 7) + (2k + 1) \Rightarrow 4(k+1) < (k^2 + 2k + 1) - 7 = (k+1)^2 - 7.$$

Therefore, by the principle of finite induction, $S(n)$ is true for all $n \geq 6$. □

Example 4.6 For $n \geq 0$ let $A_n \subset \mathbf{R}$, where $|A_n| = 2^n$ and the elements of A_n are listed in ascending order. If $r \in \mathbf{R}$, prove that in order to determine whether $r \in A_n$ (by the procedure developed below), we must compare r with no more than $n + 1$ elements in A_n.

When $n = 0$, $A_0 = \{a\}$ and only one comparison is needed. So the result is true for $n = 0$. For $n = 1$, $A_1 = \{a_1, a_2\}$ with $a_1 < a_2$. To determine whether $r \in A_1$, at most two comparisons must be made. Hence the result follows when $n = 1$. Now if $n = 2$, we write $A_2 = \{b_1, b_2, c_1, c_2\} = B_1 \cup C_1$, where $b_1 < b_2 < c_1 < c_2$, $B_1 = \{b_1, b_2\}$, $C_1 = \{c_1, c_2\}$. Comparing r with b_2 we determine which of the possibilities $r \in B_1$ or $r \in C_1$ can occur. Since $|B_1| = |C_1| = 2^1 = 2$, either possibility requires at most two more comparisons (from the prior case where $n = 1$). Consequently, we can determine whether $r \in A_2$ by making no more than $2 + 1 = n + 1$ comparisons.

We now argue in general. Assume the result true for some $k \geq 0$ and consider the case for A_{k+1}, where $|A_{k+1}| = 2^{k+1}$. Let $A_{k+1} = B_k \cup C_k$, where $|B_k| = |C_k| = 2^k$, and the elements of B_k, C_k are in ascending order with the largest element x in B_k smaller than the least element in C_k. Let $r \in \mathbf{R}$. In order to determine whether $r \in A_{k+1}$ we consider whether $r \in B_k$ or $r \in C_k$.

a) First we compare r and x. (One comparison)

b) If $r \leq x$, then because $|B_k| = 2^k$, it follows by the induction hypothesis that we can determine whether $r \in B_k$ by making no more than $k + 1$ additional comparisons. Consequently, at most $(k + 1) + 1$ comparisons are made in all.

c) If $r > x$, we do likewise with the elements in C_k. We make at most $k + 1$ additional comparisons to see whether $r \in C_k$.

The general result now follows by the principle of finite induction.   □

Example 4.7   One of our first concerns when we evaluate the quality of a computer program is whether the program does what it is supposed to do. Just as we cannot prove a theorem by checking specific cases, so we cannot establish the correctness of a program simply by testing various sets of data. (Furthermore, doing this would be quite difficult if our program were to become a part of a larger software package.) Since software development now places a great deal of emphasis on structured programming, this has brought about the need for *program verification*. Here the programmer or the programming team must prove that the program being developed is correct *regardless* of the data set supplied. The effort invested at this stage reduces considerably the time that must be spent in debugging the program (or software package). One of the methods that can play a major role in such program verification is mathematical induction. Let us see how.

The Pascal program segment shown in Fig. 4.4 is supposed to produce the answer $x(y^n)$ for real variables x, y with n a nonnegative integer. (The values for these three variables are assigned earlier in the program by the user.) We shall verify the correctness of this program segment by mathematical induction for the statement $S(n)$:   For any $x, y \in \mathbf{R}$, if the program reaches the top of the While loop with $n \in \mathbf{Z}$, $n \geq 0$, then after the loop is executed (or bypassed), the value of the real variable Answer is $x(y^n)$.

```
While n <> 0 do
   Begin
      x := x*y;
      n := n - 1
   End;
Answer := x;
```

Figure 4.4

The flowchart for this program segment is shown in Fig. 4.5. Referring to it will help us as we develop our proof.

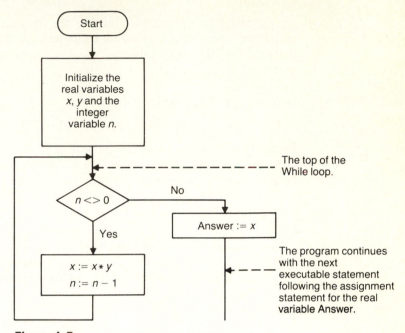

Figure 4.5

First consider $S(0)$, the statement for the case where $n = 0$. Here the program reaches the top of the While loop, but since $n = 0$ it follows the No branch in the flowchart and assigns the value $x = x(1) = x(y^0)$ to Answer, so $S(0)$ is true.

Now we assume $S(k)$, the induction hypothesis: For *any* $x, y \in \mathbf{R}$, if the program reaches the top of the While loop with $k \in \mathbf{Z}$, $k \geq 0$, then after the loop is executed (or bypassed), the value of the real variable Answer is $x(y^k)$.

To establish $S(k + 1)$, we note that because $k + 1 \geq 1$, the program will not simply follow the No branch and bypass the While loop. The While loop will be executed at least once. When the program reaches the top of the While loop for the first time, $n = k + 1 > 0$, so the loop instructions are executed and the program returns to the top of the While loop where now we find that

- The value of y is unchanged.
- The value of x is $x_1 = x(y^1) = xy$.
- The value of n is $(k + 1) - 1 = k$.

But now, by our induction hypothesis (applied to the real numbers x_1, y), we know that after the While loop for x_1, y and $n = k$ is executed (or bypassed), the value assigned to the real variable Answer is

$$x_1(y^k) = (xy)(y^k) = x(y^{k+1}).$$

Therefore $S(n)$ is true for all $n \geq 0$, and we have verified the correctness of the given Pascal program segment by the principle of mathematical induction. ☐

Our next two examples illustrate the idea of an inductive (or recursive) definition. These examples demonstrate how we can extend the associative laws that exist for logic and set theory. The first will settle the problem we mentioned at the end of Section 2.1.

Example 4.8   If p_1, p_2, p_3 are statements, we know that

$$p_1 \wedge (p_2 \wedge p_3) \Leftrightarrow (p_1 \wedge p_2) \wedge p_3,$$

so there is no ambiguity in simply writing $p_1 \wedge p_2 \wedge p_3$. The truth value for the conjunction of three statements does not depend on the way parentheses might be introduced to direct the order of forming the conjunctions of pairs of (given or resultant) statements.

The conjunction of statements was defined for only two statements (at a time). How, then, do we deal with an expression such as $p_1 \wedge p_2 \wedge p_3 \wedge p_4$? To answer this question, we introduce the idea of an *inductive* (or *recursive*) *definition*, wherein a concept at a certain $[(n + 1)\text{st}]$ stage is developed from the comparable concept at an earlier $[n\text{th}]$ stage. The following discussion illustrates the idea.

For any statements $p_1, p_2, \ldots, p_n, p_{n+1}$, we define (1) the conjunction of p_1, p_2 by $p_1 \wedge p_2$ and (2) the conjunction of $p_1, p_2, \ldots, p_n, p_{n+1}$ by

$$p_1 \wedge p_2 \wedge \cdots \wedge p_n \wedge p_{n+1} \Leftrightarrow (p_1 \wedge p_2 \wedge \cdots \wedge p_n) \wedge p_{n+1}.$$

[Note that the result on the right-hand side in (2) is the conjunction of *two* statements: p_{n+1} and the previously determined statement $(p_1 \wedge p_2 \wedge \cdots \wedge p_n)$.]

Therefore, we define the conjunction of p_1, p_2, p_3, p_4 by

$$p_1 \wedge p_2 \wedge p_3 \wedge p_4 \Leftrightarrow (p_1 \wedge p_2 \wedge p_3) \wedge p_4.$$

Then, by the associative law of $\wedge$, we find that

$$
\begin{aligned}
(p_1 \wedge p_2 \wedge p_3) \wedge p_4 &\Leftrightarrow [(p_1 \wedge p_2) \wedge p_3] \wedge p_4 \\
&\Leftrightarrow (p_1 \wedge p_2) \wedge (p_3 \wedge p_4) \\
&\Leftrightarrow p_1 \wedge [p_2 \wedge (p_3 \wedge p_4)] \\
&\Leftrightarrow p_1 \wedge (p_2 \wedge p_3 \wedge p_4).
\end{aligned}
$$

These logical equivalences show that the truth value for the conjunction of four statements is also independent of the way parentheses might be introduced to indicate how to associate the given statements.

Using the above definition, we now extend our results to the following "Generalized Associative Law for $\wedge$."

$S(n)$:   For any $n \in \mathbf{Z}^+, n \geq 3$, let $r \in \mathbf{Z}^+$, where $1 \leq r < n$.

Then for any statements $p_1, p_2, \ldots, p_r, p_{r+1}, \ldots, p_n$,

$$(p_1 \wedge p_2 \wedge \cdots \wedge p_r) \wedge (p_{r+1} \wedge \cdots \wedge p_n) \Leftrightarrow p_1 \wedge p_2 \wedge \cdots \wedge p_r \wedge p_{r+1} \wedge \cdots \wedge p_n.$$

Proof: The truth of $S(3)$ follows from the associative law for $\wedge$.

Now assume the result true for any $k \geq 3$ and any $1 \leq r < k$. That is,

$$(p_1 \wedge p_2 \wedge \cdots \wedge p_r) \wedge (p_{r+1} \wedge \cdots \wedge p_k) \Leftrightarrow p_1 \wedge p_2 \wedge \cdots \wedge p_r \wedge p_{r+1} \wedge \cdots \wedge p_k.$$

When we consider $k + 1$ statements, we find that

$$(p_1 \wedge p_2 \wedge \cdots \wedge p_r) \wedge (p_{r+1} \wedge \cdots \wedge p_k \wedge p_{k+1})$$
$$\Leftrightarrow (p_1 \wedge p_2 \wedge \cdots \wedge p_r) \wedge [(p_{r+1} \wedge \cdots \wedge p_k) \wedge p_{k+1}]$$
$$\Leftrightarrow [(p_1 \wedge p_2 \wedge \cdots \wedge p_r) \wedge (p_{r+1} \wedge \cdots \wedge p_k)] \wedge p_{k+1}$$
$$\Leftrightarrow (p_1 \wedge p_2 \wedge \cdots \wedge p_r \wedge p_{r+1} \wedge \cdots \wedge p_k) \wedge p_{k+1}$$
$$\Leftrightarrow p_1 \wedge p_2 \wedge \cdots \wedge p_r \wedge p_{r+1} \wedge \cdots \wedge p_k \wedge p_{k+1},$$

so by the principle of mathematical induction, $S(n)$ is true for all $n \in \mathbf{Z}^+$, $n \geq 3$. $\square$

Example 4.9 In Definition 3.9 we extended the binary operations of $\cup$ and $\cap$ to an arbitrary (finite or infinite) number of subsets from a given universe $\mathcal{U}$. However, these definitions do not rely on the binary nature of the operations involved, and they do not provide a *systematic* way of determining the union or intersection of any finite number of sets.

To overcome this difficulty, we define the union of the sets $A_1, A_2, \ldots,$ A_n, A_{n+1} *inductively* as follows:

1. The union of A_1, A_2 is $A_1 \cup A_2$.
2. The union of $A_1, A_2, \ldots, A_n, A_{n+1}$ is given by $A_1 \cup A_2 \cup \cdots \cup A_n \cup A_{n+1} = (A_1 \cup A_2 \cup \cdots \cup A_n) \cup A_{n+1}$, the union of the *two* sets $A_1 \cup A_2 \cup \cdots \cup A_n$ and A_{n+1}.

From this we obtain a "Generalized Associative Law for $\cup$."

$S(n)$: For any $n, r \in \mathbf{Z}^+$, where $n \geq 3$, and $1 \leq r < n$,

$$(A_1 \cup A_2 \cup \cdots \cup A_r) \cup (A_{r+1} \cup \cdots \cup A_n)$$
$$= A_1 \cup A_2 \cup \cdots \cup A_r \cup A_{r+1} \cup \cdots \cup A_n,$$

where $A_i \subseteq \mathcal{U}$ for $1 \leq i \leq n$.

Proof: The truth of $S(3)$ follows from the associative law of $\cup$. Assuming the truth of $S(k)$ for $k \geq 3$, when dealing with $k + 1$ sets we find that

$$(A_1 \cup A_2 \cup \cdots \cup A_r) \cup (A_{r+1} \cup \cdots \cup A_k \cup A_{k+1})$$
$$= (A_1 \cup A_2 \cup \cdots \cup A_r) \cup [(A_{r+1} \cup \cdots \cup A_k) \cup A_{k+1}]$$
$$= [(A_1 \cup A_2 \cup \cdots \cup A_r) \cup (A_{r+1} \cup \cdots \cup A_k)] \cup A_{k+1}$$
$$= (A_1 \cup A_2 \cup \cdots \cup A_r \cup A_{r+1} \cup \cdots \cup A_k) \cup A_{k+1}$$
$$= A_1 \cup A_2 \cup \cdots \cup A_r \cup A_{r+1} \cup \cdots \cup A_k \cup A_{k+1},$$

and by the principle of mathematical induction, $S(n)$ is true for all integers $n \geq 3$. $\square$

We close this section wth an alternative form of Theorem 4.1. This result is sometimes referred to as *complete induction*.

THEOREM 4.2 (*Finite Induction Principle–Alternative Form*) Let $S(n)$ denote a mathematical statement (or set of statements) involving a positive integer n.

    a) If $S(1)$ is true; and

    b) If $S(k + 1)$ is true, for $k \in \mathbf{Z}^+$, whenever $S(1), S(2), \ldots, S(k)$ are true,

then $S(n)$ is true for all $n \in \mathbf{Z}^+$.

Proof The proofs of this result and its equivalence to Theorem 4.1 are left for the exercises. ∎

    When we examine the principle in this form, we see that we may need the truth of statements other than just $S(k)$ in order to prove $S(k + 1)$ true. The following examples demonstrate how this can happen, and show, like our earlier examples, that we need not always start at $n_0 = 1$.

Example 4.10 The following calculations indicate that it is possible to write (without regard to order) the integers $14, 15, 16$ using only 3's and/or 8's as summands.

$$14 = 3 + 3 + 8$$
$$15 = 3 + 3 + 3 + 3 + 3$$
$$16 = 8 + 8$$

On the basis of these three results, we make the conjecture

    $S(n)$:   For every $n \in \mathbf{Z}^+$, if $n \geq 14$, then n can be expressed as a sum of 3's and/or 8's.

Proof: It is apparent that $S(14)$ is true, as are $S(15)$ and $S(16)$. So now we assume our induction hypothesis. Here this assumption is that $S(14), S(15), \ldots, S(k-2), S(k-1), S(k)$ are all true, where $k \geq 16$. Now if $n = k + 1$, then $n \geq 17$ and $k + 1 = (k - 2) + 3$. But since $14 \leq k - 2 \leq k$, by the truth of $S(k-2)$ we know that $(k - 2)$ can be written as a sum of 3's and/or 8's, so $(k + 1) = (k - 2) + 3$ can also be written in this form. Consequently, $S(n)$ is true for all $n \geq 14$ by the alternative form of the principle of mathematical induction. □

    In Example 4.10 we saw how the truth of $S(k + 1)$ was deduced by using the truth of the one prior result $S(k - 2)$. Our next example shows that there are situations wherein the truth of more than one prior result is needed.

Example 4.11 After n months of a certain greenhouse experiment, the number $P(n)$ of plants of a particular type satisfies the equations $P(0) = 3$, $P(1) = 7$, and $P(n) = 3P(n - 1) - 2P(n - 2)$, for $n \geq 2$. [Thus, for example, $P(2) = 3P(1) - 2P(0) = 3(7) - 2(3) = 15$, and $P(3) = 3P(2) - 2P(1) = 3(15) - 2(7) = 31$.] Show by mathematical induction that $P(n) = 2^{n+2} - 1$ for all $n \in \mathbf{Z}, n \geq 0$.
Proof: Here we have the statement

    $S(n)$:   If $P(0) = 3$, $P(1) = 7$, and $P(n) = 3P(n - 1) - 2P(n - 2)$, then $P(n) = 2^{n+2} - 1$, for $n \geq 0$.

    Since $2^{0+2} - 1 = 3 = P(0)$, we have $S(0)$ true. Also, $2^{1+2} - 1 = 7 = P(1)$ so $S(1)$ is true.

We now assume the truth of $S(0), S(1), \ldots, S(k-1), S(k)$ and consider the case where $n = k + 1 \geq 2$. Then

$$
\begin{aligned}
P(k+1) &= 3P(k) - 2P(k-1) \\
&= 3[2^{k+2} - 1] - 2[2^{(k-1)+2} - 1] \\
&= 3(2^{k+2}) - 3 - 2^{(k-1)+3} + 2 \\
&= 3(2^{k+2}) - (2^{k+2}) - 1 \\
&= 2(2^{k+2}) - 1 = 2^{(k+1)+2} - 1.
\end{aligned}
$$

[Here the replacements for $P(k)$ and $P(k-1)$ are based on the induction hypothesis for the two prior results $S(k)$ and $S(k-1)$.]

Consequently, by the alternative form of the principle of mathematical induction, $S(n)$ is true for all $n \geq 0$. (In Chapter 10 we shall delve further into obtaining such a solution for $P(n)$.) □

As a result of the alternative form of the principle of mathematical induction, it is possible to have an inductive (or recursive) definition wherein the concept at the $(n+1)$st stage is developed from one or more comparable concepts at the earlier 1st, 2nd, 3rd, ..., and nth stages.

EXERCISES 4.1

1. Prove each of the following by the principle of finite induction.

   a) $1^2 + 3^2 + 5^2 + \cdots + (2n-1)^2 = (n)(2n-1)(2n+1)/3$

   b) $1 \cdot 3 + 2 \cdot 4 + 3 \cdot 5 + \cdots + n(n+2) = (n)(n+1)(2n+7)/6$

   c) $\displaystyle\sum_{i=1}^{n} \frac{1}{i(i+1)} = \frac{n}{n+1}$

   d) $\displaystyle\sum_{i=1}^{n} 2^{i-1} = \sum_{i=0}^{n-1} 2^{i} = 2^{n} - 1$

   e) $\displaystyle\sum_{i=1}^{n} i^3 = \frac{n^2(n+1)^2}{4} = \left(\sum_{i=1}^{n} i\right)^2$

2. Establish each of the following by mathematical induction.

   a) $\displaystyle\sum_{i=1}^{n} i(2^i) = 2 + (n-1)2^{n+1}$

   b) $\displaystyle\sum_{i=1}^{n} 2(3^{i-1}) = 3^n - 1$

   c) $\displaystyle\sum_{i=1}^{n} (i)(i!) = (n+1)! - 1$

3. Consider the following BASIC program:

```
10  FOR I = 1 TO 123
20      FOR J = 1 TO I
30          PRINT I*J
40      NEXT J
50  NEXT I
60  END
```

a) How many times is the PRINT statement at line 30 executed?

b) Replace I in line 20 by I**2, and answer the question in part (a).

4. A wheel of fortune has the integers from 1 to 25 placed on it in a random manner. Show that regardless of how the numbers are positioned on the wheel, there are three adjacent numbers whose sum is at least 39.

5. Evaluate each of the following: a) $\sum_{i=11}^{33} i$; b) $\sum_{i=11}^{33} i^2$.

6. a) Prove that $(\cos\theta + i\sin\theta)^2 = \cos 2\theta + i\sin 2\theta$, where $i \in \mathbf{C}$ and $i^2 = -1$.

b) Using induction, prove that for any $n \in \mathbf{Z}^+$, $(\cos\theta + i\sin\theta)^n = \cos n\theta + i\sin n\theta$. (This result is known as *DeMoivre's Theorem*.)

c) Verify that $1 + i = \sqrt{2}(\cos 45° + i\sin 45°)$, and compute $(1+i)^{100}$.

7. For $n \in \mathbf{Z}^+$, if $n > 10$, prove that

$$n - 2 < \frac{n^2 - n}{12}.$$

8. Prove that for any $n \in \mathbf{Z}^+$, $n > 3 \Rightarrow 2^n < n!$

9. Prove that for any $n \in \mathbf{Z}^+$, $n > 4 \Rightarrow n^2 < 2^n$.

10. Prove that for any $n \in \mathbf{Z}^+$, $n > 9 \Rightarrow n^3 < 2^n$.

11. Let $S(n)$ be the statement: For $n \in \mathbf{Z}^+$,

$$\sum_{i=1}^{n} i = \frac{(n + (1/2))^2}{2}.$$

Show that the truth of $S(k)$ implies the truth of $S(k+1)$ for any $k \in \mathbf{Z}^+$. Is $S(n)$ true for all $n \in \mathbf{Z}^+$?

12. Let S_1 and S_2 be two sets where $|S_1| = m$, $|S_2| = n$, for $m, n \in \mathbf{Z}^+$, and the elements of S_1, S_2 are in ascending order. It can be shown that the elements in S_1 and S_2 can be merged into ascending order by making no more than $m + n - 1$ comparisons. (See Lemma 12.1.) Use this result to establish the following.

For $n \geq 0$, let S be a set with $|S| = 2^n$. Prove that the number of comparisons needed to place the elements of S in ascending order is bounded by $n \cdot 2^n$.

13. If $n \in \mathbf{Z}^+$, prove that

a) $(\cos\theta)(\cos 2\theta)(\cos 4\theta)(\cos 8\theta)\cdots[\cos(2^{n-1}\theta)] = \dfrac{\sin(2^n\theta)}{2^n\sin\theta}$

b) $\cos\theta + \cos 3\theta + \cos 5\theta + \cdots + \cos(2n-1)\theta = \dfrac{\sin 2n\theta}{2\sin\theta}$

14. In the Pascal program segment shown in Fig. 4.6, x, y, and Answer are real variables, and n is an integer variable. Prior to execution of this While loop, the user supplies real values for x and y and a nonnegative integer value for n. Prove (by mathematical induction) that for any $x, y \in \mathbf{R}$, if the program reaches the top of the While loop with $n \in \mathbf{Z}$, $n \geq 0$, then after the loop is executed (or bypassed), the value assigned to Answer is $x + ny$.

```
While n <> 0 do
   Begin
      x := x + y;
      n := n - 1
   End;
Answer := x;
```

Figure 4.6

15. During the execution of a certain Pascal program, the user assigns to the integer variables x and n any (possibly different) positive integers. The Pascal segment shown in Fig. 4.7 immediately follows these assignments. If the program reaches the top of the While loop, state and prove (by mathematical induction) what the value assigned to Answer will be after the loop is executed (or bypassed).

```
While n <> 0 do
   Begin
      x := x*n;
      n := n - 1
   End;
Answer := x;
```

Figure 4.7

16. Develop an inductive (or recursive) definition for $n!$ where $n \in \mathbb{N}$.

17. **a)** Develop an inductive (or recursive) definition for the addition of n real numbers $x_1, x_2, \ldots, x_n$, where $n \geq 2$.

   b) For any real numbers x_1, x_2, and x_3, the associative law of addition states that $x_1 + (x_2 + x_3) = (x_1 + x_2) + x_3$. Prove that if $n, r \in \mathbb{Z}^+$, where $n \geq 3$ and $1 \leq r < n$, then

   $$(x_1 + x_2 + \cdots + x_r) + (x_{r+1} + \cdots + x_n)$$
   $$= x_1 + x_2 + \cdots + x_r + x_{r+1} + \cdots + x_n.$$

18. For any $x \in \mathbb{R}$, $|x| = \sqrt{x^2} = \begin{cases} x, & \text{if } x \geq 0 \\ -x, & \text{if } x < 0 \end{cases}$, and $-|x| \leq x \leq |x|$. Consequently, $|x + y|^2 = (x + y)^2 = x^2 + 2xy + y^2 \leq x^2 + 2|x||y| + y^2 = |x|^2 + 2|x||y| + |y|^2 = (|x| + |y|)^2$, and $|x + y|^2 \leq (|x| + |y|)^2 \Rightarrow |x + y| \leq |x| + |y|$, for any x, $y \in \mathbb{R}$.

   Prove that if $n \in \mathbb{Z}^+$, $n \geq 2$, and $x_1, x_2, \ldots, x_n \in \mathbb{R}$, then

   $$|x_1 + x_2 + \cdots + x_n| \leq |x_1| + |x_2| + \cdots + |x_n|.$$

19. Use the result of Example 4.8 to prove that if $p, q_1, q_2, \ldots, q_n$ are statements and $n \geq 2$, then

   $$p \vee (q_1 \wedge q_2 \wedge \cdots \wedge q_n) \Leftrightarrow (p \vee q_1) \wedge (p \vee q_2) \wedge \cdots \wedge (p \vee q_n).$$

20. **a)** Give an inductive definition for the disjunction of statements $p_1, p_2, \ldots, p_n, p_{n+1}, n \geq 1$.

   b) Show that if $n, r \in \mathbb{Z}^+$, with $n \geq 3$ and $1 \leq r < n$, then

   $$(p_1 \vee p_2 \vee \cdots \vee p_r) \vee (p_{r+1} \vee \cdots \vee p_n)$$
   $$\Leftrightarrow p_1 \vee p_2 \vee \cdots \vee p_r \vee p_{r+1} \vee \cdots \vee p_n.$$

c) For $n \in \mathbf{Z}^+, n \geq 2$, and statements $p, q_1, q_2, \ldots, q_n$, prove that

$$p \wedge (q_1 \vee q_2 \vee \cdots \vee q_n) \Leftrightarrow (p \wedge q_1) \vee (p \wedge q_2) \vee \cdots \vee (p \wedge q_n).$$

d) For $n \in \mathbf{Z}^+, n \geq 2$, prove that for any statements $p_1, p_2, \ldots, p_n,$

(i) $\overline{(p_1 \vee p_2 \vee \cdots \vee p_n)} \Leftrightarrow \overline{p_1} \wedge \overline{p_2} \wedge \cdots \wedge \overline{p_n}$

(ii) $\overline{(p_1 \wedge p_2 \wedge \cdots \wedge p_n)} \Leftrightarrow \overline{p_1} \vee \overline{p_2} \vee \cdots \vee \overline{p_n}$

21. a) Use the result of Example 4.9 to show that if sets $A, B_1, B_2, \ldots, B_n \subseteq \mathcal{U}$ and $n \geq 2$, then

$$A \cap (B_1 \cup B_2 \cup \cdots \cup B_n) = (A \cap B_1) \cup (A \cap B_2) \cup \cdots \cup (A \cap B_n).$$

b) Give an inductive definition for the intersection of the sets $A_1, A_2, \ldots,$ $A_n, A_{n+1} \subseteq \mathcal{U}, n \geq 1$.

c) Use the result in part (b) to show that for any $n, r \in \mathbf{Z}^+$ with $n \geq 3$ and $1 \leq r < n$, $(A_1 \cap A_2 \cap \cdots \cap A_r) \cap (A_{r+1} \cap \cdots \cap A_n) =$ $A_1 \cap A_2 \cap \cdots \cap A_r \cap A_{r+1} \cap \cdots \cap A_n$.

d) For $n \geq 2$ and any sets $A_1, A_2, \ldots, A_n \subseteq \mathcal{U}$, prove that

(i) $\overline{A_1 \cap A_2 \cap \cdots \cap A_n} = \overline{A_1} \cup \overline{A_2} \cup \cdots \cup \overline{A_n}$

(ii) $\overline{A_1 \cup A_2 \cup \cdots \cup A_n} = \overline{A_1} \cap \overline{A_2} \cap \cdots \cap \overline{A_n}$

e) For $n \in \mathbf{Z}^+, n \geq 2$, and sets $A, B_1, B_2, \ldots, B_n \subseteq \mathcal{U}$, prove that

$$A \cup (B_1 \cap B_2 \cap \cdots \cap B_n) = (A \cup B_1) \cap (A \cup B_2) \cap \cdots \cap (A \cup B_n).$$

22. a) Let $n \in \mathbf{Z}^+, n \neq 1, 3$. Prove that n can be expressed as a sum of 2's and/or 5's.

b) For any $n \in \mathbf{Z}^+$ show that if $n \geq 24$, then n can be written as a sum of 5's and/or 7's.

23. For $n \in \mathbf{Z}^+$, let $P(n) =$ the (approximate) number of bacteria in a culture at the end of n hours. If $P(1) = 1000$, $P(2) = 2000$, and $P(n) = P(n-1) + P(n-2)$, for all $n > 2$, show that

$$P(n) = \left(\frac{1000}{\sqrt{5}}\right)\left[\left(\frac{1 + \sqrt{5}}{2}\right)^{n+1} - \left(\frac{1 - \sqrt{5}}{2}\right)^{n+1}\right].$$

24. A sequence of numbers $a_1, a_2, a_3, \ldots$ is defined (inductively) by

$$a_1 = 1$$
$$a_2 = 2$$
$$a_n = a_{n-1} + a_{n-2}, \quad n \geq 3$$

a) Determine the values of a_3, a_4, a_5, a_6, and a_7.

b) Prove that for all $n \geq 1$, $a_n < (7/4)^n$.

25. Verify Theorem 4.2.

26. a) For $n \geq 2$, if $p_1, p_2, p_3, \ldots, p_n, p_{n+1}$ are statements, prove that $[(p_1 \rightarrow p_2) \wedge$ $(p_2 \rightarrow p_3) \wedge \cdots \wedge (p_n \rightarrow p_{n+1})] \Rightarrow [(p_1 \wedge p_2 \wedge p_3 \wedge \cdots \wedge p_n) \rightarrow p_{n+1}].$

b) Prove that Theorem 4.2 implies Theorem 4.1.

c) Use Theorem 4.1 to establish the following: If $\emptyset \neq S \subseteq \mathbf{Z}^+$ and $n \in S$, then S contains a least element.

d) Show that Theorem 4.1 implies Theorem 4.2.

4.2
THE DIVISION ALGORITHM: PRIME NUMBERS

Although the set $\mathbf{Z}$ is not closed under nonzero division, there are many instances where one integer divides another, for example, 2 divides 6 and 7 divides 21. Here the division is exact and there is no remainder. Thus 2 dividing 6 implies the existence of a quotient, namely 3, such that $6 = 2 \cdot 3$.

DEFINITION 4.1   If $a, b \in \mathbf{Z}$ and $b \neq 0$, we say that b *divides* a, and we write $b \mid a$ if there is an integer n such that $a = bn$. When this occurs we say that b is a *divisor* of a, or a is a *multiple* of b.   ▬

With this definition we are able to speak of division inside of $\mathbf{Z}$ without going to $\mathbf{Q}$. Furthermore, when $ab = 0$ for $a, b \in \mathbf{Z}$, then either $a = 0$ or $b = 0$, and we say that $\mathbf{Z}$ has no *proper divisors of 0*. This property enables us to *cancel* as in $2x = 2y \Rightarrow x = y$, for $x, y \in \mathbf{Z}$, because $2x = 2y \Rightarrow 2(x - y) = 0 \Rightarrow 2 = 0$ or $x - y = 0 \Rightarrow x = y$. (Note that at no time did we mention multiplying both sides of the equation $2x = 2y$ by $\frac{1}{2}$. The number $\frac{1}{2}$ is outside the system $\mathbf{Z}$.)

We now summarize some properties of this division operation. Whenever we divide by an integer a, we assume that $a \neq 0$.

THEOREM 4.3   For $a, b, c \in \mathbf{Z}$

a) $1 \mid a$ and $a \mid 0$.

b) $(a \mid b \wedge b \mid a) \Rightarrow a = \pm b$.

c) $(a \mid b \wedge b \mid c) \Rightarrow a \mid c$.

d) $a \mid b \Rightarrow a \mid bx$ for all $x \in \mathbf{Z}$.

e) If $x = y + z$, for $x, y, z \in \mathbf{Z}$, and a divides two of the three integers x, y, and z, then a divides the remaining integer.

f) $(a \mid b \wedge a \mid c) \Rightarrow a \mid (bx + cy)$, for all $x, y \in \mathbf{Z}$. (The expression $bx + cy$ is called a *linear combination* of b, c.)

g) For $1 \leq i \leq n$, let $c_i \in \mathbf{Z}$. If a divides each c_i, then $a \mid (c_1 x_1 + c_2 x_2 + \cdots + c_n x_n)$, for any $x_i \in \mathbf{Z}, 1 \leq i \leq n$.

Proof   We prove part (f) and leave the remaining parts for the reader.

If $a \mid b$ and $a \mid c$ then $b = am, c = an$, for some $m, n \in \mathbf{Z}$. So $bx + cy = (am)x + (an)y = a(mx + ny)$ (by the associative law of multiplication and the distributive law of multiplication over addition, both of which $\mathbf{Z}$ satisfies). With $bx + cy = a(mx + ny)$, it follows that $a \mid (bx + cy)$.   ■

With this operation of integer division we enter the area of *number theory*. As we examine the set $\mathbf{Z}^+$ further, we notice that for all $n \in \mathbf{Z}^+, n > 1, n$ has at least two positive divisors, namely 1 and n itself. Some numbers, such as $2, 3, 5, 7, \ldots$, have exactly two positive divisors. These integers are called *primes*. All other positive integers (greater than 1 and not prime) are called *composite*. An immediate connection between prime and composite integers is expressed in the following lemma.

LEMMA 4.1   If $n \in \mathbf{Z}^+$ and n is composite, then there is a prime p such that $p \mid n$.

Proof   If not, let S be the set of all composite numbers that have no prime divisor(s). If $S \neq \emptyset$, then by the Well-Ordering Principle, S has a least element m. But m composite $\Rightarrow m = m_1 m_2$, where $1 < m_1 < m, 1 < m_2 < m$. Since $m_1 \notin S$, m_1 is prime or divisible by a prime. Consequently, there exists a prime p such that $p \mid m$, and $S = \emptyset$. ■

Now why did we call the above result a *lemma* instead of a theorem? It had to be proved like all other theorems in the book so far. A lemma is itself a theorem, but its major role is to establish a result that will be used in proving another theorem.

In listing the primes we are inclined to believe there are infinitely many such numbers. We now verify that this is true.

THEOREM 4.4   (Euclid) There are infinitely many primes.

Proof   If not, let $p_1, p_2, \ldots, p_k$ be the finite list of all primes, and let $B = p_1 p_2 \cdots p_k + 1$. Since $B > p_i$ for $1 \leq i \leq k$, B cannot be a prime. Hence B is composite. So by Lemma 4.1 there is a prime p_j, where $1 \leq j \leq k$ and $p_j \mid B$. Since $p_j \mid B$ and $p_j \mid p_1 p_2 \cdots p_k$, by Theorem 4.3(e) it follows that $p_j \mid 1$. This contradiction arises from the assumption that there are only finitely many primes; the result follows. ■

Yes, this is the same Euclid from the fourth century B.C. whose *Elements*, written on 13 parchment scrolls, comprised the first organized coverage of the geometry we studied in high school. One finds, however, that these 13 books are also concerned with number theory. In particular, Books VII, VIII, and IX dwell on this topic.

We turn now to the major idea of this section. This result enables us to deal with nonzero division in $\mathbf{Z}$ when that division is not exact.

THEOREM 4.5   (*The Division Algorithm*) If $a, b \in \mathbf{Z}$, with $b > 0$, then there exist unique $q, r \in \mathbf{Z}$ with $a = qb + r$, $0 \leq r < b$.

Proof   If $b \mid a$ the result follows, so consider the case where $b \nmid a$ (that is, b does not divide a).
     Let $S = \{a - tb \mid t \in \mathbf{Z}, a - tb > 0\}$. If $a > 0$ and $t = 0$, $a \in S$ and $S \neq \emptyset$. For $a \leq 0$, let $t = a - 1$. Then $a - tb = a - (a - 1)b = a(1 - b) + b$, with $(1 - b) \leq 0$, because $b \geq 1$. So $a - tb > 0$ and $S \neq \emptyset$. Hence, for any $a \in \mathbf{Z}$, S is a nonempty subset of $\mathbf{Z}^+$. By the Well-Ordering Principle, S has a least element r, where $0 < r = a - qb$, for some $q \in \mathbf{Z}$. If $r = b$, then $a = (q + 1)b$ and $b \mid a$, contra-

dicting $b \nmid a$. If $r > b$, then $r = b + c$, for some $c \in \mathbf{Z}^+$, and $a - qb = r = b + c \Rightarrow c = a - (q + 1)b \in S$, contradicting r being the least element of S.

This now establishes a quotient q and remainder r, $0 \leq r < b$, for the theorem. But are there other q's and r's that also work? If so, let $q_1, q_2, r_1, r_2 \in \mathbf{Z}$ with $a = q_1 b + r_1$, $0 \leq r_1 < b$ and $a = q_2 b + r_2$, $0 \leq r_2 < b$. Then $q_1 b + r_1 = q_2 b + r_2 \Rightarrow b|q_1 - q_2| = |r_2 - r_1| < b$, because $0 \leq r_1, r_2 < b$. If $q_1 \neq q_2$, we have the contradiction $b|q_1 - q_2| < b$. Hence $q_1 = q_2$, $r_1 = r_2$, and the quotient and remainder are unique.   ∎

Using the division algorithm, we consider some results on representing integers in bases other than 10.

Example 4.12   Write 6137 in the octal system (base 8). Here we seek nonnegative integers $r_0, r_1, r_2, \ldots, r_k$, with $r_k > 0$, such that $6137 = (r_k \ldots r_2 r_1 r_0)_8$.

With $6137 = r_0 + r_1 \cdot 8 + r_2 \cdot 8^2 + \cdots + r_k \cdot 8^k = r_0 + 8(r_1 + r_2 \cdot 8 + \cdots + r_k \cdot 8^{k-1})$, r_0 is the remainder obtained in the division algorithm when 6137 is divided by 8.

Consequently, $6137 = 1 + 8 \cdot 767$, so $r_0 = 1$, and $767 = r_1 + r_2 \cdot 8 + \cdots + r_k \cdot 8^{k-1} = r_1 + 8(r_2 + r_3 \cdot 8 + \cdots + r_k \cdot 8^{k-2})$. This yields $r_1 = 7$ (the remainder when 767 is divided by 8) and $95 = r_2 + r_3 \cdot 8 + \cdots + r_k \cdot 8^{k-2}$. Continuing in this manner, we find $r_2 = 7$, $r_3 = 3$, $r_4 = 1$, and $r_i = 0$ for $i \geq 5$, so

$$6137 = 1 \cdot 8^4 + 3 \cdot 8^3 + 7 \cdot 8^2 + 7 \cdot 8 + 1 = (13771)_8 .$$

We can arrange the successive divisions by 8 as follows:

		Remainders
8	$\underline{|6137}$	
8	$\underline{|767}$	1 (r_0)
8	$\underline{|95}$	7 (r_1)
8	$\underline{|11}$	7 (r_2)
8	$\underline{|1}$	3 (r_3)
	0	1 (r_4) □

Example 4.13   In the field of computer science, the binary number system (base 2) is very important. Here the only symbols that one may use are the bits 0 and 1. In Table 4.2 we have listed the binary representations of the integers from 0 to 15.

Table 4.2

Base 10	Base 2	Base 10	Base 2
0	0 0 0 0	8	1 0 0 0
1	0 0 0 1	9	1 0 0 1
2	0 0 1 0	10	1 0 1 0
3	0 0 1 1	11	1 0 1 1
4	0 1 0 0	12	1 1 0 0
5	0 1 0 1	13	1 1 0 1
6	0 1 1 0	14	1 1 1 0
7	0 1 1 1	15	1 1 1 1

Here we have included leading zeroes and find that we need four bits because of the leading 1 in the representations for the integers from 8 to 15. With five bits we can continue up to 31 ($= 32 - 1 = 2^5 - 1$); six bits are necessary to proceed to 63 ($= 64 - 1 = 2^6 - 1$). In general, if $x \in \mathbf{Z}$ and $0 \leq x < 2^n$, $n \in \mathbf{Z}^+$, then we can write x in base 2 by using n bits. Leading zeroes appear when $0 \leq x < 2^{n-1} - 1$, and for $2^{n-1} \leq x < 2^n - 1$ the first (most significant) bit is 1.

Information is generally stored in machines in units of eight bits called bytes, so for machines with memory cells of one byte we can store any one of the binary equivalents of the integers from 0 to $2^8 - 1 = 255$ in a single cell. For a machine with two-byte cells, any of the integers from 0 to $2^{16} - 1 = 65,535$ can be stored in binary form in each cell. A machine with four-byte cells would take us up to $2^{32} - 1 = 4,294,967,295$.

When a human deals with long sequences of 0's and 1's, the job soon becomes very tedious and the chance for error increases with the tedium. Consequently, it is common (especially in the study of machine and assembly languages) to represent such long sequences of bits in another notation. One such notation is the *hexadecimal (base-16) notation*. Here there are 16 symbols, and because we have only 10 symbols in the standard base-10 system, we introduce the following six additional symbols:

A (Able)	C (Charlie)	E (Echo)
B (Baker)	D (Dog)	F (Fox)

In Table 4.3 the integers from 0 to 15 are given in terms of both the binary and the hexadecimal number systems.

Table 4.3

Base 10	Base 2	Base 16	Base 10	Base 2	Base 16
0	0 0 0 0	0	8	1 0 0 0	8
1	0 0 0 1	1	9	1 0 0 1	9
2	0 0 1 0	2	10	1 0 1 0	A
3	0 0 1 1	3	11	1 0 1 1	B
4	0 1 0 0	4	12	1 1 0 0	C
5	0 1 0 1	5	13	1 1 0 1	D
6	0 1 1 0	6	14	1 1 1 0	E
7	0 1 1 1	7	15	1 1 1 1	F

To convert from base 10 to base 16, we follow a procedure like the one outlined in Example 4.12. Here we are interested in the remainders upon successive divisions by 16. Therefore, if we want to represent the (base-10) integer 13,874,945 in the hexadecimal system, we do the following calculations:

Remainders

$$16 \; \lfloor 13{,}874{,}945$$
$$16 \; \lfloor 867{,}184 \qquad\qquad 1 \qquad\qquad (r_0)$$
$$16 \; \lfloor 54{,}199 \qquad\qquad 0 \qquad\qquad (r_1)$$
$$16 \; \lfloor 3{,}387 \qquad\qquad 7 \qquad\qquad (r_2)$$
$$16 \; \lfloor 211 \qquad\qquad 11 \; (=B) \qquad (r_3)$$
$$16 \; \lfloor 13 \qquad\qquad 3 \qquad\qquad (r_4)$$
$$0 \qquad\qquad 13 \; (=D) \qquad (r_5)$$

Consequently, $13{,}874{,}945 = (D3B701)_{16}$.

There is, however, an easier approach to converting between base 2 and base 16. For example, if we want to convert the binary (one-byte) integer 01001101 to its base-16 counterpart, we break the number into blocks of four bits:

$$\underbrace{0100}_{4} \quad \underbrace{1101}_{D}$$

We then convert each block of four bits to its base-16 representation (as shown in Table 4.3), and we have $(01001101)_2 = (4D)_{16}$. If we start with the (two-byte) number $(A13F)_{16}$ and want to convert in the other direction, we replace each hexadecimal symbol by its (four-bit) binary equivalent (also as shown in Table 4.3):

$$\underbrace{A}_{1010} \quad \underbrace{1}_{0001} \quad \underbrace{3}_{0011} \quad \underbrace{F}_{1111}$$

This results in $(A13F)_{16} = (1010000100111111)_2$. □

Example 4.14 We need negative integers in order to perform the operation of subtraction in terms of addition [that is, $(a - b) = a + (-b)$]. When we are dealing with the binary representation of integers, we can use a popular method that enables us to perform addition, subtraction, multiplication, and (integer) division: the *two's complement method*. The method's popularity rests on its implementation by only two electronic circuits—one to negate and the second to add.

Table 4.4

Two's Complement Notation	
Value Represented	Four-Bit Pattern
7	0 1 1 1
6	0 1 1 0
5	0 1 0 1
4	0 1 0 0
3	0 0 1 1
2	0 0 1 0
1	0 0 0 1
0	0 0 0 0
−1	1 1 1 1
−2	1 1 1 0
−3	1 1 0 1
−4	1 1 0 0
−5	1 0 1 1
−6	1 0 1 0
−7	1 0 0 1
−8	1 0 0 0

In Table 4.4 the integers from -8 to 7 are represented by the four-bit patterns shown. The nonnegative integers are represented as they were in Tables 4.2 and 4.3. To obtain the results for $-8 \le n \le -1$, first consider the binary representation of $|n|$, the absolute value of n . Then do the following:

1. Replace each 0(1) in the binary representation of $|n|$ by 1(0); this result is called the *one's complement* of (the given representation of) $|n|$.

2. Add 1 ($= 0001$ in this case) to the result in step 1. This result is called the *two's complement* of n.

For example, to obtain the two's complement (representation) of -6, we proceed as follows.

	6
1. Start with the binary representation of 6.	↓ 0110
2. Interchange the 0's and 1's; this result is the one's complement of 0110.	↓ 1001
3. Add 1 to the prior result.	↓ $1001 + 0001 = 1010$

We can also obtain the four-bit patterns for the values $-8 \le n \le -1$ by using the four-bit patterns for the integers from 0 to 7 and complementing (interchanging 0's and 1's) these patterns as shown by four such pairs of patterns in Table 4.4. Note in Table 4.4 that the four-bit patterns for the nonnegative integers start with 0, whereas 1 is the first bit for the negative integers in the table.   □

Example 4.15   How do we perform the subtraction $33 - 15$ in base 2, using the two's complement method with patterns of eight bits ($=$one byte)?

We want to determine $33 - 15 = 33 + (-15)$. We find that $33 = (00100001)_2$, and $15 = (00001111)_2$. Therefore we represent -15 by

$$11110000 + 00000001 = 11110001.$$

The addition of integers represented in two's complement notation is the same as ordinary binary addition, except that all results must have the same size bit patterns. This means that when two integers are added by the two's complement method, any extra bit that results on the left of the answer (by a final carry) must be discarded. We illustrate this in the following calculations.

$$
\begin{array}{r}
33 \\
-\,15 \\
\end{array}
\longrightarrow
\begin{array}{r}
00100001 \\
+\,11110001 \\
\hline
100010010 \\
\end{array}
$$

This bit is discarded.

Answer $= (00010010)_2 = 18$

↑ This bit indicates that the answer is nonnegative.

To find $15 - 33$ we use $15 = (00001111)_2$ and $33 = (00100001)_2$. Then, to calculate $15 - 33$ as $15 + (-33)$, we represent -33 by $11011110 + 00000001 = 11011111$. This gives us the results

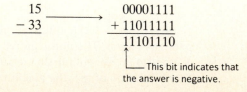

$$
\begin{array}{r}
15 \\
-\,33 \\
\end{array}
\longrightarrow
\begin{array}{r}
00001111 \\
+\,11011111 \\
\hline
11101110 \\
\end{array}
$$

This bit indicates that the answer is negative.

In order to get the positive form of the answer, we proceed as follows:

$$11101110$$

1. Take the one's complement.

$$\downarrow$$
$$00010001$$

2. Add 1 to the prior result.

$$\downarrow$$
$$00010010$$

Since $(00010010)_2 = 18$, the answer is -18.

One problem we have avoided in the two preceding calculations involves the size of the integers that we can represent by eight-bit patterns. No matter what size patterns we use, there is a limit to the size of the integers that can be represented. When we exceed this size, an *overflow error* results. For example, if we are working with eight-bit patterns and try to add 117 and 88, we obtain

$$
\begin{array}{r}
117 \\
+\ 88 \\
\end{array}
\longrightarrow
\begin{array}{r}
01110101 \\
+\ 01011000 \\
\hline
11001101 \\
\end{array}
$$

↑
└─ This bit indicates
that the answer is negative.

This result shows how we can detect an overflow error when adding two numbers. Here an overflow error is indicated: The sum of the eight-bit patterns for two positive integers has resulted in the eight-bit pattern for a negative integer. Similarly, when the addition of (the eight-bit patterns of) two negative integers results in the eight-bit pattern of a positive integer, an overflow error is detected.

□

To see why the procedure in Example 4.15 works in general, let $x, y \in \mathbf{Z}^+$ with $x > y$.

Let $2^{n-1} \le x < 2^n$. Then the binary representation for x is made up of n bits (with the leading bit 1). The binary representation for 2^n consists of $n + 1$ bits: a leading bit 1 followed by n 0's. The binary representation for $2^n - 1$ consists of n 1's.

When we subtract y from $2^n - 1$, we have

$$(2^n - 1) - y = \underbrace{11\ldots1}_{n\ 1's} - y,\ \text{the one's complement of } y.$$

Then $(2^n - 1) - y + 1$ gives us the two's complement of $-y$, and

$$x - y = x + [(2^n - 1) - y + 1] - 2^n,$$

where the final term, -2^n, results in the removal of the extra bit that arises on the left of the answer.

We close this section with one final result on composite integers. (This result will be used later in the Pascal program shown in Fig. 8.3.)

Example 4.16   If $n \in \mathbf{Z}^+$ and n is composite, then there exists a prime p such that $p \mid n$ and $p \le \sqrt{n}$.

Proof: Since n is composite, $n = n_1 n_2$, where $1 < n_1$, $n_2 < n$. We claim that one of the integers n_1, n_2 must be less than or equal to $\sqrt{n}$. If not, $n_1 > \sqrt{n}$ and $n_2 > \sqrt{n}$ give us the contradiction $n = n_1 n_2 > (\sqrt{n})(\sqrt{n}) = n$. Without loss of generality, we shall assume that $n_1 \leq \sqrt{n}$. If n_1 is prime, the result follows. If n_1 is not prime, then by Lemma 4.1 there exists a prime $p < n_1$ where $p \mid n_1$. So $p \mid n$ and $p \leq \sqrt{n}$.   □

EXERCISES 4.2

1. Verify the remaining parts of Theorem 4.3.

2. Let $a, b, c, d \in \mathbf{Z}^+$. Prove that (a) $(a \mid b \wedge c \mid d) \Rightarrow ac \mid bd$; (b) $a \mid b \Rightarrow ac \mid bc$; and (c) $ac \mid bc \Rightarrow a \mid b$.

3. If p, q are primes, prove that $p \mid q$ if and only if $p = q$.

4. If $a, b, c \in \mathbf{Z}^+$ and $a \mid bc$, does it follow that $a \mid b$ or $a \mid c$?

5. Let $a, b \in \mathbf{Z}^+$. If $b \mid a$ and $b \mid (a + 2)$, prove that $b = 1$ or $b = 2$.

6. If $n \in \mathbf{Z}^+$, and n is odd, prove that $8 \mid (n^2 - 1)$.

7. If $a, b \in \mathbf{Z}^+$, and both are odd, prove that $2 \mid (a^2 + b^2)$ but $4 \nmid (a^2 + b^2)$.

8. Write each of the following base-10 numbers in base 2, base 4, and base 8: (a) 137; (b) 6243; (c) 12,345.

9. Write each of the following (base-10) integers in base 2 and base 16.

   a) 22   **b)** 527   **c)** 1234   **d)** 6923

10. Convert each of the following hexadecimal numbers to base 2 and base 10.

    a) A7   **b)** 4C2   **c)** 1C2B   **d)** A2DFE

11. Convert each of the following binary numbers to base 10 and base 16.
    a) 11001110   **b)** 00110001
    c) 11110000   **d)** 01010111

12. Write each of the following integers in two's complement representations. Here the results are eight-bit patterns.
    a) 15   **b)** −15   **c)** 100
    d) −65   **e)** 127   **f)** −128

13. If a machine stores integers by the two's complement method, what are the largest and smallest integers that it can store if it uses bit patterns of (a) 4 bits? (b) 8 bits? (c) 16 bits? (d) 32 bits? (e) 2^n bits, $n \in \mathbf{Z}^+$?

14. In each of the following problems, we are using four-bit patterns for the two's complement representations of the integers from −8 to 7. Solve each problem (if possible), and then convert the results to base 10 to check your answers. Watch for any overflow errors.

    a)   0101   **b)**   1101   **c)**   0111
            + 0001   + 1110   + 1000

d) $\quad$ 1101 $\qquad$ **e)** $\quad$ 1011 $\qquad$ **f)** $\quad$ 0101
$\quad\quad \underline{+\ 1010} \qquad\qquad\quad \underline{+\ 0101} \qquad\qquad\quad \underline{+\ 0100}$

15. If $a, x, y \in \mathbf{Z}$, and $a \neq 0$, prove that $ax = ay \Rightarrow x = y$.

16. Write a computer program (or develop an algorithm) to convert a positive integer in base 10 to base b, where $2 \leq b \leq 9$.

17. The Division Algorithm can be generalized as follows: For $a, b \in \mathbf{Z}$, $b \neq 0$, there exist unique $q, r \in \mathbf{Z}$ with $a = qb + r, 0 \leq r < |b|$. Using Theorem 4.5 verify this generalized form of the algorithm for $b < 0$.

18. Write a computer program (or develop an algorithm) to output q and r for the generalized Division Algorithm of the previous exercise.

19. For $n \in \mathbf{Z}^+$, write a computer program (or develop an algorithm) that prints out all positive divisors of n.

20. Write a computer program (or develop an algorithm) to convert a positive integer in base 10 to base 16.

21. Let $n \in \mathbf{Z}^+$ with $n = r_k \cdot 10^k + \cdots + r_2 \cdot 10^2 + r_1 \cdot 10 + r_0$ (the base 10 representation of n). Prove that (a) $2|n$ if and only if $2|r_0$; (b) $4|n$ if and only if $4|(r_1 \cdot 10 + r_0)$; and (c) $8|n$ if and only if $8|(r_2 \cdot 10^2 + r_1 \cdot 10 + r_0)$.

State a general theorem suggested by these results.

4.3
THE GREATEST COMMON DIVISOR: THE EUCLIDEAN ALGORITHM

Continuing with the division operation developed in Section 4.2, we turn our attention to the divisors of a pair of integers.

DEFINITION 4.2 $\quad$ For $a, b \in \mathbf{Z}$, a positive integer c is said to be a *common divisor of* a, b if $c|a$ and $c|b$.

Example 4.17 $\quad$ The common divisors of 42 and 70 are 1, 2, 7, and 14, and 14 is the *greatest* of the common divisors. $\quad\square$

DEFINITION 4.3 $\quad$ Let $a, b \in \mathbf{Z}$, where at least one of $a, b \neq 0$. Then $c \in \mathbf{Z}^+$ is called a *greatest common divisor* (g.c.d.) of a, b if

a) $c|a, c|b$ (that is, c is a common divisor of a, b), and

b) for any common divisor d of a, b, we have $d|c$.

The result in Example 4.17 satisfies these conditions. However, this example deals with two small integers. What would we do with two integers each having 20 digits? We consider the following questions.

1. Given $a, b \in \mathbf{Z}$, does a greatest common divisor of a, b always exist? If so, how does one find such an integer?
2. How many greatest common divisors can a pair of integers have?

In dealing with these questions, we concentrate on $a, b \in \mathbf{Z}^+$.

THEOREM 4.6   For any $a, b \in \mathbf{Z}^+$, there exists a unique $c \in \mathbf{Z}^+$ that is the greatest common divisor of a, b.

Proof   Given $a, b \in \mathbf{Z}^+$, let $S = \{as + bt \mid s, t \in \mathbf{Z}, as + bt > 0\}$. Since $S \neq \emptyset$, by the Well-Ordering Principle S has a least element c. We claim that c is a greatest common divisor of a, b.

Since $c \in S$, $c = ax + by$, for some $x, y \in \mathbf{Z}$. Consequently, if $d \in \mathbf{Z}$ and $d \mid a$ and $d \mid b$, then by Theorem 4.3(f) $d \mid (ax + by)$, so $d \mid c$.

If $c \nmid a$, then by the division algorithm $a = qc + r$, with $0 < r < c$. Then $r = a - qc = a - q(ax + by) = (1 - qx)a + (-qy)b$, so $r \in S$, contradicting the choice of c as the least element of S. Consequently, $c \mid a$, and by a similar argument, $c \mid b$.

Hence any $a, b \in \mathbf{Z}^+$ have a greatest common divisor. If c_1, c_2 both satisfy the two conditions of Definition 4.3, then with c_1 as a greatest common divisor, and c_2 as a common divisor, it follows that $c_2 \mid c_1$. Reversing roles, we find that $c_1 \mid c_2$, and because $c_1, c_2 \in \mathbf{Z}^+$, $c_1 = c_2$. ∎

We now know that for $a, b \in \mathbf{Z}^+$, the g.c.d. of a, b exists and is unique. We denote this number by (a, b). Here $(a, b) = (b, a)$, and for $a \in \mathbf{Z}$, $a \neq 0$, $(a, 0) = |a|$. Also, for $a, b \in \mathbf{Z}^+$, $(-a, b) = (a, -b) = (-a, -b) = (a, b)$. Finally $(0, 0)$ is not defined and is of no interest to us.

From Theorem 4.6 we see that not only does (a, b) exist but also that (a, b) is the *smallest positive integer* we can write as a *linear combination* of a and b. Integers a and b are called *relatively prime* when $(a, b) = 1$—that is, when there exist $x, y \in \mathbf{Z}$ with $ax + by = 1$.

Example 4.18   Since $(42, 70) = 14$, we can find $x, y \in \mathbf{Z}$ with $42x + 70y = 14$, or $3x + 5y = 1$. By inspection $x = 2, y = -1$ is a solution; $3(2) + 5(-1) = 1$. But for $k \in \mathbf{Z}$, $1 = 3(2 - 5k) + 5(-1 + 3k)$, so $14 = 42(2 - 5k) + 70(-1 + 3k)$, and the solutions for x, y are not unique.

In general, if $(a, b) = d$, then $((a/d), (b/d)) = 1$. (Verify this!) If $(a/d)x_0 + (b/d)y_0 = 1$, then $1 = (a/d)(x_0 - (b/d)k) + (b/d)(y_0 + (a/d)k)$, for any $k \in \mathbf{Z}$. So $d = a(x_0 - (b/d)k) + b(y_0 + (a/d)k)$, yielding infinitely many solutions to $ax + by = d$. □

This example and the prior observations work well enough when a, b are fairly small. But how does one find (a, b) for some arbitrary $a, b \in \mathbf{Z}^+$? For this we turn to the following result, which we owe to Euclid.

THEOREM 4.7   (*Euclidean Algorithm*) If $a, b \in \mathbf{Z}^+$, we apply the division algorithm as follows:

$$a = q_1 b + r_1, \qquad\qquad 0 < r_1 < b$$
$$b = q_2 r_1 + r_2, \qquad\qquad 0 < r_2 < r_1$$
$$r_1 = q_3 r_2 + r_3, \qquad\qquad 0 < r_3 < r_2$$

$$\cdot \quad \cdot \quad \cdot \quad \cdot \quad \cdot \quad \cdot \quad \cdot \quad \cdot$$

$$r_i = q_{i+2} r_{i+1} + r_{i+2}, \qquad\qquad 0 < r_{i+2} < r_{i+1}$$

$$\cdot \quad \cdot \quad \cdot \quad \cdot \quad \cdot \quad \cdot \quad \cdot \quad \cdot$$

$$r_{k-3} = q_{k-1} r_{k-2} + r_{k-1}, \qquad\qquad 0 < r_{k-1} < r_{k-2}$$
$$r_{k-2} = q_k r_{k-1} + r_k, \qquad\qquad 0 < r_k < r_{k-1}$$
$$r_{k-1} = q_{k+1} r_k.$$

Then r_k, the last nonzero remainder, equals (a, b).

Proof   To verify that $r_k = (a, b)$ we establish the conditions of Definition 4.3.

Start with the first division process listed above. If $c \mid a$ and $c \mid b$, then as $a = q_1 b + r_1$, it follows that $c \mid r_1$. Next $c \mid b, c \mid r_1 \Rightarrow c \mid r_2$, because $b = q_2 r_1 + r_2$. Continuing down through the division processes, we get to where $c \mid r_{k-2}$ and $c \mid r_{k-1}$. From the next-to-last equation, we conclude that $c \mid r_k$, and this verifies condition (b) in Definition 4.3.

To establish condition (a) we go in the reverse order. From the last equation, $r_k \mid r_{k-1}$, and so $r_k \mid r_{k-2}$, because $r_{k-2} = q_k r_{k-1} + r_k$. Continuing up through the equations, we get to where $r_k \mid r_2$ and $r_k \mid r_3$, so $r_k \mid r_1$. Then $r_k \mid r_2, r_1 \Rightarrow r_k \mid b$, and finally $r_k \mid r_1, b \Rightarrow r_k \mid a$. Hence $r_k = (a, b)$. ∎

We have now used the word "algorithm" in describing the procedures set forth in Theorems 4.5 and 4.7. This term will recur frequently throughout other chapters of this text, so it may be a good idea to consider just what it connotes. First and foremost, an *algorithm* is a list of *precise* instructions designed to solve a particular *type* of problem, not just one special case. In Theorem 4.5 the situation called for evaluation of the quotient and remainder when an integer a is divided by a positive integer b. Here the two integers a, b are provided as input, and the algorithm generates, as output, the quotient and remainder. In Theorem 4.7 the problem is the determination of the g.c.d. of any two positive integers. Hence this algorithm receives the two positive integers a, b as its input and generates their g.c.d. as the output.

In general, we expect all algorithms to receive *input* and provide the needed answer(s) as *output*. Also, the algorithm should provide the same answer whenever we repeat the values for the input. This happens when the list of instructions is such that the intermediate results that arise at the execution of each instruction are unique, depending only on the (initial) input and on any results that may have been derived at any preceding instructions. Finally, our algorithms cannot go on indefinitely. They must terminate after the execution of a *finite* number of instructions.

We now apply the Euclidean algorithm (Theorem 4.7) in the following examples.

Example 4.19   Find the g.c.d. of 250 and 111, and express the result as a linear combination of these integers.

$$250 = 2(111) + 28, \qquad 0 < 28 < 111$$
$$111 = 3(28) + 27, \qquad 0 < 27 < 28$$
$$28 = 1(27) + 1, \qquad 0 < 1 < 27$$
$$27 = 27(1) + 0.$$

So 1 is the last nonzero remainder. Therefore $(250, 111) = 1$, and 250 and 111 are relatively prime. Working backwards from the third equation, we have $1 = 28 - 1(27) = 28 - 1[111 - 3(28)] = (-1)(111) + 4(28) = (-1)(111) + 4[250 - 2(111)] = 4(250) - 9(111)$, a linear combination of 250 and 111.

This expression of 1 as a linear combination of 250 and 111 is not unique, for $1 = 250[4 - 111k] + 111[-9 + 250k]$, for any $k \in \mathbf{Z}$.

We also have $(-250, 111) = (250, -111) = (-250, -111) = (250, 111) = 1$.

□

Example 4.20   Having determined a g.c.d. in Example 4.19, we now use the Euclidean algorithm to write a computer program that will find (a, b) for any $a, b \in \mathbf{Z}^+$. The output of this program is illustrated for finding $(456, 624)$ and $(116, 641)$.

The Pascal program in Fig. 4.8 employs the binary operation Mod, where for $x, y \in \mathbf{Z}^+$, x Mod y = the remainder after x is divided by y. For example, 7 Mod 3 is 1, and 18 Mod 5 is 3. (We will deal with "the arithmetic of remainders" in more detail in Chapter 14.)   □

Example 4.21   Assisting students in programming classes, Brian finds that on the average he can help a student debug a Pascal program in 6 minutes, but it takes 10 minutes to debug a program written in APL. If he works continuously for 104 minutes and doesn't waste any time, how many programs can he debug in each language?

Here we seek integers $x, y \geq 0$, where $6x + 10y = 104$, or $3x + 5y = 52$. As $(3, 5) = 1$, we can write $1 = 3(2) + 5(-1)$, so $52 = 3(104) + 5(-52) = 3(104 - 5k) + 5(-52 + 3k), k \in \mathbf{Z}$. In order to obtain $0 \leq x = 104 - 5k, 0 \leq y = -52 + 3k$, we must have $(52/3) \leq k \leq (104/5)$. So $k = 18, 19, 20$ and there are three possible solutions:

a) $(k = 18)$: $x = 14$, $y = 2$ **b)** $(k = 19)$: $x = 9$, $y = 5$
c) $(k = 20)$: $x = 4$, $y = 8$ □

The equation in Example 4.21 is an example of a *Diophantine equation*: a linear equation requiring integer solutions. This type of equation was first investigated by the Greek algebraist Diophantus, who lived in the third century A.D.

Having solved one such equation, we seek to discover when a Diophantine equation has a solution. The proof is left to the reader.

```
Program EuclideanAlgorithm (input,output);
Var
        a,b,c,d,r: integer;
Begin
        Writeln ('We wish to determine the greatest common ');
        Writeln ('divisor of two positive integers a,b.');
        Write ('a = ');
        Read (a);
        Write ('b = ');
        Read (b);
        r := a Mod b;
        d := b;
        While r > 0 do
                Begin
                        c := d;
                        d := r;
                        r := c Mod d
                End;
        Writeln ('The greatest common divisor of ',a:0,' and ');
        Write (b:0,' is ',d:0)
End.

We wish to determine the greatest common
divisor of two positive integers a,b.
a = 456
b = 624
The greatest common divisor of 456 and
624 is 24

We wish to determine the greatest common
divisor of two positive integers a,b.
a = 116
b = 641
The greatest common divisor of 116 and
641 is 1
```

Figure 4.8

THEOREM 4.8 If $a, b, c \in \mathbf{Z}^+$, the Diophantine equation $ax + by = c$ has an integer solution $x = x_0, y = y_0$ if and only if $(a, b) | c$.

We close this section with a concept that is related to the greatest common divisor.

DEFINITION 4.4 For $a, b, c \in \mathbf{Z}^+$, c is called a *common multiple* of a, b if c is a multiple of both a and b. Furthermore, c is the *least common multiple* (l.c.m.) of a, b if it is the smallest of all positive integers that are common multiples of a, b. We denote c by $[a, b]$.

THEOREM 4.9 Let $a, b, c \in \mathbf{Z}^+$, with $c = [a, b]$. If d is a common multiple of a and b, then $c | d$.

Proof   If not, by the division algorithm $d = qc + r, 0 < r < c$. Since $c = [a, b]$, it follows that $c = ma$ for some $m \in \mathbf{Z}^+$. Also, $d = na$ for some $n \in \mathbf{Z}^+$, because d is a multiple of a. Consequently, $na = qma + r \Rightarrow (n - qm)a = r > 0$, and r is a multiple of a. In a similar way r is seen to be a multiple of b, so r is a common multiple of a, b. But with $0 < r < c$, we contradict the claim that c is the least common multiple. Hence $c \mid d$. ∎

Our last result for this section ties together the g.c.d. and the l.c.m. The proof is left to the reader.

THEOREM 4.10   For $a, b \in \mathbf{Z}^+, ab = [a, b](a, b)$.

EXERCISES 4.3

1. For each pair $a, b \in \mathbf{Z}^+$, determine (a, b) and express it as a linear combination of a, b.
   a) 231, 1820
   b) 1369, 2597
   c) 2689, 4001
   d) 7982, 7983

2. For each pair a, b in Exercise 1, find $[a, b]$.

3. For any $n \in \mathbf{Z}^+$, what is $(n, n + 1)$? What is $[n, n + 1]$?

4. For $a, b \in \mathbf{Z}^+, s, t \in \mathbf{Z}$, what can we say about (a, b) if
   a) $as + bt = 2$?
   b) $as + bt = 3$?
   c) $as + bt = 4$?
   d) $as + bt = 6$?

5. For $a, b \in \mathbf{Z}^+$ and $d = (a, b)$, prove that $(a/d, b/d) = 1$.

6. For $a, b, n \in \mathbf{Z}^+$, prove that $(na, nb) = n(a, b)$.

7. For $a, b, c, d \in \mathbf{Z}^+$, prove that if $d = a + bc$, then $(b, d) = (a, b)$.

8. Let $a, b, c \in \mathbf{Z}^+$ with $(a, b) = 1$. If $a \mid c$ and $b \mid c$, prove that $ab \mid c$. Does the result follow if $(a, b) \neq 1$?

9. Determine those values of $c \in \mathbf{Z}^+, 10 < c < 20$, for which the Diophantine equation $84x + 990y = c$ has no solution. Determine the solutions for the remaining values of c.

10. If a, b are relatively prime, prove that $(a - b, a + b) = 1$ or 2.

11. Let $a, b, c \in \mathbf{Z}^+$ with $(a, b) = 1$. If $a \mid bc$, prove that $a \mid c$.

12. An executive buys \$249.00 worth of toys for the children of her employees. For each girl she gets a doll costing \$3.30; each boy receives a set of soldiers costing \$2.90. How many toys of each type did she buy?

13. Verify Theorem 4.8.

14. Verify Theorem 4.10.

4.4
THE FUNDAMENTAL THEOREM OF ARITHMETIC

In this section we extend Lemma 4.1 and show that for any $n \in \mathbf{Z}^+$, $n > 1$, either n is prime or n can be written as a product of primes, where the representation is unique up to order. This result, known as the *Fundamental Theorem of Arithmetic*, can be found in an equivalent form in Book IX of Euclid's *Elements*.

The following results prove useful.

LEMMA 4.2   If $a, b \in \mathbf{Z}^+$ and p is a prime, then $p \mid ab \Rightarrow p \mid a$ or $p \mid b$.

Proof   If $p \mid a$, then we are finished. If not, then because p is prime, it follows that $(p, a) = 1$, and there are integers x, y with $1 = px + ay$. Then $b = p(bx) + (ab)y$, where $p \mid p$ and $p \mid ab$. So by Theorem 4.3(e), $p \mid b$.   ■

LEMMA 4.3   Let $a_i \in \mathbf{Z}^+$, $1 \le i \le n$. If p is prime and $p \mid a_1 a_2 \cdots a_n$, then $p \mid a_i$ for some $1 \le i \le n$.

Proof   We leave the proof of this result to the reader.   ■

Example 4.22   In Example 2.23 we used the method of proof by contradiction to show that $\sqrt{2}$ is irrational. Here we shall generalize that earlier result and prove that $\sqrt{p}$ is irrational for any prime p.

If not, we can write $\sqrt{p} = a/b$, where $a, b \in \mathbf{Z}^+$, and $(a, b) = 1$. Then $\sqrt{p} = a/b \Rightarrow pb^2 = a^2 \Rightarrow p \mid a^2 \Rightarrow p \mid a$. (Why?) Also, $p \mid a \Rightarrow a = pa_1$, for some $a_1 \in \mathbf{Z}$, so $p^2 a_1^2 = a^2 = pb^2$ and $b^2 = pa_1^2$. But then $p \mid b^2 \Rightarrow p \mid b$, and $(a, b) \ge p$, contradicting $(a, b) = 1$.   □

We turn now to the main result of the section.

THEOREM 4.11   Any integer $n > 1$ can be written as a product of primes uniquely, up to the order of the primes. (Here a single prime is considered a product of one factor.)

Proof   The proof consists of two parts: The first covers existence and the second deals with uniqueness.

If not, let $m > 1$ be the smallest integer not expressible as a product of primes. Since m is not a prime, we are able to write $m = m_1 m_2$, where $1 < m_1 < m$, $1 < m_2 < m$. But then m_1, m_2 can be written as products of primes, because they are less than m. Consequently, with $m = m_1 m_2$ we can obtain a prime factorization of m.

For the integer 2, we have a unique prime factorization, and assuming uniqueness of representation for $3, 4, 5, \ldots, n - 1$, we suppose that $n = p_1^{s_1} p_2^{s_2} \cdots p_k^{s_k} = q_1^{t_1} q_2^{t_2} \cdots q_r^{t_r}$, where each p_i, $1 \le i \le k$, and each q_j, $1 \le j \le r$, is a prime. Also $p_1 < p_2 < \cdots < p_k$, and $q_1 < q_2 < \cdots < q_r$, and $s_i > 0$ for $1 \le i \le k$, $t_j > 0$ for $1 \le j \le r$.

Since $p_1 \mid n$, we have $p_1 \mid q_1^{t_1} q_2^{t_2} \cdots q_r^{t_r}$. By Lemma 4.3, $p_1 \mid q_j$ for some $1 \le j \le r$. With p_1, q_j primes we have $p_1 = q_j$. In fact $j = 1$, for otherwise $q_1 \mid n \Rightarrow q_1 = p_e$ for some $1 < e \leq k$ and $p_1 < p_e = q_1 < q_j = p_1$. With $p_1 = q_1$, $n_1 = n / p_1 = p_1^{s_1 - 1} p_2^{s_2} \cdots p_k^{s_k}$ $= q_1^{t_1 - 1} q_2^{t_2} \cdots q_r^{t_r}$. Since $n_1 < n$, by the induction hypothesis (Theorem 4.2) it follows that $k = r$, $p_i = q_i$ for $1 \leq i \leq k$, $s_1 - 1 = t_1 - 1$ (so $s_1 = t_1$), and $s_i = t_i$ for $2 \leq i \leq k$. Hence the prime factorization of n is unique. ∎

This result is now used in the following two examples.

Example 4.23 For $n \in \mathbf{Z}^+$, we want to count the number of positive divisors of n. For example, the number 2 has two positive divisors: 1 and itself. Likewise, 1 and 3 are the only positive divisors of 3. In the case of 4, we find the three positive divisors 1, 2, and 4.

To determine the result for any $n \in \mathbf{Z}^+$, we use Theorem 4.11 and write $n = p_1^{e_1} p_2^{e_2} \cdots p_k^{e_k}$, where for $1 \leq i \leq k$, p_i is a prime and $e_i > 0$. If $m \mid n$, then $m = p_1^{f_1} p_2^{f_2} \cdots p_k^{f_k}$ where $0 \leq f_i \leq e_i$ for $1 \leq i \leq k$. So by the rule of product, the number of positive divisors of n is

$$(e_1 + 1)(e_2 + 1) \cdots (e_k + 1).$$

For example, since $2{,}520{,}000 = 2^6 \, 3^2 \, 5^4 \, 7$, it has $(6 + 1)(2 + 1)(4 + 1)(1 + 1) = (7)(3)(5)(2) = 210$ positive divisors. □

Example 4.24 If $m, n \in \mathbf{Z}^+$, let $m = p_1^{e_1} p_2^{e_2} \cdots p_t^{e_t}$ and $n = p_1^{f_1} p_2^{f_2} \cdots p_t^{f_t}$, with each p_i prime and $0 \leq e_i$, $0 \leq f_i$ for $1 \leq i \leq t$. Then if $a_i = \min\{e_i, f_i\}$, $b_i = \max\{e_i, f_i\}$ for $1 \leq i \leq t$, we have

$$(m, n) = p_1^{a_1} p_2^{a_2} \cdots p_t^{a_t} = \prod_{i=1}^{t} p_i^{a_i}, \text{ and}$$

$$[m, n] = p_1^{b_1} p_2^{b_2} \cdots p_t^{b_t} = \prod_{i=1}^{t} p_i^{b_i}.$$

For example, let $m = 491{,}891{,}400 = 2^3 \, 3^3 \, 5^2 \, 7^2 \, 11^1 \, 13^2$ and let $n = 1{,}138{,}845{,}708 = 2^2 \, 3^2 \, 7^1 \, 11^2 \, 13^3 \, 17^1$. Then with $p_1 = 2$, $p_2 = 3$, $p_3 = 5$, $p_4 = 7$, $p_5 = 11$, $p_6 = 13$, and $p_7 = 17$, we find $a_1 = 2$, $a_2 = 2$, $a_3 = 0$ (the exponent of 5 in the factorization of n must be 0, because 5 does not appear in the prime factorization), $a_4 = 1$, $a_5 = 1$, $a_6 = 2$, and $a_7 = 0$, so

$$(m, n) = 2^2 \, 3^2 \, 5^0 \, 7^1 \, 11^1 \, 13^2 \, 17^0 = 468{,}468.$$

We also have

$$[m, n] = 2^3 \, 3^3 \, 5^2 \, 7^2 \, 11^2 \, 13^3 \, 17^1 = 1{,}195{,}787{,}993{,}400. \quad □$$

EXERCISES 4.4

1. Write each of the following numbers as a product of primes

$$p_1^{n_1} p_2^{n_2} \cdots p_k^{n_k}, \quad \text{where} \quad 0 < n_i \text{ for } 1 \leq i \leq k \quad \text{and} \quad p_1 < p_2 < \cdots < p_k.$$

a) 148,500   **b)** 7,114,800   **c)** 7,882,875

2. Determine the g.c.d. and the l.c.m. of each pair of numbers in Exercise 1.

3. Extending the results of Example 4.24, find the g.c.d. and the l.c.m. of the three numbers in Exercise 1.

4. Verify Lemma 4.3.

5. **a)** Prove that $\log_{10} 2$ is irrational.

   b) For any prime p, prove that $\log_{10} p$ is irrational.

6. Find the number of positive divisors for each integer in Exercise 1.

7. For any $n \in \mathbf{Z}^{+}$, prove that n is a perfect square (that is, $n = k^2$ for some $k \in \mathbf{Z}^{+}$) if and only if n has an odd number of positive divisors.

8. Write a computer program (or develop an algorithm) to find the prime factorization of an integer $n > 1$.

9. Find the smallest positive integer n for which the product $1260 \times n$ is a perfect cube.

10. In triangle ABC the length of side BC is 293. If the length of side AB is a perfect square, the length of side AC is a power of 2, and the length of side AC is twice the length of side AB, determine the perimeter of the triangle.

4.5
SUMMARY AND HISTORICAL REVIEW

According to the Prussian mathematician Leopold Kronecker (1823–1891), "God made the integers, all the rest is the work of man.... All results of the profoundest mathematical investigation must ultimately be expressible in the simple form of properties of the integers." In the spirit of this quotation, we find in this chapter how the handiwork of the Almighty has been further developed by man over the last 24 centuries.

Starting in the fourth century B.C. we find in Euclid's *Elements* not only the geometry of our high school experience but also the fundamental ideas of number theory. Propositions 1 and 2 of Euclid's Book VII include an example of an algorithm to determine the greatest common divisor of two positive integers by using an efficient technique to solve, in a *finite* number of steps, a specific problem. (The term "algorithm," like its predecessor "algorism," was unknown to Euclid. In fact, this term did not enter the vocabulary of most people until the late 1950s when the computer revolution began to make its impact on society. The word comes from the name of the famous Islamic textbook writer Abu Ja'far Mohammed ibn Mûsâ al-Khowârizmî (c. 825). The last part of his name, al-Khowârizmî, which is translated as "a man from the town of Khowârizm," gave rise to the term "algorism.")

In the century following Euclid, we find some number theory in the work of Eratosthenes. However, it is not until five centuries later that the first major new accomplishments in the field were made by Diophantus. In his work *Arithmetica*, his integer solutions of linear (and higher-order) equations stood as a mathematical beacon in number theory until the French mathematician Pierre de Fermat (1601–1665) came on the scene.

For more on these mathematicians and others who have worked in the theory of numbers, consult L. Dickson [2]. Chapter 5 in I. Niven and H. Zuckerman [8] deals with the solutions of Diophantine equations and their applications. The articles in [7] recount interesting contemporary developments in number theory.

In the work *Formulario Mathematico*, Giuseppe Peano (1858–1932) formulated the set of nonnegative integers on the basis of three undefined terms: zero, number, and successor. His formulation is as follows:

a) Zero is a number.

b) For any number n, its successor is a number.

c) No number has zero as its successor.

d) If two numbers m, n have the same successor, then $m = n$.

e) If T is a set of numbers where $0 \in T$, and where the successor of n is in T whenever n is in T, then T is the set of all nonnegative integers.

In these postulates the notion of order (successor) and the technique called mathematical induction are seen to be intimately related to the idea of number (that is, nonnegative integer).

The first European to apply the principle of induction in mathematical proofs was the Venetian scientist Francesco Maurocylus (1491–1575), whose arithmetic book, published in 1575, contains such work. In the next century, Pierre de Fermat made further improvements on the technique in his work involving "the method of infinite descent." Blaise Pascal (c. 1653), in proving such combinatorial results as $C(n, k)/C(n, k + 1) = (k + 1)/(n - k)$, $0 \le k \le n - 1$, used induction and referred to the technique as the work of Maurocylus. The actual term "mathematical induction" was not used, however, until the early nineteenth century when it appeared in the work of Augustus DeMorgan (1806–1871). (An interesting survey on this topic is found in the article by W. H. Bussey [1].)

The text by B. K. Youse [10] illustrates the many varied applications of the principle of mathematical induction in algebra, geometry, and trigonometry. For more on the relevance of this method of proof to the problems of programming and the development of algorithms, the text by M. Wand [9] (especially Chapter 2) provides ample background and examples.

More on the theory of numbers can be found in the texts by G. Hardy and E. Wright [3], W. J. LeVeque [5], [6], and I. Niven and H. Zuckerman [8]. At a level comparable to that of this chapter, Chapter 3 of V. Larney [4] provides an enjoyable introduction to this material.

REFERENCES

1. Bussey, W. H. "Origins of Mathematical Induction." *American Mathematical Monthly* 24 (1917): pp. 199–207.

2. Dickson, L. *History of the Theory of Numbers.* Washington, D.C.: Carnegie Institution of Washington, 1919. Reprinted by Chelsea, in New York, in 1950.
3. Hardy, G. H., and Wright, E. M. *An Introduction to the Theory of Numbers,* 4th ed. Oxford, England: Clarendon Press, 1960.
4. Larney, Violet Hachmeister. *Abstract Algebra: A First Course.* Boston: Prindle, Weber & Schmidt, 1975.
5. LeVeque, William J. *Elementary Theory of Numbers.* Reading, Mass.: Addison-Wesley, 1962.
6. LeVeque, William J. *Topics in Number Theory,* Vols. I and II. Reading, Mass.: Addison-Wesley, 1956.
7. LeVeque, William J., ed. *Studies in Number Theory.* MAA Studies in Mathematics, Vol. 6. Englewood Cliffs, N.J.: Prentice-Hall, 1969. Published by the Mathematical Association of America.
8. Niven, Ivan, and Zuckerman, Herbert S. *An Introduction to the Theory of Numbers,* 3rd ed. New York: Wiley, 1972.
9. Wand, Mitchell. *Induction, Recursion, and Programming.* New York: Elsevier North Holland, 1980.
10. Youse, Bevan K. *Mathematical Induction.* Englewood Cliffs, N.J.: Prentice-Hall, 1964.

MISCELLANEOUS EXERCISES

1. Let a, d be fixed integers. Determine a summation formula for $a + (a + d) + (a + 2d) + \cdots + (a + (n - 1)d)$, for $n \in \mathbf{Z}^+$. Verify your result by mathematical induction.

2. For $n \in \mathbf{Z}^+$, prove each of the following by mathematical induction:

   a) $5 \mid (n^5 - n)$ **b)** $6 \mid (n^3 + 5n)$

3. Let $S(n)$ be the statement: For all $n \in \mathbf{Z}^+$, $n^2 + n + 41$ is prime.

   a) Verify that $S(n)$ is true for all $1 \le n \le 9$.

   b) Does the truth of $S(k)$ imply that of $S(k + 1)$ for all $k \in \mathbf{Z}^+$?

4. For $n \in \mathbf{Z}^+$ define the sum s_n by the formula

$$s_n = \frac{1}{2!} + \frac{2}{3!} + \frac{3}{4!} + \cdots + \frac{(n - 1)}{n!} + \frac{n}{(n + 1)!}.$$

   a) Verify that $s_1 = \frac{1}{2}$, $s_2 = \frac{5}{6}$, and $s_3 = \frac{23}{24}$.

   b) Compute s_4, s_5, and s_6.

   c) On the basis of your results in parts (a) and (b), conjecture a formula for the sum of the terms in s_n.

   d) Verify your conjecture in part (c) for all $n \in \mathbf{Z}^+$ by the principle of finite induction.

5. For $n \in \mathbf{Z}^+$, use mathematical induction to show that

$$\sum_{i=1}^{n} i^4 = \frac{n(n + 1)(2n + 1)(3n^2 + 3n - 1)}{30}$$

6. Show that for any $n \in \mathbf{Z}^+$,

$$\sum_{i=1}^{n} (-1)^{i+1} i^2 = (-1)^{n+1} \sum_{i=1}^{n} i.$$

7. For any $n \in \mathbf{Z}$, $n \geq 0$, prove that

    a) $2^{2n+1} + 1$ is divisible by 3.

    b) $n^3 + (n+1)^3 + (n+2)^3$ is divisible by 9.

    c) $\dfrac{n^7}{7} + \dfrac{n^3}{3} + \dfrac{11n}{21}$ is an integer.

8. Verify that for any $n \in \mathbf{Z}^+$,

$$\sin\theta + \sin 3\theta + \sin 5\theta + \cdots + \sin(2n-1)\theta = \frac{\sin^2 n\theta}{\sin\theta}$$

9. If $n \in \mathbf{Z}^+$ and $n \geq 2$, prove that $2^n < \binom{2n}{n} < 4^n$.

10. If $n \in \mathbf{Z}^+$, prove that 57 divides $7^{n+2} + 8^{2n+1}$.

11. For any $n \in \mathbf{Z}^+$, show that

    a) if $n \geq 64$, then n can be written as a sum of 5's and/or 17's.

    b) if $n \geq 108$, then n can be expressed as a sum of 10's and/or 13's.

12. If n bits (no leading zeros) are needed to write the integer m in base 2, show that $n \geq \log_2(m+1)$.

13. Given $r \in \mathbf{Z}^+$, write $r = r_0 + r_1 \cdot 10 + r_2 \cdot 10^2 + \cdots + r_n \cdot 10^n$, where $0 \leq r_i \leq 9$ for $1 \leq i \leq n-1$, and $0 < r_n \leq 9$.

    a) Prove that $9 \mid r$ if and only if $9 \mid (r_n + r_{n-1} + \cdots + r_2 + r_1 + r_0)$.

    b) Prove that $3 \mid r$ if and only if $3 \mid (r_n + r_{n-1} + \cdots + r_2 + r_1 + r_0)$.

    c) If $t = 137486x225$, where x is a single digit, determine the value(s) of x such that $3 \mid t$. Which values of x make t divisible by 9?

14. Frances spends $6.20 on candy for prizes in a contest. If a 10-ounce box of this candy costs $.50 and a 3-ounce box costs $.20, how many boxes of each size did she purchase?

15. **a)** How many positive integers can we express as a product of nine primes (repetitions allowed and order not relevant) where the primes may be chosen from $\{2, 3, 5, 7, 11\}$?

    b) How many of the positive integers in part (a) have at least one occurrence of each of the five primes?

    c) How many of the results in part (a) are divisible by 4? How many of the results in part (b)?

16. Let $a, b, c, d, m, n, s, t \in \mathbf{Z}^+$ with $ad - bc = 1$, $s = am + bn$, and $t = cm + dn$.

    a) Solve for m, n in terms of s, t.

    b) Prove that $(s, t) = (m, n)$.

17. **a)** Ten students enter a locker room that contains 10 lockers. The first student opens all the lockers. The second student changes that status (from closed to open, or vice versa) of every other locker, starting with the second locker. The third student then changes the status of every third locker, starting at the third locker. In general, for $1 < k \leq 10$, the kth student changes the status of every kth locker, starting with the kth locker. After the tenth student has gone through the lockers, which lockers are left open?

    b) Answer part (a) if 10 is replaced by $n \in \mathbf{Z}^+$, $n \geq 2$.

CHAPTER 5

Relations and Functions

In this chapter we extend the set theory of Chapter 3 to include the concepts of relation and function. Algebra, trigonometry, and calculus all involve functions. Here, however, we shall study functions from a set-theoretic approach that includes finite functions and shall introduce new counting ideas in the study. Furthermore, we shall examine the concept of function complexity and its role in the study of the analysis of algorithms.

We take a path along which we shall find the answers to the following (closely related) six problems:

1. The Defense Department has seven different contracts that deal with a high-security project. Four companies can manufacture the distinct parts called for in each contract, and to keep the overall project as well protected as possible, it is best to have all four companies working on some part. In how many ways can the contracts be awarded so that every company is involved?

2. How many seven-symbol quaternary $(0, 1, 2, 3)$ sequences have at least one occurrence of each of the symbols 0, 1, 2, and 3?

3. An $m \times n$ *zero-one matrix* is a matrix A with m rows and n columns, such that in row i, $1 \le i \le m$, and column j, $1 \le j \le n$, the entry a_{ij} that appears is either 0 or 1. How many 7×4 zero-one matrices have exactly one 1 in each row and at least one 1 in each column? (The zero-one matrix is a data structure that arises in computer science. We shall learn more about it in later chapters.)

4. Seven (unrelated) people enter the lobby of a building which has four additional floors, and get on an elevator. What is the probability that the elevator must stop at every floor to let passengers off?

5. For positive integers m, n with $m < n$,

$$\sum_{k=0}^{n} (-1)^k \binom{n}{n-k}(n-k)^m = 0.$$

6. For every positive integer n,

$$n! = \sum_{k=0}^{n} (-1)^k \binom{n}{n-k} (n-k)^n.$$

Do you recognize the connection among the first four problems? The first three are the same problem in different settings. However, it is not obvious that the last two problems are related or that there is a connection between them and the first four. These identities, however, will be established using the same counting techniques developed to solve the first four problems.

5.1
CARTESIAN PRODUCTS AND RELATIONS

DEFINITION 5.1   For sets $A, B \subseteq \mathcal{U}$, the *Cartesian product*, or *cross product*, of A, B is denoted by $A \times B$ and equals $\{(a, b) \mid a \in A, b \in B\}$.

We say that the elements of $A \times B$ are *ordered pairs*. For $(a, b), (c, d) \in A \times B$, we have $(a, b) = (c, d)$ if and only if $a = c$ and $b = d$. (Unfortunately, the notation for an ordered pair is the same notation used for the greatest common divisor of two integers. The context will indicate which of the two ideas is being examined.)

If A, B are finite, it follows from the rule of product that $|A \times B| = |A| \cdot |B|$. Although we generally will not have $A \times B = B \times A$, we will have $|A \times B| = |B \times A|$. Also, although $A, B \subseteq \mathcal{U}$, it is not necessary that $A \times B \subseteq \mathcal{U}$, so $\mathcal{U}$ is not necessarily closed under this operation.

We can extend the definition of the Cartesian product, or cross product, to more than two sets. If $n \in \mathbf{Z}^+$, $n \geq 3$, and $A_1, A_2, \ldots, A_n \subseteq \mathcal{U}$, then the *(n-fold) product* of $A_1, A_2, \ldots, A_n$ is denoted by $A_1 \times A_2 \times \cdots \times A_n$ and equals $\{(a_1, a_2, \ldots, a_n) \mid a_i \in A_i, 1 \leq i \leq n\}$.† The elements of $A_1 \times A_2 \times \cdots \times A_n$ are called ordered *n-tuples*, although we generally use the term *triple* in place of 3-tuple. As with ordered pairs, if $(a_1, a_2, \ldots, a_n), (b_1, b_2, \ldots, b_n) \in A_1 \times A_2 \times \cdots \times A_n$, then $(a_1, a_2, \ldots, a_n) = (b_1, b_2, \ldots, b_n)$ if and only if $a_i = b_i$ for all $1 \leq i \leq n$.

Example 5.1   Let $\mathcal{U} = \{1, 2, 3, \ldots, 7\}, A = \{2, 3, 4\}, B = \{4, 5\}$. Then

a) $A \times B = \{(2, 4), (2, 5), (3, 4), (3, 5), (4, 4), (4, 5)\}$.

† When dealing with the Cartesian product of three or more sets, we must be careful about associativity. In the case of three sets, for example, there is a difference between any two of the sets $A_1 \times A_2 \times A_3$, $(A_1 \times A_2) \times A_3$, and $A_1 \times (A_2 \times A_3)$, because their respective elements are ordered triples (a_1, a_2, a_3), and the distinct ordered pairs $((a_1, a_2), a_3)$ and $(a_1, (a_2, a_3))$. Although such differences are important in certain instances, we shall not concentrate on them here and shall always use the non-parenthesized form $A_1 \times A_2 \times A_3$. This will also be our convention when dealing with four or more sets.

  b) $B \times A = \{(4,2),(4,3),(4,4),(5,2),(5,3),(5,4)\}$.

  c) $B^2 = B \times B = \{(4,4),(4,5),(5,4),(5,5)\}$.

  d) $B^3 = B \times B \times B = \{(a,b,c) \mid a,b,c \in B\}$; $(4,5,5) \in B^3$. $\square$

Example 5.2 With $\mathcal{U} = \mathbf{R}$, $\mathbf{R} \times \mathbf{R} = \{(x,y) \mid x,y \in \mathbf{R}\}$ is recognized as the real plane of coordinate geometry and two-dimensional calculus. The subset $\mathbf{R}^+ \times \mathbf{R}^+$ is the interior of the first quadrant of this plane. Likewise $\mathbf{R}^3$ represents Euclidean three-space, where three-dimensional spheres, two-dimensional planes, and one-dimensional lines are subsets of importance. $\square$

Example 5.3 An experiment $\mathcal{E}$ is conducted as follows: A single die is rolled and its outcome noted, and then a coin is flipped and its outcome noted. Determine a sample space $\mathcal{S}$ for $\mathcal{E}$.

  Let $\mathcal{E}_1$ denote the first part of experiment $\mathcal{E}$, and let $\mathcal{S}_1 = \{1,2,3,4,5,6\}$ be a sample space for $\mathcal{E}_1$. Likewise let $\mathcal{S}_2 = \{H, T\}$ be a sample space for $\mathcal{E}_2$, the second part of the experiment. Then $\mathcal{S} = \mathcal{S}_1 \times \mathcal{S}_2$ is a sample space for $\mathcal{E}$.

  This sample space can be represented pictorially with a *tree diagram* that exhibits all the possible outcomes of experiment $\mathcal{E}$. In Fig. 5.1 we have such a tree diagram, which proceeds from left to right. From the leftmost endpoint, six branches originate for the six outcomes of the first stage of the experiment $\mathcal{E}$. From each point, numbered $1,2,\ldots,6$, two branches indicate the outcomes in tossing the coin. The 12 ordered pairs at the right endpoints constitute the sample space $\mathcal{S}$. $\square$

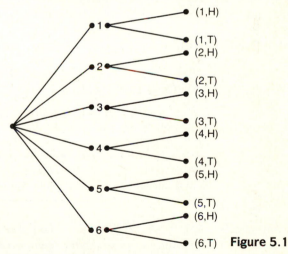

Figure 5.1

  In addition to their tie-in with Cartesian products, tree diagrams also arise in other situations.

Example 5.4 At the Wimbledon Tennis Championships, women play at most three sets in a match. The winner is the first to win two sets. If we let N and E denote the two players, the tree diagram in Fig. 5.2 indicates the six ways in which this match can be won. For example, the starred line segment (edge) indicates that player E won the first set. The double-starred edge indicates that player N has won the match by winning the first and third sets.   □

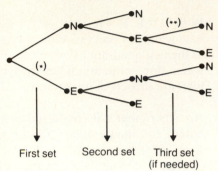

First set   Second set   Third set
(if needed)   **Figure 5.2**

Tree diagrams are examples of a general structure called a *tree*. Trees and graphs are important structures that arise in computer science and optimization theory. These will be investigated in later chapters.

Returning to the cross product of two sets, we shall find the subsets of this structure of great interest.

DEFINITION 5.2   For sets $A, B \subseteq \mathcal{U}$, any subset of $A \times B$ is called a *relation* from A to B. Any subset of $A \times A$ is called a *binary relation* on A.   ▬

Example 5.5   With $A, B, \mathcal{U}$ as in Example 5.1, the following are relations from A to B.

a) $\emptyset$ **b)** $\{(2,4)\}$ **c)** $\{(2,4),(2,5)\}$

d) $\{(2,4),(3,4),(4,4)\}$ **e)** $\{(2,4),(3,4),(4,5)\}$ **f)** $A \times B$

Since $|A \times B| = 6$, it follows from Definition 5.2 that there are 2^6 possible relations from A to B.   □

In general, for finite sets A, B with $|A| = m$ and $|B| = n$, there are 2^{mn} relations from A to B, including the empty relation as well as the relation $A \times B$ itself.

Example 5.6   Let $B = \{1,2\} \subseteq \mathbf{N}$, $\mathcal{U} = \mathcal{P}(B)$ and $A = \mathcal{U} = \{\emptyset, \{1\}, \{2\}, \{1,2\}\}$. The following is an example of a *binary relation* on A: $\mathcal{R} = \{(\emptyset, \emptyset),\ (\emptyset, \{1\}),\ (\emptyset, \{2\}),\ (\emptyset, \{1,2\}),\ (\{1\}, \{1\}),\ (\{1\}, \{1,2\}),\ (\{2\}, \{2\}),\ (\{2\}, \{1,2\}),\ (\{1,2\}, \{1,2\})\}$. We can say that the relation $\mathcal{R}$ is the *subset relation* where $(C, D) \in \mathcal{R}$ if and only if $C, D \subseteq B$ and $C \subseteq D$.   □

Example 5.7   With $A = \mathcal{U} = \mathbf{Z}^+$, we define a binary relation $\mathcal{R}$ on set A as $\{(x, y) \mid x \leq y\}$. This is the familiar "is less than or equal to" relation for the set of positive integers. It can be represented graphically as the set of points, with positive integer components, located on or above the line $y = x$ in the Euclidean plane, as partially shown in Fig. 5.3. Here we cannot list the entire relation as we did in Example 5.6, but we note, for example, that $(7, 7), (7, 11) \in \mathscr{R}$, but $(8, 2) \notin \mathscr{R}$. $(7, 11) \in \mathscr{R}$ can also be denoted by $7\mathscr{R}11$; $(8, 2) \notin \mathscr{R}$ becomes $8\not\mathscr{R}2$. $7\mathscr{R}11$ and $8\not\mathscr{R}2$ are examples of the *infix* notation for a relation. $\square$

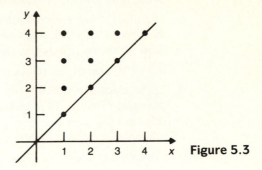

Figure 5.3

We close this section with these final observations.

1) For any set $A \subseteq \mathscr{U}$, $A \times \emptyset = \emptyset$. (If $A \times \emptyset \neq \emptyset$, let $(a, b) \in A \times \emptyset$. Then $a \in A$ and $b \in \emptyset$. Impossible!) Likewise, $\emptyset \times A = \emptyset$.

2) The Cartesian product and the binary operations of union and intersection are interrelated in the following theorem.

THEOREM 5.1   For any sets $A, B, C \subseteq \mathscr{U}$:

a) $A \times (B \cap C) = (A \times B) \cap (A \times C)$

b) $A \times (B \cup C) = (A \times B) \cup (A \times C)$

c) $(A \cap B) \times C = (A \times C) \cap (B \times C)$

d) $(A \cup B) \times C = (A \times C) \cup (B \times C)$

Proof   We prove Theorem 5.1(a) and leave the other parts for the reader. For any $a, b \in \mathscr{U}$, $(a, b) \in A \times (B \cap C) \Leftrightarrow a \in A$ and $b \in B \cap C \Leftrightarrow a \in A$ and $b \in B, C \Leftrightarrow a \in A, b \in B$, and $a \in A, b \in C \Leftrightarrow (a, b) \in A \times B$ and $(a, b) \in A \times C \Leftrightarrow (a, b) \in (A \times B) \cap (A \times C)$.   ∎

EXERCISES 5.1

1. If $\mathscr{U} = \mathbf{N}$, $A = \{1, 2, 3, 4\}$, $B = \{2, 5\}$, and $C = \{3, 4, 7\}$, determine $A \times B$; $B \times A$; $A \cup (B \times C)$; $(A \cup B) \times C$; $(A \times C) \cup (B \times C)$.

2. If $\mathscr{U} = \{1, 2, 3, 4, 5\}$, $A = \{1, 2, 3\}$, and $B = \{2, 4, 5\}$, give examples of:

a) three nonempty relations from A to B.

b) three nonempty binary relations on A.

3. For $A, B, \mathscr{U}$ as in Exercise 2, determine the following: (a) $|A \times B|$ (b) The number of relations from A to B. (c) The number of binary relations on A.

(d) The number of relations from A to B that contain $(1, 2)$ and $(1, 5)$. (e) The number of relations from A to B that contain exactly five ordered pairs. (f) The number of binary relations on A that contain at least seven elements.

4. For sets $A, B \subseteq \mathcal{U}$, when is it true that $A \times B = B \times A$?

5. For $\mathcal{U} = \mathbf{N}$, $A = \{2, 3, 4, 5, 6, 7\}$, and $B = \{10, 11, 12, 13, 14\}$, list the elements of the relation $\mathcal{R} \subseteq A \times B$ where $a \, \mathcal{R} \, b$ if a (evenly) divides b.

6. The men's final at Wimbledon is won by the first player to win three sets of the five-set match. Let C and M denote the players. Draw a tree diagram to show all the ways in which the match can be decided.

7. If $\mathcal{U} = \mathbf{R}$, sketch the relation $\{(x, y) \mid x^2 + y^2 = 4\}$. What happens if $\mathcal{U}$ is $\mathbf{R}^+$?

8. Logic chips are taken from a container, tested individually, and labeled defective or good. The testing process is continued until either two defective chips are found or five chips are tested in total. Using a tree diagram, exhibit a sample space for this process.

9. Complete the proof of Theorem 5.1.

10. A rumor is spread as follows. The originator calls two people. Each of these people phones three friends, each of whom in turn calls five associates. If no one receives more than one call, how many people now know the rumor? How many phone calls were made?

11. For $A, B, C \subseteq \mathcal{U}$, prove that $A \times (B - C) = (A \times B) - (A \times C)$.

12. Let A, B be sets with $|B| = 3$. If there are 4096 relations from A to B, what is $|A|$?

5.2
FUNCTIONS: PLAIN AND ONE-TO-ONE

In this section we concentrate on a special kind of relation called a *function*. General relations will appear again in Chapter 7.

DEFINITION 5.3   For nonempty sets A, B, a *function*, or *mapping*, *f from A to B*, denoted $f: A \to B$, is a relation from A to B in which every element of A appears exactly once as the first component of an ordered pair in the relation.   ▬

We often write $f(a) = b$ when (a, b) is an ordered pair in the function f. For $(a, b) \in f$, b is called *the image* of a under f, whereas a is a *preimage* of b. In addition, the definition suggests that f is a method for *associating* with each $a \in A$ a *unique* $b \in B$, denoting this process by $f(a) = b$. Consequently, $(a, b), (a, c) \in f$ implies $b = c$.

Example 5.8   For $A = \{1, 2, 3\}$ and $B = \{w, x, y, z\}$, $f = \{(1, w), (2, x), (3, x)\}$ is a function, and consequently a relation, from A to B. $\mathcal{R}_1 = \{(1, w), (2, x)\}$ and $\mathcal{R}_2 = \{(1, w), (2, w), (2, x), (3, z)\}$ are relations but not functions, from A to B. (Why not?)   □

DEFINITION 5.4 For the function $f: A \rightarrow B$, A is called the *domain* of f and B the *codomain* of f. The subset of B consisting of those elements that appear as second components in the ordered pairs of f is called the *range* of f and is also denoted by $f(A)$, because it is the set of images (of the elements of A) under f.

In Example 5.8, the domain of $f = \{1, 2, 3\}$, the codomain of $f = \{w, x, y, z\}$ and the range of $f = f(A) = \{w, x\}$.

A pictorial representation of these ideas appears in Fig. 5.4. This diagram suggests that a may be regarded as an *input* that is *transformed* by f into the corresponding *output*, $f(a)$. In this context, a FORTRAN compiler can be thought of as a function that transforms a source program (the input) into its corresponding object program (the output).

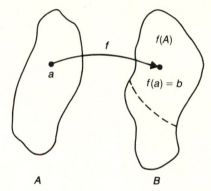

A B **Figure 5.4**

Example 5.9 Many interesting functions arise in computer science.

a) In the programming language BASIC, the greatest integer function, denoted INT, is an integer-valued function defined for the domain of all real numbers. Hence we write INT: $\mathbf{R} \rightarrow \mathbf{Z}$ where INT(X) = X, if X is an integer; INT(X) = the integer to the immediate left of X on the real number line, if X is not an integer. For example, INT(5.1) = 5 = INT(5), and INT(-7.8) = -8 = INT(-8). Here the codomain and the range are the set $\mathbf{Z}$.

b) The function trunc (for truncation) arises in Pascal and is an integer-valued function defined on $\mathbf{R}$. The function deletes the fractional part of a real number. For example, trunc $(3.78) = 3$, trunc $(5) = 5$, trunc $(-7.22) = -7$.

c) In storing a matrix in a one-dimensional array, many computer languages use the *row major* implementation. Here, if $A = (a_{ij})_{n \times n}$ is an $n \times n$ matrix, the first row of A is stored in locations $1, 2, 3, \ldots, n$ of the array if we start with a_{11} in location 1. The entry a_{21} is then found in position $n + 1$, while entry a_{34} occupies position $2n + 4$ in the array. In order to determine the location of any entry $a_{ij}, 1 \leq i, j \leq n$, from A, one defines the *access function f* from the entries of A to the positions $1, 2, 3, \ldots, n^2$ of the array. A formula for the access function here is $f(a_{ij}) = (i - 1)n + j$. $\square$

In Example 5.8 there are $2^{12} = 4096$ relations from A to B. We have examined one function among these relations, and now we wish to count the total number of functions from A to B.

> For the general case, let A, B be sets with $|A| = m$, $|B| = n$. Consequently, if $A = \{a_1, a_2, \ldots, a_m\}$, $B = \{b_1, b_2, \ldots, b_n\}$, a typical function $f: A \rightarrow B$ can be described by $\{(a_1, x_1), (a_2, x_2), \ldots, (a_m, x_m)\}$. We can select any of the n elements of B for x_1, and then do the same for x_2. (We can select any element of B for x_2 so that the same element of B may be selected for both x_1 and x_2.) We continue this selection process until one of the n elements of B is finally selected for x_m. In this way, using the rule of product, there are $n^m = |B|^{|A|}$ functions from A to B.

Therefore for A, B in Example 5.8, there are $4^3 = |B|^{|A|} = 64$ functions from A to B, and $3^4 = |A|^{|B|} = 81$ functions from B to A.

Now that we have the concept of a function as a special type of relation, we turn our attention to a special type of function.

DEFINITION 5.5   A function $f: A \rightarrow B$ is called *one-to-one*, or *injective*, if each element of B appears at most once as the image of an element of A.   ━━━

If $f: A \rightarrow B$ is one-to-one, with A, B finite, we must have $|A| \leq |B|$. For general sets A, B, if $f: A \rightarrow B$ is one-to-one, then for $a_1, a_2 \in A$, $f(a_1) = f(a_2) \Rightarrow a_1 = a_2$. Consequently, a function of this type can be characterized as one for which each element of B appears at most once as the second component of an ordered pair in f.

Example 5.10   Let $A = \{1, 2, 3\}$ and $B = \{1, 2, 3, 4, 5\}$. The function

$$f = \{(1, 1), (2, 3), (3, 4)\}$$

is a one-to-one function from A to B;

$$g = \{(1, 1), (2, 3), (3, 3)\}$$

is a function from A to B, but it fails to be one-to-one because $g(2) = g(3)$ but $2 \neq 3$. □

For A, B in Example 5.10 there are 2^{15} relations from A to B and 5^3 of these are functions. The next question we naturally want to answer is how many functions $f: A \rightarrow B$ are one-to-one. Again we argue for general finite sets.

> With $A = \{a_1, a_2, a_3, \ldots, a_m\}$, $B = \{b_1, b_2, b_3, \ldots, b_n\}$, $m \leq n$, a one-to-one function $f: A \rightarrow B$ has the form $\{(a_1, x_1), (a_2, x_2), (a_3, x_3), \ldots, (a_m, x_m)\}$, where there are n choices for x_1 (that is, any element of B), $n - 1$ choices for x_2 (that is, any element of B except the one chosen for x_1), $n - 2$ choices for x_3, and so on, finishing with $n - (m - 1) = n - m + 1$ choices for x_m. By the rule of product, the number of one-to-one functions from A to B is
>
> $$n(n - 1)(n - 2) \cdots (n - m + 1) = n!/(n - m)! = P(n, m) = P(|B|, |A|).$$

Consequently, for A, B in Example 5.10, there are $5 \cdot 4 \cdot 3 = 60$ one-to-one functions $f: A \rightarrow B$.

DEFINITION 5.6   If $f: A \rightarrow B$ and $A_1 \subseteq A$, then

$$f(A_1) = \{b \in B \mid b = f(a) \quad \text{for some } a \in A_1\},$$

and $f(A_1)$ is called the *image of A_1 under f*.

Example 5.11   Let $f: \mathbf{R} \rightarrow \mathbf{R}$ be given by $f(x) = x^2$. Then $f(\mathbf{R}) =$ the range of $f = [0, +\infty)$. The image of $\mathbf{Z}$ under f is $f(\mathbf{Z}) = \{0, 1, 4, 9, 16, \ldots\}$.   □

THEOREM 5.2   Let $f: A \rightarrow B$, with $A_1, A_2 \subseteq A$. Then

a) $f(A_1 \cup A_2) = f(A_1) \cup f(A_2)$.   b) $f(A_1 \cap A_2) \subseteq f(A_1) \cap f(A_2)$.
c) $f(A_1 \cap A_2) = f(A_1) \cap f(A_2)$ when f is injective.

Proof   We prove (b) and leave the remaining parts for the reader.
For any $b \in B$, $b \in f(A_1 \cap A_2) \Rightarrow b = f(a)$, for some $a \in A_1 \cap A_2 \Rightarrow (b = f(a)$ for some $a \in A_1)$ and $(b = f(a)$ for some (the same) $a \in A_2) \Rightarrow b \in f(A_1)$ and $b \in f(A_2) \Rightarrow b \in f(A_1) \cap f(A_2)$, so $f(A_1 \cap A_2) \subseteq f(A_1) \cap f(A_2)$.   ∎

DEFINITION 5.7   If $f: A \rightarrow B$ and $A_1 \subseteq A$, then $f|_{A_1}: A_1 \rightarrow B$ is called the *restriction of f to A_1* if $f|_{A_1}(a) = f(a)$ for all $a \in A_1$.

DEFINITION 5.8   Let $A_1 \subseteq A$ and $f: A_1 \rightarrow B$. If $g: A \rightarrow B$ and $g(a) = f(a)$ for all $a \in A_1$, then we call g an *extension of f to A*.

Example 5.12   Let $A = \{w, x, y, z\}$, $B = \{1, 2, 3, 4, 5\}$ and $A_1 = \{w, y, z\}$. Let $f: A \rightarrow B, g: A_1 \rightarrow B$ be represented by the diagrams in Fig. 5.5. Then $g = f|_{A_1}$ and f is an extension of g *from* A_1 to A. We note that for the given function $g: A_1 \rightarrow B$, there are five ways to extend g from A_1 to A.   □

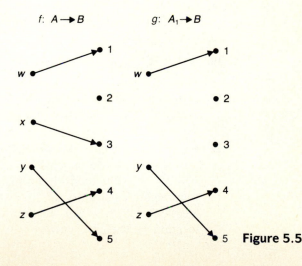

Figure 5.5

EXERCISES 5.2

1. Determine whether or not each of the following relations is a function. If a relation is a function, find its range.
   a) $\{(x,y)\,|\,x,y \in \mathbf{Z}, y = x^2 + 7\}$.
   b) $\{(x,y)\,|\,x,y \in \mathbf{R}, y^2 = x\}$.
   c) $\{(x,y)\,|\,x,y \in \mathbf{R}, y = 3x + 1\}$.
   d) $\{(x,y)\,|\,x,y \in \mathbf{Q}, x^2 + y^2 = 1\}$.
   e) $\mathscr{R}$ is a relation from A to B where $|A| = 5$, $|B| = 6$, and $|\mathscr{R}| = 6$.

2. Can the formula $f(x) = 1/(x^2 - 2)$ be used to define a function $f\colon \mathbf{R} \to \mathbf{R}$? A function $f\colon \mathbf{Z} \to \mathbf{R}$?

3. Let $A = \{1, 2, 3, 4\}$ and $B = \{x, y, z\}$. (a) List five functions from A to B. (b) How many functions $f\colon A \to B$ are there? (c) How many functions $f\colon A \to B$ are one-to-one? (d) How many functions $g\colon B \to A$ are there? (e) How many functions $g\colon B \to A$ are one-to-one? (f) How many functions $f\colon A \to B$ satisfy $f(1) = x$? (g) How many functions $f\colon A \to B$ satisfy $f(1) = f(2) = x$? (h) How many functions $f\colon A \to B$ satisfy $f(1) = x$ and $f(2) = y$?

4. If there are 2187 functions $f\colon A \to B$ and $|B| = 3$, what is $|A|$?

5. For each of the following functions, determine whether the function is one-to-one and determine its range.
   a) $f\colon \mathbf{Z} \to \mathbf{Z}, f(x) = 2x + 1$
   b) $f\colon \mathbf{Q} \to \mathbf{Q}, f(x) = 2x + 1$
   c) $f\colon \mathbf{Z} \to \mathbf{Z}, f(x) = x^3 - x$
   d) $f\colon \mathbf{R} \to \mathbf{R}, f(x) = e^x$
   e) $f\colon [-\pi/2, \pi/2] \to \mathbf{R}, f(x) = \sin x$
   f) $f\colon [0, \pi] \to \mathbf{R}, f(x) = \sin x$

6. Let $f\colon \mathbf{R} \to \mathbf{R}$ where $f(x) = x^2$. Determine $f(A)$ for the following subsets $A \subseteq \mathbf{R}$.
   a) $A = \{2, 3\}$
   b) $A = \{-3, -2, 2, 3\}$
   c) $A = (-3, 3)$
   d) $A = (-3, 2]$
   e) $A = [-7, 2]$
   f) $A = (-4, -3] \cup [5, 6]$

7. Let $A = \{1, 2, 3, 4, 5\}$, $B = \{w, x, y, z\}$, $A_1 = \{2, 3, 5\} \subseteq A$ and $g\colon A_1 \to B$. In how many ways can g be extended to a function $f\colon A \to B$?

8. Give an example of a function $f\colon A \to B$, with $A_1, A_2 \subseteq A$ and $f(A_1 \cap A_2) \neq f(A_1) \cap f(A_2)$. (Thus the inclusion in Theorem 5.2(b) may be proper.)

9. Prove parts (a) and (c) of Theorem 5.2.

10. If $A = \{1, 2, 3, 4, 5\}$ and there are 6720 injective functions $f\colon A \to B$, what is $|B|$?

11. Determine the access function $f(a_{ij})$, as described in Example 5.9(c), for a matrix $A = (a_{ij})_{m \times n}$, where (a) $m = 12, n = 12$; (b) $m = 7, n = 10$; (c) $m = 10, n = 7$.

12. For the access function developed in Example 5.9(c), the matrix $A = (a_{ij})_{m \times n}$ was stored in a one-dimensional array using the row major implementation. It is also possible to store this matrix using the column major implementation, where each entry $a_{i1}, 1 \le i \le m$, in the first column of A is stored in locations $1, 2, 3, \ldots, m$, respectively, of the array, when a_{11} is stored in location 1. Then the entries $a_{i2}, 1 \le i \le m$, of the second column of A are

stored in locations $m + 1, m + 2, m + 3, \ldots, 2m$, respectively, of the array, and so on. Find a formula for the access function $g(a_{ij})$ under these conditions.

13. **a)** Let A be an $m \times n$ matrix that is to be stored (in a contiguous manner) in a one-dimensional array of r entries. Find a formula for the access function if a_{11} is to be stored in location $k(\geq 1)$ of the array [as opposed to location 1 as in Example 5.9(c)] and we use (i) the row major implementation; (ii) the column major implementation.

b) State any conditions involving m, n, r, and k that must be satisfied in order for the results in part (a) to be valid.

5.3
ONTO FUNCTIONS: STIRLING NUMBERS OF THE SECOND KIND

The results we develop in this section will provide the answers to the first five problems stated at the beginning of this chapter. We find that the *onto* function is the key to all of the answers.

DEFINITION 5.9   A function $f: A \to B$ is called *onto*, or *surjective*, if $f(A) = B$ (that is, if for all $b \in B$ there is at least one $a \in A$ with $f(a) = b$).

Example 5.13   The function $f: \mathbf{R} \to \mathbf{R}$ defined by $f(x) = x^3$ is an onto function, but the function $g: \mathbf{R} \to \mathbf{R}$ given by $g(x) = x^2$ is not an onto function since $g(\mathbf{R}) = [0, +\infty) \subset \mathbf{R}$.   □

Example 5.14   If $A = \{1, 2, 3, 4\}$ and $B = \{x, y, z\}$, then

$$f_1 = \{(1, z), (2, y), (3, x), (4, y)\} \quad \text{and} \quad f_2 = \{(1, x), (2, x), (3, y), (4, z)\}$$

are both functions from A onto B. However, the function $g = \{(1, x), (2, x), (3, y), (4, y)\}$ is not onto, because $g(A) = \{x, y\} \subset B$.   □

If A, B are finite sets, then for any onto function $f: A \to B$ to possibly exist we must have $|A| \geq |B|$. Considering the development in the first two sections of this chapter, the reader undoubtedly feels it is time once again to use the rule of product and count the number of onto functions $f: A \to B$ where $|A| = m \geq n = |B|$. Unfortunately, the rule of product proves inadequate here. We shall obtain the needed result for some specific examples and then conjecture a general formula. In Chapter 8 we shall establish the conjecture using the Principle of Inclusion and Exclusion.

Example 5.15   If $A = \{x, y, z\}$ and $B = \{1, 2\}$, then all functions $f: A \to B$ are onto except $f_1 = \{(x, 1), (y, 1), (z, 1)\}$ and $f_2 = \{(x, 2), (y, 2), (z, 2)\}$, the *constant* functions. So there are $|B|^{|A|} - 2 = 2^3 - 2 = 6$ onto functions from A to B.   □

In general, if $|A| = m \geq 2$ and $|B| = 2$, then there are $2^m - 2$ onto functions from A to B. (Does this formula tell us anything when $m = 1$?)

Example 5.16

For $A = \{w, x, y, z\}$ and $B = \{1, 2, 3\}$, there are 3^4 functions from A to B. Considering subsets of B of size 2, there are 2^4 functions from A to $\{1, 2\}$, 2^4 functions from A to $\{2, 3\}$, and 2^4 functions from A to $\{1, 3\}$. So we have $3(2^4) = \binom{3}{2}2^4$ functions from A to B that are definitely not onto. However, before we acknowledge $3^4 - \binom{3}{2}2^4$ as the final answer, we must realize that not all of these $\binom{3}{2}2^4$ functions are distinct. For when we consider all the functions from A to $\{1, 2\}$ we are removing, among these, the function $\{(w, 2), (x, 2), (y, 2), (z, 2)\}$. Then, considering the functions from A to $\{2, 3\}$, we remove the same function: $\{(w, 2), (x, 2), (y, 2), (z, 2)\}$. Consequently, in the result $3^4 - \binom{3}{2}2^4$, we have twice removed each of the constant functions $f: A \to B$, where $f(A)$ is one of the sets $\{1\}$, $\{2\}$, or $\{3\}$. Adjusting our present result for this, we find that there are $3^4 - \binom{3}{2}2^4 + 3 = \binom{3}{3}3^4 - \binom{3}{2}2^4 + \binom{3}{1}1^4 = 36$ onto functions from A to B.

Keeping $B = \{1, 2, 3\}$, for any set A with $|A| = m \geq 3$, there are $\binom{3}{3}3^m - \binom{3}{2}2^m + \binom{3}{1}1^m$ functions from A onto B. (What result does this formula yield when $m = 1$? when $m = 2$?) $\quad\square$

The last two examples suggest a pattern that we now state, without proof, as our general formula.

For finite sets A, B with $|A| = m \geq n = |B|$, there are

$$\binom{n}{n}n^m - \binom{n}{n-1}(n-1)^m + \binom{n}{n-2}(n-2)^m - \cdots$$

$$+ (-1)^{n-2}\binom{n}{2}2^m + (-1)^{n-1}\binom{n}{1}1^m = \sum_{k=0}^{n-1}(-1)^k\binom{n}{n-k}(n-k)^m$$

$$= \sum_{k=0}^{n}(-1)^k\binom{n}{n-k}(n-k)^m$$

onto functions from A to B.

Example 5.17

Let $A = \{1, 2, 3, 4, 5, 6, 7\}$ and $B = \{w, x, y, z\}$. Applying the general formula with $m = 7$ and $n = 4$, we find that there are

$$\binom{4}{4}4^7 - \binom{4}{3}3^7 + \binom{4}{2}2^7 - \binom{4}{1}1^7 = \sum_{k=0}^{3}(-1)^k\binom{4}{4-k}(4-k)^7$$

$$= \sum_{k=0}^{4}(-1)^k\binom{4}{4-k}(4-k)^7 = 8400 \text{ functions from } A \text{ onto } B. \quad\square$$

The result in Example 5.17 is also the answer to the first three questions proposed at the start of this chapter. Once we remove the unnecessary vocabulary, we recognize that in all three cases we want to distribute seven different objects into four distinct containers with no container left empty. We can do this in terms of onto functions.

For Problem 4 we have a sample space $\mathcal{S}$ consisting of the $4^7 = 16,384$ ways in which seven people can each select one of the four floors. (Note that 4^7 is also the total number of functions $f: A \to B$ where $|A| = 7, |B| = 4$.) The event that we are concerned with contains 8400 of those selections, so the probability that the elevator must stop at every floor is $8400/16384 \doteq 0.5127$, slightly more than half of the time.

Finally, for Problem 5, since $\sum_{k=0}^{n} (-1)^k \binom{n}{n-k}(n-k)^m$ is the number of onto functions $f: A \to B$ for $|A| = m, |B| = n$, when $m < n$ there are no such functions and the summation is 0.

Problem 6 will be addressed in Section 5.6.

Before going on to anything new, however, we consider one more problem.

Example 5.18   At the CH Company, Joan, the supervisor, has a secretary, Teresa, and three other administrative assistants. If there are seven accounts to be processed, in how many ways can Joan assign the accounts so that each assistant works on at least one account and Teresa's work includes, if nothing else, the most expensive account?

First and foremost, the answer is not 8400 as in Example 5.17. Here we must consider two disjoint subcases and then apply the rule of sum.

a) If Teresa, the secretary, works only on the most expensive account, then the other six accounts can be distributed among the three administrative assistants in $\sum_{k=0}^{3} (-1)^k \binom{3}{3-k}(3 - k)^6 = 540$ ways. (540 = the number of onto functions $f: A \to B$ with $|A| = 6, |B| = 3$.)

b) If Teresa does more than just the most expensive account, the assignments can be made in $\sum_{k=0}^{4} (-1)^k \binom{4}{4-k}(4 - k)^6 = 1560$ ways. (1560 = the number of onto functions $g: C \to D$ with $|C| = 6, |D| = 4$.)

Consequently, the assignments can be given under the prescribed conditions in $540 + 1560 = 2100$ ways. (We mentioned earlier that the answer would not be 8400, but it is $(1/4)(8400) = (1/|B|)(8400)$, where 8400 is the number of onto functions $f: A \to B$, with $|A| = 7$ and $|B| = 4$. This is no coincidence, as we shall learn when we discuss Theorem 5.3.) $\square$

We now continue our discussion with the distribution of distinct objects into containers with none left empty, but now the containers become identical.

Example 5.19   If $A = \{a, b, c, d\}$ and $B = \{1, 2, 3\}$, then there are 36 onto functions from A to B or, equivalently, 36 ways to distribute four distinct objects into three distinguishable containers, with no container empty (and no regard for the location of objects in a given container). Among these 36 distributions we find the following collection of six (one of six such possible collections of six):

1. $\{a, b\}_1$ $\{c\}_2$ $\{d\}_3$ 2. $\{a, b\}_1$ $\{d\}_2$ $\{c\}_3$
3. $\{c\}_1$ $\{a, b\}_2$ $\{d\}_3$ 4. $\{c\}_1$ $\{d\}_2$ $\{a, b\}_3$
5. $\{d\}_1$ $\{a, b\}_2$ $\{c\}_3$ 6. $\{d\}_1$ $\{c\}_2$ $\{a, b\}_3,$

where, for example, the notation $\{c\}_2$ means that c is in the second container. Now if we no longer distinguish the containers, these $6 = 3!$ distributions become identical, so there are $36/(3!) = 6$ ways to distribute the distinct objects a, b, c, d among three identical containers, leaving no container empty. $\square$

Generally, for $m \geq n$ there are $\sum_{k=0}^{n}(-1)^k\binom{n}{n-k}(n-k)^m$ ways to distribute m different objects into n numbered (but otherwise identical) containers with no container left empty. Removing the numbers on the containers, so that they are now identical in appearance, we find that one distribution into these n (nonempty) identical containers corresponds with $n!$ such distributions into the numbered containers. So the number of ways in which it is possible to distribute the m distinct objects into n identical containers, with no container left empty, is

$$\frac{1}{n!}\sum_{k=0}^{n}(-1)^k\binom{n}{n-k}(n-k)^m.$$

This will be denoted by $S(m, n)$ and is called a *Stirling number of the second kind*.

We note that for $|A| = m \geq n = |B|$, there are $n! \cdot S(m, n)$ onto functions from A to B.

Table 5.1 lists some Stirling numbers of the second kind.

Table 5.1

m \ n	1	2	3	4	5	6	7	8
1	1							
2	1	1						
3	1	3	1					
4	1	7	6	1				
5	1	15	25	10	1			
6	1	31	90	65	15	1		
7	1	63	301	350	140	21	1	
8	1	127	966	1701	1050	266	28	1

Example 5.20   For $m \geq n$, $\sum_{i=1}^{n} S(m, i)$ is the number of ways in which it is possible to distribute m distinct objects into n identical containers with empty containers allowed. From the fourth row of Table 5.1 we see that there are $1 + 7 + 6 = 14$ ways to distribute the objects a, b, c, d among three identical containers, with some container(s) possibly empty. $\square$

We close this section with the derivation of an identity involving Stirling numbers. The proof is combinatorial in nature.

THEOREM 5.3   Let m, n be positive integers with $1 < n \leq m$. Then

$$S(m + 1, n) = S(m, n - 1) + nS(m, n).$$

Proof Let $A = \{a_1, a_2, \ldots, a_m, a_{m+1}\}$. Then $S(m+1, n)$ counts the number of ways in which the objects of A can be distributed among n identical containers, with no container left empty.

There are $S(m, n-1)$ ways of distributing $a_1, a_2, \ldots, a_m$ among $n-1$ identical containers, with none left empty. Then, placing a_{m+1} in the remaining empty container results in $S(m, n-1)$ of the distributions counted in $S(m+1, n)$: namely, those distributions where a_{m+1} is in a container by itself. Alternatively, distributing $a_1, a_2, \ldots, a_m$ among the n identical containers with none left empty, we have $S(m, n)$ distributions. Now, however, for each of these $S(m, n)$ distributions the n containers become distinguished by their contents. Selecting one of the n distinct containers for a_{m+1}, we have $nS(m, n)$ distributions of the total $S(m+1, n)$: namely, those where a_{m+1} is in the same container as another object from A. The result then follows by the rule of sum. ∎

To illustrate Theorem 5.3 consider the triangle shown in the table. Here the largest number corresponds with $S(m+1, n)$, for $m = 7$ and $n = 3$, and we see that $S(7+1, 3) = 966 = 63 + 3(301) = S(7, 2) + 3S(7, 3)$.

The identity in Theorem 5.3 can be used to extend Table 5.1 if necessary.

If we multiply the result in Theorem 5.3 by $(n-1)!$ we have

$$(1/n)[n!S(m+1, n)] = [(n-1)!S(m, n-1)] + [n!S(m, n)].$$

This new form of the equation relates ideas on numbers of onto functions. For if $A = \{a_1, a_2, \ldots, a_m, a_{m+1}\}$ and $B = \{b_1, b_2, \ldots, b_{n-1}, b_n\}$ with $m \geq n - 1$, then

$(1/n)$(The number of onto functions $h: A \to B$)

$\qquad = $ (The number of onto functions $f: A - \{a_{m+1}\} \to B - \{b_n\}$)
$\qquad + $ (The number of onto functions $g: A - \{a_{m+1}\} \to B$).

This is why the relationship we noted at the end of Example 5.18 is not just a coincidence.

EXERCISES 5.3

1. Give an example of finite sets A and B with $|A|, |B| \geq 4$ and a function $f: A \to B$ such that (a) f is neither one-to-one nor onto; (b) f is one-to-one but not onto; (c) f is onto but not one-to-one; (d) f is onto and one-to-one.

2. For each of the following functions $f: \mathbf{Z} \to \mathbf{Z}$, determine whether the function is one-to-one and whether it is onto. If the function is not onto, determine the range $f(\mathbf{Z})$.

   a) $f(x) = x + 7$ b) $f(x) = 2x - 3$ c) $f(x) = -x + 5$
   d) $f(x) = x^2$ e) $f(x) = x^2 + x$ f) $f(x) = x^3$

3. For each of the following functions $g: \mathbf{R} \to \mathbf{R}$, determine whether the function is one-to-one and whether it is onto. If the function is not onto, determine the range $g(\mathbf{R})$.

   a) $g(x) = x + 7$ b) $g(x) = 2x - 3$ c) $g(x) = -x + 5$
   d) $g(x) = x^2$ e) $g(x) = x^2 + x$ f) $g(x) = x^3$

4. Let $A = \{1, 2, 3, 4\}$ and $B = \{1, 2, 3, 4, 5, 6\}$.

   a) How many functions are there from A to B? How many of these are one-to-one? How many are onto?

   b) How many functions are there from B to A? How many of these are onto? How many are one-to-one?

5. Verify that $\sum_{k=0}^{n}(-1)^k \binom{n}{n-k}(n-k)^m = 0$ for $n = 5$ and $m = 2, 3, 4$.

6. A research chemist who has five laboratory assistants is engaged in a research project that calls for nine compounds that must be synthesized. In how many ways can the chemist assign these syntheses to the five assistants so that each of them is working on at least one synthesis?

7. Use the fact that every polynomial equation having real-number coefficients and odd degree has a real root in order to show that the function $f: \mathbf{R} \to \mathbf{R}$, defined by $f(x) = x^5 - 2x^2 + x$, is an onto function. Is f one-to-one?

8. Suppose we have seven different colored balls and four containers numbered I, II, III, and IV. (a) In how many ways can we distribute the balls so that no container is left empty? (b) In this collection of seven colored balls, one of them is blue. In how many ways can we distribute the balls so that no container is empty and the blue ball is in container II? (c) If we remove the numbers from the containers so that we can no longer distinguish them, in how many ways can we distribute the seven colored balls among the four identical containers, with some container(s) possibly empty?

9. Determine the next two rows ($m = 9, 10$) of Table 5.1 for the Stirling numbers $S(m, n), 1 \leq n \leq m$.

10. Write a computer program (or develop an algorithm) to compute the Stirling numbers $S(m, n)$ when $1 \leq m \leq 12$ and $1 \leq n \leq m$.

11. Let $A = \{1, 2, 3, 4, 5, 6\}$ and $B = \{1, 2, 3, 4\}$. Write a computer program (or develop an algorithm) to randomly generate a function from A to B and have it print out whether or not the generated function is onto.

5.4

SPECIAL FUNCTIONS

In Section 2 of Chapter 3 we mentioned that addition is a binary operation on the set $\mathbf{Z}^+$, whereas $\cap$ is a binary operation on $\mathcal{P}(\mathcal{U})$ for any given universe $\mathcal{U}$. We also noted in that section that "taking the minus" of an integer is a unary operation on $\mathbf{Z}$. Now it is time to make these notions of binary and unary operations more precise in terms of functions.

DEFINITION 5.10   For any set A, any function $f: A \times A \to A$ is called a *binary operation* on A. ▬

DEFINITION 5.11   A function $g: A \to A$ is called a *unary*, or *monary*, operation on A. ▬

Example 5.21   (a) The function $f: \mathbf{Z} \times \mathbf{Z} \to \mathbf{Z}$, defined by $f(a, b) = a - b$, is a binary operation on $\mathbf{Z}$. (b) If $g: \mathbf{R}^+ \to \mathbf{R}^+$ is the function where $g(a) = 1/a$, then g is a unary operation on $\mathbf{R}^+$.   □

Example 5.22   Let $\mathcal{U}$ be a universe and let $A, B \subseteq \mathcal{U}$. (a) If $f: \mathcal{P}(\mathcal{U}) \times \mathcal{P}(\mathcal{U}) \to \mathcal{P}(\mathcal{U})$ is defined by $f(A, B) = A \cup B$, then f is a binary operation on $\mathcal{P}(\mathcal{U})$. (b) The function $g: \mathcal{P}(\mathcal{U}) \to \mathcal{P}(\mathcal{U})$ defined by $g(A) = \overline{A}$ is a unary operation on $\mathcal{P}(\mathcal{U})$.   □

DEFINITION 5.12   Let $f: A \times A \to A$; that is, f is a binary operation on A.

     a) f is said to be *commutative* if $f(a, b) = f(b, a)$ for all $(a, b) \in A \times A$.

     b) f is said to be *associative* if for $a, b, c \in A$, $f(f(a, b), c) = f(a, f(b, c))$.   ▬

Example 5.23   The binary operation of Example 5.22 is commutative and associative, whereas that in Example 5.21 is neither.   □

Example 5.24   **a)** Define the binary operation $f: \mathbf{Z} \times \mathbf{Z} \to \mathbf{Z}$ by $f(a, b) = a + b - 3ab$. Since both the addition and the multiplication of integers are commutative binary operations, it follows that

$$f(a, b) = a + b - 3ab = b + a - 3ba = f(b, a),$$

so f is commutative.

To determine whether f is associative, consider $a, b, c \in \mathbf{Z}$. Then

$$f(a, b) = a + b - 3ab, \quad \text{and} \quad f(f(a, b), c) = f(a, b) + c - 3f(a, b)c$$
$$= (a + b - 3ab) + c - 3(a + b - 3ab)c$$
$$= a + b + c - 3ab - 3ac - 3bc + 9abc,$$

whereas

$$f(b, c) = b + c - 3bc, \quad \text{and}$$
$$f(a, f(b, c)) = a + f(b, c) - 3af(b, c)$$
$$= a + (b + c - 3bc) - 3a(b + c - 3bc)$$
$$= a + b + c - 3ab - 3ac - 3bc + 9abc.$$

Since $f(f(a, b), c) = f(a, f(b, c))$ for all $a, b, c \in \mathbf{Z}$, the binary operation f is associative as well as commutative.

     b) Consider the binary operation $h: \mathbf{Z} \times \mathbf{Z} \to \mathbf{Z}$, where $h(a, b) = a|b|$. Then $h(3, -2) = 3|-2| = 3(2) = 6$, but $h(-2, 3) = -2|3| = -6$. Consequently, h is *not* commutative. However, with regard to the associative property, if $a, b, c \in \mathbf{Z}$, we find that

$$h(h(a, b), c) = h(a, b)|c| = a|b||c| \quad \text{and}$$
$$h(a, h(b, c)) = a|h(b, c)| = a|b|c|| = a|b||c|,$$

so the binary operation h is associative.   □

Example 5.25   If $A = \{a, b, c, d\}$, then $|A \times A| = 16$. Consequently, there are 4^{16} functions $f: A \times A \to A$; that is, there are 4^{16} binary operations on A.

To determine the number of commutative binary operations g on A, we realize that there are four choices for each of the assignments $g(a, a)$, $g(b, b)$, $g(c, c)$, and $g(d, d)$. This then leaves us with the $4^2 - 4 = 16 - 4 = 12$ other ordered pairs (in $A \times A$) of the form $(x, y), x \neq y$. These 12 ordered pairs must be considered in sets of two in order to insure commutativity. For example, we need $g(a, b) = g(b, a)$, and may select any one of the four elements of A for $g(a, b)$. But then this choice must also be assigned to $g(b, a)$. Therefore, since there are four choices for each of these $12/2 = 6$ sets of two ordered pairs, we find that the number of commutative binary operations g on A is $4^4 \cdot 4^6 = 4^{10}$.   □

Having seen several examples of functions (in Examples 5.21, 5.22 and 5.24) where the domain is a cross product of sets, we now investigate functions where the domain is a subset of a cross product.

DEFINITION 5.13   For sets A and B, if $D \subseteq A \times B$, then $\pi_A: D \to A$, defined by $\pi_A(a, b) = a$, is called the *projection* on the first coordinate.

The function π_B is defined similarly, and we note that if $D = A \times B$, both π_A and π_B are onto.

Example 5.26   If $A = \{w, x, y\}$ and $B = \{1, 2, 3, 4\}$, let $D = \{(x, 1), (x, 2), (x, 3), (y, 1), (y, 4)\}$. Then the projection $\pi_A: D \to A$ satisfies $\pi_A(x, 1) = \pi_A(x, 2) = \pi_A(x, 3) = x$, and $\pi_A(y, 1) = \pi_A(y, 4) = y$. Since $\pi_A(D) = \{x, y\} \subset A$, this function is *not* onto.

For $\pi_B: D \to B$ we find that $\pi_B(x, 1) = \pi_B(y, 1) = 1$, $\pi_B(x, 2) = 2$, $\pi_B(x, 3) = 3$, and $\pi_B(y, 4) = 4$, so $\pi_B(D) = B$ and this projection is an onto function.   □

Example 5.27   Let $A = B = \mathbf{R}$ and consider the set $D \subseteq A \times B$ where $D = \{(x, y) | y = x^2\}$. Then D represents the subset of the Euclidean plane that contains the points on the parabola $y = x^2$.

Among the infinite number of points in D we find the point $(3, 9)$. Here $\pi_A(3, 9) = 3$, the x coordinate of $(3, 9)$, whereas $\pi_B(3, 9) = 9$, the y coordinate of the point.

For this example, $\pi_A(D) = \mathbf{R} = A$, so π_A is onto. (The projection π_A is also one-to-one.) However, $\pi_B(D) = [0, +\infty) \subset \mathbf{R}$, so π_B is *not* onto. (Nor is it one-to-one—for example, $\pi_B(2, 4) = 4 = \pi_B(-2, 4)$.)   □

We now extend the notion of projection as follows. Let $A_1, A_2, \ldots, A_n$ be sets, and $\{i_1, i_2, \ldots, i_m\} \subseteq \{1, 2, \ldots, n\}$ with $i_1 < i_2 < \cdots < i_m$ and $m \leq n$. If $D \subseteq A_1 \times A_2 \times \cdots \times A_n = \times_{i=1}^{n} A_i$, then the function $\pi: D \to A_{i_1} \times A_{i_2} \times \cdots \times A_{i_m}$ defined by $\pi(a_1, a_2, \ldots, a_n) = (a_{i_1}, a_{i_2}, \ldots, a_{i_m})$ is the projection of D on the i_1th, i_2th, ..., i_mth coordinates. The elements of D are called *n-tuples*; an element in $\pi(D)$ is an *m-tuple*.

These projections arise in a natural way in the study of *relational data bases*, a standard technique for organizing and describing large quantities of data by modern large-scale computing systems. In situations like credit card transactions, not only must existing data be organized but new data must be inserted, as when credit cards are processed for new cardholders. When bills on existing accounts are paid, or when new purchases are made on these accounts, data must be updated. Another example arises when records are searched for special considerations, as when a college admissions office searches educational records seeking, for its mailing lists, high school students who have demonstrated certain levels of mathematical achievement.

The following example demonstrates the use of projections in a method for organizing and describing data on a somewhat smaller scale.

Example 5.28   At a certain university the following sets are related for purposes of registration:

A_1 = the set of course numbers for courses offered in mathematics.

A_2 = the set of course titles offered in mathematics.

A_3 = the set of mathematics faculty.

A_4 = the set of letters of the alphabet.

Consider the *table*, or relation, $D \subseteq A_1 \times A_2 \times A_3 \times A_4$ given in Table 5.2.

Table 5.2

Course Number	Course Title	Professor	Section Letter
MA 111	Calculus I	G. Sherman	A
MA 111	Calculus I	V. Larney	B
MA 112	Calculus II	J. Kinney	A
MA 112	Calculus II	A. Schmidt	B
MA 112	Calculus II	R. Mines	C
MA 113	Calculus III	J. Kinney	A
MA 113	Calculus III	A. Schmidt	B

The sets A_1, A_2, A_3, A_4 are called the *domains of the relational data base*, and *table D* is said to have *degree* 4. Each element of D is often called a *list*.

The projection of D on $A_1 \times A_3 \times A_4$ is shown in Table 5.3. Table 5.4 shows the results for the projection of D on $A_1 \times A_2$.

Tables 5.3 and 5.4 are another way of representing the same data that appear in Table 5.2. Given Tables 5.3 and 5.4, one can recapture Table 5.2.   □

Table 5.3

Course Number	Professor	Section Letter
MA 111	G. Sherman	A
MA 111	V. Larney	B
MA 112	J. Kinney	A
MA 112	A. Schmidt	B
MA 112	R. Mines	C
MA 113	J. Kinney	A
MA 113	A. Schmidt	B

Table 5.4

Course Number	Course Title
MA 111	Calculus I
MA 112	Calculus II
MA 113	Calculus III

The theory of relational data bases is concerned with representing data in different ways and with the operations, such as projections, needed for such representations. The computer implementation of such techniques is also considered. More on this topic and additional aspects are mentioned in the exercises and chapter references.

EXERCISES 5.4

1. Each of the following functions $f\colon \mathbf{Z} \times \mathbf{Z} \to \mathbf{Z}$ is a binary operation on $\mathbf{Z}$. Determine in each case whether f is commutative and/or associative.

   a) $f(x, y) = x + y - xy$ b) $f(x, y) = \max\{x, y\}$

   c) $f(x, y) = x^y$ d) $f(x, y) = x + y - 3$

2. Let $|A| = 5$. (a) What is $|A \times A|$? (b) How many functions $f\colon A \times A \to A$ are there? (c) How many binary operations are there on A? (d) How many of these binary operations are commutative?

3. Let $A = \{a_1, a_2, a_3, \ldots, a_n\}$. (a) How many binary operations $f\colon A \times A \to A$ are there with $f(a_1, a_2) = a_3$? (b) How many of the binary operations in part (a) are commutative?

4. For a set A and a binary operation $f\colon A \times A \to A$, an element $z \in A$ is called an *identity* for f if $f(a, z) = a = f(z, a)$, for all $a \in A$. Which binary operations in Exercise 1 have an identity?

5. If $f\colon A \times A \to A$ has an identity, it is unique. For suppose that $z_1, z_2 \in A$ are both identities. Then

$$z_1 = f(z_1, z_2) = z_2$$

↙   ↓

(Since z_2 is an identity   Why?

element for f.)

6. Let $A = \{z, a, b, c, d\}$. (a) How many binary operations $f\colon A \times A \to A$ have z as the identity element? (b) How many of these binary operations are commutative?

7. Let $A = B = \mathbf{R}$. Determine $\pi_A(D)$ and $\pi_B(D)$ for each of the following sets $D \subseteq A \times B$.

   a) $D = \{(x, y) \mid x = y^2\}$

   b) $D = \{(x, y) \mid y = \sin x\}$

   c) $D = \{(x, y) \mid x^2 + y^2 = 1\}$

8. a) Let $A = \{1, 2, 3, 4, 5, 6\}$, $B = \{w, x, y, z\}$, and $D \subseteq A \times B$. If π_A and π_B are both to be onto functions, what is the minimal value that is possible for $|D|$?

   b) For A, B as in part (a), how many such different minimal sets D result in both π_A and π_B being onto?

   c) Generalize the results of parts (a) and (b).

9. Let $A_i, 1 \leq i \leq 5$, be the domains for a table $D \subseteq \times_{i=1}^{5} A_i$, where $A_1 = \{U, V, W, X, Y, Z\}$ (used as code names for different cereals in a test), and $A_2 = A_3 = A_4 = A_5 = \mathbf{Z}^+$. The table D is given as Table 5.5.

Table 5.5

Code name of cereal	Grams of sugar per 1-oz serving	% of RDA[a] of vitamin A per 1-oz serving	% of RDA of vitamin C per 1-oz serving	% of RDA of protein per 1-oz serving
U	1	25	25	6
V	7	25	2	4
W	12	25	2	4
X	0	60	40	20
Y	3	25	40	10
Z	2	25	40	10

[a] RDA = recommended daily allowance

a) What is the degree of the table?

b) Find the projection of D on $A_3 \times A_4 \times A_5$.

c) A domain of a table is called a *primary key* for the table if its value uniquely identifies each list of D. Determine the primary key(s) for this table.

10. Let $A_i, 1 \leq i \leq 5$, be the domains for a table $D \subseteq \times_{i=1}^{5} A_i$, where $A_1 = \{1, 2\}$ (used to identify the daily vitamin capsule produced by two pharmaceutical companies), $A_2 = \{A, D, E\}$, and $A_3 = A_4 = A_5 = \mathbf{Z}^+$. The table D is given as Table 5.6.

Table 5.6

Vitamin capsule	Vitamin present in capsule	Amount of vitamin in capsule in IU[a]	Dosage: capsules/day	No. of capsules per bottle
1	A	10,000	1	100
1	D	400	1	100
1	E	30	1	100
2	A	4,000	1	250
2	D	400	1	250
2	E	15	1	250

[a] IU = international units

a) What is the degree of the table?

b) What is the projection of D on $A_1 \times A_2$? on $A_3 \times A_4 \times A_5$?

c) In this table there is no primary key. (See Exercise 9.) We can, however, define a *composite primary key* as the cross product of a *minimal* number of domains of the table, whose components, taken collectively, uniquely identify each list of D. Determine some composite primary keys for this table.

5.5

THE PIGEONHOLE PRINCIPLE

It is time now for a change of pace as we introduce an interesting distribution principle. This principle may seem to have nothing in common with what we have been doing so far, but it will prove to be helpful nonetheless.

In mathematics one sometimes finds that an almost obvious idea, when applied in a rather subtle manner, is the key needed to solve a troublesome problem. On the list of such obvious ideas many would undoubtedly place the following rule, which is known as the *pigeonhole principle*.

The Pigeonhole Principle: If m pigeons occupy n pigeonholes and $m > n$, then there is at least one pigeonhole with two or more pigeons roosting in it.

Now what can pigeons roosting in pigeonholes have to do with mathematics—discrete, combinatorial, or otherwise? It turns out that this principle can be applied in various problems where we seek to establish whether or not a certain situation can actually occur. We illustrate this in the following examples. We shall find the principle useful in Section 5.6 and at other points in the text.

Example 5.29   In an office there are 13 file clerks, so at least two of them must have birthdays during the same month. Here we have 13 pigeons (the file clerks) and 12 pigeonholes (the months of the year).   □

Here is a second rather immediate application of our principle.

Example 5.30   Larry returns from the laundromat with 12 pairs of socks (each pair a different color) in a laundry bag. Drawing the socks from the bag randomly, he'll have to draw at most 13 of them in order to get a matched pair.   □

From this point on, application of the pigeonhole principle will be more subtle.

Example 5.31   Wilma operates a computer with a magnetic tape drive. One day she is given a tape that contains 500,000 "words" of four or fewer lowercase letters. (Consecutive words on the tape are separated by a blank character.) Can it be that the 500,000 words are all distinct?

From the rules of sum and product, the total number of different possible words, using four or fewer letters, is

$$26^4 + 26^3 + 26^2 + 26 = 475,254.$$

With these 475,254 words as the pigeonholes, and the 500,000 words on the tape as the pigeons, it follows that there is at least one word on the tape that is repeated.   □

Example 5.32   Let $S \subset \mathbf{Z}^+$, where $|S| = 37$. Then S contains two elements that have the same remainder upon division by 36.

Here the pigeons are the 37 positive integers in S. We know from the Division Algorithm (of Theorem 4.5) that when any positive integer n is divided by 36, there exists a unique quotient q and unique remainder r, where

$$n = 36q + r, \qquad 0 \leq r < 36.$$

The 36 possible values of r constitute the pigeonholes, and the result is now established by the pigeonhole principle. □

Example 5.33   Prove that if 101 integers are selected from the set $S = \{1, 2, 3, \ldots, 200\}$, then there are two integers such that one divides the other.

For each $x \in S$, we may write $x = 2^k y$, with $k \geq 0$, and $(2, y) = 1$. (This follows from the Fundamental Theorem of Arithmetic.) Then $y \in T = \{1, 3, 5, \ldots, 199\}$, where $|T| = 100$. Since 101 integers are selected from S, by the pigeonhole principle there are two integers of the form $a = 2^m y, b = 2^n y$. If $m \leq n$, then $a \mid b$; otherwise $b \mid a$. □

Example 5.34   Any subset of size six from the set $S = \{1, 2, 3, \ldots, 9\}$ must contain two elements whose sum is 10.

Here the numbers $1, 2, 3, \ldots, 9$ are the pigeons, and the pigeonholes are the subsets $\{1, 9\}, \{2, 8\}, \{3, 7\}, \{4, 6\}, \{5\}$. When six pigeons go to their respective pigeonholes, they must fill at least one of the two-element subsets whose members sum to 10. □

Example 5.35   Triangle ACE is equilateral with $AC = 1$. If five points are selected from the interior of the triangle, there are at least two whose distance apart is less than $1/2$.

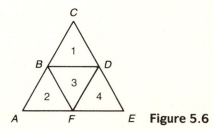

Figure 5.6

For the triangle in Fig. 5.6, the four smaller triangles are congruent equilateral triangles and $AB = 1/2$. We break up the interior of triangle ACE into the following four regions, which are mutually disjoint in pairs:

R_1:   the interior of triangle BCD together with the points on the segment BD, excluding B and D.

R_2:   the interior of triangle ABF.

R_3:   the interior of triangle BDF together with the points on the segments BF and DF, excluding B, D, and F.

R_4:   the interior of triangle FDE.

Now we apply the pigeonhole principle. Five points in the interior of triangle ACE must be such that at least two of them are in one of the four regions $R_i, 1 \leq i \leq 4$, where any two points are separated by a distance less than $1/2$.   □

Example 5.36   Let S be a set of six positive integers whose maximum is at most 14. Show that the sums of the elements in all the nonempty subsets of S cannot all be distinct.

For any nonempty subset A of S, the sum of the elements in A, denoted s_A, satisfies $1 \leq s_A \leq 9 + 10 + \cdots + 14 = 69$, and there are $2^6 - 1 = 63$ nonempty subsets of S. We should like to draw the conclusion from the pigeonhole principle by letting the possible sums, from 1 to 69, be the pigeonholes, with the 63 nonempty subsets of S as the pigeons, but then we have too few pigeons.

So instead of considering all nonempty subsets of S, we cut back to those nonempty subsets A where $|A| \leq 5$. Then $1 \leq s_A \leq 10 + 11 + \cdots + 14 = 60$. Since there are 62 nonempty subsets of S with $|A| \leq 5$, the elements of at least two of these subsets must yield the same sum.   □

Example 5.37   While on a four-week vacation, Herbert will play at least one set of tennis each day, but he won't play more than 40 sets total during this time. Prove that no matter how he distributes his sets during the four weeks, there is a consecutive span of days during which he plays exactly 15 sets.

For $1 \leq i \leq 28$, let x_i be the total number of sets Herbert has played from the start of the vacation to the end of the ith day. Then $1 \leq x_1 < x_2 < \cdots < x_{28} \leq 40$, and $x_1 + 15 < \cdots < x_{28} + 15 \leq 55$. We now have the 28 distinct numbers $x_1, x_2, \ldots, x_{28}$ and the 28 distinct numbers $x_1 + 15, x_2 + 15, \ldots, x_{28} + 15$. These 56 numbers can take on only 55 different values, so at least two of them must be equal, and we conclude that there exist $1 \leq j < i \leq 28$ with $x_i = x_j + 15$. Hence, from the start of day $j + 1$ to the end of day i, Herbert plays exactly 15 sets of tennis.   □

EXERCISES 5.5

1. In Example 5.30, what plays the roles of the pigeons and of the pigeonholes in our application of the pigeonhole principle?

2. Show that if there are eight people in a room, at least two of them have birthdays that occur on the same day of the week.

3. How many times must we roll a single die in order to get the same score (a) at least twice? (b) at least three times? (c) at least n times, for $n \geq 4$?

4. Given 8 books on Pascal, 17 FORTRAN books, 6 APL books, 12 COBOL books, and 20 BASIC books, how many of these books must we select to insure that we have 10 books dealing with the same computer language?

5. An auditorium has a seating capacity of 800. How many seats must be occupied to guarantee that at least two people seated in the auditorium have the same first and last initials?

6. a) Let $S \subset \mathbf{Z}^+$. What is the smallest value for $|S|$ that guarantees the existence of two elements $x, y \in S$ where x and y have the same remainder upon division by 1000?

b) If $S \subset \mathbf{Z}^+$, determine the smallest value for $|S|$ so that there exist three elements $x, y, z \in S$ where all three have the same remainder upon division by 1000.

c) Write a statement that generalizes the results of parts (a) and (b) and Example 5.32.

7. a) Prove that if 151 integers are selected from $\{1, 2, 3, \ldots, 300\}$, then the selection must include two integers x, y where $x \mid y$ or $y \mid x$.

b) Write a statement that generalizes the results of part (a) and Example 5.33.

8. Prove that if we select 101 integers from the set $S = \{1, 2, 3, \ldots, 200\}$, there exist m, n in the selection where $(m, n) = 1$.

9. a) Show that if any 14 integers are selected from the set $S = \{1, 2, 3, \ldots, 25\}$, there are at least two whose sum is 26.

b) Write a statement that generalizes the results of part (a) and Example 5.34.

10. Let $S = \{3, 7, 11, 15, 19, \ldots, 95, 99, 103\}$. How many elements must we select from S to insure that there will be at least two whose sum is 110?

11. a) If 11 integers are selected from $\{1, 2, 3, \ldots, 100\}$, prove that there are at least two, say x and y, such that $0 < |\sqrt{x} - \sqrt{y}| < 1$.

b) Write a statement that generalizes the result of part (a).

12. Let triangle ABC be equilateral, with $AB = 1$. Show that if we select 10 points in the interior of this triangle, there must be at least two whose distance apart is less than $1/3$.

13. Let $ABCD$ be a square with $AB = 1$. Show that if we select five points in the interior of this square, there are at least two whose distance apart is less than $1/\sqrt{2}$.

14. Let S be a set of five positive integers the maximum of which is at most 9. Prove that the sums of the elements in all the nonempty subsets of S cannot all be distinct.

15. At a certain high school there are 45 students who are training for eight different track and field events. If each student trains for only one event, and no event has more than 10 students in training for it, prove that at least three events have five or more students in training for each of them.

16. During the first six weeks of his senior year in college, Brace sends his resumé out to different companies. If he sends out at least one resumé each and every day, but no more than 60 resumés in total, show that there is a period of consecutive days during which he sends out exactly 23 resumés.

17. a) If $S \subseteq \mathbf{Z}^+$ and $|S| \geq 3$, prove that there exist $x, y \in S$ where $x + y$ is even.

b) Let $S \subseteq \mathbf{Z}^+ \times \mathbf{Z}^+$. Find the minimal value of $|S|$ that guarantees the existence of ordered pairs $(x_1, x_2), (y_1, y_2) \in S$ such that $x_1 + y_1$ and $x_2 + y_2$ are both even.

c) Extending the ideas in parts (a) and (b), consider $S \subseteq \mathbf{Z}^+ \times \mathbf{Z}^+ \times \mathbf{Z}^+$.

What size must $|S|$ be to guarantee the existence of ordered triples $(x_1, x_2, x_3), (y_1, y_2, y_3) \in S$ where $x_1 + y_1, x_2 + y_2$, and $x_3 + y_3$ are all even?

d) Generalize the results of parts (a), (b), and (c).

e) A point $P(x, y)$ in the Cartesian plane is called a *lattice point* if $x, y \in \mathbf{Z}$. Given lattice points $P_1(x_1, y_1), P_2(x_2, y_2), \ldots, P_n(x_n, y_n)$, determine the smallest value of n that guarantees the existence of $P_i(x_i, y_i)$, $P_j(x_j, y_j)$, $1 \le i < j \le n$, such that the midpoint of the line segment connecting $P_i(x_i, y_i)$ and $P_j(x_j, y_j)$ is also a lattice point.

5.6
FUNCTION COMPOSITION AND INVERSE FUNCTIONS

When computing with the elements of $\mathbf{Z}$, we find that the (binary) operation of addition provides a method for combining two integers, say a and b, into a third integer, namely $a + b$. Furthermore, for any integer c there is a second integer d where $c + d = d + c = 0$, and we call d the additive *inverse* of c. (It is also true that c is the additive *inverse* of d.)

Turning to the elements of $\mathbf{R}$ and the (binary) operation of multiplication, we have a method for combining any $r, s \in \mathbf{R}$ into their product rs. And here, for any $t \in \mathbf{R}$, if $t \ne 0$ then there is a real number u such that $ut = tu = 1$. The real number u is called the multiplicative *inverse* of t. (The real number t is also the multiplicative *inverse* of u.)

In this section we first study a method for combining two functions into a single function. Then we develop the concept of the inverse (of a function) for functions with certain properties. To accomplish these objectives, we need the following preliminary ideas.

Having examined functions that are one-to-one and those that are onto, we turn now to functions with both of these properties.

DEFINITION 5.14   If $f: A \to B$, then f is said to be *bijective*, or to be a *one-to-one correspondence*, if f is both one-to-one and onto.

Example 5.38   If $A = \{1, 2, 3, 4\}$ and $B = \{w, x, y, z\}$, then $f = \{(1, w), (2, x), (3, y), (4, z)\}$ is a one-to-one correspondence from A (on)to B, and $g = \{(w, 1), (x, 2), (y, 3), (z, 4)\}$ is a one-to-one correspondence from B (on)to A.   □

For any set A there is always a very simple but important one-to-one correspondence, as seen in the following.

DEFINITION 5.15   The function $1_A: A \to A$, defined by $1_A(a) = a$ for all $a \in A$, is called the *identity function* for A.

It should be pointed out that whenever the term "correspondence" was used in Chapter 1, the adjective "one-to-one" was implied though never stated.

DEFINITION 5.16 If $f, g: A \rightarrow B$, we say that f and g are *equal* and write $f = g$, if $f(a) = g(a)$ for all $a \in A$.

A common pitfall in dealing with the equality of functions occurs when f, g are functions with a common domain A and $f(a) = g(a)$ for all $a \in A$. It may *not* be the case that $f = g$. The pitfall results from not paying attention to the co-domains of the functions.

Example 5.39 Let $f: \mathbf{Z} \rightarrow \mathbf{Z}$, $g: \mathbf{Z} \rightarrow \mathbf{Q}$ where $f(x) = x = g(x)$, for all $x \in \mathbf{Z}$. Then f, g share the common domain $\mathbf{Z}$, have the same range $\mathbf{Z}$, and act the same on every element of $\mathbf{Z}$. Yet $f \neq g$! Here f is a one-to-one correspondence whereas g is one-to-one but not onto, so the codomains do make a difference. □

Now that we have dispensed with the necessary preliminaries, it is time to introduce an operation for combining two appropriate functions.

We have seen operations that combine integers as well as some that operate on sets. Now we introduce an operation to combine two appropriate functions.

DEFINITION 5.17 If $f: A \rightarrow B$ and $g: B \rightarrow C$, we define the *composite function*, which is denoted $g \circ f: A \rightarrow C$, by $(g \circ f)(a) = g(f(a))$, for each $a \in A$.

Example 5.40 Let $A = \{1, 2, 3, 4\}$, $B = \{a, b, c\}$ and $C = \{w, x, y, z\}$ with $f: A \rightarrow B$ and $g: B \rightarrow C$ given by $f = \{(1, a), (2, a), (3, b), (4, c)\}$ and $g = \{(a, x), (b, y), (c, z)\}$. For each element of A we find:

$$(g \circ f)(1) = g(f(1)) = g(a) = x \qquad (g \circ f)(3) = g(f(3)) = g(b) = y$$
$$(g \circ f)(2) = g(f(2)) = g(a) = x \qquad (g \circ f)(4) = g(f(4)) = g(c) = z$$

So
$$g \circ f = \{(1, x), (2, x), (3, y), (4, z)\}. \qquad \square$$

Example 5.41 Let $f: \mathbf{R} \rightarrow \mathbf{R}$, $g: \mathbf{R} \rightarrow \mathbf{R}$ be defined by $f(x) = x^2$, $g(x) = x + 5$. Then

$$(g \circ f)(x) = g(f(x)) = g(x^2) = x^2 + 5,$$

whereas

$$(f \circ g)(x) = f(g(x)) = f(x + 5) = (x + 5)^2 = x^2 + 10x + 25.$$

Here $g \circ f: \mathbf{R} \rightarrow \mathbf{R}$ and $f \circ g: \mathbf{R} \rightarrow \mathbf{R}$, but $(g \circ f)(1) = 6 \neq 36 = (f \circ g)(1)$, so even though both composites $f \circ g$ and $g \circ f$ can be formed, we do not have $f \circ g = g \circ f$. Consequently, the composition of functions is not, in general, a commutative operation. □

In the definition and examples for composite functions, it was required that codomain of f = domain of g. If range of $f \subseteq$ domain of g, this will actually be enough to yield the composite function $g \circ f: A \rightarrow C$. Also, for any $f: A \rightarrow B$, we observe that $f \circ 1_A = f = 1_B \circ f$.

An important recurring idea in mathematics is the investigation of whether the combining of two entities with a common property yields a result with this property. For example, if A and B are finite sets, then $A \cap B$ and $A \cup B$ are also finite. However, for infinite sets A and B, we have $A \cup B$ infinite but $A \cap B$ could be finite. (Give an example.)

For the composition of functions we have the following result.

THEOREM 5.4   Let $f: A \rightarrow B$ and $g: B \rightarrow C$.

a) If f, g are one-to-one, then $g \circ f$ is one-to-one.

b) If f, g are onto, then $g \circ f$ is onto.

Proof   **a)** To prove that $g \circ f: A \rightarrow C$ is one-to-one, let $a_1, a_2 \in A$ with $(g \circ f)(a_1) = (g \circ f)(a_2)$. Then $(g \circ f)(a_1) = (g \circ f)(a_2) \Rightarrow g(f(a_1)) = g(f(a_2)) \Rightarrow f(a_1) = f(a_2)$, because g is one-to-one. Also, $f(a_1) = f(a_2) \Rightarrow a_1 = a_2$, because f is one-to-one. Consequently, $g \circ f$ is one-to-one.

b) For $g \circ f: A \rightarrow C$, let $z \in C$. Since g is onto, there exists $y \in B$ with $g(y) = z$. With f onto, there exists $x \in A$ with $f(x) = y$. Hence $z = g(y) = g(f(x)) = (g \circ f)(x)$, so the range of $(g \circ f) = C = $ the codomain of $(g \circ f)$, and $g \circ f$ is onto. ∎

Although function composition is not commutative, if $f: A \rightarrow B$, $g: B \rightarrow C$, and $h: C \rightarrow D$, what can we say about the functions $(h \circ g) \circ f$ and $h \circ (g \circ f)$? That is, is function composition associative?

Before considering the general result, let us first investigate a particular example.

Example 5.42   Let $f: \mathbf{R} \rightarrow \mathbf{R}$, $g: \mathbf{R} \rightarrow \mathbf{R}$, and $h: \mathbf{R} \rightarrow \mathbf{R}$, where $f(x) = x^2$, $g(x) = x + 5$, and $h(x) = \sqrt{x^2 + 2}$.

Then $((h \circ g) \circ f)(x) = (h \circ g)(f(x)) = (h \circ g)(x^2) = h(g(x^2)) = h(x^2 + 5) = \sqrt{(x^2 + 5)^2 + 2} = \sqrt{x^4 + 10x^2 + 27}$.

On the other hand, we see that $(h \circ (g \circ f))(x) = h((g \circ f)(x)) = h(g(f(x))) = h(g(x^2)) = h(x^2 + 5) = \sqrt{(x^2 + 5)^2 + 2} = \sqrt{x^4 + 10x^2 + 27}$, as above.

So in this particular example, $(h \circ g) \circ f$ and $h \circ (g \circ f)$ are two functions with the same domain and codomain, and for all $x \in \mathbf{R}$, $((h \circ g) \circ f)(x) = \sqrt{x^4 + 10x^2 + 27} = (h \circ (g \circ f))(x)$. Consequently, $(h \circ g) \circ f = h \circ (g \circ f)$. □

In investigating our next theorem, we shall find that the result in Example 5.42 is true in general.

THEOREM 5.5   If $f: A \rightarrow B$, $g: B \rightarrow C$, and $h: C \rightarrow D$, then $(h \circ g) \circ f = h \circ (g \circ f)$.

Proof   Since the two functions have the same domain, A, and codomain, D, the result will follow by showing that for every $x \in A$, $((h \circ g) \circ f)(x) = (h \circ (g \circ f))(x)$. (See the diagram shown in Fig. 5.7.)

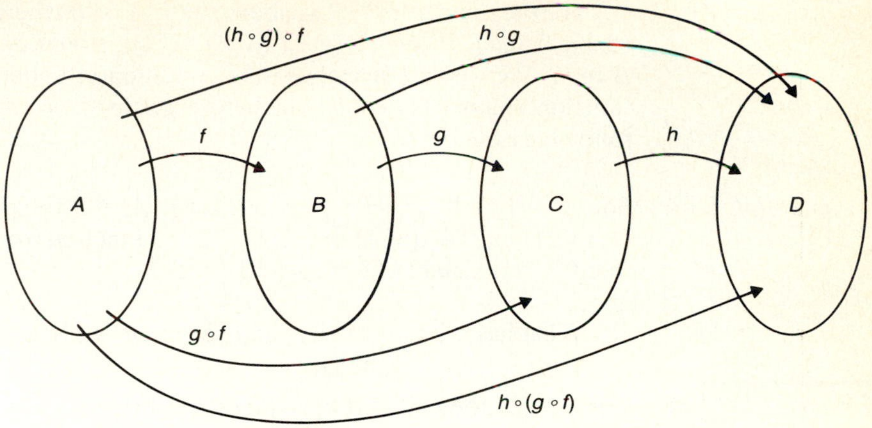

Figure 5.7

Using the definition of the composite function, we find that

$$((h \circ g) \circ f)(x) = (h \circ g)(f(x)) = h(g(f(x))),$$

whereas

$$(h \circ (g \circ f))(x) = h((g \circ f)(x)) = h(g(f(x))).$$

Consequently, the composition of functions is an associative operation. ■

By virtue of the associative property for function composition, we can write $h \circ g \circ f$, $(h \circ g) \circ f$ or $h \circ (g \circ f)$ without any problem of ambiguity. In addition, this property enables us to define powers of functions, where appropriate.

DEFINITION 5.18 If $f\colon A \to A$ we define $f^1 = f$, and for $n \in \mathbf{Z}^+$, $f^{n+1} = f \circ (f^n)$.

This definition is another example wherein the result is defined *recursively*. With $f^{n+1} = f \circ (f^n)$, we see the dependence of f^{n+1} on a previous power, namely f^n.

Example 5.43 With $A = \{1, 2, 3, 4\}$ and $f\colon A \to A$ defined by $f = \{(1,2), (2,2), (3,1), (4,3)\}$, we have $f^2 = f \circ f = \{(1,2), (2,2), (3,2), (4,1)\}$ and $f^3 = f \circ f^2 = f \circ f \circ f = \{(1,2), (2,2), (3,2), (4,2)\}$. (What are f^4, f^5?) □

We now come to the last new idea for this section: the existence of the invertible function and some of its properties.

DEFINITION 5.19 For sets $A, B \subseteq \mathscr{U}$, if $\mathscr{R}$ is a relation from A to B, then the *converse* of $\mathscr{R}$, denoted $\mathscr{R}^c$, is the relation from B to A defined by $\mathscr{R}^c = \{(b, a) \mid (a, b) \in \mathscr{R}\}$.

To get $\mathscr{R}^c$ from $\mathscr{R}$ we simply interchange the components of each ordered pair in $\mathscr{R}$. So if $A = \{1, 2, 3, 4\}$, $B = \{w, x, y\}$, and $\mathscr{R} = \{(1, w), (2, w), (3, x)\}$, then $\mathscr{R}^c = \{(w, 1), (w, 2), (x, 3)\}$, a relation from B to A.

For the same sets A, B as above, let $f\colon A \to B$ where $f = \{(1, w), (2, x), (3, y),$ $(4, x)\}$. Then $f^c = \{(w, 1), (x, 2), (y, 3), (x, 4)\}$, a relation, but not a function, from B to A. We wish to investigate the condition(s) under which the converse of a function yields a function, but before getting too abstract let us consider the following example.

Example 5.44   For $A = \{1, 2, 3\}$ and $B = \{w, x, y\}$, let $f\colon A \to B$ be given by $f = \{(1, w), (2, x), (3, y)\}$. Then $f^c = \{(w, 1), (x, 2), (y, 3)\}$ is a function from B to A, and we observe that $f^c \circ f = 1_A$ and $f \circ f^c = 1_B$. $\square$

What has happened in this finite example leads us to the following definition.

DEFINITION 5.20   If $f\colon A \to B$, then f is said to be *invertible* if there is a function $g\colon B \to A$ such that $g \circ f = 1_A$ and $f \circ g = 1_B$.   ▬

Note that the function g in Definition 5.20 is also invertible.

Example 5.45   Let $f, g\colon \mathbf{R} \to \mathbf{R}$ be defined by $f(x) = 2x + 5, g(x) = (1/2)(x - 5)$. Then $(g \circ f)(x)$ $= g(f(x)) = g(2x + 5) = (1/2)[(2x + 5) - 5] = x$,   and $(f \circ g)(x) = f(g(x)) =$ $f((1/2)(x - 5)) = 2[(1/2)(x - 5)] + 5 = x$,   so $f \circ g = 1_{\mathbf{R}}$ and $g \circ f = 1_{\mathbf{R}}$. Consequently f and g are both invertible functions. $\square$

Having seen some examples of invertible functions, we now wish to show that the function g of Definition 5.20 is unique. Then we shall find the means to identify an invertible function.

THEOREM 5.6   If a function $f\colon A \to B$ is invertible and a function $g\colon B \to A$ satisfies $g \circ f = 1_A$ and $f \circ g = 1_B$, then this function g is unique.

Proof   If g is not unique then there is another function $h\colon B \to A$ with $h \circ f = 1_A$ and $f \circ h = 1_B$. Consequently, $h = h \circ 1_B = h \circ (f \circ g) = (h \circ f) \circ g = 1_A \circ g = g$.   ∎

As a result of this theorem we shall call the function g *the inverse* of f and shall adopt the notation $g = f^{-1}$. Theorem 5.6 also implies that $f^{-1} = f^c$.

We also see that whenever f is an invertible function, so is the function f^{-1}, and $(f^{-1})^{-1} = f$, again by the uniqueness in Theorem 5.6. But we still do not know what conditions on f insure that f is invertible.

Before stating our next theorem we note that the invertible functions of Examples 5.44 and 5.45 are both bijective. Consequently, these examples provide some evidence for the following result.

THEOREM 5.7   A function $f\colon A \to B$ is invertible if and only if it is one-to-one and onto.

Proof   Assuming that $f\colon A \to B$ is invertible, we have a unique function $g\colon B \to A$ with $g \circ f = 1_A$, $f \circ g = 1_B$. If $a_1, a_2 \in A$ with $f(a_1) = f(a_2)$, then $g(f(a_1)) = g(f(a_2))$, or $(g \circ f)(a_1) = (g \circ f)(a_2)$. With $g \circ f = 1_A$ it follows that $a_1 = a_2$, so f is one-to-one.

For the onto property, let $b \in B$. Then $g(b) \in A$, so we can talk about $f(g(b))$. Since $f \circ g = 1_B$ we have $b = 1_B(b) = (f \circ g)(b) = f(g(b))$, so f is onto.

Conversely, suppose $f: A \to B$ is bijective. Since f is onto, for each $b \in B$ there is an $a \in A$ with $f(a) = b$. Consequently we define a function $g: B \to A$ by $g(b) = a$, where $f(a) = b$. This definition yields a unique function. The only problem that could arise is if $g(b) = a_1 \neq a_2 = g(b)$ because $f(a_1) = b = f(a_2)$. However, this situation cannot arise because f is one-to-one. Our definition of g is such that $g \circ f = 1_A$ and $f \circ g = 1_B$, so we find that f is invertible, with $g = f^{-1}$. ∎

Example 5.46   From Theorem 5.7 the function $f_1: \mathbf{R} \to \mathbf{R}$ defined by $f_1(x) = x^2$ is not invertible (it is neither one-to-one nor onto), but $f_2: [0, +\infty) \to [0, +\infty)$ defined by $f_2(x) = x^2$ is invertible with $f_2^{-1}(x) = \sqrt{x}$. □

The next result combines the ideas of function composition and inverse functions. The proof is left to the reader.

THEOREM 5.8   If $f: A \to B, g: B \to C$ are invertible functions, then $g \circ f: A \to C$ is invertible and $(g \circ f)^{-1} = f^{-1} \circ g^{-1}$.

Having seen some examples of functions and their inverses, one might wonder whether there is an algebraic method to determine the inverse of an invertible function. If the function is finite, we simply interchange the components of the given ordered pairs. But what if the function is defined by a formula, as in Example 5.46? Fortunately, the algebraic manipulations prove to be little more than a careful analysis of "interchanging the components of the ordered pairs." This is demonstrated in the following examples.

Example 5.47   For $m, b \in \mathbf{R}, m \neq 0$, the function $f: \mathbf{R} \to \mathbf{R}$ defined by $f = \{(x, y) \mid y = mx + b\}$ is an invertible function, because it is one-to-one and onto.

To get f^{-1} we note that

$$f^{-1} = \{(x, y) \mid y = mx + b\}^c = \{(y, x) \mid y = mx + b\}$$
$$= \underbrace{\{(x, y) \mid x = my + b\}}_{} = \{(x, y) \mid y = (1/m)(x - b)\}.$$

This is where we wish to change the components of the ordered pairs of f.

So $f: \mathbf{R} \to \mathbf{R}$ is defined by $f(x) = mx + b$, and $f^{-1}: \mathbf{R} \to \mathbf{R}$ is defined by $f^{-1}(x) = (1/m)(x - b)$. □

Example 5.48   Let $f: \mathbf{R} \to \mathbf{R}^+$ be defined by $f(x) = e^x$, where $e \doteq 2.7183$, the base for the natural logarithm. From the graph in Fig. 5.8 we see that f is one-to-one and onto, so $f^{-1}: \mathbf{R}^+ \to \mathbf{R}$ does exist, and $f^{-1} = \{(x, y) \mid y = e^x\}^c = \{(x, y) \mid x = e^y\} = \{(x, y) \mid y = \ln x\}$, so $f^{-1}(x) = \ln x$.

We should note that what happens in Fig. 5.8 happens in general. That is, the graphs of f and f^{-1} are symmetric about the line $y = x$. For example, the line segment connecting the points $(1, e)$ and $(e, 1)$ is bisected by the line $y = x$. This is true for any corresponding pair of points $(x, f(x))$ and $(f(x), f^{-1}(f(x)))$.

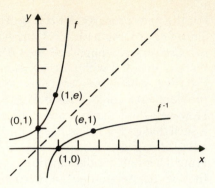

Figure 5.8

This example also yields the following formulas:

$$x = 1_\mathbf{R}(x) = (f^{-1} \circ f)(x) = \ln(e^x), \quad \text{for all } x \in \mathbf{R}.$$
$$x = 1_{\mathbf{R}^+}(x) = (f \circ f^{-1})(x) = e^{\ln x}, \quad \text{for all } x > 0. \quad \square$$

The result $x = e^{\ln x}$, for $x > 0$, is quite useful. In the standard implementation of Pascal, there is no exponentiation. To determine 2^3 one can resort to repeated multiplication, but this is futile when dealing with the evaluation of a number like $(5.73)^{4.32}$. Since exp and ln are defined functions in Pascal, we can determine $(5.73)^{4.32}$ by rewriting it as $e^{4.32 \ln(5.73)}$, since $5.73 = e^{\ln(5.73)}$ from the formula above. This becomes exp(4.32 * ln(5.73)) in Pascal.

Even when a function $f: A \to B$ is not invertible, we find use for the symbol f^{-1} in the following sense.

DEFINITION 5.21   If $f: A \to B$ and $B_1 \subseteq B$, then $f^{-1}(B_1) = \{x \in A \mid f(x) \in B_1\}$. The set $f^{-1}(B_1)$ is called the *preimage of B_1 under f.*

Be careful! Although we have the concept of a preimage for any function, not every function has an inverse function. Consequently, we cannot assume the existence of an inverse for a function f just because we find the symbol f^{-1} being used. A little caution is needed here.

Example 5.49   **a)** Let $f: \mathbf{Z} \to \mathbf{R}$ be defined by $f(x) = x^2 + 5$. Table 5.7 lists $f^{-1}(B)$ for various subsets B of the codomain $\mathbf{R}$.

b) If $g: \mathbf{R} \to \mathbf{R}$ is defined by $g(x) = x^2 + 5$, the results in Table 5.8 show how a change in domain (from $\mathbf{Z}$ to $\mathbf{R}$) affects the preimages (in Table 5.7). $\square$

Table 5.7

B	$f^{-1}(B)$
$\{6\}$	$\{-1, 1\}$
$[6, 7]$	$\{-1, 1\}$
$[6, 10]$	$\{-2, -1, 1, 2\}$
$[-4, 5)$	$\emptyset$
$[-4, 5]$	$\{0\}$
$[5, +\infty)$	$\mathbf{Z}$

Table 5.8

B	$g^{-1}(B)$
$\{6\}$	$\{-1, 1\}$
$[6, 7]$	$[-\sqrt{2}, -1] \cup [1, \sqrt{2}]$
$[6, 10]$	$[-\sqrt{5}, -1] \cup [1, \sqrt{5}]$
$[-4, 5)$	$\emptyset$
$[-4, 5]$	$\{0\}$
$[5, +\infty]$	$\mathbf{R}$

The concept of a preimage appears in conjunction with the set operations of intersection, union, and complementation in our next result. The reader should note the difference between part (a) of this theorem and part (b) of Theorem 5.2.

THEOREM 5.9 If $f: A \to B$ and $B_1, B_2 \subseteq B$, then (a) $f^{-1}(B_1 \cap B_2) = f^{-1}(B_1) \cap f^{-1}(B_2)$, (b) $f^{-1}(B_1 \cup B_2) = f^{-1}(B_1) \cup f^{-1}(B_2)$, and (c) $f^{-1}(\overline{B_1}) = \overline{f^{-1}(B_1)}$.

Proof We prove part (b) and leave parts (a) and (c) for the reader.
For $a \in A$, $a \in f^{-1}(B_1 \cup B_2) \Leftrightarrow f(a) \in B_1 \cup B_2 \Leftrightarrow f(a) \in B_1$ or $f(a) \in B_2 \Leftrightarrow a \in f^{-1}(B_1)$ or $a \in f^{-1}(B_2) \Leftrightarrow a \in f^{-1}(B_1) \cup f^{-1}(B_2)$. ∎

Using the notation of the preimage, we see that a function $f: A \to B$ is one-to-one if and only if $|f^{-1}(b)| \leq 1$ for each $b \in B$. (Note: $f^{-1}(b)$ is a shorter way to write $f^{-1}(\{b\})$.)

Since discrete mathematics is primarily concerned with finite sets, the last result of this section demonstrates how the property of finiteness can yield results that fail to be true in general. In addition, it illustrates an application of the pigeonhole principle.

THEOREM 5.10 Let $f: A \to B$ for finite sets A and B, where $|A| = |B|$. Then the following statements are equivalent: (a) f is one-to-one; (b) f is onto; and (c) f is invertible.

Proof We have already shown in Theorem 5.7 that (c) $\Rightarrow$ (a) and (b), and that together (a), (b) $\Rightarrow$ (c). Consequently, this theorem will follow when we show that for these conditions on A, B, (a) $\Leftrightarrow$ (b). Assuming (b), if f is not one-to-one then there are elements $a_1, a_2 \in A$, with $a_1 \neq a_2$, but $f(a_1) = f(a_2)$. Then $|A| > |f(A)| = |B|$, contradicting $|A| = |B|$. Conversely, if f is not onto, then $|f(A)| < |B|$. With $|A| = |B|$ we have $|A| > |f(A)|$, and it follows from the pigeonhole principle that f is not one-to-one. ∎

Using this theorem, we now verify the combinatorial identity introduced in Problem 6 at the start of this chapter. For if $n \in \mathbf{Z}^+$ and $|A| = |B| = n$, there are $n!$ one-to-one functions from A to B and $\sum_{k=0}^{n} (-1)^k \binom{n}{n-k}(n-k)^n$ onto functions from A to B. The equality $n! = \sum_{k=0}^{n} (-1)^k \binom{n}{n-k}(n-k)^n$ is then the numerical equivalent of parts (a) and (b) of Theorem 5.10. (This is also the reason why the diagonal elements $S(n, n)$, $1 \leq n \leq 8$, shown in Table 5.1 all equal 1.)

EXERCISES 5.6

1. Let $f: A \to B, g: C \to D$. Define $h: A \times C \to B \times D$ by

$$h(a, c) = (f(a), g(c)).$$

Prove that h is bijective if and only if f and g are bijective.

2. Let $f, g, h: \mathbf{Z} \to \mathbf{Z}$ be defined by $f(x) = x - 1, g(x) = 3x$,

$$h(x) = \begin{cases} 0, & x \text{ even} \\ 1, & x \text{ odd.} \end{cases}$$

Determine (a) $f \circ g, g \circ f, g \circ h, h \circ g, f \circ (g \circ h), (f \circ g) \circ h$;
(b) $f^2, f^3, g^2, g^3, h^2, h^3, h^{500}$.

3. If $\mathcal{U}$ is a given universe with $S, T \subseteq \mathcal{U}$, define $g: \mathcal{P}(\mathcal{U}) \to \mathcal{P}(\mathcal{U})$ by $g(A) = T \cap (S \cup A)$ for $A \subseteq \mathcal{U}$. Prove that $g^2 = g$.

4. Let $g: \mathbf{N} \to \mathbf{N}$ be defined by $g(n) = 2n$. If $A = \{1, 2, 3, 4\}$ and $f: A \to \mathbf{N}$ is given by $f = \{(1, 2), (2, 3), (3, 5), (4, 7)\}$, find $g \circ f$.

5. Let $f, g: \mathbf{R} \to \mathbf{R}$, where $g(x) = 1 - x + x^2$ and $f(x) = ax + b$. If $(g \circ f)(x) = 9x^2 - 9x + 3$, determine a, b.

6. Let $f, g: \mathbf{R} \to \mathbf{R}$ where $f(x) = ax + b$ and $g(x) = cx + d$ for any $x \in \mathbf{R}$, with a, b, c, d real constants. What relationship(s) must be satisfied by a, b, c, d if $(f \circ g)(x) = (g \circ f)(x)$ for all $x \in \mathbf{R}$?

7. **a)** For $A = \{1, 2, 3, \ldots, 7\}$, how many bijective functions $f: A \to A$ satisfy $f(1) \neq 1$?

   b) Answer part (a) for $A = \{x \mid x \in \mathbf{Z}^+, 1 \leq x \leq n\}$.

8. Let $f: A \to B, g: B \to C$. Then prove that (a) $g \circ f: A \to C$ onto $\Rightarrow g$ onto; and (b) $g \circ f: A \to C$ one-to-one $\Rightarrow f$ one-to-one.

9. If $A, B \subseteq \mathcal{U}$ and $\mathcal{R}_1, \mathcal{R}_2 \subseteq A \times B$, prove that (a) $(\mathcal{R}_1 \cup \mathcal{R}_2)^c = \mathcal{R}_1^c \cup \mathcal{R}_2^c$; (b) $(\mathcal{R}_1 \cap \mathcal{R}_2)^c = \mathcal{R}_1^c \cap \mathcal{R}_2^c$; and (c) $(\mathcal{R}_1^c)^c = \mathcal{R}_1$.

10. For each of the following functions $f: \mathbf{R} \to \mathbf{R}$, determine whether or not f is invertible, and, if so, determine f^{-1}.

   a) $f = \{(x, y) \mid 2x + 3y = 7\}$

   b) $f = \{(x, y) \mid ax + by = c, b \neq 0\}$

   c) $f = \{(x, y) \mid y = x^3\}$

   d) $f = \{(x, y) \mid y = x^4 - x\}$

11. **a)** Find the inverse of the function $f: \mathbf{R} \to \mathbf{R}^+$ defined by $f(x) = e^{2x+5}$.

   b) Show that $f \circ f^{-1} = 1_{\mathbf{R}^+}$ and $f^{-1} \circ f = 1_{\mathbf{R}}$.

   c) Graph f and f^{-1} on the same set of axes.

12. Determine f^{-1} for (a) $f: \mathbf{R} \to \mathbf{R}$, $f(x) = -x$; (b) $f: \mathbf{R}^2 \to \mathbf{R}^2$, $f(x, y) = (y, x)$; (c) $f: \mathbf{R}^2 \to (\mathbf{R} \times \mathbf{R}^+)$, $f(x, y) = (5x, e^y)$.

13. Prove Theorem 5.8.

14. Let $f: \mathbf{R} \to \mathbf{R}$ be defined by $f(x) = x^2$. For each of the following subsets B of $\mathbf{R}$, find $f^{-1}(B)$.

   a) $B = \{0, 1\}$ **b)** $B = \{-1, 0, 1\}$ **c)** $B = [0, 1]$

   d) $B = [0, 1)$ **e)** $B = [-1, 1]$ **f)** $B = [0, 4]$

   g) $B = [0, 1] \cup [4, 9]$ **h)** $B = (0, 1] \cup (4, 9)$.

   Determine three infinite subsets B of $\mathbf{R}$ for which $f^{-1}(B) = \emptyset$.

15. Let $f: \mathbf{R} \to \mathbf{R}$ be defined by $f(x) =$ the greatest integer in x. This is generally denoted by $f(x) = \lfloor x \rfloor$. Hence, for any $x \in \mathbf{R}$, $\lfloor x \rfloor = x$ if $x \in \mathbf{Z}$, and for $x \notin \mathbf{Z}$, $\lfloor x \rfloor$ is the integer directly to the left of x on the real number line. Find $f^{-1}(B)$ for each of the following subsets B of $\mathbf{R}$.

   a) $B = \{0, 1\}$ **b)** $B = \{-1, 0, 1\}$ **c)** $B = [0, 1)$

   d) $B = [0, 2)$ **e)** $B = [-1, 2)$ **f)** $B = [0, 1]$

   g) $B = [-1, 2]$ **h)** $B = [-1, 0) \cup (1, 3]$.

16. If $f: \mathbf{R} \to \mathbf{R}$ is given by $f(x) = \sin x$, find $f^{-1}(B)$ for (a) $B = \{0\}$; (b) $B = \{0, 1\}$; and (c) $B = [0, 1/2]$.

17. Prove parts (a) and (c) of Theorem 5.9.

18. Prove that $f: A \to B$ is one-to-one if and only if $|f^{-1}(b)| \leq 1$ for all $b \in B$.

19. a) Give an example of a function $f: \mathbf{Z} \to \mathbf{Z}$ where f is (i) one-to-one but not onto; (ii) onto but not one-to-one.

   b) Do the examples in part (a) contradict Theorem 5.10?

20. If $|A| = |B| = 5$, how many functions $f: A \to B$ are invertible?

5.7
COMPUTATIONAL COMPLEXITY †

In Section 4.3 we introduced the concept of an algorithm, following the examples set forth by the Division Algorithm (of Section 4.2) and the Euclidean Algorithm (of Section 4.3). At that time we were concerned with certain properties of a general algorithm:

- The precision of the individual step-by-step instructions.
- The input provided to the algorithm, and the output the algorithm then provides.
- The ability of the algorithm to solve a certain type of problem, not just specific cases of the problem.
- The uniqueness of the intermediate and final results, based on the input.
- The finite nature of the algorithm in that it terminates after the execution of a finite number of instructions.

When an algorithm correctly solves a certain type of problem and satisfies these five conditions, then we may find ourselves examining it further in the following ways.

1. Can we somehow measure how long it takes the algorithm to solve a problem of a certain size? This may very well depend, for example, on the compiler being used, so we want to develop a measure that doesn't depend on such considerations as compilers, execution speeds, or other characteristics of a given computer.

---

† The material in Sections 5.7 and 5.8 may be skipped at this point. It will not be used very much until Chapter 10. The only place where this material appears before Chapter 10 is in Example 7.12, but that example can be omitted without any loss of continuity.

If we want to compute $n!$ for $n \in \mathbf{Z}^+$, for example, how fast can we do it? If we want to compute a^n for $a \in \mathbf{R}$ and $n \in \mathbf{Z}^+$, is there some "function of n" that can describe how fast a given algorithm for such exponentiation accomplishes this?

2. Suppose we can answer questions such as those set forth in item 1. Then if we have two (or more) algorithms that solve a given problem, is there perhaps a way to determine whether one algorithm is "better" than another?

In particular, suppose we consider the problem of determining whether a certain real number x is present in the list of n real numbers $a_1, a_2, \ldots, a_n$. Here we have a problem of size n.

If there is an algorithm that solves this problem, how long does it take to do so? To measure this we seek a function $f(n)$, called the *time complexity function*† of the algorithm. We expect (both here and in general) that the value of $f(n)$ will increase as n increases. Also, our major concern in dealing with any algorithm is how the algorithm performs for *large* values of n.

In order to study what has now been described in a somewhat general manner, we need to introduce the following fundamental idea.

DEFINITION 5.22   Let $f, g: \mathbf{Z}^+ \to \mathbf{R}$. We say that g *dominates* f (or f is *dominated* by g) if there exist constants $m \in \mathbf{R}^+$ and $k \in \mathbf{Z}^+$ such that $|f(n)| \le m|g(n)|$ for all $n \in \mathbf{Z}^+$, where $n \ge k$.

Note that as we consider the values of $f(1), g(1), f(2), g(2), \ldots$, there is a point (namely k) after which the size of $f(n)$ is bounded above by a positive multiple (m) of the size of $g(n)$. Also, when g dominates f, $|f(n)/g(n)| \le m$ (that is, the size of the quotient $f(n)/g(n)$ is bounded by m), for those $n \in \mathbf{Z}^+$ where $n \ge k$ and $g(n) \ne 0$.

When f is dominated by g, we use what is called "big-Oh" notation to designate this, writing $f \in O(g)$, where $O(g)$ is read "order g" or "big-Oh of g." As suggested by the notation "$f \in O(g)$," $O(g)$ represents the set of all functions with domain $\mathbf{Z}^+$ and codomain $\mathbf{R}$ that are dominated by g. These ideas are demonstrated in the following examples.

Example 5.50   Let $f, g: \mathbf{Z}^+ \to \mathbf{R}$ be given by $f(n) = 5n, g(n) = n^2$, for $n \in \mathbf{Z}^+$. If we compute $f(n)$ and $g(n)$ for $1 \le n \le 4$, we find that $f(1) = 5, g(1) = 1; f(2) = 10, g(2) = 4; f(3) = 15, g(3) = 9$; and $f(4) = 20, g(4) = 16$. However, $n \ge 5 \Rightarrow n^2 \ge 5n$ and we have $|f(n)| = 5n \le n^2 = |g(n)|$. So with $m = 1$ and $k = 5$, we find that for $n \ge k$, $|f(n)| \le m|g(n)|$. Consequently, g dominates f and $f \in O(g)$. (Note how $|f(n)/g(n)|$ is bounded by 1 for all $n \ge 5$.)

We also realize that for any $n \in \mathbf{Z}^+$, $|f(n)| = 5n \le 5n^2 = 5|g(n)|$. So the dominance of f by g is shown here with $k = 1$ and $m = 5$. This is enough to demonstrate that the constants k and m of Definition 5.22 need *not* be unique.

Furthermore, we can generalize this result if we now consider functions

---

† We could also study the *space complexity function* of an algorithm, which we need when we attempt to measure the amount of memory required for the execution of an algorithm on a problem of size n. In this text, however, we limit our study to the time complexity function.

$f_1, g_1 \colon \mathbf{Z}^+ \to \mathbf{R}$ defined by $f_1(n) = an$, $g_1(n) = bn^2$, where a, b are nonzero real numbers. For if $m \in \mathbf{R}^+$ with $m|b| \geq |a|$, then for all $n \geq 1\ (= k)$, $|f_1(n)| = |an| = |a|n \leq m|b|n \leq m|b|n^2 = m|bn^2| = m|g_1(n)|$, and so $f_1 \in O(g_1)$. $\square$

In Example 5.50 we observed that $f \in O(g)$. Taking a second look at the functions f and g, we now want to show that $g \notin O(f)$.

Example 5.51   Once again let $f, g \colon \mathbf{Z}^+ \to \mathbf{R}$ be defined by $f(n) = 5n$, $g(n) = n^2$, for $n \in \mathbf{Z}^+$.
If $g \in O(f)$, then in terms of quantifiers, we would have

$$\exists m \in \mathbf{R}^+\ \exists k \in \mathbf{Z}^+\ \forall n \in \mathbf{Z}^+\ [(n \geq k) \Rightarrow |g(n)| \leq m|f(n)|].$$

Consequently, to show that $g \notin O(f)$, we need to verify that

$$\forall m \in \mathbf{R}^+\ \forall k \in \mathbf{Z}^+\ \exists n \in \mathbf{Z}^+\ [(n \geq k) \wedge (|g(n)| > m|f(n)|)].$$

To accomplish this, we first should realize that m and k are arbitrary, so we have no control over their values. The only number over which we have control is the positive integer n that we select. Now no matter what the values of m and k happen to be, we can select $n \in \mathbf{Z}^+$ such that $n > \max\{5m, k\}$. Then $n \geq k$ (actually $n > k$) and $n > 5m \Rightarrow n^2 > 5mn$, so $|g(n)| = n^2 > 5mn = m|5n| = m|f(n)|$ and $g \notin O(f)$. $\square$

Example 5.52   **a)** Let $f, g \colon \mathbf{Z}^+ \to \mathbf{R}$ with $f(n) = 5n^2 + 3n + 1$ and $g(n) = n^2$. Then $|f(n)| = |5n^2 + 3n + 1| = 5n^2 + 3n + 1 \leq 5n^2 + 3n^2 + n^2 = 9n^2 = 9|g(n)|$. Hence for all $n \geq 1\ (=k)$, $|f(n)| \leq m|g(n)|$ for any $m \geq 9$, and $f \in O(g)$. We can also write $f \in O(n^2)$ in this case.
In addition, $|g(n)| = n^2 \leq 5n^2 \leq 5n^2 + 3n + 1 = |f(n)|$ for all $n \geq 1$. So $|g(n)| \leq m|f(n)|$ for any $m \geq 1$ and all $n \geq k \geq 1$. Consequently $g \in O(f)$. (In fact $O(g) = O(f)$; that is, any function from $\mathbf{Z}^+$ to $\mathbf{R}$ that is dominated by one of f, g is also dominated by the other. We shall examine this result for the general case in the section exercises.)

b) Now consider $f, g \colon \mathbf{Z}^+ \to \mathbf{R}$ with $f(n) = 3n^3 + 7n^2 - 4n + 2$ and $g(n) = n^3$. Here we have $|f(n)| = |3n^3 + 7n^2 - 4n + 2| \leq |3n^3| + |7n^2| + |-4n| + |2| \leq 3n^3 + 7n^3 + 4n^3 + 2n^3 = 16n^3 = 16|g(n)|$, for all $n \geq 1$. So with $m = 16$ and $k = 1$, we find that f is dominated by g, and $f \in O(g)$, or $f \in O(n^3)$.
Since $7n - 4 > 0$ for all $n \geq 1$, we can write $n^3 \leq 3n^3 \leq 3n^3 + (7n - 4)n + 2$ whenever $n \geq 1$. Then $|g(n)| \leq |f(n)|$ for all $n \geq 1$, and $g \in O(f)$. (As in part (a), we also have $O(f) = O(g) = O(n^3)$ in this case.) $\square$

We generalize the results of Example 5.52 as follows. Let $f \colon \mathbf{Z}^+ \to \mathbf{R}$ where $f(n) = a_t n^t + a_{t-1} n^{t-1} + \cdots + a_2 n^2 + a_1 n + a_0$, for $a_t, a_{t-1}, \ldots, a_2, a_1, a_0 \in \mathbf{R}$, $a_t \neq 0$, $t \in \mathbf{N}$. Then

$$
\begin{aligned}
|f(n)| &= |a_t n^t + a_{t-1} n^{t-1} + \cdots + a_2 n^2 + a_1 n + a_0| \\
&\leq |a_t n^t| + |a_{t-1} n^{t-1}| + \cdots + |a_2 n^2| + |a_1 n| + |a_0| \\
&= |a_t| n^t + |a_{t-1}| n^{t-1} + \cdots + |a_2| n^2 + |a_1| n + |a_0| \\
&\leq |a_t| n^t + |a_{t-1}| n^t + \cdots + |a_2| n^t + |a_1| n^t + |a_0| n^t \\
&= (|a_t| + |a_{t-1}| + \cdots + |a_2| + |a_1| + |a_0|) n^t.
\end{aligned}
$$

In Definition 5.22, let $m = |a_t| + |a_{t-1}| + \cdots + |a_2| + |a_1| + |a_0|$ and $k = 1$, and let $g\colon \mathbf{Z}^+ \to \mathbf{R}$ be given by $g(n) = n^t$. Then $|f(n)| \le m|g(n)|$ for all $n \ge k$, so f is dominated by g, or $f \in O(n^t)$.

It is also true that $g \in O(f)$ and that $O(f) = O(g) = O(n^t)$.

This generalization provides the following special results on summations.

Example 5.53

a) Let $f\colon \mathbf{Z}^+ \to \mathbf{R}$ be given by $f(n) = 1 + 2 + 3 + \cdots + n$. Then (from Examples 1.35 and 4.1) $f(n) = (\frac{1}{2})(n)(n+1) = (\frac{1}{2})n^2 + (\frac{1}{2})n$, so $f(n) = \sum_{i=1}^{n} i \in O(n^2)$.

b) If $g\colon \mathbf{Z}^+ \to \mathbf{R}$ with $g(n) = 1^2 + 2^2 + 3^2 + \cdots + n^2 = (\frac{1}{6})(n)(n+1)(2n+1)$ (from Example 4.3), then $g(n) = (\frac{1}{3})n^3 + (\frac{1}{2})n^2 + (\frac{1}{6})n \in O(n^3)$.

c) If $t \in \mathbf{Z}^+$, and $h\colon \mathbf{Z}^+ \to \mathbf{R}$ is defined by $h(n) = \sum_{i=1}^{n} i^t$, then $h(n) = 1^t + 2^t + 3^t + \cdots + n^t \le n^t + n^t + n^t + \cdots + n^t = n(n^t) = n^{t+1}$, so $h(n) \in O(n^{t+1})$. $\square$

Now that we have examined several examples of function dominance, we shall close this section with two final observations. In the next section we shall apply the idea of function dominance in the analysis of algorithms.

1. When dealing with the concept of function dominance, we seek the best (or tightest) bound in the following sense. Suppose that $f, g, h\colon \mathbf{Z}^+ \to \mathbf{R}$, where $f \in O(g)$ and $g \in O(h)$. Then we also have $f \in O(h)$. (A proof for this is requested in the section exercises.) If $h \notin O(g)$, however, the statement $f \in O(g)$ provides a "better" bound on $|f(n)|$ than the statement $f \in O(h)$. For example, if $f(n) = 5$, $g(n) = 5n$, and $h(n) = n^2$, for all $n \in \mathbf{Z}^+$, then $f \in O(g)$, $g \in O(h)$, and $f \in O(h)$, but $h \notin O(g)$. Therefore, we are more concerned with the statement $f \in O(g)$ than with the statement $f \in O(h)$.

2. Certain orders, such as $O(n)$ and $O(n^2)$, occur often when we deal with function dominance. Therefore they have come to be designated by special names. Some of the most important of these orders are listed in Table 5.9.

Table 5.9

Big-Oh Form	Name
$O(1)$	Constant
$O(\log_2 n)$	Logarithmic
$O(n)$	Linear
$O(n \log_2 n)$	$n \log_2 n$
$O(n^2)$	Quadratic
$O(n^3)$	Cubic
$O(n^m), m = 0, 1, 2, 3, \ldots$	Polynomial
$O(c^n), c > 1$	Exponential
$O(n!)$	Factorial

EXERCISES 5.7

1. Use the results of Table 5.9 to determine the best "big-Oh" form for each of the following functions $f\colon \mathbf{Z}^+ \to \mathbf{R}$.

   a) $f(n) = 3n + 7$

    b) $f(n) = 3 + \sin(1/n)$

    c) $f(n) = n^3 - 5n^2 + 25n - 165$

    d) $f(n) = 5n^2 + 3n \, \log_2 n$

    e) $f(n) = n^2 + (n-1)^3$

    f) $f(n) = (n)(n+1)(n+2)/(n+3)$

    g) $f(n) = 2 + 4 + 6 + \cdots + 2n$

2. Let $f, g: \mathbf{Z}^+ \to \mathbf{R}$, where $f(n) = n$ and $g(n) = n + (1/n)$, for $n \in \mathbf{Z}^+$. Use Definition 5.22 to show that $f \in O(g)$ and $g \in O(f)$.

3. In each of the following, $f, g: \mathbf{Z}^+ \to \mathbf{R}$. Use Definition 5.22 to show that g dominates f.

    a) $f(n) = 100 \, \log_2 n, \ g(n) = (\frac{1}{2})n$

    b) $f(n) = 2^n, \ g(n) = 2^{2n} - 1000$

    c) $f(n) = 3n^2, \ g(n) = 2^n + 2n$

4. Let $f, g: \mathbf{Z}^+ \to \mathbf{R}$ be defined by $f(n) = n + 100$, $g(n) = n^2$. Use Definition 5.22 to show that $f \in O(g)$ but $g \notin O(f)$.

5. Let $f, g: \mathbf{Z}^+ \to \mathbf{R}$, where $f(n) = n^2 + n$ and $g(n) = (\frac{1}{2})n^3$, for $n \in \mathbf{Z}^+$. Use Definition 5.22 to show that $f \in O(g)$ but $g \notin O(f)$.

6. Let $f, g: \mathbf{Z}^+ \to \mathbf{R}$ be defined as follows:

$$f(n) = \begin{cases} n, & \text{for } n \text{ odd} \\ 1, & \text{for } n \text{ even} \end{cases}$$

$$g(n) = \begin{cases} 1, & \text{for } n \text{ odd} \\ n, & \text{for } n \text{ even} \end{cases}$$

Verify that $f \notin O(g)$ and $g \notin O(f)$.

7. Let $f, g: \mathbf{Z}^+ \to \mathbf{R}$ where $f(n) = n$ and $g(n) = \log_2 n$, for $n \in \mathbf{Z}^+$. Show that $g \in O(f)$ but $f \notin O(g)$. (Hint: $\lim\limits_{n \to \infty} \dfrac{n}{\log_2 n} = +\infty$. This requires the use of calculus.)

8. Let $f, g, h: \mathbf{Z}^+ \to \mathbf{R}$ where $f \in O(g)$ and $g \in O(h)$. Prove that $f \in O(h)$.

9. If $g: \mathbf{Z}^+ \to \mathbf{R}$ and $c \in \mathbf{R}$, we define the function $cg: \mathbf{Z}^+ \to \mathbf{R}$ by $(cg)(n) = c(g(n))$, for each $n \in \mathbf{Z}^+$. Prove that if $f, g: \mathbf{Z}^+ \to \mathbf{R}$ with $f \in O(g)$, then $f \in O(cg)$ for any $c \in \mathbf{R}$, $c \neq 0$.

10. a) Prove that $f \in O(f)$ for any $f: \mathbf{Z}^+ \to \mathbf{R}$.

    b) Let $f, g: \mathbf{Z}^+ \to \mathbf{R}$. If $f \in O(g)$ and $g \in O(f)$, prove that $O(f) = O(g)$. That is, prove that for any $h: \mathbf{Z}^+ \to \mathbf{R}$, if h is dominated by f, then h is dominated by g, and conversely.

    c) If $f, g: \mathbf{Z}^+ \to \mathbf{R}$, prove that if $O(f) = O(g)$, then $f \in O(g)$ and $g \in O(f)$.

5.8

ANALYSIS OF ALGORITHMS

Now that we have been introduced to the concept of function dominance, it is time to see how this idea is used in the study of algorithms. In this section we present our algorithms as Pascal program segments. (We shall also present algorithms as lists of instructions. The reader will find this to be the case in later chapters.)

We start with a program that computes $n!$ for $n \in \mathbf{Z}^+$.

Example 5.54   In Fig. 5.9 we have reproduced a Pascal program segment that implements an algorithm for computing $n!$ when $n \in \mathbf{Z}^+$. Here the user supplies the value of n, the input for the program. The variables i and Factorial (declared earlier in the program) are integer variables.

```
Begin
    i := 1;              {Initializes the Counter}
    Factorial := 1;      {Initializes the value of Factorial}

    While i <= n do
        Begin
            Factorial := i*Factorial;
            i := i + 1
        End;

    Writeln ('The value of ', n, ' factorial is ',
             Factorial, ' .')
End;
```

Figure 5.9

Our objective is to count (measure) the total number of operations (such as assignments, additions, multiplications, and comparisons) involved in computing $n!$ by this program. We shall let $f(n)$ denote the total number of these operations. (Then $f\colon \mathbf{Z}^+ \to \mathbf{R}$.)

There are two assignment statements at the start of the program, where the values of the integer variables i and Factorial are initialized. Then we come to the While loop, which is executed n times. Each of these executions involves the following five operations:

1. Comparing the present value of the counter i with n.
2. Increasing the present value of Factorial to $i *$ Factorial; this involves one multiplication and one assignment.
3. Incrementing the value of the counter by 1; this involves one addition and one assignment.

Finally, there is one more comparison. This is made when $i = n + 1$, so the While loop is terminated and the other four operations (in steps 2 and 3) are not performed.

Therefore, $f(n) = 2 + 5n + 1 = 5n + 3$, so $f \in O(n)$. Consequently, we say that this program implements an $O(n)$ algorithm or that the algorithm has *linear time complexity*. Assuming that all of the different operations involved take the same amount of time for execution, we can see how the function f "measures" the time for execution. If all we knew, however, was that $f \in O(n)$, then we would know that (1) the *dominant* term in f was n and (2) the time for execution was approximately cn, where c is a constant that depends on such considerations as the specific characteristics of the computer system.

Our major concern here is that the dominant term in f is n, and consequently $f \in O(n)$. For as n gets larger, the "order of magnitude" of $5n + 3$ depends primarily on the value n, the number of times the While loop is executed. Therefore, we could have obtained $f \in O(n)$ by simply counting the number of times the While loop was executed. Such short cuts will be used in our calculations for the remaining examples. ☐

Our next example introduces us to a situation where three types of complexity are determined. These measures are called the *best-case* complexity, the *worst-case* complexity, and the *average-case* complexity.

Example 5.55   In this example we examine a typical *searching* process. Here an array of n (≥ 1) integers $A[1], A[2], A[3], \ldots, A[n]$, is to be searched for the presence of an integer called Key. If the integer is found, we shall print out its first location in the array; if it is not found, an appropriate message will be given.

We cannot assume that the entries in the array are in any particular order. (If they were, the problem would be easier and a more efficient algorithm could be developed.) The input for this algorithm consists of the array (which is read in by the user or provided, perhaps, as a file from an external source), along with the number n of elements in the array, and the value of the integer Key.

The algorithm is implemented in the Pascal program segment of Fig. 5.10. Here the integer variable i is used as a counter, and the Boolean variable Found is used to record (with value true) the presence of Key in the array.

We shall define the complexity function $f(n)$ for this algorithm to be the number of elements in the array that are examined until the value Key is found or the array is exhausted (that is, the number of times the While loop is executed).

What is the best thing that can happen in our search for Key? If Key $= A[1]$, we find that Key is the first entry of the array, and we had to compare Key with only one element of the array. In this case we have $f(n) = 1$, and we say that the *best-case complexity* for our algorithm is $O(1)$ (that is, it is constant and independent of the size of the array). Unfortunately, we cannot expect such a situation to occur very often.

From the best situation we turn now to the worst. We see that we have to examine all n entries of the array if (1) the first occurrence of Key is $A[n]$ or (2) Key is not found in the array. In either case we have $f(n) = n$, and the *worst-case complexity* here is $O(n)$. (The worst-case complexity will typically be considered throughout the text.)

```
Begin

    i := 1;                    {Initializes the counter}
    Found := false;            {Changes to true if Key is found}

    While (i <= n) and (not Found) do
        If Key = A[i] then
           Found := true
        Else i = i + 1;        {The counter is incremented}
                               {only if Key is not found}

    If Found = true then
       Writeln ('The value ', Key, ' is located in
                position ', i, ' .')
    Else
       Writeln ('The value ', Key, ' does not occur in
                the array.')

End;
```

Figure 5.10

Finally, we wish to obtain an estimate of the average number of array entries examined. We shall assume that the n entries of the array are distinct and are all equally likely (with probability p) to contain the value Key, and that the probability that Key is not in the array is equal to q. Consequently, we have $np + q = 1$ and $p = (1 - q)/n$.

For any $1 \leq i \leq n$, if Key occurs at $A[i]$, then i elements of the array have been examined. If Key is not in the array, then all n array elements are examined. Therefore, the *average-case complexity* is determined by the average number of array elements examined, which is

$$f(n) = (1 \cdot p + 2 \cdot p + 3 \cdot p + \cdots + n \cdot p) + n \cdot q = p(1 + 2 + 3 + \cdots + n) + nq$$

$$= pn(n + 1)/2 + nq.$$

If $q = 0$, then Key is in the array, $p = 1/n$ and $f(n) = (n + 1)/2 \in O(n)$. For $q = 1/2$, we have an even chance that Key is in the array and $f(n) = (1/2n)[n(n + 1)/2] + (n/2) = (n + 1)/4 + (n/2) \in O(n)$. [In general, for any $0 \leq q \leq 1$, we have $f(n) \in O(n)$.] □

Early in the discussion of the previous section, we mentioned how we might want to compare two algorithms that both correctly solve a given type of problem. Such a comparison can be accomplished by using the time complexity functions for the algorithms. We demonstrate this in the results of the next two examples.

Example 5.56   In some computer languages, such as standard Pascal and C, there is no built-in function for exponentiation. The algorithm implemented in the Pascal program segment of Fig. 5.11, outputs the value of a^n where a is a real number and n is a positive integer. Here the values of a and n are determined prior to the execution of the program segment. The real variable x is initialized as 1.0 and then used to store the values $a, a^2, a^3, \ldots, a^n$ during execution of the For loop. Here the time complexity $f(n)$ for the algorithm is determined by the number of multiplications that occur in the For loop. Hence $f(n) = n \in O(n)$. □

```
Begin

    x := 1.0;

    For i := 1 to n do
        x := x*a;

    Writeln ('The value of ', a, ' raised to the power ',
             n, ' is ', x, ' .')

End;
```

Figure 5.11

Example 5.57 In Fig. 5.12 we have a second Pascal program segment to evaluate a^n for any $a \in \mathbf{R}$, $n \in \mathbf{Z}^+$. Here we are using the (predefined) nonzero integer division operation Div. The value of $(c \text{ Div } d)$ is simply the integer part (quotient) of c/d when $c, d \in \mathbf{Z}$, and $d \neq 0$. For instance, 7 Div 3 is 2, and 3 Div 4 is 0.

```
Begin

    x := 1.0;
    i := n;

    While i > 0 do
        Begin
            If i <> 2*(i Div 2) then              {i is odd}
                x := x*a;
            i := i Div 2;
            If i > 0 then
                a := a*a
        End;

Writeln ('The value of ', a, ' raised to the power ',
         n, ' is ', x, ' .')

End;
```

Figure 5.12

For this program segment the output is a^n, where the real variable a and the (positive) integer variable n have been assigned values earlier in the program. The real variable x is initialized as 1.0 and then used to store the appropriate powers of a until it contains the value of a^n. The results shown in Fig. 5.13

demonstrate what is happening to x (and a) for the cases where $n = 7$ and 8. The numbers 1, 2, 3, and 4 indicate the first, second, third, and fourth times the statements in the While loop (in particular, the statement i := i Div 2) are executed. If $n = 7$, then because $2^2 < 7 < 2^3$, we have $2 < \log_2 7 < 3$. Here the While loop is executed three times and

$$3 = \lfloor \log_2 7 \rfloor + 1 < \log_2 7 + 1,$$

where $\lfloor \log_2 7 \rfloor$ denotes the greatest integer in $\log_2 7$, which is 2. Also, when $n = 8$, the number of times the While loop is executed is

$$4 = \lfloor \log_2 8 \rfloor + 1 = \log_2 8 + 1.$$

since $\log_2 8 = 3$.

```
     n  =  7                        n  =  8
     x  :=  1.0                     x  :=  1.0
     i  :=  7                       i  :=  8

     ⎧ x  :=  x*a    {x = a}      ⎧ i  :=  4
    1⎨ i  :=  3                  1⎨ a  :=  a*a
     ⎩ a  :=  a*a

                                   ⎧ i  :=  2
     ⎧ x  :=  x*a    {x = a³}    2⎨ a  :=  a*a
    2⎨ i  :=  1
     ⎩ a  :=  a*a
                                   ⎧ i  :=  1
                                  3⎨ a  :=  a*a

     ⎧ x  :=  x*a    {x = a⁷}      ⎧ x  :=  x*a    {x = a⁸}
    3⎨ i  :=  0                  4⎨ i  :=  0

     [x = a⁷ = a · a² · a⁴]         [x = (((a)²)²)²]
```

Figure 5.13

We shall define the time complexity function $g(n)$ for (the implementation of) this exponentiation algorithm as the number of times the While loop is executed. This is also the number of times the statement i := i Div 2 is executed. (Here we assume that each Div operation is performed within a constant time interval. That is, the time interval for each call to Div is independent of the magnitude of i.) On the basis of the foregoing two observations, we want to establish that for all $n \geq 1$, $g(n) \leq \log_2 n + 1 \in O(\log_2 n)$. We shall establish this by mathematical induction (the Alternative Form—Theorem 4.2) on the value of n.

When $n = 1$, we see in Fig. 5.12 that i is odd, x is assigned the value of $a = a^1$, and a^1 is determined after only $1 = \log_2 1 + 1$ execution of the While loop. So $g(1) = 1 \leq \log_2 1 + 1$.

Now assume that for all $1 \leq n \leq k$, $g(n) \leq \log_2 n + 1$. Then for $n = k + 1$,

during the first pass through the While loop the value of i is changed to $\left\lfloor \dfrac{k+1}{2} \right\rfloor$.

Since $1 \leq \left\lfloor \dfrac{k+1}{2} \right\rfloor \leq k$, by the induction hypothesis we shall execute the While

loop $g\left(\left\lfloor \dfrac{k+1}{2} \right\rfloor\right)$ more times, where $g\left(\left\lfloor \dfrac{k+1}{2} \right\rfloor\right) \leq \log_2 \left\lfloor \dfrac{k+1}{2} \right\rfloor + 1$.

Therefore $g(k+1) \leq 1 + \left[\log_2 \left\lfloor \dfrac{k+1}{2} \right\rfloor + 1\right] \leq 1 + \left[\log_2 \left(\dfrac{k+1}{2}\right) + 1\right] = 1 +$

$[\log_2 (k+1) - \log_2 2 + 1] = \log_2 (k+1) + 1$.

For the time complexity function of Example 5.56, we found that $f(n) \in O(n)$. Here we have $g(n) \in O(\log_2 n)$. It can be verified that g is dominated by f but f is not dominated by g. Therefore, *for large n*, this second algorithm is considered more efficient than the first algorithm (of Example 5.56). (However, note how much easier the coding of the program in Fig. 5.11 is than that of the program in Fig. 5.12.) $\square$

In closing this section, we shall summarize what we have learned by making the following observations.

1. The results we established in Examples 5.54 through 5.57 are useful when we are dealing with moderate to large values of n. For small values of n, such considerations about time complexity functions have little purpose.

2. Suppose that algorithms A_1 and A_2 have time complexity functions $f(n)$ and $g(n)$, respectively, where $f(n) \in O(n)$ and $g(n) \in O(n^2)$. We must be cautious here. We might expect an algorithm with linear complexity to be "perhaps more efficient" than one with quadratic complexity. But we really need more information. If $f(n) = 1000n$ and $g(n) = n^2$, then algorithm A_2 is fine until the problem size n exceeds 1000. If the problem size is such that we never exceed 1000, then algorithm A_2 is the better choice. However as we mentioned in observation 1, as n grows larger, the algorithm of linear complexity becomes the better alternative.

3. In Fig. 5.14 we have graphed a log-linear plot for the functions associated with some of the orders given in Table 5.9. (Here we have replaced the (discrete) integer variable n by the (continuous) real variable n.) This should help us to develop some feeling for their relative growth rates (especially for large values of n).

The data in Table 5.10 provide estimates of the running times of algorithms for certain orders of complexity. Here we have the problem sizes $n = 2$, 16, and 64, and we assume that the computer can perform one operation every 10^{-6} second = 1 microsecond (on the average). The entries in the table then estimate the running times in microseconds. For example, when the problem size is 16 and the order of complexity is $n \log_2 n$, then the running time is a very brief $16 \log_2 16 = 16 \cdot 4 = 64$ microseconds; for the order of complexity 2^n, the running time is 6.5×10^4 microseconds = 0.065 seconds. Since both of these time intervals are so short, it is difficult for a human to observe much of a difference in execution times. Results appear to be instantaneous in either case.

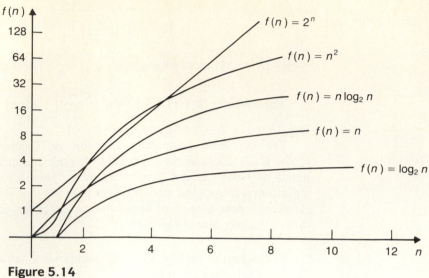

Figure 5.14

Table 5.10

Problem size n	Order of Complexity					
	$\log_2 n$	n	$n \log_2 n$	n^2	2^n	$n!$
2	1	2	2	4	4	2
16	4	16	64	256	6.5×10^4	2.1×10^{13}
64	6	64	384	4096	1.84×10^{19}	$>10^{89}$

However, such estimates can grow rather rapidly. For instance, suppose we run a program for which the input is an array A of n different integers. The results from this program are generated in two parts:

1. First the program implements an algorithm that determines the subsets of A of size 1. There are n such subsets.

2. Then a second algorithm is implemented to determine all the subsets of A. There are 2^n such subsets.

Let us assume that we have a supercomputer that can determine each subset of A in a microsecond. For the case where $|A| = 64$, the first part of the output is executed almost instantaneously—in approximately 64 microseconds. For the second part, however, Table 5.10 indicates that the amount of time needed to determine all the subsets of A will be about 1.84×10^{19} microseconds. We cannot be too content with this result, however, since

$$1.84 \times 10^{19} \text{ microseconds} \doteq 2.14 \times 10^8 \text{ days} \doteq 5845 \text{ centuries.}$$

EXERCISES 5.8

1. In each of the following Pascal program segments, the integer variables i, j, n, and sum are declared earlier in the program. The value of n (a positive integer) is supplied by the user prior to execution of the segment. In each case we define the time complexity function $f(n)$ to be the number of times the statement sum := sum + 1 is executed. Determine the best "big-Oh" form for f.

a)
```
Begin
    sum := 0;
    For i := 1 to n do
        For j := 1 to n do
            sum := sum + 1
End;
```

b)
```
Begin
    sum := 0;
    For i := 1 to n do
        For j := 1 to n*n do
            sum := sum + 1
End;
```

c)
```
Begin
    sum := 0;
    For i := 1 to n do
        For j := i to n do
            sum := sum + 1
End;
```

d)
```
Begin
    sum := 0;
    i := n;
    While i > 0 do
        Begin
            sum := sum + 1;
            i := i Div 2
        End
End;
```

e)
```
Begin
    sum := 0;
    For i := 1 to n do
        Begin
            j := n;
            While j > 0 do
                Begin
                    sum := sum + 1;
                    i := i Div 2
                End
        End
End;
```

2. The following Pascal program segment implements an algorithm for determining the maximum value in an array $A[1], A[2], A[3], \ldots, A[n]$ of integers. The array and the value of n (≥ 2) are supplied earlier in the program; the integer variables i and Max are declared at the start of the program.

```
Begin

    Max := A[1];

    For i := 2 to n do
        If A[i] > Max then
            Max := A[i]

End;
```

   a) If the worst-case complexity function $f(n)$ for this segment is determined by the number of times the comparison $A[i] > $ Max is executed, find the appropriate "big-Oh" form for f.

   b) What can we say about the best-case and average-case complexities for this implementation?

3. Let A_1 and A_2 be algorithms with worst-case complexity functions $f_1(n)$ and $f_2(n)$, respectively. Algorithm A_3 is implemented by having its input processed by A_1, and the resulting output is then processed by A_2. Find an appropriate "big-Oh" form for the worst-case complexity function $f_3(n)$ of A_3 when

   a) $f_1 \in O(n)$, $f_2 \in O(n^2)$

   b) $f_1 \in O(1)$, $f_2 \in O(\log_2 n)$

   c) $f_1 \in O(n^2)$, $f_2 \in O(n \log_2 n)$

   d) $f_1 \in O(n^2)$, $f_2 \in O(n^2)$

   e) $f_1 \in O(n^2)$, $f_2 \in O(2^n)$

4. Revise the Pascal program segment of Example 5.55 so that if Key appears in the array $A[1], A[2], A[3], \ldots, A[n]$, then all of its positions in the array are printed out.

5.9

SUMMARY AND HISTORICAL REVIEW

In this chapter we developed the function concept, which is of great importance in all areas of mathematics. Although we were primarily concerned with finite, or discrete, functions, the definition applies equally well to infinite sets and includes the functions of trigonometry and calculus. However, we did emphasize the role of a discrete function as one that transforms a finite set into a finite set. In this setting, computer output can be thought of as a function of computer input, and a compiler can be regarded as a function that transforms a (source) program into a set of machine-language instructions (object program).

The actual word "function," in its Latin form, was introduced in 1694 by Gottfried Wilhelm Leibniz (1646–1716) to denote a quantity associated with a curve (such as the slope of the curve or the coordinates of a point of the curve). By 1718, under the direction of Johann Bernoulli (1667–1748), a function was regarded as an algebraic expression made up of constants and a variable. Equations or formulas involving constants and variables came later with Leonhard Euler (1707–1783). His is the definition of "function" generally found in high school mathematics. Also, in about 1734, we find in the work of Euler and Alexis Clairaut (1713–1765) the notation $f(x)$, which is still in use today.

Euler's idea remained intact until the time of Joseph B. J. Fourier (1768–1830), who found the need for a more general type of function in his investigation of trigonometric series. In 1837, Peter Gustav Lejeune Dirichlet (1805–1859) set down a more rigorous formulation of the concepts of variable, function, and the correspondence between the independent variable x and the dependent variable y, when $y = f(x)$. Dirichlet's work emphasized the relationship between two sets of numbers and did not call for the existence of a formula or expression connecting the two sets. With the developments in set theory during the nineteenth and twentieth centuries came the generalization of the function as a particular type of relation.

In addition to his fundamental work on the definition of a function, Dirichlet was also quite active in applied mathematics and in number theory, where he found need for the pigeonhole principle, which is sometimes referred to as the Dirichlet drawer principle.

The nineteenth and twentieth centuries saw the use of the special function, one-to-one correspondence, in the study of the infinite. In about 1888, Richard Dedekind (1831–1916) defined an infinite set as one that can be placed into a one-to-one correspondence with a proper subset of itself. (Galileo (1564–1642) had observed this for the set $\mathbf{Z}^+$.) Two infinite sets that could be placed in a one-to-one correspondence with each other were said to have the same *transfinite cardinal number*. In a series of articles, Georg Cantor (1845–1918) developed the idea of levels of infinity and showed that $|\mathbf{Z}| = |\mathbf{Q}|$ but $|\mathbf{Z}| < |\mathbf{R}|$. A set A with $|A| = |\mathbf{Z}|$ is called *countable*, or *denumerable*, and we write $|\mathbf{Z}| = \aleph_0$ as Cantor did, using the Hebrew letter aleph, with the subscripted 0, to denote the first level of infinity. To show that $|\mathbf{Z}| < |\mathbf{R}|$, or that the real numbers were *uncountable*, Cantor devised a technique now referred to as the Cantor diagonal method.

The Stirling numbers of Section 5.3 are named in honor of James Stirling (1692–1770), a pioneer in the development of generating functions, a topic we will investigate later in the text. He was an associate of Sir Isaac Newton (1642–1727) and was using the Maclaurin series in his work 25 years before Colin Maclaurin (1698–1746). However, although his name is not attached to this infinite series, it appears in the approximation known as Stirling's formula: $n! \doteq (2\pi n)^{1/2} e^{-n} n^n$, which, as justice would have it, was actually developed by Abraham DeMoivre (1667–1754).

Using the counting principles developed in Section 5.3, the results in Table 5.11 extend the ideas that were summarized in Table 1.6. Here we count the number of ways it is possible to distribute m objects into n containers, under the conditions prescribed in the first three columns of the table. (The cases wherein neither the objects nor the containers are distinct will be covered in Chapter 9.)

Table 5.11

Objects Are Distinct	Containers Are Distinct	Some Container(s) May Be Empty	Number of Distributions
Yes	Yes	Yes	n^m
Yes	Yes	No	$n!\,S(m,n)$
Yes	No	Yes	$S(m,1) + S(m,2) + \cdots + S(m,n)$
Yes	No	No	$S(m,n)$
No	Yes	Yes	$\dbinom{n+m-1}{m}$
No	Yes	No	$\dbinom{n+(m-n)-1}{(m-n)} = \dbinom{m-1}{m-n}$ $= \dbinom{m-1}{n-1}$

Finally, the "big-Oh" notation of Section 5.7 was introduced by Paul Gustav Heinrich Bachmann (1837–1920) in his book *Analytische Zahlentheorie*, published in 1892. This notation has become prominent in approximation work, in such areas as numerical analysis and the analysis of algorithms. In general, the notation $f \in O(g)$ denotes that we do not know the function f explicitly, but do know a bound on its order of magnitude.

For more on infinite sets and the work of Cantor, consult Chapter 8 of H. Eves and C. V. Newsom [4] or Chapter IV of R. L. Wilder [6]. A great deal of research has been devoted to generalizations of the pigeonhole principle during this century, culminating in the subject of Ramsey theory, named for Frank Plumpton Ramsey (1903–1930). An interesting introduction to Ramsey theory can be found in Chapter 5 of D. I. A. Cohen [2]. The text by R. L. Graham, B. L. Rothschild, and J. H. Spencer [5] provides further worthwhile information. Extensive coverage on the topic of relational data bases can be found in the work of C. J. Date [3]. Finally, the text by S. Baase [1] is an excellent place to continue the study of the analysis of algorithms.

REFERENCES

1. Baase, Sara. *Computer Algorithms: Introduction to Design and Analysis,* 2nd ed. Reading, Mass.: Addison-Wesley, 1988.
2. Cohen, Daniel I. A. *Basic Techniques of Combinatorial Theory.* New York: Wiley, 1978.
3. Date, C. J. *An Introduction to Database Systems,* 3rd ed. Reading, Mass.: Addison-Wesley, 1982.

4. Eves, Howard, and Newsom, Carroll V. *An Introduction to the Foundations and Fundamental Concepts of Mathematics,* rev. ed. New York: Holt, 1965.

5. Graham, Ronald L., Rothschild, Bruce L., and Spencer, Joel H. *Ramsey Theory.* New York: Wiley, 1980.

6. Wilder, Raymond L. *Introduction to the Foundations of Mathematics,* 2nd ed. New York: Wiley, 1965.

MISCELLANEOUS EXERCISES

1. a) For sets A, B, C, and D, prove that $(A \subseteq C \wedge B \subseteq D) \Rightarrow A \times B \subseteq C \times D$.

    b) Does $A \times B \subseteq C \times D$ necessarily imply that $A \subseteq C$ and $B \subseteq D$?

2. Determine whether each of the following statements is true or false. For each false statement give a counterexample.

    a) If $f: A \to B$ and $(a, b), (a, c) \in f$, then $b = c$.

    b) If $f: A \to B$ is a one-to-one correspondence and A, B are finite, then $A = B$.

    c) If $f: A \to B$ is one-to-one, then f is invertible.

    d) If $f: A \to B$ is invertible, then f is one-to-one.

    e) If $f: A \to B$ is one-to-one and $g, h: B \to C$ with $g \circ f = h \circ f$, then $g = h$.

    f) If $f: A \to B$ and $A_1, A_2 \subseteq A$, then $f(A_1 \cap A_2) = f(A_1) \cap f(A_2)$.

    g) If $f: A \to B$ and $B_1, B_2 \subseteq B$, then $f^{-1}(B_1 \cap B_2) = f^{-1}(B_1) \cap f^{-1}(B_2)$.

3. With $\mathcal{U} = \mathbf{Z}$, let $A, B \subseteq \mathcal{U}$ where $A = \{2, 3, 5\}$ and $B = \{1, 2, 5, 6\}$.

    a) Diagram $A \times B$ as a subset of the Euclidean plane.

    b) If $\mathcal{R}$ is the relation from A to B defined as $\{(a, b) \mid a + b \text{ is odd}\}$, diagram $\mathcal{R}$ as a subset of the Euclidean plane.

    c) How many relations from A to B are not functions from A to B?

4. Let $\mathcal{U} = \mathbf{N}$ and let $A, B \subseteq \mathcal{U}$ with $1 < |A| < |B|$. If there are 262,144 relations from A to B, determine all possibilities for $|A|$ and $|B|$.

5. If $\mathcal{U}_1, \mathcal{U}_2$ are universal sets with $A, B \subseteq \mathcal{U}_1$, and $C, D \subseteq \mathcal{U}_2$, prove that

    a) $(A \cap B) \times (C \cap D) = (A \times C) \cap (B \times D)$

    b) $(A \cup B) \times (C \cup D) = (A \times C) \cup (B \times D) \cup (A \times D) \cup (B \times C)$ (so in general, $(A \cup B) \times (C \cup D) \supseteq (A \times C) \cup (B \times D)$).

6. Let $A = \{1, 2, 3, 4, 5\}$ and $B = \{1, 2, 3, 4, 5, 6\}$. How many injective functions $f: A \to B$ satisfy (a) $f(1) = 3$? (b) $f(1) = 3, f(2) = 6$?

7. Ackermann's function, $A(m, n)$, is defined *recursively* (that is, for $m > 0$, the value of $A(m, n)$ depends on the value(s) of A for one or more nonnegative integers smaller than m or n) for integers $m, n > 0$ by:

    $$A(0, n) = n + 1, \quad n \geq 0; \qquad A(m, 0) = A(m - 1, 1), \quad m > 0;$$

    $$A(m, n) = A(m - 1, A(m, n - 1)), \quad m > 0, \quad n > 0.$$

    Calculate $A(2, 3)$.

8. Let A_1, A and B be sets with $\{1, 2, 3, 4, 5\} = A_1 \subset A$, $B = \{s, t, u, v, w, x\}$, and $f: A_1 \to B$. If f can be extended to A in 216 ways, what is $|A|$?

9. Let $A = \{1, 2, 3, 4, 5\}$ and $B = \{t, u, v, w, x, y, z\}$. (a) If a function $f: A \rightarrow B$ is randomly generated, what is the probability that it is one-to-one? (b) Write a computer program (or develop an algorithm) to generate functions $f: A \rightarrow B$ and have it print out how many functions it generates until it generates one that is one-to-one.

10. Let S be a set of seven positive integers the maximum of which is at most 24. Prove that the sums of the elements in all the nonempty subsets of S cannot be distinct.

11. In a ten-day period a secretary typed 84 letters to different clients. She typed 12 of these letters on the first day, 7 on the second day, and 3 on the ninth day, and she finished the last 8 on the tenth day. Show that for a period of three consecutive days she typed at least 25 letters.

12. If $\{x_1, x_2, \dots, x_7\} \subseteq \mathbf{Z}^+$, show that for some $i \neq j$, either $x_i + x_j$ or $x_i - x_j$ is divisible by 10.

13. Let $n \in \mathbf{Z}^+, n$ odd. If $i_1, i_2, \dots, i_n$ is a permutation of the integers $1, 2, \dots, n$, prove that $(1 - i_1)(2 - i_2) \cdots (n - i_n)$ is an even integer. (Which counting principle is at work here?)

14. With both of their parents working, Thomas, Stuart, and Craig must handle ten weekly chores among themselves. (a) In how many ways can they divide up the work so that everyone is responsible for at least one chore? (b) In how many ways can the chores be assigned if Thomas, as the eldest, must mow the lawn (one of the ten weekly chores) and no one is allowed to be idle?

15. Let $n \in \mathbf{N}, n \geq 2$. Show that $S(n, 2) = 2^{n-1} - 1$.

16. Mrs. Blasi has five sons (Michael, Rick, David, Kenneth, and Donald) who enjoy reading books about sports. With Christmas approaching she visits a bookstore where she finds 12 different books on sports.

    a) In how many ways can she select nine of these books?

    b) Having made her purchase, in how many ways can she distribute the books among her sons so that each of them gets at least one book?

    c) State a problem for the situation presented here, where the answer is the product of the answers in parts (a) and (b).

    d) Two of the nine books Mrs. Blasi purchased deal with basketball, Donald's favorite sport. In how many ways can she distribute the books among her sons so that Donald gets at least the two books on basketball?

17. Let $m, n \in \mathbf{Z}^+$ with $n \geq m$. (a) In how many ways can one distribute n distinct objects among m different containers with no container left empty? (b) In the expansion of $(x_1 + x_2 + \cdots + x_m)^n$, what is the sum of all the multinomial coefficients $\binom{n}{n_1, n_2, \dots, n_m}$ where $n_1 + n_2 + \cdots + n_m = n$ and $n_i > 0$ for $1 \leq i \leq m$.

18. If $n \in \mathbf{Z}^+$ with $n \geq 4$, verify that $S(n, n - 2) = \binom{n}{3} + 3\binom{n}{4}$.

19. If $f: A \rightarrow A$ is any function, prove that for all $m, n \in \mathbf{Z}^+$, $f^m \circ f^n = f^n \circ f^m$. (First let $m = 1$ and induct on n. Then induct on m. This technique is known as *double induction*.)

20. Let $f: X \rightarrow Y$, and for each $i \in I$, let $A_i \subseteq X$. Prove that

    a) $f(\bigcup_{i \in I} A_i) = \bigcup_{i \in I} f(A_i)$

    b) $f(\bigcap_{i \in I} A_i) \subseteq \bigcap_{i \in I} f(A_i)$

    c) $f(\bigcap_{i \in I} A_i) = \bigcap_{i \in I} f(A_i)$,   for f injective

21. If $f\colon \mathbf{R} \to \mathbf{R}$ with $f(x) = x^n$, for which $n \in \mathbf{Z}^+$ is f invertible?

22. Let A be a set with $|A| = n$.

   a) How many binary operations are there on A?

   b) A ternary (3-ary) operation on A is a function $f\colon A \times A \times A \to A$. How many ternary operations are there on A?

   c) A k-ary operation on A is a function $f\colon A_1 \times A_2 \times \cdots \times A_k \to A$, where $A_i = A$, for all $1 \le i \le k$. How many k-ary operations are there on A?

   d) A k-ary operation for A is called *commutative* if

$$f(a_1, a_2, \ldots, a_k) = f(\pi(a_1), \pi(a_2), \ldots, \pi(a_k)),$$

   where $a_1, a_2, \ldots, a_k \in A$ (repetitions allowed), and $\pi(a_1), \pi(a_2), \ldots, \pi(a_k)$ is any rearrangement of $a_1, a_2, \ldots, a_k$. How many of the k-ary operations on A are commutative?

23. A function $f\colon \mathbf{R} \to \mathbf{R}$ is said to be *increasing* if for real numbers x, y, we have $x < y \Rightarrow f(x) < f(y)$. Prove that if $f, g\colon \mathbf{R} \to \mathbf{R}$ are increasing functions, then $g \circ f\colon \mathbf{R} \to \mathbf{R}$ is increasing.

24. If $\mathscr{U}$ is a universe and $A \subseteq \mathscr{U}$, we define the *characteristic function* of A by $\chi_A\colon \mathscr{U} \to \{0, 1\}$, where

$$\chi_A(x) = \begin{cases} 1, & x \in A \\ 0, & x \notin A \end{cases}$$

For sets $A, B \subseteq \mathscr{U}$, prove each of the following:

   a) $\chi_{A \cap B} = \chi_A \cdot \chi_B$ where $(\chi_A \cdot \chi_B)(x) = \chi_A(x) \cdot \chi_B(x)$

   b) $\chi_{A \cup B} = \chi_A + \chi_B - \chi_{A \cap B}$

   c) $\chi_{\overline{A}} = 1 - \chi_A$, where $(1 - \chi_A)(x) = 1(x) - \chi_A(x) = 1 - \chi_A(x)$

   (For $\mathscr{U}$ finite, placing the elements of $\mathscr{U}$ in a fixed order results in a one-to-one correspondence between subsets A of $\mathscr{U}$ and the arrays of 0's and 1's obtained as the images of $\mathscr{U}$ under χ_A. These arrays can then be used for the computer storage and manipulation of certain subsets of $\mathscr{U}$.)

25. With $A = \{x, y, z\}$, let $f, g\colon A \to A$ be given by $f = \{(x, y), (y, z), (z, x)\}, g = \{(x, y), (y, x), (z, z)\}$. Determine each of the following: $f \circ g, g \circ f, f^{-1}, g^{-1}, (g \circ f)^{-1}, f^{-1} \circ g^{-1}$, and $g^{-1} \circ f^{-1}$.

26. Let $f\colon \mathbf{R} \to \mathbf{R}$ be defined by $f(x) = (2 + \sin(x + 1))^3, x \in \mathbf{R}$. Find four functions $f_i\colon \mathbf{R} \to \mathbf{R}, 1 \le i \le 4$, so that $f = f_4 \circ f_3 \circ f_2 \circ f_1$.

27. For $A = \{1, 2, 3\}$ find the number of functions $f\colon A \to A$ where (a) $f^2 = 1_A$; (b) $f^2 = f$.

28. a) If $f\colon \mathbf{R} \to \mathbf{R}$ is defined by $f(x) = 5x + 3$, find $f^{-1}(8)$.

   b) If $g\colon \mathbf{R} \to \mathbf{R}$, where $g(x) = |x^2 + 3x + 1|$, find $g^{-1}(1)$.

   c) For $h\colon \mathbf{R} \to \mathbf{R}$, given by

$$h(x) = \left| \frac{x}{x + 2} \right|,$$

   find $h^{-1}(4)$.

29. If $f\colon X \to Y$, show that (a) for $A \subseteq X$, $A \subseteq f^{-1}(f(A))$; and (b) for $B \subseteq Y$,

$B \supseteq f(f^{-1}(B))$. State any needed condition(s) that f must satisfy in order to have equality in (a); in (b).

30. Let $f: X \to Y$ and let I be an index set, where for each $i \in I$, $B_i \subseteq Y$. Prove that

    a) $f^{-1}(\bigcup_{i \in I} B_i) = \bigcup_{i \in I} f^{-1}(B_i)$

    b) $f^{-1}(\bigcap_{i \in I} B_i) = \bigcap_{i \in I} f^{-1}(B_i)$

31. In the programming language Pascal, the functions pred and succ (for predecessor and successor, respectively) are functions from $\mathbf{Z}$ to $\mathbf{Z}$ where $\mathrm{pred}(x) = \pi(x) = x - 1$ and $\mathrm{succ}(x) = \sigma(x) = x + 1$.

    a) Determine $(\pi \circ \sigma)(x)$, $(\sigma \circ \pi)(x)$.

    b) Determine $\pi^2, \pi^3, \pi^n (n \geq 2), \sigma^2, \sigma^3, \sigma^n (n \geq 2)$.

    c) Determine $\pi^{-2}, \pi^{-3}, \pi^{-n}(n \geq 2), \sigma^{-2}, \sigma^{-3}, \sigma^{-n}(n \geq 2)$, where, for example, $\sigma^{-2} = \sigma^{-1} \circ \sigma^{-1} = (\sigma \circ \sigma)^{-1} = (\sigma^2)^{-1}$. (See Exercise 32.)

32. Let $f: A \to A$ be an invertible function. For $n \in \mathbf{Z}^+$ prove that $(f^n)^{-1} = (f^{-1})^n$. (This result can be used to define f^{-n} as either $(f^n)^{-1}$ or $(f^{-1})^n$.)

33. For $n \in \mathbf{Z}^+$, define $\tau: \mathbf{Z}^+ \to \mathbf{Z}^+$ by $\tau(n) =$ the number of positive-integer divisors of n.

    a) Let $n = p_1^{e_1} p_2^{e_2} p_3^{e_3} \cdots p_k^{e_k}$, where $p_1, p_2, p_3, \ldots, p_k$ are distinct primes and e_i is a positive integer for all $1 \leq i \leq k$. What is $\tau(n)$?

    b) Determine the three smallest values of $n \in \mathbf{Z}^+$ for which $\tau(n) = k$, where $k = 2, 3, 4, 5, 6$.

    c) For any $k \in \mathbf{Z}^+, k > 1$, prove that $\tau^{-1}(k)$ is infinite.

34. Let $a, b, c \in \mathbf{R}$ with $a^2 + bc \neq 0$. If $f(x) = (ax + b)/(cx - a)$, verify that $(f \circ f)(x) = x$ for all $x \in \mathbf{R}, x \neq a/c$.

35. a) How many subsets $A = \{a, b, c, d\} \subseteq \mathbf{Z}^+, a, b, c, d > 1$, satisfy the property $a \cdot b \cdot c \cdot d = 2 \cdot 3 \cdot 5 \cdot 7 \cdot 11 \cdot 13 \cdot 17 \cdot 19$?

    b) How many subsets $A = \{a_1, a_2, \ldots, a_m\} \subseteq \mathbf{Z}^+, a_i > 1, 1 \leq i \leq m$, satisfy the property $\prod_{i=1}^m a_i = \prod_{j=1}^n p_j$, where the $p_j, 1 \leq j \leq n$, are distinct primes and $n \geq m$?

36. Give an example of a function $f: \mathbf{Z}^+ \to \mathbf{R}$ where $f \in O(1)$ and f is one-to-one. (Hence f is not constant.)

37. Let $f, g: \mathbf{Z}^+ \to \mathbf{R}$ where

$$f(n) = \begin{cases} 2, & \text{for } n \text{ even} \\ 1, & \text{for } n \text{ odd} \end{cases} \qquad g(n) = \begin{cases} 3, & \text{for } n \text{ even} \\ 4, & \text{for } n \text{ odd} \end{cases}$$

Prove or disprove each of the following: (a) $f \in O(g)$; and (b) $g \in O(f)$.

38. For $f, g: \mathbf{Z}^+ \to \mathbf{R}$ we define $f + g: \mathbf{Z}^+ \to \mathbf{R}$ by $(f + g)(n) = f(n) + g(n)$, for $n \in \mathbf{Z}^+$. (Note that the plus sign in $f + g$ is for the addition of the functions f and g, where the plus sign in $f(n) + g(n)$ is for the addition of the real numbers $f(n)$ and $g(n)$.)

    a) Let $f_1, g_1: \mathbf{Z}^+ \to \mathbf{R}$ with $f \in O(f_1)$ and $g \in O(g_1)$. If $f_1(n) \geq 0, g_1(n) \geq 0$, for all $n \in \mathbf{Z}^+$, prove that $(f + g) \in O(f_1 + g_1)$.

    b) If the conditions $f_1(n) \geq 0$, $g_1(n) \geq 0$, for all $n \in \mathbf{Z}^+$, are not satisfied, as in part (a), provide a counterexample to show that $f \in O(f_1), g \in O(g_1) \not\Rightarrow (f + g) \in O(f_1 + g_1)$.

39. Let $a, b \in \mathbf{R}^+$, with $a, b > 1$. Let $f, g: \mathbf{Z}^+ \to \mathbf{R}$ be defined by $f(n) = \log_a n$, $g(n) = \log_b n$. Prove that $f \in O(g)$ and $g \in O(f)$. [Hence $O(\log_a n) = O(\log_b n)$.]

CHAPTER 6

Languages: Finite State Machines

In this era of computers and telecommunications, we find ourselves confronted every day with input–output situations. In purchasing a soft drink from a vending machine, we *input* some coins and then press a button to get our expected *output*, the soft drink desired. The first coin that we input sets the machine in motion. Although we usually don't care about what happens inside the machine (unless there is some kind of breakdown and we suffer a loss), we should realize that somehow the machine is keeping track of the coins we insert, until the correct total has been inserted. It is only then, and not before, that the vending machine should output the desired soft drink. Consequently, for the vendor to make the expected profit per soft drink, the machine must *internally remember*, as each coin is inserted, what sum of money has been deposited.

A computer is another example of an input–output device. Here the input is generally some type of information and the output is the result obtained after processing this information. How the input is processed depends on the internal workings of the computer; it must have the ability to remember past information as it works on the information it is presently processing.

Using the concepts developed on sets and functions, in this chapter we investigate an abstract model called a *finite state machine*, or *sequential circuit*. Such circuits are one of two basic types of control circuits found in digital computers. (The other type is a *combinational circuit* or *gating network*, which is examined in Chapter 15.) They are also found in other systems such as our vending machine, as well as in the controls for elevators and in traffic-light systems.

As the name indicates, a finite state machine has a finite number of internal states where the machine remembers certain information, when it is in that particular state. However, before getting into this concept we need some set-theoretic material in order to talk about what constitutes valid input for such a machine.

6.1
LANGUAGE: THE SET THEORY OF STRINGS

Sequences of symbols, or characters, play a key role in the processing of information by a computer. Inasmuch as computer programs are representable in terms of finite sequences of characters, some algebraic way is needed for handling such finite sequences, or *strings*.

Throughout this section we use Σ to denote a nonempty *finite* set of symbols, collectively called an *alphabet*. For example, we may have $\Sigma = \{0, 1\}$ or $\Sigma = \{a, b, c, d, e\}$.

In any alphabet Σ, we do not list elements that can be formed from other elements of Σ by juxtaposition (that is, if $a, b \in \Sigma$, then the string ab is the juxtaposition of the symbols a and b). As a result of this convention, alphabets such as $\Sigma = \{0, 1, 2, 11, 12\}$ and $\Sigma = \{a, b, c, ba, aa\}$ are not considered. (In addition, this convention will help us later in Definition 6.5, when we talk about the length of a string.)

Using an alphabet Σ as the starting place, we can construct strings from the symbols of Σ in a systematic manner by using the following idea.

DEFINITION 6.1   If Σ is an alphabet and $n \in \mathbf{Z}^+$, we define the *powers of* Σ recursively as follows:

1. $\Sigma^1 = \Sigma$, and
2. $\Sigma^{n+1} = \{xy \mid x \in \Sigma, y \in \Sigma^n\}$, where xy denotes the juxtaposition of x and y.

Example 6.1   Let Σ be any alphabet.

If $n = 2$, then $\Sigma^2 = \{xy \mid x, y \in \Sigma\}$. For instance, with $\Sigma = \{0, 1\}$ we find $\Sigma^2 = \{00, 01, 10, 11\}$.

When $n = 3$, the elements of Σ^3 have the form uv, where $u \in \Sigma$ and $v \in \Sigma^2$. But since we know the form of the elements in Σ^2, we may also regard the strings in Σ^3 as sequences of the form uxy, where $u, x, y \in \Sigma$. As an example for this case, suppose that $\Sigma = \{a, b, c, d, e\}$. Then Σ^3 would contain $5^3 = 125$ three-symbol strings—among them aaa, acb, ace, cdd, and eda.

In general, for all $n \in \mathbf{Z}^+$ we find that $|\Sigma^n| = |\Sigma|^n$ because we are dealing with arrangements (of size n) where we are allowed to repeat any of the $|\Sigma|$ objects.

$\square$

Now that we have examined Σ^n for $n \in \mathbf{Z}^+$, we shall look into one more power of Σ.

DEFINITION 6.2   For any alphabet Σ we define $\Sigma^0 = \{\lambda\}$, where λ denotes the *empty string*—that is, the string consisting of *no* symbols taken from Σ.

The symbol λ is never an element in our alphabet Σ, and we should not mistake it for the blank (space) that is found in many alphabets.

However, although $\lambda \notin \Sigma$, we do have $\emptyset \subseteq \Sigma$, so we need to be cautious here.

We observe that

1. $\{\lambda\} \not\subseteq \Sigma$ since $\lambda \notin \Sigma$, and
2. $\{\lambda\} \neq \emptyset$, because $|\{\lambda\}| = 1 \neq 0 = |\emptyset|$.

In order to speak collectively about the sets $\Sigma^0, \Sigma^1, \Sigma^2, \ldots$, we introduce the following unions.

DEFINITION 6.3   If Σ is any alphabet, then

a) $\Sigma^+ = \bigcup_{n=1}^{\infty} \Sigma^n = \bigcup_{n \in Z^+} \Sigma^n$, and

b) $\Sigma^* = \bigcup_{n=0}^{\infty} \Sigma^n$. ▬

We see that the only difference between the sets Σ^+ and Σ^* is the element λ, because $\lambda \in \Sigma^n$ only when $n = 0$. Also $\Sigma^* = \Sigma^+ \cup \Sigma^0$.

In addition to using the term "string," we shall also refer to the elements of Σ^+ or Σ^* as *words* and sometimes as *sentences*. For $\Sigma = \{0, 1, 2\}$, we find such words as 0, 01, 102, and 1112 in both Σ^+ and Σ^*.

Finally, we note that even though the sets Σ^+ and Σ^* are *infinite*, the elements of these sets are *finite* strings of symbols.

Example 6.2   For $\Sigma = \{0, 1\}$ the set Σ^* consists of all finite strings of 0's and 1's together with the empty string. For n reasonably small, we could actually list all strings in Σ^n.

If $\Sigma = \{\beta, 0, 1, 2, \ldots, 9, +, -, \times, /, (,)\}$, where β denotes the blank (or space), it is harder to describe Σ^* and for $n > 2$ there are many strings to list in Σ^n. Here in Σ^* we find familiar arithmetic expressions such as $(7 + 5)/(2 \times (3 - 10))$ as well as such gibberish as $+)((7/\times + 3/(.$ □

And now we come to a recurring situation. As with statements (Chapter 2), sets (Chapter 3), and functions (Chapter 5), once again we need to be able to decide when two objects under study—in this case strings—are to be considered equal. We investigate this issue in the next result.

DEFINITION 6.4   If $w_1, w_2 \in \Sigma^+$, then we may write $w_1 = x_1 x_2 \ldots x_m$ and $w_2 = y_1 y_2 \ldots y_n$, for $m, n \in Z^+$, and $x_1, x_2, \ldots, x_m, y_1, y_2, \ldots, y_n \in \Sigma$. We say that the strings w_1 and w_2 are *equal*, and we write $w_1 = w_2$, if $m = n$, and $x_i = y_i$ for all $1 \leq i \leq m$. ▬

It follows from this definition that two strings in Σ^+ are *equal* only when each is formed from the same number of symbols from Σ and the corresponding symbols in the two strings match identically.

We also need the number of symbols in a string to define another property for strings.

DEFINITION 6.5   If $w \in \Sigma^+$, then $w = x_1 x_2 \ldots x_n$, where $x_i \in \Sigma$ for each $1 \leq i \leq n$. We define the *length* of w, which is denoted by $\|w\|$, as the value n. For the case of λ, we have $\|\lambda\| = 0$. ▬

As a result of Definition 6.5, we find that for any alphabet Σ, if $w \in \Sigma^*$ and $\|w\| \geq 1$, then $w \in \Sigma^+$, and conversely. In addition, for $x \in \Sigma$ we find that $\|x\| = \|x\lambda\| = 1$, whereas $\|xx\| = 2$. Also, for any $y \in \Sigma^*$, $\|y\| = 1$ if and only if $y \in \Sigma$. Should Σ contain the symbol β (for the blank), it is still the case that $\|\beta\| = 1$.

If we use a particular alphabet, say $\Sigma = \{0, 1, 2\}$, and examine the elements $x = 01$, $y = 212$, and $z = 01212$ (in Σ^*), we find that

$$\|z\| = \|01212\| = 5 = 2 + 3 = \|01\| + \|212\| = \|x\| + \|y\|.$$

In order to continue our study of the properties of strings and alphabets, we need to extend the idea of juxtaposition a little further.

DEFINITION 6.6   Let $x, y \in \Sigma^*$ with $x = x_1 x_2 \ldots x_m$ and $y = y_1 y_2 \ldots y_n$, so that each x_i, for $1 \leq i \leq m$, and each y_j, for $1 \leq j \leq n$, is in Σ. The *concatenation* of x and y, which we write as xy, is the string $x_1 x_2 \ldots x_m y_1 y_2 \ldots y_n$.

Here we have a binary operation on Σ^* (and Σ^+). This operation is associative but not commutative, and since $x\lambda = \lambda x = x$ for any $x \in \Sigma^*$, the element $\lambda \in \Sigma^*$ is the identity for the operation of concatenation. (The concept of the identity was introduced in Exercises 4 and 5 of Section 5.4.) Finally, the ideas embodied in the last two definitions (the length of a string and the operation of concatenation) are interrelated in the result

$$\|xy\| = \|x\| + \|y\|, \text{ for any } x, y \in \Sigma^*.$$

The binary operation of concatenation now leads us to another recursive definition. Earlier we looked at powers of an alphabet Σ. Now we examine powers of strings.

DEFINITION 6.7   For any $x \in \Sigma^*$, we define the *powers* of x by $x^0 = \lambda$, $x^1 = x$, $x^2 = xx$, $x^3 = xx^2, \ldots$, $x^{n+1} = xx^n$, $\ldots$, where $n \in \mathbf{N}$.

In this definition we have another illustration of how a mathematical entity is given in an inductive or recursive manner: The mathematical entity we presently seek is derived from previously derived (comparable) entities. Furthermore, the definition provides a way for us to deal with an n-fold concatenation (an $(n + 1)$st power) of a string with itself and includes the special case where the string is just one symbol.

Example 6.3   If $\Sigma = \{0, 1\}$ and $x = 01$, then $x^0 = \lambda$, $x^1 = 01$, $x^2 = 0101$, and $x^3 = 010101$. For any $n > 0$, x^n consists of a string of n 0's and n 1's where the first symbol is 0 and the symbols alternate. Here $\|x^2\| = 4 = 2\|x\|$, $\|x^3\| = 6 = 3\|x\|$, and for all $n \in \mathbf{N}$, $\|x^n\| = n\|x\|$. □

We are just about ready to tackle the major theme of this section, the concept of a language. Before we do so, however, there are three other ideas we need to inquire about. These ideas involve special subsections of strings.

DEFINITION 6.8   If $x, y \in \Sigma^*$ and $w = xy$, then the string x is called a *prefix* of w, and if $y \neq \lambda$, then x is said to be a *proper prefix*. Similarly, the string y is called a *suffix* of w; it is a *proper suffix* when $x \neq \lambda$.   ▬

Example 6.4   Let $\Sigma = \{a, b, c\}$ and consider the string $w = abbcc$. Then each of the strings λ, a, $ab, abb, abbc$, and $abbcc$ is a prefix of w. On the other hand, each of the strings λ, $c, cc, bcc, bbcc$, and $abbcc$ is a suffix of w.   □

Example 6.5   If $\|x\| = 5$, $\|y\| = 4$ and $w = xy$, then w has x as a proper prefix and y as a proper suffix. In all, w has nine proper prefixes and nine proper suffixes, because λ is both a proper prefix and proper suffix for every string in Σ^+.   □

Example 6.6   If $w, a, b, c, d \in \Sigma^*$, then $(ab = w = cd) \Leftrightarrow (b$ is a suffix of d, or d is a suffix of $b) \Leftrightarrow (a$ is a prefix of c, or c is a prefix of $a)$.   □

DEFINITION 6.9   If $x, y, z \in \Sigma^*$ and $w = xyz$, then y is called a *substring* of w. When at least one of x and z is different from λ, we call y a *proper substring*.   ▬

Example 6.7   For $\Sigma = \{0, 1\}$, let $w = 00101110 \in \Sigma^*$. We find the following substrings in w:

1. 1011: This arises in only one way—when $w = xyz$, with $x = 00$, $y = 1011$, and $z = 10$.
2. 10: This example comes about in two ways:
   a) $w = xyz$ where $x = 00$, $y = 10$, and $z = 1110$, and
   b) $w = xyz$ for $x = 001011$, $y = 10$, and $z = \lambda$.

In case (b) the substring is also a (proper) suffix of w.   □

Now that we are familiar with the necessary definitions, it is time to think about the concept of language.

When we consider the standard alphabet, including the blank, there are many strings such as *qxio*, *the wxxy red atzl*, and *aeytl* that do not represent words or parts of sentences that appear in the English language, even though they are elements of Σ^*. Consequently, in order to consider only those words and expressions that make sense in the English language, we concentrate on a subset of Σ^*. This leads us to the following generalization.

DEFINITION 6.10   For a given alphabet Σ, any subset of Σ^* is called a *language* over Σ. This includes the subset $\emptyset$, which we call the *empty language*.   ▬

Example 6.8   With $\Sigma = \{0, 1\}$, the sets $A = \{0, 01, 001\}$ and $B = \{0, 01, 001, 0001, \ldots\}$ are examples of languages over Σ.   □

Example 6.9   With Σ the alphabet of 26 letters, 10 digits, and the special symbols used in a given implementation of Pascal, the collection of executable programs for that implementation constitutes a language. In the same situation, each executable program could be considered a language, as could a particular finite set of such programs. ☐

Since languages are sets we can form the union, intersection, and symmetric difference of two languages. However, for the work here, an extension of the binary operation defined (in Definition 6.6) for strings is more useful.

DEFINITION 6.11　For an alphabet Σ and languages $A, B \subseteq \Sigma^*$, the *concatenation* of A, B, denoted AB, is $\{ab \mid a \in A, b \in B\}$.

We can think of concatenation as the cross product wherein we remove the comma and parentheses in each ordered pair, so that (a, b) becomes ab. We shall see that just as $A \times B \neq B \times A$ in general, we also have $AB \neq BA$ in general. For A, B finite we did have $|A \times B| = |B \times A|$, but here $|AB| \neq |BA|$ is possible for finite languages.

Example 6.10　Let $\Sigma = \{x, y, z\}$ and let A, B be the finite languages $A = \{x, xy, z\}, B = \{\lambda, y\}$. Then $AB = \{x, xy, z, xyy, zy\}$ and $BA = \{x, xy, z, yx, yxy, yz\}$, so

　1. $|AB| = 5 \neq 6 = |BA|$, and
　2. $|AB| = 5 \neq 6 = 3 \cdot 2 = |A\|B|$.

The differences arise because there are two ways to represent xy: (1) xy for $x \in A, y \in B$ and (2) $xy\lambda$ where $xy \in A$ and $\lambda \in B$. (The concept of uniqueness of representation is something we cannot take for granted. Although it does not hold here, it is a key to the success of many mathematical ideas. We saw this in the Fundamental Theorem of Arithmetic (Theorem 4.11) and shall come across it again in Chapter 15.) ☐

The example above suggests that for finite languages A and B, $|AB| \leq |A\|B|$. This can be shown to be true in general.

The following theorem deals with some of the properties satisfied by the concatenation of languages.

THEOREM 6.1　For an alphabet Σ, let $A, B, C \subseteq \Sigma^*$. Then

　a) $A\{\lambda\} = \{\lambda\}A = A$　　　　　　　**b)** $(AB)C = A(BC)$
　c) $A(B \cup C) = AB \cup AC$　　　　**d)** $(B \cup C)A = BA \cup CA$
　e) $A(B \cap C) \subseteq AB \cap AC$　　　　**f)** $(B \cap C)A \subseteq BA \cap CA$

Proof　We prove (d) and (f) and leave the other parts for the reader.
　　　(d) With $x \in \Sigma^*$, $x \in (B \cup C)A \Rightarrow x = yz$ for $y \in B \cup C$ and $z \in A \Rightarrow$ $(x = yz$ for $y \in B, z \in A)$ or $(x = yz$ for $y \in C, z \in A) \Rightarrow x \in BA$ or $x \in CA$, so $(B \cup C)A \subseteq BA \cup CA$. Conversely, $x \in BA \cup CA \Rightarrow x \in BA$ or $x \in CA \Rightarrow x = ba_1$ where $b \in B$ and $a_1 \in A$ or $x = ca_2$ where $c \in C$ and $a_2 \in A$. Assume $x = ba_1$ for $b \in B, a_1 \in A$. Since $B \subseteq B \cup C$, we have $x = ba_1$, where $b \in B \cup C$ and $a_1 \in A$. Then $x \in (B \cup C)A$, so $BA \cup CA \subseteq (B \cup C)A$. (The argument is sim-

ilar if $x = ca_2$.) With both inclusions established, it follows that $(B \cup C)A = BA \cup CA$.

(f) For $x \in \Sigma^*$, $x \in (B \cap C)A \Rightarrow x = yz$ where $y \in B \cap C$ and $z \in A \Rightarrow$ $(x = yz$ for $y \in B$ and $z \in A)$ and $(x = yz$ for $y \in C$ and $z \in A) \Rightarrow x \in BA \cap CA$, so $(B \cap C)A \subseteq BA \cap CA$.

With $\Sigma = \{x, y, z\}$, let $B = \{x, xx, y\}$, $C = \{y, xy\}$, and $A = \{y, yy\}$. Then $xyy \in BA \cap CA$ but $xyy \notin (B \cap C)A$. Consequently $(B \cap C)A \subset BA \cap CA$. ■

Comparable to the concepts of Σ^n, Σ^*, Σ^+, the following definitions are given for an arbitrary language $A \subseteq \Sigma^*$.

DEFINITION 6.12   For any language $A \subseteq \Sigma^*$ we can construct other languages as follows:

a) $A^0 = \{\lambda\}$, $A^1 = A$, and for any $n \in \mathbf{Z}^+$, $A^{n+1} = \{ab \mid a \in A, b \in A^n\}$

b) $A^+ = \bigcup_{n \in \mathbf{Z}^+} A^n$, the *positive closure* of A

c) $A^* = A^+ \cup \{\lambda\}$. A^* is called the *Kleene closure* of A, in honor of the American logician Stephen Cole Kleene (1909–   ).

Example 6.11   If $\Sigma = \{x, y, z\}$ and $A = \{x\}$, then (1) $A^0 = \{\lambda\}$; (2) $A^n = \{x^n\}$, for any $n \in \mathbf{N}$; (3) $A^+ = \{x^n \mid n \geq 1\}$; and (4) $A^* = \{x^n \mid n \geq 0\}$. □

Example 6.12   Let $\Sigma = \{x, y\}$.

a) If $A = \{xx, xy, yx, yy\} = \Sigma^2$, then A^* is the language of all strings w in Σ^* where the length of w is even.

b) With A as in (a) and $B = \{x, y\}$, the language BA^* contains all the strings in Σ^* of odd length, and we also find in this case that $BA^* = A^*B$ and that $\Sigma^* = A^* \cup BA^*$.

c) The language $\{x\}\{x, y\}^*$ (the concatenation of the languages $\{x\}$ and $\{x, y\}^*$) contains every string in Σ^* for which x is a prefix.
The language containing all strings in Σ^* for which yy is a suffix can be defined by $\{x, y\}^*\{yy\}$.
Every string in the language $\{x, y\}^*\{xxy\}\{x, y\}^*$ has xxy as a substring.

d) Any string in the language $\{x\}^*\{y\}^*$ consists of a finite number (possibly zero) of x's followed by a finite number (also possibly zero) of y's. And although $\{x\}^*\{y\}^* \subseteq \{x, y\}^*$, the string $w = xyx$ is in $\{x, y\}^*$ but not in $\{x\}^*\{y\}^*$. Hence $\{x\}^*\{y\}^* \subset \{x, y\}^*$. □

Example 6.13   In the algebra of real numbers, if $a, b \in \mathbf{R}$ and $a, b > 0$, then $a^2 = b^2 \Rightarrow a = b$. However, in the case of languages, if $\Sigma = \{x, y\}$, $A = \{\lambda, x, x^3, x^4, \ldots\} = \{x^n \mid n \geq 0\} - \{x^2\}$ and $B = \{x^n \mid n \geq 0\}$, then $A^2 = B^2 (= B)$, but $A \neq B$. (Note: We never have $\lambda \in \Sigma$, but it is possible to have $\lambda \in A$.) □

We close this section with a lemma and a second theorem that deal with properties of languages. The lemma provides another instance where we use mathematical induction.

LEMMA 6.1.   Let Σ be an alphabet, with languages $A, B \subseteq \Sigma^*$. If $A \subseteq B$, then $A^n \subseteq B^n$, for all $n \in \mathbf{Z}^+$.

Proof   Here $S(1)$† follows, because $A^1 = A \subseteq B = B^1$. Assuming $S(k)$, we have $A^k \subseteq B^k$ for $k \in \mathbf{Z}^+$. Now consider a string $x \in A^{k+1}$. From Definition 6.12, $x = x_1 x_k$, where $x_1 \in A, x_k \in A^k$. Since $A \subseteq B$ and $A^k \subseteq B^k$ (by the induction hypothesis), we have $x_1 \in B$ and $x_k \in B^k$. Consequently, $x = x_1 x_k \in BB^k = B^{k+1}$. Therefore $S(k+1)$ is true, and by the principle of finite induction, it follows that if $A \subseteq B$, then $A^n \subseteq B^n$ for all $n \in \mathbf{Z}^+$.   ■

THEOREM 6.2   For an alphabet Σ and languages $A, B \subseteq \Sigma^*$,

a) $A \subseteq AB^*$

b) $A \subseteq B^*A$

c) $A \subseteq B \Rightarrow A^+ \subseteq B^+$

d) $A \subseteq B \Rightarrow A^* \subseteq B^*$

e) $AA^* = A^*A = A^+$

f) $A^*A^* = A^* = (A^*)^* = (A^*)^+ = (A^+)^*$

g) $(A \cup B)^* = (A^* \cup B^*)^* = (A^*B^*)^*$

Proof   We provide the proofs for parts (c) and (g).
    (c) Let $A \subseteq B$ and $x \in A^+$. Then $x \in A^+ \Rightarrow x \in A^n$, for some $n \in \mathbf{Z}^+$. From Lemma 6.1 it then follows that $x \in B^n \subseteq B^+$, and we have shown that $A^+ \subseteq B^+$.
    (g) $[(A \cup B)^* = (A^* \cup B^*)^*]$. $A \subseteq A^*, B \subseteq B^* \Rightarrow (A \cup B) \subseteq (A^* \cup B^*) \Rightarrow (A \cup B)^* \subseteq (A^* \cup B^*)^*$ (by part d). Conversely, $A, B \subseteq A \cup B \Rightarrow A^*, B^* \subseteq (A \cup B)^*$ (by part d) $\Rightarrow (A^* \cup B^*) \subseteq (A \cup B)^* \Rightarrow (A^* \cup B^*)^* \subseteq (A \cup B)^*$ (by parts d and f). From both inclusions we have $(A \cup B)^* = (A^* \cup B^*)^*$.
    $[(A^* \cup B^*)^* = (A^*B^*)^*]$. $A^*, B^* \subseteq A^*B^*$ (by parts a and b) $\Rightarrow (A^* \cup B^*) \subseteq A^*B^* \Rightarrow (A^* \cup B^*)^* \subseteq (A^*B^*)^*$ (by part d). Conversely, if $xy \in A^*B^*$ where $x \in A^*$ and $y \in B^*$, then $x, y \in A^* \cup B^*$, so $xy \in (A^* \cup B^*)^*$, and $A^*B^* \subseteq (A^* \cup B^*)^*$. Using parts (d) and (f) again, $(A^*B^*)^* \subseteq (A^* \cup B^*)^*$, and the result follows.   ■

EXERCISES 6.1

1. Let $\Sigma = \{a, b, c, d, e\}$. (a) What is $|\Sigma^2|$? $|\Sigma^3|$? (b) How many strings in Σ^* have length at most five?

2. For $\Sigma = \{w, x, y, z\}$ determine the number of strings in Σ^* of length five (a) that start with w; (b) with precisely two w's; (c) with no w's; (d) with an even number of w's.

3. If $x \in \Sigma^*$ and $\|x^3\| = 36$, what is $\|x\|$?

4. Let $\Sigma = \{\beta, x, y, z\}$ where β denotes a blank, so $x\beta \neq x$, $\beta\beta \neq \beta$, and $x\beta y \neq xy$ but $x\lambda y = xy$. Compute each of the following

a) $\|\lambda\|$

b) $\|\lambda\lambda\|$

c) $\|\beta\|$

d) $\|\beta\beta\|$

e) $\|\beta^3\|$

f) $\|x\beta\beta y\|$

g) $\|\beta\lambda\|$

h) $\|\lambda^{10}\|$

† We use the notation of Section 4.1 here. $S(n)$ is the statement: $A \subseteq B \Rightarrow A^n \subseteq B^n$.

5. Let $\Sigma = \{v, w, x, y, z\}$ and $A = \bigcup_{n=1}^{6} \Sigma^n$. How many strings in A have xy as a proper prefix?

6. Let Σ be an alphabet. Let $x_i \in \Sigma$ for $1 \le i \le 100$, where $x_i \ne x_j$ for any $1 \le i < j \le 100$. How many nonempty substrings are there for the string $s = x_1 x_2 \ldots x_{100}$?

7. Given a nonempty language $A \subseteq \Sigma^*$, prove that if $A^2 = A$, then $\lambda \in A$.

8. Let Σ be an alphabet with $a \in \Sigma$. Define the functions $p_a, s_a, r: \Sigma^* \to \Sigma^*$ and the function $d: \Sigma^+ \to \Sigma^*$ as follows:

   (i)   The prefix (by a) function: $p_a(x) = ax, x \in \Sigma^*$.

   (ii) The suffix (by a) function: $s_a(x) = xa, x \in \Sigma^*$.

   (iii) The reversal function: $r(\lambda) = \lambda$; for $x \in \Sigma^+$, if $x = x_1 x_2 \ldots x_{n-1} x_n$, where $x_i \in \Sigma$ for $1 \le i \le n$, then $r(x) = x_n x_{n-1} \ldots x_2 x_1$.

   (iv) The front deletion function: for $x \in \Sigma^+$, if $x = x_1 x_2 x_3 \ldots x_n$, then $d(x) = x_2 x_3 \ldots x_n$.

   a) Which of these four functions is (or are) one-to-one?

   b) Determine which of these four functions is (or are) onto. If a function is not onto, determine its range.

   c) Are any of these four functions invertible? If so, determine their inverse functions.

   d) Suppose that $\Sigma = \{a, e, i, o, u\}$. How many words x in Σ^4 satisfy $r(x) = x$? How many in Σ^5? How many in Σ^n, where $n \in \mathbf{N}$?

   e) For $x \in \Sigma^*$, determine $(d \circ p_a)(x)$ and $(r \circ d \circ r \circ s_a)(x)$.

   f) Verify that $r \circ p_a = s_a \circ r$.

   g) If $\Sigma = \{a, e, i, o, u\}$ and $B = \{ae, ai, ao, oo, eio, eiouu\} \subseteq \Sigma^*$, find $r^{-1}(B)$, $p_a^{-1}(B), s_a^{-1}(B)$, and $|d^{-1}(B)|$.

9. If A, B, C, and D are languages over Σ, prove that (a) $(A \subseteq B \wedge C \subseteq D) \Rightarrow AC \subseteq BD$; and (b) $A\emptyset = \emptyset A = \emptyset$.

10. For $\Sigma = \{x, y, z\}$, let $A, B \subseteq \Sigma^*$ be given by $A = \{xy\}$ and $B = \{\lambda, x\}$. Determine (a) AB; (b) BA; (c) B^3; (d) B^+; (e) A^*.

11. If $A (\ne \emptyset)$ is a language and $A^2 = A$, prove that $A = A^*$.

12. Provide the proofs for the remaining parts of Theorems 6.1 and 6.2.

13. Prove that for any finite languages $A, B \subseteq \Sigma^*$, $|AB| \le |A| \|B|$.

14. If A and B are languages over Σ, and $A \subseteq B^*$, prove that $A^* \subseteq B^*$.

15. For a given alphabet Σ, let I denote an index set, where for each $i \in I$, $B_i \subseteq \Sigma^*$. If $A \subseteq \Sigma^*$, prove that (a) $A(\bigcup_{i \in I} B_i) = \bigcup_{i \in I} AB_i$; and (b) $(\bigcup_{i \in I} B_i)A = \bigcup_{i \in I} B_i A$. (These results generalize parts (c) and (d) of Theorem 6.1.)

16. For $\Sigma = \{x, y\}$, use finite languages from Σ^* (as in Example 6.12), together with set operations, to describe the set of strings in Σ^* that (a) contain exactly one occurrence of x; (b) contain exactly two occurrences of x; (c) ·

begin with x; (d) end in yxy; (e) begin with x or end in yxy or both; (f) begin with x or end in yxy but not both.

6.2

FINITE STATE MACHINES: A FIRST ENCOUNTER

We return now to the vending machine mentioned at the start of this chapter and analyze it in the following circumstance.

At a metropolitan office, a vending machine dispenses two kinds of soft drinks in cans: cola (C) and root beer (RB). The cost of a can of either soft drink is 20¢. The machine accepts nickels, dimes, and quarters and returns the necessary change. One day Mary Jo decides to have a can of root beer. She goes to the vending machine, inserts two nickels and a dime, in that order, and presses the white button, denoted W. Out comes her can of root beer. (To get a can of cola one presses the black button, denoted B.)

What Mary Jo has done, in making her purchase, can be represented as shown in Table 6.1, where t_0 is the initial time, when she inserts her first nickel, and t_1, t_2, t_3, t_4 are later moments in time, with $t_1 < t_2 < t_3 < t_4$.

Table 6.1

	t_0	t_1	t_2	t_3	t_4
State	(1) s_0	(4) s_1 (5¢)	(7) s_2 (10¢)	(10) s_3 (20¢)	(13) s_0
Input	(2) 5¢	(5) 5¢	(8) 10¢	(11) W	
Output	(3) Nothing	(6) Nothing	(9) Nothing	(12) RB	

The numbers (1), (2), ..., (12), (13) in this table indicate the order of events in the purchase of Mary Jo's root beer. For each input at time t_i, $0 \le i \le 3$, there is *at that time* a corresponding output and then a change in state. The new state at time t_{i+1} depends on both the input and the (present) state at time t_i.

The machine is in a state of readiness at state s_0. It waits for a customer to start inserting coins that will total 20¢ or more and then press a button to get a soft drink. If at any time the total of the coins inserted exceeds 20¢, the machine provides the needed change (before the customer presses the button to get the soft drink).

At time t_0 Mary Jo provides the machine with her first input, 5¢. She receives nothing at this time but at the later time t_1, the machine is in state s_1, where it *remembers* her total of 5¢ and waits for her second input (of 5¢ at time t_1). The machine again (at time t_1) provides no output but at the next time, t_2, it is in state s_2, remembering a total of 10¢ = 5¢ (remembered at state s_1) + 5¢ (inserted at time t_1). Providing her dime (at time t_2) as the next input to the machine, Mary Jo receives no soft drink yet, because the machine doesn't "know" which type Mary Jo prefers, but it does "know" now (t_3) that she has inserted the necessary total of 20¢ = 10¢ (remembered at state s_2) + 10¢ (inserted at time t_2). At last Mary Jo presses the white button and at time t_3 the machine dispenses the output (her can of root beer) and then returns, at time t_4, to the starting state s_0, just in time for Mary Jo's friend Rizzo to deposit a quarter, receive her nickel change, press the black button and obtain the can of cola she desires. The purchase made by Rizzo is analyzed in Table 6.2.

Table 6.2

	t_0		t_1		t_2	
State	(1)	s_0	(4)	s_3 (20¢)	(7)	s_0
Input	(2)	25¢	(5)	B		
Output	(3)	5¢ change	(6)	C		

What has happened in the case of this vending machine can be abstracted to help in the analysis of certain aspects of digital computers and telephone communication systems.

The major features of such a machine are as follows:

1. The machine can be in only one of *finitely many states* at a given time. These states are called the *internal states* of the machine, and at a given time the total memory available to the machine is the knowledge of which internal state it is in at that moment.

2. The machine will accept as *input* only a finite number of symbols, and collectively these are referred to as the *input alphabet* $\mathscr{I}$. In the vending machine example, the input alphabet is {nickel, dime, quarter, W, B}, each item of which is recognized by each internal state.

3. An *output* and *next state* are determined by each combination of inputs and internal states. The finite set of all possible outputs constitutes the *output alphabet* $\mathbb{O}$ for the machine.

4. We assume that the sequential processings of the machine are *synchronized* by separate and distinct clock pulses and that the machine operates in a *deterministic* manner, where the output is completely determined by the total input provided and the starting state of the machine.

These observations lead us to the following definition.

DEFINITION 6.13 A *finite state machine* is a five-tuple $M = (S, \mathscr{I}, \mathbb{O}, v, \omega)$ where $S =$ the set of internal states for M; $\mathscr{I} =$ the input alphabet for M; $\mathbb{O} =$ the output alphabet for M; $v: S \times \mathscr{I} \to S$ is the *next state function*; and $\omega: S \times \mathscr{I} \to \mathbb{O}$ is the *output function*.

Using the notation of this definition, if the machine is in state s at time t_i and we input x at this time, then the output at time t_i is $\omega(s, x)$. This output is followed by a transition of the machine at time t_{i+1} to the next internal state given by $v(s, x)$.

We assume that when a finite state machine receives its first input, we are at time $t_0 = 0$ and the machine is in a designated starting state denoted by s_0. Our development will concentrate primarily on the output and state transitions that take place sequentially, with little or no reference to the sequence of clock pulses at times $t_0, t_1, t_2, \ldots$.

Since the sets S, $\mathcal{I}$, and $\mathcal{O}$ are finite, it is possible to represent v and ω, for a given finite state machine, by means of a table that lists $v(s, x)$ and $\omega(s, x)$ for all $s \in S$ and $x \in \mathcal{I}$. Such a table is referred to as the *state table* or *transition table* for the given machine. A second representation of the machine is made by means of a *state diagram*.

We demonstrate the state table and state diagram in the following examples.

Example 6.14   Consider the finite state machine $M = (S, \mathcal{I}, \mathcal{O}, v, \omega)$ where $S = \{s_0, s_1, s_2\}$, $\mathcal{I} = \mathcal{O} = \{0, 1\}$, and v, ω are given by the *state table* in Table 6.3. The first column of the table lists the (*present*) states for the machine. The entries in the second row are the elements of the input alphabet $\mathcal{I}$, listed once under v and then again under ω. The six numbers in the last two columns (and last three rows) are elements of the output alphabet $\mathcal{O}$.

Table 6.3

	v		ω	
	0	1	0	1
s_0	s_0	s_1	0	0
s_1	s_2	s_1	0	0
s_2	s_0	s_1	0	1

To calculate $v(s_1, 1)$, for example, we find s_1 in the column of present states and proceed horizontally over from s_1 until we are below the entry 1 in the section of the table for v. This entry gives $v(s_1, 1) = s_1$. In the same way we find $\omega(s_1, 1) = 0$.

With s_0 designated as the starting state, if the input provided to M is the string 1010, then the output is 0010, as demonstrated in Table 6.4. Here the machine is left in state s_2, so that if we had another input string, we would provide the first character of that string, here 0, at state s_2 unless the machine is *reset* to start once again at s_0.   □

Table 6.4

State	s_0	$v(s_0, 1) = s_1$	$v(s_1, 0) = s_2$	$v(s_2, 1) = s_1$	$v(s_1, 0) = s_2$
Input	1	0	1	0	0
Output	$\omega(s_0, 1) = 0$	$\omega(s_1, 0) = 0$	$\omega(s_2, 1) = 1$	$\omega(s_1, 0) = 0$	

Since we are primarily interested in the output, not in the sequence of transition states, the same machine can be represented by means of a *state diagram*. Here we can obtain the output string without actually listing the transition states. In such a diagram each internal state s is represented by a circle with s inside of it.

For states s_i and s_j, if $v(s_i, x) = s_j$ for $x \in \mathcal{I}$, and $\omega(s_i, x) = y$ for $y \in \mathcal{O}$, we represent this in the state diagram by drawing a *directed edge* (or *arc*) from the circle for s_i to the circle for s_j and labeling the arc with the input x and output y as shown in Fig. 6.1.

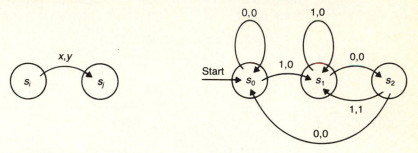

Figure 6.1   **Figure 6.2**

With these conventions, the state diagram for the machine M of Table 6.3 is shown in Fig. 6.2. Although the table is more compact, the diagram enables us to follow an input string through each transition state it determines, picking up each of the corresponding output symbols before each transition. Here if the input string is 00110101, then starting at state s_0, the first input of 0 yields an output of 0 and returns us to s_0. The next input of 0 yields the same result, but for the third input, 1, the output is 0 and we are now in state s_1. Continuing in this manner, we arrive at the output string 00000101 and finish in state s_1. (We note that the input string 00110101 is an element of $\mathcal{I}^*$, the Kleene closure of $\mathcal{I}$, and that the output string is in $\mathcal{O}^*$, the Kleene closure of $\mathcal{O}$.)

Starting at s_0, what is the output for the input string 1100101101?

Example 6.15 For the vending machine described earlier in this section, we have the state table, Table 6.5, with

1. $S = \{s_0, s_1, s_2, s_3, s_4\}$, where at state $s_k, 0 \le k \le 4$, the machine remembers retaining $5k$ cents.

Table 6.5

	v					ω				
	5¢	10¢	25¢	B	W	5¢	10¢	25¢	B	W
s_0	s_1	s_2	s_4	s_0	s_0	n	n	5¢	n	n
s_1	s_2	s_3	s_4	s_1	s_1	n	n	10¢	n	n
s_2	s_3	s_4	s_4	s_2	s_2	n	n	15¢	n	n
s_3	s_4	s_4	s_4	s_3	s_3	n	5¢	20¢	n	n
s_4	s_4	s_4	s_4	s_0	s_0	5¢	10¢	25¢	C	RB

2. $\mathcal{I} = \{5¢, 10¢, 25¢, B, W\}$, where B denotes the black button one presses for a cola, and W the white button for root beer.

3. $\mathcal{O} = \{n(\text{nothing}), RB(\text{root beer}), C(\text{cola}), 5¢, 10¢, 15¢, 20¢, 25¢\}$. □

As we observed at the end of Example 6.14, for a general finite state machine $M = (S, \mathcal{I}, \mathcal{O}, v, \omega)$, the input can be realized as an element of $\mathcal{I}^*$, with the output from $\mathcal{O}^*$. Consequently, it is to our advantage to extend the domains of v and ω from $S \times \mathcal{I}$ to $S \times \mathcal{I}^*$. For ω we enlarge the codomain to $\mathcal{O}^*$, recalling, should the need arise, that both $\mathcal{I}^*$ and $\mathcal{O}^*$ contain an empty string, λ. With these extensions, if $x_1 x_2 \ldots x_k \in \mathcal{I}^*$, for $k \in \mathbf{Z}^+$, then staring at any state $s_1 \in S$, we have

$$v(s_1, x_1) = s_2\dagger$$
$$v(s_1, x_1 x_2) = v(v(s_1, x_1), x_2) = v(s_2, x_2) = s_3$$
$$v(s_1, x_1 x_2 x_3) = v(v(\underbrace{v(s_1, x_1)}_{s_2}, x_2), x_3) = s_4$$
$$v(s_2, x_2) = s_3$$

$$\cdots \cdots \cdots$$

$$v(s_1, x_1 x_2 \ldots x_k) = v(s_k, x_k) = s_{k+1}, \quad \text{and}$$
$$\omega(s_1, x_1) = y_1$$
$$\omega(s_1, x_1 x_2) = \omega(s_1, x_1)\omega(v(s_1, x_1), x_2) = \omega(s_1, x_1)\omega(s_2, x_2) = y_1 y_2$$
$$\omega(s_1, x_1 x_2 x_3) = \omega(s_1, x_1)\omega(s_2, x_2)\omega(s_3, x_3) = y_1 y_2 y_3$$

$$\cdots \cdots \cdots$$

$$\omega(s_1, x_1 x_2 \ldots x_k) = \omega(s_1, x_1)\omega(s_2, x_2) \ldots \omega(s_k, x_k) = y_1 y_2 \ldots y_k \in \mathcal{O}^*$$

Also, $v(s_1, \lambda) = s_1$ for any $s_1 \in S$.

(We shall use these extensions again in Chapter 7.)

We close this section with an example that is relevant in computer science.

Example 6.16   Let $x = x_5 x_4 x_3 x_2 x_1 = 00111$ and $y = y_5 y_4 y_3 y_2 y_1 = 01101$ be binary numbers where x_1 and y_1 are the least significant bits. The leading 0's in x and y are there to make the strings for x and y of equal length and to guarantee enough places to complete the sum. A *serial binary adder* is a finite state machine that we can use to obtain $x + y$. The diagram in Fig. 6.3 illustrates this, where $z = z_5 z_4 z_3 z_2 z_1$ has the least significant bit z_1.

In the addition $z = x + y$, we have

$$
\begin{array}{rccccc}
x = & 0 & 0 & 1 & 1 & 1 \\
+ y = +\!\! & 0 & 1 & 1 & 0 & 1 \\
\hline
z = & 1 & 0 & 1 & 0 & 0 \\
\end{array}
$$

third addition   first addition

$x = x_5 x_4 x_3 x_2 x_1 \longrightarrow$ | Serial binary adder | $\longrightarrow z = z_5 z_4 z_3 z_2 z_1$

$y = y_5 y_4 y_3 y_2 y_1 \longrightarrow$

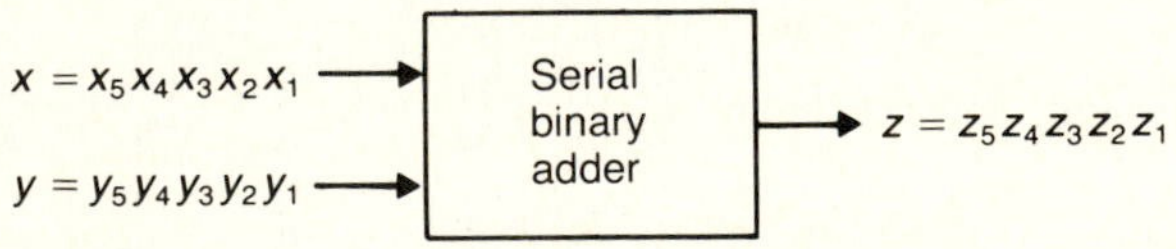

Figure 6.3

† The state s_2 is determined by s_1 and x_1. It is not simply the second in a predetermined list of states.

We note that for the first addition $x_1 = y_1 = 1$ and $z_1 = 0$, whereas for the third addition we have $x_3 = y_3 = 1$ and $z_3 = 1$ because of a *carry* from the addition of x_2 and y_2 (and the *carry* from $x_1 + y_1$). Consequently each output depends on the sum of two inputs and the ability to *remember* a carry of 0 or 1, which is crucial when it is 1.

The serial binary adder is modeled by a finite state machine $M = (S, \mathcal{I}, \mathcal{O}, \nu, \omega)$ as follows. The set $S = \{s_0, s_1\}$, where s_i indicates a carry of i; $\mathcal{I} = \{00, 01, 10, 11\}$, so there is a pair of inputs depending on whether we are seeking $0 + 0$, $0 + 1$, $1 + 0$, or $1 + 1$, respectively; and $\mathcal{O} = \{0, 1\}$. The functions ν, ω are given in the state table (Table 6.6) and the state diagram (Fig. 6.4).

Table 6.6

	ν				ω			
	00	01	10	11	00	01	10	11
s_0	s_0	s_0	s_0	s_1	0	1	1	0
s_1	s_0	s_1	s_1	s_1	1	0	0	1

In Table 6.6 we find, for example, that $\nu(s_1, 01) = s_1$ and $\omega(s_1, 01) = 0$, because s_1 indicates a carry of 1 from the addition of the previous bits. The 01 input indicates that we are adding 0 and 1 (and carrying a 1). Hence the sum is 10 and $\omega(s_1, 01) = 0$ for the 0 in 10. The carry is again remembered in $s_1 = \nu(s_1, 01)$.

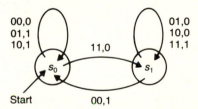

Figure 6.4

From the state diagram (Fig. 6.4) we see that the starting state must be s_0, because there is no carry prior to the addition of the least significant bits. ☐

The state diagrams in Figs. 6.2 and 6.4 are examples of *labeled directed graphs*. We shall see more about graph theory throughout the text, for it has applications not only in computer science and electrical engineering but also in coding theory (prefix codes) and optimization (transport networks).

EXERCISES 6.2

1. Using the finite state machine of Example 6.14, find the output for each of the following input strings $x \in \mathcal{I}^*$, and determine the last internal state in the transition process. (Assume that we always start at s_0.)

   a) $x = 1010101$ **b)** $x = 1001001$ **c)** $x = 101001000$

2. For the finite state machine of Example 6.14, an input string x, starting at state s_0, produces the output string 00101. Determine x.

3. Let $M = (S, \mathcal{I}, \mathbb{O}, v, \omega)$ be a finite state machine where $S = \{s_0, s_1, s_2, s_3\}$, $\mathcal{I} = \{a, b, c\}$, $\mathbb{O} = \{0, 1\}$, and v, ω are determined by Table 6.7.

Table 6.7

	v			ω		
	a	b	c	a	b	c
s_0	s_0	s_3	s_2	0	1	1
s_1	s_1	s_1	s_3	0	0	1
s_2	s_1	s_1	s_3	1	1	0
s_3	s_2	s_3	s_0	1	0	1

a) Starting at s_0, what is the output for the input string $abbccc$?

b) Draw the state diagram for this finite state machine.

4. Give the state table for the vending machine of Example 6.15 if the cost of a can of cola or root beer is increased to 25¢.

5. A finite state machine $M = (S, \mathcal{I}, \mathbb{O}, v, \omega)$ has $\mathcal{I} = \mathbb{O} = \{0, 1\}$ and is determined by the state diagram shown in Fig. 6.5.

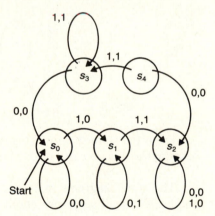

Figure 6.5

a) Determine the output string for the input string 110111, starting at s_0. What is the last transition state?

b) Answer part (a) for the same string but with s_1 as the starting state. What about s_2 and s_3 as starting states?

c) Find the state table for this machine.

d) In which state should we start so that the input string 10010 produces the output 10000?

e) Determine an input string $x \in \mathcal{I}^*$ of minimal length, such that $v(s_4, x) = s_1$. Is x unique?

6. Machine M has $\mathcal{I} = \{0, 1\} = \mathcal{O}$ and is determined by the state diagram shown in Fig. 6.6.

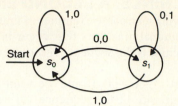

Figure 6.6

a) Describe in words what this finite state machine does.

b) What must state s_1 remember?

c) Find two languages $A, B \subseteq \mathcal{I}^*$ such that for every $x \in AB$, $\omega(s_0, x)$ has 1 as a suffix.

7. a) If S, $\mathcal{I}$, and $\mathcal{O}$ are finite sets, with $|S| = 3$, $|\mathcal{I}| = 5$, and $|\mathcal{O}| = 2$, determine

   (i) $|S \times \mathcal{I}|$;

   (ii) the number of functions $\nu \colon S \times \mathcal{I} \to S$; and

   (iii) the number of functions $\omega \colon S \times \mathcal{I} \to \mathcal{O}$.

b) For S, $\mathcal{I}$, and $\mathcal{O}$ in part (a), how many finite state machines do they determine?

8. Let $M = (S, \mathcal{I}, \mathcal{O}, \nu, \omega)$ be a finite state machine with $\mathcal{I} = \mathcal{O} = \{0, 1\}$ and S, ν, and ω determined by the state diagram shown in Fig. 6.7.

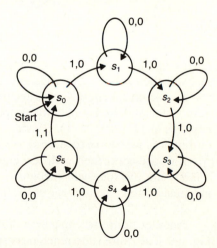

Figure 6.7

a) Find the output for the input string $x = 0110111011$.

b) Give the transition table for this finite state machine.

c) Starting in state s_0, if the output for an input string x is 0000001, determine all possibilities for x.

d) Describe in words what this finite state machine does.

6.3

FINITE STATE MACHINES: A SECOND ENCOUNTER

Having seen some examples of finite state machines, we turn to the study of some additional machines that occur in computer hardware. One important type of machine is the *sequence recognizer*.

Example 6.17

Here $\mathcal{I} = \mathcal{O} = \{0, 1\}$ and we want to construct a machine that recognizes each occurrence of the sequence 111 as it is encountered in any input string $x \in \mathcal{I}^*$. For example, if $x = 1110101111$, then the corresponding output should be 0010000011, where a 1 in the ith position of the output indicates that a 1 can be found in positions i, $i - 1$, and $i - 2$ of x. Here overlapping of sequences of 111 can occur, so some characters in the input string can be thought of as characters in more than one triple of 1's.

Letting s_0 denote the starting state, we realize that we must have a state to remember 1 (the possible start of 111) and a state to remember 11. In addition, anytime our input symbol is 0, we go back to s_0 and start the search for three successive 1's over again.

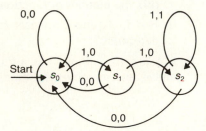

Figure 6.8

In Fig. 6.8, s_1 remembers a single 1, and s_2 remembers the string 11. If s_2 is reached, then a third "1" indicates the occurrence of the triple in the input string, and the output 1 *recognizes* this occurrence. But this third "1" also means that we have the first two 1's of another possible triple coming up in the string (as happens in 11101011"1"1). So after recognizing the occurrence of 111 with an output of 1, we return to state s_2 to remember the two inputs of 1"1".

If we are concerned with recognizing all strings that end in 111, then for any $x \in \mathcal{I}^*$, the machine will recognize such a sequence with final output 1. This machine is then a recognizer of the language $A = \{0, 1\}^*\{111\}$. □

Another finite state machine that recognizes the same triple 111 is shown in Fig. 6.9. The finite state machines represented by the state diagrams in Figs. 6.8 and 6.9 perform the same task and are said to be *equivalent*. The state diagram in Fig. 6.9 has one more state than that in Fig. 6.8, but at this stage we are not overly concerned with getting a finite state machine with a minimal number of states. In Chapter 7 we shall develop a technique to take a given finite state machine M and find one that is equivalent to it and has the smallest number of internal states needed.

The next example is a bit more selective.

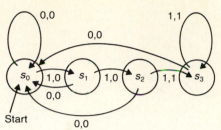

Start

Figure 6.9

Example 6.18   Now we want not only to recognize the occurrence of 111 but we want to recognize only those occurrences that end in a position that is a multiple of three. Consequently, with $\mathcal{I} = \mathbb{O} = \{0, 1\}$, if $x \in \mathcal{I}^*$, where $x = 1110111$, then we want $\omega(s_0, x) = 0010000$, not 0010001. In addition, for $x \in \mathcal{I}^*$, where $x = 111100111$, the output $\omega(s_0, x)$ is to be 001000001, not 001100001, for here, because of length considerations, overlapping of sequences of 111 is not allowed.

Again we start at s_0 (Fig. 6.10), but now s_1 must remember a first 1 only if it occurs in x in position $1, 4, 7, \ldots$. If the input at s_0 is 0, we cannot simply return to s_0 as in Example 6.17. We must remember that this 0 is the *first* of three symbols of no interest. Hence from s_0 we go to s_3 and then to s_4, processing any triple of the form $0yz$ where 0 occurs in x in position $3k + 1, k \geq 0$. The same type of situation happens at s_1 if the input is 0. Finally, at s_2 the sequence 111 is recognized with an output of 1, if it occurs. The machine then returns to s_0 to input the next symbol of the input string. $\square$

S_0 – prev. λ 0
S_1 – prev one !
S_2 – prev two !
S_3 →

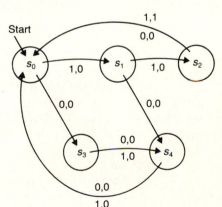

Figure 6.10

Example 6.19   Figure 6.11 shows the state diagrams for finite state machines that will recognize the occurrence of the sequence 0101 in an input string $x \in \mathcal{I}^*$, where $\mathcal{I} = \mathbb{O} = \{0, 1\}$. The machine in Fig. 6.11(a) recognizes with an output of 1 each occurrence of 0101 in an input string, regardless of where it occurs. In Fig. 6.11(b) the machine recognizes with an output of 1 only those prefixes of x whose length is a multiple of four and that end in 0101. (Hence no overlapping is allowed here.) Consequently, for $x = 01010100101$, $\omega(s_0, x) = 00010100001$ for (a), whereas for (b), $\omega(s_0, x) = 00010000000$. $\square$

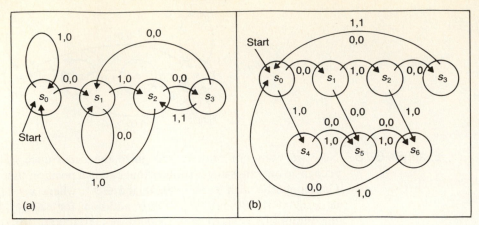

Figure 6.11

Now that we have examined some finite state machines that serve as sequence recognizers, it is only fair to consider a set of sequences that *cannot* be recognized by any finite state machine. This example gives us another opportunity to apply the pigeonhole principle.

Example 6.20　Let $\mathcal{I} = \mathcal{O} = \{0, 1\}$. Can we construct a finite state machine that recognizes precisely those sequences in the language $A = \{01, 0011, 000111, \dots\} = \{0^i 1^i | i \in \mathbf{Z}^+\}$? If so, we want $\omega(01) = 01$, $\omega(0011) = 0001$, and, in general, $\omega(0^i 1^i) = 0^{2i-1} 1$, for any $i \in \mathbf{Z}^+$.

Suppose that there is a finite state machine $M = (S, \mathcal{I}, \mathcal{O}, \nu, \omega)$ that can recognize the sequences in A. Let $|S| = n \geq 1$, where $s_0 \in S$, and s_0 is the starting state. Now consider the string $0^{n+1} 1^{n+1}$ in A. We want $\omega(s_0, 0^{n+1} 1^{n+1}) = 0^{2n+1} 1$. Therefore, we see in Table 6.8 how the finite state machine will process the $n + 1$ 0's, starting at state s_0, then continuing at states $s_1 = \nu(s_0, 0)$, $s_2 = \nu(s_1, 0), \dots, s_n = \nu(s_{n-1}, 0)$. Since $|S| = n$, by applying the pigeonhole principle to the $n + 1$ states $s_0, s_1, s_2, \dots, s_{n-1}, s_n$, we determine that there are two states s_i and s_j where $i < j$ but $s_i = s_j$.

Table 6.8

State	s_0	s_1	s_2	$\cdots$	s_{n-1}	s_n	s_{n+1}	$\cdots$	s_{2n+1}
Input	0	0	0	$\cdots$	0	0	1	$\cdots$	1
Output	0	0	0	$\cdots$	0	0	0	$\cdots$	1

Now from Table 6.9 we see how the removal of the $j - i$ columns for states $s_{i+1}, \dots, s_j$ results in Table 6.10. This table shows us how the finite state machine M recognizes the sequence $x = 0^{(n+1)-(j-i)} 1^{n+1}$, where $(n + 1) - (j - i) < n + 1$. Unfortunately $x \notin A$, so M recognizes a sequence that it is *not* supposed to recognize. This demonstrates that we cannot construct a finite state machine that recognizes precisely those sequences in the language $\{0^i 1^i | i \in \mathbf{Z}^+\}$.　□

Table 6.9

State	s_0	s_1	s_2	...	s_i	s_{i+1}	...	s_j	s_{j+1}	...	s_n	s_{n+1}	...	s_{2n+1}
Input	0	0	0	...	0	0	...	0	0	...	0	1	...	1
Output	0	0	0	...	0	0	...	0	0	...	0	0	...	1

Table 6.10

State	s_0	s_1	s_2	...	s_i	s_{i+1}	...	s_n	s_{n+1}	...	s_{2n+1}
Input	0	0	0	...	0	0	...	0	1	...	1
Output	0	0	0	...	0	0	...	0	0	...	1

A class of finite state machines that is important in the design of digital devices consists of the *k-unit delay machines*, where $k \in \mathbf{Z}^+$. For $k = 1$, we want to construct a machine M such that if $x = x_1 x_2 \ldots x_{m-1} x_m$, then for starting state s_0, $\omega(s_0, x) = 0 x_1 x_2 \ldots x_{m-1}$, so that the output is the input delayed one time unit (clock pulse). (The use of 0 as the first symbol in $\omega(s_0, x)$ is conventional.)

Example 6.21 Let $\mathcal{I} = \mathbb{O} = \{0, 1\}$. With starting state s_0, $\omega(s_0, x) = 0$ for $x = 0$ or 1, because the first output is 0; the states s_1 and s_2 remember a prior input of 0 and 1, respectively. In Fig. 6.12, we label, for example, the arc from s_1 to s_2 with 1, 0 because with an input of 1 we need to go to s_2 where inputs of 1 at time t_i are remembered so that they can become outputs of 1 at time t_{i+1}. The 0 in the label 1, 0 is the output, because starting in s_1 indicates that the prior input was 0, which becomes the present output. The labels on the other arcs are obtained by the same type of reasoning. □

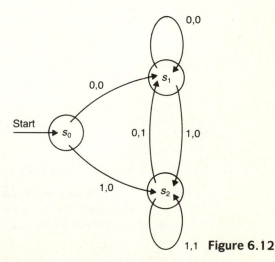

1,1 **Figure 6.12**

Example 6.22 Observing the structure for a one-unit delay, we extend our ideas to the two-unit delay machine shown in Fig. 6.13. If $x \in \mathcal{I}^*$, let $x = x_1 x_2 \ldots x_m$ where $m > 2$; if s_0 is the starting state, then $\omega(s_0, x) = 00 x_1 \ldots x_{m-2}$. For states s_0, s_1, s_2 the output is 0 for all possible inputs. States s_3, s_4, s_5, and s_6 must remember the two prior inputs

00, 01, 10, and 11, respectively. To get the other arcs in the diagram, we shall consider one such arc and then use similar reasoning for the others. For the arc from s_5 to s_3 in Fig. 6.13(a), let the input be 0. Since the prior input to s_5 from s_2 is 0, we must go to the state that remembers the two prior inputs 00. This is state s_3. Going back two states from s_5 to s_2 to s_0, we see that the input is 1 (from s_0 to s_2). This then becomes the output (delayed two places) for the arc from s_5 to s_3. The complete machine is shown in part (b) of Fig. 6.13.   □

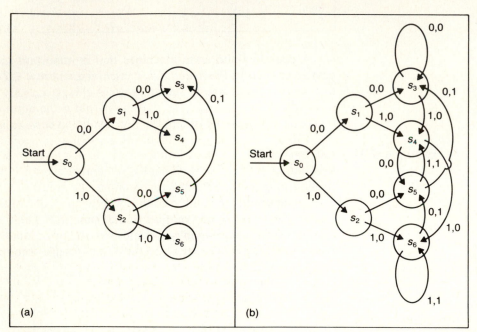

Figure 6.13

We turn now to some additional properties that arise in the study of finite state machines. The machine in Fig. 6.14 will be used for examples of the terms defined.

DEFINITION 6.14   Let $M = (S, \mathcal{I}, \mathcal{O}, v, \omega)$ be a finite state machine.

   a) For $s_i, s_j \in S$, s_j is said to be *reachable* from s_i if $s_i = s_j$ or if there is an input string $x \in \mathcal{I}^+$ such that $v(s_i, x) = s_j$. (In Fig. 6.14, state s_3 is reachable from s_0, s_1, s_2, and s_3 but not from s_4, s_5, s_6 or s_7. No state is reachable from s_3 except s_3 itself.)

   b) A state $s \in S$ is said to be *transient* if $v(s, x) = s$ for $x \in \mathcal{I}^*$ implies $x = \lambda$; that is, there is no $x \in \mathcal{I}^+$ with $v(s, x) = s$. (For the machine in Fig. 6.14, s_2 is the only transient state.)

   c) A state $s \in S$ is called a *sink*, or *sink state*, if $v(s, x) = s$, for all $x \in \mathcal{I}^*$. (s_3 is the only sink in Fig. 6.14.)

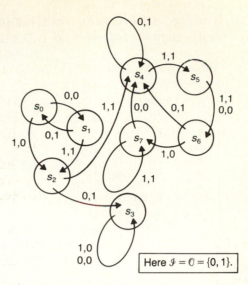

Here $\mathcal{I} = \mathbb{O} = \{0, 1\}$.

Figure 6.14

d) Let $S_1 \subseteq S$, $\mathcal{I}_1 \subseteq \mathcal{I}$. If $\nu_1 = \nu|_{s_1 \times \mathcal{I}_1} \colon S_1 \times \mathcal{I}_1 \to S_1$ (that is, the restriction of ν to $S_1 \times \mathcal{I}_1 \subseteq S \times \mathcal{I}$) has its range within S_1, then with $\omega_1 = \omega|_{s_1 \times \mathcal{I}_1}$, $M_1 = (S_1, \mathcal{I}_1, \mathbb{O}, \nu_1, \omega_1)$ is called a *submachine* of M. (With $S_1 = \{s_4, s_5, s_6, s_7\}$, and $\mathcal{I}_1 = \{0, 1\}$, we get a submachine M_1 of the machine M in Fig. 6.14.)

e) A machine is said to be *strongly connected* if for any states $s_i, s_j \in S$, s_j is reachable from s_i. (The machine in Fig. 6.14 is not strongly connected, but the submachine M_1 in (d) has this property.) ▄▄▄

We close this section with a result that uses a tree diagram.

DEFINITION 6.15   For a finite state machine M, let s_i, s_j be two distinct states in S. A shortest input string $x \in \mathcal{I}^+$ is called a *transfer* (or *transition*) *sequence* from s_i to s_j if

a) $\nu(s_i, x) = s_j$, and

b) $y \in \mathcal{I}^+$ with $\nu(s_i, y) = s_j \Rightarrow \|y\| \geq \|x\|$. ▄▄▄

There can be more than one such sequence for two states s_i, s_j.

Example 6.23   For the finite state machine M given by the state table in Table 6.11, where $\mathcal{I} = \mathbb{O} = \{0, 1\}$, find a transfer sequence from state s_0 to state s_2.

In constructing the tree diagram of Fig. 6.15, we start at state s_0 and find those states that can be reached from s_0 by using strings of length one. Here we find s_1 and s_6. Then we do the same thing with s_1 and s_6, finding, as a result, those states reachable from s_0 with input strings of length two. Continuing to expand the tree from left to right, we get to a vertex labeled with the desired state, s_2. Each time we reach a vertex labeled with a state used previously, we terminate that part of the expansion because we cannot reach any new states. After we arrive at the state we want, we backtrack to s_0 and use the state table to label the branches, as shown in Fig. 6.15. Hence, for $x = 0000$, $\nu(s_0, x) = s_2$ with $\omega(s_0, x) =$ 0100. (Here x is unique.) □

Table 6.11

	ν		ω	
	0	1	0	1
s_0	s_6	s_1	0	1
s_1	s_5	s_0	0	1
s_2	s_1	s_2	0	1
s_3	s_4	s_0	0	1
s_4	s_2	s_1	0	1
s_5	s_3	s_5	1	1
s_6	s_3	s_6	1	1

Figure 6.15

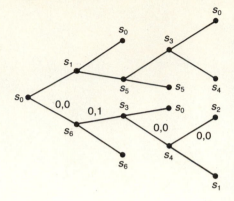

EXERCISES 6.3

1. Let $\mathcal{I} = \mathbb{O} = \{0, 1\}$. (a) Construct a state diagram for a finite state machine that recognizes each occurrence of 0000 in a string $x \in \mathcal{I}^*$. (Here overlapping is allowed.) (b) Construct a state diagram for a finite state machine that recognizes each string $x \in \mathcal{I}^*$ that ends in 0000 and has length $4k, k \in \mathbf{Z}^+$. (Here overlapping is not permitted.)

2. Answer Exercise 1 for the sequences 0110 and 1010.

3. For $k \in \mathbf{Z}^+$, k fixed, draw a state diagram for a finite state machine that recognizes a string of k consecutive 0's. (Overlapping is permitted here.)

4. Table 6.12 defines ν and ω for a finite state machine M where $\mathcal{I} = \mathbb{O} = \{0, 1\}$.

Table 6.12

	ν		ω	
	0	1	0	1
s_0	s_0	s_1	0	0
s_1	s_0	s_1	1	1

a) Draw the state diagram for M.

b) Determine the output for the following input sequences, starting at s_0 in each case: (i) $x = 111$; (ii) $x = 1010$; (iii) $x = 00011$.

c) Describe in words what machine M does.

d) How is this machine related to that shown in Fig. 6.12?

5. Verify the labels on all of the arcs in the state diagram for the two-unit delay machine shown in Fig. 6.13(b).

6. Show that it is not possible to construct a finite state machine that recognizes precisely those sequences in the language $A = \{0^i 1^j \mid i, j \in \mathbf{Z}^+, i > j\}$. (Here the alphabet for A is $\Sigma = \{0, 1\}$.)

7. For each of the machines in Table 6.13, determine the transient states, sink states, submachines, and strongly connected submachines.

Table 6.13

	ν		ω	
	0	1	0	1
s_0	s_4	s_1	0	0
s_1	s_4	s_2	0	1
s_2	s_3	s_5	0	0
s_3	s_2	s_5	1	0
s_4	s_4	s_4	1	1
s_5	s_2	s_3	0	1

(a)

	ν		ω	
	0	1	0	1
s_0	s_0	s_1	1	0
s_1	s_0	s_1	0	1
s_2	s_1	s_3	0	0
s_3	s_0	s_4	0	0
s_4	s_4	s_4	1	1

(b)

	ν		ω	
	0	1	0	1
s_0	s_1	s_2	0	1
s_1	s_0	s_2	1	1
s_2	s_2	s_3	1	1
s_3	s_6	s_4	0	0
s_4	s_5	s_5	1	0
s_5	s_3	s_4	1	0
s_6	s_6	s_6	0	0

(c)

8. Determine a transfer sequence from state s_2 to state s_5 in finite state machine (c) of Exercise 7. Is your sequence unique?

6.4

SUMMARY AND HISTORICAL REVIEW

In this chapter we have been introduced to the theory of languages and to a discrete structure called a *finite state machine*. Using our prior development of elementary set theory and finite functions, we were able to combine some abstract notions and to model digital devices such as sequence recognizers and delays. Comparable coverage appears in Chapter 1 of L. Dornhoff and F. Hohn [2] and in Chapter 2 of D. Stanat and D. McAllister [11].

The finite state machine we developed is based on the model put forth in 1955 by G. H. Mealy in [8] and is consequently referred to as the "Mealy machine." The model is based on earlier concepts found in the work of D. A. Huffman [6] and E. F. Moore [9]. For further reading on the pioneering work dealing with various aspects and applications of the finite state machine, consult the material edited by E. F. Moore [10]. Additional information on the actual synthesis of such machines and on related hardware considerations, along with an extensive coverage of many related ideas, can be found in Chapters 9–15 of Z. Kohavi [7].

For more on languages and their relation to finite state machines, one should look into the UMAP module by William J. Barnier [1], Chapters 7–10 of J. Gersting [3], and Chapters 7 and 8 of A. Gill [4]. A comprehensive coverage of these (and related) topics is given in the text by J. E. Hopcroft and J. D. Ullman [5].

REFERENCES

1. Barnier, William J. "Finite-State Machines as Recognizers" (UMAP Module 671). *The UMAP Journal* 7, no. 3, (1986): pp. 209–232.
2. Dornhoff, Larry L., and Hohn, Franz E. *Applied Modern Algebra.* New York: Macmillan, 1978.
3. Gersting, Judith L. *Mathematical Structures for Computer Science.* San Francisco: W. H. Freeman, 1982.
4. Gill, Arthur. *Applied Algebra for the Computer Sciences,* Prentice-Hall Series in Automatic Computation. Englewood Cliffs, N.J.: Prentice-Hall, 1976.
5. Hopcroft, John E., and Ullman, Jeffrey D. *Introduction to Automata Theory, Languages, and Computation.* Reading, Mass.: Addison-Wesley, 1979.
6. Huffman, D. A. "The Synthesis of Sequential Switching Circuits." *Journal of the Franklin Institute* 257, (March 1954): pp. 161–190, (April 1954): pp. 275–303. Reprinted in Moore [10].
7. Kohavi, Zvi. *Switching and Finite Automata Theory,* 2nd ed. New York: McGraw-Hill, 1978.
8. Mealy, G. H. "A Method for Synthesizing Sequential Circuits." *Bell System Technical Journal* 34, (September 1955): pp. 1045–1079.
9. Moore, E. F. "Gedanken-experiments on Sequential Machines." *Automata Studies, Annals of Mathematical Studies,* no. 34: pp. 129–153. Princeton, N.J.: Princeton University Press, 1956.
10. Moore, E. F., ed. *Sequential Machines: Selected Papers.* Reading, Mass.: Addison-Wesley, 1964.
11. Stanat, Donald F., and McAllister, David F. *Discrete Mathematics in Computer Science.* Englewood Cliffs, N.J.: Prentice-Hall, 1977.

MISCELLANEOUS EXERCISES

1. Let $\Sigma_1 = \{w, x, y\}$ and $\Sigma_2 = \{x, y, z\}$ be alphabets. If $A_1 = \{x^i y^j | i, j \in \mathbf{Z}^+, j > i \geq 1\}$, and $A_2 = \{w^i y^j | i, j \in \mathbf{Z}^+, i > j \geq 1\}$, and $A_3 = \{w^i x^j y^i z^j | i, j \in \mathbf{Z}^+, j > i \geq 1\}$, and $A_4 = \{z^j (wz)^i w^j | i, j \in \mathbf{Z}^+, i \geq 1, j \geq 2\}$, determine whether each of the following statements is true or false.

   a) A_1 is a language over Σ_1.

   b) A_1 is a language over Σ_2.

   c) A_2 is a language over Σ_1.

   d) A_2 is a language over Σ_2.

   e) A_3 is a language over $\Sigma_1 \cup \Sigma_2$.

   f) A_1 is a language over $\Sigma_1 \cap \Sigma_2$.

   g) A_4 is a language over $\Sigma_1 \triangle \Sigma_2$.

   h) $A_1 \cup A_2$ is a language over Σ_1.

   i) $A_1 \cup A_4$ is a language over Σ_1.

   j) $A_2 \cap A_3$ is a language over Σ_2.

2. For languages $A, B \subseteq \Sigma^*$, does $A^* \subseteq B^* \Rightarrow A \subseteq B$?

3. Give an example of a language A over an alphabet Σ, where $(A^2)^* \neq (A^*)^2$.

4. For a given alphabet Σ and index set I, let $B_i \subseteq \Sigma^*$ for each $i \in I$. If $A \subseteq \Sigma^*$, prove that (a) $A(\bigcap_{i \in I} B_i) \subseteq \bigcap_{i \in I} AB_i$; and (b) $(\bigcap_{i \in I} B_i)A \subseteq \bigcap_{i \in I} B_i A$. [Here, for example, $A(\bigcap_{i \in I} B_i)$ denotes the concatenation of the languages A and $\bigcap_{i \in I} B_i$.]

5. Let M be the finite state machine shown in Fig. 6.16. For states $s_i, s_j, 0 \le i, j \le 2$, let $\mathbb{O}_{ij}$ denote the set of all nonempty output strings that M can produce as it goes from state s_i to state s_j. If $i = 2, j = 0$, for example, $\mathbb{O}_{20} = \{0\}\{1, 00\}^*$.
    Find $\mathbb{O}_{02}$, $\mathbb{O}_{22}$, $\mathbb{O}_{11}$, $\mathbb{O}_{00}$, and $\mathbb{O}_{10}$.

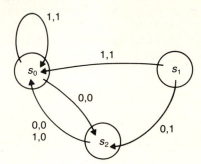

Figure 6.16

6. Let M be the finite state machine in Fig. 6.17.
    a) Find the state table for this machine.
    b) Explain what this machine does.
    c) How many distinct input strings x are there such that $\|x\| = 8$ and $\nu(s_0, x) = s_0$? How many are there with $\|x\| = 12$?

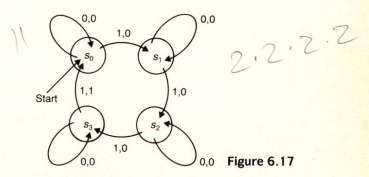

Figure 6.17

7. A finite state machine M is called *simply minimal* if, in its state table, no two output rows are identical. For $|S| = n$, $|\mathcal{I}| = m$, and $|\mathbb{O}| = p$, show that there are

$$(n^{nm})[(p^m)(p^m - 1)(p^m - 2) \cdots (p^m - (n - 1))] = \frac{(n^{nm})(p^m)!}{(p^m - n)!}$$

simply minimal machines possible.

8. Let $M = (S, \mathcal{I}, \mathbb{O}, \nu, \omega)$ be a finite state machine with $|S| = n$, and let $0 \in \mathcal{I}$.
    a) Show that for the input string $0000\ldots$, the output is eventually periodic.
    b) What is the maximum number of 0's we can input before the periodic output starts?
    c) What is the length of the maximum period that can occur?

9. For the finite state machine in part (c) of Table 6.13, determine a transfer sequence from state s_0 to state s_5. Is your sequence unique?

10. Draw the state diagram for a finite state machine $M = (S, \mathcal{I}, \mathcal{O}, \nu, \omega)$, where $\mathcal{I} = \mathcal{O} = \{0, 1\}$, if for any $x \in \mathcal{I}^+$, M puts out its first 1 when it recognizes the substring 1111 and then puts out its second 1 when it recognizes the substring 0000, after which its output is constantly 0.

11. Let $\mathcal{I} = \mathcal{O} = \{0, 1\}$. Construct a state diagram for a finite state machine that reverses (from 0 to 1 or from 1 to 0) the symbols appearing in the 4th, in the 8th, in the 12th, ..., positions of an input string $x \in \mathcal{I}^+$. For example, $\omega(0000) = 0001$, $\omega(000111) = 000011$, and $\omega(000000111) = 000100101$.

12. For $\mathcal{I} = \mathcal{O} = \{0, 1\}$, let M be the finite state machine given in Table 6.14.

Table 6.14

	ν		ω	
	0	1	0	1
s_1	s_4	s_3	0	0
s_2	s_2	s_4	0	1
s_3	s_1	s_2	1	0
s_4	s_1	s_4	1	1

If the starting state for M is *not* s_1, find an input string x (of smallest length) such that $\nu(s_i, x) = s_1$, for $i = 2, 3, 4$. (Hence x gets the machine M to state s_1 regardless of the starting state.)

13. Show that we cannot construct a finite state machine that recognizes precisely those sequences in the language $A = \{0^i 1^j \mid i, j \in \mathbf{Z}^+, i < j\}$. (The alphabet for A is $\Sigma = \{0, 1\}$.)

14. With $\mathcal{I} = \mathcal{O} = \{0, 1\}$, let M be the finite state machine given in Table 6.15. Here s_0 is the starting state.

Table 6.15

	ν		ω	
	0	1	0	1
s_0	s_1	s_2	1	0
s_1	s_2	s_1	0	1
s_2	s_2	s_3	0	1
s_3	s_1	s_0	1	0

Let $A \subseteq \mathcal{I}^+$ where $x \in A$ if and only if the last symbol in $\omega(s_0, x)$ is 1. (There may be more than one 1 in the output string $\omega(s_0, x)$.) Construct a finite state machine wherein the last symbol of the output string is 1 for all $y \in \mathcal{I}^+ - A$.

15. Let $\mathcal{I} = \mathcal{O} = \{0, 1\}$ for the two finite state machines M_1 and M_2, given in Tables 6.16 and 6.17, respectively. The starting state for M_1 is s_0, whereas s_3 is the starting state for M_2.

Table 6.16

	ν_1		ω_1	
	0	1	0	1
s_0	s_0	s_1	1	0
s_1	s_1	s_2	0	0
s_2	s_2	s_0	0	1

Table 6.17

	ν_2		ω_2	
	0	1	0	1
s_3	s_3	s_4	1	1
s_4	s_4	s_3	1	0

We connect these machines as shown in Fig. 6.18. Here each output symbol from M_1 becomes an input symbol for M_2. For example, if we input 0 to M_1, then $\omega_1(s_0, 0) = 1$ and $\nu_1(s_0, 0) = s_0$. As a result, we then input 1 $(= \omega_1(s_0, 0))$ to M_2 to get $\omega_2(s_3, 1) = 1$ and $\nu_2(s_3, 1) = s_4$.

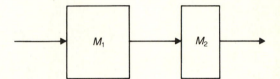

Figure 6.18

We construct a machine $M = (S, \mathcal{I}, \mathcal{O}, \nu, \omega)$ that represents this connection of M_1 and M_2 as follows:

$\mathcal{I} = \mathcal{O} = \{0, 1\}$.
$S = S_1 \times S_2$, where S_i is the set of internal states for $M_i, i = 1, 2$.
$\nu: S \times \mathcal{I} \rightarrow S$, where
$\nu((s, t), x) = (\nu_1(s, x), \nu_2(t, \omega_1(s, x)))$, for $s \in S_1, t \in S_2,$ and $x \in \mathcal{I}$.
$\omega: S \times \mathcal{I} \rightarrow \mathcal{O}$, where
$\omega((s, t), x) = \omega_2(t, \omega_1(s, x))$, for $s \in S_1, t \in S_2,$ and $x \in \mathcal{I}$.

a) Find a state table for machine M.

b) Determine the output string for the input string 1101. In what state is each of machines M_1 and M_2 after this string is processed?

16. Although the state diagram seems more convenient than the state table when we are dealing with a finite state machine $M = (S, \mathcal{I}, \mathcal{O}, \nu, \omega)$, as the input strings get longer and the sizes of $S, \mathcal{I}, \mathcal{O}$ increase, the state table proves useful when simulating the machine on a computer. The block form of the table suggests the use of a matrix or two-dimensional array for storing ν, ω. Use this observation to write a program (or develop an algorithm) that will simulate the machine in Table 6.18.

Table 6.18

	ν		ω	
	0	1	0	1
s_1	s_2	s_1	0	0
s_2	s_3	s_1	0	0
s_3	s_3	s_1	1	1

CHAPTER 7

Relations: The Second Time Around

In Chapter 5 we introduced the concept of relation and then concentrated on special kinds of relations called functions. Returning to relations in this chapter, we shall emphasize relations on a set A—that is, subsets of $A \times A$. Within the theory of languages and finite state machines from Chapter 6, we find many examples of relations on a set A, where A represents a set of strings from a given alphabet or a set of internal states from a finite state machine. Various properties of relations are developed, along with ways to represent finite relations for computer manipulation. Directed graphs reappear as a second way to represent such relations. Finally, among the many relations possible on a set A, two types are of particular importance: equivalence relations and partial orders. Equivalence relations arise in many areas in mathematics. For the present we use an equivalence relation on the set of internal states in a finite state machine M to find a machine M_1, with as few internal states as possible, that performs whatever tasks M is capable of performing. The procedure is known as the minimization process.

7.1
RELATIONS REVISITED: PROPERTIES OF RELATIONS

We start by recalling some ideas we have considered before.

DEFINITION 7.1   If A, B are sets, a *relation from A to B* is any subset of $A \times B$. Subsets of $A \times A$ are called *relations on A*.

Example 7.1   **a)** Define the relation $\mathcal{R}$ on the set $\mathbf{Z}$ by $a \mathcal{R} b$, or $(a, b) \in \mathcal{R}$, if $a \leq b$. This is the ordinary "is less than or equal to" relation on the set $\mathbf{Z}$ and can also be given on $\mathbf{Q}$ or $\mathbf{R}$, but not on $\mathbf{C}$.

    b) Let $n \in \mathbf{Z}^+$. For $x, y \in \mathbf{Z}$, the *modulo n relation* $\mathcal{R}$ is defined by $x \mathcal{R} y$ if $x - y$ is a multiple of n. With $n = 7$, we find $9 \mathcal{R} 2$, $-3 \mathcal{R} 11$, $(14, 0) \in \mathcal{R}$, but $3 \not\mathcal{R} 7$.   □

Example 7.2   Let $\mathbf{\Sigma}$ be an alphabet, with language $A \subseteq \mathbf{\Sigma}^*$. For $x, y \in A$, define $x \mathcal{R} y$ if x is a prefix of y. Other relations can be defined on A by replacing "prefix" with either "suffix" or "substring."   □

Example 7.3   Consider a finite state machine $M = (S, \mathcal{I}, \mathcal{O}, \nu, \omega)$.

    a) For $s_1, s_2 \in S$, define $s_1 \mathcal{R} s_2$ if $\nu(s_1, x) = s_2$, for some $x \in \mathcal{I}$. Relation $\mathcal{R}$ establishes the *first level of reachability*.

    b) The relation for the *second level of reachability* can also be given for S. Here $s_1 \mathcal{R} s_2$ if $\nu(s_1, x_1 x_2) = s_2$, for $x_1 x_2 \in \mathcal{I}^2$. This can be extended to higher levels if the need arises. For the general *reachability* relation we have $\nu(s_1, y) = s_2$, for $y \in \mathcal{I}^*$.

    c) Given $s_1, s_2 \in S$ the relation of *1-equivalence*, which is denoted by $s_1 E_1 s_2$ and is read "s_1 is 1-equivalent to s_2," is defined when $\omega(s_1, x) = \omega(s_2, x)$ for all $x \in \mathcal{I}$. Consequently, $s_1 E_1 s_2$ indicates that if machine M starts in either state s_1 or s_2, the output is the same for each element of $\mathcal{I}$. This idea can be extended to states being *k-equivalent*, where we write $s_1 E_k s_2$ if $\omega(s_1, y) = \omega(s_2, y)$, for all $y \in \mathcal{I}^k$. Here the same output is obtained for any string in $\mathcal{I}^k$, if we start at either s_1 or s_2.

         If two states are k-equivalent for all $k \in \mathbf{Z}^+$, then they are called *equivalent*. We shall look further into this idea later in the chapter.   □

We now proceed to an investigation of some of the properties a relation can satisfy.

DEFINITION 7.2   A relation $\mathcal{R}$ on a set A is called *reflexive* if for all $x \in A$, $(x, x) \in \mathcal{R}$.   ▬

        We say that we want each element x of A to be related to itself, for $\mathcal{R}$ to be reflexive. All the relations in Examples 7.1 and 7.2 are reflexive. The general reachability relation in Example 7.3(b) and all of the relations mentioned in part (c) of that example are also reflexive. (What goes wrong with the relations for the first and second levels of reachability given in parts (a) and (b) of Example 7.3?)

Example 7.4   For $A = \{1, 2, 3, 4\}$, a relation $\mathcal{R} \subseteq A \times A$ will be reflexive if $\mathcal{R} \supseteq \{(1, 1), (2, 2), (3, 3), (4, 4)\}$. Consequently, $\mathcal{R}_1 = \{(1, 1), (2, 2), (3, 3)\}$ is not a reflexive relation on A, whereas $\mathcal{R}_2 = \{(x, y) \mid x, y \in A, x \leq y\}$ is reflexive on A.   □

Example 7.5   Given a finite set A with $|A| = n$, we have $|A \times A| = n^2$, so there are 2^{n^2} relations on A. How many of these are reflexive?

        If $A = \{a_1, a_2, \ldots, a_n\}$, a relation $\mathcal{R}$ on A is reflexive if $\{(a_i, a_i) \mid 1 \leq i \leq n\} \subseteq \mathcal{R}$. Considering the other $n^2 - n$ ordered pairs in $A \times A$ (those of the form (a_i, a_j), where $i \neq j$ for $1 \leq i, j \leq n$) as we construct a reflexive relation $\mathcal{R}$ on A, we either include or exclude each of these ordered pairs, so by the rule of product there are $2^{(n^2-n)}$ reflexive relations on A. $\square$

DEFINITION 7.3   Relation $\mathcal{R}$ on set A is called *symmetric* if $(x, y) \in \mathcal{R} \Rightarrow (y, x) \in \mathcal{R}$, for any $x, y \in A$.

Example 7.6   With $A = \{1, 2, 3\}$, we have:

a) $\mathcal{R}_1 = \{(1, 2), (2, 1), (1, 3), (3, 1)\}$, a symmetric, but not reflexive, relation on A

b) $\mathcal{R}_2 = \{(1, 1), (2, 2), (3, 3), (2, 3)\}$, a reflexive, but not symmetric, relation on A

c) $\mathcal{R}_3 = \{(1, 1), (2, 2), (3, 3)\}$ and $\mathcal{R}_4 = \{(1, 1), (2, 2), (3, 3), (2, 3), (3, 2)\}$, reflexive and symmetric relations on A

d) $\mathcal{R}_5 = \{(1, 1), (2, 3), (3, 3)\}$, a relation on A that is neither reflexive nor symmetric $\square$

To count the symmetric relations on $A = \{a_1, a_2, \ldots, a_n\}$, we write $A \times A$ as $A_1 \cup A_2$, where $A_1 = \{(a_i, a_i) \mid 1 \leq i \leq n\}$ and $A_2 = \{(a_i, a_j) \mid 1 \leq i, j \leq n, i \neq j\}$, so that every ordered pair in $A \times A$ is in exactly one of A_1, A_2. For A_2, $|A_2| = |A \times A| - |A_1| = n^2 - n = n(n-1)$, an even integer. The set A_2 contains $(1/2)(n^2 - n)$ subsets S_{ij} of the form $\{(a_i, a_j), (a_j, a_i)\}$ where $1 \leq i < j \leq n$. In constructing a symmetric relation $\mathcal{R}$ on A, for each ordered pair in A_1 we have our usual choice of exclusion or inclusion. For each of the $(1/2)(n^2 - n)$ subsets S_{ij} $(1 \leq i < j \leq n)$ in A_2 we have the same two choices. So by the rule of product there are $2^n \cdot 2^{(1/2)(n^2-n)} = 2^{(1/2)(n^2+n)}$ symmetric relations on A.

In counting those relations on A that are both reflexive and symmetric, we have only one choice for each ordered pair in A_1. So we have $2^{(1/2)(n^2-n)}$ relations on A that are both reflexive and symmetric.

DEFINITION 7.4   For a set A, a relation $\mathcal{R}$ on A is called *transitive* if $(x, y), (y, z) \in \mathcal{R} \Rightarrow (x, z) \in \mathcal{R}$. (So if x "is related to" y, and y "is related to" z, we want x "related to" z, with y playing the role of "intermediary.")

Example 7.7   All the relations in Examples 7.1 and 7.2 are transitive, as are the relations in Example 7.3(c). $\square$

Example 7.8   Define the relation $\mathcal{R}$ on the set $\mathbf{Z}^+$ by $a \mathcal{R} b$ if a divides b, i.e., for some $c \in \mathbf{Z}^+$, $b = ca$. Now if $x \mathcal{R} y$ and $y \mathcal{R} z$, do we have $x \mathcal{R} z$? $x \mathcal{R} y \Rightarrow y = sx$ for some $s \in \mathbf{Z}^+$; $y \mathcal{R} z \Rightarrow z = ty$ where $t \in \mathbf{Z}^+$. Consequently, $z = ty = t(sx) = (ts)x$ for $ts \in \mathbf{Z}^+$, so $x \mathcal{R} z$ and $\mathcal{R}$ is transitive. In addition, $\mathcal{R}$ is reflexive, but not symmetric, because, for example, $2 \mathcal{R} 6$ but $6 \not{\mathcal{R}} 2$. $\square$

Example 7.9   If $A = \{1, 2, 3, 4\}$, then $\mathcal{R}_1 = \{(1, 1), (2, 3), (3, 4), (2, 4)\}$ is a transitive relation on A, whereas $\mathcal{R}_2 = \{(1, 3), (3, 2)\}$ is not transitive because $(1, 3), (3, 2) \in \mathcal{R}_2$ but $(1, 2) \notin \mathcal{R}_2$. $\square$

At this point the reader is probably ready to start counting the number of transitive relations on a finite set. This is harder than the previous counting arguments, but at a later point in this chapter we shall have the necessary ideas to count the relations $\mathcal{R}$ on a finite set, where $\mathcal{R}$ is reflexive, symmetric, and transitive.

We consider one last property for relations.

DEFINITION 7.5   Given a relation $\mathcal{R}$ on a set A, $\mathcal{R}$ is called *antisymmetric* if $(a\,\mathcal{R}\,b$ and $b\,\mathcal{R}\,a)\Rightarrow$ $a = b$. (Here the only way we can have both a "related to" b and b "related to" a is if a and b are one and the same element from A.)

Example 7.10   For a given universe $\mathcal{U}$, define the relation $\mathcal{R}$ on $\mathcal{P}(\mathcal{U})$ by $(A, B) \in \mathcal{R}$ if $A \subseteq B$, for $A, B \subseteq \mathcal{U}$. So $\mathcal{R}$ is the subset relation of Chapter 3 and if $A\,\mathcal{R}\,B$ and $B\,\mathcal{R}\,A$, then we have $A \subseteq B$ and $B \subseteq A$, which gives us $A = B$. Consequently, this relation is antisymmetric, as well as reflexive and transitive, but it is not symmetric. □

Before we are led astray into thinking that "not symmetric" is synonymous with "antisymmetric," let us consider the following.

Example 7.11   For $A = \{1, 2, 3\}$, the relation $\mathcal{R}$ on A given by $\mathcal{R} = \{(1, 2), (2, 1), (2, 3)\}$ is not symmetric because $(3, 2) \notin \mathcal{R}$, and it is not antisymmetric either, because $(1, 2), (2, 1) \in \mathcal{R}$, but $1 \neq 2$. The relation $\mathcal{R}_1 = \{(1, 1), (2, 2)\}$ is both symmetric and antisymmetric.

How many relations on A are antisymmetric? Writing

$$A \times A = \{(1, 1), (2, 2), (3, 3)\} \cup \{(1, 2), (2, 1), (1, 3), (3, 1), (2, 3), (3, 2)\},$$

we make two observations as we try to construct an antisymmetric relation $\mathcal{R}$ on A.

1. Any element $(x, x) \in A \times A$ can be either included or excluded with no concern about whether or not $\mathcal{R}$ is antisymmetric.

2. For an element of the form $(x, y), x \neq y$, we must consider both (x, y) and (y, x) and we note that for $\mathcal{R}$ to remain antisymmetric we have three alternatives: (a) place (x, y) in $\mathcal{R}$; (b) place (y, x) in $\mathcal{R}$; or (c) place neither (x, y) nor (y, x) in $\mathcal{R}$. (What happens if we place both (x, y) and (y, x) in $\mathcal{R}$?)

So by the rule of product, the number of antisymmetric relations on A is $(2^3)(3^3) = (2^3)(3^{(3^2-3)/2})$. If $|A| = n > 0$, then there are $(2^n)(3^{(n^2-n)/2})$ antisymmetric relations on A.   □

For our next example we return to the concept of function dominance, which we first defined in Section 5.7.

Example 7.12   Let $\mathcal{F}$ denote the set of all functions with domain $\mathbf{Z}^+$ and codomain $\mathbf{R}$; that is, $\mathcal{F} = \{f \mid f: \mathbf{Z}^+ \to \mathbf{R}\}$. For $f, g \in \mathcal{F}$, define the relation $\mathcal{R}$ on $\mathcal{F}$ by $f\,\mathcal{R}\,g$ if f is dominated by g (or $f \in O(g)$). Then $\mathcal{R}$ is reflexive and transitive.

If $f, g: \mathbf{Z}^+ \to \mathbf{R}$ are defined by $f(n) = n$ and $g(n) = n + 5$, then $f \mathcal{R} g$ and $g \mathcal{R} f$ but $f \neq g$, so $\mathcal{R}$ is *not antisymmetric*. In addition, if $h: \mathbf{Z}^+ \to \mathbf{R}$ is given by $h(n) = n^2$, then $(f, h), (g, h) \in \mathcal{R}$, but neither (h, f) nor (h, g) is in $\mathcal{R}$. Consequently the relation $\mathcal{R}$ is also *not symmetric*. □

At this point we have seen the four major properties that arise in the study of relations. Before closing this section we define two more notions, each of which involves three of these four properties.

DEFINITION 7.6   A relation $\mathcal{R}$ on a set A is called a *partial order*, or a *partial ordering relation*, if $\mathcal{R}$ is reflexive, antisymmetric, and transitive.

Example 7.13   The relation in Example 7.1(a) is a partial order, but the relation in part (b) is not, because it is not antisymmetric. All the relations of Example 7.2 are partial orders, as is the subset relation of Example 7.10. □

DEFINITION 7.7   An *equivalence relation* $\mathcal{R}$ on a set A is a relation that is reflexive, symmetric, and transitive.

Example 7.14   The relation in Example 7.1(b) and all the relations in Example 7.3(c) are equivalence relations.

For any set A, $A \times A$ is an equivalence relation on A, and if $A = \{a_1, a_2, \ldots, a_n\}$, then $\mathcal{R} = \{(a_i, a_i) | 1 \leq i \leq n\}$ is the smallest equivalence relation on A.

If $\mathcal{R}$ is a relation on a set A, then $\mathcal{R}$ is both an equivalence relation and a partial order on A if and only if $\mathcal{R}$ is the equality relation on A. □

EXERCISES 7.1

1. If $A = \{1, 2, 3, 4\}$, give an example of a relation $\mathcal{R}$ on A that is
   a) reflexive and symmetric, but not transitive
   b) reflexive and transitive, but not symmetric
   c) symmetric and transitive, but not reflexive

2. For relation (b) in Example 7.1, determine five values of x for which $(x, 5) \in \mathcal{R}$.

3. For the relation $\mathcal{R}$ in Example 7.12, let $f: \mathbf{Z}^+ \to \mathbf{R}$ where $f(n) = n$.
   a) Find three elements $f_1, f_2, f_3 \in \mathcal{F}$ such that $f_i \mathcal{R} f$ and $f \mathcal{R} f_i$, for all $1 \leq i \leq 3$.
   b) Find three elements $g_1, g_2, g_3 \in \mathcal{F}$ such that $g_i \mathcal{R} f$ but $f \not\mathcal{R} g_i$, for all $1 \leq i \leq 3$.

4. a) Rephrase the definitions for the reflexive, symmetric, transitive, and antisymmetric properties of a relation $\mathcal{R}$ (on a set A), using quantifiers.
   b) Use the results of part (a) to specify when a relation $\mathcal{R}$ (on a set A) is (i) *not* reflexive; (ii) *not* symmetric; (iii) *not* transitive; (iv) *not* antisymmetric.

5. For each of the following relations, determine whether the relation is reflexive, symmetric, antisymmetric, or transitive.

   a) $\mathcal{R} \subseteq \mathbf{Z}^+ \times \mathbf{Z}^+$ where $a\,\mathcal{R}\,b$ if $a\,|\,b$ (read "a divides b," as defined in Chapter 4).

   b) $\mathcal{R}$ is the relation on $\mathbf{Z}$ where $a\,\mathcal{R}\,b$ if $a\,|\,b$.

   c) For a given universe $\mathcal{U}$ and $C \subseteq \mathcal{U}$, define $\mathcal{R}$ on $\mathcal{P}(\mathcal{U})$ as follows: For $A, B \subseteq \mathcal{U}$, we have $A\,\mathcal{R}\,B$ if $A \cap C = B \cap C$.

   d) On the set A of all lines in $\mathbf{R}^2$, define the relation $\mathcal{R}$ for two lines ℓ_1, ℓ_2 by $\ell_1\,\mathcal{R}\,\ell_2$ if ℓ_1 is perpendicular to ℓ_2.

   e) $\mathcal{R}$ is the relation on $\mathbf{Z}$ where $x\,\mathcal{R}\,y$ if $x + y$ is even (odd).

   f) $\mathcal{R}$ is the relation on $\mathbf{Z}$ where $x\,\mathcal{R}\,y$ if $x - y$ is even (odd).

   g) $\mathcal{R}$ is the relation on $\mathbf{Z}^+$ where $a\,\mathcal{R}\,b$ if $(a, b) = 1$—that is, if a and b are relatively prime.

   h) Let T be the set of all triangles in $\mathbf{R}^2$. Define $\mathcal{R}$ on T by $t_1\,\mathcal{R}\,t_2$ if t_1 and t_2 have an angle of the same measure.

   i) $\mathcal{R}$ is the relation on $\mathbf{Z} \times \mathbf{Z}$ where $(a, b)\,\mathcal{R}\,(c, d)$ if $a \leq c$. (*Note:* $\mathcal{R} \subseteq (\mathbf{Z} \times \mathbf{Z}) \times (\mathbf{Z} \times \mathbf{Z})$.)

6. Which relations in Exercise 5 are partial orders? Which are equivalence relations?

7. a) Let $\mathcal{R}_1, \mathcal{R}_2$ be relations on a set A. Prove or disprove that $\mathcal{R}_1, \mathcal{R}_2$ reflexive $\Rightarrow \mathcal{R}_1 \cap \mathcal{R}_2$ reflexive.

   b) Answer part (a) when each occurrence of "reflexive" is replaced by (i) symmetric; (ii) antisymmetric; (iii) transitive.

8. Answer Exercise 7, replacing each occurrence of $\cap$ by $\cup$.

9. For each of the following statements about relations on a set A, where $|A| = n$, determine whether the statement is true or false. If it is false, give a counterexample.

   a) If $\mathcal{R}$ is a reflexive relation on A, then $|\mathcal{R}| \geq n$.

   b) If $\mathcal{R}$ is a relation on A and $|\mathcal{R}| \geq n$, then $\mathcal{R}$ is reflexive.

   c) If $\mathcal{R}_1, \mathcal{R}_2$ are relations on A and $\mathcal{R}_2 \supseteq \mathcal{R}_1$, then $\mathcal{R}_1$ reflexive (symmetric, antisymmetric, transitive) $\Rightarrow \mathcal{R}_2$ reflexive (symmetric, antisymmetric, transitive).

   d) If $\mathcal{R}_1, \mathcal{R}_2$ are relations on A and $\mathcal{R}_2 \supseteq \mathcal{R}_1$, then $\mathcal{R}_2$ reflexive (symmetric, antisymmetric, transitive) $\Rightarrow \mathcal{R}_1$ reflexive (symmetric, antisymmetric, transitive).

   e) If $\mathcal{R}$ is an equivalence relation on A, then $n \leq |\mathcal{R}| \leq n^2$.

10. If $A = \{w, x, y, z\}$, determine the number of relations on A that are (a) reflexive; (b) symmetric; (c) reflexive and symmetric; (d) reflexive and contain (x, y); (e) symmetric and contain (x, y); (f) antisymmetric; (g) antisymmetric and contain (x, y); (h) symmetric and antisymmetric; (i) reflexive, symmetric, and antisymmetric.

11. What is wrong with the following argument?

    Let A be a set with $\mathcal{R}$ a relation on A. If $\mathcal{R}$ is symmetric and transitive, then $\mathcal{R}$ is reflexive.

    Proof: Let $(x, y) \in \mathcal{R}$. By the symmetric property, $(y, x) \in \mathcal{R}$. Then with $(x, y), (y, x) \in \mathcal{R}$, it follows by the transitive property that $(x, x) \in \mathcal{R}$. Consequently, $\mathcal{R}$ is reflexive.

12. Let A be a set with $|A| = n$, and let $\mathcal{R}$ be a relation on A that is antisymmetric. What is the maximum value for $|\mathcal{R}|$? How many antisymmetric relations can have this size?

13. Let A be a set with $|A| = n$, and let $\mathcal{R}$ be an equivalence relation on A with $|\mathcal{R}| = r$. Why is $r - n$ always even?

14. A relation $\mathcal{R}$ on a set A is called *irreflexive* if for all $a \in A$, $(a, a) \notin \mathcal{R}$.

    a) Give an example of a relation $\mathcal{R}$ on **Z** where $\mathcal{R}$ is irreflexive and transitive but not symmetric.

    b) Let $\mathcal{R}$ be a nonempty relation on a set A. Prove that if $\mathcal{R}$ satisfies any two of the following properties—irreflexive, symmetric, and transitive—then it cannot satisfy the third.

    c) If $\mathcal{R}_1, \mathcal{R}_2$ are both irreflexive relations on a set A, what can we say about the relations $\mathcal{R}_1 \cap \mathcal{R}_2$ and $\mathcal{R}_1 \cup \mathcal{R}_2$ on A?

    d) If $|A| = n \geq 1$, how many different relations on A are irreflexive? How many are neither reflexive nor irreflexive?

7.2
COMPUTER RECOGNITION: ZERO-ONE MATRICES AND DIRECTED GRAPHS

Since our interest in relations is focused on those for finite sets, we are concerned with ways of representing such relations so that the properties of Section 7.1 can be identified. For this reason we now develop the necessary tools: relation composition, zero-one matrices, and directed graphs.

In a manner analogous to the composition of functions, relations can be combined in the following circumstances.

DEFINITION 7.8   If A, B, and C are sets with $\mathcal{R}_1 \subseteq A \times B$ and $\mathcal{R}_2 \subseteq B \times C$, then the *composite relation* $\mathcal{R}_1 \circ \mathcal{R}_2$ is a relation from A to C defined by $\mathcal{R}_1 \circ \mathcal{R}_2 = \{(x, z) \mid x \in A, z \in C,$ and there exists $y \in B$ with $(x, y) \in \mathcal{R}_1, (y, z) \in \mathcal{R}_2\}$. ——

Beware! The composition of two relations is written in an order opposite to that for function composition. We shall see why shortly.

Example 7.15   Let $A = \{1, 2, 3, 4\}$, $B = \{w, x, y, z\}$, and $C = \{5, 6, 7\}$. If $\mathcal{R}_1 = \{(1, x), (2, x), (3, y), (3, z)\}$ is a relation from A to B and $\mathcal{R}_2 = \{(w, 5), (x, 6)\}$ is a relation from B to C, then $\mathcal{R}_1 \circ \mathcal{R}_2 = \{(1,6), (2,6)\}$, and this is a relation from A to C. If $\mathcal{R}_3 = \{(w,5), (w,6)\}$ is another relation from B to C, then $\mathcal{R}_1 \circ \mathcal{R}_3 = \emptyset$.   □

Example 7.16   Let A be the set of employees at a computing center, while B denotes a set of high-level programming languages, and C is a list of projects $\{p_1, p_2, \ldots, p_8\}$ for which managers must make work assignments using the people in A. Consider $\mathcal{R}_1 \subseteq A \times B$, where an ordered pair of the form (L. Alldredge, Pascal) indicates that employee L. Alldredge is proficient in Pascal (and perhaps other programming languages). The relation $\mathcal{R}_2 \subseteq B \times C$ consists of ordered pairs of the form (Pascal, p_2) where Pascal is considered an essential language needed by anyone who works on project p_2. In the composite relation $\mathcal{R}_1 \circ \mathcal{R}_2$ we find (L. Alldredge, p_2). If no other ordered pair in $\mathcal{R}_2$ has p_2 as its second component, we know that if L. Alldredge was assigned to p_2 it was solely on the basis of his proficiency in Pascal. (Here $\mathcal{R}_1 \circ \mathcal{R}_2$ has been used to set up a matching process between employees and projects on the basis of employee knowledge of specific programming languages.)   □

Comparable to the associative law for function composition, we find the following result for relations.

THEOREM 7.1   Let A, B, C, and D be sets with $\mathcal{R}_1 \subseteq A \times B$, $\mathcal{R}_2 \subseteq B \times C$, and $\mathcal{R}_3 \subseteq C \times D$. Then $\mathcal{R}_1 \circ (\mathcal{R}_2 \circ \mathcal{R}_3) = (\mathcal{R}_1 \circ \mathcal{R}_2) \circ \mathcal{R}_3$.

Proof   Since both $\mathcal{R}_1 \circ (\mathcal{R}_2 \circ \mathcal{R}_3)$ and $(\mathcal{R}_1 \circ \mathcal{R}_2) \circ \mathcal{R}_3$ are relations from A to D, there is some reason to believe they are equal. If $(a,d) \in \mathcal{R}_1 \circ (\mathcal{R}_2 \circ \mathcal{R}_3)$, then there is an element $b \in B$ with $(a,b) \in \mathcal{R}_1$ and $(b,d) \in (\mathcal{R}_2 \circ \mathcal{R}_3)$. Also, $(b,d) \in (\mathcal{R}_2 \circ \mathcal{R}_3) \Rightarrow (b,c) \in \mathcal{R}_2$ and $(c,d) \in \mathcal{R}_3$ for some $c \in C$. Then $(a,b) \in \mathcal{R}_1$ and $(b,c) \in \mathcal{R}_2 \Rightarrow (a,c) \in \mathcal{R}_1 \circ \mathcal{R}_2$. Finally, $(a,c) \in \mathcal{R}_1 \circ \mathcal{R}_2$ and $(c,d) \in \mathcal{R}_3 \Rightarrow (a,d) \in (\mathcal{R}_1 \circ \mathcal{R}_2) \circ \mathcal{R}_3$, and $\mathcal{R}_1 \circ (\mathcal{R}_2 \circ \mathcal{R}_3) \subseteq (\mathcal{R}_1 \circ \mathcal{R}_2) \circ \mathcal{R}_3$. The opposite inclusion follows by similar reasoning.   ■

As a result of this theorem no ambiguity arises when we write $\mathcal{R}_1 \circ \mathcal{R}_2 \circ \mathcal{R}_3$ for either of the relations in Theorem 7.1. In addition, we can now define the powers of a relation $\mathcal{R}$ on a set.

DEFINITION 7.9   Given a set A and a relation $\mathcal{R}$ on A, we define the *powers of* $\mathcal{R}$ recursively by (a) $\mathcal{R}^1 = \mathcal{R}$; (b) for $n \in \mathbf{Z}^+$, $\mathcal{R}^{n+1} = \mathcal{R} \circ \mathcal{R}^n$.   ▬

Note that for $n \in \mathbf{Z}^+$, $\mathcal{R}^n$ is a relation on A.

Example 7.17   If $A = \{1, 2, 3, 4\}$ and $\mathcal{R} = \{(1,2), (1,3), (2,4), (3,2)\}$, then $\mathcal{R}^2 = \{(1,4), (1,2), (3,4)\}$, $\mathcal{R}^3 = \{(1,4)\}$, and for $n \geq 4$, $\mathcal{R}^n = \emptyset$.   □

As the set A and the relation $\mathcal{R}$ on A grow larger, calculations such as those in Example 7.17 become tedious. We seek a way to avoid this tedium. The tool we need is the computer, once a way can be found to tell the machine about the set A and the relation $\mathcal{R}$ on A.

DEFINITION 7.10   An $m \times n$ *zero-one matrix* $E = (e_{ij})_{m \times n}$ is a rectangular array of numbers arranged in m rows and n columns, where each e_{ij}, for $1 \le i \le m$ and $1 \le j \le n$, denotes the entry in the ith row and jth column of E, and each such entry is 0 or 1. (We can also write $(0, 1)$-matrix for this type of matrix.)

Example 7.18   The matrix

$$E = \begin{bmatrix} 1 & 0 & 0 & 1 \\ 0 & 1 & 0 & 1 \\ 1 & 0 & 0 & 0 \end{bmatrix}$$

is a 3×4 $(0, 1)$-matrix where, for example, $e_{11} = 1$, $e_{23} = 0$, and $e_{31} = 1$.   □

In working with these matrices, we use the standard operations of matrix addition and multiplication *with the stipulation that* $1 + 1 = 1$. (Hence the addition is called Boolean.)

Example 7.19   Consider the sets A, B, and C and the relations $\mathcal{R}_1$, $\mathcal{R}_2$ of Example 7.15. With the orders of A, B, and C fixed as in that example, we define the *relation matrices* for $\mathcal{R}_1$, $\mathcal{R}_2$ as follows:

$$
M(\mathcal{R}_1) = \begin{array}{c} \\ (1) \\ (2) \\ (3) \\ (4) \end{array}
\begin{array}{cccc} (w) & (x) & (y) & (z) \\ \left[\begin{array}{cccc} 0 & 1 & 0 & 0 \\ 0 & 1 & 0 & 0 \\ 0 & 0 & 1 & 1 \\ 0 & 0 & 0 & 0 \end{array}\right] \end{array},
\qquad
M(\mathcal{R}_2) = \begin{array}{c} \\ (w) \\ (x) \\ (y) \\ (z) \end{array}
\begin{array}{ccc} (5) & (6) & (7) \\ \left[\begin{array}{ccc} 1 & 0 & 0 \\ 0 & 1 & 0 \\ 0 & 0 & 0 \\ 0 & 0 & 0 \end{array}\right] \end{array}
$$

In finding $M(\mathcal{R}_1)$, we are dealing with a relation from A to B, so the elements of A are used to mark the rows of $M(\mathcal{R}_1)$ and the elements of B designate the columns. Then to denote, for example, that $(2, x) \in \mathcal{R}_1$, we place a 1 in the row marked (2) and the column marked (x). Each 0 in this matrix indicates an ordered pair in $A \times B$ that is missing from $\mathcal{R}_1$. For example, since $(3, w) \notin \mathcal{R}_1$, there is a 0 for the entry in row (3) and column (w) of the matrix $M(\mathcal{R}_1)$. The same process is used to obtain $M(\mathcal{R}_2)$.

Multiplying these matrices, we find that

$$
M(\mathcal{R}_1) \cdot M(\mathcal{R}_2) = \begin{bmatrix} 0 & 1 & 0 & 0 \\ 0 & 1 & 0 & 0 \\ 0 & 0 & 1 & 1 \\ 0 & 0 & 0 & 0 \end{bmatrix} \begin{bmatrix} 1 & 0 & 0 \\ 0 & 1 & 0 \\ 0 & 0 & 0 \\ 0 & 0 & 0 \end{bmatrix} = \begin{bmatrix} 0 & 1 & 0 \\ 0 & 1 & 0 \\ 0 & 0 & 0 \\ 0 & 0 & 0 \end{bmatrix} = M(\mathcal{R}_1 \circ \mathcal{R}_2),
$$

and in general we have: If $\mathcal{R}_1$ is a relation from A to B and $\mathcal{R}_2$ is a relation from B to C, then $M(\mathcal{R}_1) \cdot M(\mathcal{R}_2) = M(\mathcal{R}_1 \circ \mathcal{R}_2)$. That is, the product of the relation matrices for $\mathcal{R}_1$, $\mathcal{R}_2$, in that order, equals the relation matrix of the composite relation $\mathcal{R}_1 \circ \mathcal{R}_2$. (This is why the composition of two relations was written in the order specified in Definition 7.8.)   □

The reader will be asked to prove the general result of Example 7.19, along with some results from our next example, in Exercises 9 and 10 at the end of this section.

Further properties of relation matrices are exhibited in the following example.

Example 7.20   Let $A = \{1, 2, 3, 4\}$ and $\mathcal{R} = \{(1, 2), (1, 3), (2, 4), (3, 2)\}$, as in Example 7.17. Keeping the order of the elements in A fixed, we define the *relation matrix* for $\mathcal{R}$ as follows: $M(\mathcal{R})$ is the 4×4 $(0, 1)$-matrix whose entries m_{ij}, for $1 \leq i, j \leq 4$, are given by

$$m_{ij} = \begin{cases} 1, & \text{if } (i, j) \in \mathcal{R}, \\ 0, & \text{otherwise.} \end{cases}$$

In this case we find that

$$M(\mathcal{R}) = \begin{bmatrix} 0 & 1 & 1 & 0 \\ 0 & 0 & 0 & 1 \\ 0 & 1 & 0 & 0 \\ 0 & 0 & 0 & 0 \end{bmatrix}.$$

Now how can this be of any use? If we compute $(M(\mathcal{R}))^2$ using the convention that $1 + 1 = 1$, then we find that

$$(M(\mathcal{R}))^2 = \begin{bmatrix} 0 & 1 & 0 & 1 \\ 0 & 0 & 0 & 0 \\ 0 & 0 & 0 & 1 \\ 0 & 0 & 0 & 0 \end{bmatrix},$$

which happens to be the relation matrix for $\mathcal{R} \circ \mathcal{R} = \mathcal{R}^2$. (Check Example 7.17). Furthermore,

$$(M(\mathcal{R}))^4 = \begin{bmatrix} 0 & 0 & 0 & 0 \\ 0 & 0 & 0 & 0 \\ 0 & 0 & 0 & 0 \\ 0 & 0 & 0 & 0 \end{bmatrix},$$

the relation matrix for $\mathcal{R}^4 = \emptyset$.   □

What has happened here carries over to the general situation. We now state some results about relation matrices and their use in studying relations.

Let A be a set with $|A| = n$ and $\mathcal{R}$ a relation on A. If $M(\mathcal{R})$ is the relation matrix for $\mathcal{R}$, then

a) $M(\mathcal{R}) = \mathbf{0}$ (the matrix of all 0's) if and only if $\mathcal{R} = \emptyset$

b) $M(\mathcal{R}) = \mathbf{1}$ (the matrix of all 1's) if and only if $\mathcal{R} = A \times A$

c) $M(\mathcal{R}^m) = [M(\mathcal{R})]^m$, for $m \in \mathbf{Z}^+$

Using the $(0, 1)$-matrix for a relation, we turn to the recognition of the reflexive, symmetric, antisymmetric, and transitive properties. To accomplish this we need the concepts introduced in the following three definitions.

DEFINITION 7.11   Let $E = (e_{ij})_{m \times n}$, $F = (f_{ij})_{m \times n}$ be two $(0, 1)$-matrices. We say that *E precedes*, or *is less than*, *F*, and we write $E \leq F$, if $e_{ij} \leq f_{ij}$, for all $1 \leq i \leq m, 1 \leq j \leq n$. ▬

Example 7.21   With $E = \begin{bmatrix} 1 & 0 & 1 \\ 0 & 0 & 1 \end{bmatrix}$ and $F = \begin{bmatrix} 1 & 0 & 1 \\ 0 & 1 & 1 \end{bmatrix}$, we have $E \leq F$. In fact there are eight $(0, 1)$-matrices G for which $E \leq G$.   □

DEFINITION 7.12   For $n \in \mathbf{Z}^+$, $I_n = (\delta_{ij})_{n \times n}$ is the $n \times n$ $(0, 1)$-matrix where

$$\delta_{ij} = \begin{cases} 1, & \text{if } i = j \\ 0, & \text{if } i \neq j. \end{cases}$$

▬

DEFINITION 7.13   Let $A = (a_{ij})_{m \times n}$ be a $(0, 1)$-matrix. The *transpose* of A, written A^{tr}, is the matrix $(a_{ji}^*)_{n \times m}$ where $a_{ji}^* = a_{ij}$, for all $1 \leq j \leq n, 1 \leq i \leq m$. ▬

Example 7.22   For $A = \begin{bmatrix} 0 & 1 \\ 0 & 0 \\ 1 & 1 \end{bmatrix}$, $A^{\text{tr}} = \begin{bmatrix} 0 & 0 & 1 \\ 1 & 0 & 1 \end{bmatrix}$.

As this example demonstrates, the *i*th row (column) of A equals the *i*th column (row) of A^{tr}. This indicates a way to obtain the matrix A^{tr} from the matrix A.   □

THEOREM 7.2   Given a set A with $|A| = n$, and a relation $\mathcal{R}$ on A, let M denote the relation matrix for $\mathcal{R}$. Then

a) $\mathcal{R}$ is reflexive if and only if $I_n \leq M$.

b) $\mathcal{R}$ is symmetric if and only if $M = M^{\text{tr}}$.

c) $\mathcal{R}$ is transitive if and only if $M \cdot M = M^2 \leq M$.

d) $\mathcal{R}$ is antisymmetric if and only if $M \cap M^{\text{tr}} \leq I_n$. (The matrix $M \cap M^{\text{tr}}$ is formed by operating on corresponding entries in M and M^{tr} according to the rules $0 \cap 0 = 0 \cap 1 = 1 \cap 0 = 0$ and $1 \cap 1 = 1$.)

Proof   The results follow from the definitions of the relation properties and the $(0, 1)$-matrix. We demonstrate this for part (c), using the elements of A to designate the rows and columns in M, as in Examples 7.19 and 7.20.

Let $M^2 \leq M$. If $(x, y), (y, z) \in \mathcal{R}$, then there are 1's in row (x), column (y) and in row (y), column (z) of M. Consequently, in row (x), column (z) of M^2 there is a 1. This 1 must also occur in row (x), column (z) of M, because $M^2 \leq M$. Hence $(x, z) \in \mathcal{R}$ and $\mathcal{R}$ is transitive.

Conversely, if $\mathcal{R}$ is transitive and M is the relation matrix for $\mathcal{R}$, let s_{xz} be the entry in row (x) and column (z) of M^2, with $s_{xz} = 1$. For s_{xz} to equal 1 in M^2, there must exist at least one $y \in A$ where $m_{xy} = m_{yz} = 1$ in M. This happens only if $x \mathcal{R} y$

and $y \mathcal{R} z$. With $\mathcal{R}$ transitive, it then follows that $x \mathcal{R} z$. So $m_{xz} = 1$ and $M^2 \leq M$. The proofs of the remaining parts are left to the reader. ∎

The relation matrix is a useful tool for the computer recognition of certain properties of relations. Storing information as described here, this matrix is an example of a *data structure*. Also of interest is how the relation matrix is used in the study of graph theory† and how graph theory is used in the recognition of certain properties of relations.

DEFINITION 7.14
Let V be a finite nonempty set. A *directed graph* (or *digraph*) G on V is made up of the elements of V, called the *vertices* or *nodes* of G, and a subset E of $V \times V$, called the *(directed) edges*, or *arcs*, of G. If $a, b \in V$ and $(a, b) \in E$‡, then there is an edge *from a to b*. Vertex a is called the *origin* or *source* of the edge, with b the *terminus*, or *terminating vertex*, and we say that b is *adjacent from a* and that a is *adjacent to b*. In addition, if $a \neq b$, then $(a, b) \neq (b, a)$. An edge of the form (a, a) is called a *loop* (at a).

Example 7.23
For $V = \{1, 2, 3, 4, 5\}$, the diagram in Fig. 7.1 is a directed graph G on V with edge set $\{(1, 1), (1, 2), (1, 4), (3, 2)\}$. Vertex 5 is a part of this graph even though it is not the origin or terminus of an edge. It is referred to as an *isolated* vertex. As we see here, edges need not be straight line segments, and there is no concern about the length of an edge. □

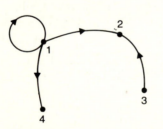

Figure 7.1

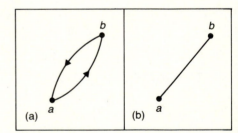

Figure 7.2

When we develop a *flowchart* to study a computer program or algorithm, we deal with a special type of directed graph where the shapes of the vertices are important in the algorithm analysis. Road maps are directed graphs, where the cities and towns are represented by vertices, and the highways linking any two localities are given by edges. In road maps, an edge is often directed in both directions. Consequently, if G is a directed graph and $a, b \in V$, with $a \neq b$, and both $(a, b), (b, a) \in E$, then the single undirected edge $\{a, b\} = \{b, a\}$, in Fig. 7.2(b) is used to represent the two directed edges shown in Fig. 7.2(a). In this

---

† Since the terminology of graph theory is not standardized, the reader may find some differences between definitions given here and in other texts.

‡ In this chapter we allow only one edge from a to b. Situations where multiple edges occur are called *multigraphs*. These are discussed in Chapter 11.

case, *a* and *b* are called *adjacent* vertices. (Directions may also be disregarded for loops.)

How are directed graphs used in the study of relations? For a set *A* and relation $\mathcal{R}$ on *A*, we construct a directed graph *G* with vertex set *A* and edge set $E \subseteq A \times A$, where $(a, b) \in E$ if $a, b \in A$ and $a\,\mathcal{R}\,b$. This is demonstrated in the following example.

Example 7.24   For $A = \{1, 2, 3, 4\}$, let $\mathcal{R} = \{(1, 1), (1, 2), (2, 3), (3, 2), (3, 3), (3, 4), (4, 2)\}$ be a relation on *A*. The directed graph associated with $\mathcal{R}$ is shown in Fig. 7.3(a). If the directions are ignored, we get the *associated undirected graph* shown in part (b) of the figure. Here we see that the graph is *connected* in the sense that for any two vertices *x*, *y*, with $x \neq y$, there is a *path* starting at *x* and ending at *y*. Such a path consists of a *finite sequence of undirected edges*, so the edges $\{1, 2\}, \{2, 4\}$ provide a path from 1 to 4, and the edges $\{3, 4\}, \{4, 2\}$, and $\{2, 1\}$ provide a path from 3 to 1. The sequence of edges $\{3, 4\}, \{4, 2\}$, and $\{2, 3\}$ provides a path from 3 to 3. Such a *closed* path is called a *cycle*. This is an example of an undirected cycle of *length* three, because it has three edges in it.

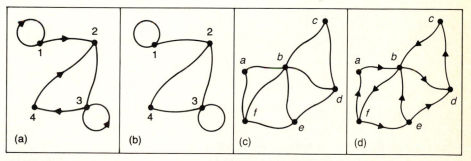

Figure 7.3

When we are dealing with paths (in both directed and undirected graphs), no vertex may be repeated. Therefore, the sequence of edges $\{a, b\}, \{b, e\}, \{e, f\}, \{f, b\}, \{b, d\}$ in Fig. 7.3(c) is *not* considered to be a path (from *a* to *d*) because we pass through the vertex *b* more than once. In the case of cycles, the path starts and terminates at the same vertex and has *at least three edges*. The sequence of edges $(b, f), (f, e), (e, d), (d, c), (c, b)$ provides a *directed cycle* of length five in Fig. 7.3(d). The six edges $(b, f), (f, e), (e, b), (b, d), (d, c), (c, b)$ do *not* yield a directed cycle in the figure because of the repetition of vertex *b*. If their directions are ignored, the corresponding six edges, in part (c) of the figure, likewise pass through vertex *b* more than once. Consequently, these edges are not considered to form a cycle for the undirected graph in Fig. 7.3(c).

Now since we require a cycle to have *length* at least three, we shall not consider loops to be cycles. We also note that loops have no bearing on graph connectivity.   □

DEFINITION 7.15 If G is a directed graph on V, then G is said to be *strongly connected* if for all $x, y \in V, x \neq y$, there is a path of *directed* edges from x to y.

It is in this sense that we talked about strongly connected machines in Chapter 6. The graph in Fig. 7.3(a) is connected but not strongly connected. For example, there is no directed path from 3 to 1. In Fig. 7.4 the directed graph on $V = \{1, 2, 3, 4\}$ is strongly connected and *loop-free*. This is also true of the directed graph in Fig. 7.3(d).

Example 7.25 For $A = \{1, 2, 3, 4\}$, consider the relations $\mathcal{R}_1 = \{(1, 1), (1, 2), (2, 1), (2, 2), (3, 3), (3, 4), (4, 3), (4, 4)\}$ and $\mathcal{R}_2 = \{(2, 4), (2, 3), (3, 2), (3, 3), (3, 4)\}$. As Fig. 7.5 illustrates, the graphs of these relations are *disconnected*. However, each graph is the union of two connected pieces called the *components* of the graph. For $\mathcal{R}_1$ the graph is made up of two strongly connected components. For $\mathcal{R}_2$, one component consists of an isolated vertex, and the other component is connected but not strongly connected. □

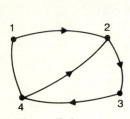

Figure 7.4

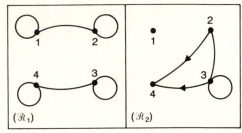

Figure 7.5

Example 7.26 The graphs in Fig. 7.6 are examples of undirected graphs that are loop-free and have an edge for every pair of distinct vertices. These graphs are called the *complete graphs* on n vertices and are denoted by K_n. In Fig. 7.6 we have examples of the complete graphs on three, four, and five vertices, respectively. The complete graph K_2 consists of two points x, y and an edge connecting them, whereas K_1 consists of one point and no edges, because loops are not allowed.

In K_5 two edges cross, namely $\{3, 5\}$ and $\{1, 4\}$. However, there is no point of intersection creating a new vertex. If we try to avoid the crossing of edges by

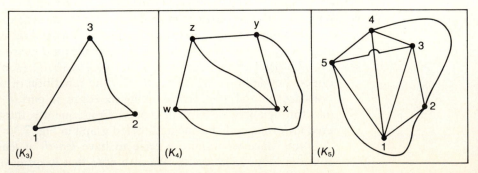

Figure 7.6

drawing the graph differently, we run into the same problem all over again. This difficulty will be examined in Chapter 11 when we deal with the planarity of graphs.   □

For a graph G on a vertex set V, the graph gives rise to a relation $\mathcal{R}$ on V where $x \mathcal{R} y$, if (x, y) is an edge in G. Consequently, there is a $(0, 1)$-matrix for G, and since this relation matrix comes about from the adjacencies of pairs of vertices, it is referred to as the *adjacency matrix* for G as well as the relation matrix for $\mathcal{R}$.

At this point we tie together the properties of relations and the structure of directed graphs.

Example 7.27   If $A = \{1, 2, 3\}$ and $\mathcal{R} = \{(1, 1), (1, 2), (2, 2), (3, 3), (3, 1)\}$, then $\mathcal{R}$ is a reflexive relation on A, but it is not symmetric, antisymmetric, or transitive. The directed graph associated with $\mathcal{R}$ consists of five edges. Three of these edges are loops that result from the reflexive property of $\mathcal{R}$. (See Fig. 7.7.) In general, if $\mathcal{R}$ is a relation on a finite set A, then $\mathcal{R}$ is reflexive if and only if its directed graph contains a loop at each vertex (element of A).   □

Example 7.28   The relation $\mathcal{R} = \{(1, 1), (1, 2), (2, 1), (2, 3), (3, 2)\}$ is symmetric on $A = \{1, 2, 3\}$ but it is not reflexive, antisymmetric, or transitive. The directed graph for $\mathcal{R}$ is found in Fig. 7.8. In general, a relation $\mathcal{R}$ on a finite set A is symmetric if and only if its directed graph contains only loops and undirected edges.   □

Example 7.29   For $A = \{1, 2, 3\}$, consider $\mathcal{R} = \{(1, 1), (1, 2), (2, 3), (1, 3)\}$. The directed graph for $\mathcal{R}$ is shown in Fig. 7.9. Here $\mathcal{R}$ is transitive and antisymmetric, but not reflexive or symmetric. The directed graph indicates that a relation on a set A is transitive if and only if its directed graph satisfies the following: For any $x, y \in A$, if there is a (directed) path from x to y in the associated graph, then there is an edge (x, y) also. (Here $(1, 2), (2, 3)$ is a (directed) path from 1 to 3, and we also have the edge $(1, 3)$ for transitivity.)

The relation $\mathcal{R}$ is antisymmetric because there are no ordered pairs in $\mathcal{R}$ of the form (x, y) and (y, x), with $x \neq y$. To use the directed graph of Fig. 7.9 to characterize antisymmetry, we observe that for any two vertices x, y, with $x \neq y$, the graph contains at most one of the edges (x, y) or (y, x). Hence there are no undirected edges aside from loops.   □

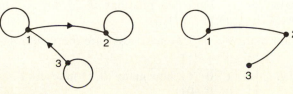

Figure 7.7   **Figure 7.8**

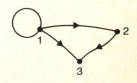

Figure 7.9

Our final example deals with equivalence relations.

Example 7.30   For $A = \{1, 2, 3, 4, 5\}$, the following are equivalence relations on A:

$$\mathcal{R}_1 = \{(1,1), (1,2), (2,1), (2,2), (3,3), (3,4), (4,3), (4,4), (5,5)\},$$
$$\mathcal{R}_2 = \{(1,1), (1,2), (1,3), (2,1), (2,2), (2,3), (3,1), (3,2), (3,3),$$
$$(4,4), (4,5), (5,4), (5,5)\}.$$

Their associated graphs are shown in Fig. 7.10. If we ignore the loops in each graph, we find the graph decomposed into components such as K_1, K_2, and K_3. In general, a relation on a finite set A is an equivalence relation if and only if its associated graph consists of the disjoint union of complete graphs augmented by loops at every vertex. $\square$

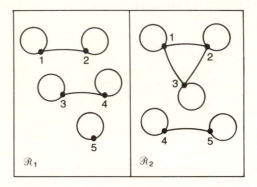

Figure 7.10

EXERCISES 7.2

1. For $A = \{1, 2, 3, 4\}$, let $\mathcal{R}$ and $\mathcal{S}$ be the relations on A defined by $\mathcal{R} = \{(1,2), (1,3), (2,4), (4,4)\}$ and $\mathcal{S} = \{(1,1), (1,2), (1,3), (2,3), (2,4)\}$. Find $\mathcal{R} \circ \mathcal{S}$, $\mathcal{S} \circ \mathcal{R}$, $\mathcal{R}^2$, $\mathcal{R}^3$, $\mathcal{S}^2$, and $\mathcal{S}^3$.

2. If $\mathcal{R}$ is a reflexive relation on a set A, prove that $\mathcal{R}^2$ is also reflexive on A.

3. Provide the proof for the opposite inclusion in Theorem 7.1.

4. For sets A, B, and C, consider relations $\mathcal{R}_1 \subseteq A \times B$, $\mathcal{R}_2 \subseteq B \times C$, and $\mathcal{R}_3 \subseteq B \times C$. Prove that (a) $\mathcal{R}_1 \circ (\mathcal{R}_2 \cup \mathcal{R}_3) = (\mathcal{R}_1 \circ \mathcal{R}_2) \cup (\mathcal{R}_1 \circ \mathcal{R}_3)$; and (b) $\mathcal{R}_1 \circ (\mathcal{R}_2 \cap \mathcal{R}_3) \subseteq (\mathcal{R}_1 \circ \mathcal{R}_2) \cap (\mathcal{R}_1 \circ \mathcal{R}_3)$. (Provide an example to show that a proper inclusion can occur in (b).)

5. For a relation $\mathcal{R}$ on a set A, define $\mathcal{R}^0 = \{(a,a) \mid a \in A\}$. If $|A| = n$, prove that there exist $s, t \in \mathbf{N}$ with $0 \leq s < t \leq 2^{n^2}$ such that $\mathcal{R}^s = \mathcal{R}^t$.

6. With $A = \{1, 2, 3, 4\}$, let $\mathcal{R} = \{(1,1), (1,2), (2,3), (3,3), (3,4), (4,4)\}$ be a relation on A. Find two relations $\mathcal{S}$, $\mathcal{T}$ on A where $\mathcal{S} \neq \mathcal{T}$ but $\mathcal{R} \circ \mathcal{S} = \mathcal{R} \circ \mathcal{T} = \{(1,1), (1,2), (1,4)\}$.

7. How many 6×6 $(0,1)$-matrices A are there with $A = A^{tr}$?

8. If $E = \begin{bmatrix} 1 & 0 & 1 & 1 \\ 0 & 1 & 0 & 1 \\ 1 & 0 & 0 & 0 \end{bmatrix}$, how many $(0,1)$-matrices F satisfy $E \leq F$?

9. Consider the sets $A = \{a_1, a_2, \ldots, a_m\}$, $B = \{b_1, b_2, \ldots, b_n\}$, and $C = \{c_1, c_2, \ldots, c_p\}$, where the elements in each set remain fixed in the order given here. Let $\mathcal{R}_1$ be a relation from A to B, and let $\mathcal{R}_2$ be a relation from B to C.

The relation matrix for $\mathcal{R}_i$ is $M(\mathcal{R}_i)$, where $i = 1, 2$. The rows and columns of these matrices are indexed by the elements from the appropriate sets A, B, and C according to the orders prescribed above. The matrix for $\mathcal{R}_1 \circ \mathcal{R}_2$ is the $m \times p$ matrix $M(\mathcal{R}_1 \circ \mathcal{R}_2)$, where the elements of A (in the order given) index the rows, while the columns are indexed by the elements of C (also in the order given).

Show that for all $1 \leq i \leq m$ and $1 \leq j \leq p$, the entries in the ith row and jth column of $M(\mathcal{R}_1) \cdot M(\mathcal{R}_2)$ and $M(\mathcal{R}_1 \circ \mathcal{R}_2)$ are equal. (Hence $M(\mathcal{R}_1) \cdot M(\mathcal{R}_2) = M(\mathcal{R}_1 \circ \mathcal{R}_2)$.)

10. Let A be a set with $|A| = n$, and consider the order for the listing of its elements as fixed. For $\mathcal{R} \subseteq A \times A$, let $M(\mathcal{R})$ denote the corresponding relation matrix.

   a) Prove that $M(\mathcal{R}) = \mathbf{0}$ (the $n \times n$ matrix of all 0's) if and only if $\mathcal{R} = \emptyset$.

   b) Prove that $M(\mathcal{R}) = \mathbf{1}$ (the $n \times n$ matrix of all 1's) if and only if $\mathcal{R} = A \times A$.

   c) Use the result of Exercise 9, along with the principle of mathematical induction, to prove that $M(\mathcal{R}^m) = [M(\mathcal{R})]^m$, for all $m \in \mathbf{Z}^+$.

11. Provide the proofs for Theorem 7.2 (a), (b), and (d).

12. Use Theorem 7.2 to write a computer program (or develop an algorithm) for the recognition of equivalence relations on a finite set.

13. For $A = \{1, 2, 3, 4\}$, let $\mathcal{R} = \{(1, 1), (1, 2), (2, 3), (3, 3), (3, 4)\}$ be a relation on A. Draw the directed graph G on A that is associated with $\mathcal{R}$. Do likewise for $\mathcal{R}^2$, $\mathcal{R}^3$, and $\mathcal{R}^4$.

14. For $|A| = 5$, how many relations $\mathcal{R}$ on A are there? How many of these relations are symmetric?

15. Let $|A| = 5$. (a) How many directed graphs can one construct on A? (b) How many of the graphs in (a) are actually undirected?

16. How many (undirected) edges are there in the complete graphs K_6, K_7, and K_n, where $n \in \mathbf{Z}^+$?

17. a) Keeping the order of the elements fixed as $1, 2, 3, 4, 5$, determine the $(0, 1)$ relation matrices for the equivalence relations in Example 7.30.

   b) Do the results of part (a) lead to any generalization?

18. a) Let $\mathcal{R}$ be the relation on $A = \{1, 2, 3, 4, 5, 6, 7\}$, where the directed graph associated with $\mathcal{R}$ consists of the components, each a directed cycle, shown in Fig. 7.11. Find the smallest integer $n > 1$, such that $\mathcal{R}^n = \mathcal{R}$. What is the smallest value of $n > 1$ for which the graph of $\mathcal{R}^n$ contains some loops? Does it ever happen that the graph of $\mathcal{R}^n$ consists of only loops?

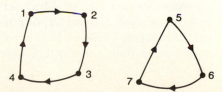

Figure 7.11

b) Answer the same questions for the relation $\mathcal{R}$ on $A = \{1, 2, 3, \dots, 9, 10\}$, if the directed graph associated with $\mathcal{R}$ is as shown in Fig. 7.12.

c) Do the results of (a) and (b) indicate anything in general?

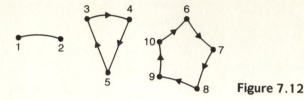

Figure 7.12

7.3
PARTIAL ORDERS: HASSE DIAGRAMS

If you ask children to recite the numbers they know, you'll hear a uniform response of "1, 2, 3, ..." Without paying attention to it, they list these numbers in increasing order. In this section we take a closer look at this idea of order, something we may have taken for granted. We start by observing something about the sets $\mathbf{N}$, $\mathbf{Z}$, $\mathbf{Q}$, $\mathbf{R}$, and $\mathbf{C}$.

The set $\mathbf{N}$ is closed under the binary operations of (ordinary) addition and multiplication, but if we seek an answer to the equation $x + 5 = 2$, we find that no element of $\mathbf{N}$ provides a solution. So we enlarge $\mathbf{N}$ to $\mathbf{Z}$, where we can perform subtraction as well as addition and multiplication. However, we soon run into trouble trying to solve the equation $2x + 3 = 4$. Enlarging to $\mathbf{Q}$, we can perform nonzero division in addition to the other operations. Yet this soon proves to be inadequate; the equation $x^2 - 2 = 0$ necessitates the introduction of the real but irrational numbers $\pm\sqrt{2}$. Even after we expand from $\mathbf{Q}$ to $\mathbf{R}$, more trouble arises when we try to solve $x^2 + 1 = 0$. Finally we arrive at $\mathbf{C}$, the complex number system, where any polynomial equation of the form $c_n x^n + c_{n-1} x^{n-1} + \cdots + c_2 x^2 + c_1 x + c_0 = 0$, where $c_i \in \mathbf{C}, 0 \leq i \leq n$, can be solved. (This result is known as the Fundamental Theorem of Algebra. Its proof requires material on functions of a complex variable, so no proof is given here.) As we kept building up from $\mathbf{N}$ to $\mathbf{C}$, gaining more ability to solve polynomial equations, something was lost when we went from $\mathbf{R}$ to $\mathbf{C}$. In $\mathbf{R}$, given numbers r_1, r_2, with $r_1 \neq r_2$, we can always decide whether $r_1 < r_2$ or $r_2 < r_1$. However, in $\mathbf{C}$, $(2 + i) \neq (1 + 2i)$, but what meaning can we attach to a statement such as "$(2 + i) < (1 + 2i)$"? We have lost the ability to "order" the elements of our number system!

As in Section 7.1, let A be a set and $\mathcal{R}$ a relation on A. The pair $(A, \mathcal{R})$ is called a *partially ordered set*, or *poset*, if relation $\mathcal{R}$ on A is a partial order, or partial ordering relation. If A is called a poset, we understand that there is a partial order $\mathcal{R}$ on A that makes A into this poset. Examples 7.1(a), 7.2, and 7.10 are posets.

Example 7.31   Let A be the set of courses offered at a college. Define relation $\mathscr{R}$ on A by $x \mathscr{R} y$, if x, y are the same course, or if x is a prerequisite for y. Then $\mathscr{R}$ makes A into a poset.   □

Example 7.32   Define $\mathscr{R}$ on $A = \{1, 2, 3, 4\}$ by $x \mathscr{R} y$ if $x \mid y$—that is, x divides y. Then $\mathscr{R} = \{(1, 1), (2, 2), (3, 3), (4, 4), (1, 2), (1, 3), (1, 4), (2, 4)\}$ is a partial order, and $(A, \mathscr{R})$ is a poset.   □

Example 7.33   In the construction of a house there are certain jobs, such as digging the foundation, that must be performed before other phases of the construction can be undertaken. If A is a set of tasks that must be performed in building a house, we can define a relation $\mathscr{R}$ on A by $x \mathscr{R} y$ if x, y denote the same task or if task x must be performed before the start of task y. In this way we place an order on the elements of A, making it into a poset that is sometimes referred to as a PERT (Program Evaluation and Review Technique) network.   □

Consider the diagrams given in Fig. 7.13. If part (a) were part of the directed graph associated with a relation $\mathscr{R}$, then because $(1, 2), (2, 1) \in \mathscr{R}$ with $1 \neq 2$, $\mathscr{R}$ could not be antisymmetric. For part (b), if the diagram were part of the graph of a transitive relation $\mathscr{R}$, then $(1, 2), (2, 3) \in \mathscr{R} \Rightarrow (1, 3) \in \mathscr{R}$. Since $(3, 1) \in \mathscr{R}$ and $1 \neq 3$, $\mathscr{R}$ is not antisymmetric, so it cannot be a partial order.

From these observations, given a relation $\mathscr{R}$ on a set A, let G be the directed graph associated with $\mathscr{R}$.

(i) If G contains a pair of edges of the form $(a, b), (b, a)$, for $a, b \in A$ with $a \neq b$, or

(ii) If $\mathscr{R}$ is transitive and G contains a directed cycle (of length greater than or equal to three),

then the relation $\mathscr{R}$ cannot be antisymmetric, so $(A, \mathscr{R})$ fails to be a partial order.

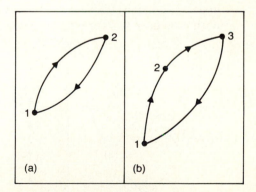

Figure 7.13

Example 7.34   Consider the directed graph for the partial order in Example 7.32. Figure 7.14(a) is the graphical representation of $\mathscr{R}$. In part (b) of the figure, we have a somewhat simpler diagram, which is called the *Hasse diagram* for $\mathscr{R}$.

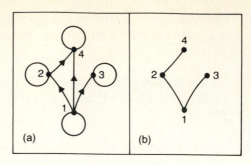

(a)   (b)

Figure 7.14

When we know that a relation $\mathscr{R}$ is a partial order on a set A, we can eliminate the loops at the vertices of its directed graph. Since $\mathscr{R}$ is also transitive, having the edges $(1, 2)$ and $(2, 4)$ is enough to insure the existence of edge $(1, 4)$, so we need not include that edge. In this way we obtain the diagram in Fig. 7.14(b). □

In general, if $\mathscr{R}$ is a partial order on a set A, we construct a Hasse diagram for $\mathscr{R}$ on A by drawing a line segment from x up to y, if $x, y \in A$ with $x \mathscr{R} y$ and, most important, if there is no other element $z \in A$ such that $x \mathscr{R} z$ and $z \mathscr{R} y$. (So there is nothing "in between" x and y.) If we adopt the convention of reading the diagram from bottom to top, then it is not necessary to direct any edges.

Example 7.35   In Fig. 7.15 we have the Hasse diagrams for the following four posets. (a) With $\mathscr{U} = \{1, 2, 3\}$ and $A = \mathscr{P}(\mathscr{U})$, $\mathscr{R}$ is the subset relation on A. (b) Here $\mathscr{R}$ is the "divides" relation applied to $A = \{1, 2, 4, 8\}$. (c) and (d) Here the same relation as in part (b) is applied to $\{2, 3, 5, 7\}$ in part (c) and to $\{2, 3, 5, 6, 7, 11, 12, 35, 385\}$ in part (d). In part (c) we note that a Hasse diagram can have all isolated vertices; it can also have two (or more) connected pieces, as shown in part (d). □

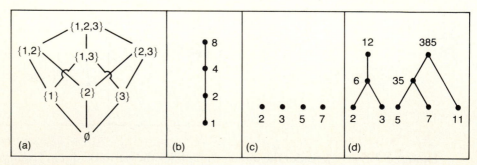

(a)   (b)   (c)   (d)

Figure 7.15

Example 7.36   Let $A = \{1, 2, 3, 4, 5\}$. The relation $\mathcal{R}$ on A, defined by $x\,\mathcal{R}\,y$ if $x \leq y$, is a partial order. This makes A into a poset that we can denote by $(A, \leq)$. If $B = \{1, 2, 4\} \subset A$, then the set $(B \times B) \cap \mathcal{R} = \{(1,1), (2,2), (4,4), (1,2), (1,4), (2,4)\}$ is a partial order on B.

In general if $\mathcal{R}$ is a partial order on A, then for any subset B of A, $(B \times B) \cap \mathcal{R}$ makes B into a poset where the partial order on B is induced from $\mathcal{R}$. $\quad\square$

We turn now to a special type of partial order.

DEFINITION 7.16   If $(A, \mathcal{R})$ is a poset, we say that A is *totally ordered* if for all $x, y \in A$ either $x\,\mathcal{R}\,y$ or $y\,\mathcal{R}\,x$.

In this case $\mathcal{R}$ is called a *total order*.   ——

Example 7.37   (a) On the set **N**, the relation $\mathcal{R}$ defined by $x\,\mathcal{R}\,y$ if $x \leq y$ is a total order. (b) The subset relation applied to $A = \mathcal{P}(\mathcal{U})$, where $\mathcal{U} = \{1, 2, 3\}$, is a partial, but not total, order: $\{1, 2\}, \{1, 3\} \in A$ but we have neither $\{1, 2\} \subseteq \{1, 3\}$ nor $\{1, 3\} \subseteq \{1, 2\}$. (c) The Hasse diagram in part (b) of Fig. 7.15 shows a total order. $\quad\square$

Could these notions of partial and total order ever arise in an industrial problem?

Say a toy manufacturer is about to market a new product and must include a set of instructions for its assembly. In order to assemble the new toy, there are seven tasks, denoted $A, B, C, \ldots, G$, that one must perform in the partial order given by the Hasse diagram of Fig. 7.16. Here we see, for example, that B, A, and E must be completed before we can work on task C. Since the set of instructions is to consist of a listing of these tasks, numbered $1, 2, 3, \ldots, 7$, how can the manufacturer write the listing and make sure that the partial order of the Hasse diagram is maintained?

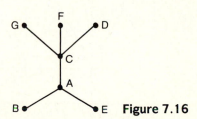

Figure 7.16

Forgetting the toy and the assembly problem, what we are really asking for here is whether we can take the partial order $\mathcal{R}$, given by the Hasse diagram, and find a total order $\mathcal{T}$ on these tasks for which $\mathcal{R} \subseteq \mathcal{T}$. The answer is yes, and the technique is known as *topological sorting*.

Topological Sorting Algorithm

(for a partial order $\mathscr{R}$ on a set A with $|A| = n$)

Step 1: Set $k = 1$. Let H_1 be the Hasse diagram of the partial order.

Step 2: Select a vertex v_k in H_k such that no (implicitly directed) edge in H_k starts at v_k.

Step 3: If $k = n$, the process is completed and we have a total order

$$\mathscr{T}: v_n < v_{n-1} < \cdots < v_2 < v_1,$$

that contains $\mathscr{R}$.

If $k < n$, then remove from H_k the vertex v_k and all (implicitly directed) edges of H_k that terminate at v_k. Call the result H_{k+1}. Increase k by 1 and return to step 2.

Here we have presented our algorithm as a precise list of instructions, making no reference to its implementation in a particular computer language.

Before we apply this algorithm to the problem at hand, we should observe the "a" before the word "vertex" in step 2. This implies that the selection need not be unique and that we can get several different total orders $\mathscr{T}$ containing $\mathscr{R}$. Also, in step 3, for vertices v_{i-1} where $2 \leq i \leq n$, the notation $v_i < v_{i-1}$ is used because it is more suggestive of "v_i before v_{i-1}" than is the notation $v_i \mathscr{T} v_{i-1}$.

In Fig. 7.17, we show the Hasse diagrams that evolve as we apply the algorithm to the partial order in Fig. 7.16. Below each diagram, the total order is listed as it evolves.

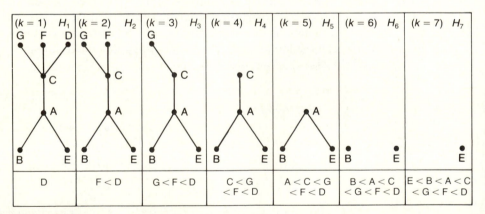

Figure 7.17

If the toy manufacturer writes the instructions in a list as 1-E, 2-B, 3-A, 4-C, 5-G, 6-F, 7-D, he or she will have a total order that preserves the partial order needed for correct assembly. This total order is one of 12 possible answers.

As is typical in discrete and combinatorial mathematics, this algorithm provides a procedure that reduces the size of our problem with each successive application.

In the topological sorting algorithm, we saw how the Hasse diagram was used in determining a total order for a poset $(A, \mathcal{R})$. This algorithm now prompts us to examine further properties of a partial order. At the start, particular emphasis will be given to the vertex v_k in step 2 of the algorithm.

What interplay exists between the partial order $(A, \mathcal{R})$ and its Hasse diagram so that we can describe a vertex such as v_k in terms of $\mathcal{R}$? This question directs us to the following results.

DEFINITION 7.17   If $(A, \mathcal{R})$ is a poset, then an element $x \in A$ is called a *maximal* element of A if for all $a \in A$, $a \neq x \Rightarrow x \, \mathcal{R} \, a$. An element $y \in A$ is called a *minimal* element of A if whenever $b \in A$ and $b \neq y$, then $b \, \mathcal{R} \, y$.

If we use the contrapositive of the first statement in Definition 7.17, then we can state that $x(\in A)$ is a maximal element if for any $a \in A$, $x \, \mathcal{R} \, a \Rightarrow x = a$. In a similar manner, for $y \in A$ we have y minimal if for any $b \in A$, $b \, \mathcal{R} \, y \Rightarrow b = y$.

Example 7.38   Let $\mathcal{U} = \{1, 2, 3\}$ and $A = \mathcal{P}(\mathcal{U})$.

a) Let $\mathcal{R}$ be the subset relation on A. Then $\mathcal{U}$ is maximal and $\emptyset$ is minimal for the poset $(A, \subseteq)$.

b) For B, the collection of proper subsets of $\{1, 2, 3\}$, let $\mathcal{R}$ be the subset relation on B. In the poset $(B, \subseteq)$, the sets $\{1, 2\}$, $\{1, 3\}$, and $\{2, 3\}$ are all maximal elements; $\emptyset$ is still the only minimal element.   □

Example 7.39   With $\mathcal{R}$ as the "is less than or equal to" relation on the set $\mathbf{Z}$, $(\mathbf{Z}, \leq)$ is a poset with neither a maximal nor a minimal element. The poset $(\mathbf{N}, \leq)$, however, has minimal element 0 but no maximal element.   □

Example 7.40   When we look back at the partial orders in parts (b), (c), and (d) of Example 7.35, the following observations come to light.

1. The partial order in part (b) has the unique maximal element 8 and the unique minimal element 1.

2. Each of the four elements—2, 3, 5, and 7—is both a maximal element and a minimal element for the poset in part (c) of Example 7.35.

3. In part (d) the elements 12 and 385 are both maximal. Each of the elements 2, 3, 5, 7, and 11 is a minimal element for this partial order.   □

Are there any conditions for which a poset must have a maximal or minimal element?

THEOREM 7.3   If $(A, \mathcal{R})$ is a poset and A is finite, then A has both a maximal and a minimal element.

Proof   Let $a_1 \in A$. If there is no element $a \in A$ where $a \neq a_1$ and $a_1 \, \mathcal{R} \, a$, then a_1 is maximal. Otherwise there is an element $a_2 \in A$ with $a_2 \neq a_1$ and $a_1 \, \mathcal{R} \, a_2$. If no element $a \in A$, $a \neq a_2$, satisfies $a_2 \, \mathcal{R} \, a$, then a_2 is maximal. Otherwise we can find

$a_3 \in A$ so that $a_3 \neq a_2$, $a_3 \neq a_1$ (Why?) while $a_1 \mathcal{R} a_2$ and $a_2 \mathcal{R} a_3$. Continuing in this manner, since A is finite we get to an element $a_n \in A$ with $a_n \mathcal{R} a$ for any $a \in A$, $a \neq a_n$, so a_n is maximal.

The proof for a minimal element follows in a similar way. ∎

Returning now to the topological sorting algorithm, we see that in each iteration of step 2 of the algorithm, we are selecting a maximal element from the original poset $(A, \mathcal{R})$, or a poset of the form $(B, \mathcal{R}')$ where $\emptyset \neq B \subset A$ and $\mathcal{R}' = (B \times B) \cap \mathcal{R}$. At least one such element exists (in each iteration) by virtue of Theorem 7.3. Then in the second part of step 3, if x is the maximal element selected (in step 2), we remove from the present poset all elements of the form (a, x). This results in a smaller poset.

We turn now to the study of some additional ideas on posets.

Among maximal and minimal elements for a poset, the following, when they exist, are of special concern.

DEFINITION 7.18   If $(A, \mathcal{R})$ is a poset, then an element $x \in A$ is called a *least* element if $x \mathcal{R} a$ for all $a \in A$. Element $y \in A$ is called a *greatest* element if $a \mathcal{R} y$ for all $a \in A$. ▬

Example 7.41   Let $\mathcal{U} = \{1, 2, 3\}$, and let $\mathcal{R}$ be the subset relation.

a) With $A = \mathcal{P}(\mathcal{U})$, the poset $(A, \subseteq)$ has $\emptyset$ as a least element and $\mathcal{U}$ as a greatest element.

b) For $B =$ the collection of nonempty subsets of $\mathcal{U}$, the poset $(B, \subseteq)$ has $\mathcal{U}$ as a greatest element. There is no least element here, but there are three minimal elements. □

Example 7.42   For the partial orders in parts (b), (c), and (d) of Example 7.35, we find that

1. The partial order in part (b) has greatest element 8 and least element 1.
2. There is no greatest element or least element for the poset in part (c).
3. No greatest element or least element exists for the partial order in part (d). □

For a poset $(A, \mathcal{R})$ we have seen that it is possible to have several maximal and minimal elements. What about least and greatest elements?

THEOREM 7.4   If the poset $(A, \mathcal{R})$ has a greatest (least) element, then that element is unique.

Proof   Suppose that $x, y \in A$ and that both are greatest elements. Since x is a greatest element, $y \mathcal{R} x$. Likewise, $x \mathcal{R} y$ because y is a greatest element. As $\mathcal{R}$ is antisymmetric, $x = y$.

The proof for the least element is similar. ∎

DEFINITION 7.19   Let $(A, \mathcal{R})$ be a poset with $B \subseteq A$. An element $x \in A$ is called a *lower bound* of B if $x \mathcal{R} b$ for all $b \in B$. If $y \in A$ and $b \mathcal{R} y$ for all $b \in B$, y is called an *upper bound* of B.

An element $x' \in A$ is called a *greatest lower bound* (glb) of B if it is a lower bound of B and if for all other lower bounds x'' of B, we have $x'' \mathcal{R} x'$. Similarly $y' \in A$ is a *least upper bound* (lub) of B if it is an upper bound of B and if $y' \mathcal{R} y''$ for all other upper bounds y'' of B.

Example 7.43 Let $\mathcal{U} = \{1, 2, 3, 4\}$, with $A = \mathcal{P}(\mathcal{U})$, and let $\mathcal{R}$ be the subset relation on A. If $B = \{\{1\}, \{2\}, \{1, 2\}\}$, then $\{1, 2\}$, $\{1, 2, 3\}$, $\{1, 2, 4\}$, and $\{1, 2, 3, 4\}$ are all upper bounds for B (in $(A, \mathcal{R})$), whereas $\{1, 2\}$ is a least upper bound (and is in B). Meanwhile, a greatest lower bound for B is $\emptyset$, which is not in B. □

Example 7.44 Let $\mathcal{R}$ be the "is less than or equal to" relation for the poset $(A, \mathcal{R})$.

   a) If $A = \mathbf{R}$ and $B = [0, 1]$, then B has glb 0 and lub 1. Note that $0, 1 \in B$. For $C = (0, 1]$, C has glb 0 and lub 1, and $1 \in C$ but $0 \notin C$.

   b) Keeping $A = \mathbf{R}$, let $B = \{q \in \mathbf{Q} \mid q^2 < 2\}$. Then B has $\sqrt{2}$ as a lub and $-\sqrt{2}$ as a glb, and neither of these real numbers is in B.

   c) Now let $A = \mathbf{Q}$, with B as in part (b). Here B has no lub or glb. □

These examples lead us to the following result.

THEOREM 7.5 If $(A, \mathcal{R})$ is a poset and $B \subseteq A$, then if B has a lub (glb), it is unique.

Proof We leave the proof to the reader. ∎

We close this section with one last ordered structure.

DEFINITION 7.20 The poset $(A, \mathcal{R})$ is called a *lattice* if for any $x, y \in A$, the elements $\text{lub}\{x, y\}$ and $\text{glb}\{x, y\}$ both exist.

Example 7.45 For $A = \mathbf{N}$ and $x, y \in \mathbf{N}$, define $x \mathcal{R} y$ by $x \leq y$. Then $\text{lub}\{x, y\} = \max\{x, y\}$, $\text{glb}\{x, y\} = \min\{x, y\}$, and $(\mathbf{N}, \leq)$ is a lattice. □

Example 7.46 For the poset in Example 7.41(a), if $S, T \subseteq \mathcal{U}$, with $\text{lub}\{S, T\} = S \cup T$ and $\text{glb}\{S, T\} = S \cap T$, then $(\mathcal{P}(\mathcal{U}), \subseteq)$ is a lattice. □

Example 7.47 Consider the poset in Example 7.35(d). Here we find, for example, that

$$\text{lub}\{2, 3\} = 6, \ \text{lub}\{3, 6\} = 6, \ \text{lub}\{5, 7\} = 35, \ \text{lub}\{7, 11\} = 385,$$

and $\text{lub}\{11, 35\} = 385$,

and $\text{glb}\{3, 6\} = 3, \ \text{glb}\{2, 12\} = 2$, and $\text{glb}\{35, 385\} = 35$.

However, even though $\text{lub}\{2, 3\}$ exists, there is no glb for the elements 2 and 3. In addition, we are also lacking (among other considerations) $\text{glb}\{5, 7\}$, $\text{glb}\{11, 35\}$, $\text{glb}\{3, 35\}$, and $\text{lub}\{3, 35\}$. Consequently, this partial order is not a lattice. □

EXERCISES 7.3

1. Draw the Hasse diagram for the poset $(\mathcal{P}(\mathcal{U}), \subseteq)$, where $\mathcal{U} = \{1, 2, 3, 4\}$.

2. Let $A = \{1, 2, 3, 6, 9, 18\}$, and define $\mathcal{R}$ on A by $x \mathcal{R} y$ if $x \mid y$. Draw the Hasse diagram for the poset $(A, \mathcal{R})$.

3. Let $(A, \mathcal{R}_1), (B, \mathcal{R}_2)$ be two posets. On $A \times B$ define relation $\mathcal{R}$ by $(a, b) \mathcal{R} (x, y)$ if $a \mathcal{R}_1 x$ and $b \mathcal{R}_2 y$. Prove that $\mathcal{R}$ is a partial order.

4. If $\mathcal{R}_1, \mathcal{R}_2$ in Exercise 3 are total orders, is $\mathcal{R}$ a total order?

5. Topologically sort the Hasse diagram in part (a) of Example 7.35.

6. For $A = \{a, b, c, d, e\}$, the Hasse diagram for the poset $(A, \mathcal{R})$ is shown in Fig. 7.18.

   a) Determine the relation matrix for $\mathcal{R}$.

   b) Construct the directed graph G (on A) that is associated with $\mathcal{R}$.

   c) Topologically sort the poset $(A, \mathcal{R})$.

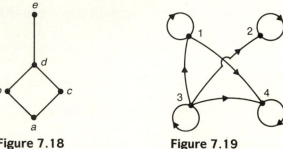

Figure 7.18   **Figure 7.19**

7. The directed graph G for a relation $\mathcal{R}$ on set $A = \{1, 2, 3, 4\}$ is shown in Fig. 7.19.

   a) Verify that $(A, \mathcal{R})$ is a poset and find its Hasse diagram.

   b) Topologically sort $(A, \mathcal{R})$.

   c) How many more directed edges are needed in Fig. 7.19 to extend $(A, \mathcal{R})$ to a total order?

8. Let $\mathcal{R}$ be a transitive relation on a set A. Prove that $\mathcal{R}$ is a partial order on A if and only if $\mathcal{R} \cap \mathcal{R}^c = \{(a, a) \mid a \in A\}$.

9. Prove that a finite poset $(A, \mathcal{R})$ has a minimal element.

10. Prove that if a poset $(A, \mathcal{R})$ has a least element, it is unique.

11. Prove Theorem 7.5.

12. Give an example of a poset with four maximal elements but no greatest element.

13. If $(A, \mathcal{R})$ is a poset but not a total order, and $\emptyset \neq B \subset A$, does it follow that $(B \times B) \cap \mathcal{R}$ makes B into a poset but not a total order?

14. If $\mathcal{R}$ is a relation on A, and G is the associated directed graph, how can one recognize from G that $(A, \mathcal{R})$ is a total order?

15. If G is the directed graph for a relation $\mathcal{R}$ on A, with $|A| = n$, and $(A, \mathcal{R})$ is a total order, how many edges (including loops) are there in G?

16. Let $M(\mathcal{R})$ be the relation matrix for relation $\mathcal{R}$ on A, with $|A| = n$. If $(A, \mathcal{R})$ is a total order, how many 1's appear in $M(\mathcal{R})$?

17. **a)** Describe the structure of the Hasse diagram for a totally ordered poset $(A, \mathcal{R})$, where $|A| = n \geq 1$.

 b) For a set A where $|A| = n \geq 1$, how many relations on A are total orders?

18. **a)** For $A = \{a_1, a_2, \ldots, a_n\}$, let $(A, \mathcal{R})$ be a poset. If $M(\mathcal{R})$ is the corresponding relation matrix, how can we recognize a minimal element of the poset from $M(\mathcal{R})$?

 b) Answer the question posed in part (a) if the adjective "minimal" is replaced by the adjective "maximal."

 c) How can one recognize the existence of a greatest or least element in $(A, \mathcal{R})$ from the relation matrix $M(\mathcal{R})$?

19. Let $\mathcal{U} = \{1, 2, 3, 4\}$, with $A = \mathcal{P}(\mathcal{U})$, and let $\mathcal{R}$ be the subset relation on A. For each of the following subsets B (of A), determine the lub and glb of B.

 a) $B = \{\{1\}, \{2\}\}$

 b) $B = \{\{1\}, \{2\}, \{3\}, \{1, 2\}\}$

 c) $B = \{\emptyset, \{1\}, \{2\}, \{1, 2\}\}$

 d) $B = \{\{1\}, \{2\}, \{1, 3\}, \{1, 2, 3\}\}$

 e) $B = \{\{1\}, \{2\}, \{3\}, \{1, 2\}, \{1, 3\}, \{2, 3\}\}$

 f) $B = \{\{1\}, \{2\}, \{3\}, \{1, 2\}, \{1, 3\}, \{2, 3\}, \{1, 2, 3\}\}$

20. Let $\mathcal{U} = \{1, 2, 3, 4, 5, 6, 7\}$, with $A = \mathcal{P}(\mathcal{U})$, and let $\mathcal{R}$ be the subset relation on A. For $B = \{\{1\}, \{2\}, \{2, 3\}\} \subseteq A$, determine each of the following.

 a) The number of upper bounds of B that contain (i) three elements of $\mathcal{U}$; (ii) four elements of $\mathcal{U}$; (iii) five elements of $\mathcal{U}$

 b) The number of upper bounds that exist for B

 c) The lub for B

 d) The number of lower bounds that exist for B

 e) The glb for B

21. Let $(A, \mathcal{R})$ be a poset. Prove or disprove each of the following statements.

 a) If $(A, \mathcal{R})$ is a lattice, then it is a total order.

 b) If $(A, \mathcal{R})$ is a total order, then it is a lattice.

22. If $(A, \mathcal{R})$ is a lattice, with A finite, prove that $(A, \mathcal{R})$ has a greatest element and a least element.

23. For $A = \{a, b, c, d, e, v, w, x, y, z\}$, consider the poset $(A, \mathcal{R})$ whose Hasse diagram is shown in Fig. 7.20. Find

 a) glb$\{b, c\}$ **b)** glb$\{b, w\}$ **c)** glb$\{e, x\}$

 d) lub$\{c, b\}$ **e)** lub$\{d, x\}$ **f)** lub$\{c, e\}$

 g) lub$\{a, v\}$

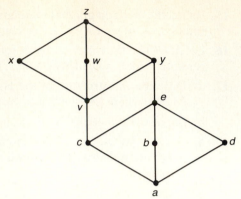

Figure 7.20

Is $(A, \mathcal{R})$ a lattice? Is there a maximal element? a minimal element? a greatest element? a least element?

24. Let $(A, \mathcal{R})$ be a totally ordered poset. If for all $\emptyset \neq B \subseteq A$, the (totally ordered) poset $(B, (B \times B) \cap \mathcal{R})$ has a least element, then $(A, \mathcal{R})$ is called *well-ordered*. (We saw this idea in Section 4.1, where we used the well-ordering of $(\mathbf{Z}^+, \leq)$ to establish the principle of mathematical induction.)

For each of the following (totally ordered) posets, determine whether or not the poset is well-ordered.

a) $(\mathbf{N}, \leq)$ **b)** $(\mathbf{Z}, \leq)$ **c)** $(\mathbf{Q}, \leq)$

d) $(\mathbf{Q}^+, \leq)$ **e)** $(P, \leq)$, where P is the set of all primes

f) $(A, \leq)$, where A is a nonempty subset of $\mathbf{Z}^+$

g) $(A, \leq)$, where $\emptyset \neq A \subset \mathbf{Z}$, and A is finite

---

7.4
EQUIVALENCE RELATIONS AND PARTITIONS

As we noted earlier, a relation $\mathcal{R}$ on a set A is an equivalence relation if it is reflexive, symmetric, and transitive. For any set $A \neq \emptyset$, the relation of equality is always an equivalence relation on A, where two elements of A are related if they are identical; equality thus establishes the property of "sameness" among the elements of A.

If we consider the relation $\mathcal{R}$ on $\mathbf{Z}$ defined by $x \mathcal{R} y$ if $x - y$ is a multiple of 2, then $\mathcal{R}$ is an equivalence relation on $\mathbf{Z}$ where all even integers are related, as are all odd integers. Here, for example, we do not have $4 = 8$, but we do have $4 \mathcal{R} 8$, for we no longer care about the size of a number but are concerned with only two properties: "evenness" and "oddness." This relation splits $\mathbf{Z}$ into two subsets consisting of the odd and even integers: $\mathbf{Z} = \{\ldots, -3, -1, 1, 3, \ldots\} \cup \{\ldots, -4, -2, 0, 2, 4, \ldots\}$. This splitting up of $\mathbf{Z}$ is an example of a partition, a concept closely related to the equivalence relation. In this section we investigate this relationship and see how it helps us count the number of equivalence relations on a finite set.

DEFINITION 7.21 Given a set A and an index set I, let $\emptyset \neq A_i \subseteq A$ for each $i \in I$. Then $\{A_i\}_{i \in I}$ is a *partition* of A if

(a) $A = \bigcup_{i \in I} A_i$ and (b) $A_i \cap A_j = \emptyset$, for all $i, j \in I$ where $i \neq j$.

Each subset A_i is called a *cell* or *block* of the partition.

Example 7.48 If $A = \{1, 2, 3, \ldots, 10\}$, then each of the following is a partition of A:

a) $A_1 = \{1, 2, 3, 4, 5\}, A_2 = \{6, 7, 8, 9, 10\}$

b) $A_1 = \{1, 3, 5, 7, 9\}, A_2 = \{2, 4, 6, 8, 10\}$

c) $A_1 = \{1, 2, 3\}, A_2 = \{4, 6, 7, 9\}, A_3 = \{5, 8, 10\}$

d) $A_i = \{i, i + 5\}, 1 \leq i \leq 5$ $\square$

Example 7.49 Let $A = \mathbf{R}$, and for each $i \in \mathbf{Z}$, let $A_i = [i, i + 1)$. Then $\{A_i\}_{i \in \mathbf{Z}}$ forms a partition of $\mathbf{R}$. $\square$

Now just how do partitions come into play with equivalence relations?

DEFINITION 7.22 Let $\mathcal{R}$ be an equivalence relation on a set A. For any $x \in A$, the *equivalence class* of x, denoted $[x]$, is defined by $[x] = \{y \in A \mid y \mathcal{R} x\}$.

Example 7.50 Define the relation $\mathcal{R}$ on $\mathbf{Z}$ by $x \mathcal{R} y$ if $4 \mid (x - y)$. For this equivalence relation we find that

$$[0] = \{\ldots, -8, -4, 0, 4, 8, 12, \ldots\} = \{4k \mid k \in \mathbf{Z}\}$$
$$[1] = \{\ldots, -7, -3, 1, 5, 9, 13, \ldots\} = \{4k + 1 \mid k \in \mathbf{Z}\}$$
$$[2] = \{\ldots, -6, -2, 2, 6, 10, 14, \ldots\} = \{4k + 2 \mid k \in \mathbf{Z}\}$$
$$[3] = \{\ldots, -5, -1, 3, 7, 11, 15, \ldots\} = \{4k + 3 \mid k \in \mathbf{Z}\}.$$

In addition, we also have, for example, $[6] = [2] = [-2]$, $[51] = [3]$, and $[17] = [1]$. Most important, $\{[0], [1], [2], [3]\}$ provides a partition of $\mathbf{Z}$. $\square$

This example leads us to the following general situation.

THEOREM 7.6 If $\mathcal{R}$ is an equivalence relation on a set A, and $x, y \in A$, then (a) $x \in [x]$; (b) $x \mathcal{R} y$ if and only if $[x] = [y]$; and (c) $[x] = [y]$ or $[x] \cap [y] = \emptyset$.

Proof a) This result follows from the reflexive property of $\mathcal{R}$.

b) If $x \mathcal{R} y$, let $w \in [x]$. Then $w \mathcal{R} x$ and because $\mathcal{R}$ is transitive, $w \mathcal{R} y$. Hence $w \in [y]$ and $[x] \subseteq [y]$. With $\mathcal{R}$ symmetric, $x \mathcal{R} y \Rightarrow y \mathcal{R} x$. So if $t \in [y]$, then $t \mathcal{R} y$ and by the transitive property, $t \mathcal{R} x$. Hence $t \in [x]$ and $[y] \subseteq [x]$. Consequently, $[x] = [y]$.

Conversely, let $[x] = [y]$. Since $x \in [x]$ by part (a), then $x \in [y]$ or $x \mathcal{R} y$.

c) This property tells us that equivalence classes can be related in only two ways. They are either identical or disjoint.

We assume that $[x] \neq [y]$ and show how it then follows that $[x] \cap [y] = \emptyset$. If $[x] \cap [y] \neq \emptyset$, then let $v \in A$ with $v \in [x]$ and $v \in [y]$. Then $v \mathcal{R} x, v \mathcal{R} y$, and, since $\mathcal{R}$ is symmetric, $x \mathcal{R} v$. Now $(x \mathcal{R} v$ and $v \mathcal{R} y) \Rightarrow x \mathcal{R} y$, by the transitive property. Also $x \mathcal{R} y \Rightarrow [x] = [y]$ by part (b). This contradicts the assumption that $[x] \neq [y]$, so we reject the supposition that $[x] \cap [y] \neq \emptyset$, and the result follows. ∎

Note that if $\mathcal{R}$ is an equivalence relation on A, then by parts (a) and (c) of Theorem 7.6 the distinct equivalence classes determined by $\mathcal{R}$ form a partition of A.

Example 7.51   If $A = \{1, 2, 3, 4, 5\}$ and $\mathcal{R} = \{(1, 1), (2, 2), (2, 3), (3, 2), (3, 3), (4, 4), (4, 5), (5, 4), (5, 5)\}$, then $\mathcal{R}$ is an equivalence relation on A. Here $[1] = \{1\}$, $[2] = \{2, 3\} = [3]$, $[4] = \{4, 5\} = [5]$, and $A = [1] \cup [2] \cup [4]$. □

Example 7.52   In ANSI FORTRAN there is a nonexecutable specification statement called the EQUIVALENCE statement that allows two or more variables in a given program to refer to the same memory location.

For example, within a program the statement

```
EQUIVALENCE (A, C, P), (UP, DOWN)
```

informs the compiler that variables A, C, and P will share one memory location while UP and DOWN will share another. Here the set of all program variables is partitioned by the equivalence relation $\mathcal{R}$, where $V_1 \mathcal{R} V_2$ if V_1 and V_2 are program variables that share the same memory location. □

Example 7.53   Having seen examples of how an equivalence relation induces a partition of a set, we now go backward. If an equivalence relation $\mathcal{R}$ on $A = \{1, 2, 3, 4, 5, 6, 7\}$ induces the partition $A = \{1, 2\} \cup \{3\} \cup \{4, 5, 7\} \cup \{6\}$, what is $\mathcal{R}$?

Consider the subset $\{1, 2\}$ of the partition. This subset implies that $[1] = \{1, 2\} = [2]$, and so $(1, 1), (2, 2), (1, 2), (2, 1) \in \mathcal{R}$. (The first two ordered pairs are necessary for the reflexive property of $\mathcal{R}$; the others preserve symmetry.)

In like manner, the subset $\{4, 5, 7\}$ implies that under $\mathcal{R}$, $[4] = [5] = [7] = \{4, 5, 7\}$ and that, as an equivalence relation, $\mathcal{R}$ must contain $\{4, 5, 7\} \times \{4, 5, 7\}$. In fact, $\mathcal{R} = (\{1, 2\} \times \{1, 2\}) \cup (\{3\} \times \{3\}) \cup (\{4, 5, 7\} \times \{4, 5, 7\}) \cup (\{6\} \times \{6\})$, and $|\mathcal{R}| = 2^2 + 1^2 + 3^2 + 1^2 = 15$. □

These examples lead us to the following.

THEOREM 7.7   If A is a set, then

a) any equivalence relation $\mathcal{R}$ on A induces a partition of A, and

b) any partition of A gives rise to an equivalence relation $\mathcal{R}$ on A.

Proof   Part (a) follows from parts (a) and (c) of Theorem 7.6. For part (b), given a partition $\{A_i\}_{i \in I}$ of A, define relation $\mathcal{R}$ on A by $x \mathcal{R} y$, if x and y are in the same cell of the partition. We leave to the reader the details of verifying that $\mathcal{R}$ is an equivalence relation.   ■

On the basis of this theorem and the examples we have examined, we state the next result. A proof for it is outlined in Exercise 12 at the end of the section.

THEOREM 7.8   For any set A, there is a one-to-one correspondence between the equivalence relations on A and the partitions of A.

We are primarily concerned with using this result for finite sets.

Example 7.54   **a)**   If $A = \{1, 2, 3, 4, 5, 6\}$, how many relations on A are equivalence relations?
     We solve this problem by counting the partitions of A, realizing that a partition of A is a distribution of the (distinct) elements of A into identical containers, with no container left empty. From Section 5.3 we know, for example, that there are $S(6, 2)$ partitions of A into two identical nonempty containers. Using the Stirling numbers of the second kind, as the number of containers varies from 1 to 6, we have $\sum_{i=1}^{6} S(6, i) = 203$ different partitions of A. Consequently, there are 203 equivalence relations on A.

b)   How many of the equivalence relations in part (a) satisfy $1, 2 \in [4]$?
     Identifying 1, 2, and 4 as the "same" element under these equivalence relations, we count as in part (a) for the set $B = \{1, 3, 5, 6\}$ and find that there are $\sum_{i=1}^{4} S(4, i) = 15$ equivalence relations on A for which $[1] = [2] = [4]$.   □

We close by noting that if A is a finite set with $|A| = n$, then for any $n \leq r \leq n^2$, there is an equivalence relation $\mathcal{R}$ on A with $|\mathcal{R}| = r$ if and only if there exist $n_1, n_2, \ldots, n_k \in \mathbf{Z}^+$ with $\sum_{i=1}^{k} n_i = n$ and $\sum_{i=1}^{k} n_i^2 = r$.

EXERCISES 7.4

1. If $A = \{1, 2, 3, 4, 5\}$ and $\mathcal{R}$ is the equivalence relation on A that induces the partition $A = \{1, 2\} \cup \{3, 4\} \cup \{5\}$, what is $\mathcal{R}$?

2. For $A = \{1, 2, 3, 4, 5, 6\}$, $\mathcal{R} = \{(1, 1), (1, 2), (2, 1), (2, 2), (3, 3), (4, 4), (4, 5), (5, 4), (5, 5), (6, 6)\}$ is an equivalence relation on A.

   a) What are $[1]$, $[2]$, and $[3]$ under this equivalence relation?

   b) What partition of A does $\mathcal{R}$ induce?

3. If $A = A_1 \cup A_2 \cup A_3$, where $A_1 = \{1, 2\}$, $A_2 = \{2, 3, 4\}$, and $A_3 = \{5\}$, define relation $\mathcal{R}$ on A by $x \mathcal{R} y$ if x and y are in the same subset A_i, for $1 \leq i \leq 3$. Is $\mathcal{R}$ an equivalence relation?

4. For $A = \mathbf{R}^2$, define $\mathcal{R}$ on A by $(x_1, y_1) \mathcal{R} (x_2, y_2)$ if $x_1 = x_2$.

   a) Verify that $\mathcal{R}$ is an equivalence relation on A.

b) Describe geometrically the equivalence classes and partition of A induced by $\mathcal{R}$.

5. Let $A = \{1, 2, 3, 4, 5\} \times \{1, 2, 3, 4, 5\}$, and define $\mathcal{R}$ on A by $(x_1, y_1) \mathcal{R} (x_2, y_2)$ if $x_1 + y_1 = x_2 + y_2$.

  a) Verify that $\mathcal{R}$ is an equivalence relation on A.

  b) Determine the equivalence classes $[(1, 3)]$, $[(2, 4)]$, and $[(1, 1)]$.

  c) Determine the partition of A induced by $\mathcal{R}$.

6. If $A = \{1, 2, 3, 4, 5, 6, 7\}$, define $\mathcal{R}$ on A by $(x, y) \in \mathcal{R}$ if $x - y$ is a multiple of 3.

  a) Show that $\mathcal{R}$ is an equivalence relation on A.

  b) Determine the equivalence classes and partition of A induced by $\mathcal{R}$.

7. Let A be a nonempty set and fix the set B, where $B \subseteq A$. Define the relation $\mathcal{R}$ on $\mathcal{P}(A)$ by $X \mathcal{R} Y$, for $X, Y \subseteq A$, if $B \cap X = B \cap Y$.

  a) Verify that $\mathcal{R}$ is an equivalence relation on $\mathcal{P}(A)$.

  b) If $A = \{1, 2, 3\}$ and $B = \{1, 2\}$, find the partition of $\mathcal{P}(A)$ induced by $\mathcal{R}$.

  c) If $A = \{1, 2, 3, 4, 5\}$ and $B = \{1, 2, 3\}$, find $[X]$ if $X = \{1, 3, 5\}$. .

  d) For $A = \{1, 2, 3, 4, 5\}$ and $B = \{1, 2, 3\}$, how many equivalence classes are in the partition induced by $\mathcal{R}$?

8. Let $A = \{v, w, x, y, z\}$. Determine the number of relations on A that are (a) reflexive and symmetric; (b) equivalence relations; (c) reflexive and symmetric but not transitive; (d) equivalence relations that determine exactly two equivalence classes; (e) equivalence relations where $w \in [x]$; (f) equivalence relations where $v, w \in [x]$; (g) equivalence relations where $w \in [x]$ and $y \in [z]$; and (h) equivalence relations where $w \in [x]$, $y \in [z]$, and $[x] \neq [z]$.

9. If $|A| = 30$ and the equivalence relation $\mathcal{R}$ on A partitions A into equivalence classes A_1, A_2, and A_3, where $|A_1| = |A_2| = |A_3|$, what is $|\mathcal{R}|$?

10. Let $A = \{1, 2, 3, 4, 5, 6, 7\}$. For each of the following values of r, determine an equivalence relation $\mathcal{R}$ on A with $|\mathcal{R}| = r$, or explain why no such relation exists. (a) $r = 6$; (b) $r = 7$; (c) $r = 8$; (d) $r = 9$; (e) $r = 11$; (f) $r = 22$; (g) $r = 23$; (h) $r = 30$; (i) $r = 31$.

11. Provide the details for the proof of part (b) of Theorem 7.7.

12. For any set $A \neq \emptyset$, let $P(A)$ denote the set of all partitions of A, and let $E(A)$ denote the set of all equivalence relations on A. Define the function f: $E(A) \rightarrow P(A)$ as follows: If $\mathcal{R}$ is an equivalence relation on A, then $f(\mathcal{R})$ is the partition of A induced by $\mathcal{R}$. Prove that f is one-to-one and onto, thus establishing Theorem 7.8.

7.5
FINITE STATE MACHINES:
THE MINIMIZATION PROCESS

In Section 6.3 we encountered two finite state machines that performed the same task but had different numbers of internal states. (See Figs. 6.8 and 6.9.) The machine with the larger number of internal states contains *redundant* states— states that can be eliminated because other states will perform their functions. Since minimization of the number of states in a machine reduces its complexity and cost, we seek a process for transforming a given machine into one that has no redundant internal states. This process is known as the *minimization process*, and its development relies on the concepts of equivalence relation and partition.

Starting with a given finite state machine $M = (S, \mathcal{I}, \mathcal{O}, v, \omega)$, we define the relation E_1 on S, by $s_1 E_1 s_2$ if $\omega(s_1, x) = \omega(s_2, x)$, for all $x \in \mathcal{I}$. This relation E_1 is an equivalence relation on S, and it partitions S into subsets such that two states are in the same subset if they produce the same output for each $x \in \mathcal{I}$. Here the states s_1, s_2 are called *1-equivalent*.

For any $k \in \mathbf{Z}^+$, we say that the states s_1, s_2 are *k-equivalent* if $\omega(s_1, x) = \omega(s_2, x)$ for all $x \in \mathcal{I}^k$. Here ω is the extension of the given output function to $S \times \mathcal{I}^*$. The relation of k-equivalence is also an equivalence relation on S; it partitions S into subsets of k-equivalent states. We write $s_1 E_k s_2$ to denote that s_1 and s_2 are k-equivalent.

Finally, if $s_1, s_2 \in S$ and s_1, s_2 are k-equivalent for all $k \geq 1$, then we call s_1 and s_2 *equivalent* and write $s_1 E s_2$. When this happens, we find that if we keep s_1 in our machine, s_2 will be redundant and can be removed. Hence our objective is to determine the partition of S induced by E and to select one state for each equivalence class. Then we shall have the minimal realization of the given machine.

In order to accomplish this, let us start with the following observations.

a) If two states in a machine are not 2-equivalent, could they possibly be 3-equivalent? (or k-equivalent, for $k \geq 4$?)

The answer is no. If $s_1, s_2 \in S$ and $s_1 \not{E}_2 s_2$ (that is, s_1 and s_2 are not 2-equivalent), then there is at least one string $xy \in \mathcal{I}^2$ such that $\omega(s_1, xy) = v_1 v_2 \neq w_1 w_2 = \omega(s_2, xy)$, where $v_1, v_2, w_1, w_2 \in \mathcal{O}$. So with regard to E_3, we find that $s_1 \not{E}_3 s_2$ because for any $z \in \mathcal{I}$, $\omega(s_1, xyz) = v_1 v_2 v_3 \neq w_1 w_2 w_3 = \omega(s_2, xyz)$.

In general, to find states that are $(k+1)$-equivalent, we look at states that are k-equivalent.

b) Now suppose that $s_1, s_2 \in S$ and $s_1 E_2 s_2$. We wish to determine whether $s_1 E_3 s_2$. That is, does $\omega(s_1, x_1 x_2 x_3) = \omega(s_2, x_1 x_2 x_3)$ for all strings $x_1 x_2 x_3 \in \mathcal{I}^3$? Consider what happens. First we get $\omega(s_1, x_1) = \omega(s_2, x_1)$, because $s_1 E_2 s_2 \Rightarrow s_1 E_1 s_2$. Then there is a transition to the states $v(s_1, x_1)$ and $v(s_2, x_1)$. Consequently, $\omega(s_1, x_1 x_2 x_3) = \omega(s_2, x_1 x_2 x_3)$ if $\omega(v(s_1, x_1), x_2 x_3) = \omega(v(s_2, x_1), x_2 x_3)$ (that is, if $v(s_1, x_1) E_2 v(s_2, x_1)$).

In general, for $s_1, s_2 \in S$ we have $s_1 E_{k+1} s_2$ if (a) $s_1 E_k s_2$ and (b) $v(s_1, x) E_k v(s_2, x)$ for all $x \in \mathcal{I}$.

With these observations to guide us, we now present an algorithm for the minimization of a finite state machine M.

Step 1: Set $k = 1$. We determine the states that are 1-equivalent by examining the rows in the state table for M. For $s_1, s_2 \in S$ it follows that $s_1 \, E_1 \, s_2$ when s_1, s_2 have the same output rows. Let P_1 be the partition of S induced by E_1.

Step 2: Having determined P_k, we obtain P_{k+1} by noting that $s_1 \, E_{k+1} \, s_2$ if $s_1 \, E_k \, s_2$ and $\nu(s_1, x) \, E_k \, \nu(s_2, x)$ for all $x \in \mathcal{I}$. We have $s_1 \, E_k \, s_2$ if s_1, s_2 are in the same subset of the partition P_k. Likewise, $\nu(s_1, x) \, E_k \, \nu(s_2, x)$ for any $x \in \mathcal{I}$, if $\nu(s_1, x)$ and $\nu(s_2, x)$ are in the same cell of the partition P_k. In this way P_{k+1} is obtained from P_k.

Step 3: If $P_{k+1} = P_k$, the process is complete. We select one state from each equivalence class and these yield a minimal realization of M.
    If $P_{k+1} \neq P_k$, we increase k by 1 and return to step 2.

We illustrate the algorithm in the following example.

Example 7.55   With $\mathcal{I} = \mathcal{O} = \{0, 1\}$, let M be given by the state table shown in Table 7.1. Looking at the output rows, we see that s_3 and s_4 are 1-equivalent, as are s_2, s_5, and s_6. Here E_1 partitions S as follows:

$$P_1: \{s_1\}, \{s_2, s_5, s_6\}, \{s_3, s_4\}.$$

For any $s \in S$ and any $k \in \mathbf{Z}^+$, $s \, E_k \, s$, so as we continue this process to determine P_2, we do not concern ourselves with equivalence classes of only one state.

Table 7.1

	ν		ω	
	0	1	0	1
s_1	s_4	s_3	0	1
s_2	s_5	s_2	1	0
s_3	s_2	s_4	0	0
s_4	s_5	s_3	0	0
s_5	s_2	s_5	1	0
s_6	s_1	s_6	1	0

Since $s_3 \, E_1 \, s_4$, there is a chance that we could have $s_3 \, E_2 \, s_4$. Here $\nu(s_3, 0) = s_2$, $\nu(s_4, 0) = s_5$ with $s_2 \, E_1 \, s_5$, and $\nu(s_3, 1) = s_4, \nu(s_4, 1) = s_3$ with $s_4 \, E_1 \, s_3$. Hence $\nu(s_3, x) \, E_1 \, \nu(s_4, x)$, for all $x \in \mathcal{I}$, and $s_3 \, E_2 \, s_4$. Similarly, $\nu(s_2, 0) = s_5, \nu(s_5, 0) = s_2$ with $s_5 \, E_1 \, s_2$, and $\nu(s_2, 1) = s_2, \nu(s_5, 1) = s_5$ with $s_2 \, E_1 \, s_5$. Thus $s_2 \, E_2 \, s_5$. Finally, $\nu(s_5, 0) = s_2$ and $\nu(s_6, 0) = s_1$, but $s_2 \, \cancel{E_1} \, s_1$, so $s_5 \, \cancel{E_2} \, s_6$. (Why don't we investigate the possibility of $s_2 \, E_2 \, s_6$?) Equivalence relation E_2 partitions S as follows:

$$P_2: \{s_1\}, \{s_2, s_5\}, \{s_3, s_4\}, \{s_6\}.$$

Since $P_2 \neq P_1$, we continue the process to get P_3. In determining whether or not $s_2 \, E_3 \, s_5$, we see that $\nu(s_2, 0) = s_5$, $\nu(s_5, 0) = s_2$, and $s_5 \, E_2 \, s_2$. Also, $\nu(s_2, 1) = s_2$,

$v(s_5, 1) = s_5$, and $s_2 \, E_2 \, s_5$. With $s_2 \, E_2 \, s_5$ and $v(s_2, x) \, E_2 \, v(s_5, x)$ for all $x \in \mathcal{I}$, we have $s_2 \, E_3 \, s_5$. For s_3, s_4, $(v(s_3, 0) = s_2) \, E_2 \, (s_5 = v(s_4, 0))$ and $(v(s_3, 1) = s_4) \, E_2 \, (s_3 = v(s_4, 1))$, so $s_3 \, E_3 \, s_4$ and E_3 induces the partition P_3: $\{s_1\}, \{s_2, s_5\}, \{s_3, s_4\}, \{s_6\}$.

Now $P_2 = P_3$, so the process is completed, as indicated in step 3 of the algorithm. We find that s_5 and s_4 are redundant states. Removing them from the table, and replacing any further occurrence of them by s_2 and s_3, respectively, we arrive at Table 7.2. This is the minimal machine that performs the same tasks as the machine given in Table 7.1.

If we do not want states that skip a subscript, we can always relabel the states in the minimal machine. Here we would have $s_1, s_2, s_3, s_4(=s_6)$, but this s_4 is not the same s_4 we started with in Table 7.1.   □

Table 7.2

	v		ω	
	0	1	0	1
s_1	s_3	s_3	0	1
s_2	s_2	s_2	1	0
s_3	s_2	s_3	0	0
s_6	s_1	s_6	1	0

You may be wondering how we knew that we could stop the process when $P_2 = P_3$. For after all, couldn't it happen that perhaps $P_4 \neq P_3$, or that $P_4 = P_3$ but $P_5 \neq P_4$? To prove that this never occurs, we define the following idea.

DEFINITION 7.23   If P_1, P_2 are any partitions of a set A, then P_2 is called a *refinement* of P_1, and we write $P_2 \leq P_1$, if every cell of P_2 is contained in a cell of P_1.   ▬

In the minimization process of Example 7.55, we had $P_3 = P_2 \leq P_1$. Whenever we apply the algorithm, as we get P_{k+1} from P_k, we always find that $P_{k+1} \leq P_k$, because $(k + 1)$-equivalence implies k-equivalence. So each successive partition refines the preceding partition.

THEOREM 7.9   In applying the minimization process, if $k \geq 1$ and P_k and P_{k+1} are partitions with $P_{k+1} = P_k$, then $P_r = P_{r+1}$ for any $r \geq k + 1$.

Proof   If not, let $r(\geq k + 1)$ be the smallest subscript such that $P_r \neq P_{r+1}$. Then $P_{r+1} < P_r$, so there exist $s_1, s_2 \in S$ with $s_1 \, E_r \, s_2$ but $s_1 \, E̸_{r+1} \, s_2$. But $s_1 \, E_r \, s_2 \Rightarrow v(s_1, x) \, E_{r-1} \, v(s_2, x)$, for all $x \in \mathcal{I}$ and with $P_r = P_{r-1}$, we then find that $v(s_1, x) \, E_r \, v(s_2, x)$, for all $x \in \mathcal{I}$, so $s_1 \, E_{r+1} \, s_2$. Consequently, $P_r = P_{r+1}$.   ■

We close this section with the following related idea. Let M be a finite state machine with $s_1, s_2 \in S$, and s_1, s_2 not equivalent. Then there must be a smallest integer $k \geq 0$ such that $s_1 \, E_k \, s_2$ but $s_1 \, E̸_{k+1} \, s_2$. For $k = 0$, we have s_1 and s_2 producing different output rows in the state table for M. In this case it is easy to find an $x \in \mathcal{I}$ such that $\omega(s_1, x) \neq \omega(s_2, x)$, and this x distinguishes these nonequivalent states. For $k \geq 1$, s_1 and s_2 produce the same output rows in the table. Now if we are to distinguish these states, we need to find a string $x = x_1 x_2 \ldots x_k x_{k+1} \in \mathcal{I}^{k+1}$ such that $\omega(s_1, x) \neq \omega(s_2, x)$, even though $\omega(s_1, x_1 x_2 \ldots x_k) = \omega(s_2, x_1 x_2 \ldots x_k)$. Such a string x is called a *distinguishing string* for the states s_1 and s_2. There may be more than one such string, but each has the (minimal) length $k + 1$.

Before we attempt to develop an algorithm for finding a distinguishing string for two nonequivalent states $s_1, s_2 \in S$, let us examine the major idea at play here. If $k \in \mathbf{Z}^+$ and $s_1 \, \mathrm{E}_k \, s_2$ but $s_1 \, \not{\mathrm{E}}_{k+1} \, s_2$, what can we conclude? We find that

$$
\begin{aligned}
s_1 \, \not{\mathrm{E}}_{k+1} \, s_2 &\Rightarrow \exists \, x_1 \in \mathcal{I} \; [\nu(s_1, x_1) \, \not{\mathrm{E}}_k \, \nu(s_2, x_1)] \\
&\Rightarrow \exists \, x_2 \in \mathcal{I} \; [\nu(\nu(s_1, x_1), x_2) \, \not{\mathrm{E}}_{k-1} \, \nu(\nu(s_2, x_1), x_2)], \\
\text{or,} \quad &\exists \, x_2 \in \mathcal{I} \; [\nu(s_1, x_1 x_2) \, \not{\mathrm{E}}_{k-1} \, \nu(s_2, x_1 x_2)], \\
&\Rightarrow \exists \, x_3 \in \mathcal{I} \; [\nu(s_1, x_1 x_2 x_3) \, \not{\mathrm{E}}_{k-2} \, \nu(s_2, x_1 x_2 x_3)] \\
&\Rightarrow \cdots \\
&\Rightarrow \exists \, x_i \in \mathcal{I} \; [\nu(s_1, x_1 x_2 \ldots x_i) \, \not{\mathrm{E}}_{k+1-i} \, \nu(s_2, x_1 x_2 \ldots x_i)] \\
&\Rightarrow \cdots \\
&\Rightarrow \exists \, x_k \in \mathcal{I} \; [\nu(s_1, x_1 x_2 \ldots x_k) \, \not{\mathrm{E}}_1 \, \nu(s_1, x_1 x_2 \ldots x_k)].
\end{aligned}
$$

This last statement about the states $\nu(s_1, x_1 x_2 \ldots x_k)$, $\nu(s_2, x_1 x_2 \ldots x_k)$ not being 1-equivalent implies that we can find $x_{k+1} \in \mathcal{I}$ where

$$
\omega(\nu(s_1, x_1 x_2 \ldots x_k), x_{k+1}) \neq \omega(\nu(s_2, x_1 x_2 \ldots x_k), x_{k+1}). \tag{1}
$$

That is, these *single* output symbols from $\mathcal{O}$ are different.

The result denoted by Equation (1) also implies that

$$
\omega(s_1, x_1 x_2 \ldots x_k x_{k+1}) \neq \omega(s_2, x_1 x_2 \ldots x_k x_{k+1}).
$$

In this case we have two output strings of length $k + 1$ that agree for the first k symbols and differ in the $(k + 1)$st symbol.

We shall use the above observations in the following development. In this algorithm we have $s_1, s_2 \in S$ with $s_1 \, \mathrm{E}_k \, s_2$ but $s_1 \, \not{\mathrm{E}}_{k+1} \, s_2$, for some (fixed) $k \in \mathbf{Z}^+$.

To find a minimal distinguishing string we apply the following algorithm, which uses the partitions $P_1, P_2, \ldots, P_k, P_{k+1}$ of the minimization process.

Step 1: Since $s_1 \, \mathrm{E}_k \, s_2$ but $s_1 \, \not{\mathrm{E}}_{k+1} \, s_2$, select $x_1 \in \mathcal{I}$ such that $\nu(s_1, x_1) \, \not{\mathrm{E}}_k \, \nu(s_2, x_1)$. (If no such x_1 can be found, then $s_1 \, \mathrm{E}_k \, s_2$ and $\nu(s_1, x) \, \mathrm{E}_k \, \nu(s_2, x)$ for all $x \in \mathcal{I}$, so $s_1 \, \mathrm{E}_{k+1} \, s_2$.)

Step 2: If $k = 1$, then $\nu(s_1, x_1) \, \not{\mathrm{E}}_1 \, \nu(s_2, x_1) \Rightarrow$ the states $\nu(s_1, x_1)$ and $\nu(s_2, x_1)$ are in different cells of P_1, so they determine different output rows. Find $x_2 \in \mathcal{I}$ with $\omega(\nu(s_1, x_1), x_2) \neq \omega(\nu(s_2, x_1), x_2)$—that is, $\omega(s_1, x_1 x_2) \neq \omega(s_2, x_1 x_2)$—and $x = x_1 x_2$ is a minimal distinguishing string for s_1 and s_2.

If $k > 1$, then set the counter $i = k$ and proceed to step 3.

Step 3: While $i > 1$, perform the following tasks a total of $k - 1$ times. This will provide the symbols $x_2, x_3, \ldots, x_k$ (from $\mathcal{I}$) for a minimal distinguishing string $x_1 x_2 x_3 \ldots x_k x_{k+1}$ (for the states s_1 and s_2).

a) Since $v(s_1, x_1 x_2 \ldots x_{k+1-i}) \not\mathrel{E}_i v(s_2, x_1 x_2 \ldots x_{k+1-i})$, use the partition P_i to determine x_{k+2-i} so that

$$v(s_1, x_1 x_2 \ldots x_{k+1-i} x_{k+2-i}) \not\mathrel{E}_{i-1} v(s_2, x_1 x_2 \ldots x_{k+1-i} x_{k+2-i}).$$

b) Reduce the value of the counter i by 1.

Step 4: At this point the value of the counter i is 1, and we have

$$v(s_1, x_1 x_2 \ldots x_k) \not\mathrel{E}_1 v(s_2, x_1 x_2 \ldots x_k)$$

for $k \in \mathbf{Z}^+$, $k \geq 2$.

Now use P_1 and select $x_{k+1} \in \mathcal{I}$ so that

$$\omega(v(s_1, x_1 x_2 \ldots x_k), x_{k+1}) \neq \omega(v(s_2, x_1 x_2 \ldots x_k), x_{k+1}).$$

That is,

$$\omega(s_1, x_1 x_2 \ldots x_k x_{k+1}) \neq \omega(s_2, x_1 x_2 \ldots x_k x_{k+1}).$$

Then $x_1 x_2 \ldots x_k x_{k+1}$ is a minimal distinguishing string for the states s_1 and s_2.

The algorithm is demonstrated in the following examples.

Example 7.56 From Example 7.55 we have the partitions shown below. Here $s_2 \mathrel{E}_1 s_6$ but $s_2 \not\mathrel{E}_2 s_6$. So we seek an input string x of length two such that $\omega(s_2, x) \neq \omega(s_6, x)$. Using step 1 of the algorithm, we start at P_2, where for s_2, s_6, we find that $\omega(s_2, 0) = 1 = \omega(s_6, 0)$, but $v(s_2, 0) = s_5$, $v(s_6, 0) = s_1$ are in different cells of P_1— that is, $v(s_2, 0) \not\mathrel{E}_1 v(s_6, 0)$. (The input 0 and the output 1 provide the labels for the arrows going from the cells of P_2 to those of P_1.) By step 2 of the algorithm, $\omega(v(s_2, 0), 0) = 1 \neq 0 = \omega(v(s_6, 0), 0)$. Hence $x = 00$ is a minimal distinguishing string for s_2 and s_6, because $\omega(s_2, 00) = 11 \neq 10 = \omega(s_6, 00)$. $\square$

$$P_2: \{s_1\}, \{s_2, s_5\}, \{s_3, s_4\}, \{s_6\}$$

0, 1 \qquad\qquad 0, 1

$$P_1: \{s_1\}, \{s_2, s_5, s_6\}, \{s_3, s_4\}$$

0, 0 \qquad 0, 1

Example 7.57 Applying the minimization process to the machine given by the state table in part (a) of Table 7.3, we obtain the partitions in part (b) of the table. (Here $P_4 = P_3$.) Applying the algorithm to states s_1 and s_4, which are 2-equivalent but not 3-equivalent, we find the following:

1. Since $s_1 \not\mathrel{E}_3 s_4$, in step 1 of our algorithm we use the partition P_3 in order to find $x_1 \in \mathcal{I}$ (namely $x_1 = 1$), so that $(v(s_1, 1) = s_2) \not\mathrel{E}_2 (s_5 = v(s_4, 1))$.

2. Here $k = 2 > 1$, so we set the counter $i = 2$ in step 2 and then proceed to step 3.

3. $v(s_1, 1) \not\!\!E_2 \, v(s_4, 1) \Rightarrow \exists \, x_2 \in \mathscr{I}$ (here $x_2 = 1$) with $(v(s_1, 1), 1) \not\!\!E_1 (v(s_4, 1), 1)$, or $v(s_1, 11) \not\!\!E_1 \, v(s_4, 11)$. We used the partition P_2 to obtain $x_2 = 1$.

4. Reducing the value of the counter by 1, we now arrive at $i = 1$. So we are finished with step 3 and proceed to step 4. Now we use the partition P_1 where we find that for $x_3 = 1 \in \mathscr{I}$,

$$\omega(v(s_1, 11), 1) = 0 \neq 1 = \omega(v(s_4, 11), 1), \qquad \text{or}$$

$$\omega(s_1, 111) = 100 \neq 101 = \omega(s_4, 111).$$

In part (b) of Table 7.3, we see how we arrived at the minimal distinguishing string $x = 111$ for these states. (Also note how this part of the table indicates that 11 is a minimal distinguishing string for the states s_2 and s_5, which are 1-equivalent but not 2-equivalent.)   □

Table 7.3

	ν		ω		P_3: $\{s_1, s_3\}, \{s_2\}, \{s_4\}, \{s_5\}$
	0	1	0	1	1,1 1,1
s_1	s_4	s_2	0	1	P_2: $\{s_1, s_3, s_4\}, \{s_2\}, \{s_5\}$
s_2	s_5	s_2	0	0	1,0 1,0
s_3	s_4	s_2	0	1	P_1: $\{s_1, s_3, s_4\}, \{s_2, s_5\}$
s_4	s_3	s_5	0	1	1,1 1,0
s_5	s_2	s_3	0	0	

    (a)   (b)

There is still a great deal more that can be done with finite state machines. Among other omissions, we have avoided offering any rigorous explanation or proof of why the minimization process works. The interested reader should consult the chapter references for more on this topic.

EXERCISES 7.5

1. Apply the minimization process to each machine in Table 7.4.

Table 7.4

	ν		ω	
	0	1	0	1
s_1	s_4	s_1	0	1
s_2	s_3	s_3	1	0
s_3	s_1	s_4	1	0
s_4	s_1	s_3	0	1
s_5	s_3	s_3	1	0

(a)

	ν		ω	
	0	1	0	1
s_1	s_6	s_3	0	0
s_2	s_5	s_4	0	1
s_3	s_6	s_2	1	1
s_4	s_4	s_3	1	0
s_5	s_2	s_4	0	1
s_6	s_4	s_6	0	0

(b)

	ν		ω	
	0	1	0	1
s_1	s_6	s_3	0	0
s_2	s_3	s_1	0	0
s_3	s_2	s_4	0	0
s_4	s_7	s_4	0	0
s_5	s_6	s_7	0	0
s_6	s_5	s_2	1	0
s_7	s_4	s_1	0	0

(c)

2. For the machine in Exercise 1(c), find a distinguishing string for each given pair of states: (a) s_1, s_5; (b) s_2, s_3; (c) s_5, s_7.

3. Let M be the finite state machine given in the state diagram shown in Fig. 7.21.

a) Minimize machine M.

b) Find a distinguishing string for each given pair of states: (i) s_3, s_6; (ii) s_3, s_4; (iii) s_1, s_2.

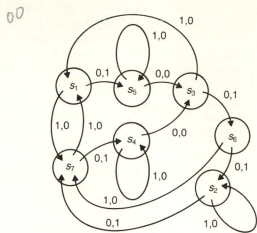

Figure 7.21

7.6
SUMMARY AND HISTORICAL REVIEW

Once again the relation concept surfaces. In Chapter 5 this idea was introduced as a generalization of the function. Here in Chapter 7 we concentrated on relations and the special properties: reflexive, symmetric, antisymmetric, and transitive. As a result we found ourselves focusing our attention on two special kinds of relations: partial orders and equivalence relations.

A relation $\mathcal{R}$ on a set A is a partial order, making A into a poset, if $\mathcal{R}$ is reflexive, antisymmetric, and transitive. Such a relation generalizes the familiar "is less than or equal to" relation on the real numbers. Try to imagine calculus, or even elementary algebra, without it! Or take a simple computer program and see what happens if the program is entered into the computer haphazardly, permuting the order of the statements. Order is with us wherever we turn. We have grown so accustomed to it that we take it for granted. For a finite poset, the Hasse diagram, a special type of directed graph, provides a pictorial representation of the order in the poset; it also proves useful when a total order, including the given partial order, is needed. The method employed to get a total order is the topological sorting algorithm that is appropriate in the solution of PERT (Program Evaluation and Review Technique) networks.

Although the equivalence relation differs from the partial order in only one property, it is quite different in structure and application. We make no attempt to trace the origin of the equivalence relation, but the ideas behind the reflexive, symmetric, and transitive properties can be found in *I Principii di Geometrica* (1889), the work of the Italian mathematician Giuseppe Peano (1858–1932). The work of Carl Friedrich Gauss (1777–1855) on *congruence*, which he developed in the 1790s, also utilizes these ideas in spirit, if not in name.

Basically, an equivalence relation $\mathcal{R}$ on a set A generalizes equality; it induces a characteristic of "sameness" among the elements of A. This "sameness" notion then causes the set A to be partitioned into subsets called *equivalence classes*. Conversely, we find that a partition of a set A induces an equivalence relation on A. The partition of a set arises in many places in mathematics and computer science. In computer science many searching algorithms rely on a technique that successively reduces the size of a given set A that is being searched. By partitioning A into smaller and smaller subsets, we apply the searching procedure in a more efficient manner. Each successive partition refines its predecessor, the key needed in the minimization process for finite state machines.

Throughout the chapter we used the interplay between relations, directed graphs, and $(0, 1)$-matrices. These matrices provide a rectangular array of information about a relation, or graph, and prove useful in certain calculations. Storing information like this, in rectangular arrays and in consecutive memory locations, has been practiced in computer science since the late 1940s and early 1950s. For more on the historical background of such considerations, consult pages 456–462 of D. Knuth [3]. Another way to store information about a graph is the *adjacency list representation*. (See Miscellaneous Exercise 17.) In the study of data structures, *linked lists* and *doubly linked lists* are prominent in implementing such storage. For more on this, consult the text by A. Aho, J. Hopcroft, and J. Ullman [1].

With regard to graph theory, we are in an area of mathematics that dates back to 1736 when the Swiss mathematician Leonhard Euler (1707–1783) solved the problem of the seven bridges of Königsberg. Since then much more has evolved in this area, especially in conjunction with data structures in computer science.

For similar coverage of some of the topics in this chapter, see Chapter 3 of D. F. Stanat and D. F. McAllister [6]. An interesting presentation of the "Equivalence Problem" can be found on pages 353–355 of D. Knuth [3] for those wanting more on the role of the computer in conjunction with the concept of the equivalence relation.

The early work on development of the minimization process can be found in the paper by E. F. Moore [5], which builds upon prior ideas of D. A. Huffman [2]. Chapter 10 of Z. Kohavi [4] covers the minimization process for different types of finite state machines and includes some hardware considerations in their design.

REFERENCES

1. Aho, Alfred V., Hopcroft, John E., and Ullman, Jeffrey D. *Data Structures and Algorithms.* Reading, Mass.: Addison-Wesley, 1983.

2. Huffman, D. A. "The Synthesis of Sequential Switching Circuits." *Journal of Franklin Institute* 257, no. 3: pp. 161–190; no. 4: pp. 275–303, 1954.

3. Knuth, Donald E. *The Art of Computer Programming,* 2nd ed., Volume 1, *Fundamental Algorithms.* Reading, Mass.: Addison-Wesley, 1973.

4. Kohavi, Zvi. *Switching and Finite Automata Theory,* 2nd ed. New York: McGraw-Hill, 1978.

5. Moore, E. F. "Gedanken-experiments on Sequential Machines." *Automata Studies, Annals of Mathematical Studies,* no. 34: pp. 129–153. Princeton, N.J.: Princeton University Press, 1956.

6. Stanat, Donald F., and McAllister, David F. *Discrete Mathematics in Computer Science.* Englewood Cliffs, N.J.: Prentice-Hall, 1977.

MISCELLANEOUS EXERCISES

1. Let A be a set and I an index set, where for each $i \in I$, $\mathcal{R}_i$ is a relation on A. Prove or disprove each of the following.

   a) $\displaystyle\bigcup_{i \in I} \mathcal{R}_i$ is reflexive on A if and only if each $\mathcal{R}_i$ is reflexive on A.

   b) $\displaystyle\bigcap_{i \in I} \mathcal{R}_i$ is reflexive on A if and only if each $\mathcal{R}_i$ is reflexive on A.

2. Repeat Exercise 1 with "reflexive" replaced by (i) symmetric; (ii) antisymmetric; (iii) transitive.

3. For a set A, let $\mathcal{R}_1$ and $\mathcal{R}_2$ be symmetric relations on A. If $\mathcal{R}_1 \circ \mathcal{R}_2 \subseteq \mathcal{R}_2 \circ \mathcal{R}_1$, prove that $\mathcal{R}_1 \circ \mathcal{R}_2 = \mathcal{R}_2 \circ \mathcal{R}_1$.

4. If $\mathcal{R}$ is a relation on a set A, prove or disprove that $\mathcal{R}^2$ reflexive $\Rightarrow \mathcal{R}$ reflexive.

5. For sets A, B, and C with relations $\mathcal{R}_1 \subseteq A \times B$ and $\mathcal{R}_2 \subseteq B \times C$, prove or disprove that $(\mathcal{R}_1 \circ \mathcal{R}_2)^c = \mathcal{R}_2^c \circ \mathcal{R}_1^c$.

6. For each of the following relations on the set specified, determine whether the relation is reflexive, symmetric, antisymmetric, or transitive. Also determine whether it is a partial order or an equivalence relation, and, if the latter, describe the partition induced by the relation.

   a) $\mathcal{R}$ is the relation on $\mathbf{Q}$ where $a \, \mathcal{R} \, b$ if $|a - b| < 1$.

   b) Let T be the set of all triangles in the plane. For $t_1, t_2 \in T$ define $t_1 \, \mathcal{R} \, t_2$ if t_1, t_2 have the same area.

   c) For T as in part (b), define $\mathcal{R}$ by $t_1 \, \mathcal{R} \, t_2$ if at least two sides of t_1 are contained within the perimeter of t_2.

   d) Let $A = \{1, 2, 3, 4, 5, 6, 7\}$. Define $\mathcal{R}$ on A by $x \, \mathcal{R} \, y$ if $xy \geq 10$.

   e) Define $\mathcal{R}$ on $\mathbf{Z}$ by $a \, \mathcal{R} \, b$ if $7 \mid (a - b)$.

   f) For $A = \{1, 2, 3, 4\} \times \{1, 2, 3, 4\}$, define $\mathcal{R}$ on A by $(x_1, y_1) \, \mathcal{R} \, (x_2, y_2)$ if $(y_1 - x_1) = \pm(y_2 - x_2)$.

   g) Define $\mathcal{R}$ on $\mathbf{R}$ by $x \, \mathcal{R} \, y$ if $\cos x = \cos y$.

   h) $\mathcal{R}$ is the relation on $\mathbf{Q}$ where $a \, \mathcal{R} \, b$ if $a - b \in \mathbf{Z}$.

7. Give an example of a poset with 5 minimal (maximal) elements but no least (greatest) element.

8. For a set A, let $C = \{P_i \mid P_i \text{ is a partition of } A\}$. Define relation $\mathcal{R}$ on C by $P_i \, \mathcal{R} \, P_j$ if $P_i \leq P_j$—that is, P_i is a refinement of P_j.

a) Verify that $\mathcal{R}$ is a partial order on C.

b) For $A = \{1, 2, 3, 4, 5\}$, let $P_i, 1 \leq i \leq 4$, be the following partitions: P_1: $\{1, 2\}$, $\{3, 4, 5\}$; P_2: $\{1, 2\}, \{3, 4\}, \{5\}$; P_3: $\{1\}, \{2\}, \{3, 4, 5\}$; P_4: $\{1, 2\}, \{3\}, \{4\}, \{5\}$. Draw the Hasse diagram for $C = \{P_i \mid 1 \leq i \leq 4\}$, where C is partially ordered by refinement.

9. If the complete graph K_n has 45 edges, what is n?

10. Let $A = \{1, 2, 3, 4, 5, 6\} \times \{1, 2, 3, 4, 5, 6\}$. Define $\mathcal{R}$ on A by $(x_1, y_1) \mathcal{R} (x_2, y_2)$, if $x_1 y_1 = x_2 y_2$.

a) Verify that $\mathcal{R}$ is an equivalence relation on A.

b) Determine the equivalence classes $[(1, 1)]$, $[(2, 2)]$, $[(3, 2)]$, and $[(4, 3)]$.

11. Let $\mathcal{F} = \{f: \mathbf{Z}^+ \rightarrow \mathbf{R}\}$. That is, $\mathcal{F}$ is the set of all functions with domain $\mathbf{Z}^+$ and codomain $\mathbf{R}$.

a) Define the relation $\mathcal{R}$ on $\mathcal{F}$ by $g \mathcal{R} h$, for $g, h \in \mathcal{F}$, if g is dominated by h and h is dominated by g. Prove that $\mathcal{R}$ is an equivalence relation on $\mathcal{F}$.

b) For $f \in \mathcal{F}$, let $[f]$ denote the equivalence class of f for the relation $\mathcal{R}$ of part (a). Let $\mathcal{F}'$ be the set of equivalence classes induced by $\mathcal{R}$. Define the relation $\mathcal{S}$ on $\mathcal{F}'$ by $[g] \mathcal{S} [h]$, for $[g], [h] \in \mathcal{F}'$, if g is dominated by h. Verify that $\mathcal{S}$ is a partial order.

c) For $\mathcal{R}$ in part (a), let $f, f_1, f_2 \in \mathcal{F}$ with $f_1, f_2 \in [f]$. If $f_1 + f_2: \mathbf{Z}^+ \rightarrow \mathbf{R}$ is defined by $(f_1 + f_2)(n) = f_1(n) + f_2(n)$, for $n \in \mathbf{Z}^+$, prove or disprove that $f_1 + f_2 \in [f]$.

12. If $A = \{2, 4, 8, 20, 28, 56, 112, 224, 336, 672, 1344, 3, 9, 15, 45, 135, 405, 675\}$, the relation $\mathcal{R}$ on A given by $x \mathcal{R} y$ if $x \mid y$ is a partial order whose Hasse diagram is structured as shown in Fig. 7.22. Label the vertices in the diagram to demonstrate the partial order. In how many different ways can we topologically sort $\mathcal{R}$ to get a total order $\mathcal{T}$ where $\mathcal{R} \subseteq \mathcal{T}$? (We sometimes say that $\mathcal{R}$ is *embedded* in $\mathcal{T}$.)

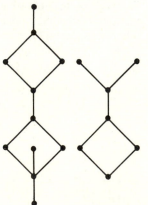

Figure 7.22

13. A lattice $(A, \mathcal{R})$ is said to be a *complete lattice* if for every nonempty subset $S \subseteq A$, lub S and glb S exist. For $A = \mathbf{Q}$, let $\mathcal{R}$ be the "is less than or equal to" relation on $\mathbf{Q}$. Then $(\mathbf{Q}, \leq)$ is a lattice. Is it a complete lattice?

14. Let M be the $(0, 1)$ adjacency matrix associated with a directed graph G. (We also have M as the $(0, 1)$-matrix for the relation associated with G.) In computing M^2, if we use regular matrix addition and multiplication so that $1 + 1 = 2$ (not Boolean, where $1 + 1 = 1$), what does "2" mean if it occurs in M^2? What about "3" in M^2?

15. For $1 \leq i \leq 9$, define the functions $f_i \colon \mathbf{Z}^+ \to \mathbf{R}$ as follows:

$$f_1(n) = 5 \qquad\qquad f_4(n) = \log_2 n^3 \qquad\qquad f_7(n) = n^2 + 5$$
$$f_2(n) = 5n - 7 \qquad\qquad f_5(n) = 5n^2 - 3n^3 \qquad\quad f_8(n) = 137$$
$$f_3(n) = 2^n \qquad\qquad f_6(n) = n^2 \log_2 n^2 \qquad\quad f_9(n) = (2/n^2) - (n^2/2).$$

Let $F = \{f_i \mid 1 \leq i \leq 9\}$. Define the relation $\mathcal{R}$ on F by $f_i \mathcal{R} f_j$, for $1 \leq i, j \leq 9$, if f_i is dominated by f_j. Determine the relation matrix associated with relation $\mathcal{R}$.

16. For $A = \{1, 2, 3, 4, 5\}$, a relation $\mathcal{R}$ on A determines the following $(0, 1)$-matrix.

$$M = \begin{bmatrix} 1 & 0 & 1 & 1 & 0 \\ 0 & 1 & 0 & 0 & 1 \\ 1 & 0 & 1 & 0 & 0 \\ 1 & 0 & 0 & 1 & 1 \\ 0 & 1 & 0 & 1 & 1 \end{bmatrix}$$

a) What is $\mathcal{R}$? What is $\mathcal{R}^2$?

b) Draw the directed graphs associated with $\mathcal{R}$ and $\mathcal{R}^2$.

c) What relational properties does $\mathcal{R}$ satisfy?

17. We have seen that the $(0, 1)$-matrix can be used to represent a graph by means of the adjacency matrix. This proves to be a rather inefficient way to represent the graph when there are many 0's present. A better method for the storage of such a graph can be accomplished using the *adjacency list representation*, which is made up of an *adjacency list* for each vertex v and an *index list*. For the graph shown in Fig. 7.23, the representation is given by the two lists in Table 7.5.

Figure 7.23

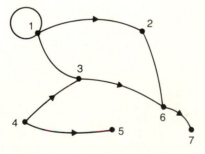

Table 7.5

Adjacency List		Index List	
1	1	1	1
2	2	2	4
3	3	3	5
4	6	4	7
5	1	5	9
6	6	6	9
7	3	7	11
8	5	8	11
9	2		
10	7		

For each vertex v in the graph, we list, preferably in numerical order, each vertex w that is adjacent from v. Hence for 1, we list $1, 2, 3$ as the first three adjacencies in our adjacency list. Next to 2 in the index list we place a 4, which tells us where to start looking in the adjacency list for the adjacencies from 2. Since there is a 5 to the right of 3 in the index list, we know that the only adjacency from 2 is 6. Likewise, the 7 to the right of 4 in the index list directs us to the seventh entry in the adjacency list, namely 3, and we find that vertex 4 is adjacent to vertices 3 (the seventh vertex in the adjacency list) and 5 (the eighth vertex in the adjacency list). We stop at vertex 5 because of the 9 to the right of vertex 5 in the index list. The 9's in the index list next to 5 and 6 indicate that no vertex is adjacent from vertex 5. In a similar way, the 11's next to 7 and 8 in the index list tell us that vertex 7 is not adjacent to any vertex in the given directed graph.

In general, this method provides an easy way to determine the vertices adjacent from a vertex v. They are listed in the positions index(v), index(v) + 1, ..., index(v + 1) − 1 of the adjacency list.

Finally, the last pair of entries in the index list, namely 8 and 11, is a "phantom" that indicates where the adjacency list would pick up from *if* there were an eighth vertex in the graph.

Represent each of the graphs in Fig. 7.24 in this manner.

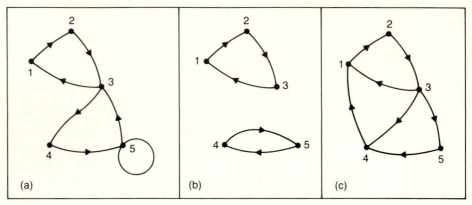

Figure 7.24

18. The adjacency list representation of a directed graph G is given by the lists in Table 7.6. Construct G from this representation.

Table 7.6

Adjacency List		Index List	
1	2	1	1
2	3	2	4
3	6	3	5
4	3	4	5
5	3	5	8
6	4	6	10
7	5	7	10
8	3	8	10
9	6		

19. Let G be an undirected graph with vertex set V. Define the relation $\mathcal{R}$ on V by $v \mathcal{R} w$ if $v = w$ or if there is a path from v to w (or from w to v, since G is undirected). (a) Prove that $\mathcal{R}$ is an equivalence relation on V. (b) What can we say about the associated partition?

20. a) For the finite state machine given in Table 7.7, determine a minimal machine that is equivalent to it.

b) Find a minimal string that distinguishes states s_4 and s_6.

Table 7.7

	ν		ω	
	0	1	0	1
s_1	s_7	s_6	1	0
s_2	s_7	s_7	0	0
s_3	s_7	s_2	1	0
s_4	s_2	s_3	0	0
s_5	s_3	s_7	0	0
s_6	s_4	s_1	0	0
s_7	s_3	s_5	1	0
s_8	s_7	s_3	0	0

21. At the computer center Maria is faced with running 10 computer programs, and, because of priorities, they are restricted by the following conditions: (a) $10 > 8, 3$; (b) $8 > 7$; (c) $7 > 5$; (d) $3 > 9, 6$; (e) $6 > 4, 1$; (f) $9 > 4, 5$; (g) $4, 5, 1 > 2$; where, for example, $10 > 8, 3$ means that program number 10 must be run before programs 8 and 3. Determine an order for running these programs so that the priorities are satisfied.

22. a) Draw the Hasse diagram for the set of positive divisors of the integer n where n is (i) 2; (ii) 4; (iii) 6; (iv) 8; (v) 12; (vi) 16; (vii) 24; (viii) 30; (ix) 32.

   b) For any $2 \leq n \leq 35$, show that the Hasse diagram for the set of positive divisors of n looks like one of the nine diagrams in part (a). (Ignore the numbers at the vertices and concentrate on the structure given by the vertices and edges.) What happens for $n = 36$?

   c) For $n \in \mathbf{Z}^+$, $\tau(n) = $ the number of positive integer divisors of n. (See Miscellaneous Exercise 33 in Chapter 5.) Let $m, n \in \mathbf{Z}^+$ and S, T be the sets of all positive integer divisors of m, n, respectively. The results of parts (a) and (b) imply that if the Hasse diagrams of S, T are structurally the same, then $\tau(m) = \tau(n)$. But is the converse true?

   d) Show that any Hasse diagram in part (a) is a lattice if we define $\mathrm{glb}\{x, y\} = (x, y)$, the g.c.d. of x, y, and $\mathrm{lub}\{x, y\} = [x, y]$, the l.c.m. of x, y.

23. For $\mathcal{I} = \{0, 1\}$, let $A = \{x_1 x_2 x_3 x_4 x_5 \mid x_i \in \mathcal{I}, 1 \leq i \leq 5\}$.

   a) What is $|A|$?

   b) For $x \in A$, define the weight of x, denoted $\mathrm{wt}(x)$, by $\mathrm{wt}(x) = \sum_{i=1}^{5} x_i$. (Hence $\mathrm{wt}(x)$ is the number of 1's that occur in x.) Define the relation $\mathcal{R}$ on A by $x \mathcal{R} y$ if $\mathrm{wt}(x) = \mathrm{wt}(y)$. Verify that $\mathcal{R}$ is an equivalence relation on A.

   c) How many equivalence classes are there in the partition of A induced by $\mathcal{R}$? How many elements of A are there in each equivalence class?

   d) Generalize the results in parts (a), (b), and (c).

24. a) For $\mathcal{U} = \{1, 2, 3\}$, let $A = \mathcal{P}(\mathcal{U})$. Define the relation $\mathcal{R}$ on A by $B \mathcal{R} C$ if $B \subseteq C$. How many ordered pairs are there in the relation $\mathcal{R}$?

   b) Answer part (a) for $\mathcal{U} = \{1, 2, 3, 4\}$.

   c) Generalize the results of parts (a) and (b).

25. For $n \in \mathbf{Z}^+$, let $\mathcal{U} = \{1, 2, 3, \ldots, n\}$. Define the relation $\mathcal{R}$ on $\mathcal{P}(\mathcal{U})$ by $A \mathcal{R} B$ if $A \not\subseteq B$ and $B \not\subseteq A$. How many ordered pairs are there in this relation?

26. Let A be a finite nonempty set with $B \subseteq A$ (B fixed), and $|A| = n$, $|B| = m$. Define the relation $\mathcal{R}$ on $\mathcal{P}(A)$ by $X \mathcal{R} Y$, for $X, Y \subseteq A$, if $X \cap B = Y \cap B$. Then $\mathcal{R}$ is an equivalence relation, as verified in Exercise 7 of Section 7.4.

   a) How many equivalence classes are in the partition of $\mathcal{P}(A)$ induced by $\mathcal{R}$?

   b) How many subsets of A are in each equivalence class of the partition induced by $\mathcal{R}$?

27. For $A \neq \emptyset$, let $(A, \mathcal{R})$ be a poset, and let $\emptyset \neq B \subseteq A$ such that $\mathcal{R}' = (B \times B) \cap \mathcal{R}$. If $(B, \mathcal{R}')$ is totally ordered, we call $(B, \mathcal{R}')$ a *chain* in $(A, \mathcal{R})$. In the case where B is finite, we may order the elements of B by $b_1 \mathcal{R}' b_2 \mathcal{R}' b_3 \mathcal{R}' \ldots \mathcal{R}' b_{n-1} \mathcal{R}' b_n$ and say that the chain has *length n*. A chain (of length n) is called *maximal* if there is no element $a \in A$ where $a \mathcal{R} b_1$, $b_n \mathcal{R} a$, or $b_i \mathcal{R} a \mathcal{R} b_{i+1}$, for any $1 \leq i \leq n-1$.

   a) Find two chains of length three for the poset given by the Hasse diagram in Fig. 7.16. Find a maximal chain for this poset. How many such maximal chains does it have?

   b) For the poset given by the Hasse diagram in Fig. 7.15(d), find two maximal chains of different lengths. What is the length of a longest (maximal) chain for this poset?

   c) Let $\mathcal{U} = \{1, 2, 3, 4\}$ and $A = \mathcal{P}(\mathcal{U})$. For the poset $(A, \subseteq)$, find two maximal chains. How many such maximal chains are there for this poset?

   d) If $\mathcal{U} = \{1, 2, 3, \ldots, n\}$, how many maximal chains are there in the poset $(\mathcal{P}(\mathcal{U}), \subseteq)$?

28. For $\emptyset \neq C \subseteq A$, let $(C, \mathcal{R}')$ be a maximal chain in the poset $(A, \mathcal{R})$, where $\mathcal{R}' = (C \times C) \cap \mathcal{R}$. If the elements of C are ordered as $c_1 \mathcal{R}' c_2 \mathcal{R}' \ldots \mathcal{R}' c_n$, prove that c_1 is a minimal element in $(A, \mathcal{R})$ and that c_n is maximal in $(A, \mathcal{R})$.

29. Let $(A, \mathcal{R})$ be a poset in which the length of a longest (maximal) chain is $n \geq 2$. Let M be the set of all maximal elements in $(A, \mathcal{R})$, and let $B = A - M$. If $\mathcal{R}' = (B \times B) \cap \mathcal{R}$, prove that the length of a longest chain in $(B, \mathcal{R}')$ is $n - 1$.

30. Let $(A, \mathcal{R})$ be a poset, and let $\emptyset \neq C \subseteq A$. If $(C \times C) \cap \mathcal{R} = \emptyset$, then for all $x, y \in C$ we have $x \not\mathcal{R} y$ and $y \not\mathcal{R} x$. The elements of C are said to form an *antichain* in the poset $(A, \mathcal{R})$.

   a) Find an antichain with three elements for the poset given in the Hasse diagram of Fig. 7.15(d). Determine a largest antichain containing the element 6. Determine a largest antichain for this poset.

   b) If $\mathcal{U} = \{1, 2, 3, 4\}$, let $A = \mathcal{P}(\mathcal{U})$. Find two different antichains for the poset $(A, \subseteq)$. How many elements occur in a largest antichain for this poset?

   c) Prove that in any poset $(A, \mathcal{R})$, the set of all maximal elements and the set of all minimal elements are antichains.

31. Let $(A, \mathcal{R})$ be a poset in which the length of a longest chain is n. Use mathematical induction to prove that the elements of A can be partitioned into n antichains $C_1, C_2, \ldots, C_n$, where $C_i \cap C_j = \emptyset$, for $1 \leq i < j \leq n$.

The Principle of Inclusion and Exclusion

We now return to the topic of enumeration as we investigate the *Principle of Inclusion and Exclusion*. Extending the ideas in the counting problems on Venn diagrams in Chapter 3, this principle will assist us in establishing the formula we conjectured in Section 5.3 for the number of onto functions $f: A \to B$, where $|A| = m \geq n = |B|$. Other applications of this principle will demonstrate its versatile nature in combinatorial and discrete mathematics as an *indirect* method for certain situations in enumeration.

8.1
THE PRINCIPLE OF INCLUSION AND EXCLUSION

In this section we develop some notation for stating our new counting principle. Then we establish the principle by a combinatorial argument. Examples will then demonstrate how this principle is applied.

Let S be a set with $N = |S|$, and let $c_1, c_2, \ldots, c_t$ be a collection of conditions or properties satisfied by some, or all, of the elements of S. Some elements of S may satisfy more than one of the conditions, whereas some may not satisfy any of them. For $1 \leq i \leq t$, $N(c_i)$ will denote the number of elements in S that satisfy condition c_i. (Elements of S are counted here when they satisfy only condition c_i, as well as when they satisfy c_i and other conditions c_j, for $j \neq i$.) For $i, j \in \{1, 2, 3, \ldots, t\}$ where $i \neq j$, $N(c_i c_j)$ will denote the number of elements in S that satisfy both of the conditions c_i, c_j, and perhaps some others. ($N(c_i c_j)$ does *not* count the elements of S that satisfy *only* c_i, c_j.) Continuing, if $1 \leq i, j, k \leq t$ are three distinct integers, then $N(c_i c_j c_k)$ denotes the number of elements in S satisfying, perhaps among others, each of the conditions c_i, c_j, and c_k.

For $1 \leq i \leq t$, $N(\overline{c}_i) = N - N(c_i)$ denotes the number of elements in S that do not satisfy condition c_i. If $1 \leq i, j \leq t$ with $i \neq j$, $N(\overline{c}_i \overline{c}_j) =$ the number of elements in S that do not satisfy either of the conditions c_i or c_j. (This is not the same as $N(\overline{c_i c_j})$.)

From the Venn diagram in Fig. 8.1, we see that if $N(c_i)$ denotes the number of elements in the left-hand circle and $N(c_j)$ denotes the number of elements in the right-hand circle, then $N(c_i c_j)$ is the number of elements in the overlap, while $N(\overline{c}_i \overline{c}_j)$ counts the elements outside the union of these circles.

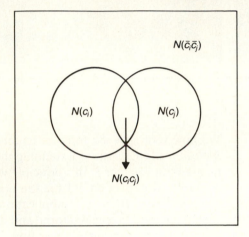

Figure 8.1

Consequently, from Fig. 8.1, $N(\overline{c}_i \overline{c}_j) = N - [N(c_i) + N(c_j)] + N(c_i c_j)$, where the last term is added on because it was eliminated twice in the term $[N(c_i) + N(c_j)]$.

In like manner, from Fig. 8.2 we find that

$$N(\overline{c}_i \overline{c}_j \overline{c}_k) = N - [N(c_i) + N(c_j) + N(c_k)]$$
$$+ [N(c_i c_j) + N(c_i c_k) + N(c_j c_k)] - N(c_i c_j c_k).$$

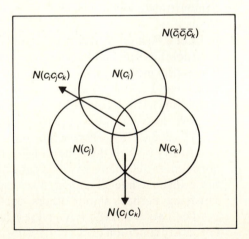

Figure 8.2

From the pattern suggested by these two cases, we state the following theorem.

THEOREM 8.1   (*The Principle of Inclusion and Exclusion*) Consider a set S, with $|S| = N$, and conditions c_i, $1 \le i \le t$, satisfied by some of the elements of S. The number of elements of S that satisfy *none* of the conditions c_i, $1 \le i \le t$, is denoted by $\overline{N} = N(\overline{c}_1 \overline{c}_2 \overline{c}_3 \ldots \overline{c}_t)$ where

$$\overline{N} = N - [N(c_1) + N(c_2) + N(c_3) + \cdots + N(c_t)]$$
$$+ [N(c_1 c_2) + N(c_1 c_3) + \cdots + N(c_1 c_t) + N(c_2 c_3) + \cdots + N(c_{t-1} c_t)]$$
$$- [N(c_1 c_2 c_3) + N(c_1 c_2 c_4) + \cdots + N(c_1 c_2 c_t) + N(c_1 c_3 c_4) + \cdots$$
$$+ N(c_1 c_3 c_t) + \cdots + N(c_{t-2} c_{t-1} c_t)] + \cdots + (-1)^t N(c_1 c_2 c_3 \ldots c_t), \qquad (1)$$

or

$$\overline{N} = N - \sum_{1 \le i \le t} N(c_i) + \sum_{1 \le i < j \le t} N(c_i c_j) - \sum_{1 \le i < j < k \le t} N(c_i c_j c_k) + \cdots$$
$$+ (-1)^t N(c_1 c_2 c_3 \ldots c_t). \qquad (2)$$

Proof   Although this result can be established by induction on t, we give a combinatorial argument here.

For each $x \in S$ we show that x contributes the same count, either 0 or 1, to each side of Eq. (2).

If x satisfies none of the conditions, then x is counted once in $\overline{N}$ and once in N, but not in any of the other terms in Eq. (2). Consequently, x contributes a count of 1 to each side of the equation.

The other possibility is that x satisfies exactly r of the conditions where $1 \le r \le t$. In this case x contributes nothing to $\overline{N}$. But on the right-hand side of Eq. (2), x is counted

1.   One time in N.

2. r times in $\sum_{1 \le i \le t} N(c_i)$. (Once for each of the r conditions.)

3. $\binom{r}{2}$ times in $\sum_{1 \le i < j \le t} N(c_i c_j)$. (Once for each pair of conditions selected from the r conditions it satisfies.)

4. $\binom{r}{3}$ times in $\sum_{1 \le i < j < k \le t} N(c_i c_j c_k)$. (Why?)

    $\cdots\cdots\cdots\cdots\cdots\cdots\cdots$

(r + 1). $\binom{r}{r} = 1$ time in $\sum N(c_{i_1} c_{i_2} \ldots c_{i_r})$, where the summation is taken over all selections of size r from the t conditions.

Consequently, on the right-hand side of Eq. (2), x is counted

$$1 - r + \binom{r}{2} - \binom{r}{3} + \cdots + (-1)^r \binom{r}{r} = [1 + (-1)]^r = 0^r = 0 \text{ times},$$

by the binomial theorem. Therefore, the two sides of Eq. (2) count the same elements of S, and the equality is verified.   ∎

An immediate corollary of this principle is given as follows:

COROLLARY 8.1   Under the hypotheses of Theorem 8.1, the number of elements in S that satisfy at least one of the conditions c_i, $1 \le i \le t$, is given by $N(c_1$ or c_2 or $\ldots$ or $c_t) = N - \overline{N}$.

Before solving some examples, we examine some further notation for simplifying the statement of Theorem 8.1.
We write

$$S_0 = N,$$
$$S_1 = [N(c_1) + N(c_2) + \cdots + N(c_t)],$$
$$S_2 = [N(c_1 c_2) + N(c_1 c_3) + \cdots + N(c_1 c_t) + N(c_2 c_3) + \cdots + N(c_{t-1} c_t)],$$

and, in general,

$$S_k = \sum N(c_{i_1} c_{i_2} \ldots c_{i_k}), \quad 1 \le k \le t,$$

where the summation is taken over all selections of size k from the collection of t conditions. Hence S_k has $\binom{t}{k}$ summands in it.

Let us look at how this principle is used now to solve certain enumeration problems.

Example 8.1   Determine the number of positive integers n where $1 \le n \le 100$ and n is *not* divisible by 2, 3, or 5.
Here $S = \{1, 2, 3, \ldots, 100\}$ and $N = 100$. For $n \in S$, n satisfies

a) condition c_1 if n is divisible by 2,

b) condition c_2 if n is divisible by 3, and

c) condition c_3 if n is divisible by 5.

Then the answer to this problem is $N(\overline{c}_1 \overline{c}_2 \overline{c}_3)$.
For $r \in \mathbf{R}$, $\lfloor r \rfloor$ denotes the *greatest integer* in r, where

$$\lfloor r \rfloor = \begin{cases} r, \text{ if } r \in \mathbf{Z} \\ \text{the largest integer smaller than } r, \text{ if } r \text{ is not an integer.} \end{cases}$$

So, for example, $\lfloor 5 \rfloor = 5$, $\lfloor \pi \rfloor = 3$ and $\lfloor -\pi \rfloor = -4$.
This function is helpful in this problem, as we find that

$$N(c_1) = \lfloor 100/2 \rfloor = 50, \qquad N(c_2) = \lfloor 100/3 \rfloor = \lfloor 33\ 1/3 \rfloor = 33,$$

$$N(c_3) = \lfloor 100/5 \rfloor = 20, \qquad N(c_1 c_2) = \lfloor 100/6 \rfloor = 16,$$

$$N(c_1 c_3) = \lfloor 100/10 \rfloor = 10, \qquad N(c_2 c_3) = \lfloor 100/15 \rfloor = 6,$$

$$\text{and } N(c_1 c_2 c_3) = \lfloor 100/30 \rfloor = 3.$$

Applying the inclusion and exclusion principle, we find that

$$N(\overline{c}_1 \overline{c}_2 \overline{c}_3) = S_0 - S_1 + S_2 - S_3 = N - [N(c_1) + N(c_2) + N(c_3)]$$
$$+ [N(c_1 c_2) + N(c_1 c_3) + N(c_2 c_3)] - N(c_1 c_2 c_3)$$
$$= 100 - [50 + 33 + 20] + [16 + 10 + 6] - 3 = 26.$$

(These 26 numbers are 1, 7, 11, 13, 17, 19, 23, 29, 31, 37, 41, 43, 47, 49, 53, 59, 61, 67, 71, 73, 77, 79, 83, 89, 91, and 97.) □

Example 8.2 In Chapter 1 we found the number of nonnegative integer solutions to the equation $x_1 + x_2 + x_3 + x_4 = 18$. We now answer the same question with the extra restriction that $x_i \le 7$, $1 \le i \le 4$.

Here S is the set of solutions of $x_1 + x_2 + x_3 + x_4 = 18$, with $0 \le x_i$ for $1 \le i \le 4$. So $|S| = N = \binom{4+18-1}{18} = \binom{21}{18}$.

We say that a solution x_1, x_2, x_3, x_4 satisfies condition c_i for $1 \le i \le 4$, if $x_i > 7$ (or $x_i \ge 8$). The answer to the problem is then $N(\bar{c}_1 \bar{c}_2 \bar{c}_3 \bar{c}_4)$.

Here $N(c_1) = N(c_2) = N(c_3) = N(c_4)$. To compute $N(c_1)$, we want the number of integer solutions of $x_1 + x_2 + x_3 + x_4 = 10$, with each $x_i \ge 0$ for $1 \le i \le 4$. Then we add 8 to the value of x_1 and get the solutions of $x_1 + x_2 + x_3 + x_4 = 18$ that satisfy condition c_1. Hence $N(c_i) = \binom{4+10-1}{10} = \binom{13}{10}$, for each $1 \le i \le 4$, and $S_1 = \binom{4}{1}\binom{13}{10}$.

Likewise, $N(c_1 c_2)$ is the number of integer solutions of $x_1 + x_2 + x_3 + x_4 = 2$, where $x_i \ge 0$ for $1 \le i \le 4$. So $N(c_1 c_2) = \binom{4+2-1}{2} = \binom{5}{2}$, and $S_2 = \binom{4}{2}\binom{5}{2}$.

Since $N(c_i c_j c_k) = 0$ for any selection of three conditions, and $N(c_1 c_2 c_3 c_4) = 0$, we have

$$N(\bar{c}_1 \bar{c}_2 \bar{c}_3 \bar{c}_4) = N - S_1 + S_2 - S_3 + S_4 = \binom{21}{18} - \binom{4}{1}\binom{13}{10} + \binom{4}{2}\binom{5}{2} - 0 + 0 = 246.$$

So of the 1330 nonnegative integer solutions of $x_1 + x_2 + x_3 + x_4 = 18$, only 246 of them satisfy $x_i \le 7$ for $1 \le i \le 4$. □

Our next example establishes the formula conjectured in Section 5.3 for counting onto functions.

Example 8.3 For finite sets A, B, with $|A| = m \ge n = |B|$, let $A = \{a_1, a_2, \ldots, a_m\}$, $B = \{b_1, b_2, \ldots, b_n\}$, and S = the set of all functions $f \colon A \to B$. Then $N = |S| = n^m$.

For $1 \le i \le n$, let c_i denote the condition on S where a function $f \colon A \to B$ satisfies c_i if b_i is *not* in the range of f. (Note the difference between c_i here and c_i in Examples 8.1 and 8.2.) Then $N(\bar{c}_i)$ is the number of functions in S that have b_i in their range, and $N(\bar{c}_1 \bar{c}_2 \ldots \bar{c}_n)$ counts the number of onto functions $f \colon A \to B$.

For any $1 \le i \le n$, $N(c_i) = (n-1)^m$, because any element of B, except b_i, can be used as the second component of an ordered pair for a function $f \colon A \to B$, where the range of f does not include b_i. Likewise, for any $1 \le i < j \le n$, there are $(n-2)^m$ functions $f \colon A \to B$ where the range contains neither b_i nor b_j. From these observations we have $S_1 = [N(c_1) + N(c_2) + \cdots + N(c_n)] = n(n-1)^m = \binom{n}{1}(n-1)^m$, and $S_2 = [N(c_1 c_2) + N(c_1 c_3) + \cdots + N(c_1 c_n) + N(c_2 c_3) + \cdots + N(c_2 c_n) + \cdots + N(c_{n-1} c_n)] = \binom{n}{2}(n-2)^m$. In general, for $1 \le k \le n$,

$$S_k = \sum_{1 \le i_1 < i_2 < \cdots < i_k \le n} N(c_{i_1} c_{i_2} \ldots c_{i_k}) = \binom{n}{k}(n-k)^m$$

It then follows by the principle of inclusion and exclusion that the number of onto functions from A to B is

$$N(\bar{c}_1\bar{c}_2\bar{c}_3\ldots\bar{c}_n) = N - S_1 + S_2 - S_3 + \cdots + (-1)^n S_n$$

$$= n^m - \binom{n}{1}(n-1)^m + \binom{n}{2}(n-2)^m - \binom{n}{3}(n-3)^m$$

$$+ \cdots + (-1)^n(n-n)^m = \sum_{i=0}^{n} (-1)^i \binom{n}{i}(n-i)^m$$

$$= \sum_{i=0}^{n} (-1)^i \binom{n}{n-i}(n-i)^m. \quad \square$$

We now solve a problem similar to those in Chapter 3 that deal with Venn diagrams.

Example 8.4   In how many ways can the 26 letters of the alphabet be permuted so that none of the patterns *car*, *dog*, *pun*, or *byte* occurs?

Let S denote the set of all permutations of the 26 letters. Then $|S| = 26!$ For $1 \le i \le 4$, a permutation in S is said to satisfy condition c_i if it contains car, dog, pun, or byte, respectively.

$$N(c_1) = N(c_2) = N(c_3) = 24!, \qquad N(c_4) = 23!$$

$$N(c_1 c_2) = N(c_1 c_3) = N(c_2 c_3) = 22!, \qquad N(c_i c_4) = 21!, \qquad i \ne 4$$

$$N(c_1 c_2 c_3) = 20!, \qquad N(c_i c_j c_4) = 19!, \qquad 1 \le i < j \le 3$$

$$N(c_1 c_2 c_3 c_4) = 17!$$

So the number of permutations in S that contain none of the given patterns is
$N(\bar{c}_1\bar{c}_2\bar{c}_3\bar{c}_4) = 26! - [3(24!) + 23!] + [3(22!) + 3(21!)] - [20! + 3(19!)] + 17! \quad \square$

Our next example deals with a number theory problem.

Example 8.5   For $n \in \mathbf{Z}^+$, let $\phi(n)$ be the number of positive integers m, where $1 \le m < n$ and $(m, n) = 1$—that is, m, n are relatively prime. This function is known as *Euler's phi function* and arises in several situations in abstract algebra involving enumeration. We find that $\phi(2) = 1$, $\phi(3) = 2$, $\phi(4) = 2$, $\phi(5) = 4$, and $\phi(6) = 2$. For any prime p, $\phi(p) = p - 1$. We would like to derive a formula for $\phi(n)$ that is related to n so that we need not make a case-by-case comparison for each m, $1 \le m < n$, against the integer n.

The derivation of our formula will use the inclusion–exclusion principle as in Example 8.1. We proceed as follows: For $n \ge 2$, write $n = p_1^{e_1} p_2^{e_2} \cdots p_t^{e_t}$, where $p_1, p_2, \ldots, p_t$ are distinct primes and $e_i \ge 1$, $1 \le i \le t$. We consider the case where $t = 4$. This will be enough to demonstrate the general idea.

With $S = \{1, 2, 3, \ldots, n\}$, $N = |S| = n$, and for $1 \le i \le 4$ we say that $k \in S$ satisfies condition c_i if k is divisible by p_i. For $1 \le k < n$, $(k, n) = 1$ if k is not divisible by any of the primes p_i, $1 \le i \le 4$. Hence $\phi(n) = N(\bar{c}_1\bar{c}_2\bar{c}_3\bar{c}_4)$.

For $1 \le i \le 4$, $N(c_i) = n/p_i$; $N(c_i c_j) = n/(p_i p_j)$, $1 \le i < j \le 4$. Also,

$N(c_i c_j c_\ell) = n/(p_i p_j p_\ell)$, $1 \le i < j < \ell \le 4$, and $N(c_1 c_2 c_3 c_4) = n/(p_1 p_2 p_3 p_4)$. So

$$\phi(n) = S_0 - S_1 + S_2 - S_3 + S_4$$

$$= n - \left[\frac{n}{p_1} + \cdots + \frac{n}{p_4}\right] + \left[\frac{n}{p_1 p_2} + \frac{n}{p_1 p_3} + \cdots + \frac{n}{p_3 p_4}\right]$$

$$- \left[\frac{n}{p_1 p_2 p_3} + \cdots + \frac{n}{p_2 p_3 p_4}\right] + \frac{n}{p_1 p_2 p_3 p_4}$$

$$= n\left[1 - \left(\frac{1}{p_1} + \cdots + \frac{1}{p_4}\right) + \left(\frac{1}{p_1 p_2} + \frac{1}{p_1 p_3} + \cdots + \frac{1}{p_3 p_4}\right)\right.$$

$$\left. - \left(\frac{1}{p_1 p_2 p_3} + \cdots + \frac{1}{p_2 p_3 p_4}\right) + \frac{1}{p_1 p_2 p_3 p_4}\right]$$

$$= \frac{n}{p_1 p_2 p_3 p_4}[p_1 p_2 p_3 p_4 - (p_2 p_3 p_4 + p_1 p_3 p_4 + p_1 p_2 p_4 + p_1 p_2 p_3)$$

$$+ (p_3 p_4 + p_2 p_4 + p_2 p_3 + p_1 p_4 + p_1 p_3 + p_1 p_2)$$

$$- (p_4 + p_3 + p_2 + p_1) + 1]$$

$$= \frac{n}{p_1 p_2 p_3 p_4}[(p_1 - 1)(p_2 - 1)(p_3 - 1)(p_4 - 1)]$$

$$= n\left[\frac{p_1 - 1}{p_1} \cdot \frac{p_2 - 1}{p_2} \cdot \frac{p_3 - 1}{p_3} \cdot \frac{p_4 - 1}{p_4}\right] = n\prod_{i=1}^{4}\left(1 - \frac{1}{p_i}\right).$$

In general, $\phi(n) = n \prod_{p|n} (1 - (1/p))$, where the product is taken over all primes p dividing n.

Consequently,

$$\phi(23,100) = \phi(2^2 \cdot 3 \cdot 5^2 \cdot 7 \cdot 11)$$
$$= (23,100)(1 - (1/2))(1 - (1/3))(1 - (1/5)) \cdot$$
$$(1 - (1/7))(1 - (1/11)) = 4800. \quad \square$$

The Pascal program in Fig. 8.3 evaluates $\phi(n)$ for $n \ge 3$. Here the Repeat-Until control structure is used to determine whether the integer i is a prime. The condition

```
j = trunc(sqrt(i))
```

for terminating execution of this loop follows from the result of Example 4.16. We use this program to find $\phi(n)$ for $n = 131$, $n = 31,500$, and $n = 198,000$.

Our final example recalls some of the graph theory of Chapter 7.

Example 8.6 In a certain area of the countryside there are five villages. An engineer is to devise a system of two-way roads so that after the system is completed, no village will be isolated. In how many ways can he do this?

```
Program EulerPhiFunction (input,output);
Var
      i,j,k,n,phi,originalvalue: integer;
Begin
      Write ('The value of n is ');
      Read (n);
      phi := n;
      originalvalue := n;
      If n Mod 2 = 0 then
              Begin
                    phi := phi Div 2;
                    While n Mod 2 = 0 do
                          n := n Div 2
              End;
      If n Mod 3 = 0 then
              Begin
                    phi := (phi * 2) Div 3;
                    While n Mod 3 = 0 do
                          n := n Div 3
              End;
      i := 5;
      While n >= 5 do
              Begin
                    j := 1;
                    Repeat
                          j := j + 1;
                          k := i Mod j
                    Until (k = 0) or (j = trunc(sqrt(i)));
                    If (k <> 0) and (n Mod i = 0) then
                            Begin
                                  phi := (phi * (i - 1)) Div i;
                                  While n Mod i = 0 do
                                        n := n Div i
                            End;
                    i := i + 2
              End;
      Write ('For n =', originalvalue:0, ' there are ', phi:0);
      Writeln (' numbers smaller than ', originalvalue:0);
      Writeln (' and relatively prime to it.')
End.

The value of n is 131
For n = 131 there are 130 numbers smaller than 131
and relatively prime to it.

The value of n is 31500
For n = 31500 there are 7200 numbers smaller than 31500
and relatively prime to it.

The value of n is 198000
For n = 198000 there are 48000 numbers smaller than 198000
and relatively prime to it.
```

Figure 8.3

Calling the villages a, b, c, d, and e, we seek the number of loop-free undirected graphs on these vertices, where no vertex is isolated. Consequently, we want to count situations such as those illustrated in parts (a) and (b) of Fig. 8.4, but not situations such as that shown in part (c).

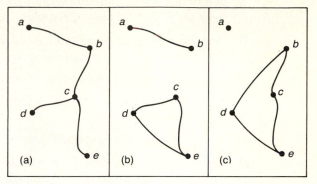

Figure 8.4

Let S be the set of loop-free undirected graphs G on $V = \{a, b, c, d, e\}$. Then $N = |S| = 2^{10}$, because there are $\binom{5}{2} = 10$ possible two-way roads for these five villages, and each road can be either included or excluded.

For $1 \leq i \leq 5$, let c_i be the condition that a system of these roads isolates villages a, b, c, d, and e, respectively. Then the answer to the problem is $N(\bar{c}_1 \bar{c}_2 \bar{c}_3 \bar{c}_4 \bar{c}_5)$. We find that $N(c_1) = 2^6$ and $S_1 = \binom{5}{1} 2^6$; $N(c_1 c_2) = 2^3$ and $S_2 = \binom{5}{2} 2^3$; $N(c_1 c_2 c_3) = 2^1$ and $S_3 = \binom{5}{3} 2^1$; $N(c_1 c_2 c_3 c_4) = 2^0$ and $S_4 = \binom{5}{4} 2^0$; and, $N(c_1 c_2 c_3 c_4 c_5) = 2^0$ and $S_5 = \binom{5}{5} 2^0$.

Consequently, $N(\bar{c}_1 \bar{c}_2 \bar{c}_3 \bar{c}_4 \bar{c}_5) = 2^{10} - \binom{5}{1} 2^6 + \binom{5}{2} 2^3 - \binom{5}{3} 2^1 + \binom{5}{4} 2^0 - \binom{5}{5} 2^0 = 768.$ □

EXERCISES 8.1

1. Determine the number of positive integers n, $1 \leq n \leq 2000$, that are
   a) not divisible by 2, 3, or 5.
   b) not divisible by 2, 3, 5, or 7.
   c) not divisible by 2, 3, or 5, but are divisible by 7.

2. Find all real numbers x such that
   a) $7 \lfloor x \rfloor = \lfloor 7x \rfloor$
   b) $\lfloor 7x \rfloor = 7$
   c) $\lfloor x + 7 \rfloor = x + 7$
   d) $\lfloor x + 7 \rfloor = \lfloor x \rfloor + 7$

3. Determine how many integer solutions there are to $x_1 + x_2 + x_3 + x_4 = 19$, if
   a) $0 \leq x_i$ for $1 \leq i \leq 4$
   b) $0 \leq x_i < 8$ for $1 \leq i \leq 4$
   c) $0 \leq x_1 \leq 5, 0 \leq x_2 \leq 6, 3 \leq x_3 \leq 7, 3 \leq x_4 \leq 8$

4. Determine the number of integer solutions to $x_1 + x_2 + x_3 + x_4 = 19$ where $-5 \leq x_i \leq 10$ for $1 \leq i \leq 4$.

5. Find the number of positive integers x that exist where $x \leq 9,999,999$ and the sum of the digits in x equals 31.

6. In how many ways can Troy select nine marbles from a bag of twelve, where three are red, three blue, three white, and three green?

7. Find the number of permutations of $a, b, c, \ldots, x, y, z$, in which none of the patterns *spin*, *game*, *path*, or *net* occurs.

8. Compute $\phi(n)$ for n equal to (a) 51; (b) 420; (c) 12300.

9. Let $n \in \mathbf{Z}^+$. (a) Determine $\phi(2^n)$. (b) Determine $\phi(2^n p)$, where p is an odd prime.

10. For $n \in \mathbf{Z}^+$, when is $\phi(n)$ odd?

11. How many positive integers n less than 6000 (a) satisfy $(n, 6000) = 1$? (b) share a common prime divisor with 6000?

12. Answer the question in Example 8.6 for the case of six villages.

13. If eight distinct dice are rolled, what is the probability that all six numbers appear?

14. How many social security numbers (nine-digit sequences) have each of the digits 1, 3, and 7 appearing at least once?

15. In how many ways can three x's, three y's, and three z's be arranged so that no consecutive triple of the same letter appears?

16. Mrs. Phillips has eight grandchildren all of whom like ice cream. In her freezer she has enough ice cream for 6 vanilla, 3 chocolate, 6 strawberry, and 5 chocolate chip cones. On her birthday all of her grandchildren come to see her, and her oldest grandchild tells her how many requests there are for each flavor. In how many ways can her grandchildren make requests that will embarrass Mrs. Phillips because she doesn't have enough of each flavor?

17. At a twelve-week conference in mathematics, Sharon met seven of her friends from college. During the conference she met each friend at lunch 35 times, every pair of them 16 times, every trio 8 times, every foursome 4 times, each set of five 2 times, each set of six once, but never all seven at once. If she had lunch every day during the 84 days of the conference, did she ever have lunch alone?

18. Extend the program in Fig. 8.3 so that it will handle the case where $n = 2$ and will print out the result so that it is grammatically correct.

19. For the program in Fig. 8.3, what purpose do the three two-line While loops serve? Why is i incremented by 2 in the line $i := i + 2$, instead of by 1? (Would we get a different result if we incremented i by 1?)

8.2
GENERALIZATIONS OF THE PRINCIPLE

Consider a set S with $|S| = N$, and conditions $c_1, c_2, \ldots, c_t$ satisfied by some of the elements of S. In Section 8.1 we saw how inclusion–exclusion provides a way to determine $N(\bar{c}_1 \bar{c}_2 \ldots \bar{c}_t)$, the number of elements in S that satisfy none of the t conditions. If $m \in \mathbf{Z}^+$ and $1 \le m \le t$, we now want to determine $E_m =$ the number of elements in S that satisfy exactly m of the t conditions. (At present we can obtain E_0.)

We can write equations such as

$$E_1 = N(c_1 \bar{c}_2 \bar{c}_3 \ldots \bar{c}_t) + N(\bar{c}_1 c_2 \bar{c}_3 \ldots \bar{c}_t) + \cdots + N(\bar{c}_1 \bar{c}_2 \bar{c}_3 \ldots \bar{c}_{t-1} c_t),$$

and

$$E_2 = N(c_1 c_2 \bar{c}_3 \ldots \bar{c}_t) + N(c_1 \bar{c}_2 c_3 \ldots \bar{c}_t) + \cdots + N(\bar{c}_1 \bar{c}_2 \bar{c}_3 \ldots \bar{c}_{t-2} c_{t-1} c_t),$$

and although these results do not assist us as much as we should like, they will be a useful starting place as we examine the Venn diagrams for the cases where $t = 3$ and 4.

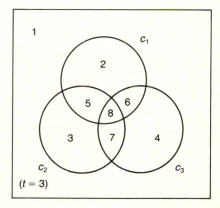

Figure 8.5

For Fig. 8.5 where $t = 3$, we place a numbered condition beside the circle representing those elements of S satisfying that particular condition. Then E_1 equals the number of elements in regions 2, 3, and 4. But we can also write

$$E_1 = N(c_1) + N(c_2) + N(c_3) - 2[N(c_1 c_2) + N(c_1 c_3) + N(c_2 c_3)] + 3N(c_1 c_2 c_3).$$

In $N(c_1) + N(c_2) + N(c_3)$ we count the elements in regions 5, 6, and 7 twice and those in region 8 three times. In the next term, the elements in regions 5, 6, and 7 are deleted twice. We remove the elements in region 8 six times in $2[N(c_1 c_2) + N(c_1 c_3) + N(c_2 c_3)]$, so we then add on the term $3N(c_1 c_2 c_3)$ and end up not counting the elements in region 8 at all. Hence we have $E_1 = S_1 - 2S_2 + 3S_3 = S_1 - \binom{2}{1}S_2 + \binom{3}{2}S_3$.

When we turn to E_2, our earlier equation indicates that we want to count the elements of S in regions 5, 6, and 7. From the Venn diagram,

$$E_2 = N(c_1 c_2) + N(c_1 c_3) + N(c_2 c_3) - 3N(c_1 c_2 c_3) = S_2 - 3S_3 = S_2 - \binom{3}{1}S_3,$$

and

$$E_3 = N(c_1 c_2 c_3) = S_3.$$

In Fig. 8.6, the conditions c_1, c_2, c_3 are associated with circular subsets of S, whereas c_4 is paired with the rather irregularly shaped area made up of regions 4, 8, 9, 11, 12, 13, 14, and 16. For $1 \le i \le 4$, E_i is determined as follows:

E_1 [regions 2, 3, 4, 5]:
$$\begin{aligned}
E_1 &= [N(c_1) + N(c_2) + N(c_3) + N(c_4)] \\
&\quad - 2[N(c_1 c_2) + N(c_1 c_3) + N(c_1 c_4) + N(c_2 c_3) + N(c_2 c_4) + N(c_3 c_4)] \\
&\quad + 3[N(c_1 c_2 c_3) + N(c_1 c_2 c_4) + N(c_1 c_3 c_4) + N(c_2 c_3 c_4)] \\
&\quad - 4N(c_1 c_2 c_3 c_4) \\
&= S_1 - 2S_2 + 3S_3 - 4S_4 = S_1 - \binom{2}{1}S_2 + \binom{3}{2}S_3 - \binom{4}{3}S_4.
\end{aligned}$$

(Note: Taking an element in region 3, we find that it is counted once in E_1 and once in S_1 (in $N(c_3)$). Taking an element in region 6, we find that it is not counted in E_1; it is counted twice in S_1 (in both $N(c_2)$ and $N(c_3)$) but removed twice in $2S_2$ (for it is counted once in S_2 in $N(c_2 c_3)$), so overall it is not counted. The reader should now consider an element from region 12 and one from region 16 and show that each contributes a count of 0 to both sides of the formula for E_1.)

E_2 [regions 6–11]:
From Fig. 8.6, $E_2 = S_2 - 3S_3 + 6S_4 = S_2 - \binom{3}{1}S_3 + \binom{4}{2}S_4$. For details on this formula we examine the results in Table 8.1, where next to each summand of S_2, S_3, and S_4 we list the regions whose elements are counted in determining that particular summand. In calculating $S_2 - 3S_3 + 6S_4$ we find the elements in regions 6–11, which are precisely those that are to be counted in E_2.

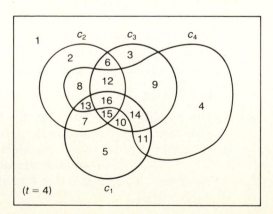

Figure 8.6

Table 8.1

S_2	S_3	S_4
$N(c_1 c_2)$: 7, 13, 15, 16	$N(c_1 c_2 c_3)$: 15, 16	$N(c_1 c_2 c_3 c_4)$: 16
$N(c_1 c_3)$: 10, 14, 15, 16	$N(c_1 c_2 c_4)$: 13, 16	
$N(c_1 c_4)$: 11, 13, 14, 16	$N(c_1 c_3 c_4)$: 14, 16	
$N(c_2 c_3)$: 6, 12, 15, 16	$N(c_2 c_3 c_4)$: 12, 16	
$N(c_2 c_4)$: 8, 12, 13, 16		
$N(c_3 c_4)$: 9, 12, 14, 16		

Finally, E_3 is made up of regions 12–15, and $E_3 = S_3 - 4S_4 = S_3 - \binom{4}{1}S_4$; E_4 consists of the elements in region 16, and $E_4 = S_4$.

These results suggest the following theorem.

THEOREM 8.2   Under the hypotheses of Theorem 8.1, for any $1 \le m \le t$, the number of elements in S that satisfy *exactly* m of the conditions $c_1, c_2, \ldots, c_t$ is given by

$$E_m = S_m - \binom{m+1}{1}S_{m+1} + \binom{m+2}{2}S_{m+2} - \cdots + (-1)^{t-m}\binom{t}{t-m}S_t. \tag{1}$$

(If $m = 0$ we obtain Theorem 8.1.)

Proof   Arguing as in Theorem 8.1, let $x \in S$ and consider the following three cases.

a) When x satisfies fewer than m conditions, it contributes a count of 0 to each of the terms $E_m, S_m, S_{m+1}, \ldots, S_t$, so it is not counted on either side of the equation.

b) When x satisfies exactly m of the conditions, it is counted once in E_m and once in S_m, but not in $S_{m+1}, \ldots, S_t$. Consequently it is included once in the count for either side of the equation.

c) Suppose x satisfies r of the conditions where $m < r \le t$. Then x contributes nothing to E_m. Yet it is counted $\binom{r}{m}$ times in S_m, $\binom{r}{m+1}$ times in $S_{m+1}, \ldots$, and $\binom{r}{r}$ times in S_r, but 0 times for any term beyond S_r. So on the right-hand side of the equation, x is counted $\binom{r}{m} - \binom{m+1}{1}\binom{r}{m+1} + \binom{m+2}{2}\binom{r}{m+2} - \cdots + (-1)^{r-m}\binom{r}{r-m}\binom{r}{r}$ times.

For $0 \le k \le r - m$,

$$\binom{m+k}{k}\binom{r}{m+k} = \frac{(m+k)!}{k!m!} \cdot \frac{r!}{(m+k)!(r-m-k)!}$$

$$= \frac{r!}{m!} \cdot \frac{1}{k!(r-m-k)!} = \frac{r!}{m!(r-m)!} \cdot \frac{(r-m)!}{k!(r-m-k)!}$$

$$= \binom{r}{m}\binom{r-m}{k}.$$

Consequently, on the right-hand side of Eq. (1), x is counted

$$\binom{r}{m}\binom{r-m}{0} - \binom{r}{m}\binom{r-m}{1} + \binom{r}{m}\binom{r-m}{2} - \cdots + (-1)^{r-m}\binom{r}{m}\binom{r-m}{r-m}$$

$$= \binom{r}{m}\left[\binom{r-m}{0} - \binom{r-m}{1} + \binom{r-m}{2} - \cdots + (-1)^{r-m}\binom{r-m}{r-m}\right]$$

$$= \binom{r}{m}[1-1]^{r-m} = \binom{r}{m} \cdot 0 = 0 \text{ times,}$$

and the formula is verified. ∎

Based on this result, if L_m denotes the number of elements of S (under the hypotheses of Theorem 8.1) that satisfy *at least* m of the t conditions, then we have the following formula.

COROLLARY 8.2 $L_m = S_m - \binom{m}{m-1}S_{m+1} + \binom{m+1}{m-1}S_{m+2} - \cdots + (-1)^{t-m}\binom{t-1}{m-1}S_t.$

Proof   A proof is outlined in the exercises at the end of this section. ∎

When $m = 1$, the result in Corollary 8.2 becomes

$$L_1 = S_1 - \binom{1}{0}S_2 + \binom{2}{0}S_3 - \cdots + (-1)^{t-1}\binom{t-1}{0}S_t$$

$$= S_1 - S_2 + S_3 - \cdots + (-1)^{t-1}S_t.$$

Comparing this with the result in Theorem 8.1, we find that

$$L_1 = N - \overline{N} = |S| - \overline{N}.$$

This is not much of a surprise, because an element x of S is counted in L_1 if it satisfies at least one of the conditions $c_1, c_2, c_3, \ldots, c_t$—that is, if $x \in S$ and x is not counted in $\overline{N} = N(\overline{c}_1 \overline{c}_2 \overline{c}_3 \ldots \overline{c}_t)$.

Example 8.7   Looking back to Example 8.6, we shall find the numbers of systems of two-way roads the engineer can devise so that, after they are built, exactly (E_2) and at least (L_2) two of the villages remain isolated.

Using the results previously calculated for this example, we have

$$E_2 = S_2 - \binom{3}{1}S_3 + \binom{4}{2}S_4 - \binom{5}{3}S_5 = 80 - 3(20) + 6(5) - 10(1) = 40,$$
$$L_2 = S_2 - \binom{2}{1}S_3 + \binom{3}{1}S_4 - \binom{4}{1}S_5 = 80 - 2(20) + 3(5) - 4(1) = 51. \quad \square$$

EXERCISES 8.2

1. For the situation in Examples 8.6 and 8.7, compute E_i for $0 \le i \le 5$ and show that $\sum_{i=0}^5 E_i = N = |S|$.

2. **a)** In how many ways can the letters in ARRANGEMENT be arranged so that there are exactly two pairs of consecutive identical letters? at least two pairs of consecutive identical letters?

   b) Answer part (a), replacing two by three.

3. In how many ways can one arrange the letters in CORRESPONDENTS so that (a) there is no pair of consecutive identical letters? (b) there are exactly two pairs of consecutive identical letters? (c) there are at least three pairs of consecutive identical letters?

4. a) Let $A = \{1, 2, 3, \ldots, 10\}$, and $B = \{1, 2, 3, \ldots, 7\}$. How many functions $f: A \to B$ satisfy $|f(A)| = 4$? How many have $|f(A)| \leq 4$?

   b) In how many ways can one distribute ten distinct prizes among four students with exactly two students getting nothing? How many ways have at least two students getting nothing?

5. a) Let $A = \{1, 2, 3, \ldots, 7\}$. A function $f: A \to A$ is said to have a *fixed point* if for some $x \in A$, $f(x) = x$. How many one-to-one functions $f: A \to A$ have at least one fixed point?

   b) In how many ways can we devise a secret code by assigning to each letter of the alphabet a different letter to represent it?

6. Zelma is having a luncheon for herself and nine of the women in her tennis league. On the morning of the luncheon she places name cards at the ten places at her table and then leaves to run a last-minute errand. Her husband, Herbert, comes home from his morning tennis match and unfortunately leaves the back door open. A gust of wind scatters the ten name cards. If Herbert replaces the ten cards at the places at the table in a random manner, in how many ways can he do so and have exactly four of the ten women sitting where Zelma wanted them? In how many ways will at least four of them be located where they are supposed to be?

7. If 13 cards are dealt from a standard deck of 52, what is the probability that these 13 cards include (a) at least one card from each suit? (b) exactly one void (for example, no clubs)? (c) exactly two voids?

8. The following provides an outline for proving Corollary 8.2. Fill in the needed details.

   a) First note that $E_t = L_t = S_t$.

   b) What is E_{t-1}, and how are L_t and L_{t-1} related?

   c) Show that $L_{t-1} = S_{t-1} - \binom{t-1}{t-2} S_t$.

   d) For any $1 \leq m \leq t - 1$, how are L_m, L_{m+1}, and E_m related?

   e) Using the results in steps (a) through (d), establish the corollary by a backward type of induction.

8.3
DERANGEMENTS: NOTHING IS IN ITS RIGHT PLACE

From elementary calculus we find that the Maclaurin series for the exponential function is given by

$$e^x = 1 + x + \frac{x^2}{2!} + \frac{x^3}{3!} + \cdots = \sum_{n=0}^{\infty} \frac{x^n}{n!},$$

so that

$$e^{-1} = \sum_{n=0}^{\infty} \frac{(-1)^n}{n!} = 1 - 1 + \frac{1}{2!} - \frac{1}{3!} + \cdots$$

To five places, $e^{-1} = 0.36788$ and $1 - 1 + (1/2!) - (1/3!) + \cdots - (1/7!) = 0.36786$. Consequently, for any $k \in \mathbf{Z}^+$, if $k \geq 7$, then e^{-1} is a very good approximation to $\sum_{n=0}^{k}((-1)^n)/n!$.

We find these ideas helpful in working the following examples.

Example 8.8   While at the racetrack, Ralph bets on each of the ten horses in a race to come in according to how they are favored. In how many ways can they reach the finish line so that he loses all of his bets?

Removing the words "horses" and "racetrack" from the problem, we really want to know in how many ways we can arrange the numbers $1, 2, 3, \ldots, 10$ so that 1 is not in first place (its natural position), 2 is not in second place (its natural position), $\ldots$, and 10 is not in tenth place (its natural position). These arrangements are called the *derangements* of $1, 2, 3, \ldots, 10$.

The principle of inclusion and exclusion provides the key. For $1 \leq i \leq 10$, an arrangement of $1, 2, 3, \ldots, 10$ is said to satisfy condition c_i if integer i is in the ith place. We obtain the number of derangements, denoted by d_{10}, as follows.

$$\begin{aligned}
d_{10} &= N(\bar{c}_1 \bar{c}_2 \bar{c}_3 \ldots \bar{c}_{10}) = 10! - \binom{10}{1}9! + \binom{10}{2}8! - \binom{10}{3}7! + \cdots + \binom{10}{10}0! \\
&= 10![1 - \binom{10}{1}(9!/10!) + \binom{10}{2}(8!/10!) - \binom{10}{3}(7!/10!) + \cdots + \binom{10}{10}(0!/10!)] \\
&= 10![1 - 1 + (1/2!) - (1/3!) + \cdots + (1/10!)] \doteq (10!)(e^{-1}).
\end{aligned}$$

The sample space here consists of the 10! ways the horses can finish. So the *probability* that Ralph will lose every bet is approximately $(10!)(e^{-1})/(10!) = e^{-1}$. This probability remains (more or less) the same if the number of horses in the race is $11, 12, \ldots$. On the other hand, for n horses, $n \geq 10$, the probability that our gambler wins at least one bet is approximately $1 - e^{-1} = 0.63212$. □

Example 8.9   At the C–H company Peggy has seven books to be reviewed, so she hires seven people to review them. She wants two reviews per book, so the first week she gives each person one book to read and then redistributes the books at the start of the second week. In how many ways can she make these two distributions so that she gets two reviews (by different people) of each book?

She can distribute the books in 7! ways the first week. Numbering both the books and the reviewers (for the first week) as $1, 2, \ldots, 7$, for the second distribution she must arrange these numbers so that none of them is in its natural position. This she can do in d_7 ways. By the rule of product, she can make the two distributions in $(7!)d_7 \doteq (7!)^2(e^{-1})$ ways. □

EXERCISES 8.3

1. In how many ways can the integers $1, 2, 3, \ldots, 10$ be arranged in a line so that no even integer is in its natural position?

2. Four applicants for a job are to be interviewed for 30 minutes each: 15 minutes with each of supervisors Nancy and Yolanda. (The interviews are in separate rooms, and interviewing starts at 9:00 A.M.) (a) In how many ways can these interviews be scheduled during a one-hour period? (b) One applicant, named Josephine, arrives at 9:00 A.M. What is the probability that she will have her two interviews one after the other? (c) Regina, another applicant, arrives at 9:00 A.M. and hopes to be finished in time to leave by 9:50 A.M. for another appointment. What is the probability that Regina will be able to do this?

3. In how many ways can Mrs. Ford distribute ten distinct books to her ten children (one book to each child) and then collect and redistribute the books so that each child has the opportunity to peruse two different books?

4. a) When n balls, numbered $1, 2, 3, \ldots, n$, are taken in succession from a container, a *rencontre* occurs if the mth ball withdrawn is numbered m, for $1 \leq m \leq n$. Find the probability of getting (i) no rencontres; (ii) exactly one rencontre; (iii) at least one rencontre; (iv) r rencontres, where $1 \leq r \leq n$.

   b) Approximate the answers to the questions in part (a).

5. Ten women attend a business luncheon. Each woman checks her coat and attaché case. Upon leaving, each woman is given a coat and case at random. (a) In how many ways can the coats and cases be distributed so that no woman gets either of her possessions? (b) In how many ways can they be distributed so that no woman gets back both of her possessions?

6. a) In how many ways can the integers $1, 2, 3, \ldots, n$ be arranged in a line so that none of the patterns $12, 23, 34, \ldots, (n-1)n$ occurs?

   b) Show that the result in part (a) equals $d_{n-1} + d_n$. (d_n = the number of derangements of $1, 2, 3, \ldots, n$.)

7. Answer part (a) of Exercise 6 if the numbers are arranged in a circle, and, as we count clockwise about the circle, none of the patterns $12, 23, 34, \ldots, (n-1)n, n1$ occurs.

8. What is the probability that the gambler in Example 8.8 wins (a) exactly five of his bets? (b) at least five of his bets?

8.4
ROOK POLYNOMIALS

Consider the six-square "chessboard" shown in Fig. 8.7. In chess there is a piece called a *rook* or *castle*, which is allowed at one turn to move horizontally or vertically over as many unoccupied spaces as it wishes. Here a rook in square 3 of the figure could move in one turn to squares 1, 2, or 4. A rook at square 5 could move to square 6 or square 2 (even though there is no square between squares 5 and 2).

For $k \in \mathbf{Z}^{+}$ we want to determine the number of ways in which k rooks can be placed on this chessboard so that no two of them can take each other—that is, no two of them are in the same row or column of the chessboard. This number is denoted by r_k, or by $r_k(C)$ if we wish to stress that we are working on chessboard C.

For any chessboard, r_1 is the number of squares on the board. Here $r_1 = 6$. Two nontaking rooks can be placed at the following pairs of positions: $(1,4)$, $(1,5)$, $(2,4)$, $(2,6)$, $(3,5)$, $(3,6)$, $(4,5)$, and $(4,6)$, so $r_2 = 8$. Continuing, we find that $r_3 = 2$, using the locations $(1,4,5)$ and $(2,4,6)$; $r_k = 0$, for $k \geq 4$.

With $r_0 = 1$, the *rook polynomial*, $r(C,x)$, for the chessboard in Fig. 8.7 is defined as $r(C,x) = 1 + 6x + 8x^2 + 2x^3$. For any $k \geq 0$, the coefficient of x^k is the number of ways we can place k nontaking rooks on chessboard C.

Figure 8.7

What we have done here (using a case-by-case analysis) soon proves tedious. As the size of the board increases, we have to consider cases wherein numbers such as r_4 and r_5 are nonzero. Consequently, we shall now make some observations that will allow us to make use of small boards and somehow break up a large board into smaller *subboards*.

The chessboard C in Fig. 8.8 is made up of 11 unshaded squares. We note that C consists of a 2×2 subboard C_1 and a seven-square subboard C_2 located in the lower right corner. These subboards are *disjoint* because they have no squares in the same row or column of C.

Figure 8.8

Calculating as we did for our first chessboard, here we find

$$r(C_1,x) = 1 + 4x + 2x^2 \qquad r(C_2,x) = 1 + 7x + 10x^2 + 2x^3$$

$$r(C,x) = 1 + 11x + 40x^2 + 56x^3 + 28x^4 + 4x^5 = r(C_1,x) \cdot r(C_2,x).$$

Hence $r(C,x) = r(C_1,x) \cdot r(C_2,x)$. But did this occur by luck or is something happening here that we should examine more closely? For example, to obtain r_3

for C, we need to know in how many ways three nontaking rooks can be placed on board C. These fall into three cases:

a) All three rooks are on subboard C_2: 2 ways.

b) Two rooks are on subboard C_2 and one is on C_1: $(10)(4) = 40$ ways.

c) One rook is on subboard C_2 and two are on C_1: $(7)(2) = 14$ ways.

Consequently, three nontaking rooks can be placed on board C in $2 + (10)(4) + (7)(2) = 56$ ways. Here we see that 56 arises just as it does as the coefficient of x^3 in the product $r(C_1, x) \cdot r(C_2, x)$.

> In general, if C is a chessboard made up of *pairwise* disjoint subboards $C_1, C_2, \ldots, C_n$, then $r(C, x) = r(C_1, x) r(C_2, x) \cdots r(C_n, x)$.

The last result for this section demonstrates the type of principle we have seen in other results in combinatorial and discrete mathematics: Given a large chessboard, break it into smaller subboards whose rook polynomials can be determined by inspection.

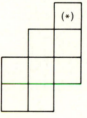

Figure 8.9

Consider chessboard C in Fig. 8.9. Let $k \geq 1$. For any square of C, such as the one designated by $(*)$, there are two possibilities to examine.

a) Place one rook on the designated square. Then we must remove, as possible locations for all of the other $k - 1$ rooks, all other squares of C in the same row or column as the designated square. We use C_s to denote the remaining smaller subboard.

b) We do not use the designated square at all. The k rooks are placed on the subboard C_e (C with the one designated square eliminated).

Since these two cases are all-inclusive and disjoint,

$$r_k(C) = r_{k-1}(C_s) + r_k(C_e).$$

From this we see that

$$r_k(C)x^k = r_{k-1}(C_s)x^k + r_k(C_e)x^k. \tag{1}$$

If $n =$ the number of squares in the chessboard (here $n = 8$), then Eq. (1) is valid for $1 \leq k \leq n$, and we write

$$\sum_{k=1}^{n} r_k(C)x^k = \sum_{k=1}^{n} r_{k-1}(C_s)x^k + \sum_{k=1}^{n} r_k(C_e)x^k. \tag{2}$$

For Eq. (2) we realize that the summations may stop before $k = n$. We have seen cases, as in Fig. 8.7, where r_n and some prior r_k's are 0. The summations start at $k = 1$, for otherwise we could find ourselves with the term $r_{-1}(C_s)x^0$ in the first summand on the right-hand side of Eq. (2).

Equation (2) may be rewritten as

$$\sum_{k=1}^{n} r_k(C)x^k = x \sum_{k=1}^{n} r_{k-1}(C_s)x^{k-1} + \sum_{k=1}^{n} r_k(C_e)x^k \tag{3}$$

or

$$1 + \sum_{k=1}^{n} r_k(C)x^k = x \cdot r(C_s, x) + \sum_{k=1}^{n} r_k(C_e)x^k + 1$$

from which it follows that

$$r(C, x) = x \cdot r(C_s, x) + r(C_e, x). \tag{4}$$

We now use this final equation to determine the rook polynomial for the chessboard shown in Fig. 8.9. Each time the idea in Eq. (4) is used, we mark the special square we are using with $(*)$. Parentheses are placed about each chessboard to denote the rook polynomial of the board.

$$= x^2(1 + 2x) + 2x(1 + 4x + 2x^2) + x(1 + 3x + x^2)$$

$$3x + 12x^2 + 7x^3 + x(1 + 2x) + (1 + 4x + 2x^2) = 1 + 8x + 16x^2 + 7x^3.$$

8.5
ARRANGEMENTS WITH FORBIDDEN POSITIONS

The rook polynomials of the previous section seem interesting on their own. Now we shall find them useful in solving the following problems.

Example 8.10 In making seating arrangements for their son's wedding reception, Grace and Nick are down to four relatives, denoted R_i, $1 \le i \le 4$, who do not get along with one another. There is a single open seat at each of the five tables T_j, $1 \le j \le 5$. Because of family differences,

a) R_1 will not sit at T_1 or T_2.   **b)** R_2 will not sit at T_2.

c) R_3 will not sit at T_3 or T_4.   **d)** R_4 will not sit at T_4 or T_5.

This situation is represented in Fig. 8.10. The number of ways we can seat these four people, and satisfy conditions (a) through (d), is the number of ways four nontaking rooks can be placed on the chessboard made up of the *unshaded* squares. However, since there are only seven shaded squares, as opposed to thirteen unshaded ones, it would be easier to work with the shaded chessboard.

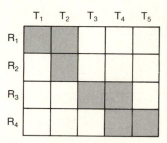

Figure 8.10

We start with the conditions that are required for us to apply inclusion and exclusion: For $1 \le i \le 4$, let c_i be the condition where a seating assignment of these four people is made with relative R_i in a forbidden (shaded) position. As usual, $|S|$ denotes the total number of ways we can place the four relatives, one to a table. Then $|S| = N = S_0 = 5!$.

To determine S_1 we consider each of the following:

- $N(c_1) = 4! + 4!$, for there are 4! ways to seat R_2, R_3, and R_4 if R_1 is in forbidden position T_1 and another 4! ways if R_1 is at table T_2, his or her other forbidden position.

- $N(c_2) = 4!$, for after placing R_2 at forbidden table T_2, we must place R_1, R_3, and R_4 at T_1, T_3, T_4, and T_5, one person to a table.

- $N(c_3) = 4! + 4!$, one summand for R_3 being in forbidden position T_3, and the other summand for R_3 being in the forbidden position T_4.

- $N(c_4) = 4! + 4!$, the two summands arising for R_4 being placed at each of the two forbidden positions T_4 and T_5.

Hence $S_1 = 7(4!)$.

Turning to S_2 we have these considerations:

- $N(c_1 c_2) = 3!$, because after we place R_1 at T_1 and R_2 at T_2, there are three tables (T_3, T_4, and T_5) where R_3 and R_4 can be seated.
- $N(c_1 c_3) = 3! + 3! + 3! + 3!$, because there are four cases where R_1 and R_3 are located at forbidden positions:

  (i) R_1 at T_1; R_3 at T_3 (ii) R_1 at T_2; R_3 at T_3

  (iii) R_1 at T_1; R_3 at T_4 (iv) R_1 at T_2; R_3 at T_4.

In a similar manner we find that $N(c_1 c_4) = 4(3!)$, $N(c_2 c_3) = 2(3!)$, $N(c_2 c_4) = 2(3!)$, and $N(c_3 c_4) = 3(3!)$. Consequently, $S_2 = 16(3!)$.

Before continuing, we make a few observations about S_1 and S_2. For S_1 we have $7(4!) = 7(5-1)!$, where 7 is the number of shaded squares in Fig. 8.10. Also, $S_2 = 16(3!) = 16(5-2)!$, where 16 is the number of ways two nontaking rooks can be placed on the shaded chessboard.

In general, for $0 \le i \le 4$, $S_i = r_i(5-i)!$, where r_i is the number of ways in which it is possible to place i nontaking rooks on the shaded chessboard shown in Fig. 8.10.

Consequently, to expedite the solution of this problem, we turn to $r(C,x)$, the rook polynomial of the chessboard. Using the decomposition of C into the disjoint subboards in the upper left and lower right corners, we find that

$$r(C,x) = (1 + 3x + x^2)(1 + 4x + 3x^2) = 1 + 7x + 16x^2 + 13x^3 + 3x^4,$$

so

$$N(\overline{c}_1 \overline{c}_2 \overline{c}_3 \overline{c}_4) = S_0 - S_1 + S_2 - S_3 + S_4 = 5! - 7(4!) + 16(3!) - 13(2!) + 3(1!)$$
$$= \sum_{i=0}^{4} (-1)^i r_i(5-i)! = 25.$$

Grace and Nick can breathe a sigh of relief. There are 25 ways in which they can seat these last four relatives at the reception and avoid any squabbling. □

Our last example demonstrates how a bit of rearranging of our chessboard can help in our calculations.

Example 8.11 We have a pair of dice; one is red, the other green. We rolled these dice six times. What is the probability that we obtained all six values on both the red die and the green die if we know that the ordered pairs $(1, 2)$, $(2, 1)$, $(2, 5)$, $(3, 4)$, $(4, 1)$, $(4, 5)$ and $(6, 6)$ did not occur? (Here an ordered pair (x, y) indicates x on red and y on green.)

Recognizing this as another problem dealing with permutations and forbidden positions, we construct the chessboard shown in Fig. 8.11(a), where the shaded squares constitute the forbidden positions. In this figure the shaded squares are scattered. Relabeling the rows and columns, we can redraw the chessboard as shown in Fig. 8.11(b), where we have taken shaded squares in the same row (or column) of the board shown in part (a) and made them adjacent. In Fig. 8.11(b), chessboard C is the union of four mutually disjoint subboards, and

$$r(C,x) = (1 + 4x + 2x^2)(1 + x)^3 = 1 + 7x + 17x^2 + 19x^3 + 10x^4 + 2x^5.$$

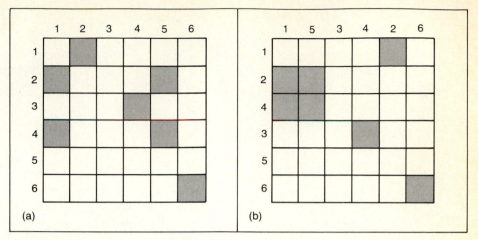

Figure 8.11

For $1 \leq i \leq 6$, define c_i as the condition where, having rolled the dice six times, we find that all six values occur on both the red die and the green die, but i on the red die is paired with one of the forbidden numbers on the green die. (Note that $N(c_5) = 0$.) Then the number of sequences of the six rolls of the dice for the event we are interested in is

$$(6!)N(\bar{c}_1 \bar{c}_2 \bar{c}_3 \bar{c}_4 \bar{c}_5 \bar{c}_6) = (6!) \sum_{i=0}^{6} (-1)^i S_i = (6!) \sum_{i=0}^{6} (-1)^i r_i \cdot (6 - i)!$$

$$= 6![6! - 7(5!) + 17(4!) - 19(3!) + 10(2!) - 2(1!) + 0(0!)]$$

$$= 6![192] = 138,240.$$

Since the sample space consists of all sequences of six ordered pairs selected with repetition from the 29 unshaded squares of the chessboard, the probability of this event is $138,240/(29)^6 \doteq 0.00023$. $\square$

EXERCISES 8.4 AND 8.5

1. Verify directly the rook polynomials for (a) the unshaded chessboards in Figs. 8.8 and 8.9, and (b) the shaded chessboards in Figs. 8.10 and 8.11(b).

2. Construct or describe a smallest (least number of squares) chessboard for which $r_{10} \neq 0$.

3. a) Find the rook polynomial for the standard 8×8 chessboard.

   b) Answer (a) with 8 replaced by n, $n \in \mathbf{Z}^+$.

4. Find the rook polynomials for the shaded chessboards in Fig. 8.12.

5. a) Find the rook polynomials for the shaded chessboards in Fig. 8.13.

   b) Generalize the chessboard (and rook polynomial) for Fig. 8.13(i).

6. Let C be a chessboard that has m rows and n columns, with $m \leq n$ (for a total

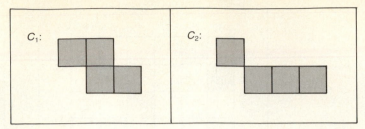

Figure 8.12

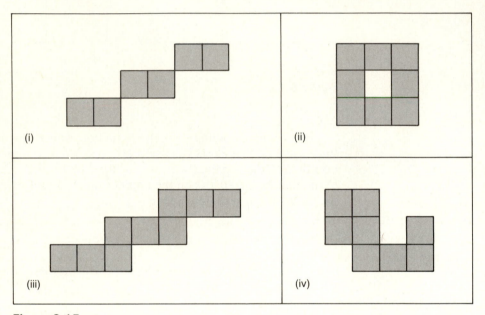

Figure 8.13

of mn squares). For $0 \leq k \leq m$, in how many ways can we arrange k (identical) nontaking rooks on C?

7. Professor Ruth has five graders to correct programs in her courses in APL, BASIC, FORTRAN, Pascal, and PL/I. Graders Jeanne and Charles both dislike FORTRAN. Sandra wants to avoid BASIC and PL/I. Paul detests APL and BASIC, and Todd refuses to work in FORTRAN and Pascal. In how many ways can Professor Ruth assign each grader to correct programs in one language, cover all five languages, and keep everybody content?

8. Why do we have 6! in the term $(6!)N(\bar{c}_1 \bar{c}_2 \ldots \bar{c}_6)$ for the solution of Example 8.11?

9. Five professors named Al, Violet, Gary, Jack, and Mary Lou are to be assigned to teach one class each from among calculus I, calculus II, calculus III, statistics, and combinatorics. Al will not teach calculus II or combinatorics, Gary cannot stand statistics, Violet and Mary Lou both refuse to teach calculus I or calculus III, and Jack detests calculus II.

a) In how many ways can the head of the mathematics department assign each of these professors one of these five courses and still keep peace in the department?

b) For the assignments in part (a), what is the probability that Violet will get to teach combinatorics?

c) For the assignments in (a), what is the probability that Gary or Mary Lou will get to teach calculus II?

10. A pair of dice, one red and the other green, is rolled six times. We know that the ordered pairs $(1, 1)$, $(1, 5)$, $(2, 4)$, $(3, 6)$, $(4, 2)$, $(4, 4)$, $(5, 1)$, and $(5, 5)$ did not come up. What is the probability that every value came up on both the red die and the green one?

11. A computer dating service wants to match each of four women with one of six men. According to the information these applicants provided when they joined the service, we can draw the following conclusions.

- Woman 1 would not be compatible with man 1, 3, or 6.
- Woman 2 would not be compatible with man 2 or 4.
- Woman 3 would not be compatible with man 3 or 6.
- Woman 4 would not be compatible with man 4 or 5.

In how many ways can the service successfully match each of the four women with a compatible partner?

8.6
SUMMARY AND HISTORICAL REVIEW

In the first and third chapters of this text we were concerned with enumeration problems in which we had to be careful of situations wherein arrangements or selections were overcounted. This became even more involved in Chapter 5 when we tried to count the number of onto functions between two finite sets.

With Venn diagrams to lead the way, in this chapter we obtained a pattern called the Principle of Inclusion and Exclusion. Using this principle, we restated each problem in terms of conditions and subsets. Using enumeration formulas on permutations and combinations that were developed earlier, we solved some simpler subproblems and let the principle manage our concern about over-counting. As a result, we were able to solve a variety of problems, some dealing with number theory and one with graph theory. We also proved the formula conjectured earlier in Section 5.3 for the number of onto functions between two finite sets.

This principle has an interesting history, being found in different manuscripts under such names as the "Sieve Method" or the "Principle of Cross Classification." A set-theoretic version of the principle, which concerned itself with set unions and intersections, is found in *Doctrine of Chances* (1718), a text on probability theory by Abraham DeMoivre (1667–1754). Somewhat earlier, in 1713,

Pierre Remond de Montmort (1678–1719) used the idea behind the principle in his solution of the problem generally known as *le probleme des rencontres* (derangements).

Credit for the way we developed and dealt with the principle belongs to James Joseph Sylvester (1814–1897). The importance of the technique was not generally appreciated, however, until somewhat later, when the publication by W. A. Whitworth [10] made mathematicians more aware of its potential and use.

For more on the use of this principle, examine Chapter 4 of C. L. Liu [4], Chapter 2 of H. J. Ryser [8], or Chapter 5 of A. Tucker [9]. More number-theoretic results related to the principle, including the Möbius inversion formula, can be found in Chapter 2 of M. Hall [1], Chapter X of C. L. Liu [5], and Chapter 16 of G. Hardy and E. Wright [3]. An extension of this formula is given in the article by G. C. Rota [7].

The article by D. Hanson, K. Seyffarth, and J. H. Weston [2] provides an interesting generalization of the derangement problem discussed in Section 8.3. The ideas behind the rook polynomials and their applications were developed in the late 1930s and during the 1940s and 1950s. Additional material on this topic is found in Chapters 7 and 8 of J. Riordan [6].

REFERENCES

1. Hall, Marshall, Jr. *Combinatorial Theory*. Waltham, Mass.: Blaisdell, 1967.
2. Hanson, Denis, Seyffarth, Karen, and Weston, J. Harley. "Matchings, Derangements, Rencontres," *Mathematics Magazine* 56, no. 4 (September 1983): pp. 224–229.
3. Hardy, Godfrey Harold, and Wright, Edward Maitland. *An Introduction to the Theory of Numbers,* 4th ed. Oxford, England: Clarendon Press, 1960.
4. Liu, C. L. *Introduction to Combinatorial Mathematics*. New York: McGraw-Hill, 1968
5. Liu, C. L. *Topics in Combinatorial Mathematics*. Mathematical Association of America, 1972.
6. Riordan, John. *An Introduction to Combinatorial Analysis*. Princeton, N.J.: Princeton University Press, 1980. (Originally published in 1958 by John Wiley & Sons.)
7. Rota, Gian Carlo. "On the Foundations of Combinatorial Theory, I. Theory of Möbius Functions," *Zeitschrift für Wahrscheinlichkeits Theorie, 2:* pp. 340–368, 1964.
8. Ryser, Herbert J. *Combinatorial Mathematics*. Carus Mathematical Monograph, No. 14. Published by the Mathematical Association of America, distributed by John Wiley & Sons, New York, 1963.
9. Tucker, Alan. *Applied Combinatorics*. New York: Wiley, 1980.
10. Whitworth, William Allen. *Choice and Chance*. Originally published at Cambridge in 1867. Reprint of the 5th ed. (1901), Hafner, New York, 1965.

MISCELLANEOUS EXERCISES

1. Determine how many $n \in \mathbf{Z}^+$ satisfy $n \leq 500$ and are not divisible by 2, 3, 5, 6, 8, or 10.

2. Determine all $x \in \mathbf{R}$ such that $\lfloor x \rfloor + \lfloor x + \frac{1}{2} \rfloor = \lfloor 2x \rfloor$.

3. Find three values for $n \in \mathbf{Z}^+$ where $\phi(n) = 16$.

4. Find the number of $n \in \mathbf{Z}^+$ such that $n \le 1000$ and n is not a perfect square, cube, or fourth power.

5. In how many ways can we arrange the integers $1, 2, 3, \ldots, 8$ in a line so that there are no occurrences of the patterns $12, 23, \ldots, 78, 81$?

6. a) If we have k different colors available, in how many ways can we paint the walls of a pentagonal room if adjacent walls are to be painted with different colors?

   b) What is the smallest value of k for which such a coloring is possible?

   c) Answer parts (a) and (b) for a hexagonal room.

7. In how many ways can five couples be seated around a circular table so that no couple is seated next to each other? (Here, as in Example 1.16, we do not distinguish between two arrangements where the first can be obtained from the second by rotating the locations of the ten people.)

8. Using the result of Theorem 8.2, prove that the number of ways we can place s different objects in n distinct containers with m containers each containing exactly r of the objects is

$$\frac{(-1)^m n! \, s!}{m!} \sum_{i=m}^{n} \frac{(-1)^i (n-i)^{s-ir}}{(i-m)!(n-i)!(s-ir)!(r!)^i}$$

9. If an arrangement of the letters in SURREPTITIOUS is selected at random, what is the probability that it contains (a) exactly three pairs of consecutive identical letters? (b) at most three pairs of consecutive identical letters?

10. Consider the list $x_1, x_1, x_2, x_2, \ldots, x_n, x_n$, consisting of n pairs of distinct symbols. In how many ways can one arrange these $2n$ symbols so that there is no pair of consecutive identical symbols?

11. In how many ways can four w's, four x's, four y's, and four z's be arranged so that there is no consecutive quadruple of the same letter?

12. Consider a list of three x's, three y's, and three z's.

   a) Show that these nine letters can be arranged in $(8!/3!3!) - (7!/3!3!)$ ways with a pair or triple of adjacent x's.

   b) Show that we can arrange these letters in $(7!/3!) - 2(6!/3!) + (5!/3!)$ ways with a pair or triple of adjacent x's and adjacent y's.

   c) Arranging these nine letters, show that there are $6! - 3(5!) + 3(4!) - 3!$ arrangements with a pair or triple of every letter occurring.

   d) In how many ways can we arrange these nine letters so that no consecutive letters are the same?

   e) Why can't we solve part (d) by considering the two ways in which we can arrange the three x's and the three y's with no consecutive letters the same, and then determine how to place the three z's so that no two are consecutive?

13. a) Given n distinct objects, in how many ways can we select r of these objects so that each selection includes some particular m of the n objects? (Here $m \le r \le n$.)

   b) Using the principle of inclusion and exclusion, prove that for $m \le r \le n$,

$$\binom{n-m}{n-r} = \sum_{i=0}^{m} (-1)^i \binom{m}{i} \binom{n-i}{r}.$$

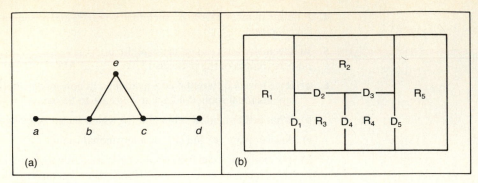

Figure 8.14

14. **a)** Let $\lambda \in \mathbf{Z}^+$. If we have λ different colors available, in how many ways can we color the vertices of the graph shown in Fig. 8.14(a) so that no adjacent vertices are the same color? This result in λ is called the *chromatic polynomial* of the graph, and the smallest value of λ for which the value of this polynomial is positive is called the *chromatic number* of the graph. What is the chromatic number of this graph? (We shall pursue this idea further in Chapter 11.)

b) If there are six colors available, in how many ways can the rooms R_i, $1 \le i \le 5$, shown in Fig. 8.14(b) be painted so that rooms with a common doorway D_j, $1 \le j \le 5$, are painted with different colors?

CHAPTER 9

Generating Functions

In this chapter and the next, we continue our study of enumeration, introducing at this time the important concept of the *generating function*.

The problem of making selections, with repetitions allowed, was studied in Chapter 1. There we sought, for example, the number of integer solutions to the equation $c_1 + c_2 + c_3 + c_4 = 25$ where $c_i \geq 0$ for $1 \leq i \leq 4$. With the principle of inclusion and exclusion, in Chapter 8 we were able to solve a more restricted version of the problem, such as $c_1 + c_2 + c_3 + c_4 = 25$ with $0 \leq c_i < 10$ for $1 \leq i \leq 4$. If, in addition, we wanted c_2 to be even and c_3 to be a multiple of 3, we could apply the results of Chapters 1 and 8 to several subcases.

The power of the generating function rests upon its ability not only to solve the kinds of problems we have considered so far but also to aid us in new situations where additional restrictions may be involved.

9.1
INTRODUCTORY EXAMPLES

Instead of defining a generating function at this point, we shall examine some examples that motivate the idea. We may find that we have dealt with this concept in a previous situation.

Example 9.1   While shopping one Saturday, Mildred buys 12 oranges for her children, Grace, Mary, and Frank. In how many ways can she distribute the oranges so that Grace gets at least four, and Mary and Frank get at least two, but Frank gets no more than five? Table 9.1 lists all the possible distributions. We see that we have all the integer solutions to the equation $c_1 + c_2 + c_3 = 12$ where $4 \leq c_1$, $2 \leq c_2$ and $2 \leq c_3 \leq 5$.

Table 9.1

G	M	F	G	M	F
4	3	5	6	2	4
4	4	4	6	3	3
4	5	3	6	4	2
4	6	2	7	2	3
5	2	5	7	3	2
5	3	4	8	2	2
5	4	3			
5	5	2			

Considering the first two cases in this table, we find the solutions $4 + 3 + 5 = 12$ and $4 + 4 + 4 = 12$. Now where in our prior algebraic experiences did anything like this happen? When multiplying polynomials we add the powers of the variable, and here, when we multiply the three polynomials,

$$(x^4 + x^5 + x^6 + x^7 + x^8)(x^2 + x^3 + x^4 + x^5 + x^6)(x^2 + x^3 + x^4 + x^5),$$

two of the ways to obtain x^{12} are as follows:

1. From the product $x^4 x^3 x^5$, where x^4 is taken from $(x^4 + x^5 + \cdots + x^8)$, x^3 from $(x^2 + x^3 + \cdots + x^6)$, and x^5 from $(x^2 + x^3 + \cdots + x^5)$.

2. From the product $x^4 x^4 x^4$, where the first x^4 is found in the first polynomial, the second x^4 in the second polynomial, and the third x^4 in the third polynomial.

Examining the product

$$(x^4 + x^5 + x^6 + x^7 + x^8)(x^2 + x^3 + x^4 + x^5 + x^6)(x^2 + x^3 + x^4 + x^5)$$

more closely, we realize that we obtain the product $x^i x^j x^k$ for every triple (i, j, k) that appears in Table 9.1. Consequently, the coefficient of x^{12} in

$$f(x) = (x^4 + x^5 + \cdots + x^8)(x^2 + x^3 + \cdots + x^6)(x^2 + x^3 + \cdots + x^5)$$

counts the number of distributions, namely 14, that we seek. The function $f(x)$ is called a *generating function* for the distributions.

But where did the factors in this product come from?

The factor $x^4 + x^5 + x^6 + x^7 + x^8$, for example, indicates that we can give Grace 4 *or* 5 *or* 6 *or* 7 *or* 8 of the oranges. Once again we make use of the interplay between the exclusive *or* and ordinary addition. The coefficient of each power of x is 1 because, considering the oranges as identical objects, there is only one way to give Grace four oranges, one way to give her five oranges, etc. Since each of Mary and Frank must receive at least two oranges, the other terms $(x^2 + x^3 + \cdots + x^6)$ and $(x^2 + x^3 + \cdots + x^5)$ start with x^2, and for Frank we stop at x^5 so that he doesn't receive more than five oranges. (Why does the term for Mary stop at x^6?)

Most of us are reasonably convinced now that the coefficient of x^{12} in $f(x)$ yields the answer. Some, however, may be a bit skeptical about this new idea. It seems that we could list the cases in Table 9.1 faster than we could multiply out

$f(x)$ or calculate the coefficient of x^{12} in $f(x)$. At present that may seem true. But as we progress to problems with more unknowns and larger quantities to distribute, the generating function will more than demonstrate its worth. (The reader may realize that the rook polynomials of Chapter 8 are examples of generating functions.) For now we consider two more examples. □

Example 9.2   If there is an unlimited number (or at least 24 of each color) of red, green, white, and black jelly beans, in how many ways can a child select 24 of these candies so that he has an even number of white beans and at least six black ones?

The polynomials associated with the jelly bean colors are as follows:

- red (green): $1 + x + x^2 + \cdots + x^{24}$, where the leading 1 is for $1x^0$, because one possibility for the red (and green) jelly beans is that none of that color is selected
- white: $(1 + x^2 + x^4 + x^6 + \cdots + x^{24})$
- black: $(x^6 + x^7 + x^8 + \cdots + x^{24})$

So the answer to the problem is the coefficient of x^{24} in the generating function

$$f(x) = (1 + x + x^2 + \cdots + x^{24})^2(1 + x^2 + x^4 + \cdots + x^{24})\cdot$$
$$(x^6 + x^7 + \cdots + x^{24}).$$

One such selection is five red, three green, eight white, and eight black jelly beans. This arises from x^5 in the first factor, x^3 in the second factor, and x^8 in the last two factors. □

One more example before closing this section!

Example 9.3   How many integer solutions are there for the equation $c_1 + c_2 + c_3 + c_4 = 25$ if $0 \le c_i$ for $1 \le i \le 4$?

We can alternatively ask in how many ways 25 (identical) pennies can be distributed among four children.

For each child the possibilities can be described by the polynomial $1 + x + x^2 + x^3 + \cdots + x^{25}$. Then the answer to this problem is the coefficient of x^{25} in the generating function

$$f(x) = (1 + x + x^2 + \cdots + x^{25})^4.$$

The answer can also be obtained as the coefficient of x^{25} in the generating function

$$g(x) = (1 + x + x^2 + x^3 + \cdots + x^{25} + x^{26} + \cdots)^4,$$

if we rephrase the question in terms of distributing, from a large (or unlimited) number of pennies, 25 pennies among four children. (Where $f(x)$ is a polynomial, $g(x)$ is a *power series* in x.) Note that the terms x^k, $k \ge 26$, are never used. So why bother with them? Because there will be times when it is easier to compute with a power series than with a polynomial. □

EXERCISES 9.1

1. For each of the following, determine a generating function and indicate the coefficient in the function that is needed to solve the problem. (Give both the polynomial and power series forms of the generating function, wherever appropriate.)

    Find the number of integer solutions for the following equations:

    a) $c_1 + c_2 + c_3 + c_4 = 20$, $0 \le c_i \le 7$, $1 \le i \le 4$

    b) $c_1 + c_2 + c_3 + c_4 = 20$, $0 \le c_i$, $1 \le i \le 4$, c_2 and c_3 even

    c) $c_1 + c_2 + c_3 + c_4 + c_5 = 30$, $2 \le c_1 \le 4$ and $3 \le c_i \le 8$, $2 \le i \le 5$

    d) $c_1 + c_2 + c_3 + c_4 + c_5 = 30$, $0 \le c_i$ for $1 \le i \le 5$, c_2 even and c_3 odd

2. Determine the generating function for the number of ways to distribute 35 pennies (from an unlimited supply) among five children if (a) there are no restrictions; (b) each child gets at least 1¢; (c) each child gets at least 2¢; (d) the oldest child gets at least 10¢; (e) the two youngest children must each get at least 10¢.

3. a) Find the generating function for the number of ways to select 10 candy bars from large supplies of six different kinds.

    b) Find the generating function for the number of ways to select, with repetitions allowed, r objects from a collection of n distinct objects.

4. a) Explain why the generating function for the number of ways to have n cents in pennies and nickels is $(1 + x + x^2 + x^3 + \cdots)(1 + x^5 + x^{10} + \cdots)$.

    b) Find the generating function for the number of ways to have n cents in pennies, nickels, and dimes.

5. Find the generating function for the number of integer solutions to the equation $c_1 + c_2 + c_3 + c_4 = 20$ where $-3 \le c_1$, $-3 \le c_2$, $-5 \le c_3 \le 5$, and $0 \le c_4$.

6. For $S = \{a, b, c\}$, consider the function $f(x) = (1 + ax)(1 + bx)(1 + cx) = 1 + ax + bx + cx + abx^2 + acx^2 + bcx^2 + abcx^3$. Here, in $f(x)$

    - The coefficient of x^0 is 1—for the subset $\emptyset$ of S.
    - The coefficient of x^1 is $a + b + c$—for the subsets $\{a\}$, $\{b\}$, and $\{c\}$ of S.
    - The coefficient of x^2 is $ab + ac + bc$—for the subsets $\{a, b\}$, $\{a, c\}$, and $\{b, c\}$ of S.
    - The coefficient of x^3 is abc—for the subset $\{a, b, c\} = S$.

    Consequently, $f(x)$ is the generating function for the subsets of S. For when we calculate $f(1)$, we obtain a sum wherein each of the eight summands corresponds with a subset of S; the summand 1 corresponds with $\emptyset$. (If we go one step further and set $a = b = c = 1$ in $f(x)$, then $f(1) = 8$, the number of subsets of S.)

    a) Give the generating function for the subsets of $S = \{a, b, c, \ldots, r, s, t\}$.

    b) Answer part (a) for selections wherein each element can be rejected, or selected as many as three times.

9.2
DEFINITION AND EXAMPLES: CALCULATIONAL TECHNIQUES

In this section we shall examine a number of formulas and examples dealing with power series. These will be used to obtain the coefficients of particular terms in a generating function.

We start with the following concept.

DEFINITION 9.1 Let $a_0, a_1, a_2, \ldots$ be a sequence of real numbers. The function

$$f(x) = a_0 + a_1 x + a_2 x^2 + \cdots = \sum_{i=0}^{\infty} a_i x^i$$

is called the *generating function* for the given sequence.

Where could this idea have come from?

Example 9.4 For any $n \in \mathbf{Z}^+$,

$$(1 + x)^n = \binom{n}{0} + \binom{n}{1}x + \binom{n}{2}x^2 + \cdots + \binom{n}{n}x^n,$$

so $(1 + x)^n$ is the generating function for the sequence

$$\binom{n}{0}, \binom{n}{1}, \binom{n}{2}, \ldots, \binom{n}{n}, 0, 0, 0, \ldots. \quad \square$$

Example 9.5 **a)** For $n \in \mathbf{Z}^+$, $(1 - x^{n+1}) = (1 - x)(1 + x + x^2 + x^3 + \cdots + x^n)$.
So $(1 - x^{n+1})/(1 - x) = (1 + x + x^2 + \cdots + x^n)$, and $(1 - x^{n+1})/(1 - x)$ is the generating function for the sequence $1, 1, 1, \ldots, 1, 0, 0, 0, \ldots$, where the first $n + 1$ terms are 1.

b) Extending the idea in part (a), we find that

$$1 = (1 - x)(1 + x + x^2 + x^3 + x^4 + \cdots),$$

so $1/(1 - x)$ is the generating function for the sequence 1, 1, 1, 1, ... $[1/(1 - x) = 1 + x + x^2 + x^3 + \cdots$ is valid for all real x where $|x| < 1$; it is for this range of values that the *geometric series* $1 + x + x^2 + \cdots$ *converges*. However, in our work with generating functions we shall be more concerned with the coefficients of the powers of x than with matters of convergence. This is not to say that the concept of convergence is not important. It is just not needed for the material we shall study in this chapter.]

c) With $1/(1 - x) = 1 + x + x^2 + x^3 + \cdots = \sum_{i=0}^{\infty} x^i$, taking the derivative of each side yields

$$\frac{d}{dx}\frac{1}{1 - x} = \frac{d}{dx}(1 - x)^{-1} = (-1)(1 - x)^{-2}(-1)$$

$$= \frac{d}{dx}(1 + x + x^2 + x^3 + \cdots) = 1 + 2x + 3x^2 + 4x^3 + \cdots$$

Consequently, $1/(1-x)^2$ is the generating function for the sequence $1, 2, 3, 4, \ldots$, while $x/(1-x)^2$ generates the sequence $0, 1, 2, 3, \ldots$.

d) Continuing from (c), $(d/dx)[x/(1-x)^2] = (d/dx)[0 + x + 2x^2 + \cdots]$, or $(x+1)/(1-x)^3 = 1 + 2^2 x + 3^2 x^2 + 4^2 x^3 + \cdots$. Hence, $(x+1)/(1-x)^3$ generates $1^2, 2^2, 3^2, \ldots$, and $x(x+1)/(1-x)^3$ generates $0^2, 1^2, 2^2, 3^2, \ldots$. ☐

Example 9.6

a) From part (b) of Example 9.5 we know that the generating function for the sequence $1, 1, 1, 1, \ldots$ is $f(x) = 1/(1-x)$. Therefore the function

$$g(x) = f(x) - x^2 = 1/(1-x) - x^2$$

is the generating function for the sequence

$$1, 1, 0, 1, 1, 1, \ldots,$$

while the function

$$h(x) = f(x) + 2x^3 = 1/(1-x) + 2x^3$$

generates the sequence

$$1, 1, 1, 3, 1, 1, \ldots.$$

b) Can we use the results of Example 9.5 to find a generating function for the sequence $0, 2, 6, 12, 20, 30, 42, \ldots$?

Here we observe that

$$a_0 = 0 = 0^2 + 0,$$
$$a_1 = 2 = 1^2 + 1,$$
$$a_2 = 6 = 2^2 + 2,$$
$$a_3 = 12 = 3^2 + 3,$$
$$a_4 = 20 = 4^2 + 4, \ldots.$$

In general, we have $a_n = n^2 + n$, for $n \geq 0$.

Using the results from parts (c) and (d) of Example 9.5, we now find that

$$\frac{x(x+1)}{(1-x)^3} + \frac{x}{(1-x)^2} = \frac{x(x+1) + x(1-x)}{(1-x)^3} = \frac{2x}{(1-x)^3}$$

is the generating function for the given sequence. (The solution here depends on our ability to recognize each a_n as the sum of n^2 and n. If we do not see this, we may be unable to answer the given question. Consequently, in Example 10.5 of the next chapter, we shall examine another technique to help us recognize the formula for a_n.) ☐

For any $n \in \mathbf{Z}^+$, the binomial theorem tells us that $(1+x)^n = \binom{n}{0} + \binom{n}{1}x + \binom{n}{2}x^2 + \cdots + \binom{n}{n}x^n$. We want to extend this idea to cases where (a) $n < 0$ and (b) n is not necessarily an integer.

With $n, r \in \mathbf{Z}^+$ and $n \geq r > 0$, we have

$$\binom{n}{r} = \frac{n!}{[r!(n-r)!]} = \frac{[n(n-1)(n-2)\cdots(n-r+1)]}{r!}$$

If $n \notin \mathbf{Z}^+$ we use $[n(n-1)(n-2)\cdots(n-r+1)]/r!$ as the definition of $\binom{n}{r}$.
  If $n \in \mathbf{Z}^+$, then

$$\binom{-n}{r} = [(-n)(-n-1)(-n-2)\cdots(-n-r+1)]/r!$$

$$= (-1)^r(n)(n+1)(n+2)\cdots(n+r-1)/r!$$

$$= (-1)^r(n+r-1)!/[(n-1)!r!] = (-1)^r\binom{n+r-1}{r}.$$

Finally, for any *real* number n, we define $\binom{n}{0} = 1$.

Example 9.7   For $n \in \mathbf{Z}^+$, the Maclaurin series expansion for $(1+x)^{-n}$ is given by

$$(1+x)^{-n} = 1 + (-n)x + (-n)(-n-1)x^2/2!$$

$$+ (-n)(-n-1)(-n-2)x^3/3! + \cdots$$

$$= 1 + \sum_{r=1}^{\infty} \frac{(-n)(-n-1)(-n-2)\cdots(-n-r+1)}{r!} x^r$$

$$= \sum_{r=0}^{\infty} (-1)^r \binom{n+r-1}{r} x^r.$$

Hence $(1+x)^{-n} = \binom{-n}{0} + \binom{-n}{1}x + \binom{-n}{2}x^2 + \cdots = \sum_{r=0}^{\infty}\binom{-n}{r}x^r$. This generalizes the binomial theorem of Chapter 1 and shows us that $(1+x)^{-n}$ is the generating function for the sequence $\binom{-n}{0}, \binom{-n}{1}, \binom{-n}{2}, \binom{-n}{3}, \ldots$.   □

Example 9.8   Find the coefficient of x^5 in $(1-2x)^{-7}$.
  With $y = -2x$, use the result of Example 9.7 to obtain $(1-2x)^{-7} = (1+y)^{-7} = \sum_{r=0}^{\infty}\binom{-7}{r}y^r = \sum_{r=0}^{\infty}\binom{-7}{r}(-2x)^r$. Consequently, the coefficient of x^5 is $\binom{-7}{5}(-2)^5 = (-1)^5\binom{7+5-1}{5}(-32) = (32)\binom{11}{5} = 14{,}784$.   □

Example 9.9   For any real number n, the Maclaurin series for $(1+x)^n$ is

$$1 + nx + n(n-1)x^2/2! + (n)(n-1)(n-2)x^3/3! + \cdots$$

$$= 1 + \sum_{r=1}^{\infty} \frac{n(n-1)(n-2)\cdots(n-r+1)}{r!}x^r.$$

Hence

$$(1+3x)^{-1/3} = 1 + \sum_{r=1}^{\infty} \frac{(-1/3)(-4/3)(-7/3)\cdots((-3r+2)/3)}{r!}(3x)^r$$

$$= 1 + \sum_{r=1}^{\infty} \frac{(-1)(-4)(-7)\cdots(-3r+2)}{r!}x^r,$$

and $(1 + 3x)^{-1/3}$ generates the sequence $1, -1, (-1)(-4)/2!, (-1)(-4)(-7)/3!,$
$\ldots, (-1)(-4)(-7) \cdots (-3r + 2)/r!, \ldots$. $\square$

Before continuing, we collect the identities shown in Table 9.2 for future
reference.

Table 9.2

For any $m, n \in \mathbf{Z}^+$, $a \in \mathbf{R}$,

1. $(1 + x)^n = \binom{n}{0} + \binom{n}{1}x + \binom{n}{2}x^2 + \cdots + \binom{n}{n}x^n$

2. $(1 + ax)^n = \binom{n}{0} + \binom{n}{1}ax + \binom{n}{2}a^2 x^2 + \cdots + \binom{n}{n}a^n x^n$

3. $(1 + x^m)^n = \binom{n}{0} + \binom{n}{1}x^m + \binom{n}{2}x^{2m} + \cdots + \binom{n}{n}x^{nm}$

4. $(1 - x^{n+1})/(1 - x) = 1 + x + x^2 + \cdots + x^n$

5. $1/(1 - x) = 1 + x + x^2 + x^3 + \cdots = \sum_{i=0}^{\infty} x^i$

6. $1/(1 + x)^n = \binom{-n}{0} + \binom{-n}{1}x + \binom{-n}{2}x^2 + \cdots$
   $\qquad = \sum_{i=0}^{\infty} \binom{-n}{i}x^i$
   $\qquad = 1 + (-1)\binom{n+1-1}{1}x + (-1)^2\binom{n+2-1}{2}x^2 + \cdots$
   $\qquad = \sum_{i=0}^{\infty}(-1)^i\binom{n+i-1}{i}x^i$

7. $1/(1 - x)^n = \binom{-n}{0} + \binom{-n}{1}(-x) + \binom{-n}{2}(-x)^2 + \cdots$
   $\qquad = \sum_{i=0}^{\infty}\binom{-n}{i}(-x)^i$
   $\qquad = 1 + (-1)\binom{n+1-1}{1}(-x) + (-1)^2\binom{n+2-1}{2}(-x)^2 + \cdots$
   $\qquad = \sum_{i=0}^{\infty}\binom{n+i-1}{i}x^i$

If $f(x) = \sum_{i=0}^{\infty} a_i x^i$, $g(x) = \sum_{i=0}^{\infty} b_i x^i$, and $h(x) = f(x)g(x)$, then $h(x) = \sum_{i=0}^{\infty} c_i x^i$,
where for any $k \geq 0$, $c_k = a_0 b_k + a_1 b_{k-1} + \cdots + a_{k-1} b_1 + a_k b_0$.

Example 9.10   Determine the coefficient of x^{15} in $f(x) = (x^2 + x^3 + x^4 + \cdots)^4$.

Since $(x^2 + x^3 + x^4 + \cdots) = x^2(1 + x + x^2 + \cdots) = x^2/(1 - x)$, the coefficient
of x^{15} in $f(x)$ is the coefficient of x^{15} in $(x^2/(1 - x))^4 = x^8/(1 - x)^4$. Hence the
coefficient sought is that of x^7 in $(1 - x)^{-4}$, namely $\binom{-4}{7}(-1)^7 = (-1)^7\binom{4+7-1}{7} \cdot (-1)^7 = \binom{10}{7} = 120$.

In general, for $n \in \mathbf{Z}^+$, the coefficient of x^n in $f(x)$ is 0, when $0 \leq n \leq 7$. For
$n \geq 8$, the coefficient of x^n in $f(x)$ is the coefficient of x^{n-8} in $(1 - x)^{-4}$, and this is
$\binom{-4}{n-8}(-1)^{n-8} = \binom{n-5}{n-8}$. $\square$

Example 9.11   In how many ways can we select, with repetitions allowed, r objects from n
distinct objects?

For each of the n distinct objects, the geometric series $1 + x + x^2 + x^3 + \cdots$
represents the possible choices for that object (namely none, one, two, ...).
Considering all of the n distinct objects, the generating function is

$$f(x) = (1 + x + x^2 + x^3 + \cdots)^n,$$

and the answer is the coefficient of x^r in $f(x)$. Now

$$(1 + x + x^2 + x^3 + \cdots)^n = \left(\frac{1}{(1-x)}\right)^n = \frac{1}{(1-x)^n} = \sum_{i=0}^{\infty} \binom{n+i-1}{i} x^i,$$

so the coefficient of x^r is

$$\binom{n+r-1}{r},$$

the result we found in Chapter 1. $\square$

Example 9.12   In how many ways can a police captain distribute 24 rifle shells to four police officers so that each officer gets at least three shells, but not more than eight?

The choices for the number of shells each officer receives are given by $x^3 + x^4 + \cdots + x^8$. There are four officers, so the resulting generating function is $f(x) = (x^3 + x^4 + \cdots + x^8)^4$.

We seek the coefficient of x^{24} in $f(x)$. With $(x^3 + x^4 + \cdots + x^8)^4 = x^{12}(1 + x + x^2 + \cdots + x^5)^4 = x^{12}((1 - x^6)/(1-x))^4$, the answer is the coefficient of x^{12} in $(1 - x^6)^4(1-x)^{-4} = [1 - \binom{4}{1}x^6 + \binom{4}{2}x^{12} - \cdots + x^{24}][\binom{-4}{0} + \binom{-4}{1}(-x) + \binom{-4}{2} \cdot (-x)^2 + \cdots]$, which is $[\binom{-4}{12}(-1)^{12} - \binom{4}{1}\binom{-4}{6}(-1)^6 + \binom{4}{2}\binom{-4}{0}] = [\binom{15}{12} - \binom{4}{1}\binom{9}{6} + \binom{4}{2}] = 125$. $\square$

Example 9.13   Verify that for all $n \in \mathbf{Z}^+$, $\binom{2n}{n} = \sum_{i=0}^{n} \binom{n}{i}^2$.

Since $(1 + x)^{2n} = [(1 + x)^n]^2$, by comparison of coefficients (of like powers of x), the coefficient of x^n in $(1 + x)^{2n}$, which is $\binom{2n}{n}$, must equal the coefficient of x^n in $[\binom{n}{0} + \binom{n}{1}x + \binom{n}{2}x^2 + \cdots + \binom{n}{n}x^n]^2$, and this is $\binom{n}{0}\binom{n}{n} + \binom{n}{1}\binom{n}{n-1} + \binom{n}{2}\binom{n}{n-2} + \cdots + \binom{n}{n}\binom{n}{0}$. With $\binom{n}{r} = \binom{n}{n-r}$, for $0 \le r \le n$, the result follows. $\square$

Example 9.14   Determine the coefficient of x^8 in $1/(x-3)(x-2)^2$.

Since $1/(x-a) = (-1/a)(1/(1-(x/a))) = (-1/a)[1 + (x/a) + (x/a)^2 + \cdots]$ for any $a \ne 0$, we could solve this problem by finding the coefficient of x^8 in $1/(x-3)(x-2)^2$ expressed as $(-1/3)[1 + (x/3) + (x/3)^2 + \cdots](1/4)[\binom{-2}{0} + \binom{-2}{1}(-x/2) + \binom{-2}{2}(-x/2)^2 + \cdots]$.

An alternative technique uses the partial fraction decomposition:

$$\frac{1}{(x-3)(x-2)^2} = \frac{A}{x-3} + \frac{B}{x-2} + \frac{C}{(x-2)^2}.$$

This decomposition implies that for all x,

$$1 = A(x-2)^2 + B(x-2)(x-3) + C(x-3),$$

or

$$1 = (A + B)x^2 + (-4A - 5B + C)x + (4A + 6B - 3C).$$

By comparing coefficients, we find that $A + B = 0$, $-4A - 5B + C = 0$, and $4A + 6B - 3C = 1$. This yields $A = 1$, $B = -1$, and $C = -1$. Hence

$$\frac{1}{(x-3)(x-2)^2} = \frac{1}{x-3} - \frac{1}{x-2} - \frac{1}{(x-2)^2}$$

$$= \left(\frac{-1}{3}\right)\frac{1}{1-(x/3)} + \left(\frac{1}{2}\right)\frac{1}{1-(x/2)} + \left(\frac{-1}{4}\right)\frac{1}{(1-(x/2))^2}$$

$$= \left(\frac{-1}{3}\right)\sum_{i=0}^{\infty}\left(\frac{x}{3}\right)^i + \left(\frac{1}{2}\right)\sum_{i=0}^{\infty}\left(\frac{x}{2}\right)^i$$

$$+ \left(\frac{-1}{4}\right)\left[\binom{-2}{0} + \binom{-2}{1}\left(\frac{-x}{2}\right) + \binom{-2}{2}\left(\frac{-x}{2}\right)^2 + \cdots\right].$$

The coefficient of x^8 is $(-1/3)(1/3)^8 + (1/2)(1/2)^8 + (-1/4)\binom{-2}{8}(-1/2)^8 = -[(1/3)^9 + 7(1/2)^{10}]$. $\square$

Example 9.15   Use generating functions to determine how many four-element subsets of $S = \{1, 2, 3, \ldots, 15\}$ contain no consecutive integers.

a) Consider one such subset (say $\{1, 3, 7, 10\}$), and write $1 \le 1 < 3 < 7 < 10 \le 15$. We see that this set of inequalities determines the differences $1 - 1 = 0$, $3 - 1 = 2$, $7 - 3 = 4$, $10 - 7 = 3$, and $15 - 10 = 5$, and these differences sum to 14. Considering another such subset (say $\{2, 5, 11, 15\}$), we write $1 \le 2 < 5 < 11 < 15 \le 15$; these inequalities yield the differences 1, 3, 6, 4, and 0, which also sum to 14.

Turning things around, we find that the nonnegative integers 0, 2, 3, 2, and 7 sum to 14 and are the differences that arise from the inequalities $1 \le 1 < 3 < 6 < 8 \le 15$ (for the subset $\{1, 3, 6, 8\}$).

These examples suggest a one-to-one correspondence between the four-element subsets to be counted and the integer solutions to $c_1 + c_2 + c_3 + c_4 + c_5 = 14$ where $0 \le c_1, c_5$, and $2 \le c_2, c_3, c_4$. The answer is the coefficient of x^{14} in

$$f(x) = (1 + x + x^2 + x^3 + \cdots)(x^2 + x^3 + x^4 + \cdots)^3 \cdot$$
$$(1 + x + x^2 + x^3 + \cdots) = x^6(1-x)^{-5}.$$

This then is the coefficient of x^8 in $(1-x)^{-5}$, which is $\binom{-5}{8}(-1)^8 = \binom{5+8-1}{8} = \binom{12}{8} = 495$.

b) Another way to look at the problem is as follows.

For the subset $\{1, 3, 7, 10\}$, we examine the strict inequalities $0 < 1 < 3 < 7 < 10 < 16$ and consider how many integers there are between any consecutive pair of these numbers. Here we get 0, 1, 3, 2, and 5: 0 because there is no integer between 0 and 1, 1 for the integer 2 between 1 and 3, 3 for the integers 4, 5, 6 between 3 and 7, etc. These 5 integers sum to 11. When we do the same thing to the subset $\{2, 5, 11, 15\}$, the strict inequalities $0 < 2 < 5 < 11 < 15 < 16$ yield the results 1, 2, 5, 3, and 0, which sum to 11.

On the other hand, we find that the nonnegative integers $0, 1, 2, 1,$ and 7 add up to 11 and arise as the numbers of distinct integers between the integers in the five consecutive strict inequalities $0 < 1 < 3 < 6 < 8 < 16$. These correspond to the subset $\{1, 3, 6, 8\}$.

These results suggest a one-to-one correspondence with the integer solutions to $b_1 + b_2 + b_3 + b_4 + b_5 = 11$ where $0 \le b_1, b_5$ and $1 \le b_2, b_3, b_4$. The number of these solutions is the coefficient of x^{11} in

$$g(x) = (1 + x + x^2 + \cdots)(x + x^2 + x^3 + \cdots)^3 (1 + x + x^2 + \cdots)$$
$$= x^3 (1 - x)^{-5}.$$

The answer is $\binom{-5}{8}(-1)^8 = 495$, as above. (The reader may now wish to look back at Miscellaneous Exercise 17 in Chapter 3.) □

In our last example for this section, we have an opportunity to use the last identity of Table 9.2.

Example 9.16   Let $f(x) = x/(1 - x)^2$. This is the generating function for the sequence $a_0, a_1, a_2, \ldots,$ where $a_k = k$ for $k \in \mathbf{N}$. The function $g(x) = x(x + 1)/(1 - x)^3$ generates the sequence $b_0, b_1, b_2, \ldots,$ for $b_k = k^2$, $k \in \mathbf{N}$.

The function $h(x) = f(x)g(x)$ consequently gives us $a_0 b_0 + (a_0 b_1 + a_1 b_0)x + (a_0 b_2 + a_1 b_1 + a_2 b_0)x^2 + \cdots$, so $h(x)$ is the generating function for the sequence $c_0, c_1, c_2, \ldots,$ where for any $k \in \mathbf{N}$,

$$c_k = a_0 b_k + a_1 b_{k-1} + a_2 b_{k-2} + \cdots + a_{k-2} b_2 + a_{k-1} b_1 + a_k b_0.$$

Here, for example, we find that

$$c_0 = 0 \cdot 0^2 = 0$$
$$c_1 = 0 \cdot 1^2 + 1 \cdot 0^2 = 0$$
$$c_2 = 0 \cdot 2^2 + 1 \cdot 1^2 + 2 \cdot 0^2 = 1$$
$$c_3 = 0 \cdot 3^2 + 1 \cdot 2^2 + 2 \cdot 1^2 + 3 \cdot 0^2 = 6$$
$$c_4 = 0 \cdot 4^2 + 1 \cdot 3^2 + 2 \cdot 2^2 + 3 \cdot 1^2 + 4 \cdot 0^2 = 20$$

and, in general, $c_k = \sum_{i=0}^{k} i(k - i)^2$. (We shall simplify this summation formula in the section exercises.)

Whenever a sequence $c_0, c_1, c_2, \ldots$ arises from two generating functions $f(x)$ [for $a_0, a_1, a_2, \ldots$] and $g(x)$ [for $b_0, b_1, b_2, \ldots$], as in this example, the sequence $c_0, c_1, c_2, \ldots$ is called the *convolution* of the sequences $a_0, a_1, a_2, \ldots$ and $b_0, b_1, b_2, \ldots$. □

EXERCISES 9.2

1. Find generating functions for the following sequences. [For example, in the case of the sequence $0, 1, 3, 9, 27, \ldots,$ the answer required is $x/(1 - 3x)$, not $\sum_{i=0}^{\infty} 3^i x^{i+1}$ or simply $0 + x + 3x^2 + 9x^3 + \cdots$.]

   a) $\binom{8}{0}, \binom{8}{1}, \binom{8}{2}, \ldots, \binom{8}{8}$

   b) $\binom{8}{1}, 2\binom{8}{2}, 3\binom{8}{3}, \ldots, 8\binom{8}{8}$

c) $1, -1, 1, -1, 1, -1, \ldots$ **d)** $0, 0, 0, 1, 1, 1, 1, 1, 1, \ldots$

e) $0, 0, 0, 6, -6, 6, -6, 6, \ldots$ **f)** $1, 0, 1, 0, 1, 0, 1, \ldots$

g) $1, 2, 4, 8, 16, \ldots$ **h)** $0, 0, 1, a, a^2, a^3, \ldots, a \neq 0$

2. Determine the sequence generated by each of the following generating functions.

a) $f(x) = (2x - 3)^3$ **b)** $f(x) = x^4/(1-x)$

c) $f(x) = x^3/(1-x^2)$ **d)** $f(x) = 1/(1+3x)$

e) $f(x) = 1/(3-x)$ **f)** $f(x) = 1/(1-x) + 3x^7 - 11$

g) $f(x) = 1/(1-2x) + 5x/(1-x)$

3. In each of the following, the function $f(x)$ is the generating function for the sequence $a_0, a_1, a_2, \ldots$, whereas the sequence $b_0, b_1, b_2, \ldots$, is generated by the function $g(x)$. Express $g(x)$ in terms of $f(x)$.

a) $b_3 = 3$ **b)** $b_3 = 3$
$b_n = a_n, \ n \in \mathbf{N}, \ n \neq 3$ $b_7 = 7$
  $b_n = a_n, \ n \in \mathbf{N}, \ n \neq 3, 7$

c) $b_1 = 1$ **d)** $b_1 = 1$
$b_3 = 3$ $b_3 = 3$
$b_n = 2a_n, \ n \in \mathbf{N}, \ n \neq 1, 3$ $b_7 = 7$
  $b_n = 2a_n + 5, \ n \in \mathbf{N}, \ n \neq 1, 3, 7$

4. Determine the constant in $(3x^2 - (2/x))^{15}$.

5. **a)** Find the coefficient of x^7 in $(1 + x + x^2 + x^3 + \cdots)^{15}$.
   b) Find the coefficient of x^7 in $(1 + x + x^2 + x^3 + \cdots)^n, n \in \mathbf{Z}^+$.

6. Find the coefficient of x^{50} in $(x^7 + x^8 + x^9 + \cdots)^6$.

7. Find the coefficient of x^{20} in $(x^2 + x^3 + x^4 + x^5 + x^6)^5$.

8. For $n \in \mathbf{Z}^+$, find in $(1 + x + x^2)(1 + x)^n$ the coefficient of (a) x^7; (b) x^8; (c) x^r for $0 \leq r \leq n + 2, r \in \mathbf{Z}$.

9. Find the coefficient of x^{15} in each of the following.

a) $x^3(1 - 2x)^{10}$ **b)** $(x^3 - 5x)/(1-x)^3$ **c)** $(1+x)^4/(1-x)^4$

10. In how many ways can two dozen identical robots be assigned to four assembly lines with (a) at least three robots assigned to each line? (b) at least three, but no more than nine, robots assigned to each line?

11. In how many ways can 3000 identical envelopes be divided, in packages of 25, among four student groups so that each group gets at least 150, but not more than 1000, of the envelopes?

12. Two cases of soft drinks, 24 bottles of one type and 24 of another, are distributed among five surveyors who are conducting taste tests. In how many ways can the 48 bottles be distributed so that each surveyor gets (a) at least two bottles of each type? (b) at least two bottles of one particular type and at least three of the other?

13. If a die is rolled 12 times, what is the probability that the sum of the rolls is 30?

14. Carol is collecting money from her cousins to have a party for her aunt. If eight of the cousins promise to give $2, $3, $4, or $5 each, and two others each give $5 or $10, what is the probability that Carol will collect exactly $40?

15. In how many ways can Traci select n marbles from a large supply of blue, red, and yellow marbles if the selection must include an even number of blue ones?

16. How can Mary split up 12 hamburgers and 16 hot dogs among her sons Richard, Peter, Christopher, and James in such a way that James gets at least one hamburger and three hot dogs, and each of his brothers gets at least two hamburgers but at most five hot dogs?

17. Verify that $(1 - x - x^2 - x^3 - x^4 - x^5 - x^6)^{-1}$ is the generating function for the number of ways the sum n, $n \in \mathbf{N}$, can be obtained when a single die is rolled an arbitrary number of times.

18. Show that $(1 - 4x)^{-1/2}$ generates the sequence $\binom{2n}{n}$, $n \in \mathbf{N}$.

19. Find a, b, and k so that $(a + bx)^k$ is the generating function for the sequence $1, 2/5, 12/25, 88/125, \dots$.

20. Find the first five terms in each of the following expansions.
    a) $(1 + 2x)^{1/3}$
    b) $(1 - 3x)^{1/2}$
    c) $(2x - 1)^{-1/3}$
    d) $(2 - x)^{1/4}$

21. Consider part (a) of Example 9.15.
    a) Determine the differences for the inequalities that result from the subset $\{3, 6, 8, 15\}$ of S, and verify that those differences add to the correct sum.
    b) Find the subset of S that determines the differences $2, 2, 3, 7$, and 0.
    c) Find the subset of S that determines the differences a, b, c, d, and e, where $0 \le a, e$, and $2 \le b, c, d$.

22. In how many ways can we select seven nonconsecutive integers from $\{1, 2, 3, \dots, 50\}$?

23. Use the following summation formulas (these appear in the material and exercises of Section 4.1) to simplify the expression for c_k in Example 9.16.

$$\sum_{i=0}^{k} i = \sum_{i=1}^{k} i = k(k + 1)/2,$$

$$\sum_{i=0}^{k} i^2 = \sum_{i=1}^{k} i^2 = k(k + 1)(2k + 1)/6, \text{ and}$$

$$\sum_{i=0}^{k} i^3 = \sum_{i=1}^{k} i^3 = k^2(k + 1)^2/4.$$

24. a) Find the first four terms $c_0, c_1, c_2,$ and c_3 of the convolutions for each of the following pairs of sequences.

    (i) $a_n = 1, b_n = 1$, for all $n \in \mathbf{N}$.

    (ii) $a_n = 1, b_n = 2^n$, for all $n \in \mathbf{N}$.

    (iii) $a_0 = a_1 = a_2 = a_3 = 1; a_n = 0, n \in \mathbf{N}, n \neq 0, 1, 2, 3; b_n = 1$, for all $n \in \mathbf{N}$.

b) Find a general formula for c_n in each of the results of part (a).

9.3
PARTITIONS OF INTEGERS

In the study of number theory, we are confronted with partitioning a positive integer n into positive summands and seeking the number of such partitions, without regard to order. This number is denoted by $p(n)$. For example,

$$p(1) = 1: \quad 1$$
$$p(2) = 2: \quad 2 = 1 + 1$$
$$p(3) = 3: \quad 3 = 2 + 1 = 1 + 1 + 1$$
$$p(4) = 5: \quad 4 = 3 + 1 = 2 + 2 = 2 + 1 + 1 = 1 + 1 + 1 + 1.$$

We should like to obtain $p(n)$ for a given n without writing all the partitions. We need a tool to keep track of the numbers of 1's, 2's, $\ldots$, n's that are used as summands for n.

If $n \in \mathbf{Z}^+$, the number of 1's we can use is 0 or 1 or 2 or $\ldots$. The power series $1 + x + x^2 + x^3 + x^4 + \cdots$ keeps account of this for us. In like manner, $1 + x^2 + x^4 + x^6 + \cdots$ keeps track of the number of 2's in the partition of n. For example, to determine $p(10)$ we want the coefficient of x^{10} in $f(x) = (1 + x + x^2 + x^3 + \cdots) \cdot (1 + x^2 + x^4 + x^6 + \cdots) \cdots (1 + x^{10} + x^{20} + \cdots)$ or in $g(x) = (1 + x + x^2 + x^3 + \cdots + x^{10})(1 + x^2 + x^4 + \cdots + x^{10}) \cdots (1 + x^{10})$.

We prefer to work with $f(x)$, because it can be written in the more compact form

$$f(x) = \frac{1}{(1-x)} \frac{1}{(1-x^2)} \cdots \frac{1}{(1-x^{10})} = \prod_{i=1}^{10} \frac{1}{(1-x^i)}.$$

If this product is extended beyond $i = 10$, we get $P(x) = \prod_{i=1}^{\infty} [1/(1-x^i)]$, which generates the sequence $p(0), p(1), p(2), p(3), \ldots$, where we define $p(0) = 1$.

Though $P(x)$ is theoretically complete, it is difficult to calculate an infinite product of terms. If we consider only $\prod_{i=1}^{r} [1/(1-x^i)]$ for some fixed r, then the coefficient of x^n is the number of partitions of n into summands that do not exceed r.

Despite the difficulty in calculating $p(n)$ from $P(x)$ for large values of n, the idea of the generating function will be useful in studying certain kinds of partitions.

Example 9.17   Find the generating function for the number of ways an advertising agent can purchase n minutes $(n \in \mathbf{Z}^+)$ of air time if time slots for commercials come in blocks of 30, 60, or 120 seconds.

Let 30 seconds represent one time unit. Then the answer is the number of integer solutions to the equation $a + 2b + 4c = 2n$ with $0 \leq a, b, c$.

The associated generating function is

$$f(x) = (1 + x + x^2 + \cdots)(1 + x^2 + x^4 + \cdots)(1 + x^4 + x^8 + \cdots)$$
$$= \frac{1}{1-x} \frac{1}{1-x^2} \frac{1}{1-x^4},$$

and the coefficient of x^{2n} is the number of partitions of $2n$ into 1's, 2's, and 4's, the answer to the problem.   □

Example 9.18   Find the generating function for $p_d(n)$, the number of partitions of a positive integer n into *distinct* summands.

Before we start, let us consider the 11 partitions of 6:

(1) $1 + 1 + 1 + 1 + 1 + 1$ **(2)** $1 + 1 + 1 + 1 + 2$

(3) $1 + 1 + 1 + 3$ **(4)** $1 + 1 + 4$

(5) $1 + 1 + 2 + 2$ **(6)** $1 + 5$

(7) $1 + 2 + 3$ **(8)** $2 + 2 + 2$

(9) $2 + 4$ **(10)** $3 + 3$

(11) 6

Partitions (6), (7), (9), and (11) have distinct summands, so $p_d(6) = 4$.

In calculating $p_d(n)$, for any $k \in \mathbf{Z}^+$ there are two choices: Either k is not used as one of the summands of n, or it is. This can be accounted for by the polynomial $1 + x^k$, and consequently, the generating function for these partitions is

$$P_d(x) = (1 + x)(1 + x^2)(1 + x^3) \cdots = \prod_{i=1}^{\infty} (1 + x^i).$$

For any $n \in \mathbf{Z}^+$, $p_d(n)$ is the coefficient of x^n in $(1 + x)(1 + x^2) \cdots (1 + x^n)$. (We define $p_d(0) = 1$.) When $n = 6$, the coefficient of x^6 in $(1 + x)(1 + x^2) \cdots (1 + x^6)$ is 4.   □

Example 9.19   Considering the partitions in Example 9.18, we see that the number of partitions of 6 into odd summands is 4, which is also $p_d(6)$. Is this a coincidence?

Let $p_o(n)$ denote the number of partitions of n into odd summands, when $n \geq 1$. We define $p_o(0) = 1$. The generating function for the sequence $p_o(0), p_o(1), p_o(2), \ldots$ is given by

$$P_o(x) = (1 + x + x^2 + x^3 + \cdots)(1 + x^3 + x^6 + \cdots)(1 + x^5 + x^{10} + \cdots) \cdot$$

$$(1 + x^7 + \cdots) \cdots = \frac{1}{1-x} \frac{1}{1-x^3} \frac{1}{1-x^5} \frac{1}{1-x^7} \cdots.$$

Now because

$$1 + x = \frac{1-x^2}{1-x}, \qquad 1 + x^2 = \frac{1-x^4}{1-x^2}, \qquad 1 + x^3 = \frac{1-x^6}{1-x^3}, \qquad \ldots,$$

we have

$$\begin{aligned} P_d(x) &= (1+x)(1+x^2)(1+x^3)(1+x^4)\cdots \\ &= \frac{1-x^2}{1-x}\frac{1-x^4}{1-x^2}\frac{1-x^6}{1-x^3}\frac{1-x^8}{1-x^4}\cdots = \frac{1}{1-x}\frac{1}{1-x^3}\cdots = P_o(x). \end{aligned}$$

From the equality of the generating functions, $p_d(n) = p_o(n)$, for all $n \geq 0$.
$\square$

We close this section with an idea called the *Ferrer's graph*. This graph uses rows of dots to represent a partition of an integer where the number of dots per row does not increase as we go from any row to the one below it.

In Fig. 9.1 we find the Ferrer's graphs for two partitions of 14: (a) $4 + 3 + 3 + 2 + 1 + 1$ and (b) $6 + 4 + 3 + 1$. The graph in part (b) is said to be the *transposition* of the graph in part (a), and vice versa, because one graph can be obtained from the other by interchanging rows and columns.

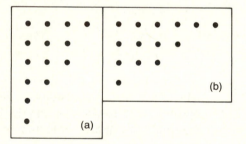

(b)

(a)

Figure 9.1

These graphs often suggest results about partitions. Here we see a partition of 14 into summands, where 4 is the largest summand, and a second partition of 14 into exactly four summands. There is a one-to-one correspondence between a Ferrer's graph and its transposition, so this example demonstrates a particular instance of the general result: The number of partitions of an integer n into m parts is equal to the number of partitions of n into summands where m is the largest summand.

EXERCISES 9.3

1. Find all partitions of 5 and 7.

2. Determine the generating function for the sequence $a_0, a_1, a_2, \ldots$, where a_n is the number of partitions of the integer n into (a) even summands; (b) distinct even summands; (c) distinct odd summands.

3. In $f(x) = [1/(1-x)][1/(1-x^2)][1/(1-x^3)]$, the coefficient of x^6 is 7. Interpret this result in terms of partitions of 6.

4. Find the generating function for the number of integer solutions of
   a) $2w + 3x + 5y + 7z = n,$ $\quad 0 \le w, x, y, z$
   b) $2w + 3x + 5y + 7z = n,$ $\quad 0 \le w, \quad 4 \le x, y, \quad 5 \le z$
   c) $2w + 3x + 5y + 7z = n,$ $\quad 2 \le w \le 4 \le x \le 7 \le y \le 10 \le z$

5. Find a generating function for the number of partitions of the nonnegative integer n into summands where (a) each summand must appear an even number of times; (b) each summand must be even.

6. Find a generating function for the number of partitions of $n \in \mathbf{N}$ into summands that (a) cannot occur more than five times; and, (b) cannot exceed 12 and cannot occur more than five times.

7. Show that the number of partitions of a positive integer n where no summand appears more than twice equals the number of partitions of n where no summand is divisible by 3.

8. Show that the number of partitions of $n \in \mathbf{Z}^+$ where no summand is divisible by 4 equals the number of partitions of n where no even summand is repeated (although odd summands may or may not be repeated).

9. Determine the generating function for the number of ways to have n cents, $0 \le n \le 99$, in pennies, nickels, dimes, quarters, and half dollars.

10. Verify the identity

$$\frac{1}{1-x} = (1 + x + x^2 + \cdots + x^9)(1 + x^{10} + x^{20} + \cdots + x^{90}) \cdot$$
$$(1 + x^{100} + x^{200} + \cdots + x^{900}) \cdots,$$

   and interpret the result in terms of partitions of an integer $n \ge 0$.

11. Using a Ferrer's graph, show that the number of partitions of an integer n into summands not exceeding m is equal to the number of partitions of n into at most m summands.

12. Using a Ferrer's graph, show that the number of partitions of n is equal to the number of partitions of $2n$ into n summands.

9.4
THE EXPONENTIAL GENERATING FUNCTION

The generating function we have been dealing with is often referred to as the *ordinary* generating function of a given sequence. This function arose in selection problems, where order was not relevant. However, turning now to problems of arrangement, where order is crucial, we seek a comparable tool. To find such a tool we return to the binomial theorem.

For any $n \in \mathbf{Z}^+$, $(1 + x)^n = \binom{n}{0} + \binom{n}{1}x + \binom{n}{2}x^2 + \cdots + \binom{n}{n}x^n$, so $(1 + x)^n$ is the (ordinary) generating function for the sequence $\binom{n}{0}, \binom{n}{1}, \binom{n}{2}, \ldots, \binom{n}{n}, 0, 0, \ldots$. When dealing with this idea in Chapter 1, we also wrote $\binom{n}{r} = C(n, r)$ when we wanted to emphasize that $\binom{n}{r}$ represented the number of combinations of n objects taken r at a time, with $0 \le r \le n$. Consequently, $(1+x)^n$ generates the sequence $C(n,0), C(n,1), C(n,2), \ldots, C(n,n), 0, 0, \ldots$.

Now for any $0 \le r \le n$,

$$C(n,r) = \frac{n!}{r!(n-r)!} = \left(\frac{1}{r!}\right) P(n,r),$$

so

$$(1+x)^n = C(n,0) + C(n,1)x + C(n,2)x^2 + C(n,3)x^3 + \cdots + C(n,n)x^n$$
$$= P(n,0) + P(n,1)x + P(n,2)\frac{x^2}{2!} + P(n,3)\frac{x^3}{3!} + \cdots + P(n,n)\frac{x^n}{n!}.$$

Hence, if in $(1+x)^n$ we consider the coefficients of $x^r/r!$, with $0 \le r \le n$, we obtain $P(n,r)$. On the basis of this observation, we have the following definition.

DEFINITION 9.2   For a sequence $a_0, a_1, a_2, a_3 \ldots$ of real numbers,

$$f(x) = a_0 + a_1 x + a_2 \frac{x^2}{2!} + a_3 \frac{x^3}{3!} + \cdots = \sum_{i=0}^{\infty} a_i \frac{x^i}{i!}$$

is called the *exponential generating function* for the given sequence.

Example 9.20   Examining the expansion for e^x, we find

$$e^x = 1 + x + \frac{x^2}{2!} + \frac{x^3}{3!} + \frac{x^4}{4!} + \cdots = \sum_{i=0}^{\infty} \frac{x^i}{i!},$$

so e^x is the exponential generating function for the sequence $1, 1, 1, \ldots$. (e^x is the ordinary generating function for the sequence $1, 1, 1/2!, 1/3!, 1/4!, \ldots$.) □

Our next example shows how this idea can help us count certain types of arrangements.

Example 9.21   In how many ways can four of the letters in ENGINE be arranged?

Table 9.3

E E N N	4!/(2!2!)		E G N N	4!/2!
E E G N	4!/2!		E I N N	4!/2!
E E I N	4!/2!		G I N N	4!/2!
E E G I	4!/2!		E I G N	4!

In Table 9.3 we list the possible selections of size 4 from the letters E, N, G, I, N, E, along with the number of arrangements those four letters determine.

We now obtain the answer by means of an exponential generating function. For the letter E, we use $[1 + x + (x^2/2!)]$, because there are 0, 1, or 2 E's to arrange. Note that the coefficient of $x^2/2!$ is 1, the number of distinct ways to arrange two E's. In like manner, we have $[1 + x + (x^2/2!)]$ for the arrangements of 0, 1, or 2 N's. The arrangements for each of the letters G and I are represented by $(1 + x)$.

Consequently, we find that the exponential generating function is $f(x) = [1 + x + (x^2/2!)]^2 (1 + x)^2$, and the answer is the coefficient of $x^4/4!$ in $f(x)$. In the computation of $f(x)$, the term involving x^4 is

$$\left(\frac{x^4}{2!2!} + \frac{x^4}{2!} + \frac{x^4}{2!} + \frac{x^4}{2!} + \frac{x^4}{2!} + \frac{x^4}{2!} + \frac{x^4}{2!} + x^4 \right)$$

$$= \left[\left(\frac{4!}{2!2!} \right) + \left(\frac{4!}{2!} \right) + \left(\frac{4!}{2!} \right) + \left(\frac{4!}{2!} \right) + \left(\frac{4!}{2!} \right) + \left(\frac{4!}{2!} \right) + \left(\frac{4!}{2!} \right) + 4! \right] \left(\frac{x^4}{4!} \right),$$

and the coefficient of $x^4/4!$ is the answer (102 arrangements) produced by the table.   □

Example 9.22   Consider the expansions of e^x and e^{-x}.

$$e^x = 1 + x + \frac{x^2}{2!} + \frac{x^3}{3!} + \frac{x^4}{4!} + \cdots \qquad e^{-x} = 1 - x + \frac{x^2}{2!} - \frac{x^3}{3!} + \frac{x^4}{4!} - \cdots$$

Adding these series, we find that

$$e^x + e^{-x} = 2 \left(1 + \frac{x^2}{2!} + \frac{x^4}{4!} + \cdots \right),$$

or

$$\frac{e^x + e^{-x}}{2} = 1 + \frac{x^2}{2!} + \frac{x^4}{4!} + \cdots .$$

Subtracting e^{-x} from e^x yields

$$\frac{e^x - e^{-x}}{2} = x + \frac{x^3}{3!} + \frac{x^5}{5!} + \cdots .   □$$

Example 9.23   A ship carries 48 flags, 12 each of red, white, blue, and black. Twelve of these flags are placed on a vertical pole in order to communicate a signal to other ships. How many of these signals use an even number of blue flags and an odd number of black flags?

The exponential generating function

$$f(x) = \left(1 + x + \frac{x^2}{2!} + \frac{x^3}{3!} + \cdots \right)^2 \left(1 + \frac{x^2}{2!} + \frac{x^4}{4!} + \cdots \right) \left(x + \frac{x^3}{3!} + \frac{x^5}{5!} + \cdots \right)$$

considers all such signals made up of n flags, where $n \geq 1$. The last two factors in $f(x)$ restrict the signals to an even number of blue flags and an odd number of black flags, respectively.

Since

$$f(x) = (e^x)^2 \left(\frac{e^x + e^{-x}}{2} \right) \left(\frac{e^x - e^{-x}}{2} \right) = \left(\frac{1}{4} \right)(e^{2x})(e^{2x} - e^{-2x}) = \frac{1}{4}(e^{4x} - 1)$$

$$= \frac{1}{4} \left(\sum_{i=0}^{\infty} \frac{(4x)^i}{i!} - 1 \right) = \left(\frac{1}{4} \right) \sum_{i=1}^{\infty} \frac{(4x)^i}{i!},$$

the coefficient of $x^{12}/12!$ in $f(x)$ yields $(1/4)(4^{12}) = 4^{11}$ signals made up of 12 flags with an even number of blue flags and an odd number of black flags. □

Our final example is reminiscent of past results.

Example 9.24   A company hires 11 new employees. Each of these employees is to be assigned to one of four subdivisions, and each subdivision gets at least one new employee. In how many ways can these assignments be made?

Calling the subdivisions A, B, C, and D, we can equivalently count the number of 11-letter sequences in which there is at least one occurrence of each of the letters A, B, C, and D. The exponential generating function for these arrangements is

$$f(x) = \left(x + \frac{x^2}{2!} + \frac{x^3}{3!} + \frac{x^4}{4!} + \cdots \right)^4 = (e^x - 1)^4 = e^{4x} - 4e^{3x} + 6e^{2x} - 4e^x + 1.$$

The answer then is the coefficient of $x^{11}/11!$ in $f(x)$. This is

$$4^{11} - 4(3^{11}) + 6(2^{11}) - 4(1^{11}) = \sum_{i=0}^{4} (-1)^i \binom{4}{i}(4 - i)^{11}.$$

This form of the answer should bring to mind some of the enumeration problems in Chapter 5. Once the vocabulary is set aside, we are counting the number of onto functions $g: X \rightarrow Y$ where $|X| = 11, |Y| = 4$. □

EXERCISES 9.4

1. Find the exponential generating function for each of the following sequences.

   a) $1, -1, 1, -1, 1, -1, \ldots$
   b) $1, 2, 2^2, 2^3, 2^4, \ldots$
   c) $1, -a, a^2, -a^3, a^4, \ldots, a \in \mathbf{R}$
   d) $1, a^2, a^4, a^6, \ldots, a \in \mathbf{R}$
   e) $a, a^3, a^5, a^7, \ldots, a \in \mathbf{R}$
   f) $0, 1, 2(2), 3(2^2), 4(2^3), \ldots$

2. Determine the sequence generated by each of the following exponential generating functions.

   a) $f(x) = 3e^{3x}$
   b) $f(x) = 6e^{5x} - 3e^{2x}$
   c) $f(x) = e^x + x^2$
   d) $f(x) = e^{2x} - 3x^3 + 5x^2 + 7x$
   e) $f(x) = 1/(1 - x)$
   f) $f(x) = 3/(1 - 2x) + e^x$

3. In each of the following, the function $f(x)$ is the exponential generating function for the sequence $a_0, a_1, a_2, \ldots$, whereas the function $g(x)$ is the exponential generating function for the sequence $b_0, b_1, b_2, \ldots$. Express $g(x)$ in terms of $f(x)$ if

a) $b_3 = 3$

   $b_n = a_n, \quad n \in \mathbf{N}, n \neq 3$

b) $a_n = 5^n, \quad n \in \mathbf{N}$

   $b_3 = -1$

   $b_n = a_n, \quad n \in \mathbf{N}, n \neq 3$

c) $b_1 = 2$

   $b_2 = 4$

   $b_n = 2a_n, \quad n \in \mathbf{N}, n \neq 1, 2$

d) $b_1 = 2$

   $b_2 = 4$

   $b_3 = 8$

   $b_n = 2a_n + 3, \quad n \in \mathbf{N}, n \neq 1, 2, 3$

4. a) For the ship in Example 9.23, how many signals use at least one flag of each color? (Solve this with an exponential generating function.)

   b) Restate part (a) in an alternative way that uses the concept of an onto function.

   c) How many signals are there in Example 9.23 where the total number of blue and black flags is even?

5. Find the exponential generating function for the number of ways to arrange n letters, $n \geq 0$, selected from each of the following words.

   a) HAWAII   b) MISSISSIPPI   c) ISOMORPHISM

6. For part (b) of Exercise 5, what is the exponential generating function if the arrangement must contain at least two I's?

7. Say the company in Example 9.24 hires 25 new employees. Give the exponential generating function for the number of ways to assign these people to the four subdivisions so that each subdivision receives at least 3, but no more than 10, new people.

8. Given the sequences $a_0, a_1, a_2, \ldots$ and $b_0, b_1, b_2, \ldots$, with exponential generating functions $f(x), g(x)$, respectively, show that if $h(x) = f(x)g(x)$, then $h(x)$ is the exponential generating function of the sequence $c_0, c_1, c_2, \ldots$, where $c_n = \sum_{i=0}^{n} \binom{n}{i} a_i b_{n-i}$, for each $n \geq 0$.

9. If a 20-digit ternary $(0, 1, 2)$ sequence is randomly generated, what is the probability that:

   a) It has an even number of 1's?

   b) It has an even number of 1's and an even number of 2's?

   c) It has an odd number of 0's?

   d) The total number of 0's and 1's is odd?

   e) The total number of 0's and 1's is even?

10. How many 20-digit quaternary $(0, 1, 2, 3)$ sequences are there where:

   a) There is at least one 2 and an odd number of 0's?

   b) No symbol occurs exactly twice?

   c) No symbol occurs exactly three times?

   d) There are exactly two 3's or none at all?

11. Twenty-five contracts (for 25 different shuttle components) are to be distributed among five companies, $c_1, c_2, \ldots, c_5$. Find the exponential generating function for the number of ways in which it is possible to award these contracts so that (a) company c_1 gets at least five contracts and all other companies get at least two; (b) every company gets at least one contract, company c_2 gets more contracts than company c_1, and company c_2 gets at most five contracts.

9.5
THE SUMMATION OPERATOR

This final section introduces a technique that helps us go from the (ordinary) generating function for the sequence $a_0, a_1, a_2, \ldots$ to the generating function for the sequence $a_0, a_0 + a_1, a_0 + a_1 + a_2, \ldots$.

For $f(x) = a_0 + a_1 x + a_2 x^2 + a_3 x^3 + \cdots$, consider the function $f(x)/(1-x)$.

$$
\begin{aligned}
f(x)/(1-x) &= [a_0 + a_1 x + a_2 x^2 + a_3 x^3 + \cdots][1 + x + x^2 + x^3 + \cdots] \\
&= a_0 + (a_0 + a_1)x + (a_0 + a_1 + a_2)x^2 \\
&\quad + (a_0 + a_1 + a_2 + a_3)x^3 + \cdots,
\end{aligned}
$$

so $f(x)/(1-x)$ generates the sequence of sums $a_0,\ a_0 + a_1,\ a_0 + a_1 + a_2,\ a_0 + a_1 + a_2 + a_3,\ \ldots$. This is the convolution of the sequence $a_0, a_1, a_2, \ldots$ and the sequence $b_0, b_1, b_2, \ldots$, where $b_n = 1$ for all $n \in \mathbf{N}$.

We find this technique handy in the following example.

Example 9.25 Find a formula to express $0^2 + 1^2 + 2^2 + \cdots + n^2$ as a function of n.

As in Section 9.2, we start with $g(x) = 1/(1-x) = 1 + x + x^2 + \cdots$. Then

$$
(-1)(1-x)^{-2}(-1) = \frac{1}{(1-x)^2} = \frac{dg(x)}{dx} = 1 + 2x + 3x^2 + 4x^3 + \cdots,
$$

so $x/(1-x)^2$ is the generating function for $0, 1, 2, 3, 4, \ldots$. Repeating this technique, we find that

$$
x\frac{d}{dx}\left[x\left(\frac{dg(x)}{dx}\right)\right] = \frac{x(1+x)}{(1-x)^3} = x + 2^2 x^2 + 3^2 x^3 + \cdots,
$$

so $x(1+x)/(1-x)^3$ generates $0^2, 1^2, 2^2, 3^2, \ldots$. As a consequence of our earlier observation,

$$
\frac{x(1+x)}{(1-x)^3}\frac{1}{(1-x)} = \frac{x(1+x)}{(1-x)^4}
$$

is the generating function for $0^2, 0^2 + 1^2, 0^2 + 1^2 + 2^2, 0^2 + 1^2 + 2^2 + 3^2, \ldots$. Hence the coefficient of x^n in $x(1 + x)/(1 - x)^4$ is $\sum_{i=0}^{n} i^2$. But the coefficient of x^n in $x(1 + x)/(1 - x)^4$ can also be calculated as follows:

$$\frac{x(1+x)}{(1-x)^4} = (x + x^2)(1 - x)^{-4} = (x + x^2)\left[\binom{-4}{0} + \binom{-4}{1}(-x) + \binom{-4}{2}(-x)^2 + \cdots\right],$$

where the coefficient of x^n is

$$\binom{-4}{n-1}(-1)^{n-1} + \binom{-4}{n-2}(-1)^{n-2}$$

$$= (-1)^{n-1}\binom{4 + (n-1) - 1}{n-1}(-1)^{n-1}$$

$$+ (-1)^{n-2}\binom{4 + (n-2) - 1}{n-2}(-1)^{n-2}$$

$$= \binom{n+2}{n-1} + \binom{n+1}{n-2} = \frac{(n+2)!}{3!(n-1)!} + \frac{(n+1)!}{3!(n-2)!}$$

$$= \tfrac{1}{6}[(n+2)(n+1)(n) + (n+1)(n)(n-1)]$$

$$= \tfrac{1}{6}(n)(n+1)[(n+2) + (n-1)] = (n)(n+1)(2n+1)/6. \quad \square$$

EXERCISES 9.5

1. Continue the development of the ideas set forth in Example 9.25 and derive the formula $\sum_{i=0}^{n} i^3 = [n(n+1)/2]^2$.

2. a) Find the generating function for the sequence of products $0 \cdot (-1)$, $1 \cdot 0, 2 \cdot 1, 3 \cdot 2, 4 \cdot 3, \ldots, i \cdot (i-1), \ldots$.

   b) Use the result from part (a) to obtain a formula for $\sum_{i=0}^{n} i(i-1)$.

3. Let $f(x)$ be the generating function for the sequence $a_0, a_1, a_2, \ldots$. For what sequence is $(1 - x)f(x)$ the generating function?

4. Let $f(x) = \sum_{i=0}^{\infty} a_i x^i$ with $f(1) = \sum_{i=0}^{\infty} a_i$, a finite number. Verify that the quotient $[f(x) - f(1)]/(x - 1)$ is the generating function for the sequence $s_0, s_1, s_2, \ldots$, where $s_n = \sum_{i=n+1}^{\infty} a_i$.

5. Find the generating function for the sequence $a_0, a_1, a_2, \ldots$, where $a_n = \sum_{i=0}^{n} (1/i!)$, $n \in \mathbf{N}$.

9.6
SUMMARY AND HISTORICAL REVIEW

In the early thirteenth century the Italian mathematician Leonardo of Pisa (c. 1175–1250), in his *Liber Abaci*, originated the study of the sequence $0, 1, 1, 2, 3, 5, 8, 13, 21, \ldots$, which can be given recursively by $F_0 = 0, F_1 = 1$,

$F_{n+2} = F_{n+1} + F_n$, $n \geq 0$. Since Leonardo was the son of Bonaccio, the sequence has come to be called the Fibonacci numbers. (Filius Bonaccii is the Latin form for "son of Bonaccio.")

If we consider the formula

$$F_n = \frac{1}{\sqrt{5}}\left[\left(\frac{1+\sqrt{5}}{2}\right)^n - \left(\frac{1-\sqrt{5}}{2}\right)^n\right], \qquad n \geq 0,$$

we find $F_0 = 0$, $F_1 = 1$, $F_2 = 1$, $F_3 = 2$, $F_4 = 3$, ... Yes, this formula determines each Fibonacci number as a function of n. (Here we have the general solution for the recursive Fibonacci relation. We shall learn more about this in the next chapter.) This formula was not derived, however, until 1718, when Abraham DeMoivre (1667–1754) obtained the result from the generating function

$$f(x) = \frac{x}{1-x-x^2} = \frac{1}{\sqrt{5}}\left[\frac{1}{1-\left(\frac{1+\sqrt{5}}{2}\right)x} - \frac{1}{1-\left(\frac{1-\sqrt{5}}{2}\right)x}\right].$$

Extending the existing techniques of the generating function, Leonhard Euler (1707–1783) advanced the study of the partitions of integers in his 1748 two-volume opus, *Introductio in Analysin Infinitorum*. With

$$P(x) = \frac{1}{1-x}\frac{1}{1-x^2}\frac{1}{1-x^3}\cdots = \prod_{i=1}^{\infty}\frac{1}{1-x^i},$$

we have the generating function for $p(0), p(1), p(2), \ldots$ where $p(n)$ is the number of partitions of n into positive summands and $p(0)$ is defined to be 1.

In the latter part of the eighteenth century, further developments on generating functions arose in conjunction with ideas in probability theory, especially with what is now called the "moment generating function." These related notions were presented in their first complete treatment by the great scholar Pierre-Simon de Laplace (1749–1827) in his 1812 publication *Théorie Analytique des Probabilités*.

Finally, we mention Norman Macleod Ferrers (1829–1903), after whom the diagram we called the Ferrer's graph is named.

For us the study of the ordinary and exponential generating functions provided a powerful technique that unified ideas found in Chapters 1, 5, and 8. Extending our prior experience with polynomials to power series, and extending the binomial theorem to $(1+x)^n$ for the cases where n need not be positive or even an integer, we found the necessary tools to compute the coefficients in these generating functions. This was more than worth the effort, because the algebraic calculations we performed took into account all of the selection processes we were trying to consider. We also found that we had seen some generating functions in a prior chapter and saw how they arose in the study of partitions.

The concept of a partition of a positive integer now enables us to complete the summaries of our earlier discussions on distributions, as given in Tables 1.6 and 5.11. Here we can now deal with the distributions of m objects into $n(\leq m)$ containers for the cases where neither the objects nor the containers are distinct.

These are covered by the entries in the second and fourth rows of Table 9.4. The notation $p(m, n)$, which appears in the last column for these entries, is used to denote the number of partitions of the positive integer m into exactly n (positive) summands. (This idea will be examined further in Miscellaneous Exercise 3 of the next chapter.) The types of distributions in the first and third rows of this table were also listed in Table 5.11. We include them here a second time for the sake of comparison and completeness.

Table 9.4

Objects Are Distinct	Containers Are Distinct	Some Container(s) May Be Empty	Number of Distributions
No	Yes	Yes	$\binom{n+m-1}{m}$
No	No	Yes	(1) $p(m)$, for $n = m$ (2) $p(m, 1) + p(m, 2) + \cdots + p(m, n)$, for $n < m$
No	Yes	No	$\binom{n+(m-n)-1}{(m-n)} = \binom{m-1}{m-n} = \binom{m-1}{n-1}$
No	No	No	$p(m, n)$

For comparable coverage of the material presented in this chapter, the interested reader should consult Chapter 2 of C. Liu [3] and Chapter 3 of A. Tucker [7]. The text by J. Riordan [5] has extensive coverage of ordinary and exponential generating functions. An interesting survey article on generating functions, written by Richard P. Stanley, can be found in the text edited by G. Rota [6].

The reader interested in learning more about the theory of partitions should consult Chapter 10 of I. Niven and H. Zuckerman [4].

Finally, a great deal about the moment generating function and its use in probability theory can be found in Chapter 3 of H. Larson [2] and in Chapter XI of the comprehensive work by W. Feller [1].

REFERENCES

1. Feller, William. *An Introduction to Probability Theory and Its Applications,* Vol. I, 3rd ed. New York: Wiley, 1968.
2. Larson, Harold J. *Introduction to Probability Theory and Statistical Inference,* 2nd ed. New York: Wiley, 1969.
3. Liu, C. L. *Introduction to Combinatorial Mathematics.* New York: McGraw-Hill, 1968.
4. Niven, Ivan, and Zuckerman, Herbert. *An Introduction to the Theory of Numbers,* 3rd ed. New York: Wiley, 1960.
5. Riordan, John. *An Introduction to Combinatorial Analysis.* Princeton, N.J.: Princeton University Press, 1980. (Originally published in 1958 by John Wiley & Sons.)
6. Rota, Gian-Carlo, ed. *Studies in Combinatorics,* Studies in Mathematics, Volume 17. Washington, D.C.: The Mathematical Association of America, 1978.
7. Tucker, Alan. *Applied Combinatorics.* New York: Wiley, 1980.

MISCELLANEOUS EXERCISES

1. Find the generating function for each of the following sequences.

   a) $7, 8, 9, 10, \ldots$ b) $1, a, a^2, a^3, a^4, \ldots, \quad a \in \mathbf{R}$

   c) $1, (1 + a), (1 + a)^2, (1 + a)^3, \ldots, \quad a \in \mathbf{R}$

   d) $2, 1 + a, 1 + a^2, 1 + a^3, \ldots, \quad a \in \mathbf{R}$

2. Let $S = \{a, b, c, d, e, f, g\}$.

   a) Find the generating function for the subsets of S. How many summands are in the coefficient of x^2 in this function?

   b) Find the generating function for the number of ways to select n objects, $0 \le n \le 14$, from S, where each object can be selected at most twice. How many summands are in the coefficient of x^2 in this function?

3. Use a partial fraction decomposition to verify that

$$\frac{x}{1 - x - x^2} = \frac{1}{\sqrt{5}} \left[\frac{1}{1 - \left(\frac{1 + \sqrt{5}}{2} \right)x} - \frac{1}{1 - \left(\frac{1 - \sqrt{5}}{2} \right)x} \right].$$

4. Find the coefficient of x^{83} in $f(x) = (x^5 + x^8 + x^{11} + x^{14} + x^{17})^{10}$.

5. Sergeant Bueti must distribute 40 bullets (20 for rifles and 20 for hand guns) among four police officers so that each officer gets at least two, but no more than seven, bullets of each type. In how many ways can he do this?

6. Find a generating function for the number of ways to partition a positive integer n into positive integer summands, where each summand appears an odd number of times or not at all.

7. For $n \in \mathbf{Z}^+$, show that the number of partitions of n in which no even summand is repeated (an odd summand may or may not be repeated) is the same as the number of partitions of n where no summand occurs more than three times.

8. Show that for any $x \in \mathbf{R}$, $\left(\sum_{n=0}^{\infty} x^n / n! \right)^{-1} = \sum_{n=0}^{\infty} (-1)^n x^n / n!$.

9. How many 10-digit telephone numbers use only the digits 1, 3, 5, and 7, with each digit appearing at least twice or not at all?

10. Use the first five terms of the binomial expansion of $(1 + x)^{1/3}$ to approximate the cube root of 2.

11. a) For what sequence of numbers is $g(x) = (1 - 2x)^{-5/2}$ the exponential generating function?

    b) Find a and b so that $(1 - ax)^b$ is the exponential generating function for the sequence $1, 7, 7 \cdot 11, 7 \cdot 11 \cdot 15, \ldots$.

12. For integers $n, k \ge 0$ let

    - P_1 be the number of partitions of n.
    - P_2 be the number of partitions of $2n + k$, where $n + k$ is the greatest summand.
    - P_3 be the number of partitions of $2n + k$ into precisely $n + k$ summands.

    Using the concept of the Ferrer's graph, prove that $P_1 = P_2$ and $P_2 = P_3$, thus concluding that the number of partitions of $2n + k$ into precisely $n + k$ summands is the same for all k.

13. In a rural area there are 12 mailboxes located at a general store.

   a) If a newsboy has 20 identical fliers, in how many ways can he distribute the fliers so that each mailbox gets at least one flier?

   b) If the mailboxes are in two rows of six, what is the probability that a distribution from part (a) will have 10 fliers distributed to the top six boxes and 10 to the bottom six?

14. Let S be a set containing n distinct objects. Verify that $e^x/(1-x)^k$ is the exponential generating function for the number of ways to choose m of the objects in S, for $0 \le m \le n$, and distribute these objects among k distinct containers, with the order of the objects in any container relevant for the distribution.

CHAPTER
10

Recurrence Relations

In various sections of the text we have seen a number of recursive definitions and constructions, where mathematical concepts of size $n + 1$ were obtained from comparable concepts of size n once we were able to establish such a concept for a first value of n, such as 0 or 1. These ideas referred back to prior results in order to obtain present and future ones. Now we shall find ourselves doing the same thing again. We shall investigate numeric functions $a(n)$, preferably written as a_n (for $n \geq 0$), where a_n depends on some of the prior terms $a_{n-1}, a_{n-2}, \ldots, a_1, a_0$. This study of what are called either *recurrence relations* or *difference equations* is the discrete counterpart to ideas applied in ordinary differential equations.

Our development will not employ any ideas from differential equations but will start with the notion of a geometric series. As further ideas are developed, we shall see some of the many applications that make this topic so important.

10.1
THE FIRST-ORDER LINEAR RECURRENCE RELATION

A *geometric series* is an infinite sequence of numbers, such as $5, 15, 45, 135, \ldots$, where the division of any term, other than the first, by its immediate predecessor is a constant, called the *common ratio*. For our sequence this common ratio is 3: $15 = 3(5), 45 = 3(15)$, and so on. If $a_0, a_1, a_2, \ldots$ is a geometric series, then $a_1/a_0 = a_2/a_1 = \cdots = a_{n+1}/a_n = \cdots = r$, the common ratio. In this particular geometric series we have $a_{n+1} = 3a_n, n \geq 0$.

The *recurrence relation* $a_{n+1} = 3a_n, n \geq 0$, does not define a unique geometric series. The sequence $7, 21, 63, 189, \ldots$ also satisfies the relation. To pinpoint the particular sequence described by $a_{n+1} = 3a_n$, we need to know one of the terms of that sequence. Hence,

$$a_{n+1} = 3a_n, \qquad n \geq 0, \qquad a_0 = 5,$$

uniquely defines the sequence $5, 15, 45, \ldots$, whereas

$$a_{n+1} = 3a_n, \qquad n \geq 0, \qquad a_1 = 21,$$

identifies $7, 21, 63, \ldots$ as the geometric series under study.

We call the equation $a_{n+1} = 3a_n, n \geq 0$, a recurrence relation because the value of a_{n+1} (the present consideration) is dependent on a_n (a prior consideration). Since a_{n+1} depends only on its immediate predecessor, the relation is said to be of *first order*. In particular, this is a *first-order linear homogeneous* recurrence relation *with constant coefficients*. (We'll say more about these ideas later.) The general form of such an equation can be written $a_{n+1} - ca_n = 0, n \geq 0$, where c is a constant.

When values such as a_0 or a_1 are given in addition to the recurrence relations, we call these values *boundary conditions*. The expression $a_0 = k$, where k is a constant, is also referred to as an *initial condition*. Our examples show the importance of the boundary condition in determining the unique solution.

Let us return now to the recurrence relation

$$a_{n+1} = 3a_n, \qquad n \geq 0, \qquad a_0 = 5.$$

Writing out the first four terms of this sequence, we find that

$$a_0 = 5,$$
$$a_1 = 3a_0 = 3(5),$$
$$a_2 = 3a_1 = 3(3a_0) = 3^2(5)$$
$$a_3 = 3a_2 = 3(3^2(5)) = 3^3(5).$$

Consequently, these results suggest that for any $n \geq 0$, $a_n = 5(3^n)$. This is called the *general solution* of the given recurrence relation. In the general solution, the value of a_n is a function of n and there is no longer any dependence on prior terms of the sequence. If we wish to compute a_{10}, we simply calculate $5(3^{10}) = 295,245$; there is no need to start at a_0 and build up to a_9 in order to obtain a_{10}.

From this example we are directed to the following. (This result can be established by mathematical induction.)

The general solution of the recurrence relation

$$a_{n+1} = ca_n, \text{ where } n \geq 0 \text{ and } c \text{ is a constant,}$$

is unique and is given by

$$a_n = a_0 c^n, \qquad n \geq 0.$$

Thus the solution $a_n = a_0 c^n, n \geq 0$, defines a discrete function whose domain is the set $\mathbf{N}$ of all nonnegative integers.

Example 10.1 Solve the recurrence relation $a_n = 7a_{n-1}$, where $n \geq 1$ and $a_2 = 98$.

This is just an alternative form of the relation $a_{n+1} = 7a_n$ for $n \geq 0$ and $a_2 = 98$. Hence the general solution has the form $a_n = a_0(7^n)$. Since $a_2 = 98 = a_0(7^2)$, it follows that $a_0 = 2$, and $a_n = 2(7^n)$, $n \geq 0$, is the unique solution. □

Example 10.2 A bank pays 6% (annual) interest on savings, compounding the interest monthly. If Bonnie deposits $1000 on the first day of May, how much will this deposit be worth a year later?

The annual interest rate is 6%, so the monthly rate is $6\%/12 = 0.5\% = 0.005$. For $0 \leq n \leq 12$, let p_n denote the value of Bonnie's deposit at the end of n months. Then $p_{n+1} = p_n + 0.005 p_n$, where $0.005 p_n$ is the interest earned on p_n during month $n + 1$, for $0 \leq n \leq 11$, and $p_0 = \$1000$.

The relation $p_{n+1} = (1.005)p_n$, $p_0 = \$1000$, has the solution $p_n = p_0(1.005)^n = \$1000(1.005)^n$. Consequently, at the end of one year, Bonnie's deposit is worth $\$1000(1.005)^{12} = \$1061.68$. □

The recurrence relation $a_{n+1} - ca_n = 0$ is called *linear* because each subscripted term appears to the first power (as x and y do in the equation of a line in the plane). Sometimes a nonlinear recurrence relation can be made into a linear one by an algebraic substitution.

Example 10.3 Find a_{12} if $a_{n+1}^2 = 5a_n^2$, where $a_n > 0$ for $n \geq 0$, and $a_0 = 2$.

Although this recurrence relation is not linear in a_n, if we let $b_n = a_n^2$, then the new relation $b_{n+1} = 5b_n$ for $n \geq 0$, and $b_0 = 4$, is a linear relation whose solution is $b_n = 4 \cdot 5^n$. Therefore, $a_n = 2(\sqrt{5})^n$ for $n \geq 0$, and $a_{12} = 31{,}250$. □

The general first-order linear recurrence relation with constant coefficients has the form $a_{n+1} + ca_n = f(n)$, $n \geq 0$, where c is a constant and $f(n)$ is a function on the set $\mathbf{N}$ of nonnegative integers.

When $f(n) = 0$ for all $n \in \mathbf{N}$, the relation is called *homogeneous*; otherwise the relation is called *nonhomogeneous*. So far we have dealt only with homogeneous relations. Now we shall solve a nonhomogeneous relation. We shall develop specific techniques that work for all linear homogeneous recurrence relations with constant coefficients. However, many different techniques prove useful when we deal with a nonhomogeneous problem, although none allows us to solve everything that can arise. But if we should recognize some pattern from our past experiences, we are more likely to succeed.

Example 10.4 Perhaps the most popular, though not the most efficient, method of sorting numeric data is a technique called the *bubble sort*. Here the input is an array $A[1], A[2], \ldots, A[n]$ of n real numbers that are to be sorted into ascending order.

The Pascal program segment shown in Fig. 10.1 provides an implementation for this type of algorithm. Here the integer variable i is the counter for the outer

For-loop, whereas the integer variable j is the counter for the inner For-loop. Finally, the real variable *temp* is used for storage that is needed when an exchange takes place.

```
Begin
  For i := 1 to n − 1 do
    For j := n downto i + 1 do
      If A[j] < A[j − 1] then
        Begin
          temp := A[j − 1];
          A[j − 1] := A[j];
          A[j] := temp
        End
End;
```

Figure 10.1

Starting with the last entry, $A[n]$, in the given array, we compare it with its immediate predecessor, $A[n-1]$. If $A[n] < A[n-1]$, we interchange the values stored in $A[n-1]$ and $A[n]$. Then we compare $A[n-1]$ (which may have originally been $A[n]$) with its immediate predecessor, $A[n-2]$. (If the original $A[n-1]$ and $A[n]$ are such that $A[n-1] \le A[n]$, then we are simply comparing the original value of $A[n-1]$ with $A[n-2]$.) If $A[n-1] < A[n-2]$, we interchange them. We continue the process. After $n-1$ such comparisons, the smallest number in the list is stored in $A[1]$. We then repeat this process for the $n-1$ numbers now stored in the (smaller) array $A[2], A[3], \ldots, A[n]$. In this way, each time (counted by i) this process is carried out, the smallest number in the remaining sublist "bubbles up" to the front of that sublist.

A small example wherein $n = 5$ and $A[1] = 7$, $A[2] = 9$, $A[3] = 2$, $A[4] = 5$, and $A[5] = 8$ is given in Fig. 10.2 to show how the bubble sort of Fig. 10.1 places a given sequence in ascending order. In this figure each comparison that leads to an interchange is denoted by the symbol $\rangle$; the symbol $\}$ indicates a comparison that results in no interchange.

In order to determine the time complexity function $f(n)$ when this algorithm is used on an input (array) of size $n \ge 1$, we count the total number of *comparisons* made in order to sort the n given numbers into ascending order.

If a_n denotes the number of comparisons needed to sort n numbers this way, then we get the following recurrence relation:

$$a_n = a_{n-1} + (n-1), \qquad n \ge 2, \qquad a_1 = 0.$$

This arises as follows. Given a list of n numbers, we make $n-1$ comparisons to bubble the smallest number up to the start of the list. The remaining sublist of $n-1$ numbers then requires a_{n-1} comparisons in order to be completely sorted.

This relation is a linear first-order relation with constant coefficients, but the term $n-1$ makes it nonhomogeneous. Since we have no technique for attacking

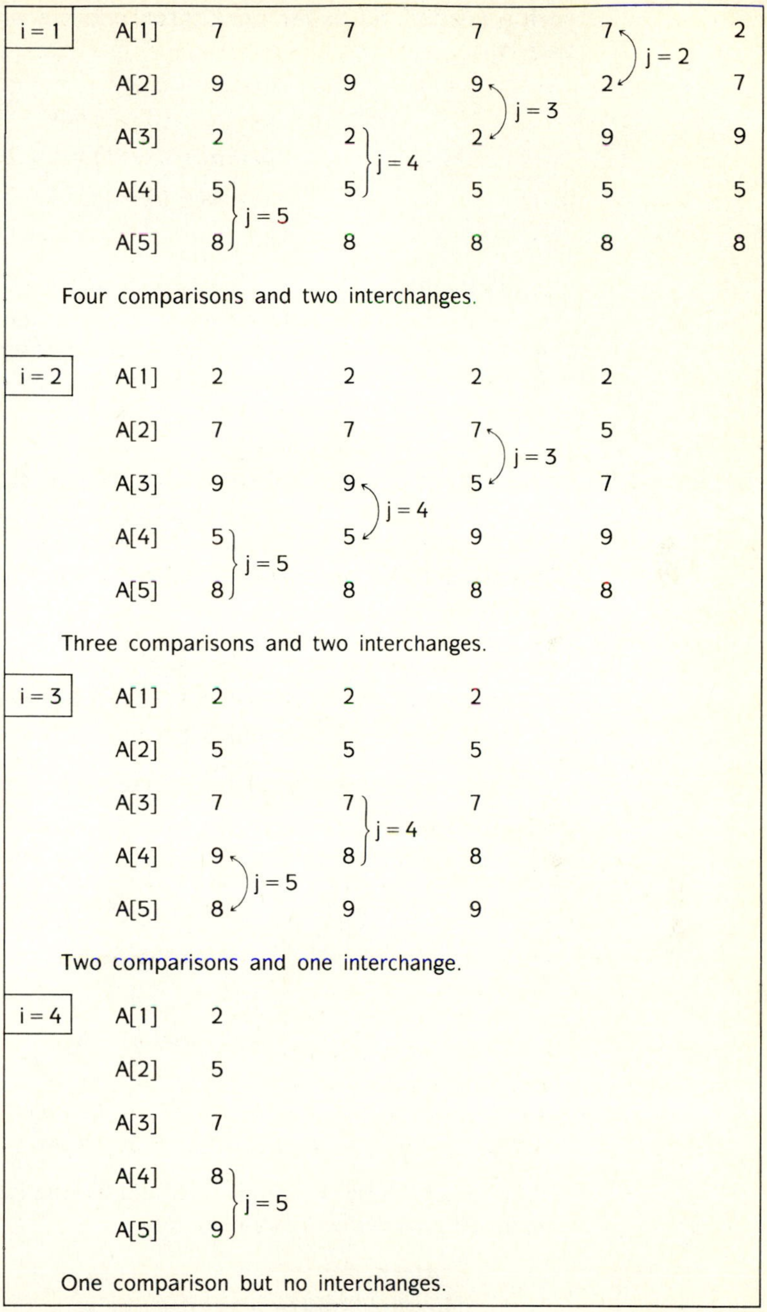

i = 1

A[1]	7	7	7	7	2
A[2]	9	9	9	2	7
A[3]	2	2	2	9	9
A[4]	5	5	5	5	5
A[5]	8	8	8	8	8

($j = 5$, $j = 4$, $j = 3$, $j = 2$)

Four comparisons and two interchanges.

i = 2

A[1]	2	2	2	2
A[2]	7	7	7	5
A[3]	9	9	5	7
A[4]	5	5	9	9
A[5]	8	8	8	8

($j = 5$, $j = 4$, $j = 3$)

Three comparisons and two interchanges.

i = 3

A[1]	2	2	2
A[2]	5	5	5
A[3]	7	7	7
A[4]	9	8	8
A[5]	8	9	9

($j = 5$, $j = 4$)

Two comparisons and one interchange.

i = 4

A[1]	2
A[2]	5
A[3]	7
A[4]	8
A[5]	9

($j = 5$)

One comparison but no interchanges.

Figure 10.2

such a relation, let us list some terms and see whether there is a recognizable pattern.

$$a_1 = 0$$
$$a_2 = a_1 + (2-1) = 1$$
$$a_3 = a_2 + (3-1) = 1+2$$
$$a_4 = a_3 + (4-1) = 1+2+3$$
$$\cdots \qquad \cdots \qquad \cdots \qquad \cdots$$

In general $a_n = 1 + 2 + \cdots + (n-1) = [(n-1)n]/2 = (n^2 - n)/2$.

As a result, the bubble sort determines the function $f: \mathbf{Z}^+ \to \mathbf{R}$ given by $f(n) = a_n = (n^2 - n)/2$. Consequently, as a measure of the running time for the algorithm, we write $f \in O(n^2)$. Hence the bubble sort is said to require $O(n^2)$ comparisons. □

Example 10.5   In part (b) of Example 9.6 we sought the generating function for the sequence $0, 2, 6, 12, 20, 30, 42, \ldots$, and our solution rested on our ability to recognize that $a_n = n^2 + n$ for each $n \in \mathbf{N}$. If we fail to see this, perhaps we can examine the given sequence and determine whether there is some other pattern that will help us.

Here $a_0 = 0$, $a_1 = 2$, $a_2 = 6$, $a_3 = 12$, $a_4 = 20$, $a_5 = 30$, $a_6 = 42$, and

$$a_1 - a_0 = 2 \qquad\qquad a_4 - a_3 = 8$$
$$a_2 - a_1 = 4 \qquad\qquad a_5 - a_4 = 10$$
$$a_3 - a_2 = 6 \qquad\qquad a_6 - a_5 = 12.$$

These calculations suggest the recurrence relation

$$a_n - a_{n-1} = 2n, \qquad n \geq 1, \qquad a_0 = 0.$$

To solve this relation, we proceed in a slightly different manner from the method we used in Example 10.4. Consider the following n equations:

$$a_1 - a_0 = 2$$
$$a_2 - a_1 = 4$$
$$a_3 - a_2 = 6$$
$$\vdots \qquad \vdots \qquad \vdots$$
$$a_n - a_{n-1} = 2n.$$

When we add these equations, the sum for the left-hand side will contain a_i and $-a_i$ for all $1 \leq i \leq n-1$. So we obtain

$$a_n - a_0 = 2 + 4 + 6 + \cdots + 2n = 2(1 + 2 + 3 + \cdots + n)$$
$$= 2[n(n+1)/2] = n^2 + n.$$

Since $a_0 = 0$, it follows that $a_n = n^2 + n$ for all $n \in \mathbf{N}$, as we found earlier in part (b) of Example 9.6. □

At this point we shall examine a recurrence relation with a variable coefficient.

Example 10.6   Solve the relation $a_n = n \cdot a_{n-1}$, where $n \geq 1$ and $a_0 = 1$.

Writing the first five terms defined by the relation, we have

$$a_0 = 1 \qquad\qquad\qquad a_3 = 3 \cdot a_2 = 3 \cdot 2 \cdot 1$$
$$a_1 = 1 \cdot a_0 = 1 \qquad\qquad a_4 = 4 \cdot a_3 = 4 \cdot 3 \cdot 2 \cdot 1$$
$$a_2 = 2 \cdot a_1 = 2 \cdot 1 \qquad \cdots \quad\cdots\quad \cdots$$

Therefore, $a_n = n!$ and the solution is the discrete numeric function $a_n, n \geq 0$, which counts the number of permutations of n objects.   □

While on the subject of permutations, we shall examine a recursive algorithm for generating the permutations of $\{1, 2, 3, \ldots, n - 1, n\}$ from those for $\{1, 2, 3, \ldots, n - 1\}$.†

There is only one permutation of $\{1\}$. Examining the permutations of $\{1, 2\}$,

$$
\begin{array}{cc}
1 & 2 \\
2 & 1
\end{array}
$$

we see that after writing the permutation 1 twice, we intertwine the number 2 about 1 to get the permutations listed. Writing each of these two permutations three times, we intertwine the number 3 and obtain

$$
\begin{array}{ccc}
  & 1 &   & 2 & 3 \\
  & 1 & 3 & 2 &   \\
3 & 1 &   & 2 &   \\
3 & 2 &   & 1 &   \\
  & 2 & 3 & 1 &   \\
  & 2 &   & 1 & 3
\end{array}
$$

We see here that the first permutation is 123 and that we obtain each of the next two permutations from its immediate predecessor by interchanging two numbers: 3 and the integer to its left. When 3 reaches the left side of the permutation, we examine the remaining numbers and permute them according to the list of permutations we generated for $\{1, 2\}$. (This makes the procedure recursive.) After that we interchange 3 with the integer on its right until 3 is on the right side of the permutation. We note that if we interchange 1 and 2 in the last permutation, we get 123, the first permutation listed.)

Continuing for $S = \{1, 2, 3, 4\}$, we first list each of the six permutations of $\{1, 2, 3\}$ four times. Starting with the permutation 1234, we intertwine the 4 throughout the remaining 23 permutations as indicated in Table 10.1. The only new idea here develops as follows. When progressing from permutation (5) to (6) to (7) to (8), we interchange 4 with the integer to its right. At permutation (8), where 4 has reached the right side, we obtain permutation (9) by keeping the location of 4 fixed and replacing the permutation 132 by 312 from the list of permutations of $\{1, 2, 3\}$. After that we continue as for the first eight permuta-

---

† The material from here to the end of this section is a digression that uses the idea of recursion. It does not deal with methods for solving recurrence relations and may be omitted with no loss of continuity.

tions until we reach permutation (16), where 4 is again on the right. We then permute 321 to obtain 231 and continue intertwining 4 until all 24 permutations have been generated. Once again, if 1 and 2 are interchanged in the last permutation, we obtain the first permutation in our list.

Table 10.1

(1)		1	2	3	4
(2)		1	2	4	3
(3)		1	4	2	3
(4)	4	1	2	3	
(5)	4	1	3	2	
(6)		1	4	3	2
(7)		1	3	4	2
(8)		1	3	2	4
(9)		3	1	2	4
(10)		3	1	4	2
(11)		3	4	1	2
...		...	...	...	
(15)		3	2	4	1
(16)		3	2	1	4
(17)		2	3	1	4
...		...	...	...	
(22)		2	4	1	3
(23)		2	1	4	3
(24)		2	1	3	4

The chapter references provide more information on recursive procedures in generating permutations and combinations.

We shall close this first section by returning to an earlier idea—the greatest common divisor of two positive integers.

Example 10.7 Recursive methods are fundamental in the areas of discrete mathematics and the analysis of algorithms. Such methods arise when we want to solve a given problem by breaking it down, or referring it, to smaller similar problems. In the programming language Pascal, this can be implemented by the use of recursive functions and procedures, which are permitted to invoke themselves. This example will provide one such function.

In computing $(333, 84)$, the g.c.d. of 333 and 84, we obtain the following calculations when we use the Euclidean Algorithm (presented in Section 4.3).

$$333 = 3(84) + 81 \qquad 0 < 81 < 84 \tag{1}$$

$$84 = 1(81) + 3 \qquad 0 < 3 < 81 \tag{2}$$

$$81 = 27(3) + 0. \tag{3}$$

Since 3 is the last nonzero remainder, the Euclidean Algorithm tells us that $(333, 84) = 3$. However, if we use only the calculations in Eqs. (2) and (3), then we find that $(84, 81) = 3$. And Eq. (3) alone implies $(81, 3) = 3$ because 3 divides 81. Consequently,

$$(333, 84) = (84, 81) = (81, 3) = 3,$$

where the integers involved in the successive calculations get smaller as we go from Eq. (1) to Eq. (2) to Eq. (3).

We also observe that

$$81 = 333 \text{ Mod } 84 \text{ and } 3 = 84 \text{ Mod } 81.$$

Therefore it follows that

$$(333, 84) = (84, 333 \text{ Mod } 84) = (333 \text{ Mod } 84, 84 \text{ Mod } (333 \text{ Mod } 84)).$$

These results provide the following recursive method for computing (a, b), where $a, b \in \mathbf{Z}^+$.

Say we have the input $a, b \in \mathbf{Z}^+$.

Step 1. If $b \mid a$ (or a Mod $b = 0$), then $(a, b) = b$.

Step 2. If $b \nmid a$, then perform the following tasks in the order specified.

- **(i)** Set $a = b$.
- **(ii)** Set $b = a$ Mod b, where the value of a for this assignment is the *old* value of a.
- **(iii)** Return to step 1.

These ideas are used in the Pascal program in Fig. 10.3, where the shaded segment is a *recursive function* that determines (a, b). (The reader may wish to compare this program with the one given in Fig. 4.8.) □

```
Program  EuclideanAlgorithm2 (input,output);

Var
   p,q: integer;

Function gcd (a,b: integer): integer;

Begin
   If  a Mod b=0  then
     gcd := b
   Else  gcd := gcd (b,a Mod b)
End;

Begin
   Writeln ('This program is designed to find');
   Writeln ('the gcd of two positive integers');
   Writeln ('p and q.');
   Write ('The first positive integer p is ');
   Readln (p);
   Write ('The second positive integer q is ');
   Readln (q);

   Writeln ('The greatest common divisor of ');
   Writeln (p,' and ', q,' is ', gcd (p,q), '.')
End.
```

Figure 10.3

EXERCISES 10.1

1. Find the recurrence relation, with initial condition, that uniquely determines each of the following geometric series.

   a) $2, 10, 50, 250, \ldots$ **b)** $6, -18, 54, -162, \ldots$

   c) $1, 1/3, 1/9, 1/27, \ldots$ **d)** $7, 14/5, 28/25, 56/125, \ldots$

2. Find the general solution of each of the following recurrence relations.

   a) $a_{n+1} - 1.5a_n = 0, n \geq 0$ **b)** $4a_n - 5a_{n-1} = 0, n \geq 1$

   c) $3a_{n+1} - 4a_n = 0, n \geq 0, a_1 = 5$ **d)** $2a_n - 3a_{n-1} = 0, n \geq 1, a_4 = 81$

3. If a_n, $n \geq 0$, is a solution of the recurrence relation $a_{n+1} - ca_n = 0$, and $a_3 = 153/49$, $a_5 = 1377/2401$, what is c?

4. The number of bacteria in a culture is 1000 (approximately), and this number increases 250% every 2 hours. Use a recurrence relation to determine the number of bacteria present after one day.

5. If Laura invests $100 at 6% interest compounded quarterly, how many months must she wait for her money to double? (She cannot withdraw the money before the quarter is up.)

6. Paul invested the stock profits he received 15 years ago in an account that paid 8% interest compounded quarterly. If his account now has $7218.27 in it, what was his initial investment?

7. Find a_{10} if $a_n^3 = 7a_{n-1}^3$, where $n \geq 1$ and $a_0 = 3$.

8. Solve for a_n, $n \geq 2$, if $5na_n + 2na_{n-1} = 2a_{n-1}$, where $n \geq 3$ and $a_3 = -30$.

9. Let $x_1, x_2, \ldots, x_{20}$ be a list of distinct real numbers to be sorted by the bubble sort technique of Example 10.4. (a) After how many comparisons will the 10 smallest numbers of the original list be arranged in ascending order? (b) How many more comparisons are needed to finish this sorting job?

10. Complete the list of permutations shown in Table 10.1.

11. Say the permutations of $\{1, 2, 3, 4, 5\}$ are generated by the procedure developed after Example 10.6. (a) What is the last permutation in the list? (b) What two permutations precede 25134? (c) What three permutations follow 25134? (d) Where do the permutations 21543, 35421, and 31524 occur in the list of 120 permutations?

12. For the implementation of the bubble sort given in Fig. 10.1, the outer For-loop is executed $n - 1$ times. This occurs regardless of whether any interchanges take place during the execution of the inner For-loop. Consequently, for $i = k$, where $1 \leq k \leq n - 2$, if the execution of the inner For-loop results in no interchanges, then the list is in ascending order. So the execution of the outer For-loop for $k + 1 \leq i \leq n - 1$ is not needed.

    a) For the situation described here, how many unnecessary comparisons are made if the execution of the inner For-loop for $i = k$ ($1 \leq k \leq n - 2$) results in no interchanges?

    b) Write an improved version of the bubble sort shown in Fig. 10.1. (Your result should eliminate the unnecessary comparisons discussed at the start of this exercise.)

c) Using the number of comparisons as a measure of its running time, determine the best-case and the worst-case time complexities for the algorithm implemented in part (b).

13. Let $x_1, x_2, x_3, \ldots, x_n$ be n real numbers, with $x_1 \leq x_2 \leq x_3 \leq \cdots \leq x_n$. The *median* for this set of n numbers is defined as

$$x_{(n+1)/2}, \text{ for } n \text{ odd}$$
$$(1/2)[x_{n/2} + x_{(n/2)+1}], \text{ for } n \text{ even}.$$

a) Suppose that $A[1], A[2], A[3], \ldots, A[n]$ is an array of n real numbers, *not* necessarily in ascending or descending order. Write a computer program (or develop an algorithm) to determine the median for the set of real numbers listed in this array.

b) Discuss the worst-case time complexity for the program you wrote in part (a).

14. For the program in Fig. 10.3, is it necessary that $a \geq b$?

10.2

THE SECOND-ORDER LINEAR HOMOGENEOUS RECURRENCE RELATION WITH CONSTANT COEFFICIENTS

Let $k \in \mathbf{Z}^+$ and $c_n(\neq 0), c_{n-1}, c_{n-2}, \ldots, c_{n-k}(\neq 0)$ be real numbers. If $a_n, n \geq 0$, is a discrete numeric function, then

$$c_n a_n + c_{n-1} a_{n-1} + c_{n-2} a_{n-2} + \cdots + c_{n-k} a_{n-k} = f(n), \qquad n \geq k,$$

is a linear recurrence relation (with constant coefficients) of *order k*. When $f(n) = 0$ for all $n \geq 0$, the relation is called *homogeneous*, otherwise it is *nonhomogeneous*.

In this section we shall concentrate on the homogeneous relation of order two: $c_n a_n + c_{n-1} a_{n-1} + c_{n-2} a_{n-2} = 0, n \geq 2$. On the basis of our work in Section 10.1, we seek a solution of the form $a_n = cr^n$, where $c \neq 0$ and $r \neq 0$.

Substituting $a_n = cr^n$ into $c_n a_n + c_{n-1} a_{n-1} + c_{n-2} a_{n-2} = 0$, we obtain

$$c_n cr^n + c_{n-1} cr^{n-1} + c_{n-2} cr^{n-2} = 0.$$

With $c, r \neq 0$, this becomes $c_n r^2 + c_{n-1} r + c_{n-2} = 0$, a quadratic equation. Hence the roots r_1, r_2 of this equation, which is called the *characteristic equation*, fall into the following three cases: (a) r_1, r_2 are distinct real numbers; (b) r_1, r_2 form a complex conjugate pair; or (c) r_1, r_2 are real, but $r_1 = r_2$. In all cases r_1 and r_2 are called the *characteristic roots*.

Case (A): (Distinct Real Roots)

Example 10.8   Solve the recurrence relation $a_n + a_{n-1} - 6a_{n-2} = 0$, where $n \geq 2$ and $a_0 = 1, a_1 = 2$.
If $a_n = cr^n$ with $c, r \neq 0$, we obtain $cr^n + cr^{n-1} - 6cr^{n-2} = 0$ from which the characteristic equation $r^2 + r - 6 = 0$ follows.

$$0 = r^2 + r - 6 = (r + 3)(r - 2) \Rightarrow r = 2, -3.$$

Since we have two distinct real roots, $a_n = 2^n$ and $a_n = (-3)^n$ are both solutions. They are *linearly independent solutions* because one is not a multiple of the other; that is, there is no real constant k such that $(-3)^n = k(2^n)$ for *all* $n \in \mathbf{N}$.† So $a_n = c_1(2)^n + c_2(-3)^n$ denotes the general solution, where c_1, c_2 are arbitrary constants.

With $a_0 = 1$ and $a_1 = 2$, c_1 and c_2 are determined as follows:

$$1 = a_0 = c_1(2)^0 + c_2(-3)^0 = c_1 + c_2$$
$$2 = a_1 = c_1(2)^1 + c_2(-3)^1 = 2c_1 - 3c_2.$$

Solving this system of equations, one finds $c_1 = 1, c_2 = 0$. Therefore $a_n = 2^n$ is the *unique* solution of the given recurrence relation.   □

An interesting second-order homogeneous recurrence relation is the *Fibonacci relation*. (This was mentioned earlier in Section 9.6.)

Example 10.9   Solve the recurrence relation $F_{n+2} = F_{n+1} + F_n$, where $n \geq 0$ and $F_0 = 0, F_1 = 1$.
As in the previous example, let $F_n = cr^n$, for $c, r \neq 0, n \geq 0$. Upon substitution we get $cr^{n+2} = cr^{n+1} + cr^n$. This gives the characteristic equation $r^2 - r - 1 = 0$. The characteristic roots are $r = (1 \pm \sqrt{5})/2$, so the general solution is $F_n = c_1[(1 + \sqrt{5})/2]^n + c_2[(1 - \sqrt{5})/2]^n$.
To solve for c_1, c_2, we use the given initial values and write $0 = F_0 = c_1 + c_2$, $1 = F_1 = c_1[(1 + \sqrt{5})/2] + c_2[(1 - \sqrt{5})/2]$. Since $-c_1 = c_2$, we have $2 = c_1(1 + \sqrt{5}) - c_1(1 - \sqrt{5})$ and $c_1 = 1/\sqrt{5}$. The general solution is given by

$$F_n = \frac{1}{\sqrt{5}}\left[\left(\frac{1 + \sqrt{5}}{2}\right)^n - \left(\frac{1 - \sqrt{5}}{2}\right)^n\right].   □$$

Example 10.10   For $n \geq 0$, let $S = \{1, 2, 3, \ldots, n\}$ (when $n = 0, S = \emptyset$), and let a_n denote the number of subsets of S that contain no consecutive integers. Find and solve a recurrence relation for a_n.
For $0 \leq n \leq 4$, we have $a_0 = 1, a_1 = 2, a_2 = 3, a_3 = 5$, and $a_4 = 8$. (For example, $a_3 = 5$ because $S = \{1, 2, 3\}$ has $\emptyset$, $\{1\}$, $\{2\}$, $\{3\}$, and $\{1, 3\}$ as subsets with no consecutive integers.) These first five terms are reminiscent of the Fibonacci sequence. But do things change as we continue?
Let $n \geq 2$ and $S = \{1, 2, 3, \ldots, n - 2, n - 1, n\}$. If $A \subseteq S$ and A is to be counted in a_n, there are two possibilities:

a) $n \in A$: When this happens $(n - 1) \notin A$, and $A - \{n\}$ would be counted in a_{n-2}.

---

† We can also call the solutions $a_n = 2^n$ and $a_n = (-3)^n$ *linearly independent* when the following condition is satisfied: For $k_1, k_2 \in \mathbf{R}$, if $k_1(2^n) + k_2(-3)^n = 0$ for all $n \in \mathbf{N}$, then $k_1 = k_2 = 0$.

b) $n \notin A$: For this case A would be counted in a_{n-1}.

These two cases are exhaustive and mutually disjoint, so we conclude that $a_n = a_{n-1} + a_{n-2}$, where $n \geq 2$ and $a_0 = 1, a_1 = 2$, is the recurrence relation for the problem. Now we could solve for a_n, but if we notice that $a_n = F_{n+2}, n \geq 0$, then the result of Example 10.9 implies that

$$a_n = \frac{1}{\sqrt{5}}\left[\left(\frac{1+\sqrt{5}}{2}\right)^{n+2} - \left(\frac{1-\sqrt{5}}{2}\right)^{n+2}\right], \qquad n \geq 0. \quad \square$$

Let us examine a comparable relation in a computer science application.

Example 10.11 In many programming languages one may consider those legal arithmetic expressions, *without parentheses*, that are made up of the digits $0, 1, 2, \ldots, 9$ and the binary operation symbols $+$, $*$, $/$. For example, $3 + 4$ and $2 + 3 * 5$ are legal arithmetic expressions; $8 + * 9$ is not. Here $2 + 3 * 5 = 17$, since there is a hierarchy of operations: multiplication and division are performed before addition. Operations at the same level are performed in their order of appearance as the expression is scanned from left to right.

For $n \in \mathbf{Z}^+$, let a_n be the number of these (legal) arithmetic expressions that are made up of n symbols. Then $a_1 = 10$, since the arithmetic expressions of one symbol are the 10 digits. Next $a_2 = 100$. This accounts for the expressions $00, 01, \ldots, 09, 10, 11, \ldots, 99$. (There are no unnecessary leading plus signs.) When $n \geq 3$, we consider two cases for a_n:

1. If x is an arithmetic expression of $n - 1$ symbols, the last symbol must be a digit. Adding one more digit to the right of x, we get $10a_{n-1}$ arithmetic expressions of n symbols where the last two symbols are digits.

2. Now let y be an expression of $n - 2$ symbols. To obtain an expression with n symbols (that is not counted in Case 1), we adjoin to the right of y one of the 29 two-symbol expressions $+1, \ldots, +9, +0, *1, \ldots, *9, *0, /1, \ldots, /9$.

    From these two cases we have $a_n = 10a_{n-1} + 29a_{n-2}$, where $n \geq 2$ and $a_1 = 10$, $a_2 = 100$. Here the characteristic roots are $5 \pm 3\sqrt{6}$ and the solution is $a_n = (5/3\sqrt{6})[(5 + 3\sqrt{6})^n - (5 - 3\sqrt{6})^n]$. (Verify this.) $\square$

Now we turn to an application in the physical sciences.

Example 10.12 Consider the linear network in Fig. 10.4, where there are k one-ohm and k three-ohm resistors connected by wires to a generator that supplies a voltage V. Find a formula for $v_n, 0 \leq n \leq k$, that gives the voltage at each junction point as a function of n. (This is also the voltage drop across the one-ohm resistor below that junction point.)

We know that $v_k = V$, the voltage supplied by the generator. To find a recurrence relation we apply the following principles.

1. *Kirchhoff's Law*: At any junction point, the sum of the currents flowing into the junction point equals the sum of the currents leaving that point.

2. *Ohm's Law*: If the voltage drop across a resistor of R ohms is $V_2 - V_1$, then the current in that resistor is $(V_2 - V_1)/R$.

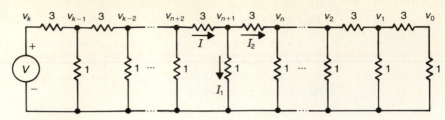

Figure 10.4

Consider the current at the junction point labeled v_{n+1}. By Kirchhoff's law, $I = I_1 + I_2$. By Ohm's law, $I = (v_{n+2} - v_{n+1})/3$, $I_1 = v_{n+1}/1$, and $I_2 = (v_{n+1} - v_n)/3$. Consequently,

$$\frac{v_{n+2} - v_{n+1}}{3} = \frac{v_{n+1}}{1} + \frac{v_{n+1} - v_n}{3}, \qquad 0 \le n \le k - 2,$$

or $v_{n+2} - 5v_{n+1} + v_n = 0$.

Let $v_n = cr^n$, for $c, r \ne 0$ and $0 \le n \le k$. The characteristic equation is $r^2 - 5r + 1 = 0$ with roots $\alpha = (5 + \sqrt{21})/2$, $\beta = (5 - \sqrt{21})/2$. From this, $v_n = c_1 \alpha^n + c_2 \beta^n$.

Now whatever v_0 is, by the two laws we have $(v_1 - v_0)/3 = v_0/1$, or $v_1 = 4v_0$. So $v_0 = c_1 + c_2$ and $4v_0 = v_1 = c_1 \alpha + c_2 \beta$. From this it follows that $c_1 = v_0(\beta - 4)/(\beta - \alpha)$, $c_2 = v_0(\alpha - 4)/(\alpha - \beta)$, and

$$v_n = v_0 \left[\frac{(\beta - 4)}{(\beta - \alpha)} \alpha^n + \frac{(\alpha - 4)}{(\alpha - \beta)} \beta^n \right].$$

Since

$$V = v_k = v_0 \left[\frac{(\beta - 4)\alpha^k - (\alpha - 4)\beta^k}{(\beta - \alpha)} \right],$$

we find $v_0 = (\beta - \alpha)V/[(\beta - 4)\alpha^k - (\alpha - 4)\beta^k]$ and this yields

$$v_n = V \left[\frac{(\beta - 4)\alpha^n - (\alpha - 4)\beta^n}{(\beta - 4)\alpha^k - (\alpha - 4)\beta^k} \right]. \qquad \square$$

The next example demonstrates how auxiliary variables may be helpful.

Example 10.13   Find a recurrence relation for the number of binary sequences of length n that have no consecutive 0's.

For $n \ge 0$, let a_n be the number of such sequences of length n. Let $a_n^{(0)}$ count those that end in 0, and $a_n^{(1)}$ those that end in 1. Hence $a_n = a_n^{(0)} + a_n^{(1)}$, and

$$a_n = 2 \cdot a_{n-1}^{(1)} + 1 \cdot a_{n-1}^{(0)}$$

The nth position can be 0 or 1.   The nth position can only be 1.

Consequently, $a_n = a_{n-1}^{(1)} + [a_{n-1}^{(1)} + a_{n-1}^{(0)}] = a_{n-1}^{(1)} + a_{n-1} = a_{n-1} + a_{n-2}$. (Why does $a_{n-1}^{(1)} = a_{n-2}$?) Therefore the recurrence relation for this problem is $a_n = a_{n-1} +$

a_{n-2}, where $n \geq 2$ and $a_0 = 1, a_1 = 2$. (We leave the details of the solution for the reader.) □

The last example for case (A) shows how to extend the results for second-order recurrence relations to those of higher order.

Example 10.14   Solve the recurrence relation

$$2a_{n+3} = a_{n+2} + 2a_{n+1} - a_n, \qquad n \geq 0, \quad a_0 = 0, \quad a_1 = 1, \quad a_2 = 2.$$

Letting $a_n = cr^n$ for $c, r \neq 0$ and $n \geq 0$, we obtain the characteristic equation $2r^3 - r^2 - 2r + 1 = 0 = (2r - 1)(r - 1)(r + 1)$. The characteristic roots are $1/2$, 1, and -1, so the solution is $a_n = c_1(1)^n + c_2(-1)^n + c_3(1/2)^n = c_1 + c_2(-1)^n + c_3(1/2)^n$. (The solutions 1, $(-1)^n$, and $(1/2)^n$ are called linearly independent because it is impossible to get any one of them as a linear combination of the other two.†) From $0 = a_0$, $1 = a_1$, and $2 = a_2$, we derive $c_1 = 5/2$, $c_2 = 1/6$, $c_3 = -8/3$. □

Case (B): (Complex Roots)

Before getting into the case of complex roots, we recall DeMoivre's Theorem: $(\cos \theta + i \sin \theta)^n = \cos n\theta + i \sin n\theta, n \geq 0$. (This is part (b) of Exercise 6 of Section 4.1.)

If $z = x + iy \in \mathbf{C}$, we can write $z = r(\cos \theta + i \sin \theta)$, where $r = \sqrt{x^2 + y^2}$ and $(y/x) = \tan \theta$. Then by DeMoivre's Theorem, $z^n = r^n(\cos n\theta + i \sin n\theta)$, for $n \geq 0$.

Example 10.15   Determine $(1 + \sqrt{3}\,i)^{10}$.

Fig. 10.5 shows a geometric way to represent the complex number $1 + \sqrt{3}\,i$ as the point $(1, \sqrt{3})$ in the xy plane. Here $r = \sqrt{1^2 + (\sqrt{3})^2} = 2$, and $\theta = \pi/3$.

So $1 + \sqrt{3}\,i = 2(\cos(\pi/3) + i \sin(\pi/3))$, and $(1 + \sqrt{3}\,i)^{10} = 2^{10}(\cos(10\pi/3) + i \sin(10\pi/3)) = 2^{10}(\cos(4\pi/3) + i \sin(4\pi/3)) = 2^{10}((-1/2) - (\sqrt{3}/2)\,i) = (-2^9)(1 + \sqrt{3}\,i)$. □

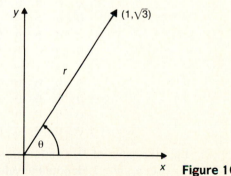

Figure 10.5

---

† Alternately, the solutions 1, $(-1)^n$, and $(-1/2)^n$ are linearly independent, because if a, b, c are real numbers, and $a(1) + b(-1)^n + c(-1/2)^n = 0$ for all $n \in \mathbf{N}$, then $a = b = c = 0$.

We'll use such results in the following examples.

Example 10.16 Solve the recurrence relation $a_n = 2(a_{n-1} - a_{n-2})$, where $n \geq 2$ and $a_0 = 1, a_1 = 2$.

Letting $a_n = cr^n$, for $c, r \neq 0$, we obtain the characteristic equation $r^2 - 2r + 2 = 0$, whose roots are $1 \pm i$. Consequently the general solution has the form $c_1(1 + i)^n + c_2(1 - i)^n$, where c_1 and c_2 presently denote arbitrary *complex* constants. [As in case (A), there are two independent solutions: $(1 + i)^n$ and $(1 - i)^n$.]

$$1 + i = \sqrt{2}(\cos(\pi/4) + i \, \sin(\pi/4))$$

and

$$1 - i = \sqrt{2}(\cos(-\pi/4) + i \, \sin(-\pi/4)) = \sqrt{2}(\cos(\pi/4) - i \, \sin(\pi/4)).$$

This yields

$$a_n = c_1(\sqrt{2})^n(\cos(n\pi/4) + i \, \sin(n\pi/4)) + c_2(\sqrt{2})^n(\cos(n\pi/4) - i \, \sin(n\pi/4))$$
$$= (\sqrt{2})^n[k_1 \cos(n\pi/4) + k_2 \sin(n\pi/4)],$$

where $k_1 = c_1 + c_2$ and $k_2 = (c_1 - c_2)i$.

$$1 = a_0 = [k_1 \cos 0 + k_2 \sin 0] = k_1$$
$$2 = a_1 = \sqrt{2}[1 \cdot \cos(\pi/4) + k_2 \sin(\pi/4)], \text{ or } 2 = 1 + k_2, \text{ and } k_2 = 1.$$

The solution for the given initial conditions is then given by $a_n = (\sqrt{2})^n \cdot [\cos(n\pi/4) + \sin(n\pi/4)], n \geq 0$. (Note that this solution contains no complex numbers. A small point may bother the reader here. How did we start with c_1, c_2 complex and end up with $k_1 = c_1 + c_2$ and $k_2 = (c_1 - c_2)i$ real? This happens if c_1, c_2 are complex conjugates.) $\square$

Let us now examine an application from linear algebra.

Example 10.17 For $b \in \mathbf{R}^+$, consider the $n \times n$ determinant D_n given by

$$
\begin{vmatrix}
b & b & 0 & 0 & 0 & 0 & \cdots & 0 & 0 & 0 & 0 & 0 \\
b & b & b & 0 & 0 & 0 & \cdots & 0 & 0 & 0 & 0 & 0 \\
0 & b & b & b & 0 & 0 & \cdots & 0 & 0 & 0 & 0 & 0 \\
0 & 0 & b & b & b & 0 & \cdots & 0 & 0 & 0 & 0 & 0 \\
0 & 0 & 0 & b & b & b & \cdots & 0 & 0 & 0 & 0 & 0 \\
\cdot & \cdot & \cdot & \cdot & \cdot & \cdot & \cdots & \cdot & \cdot & \cdot & \cdot & \cdot \\
0 & 0 & 0 & 0 & 0 & 0 & \cdots & b & b & b & 0 & 0 \\
0 & 0 & 0 & 0 & 0 & 0 & \cdots & 0 & b & b & b & 0 \\
0 & 0 & 0 & 0 & 0 & 0 & \cdots & 0 & 0 & b & b & b \\
0 & 0 & 0 & 0 & 0 & 0 & \cdots & 0 & 0 & 0 & b & b \\
\end{vmatrix}.
$$

Find the value of D_n as a function of n.

Let a_n, $n \geq 1$, denote the value of the $n \times n$ determinant D_n. Then

$$a_1 = |b| = b \quad \text{and} \quad a_2 = \begin{vmatrix} b & b \\ b & b \end{vmatrix} = 0.$$

Expanding D_n by its first row, we have $D_n =$

$$b \begin{vmatrix} b & b & 0 & 0 & 0 & \cdots & 0 & 0 & 0 & 0 \\ b & b & b & 0 & 0 & \cdots & 0 & 0 & 0 & 0 \\ 0 & b & b & b & 0 & \cdots & 0 & 0 & 0 & 0 \\ 0 & 0 & b & b & b & \cdots & 0 & 0 & 0 & 0 \\ \cdot & \cdot & \cdot & \cdot & \cdot & \cdots & \cdot & \cdot & \cdot & \cdot \\ 0 & 0 & 0 & 0 & 0 & \cdots & b & b & b & 0 \\ 0 & 0 & 0 & 0 & 0 & \cdots & 0 & b & b & b \\ 0 & 0 & 0 & 0 & 0 & \cdots & 0 & 0 & b & b \end{vmatrix}$$

(This is D_{n-1}.)

$$-b \begin{vmatrix} b & b & 0 & 0 & 0 & \cdots & 0 & 0 & 0 & 0 \\ 0 & b & b & 0 & 0 & \cdots & 0 & 0 & 0 & 0 \\ 0 & b & b & b & 0 & \cdots & 0 & 0 & 0 & 0 \\ 0 & 0 & b & b & b & \cdots & 0 & 0 & 0 & 0 \\ \cdot & \cdot & \cdot & \cdot & \cdot & \cdots & \cdot & \cdot & \cdot & \cdot \\ 0 & 0 & 0 & 0 & 0 & \cdots & b & b & b & 0 \\ 0 & 0 & 0 & 0 & 0 & \cdots & 0 & b & b & b \\ 0 & 0 & 0 & 0 & 0 & \cdots & 0 & 0 & b & b \end{vmatrix}$$

When we expand the second determinant by its first column, we find that $D_n = bD_{n-1} - (b)(b)D_{n-2} = bD_{n-1} - b^2 D_{n-2}$. This translates into the relation $a_n = ba_{n-1} - b^2 a_{n-2}$, for $n \geq 3$, $a_1 = b, a_2 = 0$.

If we let $a_n = cr^n$ for $c, r \neq 0$ and $n \geq 1$, the characteristic equation produces the roots $b[(1/2) \pm i\sqrt{3}/2]$.

Hence

$$a_n = c_1 [b((1/2) + i\sqrt{3}/2)]^n + c_2 [b((1/2) - i\sqrt{3}/2)]^n$$
$$= b^n [c_1 (\cos(\pi/3) + i \sin(\pi/3))^n + c_2 (\cos(\pi/3) - i \sin(\pi/3))^n]$$
$$= b^n [k_1 \cos(n\pi/3) + k_2 \sin(n\pi/3)].$$

$b = a_1 = b[k_1 \cos(\pi/3) + k_2 \sin(\pi/3)]$, so $1 = k_1(1/2) + k_2(\sqrt{3}/2)$, or $k_1 + \sqrt{3} k_2 = 2$.

$0 = a_2 = b^2 [k_1 \cos(2\pi/3) + k_2 \sin(2\pi/3)]$, so $0 = (k_1)(-1/2) + k_2(\sqrt{3}/2)$, or $k_1 = \sqrt{3} k_2$.

Hence $k_1 = 1, k_2 = 1/\sqrt{3}$ and the value of D_n is

$$b^n [\cos(n\pi/3) + (1/\sqrt{3}) \sin(n\pi/3)]. \quad \square$$

On to the problem of multiple roots.

Case (C): (Repeated Real Roots)

Example 10.18 Solve the recurrence relation $a_{n+2} = 4a_{n+1} - 4a_n$, where $n \geq 0$ and $a_0 = 1, a_1 = 3$.

As in the other two cases, we let $a_n = cr^n$, where $c, r \neq 0$ and $n \geq 0$. Then the characteristic equation is $r^2 - 4r + 4 = 0$ and the characteristic roots are both $r = 2$. (So $r = 2$ is called "a root of multiplicity 2.") Unfortunately, we now lack two independent solutions; 2^n and 2^n are definitely multiples of each other. We need one more (independent) solution. Let us try $n(2^n)$.

Substituting $a_n = n(2^n)$ in the given relation yields $4a_{n+1} - 4a_n = 4(n + 1) \cdot 2^{n+1} - 4n2^n = 2(n + 1)2^{n+2} - n2^{n+2} = [2n + 2 - n]2^{n+2} = (n + 2)2^{n+2} = a_{n+2}$, so $n(2^n)$ is the second independent solution. (It is independent because it is impossible to have $n2^n = k2^n$ for all $n \geq 0$ if k is a constant.)

The general solution is of the form $a_n = c_1(2^n) + c_2 n(2^n)$. With $a_0 = 1, a_1 = 3$ we find $a_n = 2^n + (1/2)n(2^n) = 2^n + n(2^{n-1}), n \geq 0$.   □

In general, if $c_n a_n + c_{n-1} a_{n-1} + \cdots + c_{n-k} a_{n-k} = 0$ with $c_n(\neq 0), c_{n-1}, \ldots, c_{n-k}(\neq 0)$ real constants, and r a characteristic root of multiplicity m, where $2 \leq m \leq k$, then the part of the general solution that involves the root r has the form

$$A_0 r^n + A_1 nr^n + A_2 n^2 r^n + \cdots + A_{m-1} n^{m-1} r^n$$
$$= (A_0 + A_1 n + A_2 n^2 + \cdots + A_{m-1} n^{m-1})r^n,$$

where $A_0, A_1, A_2, \ldots, A_{m-1}$ are arbitrary constants.

Our last example involves a little probability.

Example 10.19 If a first case of measles is recorded in a certain school system, let p_n denote the probability that at least one more case is reported during the nth week after the first recorded case. School records provide evidence that $p_n = p_{n-1} - (0.25)p_{n-2}$, where $n \geq 2$. Since $p_0 = 0$ and $p_1 = 1$, if the first case (of a new outbreak) is recorded on Monday, March 1, 1982, when did the probability for the occurrence of a new case decrease to less than 0.01 for the first time?

With $p_n = cr^n$ for $c, r \neq 0$, the characteristic equation for the recurrence relation is $r^2 - r + (1/4) = 0 = (r - (1/2))^2$. The general solution has the form $p_n = (c_1 + c_2 n)(1/2)^n$, $n \geq 0$. For $p_0 = 0$, $p_1 = 1$, we get $c_1 = 0$, $c_2 = 2$, so $p_n = n2^{-n+1}$, $n \geq 0$.

The first integer n for which $p_n < 0.01$ is 12. Hence, it was not until the week of May 16, 1982, that the probability of another new case occurring was less than 0.01.   □

EXERCISES 10.2

1. Solve the following recurrence relations. (No final answer should involve complex numbers.)

   a) $a_n = 5a_{n-1} + 6a_{n-2}$, $n \geq 2$, $a_0 = 1$, $a_1 = 3$

   b) $2a_{n+2} - 11a_{n+1} + 5a_n = 0$, $n \geq 0$, $a_0 = 2$, $a_1 = -8$

   c) $3a_{n+1} = 2a_n + a_{n-1}$, $n \geq 1$, $a_0 = 7$, $a_1 = 3$

   d) $a_{n+2} + a_n = 0$, $n \geq 0$, $a_0 = 0$, $a_1 = 3$

   e) $a_{n+2} + 4a_n = 0$, $n \geq 0$, $a_0 = a_1 = 1$

   f) $a_n - 6a_{n-1} + 9a_{n-2} = 0$, $n \geq 2$, $a_0 = 5$, $a_1 = 12$

   g) $a_n + 2a_{n-1} + 2a_{n-2} = 0$, $n \geq 2$, $a_0 = 1$, $a_1 = 3$

h) $a_n = 7a_{n-1} - 10a_{n-2}, \qquad n \geq 2, \quad a_0 = 3, \quad a_1 = 15$

i) $9a_{n+2} + 12a_{n+1} + 4a_n = 0, \qquad n \geq 0, \quad a_0 = 1, \quad a_1 = 4$

j) $a_n = 3a_{n-2} + 2a_{n-3}, \qquad n \geq 3, \quad a_0 = 1, \quad a_1 = 3, \quad a_2 = 7$

2. a) Verify the final solutions in Examples 10.11 and 10.18.

  b) Solve the recurrence relation in Example 10.13.

3. If $a_0 = 0$, $a_1 = 1$, $a_2 = 4$, and $a_3 = 37$ satisfy the recurrence relation $a_{n+2} + ba_{n+1} + ca_n = 0, n \geq 0$, where b and c are constants, solve for a_n.

4. Find and solve a recurrence relation for the number of ways to park motorcycles and compact cars in a row of n spaces if each cycle requires one space and each compact needs two. (All cycles are identical in appearance, as are the cars, and we want to use up all the n spaces.)

5. In the Fibonacci sequence, $F_n = F_{n+2} - F_{n+1}$, for $n \geq 0$. Hence, upon adding the equations

$$F_0 = F_2 - F_1$$
$$F_1 = F_3 - F_2$$
$$F_2 = F_4 - F_3$$
$$\cdots \quad \cdots \quad \cdots$$
$$F_n = F_{n+2} - F_{n+1},$$

we obtain

$$\sum_{i=0}^{n} F_i = F_{n+2} - F_1 = F_{n+2} - 1.$$

This is one of many such properties discovered by François Lucas (1842–1891). Prove the following additional properties.

a) $F_1 + F_3 + F_5 + \cdots + F_{2n-1} = F_{2n}, \qquad n \geq 1$

b) $F_0 + F_2 + F_4 + \cdots + F_{2n} = F_{2n+1} - 1, \qquad n \geq 0$

c) $F_0 + F_1 - F_2 + F_3 - F_4 + \cdots + F_{2n-1} - F_{2n} = -F_{2n-1} + 1, \qquad n \geq 1$

6. a) Prove that

$$\lim_{n \to \infty} \frac{F_{n+1}}{F_n} = \frac{1 + \sqrt{5}}{2}.$$

(This limit has come to be known as the *golden ratio*.)

  b) Consider a regular pentagon $ABCDE$ inscribed in a circle, as shown in Fig. 10.6.

  (i) Use the law of sines and the double angle formula for the sine to show that $AC/AX = 2\cos 36°$.

  (ii) As $\cos 18° = \sin 72° = 4\sin 18° \cos 18°(1 - 2\sin^2 18°)$ (Why?), show that $\sin 18°$ is a root of the polynomial equation $8x^3 - 4x + 1 = 0$, and deduce that $\sin 18° = (\sqrt{5} - 1)/4$.

  c) Verify that $AC/AX = (1 + \sqrt{5})/2$.

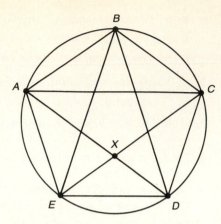

Figure 10.6

7. For $n \geq 0$, let a_n count the number of ways a sequence of 1's and 2's will sum to n. For example, $a_3 = 3$ because (1) $1, 1, 1$; (2) $1, 2$; and (3) $2, 1$ sum to 3. Find and solve a recurrence relation for a_n.

8. Find and solve a recurrence relation for the number of n-digit ternary $(0, 1, 2)$ sequences with no pair of consecutive 0's.

9. Find and solve a recurrence relation for the number of ways to stack n red, white, green, and blue poker chips so that there are no consecutive blue chips.

10. For $n \geq 1$, let a_n be the number of ways to write n as an ordered sum of positive integers where each summand is at least 2. (For example, $a_5 = 3$ because here we may represent 5 by 5, by $2 + 3$, and by $3 + 2$.) Find and solve a recurrence relation for a_n.

11. A particle moves horizontally to the right. For $n \in \mathbf{Z}^+$, the distance the particle travels in the $(n + 1)$st second is equal to twice the distance it travels during the nth second. If $x_n, n \geq 0$, denotes the position of the particle at the start of the $(n + 1)$st second, find and solve a recurrence relation for x_n, where $x_0 = 1$ and $x_1 = 5$.

12. For $n \geq 1$, let D_n be the following $n \times n$ determinant.

$$
\begin{vmatrix}
2 & 1 & 0 & 0 & 0 & \cdots & 0 & 0 & 0 & 0 \\
1 & 2 & 1 & 0 & 0 & \cdots & 0 & 0 & 0 & 0 \\
0 & 1 & 2 & 1 & 0 & \cdots & 0 & 0 & 0 & 0 \\
0 & 0 & 1 & 2 & 1 & \cdots & 0 & 0 & 0 & 0 \\
\cdot\cdot & \cdots & \cdots & \cdots & \cdots & \cdots & \cdots & \cdots & \cdots & \cdot\cdot \\
0 & 0 & 0 & 0 & 0 & \cdots & 1 & 2 & 1 & 0 \\
0 & 0 & 0 & 0 & 0 & \cdots & 0 & 1 & 2 & 1 \\
0 & 0 & 0 & 0 & 0 & \cdots & 0 & 0 & 1 & 2
\end{vmatrix}
$$

Find and solve a recurrence relation for the value of D_n.

13. Solve the recurrence relation $a_{n+2}^2 - 5a_{n+1}^2 + 4a_n^2 = 0$, where $n \geq 0$ and $a_0 = 4$, $a_1 = 13$.

14. Determine the constants b and c if $a_n = c_1 + c_2(7^n)$, $n \geq 0$, is the general solution of the relation $a_{n+2} + ba_{n+1} + ca_n = 0$, $n \geq 0$.

15. Prove that any two consecutive Fibonacci numbers are relatively prime.

16. Write a computer program (or develop an algorithm) to determine whether a given nonnegative integer is a Fibonacci number.

10.3

THE NONHOMOGENEOUS RECURRENCE RELATION

We turn now to the recurrence relations

$$a_n + ca_{n-1} = f(n), \qquad n \geq 1 \tag{1}$$

$$a_n + ba_{n-1} + ca_{n-2} = f(n), \qquad n \geq 2, \tag{2}$$

where b and c are constants, $c \neq 0$, and $f(n)$ is not identically 0.

Although there is no general method for solving all nonhomogeneous relations, when the function $f(n)$ has a certain form we shall find a successful technique.

We start with the special case of Eq. (1) when $c = -1$.

For the nonhomogeneous relation $a_n - a_{n-1} = f(n)$, we have

$$a_1 = a_0 + f(1)$$
$$a_2 = a_1 + f(2) = a_0 + f(1) + f(2)$$
$$a_3 = a_2 + f(3) = a_0 + f(1) + f(2) + f(3)$$
$$\vdots$$
$$a_n = a_0 + f(1) + \cdots + f(n) = a_0 + \sum_{i=1}^{n} f(i).$$

We can solve this type of relation in terms of $\sum_{i=1}^{n} f(i)$, if we can find a suitable summation formula from some of our earlier work.

Example 10.20   Solve the relation $a_n - a_{n-1} = 3n^2$ where $n \geq 1$ and $a_0 = 7$.

Here $f(n) = 3n^2$, so the general solution is

$$a_n = a_0 + \sum_{i=1}^{n} f(i) = 7 + 3\sum_{i=1}^{n} i^2 = 7 + \frac{1}{2}(n)(n+1)(2n+1). \quad \square$$

When a formula for the summation is not known the following procedure will handle Eq. (1), regardless of the value of c ($\neq 0$). It also works for the second-order nonhomogeneous relation in Eq. (2) for certain functions $f(n)$. Known as the *method of undetermined coefficients*, it relies on the associated homogeneous relation obtained when $f(n)$ is replaced by 0.

For either of Eq. (1) or Eq. (2), we let $a_n^{(h)}$ denote the general solution of the associated homogeneous relation, and we let $a_n^{(p)}$ be a solution of the given nonhomogeneous relation. The term $a_n^{(p)}$ is called a *particular* solution. Then $a_n = a_n^{(h)} + a_n^{(p)}$ is the general solution of the given relation. To get $a_n^{(p)}$ we use the form of $f(n)$ to suggest a form for $a_n^{(p)}$.

Example 10.21   Solve the recurrence relation $a_n - 3a_{n-1} = 5(7^n)$, where $n \geq 1$ and $a_0 = 2$.

The solution of the associated homogeneous relation is $a_n^{(h)} = c(3^n)$. Since $f(n) = 5(7^n)$, we seek a particular solution $a_n^{(p)}$ of the form $A(7^n)$. And because $a_n^{(p)}$ is to be a solution of the given nonhomogeneous relation, we place $a_n^{(p)} = A(7^n)$ into the given relation and find that $A(7^n) - 3A(7^{n-1}) = 5(7^n), n \geq 1$. Dividing by 7^{n-1}, we find that $7A - 3A = 5(7)$, so $A = 35/4$, and $a_n^{(p)} = (35/4)7^n = (5/4)7^{n+1}$, $n \geq 0$. The general solution is $a_n = c(3^n) + (5/4)7^{n+1}$. With $2 = a_0 = c + (5/4)(7)$, $c = -27/4$ and $a_n = (5/4)(7^{n+1}) - (1/4)(3^{n+3}), n \geq 0$. $\square$

Example 10.22   Solve the relation $a_n - 3a_{n-1} = 5(3^n)$, where $n \geq 1$ and $a_0 = 2$.

As in Example 10.21, $a_n^{(h)} = c(3^n)$, but here $a_n^{(h)}$ and $f(n)$ are not linearly independent. As a result we consider a particular solution $a_n^{(p)}$ of the form $Bn(3^n)$. (What happens if we substitute $a_n^{(p)} = B(3^n)$ into the given relation?)

Substituting $a_n^{(p)} = Bn3^n$ into the given relation yields

$$Bn(3^n) - 3B(n-1)(3^{n-1}) = 5(3^n), \quad \text{or} \quad Bn - B(n-1) = 5, \quad \text{so} \quad B = 5.$$

Hence $a_n = a_n^{(h)} + a_n^{(p)} = (c + 5n)3^n, n \geq 0$. With $a_0 = 2$, the general solution is $a_n = (2 + 5n)(3^n)$. $\square$

From these two examples we generalize as follows.

If $f(n) = k(r^n)$, k a constant, $n \geq 0$, in the first-order relation, then $a_n^{(p)} = Ar^n$ if r^n is not a solution of the associated homogeneous relation. When it is, $a_n^{(p)} = Bnr^n$.

For the second-order relation, if $f(n) = k(r^n)$, then

a) $a_n^{(p)} = Ar^n$ if r^n is not a homogeneous solution

b) $a_n^{(p)} = Bnr^n$, if $a_n^{(h)} = c_1 r^n + c_2 r_1^n$, where $r_1 \neq r$

c) $a_n^{(p)} = Cn^2 r^n$, if $a_n^{(h)} = (c_1 + c_2 n)r^n$

Example 10.23   (The Towers of Hanoi) Consider n circular disks with holes in their centers. These disks can be stacked on any of the pegs shown in Fig. 10.7. In the figure, $n = 5$ and the disks are stacked on peg 1 with no disk resting upon a smaller one. The objective is to transfer the disks one at a time so that we end up with the original stack on peg 3. Peg 2 may be used as a temporary location for any disk(s), but at no time are we allowed to have a larger disk on top of a smaller one on any peg. How many moves are needed to do this for n disks?

For $n \geq 0$, let a_n be the number of moves it takes to transfer n disks from peg 1 to peg 3 in the manner described. Then for $n + 1$ disks we do the following:

a) Transfer the top n disks from peg 1 to peg 2, using peg 3 as a temporary location. This takes a_n steps.

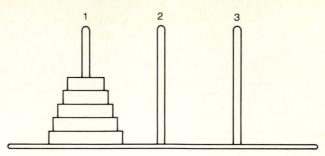

Figure 10.7

b) Transfer the largest disk from peg 1 to peg 3. This takes one step.

c) Finally, using peg 1 as a temporary location, transfer the n disks on peg 2 onto the largest disk now on peg 3. This requires another a_n moves.

This results in the relation $a_{n+1} = 2a_n + 1$, where $n \geq 0$ and $a_0 = 0$.

For $a_{n+1} - 2a_n = 1$, $a_n^{(h)} = c(2^n)$. Since $f(n) = 1$ is not a solution of $a_{n+1} - 2a_n = 0$, we set $a_n^{(p)} = A(1)^n = A$ and find from the given relation that $A = 2A + 1$, so $A = -1$ and $a_n = c(2^n) - 1$. From $a_0 = 0 = c - 1$, we conclude that $c = 1$ and $a_n = 2^n - 1$, $n \geq 0$. □

The next example arises from the mathematics of finance.

Example 10.24   Pauline takes out a loan of S dollars that is to be paid back in T periods of time. If i is the interest rate per period for the loan, what (constant) payment P must she make at the end of each period?

We let a_n denote the amount still owed on the loan at the end of the nth period (following the nth payment). Then

$$a_{n+1} = a_n + ia_n - P, \qquad 0 \leq n \leq T-1, \quad a_0 = S, \quad a_T = 0.$$

For this relation $a_n^{(h)} = c(1+i)^n$, while $a_n^{(p)} = A$ since $-P$ is not a solution of the associated homogeneous relation. With $a_n^{(p)} = A$ we find $A - (1+i)A = -P$, so $A = P/i$. From $a_0 = S$, we obtain $a_n = (S - (P/i))(1+i)^n + (P/i), 0 \leq n \leq T$.

Since $0 = a_T = (S - (P/i))(1+i)^T + (P/i)$, it follows that $(P/i) = ((P/i) - S)(1+i)^T$ and $P = (Si)[1 - (1+i)^{-T}]^{-1}$. □

We now consider a problem in the analysis of algorithms.

Example 10.25   For $n \geq 1$, let S be a set containing 2^n real numbers.

The following procedure is used to determine the maximum and minimum elements of S. We wish to determine the number of comparisons made between pairs of elements in S during the execution of this procedure.

If a_n denotes the number of needed comparisons, then $a_1 = 1$. When $n = 2$, $|S| = 2^2 = 4$, so $S = \{x_1, x_2, y_1, y_2\} = S_1 \cup S_2$ where $S_1 = \{x_1, x_2\}$, $S_2 = \{y_1, y_2\}$. Since $a_1 = 1$, it takes one comparison to determine the maximum and minimum elements in each of S_1, S_2. Comparing the minimum elements of S_1 and S_2 and then their maximum elements, we learn the maximum and minimum elements in S and find that $a_2 = 4 = 2a_1 + 2$. In general, if $|S| = 2^{n+1}$, we write $S = S_1 \cup S_2$ where $|S_1| = |S_2| = 2^n$. To determine the maximum and minimum elements in each of S_1 and S_2 requires a_n comparisons. Comparing the maximum (minimum) elements of S_1 and S_2 requires one more comparison; consequently, $a_{n+1} = 2a_n + 2, n \geq 1$.

Here $a_n^{(h)} = c(2^n)$ and $a_n^{(p)} = A$, a constant. Substituting $a_n^{(p)}$ into the relation, we find $A = 2A + 2$, or $A = -2$. So $a_n = c2^n - 2$, and with $a_1 = 1 = 2c - 2$, we obtain $c = 3/2$. Therefore $a_n = (3/2)(2^n) - 2$.

A note of caution! The existence of this procedure, which requires $(3/2)(2^n) - 2$ comparisons, does *not* exclude the possibility that we could achieve the same results via another, remarkably clever method that requires fewer comparisons. $\square$

Our last example deals with a second-order relation.

Example 10.26   Environmental records show that for a certain lake the population of a specific species of snail increases at a rate three times that of the prior year. Starting with 3000 such snails, and finding 3500 of them the following year, we remove 200 of them from this lake to increase their numbers in other lakes. We continue to remove 200 of the snails at the end of each year. If a_n represents the snail population in the original lake after n years, find and solve a recurrence relation for $a_n, n \geq 0$.

Here $(a_{n+2} - a_{n+1}) = 3(a_{n+1} - a_n) - 200$, $n \geq 0$. This yields the second-order relation $a_{n+2} - 4a_{n+1} + 3a_n = -200$, for which $a_n^{(h)} = c_1(3^n) + c_2(1^n) = c_1(3^n) + c_2$. Since $f(n) = -200 = -200(1^n)$ is a solution of the associated homogeneous relation, here $a_n^{(p)} = An$ for some constant A. This leads us to

$$A(n + 2) - 4A(n + 1) + 3An = -200, \quad \text{so} \quad -2A = -200, \quad A = 100.$$

Hence $a_n = c_1(3^n) + c_2 + 100n$. With $a_0 = 3000$ and $a_1 = 3500 - 200 = 3300$, we have $a_n = 100(3^n) + 2900 + 100n$, $n \geq 0$. $\square$

We close this section with a summary that includes the results of Examples 10.21 through 10.26.

Given a linear nonhomogeneous recurrence relation (with constant coefficients) of the form $c_n a_n + c_{n-1} a_{n-1} + \cdots + c_{n-k} a_{n-k} = f(n)$, where $c_n, c_{n-k} \neq 0$, let $a_n^{(h)}$ denote the homogeneous part of the solution a_n.

1. If $f(n)$ is a constant multiple of one of the forms in Table 10.2 and is not a solution of the associated homogeneous relation, then $a_n^{(p)}$ has the form shown in Table 10.2. (Here $A, B, A_0, A_1, A_2, \ldots, A_{t-1}, A_t$ are constants determined by substituting $a_n^{(p)}$ into the given relation; t, r, and α are also constants.)

2. When $f(n)$ comprises a sum of constant multiples of terms such as those shown in the table for item (1.) above, then $a_n^{(p)}$ is made up of the sum of the

corresponding terms in the column headed by $a_n^{(p)}$. For example, if $f(n) = n^2 + 3 \sin 2n$ and no summand of $f(n)$ is a solution of the associated homogeneous relation, then $a_n^{(p)} = (A_2 n^2 + A_1 n + A_0) + (A \sin 2n + B \cos 2n)$.

Table 10.2

$f(n)$	$a_n^{(p)}$
c, a constant	A, a constant
n	$A_1 n + A_0$
n^2	$A_2 n^2 + A_1 n + A_0$
$n^t, t \in \mathbf{Z}^+$	$A_t n^t + A_{t-1} n^{t-1} + \cdots + A_1 n + A_0$
$r^n, r \in \mathbf{R}$	Ar^n
$\sin \alpha n$	$A \sin \alpha n + B \cos \alpha n$
$\cos \alpha n$	$A \sin \alpha n + B \cos \alpha n$
$n^t r^n$	$r^n (A_t n^t + A_{t-1} n^{t-1} + \cdots + A_1 n + A_0)$
$r^n \sin \alpha n$	$Ar^n \sin \alpha n + Br^n \cos \alpha n$
$r^n \cos \alpha n$	$Ar^n \sin \alpha n + Br^n \cos \alpha n$

3. Things get trickier if a summand $f_1(n)$ of $f(n)$ is a constant multiple of a solution of the associated homogeneous relation. This happens, for example, when $f(n)$ contains summands such as cr^n or $(c_1 + c_2 n)r^n$ and r is a characteristic root. If $f_1(n)$ causes this problem, we multiply the particular solution $a_{n_1}^{(p)}$ corresponding to $f_1(n)$ by the smallest power of n, say n^s, for which no summand of $n^s f_1(n)$ is a solution of the associated homogeneous relation. Then $n^s a_{n_1}^{(p)}$ is the corresponding part of $a_n^{(p)}$.

EXERCISES 10.3

1. Solve each of the following recurrence relations.

   a) $a_{n+1} - a_n = 2n + 3$, $n \geq 0$, $a_0 = 1$

   b) $a_{n+1} - a_n = 3n^2 - n$, $n \geq 0$, $a_0 = 3$

   c) $a_{n+1} - 2a_n = 5$, $n \geq 0$, $a_0 = 1$

   d) $a_{n+1} - 2a_n = 2^n$, $n \geq 0$, $a_0 = 1$

2. Use a recurrence relation to derive the formula for $\sum_{i=0}^{n} i^2$.

3. a) Let n lines be drawn in the plane such that each line intersects every other line but no three lines are ever coincident. For $n \geq 0$, let a_n count the number of regions into which the plane is separated by the n lines. Find and solve a recurrence relation for a_n.

   b) For the situation in part (a), let b_n count the number of infinite regions that result. Find and solve a recurrence relation for b_n.

4. On the first day of a new year, Joseph deposits $1000 in an account that pays 6% interest compounded monthly. At the beginning of each month he adds $200 to his account. If he continues to do this for the next four years (so that he makes 47 additional deposits of $200), how much will his account be worth exactly four years after he opened it?

5. Solve the following recurrence relations.
   a) $a_{n+2} + 3a_{n+1} + 2a_n = 3^n$, $n \geq 0$, $a_0 = 0$, $a_1 = 1$
   b) $a_{n+2} + 4a_{n+1} + 4a_n = 7$, $n \geq 0$, $a_0 = 1$, $a_1 = 2$
   c) $a_{n+2} + 4a_{n+1} + 4a_n = n^2$, $n \geq 0$, $a_0 = 0$, $a_1 = 2$
   d) $a_{n+2} - a_n = \sin(n\pi/2)$, $n \geq 0$, $a_0 = 1$, $a_1 = 1$

6. Solve the recurrence relation $a_{n+2} - 6a_{n+1} + 9a_n = 3(2^n) + 7(3^n)$, where $n \geq 0$ and $a_0 = 1, a_1 = 4$.

7. Solve the recurrence relation $a_{n+3} - 3a_{n+2} + 3a_{n+1} - a_n = 3 + 5n$, $n \geq 0$.

8. Determine the number of n-digit quaternary $(0, 1, 2, 3)$ sequences where there is never a 3 anywhere to the right of a 0.

9. Meredith borrows $2500, at 12% compounded monthly, to buy a computer. If the loan is to be paid back over two years, what is his monthly payment?

10. The general solution of the recurrence relation $a_{n+2} + b_1 a_{n+1} + b_2 a_n = b_3 n + b_4$, $n \geq 0$, with b_i constant for $1 \leq i \leq 4$, is $c_1 2^n + c_2 3^n + n - 7$. Find b_i for each $1 \leq i \leq 4$.

11. Solve the following recurrence relations.
    a) $a_{n+2}^2 - 5a_{n+1}^2 + 6a_n^2 = 7n$, $n \geq 0$, $a_0 = a_1 = 1$
    b) $a_n + na_{n-1} = n!$, $n \geq 1$, $a_0 = 1$
    c) $a_n^2 - 2a_{n-1} = 0$, $n \geq 1$, $a_0 = 2$ (Let $b_n = \log_2 a_n$, $n \geq 0$.)

12. a) For $n \geq 1$, the nth *triangular number* t_n is defined by $t_n = 1 + 2 + \cdots + n = n(n+1)/2$. Find and solve a recurrence relation for $s_n, n \geq 1$, where $s_n = t_1 + t_2 + \cdots + t_n$, the sum of the first n triangular numbers.

   b) In an organic laboratory, Kelsey synthesizes a crystalline structure that is made up of 10,000,000 triangular layers of atoms. The first layer of the structure has one atom, the second layer has three atoms, the third layer has six atoms, and, in general, the nth layer has $1 + 2 + \cdots + n = t_n$ atoms. (We may consider each layer as though it were placed upon the spaces that result among the neighboring atoms of the succeeding layer. See Fig. 10.8.)

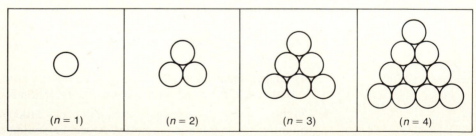

$(n = 1)$ $(n = 2)$ $(n = 3)$ $(n = 4)$

Figure 10.8

   (i) How many atoms are there in one of these crystalline structures?
   (ii) How many atoms are packed (strictly) between the 10,000th and 100,000th layer?

13. Write a computer program (or develop an algorithm) to solve the problem of the Towers of Hanoi. For $n \in \mathbf{Z}^+$, the program should provide the necessary steps for transferring the n disks from peg 1 to peg 3 under the restrictions specified in Example 10.23.

10.4
THE METHOD OF GENERATING FUNCTIONS

With all the different cases we had to consider for the nonhomogeneous linear recurrence relation, we now get some assistance from the generating function. This technique will find both the homogeneous and the particular solutions for a_n, and it will incorporate the given initial conditions as well. Furthermore, we'll be able to do even more with this method.

We demonstrate this technique in the following examples.

Example 10.27   Solve the relation $a_n - 3a_{n-1} = n$, $n \geq 1$, $a_0 = 1$.

This relation represents an infinite set of equations:

$$
\begin{array}{ll}
(n = 1) & a_1 - 3a_0 = 1 \\
(n = 2) & a_2 - 3a_1 = 2 \\
(n = 3) & a_3 - 3a_2 = 3 \\
\quad\vdots & \quad\vdots \quad \vdots
\end{array}
$$

Multiplying the first of these equations by x, the second by x^2, the third by x^3, and so on, we obtain

$$
\begin{array}{ll}
(n = 1) & a_1 x^1 - 3a_0 x^1 = 1x^1 \\
(n = 2) & a_2 x^2 - 3a_1 x^2 = 2x^2 \\
(n = 3) & a_3 x^3 - 3a_2 x^3 = 3x^3 \\
\quad\vdots & \quad\vdots \quad \vdots
\end{array}
$$

Adding this second set of equations, we find that

$$\sum_{n=1}^{\infty} a_n x^n - 3 \sum_{n=1}^{\infty} a_{n-1} x^n = \sum_{n=1}^{\infty} n x^n. \tag{1}$$

We want to solve for a_n in terms of n. To accomplish this, let $f(x) = \sum_{n=0}^{\infty} a_n x^n$ be the (ordinary) generating function for the sequence $a_0, a_1, a_2, \ldots$. Then Eq. (1) can be rewritten as

$$(f(x) - a_0) - 3x \sum_{n=1}^{\infty} a_{n-1} x^{n-1} = \sum_{n=1}^{\infty} n x^n \left(= \sum_{n=0}^{\infty} n x^n \right). \tag{2}$$

Since $\sum_{n=1}^{\infty} a_{n-1} x^{n-1} = \sum_{n=0}^{\infty} a_n x^n = f(x)$, the left-hand side of Eq. (2) becomes $(f(x) - 1) - 3xf(x)$.

Before we can proceed, we need the generating function for the sequence $0, 1, 2, 3, \ldots$. Recall from part (c) of Example 9.5 that

$$\frac{1}{(1-x)^2} = \frac{d}{dx}\left(\frac{1}{1-x}\right) = 1 + 2x + 3x^2 + \cdots$$

Consequently, $x/(1-x)^2 = x + 2x^2 + 3x^3 + \cdots$, and

$$(f(x) - 1) - 3xf(x) = \frac{x}{(1-x)^2}, \quad \text{so} \quad f(x) = \frac{1}{(1-3x)} + \frac{x}{(1-x)^2(1-3x)}.$$

Using a partial fraction decomposition, we find that

$$\frac{x}{(1-x)^2(1-3x)} = \frac{A}{(1-x)} + \frac{B}{(1-x)^2} + \frac{C}{(1-3x)},$$

or

$$x = A(1-x)(1-3x) + B(1-3x) + C(1-x)^2.$$

From the following assignments for x, we get

$$(x = 1): \quad 1 = B(-2), \qquad B = -\frac{1}{2}.$$

$$\left(x = \frac{1}{3}\right): \quad \frac{1}{3} = C\left(\frac{2}{3}\right)^2, \qquad C = \frac{3}{4}.$$

$$(x = 0): \quad 0 = A + B + C, \qquad A = -(B + C) = -\frac{1}{4}.$$

Therefore,

$$f(x) = \frac{1}{1-3x} + \frac{(-1/4)}{(1-x)} + \frac{(-1/2)}{(1-x)^2} + \frac{(3/4)}{(1-3x)}$$

$$= \frac{(7/4)}{(1-3x)} + \frac{(-1/4)}{(1-x)} + \frac{(-1/2)}{(1-x)^2}.$$

We find a_n by determining the coefficient of x^n in each of the three summands.

a) $(7/4)/(1-3x) = (7/4)[1/(1-3x)]$
$= (7/4)[1 + (3x) + (3x)^2 + (3x)^3 + \cdots]$, and the coefficient of x^n is $(7/4)3^n$.

b) $(-1/4)/(1-x) = (-1/4)[1 + x + x^2 + \cdots]$, and the coefficient of x^n here is $(-1/4)$.

c) $(-1/2)/(1-x)^2 = (-1/2)(1-x)^{-2}$
$= (-1/2)[\binom{-2}{0} + \binom{-2}{1}(-x) + \binom{-2}{2}(-x)^2 + \binom{-2}{3}(-x)^3 + \cdots]$
and the coefficient of x^n is given by $(-1/2)\binom{-2}{n}(-1)^n = (-1/2)(-1)^n\binom{2+n-1}{n} \cdot (-1)^n = (-1/2)(n+1)$.

Therefore $a_n = (7/4)3^n - (1/2)n - (3/4)$, $n \geq 0$. (Note that there is no special concern here with $a_n^{(p)}$.) □

We consider a second example, which has a familiar result.

Example 10.28 Let $n \in \mathbf{N}$. For $r \geq 0$, let $a(n, r)$ = the number of ways we can select, with repetition, r objects from a set of n distinct objects.

For $n \geq 1$, let $\{b_1, b_2, \ldots, b_n\}$ be the set of these objects, and consider object b_1. Exactly two things can happen.

a) The object b_1 is never selected. Hence the r objects are selected from $\{b_2, \ldots, b_n\}$. This we can do in $a(n-1, r)$ ways.

b) The object b_1 is selected at least once. Then we must select $r-1$ objects from $\{b_1, b_2, \ldots, b_n\}$, so we can continue to select b_1 in addition to the one selection of it we've already made. There are $a(n, r-1)$ ways to accomplish this.

Then $a(n, r) = a(n-1, r) + a(n, r-1)$, because these two cases cover all possibilities and are mutually disjoint.

Let $f_n = \sum_{r=0}^{\infty} a(n, r)x^r$ be the generating function for the sequence $a(n, 0)$, $a(n, 1), a(n, 2), \ldots$. From $a(n, r) = a(n-1, r) + a(n, r-1)$, where $n \geq 1$ and $r \geq 1$, it follows that $a(n, r)x^r = a(n-1, r)x^r + a(n, r-1)x^r$ and

$$\sum_{r=1}^{\infty} a(n, r)x^r = \sum_{r=1}^{\infty} a(n-1, r)x^r + \sum_{r=1}^{\infty} a(n, r-1)x^r.$$

Realizing that $a(n, 0) = 1$ for $n \geq 0$ and $a(0, r) = 0$ for $r > 0$, we write

$$f_n - a(n, 0) = f_{n-1} - a(n-1, 0) + x \sum_{r=1}^{\infty} a(n, r-1)x^{r-1},$$

so $f_n - 1 = f_{n-1} - 1 + xf_n$. Therefore, $f_n - xf_n = f_{n-1}$, or $f_n = f_{n-1}/(1-x)$.

If $n = 5$, for example, then

$$f_5 = \frac{f_4}{(1-x)} = \frac{1}{(1-x)} \cdot \frac{f_3}{(1-x)} = \frac{f_3}{(1-x)^2} = \frac{f_2}{(1-x)^3} = \frac{f_1}{(1-x)^4}$$

$$= \frac{f_0}{(1-x)^5} = \frac{1}{(1-x)^5},$$

since $f_0 = a(0, 0) + a(0, 1)x + a(0, 2)x^2 + \cdots = 1 + 0 + 0 + \cdots$.

In general, $f_n = 1/(1-x)^n = (1-x)^{-n}$, so $a(n, r)$ is the coefficient of x^r in $(1-x)^{-n}$, which is $\binom{-n}{r}(-1)^r = \binom{n+r-1}{r}$.

(Here we dealt with a recurrence relation for $a(n, r)$, a discrete function of the two (integer) variables $n, r \geq 0$.) □

Our last example shows how generating functions may be used to solve a system of recurrence relations.

Example 10.29   This example provides an approximate model for the propagation of high- and low-energy neutrons as they strike the nuclei of fissionable material (such as uranium) and are absorbed. Here we deal with a fast reactor where there is no moderator (such as water). (In reality, all the neutrons have fairly high energy and there are not just two energy levels. There is a continuous spectrum of energy levels, and those neutrons at the upper end of the spectrum are called the high-energy neutrons. The higher-energy neutrons tend to produce more new neutrons than the lower-energy ones.)

Consider the reactor at time 0 and suppose one high-energy neutron is injected into the system. During each time interval thereafter (about 1 microsecond, or 10^{-6} second) the following events occur.

a) When a high-energy neutron interacts with a nucleus (of fissionable material), upon absorption this results (one microsecond later) in two new high-energy neutrons and one low-energy one.

b) For interactions involving a low-energy neutron, only one neutron of each energy level is produced.

Assuming that all free neutrons interact with nuclei one microsecond after their creation, find functions of n such that

$$a_n = \text{the number of high-energy neutrons,}$$
$$b_n = \text{the number of low-energy neutrons,}$$

in the reactor after n microseconds, $n \geq 0$.

Here we have $a_0 = 1, b_0 = 0$ and the *system* of recurrence relations

$$a_{n+1} = 2a_n + b_n \tag{1}$$

$$b_{n+1} = a_n + b_n. \tag{2}$$

Let $f(x) = \sum_{n=0}^{\infty} a_n x^n$, $g(x) = \sum_{n=0}^{\infty} b_n x^n$ be the generating functions for the sequences $\{a_n \mid n \geq 0\}$, $\{b_n \mid n \geq 0\}$, respectively.

From Eqs. (1) and (2), when $n \geq 0$

$$a_{n+1} x^{n+1} = 2a_n x^{n+1} + b_n x^{n+1} \tag{1}'$$

$$b_{n+1} x^{n+1} = a_n x^{n+1} + b_n x^{n+1}. \tag{2}'$$

Summing Eq. (1)$'$ over all $n \geq 0$, we have

$$\sum_{n=0}^{\infty} a_{n+1} x^{n+1} = 2x \sum_{n=0}^{\infty} a_n x^n + x \sum_{n=0}^{\infty} b_n x^n. \tag{1}''$$

In similar fashion, Eq. (2)$'$ yields

$$\sum_{n=0}^{\infty} b_{n+1} x^{n+1} = x \sum_{n=0}^{\infty} a_n x^n + x \sum_{n=0}^{\infty} b_n x^n. \tag{2}''$$

Introducing the generating functions at this point, we get

$$f(x) - a_0 = 2xf(x) + xg(x) \tag{1}''$$

$$g(x) - b_0 = xf(x) + xg(x), \tag{2}''$$

a system of equations relating the generating functions. Solving this system, we find that

$$f(x) = \frac{1-x}{x^2 - 3x + 1} = \left(\frac{5+\sqrt{5}}{10}\right)\left(\frac{1}{\alpha - x}\right) + \left(\frac{5-\sqrt{5}}{10}\right)\left(\frac{1}{\beta - x}\right) \quad \text{and}$$

$$g(x) = \frac{x}{x^2 - 3x + 1} = \left(\frac{-5-3\sqrt{5}}{10}\right)\left(\frac{1}{\alpha - x}\right) + \left(\frac{-5+3\sqrt{5}}{10}\right)\left(\frac{1}{\beta - x}\right),$$

where

$$\alpha = \frac{3+\sqrt{5}}{2}, \qquad \beta = \frac{3-\sqrt{5}}{2}.$$

Consequently,

$$a_n = \left(\frac{5+\sqrt{5}}{10}\right)\left(\frac{3-\sqrt{5}}{2}\right)^{n+1} + \left(\frac{5-\sqrt{5}}{10}\right)\left(\frac{3+\sqrt{5}}{2}\right)^{n+1} \quad \text{and}$$

$$b_n = \left(\frac{-5-3\sqrt{5}}{10}\right)\left(\frac{3-\sqrt{5}}{2}\right)^{n+1} + \left(\frac{-5+3\sqrt{5}}{10}\right)\left(\frac{3+\sqrt{5}}{2}\right)^{n+1}, \qquad n \geq 0. \quad \square$$

EXERCISES 10.4

1. Solve the following recurrence relations by the method of generating functions.

   a) $a_{n+1} - a_n = 3^n, \quad n \geq 0, \quad a_0 = 1$

   b) $a_{n+1} - a_n = n^2, \quad n \geq 0, \quad a_0 = 1$

   c) $a_{n+2} - 2a_{n+1} + a_n = 2^n, \quad n \geq 0, \quad a_0 = 1, \quad a_1 = 2$

   d) $a_{n+2} - 2a_{n+1} + a_n = n, \quad n \geq 0, \quad a_0 = 1, \quad a_1 = 2$

2. For n distinct objects, let $a(n, r)$ denote the number of ways we can select, without repetition, r of the n objects when $0 \leq r \leq n$. Here $a(n, r) = 0$ when $r > n$. Use the recurrence relation $a(n, r) = a(n-1, r-1) + a(n-1, r)$, where $n \geq 1$ and $r \geq 1$, to show that $f(x) = (1+x)^n$ generates $a(n, r), r \geq 0$.

3. Solve the following systems of recurrence relations.

   a) $a_{n+1} = -2a_n - 4b_n$
      $b_{n+1} = 4a_n + 6b_n$
      $n \geq 0, \quad a_0 = 1, \quad b_0 = 0$

   b) $a_{n+1} = 2a_n - b_n + 2$
      $b_{n+1} = -a_n + 2b_n - 1$
      $n \geq 0, \quad a_0 = 0, \quad b_0 = 1$

10.5
A SPECIAL KIND OF NONLINEAR RECURRENCE RELATION (OPTIONAL)

Thus far our study of recurrence relations has dealt with linear relations with constant coefficients. The study of nonlinear recurrence relations and of relations with variable coefficients is not a topic we shall pursue except for one special nonlinear relation that lends itself to the method of generating functions.

We shall develop the method in a counting problem on data structures. Before doing so, we observe that if $f(x) = \sum_{i=0}^{\infty} a_i x^i$ is the generating function for $a_0, a_1, a_2, \ldots$, then $[f(x)]^2$ generates $a_0 a_0, a_0 a_1 + a_1 a_0, a_0 a_2 + a_1 a_1 + a_2 a_0, \ldots$, $a_0 a_n + a_1 a_{n-1} + a_2 a_{n-2} + \cdots + a_{n-1} a_1 + a_n a_0, \ldots$, the convolution of the sequence $a_0, a_1, a_2, \ldots$, with itself.

Example 10.30   In Section 5.1 we introduced the idea of a tree diagram. In general, a *tree* is an undirected graph that is connected and has no loops or cycles. Here we examine rooted binary trees.

In Fig. 10.9 we see two such trees, where the circled vertex denotes the *root*. These trees are called *binary* because from each vertex there are at most two edges (called *branches*) descending (since a rooted tree is a directed graph) from that vertex.

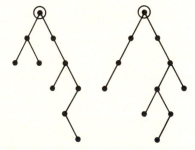

Figure 10.9

In particular these rooted binary trees are *ordered* in the sense that a left branch descending from a vertex is considered different from a right branch descending from that vertex. For the case of three vertices, the five possible ordered rooted binary trees are shown in Fig. 10.10. (If no attention were paid to order, then the last four rooted trees would be the same structure.)

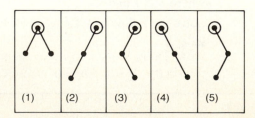

Figure 10.10

Our objective is to count, for $n \geq 0$, the number b_n of rooted ordered binary trees on n vertices. Assuming that we know the values of b_i for $0 \leq i \leq n$, in order to obtain b_{n+1} we select one vertex as the root and note, as in Fig. 10.11, that the substructures descending on the left and right of the root are smaller (rooted ordered binary) trees whose total number of vertices is n. These smaller trees are called *subtrees* of the given tree. Among these possible subtrees is the empty subtree, of which there is only $1\,(=b_0)$.

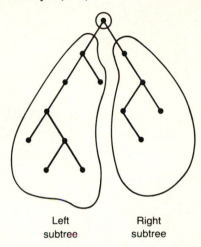

Left   Right
subtree   subtree   **Figure 10.11**

Now consider how the n vertices in these two subtrees can be divided up.

(1) 0 vertices on the left, n vertices on the right. This results in $b_0 b_n$ overall substructures to be counted in b_{n+1}.

(2) 1 vertex on the left, $n-1$ vertices on the right, yielding $b_1 b_{n-1}$ rooted ordered binary trees on $n+1$ vertices.

. . . . . .

$(i+1)$. i vertices on the left, $n-i$ on the right, for a count of $b_i b_{n-i}$ toward b_{n+1}.

. . . . . .

$(n+1)$. n vertices on the left and none on the right, contributing $b_n b_0$ of the trees.

Hence, for all $n \geq 0$,

$$b_{n+1} = b_0 b_n + b_1 b_{n-1} + b_2 b_{n-2} + \cdots + b_{n-1} b_1 + b_n b_0,$$

and

$$\sum_{n=0}^{\infty} b_{n+1} x^{n+1} = \sum_{n=0}^{\infty} (b_0 b_n + b_1 b_{n-1} + \cdots + b_{n-1} b_1 + b_n b_0) x^{n+1}. \qquad (1)$$

Now let $f(x) = \sum_{n=0}^{\infty} b_n x^n$ be the generating function for $b_0, b_1, b_2, \ldots$. We rewrite Eq. (1) as

$$(f(x) - b_0) = x \sum_{n=0}^{\infty} (b_0 b_n + b_1 b_{n-1} + \cdots + b_n b_0) x^n = x[f(x)]^2.$$

This brings us to the quadratic (in $f(x)$)

$$x[f(x)]^2 - f(x) + 1 = 0, \quad \text{so} \quad f(x) = [1 \pm \sqrt{1 - 4x}]/(2x).$$

But $\sqrt{1 - 4x} = (1 - 4x)^{1/2} = \binom{1/2}{0} + \binom{1/2}{1}(-4x) + \binom{1/2}{2}(-4x)^2 + \cdots$, where the coefficient of x^n, $n \geq 1$, is

$$\binom{1/2}{n}(-4)^n = \frac{(1/2)((1/2) - 1)((1/2) - 2) \cdots ((1/2) - n + 1)}{n!}(-4)^n$$

$$= (-1)^{n-1}\frac{(1/2)(1/2)(3/2) \cdots ((2n - 3)/2)}{n!}(-4)^n$$

$$= \frac{(-1)2^n(1)(3) \cdots (2n - 3)}{n!}$$

$$= \frac{(-1)2^n(n!)(1)(3) \cdots (2n - 3)(2n - 1)}{(n!)(n!)(2n - 1)}$$

$$= \frac{(-1)(2)(4) \cdots (2n)(1)(3) \cdots (2n - 1)}{(2n - 1)(n!)(n!)} = \frac{(-1)}{(2n - 1)}\binom{2n}{n}.$$

In $f(x)$ we select the negative radical; otherwise, we would have negative values for the b_n's. Then

$$f(x) = \frac{1}{2x}\left[1 - \left[1 - \sum_{n=1}^{\infty}\frac{1}{(2n - 1)}\binom{2n}{n}x^n\right]\right],$$

and b_n, the coefficient of x^n in $f(x)$, is half the coefficient of x^{n+1} in

$$\sum_{n=1}^{\infty}\frac{1}{(2n - 1)}\binom{2n}{n}x^n.$$

So

$$b_n = \frac{1}{2}\left[\frac{1}{2(n + 1) - 1}\right]\binom{2(n + 1)}{n + 1} = \frac{(2n)!}{(n + 1)!(n!)} = \frac{1}{(n + 1)}\binom{2n}{n}.$$

The numbers b_n are called the *Catalan numbers* after Eugene Catalan (1814–1894), who used them in determining the number of ways to parenthesize the expression $x_1 x_2 x_3 \ldots x_n$. The first seven Catalan numbers are $b_0 = 1$, $b_1 = 1$, $b_2 = 2$, $b_3 = 5$, $b_4 = 14$, $b_5 = 42$, and $b_6 = 132$. $\square$

We shall close this section with a second application of the Catalan numbers. This is based on an example given by Shimon Even. (See page 86 of reference [4].)

Example 10.31   Another important data structure that arises in computer science is the *stack*. This structure allows the storage of data items according to the following restrictions.

1. All insertions take place at *one* end of the structure. This is called the *top* of the stack, and the insertion process is referred to as the *push* procedure.

2. All deletions from the (nonempty) stack also take place from the top. We call the deletion process the *pop* procedure.

Since the *last* item inserted *in* this structure is the *first* item that can then be popped *out* of it, the stack is often referred to as a "last-in-first-out" (LIFO) structure.

Intuitive models for this data structure include a pile of poker chips on a table, a stack of trays in a cafeteria, and the discard pile used in playing certain card games, such as gin rummy. In all three of these cases, we can only (1) insert a new entry at the top of the pile or stack, or (2) take (delete) the entry at the top of the (nonempty) pile or stack.

Here we shall use this data structure, with its push and pop procedures, to help us permute the (ordered) list $1, 2, 3, \ldots, n$, for $n \in \mathbf{Z}^+$. The diagram in Fig. 10.12 shows how each integer of the input $1, 2, 3, \ldots, n$ must be pushed onto the top of the stack in the order given. However, we may pop an entry from the top of the (nonempty) stack at any time. But once an entry is popped from the stack, it may not be returned to either the top of the stack or the input left to be pushed onto the stack. The process continues until no entry is left in the stack.

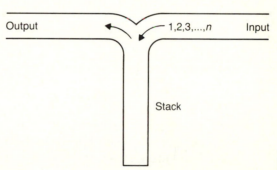

Figure 10.12

If $n = 1$, our input list consists of only the integer 1. We insert 1 at the top of the (empty) stack and then pop it out. This results in the permutation 1 (not very interesting).

For $n = 2$, there are two permutations possible for $1, 2$, and we can get both of them using the stack.

1. To get $1, 2$ we place 1 at the top of the (empty) stack and then pop it. Then 2 is placed at the top of the (empty) stack and it is popped.

2. The permutation $2, 1$ is obtained when 1 is placed at the top of the (empty) stack and 2 is then pushed onto the top of this (nonempty) stack. Upon popping first 2 from the top of the stack, and then 1, we obtain $2, 1$.

Turning to the case where $n = 3$, we find that we can obtain only five of the $3! = 6$ possible permutations of $1, 2, 3$ in this situation. For example, the permutation $2, 3, 1$ results when we take the following steps.

- Place 1 at the top of the (empty) stack.
- Push 2 onto the top of the stack (on top of 1).
- Pop 2 from the stack.
- Push 3 onto the top of the stack (on top of 1).
- Pop 3 from the stack.
- Pop 1 from the stack, leaving it empty.

The reason we fail to obtain all six permutations of $1, 2, 3$ is that we cannot generate the permutation $3, 1, 2$ using the stack. For in order to have 3 in the first position of the permutation, we must build the stack by first pushing 1 onto the (empty) stack, then pushing 2 onto the top of the stack (on top of 1), and finally pushing 3 onto the stack (on top of 2). After 3 is popped from the top of the stack, we get 3 as the first number in our permutation. But with 2 now at the top of the stack, we cannot pop 1 until after 2 has been popped, so the permutation $3, 1, 2$ cannot be generated.

When $n = 4$, there are 14 permutations of the (ordered) list $1, 2, 3, 4$ that can be generated by this stack method. We list them in the four columns of Table 10.3 according to the location of 1 in the permutation.

Table 10.3

1, 2, 3, 4	2, 1, 3, 4	2, 3, 1, 4	2, 3, 4, 1
1, 2, 4, 3	2, 1, 4, 3	3, 2, 1, 4	2, 4, 3, 1
1, 3, 2, 4			3, 2, 4, 1
1, 3, 4, 2			3, 4, 2, 1
1, 4, 3, 2			4, 3, 2, 1

1. There are five permutations with 1 in the first position, because after 1 is pushed onto and popped from the stack, there are five ways to permute $2, 3, 4$ using the stack.

2. When 1 is in the second position, 2 must be in the first position. This is because we pushed 1 onto the (empty) stack, then pushed 2 on top of it and then popped 2 and then 1. There are two permutations in column 2, because $3, 4$ can be permuted in two ways on the stack.

3. For column 3 we have 1 in position three. We note that the only numbers that can precede it are 2 and 3, which can be permuted on the stack (with 1 on the bottom) in two ways. Then 1 is popped, and we push 4 onto the (empty) stack and then pop it.

4. In the last column we obtain five permutations: After we push 1 onto the top of the (empty) stack, there are five ways to permute $2, 3, 4$ using the stack (with 1 on the bottom). Then 1 is popped from the stack to complete the permutation.

On the basis of these observations, for $1 \le i \le 4$, let a_i count the number of ways to permute the integers $1, 2, 3, \ldots, i$ (or any list of i consecutive integers) using the stack. Also, we define $a_0 = 1$ since there is only one way to permute nothing, using the stack. Then

$$a_4 = a_0 a_3 + a_1 a_2 + a_2 a_1 + a_3 a_0,$$

where

a) Each summand $a_j a_k$ satisfies $j + k = 3$.

b) The subscript j tells us that there are j integers to the left of 1 in the permutation—in particular, for $j \geq 1$, these are the integers from 2 to $j + 1$, inclusive.

c) The subscript k indicates that there are k integers to the right of 1 in the permutation—for $k \geq 1$, these are the integers from $4 - (k - 1)$ to 4.

This permutation problem can now be generalized to any $n \in \mathbf{N}$, so that

$$a_{n+1} = a_0 a_n + a_1 a_{n-1} + a_2 a_{n-2} + \cdots + a_{n-1} a_1 + a_n a_0,$$

with $a_0 = 1$. From our prior example we know that

$$a_n = \frac{1}{(n + 1)} \binom{2n}{n}. \qquad \square$$

Other examples that involve the Catalan numbers can be found in the chapter references.

EXERCISES 10.5

1. For the rooted ordered binary trees of Example 10.30, calculate b_4 and draw all of these four-vertex structures.

2. Verify that

$$\frac{1}{2}\left(\frac{1}{2n + 1}\right)\binom{2n + 2}{n + 1} = \left(\frac{1}{n + 1}\right)\binom{2n}{n},$$

for all $n \geq 0$.

3. Show that for all $n \geq 2$, $\binom{2n - 1}{n} - \binom{2n - 1}{n - 2} = \frac{1}{(n + 1)}\binom{2n}{n}$.

4. For a convex polygon of n sides, let t_n denote the number of ways the interior of the polygon can be partitioned into triangular regions by drawing nonintersecting diagonals.

   a) Show that $t_3 = 1$, $t_4 = 2$, and $t_5 = 5$.

   b) Define $t_0 = t_1 = 0$, $t_2 = 1$. Verify that $t_{n+1} = t_2 t_n + t_3 t_{n-1} + \cdots + t_{n-1} t_3 + t_n t_2$.

   c) Find t_n as a function of n.

5. For $n \geq 0$, $b_n = \left(\frac{1}{n + 1}\right)\binom{2n}{n}$ is the nth Catalan number.

   a) Show that for all $n \geq 0$, $b_{n+1} = \frac{2(2n + 1)}{(n + 2)} b_n$.

b) Use the result of part (a) to write a computer program (or develop an algorithm) that calculates the first 15 Catalan numbers.

6. For $n \geq 0$, evenly distribute $2n$ points on the circumference of a circle, and label these points cyclically with the integers $1, 2, 3, \dots, 2n$. Let a_n be the number of ways in which these $2n$ points can be paired off as n chords where no two chords intersect. (The case for $n = 3$ is shown in Fig. 10.13.) Find and solve a recurrence relation for a_n, $n \geq 0$.

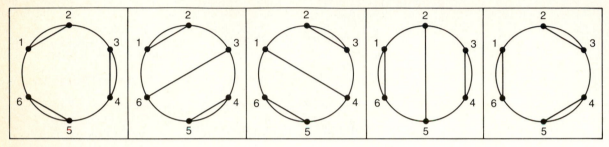

Figure 10.13

10.6
DIVIDE-AND-CONQUER ALGORITHMS (OPTIONAL)†

One of the most important and widely applicable types of efficient algorithm is based on a *divide-and-conquer* approach. Here the strategy, in general, is to solve a given problem of size n ($n \in \mathbf{Z}^+$) by

1. Solving the problem for a small value of n directly (this provides an initial condition for the resulting recurrence relation).

2. Breaking the general problem of size n into a smaller problems of the same type and (approximately) the same size—either $\lceil n/b \rceil$ or $\lfloor n/b \rfloor$,‡ where $a, b \in \mathbf{Z}^+$ with $1 \leq a < n$ and $1 < b < n$.

Then we solve the a smaller problems and use their solutions to construct a solution for the original problem of size n. We shall be especially interested in cases where n is a power of b, and $b = 2$.

---

† The material in this section may be skipped with no loss of continuity. It will be used in Section 3 of Chapter 12 to determine the time complexity function for the merge sort algorithm. However, the result there will also be obtained for a special case of the merge sort by another method that does not use the material developed in this section.

‡ For any $x \in \mathbf{R}$, $\lceil x \rceil$ denotes the *ceiling* of x and $\lfloor x \rfloor$ the *floor* of x, or *greatest integer* in x, where

    a) $\lfloor x \rfloor = \lceil x \rceil = x$, for $x \in \mathbf{Z}$

    b) $\lfloor x \rfloor =$ the integer directly to the left of x, for $x \in \mathbf{R} - \mathbf{Z}$

    c) $\lceil x \rceil =$ the integer directly to the right of x, for $x \in \mathbf{R} - \mathbf{Z}$.

We shall study those divide-and-conquer algorithms where

1. The time to solve the initial problem of size $n = 1$ is a constant $c \geq 0$, and
2. The time to break the given problem of size n into a smaller (similar) problems, together with the time to combine the solutions of these smaller problems to get a solution for the given problem, is $h(n)$, a function of n.

Our concern with time will actually be with the time complexity function $f(n)$ for these algorithms. Consequently, we shall use the notation $f(n)$ here, instead of the subscripted notation a_n that we used in the earlier sections of this chapter.

The conditions that have now been stated lead to the following recurrence relation.

$$f(1) = c,$$
$$f(n) = af(n/b) + h(n), \quad \text{for} \quad n = b^k, \quad k \geq 1.$$

We note that the domain of f is $\{1, b, b^2, b^3, \ldots\} = \{b^i \mid i \in \mathbf{N}\} \subset \mathbf{Z}^+$.

In our first result, the solution of this recurrence relation is derived for the case where $h(n)$ is the constant c.

THEOREM 10.1   Let $a, b, c \in \mathbf{Z}^+$ with $b \geq 2$, and let $f: \mathbf{Z}^+ \to \mathbf{R}$. If

$$f(1) = c, \text{ and}$$
$$f(n) = af(n/b) + c, \text{ for } n = b^k, k \geq 1,$$

then for all $n = 1, b, b^2, b^3, \ldots,$

1. $f(n) = c(\log_b n + 1)$, when $a = 1$
2. $f(n) = \dfrac{c(an^{\log_b a} - 1)}{a - 1}$, when $a \geq 2$

Proof   For $k \geq 1$ and $n = b^k$, we write the following system of k equations. (Starting with the second equation, we obtain each of these equations from its immediate predecessor by (i) replacing each occurrence of n in the prior equation by n/b and (ii) multiplying the resulting equation in (i) by a.)

$$f(n) = af(n/b) + c$$
$$af(n/b) = a^2 f(n/b^2) + ac$$
$$a^2 f(n/b^2) = a^3 f(n/b^3) + a^2 c$$
$$a^3 f(n/b^3) = a^4 f(n/b^4) + a^3 c$$
$$\vdots \qquad \vdots \qquad \vdots$$
$$a^{k-2} f(n/b^{k-2}) = a^{k-1} f(n/b^{k-1}) + a^{k-2} c$$
$$a^{k-1} f(n/b^{k-1}) = a^k f(n/b^k) + a^{k-1} c.$$

We see that each of the terms $af(n/b), a^2 f(n/b^2), a^3 f(n/b^3), \ldots, a^{k-1} f(n/b^{k-1})$ occurs one time as a summand on both the left-hand and right-hand sides of these equations. Therefore, upon adding both sides of the k equations and canceling these common summands, we obtain

$$f(n) = a^k f(n/b^k) + [c + ac + a^2 c + \cdots + a^{k-1}c].$$

Since $n = b^k$ and $f(1) = c$, we have

$$f(n) = a^k f(1) + c[1 + a + a^2 + \cdots + a^{k-1}]$$
$$= c[1 + a + a^2 + \cdots + a^{k-1} + a^k]$$

1. If $a = 1$, then $f(n) = c(k + 1)$. But $n = b^k \Leftrightarrow \log_b n = k$, so $f(n) = c(\log_b n + 1)$, for $n \in \{b^i | i \in \mathbf{N}\}$.

2. When $a \geq 2$, then $f(n) = \dfrac{c(1 - a^{k+1})}{1 - a} = \dfrac{c(a^{k+1} - 1)}{a - 1}$, from identity 4 of Table 9.2. Now $n = b^k \Leftrightarrow \log_b n = k$, so

$$a^k = a^{\log_b n} = (b^{\log_b a})^{\log_b n} = (b^{\log_b n})^{\log_b a} = n^{\log_b a},$$

and

$$f(n) = \frac{c(an^{\log_b a} - 1)}{(a - 1)}, \text{ for } n \in \{b^i | i \in \mathbf{N}\}. \quad \blacksquare$$

Example 10.32   **a)** Let $f: \mathbf{Z}^+ \to \mathbf{R}$, where

$$f(1) = 3, \text{ and}$$
$$f(n) = f(n/2) + 3, \quad \text{for} \quad n = 2^k, \ k \in \mathbf{Z}^+.$$

So by part 1 of Theorem 10.1, with $c = 3$, $b = 2$, and $a = 1$, it follows that $f(n) = 3(\log_2 n + 1)$ for $n \in \{1, 2, 4, 8, 16, \ldots\}$.

b) Suppose that $g: \mathbf{Z}^+ \to \mathbf{R}$ with

$$g(1) = 7, \text{ and}$$
$$g(n) = 4g(n/3) + 7, \quad \text{for} \quad n = 3^k, \ k \in \mathbf{Z}^+.$$

Then with $c = 7$, $b = 3$, and $a = 4$, part 2 of Theorem 10.1 implies that $g(n) = (7/3)(4n^{\log_3 4} - 1)$, when $n \in \{1, 3, 9, 27, 81, \ldots\}$. $\quad \square$

Considering Theorem 10.1, we must unfortunately realize that although we know about f for $n \in \{1, b, b^2, \ldots\}$, we cannot say anything about the value of f for the integers in $\mathbf{Z}^+ - \{1, b, b^2, \ldots\}$. So we are presently unable to deal with the concept of f as a time complexity function. To overcome this, we now generalize Definition 5.22, wherein the idea of function dominance was first introduced.

DEFINITION 10.1   Let $f, g: \mathbf{Z}^+ \to \mathbf{R}$ with S an infinite subset of $\mathbf{Z}^+$. We say that g *dominates f on S* (or f *is dominated by g on S*) if there exist constants $m \in \mathbf{R}^+$ and $k \in \mathbf{Z}^+$ such that $|f(n)| \leq m|g(n)|$ for all $n \in S$, where $n \geq k$.

Under these conditions we also say that $f \in O(g)$ *on S*.

Example 10.33 Let $f: \mathbf{Z}^+ \to \mathbf{R}$ be defined so that

$$f(n) = n, \quad \text{for} \quad n \in \{1, 3, 5, 7, \ldots\} = S_1,$$
$$f(n) = n^2, \quad \text{for} \quad n \in \{2, 4, 6, 8, \ldots\} = S_2.$$

Then $f \in O(n)$ on S_1 and $f \in O(n^2)$ on S_2. However, we *cannot* conclude that $f \in O(n)$ or $f \in O(n^2)$. ☐

Example 10.34 From Example 10.32, it now follows from Definition 10.1 that

a) $f \in O(\log_2 n)$ on $\{2^k | k \in \mathbf{N}\}$

b) $g \in O(n^{\log_3 4})$ on $\{3^k | k \in \mathbf{N}\}$ ☐

Using Definition 10.1, we now consider the following corollaries for Theorem 10.1. The first is a generalization of the particular results given in Example 10.34.

COROLLARY 10.1 Let $a, b, c \in \mathbf{Z}^+$ with $b \geq 2$, and let $f: \mathbf{Z}^+ \to \mathbf{R}$. If

$$f(1) = c, \text{ and}$$
$$f(n) = af(n/b) + c, \quad \text{for} \quad n = b^k, \quad k \geq 1,$$

then

1. $f \in O(\log_b n)$ on $\{b^k | k \in \mathbf{N}\}$, when $a = 1$, and

2. $f \in O(n^{\log_b a})$ on $\{b^k | k \in \mathbf{N}\}$, when $a \geq 2$.

Proof This proof is left as an exercise for the reader. ∎

The second corollary changes the equals signs of Theorem 10.1 to inequalities. As a result, the codomain of f must be restricted from $\mathbf{R}$ to $\mathbf{R}^+ \cup \{0\}$.

COROLLARY 10.2 For $a, b, c \in \mathbf{Z}^+$ with $b \geq 2$, let $f: \mathbf{Z}^+ \to \mathbf{R}^+ \cup \{0\}$. If

$$f(1) \leq c, \text{ and}$$
$$f(n) \leq af(n/b) + c, \quad \text{for} \quad n = b^k, \quad k \geq 1,$$

then for all $n = 1, b, b^2, b^3, \ldots,$

1. $f \in O(\log_b n)$, when $a = 1$, and

2. $f \in O(n^{\log_b a})$, when $a \geq 2$.

Proof Consider the function $g: \mathbf{Z}^+ \to \mathbf{R}^+ \cup \{0\}$, where

$$g(1) = c, \text{ and}$$
$$g(n) = ag(n/b) + c, \quad \text{for} \quad n \in \{1, b, b^2, \ldots\}.$$

By Corollary 10.1,

$$g \in O(\log_b n) \quad \text{on} \quad \{b^k | k \in \mathbf{N}\}, \quad \text{when} \quad a = 1, \text{ and}$$
$$g \in O(n^{\log_b a}) \quad \text{on} \quad \{b^k | k \in \mathbf{N}\}, \quad \text{when} \quad a \geq 2.$$

Since $f(n) \le g(n)$ for all $n \in \{1, b, b^2, \ldots\}$, it follows that $f \in O(g)$ on $\{b^k | k \in \mathbf{N}\}$, and then the corollary follows because of our earlier statement about g. ∎

Up to this point, our study of divide-and-conquer algorithms has been predominantly theoretical. It is high time we gave an example in which these ideas can be applied. The following result will confirm one of our earlier examples.

Example 10.35   For $n = 1, 2, 4, 8, 16, \ldots$, let $f(n)$ count the number of comparisons needed to find the maximum and minimum elements in a set $S \subset \mathbf{R}$, where $|S| = n$ and the procedure in Example 10.25 is used.

If $n = 1$, then the maximum and minimum elements are the same element. Therefore, no comparisons are necessary and $f(1) = 0$.

If $n > 1$, then $n = 2^k$ for some $k \in \mathbf{Z}^+$, and we partition (divide) S as $S_1 \cup S_2$ where $|S_1| = |S_2| = n/2 = 2^{k-1}$. It takes $f(n/2)$ comparisons to find the maximum M_i and the minimum m_i for each set S_i, $i = 1, 2$. For $n \ge 4$, knowing m_1, M_1, m_2, and M_2, we then compare m_1 with m_2 and M_1 with M_2 to determine the minimum and maximum elements in S. Therefore,

$$f(n) = 2f(n/2) + 1, \quad \text{when} \quad n = 2, \text{ and}$$
$$f(n) = 2f(n/2) + 2, \quad \text{when} \quad n = 4, 8, 16, \ldots.$$

Unfortunately, these results do not provide the hypotheses of Theorem 10.1. However, if we change our equations into the inequalities

$$f(1) \le 2$$
$$f(n) \le 2f(n/2) + 2, \quad \text{for} \quad n = 2^k, \quad k \ge 1,$$

then by Corollary 10.2 the time complexity function $f(n)$, measured by the number of comparisons made in this recursive procedure, satisfies $f \in O(n^{\log_2 2}) = O(n)$, for all $n = 1, 2, 4, 8, \ldots$.

We can examine the relationship between this example and Example 10.25 even further. From that earlier result, we know that if $|S| = n = 2^k$, $k \ge 1$, then the number of comparisons $f(n)$ we need (in the given procedure) to find the maximum and minimum elements in S is $(3/2)(2^k) - 2$. (*Note*: Our statement here replaces the variable n of Example 10.25 by the variable k.)

Since $n = 2^k$, we find that we can now write

$$f(1) = 0$$
$$f(n) = f(2^k) = (3/2)(2^k) - 2 = (3/2)n - 2, \quad \text{for } n = 2, 4, 8, 16, \ldots.$$

Hence $f \in O(n)$ for $n \in \{2^k | k \in \mathbf{N}\}$, just as we obtained above using Corollary 10.2. □

All of our results have required that $n = b^k$, for some $k \in \mathbf{N}$, so it is only natural to ask whether we can do anything in the case where n is allowed to be any positive integer. To find out, we introduce the following idea.

DEFINITION 10.2 A function $f: \mathbf{Z}^+ \to \mathbf{R}^+ \cup \{0\}$ is called *monotone increasing* if for all $m, n \in \mathbf{Z}^+$, $m < n \Rightarrow f(m) \leq f(n)$.

This permits us to consider results for any $n \in \mathbf{Z}^+$—under certain circumstances.

THEOREM 10.2 Let $f: \mathbf{Z}^+ \to \mathbf{R}^+ \cup \{0\}$ be monotone increasing, and let $g: \mathbf{Z}^+ \to \mathbf{R}$. For $b \in \mathbf{Z}^+$, $b \geq 2$, suppose that $f \in O(g)$ for all $n \in S = \{b^k | k \in \mathbf{N}\}$. Under these conditions,

  a) If $g \in O(\log n)$, then $f \in O(\log n)$.

  b) If $g \in O(n \log n)$, then $f \in O(n \log n)$.

  c) If $g \in O(n^r)$, then $f \in O(n^r)$, for $r \in \mathbf{R}^+ \cup \{0\}$.

Proof We shall prove parts (a) and (c) and leave part (b) for the section exercises. Before starting, we should note that the base for the logarithms in parts (a) and (b) is any positive real number greater than 1.

  a) Since $f \in O(g)$ on S, and $g \in O(\log n)$, we at least have $f \in O(\log n)$ on S. Therefore by Definition 10.1, there exist constants $m \in \mathbf{R}^+$ and $s \in \mathbf{Z}^+$ such that $f(n) = |f(n)| \leq m |\log n| = m \log n$ for all $n \in S, n \geq s$. We need to find constants $M \in \mathbf{R}^+$ and $s_1 \in \mathbf{Z}^+$ such that $f(n) \leq M \log n$ for *all* $n \geq s_1$, not just those $n \in S$.

    Let $t \in \mathbf{Z}^+$, where $s < b^k < t \leq b^{k+1}$ (and $\log s \geq 1$). Since f is monotone increasing and positive,

$$
\begin{aligned}
f(t) \leq f(b^{k+1}) &\leq m \, \log(b^{k+1}) \\
&= m[\log(b^k) + \log b] \\
&= m \, \log(b^k) + m \, \log b \\
&< m \, \log(b^k) + m \, \log b \, \log(b^k) \\
&= m(1 + \log b) \, \log(b^k) \\
&< m(1 + \log b) \, \log t.
\end{aligned}
$$

So with $M = m(1 + \log b)$ and $s_1 = b^k + 1$, we find that for all $t \in \mathbf{Z}^+$, if $t \geq s_1$ then $f(t) < M \log t$. Hence $f(t) \leq M \log t$ and $f \in O(\log n)$.

  c) Under the hypotheses for this part of the theorem, there exist constants $m \in \mathbf{R}^+, s \in \mathbf{Z}^+$ with $f(n) \leq m(n^r)$ for those $n \in S$ where $n \geq s$. So if $t \in \mathbf{Z}^+$ with $s < b^k < t \leq b^{k+1}$, then

$$
\begin{aligned}
f(t) \leq f(b^{k+1}) &\leq m(b^{k+1})^r = m[b^{(k+1)r}] \\
&= m b^r (b^k)^r \\
&< m b^r t^r.
\end{aligned}
$$

With $M = m b^r$ and $s_1 = b^k + 1$, it follows that for all $t \in \mathbf{Z}^+$, if $t \geq s_1$ then $f(t) < M t^r$ and $f \in O(n^r)$. ∎

We shall now use the result of Theorem 10.2 in determining the time complexity function $f(n)$ for a searching algorithm known as the *binary search*.

In Example 5.55 we analyzed an algorithm wherein an array $A[1]$, $A[2]$, $A[3], \ldots, A[n]$ of integers was searched for the presence of a particular integer called Key. At that time the array entries were not given in any particular order, so we simply compared the value of Key with those of the array elements $A[1], A[2], A[3], \ldots, A[n]$. But this would not be very efficient if we knew that $A[1] < A[2] < A[3] < \cdots < A[n]$. (After all, one does not search a telephone book for the telephone number of a particular person by starting at page 1 and examining every number in succession. The alphabetical ordering of the last names is used to speed up the searching process.) Let us look at a particular example.

Example 10.36   Consider the array $A[1], A[2], A[3], \ldots, A[7]$ of integers, where $A[1] = 2$, $A[2] = 4$, $A[3] = 5$, $A[4] = 7$, $A[5] = 10$, $A[6] = 17$, and $A[7] = 20$, and let Key $= 9$. We search this array as follows:

1. Compare Key with the entry at the center of the array; here it is $A[4] = 7$. Since Key $> A[4]$, we now concentrate on the remaining elements in the subarray $A[5], A[6], A[7]$.

2. Now compare Key with the center element $A[6]$. Since Key $= 9 < 17 = A[6]$, we now turn to the subarray (of $A[5], A[6], A[7]$) that consists of those elements smaller than $A[6]$. Here this is only the element $A[5]$.

3. Comparing Key with $A[5]$, we find that Key $\neq A[5]$, so Key is not present in the given array $A[1], A[2], A[3], \ldots, A[7]$.   □

From the results of Example 10.36, we make the following observations for a general (ordered) array of integers (or real numbers). Let $A[1], A[2], A[3], \ldots,$ $A[n]$ denote the given array, and let Key denote the integer (or real number) we are searching for. Unlike our array in Example 5.55, here

$$A[1] < A[2] < A[3] < \cdots < A[n].$$

1. First we compare the value of Key with the array entry at or near the center. This is the entry $A[(n + 1)/2]$ for n odd or $A[n/2]$ for n even.

   Whether n is even or odd, the array element subscripted by $m = \lfloor (n + 1)/2 \rfloor$ is the center, or near center, element. Note that at this point 1 is the value of the lower limit for the array subscripts, whereas n is the upper limit.

2. If Key is $A[m]$, we are finished. If not, then

   a) If Key exceeds $A[m]$, we search (with this dividing process) the subarray $A[m + 1], A[m + 2], \ldots, A[n]$.

   b) If Key is smaller than $A[m]$, then the dividing process is applied in searching the subarray $A[1], A[2], \ldots, A[m - 1]$.

The preceding observations have been used in the development of the Pascal program segment in Fig. 10.14. Here the input is an ordered array $A[1]$, $A[2], A[3], \ldots, A[n]$ of integers, or real numbers, in ascending order. The array and the value of the integer n are given earlier in the program, along with the value of the variable *Key*. If the array elements are integers (real numbers), then *Key* should be an integer (real number). The variables *lower* and *upper* are integer variables used for storing the lower and upper limits on the subscripts of the array or subarray being searched. The integer variable *center* stores the index for the array (subarray) element at, or near, the center of the array (subarray). In general, $center = \lfloor (lower + upper)/2 \rfloor$. Finally, the Boolean variable *flag* is assigned the value false (if *Key* is not found) or the value true (if *Key* is found).

```
Begin

 flag := false;
 lower := 1;
 upper := n;

 While (lower <= upper) and (flag = false) do
  Begin
   center := (lower + upper) Div 2;
   If Key = A[center] then
     flag := true
   Else
     If Key < A[center] then
        upper := center - 1    {The lower subarray is searched.}
      Else lower := center + 1    {The upper subarray is searched.}
  End;

 If flag = false then
  Writeln('The value ', Key, ' is not present in the array.')
 Else Writeln('The value ', Key,' is located in entry ', center,'.')

End.
```

Figure 10.14

We want to measure the (worst-case) time complexity for the algorithm implemented in Fig. 10.14. Here $f(n)$ will count the maximum number of comparisons (between Key and $A[center]$) needed to determine whether or not the given number Key appears in the ordered array $A[1], A[2], A[3], \ldots, A[n]$.

- For $n = 1$ Key is compared to $A[1]$ and $f(1) = 1$.
- When $n = 2$, in the worst case Key is compared to $A[1]$ and then to $A[2]$, so $f(2) = 2$.
- In the case of $n = 3$, $f(3) = 2$ (in the worst case).
- When $n = 4$, the worst case occurs when Key is first compared to $A[2]$ and then a binary search of $A[3], A[4]$ follows. Searching $A[3], A[4]$ requires (in the worst case) $f(2)$ comparisons. So $f(4) = 1 + f(2) = 3$.

At this point we see that $f(1) \le f(2) \le f(3) \le f(4)$, and we conjecture that f is a monotone increasing function. To verify this, we shall use the method of mathematical induction in its alternative form. Here we assume that for all $i, j \in \{1, 2, 3, \ldots, n\}, i < j \Rightarrow f(i) \leq f(j)$. Now consider the integer $n + 1$. We have two cases to examine.

1. $n + 1$ is odd: Here we write $n = 2k$ and $n + 1 = 2k + 1$, for some $k \in \mathbf{Z}^+$. In the worst case, $f(n + 1) = f(2k + 1) = 1 + f(k)$, where 1 counts the comparison of Key with $A[k + 1]$, and $f(k)$ counts the (maximum) number of comparisons needed in a binary search of the subarray $A[1], A[2], \ldots, A[k]$ or the subarray $A[k + 2], A[k + 3], \ldots, A[2k + 1]$.

   Now $f(n) = f(2k) = 1 + \max\{f(k - 1), f(k)\}$. Since $k - 1, k < n$, by the induction hypothesis we have $f(k - 1) \leq f(k)$, so $f(n) = 1 + f(k) = f(n + 1)$.

2. $n + 1$ is even: At this time we have $n + 1 = 2r$, for some $r \in \mathbf{Z}^+$, and in the worst case, $f(n + 1) = 1 + \max\{f(r - 1), f(r)\} = 1 + f(r)$, by the induction hypothesis. Therefore,

$$f(n) = f(2r - 1) = 1 + f(r - 1) \leq 1 + f(r) = f(n + 1).$$

Consequently the function f is monotone increasing.

Now it is time to determine the worst-case time complexity for the binary search algorithm, using the function $f(n)$. Since

$$f(1) = 1, \text{ and}$$
$$f(n) = f(n/2) + 1, \quad \text{for} \quad n = 2^k, \quad k \geq 1,$$

it follows from Theorem 10.1 (with $a = 1$, $b = 2$, and $c = 1$) that

$$f(n) = \log_2 n + 1, \quad \text{and} \quad f \in O(\log_2 n) \text{ for } n \in \{1, 2, 4, 8, \ldots\}.$$

But with f monotone increasing, from Theorem 10.2 it now follows that $f \in O(\log_2 n)$ (for all $n \in \mathbf{Z}^+$). Consequently, binary search is an $O(\log_2 n)$ algorithm, whereas the searching algorithm of Example 5.55 is $O(n)$. Therefore, as the value of n increases, binary search is the more efficient algorithm—but then it requires the additional condition that the array be ordered.

This section has introduced some of the basic ideas in the study of divide-and-conquer algorithms. It also extends the material first introduced on computational complexity and the analysis of algorithms in Sections 5.7 and 5.8.

The section exercises include some extensions of the results developed in this section. The reader who wants to pursue this topic further will find the chapter references both helpful and interesting.

EXERCISES 10.6

1. In each of the following, $f: \mathbf{Z}^+ \to \mathbf{R}$. Solve for $f(n)$ relative to the given set S, and determine the appropriate "big-oh" form for f on S.

   a) $f(1) = 5$
   $f(n) = 4f(n/3) + 5, \quad n = 3, 9, 27, \ldots$
   $S = \{3^i \mid i \in \mathbf{N}\}$

b) $f(1) = 7$
$f(n) = f(n/5) + 7, \quad n = 5, 25, 125, \ldots$
$S = \{5^i \mid i \in \mathbf{N}\}.$

c) $f(1) = 3$
$f(n) = 8f(n/2) + 3, \quad n = 2, 4, 8, \ldots$
$S = \{2^i \mid i \in \mathbf{N}\}$

2. Let $a, b, c \in \mathbf{Z}^+$ with $b \geq 2$, and let $d \in \mathbf{N}$. Prove that the solution for the recurrence relation

$$f(1) = d$$
$$f(n) = af(n/b) + c, \quad n = b^k, \quad k \geq 1$$

satisfies

a) $f(n) = d + c \log_b n$, for $n = b^k$, $k \in \mathbf{N}$, when $a = 1$.

b) $f(n) = dn^{\log_b a} + (c/(a-1))[n^{\log_b a} - 1]$, for $n = b^k$, $k \in \mathbf{N}$, when $a \geq 2$.

3. Determine the appropriate "big-oh" forms for f on $\{b^k \mid k \in \mathbf{N}\}$ in parts (a) and (b) of Exercise 2.

4. In each of the following, $f: \mathbf{Z}^+ \to \mathbf{R}$. Solve for $f(n)$ relative to the given set S, and determine the appropriate "big-oh" form for f on S.

a) $f(1) = 0$
$f(n) = 2f(n/5) + 3, \quad n = 5, 25, 125, \ldots$
$S = \{5^i \mid i \in \mathbf{N}\}$

b) $f(1) = 1$
$f(n) = f(n/2) + 2, \quad n = 2, 4, 8, \ldots$
$S = \{2^i \mid i \in \mathbf{N}\}$

c) $f(3) = 19$
$f(n) = 6f(n/3) + 1, \quad n = 3, 9, 27, \ldots$
$S = \{3^i \mid i \in \mathbf{N}\}.$

5. Consider a tennis tournament for n players, where $n = 2^k$, $k \in \mathbf{Z}^+$. In the first round $n/2$ matches are played, and the $n/2$ winners advance to round 2, where $n/4$ matches are played. This halving process continues until a winner is determined.

a) For $n = 2^k$, $k \in \mathbf{Z}^+$, let $f(n)$ count the total number of matches played in the tournament. Find and solve a recurrence relation for $f(n)$ of the form

$$f(1) = d$$
$$f(n) = af(n/2) + c, \quad n = 2, 4, 8, \ldots,$$

where a, c, and d are constants.

b) Show that your answer in part (a) also solves the recurrence relation

$$f(1) = d$$
$$f(n) = f(n/2) + (n/2), \quad n = 2, 4, 8, \ldots.$$

6. a) Complete the proof for Corollary 10.1.

   b) Prove part (b) of Theorem 10.2.

7. What is the best-case time complexity function for the binary search?

8. a) Modify the procedure in Example 10.35 as follows: For any $S \subset \mathbf{R}$, where $|S| = n$, partition S as $S_1 \cup S_2$, where

$$|S_1| = |S_2|, \text{ for } n \text{ even}$$
$$|S_1| = 1 + |S_2|, \text{ for } n \text{ odd}.$$

   Show that if $f(n)$ counts the number of comparisons needed (in this procedure) to find the maximum and minimum elements of S, then f is a monotone increasing function.

   b) What is the appropriate "big-oh" form for the function f of part (a)?

9. In Corollary 10.2 we were concerned with finding the appropriate "big-oh" form for a function $f \colon \mathbf{Z}^+ \to \mathbf{R}^+ \cup \{0\}$ where

$$f(1) \le c$$
$$f(n) \le af(n/b) + c, \text{ for } n = b^k, k \in \mathbf{Z}^+.$$

Here the constant c in the second inequality is interpreted as the amount of time needed to break down the given problem of size n into a smaller (similar) problems of size n/b and to combine the a solutions of these smaller problems in order to get a solution for the original problem of size n. Now we shall examine a situation wherein this amount of time is no longer constant but depends on n.

   a) Let $a, b, c \in \mathbf{Z}^+$, with $b \ge 2$. Let $f \colon \mathbf{Z}^+ \to \mathbf{R}^+ \cup \{0\}$ be a monotone increasing function, where

$$f(1) \le c$$
$$f(n) \le af(n/b) + cn, \text{ for } n = b^k, k \in \mathbf{Z}^+.$$

   Use an argument similar to the one given (for equalities) in Theorem 10.1 to show that for all $n = 1, b, b^2, b^3, \ldots$,

$$f(n) \le cn \sum_{i=0}^{k} (a/b)^i.$$

   b) Use the result of part (a) to show that $f \in O(n \log n)$, when $a = b$. (The base for the log function here is any real number greater than 1.)

   c) When $a \ne b$, show that part (a) implies that

$$f(n) \le \left(\frac{c}{a-b} \right)(a^{k+1} - b^{k+1}).$$

   d) From part (c), prove that

   (i) $f \in O(n)$ when $a < b$, and

   (ii) $f \in O(n^{\log_b a})$ when $a > b$.

   (*Note*: The "big-oh" form for f in each of parts (b) and (d) is for f on $\mathbf{Z}^+$, not just $\{b^k | k \in \mathbf{N}\}$.)

10.7
SUMMARY AND HISTORICAL REVIEW

In this chapter the recurrence relation has emerged as another tool for solving combinatorial problems. In these problems we analyze a given situation and then express the result a_n in terms of the results for certain smaller nonnegative integers. Once the recurrence relation is determined, we can solve for any value of a_n (within reason). When we have access to a computer, such relations are particularly valuable, especially if they cannot be solved explicitly.

The study of recurrence relations can be traced back to the Fibonacci relation $F_{n+2} = F_{n+1} + F_n, n \geq 0, F_0 = 0, F_1 = 1$, which was investigated by Leonardo of Pisa (c. 1175–1250) in 1202. In his *Liber Abaci*, he deals with a problem concerning the number of pairs of rabbits that result in one year if one starts with a single pair that produces another pair at the end of each month. Each new pair starts to breed similarly one month after its birth, and we assume that no rabbits die during the given year. Hence, at the end of the first month there are two pairs of rabbits; three pairs after two months; five pairs after three months, and so on. [As mentioned in the summary of Chapter 9, Abraham DeMoivre (1667–1754) obtained this result explicitly by the method of generating functions in 1718.]

This same sequence appears in the work of the German mathematician Johannes Kepler (1571–1630), who used it in his studies on how the leaves of a plant or flower are arranged about its stem. In 1844 the French mathematician Gabriel Lamé (1795–1870) used the sequence in his analysis of the efficiency of the Euclidean algorithm. Later, François Lucas (1842–1891) derived many properties of this sequence and was the first to call these numbers the Fibonacci sequence. The UMAP article by R. V. Jean [10] gives many applications of this sequence. Chapter 8 of the mathematical exposition by R. Honsberger [9] provides an interesting account of the Fibonacci numbers and of the related sequence called the Lucas numbers.

Comparable coverage of the material presented here can be found in Chapter 3 of C. Liu [14]. For more on the theoretical development of linear recurrence relations with constant coefficients, examine Chapter 9 of N. Finizio and G. Ladas [6]. This text contains an application in optics wherein a recurrence relation is used to determine the path of a ray passing through a series of equally spaced thin lenses.

Applications in probability theory dealing with recurrent events, random walks, and ruin problems can be found in Chapters XIII and XIV of the classic text by W. Feller [5]. The UMAP module by D. Sherbert [17] introduces difference equations and includes an application in economics known as the *Cobweb Theorem*. The text by S. Goldberg [8] has more on applications in the social sciences.

Recursive techniques in the generation of permutations, combinations, and partitions of integers are developed in Chapter 3 of R. Brualdi [2] and Chapter 5 of E. Page and L. Wilson [15]. The algorithm presented in Section 10.1 for the permutations of $\{1, 2, 3, \ldots, n\}$ first appeared in the work of H. Steinhaus [19] and is often referred to as the *adjacent mark ordering algorithm*. This result was rediscovered later, independently by H. Trotter [20], and S. Johnson [11]. Efficient sorting methods for permutations and other combinatorial structures are analyzed in detail in the text by D. Knuth [12]. The work of E. Reingold, J. Nievergelt, and N. Deo [16] also deals with such algorithms and their computer implementation.

For those who enjoyed the rooted ordered binary trees in Section 10.5, Chapter 3 of A. Aho, J. Hopcroft, and J. Ullman [1] should prove interesting. The basis for the example on stacks is given on page 86 of the text by S. Even [4]. The article by M. Gardner [7] provides many other examples where the Catalan numbers arise. Computational considerations in determining Catalan numbers are examined in the journal article by D. M. Campbell [3]; Chapter 9 of the work by R. Honsberger [9] discusses the connection between these numbers and the reflection principle (given in Miscellaneous Exercise 30 of Chapter 1).

Finally, the coverage on divide-and-conquer algorithms in Section 10.6 is modeled after D. F. Stanat and D. F. McAllister's presentation in Section 5.3 of [18]. Chapter 10 of the text by A. Aho, J. Hopcroft, and J. Ullman [1] provides some further information on this topic. An application of this method in a matrix multiplication algorithm appears in Chapter 10 of the text by C. L. Liu [13].

REFERENCES

1. Aho, Alfred V., Hopcroft, John E., and Ullman, Jeffrey D. *Data Structures and Algorithms.* Reading, Mass.: Addison-Wesley, 1983.
2. Brualdi, Richard A. *Introductory Combinatorics.* New York: Elsevier North-Holland, 1977.
3. Campbell, Douglas M. "The Computation of Catalan Numbers." *Mathematics Magazine* 57, no. 4 (September 1984): pp. 195–208.
4. Even, Shimon. *Graph Algorithms.* Rockville, Md.: Computer Science Press, 1979.
5. Feller, William. *An Introduction to Probability Theory and Its Applications,* Vol. I, 3rd ed. New York: Wiley, 1968.
6. Finizio, N., and Ladas, G. *An Introduction to Differential Equations (with Difference Equations, Fourier Series and Partial Differential Equations).* Belmont, Calif.: Wadsworth Publishing Company, 1982.
7. Gardner, Martin. "Mathematical Games, Catalan Numbers: An Integer Sequence that Materializes in Unexpected Places." *Scientific American* 234, no. 6 (June 1976): pp. 120–125.
8. Goldberg, Samuel. *Introduction to Difference Equations (with Illustrative Examples from Economics, Psychology, and Sociology).* New York: Wiley, 1958.
9. Honsberger, Ross. *Mathematical Gems III* (The Dolciani Mathematical Expositions, Number Nine). Washington, D.C.: The Mathematical Association of America, 1985.
10. Jean, Roger V. "The Fibonacci Sequence." *The UMAP Journal* 5, no. 1 (1984): pp. 23–47.
11. Johnson, Selmer M. "Generation of Permutations by Adjacent Transposition." *Mathematics of Computation* 17 (1963): pp. 282–285.
12. Knuth, Donald E. *The Art of Computer Programming/Volume 3 Sorting and Searching.* Reading, Mass.: Addison-Wesley, 1973.
13. Liu, C. L. *Elements of Discrete Mathematics,* 2nd ed. New York: McGraw-Hill, 1985.
14. Liu, C. L. *Introduction to Combinatorial Mathematics.* New York: McGraw-Hill, 1968.

15. Page, E. S., and Wilson, L. B. *An Introduction to Computational Combinatorics.* Cambridge: Cambridge University Press, 1979.
16. Reingold, E. M., Nievergelt, J., and Deo, N. *Combinatorial Algorithms: Theory and Practice.* Englewood Cliffs, N.J.: Prentice-Hall, 1977.
17. Sherbert, Donald R. *Difference Equations with Applications,* UMAP Module 322. Cambridge, Mass.: Birkhauser Boston, 1980.
18. Stanat, Donald F., and McAllister, David F. *Discrete Mathematics in Computer Science.* Englewood Cliffs, N.J.: Prentice-Hall, 1977.
19. Steinhaus, Hugo D. *One Hundred Problems in Elementary Mathematics.* New York: Basic Books, 1964.
20. Trotter, H. F. "ACM Algorithm 115—Permutations," *Communications of the ACM* 5 (1962): pp. 434–435.

MISCELLANEOUS EXERCISES

1. For $n \in \mathbf{Z}^+$ and $n \geq k + 1 \geq 1$, verify algebraically the recursion formula

$$\binom{n}{k+1} = \left(\frac{n-k}{k+1}\right)\binom{n}{k}$$

for binomial coefficients.

2. **a)** For $n \geq 0$, let B_n denote the number of partitions of $\{1, 2, 3, \ldots, n\}$. Set $B_0 = 1$ for the partitions of $\emptyset$. Verify that for all $n \geq 0$,

$$B_{n+1} = \sum_{i=0}^{n} \binom{n}{n-i} B_i = \sum_{i=0}^{n} \binom{n}{i} B_i.$$

(The numbers $B_i, i \geq 0$, are referred to as the *Bell numbers* after Eric Temple Bell (1883–1960).)

b) How are the Bell numbers related to the Stirling numbers of the second kind?

3. Let $n, k \in \mathbf{Z}^+$, and define $p(n, k)$ to be the number of partitions of n into exactly k summands. Prove that $p(n, k) = p(n - 1, k - 1) + p(n - k, k)$.

4. Find and solve a recurrence relation for the number of quaternary $(0, 1, 2, 3)$ sequences of length n that have no consecutive identical symbols.

5. For $n \geq 1$, let a_n count the number of ways to write n as an ordered sum of odd positive integers. (For example, $a_4 = 3$ since $4 = 3 + 1 = 1 + 3 = 1 + 1 + 1 + 1$.) Find and solve a recurrence relation for a_n.

6. For $n \geq 1$, let a_n be the number of (ordered) n-tuples $(x_1, x_2, \ldots, x_n)$ where $x_i \in \{0, 1\}$ for $1 \leq i \leq n$, and $x_1 \leq x_2 \geq x_3 \leq x_4 \geq x_5 \leq \cdots$. Find and solve a recurrence relation for a_n.

7. Let $A = \begin{bmatrix} 1 & 1 \\ 1 & 0 \end{bmatrix}$.

a) Compute A^2, A^3, and A^4.

b) Conjecture a general formula for $A^n, n \in \mathbf{Z}^+$, and establish your conjecture by mathematical induction.

8. Let $\alpha = (1 + \sqrt{5})/2$ and $\beta = (1 - \sqrt{5})/2$.

a) Verify that $\alpha^2 = \alpha + 1$ and $\beta^2 = \beta + 1$.

b) Show that $F_k = (\alpha^k - \beta^k)/(\alpha - \beta)$, for $k \geq 0$, where $\{F_k | k \geq 0\}$ is the Fibonacci sequence.

c) Prove that for any $n \geq 0$, $\sum_{k=0}^{n} \binom{n}{k} F_k = F_{2n}$.

d) Show that $\alpha^3 = 1 + 2\alpha$ and $\beta^3 = 1 + 2\beta$.

e) Prove that for any $n \geq 0$, $\sum_{k=0}^{n} \binom{n}{k} 2^k F_k = F_{3n}$.

9. For $n \in \mathbf{Z}^+$, d_n denotes the number of derangements of $\{1, 2, 3, \ldots, n\}$, as discussed in Section 8.3.

a) If $n > 2$, show that d_n satisfies the recurrence relation

$$d_n = (n-1)(d_{n-1} + d_{n-2}), \qquad d_2 = 1, \qquad d_1 = 0.$$

b) How can we define d_0 so that the result in part (a) is valid for $n \geq 2$?

c) Rewrite the result in part (a) as $d_n - nd_{n-1} = -[d_{n-1} - (n-1)d_{n-2}]$. How can $d_n - nd_{n-1}$ be expressed in terms of d_{n-2}, d_{n-3}?

d) Show that $d_n - nd_{n-1} = (-1)^n$.

e) Let $f(x) = \sum_{n=0}^{\infty} (d_n x^n)/n!$. Multiplying both sides of the equation in part (d) by $x^n/n!$ and summing for $n \geq 2$, verify that $f(x) = (e^{-x})/(1 - x)$. Hence

$$d_n = n! \left[1 - \frac{1}{1!} + \frac{1}{2!} - \frac{1}{3!} + \cdots + \frac{(-1)^n}{n!} \right].$$

10. For $n \geq 0$, let $m = \lfloor (n+1)/2 \rfloor$. Prove that $F_{n+2} = \sum_{k=0}^{m} \binom{n-k+1}{k}$. (You may want to look back at Examples 9.15 and 10.10.)

11. For $n \geq 0$, draw n ovals in the plane so that each oval intersects each of the others in exactly two points and no three ovals are coincident. If a_n denotes the number of regions in the plane that results from these n ovals, find and solve a recurrence relation for a_n.

12. a) Divide the plane into three regions by drawing two parallel lines. For $n \geq 0$, draw n additional lines so that each line intersects all other lines, including the two parallel lines with which you started. No three of these $n + 2$ lines are coincident. Find and solve a recurrence relation for the number of regions that result in the plane from these $n + 2$ lines.

b) Answer part (a) if there are k parallel lines, $k \geq 2$.

13. In a certain population model, the probability that a couple will have n children satisfies the recurrence relation $p_n = 0.6p_{n-1}, n \geq 1$. (a) Find p_n in terms of p_0. (b) Knowing that $\sum_{i=0}^{\infty} p_i = 1$, determine p_0.

14. When considering certain geographic restrictions, we replace the recurrence relation in Exercise 13 by $p_n = (1/n)p_{n-1}, n \geq 1$. Find p_n in terms of p_0 and determine p_0 for this model.

15. For $n \geq 0$, let us toss a coin $2n$ times.

a) If a_n is the number of sequences of $2n$ tosses where n heads and n tails occur, find a_n in terms of n.

b) Find constants r, s, and t so that $(r + sx)^t = f(x) = \sum_{n=0}^{\infty} a_n x^n$.

c) Let b_n denote the number of sequences of $2n$ tosses where the number of heads and tails are equal for the first time only after all $2n$ tosses have been made. (For example, if $n = 3$, then HHHTTT and HHTHTT are counted in b_n, but HTHHTT and HHTTHT are not.)

Define $b_0 = 0$ and show that for all $n \geq 1$, $a_n = a_0 b_n + a_1 b_{n-1} + \cdots + a_{n-1} b_1 + a_n b_0$.

d) Let $g(x) = \sum_{n=0}^{\infty} b_n x^n$. Show that $g(x) = 1 - 1/f(x)$, and then solve for $b_n, n \geq 1$.

16. (Gambler's Ruin)

a) When Cathy and Jill play checkers, each has probability $\frac{1}{2}$ of winning. There is never a tie, and the games are independent in the sense that no matter how many games the girls have played, each girl still has probability $\frac{1}{2}$ of winning the next game. After each game the loser gives the winner a quarter. If Cathy has $2.00 to play with and Jill has $2.50 and they play until one of them is broke, what is the probability that Cathy gets wiped out?

b) Answer part (a) if Jill is tired so that she has only probability $\frac{1}{4}$ of winning each game.

17. Let A, B be sets with $|A| = m \geq n = |B|$, and let $a(m, n)$ count the number of *onto* functions from A to B. Show that

$$a(m, 1) = 1$$
$$a(m, n) = n^m - \sum_{i=1}^{n-1} \binom{n}{i} a(m, i), \quad \text{when} \quad m \geq n > 1.$$

18. For $n, m \in \mathbf{Z}^+$, let $f(n, m)$ count the number of partitions of n where the summands form a nonincreasing sequence of positive integers and no summand exceeds m. With $n = 4$ and $m = 2$, for example, we find that $f(4, 2) = 3$, because here we are concerned with the three partitions

$$4 = 2 + 2, \qquad 4 = 2 + 1 + 1, \qquad 4 = 1 + 1 + 1 + 1.$$

a) Verify that for all $n, m \in \mathbf{Z}^+$,

$$f(n, m) = f(n - m, m) + f(n, m - 1).$$

b) Write a computer program (or develop an algorithm) to compute $f(n, m)$ for $n, m \in \mathbf{Z}^+$.

c) Write a computer program (or develop an algorithm) to compute $p(n)$, the number of partitions of any positive integer n.

19. When one examines the units digit of each Fibonacci number $F_n, n \geq 0$, one finds that these digits form a sequence that repeats after 60 terms. [This was first proved by Joseph-Louis Lagrange (1736–1813).] Write a computer program (or develop an algorithm) to calculate this sequence of 60 digits.

CHAPTER 11

An Introduction to Graph Theory

With this chapter we start to develop another major topic of this text. Unlike other areas in mathematics, the theory of graphs has a definite starting place, in a paper published in 1736 by the Swiss mathematician Leonhard Euler (1707–1783). The main idea behind this work grew out of a now-popular problem known as the seven bridges of Königsberg. We shall examine the solution of this problem, from which Euler developed the fundamental concepts for the theory of graphs.

Unlike the continuous graphs of early algebra courses, the graphs we examine here are finite in structure and can be used to analyze relationships and applications in many different settings. We have seen some examples of applications of graph theory in earlier chapters (5–8, and 10). However, the development here is independent of these prior discussions.

11.1
DEFINITIONS AND EXAMPLES

When we use a road map, we are often concerned with seeing how to get from one town to another by means of the roads indicated on the map. Consequently we are dealing with two distinct sets of objects: towns and roads. As we have seen many times before, such sets of objects can be used to define a relation. If V denotes the set of towns and E the set of roads, we can define a relation $\mathcal{R}$ on V by $a \, \mathcal{R} \, b$ if there is a system of roads that takes us from a to b. If these are all two-way roads, then if $a \, \mathcal{R} \, b$ we also have $b \, \mathcal{R} \, a$. Should all the roads under consideration be two-way, we have a symmetric relation.

One way to represent any relation is by listing the ordered pairs that are its elements. Here, however, it is more convenient to use a picture, as shown in Fig. 11.1. This figure demonstrates the possible ways of traveling among six towns using the eight roads indicated. It shows that there is at least one system of roads connecting any two towns (identical or distinct). This pictorial representation is a lot easier to work with than the 36 ordered pairs of the relation $\mathcal{R}$.

At the same time, Fig. 11.1 would be appropriate for representing six communication centers with the eight "roads" interpreted as communication links. If each link provides two-way communication, we should be quite concerned about the vulnerability of center a to such hazards as equipment breakdown or enemy attack. Without center a, neither b nor c can communicate with any of d, e, or f.

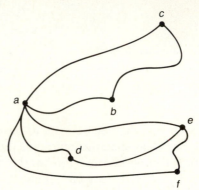

Figure 11.1

From these observations we consider the following concepts.

DEFINITION 11.1   Let V be a nonempty set, and let $E \subseteq V \times V$. The pair (V, E) is then called a *directed graph* (on V), or *digraph*† (on V), where V is the set of *vertices*, or *nodes*, and E is its set of *edges*. We write $G = (V, E)$ to denote such a graph.

Figure 11.2 provides an example of a directed graph on $V = \{a, b, c, d, e\}$ with $E = \{(a, a), (a, b), (a, d), (b, c)\}$. The direction of an edge is indicated by placing a directed arrow on the edge, as shown here. For any edge, such as (b, c), we say that the edge is *incident* with the vertices b, c; b is said to be *adjacent to* c, whereas c is *adjacent from* b. In addition, b is called the *origin*, or *source*, of the edge, and c is the *terminus*, or *terminating vertex*. The edge (a, a) is an example of a *loop*, and the vertex e that is not incident with any edge is called an *isolated* vertex.

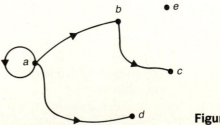

Figure 11.2

---

† Since the terminology of graph theory is not standard, the reader may find some differences between definitions used here and in other texts.

When there is no concern about the direction of an edge, the graph is called *undirected*; an undirected graph is shown in Fig. 11.3(a). This graph is a more compact way of describing the directed graph given in Fig. 11.3(b). In an undirected graph, there are undirected edges such as $\{a, b\}, \{b, c\}, \{a, c\}, \{c, d\}$ in Fig. 11.3(a). An edge such as $\{a, b\}$ stands for $\{(a, b), (b, a)\}$. Although $(a, b) = (b, a)$ only when $a = b$, we do have $\{a, b\} = \{b, a\}$ for any a, b. We write $\{a, a\}$ to denote a loop in an undirected graph; $\{a, a\}$ is considered the same as (a, a).

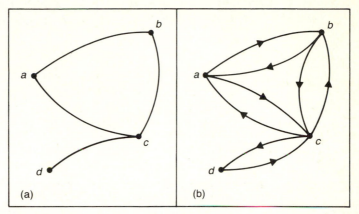

Figure 11.3

In general, if a graph G is not specified as directed or undirected, it is assumed to be undirected. If it contains no loops it is called *loop-free*.

In the next two definitions we shall not concern ourselves with any loops that may be present.

DEFINITION 11.2   Let x, y be (not necessarily distinct) vertices in an undirected graph $G = (V, E)$. An *x-y walk* in G is a finite alternating sequence

$$x = x_0, e_1, x_1, e_2, x_2, e_3, \dots, e_{n-1}, x_{n-1}, e_n, x_n = y$$

of vertices and edges from G, starting at vertex x and ending at vertex y, with the n edges $e_i = \{x_{i-1}, x_i\}, 1 \le i \le n$.

The *length* of a walk is the value of n, the number of edges in the walk. (When $n = 0$, there are no edges, $x = y$, and the walk is called *trivial*. These walks are not considered very much in our work.)

Any *x-y* walk where $x = y$ (and $n > 0$) is called a *closed walk*. Otherwise the walk is called *open*.

Note that we may repeat both vertices and edges when dealing with a walk.

Example 11.1   For the graph in Fig. 11.4 we find, for example, the following three open walks. Here we shall list either the edges or the vertices (the other is implicit).

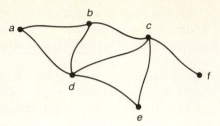

Figure 11.4

1. $\{a, b\}, \{b, d\}, \{d, c\}, \{c, e\}, \{e, d\}, \{d, b\}$: This is an a-b walk of length 6 in which we find the vertices d and b repeated, as well as the edge $\{b, d\} (= \{d, b\})$.

2. $b \rightarrow c \rightarrow d \rightarrow e \rightarrow c \rightarrow f$: Here we have a b-f walk where the length is 5 and the vertex c is repeated, but no edge appears more than once.

3. $\{f, c\}, \{c, e\}, \{e, d\}, \{d, a\}$: In this case the given f-a walk has length 4 with no repetitions of either vertices or edges.

Since the graph of Fig. 11.4 is undirected, the result in walk 1 (above) is also a b-a walk (we read the edges, if necessary, as $\{b, d\}, \{d, e\}, \{e, c\}, \{c, d\}, \{d, b\}$, and $\{b, a\}$). Similar remarks hold true for the walks in 2 and 3.

Finally the edges $\{b, c\}, \{c, d\}$, and $\{d, b\}$ provide a b-b (closed) walk. These edges are also considered c-c, or d-d, (closed) walks.   □

Now let us examine special types of walks.

DEFINITION 11.3   Consider any x-y walk in an undirected graph $G = (V, E)$.

  a) If no edge in the x-y walk is repeated, then the edges (and vertices) make up an x-y *trail*. A closed x-x trail is called a *circuit*.

  b) When no vertex of the x-y walk occurs more than once, the result is called an x-y *path*. The term *cycle* is used to describe a closed x-x path.   ▬

Convention: In dealing with circuits, we shall always understand the presence of at least one edge. When there is only one edge, then the circuit is a loop (and the graph is no longer loop-free). Circuits with two edges arise in multigraphs, a concept we shall define shortly.

The term "cycle" will always imply the presence of at least three distinct edges (from the graph).

Example 11.2   a) The b-f walk in part (2) of Example 11.1 is a b-f trail, but it is not a b-f path because of the repetition of vertex c. However, the f-a walk in part (3) of that example is both an f-a trail and an f-a path.

  b) In Fig. 11.4, the edges $\{a, b\}, \{b, d\}, \{d, c\}, \{c, e\}, \{e, d\}$, and $\{d, a\}$ provide an a-a circuit. The vertex d is repeated, so the edges do *not* give us an a-a cycle.

  c) The edges $\{a, b\}, \{b, c\}, \{c, d\}$, and $\{d, a\}$ provide a cycle (of length 4) in Fig. 11.4. We can think of this as an a-a (or b-b, c-c, d-d) cycle. This cycle is also a circuit.   □

For a directed graph we shall use the adjective "directed," as in, for example, *directed walks*, *directed paths*, and *directed cycles*.

Before continuing, we summarize (in Table 11.1) for future reference the results of Definitions 11.2 and 11.3. Each occurrence of "Yes" in the first two columns here should be interpreted as "Yes, possibly." Table 11.1 reflects the fact that a path is a trail, which in turn is an open walk. Furthermore, every cycle is a circuit, and every circuit is a closed walk.

Table 11.1

Repeated Vertex (Vertices)	Repeated Edge(s)	Open	Closed	Name
Yes	Yes	Yes		Walk (open)
Yes	Yes		Yes	Walk (closed)
Yes	No	Yes		Trail
Yes	No		Yes	Circuit
No	No	Yes		Path
No	No		Yes	Cycle

Considering how many concepts we have introduced, it is time to prove a first result in this new theory.

THEOREM 11.1   Let $G = (V, E)$ be an undirected graph, with $a, b \in V$, $a \neq b$. If there exists a trail (in G) from a to b, then there is a path (in G) from a to b.

Proof   Since there is a trail from a to b, select one of shortest length, say $\{a, x_1\}$, $\{x_1, x_2\}$, $\ldots$, $\{x_n, b\}$. If this trail is not a path, we have the situation $\{a, x_1\}$, $\{x_1, x_2\}$, $\ldots$, $\{x_{k-1}, x_k\}$, $\{x_k, x_{k+1}\}$, $\{x_{k+1}, x_{k+2}\}, \ldots$, $\{x_{m-1}, x_m\}$, $\{x_m, x_{m+1}\}, \ldots$, $\{x_n, b\}$, where $x_k = x_m$. But then $\{a, x_1\}, \{x_1, x_2\}, \ldots, \{x_{k-1}, x_k\}, \{x_m, x_{m+1}\}, \ldots, \{x_n, b\}$ is a shorter trail. ■

The notion of a path is needed in the following graph property.

DEFINITION 11.4   Let $G = (V, E)$ be an undirected graph. We call G *connected* if there is a path between any two distinct vertices of G.

Let $G = (V, E)$ be a directed graph. Its associated undirected graph is the graph obtained from G by ignoring the directions on the edges. If this associated graph is connected, we consider G connected.

A graph that is not connected is called *disconnected*.   ━━

The graphs in Figs. 11.1, 11.3, and 11.4 are connected. In Fig. 11.2 the graph is not connected because, for example, there is no path from a to e.

Example 11.3   In Fig. 11.5 we have an undirected graph on $V = \{a, b, c, d, e, f, g\}$. This graph is not connected because, for example, there is no path from a to e. However, the graph is composed of pieces that are connected, and these pieces are called the *components* of the graph. Hence a graph is connected if and only if it has only one component. □

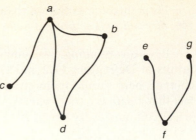

Figure 11.5

DEFINITION 11.5   For any graph $G = (V, E)$, the number of components of G is denoted by $\kappa(G)$.

Example 11.4   For the graphs in Figs. 11.1, 11.3, and 11.4, $\kappa(G) = 1$, because these graphs are connected; $\kappa(G) = 2$ for the graphs in Figs. 11.2 and 11.5.   □

Before closing this first section, we extend our concept of a graph. Thus far we have allowed at most one edge for two vertices; we now consider an extension.

DEFINITION 11.6   A graph $G = (V, E)$ is called a *multigraph* if for some $a, b \in V, a \neq b$, there are two or more edges of the form (a) (a, b) (for a directed graph); or (b) $\{a, b\}$ (for an undirected graph).

Figure 11.6 shows an example of a directed multigraph. There are three edges from a to b, so we say that the edge (a, b) has *multiplicity* 3. The edges (b, c) and (d, e) both have multiplicity 2. Also, the edge (e, d) and either one of the edges (d, e) form a (directed) circuit of length 2 in the multigraph.

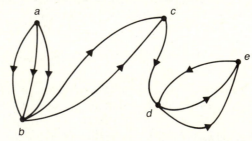

Figure 11.6

For $n \in \mathbf{Z}^+$, a multigraph is called an *n-graph* if no edge in the graph has multiplicity greater than n. The graph in Fig. 11.6 is a directed 3-graph.

We shall need the idea of a multigraph later in the chapter when we solve the problem of the seven bridges of Königsberg. Note: Whenever we are dealing with a multigraph G, we shall state explicitly that G is a multigraph.

EXERCISES 11.1

1. List three situations, different from those in this section, where a graph could prove useful.

2. For the graph in Fig. 11.7, determine (a) a walk from b to d that is not a trail; (b) a b-d trail that is not a path; (c) a path from b to d; (d) a closed walk from

b to b that is not a circuit; (e) a circuit from b to b that is not a cycle; (f) a cycle from b to b.

3. For the graph in Fig. 11.7, how many paths are there from b to f?

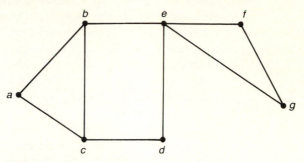

Figure 11.7

4. If a, b are vertices in a connected undirected graph G, the *distance* from a to b is defined to be the length of a shortest path from a to b. For the graph in Fig. 11.8, find the distance from d to every other vertex in G.

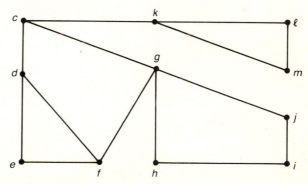

Figure 11.8

5. Seven towns $a, b, c, d, e, f,$ and g are connected by a system of highways as follows: (1) I-22 goes from a to c, passing through b; (2) I-33 goes from c to d and then passes through b as it continues to f; (3) I-44 goes from d through e to a; (4) I-55 goes from f to b, passing through g; and (5) I-66 goes from g to d.

   a) Using vertices for towns and directed edges for segments of highways between towns, draw a directed graph that models this situation.

   b) List the paths from g to a.

   c) What is the smallest number of highway segments that would have to be closed down for travel from b to d to be disrupted?

   d) Is it possible to leave town c and return there, visiting each of the other towns only once?

   e) What is the answer to part (d) if we are not required to return to c?

f) Is it possible to start at some town and drive over all of these highways exactly once? (You are allowed to visit a town more than once, and you need not return to the town from which you started.)

6. Figure 11.9 shows an undirected graph representing a section of a department store. The vertices indicate where cashiers are located; the edges denote unblocked aisles between cashiers. The department store wants to set up a security system where (plain-clothes) guards are placed at certain cashier locations so that each cashier either has a guard at his or her location or is only one aisle away from a cashier who has a guard. What is the smallest number of guards needed?

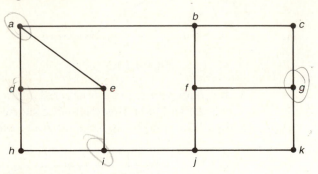

Figure 11.9

7. Let $G = (V, E)$ be a loop-free connected undirected graph, and let $\{a, b\}$ be an edge of G. Prove that $\{a, b\}$ is part of a cycle if and only if its removal (the vertices a and b are left) does not disconnect G.

8. Give an example of a connected graph G where removing any edge of G results in a disconnected graph.

9. Let G be a graph that satisfies the condition in Exercise 8.

   a) Must G be loop-free?   b) Could G be a multigraph?

   c) If G has n vertices, can we determine how many edges it has?

10. a) If $G = (V, E)$ is an undirected graph with $|V| = v$, $|E| = e$, and no loops, prove that $2e \leq v^2 - v$.

    b) State the corresponding inequality for the case when G is directed.

11. Let $G = (V, E)$ be an undirected graph. Define a relation $\mathcal{R}$ on V by $a \mathcal{R} b$ if $a = b$ or if there is a path in G from a to b. Prove that $\mathcal{R}$ is an equivalence relation. Describe the partition of V induced by $\mathcal{R}$.

12. Let $G = (V, E)$ be an undirected loop-free graph with $|V| = v$ and $|E| = e$. Determine the smallest possible value for e, in terms of v, if G is to be connected.

13. Find $\kappa(G)$ for the undirected graph $G = (V, E)$, where:

$$V = \{a, b, c, d, e, f, g, h\}, \text{ and}$$
$$E = \{\{a, b\}, \{b, c\}, \{d, e\}, \{e, f\}, \{f, d\}, \{g, h\}\}.$$

14. Let $G = (V, E)$ be a connected undirected graph, with $x, y \in V$. Prove that there exists an x-y walk in G that contains every vertex of G.

11.2
SUBGRAPHS, COMPLEMENTS, AND GRAPH ISOMORPHISM

In this section we shall focus on the following two ideas:

a) What type of substructure is there to study for a graph?

b) Is it possible to draw two graphs that appear distinct but have the same underlying structure?

To answer (a) we introduce the following definition.

DEFINITION 11.7   If $G = (V, E)$ is a graph (directed or undirected), then $G_1 = (V_1, E_1)$ is called a *subgraph* of G if $\emptyset \neq V_1 \subseteq V$ and $E_1 \subseteq E$, where each edge in E_1 is incident with vertices in V_1.

Figure 11.10(a) provides us with an undirected graph G and two of its subgraphs, G_1 and G_2. The vertices a, b are isolated in subgraph G_1. Part (b) of the figure provides a directed example.

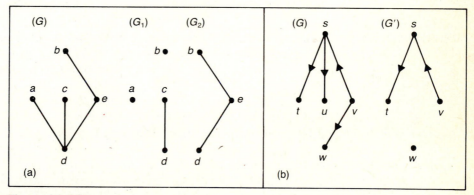

Figure 11.10

Certain special types of subgraphs arise as follows:

DEFINITION 11.8   Let $G = (V, E)$ be a graph (directed or undirected). If $\emptyset \neq U \subseteq V$, the *subgraph of G induced by U* is the subgraph with vertices in U and all edges (from G) of either the form (a) (x, y), for $x, y \in U$ (when G is directed), or (b) $\{x, y\}$, for $x, y \in U$ (when G is undirected). We denote this subgraph by $\langle U \rangle$.

A subgraph G' of a graph $G = (V, E)$ is called an *induced subgraph* if there exists $\emptyset \neq U \subseteq V$ where $G' = \langle U \rangle$.

For the subgraphs in Fig. 11.10(a), we find that G_2 is an induced subgraph of G but the subgraph G_1 is not an induced subgraph.

Example 11.5   Let $G = (V, E)$ denote the graph in Fig. 11.11(a). The subgraphs in parts (b) and (c) of the figure are induced subgraphs. For the connected subgraph in part (b), $G_1 = \langle U_1 \rangle$ for $U_1 = \{b, c, d, e\}$. In like manner, the disconnected subgraph in part (c) is $G_2 = \langle U_2 \rangle$ for $U_2 = \{a, b, e, f\}$. Finally, G_3 in part (d) of Fig. 11.11 is a subgraph of G. But it is not an induced subgraph; the vertices c, e are in G_3, but the edge $\{c, e\}$ (of G) is not present.   □

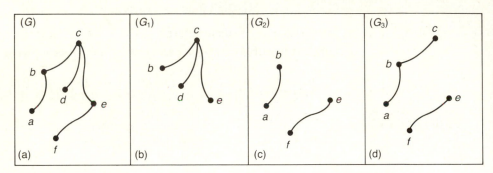

Figure 11.11

A second special type of subgraph comes about when a certain vertex or edge is deleted from the given graph. We formalize these ideas in the following definition.

DEFINITION 11.9   Let v be a vertex in a directed or an undirected graph $G = (V, E)$. The subgraph of G denoted by $G - v$ has the vertex set $V_1 = V - \{v\}$ and the edge set $E_1 \subseteq E$, where E_1 contains all the edges in E except for those that are incident with the vertex v.

In a similar way, if e is an edge of a directed or an undirected graph $G = (V, E)$, we obtain the subgraph $G - e$ of G where the set of edges $E_1 = E - \{e\}$, and the vertex set is unchanged (that is, $V_1 = V$).

Example 11.6   Let $G = (V, E)$ be the undirected graph in Fig. 11.12(a). Part (b) of this figure is the subgraph G_1 (of G), where $G_1 = G - c$. It is also the subgraph of G induced by the set of vertices $U_1 = \{a, b, d, f, g, h\}$, so $G_1 = \langle V - \{c\} \rangle = \langle U_1 \rangle$. In part (c) of Fig. 11.12 we find the subgraph G_2 of G, where $G_2 = G - e$ for e the edge $\{c, d\}$. The result in Fig. 11.12(d) shows how the ideas in Definition 11.9 can be extended to the deletion of more than one vertex (edge). We may represent this subgraph of G as $G_3 = (G - b) - f = (G - f) - b = G - \{b, f\} = \langle U_3 \rangle$, for $U_3 = \{a, c, d, g, h\}$.   □

The idea of a subgraph gives us a way to develop the complement of an undirected loop-free graph. Before doing so, however, we define a type of graph that is maximal in size for a given number of vertices.

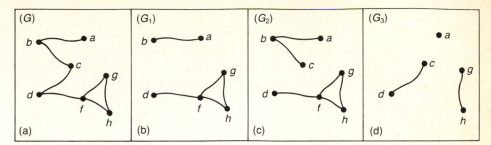

Figure 11.12

DEFINITION 11.10   Let V be a set of n vertices. The *complete graph* on V, denoted K_n, is a loop-free undirected graph where for any $a, b \in V, a \neq b$, there is an edge $\{a, b\}$. ▬

Figure 11.13 provides the complete graphs $K_n, 1 \le n \le 4$. We shall see, when we examine the idea of graph isomorphism, that these are the only possible complete graphs for the number of vertices given.

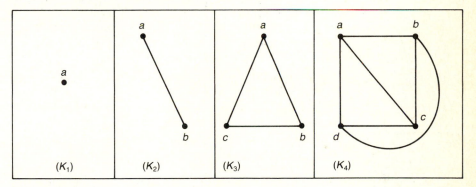

Figure 11.13

In determining the complement of a set in Chapter 3, we needed to know the universal set under consideration. The complete graph plays a role similar to a universal set.

DEFINITION 11.11   Let G be a loop-free undirected graph on n vertices. The *complement of G*, denoted $\overline{G}$, is the subgraph of K_n consisting of the n vertices in G and all edges that are not in G. (If $G = K_n$, $\overline{G}$ is a graph consisting of n vertices and no edges. Such a graph is called a *null* graph.) ▬

In Fig. 11.14(a) there is an undirected graph on four vertices. Its complement is shown in part (b) of the figure. In the complement, vertex a is isolated.

Once again we have reached a point where many new ideas have been defined. To demonstrate why some of these ideas are important, we apply them now to the solution of an interesting puzzle.

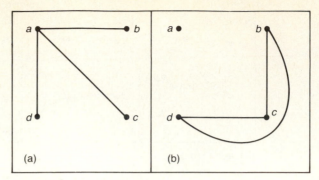

Figure 11.14

Example 11.7   (*Instant Insanity*) The game of Instant Insanity is played with four cubes. Each of the six faces on a cube is painted with one of the colors red (R), white (W), blue (B), or yellow (Y). The object of the game is to place the cubes in a column of four such that all four (different) colors appear on each of the four sides of the column.

Consider the cubes in Fig. 11.15 and number them as shown. (These cubes are only one example of this game. Many others exist.) First we shall estimate the number of arrangements that are possible here. If we place cube 1 at the bottom of the column, there are at most three different ways in which we can do this. In Fig. 11.15 cube 1 is unfolded and we see that it makes no difference whether we place the red face on the table or the opposite white face on the table. We are concerned only with the other four faces at the base of our column. With three pairs of opposite faces there will be at most three ways to place the first cube for the base of the column. Now consider cube 2. Although some colors are repeated, no pair of opposite faces has the same color. Hence we have six ways to place the second cube on top of the first. We can then rotate the second cube without changing either the face on the top of the first cube or the face on the bottom of the second cube. With four possible rotations we may place the second cube on top of the first in as many as 24 different ways. Continuing the argument, we find that there can be as many as $(3)(24)(24)(24) = 41,472$ possibilities to consider. And there may not even be a solution!

In solving the puzzle we realize that it is difficult to keep track of (1) colors on opposite faces of cubes and (2) columns of colors. A graph (actually a labeled multigraph) helps us to visualize the situation. In Fig. 11.16 we have a graph on four vertices R, W, B, and Y. As we consider each cube, we examine its three pairs of opposite faces. For example, cube 1 has a pair of opposite faces painted yellow and blue, so we draw an edge connecting Y and B and label it 1 (for cube 1). The other two edges in the figure that are labeled with 1 account for the pairs of opposite faces that are white and yellow, and red and white. Doing likewise for the other cubes, we arrive at the graph in the figure. A loop, such as the one at B, with label 3, indicates a pair of opposite faces with the same color.

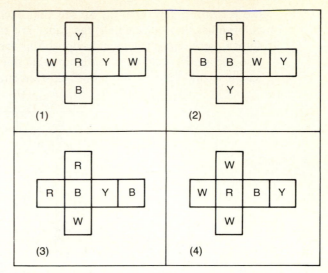

Figure 11.15

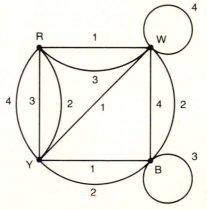

Figure 11.16

From the graph we see that there is a total of 12 edges falling into four sets of 3, according to the labels for the cubes. At each vertex the number of edges incident to (or from) the vertex counts the number of faces on the four cubes that have that color. (We count a loop twice.) Hence Fig. 11.16 tells us that for our four cubes we have five red faces, seven white ones, six blue ones, and six that are yellow.

With the four cubes stacked in a column, we examine two opposite sides of the column. This gives us four edges in the graph of Fig. 11.16, where each label appears once. Since each color is to appear only once on a side of the column, each color must appear twice as an endpoint of these four edges. If we can accomplish the same result for the other two sides of the column, we have solved the puzzle. In Fig. 11.17(a) we see that each of the two opposite sides of our column has the four colors if the cubes are arranged according to the subgraph shown there. However, to accomplish this for the other two sides of the column also, we need a second such subgraph that doesn't use any edge in part (a). In this case a second such subgraph does exist, as shown in part (b) of the figure.

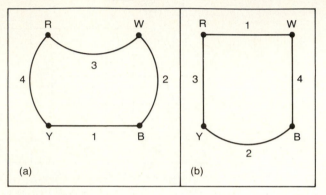

Figure 11.17

Figure 11.18 shows how to arrange the cubes as indicated by the subgraphs in Fig. 11.17.

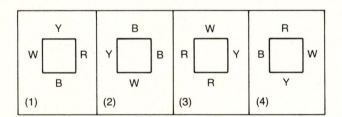

Figure 11.18

In general, for any four cubes we construct a labeled multigraph and try to find two subgraphs where (1) each subgraph contains all four vertices, and four edges, one for each label, (2) in each subgraph, each vertex is incident with exactly two edges (a loop is counted twice), and (3) no edge of the graph appears in both subgraphs.   □

Now we turn to the second question posed at the start of the section.

Figure 11.19 shows two undirected graphs on four vertices. Since straight edges and curved edges are considered the same here, each graph represents six adjacent pairs of vertices. In fact we probably feel that these graphs are both examples of the graph K_4. We make this feeling mathematically rigorous in the following definition.

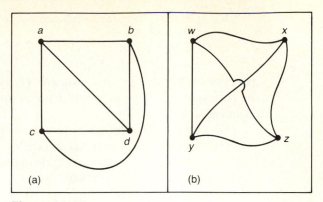

Figure 11.19

DEFINITION 11.12 Let $G_1 = (V_1, E_1)$ and $G_2 = (V_2, E_2)$ be two undirected graphs. A function $f: V_1 \rightarrow V_2$ is called a *graph isomorphism* if (a) f is one-to-one and onto, and (b) for all $a, b \in V_1$, $\{a, b\} \in E_1$ if and only if $\{f(a), f(b)\} \in E_2$. When such a function exists, G_1 and G_2 are called *isomorphic graphs*.

The vertex correspondence of a graph isomorphism preserves adjacencies. In this way the structure of the graphs is preserved.

For the graphs in Fig. 11.19 the function f defined by $f(a) = w$, $f(b) = x$, $f(c) = y$, and $f(d) = z$ provides an isomorphism. Consequently, as far as structure is concerned, these graphs are considered the same.

Let us examine this idea in a more difficult situation.

Example 11.8 In Fig. 11.20 we have two graphs, each on ten vertices. Unlike the graphs in Fig. 11.19, it is not immediate if these graphs are isomorphic.

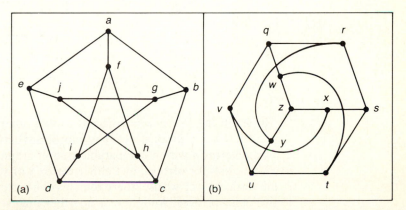

Figure 11.20

One finds that the correspondence given by

$$
\begin{array}{ccccc}
a \rightarrow q & c \rightarrow u & e \rightarrow r & g \rightarrow x & i \rightarrow z \\
b \rightarrow v & d \rightarrow y & f \rightarrow w & h \rightarrow t & j \rightarrow s
\end{array}
$$

preserves all adjacencies. For example, $\{f, h\}$ is an edge in graph (a) with $\{w, t\}$ the corresponding edge in graph (b). But how did we come up with the correspondence?

We note that because an isomorphism preserves adjacencies, it preserves graph substructures such as paths and cycles. In graph (a) the edges $\{a, f\}$, $\{f, i\}$, $\{i, d\}$, $\{d, e\}$, and $\{e, a\}$ constitute a cycle of length 5. Hence we must preserve this as we try to find an isomorphism. The corresponding edges in graph (b) are $\{q, w\}$, $\{w, z\}$, $\{z, y\}$, $\{y, r\}$, and $\{r, q\}$, which also provide a cycle of length 5. In addition, starting at vertex a in graph (a), we find a path that will "visit" each vertex only once. We express this path by $a \rightarrow f \rightarrow h \rightarrow c \rightarrow b \rightarrow g \rightarrow j \rightarrow e \rightarrow d \rightarrow i$. For the graphs to be isomorphic there must be a corresponding path in graph (b). Here the path described by $q \rightarrow w \rightarrow t \rightarrow u \rightarrow v \rightarrow x \rightarrow s \rightarrow r \rightarrow y \rightarrow z$ is the counterpart. □

These are some of the ways we can determine whether two graphs are isomorphic. Other considerations will be discussed throughout the chapter.

We close this section with one more example on graph isomorphism.

Example 11.9   Each of the two graphs in Fig. 11.21 has six vertices and nine edges. Therefore it is reasonable to ask whether they are isomorphic.

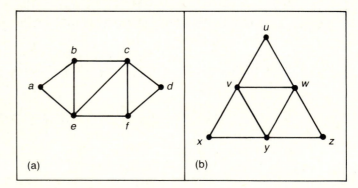

Figure 11.21

In graph (a), vertex a is adjacent to two other vertices of the graph. Consequently, if we try to construct an isomorphism between these graphs, we should associate vertex a with a comparable one in graph (b), say vertex u. A similar situation exists for vertex d and either vertex x or vertex z. But no matter which of the vertices x or z we use, there remains one vertex in graph (b) that is adjacent to two other vertices. And there is no other such vertex in graph (a) to continue our one-to-one structure preserving correspondence. Consequently, these graphs are not isomorphic.

Furthermore, in graph (b) it is possible to start at any vertex and find a circuit that includes every edge of the graph. For example, if we start at vertex u, the circuit $u \rightarrow w \rightarrow v \rightarrow y \rightarrow w \rightarrow z \rightarrow y \rightarrow x \rightarrow v \rightarrow u$ exhibits this property. This does not happen in graph (a) where the only trails that include each edge start at either b or f and then terminate at f or b, respectively. ☐

EXERCISES 11.2

1. Let G be the undirected graph in Fig. 11.22(a).

   a) How many connected subgraphs of G have four vertices and include a cycle?

   b) Describe the subgraph G_1 (of G) in part (b) of the figure as an induced subgraph and in terms of deleting a vertex of G.

   c) Describe the subgraph G_2 (of G) in part (c) of the figure as an induced subgraph and in terms of the deletion of vertices of G.

   d) Draw the subgraph of G induced by the set of vertices $U = \{b, c, d, f, i, j\}$.

   e) For the graph G, let the edge $e = \{c, f\}$. Draw the subgraph $G - e$.

   f) Let e_1, e_2 denote the edges $\{a, c\}, \{a, d\}$, respectively, in the graph G. Draw the following subgraphs of G: (i) $(G - e_1) - e_2$; (ii) $(G - e_2) - e_1$; (iii) $G - \{e_1, e_2\}$.

   g) How many connected subgraphs of G contain all nine vertices of G?

2. a) Let $G = (V, E)$ be an undirected graph, with $G_1 = (V_1, E_1)$ a subgraph of G. Under what condition(s) is G_1 *not* an induced subgraph of G?

   b) For the graph G in Fig. 11.22(a), find a subgraph that is not an induced subgraph.

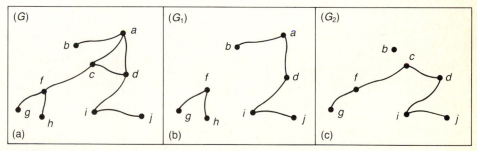

Figure 11.22

3. Let $G = (V, E)$ be an undirected graph, where $|V| \geq 2$. If every induced subgraph of G is connected, can we identify the graph G?

4. a) Find two graphs $G = (V, E)$ and $G_1 = (V_1, E_1)$, with $v \in V$ and $v_1 \in V_1$, where

   $$\kappa(G - v) = \kappa(G) \qquad \text{but} \qquad \kappa(G_1 - v_1) > \kappa(G_1).$$

   b) Find two graphs $G = (V, E)$ and $G_1 = (V_1, E_1)$, with $e \in E$ and $e_1 \in E_1$,

where
$$\kappa(G - e) = \kappa(G) \qquad \text{but} \qquad \kappa(G_1 - e_1) > \kappa(G_1).$$

5. Each of the labeled multigraphs in Fig. 11.23 arises in the analysis of a set of four blocks for the game of Instant Insanity. In each case determine a solution to the puzzle, if possible.

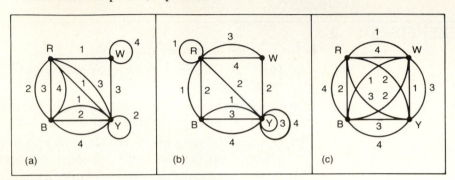

Figure 11.23

6. Find all (loop-free) nonisomorphic undirected graphs with four vertices. How many of these graphs are connected?

7. For each pair of graphs in Fig. 11.24, determine whether or not the graphs are isomorphic.

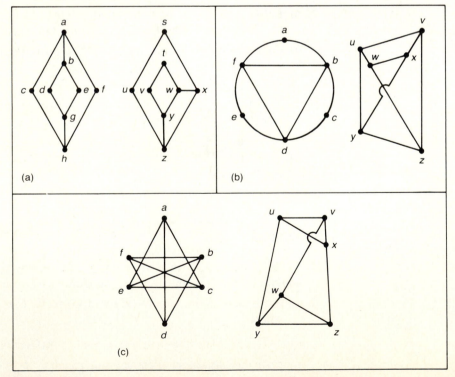

Figure 11.24

8. **a)** Coach Rizzo must make up a schedule for the five softball teams in her league. If each team plays two other teams, model a possible schedule by using a graph.

   b) Although it is possible to have different schedules in part (a), show that these schedules are the same except for a permutation of the names of the teams.

9. Let G be an undirected (loop-free) graph with v vertices and e edges. How many edges are there in $\overline{G}$?

10. Let $f: G_1 \rightarrow G_2$ be a graph isomorphism. If there is a path of length 3 between the vertices a and b in G_1, prove that in G_2 there is a path of length 3 between the vertices $f(a)$ and $f(b)$.

11. **a)** If G_1, G_2 are (loop-free) undirected graphs, prove that G_1, G_2 are isomorphic if and only if $\overline{G}_1, \overline{G}_2$ are isomorphic.

    b) Determine whether the graphs in Fig. 11.25 are isomorphic.

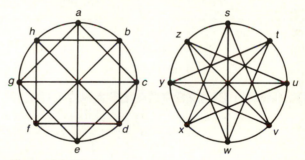

Figure 11.25

12. **a)** Let G be an undirected graph with n vertices. If G is isomorphic to its own complement $\overline{G}$, how many edges must G have? (Such a graph is called *self-complementary*.)

    b) Find an example of a self-complementary graph on four vertices and one on five vertices.

    c) If G is a self-complementary graph on n vertices, where $n > 1$, prove that $n = 4k$ or $n = 4k + 1$, for some $k \in \mathbf{Z}^+$.

13. Let G be a cycle on n vertices. Prove that G is self-complementary if and only if $n = 5$.

14. **a)** Find a graph G where both G and $\overline{G}$ are connected.

    b) If G is a graph on n vertices, $n \geq 2$, and G is not connected, prove that $\overline{G}$ is connected.

15. **a)** Extend Definition 11.12 to directed graphs.

    b) Determine, up to isomorphism, all (loop-free) directed graphs on three vertices.

    c) Determine whether the directed graphs in Fig. 11.26 are isomorphic.

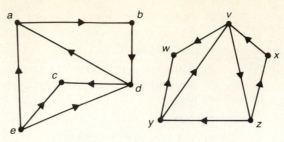

Figure 11.26

16. a) How many subgraphs $H = (V, E)$ of K_6 satisfy $|V| = 3$? (If two subgraphs are isomorphic but have different vertex sets, consider them distinct.)

  b) How many subgraphs $H = (V, E)$ of K_6 satisfy $|V| = 4$?

  c) How many subgraphs does K_6 have?

  d) For $n \geq 3$, how many subgraphs does K_n have?

11.3
VERTEX DEGREE: EULER TRAILS AND CIRCUITS

In Example 11.9 we found that the number of edges incident with a vertex could be used to show that two undirected graphs were not isomorphic. We now find this idea even more helpful.

DEFINITION 11.13   Let G be an undirected graph or multigraph. For any vertex v of G, the *degree of* v, written deg(v), is the number of edges in G that are incident with v. ▬

Example 11.10   For the graph in Fig. 11.27, $\deg(b) = \deg(d) = \deg(f) = \deg(g) = 2$, $\deg(c) = 4$, $\deg(e) = 0$, and $\deg(h) = 1$. For vertex a we have $\deg(a) = 3$, because we count a loop twice. Since h has degree 1, it is called a *pendant* vertex. □

Using the idea of vertex degree, we have the following result.

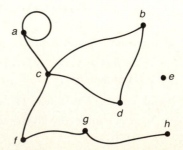

Figure 11.27

THEOREM 11.2 If $G = (V, E)$ is an undirected graph or multigraph, then $\sum_{v \in V} \deg(v) = 2|E|$.

Proof As we consider each edge $\{a, b\}$ in graph G, we find that the edge contributes a count of 1 to each of $\deg(a), \deg(b)$, and consequently a count of 2 to $\sum_{v \in V} \deg(v)$. Thus $2|E|$ accounts for $\deg(v)$, for all $v \in V$, and $\sum_{v \in V} \deg(v) = 2|E|$. ∎

This theorem provides some insight into the number of odd vertices that can exist in a graph.

COROLLARY 11.1 For any undirected graph or multigraph, the number of vertices of odd degree must be even.

Proof We leave the proof for the reader. ∎

We apply Theorem 11.2 in the following example.

Example 11.11 An undirected graph (or multigraph) where each vertex has the same degree is called a *regular* graph. Is it possible to have a regular graph where each vertex has degree 4 and there are 10 edges?

From Theorem 11.2, $2|E| = 20 = 4|V|$, so we have five vertices of degree 4. Figure 11.28 provides two nonisomorphic examples that satisfy the requirements.

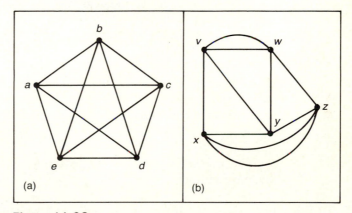

Figure 11.28

If we want each vertex to have degree 4, with 15 edges in the graph, we find that $2|E| = 30 = 4|V|$, from which it follows that no such graph is possible. □

We turn now to the reason why Euler developed the idea of the degree of a vertex: to solve the problem dealing with the seven bridges of Königsberg.

Example 11.12 (*The Seven Bridges of Königsberg*) During the eighteenth century the city of Königsberg (in East Prussia) was divided into four sections (including the island of Kneiphof) by the Pregel river. Seven bridges connected these regions, as shown in Fig. 11.29(a). It was said that residents spent their Sunday walks trying to find a starting point so that they could walk about the city, cross each bridge exactly once, and return to the starting point.

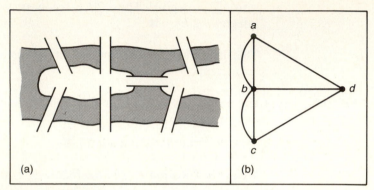

Figure 11.29

In order to determine whether or not such a circuit existed, Euler represented the four sections of the city and the seven bridges by the multigraph shown in Fig. 11.29(b). Here he found four vertices with $\deg(a) = \deg(c) = \deg(d) = 3$ and $\deg(b) = 5$. He also found that the existence of such a circuit depended on the number of vertices of odd degree in the graph. □

Before proving the general result, we give the following definition.

DEFINITION 11.14   Let $G = (V, E)$ be an undirected graph or multigraph. G is said to have an *Euler circuit* if there is a circuit in G that passes through every $v \in V$ and traverses every edge of the graph exactly once. If there is a trail from a to b in G and this trail passes through every vertex of G and traverses each edge in G exactly once, the trail is called an *Euler trail*.

The problem of the seven bridges is now settled as we characterize the graphs that have an Euler circuit.

THEOREM 11.3   Let $G = (V, E)$ be an undirected graph or multigraph. Then G has an Euler circuit if and only if G is connected and every vertex in G has even degree.

Proof   If G has an Euler circuit, then for any $a, b \in V$ there is a trail from a to b, namely that part of the circuit that starts at a and terminates at b. Therefore, it follows from Theorem 11.1 that G is connected.

Let s be the starting vertex of the Euler circuit. For any other vertex v of G, each time the circuit comes to v it then departs from the vertex. Thus the circuit has traversed either two (new) edges that are incident with v or a (new) loop at v. In either case a count of 2 is contributed to $\deg(v)$. Since v is not the starting point, a count of 2 is obtained each time the circuit passes through v, so $\deg(v)$ is even. As for the starting vertex s, the first edge of the circuit must be distinct from the last edge, and because any other visit to s results in a count of 2 for $\deg(s)$, we have $\deg(s)$ even.

Conversely, let G be connected with every vertex of even degree. If the number of edges in G is 1 or 2, then G must be as shown in Fig. 11.30. Euler circuits are immediate in these cases. We proceed now by induction and assume the result true for all situations where there are fewer than n edges. If G has n edges, select a vertex s in G as a starting point to build an Euler circuit. The graph (or multigraph) G is connected and each vertex has even degree, so we can at least construct a circuit C on s. (Verify this.) Should the circuit contain every edge of G, we are finished. If not, remove the edges of the circuit from G, making sure to remove any vertex that would become isolated. The remaining subgraph K has all vertices of even degree, but it may not be connected. However, each component of K is connected and will have an Euler circuit. (Why?) In addition, each of these Euler circuits has a vertex that is on C. Consequently, starting at s we travel on C until we arrive at a vertex s_1 that is on the Euler circuit of a component C_1 of K. Then we traverse this Euler circuit and, returning to s_1, continue on C until we reach a vertex s_2 that is on the Euler circuit of component C_2 of K. Since the graph G is finite, as we continue this process we construct an Euler circuit for G. ■

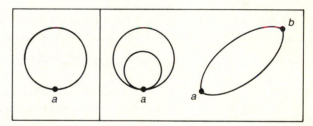

Figure 11.30

Should G be connected and not have too many vertices of odd degree, we can at least find an Euler trail in G.

COROLLARY 11.2 If G is an undirected graph or multigraph, we can construct an Euler trail in G if and only if G is connected and has only two vertices of odd degree.

Proof   If G is connected and a and b are the vertices of G that have odd degree, add an additional edge $\{a, b\}$ to G. We now have a graph G_1 that is connected and has every vertex of even degree. Hence G_1 has an Euler circuit C, and when the edge $\{a, b\}$ is removed from C, we obtain an Euler trail for G. (Thus the Euler trail starts at one of the vertices of odd degree and terminates at the other odd vertex.) We leave the details of the converse for the reader. ■

Returning now to the seven bridges of Königsberg, Figure 11.29(b) is a connected multigraph, but it has four vertices of odd degree. Consequently there is no Euler trail or circuit for this situation.

Now that we have seen how the solution of an eighteenth-century problem led to the start of graph theory, is there any more contemporary context in which we might be able to apply what we have learned?

To answer this question (in the affirmative), we shall state the directed version of Theorem 11.3. But first we need to extend our initial idea on the degree of a vertex.

DEFINITION 11.15   Let $G = (V, E)$ be a directed graph or multigraph. For any $v \in V$,

a) The *incoming*, or *in*, *degree* of v is the number of edges in G that are incident into v, and this is denoted by $\deg^-(v)$.

b) The *outgoing*, or *out*, *degree* of v is the number of edges in G that are incident from v, and this is denoted by $\deg^+(v)$.

For the case where the directed graph or multigraph contains one or more loops, each loop at a given vertex v contributes a count of 1 to each of $\deg^-(v)$ and $\deg^+(v)$.

The concepts of the in degree and the out degree for vertices now leads us to the following theorem.

THEOREM 11.4   Let $G = (V, E)$ be a directed graph or multigraph. The graph G has a directed Euler circuit if and only if G is connected and $\deg^-(v) = \deg^+(v)$ for all $v \in V$.

Proof   The proof of this theorem is left for the reader to work out in Exercise 14(a). ∎

At this time we consider an application of Theorem 11.4. This example is based on a telecommunication problem given by C. L. Liu on pages 176–178 of reference [19].

Example 11.13   In Fig. 11.31(a) we have the surface of a rotating drum that is divided into eight sectors of equal area. In part (b) of the figure we have placed conducting (shaded sectors and inner circle) and nonconducting (unshaded sectors) material on the drum. When the three terminals (shown in the figure) make contact with the three designated sectors, the nonconducting material results in no flow of current and a 1 appears on the display of a digital device. For the sectors with the conducting material, a flow of current takes place and a 0 appears on the display in each case. If the drum were rotated 45 degrees (clockwise), the screen would read 110 (from top to bottom). So we can obtain at least two (namely, 100 and 110) of the eight binary representations from 000 (for 0) to 111 (for 7). But can we represent all eight of them as the drum continues to rotate? And could we extend the problem to the 16 four-bit binary representations from 0000 through 1111, and perhaps generalize the results even further?

To answer the question for the problem in the figure, we construct a directed graph $G = (V, E)$ where $V = \{00, 01, 10, 11\}$ and E is constructed as follows: If $b_1 b_2, b_2 b_3 \in V$, draw the edge $(b_1 b_2, b_2 b_3)$. This results in the directed graph of

Fig. 11.32(a), where $|E| = 8$. We see that this graph is connected and that for all $v \in V$, $\deg^-(v) = 2 = \deg^+(v)$. Consequently, by Theorem 11.4, it has a directed Euler circuit. One such circuit is given by

$$\begin{array}{ccccccc} 100 & 000 & 001 & 010 & 101 & 011 & 111 \\ \curvearrowright 10 \longrightarrow 00 \longrightarrow 00 \longrightarrow 01 \longrightarrow 10 \longrightarrow 01 \longrightarrow 11 \longrightarrow 11 \end{array}$$

$$110$$

Here the label on each edge $e = (a, c)$, as shown in part (b) of Fig. 11.32, is the three-bit sequence $x_1 x_2 x_3$, where $a = x_1 x_2$ and $c = x_2 x_3$. Since the vertices of G are the four distinct two-bit sequences 00, 01, 10, and 11, the labels on the eight edges of G determine the eight distinct three-bit sequences. Also, any two consecutive edge labels in the Euler circuit are of the form $y_1 y_2 y_3$ and $y_2 y_3 y_4$.

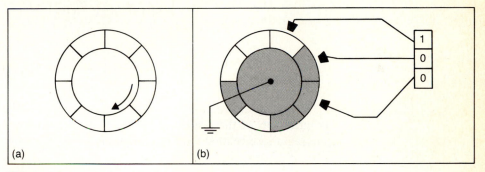

(a)   (b)

Figure 11.31

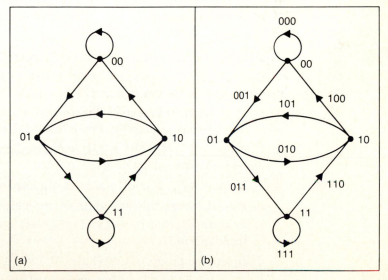

(a)   (b)

Figure 11.32

Starting with the edge label 100, in order to get the next label, 000, we concatenate the last bit in 000, namely 0, to the string 100. The resulting string 1000 then provides 100 (1000) and 000 (1000). The next edge label is 001, so we concatenate the 1 (the last bit in 001) to our present string 1000 and get 10001, which provides the three distinct three-bit sequences 100 (10001), 000 (10001), and 001 (10001). Continuing in this way, we arrive at the eight-bit sequence 10001011 (where the last 1 is *wrapped around*), and these eight bits are then arranged in the sectors of the rotating drum as in Fig. 11.33. It is from this figure that the result in Fig. 11.31(b) is obtained. And as the drum in Fig. 11.31(b) rotates, all of the eight three-bit sequences 100, 110, 111, 011, 101, 010, 001, and 000 are obtained.   □

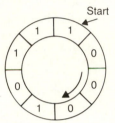

Figure 11.33

In closing this section, we wish to call reference [21] by Anthony Ralston to the attention of the reader. This article is a good source for more ideas and generalizations related to the problem discussed in Example 11.13.

EXERCISES 11.3

1. Determine $|V|$ for the following graphs or multigraphs G.

   a) G has nine edges and all vertices have degree 3.

   b) G is regular with 15 edges.

   c) G has 10 edges with two vertices of degree 4 and all others of degree 3.

2. If $G = (V, E)$ is a connected graph with $|E| = 17$ and $\deg(v) \geq 3$ for all $v \in V$, what is the maximum value for $|V|$?

3. If G is an undirected graph with n vertices and e edges, let $\delta = \min_{v \in V}\{\deg(v)\}$ and let $\Delta = \max_{v \in V}\{\deg(v)\}$. Prove that $\delta \leq 2(e/n) \leq \Delta$.

4. Let $G = (V, E)$, $H = (V', E')$ be undirected graphs with $f: V \to V'$ establishing an isomorphism between the graphs.

   a) Prove that $f^{-1}: V' \to V$ is also an isomorphism for G and H.

   b) If $a \in V$, prove that $\deg(a)$ (in G) $= \deg(f(a))$ (in H).

   c) Prove that vertex a is isolated (pendant) in G if and only if $f(a)$ is isolated (pendant) in H.

5. Complete the proofs of Corollaries 11.1 and 11.2.

6. Let $G = (V, E)$ be a loop-free undirected graph where $\deg(v) \geq k \geq 1$ for all $v \in V$. Prove that G contains a path of length k.

7. **a)** Explain why it is not possible to draw a loop-free connected undirected graph with eight vertices, where the degrees of the vertices are 1, 1, 1, 2, 3, 4, 5, and 7.

   b) Give an example of a loop-free connected undirected multigraph with eight vertices, where the degrees of the vertices are 1, 1, 1, 2, 3, 4, 5, and 7.

8. **a)** Find an Euler circuit for the graph in Fig. 11.34.

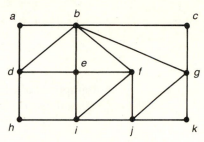

k **Figure 11.34**

   b) If the edge $\{d, e\}$ is removed from this graph, find an Euler trail for the resulting subgraph.

9. Determine the value(s) of n for which the complete graph K_n has an Euler circuit. For which n does K_n have an Euler trail but not an Euler circuit?

10. For the graph in Fig. 11.29(b), what is the smallest number of bridges that must be removed so that the resulting subgraph has an Euler trail but not an Euler circuit? Which bridge(s) should we remove?

11. When visiting a chamber of horrors, Paul and David try to figure out whether they can travel through the seven rooms and surrounding corridor of the attraction without passing through any door more than once. If they must start from the starred position in the corridor shown in Fig. 11.35, can they accomplish their goal?

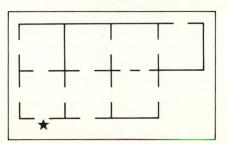

Figure 11.35

12. Let G be a loop-free undirected graph on n (≥ 3) vertices. If G has only one vertex of even degree, how many vertices in $\overline{G}$ have even degree?

13. Prove that in any directed graph or multigraph $G = (V, E)$, $\sum_{v \in V} \deg^+(v) = \sum_{v \in V} \deg^-(v)$.

14. a) Prove that a directed graph or multigraph $G = (V, E)$ has a directed Euler circuit if and only if G is connected and $\deg^+(v) = \deg^-(v)$ for all $v \in V$.

   b) Find a directed Euler circuit for the graph in Fig. 11.36.

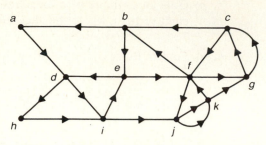

Figure 11.36

   c) A directed graph is called *strongly connected* if there is a directed path from a to b for all vertices a, b, where $a \neq b$. Prove that if a directed graph has a directed Euler circuit, then it is strongly connected. Is the converse true?

15. Let G be a directed graph on n vertices. If the associated undirected graph for G is K_n, prove that $\sum_{v \in V} [\deg^+(v)]^2 = \sum_{v \in V} [\deg^-(v)]^2$.

16. If $G = (V, E)$ is a directed graph or multigraph, prove that G has a directed Euler trail if and only if (i) G is connected; (ii) $\deg^+(v) = \deg^-(v)$ for all but two vertices x, y in V; and (iii) $\deg^+(x) = \deg^-(x) + 1, \deg^-(y) = \deg^+(y) + 1$.

17. Let $V = \{000, 001, 010, \ldots, 110, 111\}$. For any four-bit sequence $b_1 b_2 b_3 b_4$ draw an edge from the element $b_1 b_2 b_3$ to the element $b_2 b_3 b_4$ in V.

   a) Draw the graph $G = (V, E)$ as described.

   b) Find a directed Euler circuit for G.

   c) Equally space eight 0's and eight 1's around the edge of a rotating (clockwise) drum so that these 16 bits form a circular sequence where the (consecutive) subsequences of length 4 provide the binary representations of $0, 1, 2, \ldots, 14, 15$ in some order.

Quiz 6

18. Equally space nine 0's, nine 1's, and nine 2's around the edge of a rotating (clockwise) drum so that these 27 symbols form a circular sequence where the (consecutive) subsequences of length 3 provide the ternary (base 3) representations of $0, 1, 2, \ldots, 25, 26$ in some order.

19. Let $G = (V, E)$ be a loop-free connected undirected graph with $|V| \geq 2$. Prove that G contains two vertices v, w where $\deg(v) = \deg(w)$.

20. Carolyn and Richard attended a party with three other married couples. At this party a good deal of handshaking took place, but (1) no one shook hands with her or his spouse; (2) no one shook hands with herself or himself, and (3) no one shook hands with anyone more than once. Before leaving the party, Carolyn asked the other seven people how many hands she or he had shaken. She received a different answer from each of the seven. How many times did Carolyn shake hands at this party? How many times did Richard?

21. If $G = (V, E)$ is an undirected graph with $|V| = n$ and $|E| = k$, the following matrices are used to represent G.

Let $V = \{v_1, v_2, \ldots, v_n\}$. Define the *adjacency matrix* $A = (a_{ij})_{n \times n}$ where $a_{ij} = 1$ if $\{v_i, v_j\} \in E$, otherwise $a_{ij} = 0$.

If $E = \{e_1, e_2, \ldots, e_k\}$, the *incidence matrix* I is the $n \times k$ matrix $(b_{ij})_{n \times k}$ where $b_{ij} = 1$ if v_i is a vertex on the edge e_j, otherwise $b_{ij} = 0$.

a) Find the adjacency and incidence matrices associated with the graph in Fig. 11.37.

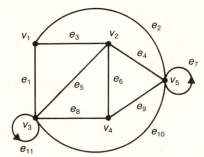

Figure 11.37

b) Calculating A^2 and using the Boolean operations where $0 + 0 = 0$, $0 + 1 = 1 + 0 = 1 + 1 = 1$, and $0 \cdot 0 = 0 \cdot 1 = 1 \cdot 0 = 0$, $1 \cdot 1 = 1$, prove that the entry in row i and column j of A^2 is 1 if and only if there is a walk of length two between the ith and jth vertices of V.

c) If we calculate A^2 using ordinary addition and multiplication, what do the entries in the matrix reveal about G?

d) What is the column sum for each column of A? Why?

e) What is the column sum for each column of I? Why?

11.4
PLANAR GRAPHS

On a road map the lines indicating the roads and highways usually intersect only at junctions or towns. But sometimes roads seem to intersect when one road is located above another, as in the case of an overpass. In this case the two roads are at different levels, or planes. This type of situation leads us to the following definition.

DEFINITION 11.16　A graph G is called *planar* if G can be drawn in the plane with its edges intersecting only at vertices of G.　━━

Example 11.14　The graphs in Fig. 11.38 are planar. The first is a 3-regular graph, because each vertex has degree 3; it is planar because no edges intersect except at the vertices. In graph (b) it appears that we have a *nonplanar* graph; the edges $\{x, z\}$ and $\{w, y\}$

overlap at a point other than a vertex. However, we can redraw this graph as shown in part (c) of the figure. Consequently K_4 is planar.   □

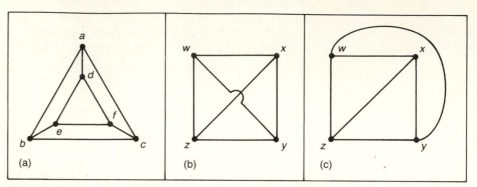

Figure 11.38

Example 11.15 Just as K_4 is planar, so are the graphs K_1, K_2, and K_3.

The construction of K_5 is shown in Fig. 11.39. Start with the pentagon in part (a) of the figure. Since a complete graph contains an edge for every pair of distinct vertices, we add edge $\{a, c\}$ as shown in part (b). This edge is contained entirely within the interior of the pentagon in part (a). (We could have drawn the edge in the exterior region determined by the pentagon. The reader will be asked in the exercises to show that the same conclusion arises in this case.) Moving to part (c), we add in the edges $\{a, d\}$, $\{c, e\}$, and $\{b, e\}$. Now we consider the vertices b and d. We need the edge $\{b, d\}$ in order to have K_5. Vertex d is inside the region formed by the cycle edges $\{a, c\}$, $\{c, e\}$, and $\{e, a\}$, whereas b is outside the region. Thus in drawing the edge $\{b, d\}$, we must intersect one of the existing edges at least once, as shown by the dotted edges in part (d). Consequently, K_5 is nonplanar. (Since this proof appeals to a diagram, it definitely lacks rigor. However, later in the section we shall prove by another method that K_5 is nonplanar.)   □

Before we can characterize all nonplanar graphs we need to examine another class of graphs.

DEFINITION 11.17 A graph $G = (V, E)$ is called *bipartite* if $V = V_1 \cup V_2$ with $V_1 \cap V_2 = \emptyset$, and every edge of G is of the form $\{a, b\}$ with one of the vertices a, b in V_1 and the other vertex in V_2. If each vertex in V_1 is joined with every vertex in V_2, we have a *complete bipartite* graph; when $|V_1| = m$, $|V_2| = n$, the graph is denoted by $K_{m,n}$.

Figure 11.40 shows two bipartite graphs. The graph in part (a) satisfies the definition for $V_1 = \{a, b\}$ and $V_2 = \{c, d, e\}$. If we add the edges $\{b, d\}$ and $\{b, c\}$, the result is the complete bipartite graph $K_{2,3}$, which is planar. Graph (b) of the figure is $K_{3,3}$. Let $V_1 = \{h_1, h_2, h_3\}$ and $V_2 = \{u_1, u_2, u_3\}$, and interpret V_1 as a set of houses and V_2 as a set of utilities. Then $K_{3,3}$ is called the *utility graph*. Can we hook up the houses with each of the utilities and avoid having overlapping utility lines? In Fig. 11.40(b) it appears that this is not possible and that $K_{3,3}$ is non-planar. (Once again we deduce the nonplanarity of a graph from a diagram. However, we shall verify that $K_{3,3}$ is nonplanar by another method, later in the section.)

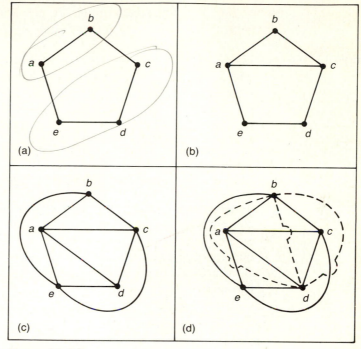

Figure 11.39

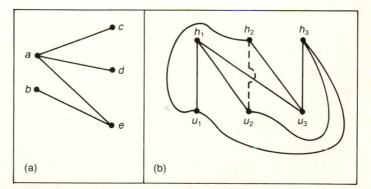

Figure 11.40

We shall see that when we are dealing with nonplanar graphs, either K_5 or $K_{3,3}$ will be the source of the problem. Before stating the general result, however, we need to develop one final new idea.

DEFINITION 11.18   Let $G = (V, E)$ be a loop-free undirected graph, where $E \neq \emptyset$. An *elementary subdivision* of G results when an edge $e = \{u, w\}$ is removed from G and then the edges $\{u, v\}, \{v, w\}$ are added to $G - e$, where $v \notin V$.

    The loop-free undirected graphs $G_1 = (V_1, E_1)$ and $G_2 = (V_2, E_2)$ are called *homeomorphic* if they are isomorphic or if they can both be obtained from the same loop-free undirected graph H by a sequence of elementary subdivisions.

Example 11.16

a) Let $G = (V, E)$ be a loop-free undirected graph with $|E| \geq 1$. If G' is obtained from G by an elementary subdivision, then the graph $G' = (V', E')$ satisfies $|V'| = |V| + 1$ and $|E'| = |E| + 1$.

b) Consider the graphs G, G_1, G_2, and G_3 in Fig. 11.41. Here G_1 is obtained from G by means of one elementary subdivision: Delete edge $\{a, b\}$ from G and then add the edges $\{a, w\}$ and $\{w, b\}$. The graph G_2 is obtained from G by two elementary subdivisions. Hence G_1 and G_2 are homeomorphic. Also, G_3 can be obtained from G by four elementary subdivisions, so G_3 is homeomorphic to both G_1 and G_2.

    However, we cannot obtain G_1 from G_2 (or G_2 from G_1) by a sequence of elementary subdivisions. Furthermore, the graph G_3 can be obtained from either G_1 or G_2 by a sequence of elementary subdivisions: six (such sequences of three elementary subdivisions) for G_1 and two for G_2. But neither G_1 nor G_2 can be obtained from G_3 by a sequence of elementary subdivisions. □

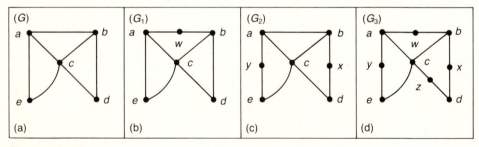

Figure 11.41

    One may think of homeomorphic graphs as being isomorphic up to vertices of degree 2. In particular, if two graphs are homeomorphic, they are simultaneously planar (or nonplanar).

    These preliminaries lead us to the following result.

THEOREM 11.5   (*Kuratowski's Theorem*) A graph is nonplanar if and only if it contains a subgraph that is homeomorphic to either K_5 or $K_{3,3}$.

Proof   (This theorem was first proved by the Polish mathematician Kasimir Kuratowski in 1930.) If a graph G has a subgraph homeomorphic to either K_5 or $K_{3,3}$, it is clear that G is nonplanar. The converse of this theorem, however, is much more difficult to prove. (A proof can be found in Chapter 8 of C. L. Liu [19].) ∎

We demonstrate the use of Kuratowski's Theorem in the following example.

Example 11.17   Figure 11.42(a) is a familiar graph called the *Petersen graph*. Part (b) of the figure provides a subgraph of the Petersen graph that is homeomorphic to $K_{3,3}$. (Figure 11.43 shows how the subgraph is obtained from $K_{3,3}$ by a sequence of four elementary subdivisions.) Hence the Petersen graph is nonplanar.   □

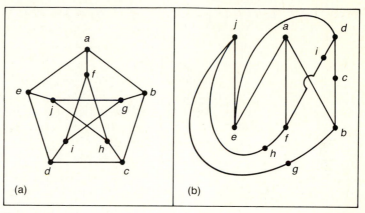

Figure 11.42

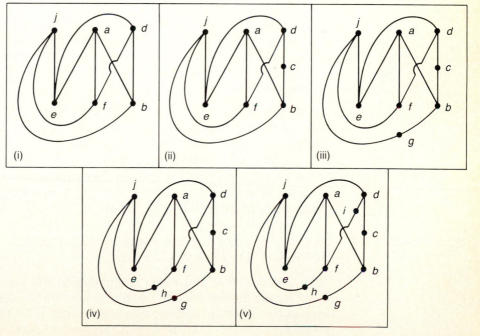

Figure 11.43

When a graph is planar and connected, we find the following relation, which was discovered by Euler.

THEOREM 11.6   Let $G = (V, E)$ be a connected planar graph with $|V| = v$ and $|E| = e$. Let r be the number of regions in the plane determined by G; one of these regions has infinite area and is called *the infinite region*. Then $v - e + r = 2$.

Proof   The proof is by induction on e. If $e = 1$, then G is isomorphic to one of the graphs in Fig. 11.44. For graph (a), $v = 1$, $e = 1$, and $r = 2$. Graph (b) has $v = 2$, $e = 1$, and $r = 1$. In both cases $v - e + r = 2$.

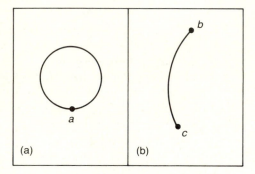

(a)   (b)   **Figure 11.44**

Assume the result is true for all connected planar graphs with k edges. Let G be connected and planar with v vertices, r regions, and $e = k + 1$ edges. We consider two cases.

Case 1: G has a pendant vertex a; that is, $\deg(a) = 1$. Let H be the subgraph of G obtained by deleting vertex a and the edge incident with it. Then H has $v - 1$ vertices, k edges, and r regions. By the induction hypothesis applied to H, $(v - 1) - k + r = 2$. Hence $[(v - 1) + 1] - (k + 1) + r = 2 = v - e + r$, so Euler's theorem holds for G.

Case 2: If G has no pendant vertex, let $\{a, b\}$ be an edge on the boundary of a finite region—a region that encloses a finite area. (Such a region must exist. If not, G is a connected graph with v vertices and e edges, where $e = v - 1$, and G has a pendant vertex.) Removing edge $\{a, b\}$ from G yields a subgraph H with v vertices, $e - 1$ edges and $r - 1$ regions. Regarding graph H, the induction hypothesis implies that $v - (e - 1) + (r - 1) = 2$, from which we have $v - e + r = 2$. Thus Euler's theorem follows for G. ∎

COROLLARY 11.3   Let $G = (V, E)$ be a loop-free connected planar graph with $|V| = v$, $|E| = e > 2$, and r regions. Then $3r \leq 2e$ and $e \leq 3v - 6$.

Proof   Since G is loop-free and is not a multigraph, the boundary of each region (including the infinite region) contains at least three edges. Each edge is used twice as part of the boundary of a region, so $2e \geq 3r$. From Euler's theorem, $2 = v - e + r \leq v - e + (2/3)e = v - (1/3)e$, so $6 \leq 3v - e$, or $e \leq 3v - 6$. ∎

We now consider what this corollary does and does not imply. If $G = (V, E)$ is a loop-free connected graph with $|E| > 2$, then if $e > 3v - 6$, it follows that G is not planar. However, if $e \leq 3v - 6$ we cannot conclude that G is planar.

Example 11.18   The graph K_5 is loop-free and connected with 10 edges and five vertices. Consequently, $3v - 6 = 15 - 6 = 9 < 10 = e$. Therefore, by Corollary 11.3, we find that K_5 is nonplanar.   □

Example 11.19   The graph $K_{3,3}$ is loop-free and connected with nine edges and six vertices. Here $3v - 6 = 18 - 6 = 12 \geq 9 = e$, so we cannot conclude that $K_{3,3}$ is planar unless we make the mistake of arguing by the converse. But we did claim earlier that $K_{3,3}$ is nonplanar.

If $K_{3,3}$ were planar, since each region in the graph is bounded by at least four edges, we have $4r \leq 2e$. (We found a similar situation in the proof of Corollary 11.3.) From Euler's theorem, $v - e + r = 2$, or $r = e - v + 2 = 9 - 6 + 2 = 5$, so $20 = 4r \leq 2e = 18$. From this contradiction we have $K_{3,3}$ being nonplanar.   □

Example 11.20   We use Euler's theorem to characterize the *Platonic solids*. (For these solids all faces are congruent and all (interior) solid angles are equal.) In Fig. 11.45 we have two of these solids. Part (a) of the figure shows the regular tetrahedron, which has four faces, each an equilateral triangle. Concentrating on the edges of the tetrahedron, we focus on its underlying framework. As we view this framework from a point directly above the center of one of the faces, we picture the planar representation in part (b). This planar graph determines four regions (corresponding to the four faces); three regions meet at each of the four vertices. Part (c) of the figure provides another Platonic solid, the cube. Its associated planar graph is given in part (d). In this graph there are six regions with three regions meeting at each vertex.

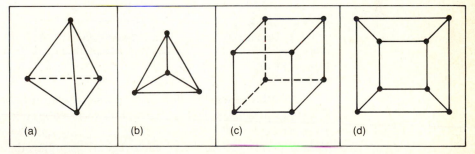

(a)   (b)   (c)   (d)

Figure 11.45

On the basis of our observations for the regular tetrahedron and the cube, we shall determine the other Platonic solids by means of their associated planar graphs. In these graphs $G = (V, E)$ we have $v = |V|$; $e = |E|$; $r =$ the number of planar regions determined by G; $m =$ the number of edges in the boundary of each region; and $n =$ the number of regions that meet at each vertex. Thus the constants $m, n \geq 3$. Since each edge is used in the boundary of two regions and there are r regions, each with m edges, it follows that $2e = mr$. Counting the endpoints of the edges we get $2e$. But all these endpoints can also be counted by considering what happens at each vertex. Since n regions meet at each vertex, n edges meet there, so there are n endpoints of edges to count at each of the v vertices. This totals nv endpoints of edges, so $2e = nv$. From Euler's theorem we have

$$0 < 2 = v - e + r = \frac{2e}{n} - e + \frac{2e}{m} = e\left(\frac{2m - mn + 2n}{mn}\right).$$

With $e, m, n > 0$, we find that $2m - mn + 2n > 0 \Rightarrow mn - 2m - 2n < 0 \Rightarrow mn - 2m - 2n + 4 < 4 \Rightarrow (m - 2)(n - 2) < 4$.

Since $m, n \geq 3$, we have $(m - 2), (n - 2) \in \mathbf{Z}^+$, and there are five cases to consider:

1. $(m - 2) = (n - 2) = 1; m = n = 3$ (The regular tetrahedron)
2. $(m - 2) = 2, (n - 2) = 1; m = 4, n = 3$ (The cube)
3. $(m - 2) = 1, (n - 2) = 2; m = 3, n = 4$ (The octahedron)
4. $(m - 2) = 3, (n - 2) = 1; m = 5, n = 3$ (The dodecahedron)
5. $(m - 2) = 1, (n - 2) = 3; m = 3, n = 5$ (The icosahedron).

The planar graphs for cases 3–5 are shown in Fig. 11.46. □

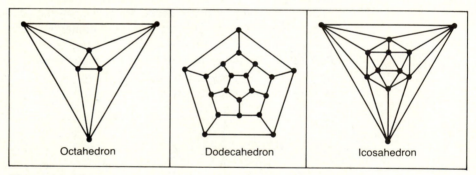

Octahedron   Dodecahedron   Icosahedron

Figure 11.46

The last idea we shall discuss for planar graphs is the notion of a *dual* graph. This concept is also valid for planar graphs with loops and for planar multigraphs. To construct a dual for a planar graph or multigraph G with $V = \{a, b, c, d, e, f\}$, place a point (vertex) inside each region, including the infinite region, determined by the graph, as in Fig. 11.47(a). For each edge shared by two regions, draw an edge connecting the vertices inside these regions. In Fig. 11.47(b), G^d is a dual for the graph $G = (V, E)$. From this example we make the following observations:

1. A vertex of degree 2 in G yields a pair of edges in G^d that connect two vertices. Hence G^d may be a multigraph. (Here vertex e provides the edges $\{a, e\}, \{e, f\}$ in G that brought about the two edges connecting v and z in G^d.)

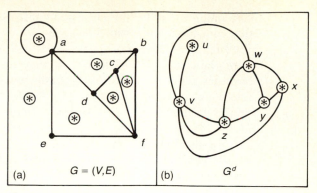

Figure 11.47

2. A loop in G yields a pendant vertex in G^d. (It is also true that a pendant vertex in G yields a loop in G^d.)

3. The degree of a vertex in G^d is the number of edges in the boundary of the region in G that contains that vertex.

(Why is G^d called *a* dual of G instead of *the* dual of G? The section exercises will show that it is possible to have isomorphic graphs G_1 and G_2 with respective duals G_1^d, G_2^d that are not isomorphic.)

In order to examine further the relationship between a graph G and a dual G^d of G, we introduce the following idea.

DEFINITION 11.19   Let $G = (V, E)$ be an undirected graph or multigraph. A subset E' of E is called a *cut-set* of G if by removing the edges (but not the vertices) in E' from G, we have $\kappa(G) < \kappa(G')$, where $G' = (V, E - E')$; but when we remove (from E) any proper subset E'' of E', we have $\kappa(G) = \kappa(G'')$, for $G'' = (V, E - E'')$. ▬

Example 11.21   For any connected graph, a cut-set is a *minimal* disconnecting set of edges. In the graph in Fig. 11.48, note that each of the sets $\{\{a, b\}, \{a, c\}\}$, $\{\{a, b\}, \{c, d\}\}$, $\{\{e, h\}, \{f, h\}, \{g, h\}\}$, and $\{\{d, f\}\}$ is a cut-set.   □

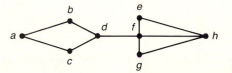

Figure 11.48

We return now to the graphs in Fig. 11.47, redrawing them as shown in Fig. 11.49 in order to emphasize relationships between their edges.

Here the edges in G are labeled $1, 2, \ldots, 10$. The numbering scheme for G^d is obtained as follows: The edge labeled 4*, for example, connects the vertices w and z in G^d. We drew this edge because edge 4 in G was a common edge of the regions containing these vertices. Likewise, edge 7 is common to the region containing x and the infinite region containing v. Hence we label the edge in G^d that connects x and v with 7^*.

In graph G the set of edges labeled $6, 7, 8$ constitutes a cycle. What about the edges labeled $6^*, 7^*, 8^*$ in G^d? If they are removed from G^d, then vertex x becomes isolated and G^d is disconnected. Since we cannot disconnect G^d by removing any proper subset of $\{6^*, 7^*, 8^*\}$, these edges form a cut-set in G^d. In similar fashion, edges $2, 4, 10$ form a cut-set in G, whereas in G^d the edges $2^*, 4^*, 10^*$ yield a cycle.

We also have the two-edge cut-set $\{3, 10\}$ in G, and we find that the edges $3^*, 10^*$ provide a two-edge circuit in G^d. Another observation: The one-edge cut-set $\{1^*\}$ in G^d comes about from edge 1, a loop in G.

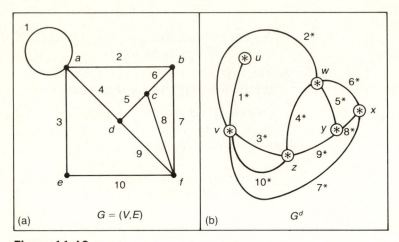

Figure 11.49

In general, there is a one-to-one correspondence between the following sets of edges in a planar graph G and a dual G^d of G.

1. Cycles (cut-sets) of n (≥ 3) edges in G correspond with cut-sets (cycles) of n edges in G^d.

2. A loop in G corresponds with a one-edge cut-set in G^d.

3. A one-edge cut-set in G corresponds with a loop in G^d.

4. A two-edge cut-set in G corresponds with a two-edge circuit in G^d.

5. If G is a planar multigraph, then each two-edge circuit in G determines a two-edge cut-set of G^d.

All these theoretical observations are interesting, but let us stop here and see how we might apply the idea of a dual.

Example 11.22   If we consider the five finite regions in Fig. 11.50(a) as countries on a map, and we construct the subgraph (because we do not use the infinite region) of a dual as shown in part (b), then we find the following relationship.

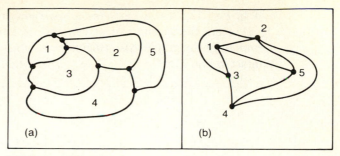

Figure 11.50

Suppose we want to color the five regions of the map in part (a) so that two countries that share a common border are colored with different colors. This idea can be translated into the dual notion of coloring the vertices in part (b) so that adjacent vertices are colored with different colors. (Such coloring problems will be examined further in Section 11.6.) □

The final result for this section provides us with an application for an electrical network. This material is based on Example 8.6 on pp. 227–230 of the text by C. L. Liu [19].

Example 11.23 In Fig. 11.51 we see an electrical network with nine contacts (switches) that control the excitation of a light. We want to construct a dual network where a second light will go on (off) whenever the light in our given network is off (on).

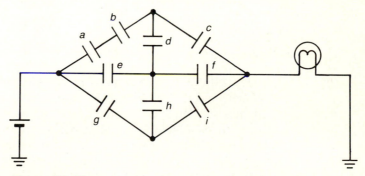

Figure 11.51

The contacts (switches) are of two types: normally open (as shown in Fig. 11.51) and normally closed. We use a and a' as in Fig. 11.52 to represent the normally open and normally closed contacts, respectively.

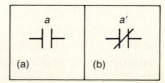

Figure 11.52

In Fig. 11.53(a) a *one-terminal-pair-graph* represents the network in Fig. 11.51. Here the special pair of vertices are labeled 1 and 2 and are called the *terminals* of the graph. Also each edge is labeled according to its corresponding contact in Fig. 11.51.

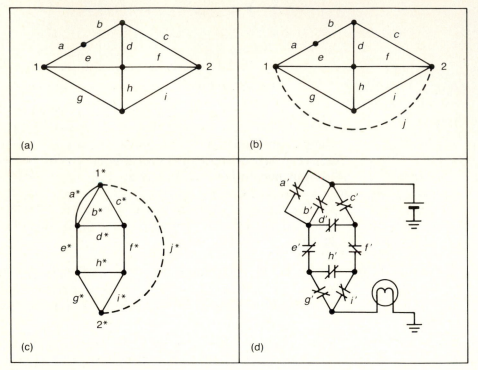

(a)   (b)

(c)   (d)

Figure 11.53

A one-terminal-pair-graph G is called a *planar-one-terminal-pair-graph* if G is planar, and the resulting graph is also planar when an edge connecting the terminals is added to G. Figure 11.53(b) shows this situation. Constructing a dual of part (b), we obtain the graph in part (c) of the figure. Removal of the dotted edge results in the terminals $1^*, 2^*$ for this dual, which is a one-terminal-pair-graph. This graph provides the dual network in Fig. 11.53(d).

We make two observations in closing.

1. When the contacts at a, b, c are closed in the original network (Fig. 11.51), the light is on. In Fig. 11.53(b) the edges a, b, c, j form a cycle that includes the terminals. In part (c) of the figure, a^*, b^*, c^*, j^* form a cut-set disconnecting the terminals $1^*, 2^*$. Finally, with a', b', c' open in part (d) of the figure, no current gets past the first level of contacts (switches) and the light is off.

2. In like manner, the edges c, f, i, j form a cut-set that separates the terminals in Fig. 11.53(b). (When the contacts at c, f, i are open in Fig. 11.51, the light is off.) Figure 11.53(c) shows how c^*, f^*, i^*, j^* form a cycle that includes $1^*, 2^*$. If c', f', i' are closed in part (d), current flows through the dual network and the light is on.   □

EXERCISES 11.4

1. Verify that the conclusion in Example 11.15 is unchanged if Fig. 11.39(b) has edge $\{a, c\}$ drawn in the exterior of the pentagon.

2. Show that when any edge is removed from K_5, the resulting subgraph is planar. Is this true for the graph $K_{3,3}$?

3. **a)** How many vertices and how many edges are there in the complete bipartite graphs $K_{4,7}$, $K_{7,11}$, and $K_{m,n}$ where $m, n \in \mathbf{Z}^+$?

   b) If the graph $K_{m,12}$ has 72 edges, what is m?

   c) Let $m, n \in \mathbf{Z}^+$. Give a combinatorial proof that

   $$\binom{m+n}{2} - mn = \binom{m}{2} + \binom{n}{2},$$

   by considering the numbers of edges in the graphs K_{m+n}, $K_{m,n}$, and $\overline{K}_{m,n}$.

4. Can a bipartite graph contain a cycle of odd length? Explain.

5. Let $G = (V, E)$ be a loop-free connected graph with $|V| = v$. If $|E| > (v/2)^2$, prove that G cannot be bipartite.

6. **a)** Find all the nonisomorphic complete bipartite graphs $G = (V, E)$ where $|V| = 6$.

   b) How many nonisomorphic complete bipartite graphs $G = (V, E)$ satisfy $|V| = n \geq 2$?

7. **a)** Let $X = \{1, 2, 3, 4, 5\}$. Construct the loop-free undirected graph $G = (V, E)$ as follows:

   - (V): Let each two-element subset of X represent a vertex in G.
   - (E): If $v_1, v_2 \in V$ correspond to subsets $\{a, b\}$ and $\{c, d\}$, respectively, of X, then draw the edge $\{v_1, v_2\}$ in G if $\{a, b\} \cap \{c, d\} = \emptyset$.

   b) To what graph is G isomorphic?

8. Determine whether the graphs in Fig. 11.54 are planar. If a given graph is planar, redraw the graph with no edges overlapping. If it is nonplanar, find a subgraph homeomorphic to either K_5 or $K_{3,3}$.

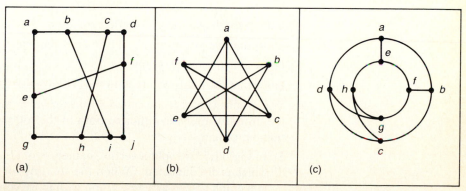

(a)   (b)   (c)

Figure 11.54

9. Let $G = (V, E)$ be a loop-free connected planar graph with $\deg(v) = 4$, for all $v \in V$. If $|E| = 16$, how many regions are there in a planar depiction of G?

10. Suppose that $G = (V, E)$ is a loop-free planar graph with $|V| = v$, $|E| = e$, and $\kappa(G) =$ the number of components of G. (a) State and prove an extension of Euler's theorem for such a graph. (b) Prove that Corollary 11.3 remains valid for G loop-free and planar but not connected.

11. If G is a loop-free connected planar graph, prove that there is a vertex v of G with $\deg(v) < 6$.

12. a) Let $G = (V, E)$ be a loop-free connected graph with $|V| \geq 11$. Prove that either G or its complement $\overline{G}$ must be nonplanar.

b) The result in part (a) is actually true for $|V| \geq 9$, but the proof for $|V| = 9$, 10, is much harder. Find a counterexample to part (a) for $|V| = 8$.

13. a) Let $k \in \mathbf{Z}^+$, $k \geq 3$. If $G = (V, E)$ is a connected planar graph with $|V| = v$, $|E| = e$, and each cycle of length at least k, prove that $e \leq \left(\dfrac{k}{k-2}\right)(v - 2)$.

b) What is the minimal cycle length in $K_{3,3}$?

c) Use parts (a) and (b) to conclude that $K_{3,3}$ is nonplanar.

d) Use part (a) to prove that the Petersen graph is nonplanar.

14. Find a dual graph for each of the planar graphs in Fig. 11.55.

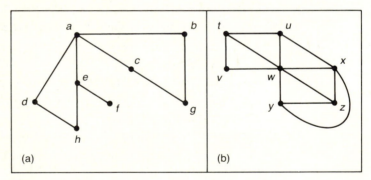

(a) (b)

Figure 11.55

15. Find duals for the planar graphs that correspond with the five Platonic solids.

16. a) Show that the graphs in Fig. 11.56 are isomorphic.

b) Draw a dual for each graph.

c) Show that the duals obtained in part (b) are not isomorphic.

d) Two graphs G and H are called *2-isomorphic* if one can be obtained from the other by applying either or both of the following procedures a finite number of times.

1. In Fig. 11.57 we split a vertex, namely r, of G and obtain the graph H, which is disconnected. (When the splitting, or separation, of a vertex in a graph results in a new graph with more components, that vertex is called a *cut-vertex* or *articulation point*.)

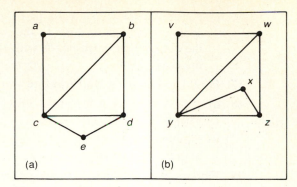

Figure 11.56

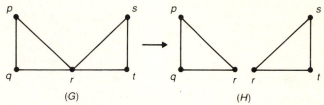

Figure 11.57

2. In Fig. 11.58 we obtain graph (d) from (a) by

   (i) first splitting the graph at two distinct vertices j and q,

   (ii) then reflecting one subgraph about the horizontal axis, and

   (iii) then joining vertex $j(q)$ in one subgraph with vertex $q(j)$ in the other subgraph.

   Prove that the dual graphs obtained in part (c) are 2-isomorphic.

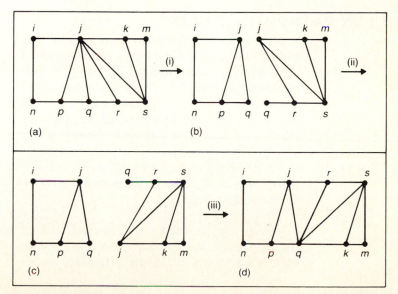

Figure 11.58

     e) For the cut-set $\{\{a, b\}, \{c, b\}, \{d, b\}\}$ in part (a) of Fig. 11.56, find the corresponding cycle in its dual. In the dual of the graph in Fig. 11.56(b), find the cut-set that corresponds with the cycle $\{w, z\}, \{z, x\}, \{x, y\}, \{y, w\}$ in the given graph.

17. For the graph in Fig. 11.58(a) find two cut-sets that contain the edge $\{k, m\}$.

18. a) If $G = (V, E)$ is a loop-free 3-regular connected graph with 8 vertices, what is $|E|$?

     b) Find two graphs G_1 and G_2 that satisfy the conditions in part (a) with G_1 planar and G_2 nonplanar.

19. Find the dual network for the electrical network shown in Fig. 11.59.

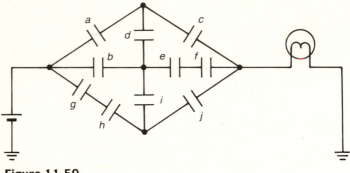

Figure 11.59

11.5
HAMILTON PATHS AND CYCLES

In 1859 the Irish mathematician Sir William Hamilton (1805–1865) developed a game that he sold to a Dublin toy manufacturer. The game consisted of a wooden regular dodecahedron with the 20 corner points (vertices) labeled with the names of prominent cities. The objective of the game was to find a cycle along the edges of the solid so that each city was on the cycle exactly once. Figure 11.60 is the planar graph for this Platonic solid; such a cycle is designated by the darkened edges. This illustration leads us to the following definition.

DEFINITION 11.20   If $G = (V, E)$ is a graph or multigraph, we say that G has a *Hamilton cycle* if there is a cycle in G that contains every vertex in V. A *Hamilton path* is a path in G that contains each vertex.

    Given a graph with a Hamilton cycle, we find that the deletion of any edge in the cycle results in a Hamilton path. It is possible, however, for a graph to have a Hamilton path without having a Hamilton cycle.

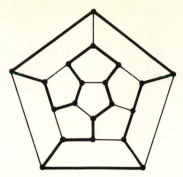

Figure 11.60

It may seem that the existence of a Hamilton cycle (path) and the existence of an Euler circuit (trail) for a graph are similar problems. The Hamilton cycle (path) is designed to visit each vertex in a graph only once; the Euler circuit (trail) traverses the graph so that each edge is traveled exactly once. Unfortunately, there is no helpful connection between the two ideas, and unlike the situation for Euler circuits (trails), there do not exist necessary and sufficient conditions on a graph G that guarantee the existence of a Hamilton cycle (path). If a graph has a Hamilton cycle, then it will at least be connected. Many theorems exist that establish either necessary or sufficient conditions for a connected graph to have a Hamilton cycle or path. We shall investigate three of these results later. When confronted with particular graphs, we shall often resort to trial and error, with a few helpful observations.

Example 11.24　If G is the graph in Fig. 11.61, the edges $\{a, b\}$, $\{b, c\}$, $\{c, f\}$, $\{f, e\}$, $\{e, d\}$, $\{d, g\}$, $\{g, h\}$, $\{h, i\}$ yield a Hamilton path for G. But does G have a Hamilton cycle?

Since G has nine vertices, if there is a Hamilton cycle in G it must contain nine edges. Let us start at vertex b and try to build a Hamilton cycle. With regard to the symmetry in the graph, it doesn't matter whether we go from b to c or to a. We'll go to c. At c we can go either to f or to i. Using symmetry again, we go to f. Then we delete edge $\{c, i\}$ from further consideration because we cannot return to vertex c. In order to include vertex i in our cycle, we must now go from f to i (to h to g). With edges $\{c, f\}$ and $\{f, i\}$ in the cycle, we delete edge $\{e, f\}$. But once we get to e we are stuck. Hence there is no Hamilton cycle for the graph. □

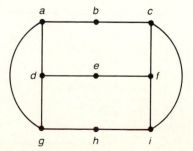

Figure 11.61

This example indicates a few helpful hints for trying to find a Hamilton cycle in a graph $G = (V, E)$.

1. If G has a Hamilton cycle, then for all $v \in V$, $\deg(v) \geq 2$.
2. If $a \in V$ and $\deg(a) = 2$, then the two edges incident with vertex a must appear in any Hamilton cycle for G.
3. If $a \in V$ and $\deg(a) > 2$, then as we try to build a Hamilton cycle, once we pass through vertex a, any unused edges incident with a are deleted from further consideration.
4. In building a Hamilton cycle for G, we cannot obtain a cycle for a subgraph of G unless it contains all the vertices of G.

Our next example provides an interesting technique for showing that a special type of graph has no Hamilton path.

Example 11.25   In Fig. 11.62(a) we have a connected graph G, and we wish to know whether G contains a Hamilton path. Part (b) of the figure provides the same graph with a set of labels x, y. This labeling is accomplished as follows: First we label vertex a with the letter x. Those vertices adjacent to a (namely b, c, and d) are then labeled with the letter y. Then we label the unlabeled vertices adjacent to b, c, and d with x. This results in the label x on the vertices e, g, and i. Finally, we label the unlabeled vertices adjacent to e, g, and i with the label y. At this point all the vertices in G are labeled. Now, since $|V| = 10$, if G is to have a Hamilton path there must be an alternating sequence of five x's and five y's. Only four vertices are labeled with x, so this is impossible. Hence G has no Hamilton path (or cycle).

But why does this argument work here? In part (c) of Fig. 11.62 we have redrawn the given graph, and we see that it is bipartite. From Exercise 4 in the previous section we know that a bipartite graph cannot have a cycle of odd length. It is also true that if a graph has no cycle of odd length, then it is bipartite. (The proof is requested of the reader in Exercise 7 of this section.) Consequently, whenever a connected graph has no odd cycle (and is bipartite), the method described above can be used to determine when the graph does not have a Hamilton path. (Exercise 8 in this section examines this idea further.)   □

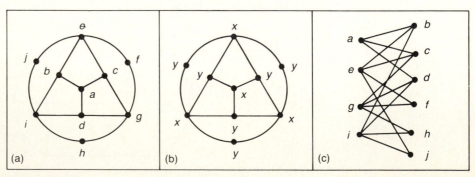

Figure 11.62

Example 11.26   Our next example provides an application that calls for Hamilton cycles in a complete graph.

At Professor Alfred's science camp, 17 students have lunch together each day at a circular table. They are trying to get to know each other better, so they make an effort to sit next to two different colleagues each afternoon. For how many afternoons can they do this? How can they arrange themselves on these occasions?

In order to solve this problem we consider the graph K_n, where $n \geq 3$ and is odd. This graph has n vertices and $\binom{n}{2} = n(n-1)/2$ edges. A Hamilton cycle in K_n has n edges, so we can have at most $(n-1)/2$ Hamilton cycles with no two having an edge in common.

Consider the circle in Fig. 11.63 and the subgraph of K_n consisting of the n vertices and the n edges $\{1,2\},\{2,3\},\ldots,\{n-1,n\},\{n,1\}$. Keep the vertices on the circumference fixed and rotate this Hamilton cycle clockwise through the angle $[1/(n-1)](2\pi)$. This gives us the Hamilton cycle (Fig. 11.64) made up of edges $\{1,3\}, \{3,5\}, \{5,2\}, \{2,7\},\ldots,\{n,n-3\}, \{n-3,n-1\}, \{n-1,1\}$. This Hamilton cycle has no edge in common with the first cycle. When $n \geq 7$ and we continue to rotate the cycle in Fig. 11.63 in this way through angles $[k/(n-1)] \cdot (2\pi)$, where $2 \leq k \leq (n-3)/2$, we obtain a set of Hamilton cycles where no two cycles have an edge in common.

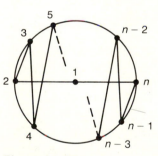

Figure 11.63

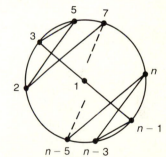

Figure 11.64

Therefore the 17 students at the science camp can dine for $[(17-1)/2] = 8$ days before each must sit next to the same student twice. Using Fig. 11.63 with $n = 17$, we can obtain eight such possible arrangements.   □

We turn now to some further results on Hamilton paths and cycles. Our first result was established in 1934 by L. Redei.

THEOREM 11.7   Let K_n^* be a complete directed graph—that is, K_n^* has n vertices and for any distinct pair x, y of vertices, exactly one of the edges (x, y) or (y, x) is in K_n^*. Such a graph (usually called a *tournament*) always contains a (directed) Hamilton path.

Proof   Let $m \geq 2$ with p_m a path containing the $m - 1$ edges $(v_1, v_2), (v_2, v_3), \ldots,$ (v_{m-1}, v_m). If $m = n$, we're finished. If not, let v be a vertex that doesn't appear in p_m.

   If (v, v_1) is an edge in K_n^* we can extend p_m by adjoining this edge. If not, then (v_1, v) must be an edge. Now suppose that (v, v_2) is in the graph. Then we have the larger path: $(v_1, v), (v, v_2), (v_2, v_3), \ldots, (v_{m-1}, v_m)$. If (v, v_2) is not an edge in K_n^*, then (v_2, v) must be. As we continue this process there are only two possibilities: (a) For some $1 \leq k \leq m - 1$ the edges $(v_k, v), (v, v_{k+1})$ are in K_n^* and we replace (v_k, v_{k+1}) with this pair of edges; or (b) (v_m, v) is in K_n^* and we add this edge to p_m. Either case results in a path p_{m+1} that includes $m + 1$ vertices and has m edges. This process can be repeated until we have such a path on n vertices.   ■

Example 11.27   In a round-robin tournament each player plays every other player exactly once. We want somehow to rank the players according to the results of the tournament. Since we could have players a, b, and c where a beats b and b beats c, but c beats a, it is not always possible to have a ranking where a player in a certain position has beaten all of the opponents in later positions. Representing the players by vertices, construct a directed graph G on these vertices by drawing edge (x, y) if x beats y. Then by Theorem 11.7, it is possible to list the players such that each has beaten the next player on the list.   □

THEOREM 11.8   Let $G = (V, E)$ be a loop-free graph with $|V| = n$. If $\deg(x) + \deg(y) \geq n - 1$ for all $x, y \in V, x \neq y$, then G has a Hamilton path.

Proof   First we prove that G is connected. If not, let C_1, C_2 be two components of G, with $x, y \in V$, and $x \in C_1$, $y \in C_2$. Let C_i have n_i vertices, $i = 1, 2$. Then $\deg(x) \leq n_1 - 1$, $\deg(y) \leq n_2 - 1$, and $\deg(x) + \deg(y) \leq (n_1 + n_2) - 2 \leq n - 2$, contradicting the condition given in the theorem. Consequently, G is connected.

   Now we build a Hamilton path for G. For $m \geq 2$, let p_m be the path $\{v_1, v_2\}, \{v_2, v_3\}, \ldots, \{v_{m-1}, v_m\}$ of length $m - 1$. (We relabel vertices if necessary.) Such a path exists, because for $m = 2$ all that is needed is one edge. If v_1 is adjacent to any vertex v other than $v_2, v_3, \ldots, v_m$, we add the edge $\{v, v_1\}$ to p_m to get p_{m+1}. The same type of procedure is carried out if v_m is adjacent to a vertex other than $v_1, v_2, \ldots, v_{m-1}$. If we are able to enlarge p_m to p_n in this way, we get a Hamilton path. Otherwise the path p_m: $\{v_1, v_2\}, \ldots, \{v_{m-1}, v_m\}$ has v_1, v_m adjacent only to vertices in p_m, and $m < n$. When this happens we claim that G contains a cycle on these vertices. If v_1 and v_m are adjacent, then the cycle is $\{v_1, v_2\}$, $\{v_2, v_3\}, \ldots, \{v_{m-1}, v_m\}, \{v_m, v_1\}$. If v_1 and v_m are not adjacent, then v_1 is adjacent to a subset S of the vertices in $\{v_2, v_3, \ldots, v_{m-1}\}$. If there is a vertex $v_t \in S$ such that v_m is adjacent to v_{t-1}, then we can get the cycle by adding $\{v_1, v_t\}, \{v_{t-1}, v_m\}$ to p_m and deleting $\{v_{t-1}, v_t\}$ as shown in Fig. 11.65. If not, let $|S| = k < m - 1$. Then $\deg(v_1) = k$ and $\deg(v_m) \leq (m - 1) - k$, and we have the contradiction $\deg(v_1) + \deg(v_m) \leq m - 1 < n - 1$. Hence there is a cycle connecting $v_1, v_2, \ldots, v_m$.

   Now consider a vertex $v \in V$ that is not found on this cycle. The graph G is connected, so there is a path from v to a first vertex v_r in the cycle, as shown in Fig. 11.66(a). Removing the edge $\{v_{r-1}, v_r\}$ (or $\{v_1, v_t\}$ if $r = t$), we get the path (longer than the original p_m) shown in Fig. 11.66(b). Repeating this process (applied to p_m) for the path in Fig. 11.66(b), we continue to increase the length of the path until it includes every vertex of G. ∎

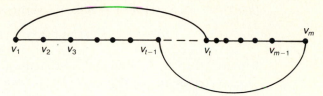

Figure 11.65

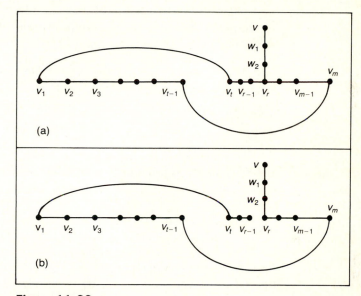

(a)

(b)

Figure 11.66

COROLLARY 11.4   Let $G = (V, E)$ be a loop-free graph with n vertices. If $\deg(v) \geq (n-1)/2$ for all $v \in V$, then G has a Hamilton path.

Proof   The proof is left as an exercise for the reader. ∎

Our last result for this section provides a sufficient condition for the existence of a Hamilton cycle in a loop-free graph. This was first proved by Oystein Ore in 1960.

THEOREM 11.9   Let $G = (V, E)$ be a loop-free undirected graph with $|V| = n \geq 3$. If $\deg(x) + \deg(y) \geq n$ for all nonadjacent $x, y \in V$, then G contains a Hamilton cycle.

Proof   Assume that G does not contain a Hamilton cycle. We add edges to G until we arrive at a subgraph H of K_n where H has no Hamilton cycle, but, for any edge e

(of K_n) not in H, $H + e$ (the subgraph of K_n obtained from H by adding the edge e) does have a Hamilton cycle.

Since $H \neq K_n$, there are vertices $a, b \in V$ where $\{a, b\}$ is not an edge of H but $H + \{a, b\}$ has a Hamilton cycle C. The graph H has no such cycle, so the edge $\{a, b\}$ is a part of cycle C. Let us list the vertices of H (and G) on cycle C as follows:

$$a \ (= v_1) \rightarrow b \ (= v_2) \rightarrow v_3 \rightarrow v_4 \rightarrow \cdots \rightarrow v_{n-1} \rightarrow v_n$$

For any $3 \leq i \leq n$, if the edge $\{b, v_i\}$ is in the graph H, then we claim that the edge $\{a, v_{i-1}\}$ cannot be an edge of H. For if both of these edges are in H, for some $3 \leq i \leq n$, then we get the Hamilton cycle

$$b \rightarrow v_i \rightarrow v_{i+1} \rightarrow \cdots \rightarrow v_{n-1} \rightarrow v_n \rightarrow a \rightarrow v_{i-1} \rightarrow v_{i-2} \rightarrow \cdots \rightarrow v_4 \rightarrow v_3$$

for the graph H (which has no Hamilton cycle). Therefore, for each $3 \leq i \leq n$, at most one of the edges $\{b, v_i\}, \{a, v_{i-1}\}$ is in H. Consequently,

$$\deg_H(a) + \deg_H(b) < n,$$

where $\deg_H(v)$ denotes the degree of vertex v in graph H. For any $v \in V$, $\deg_H(v) \geq \deg_G(v) = \deg(v)$, so we have nonadjacent (in G) vertices a, b where

$$\deg(a) + \deg(b) < n.$$

This contradicts the hypothesis that $\deg(x) + \deg(y) \geq n$ for all nonadjacent $x, y \in V$, so we reject our assumption and find that G contains a Hamilton cycle. ■

A problem that is related to the search for Hamilton cycles in a graph is the *traveling salesperson problem*. Here a traveling salesperson leaves his or her home and visits certain locations before returning. The objective is to find a way to make his or her trip most efficient (perhaps in terms of total distance traveled or total cost). The problem can be modeled with a labeled (edges have distances or costs associated with them) graph where the most efficient Hamilton cycle is sought.

The references by R. Bellman, K. L. Cooke, and J. A. Lockett [4]; M. Bellmore and G. L. Nemhauser [5]; E. A. Elsayed [11]; E. A. Elsayed and R. G. Stern [12]; and L. R. Foulds [13] should prove interesting to the reader who wants to learn more about this important optimization problem. Also, the text edited by E. L. Lawler, J. K. Lenstra, A. H. G. Rinnooy Kan, and D. B. Shmoys [18] presents twelve papers on various facets of this problem.

EXERCISES 11.5

1. Give an example of a connected graph that has:

   a) Neither an Euler circuit nor a Hamilton cycle.

   b) An Euler circuit but no Hamilton cycle.

c) A Hamilton cycle but no Euler circuit.

d) Both a Hamilton cycle and an Euler circuit.

2. Characterize the type of graph in which an Euler trail (circuit) is also a Hamilton path (cycle).

3. Find a Hamilton cycle, if one exists, for each of the graphs or multigraphs in Fig. 11.67. If the graph has no Hamilton cycle, determine whether it has a Hamilton path.

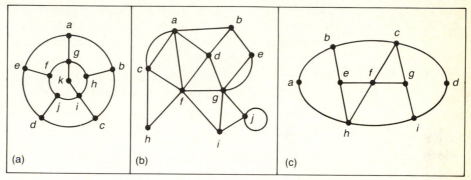

Figure 11.67

4. a) Show that the Petersen graph (Fig. 11.42(a)) contains no Hamilton cycle but that it has a Hamilton path.

b) Show that if any vertex (and the edges incident to it) is removed from the Petersen graph, then the resulting subgraph has a Hamilton cycle.

5. a) For $n \geq 3$, how many different Hamilton cycles are there in the complete graph K_n?

b) How many edge-disjoint Hamilton cycles are there in K_{21}?

c) Nineteen students in a nursery school play a game each day where they hold hands to form a circle. For how many days can they do this with no student holding hands with the same playmate twice?

6. a) For $n \in \mathbf{Z}^+, n \geq 2$, show that the number of distinct Hamilton cycles in the graph $K_{n,n}$ is $(1/2)(n-1)!\, n!$

b) How many different Hamilton paths are there for $K_{n,n}, n \geq 1$?

7. Let $G = (V, E)$ be a loop-free undirected graph. Prove that if G contains no cycle of odd length, then G is bipartite.

8. a) Let $G = (V, E)$ be a connected bipartite undirected graph with V partitioned as $V_1 \cup V_2$. Prove that if $|V_1| \neq |V_2|$, then G cannot have a Hamilton cycle.

b) Prove that if the graph G in part (a) has a Hamilton path, then $|V_1| - |V_2| = \pm 1$.

    c) Give an example of a connected bipartite undirected graph $G = (V, E)$, where V is partitioned as $V_1 \cup V_2$ and $|V_1| = |V_2| - 1$, but G has no Hamilton path.

9. a) Determine all nonisomorphic tournaments with three vertices.

   b) Find all of the nonisomorphic tournaments with four vertices. List the in degree and the out degree for each vertex, in each of these tournaments.

10. Find a counterexample to the converse of Theorem 11.8.

11. Prove Corollary 11.4.

12. Helen and Dominic invite 10 friends to dinner. In this group of 12 people everyone knows at least 6 others. Prove that the 12 can be seated around a circular table in such a way that each person is acquainted with the persons sitting on either side.

13. Let $G = (V, E)$ be a loop-free undirected graph that is 6-regular. Prove that if $|V| = 11$, then G contains a Hamilton cycle.

14. Let $n = 2^k$ for $k \in \mathbf{Z}^+$. We use the n k-bit sequences (of 0's and 1's) to represent $1, 2, 3, \ldots, n$, so that for two consecutive integers $i, i + 1$, the corresponding k-bit sequences differ in exactly one component. This representation is called a *Gray Code*.

   a) For $k = 3$, use a graph model with $V = \{000, 001, 010, \ldots, 111\}$ to find such a code for $1, 2, 3, \ldots, 8$. How is this related to the concept of a Hamilton path?

   b) Answer (a) for $k = 4$.

15. If $G = (V, E)$ is an undirected graph, a subset I of V is called *independent* if no two vertices in I are adjacent. An independent set I is called *maximal* if no vertex v can be added to I with $I \cup \{v\}$ independent. The *independence number* of G, denoted $\beta(G)$, is the size of any largest maximal independent set in G.

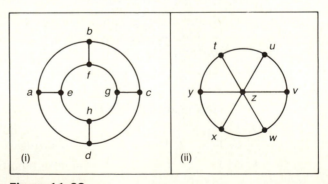

Figure 11.68

   a) For each graph in Fig. 11.68 find two maximal independent sets with different sizes.

   b) Find $\beta(G)$ for each graph in part (a).

c) Determine $\beta(G)$ for each of the following graphs: (i) $K_{1,3}$; (ii) $K_{2,3}$; (iii) $K_{3,2}$; (iv) $K_{4,4}$; (v) $K_{4,6}$; (vi) $K_{m,n}$, $m, n \in \mathbf{Z}^+$.

d) Let I be an independent set in $G = (V, E)$. What type of subgraph does I induce in $\overline{G}$?

e) If $G = (V, E)$, where $|V| = n$ and $|E| = m$, how many two-element subsets of V are independent?

16. Let $G = (V, E)$ be an undirected graph with subset I of V an independent set. For each $a \in I$ and any Hamilton cycle C for G, there will be $\deg(a) - 2$ edges in E that are incident with a and not in C. Therefore there are at least $\sum_{a \in I} [\deg(a) - 2] = \sum_{a \in I} \deg(a) - 2|I|$ edges in E that do not appear in C.

a) Why are these $\sum_{a \in I} \deg(a) - 2|I|$ edges distinct?

b) Let $v = |V|$, $e = |E|$. Prove that if $e - \sum_{a \in I} \deg(a) + 2|I| < v$, then G has no Hamilton cycle.

c) Select a suitable independent set I and use part (b) to show that the graph in Fig. 11.69 (known as the Herschel graph) has no Hamilton cycle.

Figure 11.69

11.6
GRAPH COLORING AND CHROMATIC POLYNOMIALS

At the J. & J. Chemical Company, Jeannette is in charge of the storage of chemical compounds in the company warehouse. Since certain types of compounds (such as acids and bases) should not be kept in the same vicinity, she decides to have her partner Jack partition the warehouse into separate storage areas where incompatible chemical reagents can be stored in separate compartments. How can she determine the number of storage compartments that Jack will have to build?

If this company sells 25 chemical compounds, let $\{c_1, c_2, \ldots, c_{25}\} = V$, a set of vertices. For any $1 \le i < j \le 25$, we draw the edge $\{c_i, c_j\}$ if c_i and c_j must be stored in separate compartments. This gives us an undirected graph $G = (V, E)$.

We now introduce the following concept.

DEFINITION 11.21   If $G = (V, E)$ is an undirected graph, a *proper coloring* of G occurs when we color the vertices of G so that if $\{a, b\}$ is an edge in G, then a and b are colored with different colors. (Hence adjacent vertices have different colors.) The minimum number of colors needed to properly color G is called the *chromatic number* of G and is written $\chi(G)$.

Returning to assist Jeannette at the warehouse, we find that the number of storage compartments Jack must build is equal to $\chi(G)$ for the graph we constructed on $V = \{c_1, c_2, \ldots, c_{25}\}$. But how do we compute $\chi(G)$? Before we present any work on how to determine the chromatic number of a graph, we turn to the following related idea.

In Example 11.22 we mentioned the connection between coloring the regions in a planar map (with neighboring regions having different colors) and properly coloring the vertices in an associated graph. Determining the smallest number of colors needed to color planar maps in this way has been a problem of interest for over a century.

In about 1850, Francis Guthrie (1831–1899) became interested in the general problem after showing how to color the counties on a map of England with only four colors. Shortly thereafter, he showed the "Four-color Problem" to his younger brother Frederick. At that time Frederick Guthrie (1833–1866) was a student of Augustus DeMorgan (1806–1871), who then communicated the problem (in 1852) to William Hamilton (1805–1865). The problem did not interest Hamilton and lay dormant for about 25 years. Then, in 1878, the scientific community was made aware of the problem through an announcement by Arthur Cayley (1821–1895) at a meeting of the London Mathematical Society. In 1879 Cayley stated the problem in the first volume of the *Proceedings of the Royal Geographical Society*. Shortly thereafter, the British mathematician Sir Alfred Kempe (1849–1922) devised a proof that remained unquestioned for over a decade. Then, in 1890, Percy John Heawood (1861–1955), another British mathematician, found a mistake in Kempe's work.

The problem remained unsolved until 1976, when it was finally settled by Kenneth Appel and Wolfgang Haken. Their proof employs a very intricate computer analysis of 1936 (reducible) configurations.

Although only four colors are needed to properly color the regions in a planar map, we need more than four colors to properly color the vertices of some nonplanar graphs.

We start with some small examples. Then we shall find a way to determine $\chi(G)$ from smaller subgraphs of G. We shall also obtain what is called the chromatic polynomial for G and see how it is used in computing $\chi(G)$.

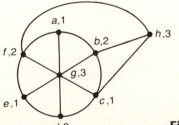

Figure 11.70

Example 11.28   For the graph G in Fig. 11.70, we start at vertex a and next to each vertex write the number of a color needed to properly color the vertices of G that have been considered up to that point. Going to vertex b, the 2 indicates the need for a second color. Proceeding alphabetically to f, we find that two colors are needed to properly color $\{a, b, c, d, e, f\}$. For vertex g a third color is needed; this third color can also be used for vertex h, because $\{g, h\}$ is not an edge in G. Thus $\chi(G) = 3$. □

Example 11.29 **a)** For any $n \geq 1$, $\chi(K_n) = n$.

b) The chromatic number of the Herschel graph (Fig. 11.69) is 2.

c) If G is the Petersen graph [see Fig. 11.42(a)], then $\chi(G) = 3$. □

Example 11.30 Let G be the graph shown in Fig. 11.71. For $U = \{b, f, h, i\}$, the induced subgraph $\langle U \rangle$ of G is isomorphic to K_4, so $\chi(G) \geq \chi(K_4) = 4$. Therefore, if we can determine a way to properly color the vertices of G with four colors, then we shall know that $\chi(G) = 4$. One way to accomplish this is to color the vertices e, f, g blue, the vertices b, j red, the vertices c, h white, and vertices a, d, i green. □

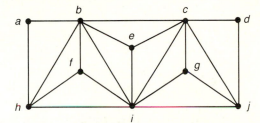

Figure 11.71

We turn now to a method for determining $\chi(G)$. Our coverage follows the development in the survey article [22] by R. C. Read.

Let G be an undirected graph, and let λ be a positive integer representing the number of colors that are available for properly coloring the vertices of G. Our objective is to find a polynomial $P(G, \lambda)$, in the variable λ, that will tell us in how many different ways we can properly color the vertices of G, using at most λ colors.

Throughout this discussion, the vertices in an undirected graph $G = (V, E)$ are distinguished by labels. Consequently, two proper colorings of such a graph will be considered different in the following sense: A proper coloring (of the vertices of G) that uses at most λ colors is a function f, with domain V and codomain $\{1, 2, 3, \ldots, \lambda\}$, where $f(u) \neq f(v)$, for adjacent vertices $u, v \in V$. Proper colorings are then different in the same way that these functions are different.

Example 11.31 **a)** If $G = (V, E)$ with $|V| = n$ and $E = \emptyset$, then G consists of n isolated points, and by the rule of product, $P(G, \lambda) = \lambda^n$.

b) If $G = K_n$, then at least n colors must be available for us to color G properly. Here, by the rule of product, $P(G, \lambda) = \lambda(\lambda - 1)(\lambda - 2) \cdots (\lambda - n + 1)$, which we denote by $\lambda^{(n)}$. For $\lambda < n$, $P(G, \lambda) = 0$ and there are no ways to properly color K_n. $P(G, \lambda) > 0$ for the first time when $\lambda = n = \chi(G)$.

c) For each path in Fig. 11.72, we consider the number of choices at each successive vertex. Proceeding alphabetically, we find that $P(G_1, \lambda) = \lambda(\lambda - 1)^3$

and $P(G_2, \lambda) = \lambda(\lambda - 1)^4$. Since $P(G_1, 1) = 0 = P(G_2, 1)$, but $P(G_1, 2) = 2 = P(G_2, 2)$, $\chi(G_1) = \chi(G_2) = 2$. If five colors are available we can properly color G_1 in $5(4)^3 = 320$ ways; G_2 can be so colored in $5(4)^4 = 1280$ ways.

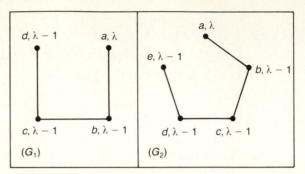

Figure 11.72

In general, if G is a path on n vertices, then $P(G, \lambda) = \lambda(\lambda - 1)^{n-1}$.

d) If G is made up of components $C_1, C_2, \ldots, C_k$, then again by the rule of product, it follows that $P(G, \lambda) = P(C_1, \lambda) \cdot P(C_2, \lambda) \cdots P(C_k, \lambda)$. ☐

As a result of Example 11.31(d), we shall concentrate on connected graphs. In many instances in discrete mathematics, methods have been employed to solve problems in large cases by breaking these down into two or more smaller cases. Once again we use this method of attack. To do so, we need the following ideas and notation.

Let $G = (V, E)$ be an undirected graph. For $e = \{a, b\} \in E$, let G_e denote the subgraph of G obtained by deleting e from G, without removing vertices a and b; that is, $G_e = G - e$ as defined in Section 11.2. From G_e a second subgraph of G is obtained by coalescing the vertices a and b. This second subgraph is denoted by G_e'.

Example 11.32   Figure 11.73 shows G_e and G_e' for graph G with the edge e as specified. Note how the coalescing of a and b in G_e' results in the coalescing of the two pairs of edges $\{d, b\}, \{d, a\}$ and $\{a, c\}, \{b, c\}$. ☐

Using these special subgraphs, we turn now to the main result.

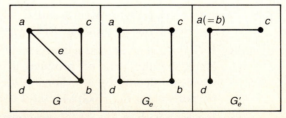

Figure 11.73

THEOREM 11.10   (*Decomposition Theorem for Chromatic Polynomials*) If $G = (V, E)$ is a connected graph and $e \in E$, then

$$P(G_e, \lambda) = P(G, \lambda) + P(G'_e, \lambda).$$

Proof   Let $e = \{a, b\}$. The number of ways to properly color the vertices in G_e with (at most) λ colors is $P(G_e, \lambda)$. Those colorings where a and b have different colors are proper colorings of G. The colorings of G_e that are not proper colorings of G occur when a and b have the same color. But each of these colorings corresponds with a proper coloring for G'_e. This partition of the $P(G_e, \lambda)$ colorings into the two disjoint subsets described results in the equation $P(G_e, \lambda) = P(G, \lambda) + P(G'_e, \lambda)$. ■

When calculating chromatic polynomials, we shall place brackets about a graph to indicate its chromatic polynomial.

Example 11.33   The following calculations yield $P(G, \lambda)$ for G a cycle of length 4.

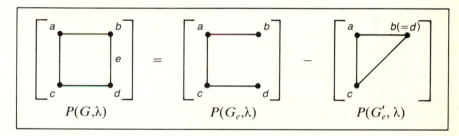

$$P(G, \lambda) \qquad\qquad P(G_e, \lambda) \qquad\qquad P(G'_e, \lambda)$$

From Example 11.31(c), $P(G_e, \lambda) = \lambda(\lambda - 1)^3$. With $G'_e = K_3$, $P(G'_e, \lambda) = \lambda^{(3)}$. Therefore,

$$P(G, \lambda) = \lambda(\lambda - 1)^3 - \lambda(\lambda - 1)(\lambda - 2) = \lambda(\lambda - 1)[(\lambda - 1)^2 - (\lambda - 2)]$$
$$= \lambda(\lambda - 1)[\lambda^2 - 3\lambda + 3] = \lambda^4 - 4\lambda^3 + 6\lambda^2 - 3\lambda.$$

Since $P(G, 1) = 0$ while $P(G, 2) = 2 > 0$, $\chi(G) = 2$.   □

Example 11.34   Here we find a second application of Theorem 11.10.

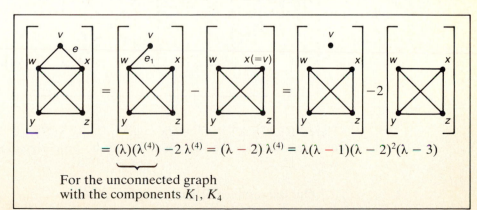

$$= (\lambda)(\lambda^{(4)}) - 2\,\lambda^{(4)} = (\lambda - 2)\,\lambda^{(4)} = \lambda(\lambda - 1)(\lambda - 2)^2(\lambda - 3)$$

For the unconnected graph
with the components K_1, K_4

For $1 \le \lambda \le 3$, $P(G, \lambda) = 0$, but $P(G, \lambda) > 0$ for $\lambda \ge 4$. Consequently the given graph has chromatic number 4. $\square$

The chromatic polynomials in Examples 11.33 and 11.34 suggest the following results.

THEOREM 11.11   For any graph G, the constant term in $P(G, \lambda)$ is 0.

Proof   For any graph G, $\chi(G) > 0$ because $V \ne \emptyset$. If $P(G, \lambda)$ has constant term a, then $P(G, 0) = a \ne 0$. This implies that there are a ways to color G properly with 0 colors. ∎

THEOREM 11.12   Let $G = (V, E)$ with $|E| > 0$. Then the sum of the coefficients in $P(G, \lambda)$ is 0.

Proof   Since $|E| \ge 1$, $\chi(G) \ge 2$. Consequently $P(G, 1) = 0 =$ the sum of the coefficients in $P(G, \lambda)$. ∎

Since the chromatic polynomial of a complete graph is easy to determine, an alternative method for finding $P(G, \lambda)$ can be obtained. Theorem 11.10 reduced the problem to smaller graphs. Here we add edges to a given graph until we reach complete graphs.

THEOREM 11.13   Let $G = (V, E)$, with $a, b \in V$ but $\{a, b\} = e \notin E$. We write G_e^+ for the graph we obtain from G by adding the edge $e = \{a, b\}$. Coalescing the vertices a and b in G gives us the subgraph G_e^{++} of G. Under these circumstances $P(G, \lambda) = P(G_e^+, \lambda) + P(G_e^{++}, \lambda)$.

Proof   This result follows from Theorem 11.10, because $P(G_e^+, \lambda) = P(G, \lambda) - P(G_e^{++}, \lambda)$. ∎

Example 11.35   Let us apply Theorem 11.13.

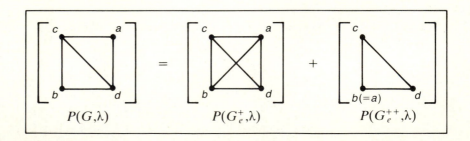

$$P(G, \lambda) \qquad\qquad P(G_e^+, \lambda) \qquad\qquad P(G_e^{++}, \lambda)$$

Here $P(G, \lambda) = \lambda^{(4)} + \lambda^{(3)} = \lambda(\lambda - 1)(\lambda - 2)^2$, so $\chi(G) = 3$. In addition, if six colors are available, the vertices in G can be properly colored in $6(5)(4)^2 = 480$ ways. $\square$

Our last result again uses complete graphs.

THEOREM 11.14 Let G be an undirected graph with subgraphs G_1, G_2. If $G = G_1 \cup G_2$ and $G_1 \cap G_2 = K_n$, then

$$P(G, \lambda) = [P(G_1, \lambda) \cdot P(G_2, \lambda)]/\lambda^{(n)}.$$

Proof Since $G_1 \cap G_2 = K_n$, it follows that K_n is a subgraph of G_1, G_2, and $\chi(G_1)$, $\chi(G_2) \geq n$. Given λ colors there are $\lambda^{(n)}$ proper colorings of K_n. For each of these $\lambda^{(n)}$ colorings there are $P(G_1, \lambda)/\lambda^{(n)}$ ways to properly color the remaining vertices in G_1. Likewise, there are $P(G_2, \lambda)/\lambda^{(n)}$ ways to properly color the remaining vertices in G_2. By the rule of product,

$$P(G, \lambda) = P(K_n, \lambda) \cdot \frac{P(G_1, \lambda)}{\lambda^{(n)}} \cdot \frac{P(G_2, \lambda)}{\lambda^{(n)}} = \frac{P(G_1, \lambda) \cdot P(G_2, \lambda)}{\lambda^{(n)}}. \quad \blacksquare$$

Example 11.36 Consider the graph in Example 11.34. Let G_1 be the subgraph induced by the vertices w, x, y, z. Let G_2 be the cycle determined by $v, w,$ and x. Then $G_1 \cap G_2$ is the edge $\{w, x\}$, so $G_1 \cap G_2 = K_2$.

Therefore

$$\begin{aligned}
P(G, \lambda) &= \frac{P(G_1, \lambda) \cdot P(G_2, \lambda)}{\lambda^{(2)}} = \frac{\lambda^{(4)} \cdot \lambda^{(3)}}{\lambda^{(2)}} \\
&= [\lambda^2 (\lambda - 1)^2 (\lambda - 2)^2 (\lambda - 3)]/[\lambda(\lambda - 1)] \\
&= \lambda(\lambda - 1)(\lambda - 2)^2(\lambda - 3),
\end{aligned}$$

as shown in Example 11.34. $\square$

Much more can be said about chromatic polynomials—in particular, there are many unanswered questions. For example, no one has found a set of conditions that indicate whether a given polynomial in λ is the chromatic polynomial for some graph. More about this topic is introduced in the article by R. C. Read [22].

EXERCISES 11.6

1. A pet-shop owner receives a shipment of tropical fish. Among the different species in the shipment are certain pairs where one species feeds on the other. These pairs must consequently be kept in different aquaria. Model this problem as a graph-coloring problem, and tell how to determine the smallest number of aquaria needed to preserve all the fish in the shipment.

2. As chairwoman of church committees, Mrs. Blasi is faced with scheduling the meeting times for 15 committees. Each committee meets for one hour each week. If someone is on more than one committee, then those committees must be scheduled at different times. Model this problem as a graph-coloring problem, and tell how to determine the least number of meeting times Mrs. Blasi has to consider for scheduling the 15 committee meetings.

3. Find the chromatic number of the following graphs.

   a) The complete bipartite graph $K_{m,n}$.

    b) A cycle on n vertices, $n \geq 3$.

    c) The graphs in Figs. 11.45(d), 11.48, and 11.68.

4. If G is a loop-free undirected graph with at least one edge, prove that G is bipartite if and only if $\chi(G) = 2$.

5. Give an example of an undirected graph $G = (V, E)$ where $\chi(G) = 3$ but no subgraph of G is isomorphic to K_3.

6. If G is a graph with components $C_1, C_2, \ldots, C_k$, how is $\chi(G)$ related to $\chi(C_i), 1 \leq i \leq k$?

7. a) Determine the chromatic polynomials for the graphs in Fig. 11.74.

    b) Find $\chi(G)$ for each graph.

    c) If five colors are available, in how many ways can the vertices of each graph be properly colored?

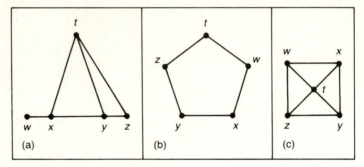

Figure 11.74

8. a) Determine whether the graphs in Fig. 11.75 are isomorphic.

    b) Find $P(G, \lambda)$ for each graph.

    c) Comment on the results found in parts (a) and (b).

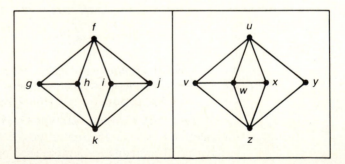

Figure 11.75

9. Let $G = (V, E)$ be a loop-free connected undirected graph. Prove or disprove: If G contains a cycle, then $P(G, \lambda)$ contains a factor of the form $(\lambda - n)$ where $n \in \mathbf{Z}^+, n \geq 2$.

10. Let G be a loop-free undirected graph where $\Delta = \max_{v \in V}\{\deg(v)\}$.

    a) Prove that $\chi(G) \le \Delta + 1$.

    b) Find two types of graphs G where $\chi(G) = \Delta + 1$.

11. For $n \ge 3$, let C_n denote the cycle of length n.

    a) What is $P(C_3, \lambda)$?

    b) If $n \ge 4$, show that $P(C_n, \lambda) = P(L_n, \lambda) - P(C_{n-1}, \lambda)$, where L_n denotes the path of length $n - 1$.

    c) Verify that $P(L_n, \lambda) = \lambda(\lambda - 1)^{n-1}$, for any $n \ge 2$.

    d) Establish the relations

$$P(C_n, \lambda) - (\lambda - 1)^n = (\lambda - 1)^{n-1} - P(C_{n-1}, \lambda), \qquad n \ge 4,$$
$$P(C_n, \lambda) - (\lambda - 1)^n = P(C_{n-2}, \lambda) - (\lambda - 1)^{n-2}, \qquad n \ge 5.$$

    e) Prove that for any $n \ge 3$, $P(C_n, \lambda) = (\lambda - 1)^n + (-1)^n(\lambda - 1)$.

12. For $n \ge 3$, the *wheel graph*, W_n, is obtained from a cycle of length n by placing a new vertex within the cycle and adding edges (*spokes*) from this new vertex to each vertex of the cycle.

    a) What relationship is there between $\chi(C_n)$ and $\chi(W_n)$?

    b) Use part (e) of Exercise 11 to show that

$$P(W_n, \lambda) = \lambda(\lambda - 2)^n + (-1)^n \lambda(\lambda - 2).$$

    c)   **(i)** If we have k different colors available, in how many ways can we paint the walls and ceiling of a pentagonal room if adjacent walls, and any wall and the ceiling, are to be painted with different colors?

        (ii) What is the smallest value of k for which such a coloring is possible?

      (iii) Answer parts (i) and (ii) for a hexagonal room. (The reader may wish to compare part (c) of this exercise with Exercise 6 in the Miscellaneous Exercises of Chapter 8.)

13. Let $G = (V, E)$ be a loop-free undirected graph with chromatic polynomial $P(G, \lambda)$ and $|V| = n$. Use Theorem 11.13 to prove that $P(G, \lambda)$ has degree n and leading coefficient 1.

14. Let $G = (V, E)$ be a loop-free undirected graph.

    a) For each such graph where $|V| \le 3$, find $P(G, \lambda)$ and show that it is a sum of consecutive powers of λ. Also show that the coefficients of these consecutive powers alternate in sign.

    b) Now consider $G = (V, E)$ where $|V| = n \ge 4$ and $|E| = k$. Prove by induction that $P(G, \lambda)$ is a sum of consecutive powers of λ and that the coefficients of these consecutive powers alternate in sign. (For the induction hypothesis, assume that the result is true for $G = (V, E)$ where either (i) $|V| = n - 1$ or (ii) $|V| = n$, but $|E| = k - 1$.)

    c) Prove that if $|V| = n$, then the coefficient of λ^{n-1} in $P(G, \lambda)$ is the negative of $|E|$.

15. Use Theorem 11.14 to find $P(G, \lambda)$ for the graph G in Fig. 11.76.

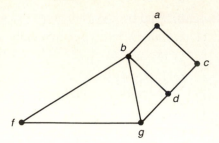

Figure 11.76

11.7
SUMMARY AND HISTORICAL REVIEW

Unlike other areas in mathematics, graph theory traces its beginnings to a definite time and place: the problem of the seven bridges of Königsberg, which was solved in 1736 by Leonhard Euler (1707–1783). In 1752 we find Euler's theorem for planar graphs. (This result was originally presented in terms of polyhedra.) However, after these developments, little was accomplished in this area for almost a century.

Then, in 1847, Gustav Kirchhoff (1824–1887) examined a special type of graph called a tree. (A *tree* is a loop-free undirected graph that is connected but contains no cycles.) Kirchhoff used this concept in applications dealing with electrical networks in his extension of Ohm's laws for electrical flow. Ten years later Arthur Cayley (1821–1895) developed this same type of graph in order to count the distinct isomers of the saturated hydrocarbons $C_n H_{2n+2}$, $n \in \mathbf{Z}^+$.

This period also saw two other major ideas come to light. The *four-color conjecture* was first investigated by Francis Guthrie (1831–1899) in about 1850. In Section 11.6 we related some of the history of this problem, which was solved via an intricate computer analysis in 1976 by Kenneth Appel and Wolfgang Haken.

The second major idea was the Hamilton cycle. This cycle is named for Sir William Hamilton (1805–1865), who used the idea in 1859 for an intriguing puzzle that used the edges on a regular dodecahedron. A solution to this puzzle is not very difficult to find, but mathematicians still search for necessary and sufficient conditions to characterize those undirected graphs that possess a Hamilton path or cycle.

Following these developments, we find little activity until after 1920. The characterization of planar graphs was solved by Kasimir Kuratowski in 1930. In 1936 we find the publication of the first book on graph theory, written by Denses König, a prominent researcher in the field. Since then there has been a great deal of activity in the area, the computer providing assistance in the last three decades. Among the many contemporary researchers (not mentioned in the chapter references) in this and related fields one finds the names Claude Berge, V. Chvátal, Paul Erdös, Laszlo Lovász, W. T. Tutte, and Hassler Whitney.

Comparable coverage of the material presented in this chapter is contained in Chapters 6, 8, and 9 of C. L. Liu [19]. More advanced work is found in the works by M. Behzad, G. Chartrand, and L. Lesniak-Foster [3], J. Bondy and U. Murty [7], N. Deo [10], and F. Harary [16]. The text by G. Chartrand and L. Lesniak [8] provides a more algorithmic approach in its presentation. A proof of Kuratowski's Theorem appears in Chapter 8 of C. L. Liu [19]. Recent developments in the field of graph theory are given in the two-volume *MAA Studies in Mathematics*, edited by D. Fulkerson [14]. The article by G. Chartrand and R. J. Wilson [9] develops many of the important concepts of graph theory by focusing on one particular graph—the Petersen graph.

Applications of graph theory in electrical networks can be found in S. Seshu and M. Reed [25]. In the text by N. Deo [10], applications in coding theory, electrical networks, operations research, computer programming, and chemistry occupy Chapters 12–15. The text by F. Roberts [23] applies the methods of graph theory to the social sciences. Recent applications of graph theory in chemistry are given in the article by D. H. Rouvray [24].

More on chromatic polynomials can be found in the survey article by R. C. Read [22] and in Chapter VI of C. L. Liu [20]. The role of Polya's theory in graphical enumeration is examined in Chapter 10 of N. Deo [10]. A thorough coverage of this topic is found in the text by F. Harary and E. Palmer [17]. We shall introduce the basic ideas behind this method of enumeration in Chapter 16.

Additional coverage on the historical development of graph theory is given in N. Biggs, E. Lloyd, and R. Wilson [6].

Many applications in graph theory involve large graphs that require the decision-making capability of a computer in conjunction with the ingenuity of mathematical methods. Chapter 11 of N. Deo [10] presents computer algorithms dealing with several of the graph-theoretic properties we have studied here. Along the same line, the text by A. Aho, J. Hopcroft, and J. Ullman [1] provides even more for the reader interested in computer science.

As mentioned at the end of Section 11.5, the problem of the traveling salesperson is closely related to the search for a Hamilton cycle in a graph. This is a graph-theoretic problem of interest in both operations research and computer science. The article by M. Bellmore and G. L. Nemhauser [5] provides a good introductory survey of results on this problem. The text by R. Bellman, K. L. Cooke, and J. A. Lockett [4] includes an algorithmic treatment of this problem along with other graph problems. A number of heuristics for obtaining an approximate solution to the problem are given in Chapter 4 of the text by L. R. Foulds [13]. In the text edited by E. L. Lawler, J. K. Lenstra, A. H. G. Rinnooy Kan, and D. B. Shmoys [18], there are 12 papers dealing with various aspects of this problem, including historical considerations as well as some results on computational complexity. Recent developments on applications, where a robot visits different locations in an automated warehouse in order to fill a given order, are examined in the articles by E. A. Elsayed [11] and by E. A. Elsayed and R. G. Stern [12].

The solution of the four-color problem can be examined further by starting with the paper by K. Appel and W. Haken [2]. An appreciation for the development of this work can be gained from a study of the earlier article by W. Haken [15]. The proof uses a computer analysis to handle a large number of cases; the article by T. Tymoczko [26] examines the role of such techniques in pure mathematics.

Finally, the article by A. Ralston [21] demonstrates the connections among coding theory, combinatorics, graph theory, and computer science.

REFERENCES

1. Aho, Alfred V., Hopcroft, John E., and Ullman, Jeffrey D. *Data Structures and Algorithms.* Reading, Mass.: Addison-Wesley, 1983.
2. Appel, Kenneth, and Haken, Wolfgang. "Every Planar Map Is Four Colorable." *Bulletin of the American Mathematical Society* 82 (1976): pp. 711–712.
3. Behzad, Mehdi, Chartrand, Gary, and Lesniak-Foster, Linda. *Graphs and Digraphs.* Belmont, Calif.: Wadsworth, 1979.
4. Bellman, R., Cooke, K. L., and Lockett, J. A. *Algorithms, Graphs, and Computers.* New York: Academic Press, 1970.
5. Bellmore, M., and Nemhauser, G. L. "The Traveling Salesman Problem: A Survey." *Operations Research* 16, 1968, pp. 538–558.
6. Biggs, N., Lloyd, E. K., and Wilson, R. J. *Graph Theory (1736–1936).* Oxford, England: Clarendon Press, 1976.
7. Bondy, J. A., and Murty, U. S. R. *Graph Theory with Applications.* New York: Elsevier North-Holland, 1976.
8. Chartrand, Gary, and Lesniak, Linda. *Graphs and Digraphs,* 2nd ed. Monterey, Calif.: Wadsworth and Brooks/Cole, 1986.
9. Chartrand, Gary, and Wilson, Robin J. "The Petersen Graph." In Frank Harary and John S. Maybee, eds., *Graphs and Applications (Proceedings of the First Colorado Symposium on Graph Theory).* New York: Wiley, 1985.
10. Deo, Narsingh. *Graph Theory with Applications to Engineering and Computer Science.* Englewood Cliffs, N.J.: Prentice-Hall, 1974.
11. Elsayed, E. A. "Algorithms for Optimal Material Handling in Automatic Warehousing Systems." *Int. J. Prod. Res.* 19, 1981, pp. 525–535.
12. Elsayed, E. A., and Stern, R. G. "Computerized Algorithms for Order Processing in Automated Warehousing Systems." *Int. J. Prod. Res.* 21, 1983, pp. 579–586.
13. Foulds, L. R. *Combinatorial Optimization for Undergraduates.* New York: Springer-Verlag, 1984.
14. Fulkerson, D. R., ed. *Studies in Graph Theory,* Parts I, II, *MAA Studies in Mathematics,* Vol. 11 and 12. The Mathematical Association of America, 1975.
15. Haken, Wolfgang. "An Attempt to Understand the Four-Color Problem." *Journal of Graph Theory* 1, no. 3 (1977).
16. Harary, Frank. *Graph Theory.* Reading, Mass.: Addison-Wesley, 1969.
17. Harary, Frank, and Palmer, Edgar M. *Graphical Enumeration.* New York: Academic Press, 1973.
18. Lawler, E. L., Lenstra, J. K., Rinnooy Kan, A. H. G., and Shmoys, D. B., eds. *The Traveling Salesman Problem.* New York: Wiley, 1986.

19. Liu, C. L. *Introduction to Combinatorial Mathematics*. New York: McGraw-Hill, 1968.

20. Liu, C. L. *Topics in Combinatorial Mathematics*. Mathematical Association of America, 1972.

21. Ralston, Anthony. "De Bruijn Sequences—A Model Example of the Interaction of Discrete Mathematics and Computer Science." *Mathematics Magazine* 55, no. 3 (May 1982): pp. 131–143.

22. Read, R. C. "An Introduction to Chromatic Polynomials." *Journal of Combinatorial Theory* 4, 1968, pp. 52–71.

23. Roberts, Fred S. *Discrete Mathematical Models*. Englewood Cliffs, N.J.: Prentice-Hall, 1976.

24. Rouvray, Dennis H. "Predicting Chemistry from Topology." *Scientific American* 255, no. 3 (September 1986): pp. 40–47.

25. Seshu, S., and Reed, M. B. *Linear Graphs and Electrical Networks*. Reading, Mass.: Addison-Wesley, 1961.

26. Tymoczko, Thomas. "Computers, Proofs and Mathematicians: A Philosophical Investigation of the Four-Color Proof." *Mathematics Magazine* 53, no. 3 (May 1980): pp. 131–138.

MISCELLANEOUS EXERCISES

1. Let G be a loop-free undirected graph on n vertices. If G has 56 edges and $\overline{G}$ has 80 edges, what is n?

2. Let G_1 and G_2 be two loop-free undirected connected graphs, each on n vertices. If every vertex in both G_1 and G_2 has degree 2, prove that G_1 and G_2 are isomorphic.

3. An n-cube is an undirected graph $G = (V, E)$ where the vertices of G are the (ordered) n-tuples of 0's and 1's. Edges in G are of the form $\{v, w\}$ where v and w differ in exactly one component.

   a) Prove that G is connected.

   b) Find $|V|$ and $|E|$.

4. Let G be an undirected graph with C a cycle in G. If e_1 and e_2 are any two edges in C, prove that G contains a cut-set S such that $C \cap S = \{e_1, e_2\}$.

5. a) If the edges of K_6 are painted either red or blue, prove that there is a red triangle or a blue triangle that is a subgraph.

   b) Prove that in any group of six people there must be three who are total strangers to one another or three who are mutual friends.

6. Let $G = (V, E)$ be a loop-free undirected graph. Recall that G is called *self-complementary* if G and $\overline{G}$ are isomorphic. If G is self-complementary with $|V| = n$,

   a) what is $|E|$?

   b) prove that G is connected.

   c) can G contain a vertex v where $\deg(v) = n - 1$?

7. a) Show that the graphs G_1 and G_2, in Fig. 11.77, are isomorphic.

   b) How many different isomorphisms $f: G_1 \to G_2$ are possible here?

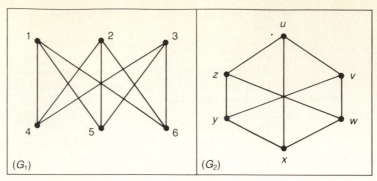

Figure 11.77

8. Let $X = \{1, 2, 3, \ldots, n\}$, where $n \geq 2$. Construct the loop-free undirected graph $G = (V, E)$ as follows:

   - (V): Each two-element subset of X determines a vertex of G.

   - (E): If $v_1, v_2 \in V$ correspond to subsets $\{a, b\}$ and $\{c, d\}$, respectively, of X, draw the edge $\{v_1, v_2\}$ in G when $\{a, b\} \cap \{c, d\} = \emptyset$.

   a) Show that G is an isolated vertex when $n = 2$ and that G is disconnected for $n = 3, 4$.

   b) Show that for $n \geq 5$, G is connected. (In fact, for any $v_1, v_2 \in V$, either $\{v_1, v_2\} \in E$ or there is a path of length 2 connecting v_1 and v_2.)

   c) Prove that G is nonplanar for $n \geq 5$.

9. If $G = (V, E)$ is an undirected graph, a subset K of V is called a *covering* of G if for every edge $\{a, b\}$ of G either a or b is in K. The set K is a *minimal covering* if $K - \{x\}$ fails to cover G for each $x \in K$. The number of vertices in a smallest (minimal) covering is called the *covering number* of G.

   a) Prove that if $I \subseteq V$, then I is an independent set in G if and only if $V - I$ is a covering of G.

   b) Verify that $|V|$ is the sum of the independence number of G and its covering number.

10. If $G = (V, E)$ is an undirected graph, a subset D of V is called a *dominating set* if for all $v \in V$, either $v \in D$ or v is adjacent to a vertex in D. If D is a dominating set and no proper subset of D has this property, D is called *minimal*. The size of any smallest minimal dominating set in G is denoted by $\gamma(G)$ and is called the *domination number* of G.

   a) If G has no isolated vertices, prove that if D is a minimal dominating set, then $V - D$ is a dominating set.

   b) If $I \subseteq V$ and I is independent, prove that I is a dominating set if and only if I is maximal independent.

   c) Show that $\gamma(G) \leq \beta(G)$, and $|V| \leq \beta(G)\chi(G)$.

11. a) Let $G = (V, E)$ be an undirected graph made up of n components $C_i, 1 \leq i \leq n$. How many different maximal independent sets are there in G under each of the following conditions?

(i) Each C_i has vertices a_i, b_i, c_i and edges $\{a_i, b_i\}, \{a_i, c_i\}, \{b_i, c_i\}, 1 \le i \le n$?

(ii) Each C_i has vertices a_i, b_i, c_i, d_i and edges $\{a_i, b_i\}, \{b_i, c_i\}, \{c_i, d_i\}, \{d_i, a_i\},$ $1 \le i \le n$?

(iii) Each C_i has vertices a_i, b_i, c_i, d_i and is isomorphic to $K_4, 1 \le i \le n$?

b) Generalize the results of part (a).

12. For $n \ge 1$, let L_n denote the path made up of n vertices and $n - 1$ edges. Let a_n be the number of independent subsets of vertices in L_n. (The empty subset is considered one of these independent subsets.) Find and solve a recurrence relation for a_n.

13. Let $G = (V, E)$ be the undirected connected "ladder graph" shown in Fig. 11.78.

a) Determine $|V|$ and $|E|$.

b) For $n \ge 0$, let a_n denote the number of ways one can select n of the edges in G so that no two edges share a common vertex. Find and solve a recurrence relation for a_n.

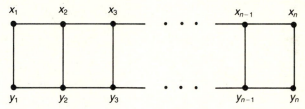

Figure 11.78

14. For $n \ge 1$, let $a_n = \binom{n}{2}$, the number of edges in K_n, and let $a_0 = 0$. Find the generating function $f(x) = \sum_{n=0}^{\infty} a_n x^n$.

15. For the graph G in Fig. 11.79, answer the following questions.

a) What are $\gamma(G)$, $\beta(G)$, and $\chi(G)$?

b) Does G have an Euler circuit or a Hamilton cycle?

c) Is G bipartite? Is it planar?

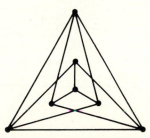

Figure 11.79

16. **a)** The complete bipartite graph $K_{m,n}$ contains 16 edges and satisfies $m \le n$. Determine m, n so that $K_{m,n}$ possesses (i) an Euler circuit but not a Hamilton cycle; (ii) both a Hamilton cycle and an Euler circuit.

b) Generalize the results of part (a).

17. If $G = (V, E)$ is an undirected graph, any subgraph of G that is a complete graph is

called a *clique* in G. The number of vertices in a largest clique in G is called the *clique number* for G and is denoted by $\omega(G)$.

a) How are $\chi(G)$ and $\omega(G)$ related?

b) If G is made up of components $C_1, C_2, \ldots, C_k$, how is $\omega(G)$ related to $\omega(C_1)$, $\omega(C_2), \ldots, \omega(C_k)$?

c) Can a graph G have 25 edges and clique number 8?

d) Is there any relationship between $\omega(G)$ and $\beta(\overline{G})$?

18. If $G = (V, E)$ is an undirected loop-free graph, the *line graph* of G, denoted $L(G)$, is a graph with the set E as vertices, where we join two vertices e_1, e_2 in $L(G)$ if and only if e_1, e_2 are adjacent edges in G.

a) Find $L(G)$ for each of the graphs in Fig. 11.80.

b) Assuming that $|V| = n$ and $|E| = e$, show that $L(G)$ has e vertices and $(1/2) \sum_{v \in V} \deg(v)[\deg(v) - 1] = [(1/2) \sum_{v \in V} [\deg(v)]^2] - e = \sum_{v \in V} \binom{\deg(v)}{2}$ edges.

c) Prove that if G has an Euler circuit, then $L(G)$ has both an Euler circuit and a Hamilton cycle.

d) If $G = K_4$, examine $L(G)$ to show that the converse of part (c) is false.

e) Prove that if G has a Hamilton cycle, then so does $L(G)$.

f) Examine $L(G)$ for the graph in Fig. 11.80(b) to show that the converse of part (e) is false.

g) Verify that $L(G)$ is nonplanar for $G = K_5$ and $G = K_{3,3}$.

h) Give an example of a graph G where G is planar but $L(G)$ is not.

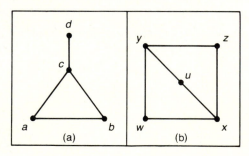

Figure 11.80

19. Explain why each of the following polynomials in λ cannot be a chromatic polynomial.

a) $\lambda^4 - 5\lambda^3 + 7\lambda^2 - 6\lambda + 3$ b) $3\lambda^3 - 4\lambda^2 + \lambda$

c) $\lambda^4 - 3\lambda^3 + 5\lambda^2 - 4\lambda$ d) $\lambda^4 - 5\lambda^2 + 4\lambda$

20. a) For any $x, y \in \mathbf{Z}^+$ prove that $x^3 y - xy^3$ is even.

b) Let $V = \{1, 2, 3, \ldots, 8, 9\}$. Construct the loop-free undirected graph $G = (V, E)$ as follows: For $m, n \in V, m \neq n$, draw the edge $\{m, n\}$ in G if 5 divides $m + n$ or $m - n$.

c) Given any three distinct positive integers, prove that there are at least two of these, say x and y, where 10 divides $x^3 y - xy^3$.

21. A *complete n-partite* graph $G = (V, E)$ is a loop-free undirected graph where

    (i) $V = V_1 \cup V_2 \cup \cdots \cup V_n$, with $V_i \neq \emptyset$ for $1 \leq i \leq n$, and
    $V_i \cap V_j = \emptyset$, for $1 \leq i < j \leq n$; and,

    (ii) if $u \in V_i, v \in V_j$, with $i \neq j$, then $\{u, v\} \in E$.

    a) If $|V_i| = 1$, $1 \leq i \leq n$, what is the resulting graph?

    b) If $|V_i| = p_i$, $1 \leq i \leq n$, determine $|V|$ and $|E|$.

    c) For $|V_i| = p_i$, $1 \leq i \leq n$, what is $\kappa(\overline{G})$? How many edges are there in $\overline{G}$?

    d) Find the values of n (>1) and $p_1, p_2, \ldots, p_n$ for which a complete n-partite graph is planar.

22. Just as we defined $\lambda^{(n)}$ when dealing with the chromatic polynomial for K_n, for $n \geq 1$, we can also define $x^{(n)} = x(x-1)(x-2) \cdots (x-n+1)$. Multiplying this out results in a polynomial in x of degree n. We write this polynomial as $\sum_{i=0}^{n} s(n, i) x^i$ and call each coefficient $s(n, i)$, $0 \leq i \leq n$, a *Stirling number of the first kind*.

    For all n, $s(n, n) = 1$ and $s(n, 0) = 0$. Also, for $i < 0$ or $i > n$, $s(n, i) = 0$. In addition $\sum_{i=0}^{n+1} s(n+1, i) x^i = x^{(n+1)} = x^{(n)}(x - n) = [\sum_{i=0}^{n} s(n, i) x^i](x - n)$. Consequently, for any $1 \leq i \leq n$, the comparison of coefficients of like powers of x results in the recurrence relation

    $$s(n+1, i) = s(n, i-1) - n s(n, i).$$

    a) Use the recurrence relation to find a table for $s(n, i)$, $1 \leq n \leq 5$, $0 \leq i \leq n$.

    b) Write a computer program (or develop an algorithm) to extend the table in part (a) to all $n \leq 15$, $0 \leq i \leq n$.

    c) For $m, n \in \mathbf{Z}^+$, we define the delta function $\delta(m, n)$ to be 1 when $m = n$ and to be 0 when $m \neq n$. If $k = \max\{m, n\}$, then the Stirling numbers of the first and second kinds are related by

    $$\sum_{i=0}^{k} s(m, i) S(i, n) = \delta(m, n) = \sum_{i=0}^{k} S(m, i) s(i, n).$$

    Verify these results for the cases where $m = n = 4$ and where $m = 5, n = 4$.

CHAPTER 12

Trees

Continuing our study of graph theory, we focus on a special type of graph called a tree. First used in 1847 by Gustav Kirchhoff (1824–1887) in his work on electrical networks, trees were later redeveloped and named by Arthur Cayley (1821–1895). In 1857 Cayley used these graphs to enumerate the different isomers of the saturated hydrocarbons $C_n H_{2n+2}, n \in \mathbf{Z}^+$.

With the advent of digital computers, many new applications were found for trees. Special types of trees are prominent in the study of data structures, sorting, and coding theory, and in the solution of certain optimization problems.

12.1
DEFINITIONS, PROPERTIES, AND EXAMPLES

DEFINITION 12.1 Let $G = (V, E)$ be a loop-free undirected graph. The graph G is called a *tree*† if G is connected and contains no cycles.

In Fig. 12.1 the graph G_1 is a tree, but the graph G_2 is not a tree because it contains the cycle $\{a, b\}, \{b, c\}, \{c, a\}$. The graph G_3 is not connected, so it cannot be a tree. However, each component of G_3 is a tree, and we call G_3 a *forest*.

When a graph is a tree we write T instead of G to emphasize this structure.

In Fig. 12.1 we see that G_1 is a subgraph of G_2 where G_1 contains all the vertices of G_2 and G_1 is a tree. In this situation G_1 is called a *spanning tree* for G_2. We think of a spanning tree as providing minimal connectivity for the graph and as a minimal skeletal framework holding the vertices together.

We now examine some properties of trees.

---

† As in the case of graphs, the terminology in the study of trees is not standard and the reader may find some differences from one textbook to another.

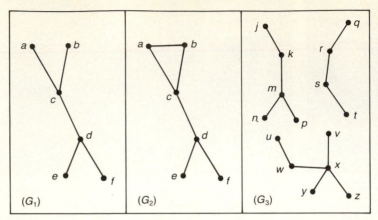

Figure 12.1

THEOREM 12.1   If a, b are vertices in a tree $T = (V, E)$, then there is a unique path that connects these vertices.

Proof   Since T is connected, there is at least one path in T that connects a and b. If there were more, from two such paths some of the edges could be combined to form a cycle. But T has no cycles.   ∎

THEOREM 12.2   If $G = (V, E)$ is an undirected graph, then G is connected if and only if G has a spanning tree.

Proof   If G has a spanning tree T, then for every pair a, b of vertices in V, a subset of the edges in T provides a (unique) path between a and b, so G is connected. Conversely, if G is connected and G is not a tree, remove all loops from G. If the resulting subgraph G_1 is not a tree, then G_1 must contain a cycle. For each such cycle, remove a distinct edge of G_1. The resulting subgraph G_2 contains all the vertices in G, is loop-free and connected, and contains no cycles. Hence G_2 is a spanning tree for G.   ∎

Figure 12.2 shows three nonisomorphic trees that exist for five vertices. Although they are not isomorphic, they all have the same number of edges, namely four. This leads us to the following general result.

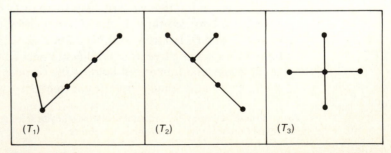

Figure 12.2

THEOREM 12.3   In any tree $T = (V, E)$, $|V| = |E| + 1$.

Proof   The proof is by induction on $|E|$. If $|E| = 0$, then the tree consists of a single isolated vertex as in Fig. 12.3(a). Here $|V| = 1 = |E| + 1$. Parts (b) and (c) of the figure verify the result for the cases where $|E| = 1$ or 2.

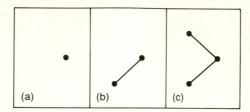

(a)   (b)   (c)   **Figure 12.3**

Assume the theorem is true for any tree that contains at most k edges, where $k \geq 0$. Now consider a tree $T = (V, E)$, as in Fig. 12.4, where $|E| = k + 1$. (The dotted edge(s) indicates that some of the tree doesn't appear in the figure.) If the edge with endpoints y, z is removed from T, we obtain two *subtrees* $T_1 = (V_1, E_1)$ and $T_2 = (V_2, E_2)$ where $|V| = |V_1| + |V_2|$ and $|E_1| + |E_2| + 1 = |E|$. (One of these subtrees could consist of just a single vertex if, for example, the edge with endpoints w, x were removed.) Since $0 \leq |E_1|, |E_2| \leq k$, by the induction hypothesis, $|E_i| + 1 = |V_i|$, $i = 1, 2$. Consequently, $|V| = |V_1| + |V_2| = (|E_1| + 1) + (|E_2| + 1) = (|E_1| + |E_2| + 1) + 1 = |E| + 1$, and the theorem follows by the alternative form of finite induction. ∎

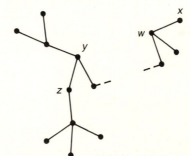

Figure 12.4

As we examine the trees in Fig. 12.2 we also see that each tree has at least two pendant vertices—that is, vertices of degree 1. This is also true in general.

THEOREM 12.4   For any tree $T = (V, E)$, if $|V| \geq 2$ then T has at least two pendant vertices.

Proof   Let $|V| = n \geq 2$. From Theorem 12.3, $|E| = n - 1$, and by Theorem 11.2, $2(n - 1) = 2n - 2 = 2|E| = \sum_{v \in V} \deg(v)$. If $\deg(v) \geq 2$ for all $v \in V$, then $\sum_{v \in V} \deg(v) \geq 2|V| = 2n$. From the contradiction $2n - 2 \geq 2n$, it follows that there are at least two vertices v for which $\deg(v) < 2$. Since there are no isolated vertices, these vertices must be pendant. ∎

Example 12.1   In Fig. 12.5 we have two trees, each with 14 vertices (labeled with C's and H's) and 13 edges. Each vertex has degree 4 (C, carbon atom) or degree 1 (H, hydrogen atom). Part (b) of the figure has a carbon atom (C) at the center of the tree. This carbon atom is adjacent to four vertices, three of which have degree 4. There is no vertex (C atom) in part (a) that possesses this property, so the two trees are not isomorphic. They serve as models for the two chemical isomers that correspond with the saturated† hydrocarbon $C_4 H_{10}$. Part (a) represents butane; part (b) represents isobutane.   □

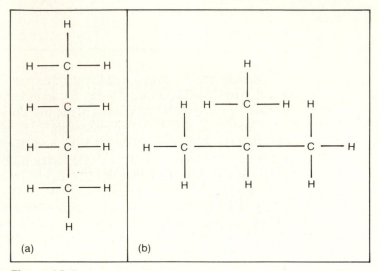

Figure 12.5

A second result from chemistry is given in the following example.

Example 12.2   If a saturated hydrocarbon has n carbon atoms, show that it has $2n + 2$ hydrogen atoms.

Considering the saturated hydrocarbon as a tree $T = (V, E)$, let k equal the number of pendant vertices, or hydrogen atoms, in the tree. Then with a total of $n + k$ vertices, where each of the n carbon atoms has degree 4, we find that

$$4n + k = \sum_{v \in V} \deg(v) = 2|E| = 2(|V| - 1) = 2(n + k - 1),$$

and

$$4n + k = 2(n + k - 1) \Rightarrow k = 2n + 2. \quad \square$$

We close this section with a theorem that provides several different ways to characterize trees.

---

† The adjective "saturated" is used here to indicate that for the number of carbon atoms present in the molecule, we have the maximum number of hydrogen atoms.

THEOREM 12.5   The following statements are equivalent for a loop-free undirected graph $G = (V, E)$.

a) G is a tree.

b) G is connected but the removal of any edge from G disconnects G into two subgraphs that are trees.

c) G contains no cycles and $|V| = |E| + 1$.

d) G is connected and $|V| = |E| + 1$.

e) G contains no cycles, and if $a, b \in V$ with $\{a, b\} \notin E$, then the graph obtained by adding edge $\{a, b\}$ to G has precisely one cycle.

Proof   We shall prove that (a) $\Rightarrow$ (b), (b) $\Rightarrow$ (c), and (c) $\Rightarrow$ (d), leaving to the reader the proofs for (d) $\Rightarrow$ (e) and (e) $\Rightarrow$ (a).

[(a) $\Rightarrow$ (b)]: If G is a tree, then G is connected. So let $e = \{a, b\}$ be any edge of G. Then if $G - e$ is connected, there are at least two paths in G from a to b. But this contradicts Theorem 12.1. Hence $G - e$ is disconnected and so the vertices in $G - e$ may be partitioned into two subsets: (1) vertex a and those vertices that can be reached from a by a path in $G - e$, and (2) vertex b and those vertices that can be reached from b by a path in $G - e$. These two connected components are trees, because if there were a loop or cycle present in either component, then that same loop or cycle would also be in G.

[(b) $\Rightarrow$ (c)]: If G contains a cycle, then let $e = \{a, b\}$ be an edge of the cycle. But then $G - e$ is connected, contradicting the hypothesis in (b).

To prove that $|V| = |E| + 1$, we shall apply the alternative form of mathematical induction on $|V|$. For $|V| = 1$, we have $|E| = 0$ (because our graph is loop-free) and the result is true. Now we assume the result true for all graphs satisfying condition (b) when $1 \le |V| \le k$. If $G = (V, E)$ satisfies condition (b) and $|V| = k + 1$, we remove an edge from G and consider the resulting connected subgraphs $G_1 = (V_1, E_1)$ and $G_2 = (V_2, E_2)$. These subgraphs are trees, and we know that (a) $\Rightarrow$ (b), so it follows that these subgraphs satisfy condition (b). Therefore, by the induction hypothesis, $|V_i| = |E_i| + 1$, $i = 1, 2$. Consequently, $|V| = |V_1| + |V_2| = (|E_1| + 1) + (|E_2| + 1) = (|E_1| + |E_2| + 1) + 1 = |E| + 1$.

[(c) $\Rightarrow$ (d)]: If $G = (V, E)$ is disconnected, let $\kappa(G) = r$ and let $C_1, C_2, \ldots, C_r$ denote the components of G. For $1 \le i \le r$, select a vertex $v_i \in C_i$ and add the $r - 1$ edges $\{v_1, v_2\}, \{v_2, v_3\}, \ldots, \{v_{r-1}, v_r\}$ to G to form the graph $G' = (V, E')$, which is a tree. Since (a) $\Rightarrow$ (c), G' being a tree implies that $|V| = |E'| + 1$. (This also follows from Theorem 12.3.) But from (c), $|V| = |E| + 1$, so $|E| = |E'|$ and $r - 1 = 0$. With $r = 1$, it follows that G is connected. ∎

EXERCISES 12.1

1. a) Draw the graphs of all nonisomorphic trees on six vertices.

   b) How many isomers does hexane ($C_6 H_{14}$) have?

2. Let $T_1 = (V_1, E_1)$, $T_2 = (V_2, E_2)$ be two trees where $|E_1| = 17$ and $|V_2| = 2|V_1|$. Determine $|V_1|$, $|V_2|$, and $|E_2|$.

3. **a)** If $G = (V, E)$ is a forest with $|V| = v$, $|E| = e$, and κ components (trees), what relationship exists among v, e, and κ?

   b) What is the smallest number of edges we must add to G in order to get a tree?

4. **a)** Verify that all trees are planar.

   b) Derive Theorem 12.3 from part (a) and Euler's theorem.

5. Give an example of an undirected graph $G = (V, E)$ where $|V| = |E| + 1$ but G is not a tree.

6. **a)** If a tree has four vertices of degree 2, one vertex of degree 3, two of degree 4, and one of degree 5, how many pendant vertices does it have?

   b) If a tree $T = (V, E)$ has v_2 vertices of degree 2, v_3 vertices of degree 3, . . . , and v_m vertices of degree m, what are $|V|$ and $|E|$?

7. If $G = (V, E)$ is a loop-free undirected graph, prove that G is a tree if there is a unique path between any two vertices of G.

8. The connected undirected graph $G = (V, E)$ has 30 edges. What is the maximum value that $|V|$ can have?

9. Let $T = (V, E)$ be a tree with $|V| = n \geq 2$. How many distinct paths are there (as subgraphs) in T?

10. Let $G = (V, E)$ be a loop-free connected undirected graph where $V = \{v_1, v_2, v_3, \dots, v_n\}$, $n \geq 2$, $\deg(v_1) = 1$, and $\deg(v_i) \geq 2$ for $2 \leq i \leq n$. Prove that G must have a cycle.

11. How many nonidentical (though isomorphic) spanning trees does the graph C_n (cycle on n vertices, $n \geq 3$) have?

12. Determine the number of nonidentical (though some may perhaps be isomorphic) spanning trees that exist for each of the graphs shown in Fig. 12.6.

13. Let $G = (V, E)$ be a loop-free connected undirected graph. Let H be a subgraph of G. The *complement of H in G* is the subgraph of G made up of those edges in G that are not in H (along with the vertices incident to these edges).

    a) If T is a spanning tree of G, prove that the complement of T in G does not contain a cut-set in G.

    b) If C is a cut-set in G, prove that the complement of C in G does not contain a spanning tree of G.

14. Complete the proof of Theorem 12.5.

15. A labeled tree is one wherein the vertices are labeled. If the tree has n vertices, $\{1, 2, 3, \dots, n\}$ is used as the set of labels. We find that two trees that are isomorphic without labels may become nonisomorphic when labeled. In Fig. 12.7 the first two trees are isomorphic as labeled trees. The third tree is isomorphic to the other two if we ignore the labels; as a labeled tree, however, it is not isomorphic to either of the other two.

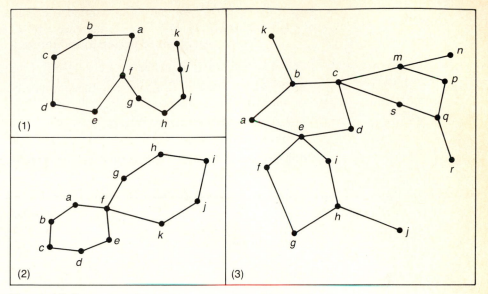

Figure 12.6

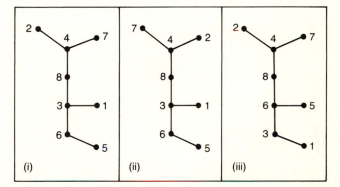

Figure 12.7

The number of nonisomorphic trees with n labeled vertices can be counted by setting up a one-to-one correspondence between these trees and the n^{n-2} sequences (with repetitions allowed) $x_1, x_2, \ldots, x_{n-2}$ from $\{1, 2, 3, \ldots, n\}$. If T is one such labeled tree, we use the following algorithm to establish the one-to-one correspondence. (Here T has at least one edge.)

Step 1: Set the counter i to 1.

Step 2: Set $T(i) = T$.

Step 3: Since a tree has at least two pendant vertices, select the pendant vertex in $T(i)$ with the smallest label y_i. Now remove the edge $\{x_i, y_i\}$ from $T(i)$ and use x_i for the ith component of the sequence.

Step 4: If $i = n - 2$, we have the sequence corresponding to the given labeled tree $T(1)$. If $i \neq n - 2$, increase i by 1, set $T(i)$ equal to the resulting subtree obtained in step 3, and return to step 3.

a) Find the six-digit sequence for trees (i) and (iii) in Fig. 12.7.

b) If v is a vertex in T, show that the number of times the label on v appears in the sequence $x_1, x_2, \ldots, x_{n-2}$ is $\deg(v) - 1$.

c) Reconstruct the labeled tree on eight vertices that is associated with the sequence $2, 6, 5, 5, 5, 5$.

d) Develop an algorithm for reconstructing a tree from a given sequence $x_1, x_2, \ldots, x_{n-2}$.

12.2
ROOTED TREES

We turn now to directed trees. We find a variety of applications for a special type of directed tree called a rooted tree.

DEFINITION 12.2   If G is a directed graph, then G is called a *directed tree* if the undirected graph associated with G is a tree. When G is a directed tree, G is called a *rooted tree* if there is a unique vertex r, called the *root*, in G with the in degree of $r = \deg^-(r) = 0$, and for all other vertices v, $\deg^-(v)$, the in degree of v, is 1.   ▬

The tree in part (a) of Fig. 12.8 is directed but not rooted; the tree in part (b) is rooted with root r.

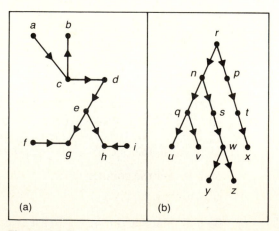

(a)   (b)

Figure 12.8

We draw rooted trees as in Fig. 12.8(b) but with the directions understood as going from the upper level to the lower level, so that the arrows aren't needed.

In a rooted tree, a vertex v with the out degree of $v = \deg^+(v) = 0$ is called a *leaf* (or *terminal vertex*). Vertices u, v, x, y, z are leaves in Fig. 12.8(b). All other vertices are called *branch nodes* (or *internal vertices*).

Consider the vertex s in this rooted tree [Fig. 12.8(b)]. The path from the root, r, to s is of length 2, so we say that s is at *level* 2 in the tree, or that s has *level number* 2. Similarly, x is at level 3, whereas y has level number 4. We call s a *child* of n, and we call n the *parent* of s. Vertices w, y, and z are considered *descendants* of s, n, and r; s, n, and r are called *ancestors* of w, y, and z. In general, if v_1 and v_2 are vertices in a rooted tree and v_1 has the smaller level number, then v_1 is an ancestor of v_2 (or v_2 is a descendant of v_1) if there is a path from v_1 to v_2. Two vertices with a common parent are referred to as *siblings*. Such is the case for vertices q and s, whose parent is vertex n. Finally, if v_1 is any vertex of the tree, the *subtree at* v_1 is the subgraph induced by the root v_1 and all of its descendants (there may be none).

Example 12.3   In Fig. 12.9(a) a rooted tree is used to represent the table of contents of a three-chapter $(C1, C2, C3)$ book. Vertices with level number 2 are for sections within a chapter; those at level 3 represent subsections within a section.   □

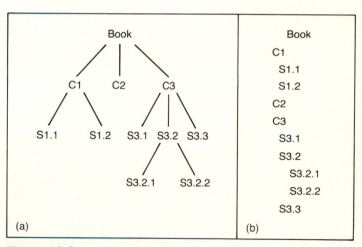

Figure 12.9

The tree in Fig. 12.9(a) suggests an order for the vertices if we examine the subtrees at $C1$, $C2$, and $C3$ from left to right. (This order will recur again in this section, in a more general context.) We now consider a second example that provides such an order.

Example 12.4   In the tree T shown in Fig. 12.10, the edges (branches) leaving each internal vertex are *ordered* from left to right. Hence T is called an *ordered rooted tree*.

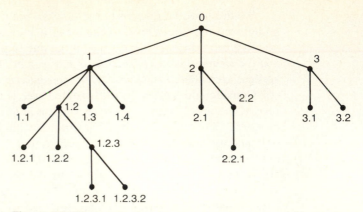

Figure 12.10

We label the vertices for this tree by the following algorithm.

Step 1: First assign the root the label (or *address*) 0.

Step 2: Next assign the positive integers $1, 2, 3, \ldots$ to the vertices at level 1, going from left to right.

Step 3: Now let v be an internal vertex at level $n \geq 1$, and let $v_1, v_2, \ldots, v_k$ denote the children of v (going from left to right). If a is the label assigned to vertex v, assign the labels $a.1, a.2, \ldots, a.k$ to the children $v_1, v_2, \ldots, v_k$, respectively.

Consequently, each vertex in T, other than the root, has a label of the form $a_1.a_2.a_3.\ldots.a_n$ if and only if that vertex has level number n. This is known as the *universal address system*.

This system provides a way to *order* all vertices in T. If u and v are two vertices in T with addresses b and c, respectively, we define $b < c$ if (a) $b = a_1.a_2.\ldots.a_m$, $c = a_1.a_2.\ldots.a_m.a_{m+1}.\ldots.a_n$, with $m < n$; or, (b) $b = a_1.a_2.\ldots.a_m.x_1.\ldots.y$, $c = a_1.a_2.\ldots.a_m.x_2.\ldots.z$, where $x_1, x_2 \in \mathbf{Z}^+$ and $x_1 < x_2$.

For the tree under consideration, this ordering yields

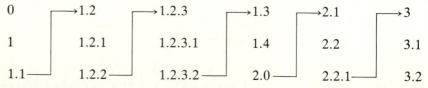

Since this resembles the alphabetical ordering in a dictionary, the order is called the *lexicographic*, or *dictionary*, *order*.   □

We consider one further application of a rooted tree in the study of computer algorithms.

Example 12.5   A rooted tree is called a *binary* rooted tree if for each vertex v, $\deg^+(v) = 0, 1$, or 2—that is, if v has at most two children. If $\deg^+(v) = 0$ or 2 for all $v \in V$, then the rooted tree is called a *complete* binary tree. Such a tree can represent a binary operation, as in parts (a) and (b) of Fig. 12.11. To avoid confusion when dealing with a noncommutative operation $\circ$, we label the root as $\circ$ and require the result to be $a \circ b$, where a is the left child, and b the right child of the root.

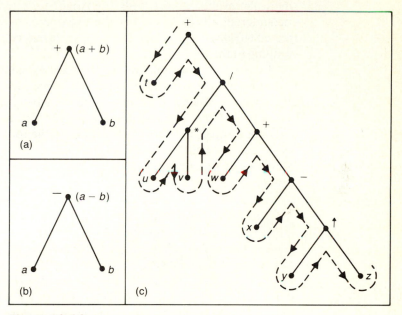

Figure 12.11

In evaluating $t + (uv)/(w + x - y^2)$ in a procedural language such as FORTRAN, we write the expression in the form $t + (u * v)/(w + x - y \uparrow z)$, where $*$ denotes multiplication and $\uparrow$ exponentiation. When the computer evaluates this expression, it performs the binary operations (within each parenthesized part) according to a hierarchy of operations whereby exponentiation precedes multiplication and division, which in turn precede addition and subtraction. In Fig. 12.12 we number the operations in the order in which they are performed by the computer. In order for the computer to evaluate this expression, it must somehow scan the expression to perform the operations in the order specified.

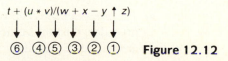

Figure 12.12

Instead of scanning back and forth continuously, however, the machine converts the expression into a notation that is independent of parentheses. This is known as Polish notation, in honor of the Polish (actually Ukrainian) logician Jan Lukasiewicz (1878–1956). Here the *infix* notation $a \circ b$ for a binary operation $\circ$ becomes $\circ ab$, the *prefix* (or Polish) notation. The advantage is that the expression in Fig. 12.12 can be rewritten without parentheses as

$$+ t / * uv + w - x \uparrow yz,$$

where the evaluation proceeds from right to left. When a binary operation is encountered, it is performed on the two operands to its right. The result is then treated as one of the operands for the next binary operation encountered as we continue to the left.

The use of Polish notation is important for compilation and can be attained by representing a given expression by a rooted tree, as shown in Fig. 12.11(c). Here each variable (or constant) is used to label a leaf of the tree. Each internal vertex is labeled by a binary operation whose left and right operands are the left and right subtrees it determines. Starting at the root, as we traverse the tree from top to bottom and left to right, as shown in Fig. 12.11(c), we find the Polish notation by writing down the labels of the vertices in the order in which they are visited.   □

These last two examples illustrate the importance of order. Several methods exist for systematically ordering the vertices in a tree. Two of the most prevalent in the study of data structures are the preorder and postorder. These are defined recursively in the following definition.

DEFINITION 12.3   Let $T = (V, E)$ be a rooted tree with root r. If T has no other vertices, then the root by itself constitutes the *preorder* and *postorder traversals* of T. If $|V| > 1$, let $T_1, T_2, T_3, \ldots, T_k$ denote the subtrees of T as we go from left to right (as in Fig. 12.13).

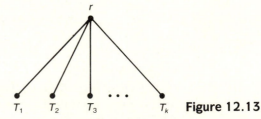

Figure 12.13

a) The *preorder traversal* of T first visits r and then visits the vertices of T_1 in preorder, then the vertices in T_2 in preorder, and so on until the vertices in T_k are visited in preorder.

b) The *postorder traversal* of T visits in postorder the vertices of the subtrees $T_1, T_2, \ldots, T_k$ and then visits the root.

We demonstrate these ideas in the following example.

Example 12.6   Consider the rooted tree shown in Fig. 12.14.

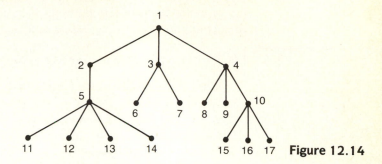

Figure 12.14

a) *Preorder:* After visiting vertex 1 we visit the subtree T_1 rooted at vertex 2. After visiting vertex 2 we proceed to the subtree rooted at vertex 5, and after visiting vertex 5 we go to the subtree rooted at vertex 11. This subtree has no other vertices, so we visit vertex 11 and then return to vertex 5 from which we visit, in succession, vertices 12, 13, and 14. Following this we *backtrack* (14 to 5 to 2 to 1) to the root and then visit the vertices in the subtree T_2 in the preorder 3, 6, 7. Finally, after returning to the root for the last time, we traverse the subtree T_3 in the preorder 4, 8, 9, 10, 15, 16, 17. Hence the preorder listing of the vertices in this tree is 1, 2, 5, 11, 12, 13, 14, 3, 6, 7, 4, 8, 9, 10, 15, 16, 17.

In this ordering we start at the root and build a path as far as we can. At each level we go to the leftmost vertex (not previously visited) at the next level, until we reach a leaf ℓ. Then we backtrack to the parent p of this leaf ℓ and visit its sibling s (and the subtree that s determines) directly on its right. If no such sibling s exists, we backtrack to the grandparent g of the leaf ℓ and visit, if it exists, a vertex u that is the sibling of p directly to its right in the tree. Eventually we generate a set of paths from the root to each of the leaves (going from left to right) in the tree.

The vertices in Figs. 12.9(a), 12.10, and 12.11(c) are visited in preorder.

b) *Postorder:* For the postorder traversal of a tree, we start at the root r and build the longest path, going to the leftmost child at each internal vertex arrived at. When we arrive at a leaf ℓ we visit this vertex and then backtrack to its parent p. However, we do not visit p until after all of its descendants are visited. The next vertex we visit is found by applying the same procedure at p that was originally applied at r in obtaining ℓ. And at no time is any vertex visited more than once or before any of its descendants.

For the given tree, the postorder traversal starts with a postorder traversal of the subtree T_1 rooted at vertex 2. This yields the listing 11, 12, 13, 14, 5, 2. We proceed to the subtree T_2, and the postorder listing continues with 6, 7, 3. Then for T_3 we find 8, 9, 15, 16, 17, 10, 4 as the postorder listing. Finally, vertex 1 is visited. Consequently, for this tree, the postorder traversal visits the vertices in the order 11, 12, 13, 14, 5, 2, 6, 7, 3, 8, 9, 15, 16, 17, 10, 4, 1.   ☐

In the case of binary rooted trees, a third type of tree traversal called the *inorder traversal* is used. Here we do *not* consider subtrees as first and second, but rather in terms of left and right. The formal definition is recursive, as were the definitions of preorder and postorder traversals.

DEFINITION 12.4   Let $T = (V, E)$ be a binary rooted tree with vertex r the root.

1. If $|V| = 1$, then the vertex r constitutes the *inorder traversal* of T.
2. When $|V| > 1$, let T_L and T_R denote the left and right subtrees of T. The *inorder traversal* of T first visits the vertices of T_L in inorder, then it visits the root r, and then it visits, in inorder, the vertices of T_R.

We realize that here a left or right subtree may be empty. Also, if v is a vertex in such a tree and $\deg^+(v) = 1$, then if w is the child of v, we must distinguish between w's being the left child and its being the right child.

Example 12.7   As a result of the previous comments, the two binary rooted trees shown in Fig. 12.15 are not considered the same. As rooted binary trees they are isomorphic. However, when we consider the additional concept of left and right children, we see that in part (a) of the figure vertex v has right child a, whereas in part (b) vertex a is the left child of v. Consequently, when the difference between left and right children is taken into consideration, these trees are no longer viewed as the same tree.

In visiting the vertices for the tree in part (a) of Fig. 12.15, we first visit in inorder the left subtree of the root r. This subtree consists of the root v and its *right* child a. (Here the left child is *null*, or nonexistent.) Since v has no left subtree, we visit in inorder vertex v and then its right subtree, namely a. Having traversed the left subtree of r, we now visit vertex r and then traverse, in inorder, the vertices in the right subtree of r. This results in our visiting first vertex b (because b has no left subtree) and then vertex c. Hence the inorder listing for the tree shown in Fig. 12.15(a) is v, a, r, b, c.

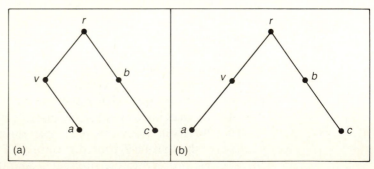

(a)　　　(b)

Figure 12.15

When we consider the tree in part (b) of the figure, once again we start by visiting, in inorder, the vertices in the left subtree of the root r. Here, however, this left subtree consists of vertex v (the root of the subtree) and its *left* child a. (In this case, the right child of v is null, or nonexistent.) Therefore this inorder traversal first visits vertex a (the left subtree of v), and then vertex v. Since v has no right subtree, we are now finished visiting the left subtree of r, in inorder. So next the root r is visited, and then the vertices of the right subtree of r are traversed, in inorder. This results in the inorder listing a, v, r, b, c for the tree shown in Fig. 12.15(b).

We should note, however, that for the preorder and postorder traversals *in this particular example*,† the same results are obtained for both trees.

- Preorder listing: r, v, a, b, c
- Postorder listing: a, v, c, b, r

It is only for the inorder traversal, with its distinctions between left and right children and between left and right subtrees, that a difference occurs. ☐

Example 12.8 If we apply the inorder traversal to the binary rooted tree shown in Fig. 12.16, we find that the inorder listing for the vertices is $p, j, q, f, c, k, g, a, d, r, b, h, s, m, e, i, t, n, u$. ☐

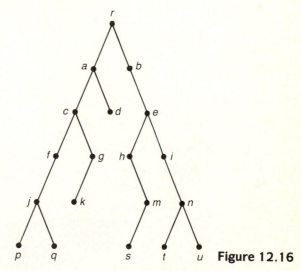

Figure 12.16

The following spanning tree for a connected graph is related to the notion of preorder.

Let $G = (V, E)$ be a connected undirected graph with $r \in V$. Starting from r we construct a path in G. If this path includes every vertex in V, then the path is a spanning tree T for G and we are finished. If not, let x and y be the last two vertices visited along this path, with y the last vertex. We then return, or *backtrack*, to the vertex x and construct a second path in G that starts at x and doesn't

---

† A note of caution! If we interchange the order of the two existing children (of a certain parent) in a binary tree, then a change results in the preorder, postorder, and inorder traversals. If one child is "null," however, then only the inorder traversal changes.

include any vertex already visited. If no such path exists, backtrack to the parent p of x and see how far it is possible to branch off from p, building a path (with no previously visited vertices) to a new leaf y_1. Should all edges from the vertex p lead to vertices already encountered, backtrack one level higher and continue the process. Since the graph is finite and connected, this technique, which is called *backtracking*, or *depth-first search*, eventually determines a spanning tree T for G, where r is regarded as the root of T. Using T, we then order the vertices of G in a preorder listing.

The depth-first search serves as a framework around which many algorithms can be designed to test for certain graph properties. One such algorithm will be examined in detail in Section 12.5.

In order to implement the depth-first search in a computer program, we must assign a fixed order to the vertices of the given graph $G = (V, E)$. Then if there are two or more vertices adjacent to a vertex v, we shall know exactly which vertex to visit first. This order now helps us to develop the foregoing description of the depth-first search as an algorithm.

Let $G = (V, E)$ be a loop-free connected undirected graph where $|V| = n$ and the vertices are ordered as $v_1, v_2, v_3, \ldots, v_n$. To find the rooted ordered depth-first spanning tree for the prescribed order, we apply the following algorithm, wherein the variable v is used to store the vertex presently being examined.

Depth-First Search Algorithm

Step 1: Assign v_1 to the variable v and initialize T as the tree consisting of just this one vertex. (The vertex v_1 will be the root of the spanning tree that develops.)

Step 2: Select the smallest subscript i, for $2 \leq i \leq n$, such that $\{v, v_i\} \in E$ and v_i has not already been visited.

If no such subscript is found, then go to step 3. Otherwise, perform the following: (1) Attach the edge $\{v, v_i\}$ to the tree T; (2) Assign v_i to v; and (3) Return to step 2.

Step 3: If $v = v_1$, the tree T is the (rooted ordered) spanning tree for the order specified.

Step 4: For $v \neq v_1$, backtrack from v. If u is the parent of the vertex assigned to v in T, then assign u to v and return to step 2.

Example 12.9   We now apply this algorithm to the graph $G = (V, E)$ shown in Fig. 12.17(a). Here the order for the vertices is alphabetic: $a, b, c, d, e, f, g, h, i, j$.

First we assign the vertex a to the variable v and initialize T as just the vertex a (the root). Going to step 2, we find that the vertex b is the first vertex such that $\{a, b\} \in E$ and b has not been visited earlier. So we attach edge $\{a, b\}$ to T, assign b to v, and return to step 2.

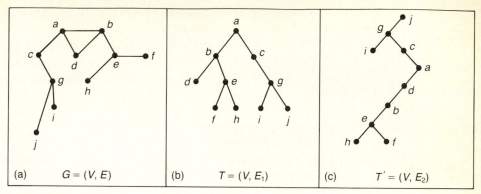

(a) $G = (V, E)$ (b) $T = (V, E_1)$ (c) $T' = (V, E_2)$

Figure 12.17

At $v = b$ we find that the first vertex (not visited earlier) that provides an edge for the spanning tree is d. Consequently the edge $\{b, d\}$ is attached to T, d is assigned to v, and we again return to step 2.

This time, however, there is no new vertex that we can obtain from d, because vertices a and b have already been visited. So we go to step 3. But here the value of v is d, not a, and we go to step 4. Now we backtrack from d, assigning the vertex b to v, and then we return to step 2. At this time we see that the edge $\{b, e\}$ can be added to T.

Continuing the process, we attach the edges $\{e, f\}$ and $\{e, h\}$ next. But now the vertex h has been assigned to v, and we must backtrack from h to e to b to a. When v is assigned the vertex a this (second) time, the new edge $\{a, c\}$ is obtained. Then we proceed to attach the edges $\{c, g\}$, $\{g, i\}$, and $\{g, j\}$. At this point all of the vertices in G have been visited, and we backtrack from j to g to c to a. With $v = a$ once again we return to step 2 and from there to step 3, where the process terminates.

The resulting tree $T = (V, E_1)$ is shown in part (b) of Fig. 12.17. Part (c) of the figure shows the tree T' that results for the vertex ordering: $j, i, h, g, f, e, d, c,$ b, a. $\square$

A second method for searching the vertices of a loop-free connected undirected graph is the *breadth-first search*. Here we designate one vertex as the root and fan out to all vertices adjacent to the root. From each child of the root we then fan out to those vertices (not previously visited) that are adjacent to one of these children. As we continue this process, we never list a vertex twice, so no cycle is constructed, and with G finite the process eventually terminates.

We actually used this technique earlier in Example 11.25 of Section 11.5.

A certain data structure proves useful in developing an algorithm for this second searching method. A *queue* is an ordered list wherein items are inserted at one end (called the *rear*) of the list and deleted at the other end (called the *front*). The *first* item inserted *in* the queue is the *first* item that can be taken *out* of it. Consequently, a queue is referred to as a "first-in, first-out" or FIFO structure.

As in the depth-first search, we need to assign an order to the vertices of our graph.

We start with a loop-free connected undirected graph $G = (V, E)$, where $|V| = n$ and the vertices are ordered as $v_1, v_2, v_3, \ldots, v_n$. The following algorithm generates the (rooted ordered) breadth-first spanning tree T of G for the given order.

Breadth-First Search Algorithm

Step 1: Insert vertex v_1 in the queue Q and initialize T as the tree made up of this one vertex v_1 (the root of the final version of T).

Step 2: Delete in turn vertices from the front of Q. As each vertex v is deleted, consider $v_i, 2 \le i \le n$. If the edge $\{v, v_i\} \in E$ and v_i has not been visited previously, attach the edge to T. If we examine *all* of the vertices previously in Q and obtain no new edges, the tree T (generated to this point) is the (rooted ordered) spanning tree for the given order.

Step 3: Insert the vertices adjacent to each v (from step 2) at the rear of the queue Q, according to the order in which they are (first) visited. Then return to step 2.

Example 12.10   We shall employ the graph of Fig. 12.17(a) with the prescribed order $a, b, c, d, e, f, g, h, i, j$ to illustrate the use of the algorithm for the breadth-first search.

Start with vertex a. Insert a in Q and initialize T as this one vertex (the root of the resulting tree).

In step 2 we delete a from Q and visit the vertices adjacent to it: b, c, d. (These vertices have not been previously visited.) This results in our attaching to T the edges $\{a, b\}, \{a, c\}$, and $\{a, d\}$.

Continuing to step 3, we insert b, c, d (in this order) in Q and then return to step 2. Now each of these vertices is deleted from Q, and the vertices adjacent to them (not previously visited) are visited according to the order given for the vertices of G. This results in the two new vertices e, g, and in the edges $\{b, e\}, \{c, g\}$ which are attached to T.

Next we go to step 3 and insert e, g in Q. Returning to step 2, we delete each of these vertices from Q and find, in order, the (previously unvisited) new vertices f, h and i, j. This results in our attaching the edges $\{e, f\}, \{e, h\}$ and $\{g, i\}, \{g, j\}$ to the tree T.

Once again we return to step 3 where the vertices f, h, i, j are inserted in Q. But now when we go to step 2 and delete the vertices from Q, we do not find any new vertices (that have not been visited). Consequently, the tree T in Fig. 12.18(a) is the breadth-first search spanning tree for G, for the order prescribed. (The tree T_1 shown in part (b) of the figure arises for the order $j, i, h, g, f, e, d, c, b, a$.) □

Let us apply these ideas on graph searching to one more example.

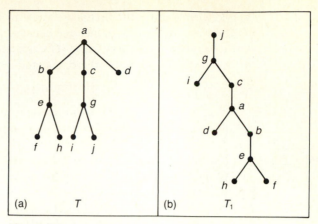

Figure 12.18

Example 12.11   Let $G = (V, E)$ be an undirected graph where the vertices are ordered as $v_1, v_2, \ldots, v_7$. If Fig. 12.19(a) is the adjacency matrix $A(G)$ for G, how can we use this representation of G to determine whether G is connected, without drawing the graph?

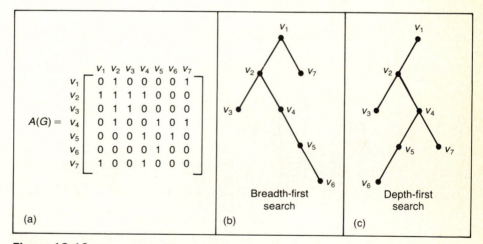

Figure 12.19

Using v_1 as the root, in part (b) of the figure we search the graph by means of its adjacency matrix, using a breadth-first search. First we visit the vertices adjacent to v_1, listing them in ascending order according to the subscripts on the v's in $A(G)$. The search continues, and as all vertices in G are reached, G is shown to be connected.

The same conclusion follows from the depth-first search in part (c). The tree here also has v_1 as its root. As the tree branches out to search the graph, it does so by listing the first vertex found adjacent to v_1 according to the row in $A(G)$ for

v_1. Likewise, from v_2 the first new vertex in this search is found from $A(G)$ to be v_3. The vertex v_3 is a leaf in this tree because no new vertex can be visited from v_3. As we backtrack to v_2, row 2 of $A(G)$ indicates that v_4 can now be visited from v_2. As this process continues, the connectedness of G follows from part (c) of the figure.   ☐

It is time now to return to our main discussion on rooted trees. The following definition generalizes the ideas that were introduced for Example 12.5.

DEFINITION 12.5   Let $T = (V, E)$ be a rooted tree and let $m \in \mathbf{Z}^+$.
   We call T an *m-ary tree* if $\deg^+(v) \leq m$ for all $v \in V$. When $m = 2$, the tree is called a *binary tree*.
   If $\deg^+(v) = 0$ or m for all $v \in V$, then T is called a *complete m-ary tree*. The special case of $m = 2$ results in a *complete binary tree*.   ▬

In a complete m-ary tree, each internal vertex has exactly m children. (Each leaf of this tree still has no children.)
   Some properties of these trees are considered in the following theorem.

THEOREM 12.6   Let $T = (V, E)$ be a complete m-ary tree with $|V| = n$. If T has ℓ leaves and i internal vertices, then (a) $n = mi + 1$; (b) $\ell = (m - 1)i + 1$; and (c) $i = (\ell - 1)/(m - 1) = (n - 1)/m$.

Proof   This proof is left for the section exercises.   ■

Example 12.12   The Wimbledon tennis championship is a single-elimination tournament wherein a player (or doubles team) is eliminated after a single loss. If 27 women compete in the singles championship, how many matches must be played to determine the number-one female player?
   Consider the tree shown in Fig. 12.20. With 27 women competing, there are 27 leaves in this complete binary tree, so from Theorem 12.6(c) the number of internal vertices (which is the number of matches) is $i = (\ell - 1)/(m - 1) = (27 - 1)/(2 - 1) = 26$.   ☐

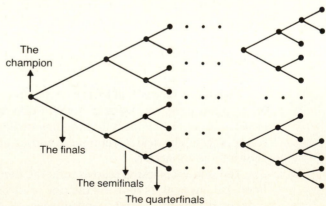

Figure 12.20

Example 12.13   A classroom contains 25 microcomputers that must be connected to a wall socket that has four outlets. Connections are made by using extension cords that have four outlets each. What is the least number of cords needed to get these computers set up for class use?

  The wall socket is considered the root of a complete m-ary tree for $m = 4$. The microcomputers are the leaves of this tree, so $\ell = 25$. Each internal vertex, except the root, corresponds with an extension cord. So by part (c) of Theorem 12.6, there are $(\ell - 1)/(m - 1) = (25 - 1)/(4 - 1) = 8$ internal vertices. Hence we need $8 - 1$ (where the 1 is subtracted for the root) $= 7$ extension cords. $\square$

DEFINITION 12.6   If $T = (V, E)$ is a rooted tree and h is the largest level number achieved by a leaf of T, then T is said to have *height h*. A rooted tree T of height h is said to be *balanced* if the level number of every leaf in T is $h - 1$ or h.   ▬

  The rooted tree shown in Fig. 12.14 is a balanced tree of height 3. Tree T' in Fig. 12.17(c) has height 7 but is not balanced. (Why?)

  The tree for the tournament in Example 12.12 must be balanced in order for the tournament to be as fair as possible. If it is not balanced, some competitor will receive more than one bye (an opportunity to advance without playing a match).

  Before stating our next theorem, let us recall that for any $x \in \mathbf{R}$, $\lfloor x \rfloor$ denotes the greatest integer in x, whereas $\lceil x \rceil$, *the ceiling of x*, is defined by $\lceil x \rceil = \lfloor x \rfloor = x$, for $x \in \mathbf{Z}$; $\lceil x \rceil = \lfloor x \rfloor + 1$, $x \notin \mathbf{Z}$.

THEOREM 12.7   Let $T = (V, E)$ be a complete m-ary tree of height h. If T has ℓ leaves, then $\ell \leq m^h$, and $h \geq \lceil \log_m \ell \rceil$.

Proof   The proof that $\ell \leq m^h$ will be established by induction on h. When $h = 1$, T is a tree with a root and m children. In this case $\ell = m = m^h$, and the result is true. Assume the result for any tree of height $< h$, and consider a tree T with height h and ℓ leaves. (The level numbers that are possible for these leaves are $1, 2, \ldots, h$, with at least m of the leaves at level h.) The ℓ leaves of T are also the ℓ leaves (total) for the m subtrees T_i, $1 \leq i \leq m$, of T rooted at each of the children of the root. For $1 \leq i \leq m$, let ℓ_i be the number of leaves in subtree T_i. (In the case where leaf and root coincide, $\ell_i = 1$. But since $m \geq 1$ and $h - 1 \geq 0$, we have $m^{h-1} \geq 1 = \ell_i$.) By the induction hypothesis, $\ell_i \leq m^{h(T_i)} \leq m^{h-1}$, where $h(T_i)$ denotes the height of the subtree T_i, and so $\ell = \ell_1 + \ell_2 + \cdots + \ell_m \leq m(m^{h-1}) = m^h$.

  With $\ell \leq m^h$, $\log_m \ell \leq \log_m(m^h) = h$, and since $h \in \mathbf{Z}^+$, it follows that $h \geq \lceil \log_m \ell \rceil$.   ∎

COROLLARY 12.1   Let T be a complete m-ary tree with ℓ leaves and height h. If T is balanced, then $h = \lceil \log_m \ell \rceil$.

Proof   This proof is left as an exercise.   ∎

  We close this section with an application that uses a complete ternary $(m = 3)$ tree.

Example 12.14   (*Decision Trees*) Given eight coins (identical in appearance) and a pan balance, if exactly one of these coins is counterfeit and heavier than the other seven, determine the counterfeit coin.

Let the coins be labeled $1, 2, 3, \ldots, 8$. In using the pan balance there are three outcomes to consider: (a) the two sides balance to indicate that the coins in the two pans are not counterfeit; (b) the left pan of the balance goes down, indicating that the counterfeit coin is in the left pan; or (c) the right pan goes down, indicating that it holds the counterfeit coin.

In Fig. 12.21(a), we search for the counterfeit coin by first balancing coins $1, 2, 3, 4$ against $5, 6, 7, 8$. If the balance tips to the right, we follow the right branch from the root to then analyze coins $5, 6$ against $7, 8$. If the balance tips to the left, we test coins $1, 2$ against $3, 4$. At each successive level, we have half as many coins to test, so at level 3 the heavier counterfeit coin has been identified.

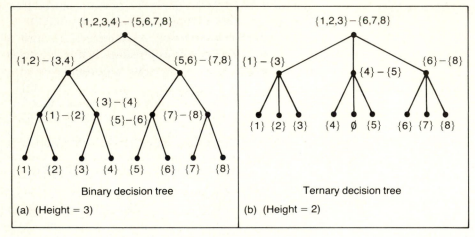

Binary decision tree

(a) (Height = 3)

Ternary decision tree

(b) (Height = 2)

Figure 12.21

The tree in part (b) of the figure finds the heavier coin in two weighings. The first weighing balances coins $1, 2, 3$ against $6, 7, 8$. Three possible outcomes can occur: (i) the balance tips to the right, indicating that the heavier coin is 6, 7, or 8, and we follow the right branch from the root; (ii) the balance tips to the left and we follow the left branch to find which of 1, 2, 3 is the heavier, or, (iii) the pans balance and we follow the center branch to find which of 4, 5 is heavier. At each internal vertex the label indicates which coins are being compared. Unlike part (a), a conclusion may be deduced in part (b) when a coin is not included in a weighing. Finally, when comparing coins 4 and 5, because equality cannot take place we label the center leaf with $\emptyset$.

In this particular problem, the height of the complete ternary tree used must be at least 2. With eight coins involved, the tree will have at least eight leaves. Consequently, with $\ell \geq 8$, it follows from Theorem 12.7 that $h \geq \lceil \log_3 \ell \rceil \geq \lceil \log_3 8 \rceil = 2$, so at least two weighings are needed. If n coins are involved, the complete ternary tree will have ℓ leaves where $\ell \geq n$, and its height h satisfies $h \geq \lceil \log_3 n \rceil$. $\square$

EXERCISES 12.2

1. Answer the following questions for the tree shown in Fig. 12.22.
   a) Which vertices are the leaves?
   b) Which vertex is the root?
   c) Which vertex is the parent of g?
   d) Which vertices are the descendants of c?
   e) Which vertices are the siblings of s?
   f) What is the level number of vertex f?
   g) Which vertices have level number 4?
   h) What is the height of the tree?
   i) Is the tree balanced?

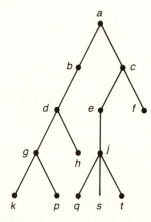

Figure 12.22

2. Let $T = (V, E)$ be a rooted tree ordered by a universal address system.
   a) If vertex v in T has address 2.1.3.6, what is the smallest number of siblings that v must have?
   b) For the vertex v in part (a), find the address of its parent.
   c) How many ancestors does the vertex v in part (a) have?
   d) With the presence of v in T, what other addresses must there be in the system?

3. a) Write the expression $(w + x - y)/(\pi * z^3)$ in Polish notation, using a rooted tree.
   b) What is the value of the expression (in Polish notation)
      $/ \uparrow a - bc + d * ef$, if $a = c = d = e = 2, b = f = 4$?

    c) Extend the structure of the rooted tree in Fig. 12.11(c) to include monary (or unary) operations such as the trigonometric functions and multiplicative inverse. Find the rooted trees for the following.

      (i) $(\sin \pi x) + (y - z)^{-1}$ **(ii)** $\sin(\pi x + y) - z^{-1}$

4. List the vertices in the tree shown in Fig. 12.23 when they are visited in a preorder traversal and in a postorder traversal.

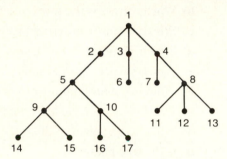

Figure 12.23

5. For the tree shown in Fig. 12.24, list the vertices according to a preorder traversal, an inorder traversal, and a postorder traversal.

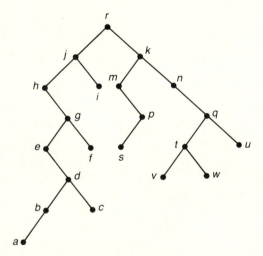

Figure 12.24

6. Determine whether each of the following statements is true or false. If the statement is false, provide a counterexample.

    a) $\lfloor a \rfloor = \lceil a \rceil$ for all $a \in \mathbf{Z}$.
    b) $\lfloor a \rfloor = \lceil a \rceil$ for all $a \in \mathbf{R}$.
    c) $\lfloor a \rfloor = \lceil a \rceil - 1$ for all $a \in \mathbf{R} - \mathbf{Z}$.
    d) $\lfloor a \rfloor \lceil b \rceil = \lceil a \rceil \lfloor b \rfloor$ for all $a, b \in \mathbf{R}$.
    e) $-\lceil a \rceil = \lceil -a \rceil$ for all $a \in \mathbf{R}$.
    f) $-\lceil a \rceil = \lfloor -a \rfloor$ for all $a \in \mathbf{R}$.
    g) $-\lfloor a \rfloor = \lceil -a \rceil$ for all $a \in \mathbf{R}$.

7. The complete binary tree $T = (V, E)$ has $V = \{a, b, c, \ldots, i, j, k\}$. The postorder listing of V yields $d, e, b, h, i, f, j, k, g, c, a$. From this information draw T if (a) the height of T is 3; (b) the height of the left subtree of T is 3.

8. **a)** Let $T = (V, E)$ be a rooted tree, and let preorder (v) denote the position of vertex v in a preorder listing of V. Also, let desc(v) denote the number of descendants of v. For $v_1, v_2 \in V$, prove that v_1 is a descendant of v_2 if and only if preorder$(v_2) <$ preorder$(v_1) \leq$ preorder(v_2) + desc(v_2).

   b) State and prove a comparable result, using postorder in place of preorder.

9. **a)** Find the depth-first spanning tree for the graph shown in Fig. 11.55(a) if the order of the vertices is given as (i) a, b, c, d, e, f, g, h; (ii) h, g, f, e, d, c, b, a; (iii) a, b, c, d, h, g, f, e.

   b) Repeat part (a) for the graph shown in Fig. 11.68(i).

10. Find the breadth-first spanning trees for the graphs and prescribed orders given in Exercise 9.

11. Let $G = (V, E)$ be an undirected graph with adjacency matrix $A(G)$ as shown here.

	v_1	v_2	v_3	v_4	v_5	v_6	v_7	v_8
v_1	0	1	0	0	0	0	1	0
v_2	1	1	0	1	1	0	1	0
v_3	0	0	0	1	0	1	0	1
v_4	0	1	1	0	0	0	0	0
v_5	0	1	0	0	0	0	1	0
v_6	0	0	1	0	0	1	0	0
v_7	1	1	0	0	1	0	0	0
v_8	0	0	1	0	0	0	0	0

Use a breadth-first search based on $A(G)$ to determine whether G is connected.

12. **a)** For the graph shown in Fig. 12.25, find the breadth-first spanning tree when the vertices are ordered as (i) a, b, c, d, e, f; (ii) a, b, e, c, d, f; (iii) a, c, d, b, e, f.

   b) How many different breadth-first spanning trees are rooted at vertex a?

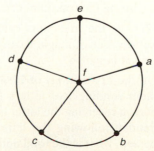

Figure 12.25

13. Prove Theorem 12.6 and Corollary 12.1.

14. With m, n, i, ℓ as in Theorem 12.6, prove that

    a) $n = (m\ell - 1)/(m - 1)$; and   b) $\ell = [(m - 1)n + 1]/m$.

15. For $m \geq 3$, a complete m-ary tree can be transformed into a complete binary tree as shown in Fig. 12.26.

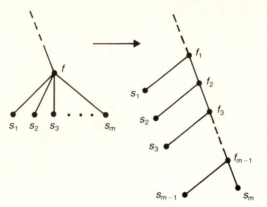

Figure 12.26

a) Use this technique to transform the complete ternary decision tree shown in Fig. 12.21(b).

b) If T is a complete quaternary tree of height 3, what is the height of T after it is transformed into a complete binary tree?

c) Answer part (b) if T is a complete m-ary tree of height h.

16. a) At a men's singles tennis tournament, each of 25 players brings a can of tennis balls. When a match is played, one can of balls is opened and used, then kept by the loser. The winner takes the unopened can on to his next match. How many cans of tennis balls will be opened during this tournament? How many matches are played in the tournament?

    b) In how many matches did the tournament champion play?

    c) If a match is won by the first opponent to win three sets, what is the maximum number of sets that could have been played (by all entrants) during the tournament?

17. What is the maximum number of internal vertices that a complete quaternary tree of height 8 can have? What is the number for a complete m-ary tree of height h?

18. On the first Sunday of 1984 Rizzo and Frenchie start a chain letter, each of them sending five letters (to ten different friends between them). Each person receiving the letter is to send out five copies to five new people on the Sunday following the letter's arrival. After the first seven Sundays have passed, what is the total number of chain letters that have been mailed? How many were mailed on the last three Sundays?

19. Use a complete ternary decision tree to repeat Example 12.14 for a set of 12 coins, where exactly one of the coins is a heavier counterfeit.

12.3
TREES AND SORTING

In Example 10.4 (Chapter 10), the bubble sort was introduced in the study of linear recurrence relations. There we found that the number of comparisons needed to sort a list of n items is $n(n-1)/2$. Consequently, this algorithm determines a function $f: \mathbf{Z}^+ \to \mathbf{R}$ defined by $f(n) = n(n-1)/2$. This is the (worst-case) time complexity function for the algorithm, and we often express this by writing $f \in O(n^2)$. Consequently, the bubble sort is said to require $O(n^2)$ comparisons. We interpret this to mean that for large n, the number of comparisons equals cn^2, where c is a proportionality constant that is generally not specified because it depends on such factors as the compiler and the computer that are used.

In this section we shall study a second method for sorting a given list of n items into ascending order. The method is called the *merge sort*, and we shall find that the order of its worst-case time complexity function is $O(n \log_2 n)$. This will be accomplished in the following manner:

1. First we shall measure the number of comparisons needed when n is a power of 2. Our method will employ a pair of balanced complete binary trees.

2. Then we shall cover the case for general n by using the optional material on divide-and-conquer algorithms in Section 10.6.

For the case where n is an arbitrary positive integer, we start by considering the following procedure.

Given a list of n items to sort into ascending order, the *merge sort* recursively splits the given list and all subsequent sublists in half (or as close as possible to half) until each sublist contains a single element. Then the procedure merges these sublists in ascending order until the original n items have been so sorted. The splitting and merging processes can best be described by a pair of balanced complete binary trees, as in the next example.

Example 12.15   (*Merge Sort*) Using the merge sort, Fig. 12.27 sorts the list $6, 2, 7, 3, 4, 9, 5, 1, 8$. The tree at the top of the figure shows how the process first splits the given list into sublists of size 1. The merging process is then outlined by the tree at the bottom of the figure.   □

To compare the merge sort to the bubble sort, we want to determine its complexity function. The following lemma will be needed for this task.

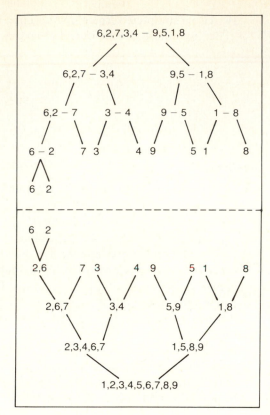

Figure 12.27

LEMMA 12.1   Let L_1 and L_2 be two sorted lists of ascending numbers, where L_i contains n_i elements, $i = 1, 2$. Then L_1 and L_2 can be merged into one ascending list L using at most $n_1 + n_2 - 1$ comparisons.

Proof   To merge L_1, L_2 into list L, we perform the following algorithm.

> **Step 1:** Set L equal to the empty list $\emptyset$.
>
> **Step 2:** Compare the first elements in L_1, L_2. Remove the smaller of the two from the list it is in and place it at the end of L.
>
> **Step 3:** For the *present* lists L_1, L_2 (one change is made in one of these lists each time step 2 is executed), there are two considerations.
>
> **a)** If either of L_1, L_2 is empty, then the other list is concatenated to the end of L. This completes the merging process.
>
> **b)** If not, return to step 2.

Each comparison of a number from L_1 with one from L_2 results in an element being added to list L, so there cannot be more than $n_1 + n_2$ comparisons. When one of the lists L_1 or L_2 becomes empty, no further comparisons are needed, so the maximum number of comparisons needed is $n_1 + n_2 - 1$. ∎

To determine the complexity function of the merge sort, consider a list of n elements. For the moment, we do not treat the general problem, assuming here that $n = 2^h$. † In the splitting process, the list of 2^h elements is first split into two sublists of size 2^{h-1}. (These are the level 1 vertices in the tree representing the splitting process.) As the process continues, each successive list of size 2^{h-k}, $h > k$, is at level k and splits into two sublists of size $(1/2)(2^{h-k}) = 2^{h-k-1}$. At level h the sublists each contain $2^{h-h} = 1$ element.

Reversing the process, we first merge the $n = 2^h$ leaves into 2^{h-1} ordered sublists of size 2. These sublists are at level $h - 1$ and require $(1/2)(2^h) = 2^{h-1}$ comparisons (one per pair). As this merging process continues, at each of the 2^k vertices at level k, $1 \leq k < h$, there is a sublist of size 2^{h-k}, obtained from merging the two sublists of size 2^{h-k-1} at its children (on level $k + 1$). From Lemma 12.1, this merging requires at most $2^{h-k-1} + 2^{h-k-1} - 1 = 2^{h-k} - 1$ comparisons. When the children of the root are reached, there are two sublists of size 2^{h-1} (at level 1). To merge these sublists into the final list requires at most $2^{h-1} + 2^{h-1} - 1 = 2^h - 1$ comparisons.

Consequently, for $1 \leq k \leq h$, at level k there are 2^{k-1} pairs of vertices. At each of these vertices is a sublist of size 2^{h-k}, so it takes at most $2^{h-k+1} - 1$ comparisons to merge each pair of sublists. With 2^{k-1} pairs of vertices at level k, the total number of comparisons at level k is at most $2^{k-1}(2^{h-k+1} - 1)$. When we sum over all levels k where $1 \leq k \leq h$, we find that the total number of comparisons is at most

$$\sum_{k=1}^{h} 2^{k-1}(2^{h-k+1} - 1) = \sum_{k=0}^{h-1} 2^k(2^{h-k} - 1) = \sum_{k=0}^{h-1} 2^h - \sum_{k=0}^{h-1} 2^k = h \cdot 2^h - (2^h - 1).$$

With $n = 2^h$, we have $h = \log_2 n$ and

$$h \cdot 2^h - (2^h - 1) = n \log_2 n - (n - 1) = n \log_2 n - n + 1,$$

where $n \log_2 n$ is the dominating term for large n. Thus the (worst-case) time complexity function for this sorting procedure is $g(n) = n \log_2 n - n + 1$ and $g \in O(n \log_2 n)$, for $n = 2^h$, $h \in \mathbf{Z}^+$. Hence the number of comparisons needed to merge sort a list of n items is $dn \log_2 n$ for some proportionality constant d, and for all $n \geq n_0$, where n_0 is some particular (large) positive integer.

To show that the order of the merge sort is $O(n \log_2 n)$ for all $n \in \mathbf{Z}^+$, our second approach will use the result of Exercise 9 from Section 10.6. We state that now:

Let $a, b, c \in \mathbf{Z}^+$, with $b \geq 2$. If $g: \mathbf{Z}^+ \to \mathbf{R}^+ \cup \{0\}$ is a monotone increasing function, where

$$g(1) \leq c,$$
$$g(n) \leq ag(n/b) + cn, \qquad \text{for } n = b^h, h \in \mathbf{Z}^+,$$

---

† The result obtained here for $n = 2^h$, $h \in \mathbf{N}$, is actually true for all $n \in \mathbf{Z}^+$. However, the derivation for general n requires the optional material in Section 10.6. That is why this counting argument is included here—for the benefit of those readers who did not cover Section 10.6.

then for the case where $a = b$, we have $g \in O(n \log n)$, for *all* $n \in \mathbf{Z}^+$. (The base for the log function may be any real number greater than 1. Here we shall use the base 2.)

Before we can apply this result to the merge sort, we wish to formulate this sorting process (illustrated in Fig. 12.27) as a precise algorithm. To do so, we call the procedure outlined in Lemma 12.1 the "merge" algorithm. Then we shall write "merge (L_1, L_2)" in order to represent the application of that procedure to the lists L_1, L_2, which are in ascending order.

The algorithm for merge sort is a recursive procedure, because it may invoke itself. Here the input is an array (called List) of n items, such as real numbers.

The MergeSort Algorithm:

Step 1: If $n = 1$, then List is already sorted and the process terminates. If $n > 1$, then go to step 2.

Step 2: (Divide the array and sort the subarrays.) Perform the following:

(1) Assign m the value $\lfloor n/2 \rfloor$.

(2) Assign to List1 the subarray

$$\text{List}[1], \text{List}[2], \ldots, \text{List}[m].$$

(3) Assign to List2 the subarray

$$\text{List}[m+1], \text{List}[m+2], \ldots, \text{List}[n].$$

(4) Apply MergeSort to List1 (and m) and to List2 (and $n - m$).

Step 3: Merge (List1, List2).

The function $g: \mathbf{Z}^+ \rightarrow \mathbf{R}^+ \cup \{0\}$ will measure the (worst-case) time complexity for this algorithm by counting the maximum number of comparisons needed to merge sort an array of n items. For $n = 2^h$, $h \in \mathbf{Z}^+$, we have

$$g(n) = 2g(n/2) + [(n/2) + (n/2) - 1].$$

The term $2g(n/2)$ results from step 2 of the MergeSort algorithm, and the summand $[(n/2) + (n/2) - 1]$ follows from step 3 of the algorithm and Lemma 12.1.

With $g(1) = 0$, the above equation provides the inequalities

$$g(1) = 0 \le 1,$$
$$g(n) = 2g(n/2) + (n - 1) \le 2g(n/2) + n, \qquad \text{for } n = 2^h, h \in \mathbf{Z}^+.$$

We also observe that $g(1) = 0$, $g(2) = 1$, $g(3) = 3$, and $g(4) = 5$, so $g(1) \le g(2) \le g(3) \le g(4)$. Consequently, it appears that g may be a monotone increasing function. The proof that it is monotone increasing is similar to that given for the time complexity function of the binary search. This follows Example 10.35 in

Section 10.6, so we leave the details showing that g is monotone increasing to the section exercises.

Now with $a = b = 2$ and $c = 1$, the result stated earlier implies that $g \in O(n \log_2 n)$ for *all* $n \in \mathbf{Z}^+$.

Although $n \log_2 n \leq n^2$ for all $n \in \mathbf{Z}^+$, it does *not* follow that because the bubble sort is $\mathbb{O}(n^2)$ and the merge sort is $\mathbb{O}(n \log_2 n)$, the merge sort is more efficient than the bubble sort for all $n \in \mathbf{Z}^+$. The bubble sort requires less programming effort and generally takes less time than the merge sort for small values of n (depending on factors such as the programming language, the compiler, and the computer). However, as n increases, the ratio of the running times, as measured by $(cn^2)/(dn \log_2 n) = (c/d)(n/\log_2 n)$, gets arbitrarily large. Consequently, as the input list increases in size, the $\mathbb{O}(n^2)$ algorithm (bubble sort) takes significantly more time than the $\mathbb{O}(n \log_2 n)$ algorithm (merge sort).

For more on sorting algorithms and their time complexity functions, the reader should examine [1], [2], [5], and [6] in the chapter references.

EXERCISES 12.3

1. a) Give an example of two lists L_1, L_2, each of which is in ascending order and contains five elements, and where nine comparisons are needed to merge L_1, L_2 by the algorithm given in Lemma 12.1.

   b) Generalize the result in part (a) to the case where L_1 has m elements, L_2 has n elements, and $m + n - 1$ comparisons are needed to merge the two lists.

2. Apply the merge sort to each of the following lists. Draw the splitting and merging trees for each application of the procedure.

   a) $-1, 0, 2, -2, 3, 6, -3, 5, 1, 4$

   b) $-1, 7, 4, 11, 5, -8, 15, -3, -2, 6, 10, 3$

3. Related to the merge sort is a somewhat more efficient procedure called the *quick sort*. Here we start with a list $L: a_1, a_2, \ldots, a_n$, and use a_1 as a pivot to develop two sublists L_1 and L_2 as follows. For $i > 1$, if $a_i < a_1$, place a_i at the end of the first list being developed (this is L_1 at the end of the process); otherwise, place a_i at the end of the second list L_2.

   After all $a_i, i > 1$, have been processed, place a_1 at the end of the first list. Now apply quick sort recursively to each of the lists L_1 and L_2 to obtain sublists L_{11}, L_{12}, L_{21}, and L_{22}. Continue the process until each of the resulting sublists contains one element. The sublists are then ordered, and their concatenation gives the ordering sought for the original list L.

   Apply quick sort to each list in Exercise 2.

4. Prove that the function g used in the second method to analyze the time complexity of the merge sort is monotone increasing.

12.4
WEIGHTED TREES AND PREFIX CODES

Among the many applications of discrete mathematics, the topic of coding theory is one area wherein different finite structures play a major role. These structures enable us to represent and transmit information that is coded in terms of the symbols in a given alphabet. For instance, the way we most often code, or represent, characters internally in a computer is by means of strings of fixed length, using the symbols 0 and 1.

The codes developed in this section, however, will use strings of different lengths. Why a person should want to develop such a coding scheme and how the scheme can be constructed will be our major concerns in this section.

Suppose we wish to develop a way to represent the letters of the alphabet using strings of 0's and 1's. Since there are 26 letters, we should be able to encode these symbols in terms of sequences of five bits, since $2^4 < 26 < 2^5$. However, in the English (or any other) language, not all letters occur with the same frequency. Consequently, it would be more efficient to use binary sequences of different lengths, with the most frequently occurring letters (such as e, i, t) represented by the shortest possible sequences. For example, consider $S = \{a, e, n, r, t\}$, a subset of the alphabet. Represent the elements of S by the binary sequences

$$a: 01 \qquad e: 0 \qquad n: 101 \qquad r: 10 \qquad t: 1.$$

If the message "*ata*" is to be transmitted, the binary sequence 01101 is sent. Unfortunately, this sequence is also transmitted for the messages "*etn*", "*atet*", and "*an*".

Consider a second encoding scheme, one given by

$$a: 111 \qquad e: 0 \qquad n: 1100 \qquad r: 1101 \qquad t: 10.$$

Here the message "*ata*" is represented by the sequence 11110111 and there are no other possibilities to confuse the situation. What's more, the labeled complete binary tree shown in Fig. 12.28 can be used to decode the sequence 11110111. Starting at the root, traverse the edge labeled 1 to the right child (of the root). Continuing along the next two edges labeled with 1, we arrive at the leaf labeled *a*. Hence the unique path from the root to the vertex at *a* is unambiguously determined by the first three 1's in the sequence 11110111. After we return to the root, the next two symbols in the sequence, namely 10, determine the unique path along the edge from the root to its right child, followed by the edge from that child to its left child. This terminates at the vertex labeled *t*. Again returning to the root, the final three bits of the sequence determine the letter *a* for a second time. Hence the tree "decodes" 11110111 as *ata*.

Why did the second encoding scheme work out so readily when the first led to ambiguities? In the first scheme, *r* is represented as 10 and *n* as 101. If we encounter the symbols 10, how can we determine whether the symbols represent *r* or the first two symbols of 101, which represent *n*? The problem is that the sequence for r is a prefix of the sequence for n. This dilemma does not occur in the second encoding scheme, which suggests the following definition.

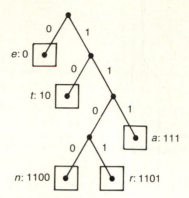

Figure 12.28

DEFINITION 12.7   A set P of binary sequences (representing a set of symbols) is called a *prefix code* if no sequence in P is the prefix of any other sequence in P.

Consequently, the binary sequences $111, 0, 1100, 1101, 10$ constitute a prefix code for the letters a, e, n, r, t, respectively. But how did the complete binary tree of Fig. 12.28 come about? To deal with this problem, we need the following concept.

DEFINITION 12.8   If T is a complete binary tree of height h, then T is called a *full* binary tree if all the leaves in T are at level h.

Example 12.16   For the prefix code $P = \{111, 0, 1100, 1101, 10\}$, the longest binary sequence has length 4. Draw the labeled full binary tree of height 4, as shown in Fig. 12.29. The elements of P are assigned to the vertices of this tree as follows. For example, the sequence 10 traces the path from the root r to its right child c_R. Then it continues to the left child of c_R where the box (marked with the asterisk) indicates completion of the sequence. Returning to the root, the other four sequences are traced out in similar fashion, resulting in the five boxed vertices. For each boxed vertex remove the subtree (except for the root) that it determines. The resulting pruned tree is the tree of Fig. 12.28 where no "box" is an ancestor of another "box."

□

We turn now to a method of determining a labeled tree, where the frequency of occurrence of each symbol is taken into account. Once again a prefix code is needed wherein short sequences are used for the more frequently occurring symbols. If there are many symbols, such as all 26 letters of the alphabet, a trial-and-error method for constructing such a tree is not efficient. An elegant construction developed by David A. Huffman provides a recursive technique for constructing such trees.

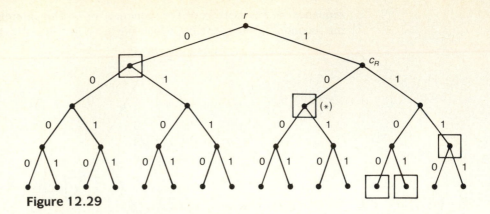

Figure 12.29

The general problem now arises as follows.

Let $w_1, w_2, \ldots, w_n$ be a set of positive numbers called *weights*, where $w_1 \le w_2 \le \cdots \le w_n$. If $T = (V, E)$ is a complete binary tree with n leaves, assign these weights (in any one-to-one manner) to the n leaves. The result is called a *complete binary tree for the weights* $w_1, w_2, \ldots, w_n$. The *weight of the tree*, denoted $W(T)$, is defined as $\sum_{i=1}^{n} w_i \ell(w_i)$ where, for $1 \le i \le n$, $\ell(w_i)$ is the level number of the leaf assigned the weight w_i. The objective is to assign the weights so that $W(T)$ is as small as possible. A complete binary tree T_1 for these weights is said to be an *optimal* tree if $W(T_1) \le W(T)$ for any other complete binary tree T for the weights.

Figure 12.30 shows two complete binary trees for the weights 3, 5, 6, and 9. For tree T_1, $W(T_1) = \sum_{i=1}^{4} w_i \ell(w_i) = (3 + 9 + 5 + 6) \cdot 2 = 46$, because each leaf has level number 2. In the case of T_2, $W(T_2) = 3 \cdot 3 + 5 \cdot 3 + 6 \cdot 2 + 9 \cdot 1 = 45$, which we shall find is optimal.

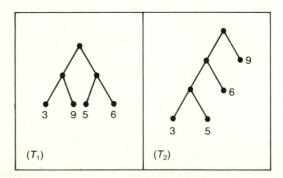

Figure 12.30

The major idea behind Huffman's construction is that in order to obtain an optimal tree T for the n weights $w_1, w_2, w_3, \ldots, w_n$, one considers an optimal tree T_1 for the $n - 1$ weights $w_1 + w_2, w_3, \ldots, w_n$. (It cannot be assumed that $w_1 + w_2 \le w_3$.) In particular, the tree T_1 is transformed into T by replacing the leaf with weight $w_1 + w_2$ by a tree of height 1 with left child of weight w_1 and right child of weight w_2. If the tree T_2 in Fig. 12.30 is optimal for the four weights

$1 + 2, 5, 6, 9$, then the tree in Fig. 12.31 will be optimal for the five weights 1, 2, 5, 6, 9.

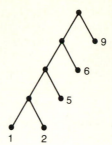

Figure 12.31

To establish these claims, we need the following lemma.

LEMMA 12.2   If T is an optimal tree for the n weights $w_1 \le w_2 \le \cdots \le w_n$, then there exists an optimal tree T' in which the leaves of weights w_1 and w_2 are siblings.

Proof   Let v be an internal vertex of T where the level number of v is maximal for all internal vertices. Let w_x and w_y be the weights assigned to the children x, y of vertex v, with $w_x \le w_y$. By the choice of vertex v, $\ell(w_x) = \ell(w_y) \ge \ell(w_1), \ell(w_2)$. Consider the case of $w_1 < w_x$. (If $w_1 = w_x$, then w_1 and w_x can be interchanged and we would consider the case of $w_2 < w_y$. Applying the following proof to this case, we would find that w_y and w_2 can be interchanged.)

If $\ell(w_x) > \ell(w_1)$, let $\ell(w_x) = \ell(w_1) + j$, for some $j \in \mathbf{Z}^+$. Then $w_1 \ell(w_1) + w_x \ell(w_x) = w_1 \ell(w_1) + w_x [\ell(w_1) + j] = w_1 \ell(w_1) + w_x j + w_x \ell(w_1) > w_1 \ell(w_1) + w_1 j + w_x \ell(w_1) = w_1 \ell(w_x) + w_x \ell(w_1)$. So $W(T) = w_1 \ell(w_1) + w_x \ell(w_x) + \sum_{i \ne 1, x} w_i \ell(w_i) > w_1 \ell(w_x) + w_x \ell(w_1) + \sum_{i \ne 1, x} w_i \ell(w_i)$. Consequently, by interchanging the locations of the weights w_1 and w_x, we obtain a tree of smaller weight. But this contradicts the choice of T as an optimal tree. Therefore $\ell(w_x) = \ell(w_1) = \ell(w_y)$. In a similar manner, it can be shown that $\ell(w_y) = \ell(w_2)$, so $\ell(w_x) = \ell(w_y) = \ell(w_1) = \ell(w_2)$. Interchanging the locations of the pair w_1, w_x, and the pair w_2, w_y, we obtain an optimal tree T_1, where w_1, w_2 are siblings. ∎

(From this lemma we see that small weights will be at the bottom level in an optimal tree.)

THEOREM 12.8   Let T be an optimal tree for the weights $w_1 + w_2, w_3, \ldots, w_n$, where $w_1 \le w_2 \le w_3 \le \cdots \le w_n$. At the leaf with weight $w_1 + w_2$ place a (complete) binary tree of height 1 and assign the weights w_1, w_2 to the children (leaves) of this former leaf. The new binary tree T_1 so constructed is then optimal for the weights $w_1, w_2, w_3, \ldots, w_n$.

Proof   Let T_2 be an optimal tree for the weights $w_1, w_2, \ldots, w_n$, where the leaves for weights w_1, w_2 are siblings. Remove the leaves of weights w_1, w_2 and assign the weight $w_1 + w_2$ to their parent (now a leaf). This complete binary tree is denoted T_3 and $W(T_2) = W(T_3) + w_1 + w_2$. Also, $W(T_1) = W(T) + w_1 + w_2$. Since T is optimal, $W(T) \le W(T_3)$. If $W(T) < W(T_3)$, then $W(T_1) < W(T_2)$, contradicting the choice of T_2 as optimal. Hence $W(T_1) = W(T_2)$ and T_1 is optimal for the weights $w_1, w_2, \ldots, w_n$. ∎

Remark. The proof above started with an optimal tree T_2 whose existence rests on the fact that there is only a finite number of ways in which we can assign n weights to a complete binary tree with n leaves. Consequently, with a finite number of assignments there is at least one where $W(T)$ is minimal. But finite numbers can be large. This proof establishes the existence of an optimal tree for a set of weights and develops a recursive construction for such a tree. The following example uses such a construction.

Example 12.17 Construct an optimal prefix code for the symbols a, o, q, u, y, z that occur (in a given sample) with frequencies $20, 28, 4, 17, 12, 7$, respectively.

Figure 12.32 shows the construction that follows Huffman's recursive procedure. In part (a), weights 4 and 7 are combined so that we then consider the construction for the weights $11, 12, 17, 20, 28$. At each step we create a tree with subtrees rooted at the two smallest weights. These two smallest weights belong to vertices each of which is originally either isolated (a tree with just a root) or the root of a tree obtained earlier in the construction. From the last result, the prefix code is determined as

a: 11 o: 01 q: 0000 u: 10 y: 001 z: 0001   □

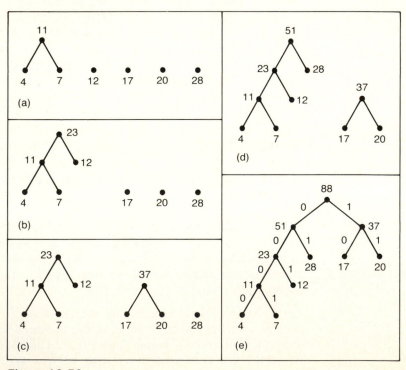

Figure 12.32

EXERCISES 12.4

1. For the prefix code given in Fig. 12.28, decode the sequences (a) 1001111101; (b) 10111100110001101; (c) 1101111110010.

2. A code for $\{a, b, c, d, e\}$ is given by a: 00 b: 01 c: 101 d: x10 e: yz1, where $x, y, z \in \{0, 1\}$. Determine x, y, and z so that the given code is a prefix code.

3. Construct an optimal prefix code for the symbols $a, b, c, \ldots, i, j$ that occur (in a given sample) with respective frequencies 78, 16, 30, 35, 125, 31, 20, 50, 80, 3.

4. How many leaves does a full binary tree have if its height is (a) 3? (b) 7? (c) 12? (d) h?

5. Let $T = (V, E)$ be a complete m-ary tree of height h. This tree is called a *full* m-ary tree if all of its leaves are at level h. If T is a full m-ary tree with height 7 and 279,936 leaves, how many internal vertices are there in T?

6. Let $L_i, 1 \le i \le 4$, be four lists of numbers, each sorted in ascending order. The numbers of entries in these lists are 75, 40, 110, and 50, respectively.

   a) How many comparisons are needed to merge these four lists by merging L_1 and L_2, merging L_3 and L_4, and then merging the two resulting lists?

   b) How many comparisons are needed if we first merge L_1 and L_2, then merge the result with L_3, and finally merge this result with L_4?

   c) In order to minimize the total number of comparisons in this merging of the four lists, what order should the merging follow?

   d) Extend the result in part (c) to n sorted lists $L_1, L_2, \ldots, L_n$.

12.5
BICONNECTED COMPONENTS AND ARTICULATION POINTS

Let $G = (V, E)$ be the loop-free connected undirected graph shown in Fig. 12.33(a), where each vertex represents a communication center. Here an edge $\{x, y\}$ indicates the existence of a communication link between the centers at x and y.

By splitting the vertices at c and f, we obtain the collection of subgraphs in part (b) of the figure. These vertices are examples of the following.

DEFINITION 12.9   A vertex v in a loop-free undirected graph $G = (V, E)$ is called an *articulation point* if $\kappa(G - v) > \kappa(G)$; that is, the subgraph $G - v$ has more components than the given graph G.

A graph with no articulation points is called *biconnected*.

A *biconnected component* of a graph is a maximal biconnected subgraph—a biconnected subgraph that is not a subgraph of a larger biconnected subgraph.

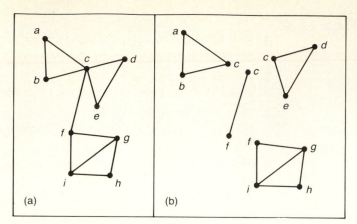

Figure 12.33

The graph shown in Fig. 12.33(a) has the two articulation points c and f, and its four biconnected components are shown in part (b) of the figure.

In terms of communication centers and links, the articulation points of the graph indicate where the system is most vulnerable. Without articulation points, such a system is more likely to survive disruptions at a communication center, whether by the breakdown of a technical device or by external forces.

The problem of finding the articulation points in a connected graph provides an application for the depth-first spanning tree. The objective here is the development of an algorithm that determines the articulation points of a loop-free connected undirected graph. If no such points exist, then the graph is biconnected. Should such vertices exist, the resulting biconnected components provide information about such properties as the planarity and chromatic number of the given graph.

The following preliminaries are needed for developing this algorithm.

LEMMA 12.3   Let $G = (V, E)$ be a loop-free connected undirected graph with $T = (V, E')$ a depth-first spanning tree of G. If $\{a, b\} \in E$ but $\{a, b\} \notin E'$, then a is either an ancestor or a descendant of b in the tree T.

Proof   From the depth-first spanning tree T, we obtain a preorder listing for the vertices in V. For any $v \in V$, let dfi(v) denote the depth-first index of vertex v—that is, the position of v in the preorder listing. Assume that dfi(a) < dfi(b). Consequently, a is encountered before b in the preorder traversal of T, so a cannot be a descendant of b. If, in addition, vertex a is not an ancestor of b, then b is not in the subtree T_a of T rooted at a. But when we backtrack (through T_a) to a, we find that because $\{a, b\} \in E$, it should have been possible for the depth-first search to go from a to b and to use the edge $\{a, b\}$ in T. ■

If $G = (V, E)$ is a loop-free connected undirected graph, let $T = (V, E')$ be a depth-first spanning tree for G, as shown in Fig. 12.34. By Lemma 12.3, the dotted edge $\{a, b\}$, which is not part of T, indicates an edge that may exist in G. Such an edge is called a *back edge* of T, and here a is an ancestor of b. (Here $\mathrm{dfi}(a) = 3$, whereas $\mathrm{dfi}(b) = 6$.) A dotted edge, such as $\{b, c\}$ in the figure, cannot exist in G, also because of Lemma 12.3. Thus all edges of G are either edges in T or back edges of T.

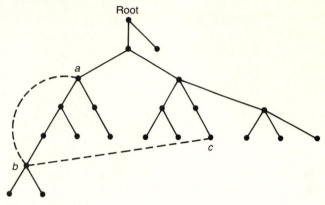

Root

a

b

c

Figure 12.34

Let $x \in V$. For G as above, let $T_{x,c}$ be the subtree consisting of edge $\{x, c\}$ (c a child of x) together with the tree T_c rooted at c. If there is no back edge from a descendant of x in $T_{x,c}$ to an ancestor of x, then the splitting of vertex x results in the separation of $T_{x,c}$ from G, and x is an articulation point. If no other articulation points of G occur in $T_{x,c}$, then the addition to $T_{x,c}$ of all other edges in G determined by the vertices in $T_{x,c}$ (the subgraph of G induced by the vertices in $T_{x,c}$) results in a biconnected component of G. (A root has no ancestors, so the root is an articulation point if and only if it has more than one child.)

The depth-first spanning tree preorders the vertices of G. Let $\mathrm{dfi}(x)$ denote the depth-first index of x in that preorder. If y is a descendant of x, then $\mathrm{dfi}(x) < \mathrm{dfi}(y)$. For y an ancestor of x, $\mathrm{dfi}(x) > \mathrm{dfi}(y)$. Define $\mathrm{low}(x) = \min\{\mathrm{dfi}(y) \,|\, y$ is connected in G to either x or a descendant of $x\}$. If z is the parent of x (in T), then there are two possibilities to consider:

1. $\mathrm{low}(x) = \mathrm{dfi}(z)$: In this case T_x, the subtree rooted at x, contains no vertex that is adjacent to an ancestor of z by means of a back edge of T. Hence z is an articulation point of G. If T_x contains no articulation points, then T_x together with edge $\{z, x\}$ spans a biconnected component of G (that is, the subgraph of G induced by vertex z and the vertices in T_x is a biconnected component of G). Now remove T_x and the edge $\{z, x\}$ from T, and apply this idea to the remaining subtree of T.

2. $\mathrm{low}(x) < \mathrm{dfi}(z)$: Here there is a descendant of z that can be joined (by a back edge) to an ancestor of z.

To deal in an efficient manner with these ideas, we develop the following algorithm. Let $G = (V, E)$ be a loop-free connected undirected graph defined in terms of its adjacency matrix $A(G)$.

Step 1: Find the depth-first spanning tree T for G, for the vertices ordered by the rows (or columns) of $A(G)$. Let $y_1, y_2, \ldots, y_n$ be the vertices of G pre-ordered by T. Then dfi(y_j) = j, for all $1 \le j \le n$.

Step 2: Start with y_n and continue back to $y_{n-1}, y_{n-2}, \ldots, y_4, y_3$, determining low($y_j$), $3 \le j \le n$, recursively as follows:

a) low$'(y_j) = \min\{\text{dfi}(z) \mid z$ is adjacent in G to $y_j\}$.

b) If $c_1, c_2, \ldots, c_m$ are the children of y_j, then low(y_j) = min{low$'(y_j)$, low(c_1), low(c_2), . . . , low(c_m)}. (No problem arises here, for the vertices are examined in the reverse order to the given preorder. Consequently, if c is a child of p, then low(c) is determined before low(p).)

Step 3: Let w_j be the parent of y_j. If low(y_j) = dfi(w_j) then w_j is an articulation point of G (and T). Moreover, the subtree rooted at y_j together with the edge $\{w_j, y_j\}$ is part of a biconnected component of G.

Example 12.18   We apply this algorithm to the graph $G = (V, E)$ given by the adjacency matrix $A(G)$ shown here. The vertices are considered ordered as $x_1, x_2, x_3, \ldots, x_{10}$.

	x_1	x_2	x_3	x_4	x_5	x_6	x_7	x_8	x_9	x_{10}	
x_1	0	1	1	1	1	1	0	0	1	0	y_1
x_2	1	0	0	0	0	0	0	0	0	0	y_2
x_3	1	0	0	0	1	0	0	0	0	0	y_3
x_4	1	0	0	0	0	1	1	1	0	0	y_5
x_5	1	0	1	0	0	0	0	0	0	0	y_4
x_6	1	0	0	1	0	0	0	0	0	0	y_6
x_7	0	0	0	1	0	0	0	1	0	0	y_7
x_8	0	0	0	1	0	0	1	0	1	1	y_8
x_9	1	0	0	0	0	0	0	1	0	0	y_9
x_{10}	0	0	0	0	0	0	0	1	0	0	y_{10}
	y_1	y_2	y_3	y_5	y_4	y_6	y_7	y_8	y_9	y_{10}	

The 10 vertices of G may be traversed in the preorder listing given in Figure 12.35(a). Part (b) of that figure provides a re-indexing of these vertices so that for the depth-first spanning tree shown, dfi(y_j) = j, for all $1 \le j \le 10$. These are the y_j's, $1 \le j \le 10$, that appear at the ends of the rows and columns in the adjacency matrix $A(G)$. Starting with vertex y_{10}, from $A(G)$ we find low$'(y_{10}) = 8$, because y_8 is adjacent to y_{10}, but for $1 \le i \le 7$, y_i is not adjacent to y_{10}. With y_{10} a leaf, it follows that low(y_{10}) = 8. Since low(y_{10}) = dfi(y_8), and y_8 is the parent of y_{10}, y_8 is an articulation point of G; the edge $\{y_8, y_{10}\}$ is the biconnected component. In Fig. 12.36(a), for each y_j, $3 \le j \le 10$, the ordered pair next to y_j is (low$'(y_j)$, low(y_j)). These are computed by the algorithm as we proceed backwards from y_{10} to y_9, to y_8, and so forth, to y_3. In computing y_8, for example, from $A(G)$ we find that low$'(y_8) = 5$, because y_5 is adjacent to y_8 and for any $1 \le i \le 4$, y_i is not adjacent to y_8. Since y_9 is a child of y_8 and low(y_9) = 1, it follows that low(y_8) = 1. For both y_3 and y_5, low(y_3) = low(y_5) = 1 = dfi(y_1), with y_1 the parent of both y_3 and y_5. Hence y_1 is an articulation point and the following subtrees give rise to two more biconnected components: (a) the subtree rooted at y_3 together with the edge $\{y_1, y_3\}$; and (b) the subtree rooted at y_5, with the edge $\{y_8, y_{10}\}$ deleted and the edge $\{y_1, y_5\}$ adjoined. Finally, because y_2 is a leaf, the edge $\{y_1, y_2\}$ constitutes one more biconnected component.

Figure 12.36(b) shows the articulation points y_1 and y_8 for G. These determine the four biconnected components of G shown in the figure. □

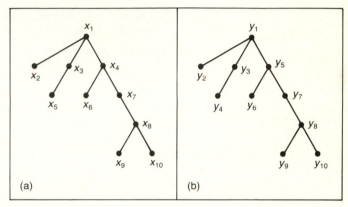

Figure 12.35

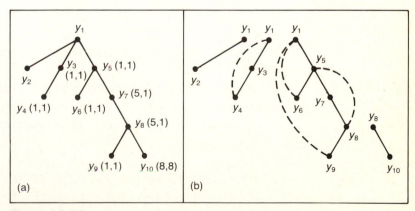

Figure 12.36

EXERCISES 12.5

1. Find the articulation points and biconnected components for the graph shown in Fig. 12.37.

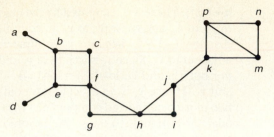

Figure 12.37

2. Let $G = (V, E)$ be a loop-free connected undirected graph. If $z \in V$, prove that z is an articulation point of G if and only if there exist $x, y \in V$ such that every path connecting x and y contains the vertex z.

3. Let $T = (V, E)$ be a tree with $|V| = n \geq 2$.

a) What are the smallest and the largest numbers of articulation points that T can have? Describe the trees for each of these cases.

b) How many biconnected components does T have in each case?

4. a) Let $T = (V, E)$ be a tree. If $v \in V$, prove that v is an articulation point of T if and only if $\deg(v) > 1$.

b) Let $G = (V, E)$ be a loop-free connected undirected graph with $|E| \geq 1$. Prove that G has at least two vertices that are not articulation points.

5. If $B_1, B_2, \ldots, B_k$ are the biconnected components of a loop-free connected undirected graph G, how is $\chi(G)$ related to $\chi(B_i), 1 \leq i \leq k$?

6. For a loop-free connected undirected graph G, explain why the planarity of G depends on the planarity of its biconnected components.

7. A loop-free connected undirected graph G has biconnected components B_1, B_2, and B_3. A second loop-free connected undirected graph G' has biconnected components B'_1, B'_2, and B'_3. If B_i is isomorphic to B'_i, for all $1 \leq i \leq 3$, does it follow that G and G' are isomorphic?

8. Each of the following adjacency matrices represents a loop-free connected undirected graph G. In each case:

a) Determine a depth-first spanning tree for G rooted at x_1.

b) Using the result in part (a), apply the algorithm developed in this section to find the articulation points and biconnected components of G.

	x_1	x_2	x_3	x_4	x_5	x_6	x_7	x_8
x_1	0	1	1	1	1	0	0	0
x_2	1	0	0	1	1	0	0	0
x_3	1	0	0	0	0	1	1	0
x_4	1	1	0	0	1	0	0	0
x_5	1	1	0	1	0	0	0	0
x_6	0	0	1	0	0	0	1	0
x_7	0	0	1	0	0	1	0	1
x_8	0	0	0	0	0	0	1	0

$$
\begin{array}{c c c c c c c c c}
 & x_1 & x_2 & x_3 & x_4 & x_5 & x_6 & x_7 & x_8 \\
x_1 & 0 & 1 & 1 & 1 & 1 & 0 & 0 & 0 \\
x_2 & 1 & 0 & 1 & 1 & 0 & 0 & 0 & 0 \\
x_3 & 1 & 1 & 0 & 1 & 0 & 1 & 0 & 0 \\
x_4 & 1 & 1 & 1 & 0 & 0 & 0 & 0 & 0 \\
x_5 & 1 & 0 & 0 & 0 & 0 & 0 & 1 & 1 \\
x_6 & 0 & 0 & 1 & 0 & 0 & 0 & 0 & 0 \\
x_7 & 0 & 0 & 0 & 0 & 1 & 0 & 0 & 1 \\
x_8 & 0 & 0 & 0 & 0 & 1 & 0 & 1 & 0
\end{array}
$$

9. In step 2 of the algorithm for articulation points, why wasn't it necessary to compute $\text{low}(y_1)$ and $\text{low}(y_2)$?

10. Let $G = (V, E)$ be a loop-free connected undirected graph with $v \in V$.

   a) Prove that $\overline{G - v} = \overline{G} - v$.

   b) If v is an articulation point of G, prove that v cannot be an articulation point of $\overline{G}$.

12.6
SUMMARY AND HISTORICAL REVIEW

The structure now called a tree first appeared in 1847 in the work of Gustav Kirchhoff (1824–1877) on electrical networks. The concept also appeared at this time in the book *Geometrie die Lage* by Karl von Staudt (1798–1867). In 1857 trees were rediscovered by Arthur Cayley (1821–1895), who was unaware of these earlier developments. Cayley was the first to call the structure a "tree," and he used it in applications dealing with chemical isomers. He also investigated the enumeration of certain classes of trees. In his first work on trees, Cayley enumerated unlabeled rooted trees. This was then followed by the enumeration of unlabeled ordered trees. Two of Cayley's contemporaries who also studied trees were Carl Borchardt (1817–1880) and Marie Ennemond Jordan (1838–1922).

The formula n^{n-2} for the number of labeled trees on n vertices (Exercise 15 at the end of Section 12.1) was discovered in 1860 by Carl Borchardt. Cayley later gave an independent development of the formula in 1889. Since then, there have been other derivations. These are surveyed in the book by J. W. Moon [8].

The paper by G. Polya [9] is a pioneering work on the enumeration of trees and other combinatorial structures. Polya's theory of enumeration, which we shall see in Chapter 16, was developed in this work. For more on the enumeration of trees, the reader should see Chapter 15 of F. Harary [3].

The high-speed digital computer has proved to be a constant impetus for the discovery of new applications of trees. The first application of these structures was in the manipulation of algebraic formulae. This dates back to 1951 in the work of Grace Hopper. Since then, computer applications of trees have been widely investigated. In the beginning, particular results appeared only in the documentation of specific algorithms. The first general survey of the applications of trees was made in 1961 by Kenneth Iverson as part of a broader survey on data structures. Such ideas as preorder and postorder can be traced to the early 1960s, as evidenced in the work of Zdzislaw Pawlak, Lyle Johnson, and Kenneth Iverson. Additional material on these orders and the procedures for their implementation on a computer can be found in Chapter 3 of the text by A. Aho, J. Hopcroft, and J. Ullman [1].

If $G = (V, E)$ is a loop-free undirected graph, then the depth-first search and the breadth-first search (given in Section 12.2) provide ways to determine whether the given graph is connected. The algorithms developed for these searching procedures are also important in developing other algorithms. For example, the depth-first search arises in the algorithm for finding the articulation points and biconnected components of a loop-free connected undirected graph. If $|V| = n$ and $|E| = e$, then it can be shown that both the depth-first search and the breadth-first search are $\mathbb{O}(\max\{n, e\})$. For most graphs $e > n$, so the algorithms are generally labeled as $\mathbb{O}(e)$. But since $e \le \binom{n}{2} = (1/2)(n)(n-1)$, the order for each of these algorithms sometimes appears as $\mathbb{O}(n^2)$. Hence the algorithms are said to have linear order in e but quadratic order in n. These ideas are developed in great detail on pages 172–191 of the text by S. Baase [2], where the coverage also includes an analysis of the time complexity function for the algorithm (of Section 12.5) that determines articulation points (and biconnected components). Chapter 6 of the text by A. Aho, J. Hopcroft, and J. Ullman [1] also deals with the depth-first search, whereas Chapter 7 covers the breadth-first search and the algorithm for articulation points.

More on the properties and computer applications of trees is given in Section 3 of Chapter 2 in the work by D. Knuth [5]. Sorting techniques and their use of trees can be further studied in Chapter 11 of A. Aho, J. Hopcroft, and J. Ullman [1]. An extensive investigation will warrant the coverage found in volume 3 of D. Knuth [6].

The technique in Section 12.4 for designing prefix codes is based on methods developed by D. Huffman [4].

Finally, Chapter 7 of C. L. Liu [7] deals with trees, cycles, cut-sets, and the vector spaces associated with these ideas. The reader with a background in linear or abstract algebra should find this material of interest.

REFERENCES

1. Aho, Alfred V., Hopcroft, John E., and Ullman, Jeffrey D. *Data Structures and Algorithms.* Reading, Mass.: Addison-Wesley, 1983.
2. Baase, Sara. *Computer Algorithms: Introduction to Design and Analysis,* 2nd ed. Reading, Mass.: Addison-Wesley, 1988.
3. Harary, Frank. *Graph Theory.* Reading, Mass.: Addison-Wesley, 1969.
4. Huffman, David A. "A Method for the Construction of Minimum Redundancy Codes." *Proceedings of the IRE,* Vol. 40, 1952, pp. 1098–1101.
5. Knuth, Donald E. *The Art of Computer Programming,* Vol. 1, 2nd ed. Reading, Mass.: Addison-Wesley, 1973.

6. Knuth, Donald E. *The Art of Computer Programming,* Vol. 3. Reading, Mass.: Addison-Wesley, 1973.

7. Liu, C. L. *Introduction to Combinatorial Mathematics.* New York: McGraw-Hill, 1968.

8. Moon, John Wesley. *Counting Labelled Trees.* Canadian Mathematical Congress, Montreal, Canada, 1970.

9. Polya, George. "Kombinatorische Anzahlbestimmungen für Gruppen, Graphen und Chemische Verbindungen." *Acta Mathematica,* Vol. 68, 1937, pp. 145–234.

MISCELLANEOUS EXERCISES

1. **a)** Let $G = (V, E)$ be a loop-free undirected graph with $|V| = n$. Prove that G is a tree if and only if $P(G, \lambda) = \lambda(\lambda - 1)^{n-1}$.

   b) Prove that $\chi(G) = 2$ for any tree with two or more vertices.

   c) If $G = (V, E)$ is a connected undirected graph with $|V| = n$, prove that for any integer $\lambda \geq 0$, $P(G, \lambda) \leq \lambda(\lambda - 1)^{n-1}$.

2. The forest $G = (V, E)$ contains 26 vertices and 21 edges. How many components does G have?

3. Let $T = (V, E)$ be an ordered rooted tree, as in Example 12.4. Assume vertex v has address $a_1.a_2.a_3. \ldots .a_{10}$.

   a) At what level is vertex v in the tree?

   b) What are the addresses of the vertices on the path from the root of T to vertex v?

4. Let $G = (V, E)$ be an undirected graph where $V = \{v_2, v_3, v_4, \ldots, v_{30}\}$; $E = \{\{v_i, v_j\} \mid 2 \leq i < j \leq 30$ and i, j have a common divisor greater than 1$\}$.

   a) Determine $\kappa(G)$.

   b) Determine a spanning forest for G.

5. Let $G = (V, E)$ be a connected undirected graph. If $H = (V, E')$ is a subgraph of G, prove that H is a Hamilton path in G if and only if H is a spanning tree of G where (in H) each vertex has degree at most 2.

6. A telephone communication system is set up at a company where 125 executives are employed. The system is initialized by the president, who calls her four vice-presidents. Each vice-president then calls four other executives, who in turn call four others, and so on.

   a) How many calls are made in reaching all 125 executives?

   b) How many executives, aside from the president, are required to make calls?

7. If T is a complete m-ary tree, the *total path length* of T is the sum of the lengths of all paths in the tree from the root to each of its vertices.

   a) For any nonnegative integer h, let x_h denote the minimal total path length for a complete m-ary tree of height h. Show that x_h satisfies the recurrence relation $x_{h+1} = x_h + m(h + 1)$, with initial condition $x_0 = 0$.

   b) Solve the recurrence relation in part (a) for x_h.

8. In a complete m-ary tree T, the *external path length* of T is the sum of the lengths of all paths in T from the root to each of the leaves.

a) If $h \geq 0$, let y_h denote the minimal external path length for a complete m-ary tree of height h. Show that y_h satisfies the recurrence relation $y_{h+1} = y_h - h + m(h+1)$, with the initial condition $y_0 = 0$.

b) Solve the recurrence relation in part (a) for y_h.

9. Let T be a complete binary tree with the vertices of T ordered by a preorder traversal. This traversal assigns the label 1 to all internal vertices of T and the label 0 to each leaf. The sequence of 0's and 1's that results from the preorder traversal of T is called the tree's *characteristic sequence*.

a) Find the characteristic sequence for the complete binary tree shown in Fig. 12.11(c).

b) Determine the complete binary trees for the characteristic sequences
(i) 1011001010100 and (ii) 10111100001011000.

c) What are the last two symbols in the characteristic sequence for all complete binary trees? Why?

10. For $k \in \mathbf{Z}^+$, let $n = 2^k$, and consider the list $L: a_1, a_2, a_3, \ldots, a_n$. To sort L in ascending order, first compare the entries a_i and $a_{i+(n/2)}, 1 \leq i \leq n/2$. For the resulting 2^{k-1} ordered pairs, merge sort the ith and $(i + (n/4))$-th ordered pairs, $1 \leq i \leq n/4$. Now do a merge sort on the ith and $(i + (n/8))$-th ordered quadruples, $1 \leq i \leq n/8$. Continue the process until the elements of L are in ascending order.

a) Apply this sorting procedure to the list

$$L: 11, 3, 4, 6, -5, 7, 35, -2, 1, 23, 9, 15, 18, 2, -10, 5.$$

b) If $n = 2^k$, how many comparisons at most does this procedure require?

11. a) If T is a full binary tree of height 5, how many leaves does T have? How many internal vertices? How many edges (branches)?

b) Answer part (a) for a full binary tree of height h, where $h \in \mathbf{Z}^+$.

12. Let T be a full binary tree of height h. If $v_1, v_2, \ldots, v_m$ are the leaves of T, evaluate $\sum_{i=1}^{m} 2^{-d_i}$, where d_i is the level number of leaf $v_i, 1 \leq i \leq m$.

13. Let $G = (V, E)$ be a loop-free undirected graph. If $\deg(v) \geq 2$ for all $v \in V$, prove that G must contain a cycle.

14. Let $T = (V, E)$ be a rooted tree with root r. Define the relation $\mathcal{R}$ on V by $x \mathcal{R} y$, for $x, y \in V$, if $x = y$ or if x is on the path from r to y. Prove that $\mathcal{R}$ is a partial order.

15. Let $T = (V, E)$ be a tree with $V = \{v_1, v_2, \ldots, v_n\}, n \geq 2$. Prove that the number of pendant vertices in T is equal to

$$2 + \sum_{\deg(v_i) \geq 3} (\deg(v_i) - 2).$$

16. Let $T = (V, E)$ be a complete m-ary tree of height h. If i is the number of internal nodes of T and $|V| = v$, show that the ratio i/v approaches $1/m$ as h increases.

17. Let $G = (V, E)$ be a loop-free connected undirected graph. If $e \in E$, under what condition(s) will e be an edge in every spanning tree of G?

18. Let $G = (V, E)$ be a loop-free undirected graph. Define the relation $\mathcal{R}$ on E as follows: If $e_1, e_2 \in E$, then $e_1 \mathcal{R} e_2$ if $e_1 = e_2$ or if e_1 and e_2 are edges of a cycle C in G.

a) Verify that $\mathcal{R}$ is an equivalence relation on E.

b) Describe the partition of E induced by $\mathcal{R}$.

19. If $G = (V, E)$ is a loop-free connected undirected graph and $a, b \in V$, then we define the *distance* from a to b (or from b to a), denoted $d(a, b)$, as the length of a shortest path (in G) connecting a and b. (This is the number of edges in a shortest path connecting a and b.)

For any loop-free connected undirected graph $G = (V, E)$, the *square* of G, denoted G^2, is the graph with vertex set V (the same as G) and edge set defined as follows: For any $a, b \in V$, $\{a, b\}$ is an edge in G^2 if $d(a, b) \le 2$ (in G). In parts (a) and (b) of Fig. 12.38, we have a graph G and its square.

a) Find the square of the graph in part (c) of the figure.

b) Find G^2 if G is the graph $K_{1,3}$.

c) If G is the graph $K_{1,n}$, for $n \ge 4$, how many edges are added to G in order to construct G^2?

d) For any loop-free connected undirected graph G, prove that G^2 has no articulation points.

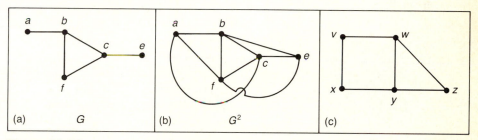

Figure 12.38

20. Let $G = (V, E)$ be the undirected connected "ladder graph" shown in Fig. 12.39. For $n \ge 0$, let a_n be the number of spanning trees of G, whereas b_n is the number of these spanning trees that contain the edge $\{x_1, y_1\}$.

a) Explain why $a_n = a_{n-1} + b_n$.

b) Find an equation that expresses b_n in terms of a_{n-1} and b_{n-1}.

c) Use the results in parts (a) and (b) to set up and solve a recurrence relation for a_n.

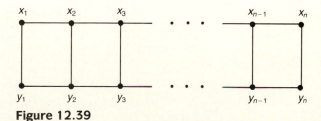

Figure 12.39

21. Let $G = (V, E)$ be a loop-free connected undirected graph where $|V| = v$ and $|E| = e$. The graph G is called *graceful* if it is possible to assign the labels $\{1, 2, 3, \ldots, v\}$ to the vertices of G in such a manner that the induced edge labeling—where each edge $\{i, j\}$ is assigned the label $|i - j|$, for $i, j \in \{1, 2, 3, \ldots, v\}$, $i \neq j$—results in the e edges being labeled by $1, 2, 3, \ldots, e$.

a) Prove that every path on n vertices, $n \geq 2$, is graceful.

b) For $n \in \mathbf{Z}^+$, $n \geq 2$, show that $K_{1,n}$ is graceful.

c) If $T = (V, E)$ is a tree with $4 \leq |V| \leq 6$, show that T is graceful. (It has been conjectured that every tree is graceful.)

13
Optimization and Matching

Using the structures of trees and graphs, the final chapter for this part of the text introduces techniques that arise in the area of mathematics called *operations research*. These techniques optimize certain results by considering graphs and multigraphs that have a nonnegative real number, called a weight, associated with each edge of the graph. These numbers relate information such as the distance between the vertices that are the endpoints of the edge, or perhaps the amount of material that can be shipped from one vertex to another along an edge that represents a highway or air route. With the graphs providing the framework, the optimization methods are developed in an algorithmic manner to facilitate their implementation on a computer. Among the problems we analyze are the determinations of

1. The shortest distance between a designated vertex v_0 and each of the other vertices in a special type of loop-free connected directed graph.

2. A spanning tree for a given graph or multigraph where the sum of the weights of the edges in the tree is minimal.

3. The maximum amount of material that can be transported from a starting point (the source) to a terminating point (the sink), where the weight of an edge indicates an allowable capacity for the material being transported.

13.1

DIJKSTRA'S SHORTEST-PATH ALGORITHM

We start with a loop-free connected directed graph $G = (V, E)$. Now to each edge $e = (a, b)$ of this graph, we assign a nonnegative real number called the *weight* of e. This is denoted by wt(e), or wt(a, b). If $x, y \in V$ but $(x, y) \notin E$, we define wt(x, y) = ∞.

For any $e = (a, b) \in E$, wt(e) may represent (1) the length of a road from a to b, (2) the time it takes to travel on this road from a to b, or (3) the cost of traveling from a to b on this road.

Whenever such a graph $G = (V, E)$ is given with the weight assignments described here, the graph is referred to as a *weighted* graph.

Example 13.1   In Figure 13.1 the weighted graph $G = (V, E)$ represents travel routes between certain pairs of cities. Here the weight of each edge (x, y) indicates the approximate flying time for a direct flight from city x to city y.

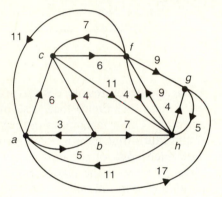

Figure 13.1

In this directed graph there are situations where wt$(x, y) \neq$ wt(y, x) for certain edges (x, y) and (y, x) in G. For example wt$(c, f) = 6 \neq 7 =$ wt(f, c). Perhaps this is due to tailwinds. As a plane flies from c to f, the plane may be assisted by tailwinds that, in turn, slow it down when it is flying in the opposite direction (from f to c).

We see that $c, g \in V$ but $(c, g), (g, c) \notin E$, so wt$(g, c) =$ wt$(c, g) = \infty$. This is also true for other pairs of vertices. On the other hand, for certain pairs of vertices such as a, f, we have wt$(a, f) = \infty$ whereas wt$(f, a) = 11$, a finite number.
□

Our objective in this section has two parts. Given a weighted graph $G = (V, E)$, for each $e = (a, b) \in E$, we shall interpret wt(e) as the length of a direct route (whether by automobile, plane, or boat) from a to b. We write $d(a, b)$ for the (shortest) distance from a to b. This is the length of a shortest directed path (in G) from a to b. If no such path exists (in G) from a to b, then we define $d(a, b) = \infty$. And for any $a \in V$, $d(a, a) = 0$. Consequently, we have the distance function $d: V \times V \to \mathbf{R}^+ \cup \{0, \infty\}$.

Now fix $v_0 \in V$. Then for all $v \in V$, we shall determine

1. $d(v_0, v)$

2. a directed path from v_0 to v, for those $v \in V$ where $d(v_0, v)$ is finite

To accomplish these objectives, we shall introduce a version of the algorithm that was developed by Edsger Wybe Dijkstra in 1959. This procedure is an example of a *greedy* algorithm, for what we do to obtain the best result *locally* (for vertices "close" to v_0) turns out to be the best result *globally* (for all vertices of the graph).

Before we state the algorithm, we wish to examine some properties of the distance function d. These properties will help us understand why the algorithm works.

With $v_0 \in V$ fixed (as it was earlier), let $S \subseteq V$ with $v_0 \in S$, and $\overline{S} = V - S$. Then we define the *distance from v_0 to $\overline{S}$* by

$$d(v_0, \overline{S}) = \min_{v \in \overline{S}} \{d(v_0, v)\}.$$

So $d(v_0, \overline{S})$ is the length of a shortest directed path from v_0 to a vertex in $\overline{S}$. When $d(v_0, \overline{S}) < \infty$ there will be at least one vertex v_{m+1} in $\overline{S}$ with $d(v_0, \overline{S}) = d(v_0, v_{m+1})$, so v_{m+1} is the vertex at the end of a minimal path. In addition, if P: (v_0, v_1), $(v_1, v_2), \ldots, (v_{m-1}, v_m), (v_m, v_{m+1})$ is a shortest directed path (in G) from v_0 to v_{m+1}, then we find that

1. $v_0, v_1, v_2, \ldots, v_m \in S$, and
2. P': $(v_0, v_1), (v_1, v_2), \ldots, (v_{k-1}, v_k)$ is a shortest directed path (in G) from v_0 to v_k, for each $1 \le k \le m$.

(The proofs for these two results are requested in the first exercise for this section.)

From these observations it follows that

$$d(v_0, \overline{S}) = \min\{d(v_0, u) + \mathrm{wt}(u, w)\},$$

where the minimum is evaluated over all $u \in S, w \in \overline{S}$. If a minimum occurs for $u = x$ and $w = y$, then

$$d(v_0, y) = d(v_0, x) + \mathrm{wt}(x, y)$$

is the (shortest) distance from v_0 to y.

The formula for $d(v_0, \overline{S})$ is the cornerstone of the algorithm. We start with the set $S_0 = \{v_0\}$ and then determine

$$d(v_0, \overline{S}_0) = \min_{\substack{u \in S_0 \\ w \in \overline{S}_0}} \{d(v_0, u) + \mathrm{wt}(u, w)\}.$$

This gives us $d(v_0, \overline{S}_0) = \min_{w \in \overline{S}_0}\{\mathrm{wt}(v_0, w)\}$, since $S_0 = \{v_0\}$. If $v_1 \in \overline{S}_0$ and $d(v_0, \overline{S}_0) = \mathrm{wt}(v_0, v_1)$, then we replace S_0 by $S_1 = S_0 \cup \{v_1\}$ and determine

$$d(v_0, \overline{S}_1) = \min_{\substack{u \in S_1 \\ w \in \overline{S}_1}} \{d(v_0, u) + \mathrm{wt}(u, w)\}.$$

This leads us to a vertex v_2 in $\overline{S}_1$ with $d(v_0, \overline{S}_1) = d(v_0, v_2)$. Continuing the process, if $S_i = \{v_0, v_1, v_2, \ldots, v_i\}$ has been determined and $v_{i+1} \in \overline{S}_i$ with $d(v_0, v_{i+1}) = d(v_0, \overline{S}_i)$, then we extend S_i to S_{i+1}. We stop when we reach $\overline{S}_{n-1} = \emptyset$ (where $n = |V|$), or when $d(v_0, \overline{S}_i) = \infty$ for some $0 \le i \le n - 2$.

Throughout this process, various labels will be placed on each vertex $v \in V$, $v \neq v_0$. The last set of labels will have the form $(L(v), u)$, where $L(v) = d(v_0, v)$, the distance from v_0 to v, and u is the vertex (if one exists) that precedes v along a shortest path from v_0 to v. That is, (u, v) is the last edge in a directed path from v_0 to v, and this path determines $d(v_0, v)$. At first we label v_0 with $(0, -)$ and all of the other vertices v with the label $(\infty, -)$. As we apply the algorithm, the label on each $v \neq v_0$ will change (sometimes more than once) from $(\infty, -)$ to the final label $(L(v), u) = (d(v_0, v), u)$, unless $d(v_0, v) = \infty$.

Now that these preliminaries are behind us, it is time to formally state the algorithm.

Let $G = (V, E)$ be a weighted graph, with $|V| = n$. To find the shortest distance from a fixed vertex v_0 to all other vertices in G, as well as a shortest directed path for each of these vertices, we apply the following algorithm.

Dijkstra's Shortest-Path Algorithm

Step 1: Set the counter $i = 0$ and $S_0 = \{v_0\}$. Label v_0 with $(0, -)$ and each $v \neq v_0$ with $(\infty, -)$.

   If $n = 1$, then $V = \{v_0\}$ and the problem is solved.

   If $n > 1$, continue to step 2.

Step 2: For each $v \in \overline{S}_i$ replace, when possible, the label on v by the new label $(L(v), y)$ where

$$L(v) = \min_{u \in S_i}\{L(v), L(u) + \text{wt}(u, v)\},$$

and y is a vertex in S_i that produces the minimum $L(v)$. [When a replacement does take place, it is due to the fact that we can go from v_0 to v and travel a shorter distance by going along a path that includes the edge (y, v).]

Step 3: If every vertex in $\overline{S}_i$ (for some $0 \leq i \leq n - 2$) has the label $(\infty, -)$, then the labeled graph contains the information we are seeking.

   If not, then there is at least one vertex $v \in \overline{S}_i$ that is not labeled by $(\infty, -)$, and we perform the following tasks:

1. Select a vertex v_{i+1} where $L(v_{i+1})$ is a minimum (for all such v). There may be more than one such vertex, in which case we are free to choose among the possible candidates. The vertex v_{i+1} is an element of $\overline{S}_i$ that is closest to v_0 (and S_i).

2. Assign $S_i \cup \{v_{i+1}\}$ to S_{i+1}.

3. Increase the counter i by 1.
   If $i = n - 1$, the labeled graph contains the information we want. If $i < n - 1$, return to step 2.

We now apply this algorithm in the following example.

Example 13.2   Apply Dijkstra's algorithm to the weighted graph $G = (V, E)$ shown in Fig. 13.1 in order to find the shortest distance from vertex c ($= v_0$) to each of the other five vertices in G.

Initialization: ($i = 0$). Set $S_0 = \{c\}$. Label c with $(0, -)$ and all other vertices in G with $(\infty, -)$.

First Iteration: ($\bar{S}_0 = \{a, b, f, g, h\}$). Here $i = 0$ in step 2 and we find, for example, that
$$L(a) = \min\{L(a), L(c) + \text{wt}(c, a)\}$$
$$= \min\{\infty, 0 + \infty\} = \infty, \text{ whereas}$$
$$L(f) = \min\{L(f), L(c) + \text{wt}(c, f)\}$$
$$= \min\{\infty, 0 + 6\} = 6.$$

Similar calculations yield $L(b) = L(g) = \infty$ and $L(h) = 11$. So we label the vertex f with $(6, c)$ and the vertex h with $(11, c)$. The other vertices in $\bar{S}_0$ remain labeled by $(\infty, -)$. [See Fig. 13.2(a).] In step 3 we see that f is the vertex v_1 in $\bar{S}_0$ closest to v_0 (and S_0), so we assign to S_1 the set $S_0 \cup \{f\} = \{c, f\}$ and increase the counter i to 1. Since $i = 1 < 5 (= 6 - 1)$, we return to step 2.

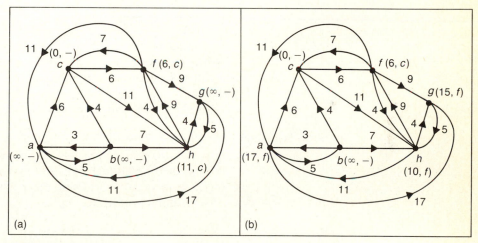

(a)   (b)

Figure 13.2

Second Iteration: ($\bar{S}_1 = \{a, b, g, h\}$). Now $i = 1 > 0$ in step 2, so for each $v \in \bar{S}_1$ we set
$$L(v) = \min_{u \in S_1}\{L(v), L(u) + \text{wt}(u, v)\}.$$

This yields
$$L(a) = \min\{L(a), L(c) + \text{wt}(c, a), L(f) + \text{wt}(f, a)\}$$
$$= \min\{\infty, 0 + \infty, 6 + 11\} = 17,$$

so vertex a is labeled $(17, f)$. In a similar manner, we find

$$L(b) = \min\{\infty, 0 + \infty, 6 + \infty\} = \infty,$$
$$L(g) = \min\{\infty, 0 + \infty, 6 + 9\} = 15, \text{ and,}$$
$$L(h) = \min\{\infty, 0 + \infty, 6 + 4\} = 10.$$

[These results provide the labeling in Fig. 13.2(b).] In step 3 we find that the vertex v_2 is h, because $h \in \overline{S}_1$ and $L(h)$ is a minimum. Then S_2 is assigned $S_1 \cup \{h\} = \{c, f, h\}$, the counter is increased to 2, and since $2 < 5$, the algorithm directs us back to step 2.

Third Iteration: $(\overline{S}_2 = \{a, b, g\})$. With $i = 2 > 0$, in step 2 the following are computed:

$$L(a) = \min_{u \in S_2}\{L(a), L(u) + \text{wt}(u, a)\}$$
$$= \min\{17, 0 + \infty, 6 + 11, 10 + 11\} = 17$$

(so the label on a is not changed);

$$L(b) = \min\{\infty, 0 + \infty, 6 + \infty, 10 + \infty\} = \infty$$

(so the label on b remains ∞); and

$$L(g) = \min\{15, 0 + \infty, 6 + 9, 10 + 4\} = 14 < 15,$$

so the label on g is changed to $(14, h)$ because $14 = L(h) + \text{wt}(h, g)$. Among the vertices in $\overline{S}_2$, g is the closest to v_0 (and S_2) since $L(g)$ is a minimum. In step 3, vertex v_3 is defined as g and $S_3 = S_2 \cup \{g\} = \{c, f, h, g\}$. Then the counter i is increased to $3 < 5$, and we return to step 2.

Fourth Iteration: $(\overline{S}_3 = \{a, b\})$. With $i = 3 > 0$, the following are determined in step 2: $L(a) = 17$; $L(b) = \infty$. (Thus no labels are changed during this iteration.) We set $v_4 = a$ and $S_4 = S_3 \cup \{a\} = \{c, f, h, g, a\}$ in step 3. Then the counter i is increased to $4 (< 5)$, and we return to step 2.

Fifth Iteration: $(\overline{S}_4 = \{b\})$. Here $i = 4 > 0$ and in step 2 we find $L(b) = L(a) + \text{wt}(a, b) = 17 + 5 = 22$. Now the label on b is changed to $(22, a)$. Then $v_5 = b$ in step 3, S_5 is set to $\{c, f, h, g, a, b\}$, and i is incremented to 5. But now that $i = 5 = |V| - 1$, the process terminates. We reach the labeled graph shown in Fig. 13.3.

From the labels in Fig. 13.3 we have the following shortest distances from c to the other five vertices in G:

1. $d(c, f) = L(f) = 6$ **2.** $d(c, h) = L(h) = 10$

3. $d(c, g) = L(g) = 14$ **4.** $d(c, a) = L(a) = 17$

5. $d(c, b) = L(b) = 22$

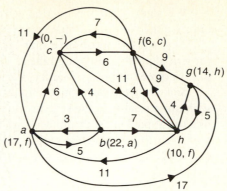

Figure 13.3

To determine, for example, a shortest directed path from c to b, we start at vertex b, which is labeled $(22, a)$. Hence a is the predecessor of b on this shortest path. The label on a is $(17, f)$, so f precedes a on the path. Finally, the label on f is $(6, c)$, so we are back at vertex c, and the shortest directed path from c to b determined by the algorithm is given by the edges (c, f), (f, a), and (a, b). □

Now that we have demonstrated one application of this algorithm, our next concern is the order of its worst-case time complexity function $f(n)$, where $n = |V|$ in the weighted graph $G = (V, E)$. We shall measure the complexity in terms of the number of additions and comparisons that are made in steps 2 and 3 during execution of the algorithm.

Following the initialization process in step 1, there are at most $n - 1$ iterations, because each iteration determines the next closest vertex to v_0, and $n - 1 = |V - \{v_0\}|$.

If $0 \le i \le n - 2$, then in step 2 for that iteration (the $(i + 1)$st), we find that the following takes place for each $v \in \overline{S}_i$:

1. When $0 \le i \le n - 2$, we perform at most $n - 1$ additions to calculate

$$L(v) = \min_{u \in S_i}\{L(v), L(u) + \text{wt}(u, v)\}$$

—one addition for each $u \in S_i$.

2. We compare the present value of $L(v)$ with each of the (possibly infinite) numbers $L(u) + \text{wt}(u, v)$—one for each $u \in S_i$—in order to determine the updated value of $L(v)$. This requires at most $n - 1$ comparisons. Therefore, before we get to step 3 we have performed at most $2(n - 1)$ steps for *each* $v \in \overline{S}_i$—a total of at most $2(n - 1)^2$ steps for *all* $v \in \overline{S}_i$.

   Continuing to step 3, we now must select the minimum from among at most $n - 1$ numbers $L(v)$, where $v \in \overline{S}_i$. This requires $n - 2$ additional comparisons—in the worst case.

   Consequently, each iteration needs no more than $2(n - 1)^2 + (n - 2)$ steps in all. It is possible to have as many as $n - 1$ iterations, so it follows that

$$f(n) \le (n - 1)[2(n - 1)^2 + (n - 2)] \in O(n^3).$$

We shall close this section with some observations that can be used to improve the worst-case time complexity of this algorithm.

First we should observe that for $0 \leq i \leq n - 2$, the $(i + 1)$st iteration of our present algorithm generated the $(i + 1)$st closest vertex to v_0. This was the vertex v_{i+1}. In our example we found $v_1 = f$, $v_2 = h$, $v_3 = g$, $v_4 = a$, and $v_5 = b$.

Second, note how much duplication we had when computing $L(v)$. This is seen quite readily in the second and third iterations of Example 13.2. We should like to cut back on such unnecessary calculations, so let us try a slightly different approach to our shortest-path problem. Once again we start with a weighted graph $G = (V, E)$ with $|V| = n$ and $v_0 \in V$. We shall now let v_i denote the ith closest vertex to v_0, where $0 \leq i \leq n - 1$, $S_i = \{v_0, v_1, \ldots, v_i\}$, and $\overline{S}_i = V - S_i$. At the start we assign to each $v \in V$ the number $L_0(v)$ as follows:

$$L_0(v_0) = 0, \text{ because } d(v_0, v_0) = 0 \text{ and}$$
$$L_0(v) = \infty, \text{ for } v \neq v_0.$$

Then for $i \geq 0$ and $v \in \overline{S}_i$, we define

$$L_{i+1}(v) = \min\{L_i(v), L_i(v_i) + \text{wt}(v_i, v)\},$$

where v_i is a vertex for which $L_i(v_i)$ is minimal: a vertex that is ith closest to v_0. We find that

$$L_{i+1}(v) = \min_{1 \leq j \leq i}\{d(v_0, v_j) + \text{wt}(v_j, v)\}.$$

Now let us see what happens at each of the (at most) $n - 1$ iterations when we employ the definition of $L_{i+1}(v)$ that uses the vertex v_i.

For any $v \in S_i$ we need only one addition [namely, $L_i(v_i) + \text{wt}(v_i, v)$] and one comparison [between $L_i(v)$ and $L_i(v_i) + \text{wt}(v_i, v)$] in order to compute $L_{i+1}(v)$. Since there are at most $n - 1$ vertices in $\overline{S}_i$, this necessitates at most $2(n - 1)$ steps to obtain $L_{i+1}(v)$ for *all* $v \in \overline{S}_i$. Finding the minimum of $\{L_{i+1}(v) \,|\, v \in \overline{S}_i\}$ requires at most $n - 2$ comparisons, so at each iteration we can obtain v_{i+1}—a vertex $v \in \overline{S}_i$ where $L_{i+1}(v)$ is a minimum—in at most $2(n - 1) + (n - 2) = 3n - 4$ steps. We perform at most $n - 1$ iterations, so we find that for this version of Dijkstra's algorithm, the worst-case time complexity function is $O(n^2)$.

In order to find a shortest path from v_0 to each $v \in V, v \neq v_0$, we see that whenever $L_{i+1}(v) < L_i(v)$, for any $0 \leq i \leq n - 2$, we need to keep track of the vertex $y \in S_i$ for which $L_{i+1}(v) = d(v_0, y) + \text{wt}(y, v)$.

EXERCISES 13.1

1. Let $G = (V, E)$ be a weighted graph where for each edge $e = (a, b)$ in E, $\text{wt}(a, b)$ equals the distance from a to b along edge e. If $(a, b) \notin E$, then $\text{wt}(a, b) = \infty$.

Fix $v_0 \in V$ and let $S \subseteq V$, with $v_0 \in S$. Then for $\overline{S} = V - S$ we define $d(v_0, \overline{S}) = \min_{v \in \overline{S}}\{d(v_0, v)\}$. If $v_{m+1} \in \overline{S}$ and $d(v_0, \overline{S}) = d(v_0, v_{m+1})$, then P: $(v_0, v_1), (v_1, v_2), \ldots, (v_{m-1}, v_m), (v_m, v_{m+1})$ is a shortest directed path (in G) from v_0 to v_{m+1}. Prove that

a) $v_0, v_1, v_2, \ldots, v_{m-1}, v_m \in S$

b) $P': (v_0, v_1), (v_1, v_2), \ldots, (v_{k-1}, v_k)$ is a shortest directed path (in G) from v_0 to v_k, for any $1 \le k \le m$.

2. a) Apply Dijkstra's algorithm to the weighted graph $G = (V, E)$ in Fig. 13.4, and determine the shortest distance from vertex a to each of the other six vertices in G. Here $\text{wt}(e) = \text{wt}(x, y) = \text{wt}(y, x)$ for each edge $e = \{x, y\}$ in E.

b) Determine a shortest path from vertex a to each of the vertices c, f, and i.

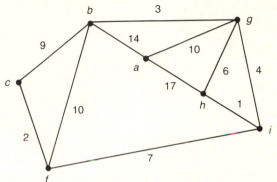

Figure 13.4

3. a) Apply Dijkstra's algorithm to the graph shown in Fig. 13.1, and determine the shortest distance from vertex a to each of the other vertices in the graph.

b) Find a shortest path from vertex a to each of the vertices f, g, and h.

4. Use the ideas developed at the end of the section to confirm the results obtained in (a) Example 13.2 and (b) part (a) of Exercise 2.

5. Prove or disprove the following for a weighted graph $G = (V, E)$, where $V = \{v_0, v_1, v_2, \ldots, v_n\}$ and $e_1 \in E$ with $\text{wt}(e_1) < \text{wt}(e)$ for all $e \in E, e \ne e_1$. If Dijkstra's algorithm is applied to G, and the shortest distance $d(v_0, v_i)$ is computed for each vertex $v_i, 1 \le i \le n$, then there exists a vertex $v_j, 1 \le j \le n$, where the edge e_1 is used in the shortest path from v_0 to v_j.

13.2
MINIMAL SPANNING TREES: THE ALGORITHMS OF KRUSKAL AND PRIM

A loosely coupled computer network is to be set up for a system of seven computers. The graph G in Fig. 13.5 models the situation. The computers are represented by the vertices in the graph; the edges represent transmission lines that are being considered for linking certain pairs of computers. Associated with each edge e in G is a nonnegative real number $\text{wt}(e)$, the weight of e. Here the weight of an edge indicates the projected cost for constructing that particular transmission line. The objective is to link all the computers while minimizing the total cost of construction. To do so requires a spanning tree T, where the sum of the weights of the edges in T is minimal. The construction of such an *optimal* spanning tree can be accomplished by algorithms developed by Joseph Kruskal and Robert Prim.

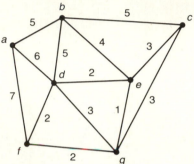

Figure 13.5

Like Dijkstra's algorithm, these algorithms are *greedy* algorithms; when each of these algorithms is used, at each step of the process an optimal (here minimal) choice is made from the remaining available data. Once again, if what appears to be the best choice *locally* (for example, for a vertex c and the vertices near c) turns out to be the best choice *globally* (for all vertices of the graph), then the greedy algorithm will lead to an optimal solution.

We first consider Kruskal's algorithm. This algorithm is given as follows.

Let $G = (V, E)$ be an undirected connected graph where $|V| = n$ and each edge e is assigned a nonnegative real number wt(e). To find an optimal (minimal) spanning tree for G, apply the following algorithm.

Kruskal's Algorithm

Step 1: Set the counter $i = 1$ and select an edge $e_1 \in E$ where wt(e_1) is as small as possible.

Step 2: For $1 \leq i \leq n - 2$, if edges $e_1, e_2, \ldots, e_i$ have been selected, then select edge e_{i+1} from the remaining edges in G so that (a) wt(e_{i+1}) is as small as possible, and (b) the subgraph of G determined by the edges $e_1, e_2, \ldots, e_i, e_{i+1}$ (and the vertices they are incident with) contains no cycles.

Step 3: Replace i by $i + 1$.

If $i = n - 1$, the subgraph of G determined by edges $e_1, e_2, \ldots, e_{n-1}$ is connected with n vertices and $n - 1$ edges, and is an optimal spanning tree for G.

If $i < n - 1$, return to step 2.

Before establishing the validity of the algorithm, we consider the following example.

Example 13.3 Apply Kruskal's algorithm to the graph shown in Fig. 13.5.

Initialization: ($i = 1$) Since there is a unique edge, namely $\{e, g\}$, of weight 1, start with $T = \{\{e, g\}\}$. (T starts as a tree with one edge, and after each iteration it grows into a larger tree or forest. After the last iteration the subgraph T is an optimal spanning tree for the given graph G.)

First Iteration: Among the remaining edges in G, three have weight 2. Select $\{d, f\}$, which satisfies the conditions in step 2. Now T is the forest $\{\{e, g\}, \{d, f\}\}$, and i is increased to 2. With $i = 2 < 6$, return to step 2.

Second Iteration: Two remaining edges have weight 2. Select $\{d, e\}$. Now T is the tree $\{\{e, g\}, \{d, f\}, \{d, e\}\}$, and i increases to 3. But because $3 < 6$, the algorithm directs us back to step 2.

Third Iteration: Among the edges of G that are not in T, edge $\{f, g\}$ has minimal weight 2. However, if this edge is added to T, the result contains a cycle, which destroys the tree structure being sought. Consequently, the edges $\{c, e\}$, $\{c, g\}$, and $\{d, g\}$ are considered. Edge $\{d, g\}$ brings about a cycle, but either $\{c, e\}$ or $\{c, g\}$ satisfies the conditions in step 2. Select $\{c, e\}$. T grows to $\{\{e, g\}, \{d, f\}, \{d, e\}, \{c, e\}\}$ and i is increased to 4. Returning to step 2, we find that the fourth and fifth iterations provide the following.

Fourth Iteration: $T = \{\{e, g\}, \{d, f\}, \{d, e\}, \{c, e\}, \{b, e\}\}$; i increases to 5.

Fifth Iteration: $T = \{\{e, g\}, \{d, f\}, \{d, e\}, \{c, e\}, \{b, e\}, \{a, b\}\}$. The counter i now becomes $6 = $ (number of vertices in G) $- 1$. So T is an optimal tree for graph G and has weight $1 + 2 + 2 + 3 + 4 + 5 = 17$.

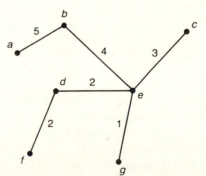

Figure 13.6

Figure 13.6 shows this spanning tree of minimal weight. □

Example 13.3 demonstrates that Kruskal's algorithm does generate a spanning tree. This follows from parts (a) and (c) of Theorem 12.5, because the resulting subgraph has n ($= |V|$) vertices, $n - 1$ edges, and no cycles. Also, the algorithm is greedy; it selects from the remaining edges an edge of minimal weight that doesn't create a cycle. The following result guarantees that the spanning tree obtained is optimal.

THEOREM 13.1   Let $G = (V, E)$ be a weighted connected undirected graph. Any spanning tree for G that is obtained by Kruskal's algorithm is optimal.

Proof   Let $|V| = n$, and let T be a spanning tree for G obtained by Kruskal's algorithm. The edges in T are labeled as $e_1, e_2, \ldots, e_{n-1}$, according to the order in which they are generated by the algorithm. For any optimal tree T' of G, define $d(T') = k$ if k is the smallest positive integer such that T and T' both contain $e_1, e_2, \ldots, e_{k-1}$, but $e_k \notin T'$.

Let T_1 be an optimal tree for which $d(T_1) = r$ is maximal. If $r = n$, then $T = T_1$ and the result follows. Otherwise, $r \leq n - 1$ and adding edge e_r (of T) to T_1 produces the cycle C, where there exists an edge e_r' of C that is in T_1 but not in T.

Start with tree T_1. Adding e_r to T_1 and deleting e_r', we obtain a connected graph with n vertices and $n - 1$ edges. This graph is a tree, T_2. The weights of T_1 and T_2 satisfy $\text{wt}(T_2) = \text{wt}(T_1) + \text{wt}(e_r) - \text{wt}(e_r')$.

Following the selection of $e_1, e_2, \ldots, e_{r-1}$ in Kruskal's algorithm, the edge e_r is chosen so that $\text{wt}(e_r)$ is minimal and no cycle results when e_r is added to the subgraph H of G determined by $e_1, e_2, \ldots, e_{r-1}$. Since e_r' produces no cycle when added to the subgraph H, by the minimality of $\text{wt}(e_r)$, $\text{wt}(e_r') \geq \text{wt}(e_r)$. Hence $\text{wt}(e_r) - \text{wt}(e_r') \leq 0$, so $\text{wt}(T_2) \leq \text{wt}(T_1)$. But with T_1 optimal, $\text{wt}(T_2) = \text{wt}(T_1)$, so T_2 is optimal.

The tree T_2 is optimal and has the edges $e_1, e_2, \ldots, e_{r-1}, e_r$ in common with T, so $d(T_2) \geq r + 1 > r = d(T_1)$, contradicting the choice of T_1. Consequently, $T_1 = T$ and Kruskal's algorithm produces the optimal tree T.   ■

We measure the worst-case time complexity function f for Kruskal's algorithm by making the following observations. Given a weighted connected undirected graph $G = (V, E)$, where $|V| = n$ and $|E| = m \geq 2$, we can use the merge sort of Section 12.3 to list (and relabel, if necessary) the edges in E as $e_1, e_2, \ldots, e_m$, where $\text{wt}(e_1) \leq \text{wt}(e_2) \leq \cdots \leq \text{wt}(e_m)$. The number of comparisons needed to do this is $O(m \log_2 m)$. Then once we have the edges of G listed in this order (of nondecreasing weights), step 2 of the algorithm is carried out for at most the $m - 2$ edges $e_2, e_3, \ldots, e_{m-1}$. (If these edges do not provide a minimal spanning tree, then we simply add in the final edge e_m to obtain the structure.)

For each edge $e_i, 2 \leq i \leq m - 1$, we must determine whether e_i causes the formation of a cycle in the tree, or forest, that we have developed (after considering the edges $e_1, e_2, \ldots, e_{i-1}$). This cannot be done in a constant [that is, $O(1)$] amount of time for each edge. However, it does turn out that *all* of the work needed for cycle detection can be carried out in at most $O(n \log_2 n)$ steps.†

Consequently, we shall define the worst-case time complexity function f, for $m \geq 2$, as the sum of the following:

---

† For more on the analysis of the segment dealing with cycle detection, we refer the reader to Chapter 8 of the text by S. Baase [2] and to Chapter 4 of the text by E. Horowitz and S. Sahni [14].

1. The total number of comparisons needed to sort the edges of G into non-decreasing order, and

2. The total number of steps that are carried out in step 2 in order to detect the formation of a cycle.

Unless G is a tree, it follows that $n \leq m$ because G is connected. As a result, $n \log_2 n \leq m \log_2 m$ and $f \in O(m \log_2 m)$.

A measure in terms of n, the number of vertices in G, can also be given. Here $n - 1 \leq m$ because the graph is connected, and $m \leq \binom{n}{2} = (1/2)(n)(n-1)$, the number of edges in K_n. Consequently $m \log_2 m \leq n^2 \log_2 n^2 = 2n^2 \log_2 n$, and we can express the complexity of Kruskal's algorithm as $O(n^2 \log_2 n)$.

A second technique for constructing an optimal tree was developed by Robert Prim. In this greedy algorithm, the vertices in the graph are partitioned into two sets: processed and not processed. At first only one vertex is in the set P of processed vertices, and all other vertices are in the set N of vertices to be processed. Each iteration of the algorithm increases the set P by one vertex while the size of set N decreases by one. The algorithm is summarized as follows.

Let $G = (V, E)$ be a weighted connected undirected graph. To obtain an optimal tree T for G, apply the following procedure.

Prim's Algorithm

Step 1: Set the counter $i = 1$ and place a vertex $v_1 \in V$ into set P. Define $N = V - \{v_1\}$ and $T = \emptyset$.

Step 2: For $1 \leq i \leq n - 1$, let $P = \{v_1, v_2, \ldots, v_i\}$, $T = \{e_1, e_2, \ldots, e_{i-1}\}$, and $N = V - P$. Add to T the shortest edge (the edge of minimal weight) in G that connects a vertex x in P with a vertex y $(= v_{i+1})$ in N. Place y in P and delete it from N.

Step 3: Increase the counter by 1.
    If $i = n$, the subgraph of G determined by the edges $e_1, e_2, \ldots, e_{n-1}$ is connected with n vertices and $n - 1$ edges and is an optimal tree for G.
    If $i < n$, go back to step 2.

We use this algorithm to find an optimal tree for the graph in Fig. 13.5.

Example 13.4   Prim's algorithm generates an optimal tree as follows.

Initialization: $i = 1$; $P = \{a\}$; $N = \{b, c, d, e, f, g\}$; $T = \emptyset$.

First Iteration: $T = \{\{a, b\}\}$; $P = \{a, b\}$; $N = \{c, d, e, f, g\}$; $i = 2$.

Second Iteration: $T = \{\{a, b\}, \{b, e\}\}$; $P = \{a, b, e\}$; $N = \{c, d, f, g\}$; $i = 3$.

Third Iteration: $T = \{\{a, b\}, \{b, e\}, \{e, g\}\}$; $P = \{a, b, e, g\}$; $N = \{c, d, f\}$; $i = 4$.

Fourth Iteration: $T = \{\{a,b\},\{b,e\},\{e,g\},\{d,e\}\}$; $P = \{a,b,e,g,d\}$;
$N = \{c,f\}$; $i = 5$.

Fifth Iteration: $T = \{\{a,b\},\{b,e\},\{e,g\},\{d,e\},\{f,g\}\}$; $P = \{a,b,e,g,d,f\}$;
$N = \{c\}$; $i = 6$.

Sixth Iteration: $T = \{\{a,b\},\{b,e\},\{e,g\},\{d,e\},\{f,g\},\{c,g\}\}$;
$P = \{a,b,e,g,d,f,c\} = V$; $N = \emptyset$; $i = 7 = |V|$. Hence T is an optimal spanning tree of weight 17 for G, as seen in Fig. 13.7. $\square$

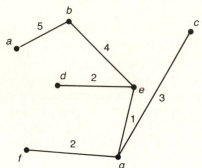

Figure 13.7

We shall only state the following theorem, which establishes the validity of Prim's algorithm. The proof is left for the reader.

THEOREM 13.2   Let $G = (V, E)$ be a weighted connected undirected graph. Any spanning tree for G that is obtained by Prim's algorithm is optimal.

We conclude this section with a few words and references about the worst-case time complexity function for Prim's algorithm. When the algorithm is applied to a weighted connected undirected graph $G = (V, E)$, where $|V| = n$ and $|E| = m$, the typical implementations require $O(n^2)$ steps. (This can be found in Chapter 7 of A. V. Aho, J. E. Hopcroft, and J. D. Ullman [1]; in Chapter 4 of S. Baase [2]; and in Chapter 4 of E. Horowitz and S. Sahni [14].) More recent implementations of the algorithm have improved the situation so that it now requires $O(m \log_2 n)$ steps. (This is discussed in the articles by R. Graham and P. Hell [13]; by D. B. Johnson [15]; and by A. Kershenbaum and R. Van Slyke [16].)

EXERCISES 13.2

1. Apply Kruskal's and Prim's algorithms to determine minimal spanning trees for the graph shown in Fig. 13.8.

2. Table 13.1 provides information on the distance (in miles) between pairs of cities in the state of Indiana.

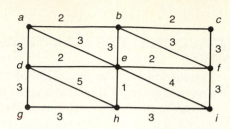

Figure 13.8

Table 13.1

	Bloomington	Evansville	Fort Wayne	Gary	Indianapolis	South Bend
Evansville	119	—	—	—	—	—
Fort Wayne	174	290	—	—	—	—
Gary	198	277	132	—	—	—
Indianapolis	51	168	121	153	—	—
South Bend	198	303	79	58	140	—
Terre Haute	58	113	201	164	71	196

A system of highways that connects these seven cities is to be constructed. Determine which highways should be constructed so that the cost of construction is minimal. (Assume that the cost of construction of a mile of highway is the same between any pair of cities.)

3. **a)** Answer Exercise 2 under the additional requirement that the system includes a highway directly linking Evansville and Indianapolis.

   b) If there must be a direct link between Fort Wayne and Gary in addition to the one connecting Evansville and Indianapolis, find the smallest number of miles of highway that must be constructed.

4. Modify Kruskal's algorithm to find a minimal spanning tree that includes one or more prescribed edges.

5. **a)** Modify Kruskal's algorithm to determine an optimal tree of *maximal* weight.

   b) Interpret the information of Exercise 2 in terms of the number of calls that can be placed between pairs of cities via the adoption of certain new telephone transmission lines. (Cities that are not directly linked must communicate through one or more intermediate cities.) How can the seven cities be minimally connected and allow a maximum number of calls to be placed?

6. Prove Theorem 13.2.

7. Let $G = (V, E)$ be a weighted connected undirected graph where for distinct edges $e_1, e_2 \in E$, $wt(e_1) \neq wt(e_2)$. Prove that G has only one minimal spanning tree.

13.3

TRANSPORT NETWORKS: THE MAX-FLOW MIN-CUT THEOREM

This section provides an application for weighted directed graphs to the flow of a commodity from a source to a prescribed destination. Such commodities may be gallons of oil that flow through pipelines or numbers of telephone calls transmitted in a communication system. In modeling such situations, we interpret the weight of an edge in the directed graph as a capacity that places an upper limit, for example, on the amount of oil that can flow through a certain part of a system of pipelines. These ideas are expressed formally in the following definition.

DEFINITION 13.1   Let $N = (V, E)$ be a loop-free connected directed graph. Then N is called a *network*, or *transport network*, if the following conditions are satisfied:

   a) There exists a unique vertex $a \in V$ with $\deg^-(a)$, the in degree of a, equal to 0. This vertex a is called the *source*.

   b) There is a unique vertex $z \in V$, called the *sink*, where $\deg^+(z)$, the out degree of z, equals 0.

   c) The graph N is weighted, so there is a function from E to the set of non-negative real numbers that assigns to each edge $e = (v, w) \in E$ a *capacity*, denoted by $c(e) = c(v, w)$.   ■

Example 13.5   The graph in Fig. 13.9 is a transport network. Here vertex a is the source; the sink is at vertex z. Since $c(a, b) + c(a, g) = 5 + 7 = 12$, the amount of the commodity being transported from a to z cannot exceed 12. With $c(d, z) + c(h, z) = 5 + 6 = 11$, the amount is further restricted to be no greater than 11. To determine the maximum amount that can be transported from a to z, we must consider the capacities of all edges in the network.   □

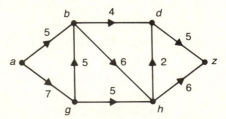

Figure 13.9

The following definition is introduced to assist us in solving this problem.

DEFINITION 13.2   If $N = (V, E)$ is a transport network, a function f from E to the nonnegative real numbers is called a *flow* for N if

   a) $f(e) \leq c(e)$ for each edge $e \in E$, and

   b) for any $v \in V$, other than the source a or the sink z, $\sum_{w \in V} f(w, v) = \sum_{w \in V} f(v, w)$. (If there is no edge (v, w), $f(v, w) = 0$.)   ■

The first condition specifies that the amount of material transported along a given edge cannot exceed the capacity of that edge. Condition (b) requires that the amount of material flowing into a vertex v must equal the amount that flows out from this vertex. This is so for all vertices except the source and sink.

Example 13.6 For each network in Fig. 13.10, the label x, y on edge e is determined such that $x = c(e)$ and y is the value assigned for a possible flow f. The label on each edge e satisfies $f(e) \le c(e)$. In part (a) of the figure, the "flow" into vertex g is 5, but the "flow" out from that vertex is $2 + 2 = 4$. Hence the function f is not a flow in this case. The function f for part (b) does satisfy both conditions, so it is a flow for the given network. $\square$

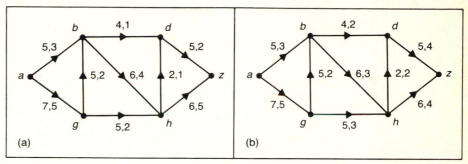

Figure 13.10

DEFINITION 13.3 Let f be a flow for a transport network $N = (V, E)$.

a) An edge e of the network is called *saturated* if $f(e) = c(e)$. When $f(e) < c(e)$, the edge is called *unsaturated*.

b) If a is the source of N, then $\text{val}(f) = \sum_{v \in V} f(a, v)$ is called the *value of the flow*.

Example 13.7 For the network in Fig. 13.10(b), only the edge (h, d) is saturated. All other edges are unsaturated. The value of the flow in this network is

$$\text{val}(f) = \sum_{v \in V} f(a, v) = f(a, b) + f(a, g) = 3 + 5 = 8.$$

But is there another flow f_1 such that $\text{val}(f_1) > 8$? The determination of a *maximal flow* (a flow that achieves the greatest possible value) is the objective of the remainder of this section. To accomplish this, we observe that in the network of Fig. 13.10(b),

$$\sum_{v \in V} f(a, v) = 8 = 4 + 4 = f(d, z) + f(h, z) = \sum_{v \in V} f(v, z).$$

Consequently, the total flow leaving the source a equals the total flow into the sink z. $\square$

The last remark in Example 13.7 seems like a reasonable circumstance, but will it occur in general? To prove the result for any network, we need the following special type of cut-set.

DEFINITION 13.4   If $N = (V, E)$ is a transport network and C is a cut-set for the undirected graph associated with N, then C is called a *cut*, or an *a-z cut*, if the removal of the edges in C from the network results in the separation of a and z. ▬

Example 13.8   Each of the dotted curves in Fig. 13.11 indicates a cut for the given network. The cut C_1 consists of the undirected edges $\{a, g\}$, $\{b, d\}$, $\{b, g\}$, and $\{b, h\}$. This cut partitions the vertices of the network into the two sets $P = \{a, b\}$ and its complement $\bar{P} = \{d, g, h, z\}$, so C_1 is denoted as $(P, \bar{P})$. The *capacity of a cut*, denoted $c(P, \bar{P})$, is defined by

$$c(P, \bar{P}) = \sum_{\substack{v \in P \\ w \in \bar{P}}} c(v, w),$$

the sum of the capacities of all edges (v, w), where $v \in P$ and $w \in \bar{P}$. In this example, $c(P, \bar{P}) = c(a, g) + c(b, d) + c(b, h) = 7 + 4 + 6 = 17$. [Considering the *directed* edges of the cut $C_1 = (P, \bar{P})$, namely $(a, g), (b, d), (b, h)$, we find that the removal of these edges does not result in a subgraph with two components. However, these three edges are minimal in the sense that their removal eliminates any possible directed path from a to z.]

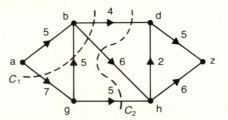

Figure 13.11

The cut C_2 induces the vertex partition $Q = \{a, b, g\}$, $\bar{Q} = \{d, h, z\}$ and has capacity $c(Q, \bar{Q}) = c(b, d) + c(b, h) + c(g, h) = 4 + 6 + 5 = 15$.

A third cut of interest is the one that induces the vertex partition $S = \{a, b, d, g, h\}$, $\bar{S} = \{z\}$. (What are the edges in this cut?) Its capacity is 11. ☐

Using the idea of the capacity of a cut, this next result provides an upper bound for the value of a flow in a network.

THEOREM 13.3   Let f be a flow in a network $N = (V, E)$. If $C = (P, \bar{P})$ is any cut in N, then val(f) cannot exceed $c(P, \bar{P})$.

Proof   Let vertex a be the source in N and vertex z the sink. Since deg$^-(a) = 0$, for any $w \in V$, $f(w, a) = 0$. Consequently,

$$\text{val}(f) = \sum_{v \in V} f(a, v) = \sum_{v \in V} f(a, v) - \sum_{w \in V} f(w, a).$$

By condition (b) in the definition of a flow, for any $x \in P, x \neq a, \sum_{v \in V} f(x, v) - \sum_{w \in V} f(w, x) = 0$.

Adding the results in the above equations yields

$$\text{val}(f) = \left[\sum_{v \in V} f(a, v) - \sum_{w \in V} f(w, a) \right] + \sum_{\substack{x \in P \\ x \neq a}} \left[\sum_{v \in V} f(x, v) - \sum_{w \in V} f(w, x) \right]$$

$$= \sum_{\substack{x \in P \\ v \in V}} f(x, v) - \sum_{\substack{x \in P \\ w \in V}} f(w, x)$$

$$= \left[\sum_{\substack{x \in P \\ v \in P}} f(x, v) + \sum_{\substack{x \in P \\ v \in \overline{P}}} f(x, v) \right] - \left[\sum_{\substack{x \in P \\ w \in P}} f(w, x) + \sum_{\substack{x \in P \\ w \in \overline{P}}} f(w, x) \right].$$

Since

$$\sum_{\substack{x \in P \\ v \in P}} f(x, v) \qquad \text{and} \qquad \sum_{\substack{x \in P \\ w \in P}} f(w, x)$$

are summed over the same set of all ordered pairs in $P \times P$, these summations are equal. Consequently,

$$\text{val}(f) = \sum_{\substack{x \in P \\ v \in \overline{P}}} f(x, v) - \sum_{\substack{x \in P \\ w \in \overline{P}}} f(w, x).$$

For all $x, w \in V, f(w, x) \geq 0$, so

$$\sum_{\substack{x \in P \\ w \in \overline{P}}} f(w, x) \geq 0 \qquad \text{and} \qquad \text{val}(f) \leq \sum_{\substack{x \in P \\ v \in \overline{P}}} f(x, v) \leq \sum_{\substack{x \in P \\ v \in \overline{P}}} c(x, v) = c(P, \overline{P}). \qquad \blacksquare$$

From Theorem 13.3 we find that in a network N, the value for *any* flow is less than or equal to the capacity of *any* cut in that network. Hence the value of the maximum flow cannot exceed the minimum capacity over all cuts in a network. For the network in Fig. 13.11, it can be shown that the cut consisting of edges (d, z) and (h, z) has minimum capacity 11. Consequently, the maximum flow f for the network satisfies $\text{val}(f) \leq 11$. It will turn out that the value of the maximum flow is 11. How to construct such a flow and why its value equals the minimum capacity among all cuts will be dealt with in the main theorem of this section.

However, before we deal with this construction, let us note that in the proof of Theorem 13.3, the value of a flow is given by

$$\text{val}(f) = \sum_{\substack{x \in P \\ v \in \overline{P}}} f(x, v) - \sum_{\substack{x \in P \\ w \in \overline{P}}} f(w, x),$$

where $(P, \overline{P})$ is *any* cut in N. Therefore, once a flow is constructed in a network, then for any cut $(P, \overline{P})$ in the network, the value of the flow equals the sum of the flows in the edges directed from the vertices in P to those in $\overline{P}$ minus the sum of the flows in the edges directed from the vertices in $\overline{P}$ to those in P.

This observation leads to the following result.

COROLLARY 13.1   If f is a flow in a transport network $N = (V, E)$, then the value of the flow from the source a is equal to the value of the flow into the sink z.

Proof   Let $P = \{a\}$, $\overline{P} = V - \{a\}$, and $Q = V - \{z\}$, $\overline{Q} = \{z\}$. From the above observation,

$$\sum_{\substack{x \in P \\ v \in \overline{P}}} f(x, v) - \sum_{\substack{x \in P \\ w \in \overline{P}}} f(w, x) = \mathrm{val}(f) = \sum_{\substack{y \in Q \\ v \in \overline{Q}}} f(y, v) - \sum_{\substack{y \in Q \\ w \in \overline{Q}}} f(w, y).$$

With $P = \{a\}$ and $\deg^-(a) = 0$, $\sum_{x \in P, w \in \overline{P}} f(w, x) = \sum_{w \in \overline{P}} f(w, a) = 0$. Similarly, for $\overline{Q} = \{z\}$ and $\deg^+(z) = 0$, $\sum_{y \in Q, w \in \overline{Q}} f(w, y) = \sum_{y \in Q} f(z, y) = 0$. Consequently,

$$\sum_{\substack{x \in P \\ v \in \overline{P}}} f(x, v) = \sum_{v \in \overline{P}} f(a, v) = \mathrm{val}(f) = \sum_{\substack{y \in Q \\ v \in \overline{Q}}} f(y, v) = \sum_{y \in Q} f(y, z),$$

and this establishes the corollary.   ■

We turn now to the main result of the section.

THEOREM 13.4   (*The Max-Flow Min-Cut Theorem*) For a transport network $N = (V, E)$, the maximum flow that can be attained in N is equal to the minimum capacity over all cuts in the network.

Proof   By Theorem 13.3, if $(P, \overline{P})$ is a cut of minimum capacity in N, then the value of any flow f in N satisfies $\mathrm{val}(f) \leq c(P, \overline{P})$. To verify the existence of a flow f for which $\mathrm{val}(f) = c(P, \overline{P})$, the following iterative algorithm, called the *labeling procedure*, provides the necessary steps.

The Labeling Procedure

Step 1: Given a network N, define an initial flow f in N by $f(e) = 0$ for every e in E. (This function f satisfies the conditions in the definition of a flow.)

Step 2: Label the source a with $(-, \infty)$. (This label indicates that from the source a as much material is available as is needed to achieve a maximum flow.)

Step 3: For any vertex x that is adjacent from a, label x as follows:

a) If $c(a, x) - f(a, x) > 0$, define $\Delta(x) = c(a, x) - f(a, x)$ and label vertex x with $(a^+, \Delta(x))$.

b) If $c(a, x) - f(a, x) = 0$, leave vertex x unlabeled.

(The label $(a^+, \Delta(x))$ indicates that the present flow from a to x can be increased by the amount $\Delta(x)$, with the $\Delta(x)$ additional units supplied from the source a.)

Step 4: Let $x \in V$, $x \neq a$, and suppose that x is labeled.
     If $(x, y) \in E$ and y is not labeled, label vertex y as follows:

a) If $c(x, y) - f(x, y) > 0$, define $\Delta(y) = \min\{\Delta(x), c(x, y) - f(x, y)\}$, and label vertex y as $(x^+, \Delta(y))$.

b) If $c(x, y) - f(x, y) = 0$, leave vertex y unlabeled.

(The label $(x^+, \Delta(y))$ indicates that the present flow into vertex y can be increased by an amount $\Delta(y)$ taken from vertex x.)

Step 5: Let $x \in V$, with x labeled.
   If $(y, x) \in E$, and y is not labeled, label vertex y as follows:

a) If $f(y, x) > 0$, label vertex y as $(x^-, \Delta(y))$, where $\Delta(y) = \min\{\Delta(x), f(y, x)\}$.
b) If $f(y, x) = 0$, leave vertex y unlabeled.

(The label $(x^-, \Delta(y))$ tells us that by decreasing the flow from y to x, the total flow out of y to the labeled vertices can be decreased by $\Delta(y)$. These $\Delta(y)$ units may then be used to increase the total flow from y to the unlabeled vertices.)

Since a vertex y may be adjacent to, or from, more than one labeled vertex, the results of this procedure need not be unique. In addition, if x is labeled, the network could include both of the edges (x, y) and (y, x), perhaps resulting in two possible labels for y. But the procedure is designed to create a maximum flow, and there may be more than one. Nonetheless, when a vertex can be labeled in more than one way, an arbitrary choice must be made.

As we apply the labeling procedure to the vertices of the given network, steps 4 and 5 are repeated as long as possible, with regard to the current (and changing) set of labeled vertices. With each iteration there are two possible cases to consider.

Case 1: If the sink z is labeled as $(x^+, \Delta(z))$, the flow in the edge (x, z) can be increased from $f(x, z)$ to $f(x, z) + \Delta(z)$, as indicated by the label.

Considering vertex x, it can be labeled as either $(v^+, \Delta(x))$ or $(v^-, \Delta(x))$, where $\Delta(x) \geq \Delta(z)$. For the label $(v^+, \Delta(x))$, we may regard vertex v as the source for increasing the flow in edge (x, z) by the amount $\Delta(z)$. In this case we likewise increase the present flow in edge (v, x) from $f(v, x)$ to $f(v, x) + \Delta(z)$ (not $f(v, x) + \Delta(x)$). If x has the label $(v^-, \Delta(x))$, then the flow in the edge (x, v) is changed from $f(x, v)$ to $f(x, v) - \Delta(z)$ in order to provide the additional flow of Δz units from x to z.

As this process continues back to source a, each directed edge along a path from a to z has its flow increased by Δz units. When this is accomplished, all vertex labels, except for $(-, \infty)$ on the source, are removed; the process is repeated to see whether it is possible to increase the flow even more.

Case 2: If the labeling procedure is carried out as far as possible and the sink z is unlabeled, then the maximum flow has been attained. Let P be the vertices in V that are labeled, and $\overline{P} = V - P$. Since vertices in $\overline{P}$ are not labeled, the flows in the edges (x, y), where $x \in P$ and $y \in \overline{P}$, satisfy $f(x, y) = c(x, y)$. Also, for any edge (w, v) with $w \in \overline{P}$ and $v \in P$, we have $f(w, v) = 0$. Consequently, there is a flow for the given network, where the value of the flow is equal to the capacity of the cut $(P, \overline{P})$. From Theorem 13.3, this flow is a maximum. ∎

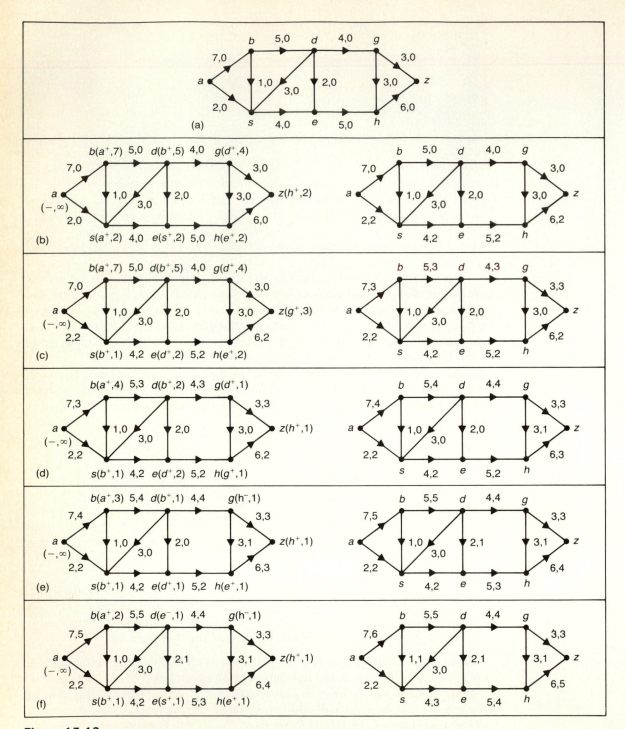

Figure 13.12

Before demonstrating the labeling procedure we state one last corollary, the proof of which is left as an exercise.

COROLLARY 13.2   Let $N = (V, E)$ be a transport network where for each $e \in E$, $c(e)$ is a positive integer. Then there is a maximum flow f for N, where $f(e)$ is a nonnegative integer for each edge e.

Example 13.9   Use the labeling procedure to find a maximum flow for the transport network in Fig. 13.12(a).

In this network, each edge is labeled with an ordered pair x, y, where x is the capacity of the edge and $y = 0$ indicates an initial flow for the network. Figure 13.12(b) demonstrates the first application of the labeling procedure. Here a choice is made in labeling the sink, z. We chose $(h^+, 2)$ as the label instead of $(g^+, 3)$. (What other label could have been used for vertex e?) Backtracking from z to h to e to s to a, and increasing the flow in each edge by $\Delta(z) = 2$, we obtain the new flow in Fig. 13.12(b). Parts (c), (d), (e), and (f) of the figure show a second, third, fourth, and fifth application of the labeling procedure. Note how vertex g has a negative label in parts (e) and (f). Also, part (f) provides a second instance of a negative label, this time for vertex d. Applying the labeling procedure one last time, the transport network is labeled as in Fig. 13.13. Here the sink z is unlabeled and the second case in the procedure is now followed. Letting $P = \{a, b\}$ and $\bar{P} = \{s, d, e, g, h, z\}$, we find that $c(P, \bar{P}) = c(b, d) + c(b, s) + c(a, s) = 5 + 1 + 2 = 8 = 5 + 3 = f(h, z) + f(g, z)$, the flow into z. The edges in N that are crossed by the dotted curve are the edges of the (undirected) cut-set associated with the cut $(P, \bar{P})$. It consists of all edges of the form $\{x, y\}$ where $x \in P, y \in \bar{P}$. □

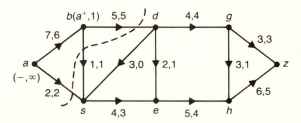

Figure 13.13

We close this section with three examples that are modeled with the concepts of the transport network. After setting up the models, the final solution of each example is left to the section exercises.

Example 13.10   Computer chips are manufactured (in units of a thousand) at three companies c_1, c_2, and c_3. These chips are then distributed to two computer manufacturers, m_1 and m_2, through the "transport network" in Fig. 13.14(a), where there are the three sources—c_1, c_2, and c_3—and the two sinks m_1 and m_2. Company c_1 can produce up to 15 units, company c_2 up to 20 units, and company c_3 up to 25 units. If each manufacturer needs 25 units, how many units should each of the companies produce so that they can meet the demand of the manufacturers, or at least supply them with as many units as the network will allow?

In order to model this example with a transport network, we introduce a source a and a sink z, as shown in Fig. 13.14(b). The manufacturing capabilities of the three companies are then used to label the edges (a, c_1), (a, c_2), and (a, c_3). For the edges (m_1, z) and (m_2, z) the demands are used as labels. To answer the question posed here, one applies the labeling procedure to this network to find the value of a maximum flow. □

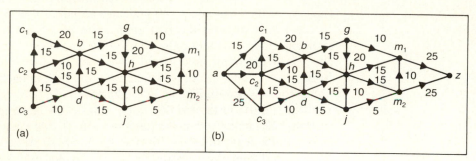

Figure 13.14

Example 13.11   The transport network shown in Fig. 13.15(a) has an added restriction, for there are capacities assigned to vertices other than the source and sink. Part (b) of the figure shows how to redraw the network to obtain one where the labeling procedure can be applied. For any vertex v other than a or z, split v into vertices v_1 and v_2. Draw an edge from v_1 to v_2 and label it with the capacity originally assigned to v. An edge of the form (v, w), where $v \neq a, w \neq z$, then becomes the edge (v_2, w_1), maintaining the capacity of (v, w). Edges of the form (a, v) become (a, v_1) with capacity $c(a, v)$. An edge such as (w, z) is replaced by the edge (w_2, z), with capacity $c(w, z)$.

The maximum flow for the given network is now determined by applying the labeling procedure to the network shown in Fig. 13.15(b). □

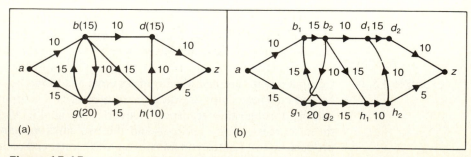

Figure 13.15

Example 13.12   During the practice of war games, messengers must deliver information from headquarters (vertex a) to a field command station (vertex z). Since certain roads may be blocked or destroyed, how many messengers should be sent out so that each travels along a path that has no edge in common with any other path taken?

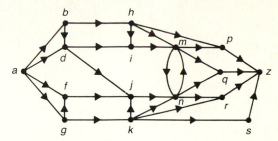

Figure 13.16

   Since the distances between vertices is not relevant here, the graph shown in Fig. 13.16 has no capacities assigned to its edges. The problem here is to determine the maximum number of edge-disjoint paths from a to z. Assigning each edge a capacity of 1 converts the problem into a maximum-flow problem, where the number of edge-disjoint paths (from a to z) equals the value of a maximum flow for the network.   □

EXERCISES 13.3

1. **a)** For the network shown in Fig. 13.17, let the capacity of each edge be 10. If each edge e in the figure is labeled by a function f, as shown, determine the values of s, t, w, x, and y so that f is a flow in the network.

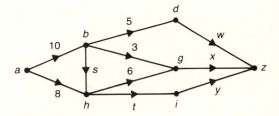

Figure 13.17

   b) What is the value of this flow?

   c) Find three cuts $(P, \bar{P})$ in this network that have capacity 30.

2. If $N = (V, E)$ is a transport network, let f be a flow in N and let $(P, \bar{P})$ be a cut. Prove that the value of the flow f equals $c(P, \bar{P})$ if and only if

   a) $f(e) = c(e)$ for each edge $e = (x, y), x \in P, y \in \bar{P}$, and

   b) $f(e) = 0$ for each edge $e = (v, w), v \in \bar{P}, w \in P$.

3. Find a maximum flow and the corresponding minimum cut for each transport network shown in Fig. 13.18.

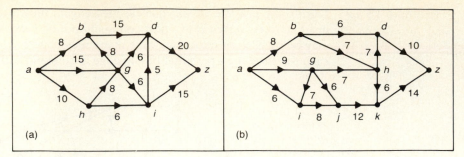

Figure 13.18

4. Apply the labeling procedure to find a maximum flow in Examples 13.10, 13.11, and 13.12.

5. Prove Corollary 13.2.

6. In each of the following "transport networks" two companies c_1 and c_2 produce a certain product that is used by two manufacturers m_1 and m_2. For the network shown in part (a) of Fig. 13.19, c_1 produces 8 units and c_2 produces 7 units; manufacturer m_1 requires 7 units and manufacturer m_2 needs 6 units. In the network shown in Fig. 13.19(b), each company produces 7 units and each manufacturer needs 6 units. In which situation(s) can the suppliers meet the manufacturers' demands?

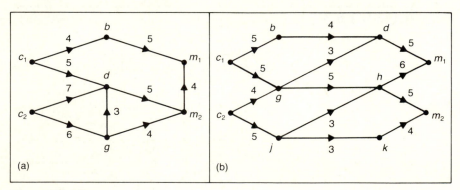

Figure 13.19

7. Find a maximum flow for the network shown in Fig. 13.20. The capacities on the undirected edges indicate that the capacity is the same in either direction.

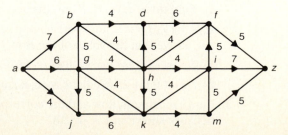

Figure 13.20

13.4
MATCHING THEORY

The Villa school district must hire four teachers to teach classes in the following subjects: mathematics (s_1), computer science (s_2), chemistry (s_3), physics (s_4), and biology (s_5). Four candidates who are interested in teaching in this district are Miss Carelli (c_1), Mr. Ritter (c_2), Ms. Camille (c_3), and Mrs. Lewis (c_4). Miss Carelli is certified in mathematics and computer science; Mr. Ritter in mathematics and physics; Ms. Camille in biology; and Mrs. Lewis in chemistry, physics, and computer science. If the district hires all four candidates, can each teacher be assigned to teach a (different) subject in which he or she is certified?

This problem is an example of a general situation called the *assignment problem*. Using the principle of inclusion and exclusion in conjunction with the rook polynomial (see Sections 8.4 and 8.5), one can determine in how many ways, if any, the four teachers may be assigned so that each teaches a different subject for which he or she is qualified. However, these techniques do not provide a means of setting up any of these assignments. In Fig. 13.21 the problem is modeled by means of a bipartite graph $G = (V, E)$, where V is partitioned as $X \cup Y$ with $X = \{c_1, c_2, c_3, c_4\}$ and $Y = \{s_1, s_2, s_3, s_4, s_5\}$. The edges $\{c_1, s_2\}$, $\{c_2, s_4\}$, $\{c_3, s_5\}$, $\{c_4, s_3\}$ demonstrate such an assignment of X into Y.

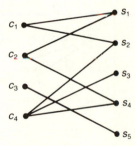

Figure 13.21

To examine this idea further, the following concepts are introduced.

DEFINITION 13.5   Let $G = (V, E)$ be a bipartite graph with V partitioned as $X \cup Y$. (Each edge of E has the form $\{x, y\}$ with $x \in X$ and $y \in Y$.)

    a) A *matching* in G is a subset of E such that no two edges share a common vertex in X or Y.

    b) A *complete matching* of X into Y is a matching in G such that every $x \in X$ is the endpoint of an edge.

    In terms of functions, a matching is a function that establishes a one-to-one correspondence between a subset of X and a subset of Y. When the matching is complete, a one-to-one function from X into Y is defined. The example in Fig. 13.21 contains such a function and a complete matching.

    For a bipartite graph $G = (V, E)$ with V partitioned as $X \cup Y$, a complete matching of X into Y requires $|X| \le |Y|$. If $|X|$ is large, then the construction of such a matching cannot be accomplished just by observation or trial and error. The following theorem, due to the English mathematician Philip Hall (1935), provides necessary and sufficient conditions for the existence of such a matching. The proof of the theorem, however, is not that given by Hall. A constructive proof that uses the material developed on transport networks is given.

THEOREM 13.5   Let $G = (V, E)$ be bipartite with V partitioned as $X \cup Y$. A complete matching of X into Y exists if and only if for every subset A of X, $|A| \leq |R(A)|$, where $R(A)$ is the subset of Y consisting of those vertices that are adjacent to the vertices in A.

Before proving the theorem, we illustrate its use in the following example.

Example 13.13   **a)** The bipartite graph shown in Fig. 13.22(a) has no complete matching. Any attempt to construct such a matching must include $\{x_1, y_1\}$ and either $\{x_2, y_3\}$ or $\{x_3, y_3\}$. If $\{x_2, y_3\}$ is included, there is no match for x_3. Likewise, if $\{x_3, y_3\}$ is included, we are not able to match x_2. If $A = \{x_1, x_2, x_3\} \subseteq X$, then $R(A) = \{y_1, y_2\}$. With $|A| = 3 > 2 = |R(A)|$, it follows from Theorem 13.5 that no complete matching can exist.

Figure 13.22

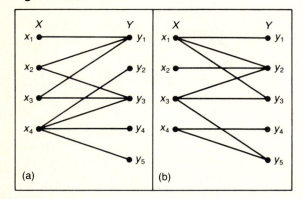

(a)   (b)

Table 13.2

A	$R(A)$	$|A|$	$|R(A)|$
$\emptyset$	$\emptyset$	0	0
$\{x_1\}$	$\{y_1, y_2, y_3\}$	1	3
$\{x_2\}$	$\{y_2\}$	1	1
$\{x_3\}$	$\{y_2, y_3, y_5\}$	1	3
$\{x_4\}$	$\{y_4, y_5\}$	1	2
$\{x_1, x_2\}$	$\{y_1, y_2, y_3\}$	2	3
$\{x_1, x_3\}$	$\{y_1, y_2, y_3, y_5\}$	2	4
$\{x_1, x_4\}$	Y	2	5
$\{x_2, x_3\}$	$\{y_2, y_3, y_5\}$	2	3
$\{x_2, x_4\}$	$\{y_2, y_4, y_5\}$	2	3
$\{x_3, x_4\}$	$\{y_2, y_3, y_4, y_5\}$	2	4
$\{x_1, x_2, x_3\}$	$\{y_1, y_2, y_3, y_5\}$	3	4
$\{x_1, x_2, x_4\}$	Y	3	5
$\{x_1, x_3, x_4\}$	Y	3	5
$\{x_2, x_3, x_4\}$	$\{y_2, y_3, y_4, y_5\}$	3	4
X	Y	4	5

b) For the graph in part (b) of the figure, consider the exhaustive listing in Table 13.2. Assuming the validity of Theorem 13.5, this listing indicates that the graph contains a complete matching. ☐

We turn now to a proof of the theorem.

Proof   With V partitioned as $X \cup Y$, let $X = \{x_1, x_2, \ldots, x_m\}$ and $Y = \{y_1, y_2, \ldots, y_n\}$. Construct a transport network N that extends graph G by introducing two new vertices a (the source) and z (the sink). For each vertex x_i, $1 \leq i \leq m$, draw edge

(a, x_i); for each vertex y_j, $1 \leq j \leq n$, draw edge (y_j, z). Each new edge is given a capacity of 1. Let M be any positive number that exceeds $|X|$. Assign each edge in G the capacity M. The original graph G and its associated network N appear as shown in Fig. 13.23. It follows that a complete matching exists in G if and only if there is a maximum flow in N that uses all edges (a, x_i), $1 \leq i \leq m$. Then the value of such a maximum flow is $m = |X|$.

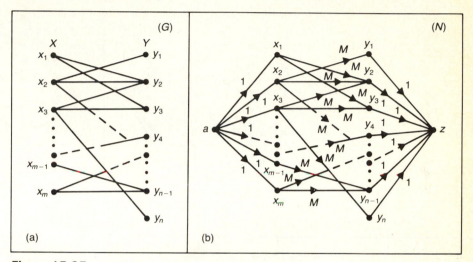

Figure 13.23

If $|A| \leq |R(A)|$ for all $A \subseteq X$, we shall prove that there is a complete matching in G by showing that for any cut $(P, \bar{P})$ in N, $c(P, \bar{P}) \geq |X|$. Let $A \subseteq X$, where $A = \{x_1, x_2, \ldots, x_i\}$, for some $0 \leq i \leq m$. (Elements of X are relabeled, if necessary, so that the subscripts on the elements of A are consecutive. When $i = 0$, $A = \emptyset$.) Let P be the source a together with the vertices in A and the set $B \subseteq Y$, as shown in Fig. 13.24(a). (Elements of Y are also relabeled if necessary.) Then $\bar{P} = (X - A) \cup (Y - B) \cup \{z\}$. If there is an edge $\{x, y\}$ with $x \in A$ and $y \in (Y - B)$, then the capacity of that edge is a summand in $c(P, \bar{P})$ and $c(P, \bar{P}) \geq M > |X|$. Should no such edge exist, then $c(P, \bar{P})$ is determined by the capacities of (1) the edges from the source a to the vertices in $X - A$ and (2) the edges from the vertices in B to the sink z. Since each of these edges has capacity 1, $c(P, \bar{P}) = |X - A| + |B| = |X| - |A| + |B|$. With $B \supseteq R(A)$, we have $|B| \geq |R(A)|$, and since $|R(A)| \geq |A|$, it follows that $|B| \geq |A|$. Consequently $c(P, \bar{P}) = |X| + (|B| - |A|) \geq |X|$. Therefore, since every cut in network N has capacity at least $|X|$, by Theorem 13.4 any maximum flow for N has value $|X|$. Such a flow will result in exactly $|X|$ edges from X to Y having flow 1, and this flow provides a complete matching of X into Y.

Conversely, let $A \subseteq X$ with $|A| > |R(A)|$. Let $(P, \bar{P})$ be the cut shown for the network in Fig. 13.24(b), with $P = \{a\} \cup A \cup R(A)$ and $\bar{P} = (X - A) \cup (Y - R(A)) \cup \{z\}$. Then $c(P, \bar{P})$ is determined by (1) the edges from the source a to the vertices in $X - A$ and (2) the edges from the vertices in $R(A)$ to the sink z. Hence $c(P, \bar{P}) = |X - A| + |R(A)| = |X| - (|A| - |R(A)|) < |X|$, since $|A| >$

$|R(A)|$. The network has a cut of capacity less than $|X|$, so once again by Theorem 13.4 it follows that any maximum flow in the network has value smaller than $|X|$. Therefore there is no complete matching from X into Y for the given bipartite graph G. ∎

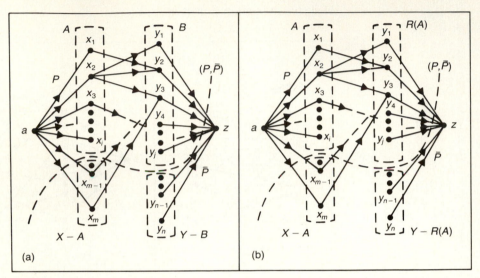

Figure 13.24

Example 13.14   Five students s_1, s_2, s_3, s_4, and s_5 are members of three committees c_1, c_2, and c_3. The bipartite graph shown in Fig. 13.25(a) indicates the committee membership. Each committee is to select a student representative to meet with the school president. Can a selection be made in such a way that each committee has a distinct representative?

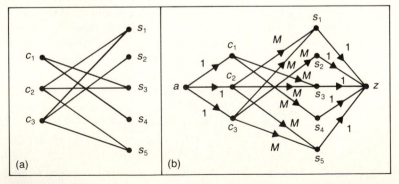

Figure 13.25

Although this problem is small enough to solve by inspection, we use the ideas developed in the proof of Theorem 13.5. Figure 13.25(b) provides the network for the given bipartite graph. In Fig. 13.26(a) the labeling procedure

(Theorem 13.4) is applied for the first time and indicates the edge (c_3, s_5) as a possible start for a complete matching. (Many edge labels are omitted in the figures in order to simplify the diagrams. Any unlabeled edge that starts at a or terminates at z has the label $1, 0$; all other unlabeled edges are labeled $M, 0$.) Two further applications of the procedure provide the network shown in Fig. 13.26(b). Here the maximum flow indicates that $(c_1, s_4), (c_2, s_3), (c_3, s_5)$ is a possible solution to the choice of representatives.

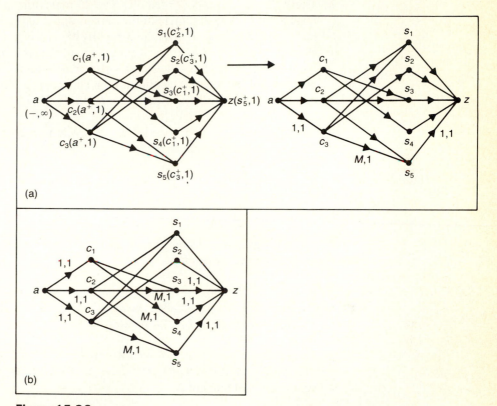

Figure 13.26

This example is a particular instance of a problem studied by Philip Hall. He considered a collection of sets $A_1, A_2, \ldots, A_n$, where the elements $a_1, a_2, \ldots, a_n$ were called a *system of distinct representatives* for the collection if (a) $a_i \in A_i$, $1 \le i \le n$; and (b) $a_i \ne a_j, 1 \le i < j \le n$. Rewording Theorem 13.5 in this context, we find that the collection $A_1, A_2, \ldots, A_n$ has a system of distinct representatives if and only if, for all $1 \le i \le n$, the union of any i of the sets $A_1, A_2, \ldots, A_n$ contains at least i elements. $\square$

Although the condition in Theorem 13.5 may be very tedious to check, the following corollary provides sufficient conditions for the existence of a complete matching.

COROLLARY 13.3 Let $G = (V, E)$ be a bipartite graph with V partitioned as $X \cup Y$. There is a complete matching of X into Y if, for some $k \in \mathbf{Z}^+$, $\deg(x) \geq k \geq \deg(y)$ for all vertices $x \in X$ and $y \in Y$.

Proof This proof is left for the section exercises. ∎

Example 13.15

a) Corollary 13.3 is applicable to the graph shown in Fig. 13.25(a). Here the appropriate value of k is 2.

b) There are 50 students (25 females and 25 males) in the senior class at Bell High School. If each female in the class is appreciated by exactly five of the males, and each male enjoys the company of exactly five of the females in the class, then it is possible for each male to go to the class party with a female he likes, and each female will attend with a male who likes her. (As a result of problems of this type, the condition in Theorem 13.5 is often referred to in the literature as *Hall's Marriage Condition*.) □

For problems such as Example 13.13(a), where a complete matching does not exist, the following type of matching is often of interest.

DEFINITION 13.6 If $G = (V, E)$ is a bipartite graph with V partitioned as $X \cup Y$, a *maximal matching* in G is one that matches as many vertices in X as possible with the vertices in Y.

To investigate the existence and construction of a maximal matching, the following new idea is presented.

DEFINITION 13.7 Let $G = (V, E)$ be a bipartite graph where V is partitioned as $X \cup Y$. If $A \subseteq X$, then $\delta(A) = |A| - |R(A)|$ is called the *deficiency of A*. The *deficiency of graph G*, denoted $\delta(G)$, is given by $\delta(G) = \max\{\delta(A) | A \subseteq X\}$.

For $\emptyset \subseteq X$, we have $R(\emptyset) = \emptyset$, so $\delta(\emptyset) = 0$ and $\delta(G) \geq 0$. If $\delta(G) > 0$, there is a subset A of X with $|A| - |R(A)| > 0$, so $|A| > |R(A)|$ and by Theorem 13.5 there is no complete matching of X into Y.

Example 13.16 The graph in Fig. 13.27(a) has no complete matching. [See Example 13.13(a).] For $A = \{x_1, x_2, x_3\}$, we find that $R(A) = \{y_1, y_3\}$ and $\delta(A) = 3 - 2 = 1$. As a result of this subset A, $\delta(G) = 1$. Removing one of the vertices from A (and the edges incident with it), we obtain the subgraph shown in part (b) of the figure. This (bipartite) subgraph provides a complete matching from $X_1 = \{x_2, x_3, x_4\}$ into Y. The edges $\{x_2, y_1\}$, $\{x_3, y_3\}$, and $\{x_4, y_4\}$ indicate one such matching that is also a maximal matching of X into Y. □

The ideas developed in Example 13.16 lead to the following theorem.

THEOREM 13.6 Let $G = (V, E)$ be bipartite with V partitioned as $X \cup Y$. The maximum number of vertices in X that can be matched with those in Y is $|X| - \delta(G)$. Moreover, a matching of size $|X| - \delta(G)$ exists.

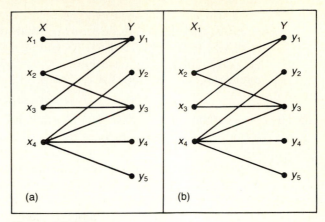

Figure 13.27

Proof We provide a constructive proof, using transport networks as in the proof of Theorem 13.5. As in Figure 13.23, let N be the network associated with the bipartite graph G. The result will follow when we show that (a) the capacity of any cut $(P, \bar{P})$ in N is greater than or equal to $|X| - \delta(G)$, and (b) there exists a cut with capacity $|X| - \delta(G)$.

Let $(P, \bar{P})$ be any cut in N, where P is made up of the source a, the vertices in $A \subseteq X$, and the vertices in $B \subseteq Y$. [See Fig. 13.24(a).] As in the proof of Theorem 13.5, the subsets A, B may be $\emptyset$.

1. If edge (x, y) is in N with $x \in A$ and $y \in Y - B$, then $c(x, y)$ is a summand in $c(P, \bar{P})$. Since $c(x, y) = M > |X|$, it follows that $c(P, \bar{P}) > |X| \geq |X| - \delta(G)$.

2. If no such edge as in (1) exists, then $c(P, \bar{P})$ is determined by the $|X - A|$ edges from a to $X - A$ and the $|B|$ edges from B to z. Since each of these edges has capacity 1, we find that $c(P, \bar{P}) = |X - A| + |B| = |X| - |A| + |B|$. No edge connects a vertex in A with a vertex in $Y - B$, so $R(A) \subseteq B$ and $|R(A)| \leq |B|$. Consequently, $c(P, \bar{P}) = (|X| - |A|) + |B| \geq (|X| - |A|) + |R(A)| = |X| - (|A| - |R(A)|) = |X| - \delta(A) \geq |X| - \delta(G)$.

Therefore, in either case, $c(P, \bar{P}) \geq |X| - \delta(G)$ for any cut $(P, \bar{P})$ in N.

To complete the proof, we must establish the existence of a cut with capacity $|X| - \delta(G)$. Since $\delta(G) = \max\{\delta(A) | A \subseteq X\}$, we can select a subset A of X with $\delta(G) = \delta(A)$. Examining Fig. 13.24(b), we let $P = \{a\} \cup A \cup R(A)$. Then $\bar{P} = (X - A) \cup (Y - R(A)) \cup \{z\}$. There is no edge between the vertices in A and those in $Y - R(A)$, so $c(P, \bar{P}) = |X - A| + |R(A)| = |X| - (|A| - |R(A)|) = |X| - \delta(A) = |X| - \delta(G)$. ∎

We close this section with two examples that deal with these concepts.

Example 13.17 Let $G = (V, E)$ be bipartite with V partitioned as $X \cup Y$. For each $x \in X$, $\deg(x) \geq 4$ and for each $y \in Y$, $\deg(y) \leq 5$. If $|X| \leq 15$, find an upper bound for $\delta(G)$.

Let $\emptyset \neq A \subseteq X$ and let $E_1 \subseteq E$ where $E_1 = \{\{a, b\} | a \in A, b \in R(A)\}$. Since $\deg(a) \geq 4$ for all $a \in A$, $|E_1| \geq 4|A|$. With $\deg(b) \leq 5$ for all $b \in R(A)$, $|E_1| \leq 5|R(A)|$. Hence $4|A| \leq 5|R(A)|$ and $\delta(A) = |A| - |R(A)| \leq |A| - (4/5)|A| = (1/5)|A|$. Since $A \subseteq X$, we have $|A| \leq 15$, so $\delta(A) \leq (1/5)(15) = 3$. Consequently, $\delta(G) = \max\{\delta(A) | A \subseteq X\} \leq 3$. ☐

Example 13.18   At a certain college, each of 110 freshmen is to be advised by a faculty advisor when she or he arrives for registration. These freshmen fall into four groups: 20 chemistry majors, 20 physics majors, 35 mathematics majors, and 35 computer science majors. Each prospective chemist may select any one of 12 faculty advisors (not necessarily the same 12 for each prospective chemist); each physics major, any one of 8 such advisors. The mathematics majors each have a choice of 5 advisors, but the future computer scientists are limited to a choice of only 3. There are 60 faculty advisors and each can handle up to 7 students at one time. On registration day, all of the students are waiting in the auditorium when the registrar arrives. If students are assigned according to their preferences as described above, what is the largest crowd of students that should have to wait for a later meeting with an advisor of their choice?

Let X be the set of 110 students and Y the set of 60 faculty members. The objective then is to find a maximum matching of X into Y. For any $x \in X, y \in Y$, draw an edge $\{x, y\}$ if student x can select faculty advisor y. This yields the bipartite graph $G = (V, E)$, where V is partitioned as $X \cup Y$. To answer the question we need to compute $\delta(G)$.

Suppose that $A \subseteq X$, where the numbers of chemistry, physics, mathematics, and computer science majors in A are n_1, n_2, n_3, and n_4, respectively. Then $\delta(A) = |A| - |R(A)| = (n_1 + n_2 + n_3 + n_4) - |R(A)|$. With the given restrictions on the number of possible advisor selections for each major and on the number of students an advisor can assist at one time, $7|R(A)| \geq 12n_1 + 8n_2 + 5n_3 + 3n_4$. Therefore, $\delta(A) \leq (n_1 + n_2 + n_3 + n_4) - (12n_1 + 8n_2 + 5n_3 + 3n_4)/7 = (-5n_1/7) + (-n_2/7) + (2n_3/7) + (4n_4/7)$. The subset A was chosen arbitrarily, so this upper bound on $\delta(A)$ is maximal when $n_1 = n_2 = 0$, $n_3 = 35$, and $n_4 = 35$. Hence $\delta(G) \leq (70 + 140)/7 = 30$, and no matter how the students are assigned (following the procedures), there should be at least $|X| - \delta(G) = 80$ who are assigned and no more than 30 who are kept waiting. ☐

EXERCISES 13.4

1. For the graph shown in Fig. 13.21, if four edges are selected at random, what is the probability that they provide a complete matching of X into Y?

2. Cathy is liked by Albert, Joseph, and Robert; Janice by Joseph and Dennis; Theresa by Albert and Joseph; Nettie by Dennis, Joseph, and Frank; and Karen by Albert, Joseph, and Robert. (a) Set up a bipartite graph to model the matching problem where each man is paired with a woman he likes. (b) Draw the associated network for the graph in part (a) and determine a

maximum flow for this network. What complete matching does this determine? (c) Is there a complete matching that pairs Janice with Dennis and Nettie with Frank? (d) Is it possible to determine two complete matchings where each man is paired with two different women?

3. At Rydell High School the senior class is represented on six school committees by Annemarie (A), Gary (G), Jill (J), Kenneth (K), Michael (M), Norma (N), Paul (P), and Rosemary (R). The senior members of these committees are $\{A, G, J, P\}$, $\{G, J, K, R\}$, $\{A, M, N, P\}$, $\{A, G, M, N, P\}$, $\{A, G, K, N, R\}$, and $\{G, K, N, R\}$.

   a) The student government calls a meeting that requires the presence of a senior member from each committee. Find a selection that maximizes the number of seniors involved.

   b) Before the meeting the finances of each committee are to be reviewed by a senior who is not on that committee. Can this be accomplished? If so, how?

4. Prove Corollary 13.3.

5. Brenda is in charge of assigning students to part-time jobs at the college where she works. She has 25 student applications and there are 25 different part-time jobs available on the campus. Each applicant is qualified for at least four of the jobs, but each job can be performed by at most four of the applicants. Can Brenda assign all the students to jobs for which they are qualified? Explain.

6. For each of the following collections of sets, determine if possible a system of distinct representatives. If no such system exists, explain why.

   a) $A_1 = \{2, 3, 4\}, A_2 = \{3, 4\}, A_3 = \{1\}, A_4 = \{2, 3\}$
   b) $A_1 = A_2 = A_3 = \{2, 4, 5\}, A_4 = A_5 = \{1, 2, 3, 4, 5\}$
   c) $A_1 = \{1, 2\}, A_2 = \{2, 3, 4\}, A_3 = \{2, 3\}, A_4 = \{1, 3\}, A_5 = \{2, 4\}$

7. a) Determine all systems of distinct representatives for the collection of sets $A_1 = \{1, 2\}, A_2 = \{2, 3\}, A_3 = \{3, 4\}, A_4 = \{4, 1\}$.

   b) Given the collection of sets $A_1 = \{1, 2\}$, $A_2 = \{2, 3\}$, $A_3 = \{3, 4\}, \ldots, A_n = \{n, 1\}$, determine how many different systems of distinct representatives exist for the collection.

8. Let $A_1, A_2, \ldots, A_n$ be a collection of sets where $A_1 = A_2 = \ldots = A_n$ and $|A_i| = k > 0, 1 \le i \le n$. (a) Prove that the given collection has a system of distinct representatives if and only if $n \le k$. (b) When $n \le k$, how many different systems exist for the collection?

9. At each of seven universities there are six graduates with training in both computer science and Japanese. Six international corporations wish to hire 10, 9, 8, 7, 7, and 7 graduates, respectively, with these qualifications. Company policy (at all six corporations) dictates that no two newly hired professionals with this training may be from the same university. Determine whether all of these graduates will be hired by these six corporations.

10. Let $G = (V, E)$ be bipartite with V partitioned as $X \cup Y$. For all $x \in X$, $\deg(x) \geq 3$, and for all $y \in Y$, $\deg(y) \leq 7$. If $|X| \leq 50$, find an upper bound on $\delta(G)$.

11. **a)** Let $G = (V, E)$ be the bipartite graph shown in Fig. 13.28, with V partitioned as $X \cup Y$. Determine $\delta(G)$ and a maximal matching of X into Y.

    b) For any bipartite graph $G = (V, E)$, with V partitioned as $X \cup Y$, if $\beta(G)$ denotes the independence number of G, show that $|Y| = \beta(G) - \delta(G)$.

    c) Determine a largest maximal independent set of vertices for the graphs shown in Fig. 13.27(a) and Fig. 13.28.

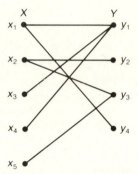

Figure 13.28

12. An aircraft company hires 100 new employees: 10 mathematicians, 10 computer scientists, 20 civil engineers, 25 mechanical engineers, and 35 electrical engineers. At present the company is undertaking 25 projects. Each mathematician has a choice of 4 projects on which to work, each civil engineer or computer scientist a choice of 5, each mechanical engineer a choice of 12, and each electrical engineer a choice of 15. If there is not enough work for more than 10 people on any one of these projects, what is the largest number of new employees that should have to wait for the start of new projects before they can begin working on a project?

13.5
SUMMARY AND HISTORICAL REVIEW

This chapter has provided us with a sample of the ways in which graph theory enters into an area of mathematics called operations research. Each topic was presented in an algorithmic manner that can be used in the computer implementation needed for solving each type of problem. Comparable coverage of this material can be found in Chapters 10 and 11 of the text by C. L. Liu [19] and in Chapter 9 of A. Tucker [25]. Chapters 4 and 5 of E. Lawler [18] offer an extensive coverage of many of the more recent developments on networks and matching. This text provides a wide variety of applications and includes references for additional reading.

In Section 13.1 we examined a shortest-path algorithm for weighted graphs. The full development of the algorithm is given in the article by E. W. Dijkstra [8].

Section 13.2 provided two techniques for finding a minimal spanning tree in a weighted undirected graph. These techniques were developed in the late 1950s by Joseph Kruskal [17] and Robert Prim [22]. Actually, however, methods for constructing minimal spanning trees can be traced back to 1926 in the work of Otakar Borůvka dealing with the construction of an electric power network. Even before this (1909–1911) the anthropologist Jan Czekanowski, in his work on various classification schemes, was very close to recognizing the minimal spanning tree problem and to providing a greedy algorithm for its solution. The survey paper by R. L. Graham and P. Hell [13] mentions the contributions made by Borůvka and Czekanowski and gives more information on the history and applications of this structure.

The computer implementation of all the techniques given in the first two sections can be found in Chapters 6 and 7 of A. Aho, J. Hopcroft, and J. Ullman [1]; in Chapter 4 of S. Baase [2]; and in Chapter 4 of E. Horowitz and S. Sahni [14]. These references also discuss the efficiency and speed of these algorithms. As we mentioned at the end of Section 13.2, the articles by R. Graham and P. Hell [13], by D. B. Johnson [15], and by A. Kershenbaum and R. Van Slyke [16] discuss some of the more recent implementations of Prim's algorithm. An interesting application of the concept of the minimal spanning tree in a physical science setting is provided in the article by D. Shier [24].

As we noted in Section 13.3, problems dealing with the allocation of resources or the shipment of goods can be modeled by means of transport networks. The fundamental work by G. Dantzig, L. Ford, and D. Fulkerson that brought about the labeling procedure for these networks and the Max-Flow Min-Cut Theorem can be found in their pioneering articles [6, 7, 9, 10]. The classic text by L. Ford and D. Fulkerson [11] provides excellent coverage of this topic. In addition, the reader may wish to examine Chapter 8 of the text by C. Berge [3] or Chapter 7 of the book by R. Busacker and T. Saaty [5]. Chapter 10 in C. L. Liu [19] includes coverage on an extension to networks wherein the flow in each edge is restricted by a lower as well as an upper capacity. For more applications the reader should examine the article by D. Fulkerson on pages 139–171 of reference [12].

The last topic discussed here dealt with matching in a bipartite graph. The theory behind this was first developed by Philip Hall in 1935, but here the ideas on transport networks were used to provide an algorithm for a solution. Chapter 7 of the text by O. Ore [21] provides a very readable introduction to this topic, along with some applications. For more on systems of representatives, the reader should examine Chapter 5 of the monograph by H. Ryser [23]. A second method for finding a maximal matching in a bipartite graph is called the *Hungarian method*. This is given in Chapter 5 of the text by J. Bondy and U. Murty [4] and in Chapter 10 of the book by C. Berge [3]. In addition to its application in solving the assignment problem, matching theory has many interesting combinatorial implications. One may learn more about these in the survey article by L. Mirsky and H. Perfect [20].

REFERENCES

1. Aho, Alfred V., Hopcroft, John E., and Ullman, Jeffrey D. *Data Structures and Algorithms*. Reading, Mass.: Addison-Wesley, 1983.

2. Baase, Sara. *Computer Algorithms, Introduction to Design and Analysis,* 2nd ed. Reading, Mass.: Addison-Wesley, 1988.

3. Berge, Claude. *The Theory of Graphs and Its Applications*. New York: Wiley, 1962.

4. Bondy, J. A., and Murty, U. S. R. *Graph Theory with Applications*. New York: Elsevier North Holland, 1976.

5. Busacker, Robert G., and Saaty, Thomas L. *Finite Graphs and Networks*. New York: McGraw-Hill, 1965.

6. Dantzig, G.B., and Fulkerson, D. R. *Computation of Maximal Flows in Networks*. The RAND Corporation, P-677, 1955.

7. Dantzig, G. B., and Fulkerson, D. R. *On the Max Flow Min Cut Theorem*. The RAND Corporation, RM-1418-1, 1955.

8. Dijkstra, Edsger W. "A Note on Two Problems in Connexion with Graphs." *Numerishe Mathematik,* Vol. 1, 1959, pp. 269–271.

9. Ford, L. R., Jr. *Network Flow Theory*. The RAND Corporation, P-923, 1956.

10. Ford, L. R., Jr., and Fulkerson, D. R. "Maximal Flow Through a Network," *Canadian Journal of Mathematics,* Vol. 8, 1956, pp. 399–404.

11. Ford, L. R., Jr., and Fulkerson, D. R. *Flows in Networks*. Princeton, N.J.: Princeton University Press, 1962.

12. Fulkerson, D. R., ed. *Studies in Graph Theory,* Part I. *MAA Studies in Mathematics,* Vol. 11, The Mathematical Association of America, 1975.

13. Graham, Ronald L., and Hell, Pavol. "On the History of the Minimum Spanning Tree Problem." *Annals of the History of Computing* 7, no. 1 (January 1985): pp. 43–57.

14. Horowitz, Ellis, and Sahni, Sartaj. *Fundamentals of Computer Algorithms*. Potomac, Md.: Computer Science Press, 1978.

15. Johnson, D. B. "Priority Queues with Update and Minimum Spanning Trees." *Information Processing Letters,* Vol. 4, 1975, pp. 53–57.

16. Kershenbaum, A., and Van Slyke, R. "Computing Minimum Spanning Trees Efficiently." *Proceedings of the Annual ACM Conference,* 1972, pp. 518–527.

17. Kruskal, Joseph B. "On the Shortest Spanning Subtree of a Graph and the Traveling Salesman Problem." *Proceedings of the AMS* 1, No. 1 (1956): pp. 48–50.

18. Lawler, Eugene. *Combinatorial Optimization: Networks and Matroids*. New York: Holt, 1976.

19. Liu, C. L. *Introduction to Combinatorial Mathematics*. New York: McGraw-Hill, 1968.

20. Mirsky, L., and Perfect, H. "Systems of Representatives." *Journal of Mathematical Analysis and Applications,* Vol. 3, 1966, pp. 520–568.

21. Ore, Oystein. *Theory of Graphs*. Providence, R.I.: American Mathematical Society, 1962.

22. Prim, Robert C. "Shortest Connection Networks and Some Generalizations." *Bell System Technical Journal,* Vol. 36, 1957, pp. 1389–1401.

23. Ryser, Herbert J. *Combinatorial Mathematics*. Carus Mathematical Monographs, Number 14, Mathematical Association of America, 1963.

24. Shier, Douglas R. "Testing for Homogeneity Using Minimum Spanning Trees." *The UMAP Journal* 3, No. 3 (1982): pp. 273–283.

25. Tucker, Alan. *Applied Combinatorics*. New York: Wiley, 1980.

MISCELLANEOUS EXERCISES

1. Apply Dijkstra's algorithm to the weighted directed multigraph shown in Fig. 13.29, and find the shortest distance from vertex a to the other seven vertices in the graph.

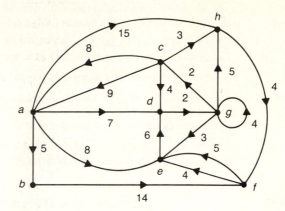

Figure 13.29

2. For her class in analysis of algorithms, Stacy writes the following algorithm to determine the shortest distance from a vertex a to a vertex b in a weighted directed graph $G = (V, E)$.

Step 1: Set Distance equal to 0, assign vertex a to the variable v, and let $T = V$.

Step 2: If $v = b$, the value of Distance is the answer to the problem. If $v \neq b$, then

1. Replace T by $T - \{v\}$ and select $w \in T$ with $\mathrm{wt}(v, w)$ minimal.
2. Set Distance equal to Distance $+ \mathrm{wt}(v, w)$.
3. Assign vertex w to the variable v and then return to step 2.

   Is Stacy's algorithm correct? If so, prove it. If not, provide a counterexample.

3. **a)** Let $G = (V, E)$ be a weighted connected undirected graph. If $e_1 \in E$ with $\mathrm{wt}(e_1) < \mathrm{wt}(e)$ for all other edges $e \in E$, prove that edge e_1 is part of every minimal spanning tree for G.

   b) With G as in part (a), suppose that there are edges $e_1, e_2 \in E$ with $\mathrm{wt}(e_1) < \mathrm{wt}(e_2) < \mathrm{wt}(e)$ for all other edges $e \in E$. Prove or disprove: Edge e_2 is part of every minimal spanning tree for G.

4. **a)** Let $G = (V, E)$ be a weighted connected undirected graph where each edge e of G is part of a cycle. Prove that if $e_1 \in E$ with $\mathrm{wt}(e_1) > \mathrm{wt}(e)$ for all other edges $e \in E$, then any spanning tree for G that contains e_1 cannot be minimal.

   b) With G as in part (a), suppose that $e_1, e_2 \in E$ with $\mathrm{wt}(e_1) > \mathrm{wt}(e_2) > \mathrm{wt}(e)$ for all other edges $e \in E$. Prove or disprove: Edge e_2 is not part of any minimal spanning tree for G.

5. Use the concept of flow in a transport network to construct a directed multigraph $G = (V, E)$, with $V = \{u, v, w, x, y\}$ and $\deg^-(u) = 1, \deg^+(u) = 3$; $\deg^-(v) = 3, \deg^+(v) = 3$; $\deg^-(w) = 3, \deg^+(w) = 4$; $\deg^-(x) = 5, \deg^+(x) = 4$; and $\deg^-(y) = 4, \deg^+(y) = 2$.

6. A set of words $\{qs, tq, ut, pqr, srt\}$ is to be transmitted using a binary code for each letter.

    a) Show that it is possible to select one letter from each word as a system of distinct representatives for these words.

    b) If a letter is selected at random from each of the five words, what is the probability that the selection is a system of distinct representatives for the words?

7. For $n \in \mathbf{Z}^+$ and for each $1 \le i \le n$, let $A_i = \{1, 2, 3, \ldots, n\} - \{i\}$. How many different systems of distinct representatives exist for the collection $A_1, A_2, A_3, \ldots, A_n$?

8. This exercise outlines a proof of the Birkhoff–von Neumann Theorem.

    a) For $n \in \mathbf{Z}^+$, an $n \times n$ matrix is called a *permutation* matrix if there is exactly one 1 in each row and column, and all other entries are 0. How many 5×5 permutation matrices are there? How many $n \times n$?

    b) An $n \times n$ matrix B is called *doubly stochastic* if $b_{ij} \ge 0$ for all $1 \le i, j \le n$, and the sum of the entries in any row or column is 1.

$$\text{If } B = \begin{bmatrix} 0.2 & 0.1 & 0.7 \\ 0.4 & 0.5 & 0.1 \\ 0.4 & 0.4 & 0.2 \end{bmatrix}, \text{ verify that } B \text{ is doubly stochastic.}$$

    c) Find four positive real numbers c_1, c_2, c_3, and c_4 and four permutation matrices P_1, P_2, P_3, and P_4 such that $c_1 + c_2 + c_3 + c_4 = 1$ and $B = c_1 P_1 + c_2 P_2 + c_3 P_3 + c_4 P_4$.

    d) Part (c) is a special case of the Birkhoff–von Neumann Theorem: If B is any $n \times n$ doubly stochastic matrix, then there exist positive real numbers $c_1, c_2, \ldots, c_k$ and permutation matrices $P_1, P_2, \ldots, P_k$ such that $\sum_{i=1}^{k} c_i = 1$ and $\sum_{i=1}^{k} c_i P_i = B$. To prove this result, proceed as follows: Construct a bipartite graph $G = (V, E)$ with V partitioned as $X \cup Y$, where $X = \{x_1, x_2, \ldots, x_n\}$ and $Y = \{y_1, y_2, \ldots, y_n\}$. The vertex $x_i, 1 \le i \le n$, corresponds with the ith row of B; the vertex $y_j, 1 \le j \le n$, corresponds with the jth column of B. The edges of G are of the form $\{x_i, y_j\}$ if and only if $b_{ij} > 0$. We claim that there is a complete matching of X into Y.

        If not, there is a subset A of X with $|A| > |R(A)|$. That is, there is a set of r rows of B having positive entries in s columns and $r > s$. What is the sum of these r rows of B? Yet the sum of these same entries, when added column by column, is less than or equal to s. (Why?) Consequently, we have a contradiction.

        As a result of the complete matching of X into Y, there are n positive entries in B that occur so that no two are in the same row or column. (Why?) If c_1 is the smallest of these entries, then we may write $B = c_1 P_1 + B_1$, where P_1 is an $n \times n$ permutation matrix wherein the 1's are located according to the positive entries in B that came about from the complete matching. What are the sums of the entries in the rows and columns of B_1?

    e) How is the proof completed?

9. Let $G = (V, E)$ be a bipartite graph with V partitioned as $X \cup Y$. If $E' \subseteq E$, and E' determines a complete matching of X into Y, what property do the vertices determined by E' in the line graph $L(G)$ have? (The line graph $L(G)$ for a loop-free undirected graph G is defined in Miscellaneous Exercise 18 of Chapter 11.)

CHAPTER 14

Rings and Modular Arithmetic

In this fourth and final section of the text, our emphasis returns to structure as we begin the investigation of sets of elements in which the structure arises from two binary operations. The concepts of structure and enumeration often reinforce each other. Here we will see this occur as ideas from Chapters 4, 5 and 8 come to the forefront again.

When we examined the set **Z** in Chapter 4, it was in conjunction with the binary operations of addition and multiplication. In this chapter we emphasize these operations by writing $(\mathbf{Z}, + , \cdot)$, instead of just **Z**. From some of the properties of $(\mathbf{Z}, + , \cdot)$ we define the algebraic structure called a *ring*. Without knowing it, we have been dealing with rings in many mathematical settings. Now we will be concerned with finite rings that arise in number theory and computer science applications. Of particular interest is the *hashing function*, which provides a means of identifying records stored in a table.

14.1

THE RING STRUCTURE: DEFINITION AND EXAMPLES

We start by defining the ring structure, realizing as we do that most abstract definitions, like theorems, come about from a study of many examples where one identifies the common idea or ideas present in what may seem to be a collection of unrelated objects.

DEFINITION 14.1   Let R be a nonempty set on which we define two binary operations, denoted by + and · (which may be quite different from the ordinary addition and multiplication we are accustomed to). Then $(R, + , \cdot)$ is a *ring* if for all $a, b, c \in R$, the following conditions are satisfied:

a) $a + b = b + a$ Commutative Law of +

b) $a + (b + c) = (a + b) + c$ Associative Law of +

c) There exists $z \in R$ such that   Existence of an identity for $+$
$a + z = z + a = a$ for every $a \in R$.

d) For each $a \in R$ there is an element   Existence of inverses under $+$
$b \in R$ with $a + b = b + a = z$.

e) $a \cdot (b \cdot c) = (a \cdot b) \cdot c$ Associative Law of $\cdot$

f) $a \cdot (b + c) = a \cdot b + a \cdot c$ Distributive Laws of $\cdot$ over $+$
$(b + c) \cdot a = b \cdot a + c \cdot a$

The binary operations here are closed by definition. We are dealing, as in Chapter 5, with two functions, which can be denoted by f, g, where $f, g: R \times R \to R$ and $f(a, b) = a + b$, $g(a, b) = a \cdot b$. (We often write ab for $a \cdot b$.)

In the next section we will show that the additive identity is unique, as is the additive inverse of each element.

Example 14.1   Under ordinary addition and multiplication, we find that **Z**, **Q**, **R**, and **C** are rings. In all of these rings the additive identity z is the integer 0, and the additive inverse of any number x is $-x$. □

Example 14.2   Let $M_2(\mathbf{Z})$ denote the set of all 2×2 matrices with integer components. (The sets $M_2(\mathbf{Q})$, $M_2(\mathbf{R})$, and $M_2(\mathbf{C})$ are defined similarly.) In $M_2(\mathbf{Z})$ two matrices are equal if their corresponding components are equal.

Here we define $+$ and $\cdot$ by

$$\begin{bmatrix} a & b \\ c & d \end{bmatrix} + \begin{bmatrix} e & f \\ g & h \end{bmatrix} = \begin{bmatrix} a+e & b+f \\ c+g & d+h \end{bmatrix}, \qquad \begin{bmatrix} a & b \\ c & d \end{bmatrix}\begin{bmatrix} e & f \\ g & h \end{bmatrix} = \begin{bmatrix} ae+bg & af+bh \\ ce+dg & cf+dh \end{bmatrix}$$

Under these operations, $M_2(\mathbf{Z})$ is a ring where $z = \begin{bmatrix} 0 & 0 \\ 0 & 0 \end{bmatrix}$ and the additive inverse of $\begin{bmatrix} a & b \\ c & d \end{bmatrix}$ is $\begin{bmatrix} -a & -b \\ -c & -d \end{bmatrix}$.

A few things happen here that do not occur in the rings of Example 14.1. For example,

$$\begin{bmatrix} 1 & 2 \\ 1 & 1 \end{bmatrix}\begin{bmatrix} 3 & 7 \\ 1 & 0 \end{bmatrix} = \begin{bmatrix} 5 & 7 \\ 4 & 7 \end{bmatrix} \neq \begin{bmatrix} 10 & 13 \\ 1 & 2 \end{bmatrix} = \begin{bmatrix} 3 & 7 \\ 1 & 0 \end{bmatrix}\begin{bmatrix} 1 & 2 \\ 1 & 1 \end{bmatrix}$$

shows that multiplication need not be commutative in a ring. That is why there are two distributive laws. Also,

$$\begin{bmatrix} 1 & -1 \\ -1 & 1 \end{bmatrix}\begin{bmatrix} 2 & 1 \\ 2 & 1 \end{bmatrix} = \begin{bmatrix} 0 & 0 \\ 0 & 0 \end{bmatrix},$$

even though neither $\begin{bmatrix} 1 & -1 \\ -1 & 1 \end{bmatrix}$ nor $\begin{bmatrix} 2 & 1 \\ 2 & 1 \end{bmatrix}$ is the additive identity. Hence a ring may contain *proper divisors of zero*. □

DEFINITION 14.2 Let $(R, +, \cdot)$ be a ring.

    a) If $ab = ba$ for all $a, b \in R$, then R is called a *commutative* ring.

    b) The ring R is said to have no *proper divisors of zero* if for any $a, b \in R$, $ab = z \Rightarrow a = z$ or $b = z$.

    c) If there is an element $u \in R$ such that $u \neq z$ and $au = ua = a$ for all $a \in R$, we call u the *unity*, or *multiplicative identity*, of R. R is then called a *ring with unity*.

It follows from part (c) of Definition 14.2 that whenever we have a ring R with unity, then R contains at least two elements.

The rings in Example 14.1 are all commutative rings whose unity is the integer 1. None of these rings has any proper divisors of zero. Meanwhile, the ring $M_2(\mathbf{Z})$ is a noncommutative ring whose unity is the matrix $\begin{bmatrix} 1 & 0 \\ 0 & 1 \end{bmatrix}$. This ring does contain proper divisors of zero.

We turn now to finite rings.

Example 14.3 Let $\mathcal{U} = \{1, 2\}$ and $R = \mathcal{P}(\mathcal{U})$. Define $+$ and $\cdot$ on the elements of R by

$$A + B = A \bigtriangleup B = \{x \mid x \in A \text{ or } x \in B, \text{ but not both}\};$$
$$A \cdot B = A \cap B = \text{the intersection of sets } A, B \subseteq \mathcal{U}.$$

We form Tables 14.1(a) and (b) for these operations.

Table 14.1

$+ (\bigtriangleup)$	$\emptyset$	$\{1\}$	$\{2\}$	$\mathcal{U}$
$\emptyset$	$\emptyset$	$\{1\}$	$\{2\}$	$\mathcal{U}$
$\{1\}$	$\{1\}$	$\emptyset$	$\mathcal{U}$	$\{2\}$
$\{2\}$	$\{2\}$	$\mathcal{U}$	$\emptyset$	$\{1\}$
$\mathcal{U}$	$\mathcal{U}$	$\{2\}$	$\{1\}$	$\emptyset$

(a)

$\cdot (\cap)$	$\emptyset$	$\{1\}$	$\{2\}$	$\mathcal{U}$
$\emptyset$	$\emptyset$	$\emptyset$	$\emptyset$	$\emptyset$
$\{1\}$	$\emptyset$	$\{1\}$	$\emptyset$	$\{1\}$
$\{2\}$	$\emptyset$	$\emptyset$	$\{2\}$	$\{2\}$
$\mathcal{U}$	$\emptyset$	$\{1\}$	$\{2\}$	$\mathcal{U}$

(b)

From results in Chapter 3, R satisfies conditions (a), (b), (e), and (f) of Definition 14.1 for these operations of "addition" and "multiplication." The table for "addition" shows that $\emptyset$ is the additive identity. For any $x \in R$, the additive inverse of x is x itself. The multiplication table is symmetric about the diagonal from the upper left to the lower right, so the operation defined by the table is commutative. The table also indicates that R has *unity* $\mathcal{U}$. So R is a finite commutative ring with unity. The elements $\{1\}, \{2\}$ provide an example of proper divisors of zero. $\square$

Example 14.4   For $R = \{a, b, c, d, e\}$ we define $+$ and $\cdot$ by Tables 14.2(a) and (b).

Table 14.2

+	a	b	c	d	e
a	a	b	c	d	e
b	b	c	d	e	a
c	c	d	e	a	b
d	d	e	a	b	c
e	e	a	b	c	d

·	a	b	c	d	e
a	a	a	a	a	a
b	a	b	c	d	e
c	a	c	e	b	d
d	a	d	b	e	c
e	a	e	d	c	b

(a)   (b)

Although we do not verify the 125 equalities needed to establish each of the associative laws and the distributive law, $(R, +, \cdot)$ is a finite commutative ring with unity, and it has no proper divisors of zero. The element a is the zero (that is, the additive identity) of R, whereas b is the unity. Here every nonzero element x has a *multiplicative inverse* y, where $xy = yx = b$, the unity. Elements c and d are multiplicative inverses of each other; b is its own inverse, as is e.   □

DEFINITION 14.3   Let R be a ring with unity u. If $a, b \in R$ and $ab = ba = u$, then b is called a *multiplicative inverse* of a and a is called a *unit* of R. (The element b is also a unit of R.)

DEFINITION 14.4   Let R be a commutative ring with unity. Then

a) R is called an *integral domain* if R has no proper divisors of zero.

b) R is called a *field* if every nonzero element of R is a unit.

The ring $(\mathbf{Z}, +, \cdot)$ is an integral domain but not a field, while $\mathbf{Q}, \mathbf{R}, \mathbf{C}$, under ordinary addition and multiplication, are both integral domains and fields. The ring in Example 14.4 is both an integral domain and a field.

It follows from part (c) of Definition 14.2 that if R is an integral domain or a field, then $|R| \geq 2$.

Example 14.5   For our last finite ring of this section we let $R = \{s, t, v, w, x, y\}$ where $+$ and $\cdot$ are given by Tables 14.3(a) and (b).

Table 14.3

+	s	t	v	w	x	y
s	s	t	v	w	x	y
t	t	v	w	x	y	s
v	v	w	x	y	s	t
w	w	x	y	s	t	v
x	x	y	s	t	v	w
y	y	s	t	v	w	x

·	s	t	v	w	x	y
s	s	s	s	s	s	s
t	s	t	v	w	x	y
v	s	v	x	s	v	x
w	s	w	s	w	s	w
x	s	x	v	s	x	v
y	s	y	x	w	v	t

(a)   (b)

From these tables we see that $(R, +, \cdot)$ is a commutative ring with unity, but it is neither an integral domain nor a field. The element t is the unity, and t and y are the units of R.

We also note that $vv = vy$, and even though v is not the zero element of R, we cannot cancel and say that $v = y$. So a general ring does *not* satisfy the cancellation law of multiplication that we take for granted. We shall look at this idea again in the next section. □

EXERCISES 14.1

1. What is the additive inverse of each element in the rings of Examples 14.4 and 14.5?

2. Determine whether or not each of the following sets of numbers is a ring under ordinary addition and multiplication.

    a) $R = $ the set of positive integers and zero

    b) $R = $ the set of all even integers

    c) $R = \{kn \mid n \in \mathbf{Z}, k \text{ a fixed positive integer}\}$

    d) $R = \{a + b\sqrt{2} \mid a, b \in \mathbf{Z}\}$

    e) $R = \{a + b\sqrt{2} + c\sqrt{3} \mid a \in \mathbf{Z}, b, c \in \mathbf{Q}\}$

3. Table 14.4(a) and (b) makes $(R, +, \cdot)$ into a ring, where $R = \{s, t, x, y\}$. (a) What is the zero for this ring? (b) What is the additive inverse of each element? (c) What is $t(s + xy)$? (d) Is R a commutative ring? (e) Does R have a unity? (f) Find a pair of zero divisors.

Table 14.4

+	s	t	x	y
s	y	x	s	t
t	x	y	t	s
x	s	t	x	y
y	t	s	y	x

(a)

·	s	t	x	y
s	y	y	x	x
t	y	y	x	x
x	x	x	x	x
y	x	x	x	x

(b)

4. For the set R in Example 14.3, keep $A \cdot B = A \cap B$, but define $A + B = A \cup B$. Is $(R, \cup, \cap)$ a ring?

5. Let $R = \{a + bi \mid a, b \in \mathbf{Z}, \ i^2 = -1\}$, with addition and multiplication defined by $(a + bi) + (c + di) = (a + c) + (b + d)i$ and $(a + bi)(c + di) = (ac - bd) + (bc + ad)i$, respectively. (a) Verify that R is an integral domain. (b) Determine all units in R.

6. **a)** Determine the multiplicative inverse of the matrix $\begin{bmatrix} 1 & 2 \\ 3 & 7 \end{bmatrix}$ in the ring $M_2(\mathbf{Z})$—that is, find a, b, c, d so that

$$\begin{bmatrix} 1 & 2 \\ 3 & 7 \end{bmatrix}\begin{bmatrix} a & b \\ c & d \end{bmatrix} = \begin{bmatrix} a & b \\ c & d \end{bmatrix}\begin{bmatrix} 1 & 2 \\ 3 & 7 \end{bmatrix} = \begin{bmatrix} 1 & 0 \\ 0 & 1 \end{bmatrix}.$$

b) Show that $\begin{bmatrix} 1 & 2 \\ 3 & 8 \end{bmatrix}$ is a unit in the ring $M_2(\mathbf{Q})$ but not a unit in $M_2(\mathbf{Z})$.

7. If $\begin{bmatrix} a & b \\ c & d \end{bmatrix} \in M_2(\mathbf{R})$, prove that $\begin{bmatrix} a & b \\ c & d \end{bmatrix}$ is a unit of this ring if and only if $ad - bc \neq 0$.

8. Give an example of a ring with eight elements. How about one with 16 elements? Generalize.

9. Define a new type of addition and multiplication, denoted by $\oplus$ and $\odot$, on the set $\mathbf{Z}$ as follows. For $x, y \in \mathbf{Z}$, $x \oplus y = x + y - 1$, $x \odot y = x + y - xy$. (a) Prove that $(\mathbf{Z}, \oplus, \odot)$ is a ring. (b) Is this ring commutative? (c) Does the ring have a unity element? What about units? (d) Is the ring an integral domain? a field?

10. For $R = \{s, t, x, y\}$, define $+$ and $\cdot$, making R into a ring, by Table 14.5(a) for $+$ and by the partial table for $\cdot$ in Table 14.5(b).

Table 14.5

+	s	t	x	y
s	s	t	x	y
t	t	s	y	x
x	x	y	s	t
y	y	x	t	s

(a)

·	s	t	x	y
s	s	s	s	s
t	s	t	?	?
x	s	t	?	y
y	s	?	s	?

(b)

a) Using the associative and distributive laws, determine the entries for the missing spaces in the multiplication table.

b) Is this ring commutative?

c) Does it have a unity? How about units?

d) Is the ring an integral domain or a field?

14.2
RING PROPERTIES AND SUBSTRUCTURES

In each ring of Section 14.1 we were concerned with *the* zero element of the ring and *the* additive inverse of each ring element. It is time now to show, along with other properties, that these elements are unique.

THEOREM 14.1   In any ring $(R, +, \cdot)$,

a) the zero element z is unique, and

b) the additive inverse of each ring element is unique.

Proof   **a)** If R has more than one additive identity, let z_1, z_2 denote two such elements. Then

$$z_1 = z_1 + z_2 = z_2.$$

Since z_2 is an   Since z_1 is an
additive identity   additive identity

b) For $a \in R$, suppose there are two elements $b, c \in R$ where $a + b = b + a = z$ and $a + c = c + a = z$. Then $b = b + z = b + (a + c) = (b + a) + c = z + c = c$. (The reader should supply the condition that establishes each equality.) ∎

As a result of the uniqueness in (b), from this point on we shall denote *the* additive inverse of $a \in R$ by $-a$. Further, we may now speak of *subtraction* in the ring, where we understand that $a - b = a + (-b)$.

From Theorem 14.1(b) we obtain the following for ring R.

THEOREM 14.2   (*The Cancellation Laws of Addition*) For all $a, b, c \in R$,

   a) $a + b = a + c \Rightarrow b = c$, and
   b) $b + a = c + a \Rightarrow b = c$.

Proof   **a)** Since $a \in R$, it follows that $-a \in R$ and we have

$$a + b = a + c \Rightarrow (-a) + (a + b) = (-a) + (a + c) \Rightarrow$$
$$[(-a) + a] + b = [(-a) + a] + c \Rightarrow$$
$$z + b = z + c \Rightarrow b = c.$$

b) We leave this similar proof for the reader. ∎

THEOREM 14.3   For any ring R and any $a \in R$, we have $az = za = z$.

Proof   If $a \in R$, then $az = a(z + z)$ because $z + z = z$. Hence $z + az = az = az + az$. (Why?) Using the cancellation law of addition, we have $z = az$.
The proof that $za = z$ is done similarly. ∎

The reader may feel that the result of Theorem 14.3 is obvious. But we are not dealing with just $\mathbf{Z}$ or $\mathbf{Q}$ or $M_2(\mathbf{Z})$. Our objective is to show that *any* ring satisfies such a result, and to get the result we may only use the conditions in the definition of a ring and whatever properties we've derived up to this point.

The uniqueness of additive inverses now implies the following theorem.

THEOREM 14.4   Given a ring $(R, +, \cdot)$, for any $a, b \in R$,

   a) $-(-a) = a$,
   b) $a(-b) = (-a)b = -(ab)$, and
   c) $(-a)(-b) = ab$.

Proof   **a)** By the convention stated after Theorem 14.1, $-(-a)$ denotes *the* additive inverse of $-a$. Since $(-a) + a = z$, a is also an additive inverse for $-a$. Consequently, by the uniqueness of such inverses, $-(-a) = a$.

b) We shall prove that $a(-b) = -(ab)$ and shall leave the other part for the reader. We know that $-(ab)$ denotes *the* additive inverse of ab. However, $ab + a(-b) = a[b + (-b)] = az = z$, by Theorem 14.3, so by the uniqueness of additive inverses, $a(-b) = -(ab)$.

c) Here we establish an idea we have used in algebra since our first encounter with *signed numbers*. "Minus times minus does indeed equal plus," and the proof follows from the properties and definition of a ring. From part (b) we have $(-a)(-b) = -[a(-b)] = -[-(ab)]$, and the result then follows from part (a). ∎

For the operation of multiplication one finds the following, which is comparable to Theorem 14.1.

THEOREM 14.5   Given a ring $(R, +, \cdot)$,

a) if R has a unity, then it is unique, and

b) if R has a unity, and x is a unit of R, then the multiplicative inverse of x is unique.

Proof   The proofs of these results are left to the reader. ∎

As a result of this theorem we shall denote *the* multiplicative inverse of any unit x in R by x^{-1}. Also, the definition of a field can now be restated: A field is a commutative ring F with unity, such that for all $x \in F$, $x \neq z \Rightarrow x^{-1} \in F$.

With this notion to assist us, we examine some further properties and relations between fields and integral domains.

THEOREM 14.6   Let $(R, +, \cdot)$ be a commutative ring with unity. Then R is an integral domain if and only if, for any $a, b, c \in R$ where $a \neq z$, $ab = ac \Rightarrow b = c$. (Hence, a commutative ring with unity that satisfies the *cancellation law of multiplication* is an integral domain.)

Proof   If R is an integral domain and $x, y \in R$, then $xy = z \Rightarrow x = z$ or $y = z$. Now if $ab = ac$, then $ab - ac = a(b - c) = z$, and because $a \neq z$, it follows that $b - c = z$ or $b = c$. Conversely, if R is commutative with unity and R satisfies multiplicative cancellation, then let $a, b \in R$ with $ab = z$. If $a = z$, we are finished. If not, as $az = z$, we can write $ab = az$ and conclude that $b = z$. So there are no proper divisors of zero and R is an integral domain. ∎

Before going on, let us realize that the cancellation law of multiplicaton does *not* imply the existence of multiplicative inverses. The integral domain $(\mathbf{Z}, +, \cdot)$ satisfies multiplicative cancellation, but it contains only two elements, namely 1 and -1, that are units. Hence, an integral domain need not be a field. But what about a field? Is it an integral domain?

THEOREM 14.7 If $(F, +, \cdot)$ is a field, then it is an integral domain.

Proof Let $a, b \in F$ with $ab = z$. If $a = z$, we are finished. If not, a has a multiplicative inverse a^{-1} because F is a field. Then

$$ab = z \Rightarrow a^{-1}(ab) = a^{-1}z \Rightarrow (a^{-1}a)b = a^{-1}z \Rightarrow ub = z \Rightarrow b = z.$$

Hence F has no proper divisors of zero and is an integral domain. ∎

In Chapter 5 we found that functions $f: A \to A$ could be one-to-one (or onto) without being onto (or one-to-one). However, if A were *finite*, such a function f was one-to-one if and only if it was onto. (See Theorem 5.10.) The same situation occurs with *finite* integral domains. An integral domain need not be a field, but when it is finite we find that this structure is a field.

THEOREM 14.8 A *finite* integral domain $(D, +, \cdot)$ is a field.

Proof Since D is finite, we list the elements of D as $\{d_1, d_2, \ldots, d_n\}$. For $d \in D$, $d \neq z$, we have $dD = \{dd_1, dd_2, \ldots, dd_n\} \subseteq D$, because D is closed under multiplication. $|D| = n$ and $dD \subseteq D$, so if we could show that dD contains n elements, we would have $dD = D$. If $|dD| < n$, then $dd_i = dd_j$, for some $1 \leq i < j \leq n$. But since D is an integral domain and $d \neq z$, we have $d_i = d_j$, when they are supposed to be distinct. So $dD = D$ and for some $1 \leq k \leq n$, $dd_k = u$, the unity of D. Then $dd_k = u \Rightarrow d$ is a unit of D, and since d was chosen arbitrarily, it follows that $(D, +, \cdot)$ is a field. ∎

In the next section we shall look at finite fields that are useful in discrete mathematics. Before closing this section, however, let us examine some special subsets of a ring.

When dealing with finite state machines in Chapter 6, we saw instances where subsets of the set of internal states gave rise to machines on their own (when the next state and output functions of the original machines were restricted). These were called submachines. Since binary operations are special kinds of functions, we encounter a similar idea in the following definition.

DEFINITION 14.5 For a ring $(R, +, \cdot)$, a nonempty subset S of R is called a *subring* of R if $(S, +, \cdot)$—that is, S under the addition and multiplication of R, restricted to S—is a ring.

Example 14.6 For any ring R, the subsets $\{z\}$ and R are always subrings of R. □

Example 14.7
a) The set of all even integers is a subring of $(\mathbf{Z}, +, \cdot)$. In fact, for any $n \in \mathbf{Z}^+$, $n\mathbf{Z} = \{nx \mid x \in \mathbf{Z}\}$ is a subring of $(\mathbf{Z}, +, \cdot)$.

b) $(\mathbf{Z}, +, \cdot)$ is a subring of $(\mathbf{Q}, +, \cdot)$, which is a subring of $(\mathbf{R}, +, \cdot)$, which is a subring of $(\mathbf{C}, +, \cdot)$. □

Example 14.8 In Example 14.5, the subsets $S = \{s, w\}$ and $T = \{s, v, x\}$ are subrings of R. □

The next result characterizes those subsets of a ring that are subrings.

THEOREM 14.9   Given a ring $(R, +, \cdot)$, a nonempty subset S of R is a subring of R if and only if

1.   for all $a, b \in S$, we have $a + b, ab \in S$ (that is, S is closed under the binary operations of addition and multiplication defined on R), and

2.   for all $a \in S$, we have $-a \in S$.

Proof   If $(S, +, \cdot)$ is a subring of R, then in its own right it satisfies all the conditions of a ring. Hence it satisfies conditions 1 and 2 of the theorem. Conversely, let S be a nonempty subset of R that satisfies conditions 1 and 2. Conditions (a), (b), (e), and (f) of the definition of a ring are inherited by the elements of S, because they are also elements of R. Thus, all we need to verify here is that S has an additive identity. $S \neq \emptyset$, so there is an element $a \in S$, and by condition 2, $-a \in S$. Then by condition 1, $z = a + (-a) \in S$.   ∎

Note that $(\mathbf{Z}^+, +, \cdot)$ satisfies condition 1 above, but not condition 2, so it is not a subring of $(\mathbf{Z}, +, \cdot)$.

The result in Theorem 14.9 can also be given as follows.

THEOREM 14.10   For any ring $(R, +, \cdot)$, if $\emptyset \neq S \subseteq R$,

a)   then $(S, +, \cdot)$ is a subring of R if and only if for all $a, b \in S$, we have $a - b$, $ab \in S$;

b)   and if S is finite, then $(S, +, \cdot)$ is a subring of R if and only if for all $a, b \in S$, we have $a + b$, $ab \in S$. (Once again, additional help comes from a finiteness condition.)

Proof   These proofs we leave for the reader.   ∎

We close with an important type of subring.

DEFINITION 14.6   A nonempty subset I of a ring R is called an *ideal* of R if for all $a, b \in I$ and all $r \in R$, we have (a) $a - b \in I$ and (b) $ar, ra \in I$.   ▬

An ideal is a subring, but the converse is not so: $(\mathbf{Z}, +, \cdot)$ is a subring of $(\mathbf{Q}, +, \cdot)$ but not an ideal, because $(1/2)9 \notin \mathbf{Z}$ for $(1/2) \in \mathbf{Q}$, $9 \in \mathbf{Z}$. Meanwhile, all the subrings in Example 14.7(a) are ideals of $(\mathbf{Z}, +, \cdot)$.

EXERCISES 14.2

1.   Complete the proofs of Theorems 14.2, 14.4, 14.5, and 14.10.

2.   If a, b, and c are any elements in a ring $(R, +, \cdot)$, prove that (a) $a(b - c) = ab - (ac)$ and (b) $(b - c)a = ba - (ca)$.

3.   a)   If R is a ring with unity and a, b are units of R, prove that ab is a unit of R and that $(ab)^{-1} = b^{-1}a^{-1}$.

   b)   For the ring $M_2(\mathbf{Z})$, find A^{-1}, B^{-1}, $(AB)^{-1}$, $(BA)^{-1}$, and $B^{-1}A^{-1}$ if

$$A = \begin{bmatrix} 4 & 7 \\ 1 & 2 \end{bmatrix}, \qquad B = \begin{bmatrix} 5 & 2 \\ 2 & 1 \end{bmatrix}.$$

4. Prove that a unit in a ring R cannot be a proper divisor of zero.

5. If a is a unit in ring R, prove that $-a$ is also a unit in R.

6. If a, b are units in a ring R, is $a + b$ necessarily a unit in R?

7. **a)** Find all subrings in the ring given in Table 14.4.

   b) Are any of the subrings in part (a) also ideals?

8. **a)** Verify that the subsets $S = \{s, w\}$ and $T = \{s, v, x\}$ are subrings of the ring R in Example 14.5. (The binary operations for the elements of S, T are those given in Table 14.3.)

   b) Are the subrings in part (a) ideals of R?

9. Let S and T be subrings of a ring R. Prove that $S \cap T$ is a subring of R.

10. Answer Exercise 9, replacing "subring" with "ideal."

11. **a)** For $R = M_2(\mathbf{Z})$, prove that

$$S = \left\{ \begin{bmatrix} a & 0 \\ 0 & 0 \end{bmatrix} \middle| a \in \mathbf{Z} \right\}$$

   is a subring of R.

   b) What is the unity of R?

   c) Does S have a unity?

   d) Does S have any properties that R does not have?

   e) Is S an ideal of R?

12. Let S and T be the following subsets of the ring $R = M_2(\mathbf{Z})$:

$$S = \left\{ \begin{bmatrix} a & 0 \\ b & c \end{bmatrix} \middle| a, b, c \in \mathbf{Z} \right\}, \qquad T = \left\{ \begin{bmatrix} 2a & 2b \\ 2c & 2d \end{bmatrix} \middle| a, b, c, d \in \mathbf{Z} \right\}.$$

   a) Verify that S is a subring of R. Is it an ideal?

   b) Verify that T is a subring of R. Is it an ideal?

13. Let $(R, +, \cdot)$ be a commutative ring, and let z denote the zero element of R. For a fixed element $a \in R$, define $N(a) = \{r \in R \mid ra = z\}$. Prove that $N(a)$ is an ideal of R.

14. Let R be a commutative ring with unity u, and let I be an ideal of R. (a) If $u \in I$, prove that $I = R$. (b) If I contains a unit of R, prove that $I = R$.

15. If R is a field, how many ideals does R have?

16. Let $(R, +, \cdot)$ be the commutative ring with unity given by Tables 14.6(a) and (b).

Table 14.6

+	z	u	a	b
z	z	u	a	b
u	u	z	b	a
a	a	b	z	u
b	b	a	u	z

(a)

·	z	u	a	b
z	z	z	z	z
u	z	u	a	b
a	z	a	b	u
b	z	b	u	a

(b)

a) Verify that R is a field.

b) Find a subring of R that is not an ideal.

c) Let x and y be unknowns. Solve the following system of linear equations in R: $bx + y = u$; $x + by = z$.

17. Let R be a commutative ring with unity u.

a) For any $a \in R$, prove that $aR = \{ar \,|\, r \in R\}$ is an ideal of R.

b) If the only ideals of R are $\{z\}$ and R, prove that R is a field.

18. Let $(S, +, \cdot)$ and $(T, +', \cdot')$ be two rings. For $R = S \times T$, define addition "$\oplus$" and multiplication "$\odot$" by

$$(s_1, t_1) \oplus (s_2, t_2) = (s_1 + s_2, t_1 +' t_2),$$
$$(s_1, t_1) \odot (s_2, t_2) = (s_1 \cdot s_2, t_1 \cdot' t_2).$$

a) Prove that under these binary operations, R is a ring.

b) If both S and T are commutative, prove that R is commutative.

c) If S has unity u_S and T has unity u_T, what is the unity of R?

d) If S and T are fields, is R also a field?

19. Let $(R, +, \cdot)$ be a ring with unity u, and $|R| = 8$. On $R^4 = R \times R \times R \times R$, define $+$ and $\cdot$ as suggested by Exercise 18. In the ring R^4, (a) how many elements have exactly two nonzero components? (b) how many elements have all nonzero components? (c) is there a unity? (d) how many units are there if R has four units?

20. Let $(R, +, \cdot)$ be a ring, with $a \in R$. Define $0a = z$, $1a = a$, and $(n + 1)a = na + a$, for any $n \in \mathbf{Z}^+$. (Here we are multiplying elements of R by elements of $\mathbf{Z}$, so we have an operation different from the multiplication in either $\mathbf{Z}$ or R.) For $n > 0$, we define $(-n)a = n(-a)$, so, for example, $(-3)a = 3(-a) = 2(-a) + (-a) = [(-a) + (-a)] + (-a) = [-(a + a)] + (-a) = -[(a + a) + a] = -[2a + a] = -(3a)$.

For any $a, b \in R$, and any $m, n \in \mathbf{Z}$, prove that

a) $ma + na = (m + n)a$ **b)** $m(na) = (mn)a$

c) $n(a + b) = na + nb$ **d)** $n(ab) = (na)b = a(nb)$

e) $(ma)(nb) = (mn)(ab) = (na)(mb)$

21. a) For ring $(R, +, \cdot)$ and any $a \in R$, we define $a^1 = a$, and $a^{n+1} = a^n a$, for any $n \in \mathbf{Z}^+$. Prove that for any $m, n \in \mathbf{Z}^+$, $(a^m)(a^n) = a^{m+n}$ and $(a^m)^n = a^{mn}$.

b) Can you suggest how we might define a^0 or a^{-n}, $n \in \mathbf{Z}^+$, including any necessary conditions that R must satisfy for these definitions to make sense?

22. Let S be the set of all functions $f: \mathbf{R} \to \mathbf{R}$. If $f, g \in S$, define $f + g$ and $f \cdot g$ as follows:

For $x \in \mathbf{R}$,

$$f + g: \ \mathbf{R} \to \mathbf{R}, \text{ where } (f + g)(x) = f(x) + g(x), \text{ and}$$
$$f \cdot g: \ \mathbf{R} \to \mathbf{R}, \text{ where } (f \cdot g)(x) = f(x) \cdot g(x).$$

(Note: $f + g$ denotes the addition of functions, and this is defined in terms of $f(x) + g(x)$, the addition of real numbers. The situation is similar for $f \cdot g$.)

a) Prove that $(S, +, \cdot)$ is a commutative ring with unity, but not an integral domain.

b) Let $T = \{f \in S \mid f(\mathbf{R}) \subseteq \mathbf{Z}\}$. Verify that $(T, +, \cdot)$ is a subring of S. Is it an ideal?

c) Let $I = \{f \in S \mid f(0) = 0\}$. Prove that I is an ideal of S.

14.3
THE INTEGERS MODULO *n*

Enough theory for a while! We shall now concentrate on the construction and use of special finite rings and fields.

DEFINITION 14.7   Let $n \in \mathbf{Z}^+$, $n > 1$. For $a, b \in \mathbf{Z}$, we say that *a is congruent to b modulo n*, and write $a \equiv b \pmod{n}$, if $n \mid (a - b)$, or $a = b + kn$ for some $k \in \mathbf{Z}$.   ▬

Example 14.9   (a) $17 \equiv 2 \pmod{5}$; (b) $-7 \equiv -49 \pmod{6}$.   □

THEOREM 14.11   Congruence modulo *n* is an equivalence relation on $\mathbf{Z}$.

Proof   The proof is left for the reader.   ■

Since an equivalence relation induces a partition, for $n \geq 2$, congruence modulo *n* partitions $\mathbf{Z}$ into the *n* equivalence classes

$$[0] = \{\ldots, -2n, -n, 0, n, 2n, \ldots\} = \{0 + nx \mid x \in \mathbf{Z}\}$$
$$[1] = \{\ldots, -2n + 1, -n + 1, 1, n + 1, 2n + 1, \ldots\} = \{1 + nx \mid x \in \mathbf{Z}\}$$
$$[2] = \{\ldots, -2n + 2, -n + 2, 2, n + 2, 2n + 2, \ldots\} = \{2 + nx \mid x \in \mathbf{Z}\}$$
$$\vdots \qquad\qquad\qquad\qquad \vdots$$
$$[n - 1] = \{\ldots, -n - 1, -1, n - 1, 2n - 1, 3n - 1, \ldots\}$$
$$= \{(n - 1) + nx \mid x \in \mathbf{Z}\}.$$

For any $t \in \mathbf{Z}$, by the division algorithm (of Section 4.2) we can write $t = qn + r$, where $0 \leq r < n$, so $t \in [r]$, or $[t] = [r]$. We use $\mathbf{Z}_n$ to denote $\{[0], [1], [2], \ldots, [n-1]\}$. (When there is no danger of ambiguity, we often replace $[a]$ by a and write $\mathbf{Z}_n = \{0, 1, 2, \ldots, n-1\}$.) Our objective now is to define addition and multiplication on $\mathbf{Z}_n$ so that we obtain a ring.

For $[a], [b] \in \mathbf{Z}_n$, define $+$ and $\cdot$ by

$$[a] + [b] = [a + b] \qquad \text{and} \qquad [a] \cdot [b] = [a][b] = [ab].$$

For example, if $n = 7$, then $[2] + [6] = [2 + 6] = [8] = [1]$, and $[2][6] = [12] = [5]$.

Before these definitions are so readily accepted, we must investigate whether or not these operations are *well-defined* in the sense that if $[a] = [c]$, $[b] = [d]$, then $[a] + [b] = [c] + [d]$ and $[a][b] = [c][d]$. Since $[a] = [c]$ can occur with $a \neq c$, do the results of our addition and multiplication depend on which representatives are chosen from the equivalence classes? We shall prove that the results are independent of such representatives and that the operations are very definitely well-defined.

First, we observe that $[a] = [c] \Rightarrow a = c + sn$, and $[b] = [d] \Rightarrow b = d + tn$. Hence

$$a + b = (c + sn) + (d + tn) = c + d + (s + t)n,$$

so $(a + b) \equiv (c + d)(\operatorname{mod} n)$ and $[a + b] = [c + d]$.
Also,

$$ab = (c + sn)(d + tn) = cd + (sd + ct + stn)n$$

and $ab \equiv cd \ (\operatorname{mod} n)$, or $[ab] = [cd]$.

This result now leads us to the following.

THEOREM 14.12 For $n \in \mathbf{Z}^+, n > 1$, under the binary operations defined above, $\mathbf{Z}_n$ is a commutative ring with unity [1].

Proof The proof is left to the reader. Verification of the ring properties follows from the definitions of addition and multiplication in $\mathbf{Z}_n$ and from the corresponding properties for the ring $(\mathbf{Z}, + , \cdot)$. ∎

Before stating any further results, let us examine two particular examples. In Tables 14.7(a) and (b) and 14.8(a) and (b), we simplify $[a]$ by writing a.

Table 14.7

$\mathbf{Z}_5$

+	0	1	2	3	4
0	0	1	2	3	4
1	1	2	3	4	0
2	2	3	4	0	1
3	3	4	0	1	2
4	4	0	1	2	3

·	0	1	2	3	4
0	0	0	0	0	0
1	0	1	2	3	4
2	0	2	4	1	3
3	0	3	1	4	2
4	0	4	3	2	1

(a) (b)

Table 14.8

$\mathbf{Z}_6$

+	0	1	2	3	4	5
0	0	1	2	3	4	5
1	1	2	3	4	5	0
2	2	3	4	5	0	1
3	3	4	5	0	1	2
4	4	5	0	1	2	3
5	5	0	1	2	3	4

·	0	1	2	3	4	5
0	0	0	0	0	0	0
1	0	1	2	3	4	5
2	0	2	4	0	2	4
3	0	3	0	3	0	3
4	0	4	2	0	4	2
5	0	5	4	3	2	1

(a) (b)

In $\mathbf{Z}_5$ every nonzero element has a multiplicative inverse, so $\mathbf{Z}_5$ is a field. For $\mathbf{Z}_6$, however, 1 and 5 are the only units and $2, 3, 4$ are divisors of zero. In $\mathbf{Z}_9$, $3 \cdot 3 = 3 \cdot 6 = 0$, so for $\mathbf{Z}_n, n > 1$, to be a field, we need more than just an odd modulus.

THEOREM 14.13 $\mathbf{Z}_n$ is a field if and only if n is a prime.

Proof   Let n be a prime, and suppose that $0 < a < n$. Then $(a, n) = 1$, so there are integers s, t with $as + tn = 1$. Thus $as \equiv 1 \pmod{n}$, or $[a][s] = [1]$, and $[a]$ is a unit of $\mathbf{Z}_n$, which is consequently a field.

Conversely, if n is not a prime, then $n = n_1 n_2$, where $1 < n_1, n_2 < n$. So $[n_1] \neq [0]$ and $[n_2] \neq [0]$ but $[n_1][n_2] = [n_1 n_2] = [0]$, and $\mathbf{Z}_n$ is not even an integral domain, so it cannot be a field.   ∎

In $\mathbf{Z}_6$, $[5]$ is a unit and $[3]$ is a zero divisor. We seek a way to recognize when $[a]$ is a unit in $\mathbf{Z}_n$, for n composite.

THEOREM 14.14   In $\mathbf{Z}_n$, $[a]$ is a unit if and only if $(a, n) = 1$.

Proof   If $(a, n) = 1$, the result follows as in Theorem 14.13. For the converse, let $[a] \in \mathbf{Z}_n$ and $[a]^{-1} = [s]$. Then $[as] = [a][s] = [1]$, so $as \equiv 1 \pmod{n}$ and $as = 1 + tn$, for some $t \in \mathbf{Z}$. But $1 = as + n(-t) \Rightarrow (a, n) = 1$.   ∎

Example 14.10   Find $[25]^{-1}$ in $\mathbf{Z}_{72}$.

Since $(25, 72) = 1$, the Euclidean algorithm leads us to

$$
\begin{aligned}
72 &= 2(25) + 22, && 0 < 22 < 25 \\
25 &= 1(22) + 3, && 0 < 3 < 22 \\
22 &= 7(3) + 1, && 0 < 1 < 3.
\end{aligned}
$$

As 1 is the last nonzero remainder, we have

$$
\begin{aligned}
1 &= 22 - 7(3) = 22 - 7[25 - 22] = (-7)(25) + (8)(22) \\
&= (-7)(25) + 8[72 - 2(25)] = 8(72) - 23(25).
\end{aligned}
$$

But

$$
1 = 8(72) - 23(25) \Rightarrow 1 \equiv (-23)(25) \equiv (-23 + 72)(25) \pmod{72},
$$

so $[1] = [49][25]$ and $[25]^{-1} = [49]$ in $\mathbf{Z}_{72}$.   □

Now $[25]$ is a unit in $\mathbf{Z}_{72}$, but is there any way of knowing how many units this ring has? From Theorem 14.14, if $1 \leq a < 72$, then $[a]^{-1}$ exists if and only if $(a, 72) = 1$. Consequently, the number of units in $\mathbf{Z}_{72}$ is the number of integers a, where $1 \leq a < 72$ and $(a, 72) = 1$. Using Euler's phi function (Example 8.5) we find that this is

$$
\phi(72) = \phi(2^3 3^2) = (72)[1 - (1/2)][1 - (1/3)] = (72)(1/2)(2/3) = 24.
$$

> In general, for any $n \in \mathbf{Z}^+$, $n > 1$, there are $\phi(n)$ units and $n - 1 - \phi(n)$ proper divisors of zero in $\mathbf{Z}_n$.

The last example for this section provides an application in information retrieval.

Example 14.11   When searching a table of records stored in a computer, each record is assigned a memory location or *address* in the computer's memory. The record itself is often made up of fields (this has nothing to do with ring structures). For instance, a college registrar keeps a record on each student, with the record containing information on the student's social security number, name, and major, for a total of three fields.

In searching for a particular student's record, we can use his or her social security number as the *key* to the record, because it uniquely identifies that record. As a result, we develop a function from the set of keys to the set of addresses in the table.

If the college is small enough, we may find that the first four digits of the social security number are enough for identification. We develop a *hashing* (or *scattering*) function h from the set of keys (still social security numbers) to the set of addresses, determined now by the first four digits of the key. For example, $h(081\text{-}37\text{-}6495)$ identifies the record at the address associated with 0813. In this way we can store the table using at most 10,000 addresses. All is well so long as h is one-to-one. Should a second student have social security number 081-39-0207, h would no longer uniquely identify a student's record. When this happens, a *collision* is said to occur. Since increasing the size of the stored table often results in more unused storage, we must balance the cost of this storage against the cost of handling such collisions. Techniques for resolving collisions have been devised. They depend on the data structures (such as vectors or linear linked lists) that are used to store the records.

Different kinds of hashing functions that have been developed include the following.

a) The *division* method: Here we restrict the number of addresses we want to use to a fixed integer n. For any key k (a positive integer), we define $h(k) = r$, where $r \equiv k \pmod{n}$ and $0 \le r < n$.

b) Often implemented is the *folding* method, where the key is split into parts and the parts are added together to give $h(\text{key})$.
For example, $h(081\text{-}37\text{-}6495) = 081 + 37 + 6495 = 6613$ utilizes folding, and if we want only three-digit addresses, suppressing the first 6, we can have $h(081\text{-}37\text{-}6495) = 613$.

The importance of choosing a pertinent hashing function cannot be emphasized enough as we try to improve efficiency in terms of greater speed and less unused storage.

Using the modular concept we can develop a hashing function h, using the same keys as above, where

$$h(x_1 x_2 x_3 \text{-} x_4 x_5 \text{-} x_6 x_7 x_8 x_9) = y_1 y_2 y_3,$$

with

$$y_1 \equiv x_1 + x_2 + x_3 \pmod{5}, \qquad 0 \le y_1 < 5$$
$$y_2 \equiv x_4 + x_5 \pmod{3}, \qquad 0 \le y_2 < 3$$
$$y_3 \equiv x_6 + x_7 + x_8 + x_9 \pmod{7}, \qquad 0 \le y_3 < 7.$$

Here, for example, $h(081\text{-}37\text{-}6495) = 413$. ☐

EXERCISES 14.3

1. List four elements in each of the following equivalence classes.
   a) $[1]$ in $\mathbf{Z}_7$ $\qquad\qquad$ b) $[2]$ in $\mathbf{Z}_{11}$ $\qquad\qquad$ c) $[10]$ in $\mathbf{Z}_{17}$

2. Complete the proofs of Theorems 14.11 and 14.12.

3. Define relation $\mathcal{R}$ on $\mathbf{Z}^+$ by $a\,\mathcal{R}\,b$, if $\tau(a) = \tau(b)$, where $\tau(a) =$ the number of positive divisors of a. For example, $2\,\mathcal{R}\,3$ and $4\,\mathcal{R}\,25$ but $5\,\mathcal{\not R}\,9$.
   a) Verify that $\mathcal{R}$ is an equivalence relation on $\mathbf{Z}^+$.
   b) For the equivalence classes $[a]$ and $[b]$ induced by $\mathcal{R}$, define operations of addition and multiplication by $[a] + [b] = [a + b]$ and $[a][b] = [ab]$. Are these operations well-defined (that is, does $a\,\mathcal{R}\,c, b\,\mathcal{R}\,d \Rightarrow (a + b)\mathcal{R}(c + d)$, $(ab)\,\mathcal{R}\,(cd)$)?

4. Find the multiplicative inverse of each element in $\mathbf{Z}_{11}, \mathbf{Z}_{13}$, and $\mathbf{Z}_{17}$.

5. Find $[a]^{-1}$ in $\mathbf{Z}_{1009}$ for (a) $a = 17$, (b) $a = 100$, and (c) $a = 777$.

6. a) Find all subrings of $\mathbf{Z}_{12}, \mathbf{Z}_{18}$, and $\mathbf{Z}_{24}$.
   b) Construct the Hasse diagram for each of these collections of subrings, where the partial order arises from set inclusion. Compare these diagrams with those for the set of positive divisors of n ($n = 12; 18; 24$), where the partial order now comes from the divisibility relation.
   c) Find the formula for the number of subrings in $\mathbf{Z}_n, n > 1$.

7. How many units and how many (proper) zero divisors are there in (a) $\mathbf{Z}_{17}$? (b) $\mathbf{Z}_{117}$? (c) $\mathbf{Z}_{1117}$?

8. Prove that in any list of n consecutive integers, one of the integers is divisible by n.

9. If three distinct integers are selected from the set $\{1, 2, 3, \dots, 1000\}$, what is the probability that their sum is divisible by 3?

10. a) For $c, d, n, m \in \mathbf{Z}$, with $n > 1$ and $m > 0$, prove that if $c \equiv d \pmod{n}$, then $mc \equiv md \pmod{n}$ and $c^m \equiv d^m \pmod{n}$.
    b) If $x_n x_{n-1} \dots x_1 x_0 = x_n \cdot 10^n + \cdots + x_1 \cdot 10 + x_0$ denotes an $(n + 1)$-digit integer, then prove that
    $$x_n x_{n-1} \dots x_1 x_0 \equiv x_n + x_{n-1} + \cdots + x_1 + x_0 \pmod{9}.$$

11. a) Prove that for all $n \in \mathbf{N}$, $10^n \equiv (-1)^n \pmod{11}$.
    b) Consider the result for mod 9 in part (b) of Exercise 10. State and prove a comparable result for mod 11.

    c) For each of the following, determine the value of the digit d so that the result is an integer divisible by 11: (i) $2d653874$; (ii) $37d64943252$.

12. Determine the unit's digit in 3^{55}.

13. For $a, b, n \in \mathbf{Z}^+$ and $n > 1$, prove that $a \equiv b \ (\mathrm{mod}\, n) \Rightarrow (a, n) = (b, n)$.

14. a) Show that for any $[a] \in \mathbf{Z}_7$, if $[a] \neq [0]$, then $[a]^6 = [1]$.

    b) Let $n \in \mathbf{Z}^+$ with $(n, 7) = 1$. Prove that $7 \,|\, (n^6 - 1)$.

15. For the hashing function at the end of Example 14.11, find
(a) $h(123\text{-}04\text{-}2275)$; (b) a social security number n such that $h(n) = 413$, thus causing a collision with the number 081-37-6495 of the example.

16. Write a computer program (or develop an algorithm) that implements the hashing function of Exercise 15.

14.4
RING HOMOMORPHISMS AND ISOMORPHISMS

In this final section we shall examine functions (between rings) that obey special properties which depend on the binary operations in the rings.

Example 14.12 Consider the rings $(\mathbf{Z}, +, \cdot)$ and $(\mathbf{Z}_6, +, \cdot)$, where addition and multiplication in $\mathbf{Z}_6$ are as defined in Section 14.3.
    Define $f\colon \mathbf{Z} \to \mathbf{Z}_6$ by $f(x) = [x]$. For example, $f(1) = [1] = [7] = f(7)$, $f(2) = f(8) = f(2 + 6k) = [2]$, $k \in \mathbf{Z}$. (So f is onto though not one-to-one.)
    For $2, 3 \in \mathbf{Z}$, $f(2) = [2]$, $f(3) = [3]$ and we have $f(2 + 3) = f(5) = [5] = [2] + [3] = f(2) + f(3)$, and $f(2 \cdot 3) = f(6) = [0] = [2][3] = f(2) \cdot f(3)$.
    In fact for any $x, y \in \mathbf{Z}$,

$$f(x + y) = [x + y] = [x] + [y] = f(x) + f(y),$$

           ↑   ↑
      Addition in **Z**   Addition in $\mathbf{Z}_6$

and

$$f(x \cdot y) = [xy] = [x][y] = f(x) \cdot f(y). \qquad \square$$

           ↑   ↑
   Multiplication in **Z**   Multiplication in $\mathbf{Z}_6$

This example suggests the following definition.

DEFINITION 14.8 Let $(R, +, \cdot)$ and $(S, \oplus, \odot)$ be rings. A function $f\colon R \to S$ is called a *ring homomorphism* if for all $a, b \in R$,

    a) $f(a + b) = f(a) \oplus f(b)$, and

    b) $f(a \cdot b) = f(a) \odot f(b)$.

This function is said to *preserve* the ring operations. Consider $f(a + b) = f(a) \oplus f(b)$. *Adding a, b in R first and then finding the image (under f) in S of this* sum, we get the same result as when we first determine the images (under f) in S of a, b, and then *add* these images *in* S. (Hence we have the function operation and the addition operations commuting with each other.) Similar remarks can be made about the multiplicative operations in the rings.

DEFINITION 14.9   Let $f: (R, +, \cdot) \rightarrow (S, \oplus, \odot)$ be a ring homomorphism. If f is one-to-one and onto, then f is called a *ring isomorphism* and we say that R and S are *isomorphic rings*.

We can think of isomorphic rings arising when the "same" ring is dealt with in two different languages. The function f then provides a dictionary for translating from one language into the other.

The terms "homomorphism" and "isomorphism" come from the Greek, where *morphe* refers to shape or structure, *homo* means similar, and *iso* means identical or same. Hence homomorphic rings may be thought of as similar in structure, while isomorphic rings are abstractly the same structure.

In Definition 11.12 we defined the concept of *graph* isomorphism. There we called the undirected graphs $G_1 = (V_1, E_1)$ and $G_2 = (V_2, E_2)$ isomorphic when we could find a function $f: V_1 \rightarrow V_2$, where

a) f is one-to-one and onto, and

b) $\{a, b\} \in E_1$ if and only if $\{f(a), f(b)\} \in E_2$.

In light of our statements about ring isomorphisms, another way to think about condition (b) here is in terms of the function f preserving the structures of the undirected graphs G_1 and G_2. When $|V_1| = |V_2|$, it is not difficult to find a function $f: V_1 \rightarrow V_2$ that is one-to-one and onto. However, for a given set V of vertices, what determines the structure of an undirected graph $G = (V, E)$ is its set of edges (where the vertex adjacencies are defined). Therefore a one-to-one correspondence $f: V_1 \rightarrow V_2$ is a graph isomorphism when it preserves the structures of G_1 and G_2 by preserving these vertex adjacencies.

Example 14.13   For the ring R in Example 14.4 and the ring $\mathbf{Z}_5$, the function $f: R \rightarrow \mathbf{Z}_5$ given by

$$f(a) = [0], \qquad f(b) = [1], \qquad f(c) = [2], \qquad f(d) = [3], \qquad f(e) = [4]$$

provides us with a ring isomorphism.

For example, $f(c + d) = f(a) = [0] = [2] + [3] = f(c) + f(d)$, while $f(be) = f(e) = [4] = [1][4] = f(b)f(e)$. (There are 25 such equalities that must be verified for the preservation of each of the binary operations.)   □

Inasmuch as there are $5! = 120$ one-to-one functions from R onto $\mathbf{Z}_5$, is there any assistance we can call upon in attempting to determine when one of these functions is an ismorphism? Suggested by Example 14.13, the following theorem provides ways of at least starting to determine when functions between rings can be homomorphisms and isomorphisms. [Parts (c) and (d) of this theorem rely on the results of Exercises 20 and 21 in Section 14.2.]

THEOREM 14.15   If $f: (R, +, \cdot) \to (S, \oplus, \odot)$ is a ring homomorphism, then

    a) $f(z_R) = z_S$, where z_R, z_S are the zero elements of R, S, respectively;

    b) $f(-a) = -f(a)$, for any $a \in R$;

    c) $f(na) = nf(a)$, for any $a \in R, n \in \mathbf{Z}$;

    d) $f(a^n) = [f(a)]^n$, for any $a \in R, n \in \mathbf{Z}^+$; and

    e) if A is a subring of R, $f(A)$ is a subring of S.

Proof   **a)** $z_S \oplus f(z_R) = f(z_R) = f(z_R + z_R) = f(z_R) \oplus f(z_R)$. (Why?) So by the cancellation law of addition in S, we have $f(z_R) = z_S$.

    b) $z_S = f(z_R) = f(a + (-a)) = f(a) \oplus f(-a)$. Since additive inverses in S are unique and $f(-a)$ is an additive inverse of $f(a)$, it follows that $f(-a) = -f(a)$.

    c) If $n = 0$, then $f(na) = f(z_R) = z_S = nf(a)$. The result is also true for $n = 1$, so we assume the truth for $n = k$. Proceeding by mathematical induction, we examine the case where $n = k + 1$. By the results of Exercise 20 of Section 14.2, we get $f((k + 1)a) = f(ka + a) = f(ka) \oplus f(a) = kf(a) \oplus f(a)$ (Why?) $= (k + 1)(f(a))$ (Why?). (Note: There are three different kinds of addition here.)

        When $n > 0$, $f(-na) = -nf(a)$, from our prior proof by induction, part (b) above, and part (b) of Theorem 14.1, since $f(-na) + f(na) = f(n(-a)) + f(na) = nf(-a) + nf(a) = n[f(-a) + f(a)] = n[-f(a) + f(a)] = nz_S = z_S$. Hence the result follows for all $n \in \mathbf{Z}$.

    d) We leave this result for the reader to prove.

    e) Since $A \neq \emptyset$, $f(A) \neq \emptyset$. If $x, y \in f(A)$, then $x = f(a)$, $y = f(b)$ for some $a, b \in A$. Then $x \oplus y = f(a) \oplus f(b) = f(a + b)$, and $x \odot y = f(a) \odot f(b) = f(ab)$, with $a + b, ab \in A$ (Why?), so $x \oplus y, x \odot y \in f(A)$. Also, for $x \in f(A)$ with $x = f(a), a \in A$, we have $f(-a) = -f(a) = -x$, and because $-a \in A$ (Why?), we have $-x \in f(A)$. Therefore $f(A)$ is a subring of S. ∎

When the homomorphism is onto, we obtain the following theorem.

THEOREM 14.16   If $f: (R, +, \cdot) \to (S, \oplus, \odot)$ is a ring homomorphism from R *onto* S, then

    a) if R has unity u_R, $f(u_R)$ is the unity of S;

    b) if R has unity u_R and a is a unit in R, then $f(a)$ is a unit in S and $f(a^{-1}) = [f(a)]^{-1}$;

    c) if R is commutative, then S is commutative; and

    d) if I is an ideal of R, then $f(I)$ is an ideal of S.

Proof   We shall prove part (d) and leave the other parts to the reader. Since I is a subring of R, it follows that $f(I)$ is a subring of S by part (e) of Theorem 14.15. To verify that $f(I)$ is an ideal, let $x \in f(I)$ and $s \in S$. Then $x = f(a)$ and $s = f(r)$, for some $a \in I, r \in R$. So $s \odot x = f(r) \odot f(a) = f(ra)$, with $ra \in I$, and we have $s \odot x \in f(I)$. Similarly, $x \odot s \in f(I)$, so $f(I)$ is an ideal of S. ∎

These theorems reinforce the way in which homomorphisms and isomorphisms preserve structure. But can we find any use for these functions, aside from using them to prove more theorems? Before calling it quits, we consider two more examples.

Example 14.14   With $\mathbf{C}$ the field of complex numbers and S the ring of 2×2 real matrices of the form $\begin{bmatrix} a & b \\ -b & a \end{bmatrix}$, define $f: \mathbf{C} \to S$ by $f(a + bi) = \begin{bmatrix} a & b \\ -b & a \end{bmatrix}$, for $a + bi \in \mathbf{C}$.

Then

$$a + bi = c + di \Leftrightarrow a = c, b = d \Leftrightarrow \begin{bmatrix} a & b \\ -b & a \end{bmatrix} = \begin{bmatrix} c & d \\ -d & c \end{bmatrix},$$

so f is a one-to-one function. It is also onto. (Why?)
  Further,

$$f((a + bi) + (x + yi)) = f((a + x) + (b + y)i)$$
$$= \begin{bmatrix} a + x & b + y \\ -(b + y) & a + x \end{bmatrix} = \begin{bmatrix} a & b \\ -b & a \end{bmatrix} + \begin{bmatrix} x & y \\ -y & x \end{bmatrix}$$
$$= f(a + bi) + f(x + yi),$$

and

$$f((a + bi)(x + yi)) = f((ax - by) + (bx + ay)i)$$
$$= \begin{bmatrix} ax - by & bx + ay \\ -(bx + ay) & ax - by \end{bmatrix} = \begin{bmatrix} a & b \\ -b & a \end{bmatrix} \begin{bmatrix} x & y \\ -y & x \end{bmatrix}$$
$$= f(a + bi)f(x + yi),$$

so f is a ring isomorphism.
  But where can we use this isomorphism?
  Suppose we need to find the value of the complex number

$$(-8.3 + 9.9i) + \frac{(5.2 - 7.1i)^7}{(1.3 + 3.7i)^4}.$$

If a computer with an implementation of BASIC that included matrix operations were available, then using this isomorphism we could *translate* the problem from complex numbers, which are not implemented in BASIC, into matrices, which the computer can handle. Hence we seek the result

$$\begin{bmatrix} -8.3 & 9.9 \\ -9.9 & -8.3 \end{bmatrix} + \begin{bmatrix} 5.2 & -7.1 \\ 7.1 & 5.2 \end{bmatrix}^7 \left(\begin{bmatrix} 1.3 & 3.7 \\ -3.7 & 1.3 \end{bmatrix}^4 \right)^{-1}.$$

Using the BASIC program in Fig. 14.1, we find the answer (to two decimal places) to the complex arithmetic problem to be $8379.98 + 15122.7i$.   □

```
10      Dim A(2,2), B(2,2), C(2,2), D(2,2), E(2,2), F(2,2),
        G(2,2), H(2,2), K(2,2), M(2,2)
20      Mat Read A, E, K
30      Data 1.3,3.7,-3.7,1.3,5.2,-7.1,7.1,5.2,-8.3,9.9,
        -9.9,-8.3
40      Mat B = A
50      For I = 1 To 3
60          Mat C = A*B
70          Mat A = C
80      Next I
90      Mat D = Inv(A)
100     Mat F = E
110     For J = 1 To 6
120         Mat G = E*F
130         Mat E = G
140     Next J
150     Mat H = D*E
160     Mat M = K + H
170     Print "The solution to the problem is"; M(1,1);
        "+"; M(1,2); "i"
180     End

The solution to the problem is 8379.98 + 15122.7 i
```

Figure 14.1

Example 14.15   Extending the idea developed in Exercise 18 of Section 14.2, let R be the ring $\mathbf{Z}_2 \times \mathbf{Z}_3 \times \mathbf{Z}_5$. Then $|R| = |\mathbf{Z}_2| \cdot |\mathbf{Z}_3| \cdot |\mathbf{Z}_5| = 30$, and the operations of addition and multiplication are defined in R as follows:

For any (a_1, a_2, a_3), $(b_1, b_2, b_3) \in R$ where $a_1, b_1 \in \mathbf{Z}_2$, $a_2, b_2 \in \mathbf{Z}_3$, and $a_3, b_3 \in \mathbf{Z}_5$, we have

$$(a_1, a_2, a_3) + (b_1, b_2, b_3) = (a_1 + b_1, a_2 + b_2, a_3 + b_3)$$

$$\begin{array}{cccc} \uparrow & \uparrow & \uparrow & \uparrow \\ \text{Addition} & \text{Addition} & \text{Addition} & \text{Addition} \\ \text{in } R & \text{in } \mathbf{Z}_2 & \text{in } \mathbf{Z}_3 & \text{in } \mathbf{Z}_5 \end{array}$$

and

$$(a_1, a_2, a_3) \cdot (b_1, b_2, b_3) = (a_1 \cdot b_1, a_2 \cdot b_2, a_3 \cdot b_3).$$

$$\begin{array}{cccc} \text{Multiplication} & \text{Multiplication} & \text{Multiplication} & \text{Multiplication} \\ \text{in } R & \text{in } \mathbf{Z}_2 & \text{in } \mathbf{Z}_3 & \text{in } \mathbf{Z}_5 \end{array}$$

Define the function $f: \mathbf{Z}_{30} \to R$ by $f(x) = (x_1, x_2, x_3)$, where

$$x \equiv x_1 \,(\text{mod}\, 2), \qquad 0 \le x_1 \le 1$$
$$x \equiv x_2 \,(\text{mod}\, 3), \qquad 0 \le x_2 \le 2$$
$$x \equiv x_3 \,(\text{mod}\, 5), \qquad 0 \le x_3 \le 4.$$

Consequently, x_1, x_2, and x_3 are the remainders that result when x is divided by 2, 3, and 5, respectively.

The results in Table 14.9 show that f is a function that is one-to-one and onto.

Table 14.9

x (in $\mathbf{Z}_{30}$)	$f(x)$ (in R)	x (in $\mathbf{Z}_{30}$)	$f(x)$ (in R)	x (in $\mathbf{Z}_{30}$)	$f(x)$ (in R)
0	$(0,0,0)$	10	$(0,1,0)$	20	$(0,2,0)$
1	$(1,1,1)$	11	$(1,2,1)$	21	$(1,0,1)$
2	$(0,2,2)$	12	$(0,0,2)$	22	$(0,1,2)$
3	$(1,0,3)$	13	$(1,1,3)$	23	$(1,2,3)$
4	$(0,1,4)$	14	$(0,2,4)$	24	$(0,0,4)$
5	$(1,2,0)$	15	$(1,0,0)$	25	$(1,1,0)$
6	$(0,0,1)$	16	$(0,1,1)$	26	$(0,2,1)$
7	$(1,1,2)$	17	$(1,2,2)$	27	$(1,0,2)$
8	$(0,2,3)$	18	$(0,0,3)$	28	$(0,1,3)$
9	$(1,0,4)$	19	$(1,1,4)$	29	$(1,2,4)$

To verify that f is an isomorphism, let $x, y \in \mathbf{Z}_{30}$. Then

$$f(x + y) = ((x + y)\bmod 2, (x + y)\bmod 3, (x + y)\bmod 5)$$
$$= (x\bmod 2, x\bmod 3, x\bmod 5) + (y\bmod 2, y\bmod 3, y\bmod 5)$$
$$= f(x) + f(y), \text{ and}$$

$$f(xy) = (xy\bmod 2, xy\bmod 3, xy\bmod 5)$$
$$= (x\bmod 2, x\bmod 3, x\bmod 5) \cdot (y\bmod 2, y\bmod 3, y\bmod 5)$$
$$= f(x)f(y),$$

so f is an isomorphism.

In examining Table 14.9 we find, for example, that

1. $f(0) = (0,0,0)$, where 0 is the zero element of $\mathbf{Z}_{30}$ and $(0,0,0)$ is the zero element of $\mathbf{Z}_2 \times \mathbf{Z}_3 \times \mathbf{Z}_5$.

2. $f(2 + 4) = f(6) = (0,0,1) = (0,2,2) + (0,1,4) = f(2) + f(4)$.

3. The element 21 is the additive inverse of 9 in $\mathbf{Z}_{30}$, whereas $f(21) = (1,0,1)$ is the additive inverse of $(1,0,4) = f(9)$ in $\mathbf{Z}_2 \times \mathbf{Z}_3 \times \mathbf{Z}_5$.

4. $\{0, 5, 10, 15, 20, 25\}$ is a subring of $\mathbf{Z}_{30}$ with $\{(0,0,0)(=f(0)), (1,2,0)(=f(5)), (0,1,0)(=f(10)), (1,0,0)(=f(15)), (0,2,0)(=f(20)), (1,1,0)(=f(25))\}$ the corresponding subring in $\mathbf{Z}_2 \times \mathbf{Z}_3 \times \mathbf{Z}_5$.

But what else can we do with this isomorphism between $\mathbf{Z}_{30}$ and $\mathbf{Z}_2 \times \mathbf{Z}_3 \times \mathbf{Z}_5$? Suppose, for example, that we need to calculate $28 \cdot 17$ in $\mathbf{Z}_{30}$. We can transfer the problem to $\mathbf{Z}_2 \times \mathbf{Z}_3 \times \mathbf{Z}_5$ and compute $f(28) \cdot f(17) = (0,1,3) \cdot (1,2,2)$, where the moduli 2, 3, and 5 are smaller than 30 and easier to work with. Since $(0,1,3) \cdot (1,2,2) = (0 \cdot 1, 1 \cdot 2, 3 \cdot 2) = (0,2,1)$ and $f^{-1}(0,2,1) = 26$, it follows that $28 \cdot 17$ (in $\mathbf{Z}_{30}$) is 26. $\square$

The isomorphism in Example 14.15 is a special case of a more general result called the *Chinese Remainder Theorem*. This we shall now state but not prove. We know from the Fundamental Theorem of Arithmetic that for any $n \in \mathbf{Z}^+$, $n > 1$, we can factor n as $p_1^{n_1} p_2^{n_2} \cdots p_k^{n_k}$, where $p_1, p_2, \ldots, p_k$ are k distinct primes, $k \geq 1$, and $n_1, n_2, \ldots, n_k \in \mathbf{Z}^+$. The Chinese Remainder Theorem then tells us that the rings $\mathbf{Z}_n$ and $\mathbf{Z}_{m_1} \times \mathbf{Z}_{m_2} \times \cdots \times \mathbf{Z}_{m_k}$ are isomorphic for $m_1 = p_1^{n_1}$, $m_2 = p_2^{n_2}, \ldots, m_k = p_k^{n_k}$. For the interested reader we mention that on pages 344–359 of the text by J. P. Tremblay and R. Manohar [6], the Chinese Remainder Theorem is studied in conjunction with applications of residue arithmetic to computers.

EXERCISES 14.4

1. If R is the ring of Example 14.5, determine an isomorphism $f: R \to \mathbf{Z}_6$.

2. Complete the proofs of Theorems 14.15 and 14.16.

3. If R, S, and T are rings, and $f: R \to S, g: S \to T$ are ring homomorphisms, prove that the composite function $g \circ f: R \to T$ is a ring homomorphism.

4. If $S = \left\{ \begin{bmatrix} a & 0 \\ 0 & a \end{bmatrix} \middle| a \in \mathbf{R} \right\}$, then S is a ring under matrix addition and multiplication. Prove that $\mathbf{R}$ is isomorphic to S.

5. a) Let $(R, +, \cdot)$ and $(S, \oplus, \odot)$ be rings with zero elements z_R and z_S, respectively. If $f: R \to S$ is a ring homomorphism, let $K = \{a \in R \mid f(a) = z_S\}$. Prove that K is an ideal of R. (K is called the *kernel* of the homomorphism f.)

   b) Find the kernel of the homomorphism in Example 14.12.

   c) Let $f, (R, +, \cdot)$ and $(S, \oplus, \odot)$ be as in part (a). Prove that f is one-to-one if and only if the kernel of f is $\{z_R\}$.

6. Use the information in Table 14.9 to compute each of the following in $\mathbf{Z}_{30}$.

   a) $(13)(23) + 18$ b) $(11)(21) - 20$

   c) $(13 + 19)(27)$ d) $(13)(29) + (24)(8)$

7. a) Construct a table (as in Example 14.15) for the isomorphism $f: \mathbf{Z}_{20} \to \mathbf{Z}_4 \times \mathbf{Z}_5$.

   b) Use the table from part (a) to compute the following in $\mathbf{Z}_{20}$.

   (i) $(17)(19) + (12)(14)$ (ii) $(18)(11) - (9)(15)$

8. Let $n, r, s \in \mathbf{Z}^+$ with $n, r, s \geq 2$, $n = rs$, and $(r, s) = 1$. If $f: \mathbf{Z}_n \to \mathbf{Z}_r \times \mathbf{Z}_s$ is a ring isomorphism with $f(a) = (1, 0)$ and $f(b) = (0, 1)$, prove that if $(m, t) \in \mathbf{Z}_r \times \mathbf{Z}_s$, then $f^{-1}(m, t) \equiv ma + tb \pmod{n}$.

9. a) How many units are there in the ring $\mathbf{Z}_8$?

   b) How many units are there in the ring $\mathbf{Z}_2 \times \mathbf{Z}_2 \times \mathbf{Z}_2$?

   c) Are $\mathbf{Z}_8$ and $\mathbf{Z}_2 \times \mathbf{Z}_2 \times \mathbf{Z}_2$ isomorphic rings?

10. a) How many units are there in $\mathbf{Z}_{15}$? How many in $\mathbf{Z}_3 \times \mathbf{Z}_5$?

    b) Are $\mathbf{Z}_{15}$ and $\mathbf{Z}_3 \times \mathbf{Z}_5$ isomorphic?

11. Are $\mathbf{Z}_4$ and the ring in Example 14.3 isomorphic?

12. If $f: R \to S$ is a ring homomorphism and J is an ideal of S, prove that $f^{-1}(J) = \{a \in R \mid f(a) \in J\}$ is an ideal of R.

13. In the solution of Example 14.14, part of the problem required the calculation of

$$\left(\begin{bmatrix} 1.3 & 3.7 \\ -3.7 & 1.3 \end{bmatrix}^4\right)^{-1}.$$

Algebraically this equals

$$\left(\begin{bmatrix} 1.3 & 3.7 \\ -3.7 & 1.3 \end{bmatrix}^{-1}\right)^4.$$

Is there any advantage, with respect to computer computation, in using one form instead of the other?

14.5
SUMMARY AND HISTORICAL REVIEW

Emphasizing structure induced by two binary operations, this chapter has introduced us to the mathematical system called a ring. Throughout the development of mathematics, the ring of integers has played a key role. In the branch of mathematics called number theory, we examine the basic properties of $(\mathbf{Z}, +, \cdot)$, as well as the finite rings $(\mathbf{Z}_n, +, \cdot)$. The matrix rings provide familiar examples of noncommutative rings.

This chapter contains the development of an abstract theory. On the basis of the definition of a ring, we established principles of elementary algebra that we have been using since our early encounters with arithmetic, signed numbers, and the manipulation of unknowns. The reader may have found some of the proofs tedious, as we justified all the steps in the derivations. Faced with the challenge of trying to prove a result in abstract mathematics, one should follow the advice given by the Roman rhetorician Marcus Fabius Quintilianus (first century A.D.), when he said "One should not aim at being possible to understand (or follow), but at being impossible to be misunderstood."

A famous problem in number theory, known as *Fermat's Last Theorem*, claims that the equation $x^n + y^n = z^n$, $n \in \mathbf{Z}^+$, $n > 1$, has no solutions in $\mathbf{Z}^+$ when $n > 2$. The French mathematician, Pierre de Fermat (1601–1665) wrote that he had proved this result but that the proof was too long to include in the margin of his manuscript. Unfortunately, to this day, there are infinitely many unresolved values of n for the theorem. Attempts to solve this problem have resulted in new mathematical ideas and theories.

In trying to prove Fermat's last theorem, the German mathematician Ernst Kummer (1810–1893) developed the foundations for the concept of the *ideal*. This was later discovered and used by his countryman Richard Dedekind (1831–1916) in his research on what are now called Dedekind domains. Use of the term "ring," however, seems to be attributable to the German mathematician David Hilbert (1862–1943).

Ring homomorphisms and their interplay with ideals were extensively developed by the German mathematician Emmy Noether (1882–1935). This great genius received little remuneration, financial or otherwise, from the governing bodies of her native land because of the sexual bias that existed in the universities at that time. Emmy Noether's talents were nonetheless recognized by her colleagues, and she was eulogized in *The New York Times* on May 3, 1935 by none other than Albert Einstein (1879–1955). In addition to enduring sexual bias, as a Jew she was forced to flee her homeland in 1933 when the Nazis came to power. She spent the last two years of her life guiding young mathematicians in the United States. For more on the life of this fascinating person, examine the biography by A. Dick [2].

The special rings called *fields* arise in the rational, real, and complex number systems. But we also found some interesting finite fields. These structures will be examined again in Chapter 17 in connection with combinatorial designs. The field theory developed by the French genius Evariste Galois (1811–1832) established results about the solutions of polynomial equations of degree > 4. These results had baffled mathematicians for centuries, and his ideas, now known as Galois theory, still comprise one of the most elegant mathematical theories ever developed. More on Galois theory appears in the text by O. Zariski and P. Samuel [8].

For supplemental reading on ring theory at the introductory level, the interested reader should examine Chapters 6, 7, and 12 of N. McCoy and T. Berger [4] and Chapter 6 of V. Larney [3]. A somewhat more advanced coverage can be found in Chapter 4 of the text by E. A. Walker [7].

The development of modular congruence, along with many related ideas, we owe primarily to Carl Friedrich Gauss (1777–1855). More on the solution of congruences can be found in the text by I. Niven and H. Zuckerman [5].

Finally, the topic of hashing, or scattering, can be further investigated in Chapter 2 of J. Tremblay and R. Manohar [6]. Chapter 4 of A. Aho, J. Hopcroft, and J. Ullman [1] includes a discussion on the efficiency of hashing functions and a probabilistic investigation of the collision problem that arises for these functions.

REFERENCES

1. Aho, Alfred V., Hopcroft, John E., and Ullman, Jeffrey D. *Data Structures and Algorithms.* Reading, Mass.: Addison-Wesley, 1983.
2. Dick, Auguste. *Emmy Noether (1882–1935),* trans. Heidi Blocher. Boston: Birkhäuser-Boston, 1981.
3. Larney, Violet Hachmeister. *Abstract Algebra: A First Course.* Boston: Prindle, Weber & Schmidt, 1975.
4. McCoy, Neal H., and Berger, Thomas R. *Algebra: Groups, Rings and Other Topics.* Boston: Allyn and Bacon, 1977.
5. Niven, Ivan, and Zuckerman, Herbert S. *An Introduction to the Theory of Numbers,* 3rd ed. New York: Wiley, 1972.
6. Tremblay, Jean-Paul, and Manohar, R. *Discrete Mathematical Structures with Applications to Computer Science.* New York: McGraw-Hill, 1975.

7. Walker, Elbert A. *Introduction to Abstract Algebra.* New York: Random House/Birkhäuser, 1987.

8. Zariski, Oscar, and Samuel, Pierre. *Commutative Algebra,* Vol. I. Princeton, N.J.: Van Nostrand, 1958.

MISCELLANEOUS EXERCISES

1. Determine whether each of the following statements is true or false. For each false statement give a counterexample.

   a) If $(R, +, \cdot)$ is a ring, and $\emptyset \neq S \subseteq R$ with S closed under $+$ and $\cdot$, then S is a subring of R.

   b) If $(R, +, \cdot)$ is a ring with unity, and S is a subring of R, then S has a unity.

   c) If $(R, +, \cdot)$ is a ring with unity u_R, and S is a subring of R with unity u_S, then $u_R = u_S$.

   d) Every field is an integral domain.

   e) No subring of an integral domain has any proper divisors of zero.

   f) Any subring of a field is a field.

   g) A field can have only two subrings.

   h) The function $f: \mathbf{Z} \to \mathbf{Z}$ defined by $f(x) = 2x$ is a ring homomorphism.

   i) Every finite field has a prime number of elements.

   j) $(\mathbf{Q}, +, \cdot)$ has an infinite number of subrings.

   k) If S and T are ideals of a ring R, then $S \cup T$ is an ideal of R.

   ℓ) If $a, b \in \mathbf{Z}_p$ (p a prime) and $a^2 = b^2$, then $a = b$.

2. Prove that a ring R is commutative if and only if $(a + b)^2 = a^2 + 2ab + b^2$, for all $a, b \in R$.

3. A ring R is called *Boolean* if $a^2 = a$ for all $a \in R$. If R is Boolean, prove that (a) $a + a = 2a = z$, for all $a \in R$, and (b) R is commutative.

4. a) In the field $\mathbf{C}$ of complex numbers, the *conjugate* of a complex number $z = x + iy$ is given by $\bar{z} = x - iy$. Hence $\overline{2 + 3i} = 2 - 3i$ and $\overline{5i} = -5i$, while $\bar{7} = 7$.
      If $z, z_1, z_2 \in \mathbf{C}$, prove that

   (i) $\overline{z_1 + z_2} = \bar{z_1} + \bar{z_2}$

   (ii) $\overline{z_1 z_2} = \bar{z_1}\,\bar{z_2}$

   (iii) $(\bar{z})^n = (\overline{z^n})$, for all $n \in \mathbf{Z}^+$

   (iv) $(\bar{z})^{-1} = (\overline{z^{-1}})$, for all $z \neq 0$

   (v) $z + \bar{z} \in \mathbf{R}$

   (vi) $z\bar{z} \in \mathbf{R}^+$, for all $z \neq 0$

   b) Let $f: \mathbf{C} \to \mathbf{C}$ be defined by $f(z) = \bar{z}$. Prove that f is an isomorphism.

5. If $(R, +, \cdot)$ is a ring, prove that $C = \{r \in R \mid ar = ra, \text{ for all } a \in R\}$ is a subring of R. (The subring C is called the *center* of R.)

6. Given a finite field F, let $M_2(F)$ denote the set of all 2×2 matrices with entries from F. As in Example 14.2, $(M_2(F), +, \cdot)$ becomes a noncommutative ring with unity.

a) Determine the number of elements in $M_2(F)$ if F is

(i) $\mathbf{Z}_2$ (ii) $\mathbf{Z}_3$ (iii) $\mathbf{Z}_p$, p a prime

b) As in Exercise 7 of Section 14.1, $A = \begin{bmatrix} a & b \\ c & d \end{bmatrix} \in M_2(\mathbf{Z}_p)$ is a unit if and only if $ad - bc \neq z$. This occurs if the first row of A does not contain all zeros and the second row is not a multiple (by an element of $\mathbf{Z}_p$) of the first. Use this observation to determine the number of units in

(i) $M_2(\mathbf{Z}_2)$ (ii) $M_2(\mathbf{Z}_3)$ (iii) $M_2(\mathbf{Z}_p)$, p a prime

7. a) Let $(D, +, \cdot)$ be an integral domain. Define relation $\mathcal{R}$ on D by $a \mathcal{R} b$ if $a = bs$, where s is a unit of D. Prove that $\mathcal{R}$ is an equivalence relation.

b) Find the partition induced by $\mathcal{R}$ if the integral domain is

(i) $(\mathbf{Z}, +, \cdot)$ (ii) $\{a + bi \mid a, b \in \mathbf{Z}, i^2 = -1\}$ (iii) $\mathbf{Z}_7$

8. Let $A = \mathbf{R}^+$. Define $\oplus$ and $\odot$ on A by $a \oplus b = ab$, the ordinary product of a, b; and $a \odot b = a^{\log_2 b}$.

a) Verify that $(A, \oplus, \odot)$ is a commutative ring with unity.

b) Is it an integral domain or field?

9. Let R be a ring with ideals A and B. Define $A + B = \{a + b \mid a \in A, b \in B\}$. Prove that $A + B$ is an ideal of R. (For any ring R, the ideals of R form a poset under set inclusion. If A and B are ideals of R, with $\mathrm{glb}\{A, B\} = A \cap B$ and $\mathrm{lub}\{A, B\} = A + B$, the poset is a lattice.)

10. a) If p is a prime, prove that p divides $\binom{p}{k}$, $0 < k < p$.

b) If $a, b \in \mathbf{Z}$, prove that $(a + b)^p \equiv a^p + b^p \pmod{p}$.

11. Given n positive integers $x_1, x_2, \ldots, x_n$, not necessarily distinct, prove that either $n \mid (x_1 + x_2 + \cdots + x_i)$, for some $1 \leq i \leq n$, or there exist $1 \leq i < j \leq n$ such that $n \mid (x_{i+1} + \cdots + x_{j-1} + x_j)$.

12. Write a computer program (or develop an algorithm) that reverses the order of the digits in a given positive integer. For example, the input 1374 results in the output 4731.

CHAPTER 15

Boolean Algebra and Switching Functions

Again we confront an algebraic system in which the structure depends primarily on two binary operations. Unlike the coverage of rings, in dealing with Boolean algebras we shall stress applications more than the abstract nature of the system. Nonetheless, we shall find results here that are quite different from those for rings. Among other things, a finite Boolean algebra cannot have just any number of elements. It must have 2^n elements, where $n \in \mathbf{Z}^+$.

In 1854 the English mathematician George Boole published his monumental work *An Investigation of the Laws of Thought*. Within this work Boole created a system of mathematical logic that he developed in terms of what is now called a Boolean algebra.

In 1938 Claude Shannon developed the algebra of switching functions and showed how its structure was related to the ideas established by Boole. As a result, an example of abstract mathematics in the nineteenth century became an applied mathematical discipline in the twentieth.

15.1
SWITCHING FUNCTIONS: DISJUNCTIVE AND CONJUNCTIVE NORMAL FORMS

An electric switch can be turned on (allowing the flow of current) or off (preventing the flow of current). Similarly, in a transistor, current is either passing (conducting) or not passing (nonconducting). These are two examples of *two-state devices*. (In Section 2.2 we saw how the electric switch was related to the two-valued logic we introduced there.)

In order to investigate such devices, we abstract these notions of "true" and "false," "on" and "off," as follows.

Let $B = \{0, 1\}$. We define addition, multiplication, and the complements for the elements of B by

a)	**or:**	$0 + 0 = 0;\quad 0 + 1 = 1 + 0 = 1 + 1 = 1.$
b)	**and:**	$0 \cdot 0 = 0 = 1 \cdot 0 = 0 \cdot 1;\quad 1 \cdot 1 = 1.$
c)	**not:**	$\bar{0} = 1;\quad \bar{1} = 0.$

A variable x is called a _Boolean variable_ if x takes on only values in $B = \{0,1\}$. Consequently, $x + x = x$ and $x^2 = x \cdot x = xx = x$ for any Boolean variable x.

If x, y are Boolean variables, then

1. $x + y = 0$ if and only if $x = y = 0$, and

2. $xy = 1$ if and only if $x = y = 1$.

If $n \in \mathbf{Z}^+$, $B^n = \{(b_1, b_2, \ldots, b_n) \mid b_i \in \{0, 1\}, 1 \le i \le n\}$. A function $f: B^n \to B$ is called a _Boolean_, or _switching_, _function_ of n variables. The n variables are emphasized by writing $f(x_1, x_2, \ldots, x_n)$, where each $x_i, 1 \le i \le n$, is a Boolean variable.

Example 15.1   Let $f: B^3 \to B$, where $f(x, y, z) = xy + z$. (We write xy for $x \cdot y$.) This Boolean function is determined by evaluating f for each of the eight possible assignments of the variables x, y, z. Table 15.1 demonstrates this.   □

Table 15.1

x	y	z	xy	$f(x, y, z) = xy + z$
0	0	0	0	0
0	0	1	0	1
0	1	0	0	0
0	1	1	0	1
1	0	0	0	0
1	0	1	0	1
1	1	0	1	1
1	1	1	1	1

DEFINITION 15.1   If $f, g: B^n \to B$, then $f = g$ if their corresponding (function) tables are exactly the same. (The tables show that $f(b_1, b_2, \ldots, b_n) = g(b_1, b_2, \ldots, b_n)$, for all $(b_1, b_2, \ldots, b_n) \in B^n$.)

DEFINITION 15.2   For $f: B^n \to B$, the _complement_ of f, denoted $\bar{f}$, is the Boolean function defined on B^n by

$$\bar{f}(b_1, b_2, \ldots, b_n) = \overline{f(b_1, b_2, \ldots, b_n)}.$$

If $g: B^n \to B$, we define $f + g$, $f \cdot g: B^n \to B$, the _sum_ and _product_ of f, g, respectively, by

$$(f + g)(b_1, b_2, \ldots, b_n) = f(b_1, b_2, \ldots, b_n) + g(b_1, b_2, \ldots, b_n)$$

and

$$(f \cdot g)(b_1, b_2, \ldots, b_n) = f(b_1, b_2, \ldots, b_n) \cdot g(b_1, b_2, \ldots, b_n).$$

Table 15.2

1. $\overline{\overline{f}} = f$	$\overline{\overline{x}} = x$	Law of the *Double Complement*
2. $\overline{f+g} = \overline{f}\,\overline{g}$ $\overline{fg} = \overline{f} + \overline{g}$	$\overline{x+y} = \overline{x}\,\overline{y}$ $\overline{xy} = \overline{x} + \overline{y}$	*DeMorgan's* Laws
3. $f + g = g + f$ $fg = gf$	$x + y = y + x$ $xy = yx$	*Commutative* Laws
4. $f + (g + h) = (f + g) + h$ $f(gh) = (fg)h$	$x + (y + z) = (x + y) + z$ $x(yz) = (xy)z$	*Associative* Laws
5. $f + gh = (f + g)(f + h)$ $f(g + h) = fg + fh$	$x + yz = (x + y)(x + z)$ $x(y + z) = xy + xz$	*Distributive* Laws
6. $f + f = f$ $ff = f$	$x + x = x$ $xx = x$	*Idempotent* Laws
7. $f + 0 = f$ $f \cdot 1 = f$	$x + 0 = x$ $x \cdot 1 = x$	*Identity* Laws
8. $f + \overline{f} = 1$ $f\overline{f} = 0$	$x + \overline{x} = 1$ $x\overline{x} = 0$	*Inverse* Laws
9. $f + 1 = 1$ $f \cdot 0 = 0$	$x + 1 = 1$ $x \cdot 0 = 0$	*Dominance* Laws
10. $f + fg = f$ $\;f \cdot 1 + f \cdot g$ $f(f + g) = f$	$x + xy = x$ $x(x + y) = x$	*Absorption* Laws

As with the laws of logic (in Chapter 2) and the laws of set theory (in Chapter 3), the properties shown in Table 15.2 are satisfied by any Boolean functions $f, g, h: B^n \rightarrow B$ and by any Boolean variables x, y, z. (We write fg for $f \cdot g$.)

The function **0** denotes the Boolean function whose value is always 0, and **1** is the function whose only value is 1. [Note: $\mathbf{0}, \mathbf{1} \notin B$.]

Once again the idea of duality appears in properties 2–10. If s denotes a theorem about Boolean functions, then s^d, the *dual* of s, is obtained by replacing in s all occurrences of $+ (\cdot)$ by $\cdot (+)$ and all occurrences of **0** (**1**) by **1** (**0**). By the principle of duality (which we shall examine in Section 15.4) the statement s^d is also a theorem. The same is true for a theorem dealing with Boolean variables, except here it is the Boolean values 0 and 1 that are replaced, not the constant functions **0** and **1**.

We can prove property 5 of Table 15.2 as follows.

Example 15.2   (The Distributive Law of $+$ over $\cdot$) The last two columns of Table 15.3 show that $f + gh = (f + g)(f + h)$. By the principle of duality, we obtain $f(g + h) = fg + fh$. We also see that $x + yz = (x + y)(x + z)$ is a special case of this property for the situation where $f, g, h: B^3 \rightarrow B$, with $f(x, y, z) = x$, $g(x, y, z) = y$, and $h(x, y, z) = z$. Hence no additional tables are needed to establish these properties for Boolean variables. $\square$

Table 15.3

f	g	h	gh	$f+g$	$f+h$	$f+gh$	$(f+g)(f+h)$
0	0	0	0	0	0	0	0
0	0	1	0	0	1	0	0
0	1	0	0	1	0	0	0
0	1	1	1	1	1	1	1
1	0	0	0	1	1	1	1
1	0	1	0	1	1	1	1
1	1	0	0	1	1	1	1
1	1	1	1	1	1	1	1

Example 15.3

a) To establish the first absorption property for Boolean variables, in place of a table we have the following:

$$
\begin{array}{ll}
& \textbf{Reasons} \\
x + xy = x1 + xy & \text{Identity Law} \\
\quad = x(1+y) & \text{Distributive Law of } \cdot \text{ over } + \\
\quad = x1 & \text{Dominance Law (and Commutative Law of } +) \\
\quad = x & \text{Identity Law}
\end{array}
$$

This result indicates that some of our properties can be derived from others. The question then is which properties we must establish with tables so that we can derive the other properties as we did here. We shall consider this later when we study the structure of a Boolean algebra.

In the meantime, let us demonstrate how the results of Table 15.2 can be used to simplify two other Boolean expressions.

b) For any Boolean variables $w, x, y,$ and z, we find that

$$
\begin{array}{ll}
& \textbf{Reasons} \\
wy + xy + wz + xz = (w+x)y + (w+x)z & \text{Commutative Law of } \cdot \\
& \text{and the Distributive Law} \\
& \text{of } \cdot \text{ over } + \\
\quad = (w+x)(y+z) & \text{Distributive Law of } \cdot \\
& \text{over } +
\end{array}
$$

c) Simplify the expression $wx + \overline{\overline{x}\overline{z}} + (y + \overline{z})$, where $w, x, y,$ and z are Boolean variables.

$$
\begin{array}{ll}
& \textbf{Reasons} \\
wx + \overline{\overline{x}\overline{z}} + (y + \overline{z}) = wx + (\overline{\overline{x}} + \overline{z}) + (y + \overline{z}) & \text{DeMorgan's Law} \\
\quad = wx + (x + \overline{z}) + (y + \overline{z}) & \text{Law of the Double} \\
& \text{Complement} \\
\quad = [(wx + x) + \overline{z}] + (y + \overline{z}) & \text{Associative Law of } + \\
\quad = (x + \overline{z}) + (y + \overline{z}) & \text{Absorption Law (and the} \\
& \text{Commutative Law of } +) \\
\quad = x + (\overline{z} + \overline{z}) + y & \text{Commutative and} \\
& \text{Associative Laws of } + \\
\quad = x + \overline{z} + y & \text{Idempotent Law of } +
\end{array}
$$

$\square$

Up to this point we have repeated for Boolean functions what we did with propositions: Given the Boolean function, we constructed its table of values. Now we consider the reverse process: Given a table of values, find a Boolean function for which it is the correct table.

Example 15.4   Given three Boolean variables x, y, z, find functions $f, g, h: B^3 \to B$ for the columns specified in Table 15.4.

For the column under f we want a result that has the value 1 in the case where $x = y = 0$ and $z = 1$. The function $f(x, y, z) = \bar{x}\bar{y}z$ is one such function. In the same way, $g(x, y, z) = x\bar{y}\bar{z}$ yields the value 1 for $x = 1$, $y = z = 0$, and is 0 in all other cases. As each of f and g has the value 1 in only one case and these cases are distinct from each other, their sum $f + g$ has the value 1 in exactly these two cases. So $h(x, y, z) = f(x, y, z) + g(x, y, z) = \bar{x}\bar{y}z + x\bar{y}\bar{z}$ has the column of values given under h.   □

Table 15.4

x	y	z	f	g	h
0	0	0	0	0	0
0	0	1	1	0	1
0	1	0	0	0	0
0	1	1	0	0	0
1	0	0	0	1	1
1	0	1	0	0	0
1	1	0	0	0	0
1	1	1	0	0	0

This example leads us to the following definition.

DEFINITION 15.3   For any $n \in \mathbf{Z}^+$, if f is a Boolean function on the n variables $x_1, x_2, \ldots, x_n$, we call

a) each term x_i or its complement $\bar{x}_i$, $1 \leq i \leq n$, a *literal*;

b) a term of the form $y_1 y_2 \cdots y_n$, where each $y_i = x_i$ or $\bar{x}_i$, $1 \leq i \leq n$, a *fundamental conjunction*; and

c) a representation of f as a sum of fundamental conjunctions a *disjunctive normal form* (d.n.f.) of f.

Although no formal proof is given here, the following examples suggest that any $f: B^n \to B$, $f \neq \mathbf{0}$, has a unique (up to the order of fundamental conjunctions) representation as a d.n.f.

Example 15.5   Find the d.n.f. for $f: B^3 \to B$, where $f(x, y, z) = xy + \bar{x}z$.

From Table 15.5, we see that the column for f contains four 1's. These indicate the four fundamental conjunctions needed in the d.n.f. of f, so $f(x, y, z) = \bar{x}\bar{y}z + \bar{x}yz + xy\bar{z} + xyz$.

Another way to solve this problem is to take each product term and somehow involve whichever variables are missing. Using the properties of these variables, we have $xy + \bar{x}z = xy(z + \bar{z}) + \bar{x}(y + \bar{y})z$ (Why?) $= xyz + xy\bar{z} + \bar{x}yz + \bar{x}\bar{y}z$.   □

$xyz + xy\bar{z} + \bar{x}yz + \bar{x}\bar{y}z$

Table 15.5

x	y	z	xy	$\bar{x}z$	f
0	0	0	0	0	0
0	0	1	0	1	1
0	1	0	0	0	0
0	1	1	0	1	1
1	0	0	0	0	0
1	0	1	0	0	0
1	1	0	1	0	1
1	1	1	1	0	1

Example 15.6   Find the d.n.f. for $g(w, x, y, z) = wx\bar{y} + wy\bar{z} + xy$.

We examine each term, as follows:

a) $wx\bar{y} = wx\bar{y}(z + \bar{z}) = wx\bar{y}z + wx\bar{y}\,\bar{z}$

b) $wy\bar{z} = w(x + \bar{x})y\bar{z} = wxy\bar{z} + w\bar{x}y\bar{z}$

c) $xy = (w + \bar{w})xy(z + \bar{z}) = wxyz + wxy\bar{z} + \bar{w}xyz + \bar{w}xy\bar{z}$

From the idempotent property of $+$, the d.n.f. of g is

$$g(w, x, y, z) = wx\bar{y}z + wx\bar{y}\,\bar{z} + wxy\bar{z} + w\bar{x}y\bar{z} + wxyz + \bar{w}xyz + \bar{w}xy\bar{z}. \quad \square$$

Consider Table 15.6. If we agree to list the Boolean variables in alphabetical order, we see that the values for x, y, z in any row determine a binary number. These binary numbers for $0, 1, 2, \ldots, 7$ arise for rows $1, 2, \ldots, 8$, respectively. Because of this, the d.n.f. of a nonzero Boolean function can be expressed more compactly. The function f in Example 15.5 can be given by $f = \sum m(1, 3, 6, 7)$, where m indicates the *minterms* at rows $2, 4, 7, 8$, where the respective binary labels are $1, 3, 6, 7$. The word "minterm" arises as follows: $m(1)$, for example, is the minterm for the row where $x = y = 0$, $z = 1$; the prefix "min" refers to the fact that the fundamental conjunction $\bar{x}\,\bar{y}z$ has value 1 a minimal number of times (without being identically **0**).

Lacking a table, we can still represent the d.n.f. of the function g of Example 15.6 as a *sum of minterms*. For each fundamental conjunction $c_1 c_2 c_3 c_4$, where $c_1 = w$ or $\bar{w}, \ldots, c_4 = z$ or $\bar{z}$, we replace each c_i, $1 \le i \le 4$, by 0 if c_i is a complemented variable, and by 1 otherwise. In this way the binary number associated with that fundamental conjunction is obtained. As a sum of minterms, $g = \sum m(6, 7, 10, 12, 13, 14, 15)$.

Table 15.6

x	y	z
0	0	0
0	0	1
0	1	0
0	1	1
1	0	0
1	0	1
1	1	0
1	1	1

Dual to the disjunctive normal form is the conjunctive normal form, which we discuss before closing this section.

Example 15.7 Let $f: B^3 \to B$ be given by Table 15.7. A term of the form $c_1 + c_2 + c_3$, where $c_1 = x$ or $\bar{x}$, $c_2 = y$ or $\bar{y}$, and $c_3 = z$ or $\bar{z}$, is called a *fundamental disjunction*. The fundamental disjunction $x + y + z$ has value 1 in all cases except where the values of x, y, z are 0. Similarly, $x + \bar{y} + z$ has value 1 except when $x = z = 0$ and $y = 1$. Since each of these fundamental disjunctions has the value 0 in only one case, and these cases do not occur simultaneously, the product $(x + y + z)(x + \bar{y} + z)$ has the value 0 in precisely the two cases listed above. Continuing in this manner, we write the function f as

$$f = (x + y + z)(x + \bar{y} + z)(\bar{x} + \bar{y} + z)$$

and call this the *conjunctive normal form* (c.n.f.) for f.

Since $x + y + z$ has the value 1 a maximum number of times (without being identically **1**), it is called a *maxterm*, especially when we use a binary row label to represent it. Using the binary numbers indexing the rows of the table, we write $f = \prod M(0, 2, 6)$, a *product of maxterms*.

Such representations exist for any $f \neq \mathbf{1}$, and they are unique up to the order of the fundamental disjunctions (or maxterms). $\square$

Table 15.7

x	y	z	f
0	0	0	0
0	0	1	1
0	1	0	0
0	1	1	1
1	0	0	1
1	0	1	1
1	1	0	0
1	1	1	1

Example 15.8 Let $g: B^4 \to B$, where $g(w, x, y, z) = (w + x + y)(x + \bar{y} + z)(w + \bar{y})$. To obtain the c.n.f. for g, we write each disjunction in the product as follows:

a) $w + x + y = w + x + y + 0 = w + x + y + z\bar{z}$
$\qquad = (w + x + y + z)(w + x + y + \bar{z})$

b) $x + \bar{y} + z = w\bar{w} + x + \bar{y} + z = (w + x + \bar{y} + z)(\bar{w} + x + \bar{y} + z)$

c) $w + \bar{y} = w + x\bar{x} + \bar{y} = (w + x + \bar{y})(w + \bar{x} + \bar{y})$
$\qquad = (w + x + \bar{y} + z\bar{z})(w + \bar{x} + \bar{y} + z\bar{z})$
$\qquad = (w + x + \bar{y} + z)(w + x + \bar{y} + \bar{z})(w + \bar{x} + \bar{y} + z)(w + \bar{x} + \bar{y} + \bar{z})$

Consequently, using the idempotent law of $\cdot$, we have $g(w, x, y, z) = (w + x + y + z)(w + x + y + \bar{z})(w + x + \bar{y} + z)(\bar{w} + x + \bar{y} + z)(w + x + \bar{y} + \bar{z}) \cdot (w + \bar{x} + \bar{y} + z)(w + \bar{x} + \bar{y} + \bar{z})$.

To obtain g as a product of maxterms, we associate with each fundamental disjunction $d_1 + d_2 + d_3 + d_4$ the binary number $b_1 b_2 b_3 b_4$, where $b_1 = 0$ if $d_1 = w$;

$b_1 = 1$ if $d_1 = \overline{w}; \ldots;$ $b_4 = 0$ if $d_4 = z;$ $b_4 = 1$ if $d_4 = \overline{z}$. As a result, $g = \prod M(0, 1, 2, 3, 6, 7, 10)$. □

EXERCISES 15.1

1. Find the value of each of the following Boolean expressions if the values of the Boolean variables $w, x, y,$ and z are 1, 1, 0, and 0, respectively.
   a) $\overline{xy} + \overline{x}\,\overline{y}$
   b) $w + \overline{x}y$
   c) $wx + \overline{y} + yz$
   d) $wx + xy + yz$
   e) $(wx + y\overline{z}) + w\overline{y} + \overline{(w + y)(\overline{x} + y)}$

2. Let $w, x,$ and y be Boolean variables where the value of x is 1. For each of the following Boolean expressions, determine, if possible, the value of the expression. If you cannot determine the value of the expression, then find the number of assignments of values for w and y that will result in the value 1 for the expression.
   a) $x + xy + w$ b) $xy + w$
   c) $\overline{x}y + xw$ d) $\overline{x}y + w$

3. a) How many rows are needed to construct the (function) table for a Boolean function of n variables?
   b) How many different Boolean functions of n variables are there?

4. Suppose that $f: B^3 \rightarrow B$ is defined by $f(x, y, z) = \overline{\overline{(x + y)} + (\overline{x}z)}$.
   a) Determine the d.n.f. and c.n.f. for f.
   b) Write f as a sum of minterms and a product of maxterms (utilizing binary indices).

5. Let F_6 denote the set of all Boolean functions $f: B^6 \rightarrow B$.
   a) What is $|F_6|$?
   b) How many fundamental conjunctions (disjunctions) are there in F_6?
   c) How many minterms (maxterms) are there in F_6?
   d) How many functions $f \in F_6$ have the value 1 when (exactly) two of its variables have the value 1? (In all other circumstances, the value of f may be 0 or 1.)
   e) How many functions $f \in F_6$ have the value 1 when at least two of its variables have the value 1? (In all other circumstances, the value of f may be 0 or 1.)
   f) Let $u, v, w, x, y,$ and z denote the six Boolean variables for the functions in F_6. How many of these functions are independent of x [that is, $f(u, v, w, x, y, z) = f(u, v, w, \overline{x}, y, z)$]? How many are independent of x, y, z?

6. Let $f: B^4 \to B$. Find the disjunctive normal form for f if
   a) $f^{-1}(1) = \{0101 \text{ (that is, } w = 0, x = 1, y = 0, z = 1), 0110, 1000, 1011\}.$ = min
   b) $f^{-1}(0) = \{0000, 0001, 0010, 0100, 1000, 1001, 0110\}.$ = max

7. Let $f: B^n \to B$. If the d.n.f. of f has m fundamental conjunctions and its c.n.f. has k fundamental disjunctions, how are m, n, and k related?

8. If x, y, and z are Boolean variables and $x + y + z = xyz$, prove that x, y, z are simultaneously 0 or 1.

9. Simplify the following Boolean expressions.
   a) $xy + (x + y)\bar{z} + y$
   b) $x + y + \overline{(\bar{x} + y + z)}$
   c) $yz + wx + z + [wz(xy + wz)]$
   d) $x_1 + \bar{x}_1 x_2 + \bar{x}_1\bar{x}_2 x_3 + \bar{x}_1\bar{x}_2\bar{x}_3 x_4 + \cdots$

10. Find the values of the Boolean variables w, x, y, z that satisfy the following system of simultaneous (Boolean) equations.
$$x + \bar{x}y = 0 \qquad \bar{x}y = \bar{x}z \qquad \bar{x}y + \bar{x}\bar{z} + zw = \bar{z}w$$

11. a) For $f, g, h: B^n \to B$, prove that $fg + \bar{f}h + gh = fg + \bar{f}h$ and that $fg + f\bar{g} + \bar{f}g + \bar{f}\bar{g} = 1$.
    b) State the dual of each result in part (a).

12. Let $f, g: B^n \to B$. Define the relation "$\leq$" on F_n, the set of all Boolean functions of n variables, by $f \leq g$ if the value of g is 1 at least whenever the value of f is 1.
    a) Prove that this relation is a partial order on F_n.
    b) Prove that $fg \leq f$ and $f \leq f + g$.
    c) For $n = 2$, draw the Hasse diagram for the 16 functions in F_2. Where are the minterms and maxterms located in the diagram? Compare this diagram with that for the power set of $\{a, b, c, d\}$ partially ordered under the subset relation.

13. For $f, g: B^n \to B$, define the binary operation $\oplus$ (Exclusive Or) by $f \oplus g = f\bar{g} + \bar{f}g$.
    a) Determine $f \oplus f$, $f \oplus \bar{f}$, $f \oplus 1$, and $f \oplus 0$.
    b) Prove or disprove each of the following:
       (i) $f \oplus g = 0 \Rightarrow f = g$
       (ii) $f \oplus (g \oplus h) = (f \oplus g) \oplus h$
       (iii) $f \oplus g = \bar{f} \oplus \bar{g}$
       (iv) $f \oplus gh = (f \oplus g)(f \oplus h)$
       (v) $f(g \oplus h) = fg \oplus fh$
       (vi) $\overline{(f \oplus g)} = \bar{f} \oplus g = f \oplus \bar{g}$
       (vii) $f \oplus g = f \oplus h \Rightarrow g = h$

15.2
GATING NETWORKS: MINIMAL SUMS OF PRODUCTS: KARNAUGH MAPS

The switching functions of Section 15.1 present an interesting mathematical theory. Their importance lies in their implementation by means of *logic gates* (devices in a digital computer that perform specified tasks in the processing of data). The electrical and mechanical components of such gates depend on the state of the art. Consequently, we shall not concern ourselves here with questions relating to hardware.

Figure 15.1 contains the logic gates for negation (complement), conjunction, and disjunction in parts (a), (b), and (c), respectively. Since the Boolean operations of $+$ and $\cdot$ are associative, we may have more than two inputs for an AND gate or an OR gate.

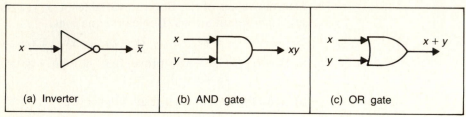

(a) Inverter	(b) AND gate	(c) OR gate

Figure 15.1

Figure 15.2 shows the *logic*, or *gating*, *network* for the expression $(w + \bar{x}) \cdot (y + xz)$. Symbols on a line to the left of a gate (or inverter) are *inputs*. When they are on line segments to the right of a gate, they are *outputs*. We have *split* the input line for x, so that x may serve as input for both an AND gate and an inverter.

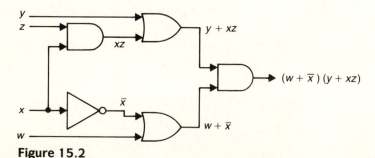

Figure 15.2

The exercises will provide practice in drawing the logic network for a Boolean expression and in going from the network to the expression. Meanwhile certain features of these networks need to be emphasized.

1. An input line may be split to provide that input to more than one gate.

2. Input and output lines come together only at gates.

3. There is no doubling back; that is, the output from a gate g cannot be used as an input for gate g, or any gate (directly or indirectly) leading into g.

4. We assume that the output of a gating network is an instantaneous function of the present inputs. There is no time dependence and we attach no importance to prior inputs, as we do with finite state machines.

With these ideas in mind, let us analyze the computer addition of binary numbers.

Example 15.9   When we add two bits (binary digits), the result consists of a sum s and a carry c. In three of four cases the carry is 0, so we shall concentrate on the computation of $1 + 1$. Examining parts (b) and (c) of Table 15.8, we consider the sum s and the carry c as Boolean functions of the variables x and y. Then $c = xy$ and $s = \bar{x}y + x\bar{y} = x \oplus y = (x + y)(\overline{xy})$.

Table 15.8

x	y	Binary sum		x	y	sum		x	y	carry
0	0	$0 + 0 = 0$		0	0	0		0	0	0
0	1	$0 + 1 = 1$		0	1	1		0	1	0
1	0	$1 + 0 = 1$		1	0	1		1	0	0
1	1	$1 + 1 = 10$		1	1	0		1	1	1
(a)				(b)				(c)		

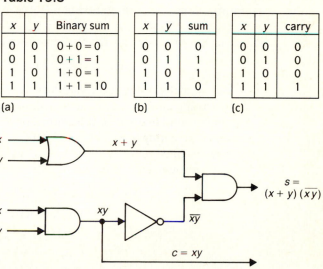

The half-adder

Figure 15.3

Figure 15.3 is a gating network with two outputs, and it is referred to as a *multiple output* network. This device, called a *half-adder*, implements the results in Tables 15.8(b) and (c). Using two half-adders and an OR gate, we construct the *full-adder* shown in Fig. 15.4(a). If $x = x_n x_{n-1} \ldots x_2 x_1 x_0$ and $y = y_n y_{n-1} \ldots y_2 y_1 y_0$, consider the process of adding the binary digits x_i and y_i in finding the sum $x + y$. Here c_{i-1} is the carry from the addition of x_{i-1} and y_{i-1} (and a possible carry c_{i-2}). The input c_{i-1}, together with the inputs x_i and y_i, produce the sum s_i and the carry c_i as shown in the figure. Finally, in Fig. 15.4(b) two full-adders and a half-adder are combined to produce the sum of the two binary numbers $x_2 x_1 x_0$ and $y_2 y_1 y_0$, whose sum is $c_2 s_2 s_1 s_0$.   □

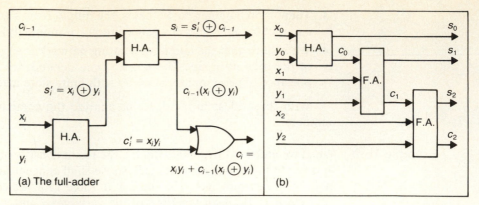

Figure 15.4

The next example introduces the main theme of this section—the minimal-sum-of-products representation of a Boolean function.

Example 15.10   Find a gating network for the Boolean function

$$f(w, x, y, z) = \Sigma m(4, 5, 7, 8, 9, 11).$$

Consider the order of the variables as w, x, y, z. We can determine the d.n.f. of f by writing each minterm number in binary notation and then finding its corresponding fundamental conjunction. For example, (a) $5 = 0101$, indicating the fundamental conjunction $\overline{w}x\overline{y}z$; and (b) $7 = 0111$, indicating $\overline{w}xyz$. Continuing in this way, we have $f(w, x, y, z) = \overline{w}x\overline{y}z + \overline{w}xyz + w\overline{x}\,\overline{y}\,\overline{z} + w\overline{x}\,\overline{y}z + w\overline{x}yz + \overline{w}x\overline{y}\,\overline{z}$.

Using properties of Boolean variables, we find that

$$f = \overline{w}xz(\overline{y} + y) + w\overline{x}\,\overline{y}(\overline{z} + z) + w\overline{x}yz + \overline{w}x\overline{y}\,\overline{z}$$
$$= \overline{w}xz + w\overline{x}\,\overline{y} + w\overline{x}yz + \overline{w}x\overline{y}\,\overline{z}$$
$$= \overline{w}x(z + \overline{y}\,\overline{z}) + w\overline{x}(\overline{y} + yz)$$
$$= \overline{w}x(z + \overline{y}) + w\overline{x}(\overline{y} + z) \text{ (Why?)}$$
$$= \overline{w}x(\overline{y} + z) + w\overline{x}(\overline{y} + z),$$

so

a) $f(w, x, y, z) = \overline{w}xz + \overline{w}x\overline{y} + w\overline{x}\,\overline{y} + w\overline{x}z$, a *minimal sum of products*; or,

b) $f(w, x, y, z) = \overline{w}x(\overline{y} + z) + w\overline{x}(\overline{y} + z)$.   □

From this point on we shall consider an input of the form $\overline{w}$, which has not passed through any gates, as the exact input, instead of regarding it as the result obtained from inputting w and passing it through an inverter.

In Fig. 15.5(a), we have a gating network implementing the d.n.f. of the function f in Example 15.10. Part (b) of the figure is the gating network for f as a minimal sum of products. This sum is *much* better than the d.n.f. and is minimal in the sense that it cannot be reduced and still consist of a sum of product terms. Figure 15.5(c) has a gating network for $f = \overline{w}x(\overline{y} + z) + w\overline{x}(\overline{y} + z)$.

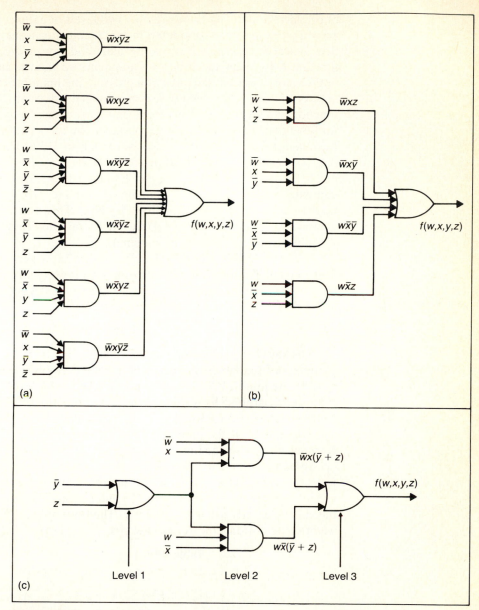

Figure 15.5

The network in part (c) has only four logic gates, whereas that in part (b) has five such devices. Consequently, we may feel that the network in part (c) is better with regard to minimizing cost, because each extra gate increases the cost of production. However, even though there are fewer inputs and fewer gates for the implementation in part (c), some of the inputs (namely $\bar{y}$ and z) must pass through three *levels of gating* before providing the output f. For the minimal sum of products in part (b), there are only two levels of gating. In the study of gating networks, outputs are considered instantaneous functions of the input. In practice, however, each level of gating adds a delay in the development of the function f. For high-speed digital equipment we want to minimize delay, so we opt for more speed at the price of increasing the manufacturing cost.

As a result of this need to maximize speed, we want to represent a Boolean function in a form called a minimal sum of products. In order to accomplish this for functions of not more than six variables, we use a pictorial method called the *Karnaugh map*, developed by M. Karnaugh in 1953. The d.n.f. of a Boolean function is a major key behind this technique.

In simplifying the d.n.f. of f in Example 15.10, we combined the two fundamental conjunctions $\overline{w}x\overline{y}z$ and $\overline{w}xyz$ into the product term $\overline{w}xz$, because $\overline{w}x\overline{y}z + \overline{w}xyz = \overline{w}xz(\overline{y} + y) = \overline{w}xz(1) = \overline{w}xz$. This indicates that if two fundamental conjunctions differ in exactly one literal, then they can be combined into a product term with that literal missing.

For $g: B^4 \rightarrow B$, where $g(w, x, y, z) = wx\overline{y}\,\overline{z} + wx\overline{y}z + wxyz + wxy\overline{z}$, each fundamental conjunction differs from its predecessor in exactly one literal. Hence we can simplify g as $g = wx\overline{y}(\overline{z} + z) + wxy(z + \overline{z}) = wx\overline{y} + wxy = wx(\overline{y} + y) = wx$. We could have also written

$$g = wx(\overline{y}\,\overline{z} + \overline{y}z + yz + y\overline{z}) = wx(y + \overline{y})(z + \overline{z}) = wx.$$

The key to this reduction process is the recognition of pairs (quadruples, ... , 2^n-tuples) of fundamental conjunctions where any two adjacent terms differ in exactly one literal. If $h: B^4 \rightarrow B$, and the d.n.f. of h has 12 terms, can we move these terms around to recognize the best reductions? The Karnaugh map organizes these terms for us.

We start with the case of two variables, w and x. Table 15.9 shows the Karnaugh maps for the functions $f(w, x) = wx$ and $g(w, x) = w + x$.

In part (a), the 1 interior to the table indicates the fundamental conjunction wx. This occurs in the row for $w = 1$ and the column for $x = 1$, the one case when $wx = 1$. In part (b), there are three 1's in the table. The top 1 is for $\overline{w}x$, which has the value 1 exactly when $w = 0$, $x = 1$. The bottom two 1's are for $w\overline{x}$ and wx, as we read the bottom row from left to right.

Table 15.9

$w \backslash x$	0	1
0		
1		1

$w \backslash x$	0	1
0		1
1	1	1

(a) wx (b) $w + x$

Table 15.9(b) represents the d.n.f. $\overline{w}x + w\overline{x} + wx$. As a result of their adjacency in the bottom row, the table indicates that $w\overline{x}$ and wx differ in only one literal and can be combined to yield w. By the idempotent law of addition, we can use wx again. The adjacency in the second column of the table indicates the combining of $\overline{w}x$ and wx to get x. (In the x column all possibilities for w, namely w and $\overline{w}$, appear. This is a way to recognize x as the result for that column.)

Example 15.11 We now consider three Boolean variables w, x, y. In Table 15.10, the first new idea we encounter is in the column headings for xy. These are not the same as the ones we had for the rows in the function tables. We see here, in going from left to right, that 00 differs from 01 in exactly one place, 01 differs from 11 in exactly one place, etc.

Table 15.10

$w\backslash xy$	00	01	11	10
0	1			1
1	1		1	

If $f(w, x, y) = \Sigma m(0, 2, 4, 7)$, then because $0 = 000\,(\overline{w}\,\overline{x}\,\overline{y})$, $2 = 010\,(\overline{w}x\overline{y})$, $4 = 100\,(w\overline{x}\,\overline{y})$, and $7 = 111\,(wxy)$, we can represent these terms by placing 1's as shown in Table 15.10. The 1 for wxy is not adjacent to any other 1, so it is *isolated*; we shall have wxy in the minimal sum of products representing f. The 1 for $\overline{w}x\overline{y}$ is not isolated, for we consider the table as wrapping around, making this 1 adjacent to the 1 for $\overline{w}\,\overline{x}\,\overline{y}$. These combine to give $\overline{w}\,\overline{y}$. Finally, the 1's in the column for $x = y = 0$ indicate a reduction of $\overline{w}\,\overline{x}\,\overline{y} + w\overline{x}\,\overline{y}$ to $\overline{x}\,\overline{y}$. Hence, as a minimal sum of products, $f = wxy + \overline{w}\,\overline{y} + \overline{x}\,\overline{y}$. □

Example 15.12 From the respective parts of Table 15.11 we have

a) $f(w, x, y) = \Sigma m(0, 2, 4, 6) = \overline{y}$, the only variable whose value does not change when the four terms designated by the 1's are considered. (The value of y is 0 here, so $f(w, x, y) = \overline{y}$.)

b) $f(w, x, y) = \Sigma m(0, 1, 2, 3) = \overline{w}$

c) $f(w, x, y) = \Sigma m(1, 2, 3, 5, 6, 7) = y + x$ □

Table 15.11

$w\backslash xy$	00	01	11	10
0	1			1
1	1			1

(a)

$w\backslash xy$	00	01	11	10
0	1	1	1	1
1				

(b)

$w\backslash xy$	00	01	11	10
0		1	1	1
1		1	1	1

(c)

Advancing to four variables, we consider the following example.

Example 15.13 Find a minimal-sum-of-products representation for the function $f(w, x, y, z) = \Sigma m(0, 1, 2, 3, 8, 9, 10)$.

The Karnaugh map for f in Table 15.12 combines the 1's in the four corners to give the term $\overline{x}\,\overline{z}$. The four 1's in the top row combine to give $\overline{w}\,\overline{x}$. (Using only the middle two 1's, we do not make use of all the available adjacencies and get the term $\overline{w}\,\overline{x}z$, which has one more literal than $\overline{w}\,\overline{x}$.) Finally, the 1 in the row

($w = 1, x = 0$) and the column ($y = 0, z = 1$) can be combined with the 1 on its left, *and* these can then be combined with the first two 1's in the top row to give $\bar{x}\bar{y}$. Hence, as a minimal sum of products, $f(w,x,y,z) = \bar{x}\bar{z} + \bar{w}\bar{x} + \bar{x}\bar{y}$.   □

Table 15.12

$wx \backslash yz$	00	01	11	10
00	1	1	1	1
01				
11				
10	1	1		1

Example 15.14   The map for $f(w,x,y,z) = \Sigma m(9,10,11,12,13)$ appears in Table 15.13. The only 1 in the table that has not been combined with another term is adjacent to a 1 on its right (this combination yields $w\bar{x}z$) and to a 1 above it (this combination yields $w\bar{y}z$). Consequently, we can represent f as a minimal sum of products in two ways: $wx\bar{y} + w\bar{x}y + w\bar{x}z$ and $wx\bar{y} + w\bar{x}y + w\bar{y}z$. This type of representation, then, is not unique. However, we should observe that the same number of product terms and the same total number of literals appear in each case.   □

Table 15.13

$wx \backslash yz$	00	01	11	10
00				
01				
11	1	1		
10		1	1	1

Example 15.15   There is a right way and there is a wrong way to use a Karnaugh map.

Let $f(w,x,y,z) = \Sigma m(3,4,5,7,9,13,14,15)$. In Table 15.14(a) we combine a block of four 1's into the term xz. But when we account for the other four 1's, we do what is shown in (b). So the result in (b) will yield f as a sum of four terms (each with three literals), whereas the method suggested in (a) adds the extra (unneeded) term xz.   □

Table 15.14

$wx \backslash yz$	00	01	11	10
00			1	
01	1	1	1	
11		1	1	1
10		1		

(a)

$wx \backslash yz$	00	01	11	10
00			1	
01	1	1	1	
11		1	1	1
10		1		

(b)

The following suggestions on the use of Karnaugh maps are based on what we have done so far. We state them now so that they may be used for larger maps.

1. Start by combining those terms in the table where there is at most one possibility for simplification.

2. Check the four corners of a table. They may contain adjacent 1's even though the 1's appear isolated.

3. In all simplifications, try to obtain the largest possible block of adjacent 1's in order to get a minimal product term. (Recall that 1's can be used more than once, if necessary, because of the idempotent law of +.)

4. If there is a choice in simplifying an entry in the table, try to use adjacent 1's that have not been used in any prior simplification.

Example 15.16   If $f(v, w, x, y, z) = \Sigma m(1, 5, 10, 11, 14, 15, 18, 26, 27, 30, 31)$, we construct two 4×4 tables, one for $v = 0$, the other for $v = 1$. (Table 15.15)

Table 15.15

$(v = 0)$ $(v = 1)$

Following the order of the variables, we write, for example, $5 = 00101$ to indicate the need for a 1 in the second row and second column of the table for $v = 0$. Filling in the other 1's, we see that the 1 in the first row, fourth column of the table for $v = 1$ can be combined with another term in only one way, yielding the product $v\bar{x}y\bar{z}$. This is also true for the two 1's in the second column of the $(v = 0)$ table. These give the product $\bar{v}\bar{w}\bar{y}z$. The block of eight 1's yields wy, and we have $f(v, w, x, y, z) = wy + \bar{v}\bar{w}\bar{y}z + v\bar{x}y\bar{z}$.   □

A function f of the six variables t, v, w, x, y, and z requires four tables for each of the cases (a) $t = 0, v = 0$; (b) $t = 0, v = 1$; (c) $t = 1, v = 1$; (d) $t = 1, v = 0$. Beyond six variables, this method becomes overly complicated. Another procedure, the *Quine–McCluskey Method*, can be used. For a large number of variables the method is tedious to perform by hand, but it is a systematic procedure suitable for computer implementation, particularly for computers possessing some type of "binary compare" command. (More about this technique is given in Chapter 7 of Reference 3.)

We close with an example involving a *minimal product of sums*.

Example 15.17   For $g(w, x, y, z) = \Pi M(1, 5, 7, 9, 10, 13, 14, 15)$, we place a 0 in each of the positions for the binary equivalents of the maxterms listed. This yields the results shown in Table 15.16.

The 0 in the lower right-hand corner can be combined only with the 0 above it, and $(\bar{w} + x + \bar{y} + z))(\bar{w} + \bar{x} + \bar{y} + z) = (\bar{w} + \bar{y} + z) + x\bar{x} = (\bar{w} + \bar{y} + z) + 0 =$

$\overline{w} + \overline{y} + z$. The block of four 0's simplifies to $\overline{x} + \overline{z}$, whereas the four 0's in the second column yield $y + \overline{z}$. So $g(w, x, y, z) = (\overline{w} + \overline{y} + z)(\overline{x} + \overline{z})(y + \overline{z})$, a minimal product of sums. □

Table 15.16

wx \ yz	00	01	11	10
00		0		
01		0	0	
11		0	0	0
10		0		0

EXERCISES 15.2

1. Using inverters, AND gates, and OR gates, construct the gates shown in Fig. 15.6.

2. Using only NAND gates (see Fig. 15.6), construct the inverter, AND gate, and OR gate.

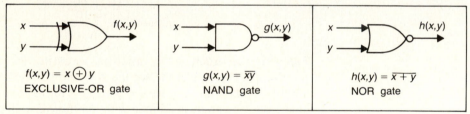

$f(x,y) = x \oplus y$
EXCLUSIVE-OR gate

$g(x,y) = \overline{xy}$
NAND gate

$h(x,y) = \overline{x + y}$
NOR gate

Figure 15.6

3. Answer Exercise 2, replacing NAND by NOR.

4. Using inverters, AND gates, and OR gates, construct gating networks for
   a) $f(x, y, z) = x\overline{z} + y\overline{z} + x$ **b)** $g(x, y, z) = (x + \overline{z})(y + \overline{z})\overline{x}$
   c) $h(x, y, z) = \overline{(xy \oplus yz)}$

5. For the network in Fig. 15.7, express f as a function of w, x, y, z.

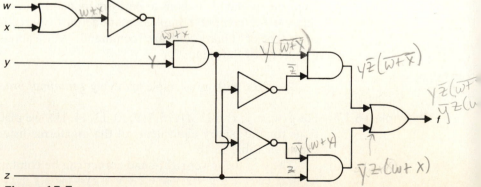

Figure 15.7

6. Implement the half-adder of Fig. 15.3 using only (a) NAND gates; (b) NOR gates.

7. For each of the following Boolean functions f, design a level-two gating network for f as a minimal sum of products.

   a) $f: B^3 \rightarrow B$, where $f(x, y, z) = 1$ if and only if exactly two of the variables have the value 1.

   b) $f: B^3 \rightarrow B$, where $f(x, y, z) = 1$ if and only if at least two of the variables have the value 1.

   c) $f: B^4 \rightarrow B$, where $f(w, x, y, z) = 1$ if and only if an odd number of variables have the value 1.

8. Find a minimal-sum-of-products representation for

   a) $f(w, x, y) = \Sigma m(1, 2, 5, 6)$

   b) $f(w, x, y) = \Pi M(0, 1, 4, 5)$

   c) $f(w, x, y, z) = \Sigma m(0, 2, 5, 7, 8, 10, 13, 15)$

   d) $f(w, x, y, z) = \Sigma m(0, 2, 5, 7, 8, 9, 10, 13, 15)$

   e) $f(w, x, y, z) = \Sigma m(5, 6, 8, 11, 12, 13, 14, 15)$

   f) $f(w, x, y, z) = \Sigma m(7, 9, 10, 11, 14, 15)$

   g) $f(v, w, x, y, z) = \Sigma m(1, 2, 3, 4, 10, 17, 18, 19, 22, 23, 27, 28, 30, 31)$

9. Obtain a minimal-product-of-sums representation for $f(w, x, y, z) = \Pi M(0, 1, 2, 4, 5, 10, 12, 13, 14)$.

10. Let $f: B^n \rightarrow B$ be a function of the Boolean variables $x_1, x_2, \ldots, x_n$. Determine n if the number of 1's needed to express x_1 in the Karnaugh map for f is (a) 2; (b) 4; (c) 8; (d) 2^k, for $k \in \mathbf{Z}^+$ with $1 \leq k \leq n - 1$.

11. If $g: B^7 \rightarrow B$ is a Boolean function of the Boolean variables $x_1, x_2, \ldots, x_7$, how many 1's are needed in the Karnaugh map of g in order to represent the product term (a) x_1; (b) $x_1 x_2$; (c) $x_1 \bar{x}_2 x_3$; (d) $x_1 x_3 x_5 x_7$?

12. In each of the following, $f: B^4 \rightarrow B$, where the Boolean variables (in order) are w, x, y, and z. Determine $|f^{-1}(0)|$ and $|f^{-1}(1)|$ if, as a minimal sum of products, f reduces to

   a) $\bar{x}$ b) wy c) $w\bar{y}z$

   d) $x + y$ e) $xy + z$ f) $xy\bar{z} + w$

15.3

FURTHER APPLICATIONS: DON'T-CARE CONDITIONS

Our objective now is to use the ideas we have developed in the first two sections in a variety of applications.

Example 15.18   As head of her church bazaar, Paula has volunteered to bake a cake to be sold. Members of the bazaar committee volunteer to donate the needed ingredients as shown in Table 15.17.

Table 15.17

	Flour	Milk	Butter	Pecans	Eggs
Sue	x		x		
Dorothy			x	x	
Bettie	x	x			
Theresa		x			x
Ruthanne		x	x	x	

Paula sends her daughter Amy to pick up the ingredients. Write a Boolean expression to help Paula determine which sets of volunteers she should consider so that Amy can collect all of the necessary ingredients (and nothing extra).

Let s, d, b, t, and r denote five Boolean variables corresponding, respectively, to the women listed in the first column of the table. To get the flour, Amy must visit Sue or Bettie. In Boolean terminology, we can say that flour determines the sum $s + b$. This term will be part of a product of sums. For the other ingredients, the following sums denote the choices.

   milk: $b + t + r$ butter: $s + d + r$ pecans: $d + r$ eggs: t

To answer the question posed here, we seek a minimal sum of products for the function $f(s, d, b, t, r) = (s + b)(b + t + r)(s + d + r)(d + r)t$. The answer can be obtained by multiplying everything out and then simplifying the result, or by using a Karnaugh map. This time we'll use the map (in Table 15.18).

Table 15.18

$db \backslash tr$	00	01	11	10
00	0	0	0	0
01	0	0	1	0
11	0	0	1	1
10	0	0	0	0

$(s = 0)$

$db \backslash tr$	00	01	11	10
00	0	0	1	0
01	0	0	1	0
11	0	0	1	1
10	0	0	1	1

$(s = 1)$

We are starting with f as a product (*not* minimal) of sums. Consequently we first fill in the 0's of the table as follows: Here $s + b$, for example, is represented by the eight 0's in the first and fourth rows of the table for $s = 0$; t requires the 16 0's in the first two columns of both tables. Filling in the 0's for the other three sums in the product, we then place a 1 in the nine remaining spaces and arrive at the table shown. Now we need a minimal sum of products for the nine 1's in the table. We find the result is $srt + sdt + brt + dbt$. (Verify this.) □

In our next application, we examine a certain property of graphs. This property was introduced earlier in Miscellaneous Exercise 10 of Chapter 11. The development here, however, does not rely on that prior presentation.

DEFINITION 15.4   Let $G = (V, E)$ denote a graph (undirected) with vertex set V and edge set E. A subset D of V is called a *dominating set* for G if for every $v \in V$, either $v \in D$ or v is adjacent to a vertex in D.

For the graph shown in Fig. 15.8, the sets $\{a, d\}$, $\{a, c, e\}$, and $\{b, d, e, f\}$ are examples of dominating sets. The set $\{a, c, e\}$ is a *minimal dominating* set, for if any of the three vertices a, c, or e is removed, the remaining two no longer dominate the graph. The set $\{a, d\}$ is also minimal, but $\{b, d, e, f\}$ is not, because $\{b, d, e\}$ dominates G.

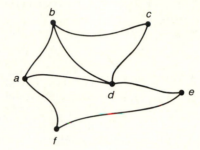

Figure 15.8

Example 15.19   For the graph shown in Fig. 15.8, let the vertices represent cities and the edges highways. We wish to build hospitals in some of these cities so that each city either has a hospital or is adjacent to a city that does. In how many ways can this be accomplished by building a minimal number of hospitals in each case?

To answer this question, we need the minimal dominating sets for G. Consider vertex a. To guarantee that a will satisfy our objective, we must build a hospital in a, or b, or d, or f. Hence we have the term $a + b + d + f$. For b to satisfy our objective, we generate the term $a + b + c + d$. Continuing with the other four locations, we find that the answer is then a minimal-sum-of-products representation for the Boolean function $g(a, b, c, d, e, f) = (a + b + d + f) \cdot (a + b + c + d)(b + c + d)(a + b + c + d + e)(d + e + f)(a + e + f)$. Using the properties of Boolean variables, we have

$g = (a + b + d + f)(b + c + d) \cdot$ $\quad (d + e + f)(a + e + f)$	Absorption Law
$= [(a + f)c + (b + d)][da + (e + f)]$	Distributive Law of $+$ over $\cdot$, and the Absorption Law
$= [ac + fc + b + d][da + e + f]$	Distributive Law of $\cdot$ over $+$
$= acda + ace + acf + fcda + fce + fcf$ $\quad + bda + be + bf + dda + de + df$	Distributive Law of $\cdot$ over $+$
$= ace + cf + ad + be + bf + de + df$	Commutative and Idempotent Laws of $\cdot$, and the Absorption Law

Consequently, in six of the cases the objective can be achieved by building only two hospitals. If a and c have the largest populations and we want to locate hospitals in each of these cities, then we would also have to construct a hospital at e. □

The final application we shall examine introduces the notion of *"don't-care" conditions*.

Example 15.20   The four input lines for the gating network shown in Fig. 15.9 provide the binary equivalents of the digits $0, 1, 2, \ldots, 9$, with each number represented as $abcd$ (d is least significant). Construct a gating network with two levels of gating such that the output function f equals 1 for the input digits $0, 3, 6, 9$ (that is, f detects digits divisible by 3).

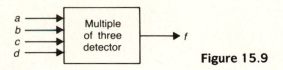

Figure 15.9

Before concluding that $f = 0$ for the other 12 cases, we examine Table 15.19, where an "x" appears for the value of f in the last six cases. These input combinations do not occur (because of certain external constraints), so we *don't care* what the value of f is in these situations. For such occurrences, the outputs are referred to as *unspecified* and f is called *incompletely specified*. We write $f = \sum m (0, 3, 6, 9) + d(10, 11, 12, 13, 14, 15)$. When seeking a minimal sum of products for f, we can use any or all of these don't-care conditions in the simplification process.

Table 15.19

a	b	c	d	f		a	b	c	d	f
0	0	0	0	1		1	0	0	0	0
0	0	0	1	0		1	0	0	1	1
0	0	1	0	0		1	0	1	0	x
0	0	1	1	1		1	0	1	1	x
0	1	0	0	0		1	1	0	0	x
0	1	0	1	0		1	1	0	1	x
0	1	1	0	1		1	1	1	0	x
0	1	1	1	0		1	1	1	1	x

From the Karnaugh map in Table 15.20, we write f as a minimal sum of products, obtaining

$$f = \overline{a}\,\overline{b}\,\overline{c}\,\overline{d} + \overline{b}cd + bc\overline{d} + ad.$$

The first summand in f is for recognition of 0; $\overline{b}cd$ provides recognition for 3 because it stands for 0011 ($\overline{a}\,bcd$), since 1011 ($ab\overline{c}d$) does not occur. Likewise $bc\overline{d}$ is needed to recognize 6, whereas ad takes care of 9. Fig. 15.10 provides the interior details (minus the inverters) of Fig. 15.9. (Note that in Table 15.20 there are some don't-care conditions that were not used.)   □

Table 15.20

ab\cd	00	01	11	10
00	①		①	
01				①
11	x	x	x	x
10		1	x	x

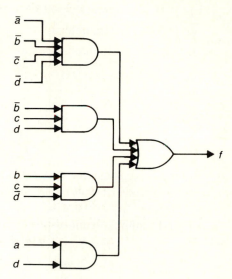

Figure 15.10

EXERCISES 15.3

1. For his tenth birthday, Mona wants to buy her son Jason some stamps for his collection. At the hobby shop she finds six different packages (which we shall call u, v, w, x, y, z). The kinds of stamps in each of these packages are as follows:

Table 15.21

	United States	European	Asian	African
u		✓		✓
v	✓		✓	
w	✓	✓		
x	✓			
y	✓			✓
z			✓	✓

Determine all minimal combinations of packages Mona can buy so that Jason will get some stamps from all four geographical locations.

2. Rework Example 15.19 using a Karnaugh map on six variables.

3. Determine whether each of the following statements is true or false. If the statement is false, provide a counterexample.

   Let $G = (V, E)$ be an undirected graph with $D_1, D_2 \subseteq V$.

   a) If D_1, D_2 are dominating sets of G, then $D_1 \cup D_2$ is likewise.

   b) If D_1, D_2 are dominating sets of G, then $D_1 \cap D_2$ is also.

   c) If D_1 is a dominating set of G and $D_1 \subseteq D_2$, then D_2 dominates G.

   d) If D_1 is a dominating set of G and $\emptyset \neq D_2 \subseteq D_1$, then D_2 dominates G.

   e) If $D_1 \cup D_2$ dominates G, then at least one of D_1, D_2 dominates G.

4. Determine all minimal dominating sets for the graph G shown in Fig. 15.11.

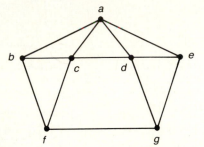

Figure 15.11

5. Find a minimal-sum-of-products representation for

   a) $f(w, x, y, z) = \Sigma m(1, 3, 5, 7, 9) + d(10, 11, 12, 13, 14, 15)$

   b) $f(w, x, y, z) = \Sigma m(0, 5, 6, 8, 13, 14) + d(4, 9, 11)$

   c) $f(v, w, x, y, z) = \Sigma m(0, 2, 3, 4, 5, 6, 12, 19, 20, 24, 28)$
   $+ d(1, 13, 16, 29, 31)$

6. The four input lines for the gating network shown in Fig. 15.12 provide the binary equivalents of the numbers $0, 1, 2, \ldots, 15$, where each number is represented as $abcd$, with d the least significant bit.

   a) Determine the d.n.f. of f, whose value is 1 for $abcd$ prime, and 0 otherwise.

   b) Draw the two-level gating network for f as a minimal sum of products.

   c) We are informed that the given network is part of a larger network and that as a result, the binary equivalents of the numbers 10 through 15 are never provided as input. Design a two-level gating network for f under these circumstances.

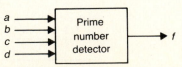

Figure 15.12

15.4

THE STRUCTURE OF A BOOLEAN ALGEBRA (OPTIONAL)

In this last section we analyze the structure of a Boolean algebra and determine those $n \in \mathbf{Z}^+$ for which there is a Boolean algebra of n elements.

DEFINITION 15.5   Let $\mathcal{B}$ be a nonempty set that contains two special elements 0 and 1 and on which we define binary operations $+$, $\cdot$, and a monary (or unary) operation $^-$. Then $(\mathcal{B}, +, \cdot, ^-, 0, 1)$ is called a *Boolean algebra* if the following conditions are satisfied for all $x, y, z \in \mathcal{B}$.

a) $x + y = y + x$	a)′ $xy = yx$	Commutative Laws
b) $x(y + z) = xy + xz$	b)′ $x + yz = (x + y)(x + z)$	Distributive Laws
c) $x + 0 = x$	c)′ $x \cdot 1 = x$	Identity Laws
d) $x + \bar{x} = 1$	d)′ $x \cdot \bar{x} = 0$	Inverse Laws
e) $0 \neq 1$		

As seen above, we often write xy for $x \cdot y$. When the operations and identity elements are known, we write $\mathcal{B}$ instead of $(\mathcal{B}, +, \cdot, ^-, 0, 1)$.

From our past experience we have the following examples.

Example 15.21   If $\mathcal{U}$ is a (finite) set, then $\mathcal{B} = \mathcal{P}(\mathcal{U})$ is a Boolean algebra where for $A, B \subseteq \mathcal{U}$, we have $A + B = A \cup B$, $AB = A \cap B$, $\bar{A} = $ the complement of A (in $\mathcal{U}$), and $0 = \emptyset$, $1 = \mathcal{U}$. □

Example 15.22   For $n \in \mathbf{Z}^+$, $F_n = \{f: B^n \to B\}$, the set of Boolean functions on n Boolean variables, is a Boolean algebra where $+$, $\cdot$, and $^-$ are as defined in Definition 15.2, and $0 = \mathbf{0}$, $1 = \mathbf{1}$. □

Let us now examine a new type of Boolean algebra.

Example 15.23   Let $\mathcal{B}$ be the set of all positive integer divisors of 30: $\mathcal{B} = \{1, 2, 3, 5, 6, 10, 15, 30\}$. For any $x, y \in \mathcal{B}$, define $x + y = [x, y]$, the l.c.m. of x, y; $xy = (x, y)$, the g.c.d. of x, y; and, $\bar{x} = 30/x$. Then with 1 as the 0 element and 30 as the 1 element, $(\mathcal{B}, +, \cdot, ^-, 0, 1)$ is a Boolean algebra.

We shall establish one of the distributive laws for this Boolean algebra and leave the other conditions for the reader to check.

For the first distributive law we want to show that $(x, [y, z]) = [(x, y), (x, z)]$, for any $x, y, z \in \mathcal{B}$. Write $x = 2^{k_1} 3^{k_2} 5^{k_3}$, $y = 2^{m_1} 3^{m_2} 5^{m_3}$, and $z = 2^{n_1} 3^{n_2} 5^{n_3}$, where for $1 \le i \le 3$ we have $0 \le k_i, m_i, n_i \le 1$.

Then $[y, z] = 2^{s_1} 3^{s_2} 5^{s_3}$ where $s_i = \max\{m_i, n_i\}$, $1 \le i \le 3$, so $(x, [y, z]) = 2^{t_1} 3^{t_2} 5^{t_3}$ where $t_i = \min\{k_i, \max\{m_i, n_i\}\}$, $1 \le i \le 3$.

Also, $(x, y) = 2^{u_1} 3^{u_2} 5^{u_3}$, where $u_i = \min\{k_i, m_i\}$, $1 \le i \le 3$, and $(x, z) = 2^{v_1} 3^{v_2} 5^{v_3}$, with $v_i = \min\{k_i, n_i\}$. So $[(x, y), (x, z)] = 2^{w_1} 3^{w_2} 5^{w_3}$, where $w_i = \max\{u_i, v_i\}$, $1 \le i \le 3$.

Therefore

$$w_i = \max\{u_i, v_i\} = \max\{\min\{k_i, m_i\}, \min\{k_i, n_i\}\}, \text{ and } t_i = \min\{k_i, \max\{m_i, n_i\}\}.$$

In order to verify the result, we need to show that $w_i = t_i$ for $1 \le i \le 3$. If $k_i = 0$, then $w_i = 0 = t_i$. If $k_i = 1$, then $w_i = \max\{m_i, n_i\} = t_i$. This exhausts all possibilities, so $w_i = t_i$ for $1 \le i \le 3$ and $(x, [y, z]) = [(x, y), (x, z)]$.

If we analyze this result further, we find that 30 can be replaced by any number $m = p_1 p_2 p_3$, where p_1, p_2, p_3 are distinct primes. In fact, the result follows for the set of all divisors of $p_1 p_2 \cdots p_n$, a product of n distinct primes. (Note that such a product is square-free; that is, there is no $k \in \mathbf{Z}^+, k > 1$, with k^2 dividing it.) □

Example 15.24 A word about the propositional calculus. If p, q are two propositions, we may feel that the collection of all statements obtained from p, q, using $\vee, \wedge$, and $^-$, should be a Boolean algebra. After all, just look at the laws of logic and the way they compare with the comparable results for set theory and Boolean functions. There is one main difference. In logic we have, for example, $p \wedge q \Leftrightarrow q \wedge p$, not $p \wedge q = q \wedge p$. To get around this we define a relation $\mathcal{R}$ on the set S of all propositions so obtained from p, q, where $s_1 \mathcal{R} s_2$ if $s_1 \Leftrightarrow s_2$. Then $\mathcal{R}$ is an equivalence relation on S and partitions S, in this case, into 16 equivalence classes. If we define $+, \cdot$, and $^-$ on these equivalence classes by $[s_1] + [s_2] = [s_1 \vee s_2]$, $[s_1][s_2] = [s_1 \wedge s_2]$, and $\overline{[s_1]} = [\bar{s}_1]$, and if we recognize $[T_0]$ as the 1 element and $[F_0]$ as the 0 element, then we get a Boolean algebra. □

In the definition of a Boolean algebra, there are nine conditions. Yet in the lists of properties for set theory, logic, and Boolean functions, we listed 19 properties. And there were even more! Undoubtedly, there is a way to get the remaining properties, and others not listed among the 19, from the ones given in the definition.

THEOREM 15.1 (*The Idempotent Laws*) For any $x \in \mathcal{B}$, a Boolean algebra, (i) $x + x = x$; and, (ii) $xx = x$.

Proof (To the right of each equality appearing in this proof, we list the letter of the condition from Definition 15.5 that verifies it.)

(i) $x = x + 0$ c)
$\quad = x + x\bar{x}$ d)'
$\quad = (x + x)(x + \bar{x})$ b)'
$\quad = (x + x) \cdot 1$ d)
$\quad = x + x$ c)'

(ii) $x = x \cdot 1$ c)'
$\quad = x(x + \bar{x})$ d)
$\quad = xx + x\bar{x}$ b)
$\quad = xx + 0$ d)'
$\quad = xx$ c) ■

In proving this theorem we can obtain the proof of (ii) from that of (i) by changing all occurrences of $+$ to $\cdot$, and vice versa, and all occurrences of 0 to 1, and vice versa. Also, the reasons verifying the corresponding steps constitute a pair of conditions in Definition 15.5. As in the past, these pairs are said to be *duals* of each other; condition (e) is called *self-dual*. This now leads us to the following result.

THEOREM 15.2 (*The Principle of Duality*) If s is a theorem about a Boolean algebra, and s can be proved from the conditions in Definition 15.5 and properties derived from these same conditions, then its dual s^d is likewise a theorem.

Proof Let s be such a theorem. Dualizing all the steps and reasons in the proof of s (as in the proof of Theorem 15.1), we obtain a proof for s^d. ∎

We now list some further properties for a Boolean algebra. We shall prove some of these properties and leave the remaining proofs for the reader.

THEOREM 15.3 For any Boolean algebra $\mathscr{B}$, if $x, y, z \in \mathscr{B}$, then

a) $x \cdot 0 = 0$	a)′ $x + 1 = 1$	Dominance Laws
b) $x(x + y) = x$	b)′ $x + xy = x$	Absorption Laws
c) $xy = xz, \bar{x}y = \bar{x}z \Rightarrow y = z$		Cancellation Laws
c)′ $x + y = x + z, \bar{x} + y = \bar{x} + z \Rightarrow y = z$		
d) $x(yz) = (xy)z$	d)′ $x + (y + z) = (x + y) + z$	Associative Laws
e) $x + y = 1, xy = 0 \Rightarrow y = \bar{x}$		Uniqueness of Inverses
f) $\bar{\bar{x}} = x$		Law of the Double Complement
g) $\overline{xy} = \bar{x} + \bar{y}$	g)′ $\overline{x + y} = \bar{x}\,\bar{y}$	DeMorgan's Laws
h) $\bar{0} = 1$	h)′ $\bar{1} = 0$	
i) $x\bar{y} = 0$ iff $xy = x$	i)′ $x + \bar{y} = 1$ iff $x + y = x$	

Proof a) $x \cdot 0 = 0 + x \cdot 0,$ by Definition 15.5(c), (a)
  $= x \cdot \bar{x} + x \cdot 0,$ by Definition 15.5(d)′
  $= x \cdot (\bar{x} + 0),$ by Definition 15.5(b)
  $= x \cdot \bar{x},$ by Definition 15.5(c)
  $= 0,$ by Definition 15.5(d)′

a)′ This follows from (a) by the Principle of Duality.

c) $y = 1 \cdot y = (x + \bar{x})y = xy + \bar{x}y = xz + \bar{x}z = (x + \bar{x})z = 1 \cdot z = z$ (Verify all equalities.)

c)′ This is the dual of (c).

d) To establish this result, we use (c)′ and arrive at the conclusion by showing that $x + [x(yz)] = x + [(xy)z]$ and $\bar{x} + [x(yz)] = \bar{x} + [(xy)z]$. Using the absorption law, we find that $x + [x(yz)] = x$. Likewise $x + [(xy)z] = [x + (xy)] \cdot (x + z) = x(x + z) = x$. Then $\bar{x} + [x(yz)] = (\bar{x} + x)(\bar{x} + yz) = 1 \cdot (\bar{x} + yz) = \bar{x} + yz$, whereas $\bar{x} + [(xy)z] = (\bar{x} + xy)(\bar{x} + z) = ((\bar{x} + x)(\bar{x} + y))(\bar{x} + z) = (1 \cdot (\bar{x} + y))(\bar{x} + z) = (\bar{x} + y) \cdot (\bar{x} + z) = \bar{x} + yz$ (Verify all equalities.)

The result now follows by the cancellation law in (c)'.

d)' Fortunately, this is the dual of (d).

e) $\bar{x} = \bar{x} + 0 = \bar{x} + xy = (\bar{x} + x)(\bar{x} + y) = 1 \cdot (\bar{x} + y) = (\bar{x} + y) \cdot 1 = (\bar{x} + y)(x + y) = \bar{x}x + y = 0 + y = y$ (Verify all equalities.)

We note that (e) is self-dual. Statement (f) is a corollary of (e) because $\bar{\bar{x}}$ and x are both complements (inverses) of $\bar{x}$.

g) This result will follow from (e) if we can show that $\bar{x} + \bar{y}$ is a complement of xy.

$$xy + (\bar{x} + \bar{y}) = (xy + \bar{x}) + \bar{y} = (x + \bar{x})(y + \bar{x}) + \bar{y}$$
$$= 1 \cdot (y + \bar{x}) + \bar{y} = (y + \bar{y}) + \bar{x} = 1 + \bar{x} = 1.$$

Also, $xy(\bar{x} + \bar{y}) = (xy\bar{x}) + (xy\bar{y}) = ((x\bar{x})y) + (x(y\bar{y})) = 0 \cdot y + x \cdot 0 = 0 + 0 = 0$.

Consequently, $\bar{x} + \bar{y}$ is a complement of xy, and by uniqueness of complements, $\overline{xy} = \bar{x} + \bar{y}$. ∎

Enough proving for a while! Now we are going to investigate how to place an order on the elements of a Boolean algebra. In fact, we shall want a partial order, and for this reason we turn now to the Hasse diagram.

Let us start by considering the Hasse diagrams for the following two Boolean algebras.

a) $(\mathcal{P}(\mathcal{U}), \cup, \cap, ^-, \emptyset, \mathcal{U})$, where $\mathcal{U} = \{1, 2, 3\}$, and the partial order is induced by the subset relation.

b) $(\mathcal{S}, +, \cdot, ^-, 1, 30)$, where $\mathcal{S} = \{1, 2, 3, 5, 6, 10, 15, 30\}$, $x + y = [x, y]$, $xy = (x, y)$, and $\bar{x} = 30/x$. Here the 0 element is the divisor 1 and the 1 element is the divisor 30. The relation $\mathcal{R}$ on $\mathcal{S}$, defined by $x \mathcal{R} y$ if x divides y, makes $\mathcal{S}$ into a poset.

Figure 15.13 shows the Hasse diagrams for these two Boolean algebras. Ignoring the labels at the vertices in each diagram, we see that the underlying structures are the same. Hence we have the suggestion of an *isomorphism of Boolean algebras*.

These examples also suggest two other ideas.

1. Can we partially order any finite Boolean algebra?
2. Looking at Fig. 15.13(a), we see that the nonzero elements just above $\emptyset$ are such that every element other than $\emptyset$ can be obtained as a Boolean sum of these three. For example, $\{1, 3\} = \{1\} \cup \{3\}$ and $\{1, 2, 3\} = \{1\} \cup \{2\} \cup \{3\}$. For part (b), the numbers 2, 3, and 5 are such that every divisor other than 1 is realized as the Boolean sum of these three. For example, $6 = [2, 3]$ and $30 = [2, [3, 5]]$.

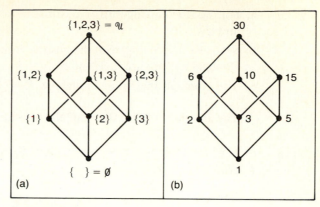

Figure 15.13

We now start to deal formally with these suggestions.

When dealing with sets in Chapter 3 we related the operations of $\cup$, $\cap$, and $^-$ to the subset relation by the equivalence of the statements: (a) $A \subseteq B$; (b) $A \cap B = A$; (c) $A \cup B = B$; and (d) $\overline{B} \subseteq \overline{A}$, where $A, B \subseteq \mathcal{U}$. We now use (a) and (b) in an attempt to partially order any Boolean algebra $\mathcal{B}$.

DEFINITION 15.6   If $x, y \in \mathcal{B}$, define $x \leq y$ if $xy = x$.   —

Hence we define a new concept, namely "$\leq$", in terms of notions we have in $\mathcal{B}$, namely $\cdot$ and the notion of equality. We can make up definitions! But does this one lead us anywhere?

THEOREM 15.4   The relation "$\leq$", defined above, is a partial order.

Proof   Since $xx = x$ for any $x \in \mathcal{B}$, we have $x \leq x$ and the relation is reflexive. To establish antisymmetry, suppose that $x, y \in \mathcal{B}$ with $x \leq y$ and $y \leq x$. Then $xy = x$ and $yx = y$. By the commutative property, $xy = yx$, so $x = y$. Finally, if $x \leq y$ and $y \leq z$, then $xy = x$ and $yz = y$, so $x = xy = x(yz) = (xy)z = xz$, and with $x = xz$, we have $x \leq z$, so the relation is transitive.   ∎

Now we can partially order any Boolean algebra. Before going on, however, let us consider the Boolean algebra consisting of the divisors of 30. How do we apply Theorem 15.4 in this example? Here the partial order is given by $x \leq y$ if $xy = x$. Since xy is (x, y), if $(x, y) = x$, then x divides y. But this was precisely the partial order we had on this Boolean algebra when we started.

Armed with this concept of partial order, we return to the observations we made earlier about the elements in the Hasse diagrams of Fig. 15.13.

DEFINITION 15.7   Let 0 denote the zero element of a Boolean algebra $\mathcal{B}$. An element $x \in \mathcal{B}, x \neq 0$, is called an *atom* of $\mathcal{B}$ if for all $y \in \mathcal{B}, 0 \leq y \leq x \Rightarrow y = 0$ or $y = x$.   —

Example 15.25   **a)** For the Boolean algebra of all subsets of $\mathcal{U} = \{1,2,3\}$, the atoms are $\{1\}$, $\{2\}$, and $\{3\}$.

b) When we are dealing with the positive integer divisors of 30, the atoms of this Boolean algebra are 2, 3, and 5.

c) The atoms in the Boolean algebra $F_n = \{f \colon B^n \to B\}$ are the minterms.   □

The atoms of a finite Boolean algebra satisfy the following properties.

THEOREM 15.5   **a)** If x is an atom in a Boolean algebra $\mathcal{B}$, then for all $y \in \mathcal{B}$, $xy = 0$ or $xy = x$.

b) If x_1, x_2 are atoms of $\mathcal{B}$ and $x_1 \neq x_2$, then $x_1 x_2 = 0$.

Proof   **a)** For any $x, y \in \mathcal{B}$, $xy \leq x$, because $(xy)x = xy$. For x an atom, $xy \leq x \Rightarrow xy = 0$ or $xy = x$.

b) This follows from part (a). The reader should supply the details.   ∎

THEOREM 15.6   If $x_1, x_2, \ldots, x_n$ are all the atoms in a Boolean algebra $\mathcal{B}$ and $x \in \mathcal{B}$ with $xx_i = 0$ for all $1 \leq i \leq n$, then $x = 0$.

Proof   If $x \neq 0$, let $S = \{y \in \mathcal{B} \mid 0 < y \leq x\}$. ($0 < y$ denotes $0 \leq y$ and $0 \neq y$.) With $x \in S$, we have $S \neq \emptyset$. Since S is finite, we can find an element z in $\mathcal{B}$ where $0 < z \leq x$ and no element of $\mathcal{B}$ is between 0 and z. Then z is an atom and $0 = xz = z > 0$. This possibility has led us to a contradiction, so we cannot have $x \neq 0$; that is, $x = 0$.   ∎

This leads us to the following result on representation.

THEOREM 15.7   Given a finite Boolean algebra $\mathcal{B}$ with atoms $x_1, x_2, \ldots, x_n$, then any $x \in \mathcal{B}$, $x \neq 0$, can be written as a sum of atoms uniquely, up to order.

Proof   Since $x \neq 0$, by Theorem 15.6, $S = \{x_i \mid xx_i \neq 0\} \neq \emptyset$. Let $S = \{x_{i_1}, x_{i_2}, \ldots, x_{i_k}\}$, and $y = x_{i_1} + x_{i_2} + \cdots + x_{i_k}$.   Then $xy = x(x_{i_1} + x_{i_2} + \cdots + x_{i_k}) = xx_{i_1} + xx_{i_2} + \cdots + xx_{i_k} = x_{i_1} + x_{i_2} + \cdots + x_{i_k}$, by Theorem 15.5(a). So $xy = y$.

Now consider $(x\bar{y})x_i$ for each $1 \leq i \leq n$. If $x_i \notin S$, then $xx_i = 0$, and $(x\bar{y})x_i = 0$. For $x_i \in S$, we have $(x\bar{y})x_i = xx_i\overline{(x_{i_1} + x_{i_2} + \cdots + x_{i_k})} = xx_i(\bar{x}_{i_1}\bar{x}_{i_2}\cdots\bar{x}_{i_k}) = x(x_i\bar{x}_i)(z)$, where z is the product of the complements of all elements in $S - \{x_i\}$. As $x_i\bar{x}_i = 0$, it follows that $(x\bar{y})x_i = 0$. So $(x\bar{y})x_i = 0$ for all x_i, $1 \leq i \leq n$. By Theorem 15.6, $x\bar{y} = 0$.

With $xy = y$ and $x\bar{y} = 0$, it follows that $x = x \cdot 1 = x(y + \bar{y}) = xy + x\bar{y} = xy + 0 = y = x_{i_1} + x_{i_2} + \cdots + x_{i_k}$, a sum of atoms.

To show that this representation is unique up to order, suppose $x = x_{j_1} + x_{j_2} + \cdots + x_{j_\ell}$.

If x_{j_1} does not appear in $x_{i_1} + x_{i_2} + \cdots + x_{i_k}$, then $x_{j_1} = x_{j_1} \cdot x_{j_1} = x_{j_1}(x_{j_1} + x_{j_2} + \cdots + x_{j_\ell})$ [By Theorem 15.5(b)] $= x_{j_1}x = x_{j_1}(x_{i_1} + x_{i_2} + \cdots + x_{i_k}) = 0$ [Again by Theorem 15.5(b)]. Hence x_{j_1} must appear as a term in $x_{i_1} + x_{i_2} + \cdots + x_{i_k}$, as must $x_{j_2}, \ldots, x_{j_\ell}$. So $\ell \leq k$. By the same reasoning, we get $k \leq \ell$ and find the representations identical, except for order.   ∎

From this result we see that if $\mathcal{B}$ is a finite Boolean algebra with atoms $x_1, x_2, \ldots, x_n$, then each $x \in \mathcal{B}$ can be uniquely written as $\sum_{i=1}^{n} c_i x_i$, with each $c_i \in \{0, 1\}$. If $c_i = 0$, this indicates that x_i is not in the representation of x; $c_i = 1$ indicates that it is. Consequently, each $x \in \mathcal{B}$ is associated with an n-tuple $(c_1, c_2, \ldots, c_n)$, and there are 2^n such n-tuples. Therefore we have proved the following result.

THEOREM 15.8   If $\mathcal{B}$ is a finite Boolean algebra with n atoms, then $|\mathcal{B}| = 2^n$.

There is one final question to resolve. If $n \in \mathbf{Z}^+$, how many different Boolean algebras of size 2^n are there? Looking at the Hasse diagrams in Fig. 15.13, we see two different pictures. But if we ignore the labels on the vertices, the underlying structures emerge as exactly the same. Hence these two Boolean algebras are abstractly identical or isomorphic.

DEFINITION 15.8   Suppose $(\mathcal{B}_1, +, \cdot, ^-, 0, 1)$ and $(\mathcal{B}_2, +, \cdot, ^-, 0, 1)$ are Boolean algebras. Then $\mathcal{B}_1, \mathcal{B}_2$ are called *isomorphic* if there is a one-to-one correspondence $f: \mathcal{B}_1 \to \mathcal{B}_2$ such that for all $x_1, y_1 \in \mathcal{B}_1$,

a) $f(x_1 + y_1) = f(x_1) + f(y_1)$
   $\quad\uparrow \qquad\qquad\quad \uparrow$
   $\text{(in } \mathcal{B}_1) \qquad\quad \text{(in } \mathcal{B}_2)$

b) $f(x_1 \cdot y_1) = f(x_1) \cdot f(y_1)$
   $\quad\uparrow \qquad\qquad\quad \uparrow$
   $\text{(in } \mathcal{B}_1) \qquad\quad \text{(in } \mathcal{B}_2)$

c) $f(\overline{x_1}) = \overline{f(x_1)}$ (In $f(\overline{x_1})$ we take the complement in $\mathcal{B}_1$, while for $\overline{f(x_1)}$ the complement is taken in $\mathcal{B}_2$.)

Such a function f *preserves* the operations of the algebraic structures.

Example 15.26   For the two Boolean algebras in Fig. 15.13, define f by

$$f: \emptyset \to 1 \qquad f: \{2\} \to 3 \qquad f: \{1, 2\} \to 6 \qquad f: \{2, 3\} \to 15$$
$$f: \{1\} \to 2 \qquad f: \{3\} \to 5 \qquad f: \{1, 3\} \to 10 \qquad f: \{1, 2, 3\} \to 30$$

Note the following:

a) The 0 elements correspond under f, as do the 1 elements.

b) $f(\{1\} \cup \{2\}) = f(\{1, 2\}) = 6 = \text{l.c.m. of } 2, 3 = [f(\{1\}), f(\{2\})]$

c) $f(\{1, 2\} \cap \{2, 3\}) = f(\{2\}) = 3 = \text{g.c.d. of } 6, 15 = (f(\{1, 2\}), f(\{2, 3\}))$

d) $f(\overline{\{2\}}) = f(\{1, 3\}) = 10 = 30/3 = \overline{3} = \overline{f(\{2\})}$

e) The image of any atom ($\{1\}, \{2\}, \{3\}$) is an atom (2, 3, 5, respectively).

This function is an isomorphism. Once we establish a correspondence between the respective 0 elements and between the respective atoms, the remaining correspondences are determined from these by Theorem 15.7 and the preservation of the operations under f. $\square$

From this example we have our final result.

THEOREM 15.9   Any finite Boolean algebra $\mathcal{B}$ is isomorphic to a Boolean algebra of sets.

Proof   Since $\mathcal{B}$ is finite, $\mathcal{B}$ has n atoms x_i, $1 \le i \le n$, and $|\mathcal{B}| = 2^n$. Let $\mathcal{U} = \{1, 2, \dots, n\}$ and $\mathcal{P}(\mathcal{U})$ be the Boolean algebra of subsets of $\mathcal{U}$.

We define $f\colon \mathcal{B} \to \mathcal{P}(\mathcal{U})$ as follows. For each $x \in \mathcal{B}$, we write $x = \sum_{i=1}^{n} c_i x_i$, where each c_i is 0 or 1. This follows from Theorem 15.7. Then we define

$$f(x) = \{i \mid 1 \le i \le n \text{ and } c_i = 1\}.$$

[For example, (1) $f(0) = \emptyset$; (2) $f(x_i) = \{i\}$ for each atom x_i, $1 \le i \le n$; (3) $f(x_1 + x_2) = \{1, 2\}$; and (4) $f(x_2 + x_4 + x_7) = \{2, 4, 7\}$.] Now consider $x, y \in \mathcal{B}$, with $x = \sum_{i=1}^{n} c_i x_i$ and $y = \sum_{i=1}^{n} d_i x_i$, where $c_i, d_i \in \{0, 1\}$ for $1 \le i \le n$.

First we find that $x + y = \sum_{i=1}^{n} s_i x_i$, where $s_i = c_i + d_i$ for $1 \le i \le n$. (Here $1 + 1 = 1$.) Consequently,

$$
\begin{aligned}
f(x + y) &= \{i \mid 1 \le i \le n \text{ and } s_i = 1\} \\
&= \{i \mid 1 \le i \le n \text{ and, } c_i = 1 \text{ or } d_i = 1\} \\
&= \{i \mid 1 \le i \le n \text{ and } c_i = 1\} \cup \{i \mid 1 \le i \le n \text{ and } d_i = 1\} \\
&= f(x) \cup f(y).
\end{aligned}
$$

In a similar way, we have

$$x \cdot y = \sum_{i=1}^{n} t_i x_i, \text{ where } t_i = c_i d_i \text{ for } 1 \le i \le n, \text{ and}$$

$$
\begin{aligned}
f(x \cdot y) &= \{i \mid 1 \le i \le n \text{ and } t_i = 1\} \\
&= \{i \mid 1 \le i \le n \text{ and, } c_i = 1 \text{ and } d_i = 1\} \\
&= \{i \mid 1 \le i \le n \text{ and } c_i = 1\} \cap \{i \mid 1 \le i \le n \text{ and } d_i = 1\} \\
&= f(x) \cap f(y).
\end{aligned}
$$

To complete the proof that f is an isomorphism, we should first observe that if $x = \sum_{i=1}^{n} c_i x_i$, then $\bar{x} = \sum_{i=1}^{n} \bar{c}_i x_i$. This follows from Theorems 15.3(e) and 15.5(b), because

$$\sum_{i=1}^{n} c_i x_i + \sum_{i=1}^{n} \bar{c}_i x_i = \sum_{i=1}^{n} (c_i + \bar{c}_i) x_i = \sum_{i=1}^{n} x_i = 1$$

(Why is this true?) and

$$\left(\sum_{i=1}^{n} c_i x_i \right) \left(\sum_{i=1}^{n} \bar{c}_i x_i \right) = \sum_{i=1}^{n} c_i \bar{c}_i x_i = \sum_{i=1}^{n} 0 x_i = 0.$$

Now we find that

$$
\begin{aligned}
f(\bar{x}) &= \{i \mid 1 \le i \le n \text{ and } \bar{c}_i = 1\} \\
&= \{i \mid 1 \le i \le n \text{ and } c_i = 0\} \\
&= \overline{\{i \mid 1 \le i \le n \text{ and } c_i = 1\}} \\
&= \overline{f(x)},
\end{aligned}
$$

so the function f preserves the operations in the Boolean algebras $\mathcal{B}$ and $\mathcal{P}(\mathcal{U})$.

We leave to the reader the details showing that f is one-to-one and onto, thus establishing that f is an isomorphism.   ∎

EXERCISES 15.4

1. Verify the second distributive law and the identity and inverse laws for Example 15.23.

2. Complete the proof of Theorem 15.3.

3. Let $\mathscr{B}$ be the set of positive integer divisors of 210, and define $+$, $\cdot$, and $^-$ for $\mathscr{B}$ by $x + y = [x, y]$, $x \cdot y = (x, y)$, and $\overline{x} = 210/x$. Determine each of the following:

   a) $30 + 5 \cdot 7$ b) $(30 + 5) \cdot (30 + 7)$ c) $\overline{(14 + 15)}$

   d) $21(2 + \overline{10})$ e) $(2 + 3) + 5$ f) $(6 + 35)(7 + 10)$

4. For a Boolean algebra $\mathscr{B}$ the relation "$\leq$" on $\mathscr{B}$, defined by $x \leq y$ if $xy = x$, was shown to be a partial order. Prove that: (a) if $x \leq y$ then $x + y = y$; and (b) if $x \leq y$ then $\overline{y} \leq \overline{x}$.

5. If $\mathscr{B}$ is a Boolean algebra, partially ordered by $\leq$, and $x, y \in \mathscr{B}$, what is the dual of the statement "$x \leq y$"?

6. Let $F_n = \{f : B^n \to B\}$ be the Boolean algebra of all Boolean functions on n variables. How many atoms does F_n have?

7. Verify Theorem 15.5(b).

8. Given a Boolean algebra $\mathscr{B}$, a nonempty subset $\mathscr{B}_1$ of $\mathscr{B}$ is called a *subalgebra* if for all $x, y \in \mathscr{B}_1$, we find that $x + y$, xy, $\overline{x} \in \mathscr{B}_1$.

   a) Prove that for a subalgebra $\mathscr{B}_1$ of $\mathscr{B}$ we have $0, 1 \in \mathscr{B}_1$. (Hence $|\mathscr{B}_1| \geq 2$).

   b) Find two examples of subalgebras in Example 15.23.

   c) For $\mathscr{U} = \{1, 2, 3\}$ and $\mathscr{B} = $ the Boolean algebra of subsets of $\mathscr{U}$, how many different subalgebras can we find?

   d) Show that the definition above can be shortened by requiring only that whenever $x, y \in \mathscr{B}_1$, $x + y$ and $\overline{x} \in \mathscr{B}_1$.

   e) If $\mathscr{B}_1, \mathscr{B}_2$ are subalgebras of $\mathscr{B}$, is $\mathscr{B}_1 \cap \mathscr{B}_2$ a subalgebra?

9. Given a Boolean algebra $\mathscr{B}$, let $\emptyset \neq \mathscr{B}_1 \subseteq \mathscr{B}$. If $0, 1 \in \mathscr{B}_1$, and $\mathscr{B}_1$ is closed under $+$ and $\cdot$, is $\mathscr{B}_1$ a subalgebra of $\mathscr{B}$?

10. If $\mathscr{B}$ is a Boolean algebra, prove that the zero element and the one element of $\mathscr{B}$ are unique.

11. Let $f : \mathscr{B}_1 \to \mathscr{B}_2$ be an isomorphism of Boolean algebras. Prove each of the following:

   a) $f(0) = 0$

   b) $f(1) = 1$

   c) If $x, y \in \mathscr{B}_1$ with $x \leq y$, then in $\mathscr{B}_2$, $f(x) \leq f(y)$.

   d) If x is an atom of $\mathscr{B}_1$, then $f(x)$ is an atom in $\mathscr{B}_2$.

   e) If $\mathscr{S}_1$ is a subalgebra of $\mathscr{B}_1$, then $f(\mathscr{S}_1)$ is a subalgebra of $\mathscr{B}_2$.

12. Let $\mathscr{B}_1$ be the Boolean algebra of all positive integer divisors of 2310, with $\mathscr{B}_2$ the Boolean algebra of subsets of $\{a, b, c, d, e\}$.

    a) Define $f: \mathscr{B}_1 \to \mathscr{B}_2$ so that $f(2) = \{a\}$, $f(3) = \{b\}$, $f(5) = \{c\}$, $f(7) = \{d\}$, $f(11) = \{e\}$. For f to be an isomorphism, what must be the images of 35, 110, 210, and 330?

    b) How many different isomorphisms can one define between $\mathscr{B}_1$ and $\mathscr{B}_2$?

13. a) If $\mathscr{B}_1, \mathscr{B}_2$ are Boolean algebras and $f: \mathscr{B}_1 \to \mathscr{B}_2$ is one-to-one, onto, and such that $f(x + y) = f(x) + f(y)$ and $f(\bar{x}) = \overline{f(x)}$, for all $x, y \in \mathscr{B}_1$, prove that f is an isomorphism.

    b) State and prove another result comparable to that in (a). (What principle is used here?)

14. Prove that the function f in Theorem 15.9 is one-to-one and onto.

15. Let $\mathscr{B}$ be a finite Boolean algebra with the n atoms $x_1, x_2, \ldots, x_n$. (So $|\mathscr{B}| = 2^n$.) Prove that

$$1 = x_1 + x_2 + \cdots + x_n.$$

15.5
SUMMARY AND HISTORICAL REVIEW

The modern concept of abstract algebra was developed by George Boole in his study of general abstract systems, as opposed to particular examples of such systems. In his 1854 publication *An Investigation of the Laws of Thought*, he formulated the mathematical structure now called a Boolean algebra. Although abstract in nature during the nineteenth century, the study of Boolean algebra was investigated in the twentieth century for its applicative value.

Starting in 1938, C. E. Shannon [8] made the first major contribution in applied Boolean algebra. He devised the algebra of switching circuits and showed its relation to the algebra of logic. Additional developments that were made in this area during the 1940s and 1950s are noted in the paper by C. E. Shannon [9] and in the report of the Harvard University Computation Laboratory [10].

We found that switching functions can be represented by their disjunctive and conjunctive normal forms. These allowed us to write such functions in a compact way using binary indices. The minimization process showed how to obtain a given Boolean function as a minimal sum of products or as a minimal product of sums. Based on the map method by E. W. Veitch [11], M. Karnaugh's modification [4] was developed here as a pictorial method for the simplification of Boolean functions. Another technique that we mentioned in the text is the tabulation algorithm known as the Quine–McCluskey method. Originally developed by W. V. Quine in references [6] and [7], this technique was modified by E. J. McCluskey, Jr. [5]. It is very useful for functions with more than six variables and lends itself to computer implementation. The interested reader can find more about Karnaugh maps in Chapter 6 of F. Hill and G. Peterson [3]. Chapter 7 of [3] provides an excellent coverage of the Quine–McCluskey method. A. Friedman and P. Menon, in reference [2], examine gating networks in the light of contemporary technology, whereas T. Booth [1] investigates more specific applications of logic design in the study of computers.

Although the major part of this chapter was applied in nature, Section 15.4 found us investigating the structure of a Boolean algebra. Unlike commutative rings with unity, which come in all possible sizes, we found that a Boolean algebra can contain only 2^n elements, where $n \in \mathbf{Z}^+$. Uniqueness of representation appeared as we found the atoms of a Boolean algebra used to build the rest of the algebra. The Boolean algebra of sets that we studied in Chapter 3 was found to represent all Boolean algebras in the sense that a finite Boolean algebra with n atoms is isomorphic to the algebra of all subsets of $\{1, 2, 3, \ldots, n\}$.

REFERENCES

1. Booth, Taylor L. *Digital Networks and Computer Systems,* 2nd ed. New York: Wiley, 1978.
2. Friedman, Arthur D., and Menon, P. R. *Theory and Design of Switching Circuits.* Woodland Hills, Calif.: Computer Science Press, 1975.
3. Hill, Frederick J., and Peterson, Gerald R. *Introduction to Switching Theory and Logical Design,* 3rd ed. New York: Wiley, 1981.
4. Karnaugh, M. "The Map Method for Synthesis of Combinational Logic Circuits," *Transactions of the AIEE,* part I, vol. 72, no. 9 (1953): pp. 593–599.
5. McCluskey, E. J., Jr. "Minimization of Boolean Functions," *Bell System Technical Journal* 35, no. 6 (November 1956): pp. 1417–1444.
6. Quine, W. V. "The Problem of Simplifying Truth Functions," *American Mathematical Monthly* 59, no. 8 (October 1952): pp. 521–531.
7. Quine, W. V. "A Way to Simplify Truth Functions," *American Mathematical Monthly* 62, no. 9 (November 1955): pp. 627–631.
8. Shannon, C. E. "A Symbolic Analysis of Relay and Switching Circuits," *Transactions of the AIEE,* vol. 57 (1938): pp. 713–723.
9. Shannon, C. E. "The Synthesis of Two-terminal Switching Circuits," *Bell System Technical Journal,* vol. 28 (1949): pp. 59–98.
10. Staff of the Computation Laboratory. *Synthesis of Electronic Computing and Control Circuits,* Annals 27. Cambridge, Mass.: Harvard University Press, 1951.
11. Veitch, E. W. "A Chart Method for Simplifying Truth Functions." *Proceedings of the ACM.* Pittsburgh, Penna. (May 1952): pp. 127–133.

MISCELLANEOUS EXERCISES

1. **a)** Let $n \geq 2$. If x_i is a Boolean variable for $1 \leq i \leq n$, prove that

   (i) $\overline{(x_1 + x_2 + \cdots + x_n)} = \bar{x}_1 \bar{x}_2 \cdots \bar{x}_n$

   (ii) $\overline{(x_1 x_2 \cdots x_n)} = \bar{x}_1 + \bar{x}_2 + \cdots + \bar{x}_n$

   (iii) $(\sum_{i=1}^{n} x_i)(\overline{\prod_{i=1}^{n} x_i}) = x_1 \bar{x}_2 + x_2 \bar{x}_3 + x_3 \bar{x}_4 + \cdots + x_{n-1} \bar{x}_n + x_n \bar{x}_1$

   b) State the dual of statement (iii) in part (a).

2. Let f, g: $B^5 \to B$ be Boolean functions, where $f = \Sigma m(1, 2, 4, 7, x)$ and $g = \Sigma m(0, 1, 2, 3, y, z, 16, 25)$. If $f \leq g$, what are x, y, z?

3. Eileen is having a party and finds herself confronted with decisions about inviting five of her friends.

   a) If she invites Margaret, she must also invite Joan.

   b) If Kathleen is invited, Nettie and Margaret must also be invited.

   c) She can invite Cathy or Joan, but definitely not both of them.

   d) Neither Cathy nor Nettie will show up if the other is not invited.

   e) Either Kathleen or Nettie or both must be invited.

   Determine which subsets of these five friends Eileen can invite to her party and still satisfy conditions (a) through (e).

4. Let f, g: $B^4 \to B$, where $f = \Sigma m(2, 4, 6, 8)$, and $g = \Sigma m(1, 2, 3, 4, 5, 6, 7, 8, 9, 11, 13, 15)$. Find a function h: $B^4 \to B$ such that $f = gh$.

5. Use a Karnaugh map to find a minimal-sum-of-products representation for

   a) $f(w, x, y, z) = \Sigma m(0, 1, 2, 3, 6, 7, 14, 15)$

   b) $g(v, w, x, y, z) = \Pi M(1, 2, 4, 6, 9, 10, 11, 14, 17, 18, 19, 20, 22, 25, 26, 27, 30)$

6. In the feedback network of Fig. 15.14, we have a unit delay (one of the finite state machines of Chapter 6). This device causes the complement of the output f at time t to be fed back into the AND gate, along with new values for the inputs x, y, at the next unit of time, or clock pulse, $t + 1$.

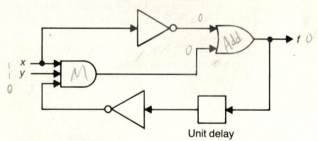

Figure 15.14

   Complete Table 15.22, which provides the output f at each time t for the values of x and y given.

Table 15.22

t	0	1	2	3	4	5	6
x	0	0	1	0	1	1	0
y	0	1	1	0	0	1	1
f	1	1	?	?	?	?	?

7. For n Boolean variables there are 2^{2^n} Boolean functions, each of which can be represented by a function table.

a) A Boolean function f on the n variables $x_1, x_2, \ldots, x_n$ is called *self-dual* if $f(x_1, x_2, \ldots, x_n) = f(\overline{x}_1, \overline{x}_2, \ldots, \overline{x}_n)$. How many Boolean functions on n variables are self-dual?

b) Let $f \colon B^3 \to B$. Then f is called *symmetric* if $f(x, y, z) = f(x, z, y) = f(y, x, z) = f(y, z, x) = f(z, x, y) = f(z, y, x)$. So the value of f is unchanged when we rearrange the three columns of values listed under x, y, and z in the table for f. How many such functions are there on three Boolean variables? How many are there on n Boolean variables?

8. The four input lines for the network in Fig. 15.15 provide the binary equivalents of the numbers $0, 1, 2, \ldots, 15$, where each number is represented as $abcd$, with d the least significant bit.

a) Find the d.n.f. of g, whose value is 1 exactly when $abcd$ is the binary equivalent of 1, 2, 4, or 8.

b) Draw the two-level gating network for g as a minimal sum of products.

c) If this network is part of a larger network and, consequently, the binary equivalents of the numbers 10 through 15 never occur as inputs, design a two-level gating network for g in this case.

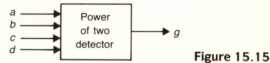

Figure 15.15

9. For (a) $n = 60$, and (b) $n = 120$, explain why the positive divisors of n do not yield a Boolean algebra. (Here $x + y = [x, y]$, $xy = (x, y)$, $\overline{x} = n/x$, 1 is the zero element, and n is the one element.)

10. Let $\mathcal{B}_1$ be the Boolean algebra of all positive integer divisors of 30030, and let $\mathcal{B}_2$ be the Boolean algebra of subsets of $\{u, v, w, x, y, z\}$. How many isomorphisms $f \colon \mathcal{B}_1 \to \mathcal{B}_2$ satisfy $f(2) = \{u\}$ and $f(6) = \{u, v\}$?

11. Let $a, b, c \in \mathcal{B}$, a Boolean algebra. Prove that $ab + c = a(b + c)$ if and only if $c \le a$.

Groups, Coding Theory, and Polya's Method of Enumeration

In the study of algebraic structures we examine properties shared by particular mathematical systems. Then we generalize our findings in order to study the underlying structure common to these particular examples.

In Chapter 14 we did this with the ring structure, which depended on two binary operations. Now we turn to a structure dependent on one binary operation. This structure is called a *group*.

Our study of groups will examine many ideas comparable to those for rings. However, here we shall dwell primarily on those aspects of the structure that are needed for applications in coding theory and for a counting method developed by George Polya.

16.1
DEFINITION, EXAMPLES, AND ELEMENTARY PROPERTIES

DEFINITION 16.1   If G is a nonempty set and $\circ$ is a binary operation on G, then $(G, \circ)$ is called a *group* if the following conditions are satisfied.

1. For all $a, b \in G$, $a \circ b \in G$. (Closure of G under $\circ$)

2. For all $a, b, c \in G$, $a \circ (b \circ c) = (a \circ b) \circ c$. (The Associative Property)

3. There exists $e \in G$ where $a \circ e = e \circ a = a$, for all $a \in G$. (The Existence of an Identity)

4. For each $a \in G$ there is an element $b \in G$ such that $a \circ b = b \circ a = e$. (Existence of Inverses)

   If, in addition, $a \circ b = b \circ a$ for all $a, b \in G$, then G is called a *commutative*, or *abelian*, group. The adjective "abelian" honors the Norwegian mathematician Niels Henrik Abel (1802–1829).

Example 16.1   Under ordinary addition, each of $\mathbf{Z}, \mathbf{Q}, \mathbf{R}, \mathbf{C}$ is an abelian group. None of these is a group under multiplication, because 0 has no multiplicative inverse. However, $\mathbf{Q}^*, \mathbf{R}^*$, and $\mathbf{C}^*$ (the nonzero elements of $\mathbf{Q}, \mathbf{R}$, and $\mathbf{C}$, respectively) are multiplicative abelian groups.

If $(R, +, \cdot)$ is a ring, then $(R, +)$ is an abelian group; the nonzero elements of a *field* form a multiplicative abelian group. $\square$

Example 16.2   For $n \in \mathbf{Z}^+, n > 1, (\mathbf{Z}_n, +)$ is an abelian group. When p is a prime, $(\mathbf{Z}_p^*, \cdot)$ is an abelian group. Tables 16.1 and 16.2 demonstrate this for $n = 6$ and $p = 7$. (Recall that in $\mathbf{Z}_n$ we often write a for $[a] = \{a + kn \mid k \in \mathbf{Z}\}$. A similar notation is used in $\mathbf{Z}_p^*$.) $\square$

Table 16.1

+	0	1	2	3	4	5
0	0	1	2	3	4	5
1	1	2	3	4	5	0
2	2	3	4	5	0	1
3	3	4	5	0	1	2
4	4	5	0	1	2	3
5	5	0	1	2	3	4

Table 16.2

·	1	2	3	4	5	6
1	1	2	3	4	5	6
2	2	4	6	1	3	5
3	3	6	2	5	1	4
4	4	1	5	2	6	3
5	5	3	1	6	4	2
6	6	5	4	3	2	1

DEFINITION 16.2   When G is a finite group, the number of elements in G is called the *order* of G and is denoted $|G|$.

Example 16.3   For any $n \in \mathbf{Z}^+, |(\mathbf{Z}_n, +)| = n$, while $|(\mathbf{Z}_p^*, \cdot)| = p - 1$ for any prime p. $\square$

From here on the group operation will be written multiplicatively, unless it is given otherwise. So $a \circ b$ now becomes ab.

We examine the following properties shared by all groups.

THEOREM 16.1   For any group G

a)   the identity of G is unique.

b)   the inverse of each element of G is unique.

c)   if $a, b, c \in G$ and $ab = ac$, then $b = c$.

d)   if $a, b, c \in G$ and $ba = ca$, then $b = c$.

e) $(ab)^2 = a^2 b^2$ for all $a, b \in G$ if and only if G is abelian.

Proof   **a)**   If e_1, e_2 are both identities in G, then $e_1 = e_1 e_2 = e_2$. (Justify each equality.)

b)   Let $a \in G$ and suppose that b, c are both inverses of a. Then $b = be = b(ac) = (ba)c = ec = c$. (Justify each equality.)

The proofs of properties (c) through (e) are left for the reader.   ∎

On the basis of the result in Theorem 16.1(b) the unique inverse of a will be designated by a^{-1}. When the group is written additively, $-a$ is used.

As in the case of multiplication in a ring, we have powers of elements in a group. We define $a^0 = e$, $a^1 = a$, $a^2 = a \cdot a$, and in general $a^{n+1} = a^n \cdot a$, $n \geq 0$. Since each group element has an inverse, here the case of a^{-n} must also be considered. For $n \in \mathbf{Z}^+$, we define $a^{-n} = (a^{-1})^n$. Then a^n is defined for all $n \in \mathbf{Z}$ and it can be shown that for all $m, n \in \mathbf{Z}$, $a^m \cdot a^n = a^{m+n}$; and $(a^m)^n = a^{mn}$.

If the group operation is addition, then multiples replace powers and for any $m, n \in \mathbf{Z}$, and any $a, b \in G$, we find that

$$ma + na = (m + n)a \qquad m(na) = (mn)a \qquad m(a + b) = ma + mb.$$

Example 16.4   Let $G = (\mathbf{Z}_6, +)$. If $H = \{0, 2, 4\}$, then H is a nonempty subset of G. Table 16.3 shows that $(H, +)$ is also a group under the binary operation of G.   □

Table 16.3

+	0	2	4
0	0	2	4
2	2	4	0
4	4	0	2

This situation gives rise to the following definition.

DEFINITION 16.3   Let G be a group and $\emptyset \neq H \subseteq G$. If H is a group under the binary operation of G, then we call H a *subgroup* of G.

Example 16.5
a) Every group G has $\{e\}$, G as subgroups. These are the *trivial* subgroups of G. All others are termed *nontrivial*, or *proper*.

b) In addition to $H = \{0, 2, 4\}$, the subset $K = \{0, 3\}$ is also a (proper) subgroup of $G = (\mathbf{Z}_6, +)$.

c) The group $(\mathbf{Z}, +)$ is a subgroup of $(\mathbf{Q}, +)$, which is a subgroup of $(\mathbf{R}, +)$. Yet $\mathbf{Z}^*$ under multiplication is not a subgroup of $(\mathbf{Q}^*, \cdot)$. (Why not?)   □

For a group G and $\emptyset \neq H \subseteq G$, the following tells us when H is a subgroup of G.

THEOREM 16.2   If H is a nonempty subset of a group G, then H is a subgroup of G if and only if (a) for all $a, b \in H$, $ab \in H$; and (b) for all $a \in H$, $a^{-1} \in H$.

Proof   If H is a subgroup of G, then by Definition 16.3 H is a group under the same binary operation. Hence it satisfies all the group conditions, including the two mentioned here. Conversely, let $\emptyset \neq H \subseteq G$ with H satisfying conditions (a) and (b). For all $a, b, c \in H$, $(ab)c = a(bc)$ in G, so $(ab)c = a(bc)$ in H. (We say that H "inherits" the associative property from G.) Finally, as $H \neq \emptyset$, let $a \in H$. By condition (b), $a^{-1} \in H$ and by condition (a), $aa^{-1} = e \in H$, so H contains the identity element and is a group.   ■

A finiteness condition improves the situation.

THEOREM 16.3 If G is a group and $\emptyset \neq H \subseteq G$, with H *finite*, then H is a subgroup of G if and only if H is closed under the binary operation of G.

Proof As in the proof of Theorem 16.2, if H is a subgroup of G, then H is closed under the binary operation of G. Conversely, let H be a finite nonempty subset of G that is so closed. If $a \in H$, then $aH = \{ah \,|\, h \in H\} \subseteq H$ because of the closure condition. By cancellation in G, $ah_1 = ah_2 \Rightarrow h_1 = h_2$, so $|aH| = |H|$. With $aH \subseteq H$ and $|aH| = |H|$, it follows from H being *finite* that $aH = H$. As $a \in H$, there exists $b \in H$ with $ab = a$. But (in G) $ab = ae$, so $b = e$ and H contains the identity. Since $e \in H = aH$, there is an element $c \in H$ such that $ac = e$. Then $(ca)(ca) = (ca)^2 = (c(ac))a = (ce)a = ca = (ca)e$, so $ca = e$, and $c = a^{-1} \in H$. Consequently, by Theorem 16.2, H is a subgroup of G. ∎

The finiteness condition in Theorem 16.3 is crucial. Both $\mathbf{Z}^+$ and $\mathbf{N}$ are nonempty closed subsets of the group $(\mathbf{Z}, +)$, yet neither has the additive inverses needed for the group structure.

The next example provides a nonabelian group.

Example 16.6 Consider the first equilateral triangle shown in Fig. 16.1(a). When we rotate this triangle counterclockwise through 120° about an axis perpendicular to its plane and passing through its center C, we obtain the second triangle shown in Fig. 16.1(a). As a result, the vertex originally labeled 1 in Fig. 16.1(a) is now in the position that was labeled 3. Likewise, 2 is now in the position originally occupied by 1, and 3 has moved to where 2 was. This can be described by the function $\pi_1: \{1, 2, 3\} \to \{1, 2, 3\}$, where $\pi_1(1) = 3, \pi_1(2) = 1, \pi_1(3) = 2$. A more compact notation, $\left(\begin{smallmatrix} 1 & 2 & 3 \\ 3 & 1 & 2 \end{smallmatrix}\right)$, where we write $\pi_1(i)$ below i for each $1 \leq i \leq 3$, emphasizes that π_1 is a permutation of $\{1, 2, 3\}$. If π_2 denotes the counterclockwise rotation through 240°, then $\pi_2 = \left(\begin{smallmatrix} 1 & 2 & 3 \\ 2 & 3 & 1 \end{smallmatrix}\right)$. For the identity π_0—that is, the rotation through $n(360°)$ for $n \in \mathbf{Z}$—we write $\pi_0 = \left(\begin{smallmatrix} 1 & 2 & 3 \\ 1 & 2 & 3 \end{smallmatrix}\right)$. These rotations are called *rigid motions* of the triangle. They are two-dimensional motions that keep the center C fixed and preserve the shape of the triangle. Hence the triangle looks the same as when we started, except for a possible rearrangement of some of its vertices.

In addition to these rotations, the triangle can be reflected along an axis passing through a vertex and the midpoint of the opposite side. For the diagonal axis that bisects the base angle on the right, the reflection gives the result in Fig. 16.1(b). This we represent by $r_1 = \left(\begin{smallmatrix} 1 & 2 & 3 \\ 2 & 1 & 3 \end{smallmatrix}\right)$. A similar reflection about the axis bisecting the left base angle yields the permutation $r_2 = \left(\begin{smallmatrix} 1 & 2 & 3 \\ 1 & 3 & 2 \end{smallmatrix}\right)$. When the triangle is reflected about its vertical axis, we have $r_3 = \left(\begin{smallmatrix} 1 & 2 & 3 \\ 3 & 2 & 1 \end{smallmatrix}\right)$. Each $r_i, 1 \leq i \leq 3$, is a three-dimensional rigid motion.

Let $G = \{\pi_0, \pi_1, \pi_2, r_1, r_2, r_3\}$, the set of rigid motions (in space) of the equilateral triangle. We make G into a group by defining the rigid motion $\alpha\beta$, for $\alpha, \beta \in G$, as that motion obtained by applying first α and then β. Hence, for example, $\pi_1 r_1 = r_3$. We can see this geometrically, but it will be easier to consider the permutations as follows: $\pi_1 r_1 = \left(\begin{smallmatrix} 1 & 2 & 3 \\ 3 & 1 & 2 \end{smallmatrix}\right)\left(\begin{smallmatrix} 1 & 2 & 3 \\ 2 & 1 & 3 \end{smallmatrix}\right)$, where, for example, $\pi_1(1) = 3$ and $r_1(3) = 3$ and we write $1 \xrightarrow{\pi_1} 3 \xrightarrow{r_1} 3$. So $1 \xrightarrow{\pi_1 r_1} 3$ in the product $\pi_1 r_1$. (Note that the order in which we write the product $\pi_1 r_1$ here is opposite from the order of their composite function as defined in Section 5.6. The notation of Section 5.6 occurs in analysis, whereas in algebra there is a tendency toward this opposite order.) Also, since $2 \xrightarrow{\pi_1} 1 \xrightarrow{r_1} 2$ and $3 \xrightarrow{\pi_1} 2 \xrightarrow{r_1} 1$, it follows that $\pi_1 r_1 = \left(\begin{smallmatrix} 1 & 2 & 3 \\ 3 & 2 & 1 \end{smallmatrix}\right) = r_3.$

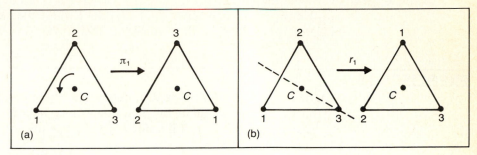

Figure 16.1

Table 16.4 verifies that under this binary operation G is closed, with identity π_0. Also $\pi_1^{-1} = \pi_2$, $\pi_2^{-1} = \pi_1$, and every other element is its own inverse. Since the elements of G are actually functions, the associative property follows from Theorem 5.5 (although in reverse order).

We computed $\pi_1 r_1$ as r_3, but from Table 16.4 we see that $r_1 \pi_1 = r_2$. With $\pi_1 r_1 = r_3 \neq r_2 = r_1 \pi_1$, G is nonabelian.

Table 16.4

$\cdot$	π_0	π_1	π_2	r_1	r_2	r_3
π_0	π_0	π_1	π_2	r_1	r_2	r_3
π_1	π_1	π_2	π_0	r_3	r_1	r_2
π_2	π_2	π_0	π_1	r_2	r_3	r_1
r_1	r_1	r_2	r_3	π_0	π_1	π_2
r_2	r_2	r_3	r_1	π_2	π_0	π_1
r_3	r_3	r_1	r_2	π_1	π_2	π_0

This group can also be obtained as the group of all permutations of the set $\{1, 2, 3\}$. Consequently it is denoted by S_3 (the *symmetric* group on three symbols). $\square$

Example 16.7   The symmetric group S_4 consists of the 24 permutations of $\{1, 2, 3, 4\}$. Here $\pi_0 = \left(\begin{smallmatrix} 1 & 2 & 3 & 4 \\ 1 & 2 & 3 & 4 \end{smallmatrix}\right)$ is the identity. If $\alpha = \left(\begin{smallmatrix} 1 & 2 & 3 & 4 \\ 2 & 1 & 4 & 3 \end{smallmatrix}\right), \beta = \left(\begin{smallmatrix} 1 & 2 & 3 & 4 \\ 3 & 1 & 2 & 4 \end{smallmatrix}\right)$, then $\alpha\beta = \left(\begin{smallmatrix} 1 & 2 & 3 & 4 \\ 1 & 3 & 4 & 2 \end{smallmatrix}\right)$ but $\beta\alpha = \left(\begin{smallmatrix} 1 & 2 & 3 & 4 \\ 4 & 2 & 1 & 3 \end{smallmatrix}\right)$, so S_4 is nonabelian. Also, $\beta^{-1} = \left(\begin{smallmatrix} 1 & 2 & 3 & 4 \\ 2 & 3 & 1 & 4 \end{smallmatrix}\right)$ and $\alpha^2 = \pi_0 = \beta^3$. Within S_4 there is a subgroup of order 8 that represents the group of rigid motions for a square. $\square$

We turn now to a construction for making larger groups out of smaller ones.

THEOREM 16.4 Let $(G, \circ)$ and $(H, *)$ be groups. Define the binary operation $\cdot$ on $G \times H$ by $(g_1, h_1) \cdot (g_2, h_2) = (g_1 \circ g_2, h_1 * h_2)$. Then $(G \times H, \cdot)$ is a group and is called the *direct product* of G and H.

Proof   The verification of the group postulates for $(G \times H, \cdot)$ is left to the reader. ∎

Example 16.8   Consider the groups $(\mathbf{Z}_2, +)$, $(\mathbf{Z}_3, +)$. On $G = \mathbf{Z}_2 \times \mathbf{Z}_3$, define $(a_1, b_1) \cdot (a_2, b_2) = (a_1 + a_2, b_1 + b_2)$. Then G is a group of order 6 where the identity is $(0, 0)$, and the inverse, for example, of the element $(1, 2)$ is $(1, 1)$. □

EXERCISES 16.1

1. For each of the following sets, determine whether or not the set is a group under the stated operation. If so, determine its identity and the inverse of each of its elements. If it is not a group, state the condition(s) of the definition that it violates.

   a) $\{-1, 1\}$ under multiplication   b) $\{-1, 1\}$ under addition

   c) $\{-1, 0, 1\}$ under addition   d) $\{10n \mid n \in \mathbf{Z}\}$ under addition

   e) The set of all functions $f: A \to A$, where $A = \{1, 2, 3, 4\}$, under function composition

   f) The set of all one-to-one functions $g: A \to A$, where $A = \{1, 2, 3, 4\}$, under function composition

   g) $\{a/2^n \mid a, n \in \mathbf{Z}, n \geq 0\}$ under addition

2. Prove parts (c) through (e) of Theorem 16.1.

3. Why is the set $\mathbf{Z}$ *not* a group under subtraction?

4. Let $G = \{q \in \mathbf{Q} \mid q \neq -1\}$. Define the binary operation $\circ$ on G by $x \circ y = x + y + xy$. Prove that $(G, \circ)$ is an abelian group.

5. Define the binary operation $\circ$ on $\mathbf{Z}$ by $x \circ y = x + y + 1$. Verify that $(\mathbf{Z}, \circ)$ is an abelian group.

6. Let $S = \mathbf{R}^* \times \mathbf{R}$. Define the binary operation $\circ$ on S by $(u, v) \circ (x, y) = (ux, vx + y)$. Prove that $(S, \circ)$ is a nonabelian group.

7. If G is a group, prove that for all $a, b \in G$,

   a) $(a^{-1})^{-1} = a$ b) $(ab)^{-1} = b^{-1} a^{-1}$

8. Prove that a group G is abelian if and only if for all $a, b \in G$, $(ab)^{-1} = a^{-1} b^{-1}$.

9. Find all subgroups in each of the following groups.

   a) $(\mathbf{Z}_{12}, +)$ b) $(\mathbf{Z}_{11}^*, \cdot)$ c) S_3

10. a) How many rigid motions (in two or three dimensions) are there for a square?

    b) Make a group table for these rigid motions like the one in Table 16.4 for the equilateral triangle. What is the identity for this group? Describe the inverse of each element geometrically.

11. **a)** How many rigid motions (in two or three dimensions) are there for a regular pentagon? Describe them geometrically.

    b) Answer part (a) for a regular n-gon, $n \geq 3$.

12. In the group S_5, let

    $$\alpha = \begin{pmatrix} 1 & 2 & 3 & 4 & 5 \\ 2 & 3 & 1 & 4 & 5 \end{pmatrix} \quad \text{and} \quad \beta = \begin{pmatrix} 1 & 2 & 3 & 4 & 5 \\ 2 & 1 & 5 & 3 & 4 \end{pmatrix}.$$

    Determine $\alpha\beta$, $\beta\alpha$, α^3, β^4, α^{-1}, β^{-1}, $(\alpha\beta)^{-1}$, $(\beta\alpha)^{-1}$, and $\beta^{-1}\alpha^{-1}$.

13. If G is a group, let $H = \{a \in G \mid ag = ga \text{ for all } g \in G\}$. Prove that H is a subgroup of G.

14. Let ω be the complex number $(1/\sqrt{2})(1 + i)$.

    a) Show that $\omega^8 = 1$ but $\omega^n \neq 1$ for $n \in \mathbf{Z}^+$, $1 \leq n \leq 7$.

    b) Verify that $\{\omega^n \mid n \in \mathbf{Z}^+, \ 1 \leq n \leq 8\}$, is an abelian group under multiplication.

15. **a)** Prove Theorem 16.4.

    b) Extending the idea developed in Theorem 16.4 and Example 16.8 to the group $\mathbf{Z}_6 \times \mathbf{Z}_6 \times \mathbf{Z}_6 = \mathbf{Z}_6^3$, answer the following.

    (i) What is the order of this group?

    (ii) Find a subgroup of $\mathbf{Z}_6^3$ of order 6, one of order 12, and one of order 36.

    (iii) Determine the inverse of each of the elements $(2, 3, 4)$, $(4, 0, 2)$, $(5, 1, 2)$.

16. If H, K are subgroups of a group G, prove that $H \cap K$ is also a subgroup of G.

16.2
HOMOMORPHISMS, ISOMORPHISMS, AND CYCLIC GROUPS

We turn our attention once again to functions that preserve structure.

Example 16.9   Let $G = (\mathbf{Z}, +)$ and $H = (\mathbf{Z}_4, +)$. Define $f: G \to H$ by

$$f(x) = [x] = \{x + 4k \mid k \in \mathbf{Z}\}.$$

For any $x, y \in G$,

$$f(x + y) = \underset{\uparrow}{[x + y]} = [x] + \underset{\uparrow}{[y]} = f(x) + f(y).$$

The operation in G The operation in H

Here f preserves the group operations and is an example of a special type of function which we shall now define. $\square$

DEFINITION 16.4   If $(G, \circ)$ and $(H, *)$ are groups and $f: G \to H$, then f is called a *group homo-morphism* if for all $a, b \in G$, $f(a \circ b) = f(a) * f(b)$.   ▬

When we know that the given structures are groups, the function f is simply called a homomorphism.

Some properties of homomorphisms are given in the following.

THEOREM 16.5   Let $(G, \circ)$, $(H, *)$ be groups with respective identities e_G, e_H. If $f: G \to H$ is a homomorphism, then

a) $f(e_G) = e_H$

b) $f(a^{-1}) = [f(a)]^{-1}$ for any $a \in G$

c) $f(a^n) = [f(a)]^n$ for all $a \in G$ and all $n \in \mathbf{Z}$

d) $f(S)$ is a subgroup of H for any subgroup S of G

Proof

a) $e_H * f(e_G) = f(e_G) = f(e_G \circ e_G) = f(e_G) * f(e_G)$, so by Theorem 16.1(d), $f(e_G) = e_H$.

b & c) The proofs of these parts are left for the reader.

d) If S is a subgroup of G, then $S \neq \emptyset$, so $f(S) \neq \emptyset$. Let $x, y \in f(S)$. Then $x = f(a), y = f(b)$, for some $a, b \in S$. Since S is a subgroup of G, it follows that $a \circ b \in S$, so $x * y = f(a) * f(b) = f(a \circ b) \in f(S)$. Finally, $x^{-1} = [f(a)]^{-1} = f(a^{-1}) \in f(S)$ because $a^{-1} \in S$ when $a \in S$. Consequently, by Theorem 16.2, $f(S)$ is a subgroup of H. ■

DEFINITION 16.5   If $f: (G, \circ) \to (H, *)$ is a homomorphism, we call f an *isomorphism* if it is one-to-one and onto. In this case G, H are said to be *isomorphic groups*.   ▬

Example 16.10   Let $f: (\mathbf{R}^+, \cdot) \to (\mathbf{R}, +)$ where $f(x) = \log_{10}(x)$. This function is both one-to-one and onto. (Verify these properties.) For any $a, b \in \mathbf{R}^+$, $f(ab) = \log_{10}(ab) = \log_{10} a + \log_{10} b = f(a) + f(b)$. Therefore, f is an isomorphism and the group of positive real numbers under multiplication is abstractly the same as the group of all real numbers under addition. Here the function f translates a problem in the multiplication of real numbers (a somewhat difficult problem without a calculator) into a problem dealing with the addition of real numbers (an easier arithmetic consideration). This was a major reason behind the use of logarithms before the advent of calculators. □

Example 16.11   Let G be the group of complex numbers $\{1, -1, i, -i\}$ under multiplication. Table 16.5 shows the multiplication table for the group. With $H = (\mathbf{Z}_4, +)$, consider $f: G \to H$ defined by

$$f(1) = [0] \qquad f(-1) = [2] \qquad f(i) = [1] \qquad f(-i) = [3].$$

Then $f((i)(-i)) = f(1) = [0] = [1] + [3] = f(i) + f(-i)$, and $f((-1)(-i)) = f(i) = [1] = [2] + [3] = f(-1) + f(-i)$.

Although we have not checked all possible cases, the function is an isomorphism. Note that the image under f of the subgroup $\{1, -1\}$ of G is $\{[0], [2]\}$, a subgroup of H.

Let us take a closer look at the group G. Here $i^1 = i$, $i^2 = -1$, $i^3 = -i$, and $i^4 = 1$, so every element of G is a power of i, and we say that i *generates* G. This is denoted by $G = \langle i \rangle$. (It is also true that $G = \langle -i \rangle$. Verify this.) □

Table 16.5

·	1	−1	i	$-i$
1	1	−1	i	$-i$
−1	−1	1	$-i$	i
i	i	$-i$	−1	1
$-i$	$-i$	i	1	−1

This example leads us to the following definition.

DEFINITION 16.6 A group G is called *cyclic* if there is an element $x \in G$ such that for all $a \in G$, $a = x^n$ for some $n \in \mathbf{Z}$.

Example 16.12 The group $H = (\mathbf{Z}_4, +)$ is cyclic. Here the operation is addition, so we have multiples instead of powers. We find that both [1] and [3] generate H. For the case of [3], we have $1 \cdot [3] = [3]$, $2 \cdot [3] = [2]$, $3 \cdot [3] = [1]$, and $4 \cdot [3] = [0]$. Hence $H = \langle [3] \rangle = \langle [1] \rangle$. □

The concept of a cyclic group leads to a related idea. Given a group G, if $a \in G$ consider the set $S = \{a^k \mid k \in \mathbf{Z}\}$. From Theorem 16.2 it follows that S is a subgroup of G. This subgroup is called the *subgroup generated by* a and is designated by $\langle a \rangle$. In Example 16.11 $\langle i \rangle = \langle -i \rangle = G$; also, $\langle -1 \rangle = \{-1, 1\}$ and $\langle 1 \rangle = \{1\}$.

DEFINITION 16.7 If G is a group and $a \in G$, the *order of* a, denoted $o(a)$, is $|\langle a \rangle|$. (If $|\langle a \rangle|$ is infinite we say that a has infinite order.)

In Example 16.11, $o(1) = 1$, $o(-1) = 2$, whereas both i and $-i$ have order 4.

Let us take a second look at the idea of order for $|\langle a \rangle|$ finite. With $a \in G$ and $\langle a \rangle = \{\ldots, a^{-1}, e, a, a^2, a^3, \ldots\}$ finite, by the pigeonhole principle there must be repetitions. So there is a *least* positive integer s where $a^s = a^t$, for $1 \le t < s$. Then $a^{s-t} = e$ and $\langle a \rangle = \{a, a^2, \ldots, a^{s-t}\}$. Therefore $o(a)$ can be defined as the *smallest positive integer n for which $a^n = e$.*

THEOREM 16.6 Let $a \in G$ with $o(a) = n$. If $k \in \mathbf{Z}$ and $a^k = e$, then $n \mid k$.

Proof By the division algorithm $k = qn + r$, for $0 \le r < n$, and so it follows that $e = a^k = a^{qn+r} = (a^n)^q a^r = (e^q)(a^r) = a^r$. If $0 < r < n$, we contradict the definition of n as $o(a)$. Hence $r = 0$ and $k = qn$. ∎

We now examine some further results on cyclic groups.

THEOREM 16.7   Let G be a cyclic group.

a) If G is infinite, then G is isomorphic to $(\mathbf{Z}, +)$.

b) If $|G| = n$, then G is isomorphic to $(\mathbf{Z}_n, +)$.

Proof   a)   For $G = \langle a \rangle = \{a^k | k \in \mathbf{Z}\}$, let $f: G \to \mathbf{Z}$ be defined by $f(a^k) = k$. (Could we have $a^k = a^t$ with $k \neq t$? If so, f would not be a function.) For $a^m, a^n \in G$, $f(a^m \cdot a^n) = f(a^{m+n}) = m + n = f(a^m) + f(a^n)$, so f is a homomorphism. We leave to the reader the verification that f is one-to-one and onto.

b)   If $G = \langle a \rangle = \{a, a^2, \dots, a^{n-1}, a^n = e\}$, then the function $f: G \to \mathbf{Z}_n$ defined by $f(a^k) = [k]$ is an isomorphism. (Verify this.)   ∎

Example 16.13   If $G = \langle g \rangle$, G is abelian because $g^m \cdot g^n = g^{m+n} = g^{n+m} = g^n \cdot g^m$ for all $m, n \in \mathbf{Z}$. The converse, however, is false. The group H of Table 16.6 is abelian, and $o(e) = 1, o(a) = o(b) = o(c) = 2$. Since no element of H has order 4, H cannot be cyclic. (The group H is the smallest noncyclic group and is known as the *Klein Four* group.)   □

Table 16.6

·	e	a	b	c
e	e	a	b	c
a	a	e	c	b
b	b	c	e	a
c	c	b	a	e

Our last result concerns the structure of subgroups in a cyclic group.

THEOREM 16.8   Any subgroup of a cyclic group is cyclic.

Proof   Let $G = \langle a \rangle$. If H is a subgroup of G, each element of H has the form a^k, for some $k \in \mathbf{Z}$. For $H \neq \{e\}$, let t be the smallest positive integer such that $a^t \in H$. (How do we know such an integer t exists?) We claim that $H = \langle a^t \rangle$. Since $a^t \in H$, by the closure property for the subgroup H, $\langle a^t \rangle \subseteq H$. For the opposite inclusion, let $b \in H$, with $b = a^s$, for some $s \in \mathbf{Z}$. By the division algorithm, $s = qt + r$, where $q, r \in \mathbf{Z}$ and $0 \leq r < t$. Consequently, $a^s = a^{qt+r}$ and so $a^r = a^{-qt}a^s = (a^t)^{-q}b$. H is a subgroup of G, so $a^t \in H \Rightarrow (a^t)^{-q} \in H$. Then with $(a^t)^{-q}, b \in H$, it follows that $a^r = (a^t)^{-q}b \in H$. But if $a^r \in H$, we contradict the minimality of t. Hence $r = 0$ and $b = a^{qt} = (a^t)^q \in \langle a^t \rangle$, so $H = \langle a^t \rangle$, a cyclic group.   ∎

EXERCISES 16.2

1. Prove parts (b) and (c) of Theorem 16.5.

2. If $f: G \to H, g: H \to K$ are homomorphisms, prove that the composite function $g \circ f: G \to K$, where $(g \circ f)(x) = g(f(x))$, is a homomorphism.

3. If $G = (\mathbf{Z}_6, +)$, $H = (\mathbf{Z}_3, +)$, and $K = (\mathbf{Z}_2, +)$, find an isomorphism for the groups $H \times K$ and G.

4. Let $f: G \rightarrow H$ be a group homomorphism onto H. If G is abelian, prove that H is abelian.

5. Find the order of each element in the group of rigid motions of (a) the equilateral triangle; and (b) the square.

6. In S_5 find an element of order n, for all $2 \leq n \leq 5$. Also determine the (cyclic) subgroups of S_5 that each of these elements generates.

7. Verify that $(\mathbf{Z}_p^*, \cdot)$ is cyclic for the primes 5, 7, and 11.

8. For a group G, prove that the function $f: G \rightarrow G$ defined by $f(a) = a^{-1}$ is an isomorphism if and only if G is abelian.

9. Prove or disprove the following statement: If G is a group where every proper subgroup of G is cyclic, then G is cyclic.

10. For $\omega = (1/\sqrt{2})(1 + i)$, let G be the multiplicative group $\{\omega^n \mid n \in \mathbf{Z}^+, 1 \leq n \leq 8\}$.

    a) Show that G is cyclic and find each element $x \in G$ such that $\langle x \rangle = G$.

    b) Prove that G is isomorphic to the group $(\mathbf{Z}_8, +)$.

11. a) Find all generators of the cyclic groups $(\mathbf{Z}_{12}, +)$, $(\mathbf{Z}_{16}, +)$, and $(\mathbf{Z}_{24}, +)$.

    b) Let $G = \langle a \rangle$ with $o(a) = n$. Prove that a^k, $k \in \mathbf{Z}^+$, generates G if and only if k and n are relatively prime.

    c) If G is a cyclic group of order n, how many distinct generators does it have?

12. Let $f: G \rightarrow H$ be a group homomorphism. If $a \in G$ with $o(a) = n$, and $o(f(a)) = k$ (in H), prove that $k \mid n$.

16.3

COSETS AND LAGRANGE'S THEOREM

In the last two sections, for any finite group G and subgroup H of G, we had $|H|$ dividing $|G|$. In this section we'll see that this was not mere chance but is true in general. To prove this we need one new idea.

DEFINITION 16.8   If H is a subgroup of G, then for any $a \in G$, the set $aH = \{ah \mid h \in H\}$ is called a *left coset* of H in G. The set $Ha = \{ha \mid h \in H\}$ is a *right coset* of H in G. ▬

If the operation in G is addition, we write $a + H$ in place of aH, where $a + H = \{a + h \mid h \in H\}$.

When the term "coset" is used in this chapter, it will refer to left coset. For abelian groups there is no need to distinguish between left and right cosets. This is not the case, however, for nonabelian groups.

Example 16.14   If G is the group of Example 16.6 and $H = \{\pi_0, \pi_1, \pi_2\}$, the coset $r_1 H = \{r_1 \pi_0, r_1 \pi_1, r_1 \pi_2\} = \{r_1, r_2, r_3\}$. Likewise we have $r_2 H = r_3 H = \{r_1, r_2, r_3\}$, whereas $\pi_0 H = \pi_1 H = \pi_2 H = H$.

We see that $|\alpha H| = |H|$ for any $\alpha \in G$ and that $G = H \cup r_1 H$ is a partition of G.

For the subgroup $K = \{\pi_0, r_1\}$, we find $r_2 K = \{r_2, \pi_2\}$ and $r_3 K = \{r_3, \pi_1\}$. Again a partition of G arises: $G = K \cup r_2 K \cup r_3 K$. (*Note:* $Kr_2 = \{\pi_0 r_2, r_1 r_2\} = \{r_2, \pi_1\} \neq r_2 K$)   □

Example 16.15   For $G = (\mathbf{Z}_{12}, +)$ and $H = \{[0], [4], [8]\}$, we find that

$$[0] + H = \{[0], [4], [8]\} = [4] + H = [8] + H = H$$
$$[1] + H = \{[1], [5], [9]\} = [5] + H = [9] + H$$
$$[2] + H = \{[2], [6], [10]\} = [6] + H = [10] + H$$
$$[3] + H = \{[3], [7], [11]\} = [7] + H = [11] + H,$$

and $H \cup ([1] + H) \cup ([2] + H) \cup ([3] + H)$ is a partition of G.   □

These examples lead us to the following results.

LEMMA 16.1   If H is a subgroup of the finite group G, then for any $a, b \in G$, (a) $|aH| = |H|$; and (b) $aH = bH$ or $aH \cap bH = \emptyset$.

Proof   **a)** Since $aH = \{ah \,|\, h \in H\}$, $|aH| \leq |H|$. If $|aH| < |H|$, we have $ah_i = ah_j$ with h_i, h_j distinct elements of H. By left cancellation in G we get the contradiction $h_i = h_j$, so $|aH| = |H|$.

b) If $aH \cap bH \neq \emptyset$, let $c = ah_1 = bh_2$, for some $h_1, h_2 \in H$. If $x \in aH$, then $x = ah$ for some $h \in H$, and so $x = (bh_2 h_1^{-1})h = b(h_2 h_1^{-1} h) \in bH$, and $aH \subseteq bH$. Similarly, $y \in bH \Rightarrow y = bh_3$ for some $h_3 \in H$

$$\Rightarrow y = (ah_1 h_2^{-1})h_3 = a(h_1 h_2^{-1} h_3) \in aH,$$

so $bH \subseteq aH$. Therefore aH and bH are either disjoint or identical.

(We observe here that if $g \in G$, then $g \in gH$ because $e \in H$. Also, by part (b), G can be partitioned into mutually disjoint cosets.)   ∎

We are now ready to prove the main result of this section.

THEOREM 16.9   (*Lagrange's Theorem*) If G is a group of order n with H a subgroup of order m, then m divides n.

Proof   If $H = G$ the result follows. Otherwise $m < n$ and there exists an element $a \in G - H$. Since $a \notin H$, it follows that $aH \neq H$, so $aH \cap H = \emptyset$. If $G = aH \cup H$, then $|G| = |aH| + |H| = 2|H|$ and the theorem follows. If not, there is an element $b \in G - (H \cup aH)$, with $bH \cap H = \emptyset = bH \cap aH$ and $|bH| = |H|$. If $G = bH \cup aH \cup H$, we have $|G| = 3|H|$. Otherwise we're back to an element $c \in G$ with $c \notin bH \cup aH \cup H$. The group G is finite, so this process terminates and we find that $G = a_1 H \cup a_2 H \cup \cdots \cup a_k H$. Therefore, $|G| = k|H|$ and m divides n.   ∎

An alternative method for proving this theorem is given in Exercise 12 for this section.

We close with the statements of two corollaries. Their proofs are requested in the section exercises.

COROLLARY 16.1   If G is a finite group and $a \in G$, then $o(a)$ divides $|G|$.

COROLLARY 16.2   Any group of prime order is cyclic.

EXERCISES 16.3

1. Let $G = S_4$. (a) For $\alpha = \left(\begin{smallmatrix} 1 & 2 & 3 & 4 \\ 2 & 3 & 4 & 1 \end{smallmatrix}\right)$, find the subgroup $H = \langle a \rangle$. (b) Determine the left cosets of H in G.

2. Answer Exercise 1 for $\beta = \left(\begin{smallmatrix} 1 & 2 & 3 & 4 \\ 2 & 3 & 1 & 4 \end{smallmatrix}\right)$.

3. If $\gamma = \left(\begin{smallmatrix} 1 & 2 & 3 & 4 \\ 2 & 1 & 4 & 3 \end{smallmatrix}\right) \in S_4$, how many cosets does $\langle \gamma \rangle$ determine?

4. For $G = (\mathbf{Z}_{24}, +)$, find the cosets determined by the subgroup $H = \langle [3] \rangle$. Do likewise for the subgroup $K = \langle [4] \rangle$.

5. Let $G = \mathbf{R} \times \mathbf{R}$, with the group operation $+$ defined by $(a, b) + (c, d) = (a + c, b + d)$, for $a, b, c, d \in \mathbf{R}$.

   a) If $H = \{(a, 0) \,|\, a \in \mathbf{R}\}$, prove that H is a subgroup of G.

   b) Give a geometric interpretation of the cosets of H in G.

6. a) Let R be a ring with unity u. Prove that the units of R form a group under the multiplication of the ring.

   b) If $R = (\mathbf{Z}_n, + , \cdot)$, how many elements are in its (multiplicative) group of units?

7. If G is a group of order n and $a \in G$, prove that $a^n = e$.

8. a) (*Fermat's Theorem*) If p is a prime, prove that $a^p \equiv a \pmod{p}$ for any $a \in \mathbf{Z}$. (How is this related to Exercise 14(a) of Section 14.3?)

   b) (*Euler's Theorem*) For any $n \in \mathbf{Z}^+$ and any $a \in \mathbf{Z}$, prove that if $(a, n) = 1$, then $a^{\phi(n)} \equiv 1 \pmod{n}$.

   c) How are the theorems in parts (a) and (b) related?

   d) Is there any connection between these two theorems and the results in Exercises 6 and 7?

9. Let p be a prime. (a) If G has order $2p$, prove that every proper subgroup of G is cyclic. (b) If G has order p^2, prove that G has a subgroup of order p.

10. Prove Corollaries 16.1 and 16.2.

11. Let H and K be subgroups of a group G, where e is the identity of G.

    a) Prove that if $|H| = 10$ and $|K| = 21$, then $H \cap K = \{e\}$.

    b) If $|H| = m$ and $|K| = n$, with $(m, n) = 1$, prove that $H \cap K = \{e\}$.

12. The following provides an alternative way to establish Lagrange's Theorem. Let G be a group of order n, and let H be a subgroup of G of order m.

a) Define the relation $\mathscr{R}$ on G as follows: If $a, b \in G$, then $a \mathscr{R} b$ if $a^{-1}b \in H$. Prove that $\mathscr{R}$ is an equivalence relation on G.

b) For $a, b \in G$, prove that $a \mathscr{R} b$ if and only if $aH = bH$.

c) If $a \in G$, prove that $[a]$, the equivalence class of a under $\mathscr{R}$, satisfies $[a] = aH$.

d) For any $a \in G$, prove that $|aH| = |H|$.

e) Now establish the conclusion of Lagrange's Theorem, namely that $|H|$ divides $|G|$.

16.4
ELEMENTS OF CODING THEORY

In this and the next four sections we introduce an area of applied mathematics called *algebraic coding theory*. This theory was inspired by the fundamental paper of Claude Shannon (1948) along with results by Marcel Golay (1949) and Richard Hamming (1950). Since that time it has become an area of great interest where algebraic structures, probability, and combinatorics all play a role.

Our coverage will be held to an introductory level as we seek to model the transmission of information represented by strings of the signals 0 and 1.

In digital communications, when information is transmitted in the form of strings of 0's and 1's, certain problems arise. As a result of "noise" in the channel, when a certain signal is transmitted a different signal may be received, thus causing a wrong decision to be made by the receiver. Hence we want to develop techniques to help us detect, and perhaps even correct, transmission errors. However, we can only improve the chances of correct transmission; there are no guarantees.

Our model uses a *binary symmetric channel*, as shown in Fig. 16.2. The adjective "binary" appears because an individual signal is represented by one of the bits 0 or 1. When a transmitter sends the signal 0 or 1 in such a channel, associated with either signal is a (constant) probability p for incorrect transmission. When that probability p is the same for both signals, the channel is called *symmetric*. Here, for example, we have probability p of sending 0 and having 1 received. The probability of sending signal 0 and having it received correctly is then $1 - p$. All possibilities are considered in Fig. 16.2.

Example 16.16 Consider the string $c = 10110$. We regard c as an element of the group $\mathbf{Z}_2^5$, formed from the direct product of five copies of $(\mathbf{Z}_2, +)$. To shorten notation we write 10110 instead of $(1, 0, 1, 1, 0)$. When sending each bit (individual signal) of c through the binary symmetric channel, we assume that the probability of incorrect transmission is $p = 0.05$.

The probability of transmitting c with no errors is $(.95)^5 \doteq 0.77$. Here, and throughout our discussion of coding theory, we assume that the transmission of any signal does not depend in any way on the transmissions of the prior signals.

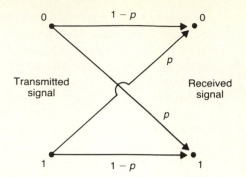

The Binary Symmetric Channel   **Figure 16.2**

Consequently, the probability of the occurrence of all of these *independent* events (in their prescribed order) is given by the product of their individual probabilities.

What is the probability that the party receiving the five-bit message receives the string $r = 00110$—that is, the original message with an error in the first position? The probability of incorrect transmission for the first bit is 0.05, so with the assumption of independent events, $(0.05)(0.95)^4 \doteq 0.041$ is the probability of sending $c = 10110$ and receiving $r = 00110$. With $e = 10000$, we can write $c + e = r$ and interpret r as the result of the sum of the original message c and the particular *error pattern* $e = 10000$. Since $c, r, e \in \mathbf{Z}_2^5$ and $-1 = 1$ in $\mathbf{Z}_2$, we also have $c + r = e$ and $r + e = c$.

In transmitting $c = 10110$, the probability of receiving $r = 00100$ is

$$(0.05)(0.95)^2(0.05)(0.95) \doteq 0.002,$$

so this multiple error is not very likely to occur.

Finally if we transmit $c = 10110$, what is the probability that r differs from c in exactly two places? To answer this we sum the probabilities for each error pattern consisting of two 1's and three 0's. Each such pattern has probability 0.002. There are $\binom{5}{2}$ such patterns, so the probability of two errors in transmission is given by

$$\binom{5}{2}(0.05)^2(0.95)^3 \doteq 0.021. \quad \square$$

These results lead us to the following theorem.

THEOREM 16.10   Let $c \in \mathbf{Z}_2^n$. For the transmission of c through a binary symmetric channel with probability p of incorrect transmission,

a) the probability of receiving $r = c + e$, where e is a *particular* error pattern consisting of k 1's and $(n - k)$ 0's, is $p^k(1 - p)^{n-k}$.

b) the probability that k errors are made in the transmission is $\binom{n}{k}p^k(1 - p)^{n-k}$.

In Example 16.16, the probability of making at most one error in the transmission of $c = 10110$ is $(0.95)^5 + \binom{5}{1}(0.05)(0.95)^4 \doteq 0.977$. Thus the chance for multiple errors in transmission will be considered negligible throughout the discussion in this chapter. Such an assumption is valid when p is small. In actuality, a binary symmetric channel is considered "good" when $p < 10^{-5}$. However, no matter what else we stipulate, we always want $p < 1/2$.

To improve the accuracy of transmission in a binary symmetric channel, certain types of coding schemes can be used where extra signals are provided.

For $m, n \in \mathbf{Z}^+$, let $n > m$. Consider $\emptyset \neq W \subseteq \mathbf{Z}_2^m$. The set W consists of the *messages* to be transmitted. We add to each $w \in W$ extra signals to form the *code word* c, where $c \in \mathbf{Z}_2^n$. This process is called *encoding* and is represented by the function $E: W \to C$. Then $E(w) = c$ and $E(W) = C \subseteq \mathbf{Z}_2^n$. To keep the code words for the messages distinct, E must be one-to-one. Upon transmission, c is transformed into $T(c)$, where $T(c) \in \mathbf{Z}_2^n$. Unfortunately, T is not a function, because $T(c)$ may be different at different transmission times. (See Fig. 16.3.)

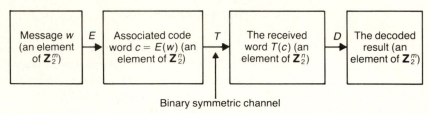

Figure 16.3

Upon receiving $T(c)$, we want to apply a decoding function $D: \mathbf{Z}_2^n \to \mathbf{Z}_2^m$ to remove the extra signals and, we hope, obtain the original message w. Ideally $D \circ T \circ E$ should be the identity function on W, with $D: C \to W$. Since this cannot be achieved, we seek functions E and D such that there is a high probability of decoding the received word $T(c)$ and recapturing the original message w. In addition, we want the ratio m/n to be as large as possible so that an excessive number of signals are not added to w in getting $E(w)$. This ratio m/n measures the *efficiency* of our scheme and is called the *rate* of the code. Finally, the functions E and D should be more than theoretical results; they must be practical in the sense that they can be implemented electronically.

In such a scheme, the functions E and D are called the *encoding* and *decoding* functions, respectively, of an (n, m) *block code*.

We illustrate these ideas in the following two examples.

Example 16.17   Consider the $(m + 1, m)$ block code for $m = 8$. Let $W = \mathbf{Z}_2^8$. For any $w = w_1 w_2 \ldots w_8 \in W$, define $E: \mathbf{Z}_2^8 \to \mathbf{Z}_2^9$ by $E(w) = w_1 w_2 \ldots w_8 w_9$, where $w_9 = \sum_{i=1}^{8} w_i$, with the addition performed modulo 2. For example, $E(11001101) = 110011011$, and $E(00110011) = 001100110$.

For all $w \in \mathbf{Z}_2^8$, $E(w)$ contains an even number of 1's. So for $w = 11010110$ and $E(w) = 110101101$, if we receive $T(c) = T(E(w))$ as 100101101, from the odd number of 1's in $T(c)$ we know that a mistake has occurred in transmission. Hence we are able to *detect* single errors in transmission. But we have no way to correct such errors.

The probability of sending the code word 110101101 and making at most one error in transmission is

$$\underbrace{(1-p)^9}_{\substack{\text{All nine bits are} \\ \text{correctly transmitted.}}} + \underbrace{\tbinom{9}{1}p(1-p)^8}_{\substack{\text{One bit is changed in} \\ \text{transmission and an error is detected.}}}.$$

For $p = 0.001$ this gives $(0.999)^9 + \binom{9}{1}(0.001)(0.999)^8 \doteq 0.999964$.

If we detect an error and we are able to relay a signal back to the transmitter to repeat the transmission of the code word, and continue this until the received word has an even number of 1's, then the probability of sending the code word 110101101 and receiving the correct transmission is approximately 0.999964.†

Should an even number of errors occur in transmission, $T(c)$ is unfortunately accepted as the correct code word and we interpret its first eight components as the original message. This scheme is called the $(m + 1, m)$ *parity-check code* and is appropriate only when multiple errors are not likely to occur.

If we send the message 11010110 through the channel, we have probability $(0.999)^8 = 0.992028$ of correct transmission. By using this parity-check code, we increase our chances of getting the correct message to (approximately) 0.999964. However, an extra signal is sent (and perhaps additional transmissions are needed) and the rate of the code has decreased from 1 to 8/9.

But suppose that instead of sending eight bits we sent 160 bits, in successive strings of length 8. The chances of receiving the correct message without any coding scheme would be $(0.999)^{160} \doteq 0.852076$. With the parity-check method we send 180 bits, but the chance for correct transmission increases to $(0.999964)^{20} \doteq 0.999280$. $\square$

Example 16.18   The $(3m, m)$ *triple repetition code* is one where we can *detect* and *correct* single errors in transmission. With $m = 8$ and $W = \mathbf{Z}_2^8$, we define $E: \mathbf{Z}_2^8 \rightarrow \mathbf{Z}_2^{24}$ by $E(w_1 w_2 \ldots w_7 w_8) = w_1 w_2 \ldots w_8 w_1 w_2 \ldots w_8 w_1 w_2 \ldots w_8$.

Hence if $w = 10110111$, $c = E(w) = 101101111011011110110111$.

The decoding function $D: \mathbf{Z}_2^{24} \rightarrow \mathbf{Z}_2^8$ is carried out by the majority rule. For example, if $T(c) = 101001110011011110110110$, then we have three errors. We decode $T(c)$ by examining the first, ninth, and seventeenth positions to see which signal appears more times. Here it is 1, so we decode the first entry in the decoded message as 1. Continuing with the entries in the second, tenth, and eighteenth positions, the result for the second entry of the decoded message is 0. As we proceed, we recapture the correct message, 10110111.

---

† The probability of sending the code word 110101101 and receiving the correct transmission under the conditions described is, for $p = 0.001$,

$$(0.999)^9 + [\tbinom{9}{1}(0.001)(0.999)^8](0.999)^9 + [\tbinom{9}{1}(0.001)(0.999)^8]^2(0.999)^9 + \cdots$$
$$= (0.999)^9(1 + x + x^2 + \cdots) = (0.999)^9(1/(1-x)),$$

where $x = \binom{9}{1}(0.001)(0.999)^8$. This probability is 0.999964 (to six decimal places).

Although we have more than one transmission error here, all is well unless two errors occur with the second error eight or sixteen spaces after the first—that is, if two or more incorrect transmissions occur for the same bit of the original message.

Now how does this scheme compare with the other methods we have? With $p = 0.001$, the probability of correctly decoding a single bit is $(0.999)^3 + \binom{3}{1} \cdot (0.001)(0.999)^2 \doteq 0.999997$. So the probability of receiving and correctly decoding the eight-bit message is $(0.999997)^8 = 0.999976$, just slightly better than the result from the parity-check method. But we transmit 24 signals for this message, so our rate is now $\frac{1}{3}$. For this increased accuracy and the ability to detect and correct single errors, we pay with an increase in transmission time.   □

EXERCISES 16.4

1. Let C be a set of code words, where $C \subseteq \mathbf{Z}_2^7$. In each of the following, two of e (error pattern), r (received word) and c (code word) are given, with $r = c + e$. Determine the third term.

     a) $c = 1010110$, $r = 1011111$

     b) $c = 1010110$, $e = 0101101$

     c) $e = 0101111$, $r = 0000111$

2. A binary symmetric channel has probability $p = 0.05$ of incorrect transmission. If the code word $c = 011011101$ is transmitted, what is the probability that (a) we receive $r = 011111101$? (b) we receive $r = 111011100$? (c) a single error occurs? (d) a double error occurs? (e) a triple error occurs? (f) three errors occur, no two of them consecutive?

3. Let $E: \mathbf{Z}_2^3 \to \mathbf{Z}_2^9$ be the encoding function for the $(9, 3)$ triple repetition code.

     a) If $D: \mathbf{Z}_2^9 \to \mathbf{Z}_2^3$ is the corresponding decoding function, apply D to decode the received words (i) 111101100; (ii) 000100011; (iii) 010011111.

     b) Find three different received words r for which $D(r) = 000$.

     c) For any $w \in \mathbf{Z}_2^3$, what is $|D^{-1}(w)|$?

4. The $(5m, m)$ five-times repetition code has encoding function $E: \mathbf{Z}_2^m \to \mathbf{Z}_2^{5m}$, where $E(w) = wwwww$. Decoding with $D: \mathbf{Z}_2^{5m} \to \mathbf{Z}_2^m$ is accomplished by the majority rule. (Here we are able to correct single and double errors made in transmission.)

     a) With $p = 0.05$, what is the probability for the transmission and correct decoding of the signal 0?

     b) Answer part (a) for the message 110.

     c) For $m = 2$, decode the received word $r = 0111001001$.

     d) If $m = 2$, find three received words r where $D(r) = 00$.

     e) For $m = 2$ and $D: \mathbf{Z}_2^{10} \to \mathbf{Z}_2^2$, what is $|D^{-1}(w)|$ for $w \in \mathbf{Z}_2^2$?

16.5
THE HAMMING METRIC

In this section we develop the general principles for discussing the error-detecting and error-correcting capabilities of a coding scheme. These ideas we owe to Richard Hamming.

We start by considering a code $C \subseteq \mathbf{Z}_2^4$, with $c_1 = 0111$, $c_2 = 1111 \in C$. Now both the transmitter and the receiver know the elements of C. So if the transmitter sends c_1 but the person receiving the code word receives $T(c_1)$ as 1111, then he or she feels that c_2 was transmitted and makes whatever decision (a wrong one) c_2 implies. Consequently, although only one transmission error was made, the results could be unpleasant. Why is this? Unfortunately we have two code words that are almost the same. They are rather *close* to each other, for they differ in only one component.

We describe this notion of closeness more precisely as follows.

DEFINITION 16.9   For any element $x = x_1 x_2 \ldots x_n \in \mathbf{Z}_2^n$, where $n \in \mathbf{Z}^+$, the *weight of x*, denoted wt(x), is the number of components x_i, of x, $1 \leq i \leq n$, where $x_i = 1$. If $y \in \mathbf{Z}_2^n$, the *distance between x and y*, denoted $d(x, y)$, is the number of components where $x_i \neq y_i$, for $1 \leq i \leq n$.   ▬

Example 16.19   For $n = 5$, let $x = 01001$ and $y = 11101$. Then wt$(x) = 2$, wt$(y) = 4$, and $d(x, y) = 2$. In addition, $x + y = 10100$, so wt$(x + y) = 2$. Is it just by chance that $d(x, y) = $ wt$(x + y)$? For $1 \leq i \leq 5$, $x_i + y_i$ contributes a count of 1 to wt$(x + y) \Leftrightarrow x_i \neq y_i \Leftrightarrow x_i, y_i$ contribute a count of 1 to $d(x, y)$. (This is true for any $n \in \mathbf{Z}^+$, so wt$(x + y) = d(x, y)$ for all $x, y \in \mathbf{Z}_2^n$.)   □

When $x, y \in \mathbf{Z}_2^n$, we write $d(x, y) = \sum_{i=1}^n d(x_i, y_i)$ where,

$$\text{for } 1 \leq i \leq n, \qquad d(x_i, y_i) = \begin{cases} 0 & \text{if } x_i = y_i \\ 1 & \text{if } x_i \neq y_i. \end{cases}$$

LEMMA 16.2   For any $x, y \in \mathbf{Z}_2^n$, wt$(x + y) \leq$ wt$(x) +$ wt(y).

Proof   We prove this lemma by examining the individual components x_i, y_i, $x_i + y_i$, $1 \leq i \leq n$, of x, y, $x + y$, respectively. Only one situation would cause this inequality to be false: if $x_i + y_i = 1$ while $x_i = 0$ and $y_i = 0$, for some $1 \leq i \leq n$. But this never occurs, because $x_i + y_i = 1$ implies that exactly one of x_i and y_i is 1.   ■

In Example 16.19 we found that

$$\text{wt}(x + y) = \text{wt}(10100) = 2 \leq 2 + 4 = \text{wt}(01001) + \text{wt}(11101) = \text{wt}(x) + \text{wt}(y).$$

THEOREM 16.11   The distance function d defined on $\mathbf{Z}_2^n \times \mathbf{Z}_2^n$ satisfies the following for all $x, y, z \in \mathbf{Z}_2^n$.

a) $d(x, y) \geq 0$

b) $d(x, y) = 0 \Leftrightarrow x = y$

c) $d(x, y) = d(y, x)$

d) $d(x, z) \leq d(x, y) + d(y, z)$

Proof   We leave the first three parts for the reader and prove part (d).

In $\mathbf{Z}_2^n$, $y + y = 0$, so $d(x, z) = \text{wt}(x + z) = \text{wt}(x + (y + y) + z) = \text{wt}((x + y) + (y + z)) \leq \text{wt}(x + y) + \text{wt}(y + z)$, by Lemma 16.2. With $\text{wt}(x + y) = d(x, y)$ and $\text{wt}(y + z) = d(y, z)$, the result follows. (This property is generally called the *Triangle Inequality*.) ∎

When a function satisfies the four properties listed in Theorem 16.11, it is called a *distance function*, and we call $(\mathbf{Z}_2^n, d)$ a *metric space*. Hence d is often referred to as the *Hamming metric*.

DEFINITION 16.10   For $n, k \in \mathbf{Z}^+$ and $x \in \mathbf{Z}_2^n$, the *sphere* of radius k centered at x is defined as $S(x, k) = \{y \in \mathbf{Z}_2^n \mid d(x, y) \leq k\}$. ━━━

Example 16.20   For $n = 3$ and $x = 110 \in \mathbf{Z}_2^3$, $S(x, 1) = \{110, 010, 100, 111\}$ and $S(x, 2) = \{110, 010, 100, 111, 000, 101, 011\}$. □

With these preliminaries we turn now to the two major results of this section.

THEOREM 16.12   Let $E: W \to C$ be an encoding function with the set of messages $W \subseteq \mathbf{Z}_2^m$ and the set of code words $E(W) = C \subseteq \mathbf{Z}_2^n$, where $m < n$. For $k \in \mathbf{Z}^+$, we can detect transmission errors of weight $\leq k$ if and only if the minimum distance between code words is at least $k + 1$.

Proof   The set C is known to both the transmitter and the receiver, so if $w \in W$ is the message and $c = E(w)$ is transmitted, let $c \neq T(c) = r$. If the minimum distance between code words is at least $k + 1$, then the transmission of c can result in as many as k errors and r will not be listed in C. Hence we can detect all errors e where $\text{wt}(e) \leq k$. Conversely, let c_1, c_2 be code words with $d(c_1, c_2) < k + 1$. Then $c_2 = c_1 + e$ where $\text{wt}(e) \leq k$. If we send c_1 and $T(c_1) = c_2$, then we would feel that c_2 had been sent, thus failing to detect an error of weight $\leq k$. ∎

What can we say about error-correcting capability?

THEOREM 16.13   With E, W, C as in Theorem 16.12, and $k \in \mathbf{Z}^+$, we can construct a decoding function $D: \mathbf{Z}_2^n \to W$ that corrects all transmission errors of weight $\leq k$ if and only if the minimum distance between code words is at least $2k + 1$.

Proof   For $c \in C$, consider $S(c, k) = \{x \in \mathbf{Z}_2^n \mid d(c, x) \leq k\}$. Define $D: \mathbf{Z}_2^n \to W$ as follows. If $r \in \mathbf{Z}_2^n$ and $r \in S(c, k)$ for some code word c, then $D(r) = w$ where $E(w) = c$. If $r \notin S(c, k)$ for any $c \in C$, then we define $D(r) = w_0$, where w_0 is some arbitrary message that remains fixed once it is chosen. The only problem we could face here is that D might not be a function. This will happen if there is an element r in $\mathbf{Z}_2^n$ with r in both $S(c_1, k)$ and $S(c_2, k)$ for distinct code words c_1, c_2. But $r \in S(c_1, k) \Rightarrow d(r, c_1) \leq k$, and $r \in S(c_2, k) \Rightarrow d(r, c_2) \leq k$, so $d(c_1, c_2) \leq d(c_1, r) + d(r, c_2) \leq k + k < 2k + 1$. Consequently, if the minimum distance between code words is at least $2k + 1$, D is a function, and it will decode all possible received words, correcting any transmission error of weight $\leq k$. Conversely, if $c_1, c_2 \in C$

and $d(c_1, c_2) \le 2k$, then c_2 can be obtained from c_1 by making at most $2k$ changes. Starting at code word c_1 we make approximately half (exactly, $\lfloor d(c_1, c_2)/2 \rfloor$) of these changes. This brings us to $r = c_1 + e_1$ with $\text{wt}(e_1) \le k$. Continuing from r, we make the remaining changes to get to c_2 and find $r + e_2 = c_2$ with $\text{wt}(e_2) \le k$. But then $r = c_2 + e_2$. Now with $c_1 + e_1 = r = c_2 + e_2$ and $\text{wt}(e_1)$, $\text{wt}(e_2) \le k$, how can one decide on the code word from which r arises? As a result, we have an error of weight $\le k$ that cannot be corrected. ∎

Example 16.21   With $W = \mathbf{Z}_2^2$ let $E: W \to \mathbf{Z}_2^6$ be given by

$$E(00) = 000000 \quad E(10) = 101010 \quad E(01) = 010101 \quad E(11) = 111111.$$

Then the minimum distance between code words is 3, so we can detect double errors and correct single ones.

With

$$S(000000, 1) = \{x \in \mathbf{Z}_2^6 | d(000000, x) \le 1\}$$
$$= \{000000, 100000, 010000, 001000, 000100, 000010, 000001\},$$

the decoding function $D: \mathbf{Z}_2^6 \to W$ gives $D(x) = 00$ for all $x \in S(000000, 1)$.

Similarly,

$$S(010101, 1) = \{x \in \mathbf{Z}_2^6 | d(010101, x) \le 1\}$$
$$= \{010101, 110101, 000101, 011101, 010001, 010111, 010100\},$$

and here $D(x) = 01$ for each $x \in S(010101, 1)$. At this point our definition of D accounts for 14 of the elements in $\mathbf{Z}_2^6$. Continuing to define D for the 14 elements in $S(101010, 1)$ and $S(111111, 1)$, there remain 36 other elements to account for. We define $D(x) = 00$ (or any other message) for these 36 other elements and have a decoding function that will correct single errors.

With regard to detection, if $c = 010101$ and $T(c) = r = 111101$, we can detect this double error. But if $T(c) = r_1 = 111111$, a triple error has occurred, so we think that $c = 111111$ and incorrectly decode r_1 as 11, instead of as the correct message 01. □

16.6
THE PARITY-CHECK AND GENERATOR MATRICES

In this section we introduce an example where encoding and decoding functions are given by matrices over $\mathbf{Z}_2$. One of these matrices will help us to locate the *nearest* code word for a given received word. This will be especially helpful when the set C of code words grows larger.

Example 16.22   Let

$$G = \begin{bmatrix} 1 & 0 & 0 & 1 & 1 & 0 \\ 0 & 1 & 0 & 0 & 1 & 1 \\ 0 & 0 & 1 & 1 & 0 & 1 \end{bmatrix}$$

be a 3×6 matrix over $\mathbf{Z}_2$. The first three columns of G form the 3×3 identity matrix I_3. Letting A denote the matrix formed from the last three columns of G, we write $G = [I_3 | A]$ to denote its structure. The matrix G is called a *generator matrix*.

We use G to define an encoding function $E: \mathbf{Z}_2^3 \rightarrow \mathbf{Z}_2^6$ as follows. For $w \in \mathbf{Z}_2^3$, $E(w) = wG$ is the element in $\mathbf{Z}_2^6$ obtained by multiplying w, considered as a three-dimensional row vector, by the matrix G on its right. Unlike the results on matrix multiplication in Chapter 7, in the calculations here we have $1 + 1 = 0$, not $1 + 1 = 1$.

(If the set W of messages is not all of $\mathbf{Z}_2^3$, we'll assume that all of $\mathbf{Z}_2^3$ is encoded. The transmitter and receiver will both know the real messages of importance and their corresponding code words.)

We find here, for example, that

$$E(110) = (110)G = [110] \begin{bmatrix} 1 & 0 & 0 & 1 & 1 & 0 \\ 0 & 1 & 0 & 0 & 1 & 1 \\ 0 & 0 & 1 & 1 & 0 & 1 \end{bmatrix} = [110101],$$

and

$$E(010) = (010)G = [010] \begin{bmatrix} 1 & 0 & 0 & 1 & 1 & 0 \\ 0 & 1 & 0 & 0 & 1 & 1 \\ 0 & 0 & 1 & 1 & 0 & 1 \end{bmatrix} = [010011].$$

Note that $E(110)$ can be obtained by adding the first two rows of G, whereas $E(010)$ is simply the second row of G.

The set of code words is

$$C = \{000000, 100110, 010011, 001101, 110101, 101011, 011110, 111000\} \subseteq \mathbf{Z}_2^6,$$

so decoding is accomplished by simply dropping the last three components of the code word. In addition, the minimum distance between code words is 3, so we can detect errors of weight ≤ 2 and correct single errors.

For any $w = w_1 w_2 w_3 \in \mathbf{Z}_2^3$, $E(w) = w_1 w_2 w_3 w_4 w_5 w_6 \in \mathbf{Z}_2^6$. Since

$$E(w) = [w_1 \, w_2 \, w_3] \begin{bmatrix} 1 & 0 & 0 & 1 & 1 & 0 \\ 0 & 1 & 0 & 0 & 1 & 1 \\ 0 & 0 & 1 & 1 & 0 & 1 \end{bmatrix}$$
$$= [w_1 \, w_2 \, w_3 (w_1 + w_3)(w_1 + w_2)(w_2 + w_3)],$$

we have $w_4 = w_1 + w_3$, $w_5 = w_1 + w_2$, $w_6 = w_2 + w_3$, and these equations are called the *parity-check equations*. Since $w_i \in \mathbf{Z}_2$ for each $1 \leq i \leq 6$, it follows that $w_i = -w_i$ and so the equations can be rewritten as

$$\begin{aligned} w_1 \quad\quad\;\; + w_3 + w_4 \quad\quad\quad\quad\quad &= 0 \\ w_1 + w_2 \quad\quad\quad\quad + w_5 \quad\quad &= 0 \\ w_2 + w_3 \quad\quad\quad\quad\; + w_6 &= 0. \end{aligned}$$

Thus we find that

$$\begin{bmatrix} 1 & 0 & 1 & 1 & 0 & 0 \\ 1 & 1 & 0 & 0 & 1 & 0 \\ 0 & 1 & 1 & 0 & 0 & 1 \end{bmatrix} \begin{bmatrix} w_1 \\ w_2 \\ w_3 \\ w_4 \\ w_5 \\ w_6 \end{bmatrix} = H \cdot (E(w))^{tr} = \begin{bmatrix} 0 \\ 0 \\ 0 \end{bmatrix},$$

where $(E(w))^{tr}$ denotes the transpose of $E(w)$. Consequently, if $r = r_1 r_2 \ldots r_6 \in \mathbf{Z}_2^6$, we can identify r as a code word if

$$H \cdot r^{tr} = \begin{bmatrix} 0 \\ 0 \\ 0 \end{bmatrix}.$$

Writing $H = [B \,|\, I_3]$, we notice that if the rows and columns of B are interchanged, then we get A. Hence $B = A^{tr}$.

From the theory developed earlier on error correction, because the minimum distance between the code words of this example is 3, we should be able to develop a decoding function that corrects single errors.

Suppose we receive $r = 110110$. If we have a long list of code words to check r against, we would be better off to examine $H \cdot r^{tr}$, which is called the *syndrome* of r. Here

$$H \cdot r^{tr} = \begin{bmatrix} 1 & 0 & 1 & 1 & 0 & 0 \\ 1 & 1 & 0 & 0 & 1 & 0 \\ 0 & 1 & 1 & 0 & 0 & 1 \end{bmatrix} \begin{bmatrix} 1 \\ 1 \\ 0 \\ 1 \\ 1 \\ 0 \end{bmatrix} = \begin{bmatrix} 0 \\ 1 \\ 1 \end{bmatrix},$$

so r is not a code word. Hence we at least detect an error. Looking back at the list of code words, we see that $d(100110, r) = 1$. For all other $c \in C$, $d(r, c) \geq 2$. Writing $r = c + e = 100110 + 010000$, we find that the transmission error (of weight 1) occurs in the second component of r. Is it a coincidence that the syndrome $H \cdot r^{tr}$ produced the second column of H? If not, we can use this to realize that if a single transmission error occurred, it took place at the second component. Changing the second component of r, we get c; the message w comprises the first three components of c.

Let $r = c + e$, where c is a code word and e is an error pattern of weight 1. Suppose that 1 is in the ith component of e, $1 \leq i \leq 6$. Then

$$H \cdot r^{tr} = H \cdot (c + e)^{tr} = H \cdot (c^{tr} + e^{tr}) = H \cdot c^{tr} + H \cdot e^{tr}.$$

With c a code word, it follows that $H \cdot c^{tr} = \mathbf{0}$, so $H \cdot r^{tr} = H \cdot e^{tr} = i$th column of matrix H.

We are primarily concerned with transmissions where multiple errors are rare, so this technique is of definite value. If we ask for more, however, we find ourselves expecting too much.

Suppose that we receive $r = 000111$. Computing the syndrome

$$H \cdot r^{\mathrm{tr}} = \begin{bmatrix} 1 & 0 & 1 & 1 & 0 & 1 \\ 1 & 1 & 0 & 0 & 1 & 0 \\ 0 & 1 & 1 & 0 & 0 & 1 \end{bmatrix} \begin{bmatrix} 0 \\ 0 \\ 0 \\ 1 \\ 1 \\ 1 \end{bmatrix} = \begin{bmatrix} 1 \\ 1 \\ 1 \end{bmatrix},$$

we obtain a result that is not one of the columns of H. Yet $H \cdot r^{\mathrm{tr}}$ can be obtained as the sum of two columns from H. If $H \cdot r^{\mathrm{tr}}$ came from the first and sixth columns of H, correcting these components in r results in the code word 100110. If we sum the third and fifth columns of H to get this syndrome, upon changing the third and fifth components of r we get a second code word, 001101. So we cannot expect H to correct multiple errors. This is no surprise since the minimum distance between code words is 3.   □

We summarize the results of Example 16.22 for the general situation. For $m, n \in \mathbf{Z}^+$ with $m < n$, the encoding function $E: \mathbf{Z}_2^m \to \mathbf{Z}_2^n$ is given by an $m \times n$ matrix G over $\mathbf{Z}_2$. This matrix G is called the generator matrix for the code and has the form $[I_m | A]$, where A is an $m \times (n - m)$ matrix. Here $E(w) = wG$ for each message $w \in \mathbf{Z}_2^m$, and the code $C = E(\mathbf{Z}_2^m) \subset \mathbf{Z}_2^n$.

The associated *parity-check matrix* H is an $(n - m) \times n$ matrix of the form $[A^{\mathrm{tr}} | I_{n-m}]$. This matrix can also be used to define the encoding function E, because if $w = w_1 w_2 \ldots w_m \in \mathbf{Z}_2^m$, then $E(w) = w_1 w_2 \ldots w_m w_{m+1} \ldots w_n$, where $w_{m+1}, \ldots, w_n$ can be determined from the set of $n - m$ (parity-check) equations that arise from $H \cdot (E(w))^{\mathrm{tr}} = \mathbf{0}$, the column vector of $n - m$ 0's.

This unique parity-check matrix H also provides a decoding scheme that corrects single errors in transmissions if:

a) H does not contain a column of 0's. (If the ith column of H had all 0's and $H \cdot r^{\mathrm{tr}} = \mathbf{0}$ for a received word r, we couldn't decide whether r was a code word or a received word whose ith component was incorrectly transmitted. We do not want to compare r with all code words when C is large.)

b) No two columns of H are the same. (If the ith and jth columns of H are the same and $H \cdot r^{\mathrm{tr}}$ equals this repeated column, how would we decide which component of r to change?)

When H satisfies these two conditions, we get the following decoding algorithm. For any $r \in \mathbf{Z}_2^n$, if $T(c) = r$, then

1. With $H \cdot r^{\mathrm{tr}} = \mathbf{0}$, we feel that the transmission was correct and that r is the code word that was transmitted. The decoded message then consists of the first m components of r.

2. With $H \cdot r^{\mathrm{tr}}$ equal to the ith column of H, we feel that there has been a single error in transmission and change the ith component of r in order to get the code word c. Here the first m components of c yield the original message.

3. If neither case 1 nor case 2 occurs, we feel that there has been more than one transmission error and we cannot provide a reliable way to decode in this situation.

We close with one final comment on the matrix H. If we start with a parity-check matrix $H = [B \,|\, I_{n-m}]$ and use it, as described above, to define the function E, then we obtain the same set of code words that is generated by the unique associated generator matrix $G = [I_m \,|\, B^{tr}]$.

EXERCISES 16.5 AND 16.6

1. For Example 16.21, list the elements in $S(101010, 1)$ and $S(111111, 1)$.

2. Decode each of the following received words for Example 16.21.

   a) 110101

   b) 101011

   c) 001111

   d) 110000

3. a) If $x \in \mathbf{Z}_2^{10}$, determine $|S(x, 1)|, |S(x, 2)|, |S(x, 3)|$.

   b) For $n, k \in \mathbf{Z}^+$ with $1 \le k \le n$, if $x \in \mathbf{Z}_2^n$, what is $|S(x, k)|$?

4. Let $E: \mathbf{Z}_2^5 \to \mathbf{Z}_2^{25}$ be an encoding function where the minimum distance between code words is 9. What is the largest value of k such that we can detect errors of weight $\le k$? If we wish to correct errors of weight $\le n$, what is the maximum value for n?

5. For each of the following encoding functions, find the minimum distance between the code words. Discuss the error-detecting and error-correcting capabilities of each code.

   a) $E: \mathbf{Z}_2^2 \to \mathbf{Z}_2^5$

$00 \to 00001$	$01 \to 01010$
$10 \to 10100$	$11 \to 11111$

   b) $E: \mathbf{Z}_2^2 \to \mathbf{Z}_2^{10}$

$00 \to 0000000000$	$01 \to 0000011111$
$10 \to 1111100000$	$11 \to 1111111111$

   c) $E: \mathbf{Z}_2^3 \to \mathbf{Z}_2^6$

$000 \to 000111$	$001 \to 001001$
$010 \to 010010$	$011 \to 011100$
$100 \to 100100$	$101 \to 101010$
$110 \to 110001$	$111 \to 111000$

   d) $E: \mathbf{Z}_2^3 \to \mathbf{Z}_2^8$

$000 \to 00011111$	$001 \to 00111010$
$010 \to 01010101$	$011 \to 01110000$
$100 \to 10001101$	$101 \to 10101000$
$110 \to 11000100$	$111 \to 11100011$

6. a) Use the parity-check matrix H of Example 16.22 to decode the following received words.

   (i) 111101　　(ii) 110101　　(iii) 001111　　(iv) 100100

   (v) 110001　　(vi) 111111　　(vii) 111100　　(viii) 010100

   b) Are all the results in part (a) uniquely determined?

7. The encoding function $E: \mathbf{Z}_2^2 \to \mathbf{Z}_2^5$ is given by the generator matrix

$$G = \begin{bmatrix} 1 & 0 & 1 & 1 & 0 \\ 0 & 1 & 0 & 1 & 1 \end{bmatrix}.$$

a) Determine all code words. What can we say about the error-detection capability of this code? What about its error-correction capability?

b) Find the associated parity-check matrix H.

c) Use H to decode each of the following received words.

    (i) 11011   (ii) 10101   (iii) 11010

    (iv) 00111   (v) 11101   (vi) 00110

8. Define the encoding function $E: \mathbf{Z}_2^3 \to \mathbf{Z}_2^6$ by means of the parity-check matrix

$$H = \begin{bmatrix} 1 & 0 & 1 & 1 & 0 & 0 \\ 1 & 1 & 0 & 0 & 1 & 0 \\ 1 & 0 & 1 & 0 & 0 & 1 \end{bmatrix}.$$

a) Determine all code words.

b) Does this code correct all single errors in transmission?

9. Find the generator and parity-check matrices for the $(9, 8)$ single parity-check coding scheme of Example 16.17.

10. a) Show that the 1×9 matrix $G = [1 \quad 1 \quad 1 \quad \ldots \quad 1]$ is the generator matrix for the $(9, 1)$ nine-times repetition code.

b) What is the associated parity-check matrix H in this case?

11. For an (n, m) code C with generator matrix $G = [I_m | A]$ and parity-check matrix $H = [A^{\text{tr}} | I_{n-m}]$, the $(n, n - m)$ code C^d with generator matrix $[I_{n-m} | A^{\text{tr}}]$ and parity-check matrix $[A | I_m]$ is called the *dual code* of C. Show that the codes in each of Exercises 9 and 10 constitute a pair of dual codes.

12. Given $n \in \mathbf{Z}^+$, let the set $M(n, k) \subseteq \mathbf{Z}_2^n$ contain the maximum number of code words of length n, where the minimum distance between code words is $2k + 1$. Prove that

$$\frac{2^n}{\sum_{i=0}^{2k} \binom{n}{i}} \leq |M(n, k)| \leq \frac{2^n}{\sum_{i=0}^{k} \binom{n}{i}}.$$

(The upper bound on $|M(n, k)|$ is called the *Hamming bound*; the lower bound is referred to as the *Gilbert bound*.)

16.7
GROUP CODES: DECODING WITH COSET LEADERS

Now that we've examined some introductory material on coding theory, it is time to see how the group structure enters the picture.

DEFINITION 16.11   Let $E: \mathbf{Z}_2^m \to \mathbf{Z}_2^n$ be an encoding function. The code $C = E(\mathbf{Z}_2^m)$ is called a *group code* if C is a subgroup of $\mathbf{Z}_2^n$.

When the code words form a group, it is easier to compute the minimum distance between code words.

THEOREM 16.14   In a group code, the minimum distance between code words is the minimum of the weights of the nonzero elements of the code.

Proof   Let $a, b, c \in C$ where $d(a, b)$ is minimum and c is nonzero with minimum weight. By closure in the group C, $a + b$ is a code word. Since $d(a, b) = \text{wt}(a + b)$, by the choice of c, $d(a, b) \geq \text{wt}(c)$. (Why is $d(a, b) > 0$?) Conversely, $\text{wt}(c) = d(c, \mathbf{0})$, where $\mathbf{0}$ is a code word because C is a group. Then $d(c, \mathbf{0}) \geq d(a, b)$ by the choice of a, b, so $\text{wt}(c) \geq d(a, b)$. Consequently, $d(a, b) = \text{wt}(c)$.   ■

If C is a set of code words and $|C| = 1024$, we have to compute $\binom{1024}{2} = 523{,}776$ distances to find the minimum distance between code words. But if we can recognize that C possesses a group structure, we need only compute the weights of the 1023 nonzero elements of C.

Is there some way to guarantee a group structure on the code words? By Theorem 16.5(d), if $E: \mathbf{Z}_2^m \to \mathbf{Z}_2^n$ is a group homomorphism, then $C = E(\mathbf{Z}_2^m)$ will be a subgroup of $\mathbf{Z}_2^n$. Our next result will use this fact to show that the codes we obtain when using a generator matrix G or a parity-check matrix H are group codes. Furthermore, the proof of this result reconfirms the observation we made (at the end of the previous section) about *the* code that arises from a generator matrix G or its associated parity-check matrix H.

THEOREM 16.15   Let $E: \mathbf{Z}_2^m \to \mathbf{Z}_2^n$ be an encoding function given by a generator matrix G or the associated parity-check matrix H. Then $C = E(\mathbf{Z}_2^m)$ is a group code.

Proof   We establish these results by proving that the function E arising from G or H is a group homomorphism.

If $x, y \in \mathbf{Z}_2^m$, then $E(x + y) = (x + y)G = xG + yG = E(x) + E(y)$. Hence E is a homomorphism and $C = E(\mathbf{Z}_2^m)$ is a group code.

For the case of H, if x is a message, then $E(x) = x_1 x_2 \ldots x_m x_{m+1} \ldots x_n$, where $x = x_1 x_2 \ldots x_m \in \mathbf{Z}_2^m$ and $H \cdot (E(x))^{\text{tr}} = \mathbf{0}$. In particular, $E(x)$ is uniquely determined by these two properties. If y is also a message, then $x + y$ is likewise, and $E(x + y)$ has $(x_1 + y_1), (x_2 + y_2), \ldots, (x_m + y_m)$ as its first m components, as does $E(x) + E(y)$. Further, $H \cdot (E(x) + E(y))^{\text{tr}} = H \cdot (E(x)^{\text{tr}} + E(y)^{\text{tr}}) = H \cdot E(x)^{\text{tr}} + H \cdot E(y)^{\text{tr}} = \mathbf{0} + \mathbf{0} = \mathbf{0}$. Since $E(x + y)$ is the unique element of $\mathbf{Z}_2^n$ with $(x_1 + y_1)$, $(x_2 + y_2), \ldots, (x_m + y_m)$ as its first m components and with $H \cdot (E(x + y))^{\text{tr}} = \mathbf{0}$, it follows that $E(x + y) = E(x) + E(y)$. So E is a group homomorphism and $C = \{c \in \mathbf{Z}_2^n \mid H \cdot c^{\text{tr}} = \mathbf{0}\}$ is a group code.   ■

Now we use the group structure of C, together with its cosets in $\mathbf{Z}_2^n$, to develop a scheme for decoding. Our example uses the code developed in Example 16.22, but the procedure applies for any group code.

Example 16.23   We develop a table for decoding as follows.

1. First list in a row the elements of the group code C, starting with the identity.

000000   100110   010011   001101   110101   101011   011110   111000.

2. Next select an element x of $\mathbf{Z}_2^6$ ($\mathbf{Z}_2^n$, in general) where x does not appear anywhere in the table developed so far and has minimum weight. Then list the elements of the coset $x + C$, with $x + c$ directly below c for each $c \in C$. For $x = 100000$ we have

000000 100110 010011 001101 110101 101011 011110 111000
100000 000110 110011 101101 010101 001011 111110 011000.

3. Repeat step 2 until the cosets provide a partition of $\mathbf{Z}_2^6$ ($\mathbf{Z}_2^n$, in general). This results in the *decoding table* shown in Table 16.7.

4. Once the decoding table is constructed, for any received word r we find the column containing r and use the first three components of the code word c at the top of the column to decode r.

Table 16.7
Decoding Table for the Code of Example 16.22

000000	100110	010011	001101	110101	101011	011110	111000
100000	000110	110011	101101	010101	001011	111110	011000
010000	110110	000011	011101	100101	111011	001110	101000
001000	101110	011011	000101	111101	100011	010110	110000
000100	100010	010111	001001	110001	101111	011010	111100
000010	100100	010001	001111	110111	101001	011100	111010
000001	100111	010010	001100	110100	101010	011111	111001
010100	110010	000111	011001	100001	111111	001010	101100

From the table we find that the code words for the received words

$$r_1 = 101001 \qquad r_2 = 111010 \qquad r_3 = 001001 \qquad r_4 = 111011$$

are

$$c_1 = 101011 \qquad c_2 = 111000 \qquad c_3 = 001101 \qquad c_4 = 101011,$$

respectively. From these results the respective messages are

$$w_1 = 101 \qquad w_2 = 111 \qquad w_3 = 001 \qquad w_4 = 101.$$

The entries in the first column of Table 16.7 are called the *coset leaders*. For the first seven rows, the coset leaders are the same in all tables, with some permutations of rows possible. However, for the last row, either 100001 or 001010 could have been used in place of 010100 because they also have minimum weight 2. So the table need not be unique.

How do the coset leaders really help us? It seems that the code words in the first row are what we used to decode r_1, r_2, r_3, and r_4 above.

Consider the received words $r_1 = 101001$ and $r_2 = 111010$ in the sixth row, where the coset leader is $x = 000010$. Computing syndromes, we find that

$$H \cdot (r_1)^{\text{tr}} = \begin{bmatrix} 0 \\ 1 \\ 0 \end{bmatrix} = H \cdot (r_2)^{\text{tr}} = H \cdot x^{\text{tr}}.$$

This is not just a coincidence. □

THEOREM 16.16   Let $C \subseteq \mathbf{Z}_2^n$ be a group code for a parity-check matrix H, and let $r_1, r_2 \in \mathbf{Z}_2^n$. For the table of cosets of C in $\mathbf{Z}_2^n$, r_1 and r_2 are in the same coset of C if and only if $H \cdot (r_1)^{\text{tr}} = H \cdot (r_2)^{\text{tr}}$.

Proof   If r_1 and r_2 are in the same coset, then $r_1 = x + c_1$ and $r_2 = x + c_2$, where x is the coset leader, and c_1 and c_2 are the code words at the tops of the respective columns for r_1 and r_2. Then $H \cdot (r_1)^{\text{tr}} = H \cdot (x + c_1)^{\text{tr}} = H \cdot x^{\text{tr}} + H \cdot (c_1)^{\text{tr}} = H \cdot x^{\text{tr}} + \mathbf{0} = H \cdot x^{\text{tr}}$, because c_1 is a code word. Likewise, $H \cdot (r_2)^{\text{tr}} = H \cdot x^{\text{tr}}$, so r_1, r_2 have the same syndrome. Conversely, $H \cdot (r_1)^{\text{tr}} = H \cdot (r_2)^{\text{tr}} \Rightarrow H \cdot (r_1 + r_2)^{\text{tr}} = \mathbf{0} \Rightarrow r_1 + r_2$ is a code word c. Hence $r_1 + r_2 = c$, so $r_1 = r_2 + c$ and $r_1 \in r_2 + C$. Since $r_2 \in r_2 + C$, we have r_1, r_2 in the same coset. ∎

In decoding received words, when Table 16.7 is used we must search through 64 elements to find a given received word. For $C \subseteq \mathbf{Z}_2^{12}$ there are 4096 strings, each with 12 bits. Such a searching process is tedious. So perhaps we should be thinking about having a computer do the searching. Presently it appears that this means storing the entire table: $6 \times 64 = 384$ bits of storage for Table 16.7; $12 \times 4096 = 49,152$ bits for $C \subseteq \mathbf{Z}_2^{12}$. We should like to improve this situation. Before things get better, however, they'll look worse as we enlarge Table 16.7, as shown in Table 16.8. This new table includes to the left of the coset leaders (the transposes of) the syndromes for each row.

Table 16.8
Decoding Table 16.7 with Syndromes

000	000000	100110	010011	001101	110101	101011	011110	111000
110	100000	000110	110011	101101	010101	001011	111110	011000
011	010000	110110	000011	011101	100101	111011	001110	101000
101	001000	101110	011011	000101	111101	100011	010110	110000
100	000100	100010	010111	001001	110001	101111	011010	111100
010	000010	100100	010001	001111	110111	101001	011100	111010
001	000001	100111	010010	001100	110100	101010	011111	111001
111	010100	110010	000111	011001	100001	111111	001010	101100

Now we can decode a received word r by the following procedure.

1. Compute the syndrome $H \cdot r^{\text{tr}}$.

2. Find the coset leader x to the right of $H \cdot r^{\text{tr}}$.

3. Add x to r to get c. (The code word c that we are seeking at the top of the column containing r satisfies $c + x = r$, or $c = x + r$.)

Consequently, all that is needed from Table 16.8 are the first two columns, which will require $(3)(8) + (6)(8) = 72$ storage bits. With 18 more storage bits for H we can store what we need for this decoding process, called *decoding by coset leaders*, in 90 storage bits, as opposed to the original estimate of 384 bits.

Applying this procedure to $r = 110110$, we find the syndrome

$$H \cdot r^{\text{tr}} = \begin{bmatrix} 0 \\ 1 \\ 1 \end{bmatrix}.$$

Since 011 is to the left of the coset leader $x = 010000$, the code word $c = x + r = 010000 + 110110 = 100110$, from which we recapture the original message, 100.

The code here is a group code where the minimum weight of the nonzero code words is 3, so we expected to be able to find a decoding scheme that corrected single errors. Here this is accomplished because the error patterns of weight 1 are all coset leaders. We cannot correct double errors; only one error pattern of weight 2 is a coset leader. All error patterns of weight 1 or 2 would have to be coset leaders before our decoding scheme could correct both single and double errors in transmission.

Unlike the situation in Example 16.22, where syndromes were also used for decoding, things here are a bit different. Once we have a complete table listing all of the cosets of C in $\mathbf{Z}_2^6$, the process of decoding by coset leaders will give us an answer for *all* received words, not just for those that are code words or have syndromes that appear among the columns of the parity-check matrix H. However, we do realize that there is still a problem here, because the last row of our table is not unique. Nonetheless, as our last result will affirm, this method provides a decoding scheme that is as good as any other.

Theorem 16.17   When we are decoding by coset leaders, if $r \in \mathbf{Z}_2^n$ is a received word and r is decoded as the code word c^* (which we then decode to recapture the message), then $d(c^*, r) \le d(c, r)$ for all code words c.

Proof   Let x be the coset leader for the coset containing r. Then $r = c^* + x$, or $r + c^* = x$, so $d(c^*, r) = \text{wt}(r + c^*) = \text{wt}(x)$. If c is any code word, $d(c, r) = \text{wt}(c + r)$, and $c + r = c + (c^* + x) = (c + c^*) + x$. Since C is a group code, $c + c^* \in C$ and $c + r$ is in the coset $x + C$. Among the elements in the coset $x + C$, the coset leader x is chosen to have minimum weight, so $\text{wt}(c + r) \ge \text{wt}(x)$. Consequently, $d(c^*, r) = \text{wt}(x) \le \text{wt}(c + r) = d(c, r)$. ∎

16.8
HAMMING MATRICES

We found the parity-check matrix H helpful in correcting single errors in transmission when (a) H had no column of 0's and (b) no two columns of H were alike. For the matrix

$$H = \begin{bmatrix} 1 & 1 & 0 & 1 & 1 & 0 & 0 \\ 1 & 0 & 1 & 1 & 0 & 1 & 0 \\ 0 & 1 & 1 & 1 & 0 & 0 & 1 \end{bmatrix}$$

we find that H satisfies these two conditions and that for the number of rows, $r = 3$, in H we have the maximum number of columns possible. If any additional column is added, H will no longer be useful for correcting single errors.

The generator matrix G associated with H is

$$G = \begin{bmatrix} 1 & 0 & 0 & 0 & 1 & 1 & 0 \\ 0 & 1 & 0 & 0 & 1 & 0 & 1 \\ 0 & 0 & 1 & 0 & 0 & 1 & 1 \\ 0 & 0 & 0 & 1 & 1 & 1 & 1 \end{bmatrix}.$$

Consequently we have a $(7,4)$ group code. The encoding function $E: \mathbf{Z}_2^4 \to \mathbf{Z}_2^7$ encodes four-bit messages into seven-bit code words. We realize that because H is determined by three parity-check equations, we have now maximized the number of bits we can have in the messages. (In addition, the columns of H, read from top to bottom, are the binary equivalents of the integers from 1 to 7.)

In general, if we start with r parity-check equations, then the parity-check matrix H can have as many as $2^r - 1$ columns and still be used to correct single errors. Under these circumstances $H = [A \,|\, I_r]$, where A is an $r \times 2^r - 1 - r$ matrix, and $G = [I_m \,|\, A^{\mathrm{tr}}]$ with $m = 2^r - 1 - r$. The parity-check matrix H associated with a $(2^r - 1, 2^r - 1 - r)$ group code in this way is called a *Hamming matrix*, and the code is referred to as a *Hamming code*.

Example 16.24 If $r = 4$, then $2^r - 1 = 15$ and $2^r - 1 - r = 11$. The one possible Hamming matrix H for $r = 4$ is

$$\begin{bmatrix} 1 & 1 & 1 & 1 & 1 & 1 & 1 & 0 & 0 & 0 & 0 & 1 & 0 & 0 & 0 \\ 1 & 1 & 1 & 1 & 0 & 0 & 0 & 1 & 1 & 1 & 0 & 0 & 1 & 0 & 0 \\ 1 & 1 & 0 & 0 & 1 & 1 & 0 & 1 & 1 & 0 & 1 & 0 & 0 & 1 & 0 \\ 1 & 0 & 1 & 0 & 1 & 0 & 1 & 1 & 0 & 1 & 1 & 0 & 0 & 0 & 1 \end{bmatrix}.$$

Once again, the columns of H contain the binary equivalents of the integers from 1 to 15 $(= 2^4 - 1)$.

This matrix H is the parity-check matrix of a Hamming $(15, 11)$ code whose rate is $11/15$. $\square$

With regard to the rate of these Hamming codes, for any $r \geq 2$, the rate m/n of such a code is given by

$$m/n = (2^r - 1 - r)/(2^r - 1) = 1 - [r/(2^r - 1)].$$

As r increases, $r/(2^r - 1)$ goes to 0 and the rate approaches 1.

We close our discussion on coding theory with one final observation. In Section 16.6 we presented G (and H) in what is called the *systematic form*. Other arrangements of the rows and columns of these matrices are also possible, and these yield *equivalent codes*. (More on this can be found in the text by L. Dornhoff and F. Hohn [2].) We mention this here because it is often common practice to list the columns in a Hamming matrix of r rows so that the binary representations of the integers from 1 to $2^r - 1$ appear as the columns of H are read from left to right. For the Hamming $(7, 4)$ code, the matrix H mentioned at the start of this section would take the (equivalent) form

$$H_1 = \begin{bmatrix} 0 & 0 & 0 & 1 & 1 & 1 & 1 \\ 0 & 1 & 1 & 0 & 0 & 1 & 1 \\ 1 & 0 & 1 & 0 & 1 & 0 & 1 \end{bmatrix}.$$

Here the identity appears in the first, second, and fourth columns instead of in the last three. Consequently, we would use these components for the parity checks and find that if we send the message $w = w_1 w_2 w_3 w_4$, then the corresponding code word $E(w)$ is $c_1 c_2 w_1 c_3 w_2 w_3 w_4$, where

$$
\begin{aligned}
c_1 &= w_1 + w_2 &&+ w_4 \\
c_2 &= w_1 &&+ w_3 + w_4 \\
c_3 &= &&w_2 + w_3 + w_4,
\end{aligned}
$$

so that $H_1 \cdot (E(w))^{\text{tr}} = \mathbf{0}$.

EXERCISES 16.7 AND 16.8

1. Let $E: \mathbf{Z}_2^8 \rightarrow \mathbf{Z}_2^{12}$ be the encoding function for a code C. How many calculations are needed to find the minimum distance between code words? How many calculations are needed if E is a group homomorphism?

2. a) Use Table 16.8 to decode the following received words.

   000011 100011 111110 100001 001100 011110 001111 111100

   b) Do any of the results in part (a) change if a different set of coset leaders is used?

3. a) Construct a decoding table (with syndromes) for the group code given by the generator matrix

   $$G = \begin{bmatrix} 1 & 0 & 1 & 1 & 0 \\ 0 & 1 & 0 & 1 & 1 \end{bmatrix}.$$

   b) Use the table from part (a) to decode the following received words.

   11110 11101 11011 10100 10011 10101 11111 01100

   c) Does this code correct single errors in transmission?

4. Let

   $$H = \begin{bmatrix} 1 & 1 & 0 & 1 & 1 & 0 & 0 \\ 1 & 0 & 1 & 1 & 0 & 1 & 0 \\ 0 & 1 & 1 & 1 & 0 & 0 & 1 \end{bmatrix}$$

   be the parity-check matrix for a Hamming $(7, 4)$ code.

   a) Encode the following messages: 1000 1100 1011 1110 1001 1111.

   b) Decode the following received words:

   1100001 1110111 0010001 0011100.

   c) Construct a decoding table consisting of the syndromes and coset leaders for this code.

   d) Use the result in part (c) to decode the received words given in part (b).

5. a) What are the dimensions of the generator matrix for the Hamming $(63, 57)$ code? What are the dimensions for the associated parity-check matrix H?

   b) What is the rate of this code?

6. Compare the rates of the Hamming $(7, 4)$ code and the $(3, 1)$ triple-repetition code.

7. a) Let $p = 0.01$ be the probability of incorrect transmission for a binary symmetric channel. If the message 1011 is sent via the Hamming $(7, 4)$ code, what is the probability of correct decoding?

   b) Answer part (a) for a 20-bit message sent in five blocks of length 4.

8. A standard practice in dealing with Hamming codes is to write the columns in normal binary order (going from left to right), as we did with the matrix H_1 at the end of Section 16.8. If we do this for the Hamming $(15, 11)$ code, in which columns do we find the identity matrix? Where is the identity for the Hamming $(31, 26)$ code? Where is it for the Hamming $(2^r - 1, 2^r - 1 - r)$ code?

16.9
COUNTING AND EQUIVALENCE: BURNSIDE'S THEOREM

In this section and the next two we shall develop a counting technique known as Polya's method of enumeration. Our development will not be very rigorous. Often we shall only state the general results of the theory as seen in the solution of a specific problem. Our first encounter with this problem is presented in the following example.

Example 16.25   We have a set of sticks, all of the same length and color, and a second set of round plastic disks. Each disk contains two holes, as shown in Fig. 16.4, and the sticks can be inserted into these holes in order to form different shapes, such as a square. If each disk is either red or white, how many distinct squares can we form?

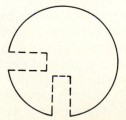

Figure 16.4

If the square is considered stationary, then the four disks are located at four distinct locations; a red or white disk is used at each location. Thus there are $2^4 = 16$ different configurations, as shown in Fig. 16.5, where a dark circle indicates a red disk. The configurations have been split into six classes, $c\ell(1)$, $c\ell(2), \ldots, c\ell(6)$, according to the number and relative location of the red disks.

Now suppose that the square is not fixed but can be moved about in space. Unless the vertices (disks) are marked somehow, certain configurations in Fig. 16.5 are indistinguishable when we move them about.

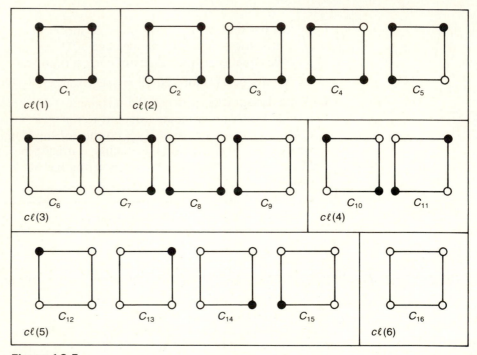

Figure 16.5

To place these notions in a more mathematical setting, we use the nonabelian group of three-dimensional rigid motions of a square to define an equivalence relation on the configurations in Fig. 16.5. Since this group will be used throughout this section and the next two sections, we now give a detailed description of its elements.

In Fig. 16.6 we have the group $G = \{\pi_0, \pi_1, \pi_2, \pi_3, r_1, r_2, r_3, r_4\}$ for the rigid motions of the square in part (a). Parts (b) through (i) of the figure show how each element of G is applied. We have expressed each group element as a permutation of $\{1, 2, 3, 4\}$ and in a new form called a *product of disjoint cycles*. For example, in part (b) we find $\pi_1 = (1234)$. The cycle (1234) indicates that if we start with the square in part (a), after applying π_1, we find that 1 has moved to the position originally occupied by 2, 2 to that of 3, 3 to that of 4, and 4 to that of 1. In general, if xy appears in a cycle, then x moves to the position originally occupied by y. Also, for a cycle where x and y appear as $(x \ldots y)$, y moves to the position originally occupied by x when the motion described by this cycle is applied. Note that $(1234) = (2341) = (3412) = (4123)$. We say that each of these cycles has *length* 4, the number of elements in the cycle. In the case of r_1 in part (f) of the figure, starting with 1 we find that r_1 sends 1 to 4, so we have $(14 \ldots)$ as the start of our first cycle in this *decomposition* of r_1. However, here r_1 sends 4 to 1, so we have completed a portion, namely (14), of the complete decomposition. We then select a vertex that has not yet appeared—for example, vertex 2. Since r_1 sends 2 to 3 and 3, in turn, to 2, we get a second cycle (23). This exhausts all vertices and so $(14)(23) = r_1$, where these cycles have no vertex in common. Here $(14)(23) = (23)(14) = (23)(41) = (32)(41)$ all provide a representation of r_1 as a product of disjoint cycles, each of length two. Last, for the group element $r_3 = (13)(2)(4)$, the cycle (2) indicates that 2 is fixed by, or *invariant* under, the permutation r_3. When the number of vertices involved is known, the permutation r_3 may also be written as $r_3 = (13)$, where the missing elements are understood to be fixed. However, we shall write all cycles in our decompositions, for this will be useful later in our discussion.

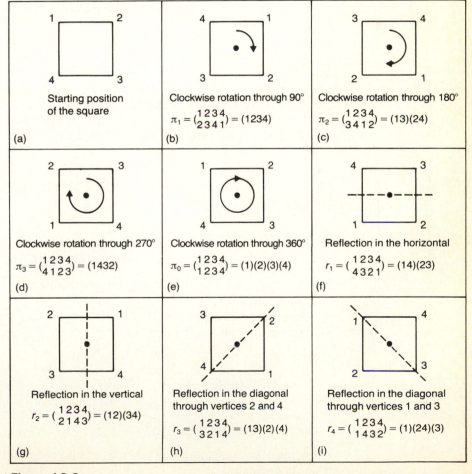

Figure 16.6

Before continuing with the main discussion, let us examine one more result on disjoint cycles.

In the group S_6 of all permutations of $\{1, 2, 3, 4, 5, 6\}$, let $\pi = \left(\begin{smallmatrix} 1 & 2 & 3 & 4 & 5 & 6 \\ 2 & 3 & 1 & 4 & 6 & 5 \end{smallmatrix}\right)$. As a product of disjoint cycles,

$$\pi = (123)(4)(56) = (56)(4)(123) = (4)(231)(65).$$

If $\sigma \in S_6$, with $\sigma = \left(\begin{smallmatrix} 1 & 2 & 3 & 4 & 5 & 6 \\ 2 & 4 & 3 & 1 & 6 & 3 \end{smallmatrix}\right)$, then

$$\sigma = (124)(356) = \begin{pmatrix} 1 & 2 & 3 & 4 & 5 & 6 \\ 2 & 4 & 3 & 1 & 5 & 6 \end{pmatrix}\begin{pmatrix} 1 & 2 & 3 & 4 & 5 & 6 \\ 1 & 2 & 5 & 4 & 6 & 3 \end{pmatrix},$$

so each cycle can be thought of as an element of S_6.

Finally, if $\alpha = (124)(3)(56)$ and $\beta = (13)(245)(6)$ are elements of S_6, then

$$\alpha\beta = (124)(3)(56)(13)(245)(6) = (143)(256),$$

whereas

$$\beta\alpha = (13)(245)(6)(124)(3)(56) = (132)(465).$$

Returning to the 16 configurations, or colorings, in Fig. 16.5, we now examine how each element of the group G, in Fig. 16.6, acts upon these configurations. For example, $\pi_1 = \left(\begin{smallmatrix} 1 & 2 & 3 & 4 \\ 2 & 3 & 4 & 1 \end{smallmatrix}\right)$ permutes the numbers $\{1, 2, 3, 4\}$ according to a 90° clockwise rotation for the square in Fig. 16.6(a), yielding the result in Fig. 16.6(b). How does such a rotation act on $S = \{C_1, C_2, \ldots, C_{16}\}$, our set of colorings? We use π_1^* to distinguish between the 90° clockwise rotation for $\{1, 2, 3, 4\}$ and the same rotation when applied to $S = \{C_1, C_2, \ldots, C_{16}\}$. We find that

$\pi_1^* =$

$$\begin{pmatrix} C_1 & C_2 & C_3 & C_4 & C_5 & C_6 & C_7 & C_8 & C_9 & C_{10} & C_{11} & C_{12} & C_{13} & C_{14} & C_{15} & C_{16} \\ C_1 & C_3 & C_4 & C_5 & C_2 & C_7 & C_8 & C_9 & C_6 & C_{11} & C_{10} & C_{13} & C_{14} & C_{15} & C_{12} & C_{16} \end{pmatrix}.$$

As a product of disjoint cycles,

$$\pi_1^* = (C_1)(C_2 C_3 C_4 C_5)(C_6 C_7 C_8 C_9)(C_{10} C_{11})(C_{12} C_{13} C_{14} C_{15})(C_{16}).$$

We note that under the action of π_1^*, no configuration is changed into one that is in another class.

As a second example, consider the reflection r_3 in Fig. 16.6(h). The action of this rigid motion on S is given by

$r_3^* =$

$$\begin{pmatrix} C_1 & C_2 & C_3 & C_4 & C_5 & C_6 & C_7 & C_8 & C_9 & C_{10} & C_{11} & C_{12} & C_{13} & C_{14} & C_{15} & C_{16} \\ C_1 & C_2 & C_5 & C_4 & C_3 & C_7 & C_6 & C_9 & C_8 & C_{10} & C_{11} & C_{14} & C_{13} & C_{12} & C_{15} & C_{16} \end{pmatrix}$$

$$= (C_1)(C_2)(C_3 C_5)(C_4)(C_6 C_7)(C_8 C_9)(C_{10})(C_{11})(C_{12} C_{14})(C_{13})(C_{15})(C_{16}).$$

Once again, no configuration is taken by r_3^* into one that is outside the class that it was in originally.

Using the idea of *the group G acting on the set S*, we define a relation $\mathcal{R}$ on S as follows. For colorings $C_i, C_j \in S$, where $1 \le i, j \le 16$, we write $C_i \mathcal{R} C_j$ if there is a permutation $\sigma \in G$ such that $\sigma^*(C_i) = C_j$. That is, as σ^* acts on the 16 configurations in S, C_i is transformed into C_j. This relation $\mathcal{R}$ is an equivalence relation, as we now verify.

a) (Reflexive Property) For all $C_i \in S, 1 \le i \le 16$, it follows that $C_i \mathcal{R} C_i$ because G contains the identity permutation. ($\pi_0^*(C_i) = C_i$ for all $1 \le i \le 16$.)

b) (Symmetric Property) If $C_i \mathcal{R} C_j$ for $C_i, C_j \in S$, then $\sigma^*(C_i) = C_j$, for some $\sigma \in G$. G is a group, so $\sigma^{-1} \in G$, and we find that $(\sigma^*)^{-1} = (\sigma^{-1})^*$. (Verify this for two choices of $\sigma \in G$.) Hence $C_i = (\sigma^{-1})^*(C_j)$, and $C_j \mathcal{R} C_i$.

c) (Transitive Property) Let $C_i, C_j, C_k \in S$ with $C_i \mathcal{R} C_j$ and $C_j \mathcal{R} C_k$. Then $C_j = \sigma^*(C_i)$ and $C_k = \tau^*(C_j)$, for some $\sigma, \tau \in G$. By closure in G, $\sigma\tau \in G$, and we find that $(\sigma\tau)^* = \sigma^*\tau^*$, where σ is applied first in $\sigma\tau$ and σ^* first in $\sigma^*\tau^*$. (Verify this for two specific permutations $\sigma, \tau \in G$.) Then $C_k = (\sigma\tau)^*(C_i)$ and $\mathcal{R}$ is transitive. (The reader may have noticed that $C_k = \tau^*(C_j) = \tau^*(\sigma^*(C_i))$ and felt that we should have written $(\sigma\tau)^* = \tau^*\sigma^*$. Once again, there has been a change in the notation for the composite function as we first defined it in Chapter 5. Here we write $\sigma^*\tau^*$ for $(\sigma\tau)^*$, and σ^* is applied first.)

Since $\mathcal{R}$ is an equivalence relation on S, $\mathcal{R}$ partitions S into equivalence classes, which are precisely the classes $c\ell(1), c\ell(2), \ldots, c\ell(6)$ of Fig. 16.5. Consequently, there are six nonequivalent configurations under the group action. So among the original 16 colorings only 6 are really distinct.

What has happened in this example generalizes as follows. With S a set of configurations, let G be a group of permutations that act on S. If the relation $\mathcal{R}$ is defined on S by $x \mathcal{R} y$ if $\pi^*(x) = y$, for some $\pi \in G$, then $\mathcal{R}$ is an equivalence relation.

With only red and white disks to connect the sticks, the answer to this example could have been determined from the results in Fig. 16.5. However, we developed quite a bit of mathematical overkill to answer the question. Referring to S as the set of 2-colorings of the vertices of a square, we start to wonder about the role of 2 and seek the number of nonequivalent configurations if the disks come in three or more colors.

In addition, we might notice that the function $f(r, w) = r^4 + r^3w + 2r^2w^2 + rw^3 + w^4$ is the generating function (of two variables) for the number of nonequivalent configurations from S. Here the coefficient of $r^i w^{4-i}$, for $0 \le i \le 4$, yields the number of distinct 2-colorings that have i red disks and $(4 - i)$ white ones. The coefficient of $r^2 w^2$ is 2 because of the two equivalence classes $c\ell(3)$ and $c\ell(4)$. Finally, $f(1, 1) = 6$, the number of equivalence classes. This generating function $f(r, w)$ is called the *pattern inventory* for the configurations. We shall examine it in more detail in the next two sections. □

For now we extend our present results in the following theorem. (A proof of this result is given on pages 136–137 of C. L. Liu [7].)

THEOREM 16.18   (*Burnside's Theorem*) Let S be a set of configurations on which a group G of permutations acts. The number of equivalence classes into which S is partitioned by the action of G is then given by

$$\frac{1}{|G|} \sum_{\pi \in G} \psi(\pi^*),$$

where $\psi(\pi^*)$ is the number of configurations in S fixed by π^*.

To convince ourselves of the truth of this theorem, we first examine two examples where we already know the answers.

Example 16.26   In Example 16.25 we find that $\psi(\pi_1^*) = 2$, because only C_1 and C_{16} are fixed, or *invariant*, under π_1^*. For $r_3 \in G$, however, $\psi(r_3^*) = 8$, because $C_1, C_2, C_4, C_{10}, C_{11}, C_{13}, C_{15},$ and C_{16} remain fixed under this group action. In like manner $\psi(\pi_2^*) = 4$, $\psi(\pi_3^*) = 2$, $\psi(\pi_0^*) = 16$, $\psi(r_1^*) = \psi(r_2^*) = 4$, and $\psi(r_4^*) = 8$. With $|G| = 8$, Burnside's theorem implies that the number of equivalence classes, or nonequivalent configurations, is $(1/8)(16 + 2 + 4 + 2 + 4 + 4 + 8 + 8) = (1/8)(48) = 6$, the original answer.   □

Example 16.27   In how many ways can six people be arranged around a circular table if two arrangements are considered equivalent when one can be obtained from the other by means of a clockwise rotation through $i \cdot 60°$, for $0 \le i \le 5$?

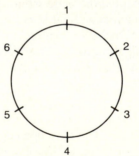

Figure 16.7

Here the six distinct people are to be placed in six chairs located at a table, as shown in Fig. 16.7. Our permutation group G consists of the clockwise rotations π_i through $i \cdot 60°$, where $0 \le i \le 5$. Here reflections are not meaningful. The situation is two-dimensional, for we can rotate the circle (representing the table) only in the plane; the circle never lifts off the plane. The total number of possible configurations is 6! We find that $\psi(\pi_0^*) = 6!$ and that $\psi(\pi_i^*) = 0$, for $1 \le i \le 5$. (It's impossible to move different people and simultaneously have them stay in a fixed location.) Consequently, the total number of nonequivalent seating arrangements is

$$\left(\frac{1}{|G|}\right) \sum_{\sigma \in G} \psi(\sigma^*) = \left(\frac{1}{6}\right)(6! + 0 + 0 + 0 + 0 + 0) = 5!,$$

as we found in Example 1.16 of Chapter 1.   □

We now examine a situation where the power of this theorem is made apparent.

Example 16.28   In how many ways can the vertices of a square be 3-colored, if the square can be moved about in three dimensions?

Now we have the sticks of Example 16.25, along with red, white, and blue disks. Considering the group in Fig. 16.6, we find the following:

$\psi(\pi_0^*) = 3^4$, because the identity fixes all 81 configurations in the set S of possible configurations.

$\psi(\pi_1^*) = \psi(\pi_3^*) = 3$, for each of π_1^*, π_3^* leaves invariant only those configurations with all vertices the same color.

$\psi(\pi_2^*) = 9$, for π_2^* can fix only those configurations where the opposite (diagonally) vertices have the same color. Consider a square like the one shown in Fig. 16.8. There are three choices for placing a colored disk at vertex 1 and then one choice for matching it at vertex 3. Likewise, there are three choices for colors at vertex 2 and then one for vertex 4. Consequently there are nine configurations invariant under π_2^*.

$\psi(r_1^*) = \psi(r_2^*) = 9$. In the case of r_1^*, for the square shown in Fig. 16.8 we have three choices for coloring each of the vertices 1 and 2, and then we must match the color of vertex 4 with the color of vertex 1, and the color of vertex 3 with that of vertex 2.

Figure 16.8

Finally, $\psi(r_3^*) = \psi(r_4^*) = 27$. For r_3^*, we have nine choices for coloring the two vertices at 2 and 4, and three choices for vertex 1. Then there is only one choice for vertex 3, because we must match the color of vertex 1.

By Burnside's theorem, the number of nonequivalent configurations is $(1/8)(3^4 + 3 + 3^2 + 3 + 3^2 + 3^2 + 3^3 + 3^3) = 21$.   □

EXERCISES 16.9

1. Consider the configurations shown in Fig. 16.5.
    a) Determine π_2^*, π_3^*, r_2^*, and r_4^*.
    b) Verify that $(\pi_1^{-1})^* = (\pi_1^*)^{-1}$ and $(r_3^{-1})^* = (r_3^*)^{-1}$.
    c) Verify that $(\pi_1 r_1)^* = \pi_1^* r_1^*$ and $(\pi_3 r_4)^* = \pi_3^* r_4^*$.

2. Express each of the following elements of S_7 as a product of disjoint cycles.

$$\alpha = \begin{pmatrix} 1 & 2 & 3 & 4 & 5 & 6 & 7 \\ 2 & 4 & 6 & 7 & 1 & 5 & 3 \end{pmatrix} \qquad \beta = \begin{pmatrix} 1 & 2 & 3 & 4 & 5 & 6 & 7 \\ 3 & 6 & 5 & 2 & 1 & 7 & 4 \end{pmatrix}$$

$$\gamma = \begin{pmatrix} 1 & 2 & 3 & 4 & 5 & 6 & 7 \\ 2 & 3 & 1 & 7 & 5 & 4 & 6 \end{pmatrix} \qquad \delta = \begin{pmatrix} 1 & 2 & 3 & 4 & 5 & 6 & 7 \\ 4 & 2 & 7 & 1 & 3 & 6 & 5 \end{pmatrix}$$

3. **a)** Determine the order of each of the elements in Exercise 2.

   b) State a general result about the order of an element in S_n in terms of the lengths of the cycles in its decomposition as a product of disjoint cycles.

4. **a)** Determine the number of distinct ways one can color the vertices of an equilateral triangle using the colors red and white, if the triangle is free to move in three dimensions.

   b) Answer part (a) if the color blue is also available.

5. Answer the questions in Exercise 4 for a regular pentagon.

6. **a)** How many distinct ways are there to paint the *edges* of a square with three different colors?

   b) Answer part (a) for the edges of a regular pentagon.

7. We make a child's bracelet by symmetrically placing four beads about a circular wire. The colors of the beads are red, white, blue, and green, and there are at least four beads of each color.

   a) How many distinct bracelets can we make in this way, if the bracelets can be rotated but not reflected?

   b) Answer part (a) if the bracelets can be rotated and reflected.

8. A baton is painted with three cylindrical bands of color (not necessarily distinct), with each band of the same length.

   a) How many distinct paintings can be made if there are three colors of paint available? How many for four colors?

   b) Answer part (a) for batons with four cylindrical bands.

   c) Answer part (a) for batons with n cylindrical bands.

   d) Answer parts (a) and (b) if adjacent cylindrical bands are to have different colors.

9. In how many ways can we 2-color the vertices of the configurations shown in Fig. 16.9 if they are free to move in (a) two dimensions? (b) three dimensions?

 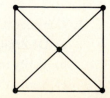

Figure 16.9

10. A pyramid has a square base and four faces that are equilateral triangles. If we can move the pyramid about (in three dimensions), how many nonequivalent ways are there to paint its five faces if we have paint of four different colors? How many if the color of the base must be different from the color(s) of the triangular faces?

11. a) In how many ways can we paint the cells of a 3×3 chessboard using red and blue paint? (The back of the chessboard is black.)

    b) In how many ways can we construct a 3×3 chessboard by joining (with paste) the edges of nine 1×1 plastic squares that are transparent and tinted red or blue? (There are nine squares of each color available.)

12. Answer Exercise 11 for a 4×4 chessboard. (Replace each "nine" in part (b) with "sixteen.")

13. In how many ways can we paint the seven (identical) horses on a carousel using black, brown, and white paint?

14. a) Let S be a set of configurations and G a group of permutations that act on S. If $x \in S$, prove that $\{\pi \in G \mid \pi^*(x) = x\}$ is a subgroup of G.

    b) Determine the respective subgroups in part (a) for the configurations C_7 and C_{15} in Fig. 16.5.

16.10
THE CYCLE INDEX

In applying Burnside's theorem we have been faced with computing $\psi(\pi^*)$ for each $\pi \in G$, where G is a permutation group acting on a set S of configurations. As the number of available colors increases and the configurations get more complex, such computations can get a bit involved. In addition, it seems that if we can determine the number of 2-colorings for a set S of configurations, we should be able to use some of the work in this case to determine the number of 3-colorings, 4-colorings, and so on. We shall now find some assistance as we return to the solution of Example 16.25. This time more attention will be paid to the representation of each permutation $\pi \in G$ as a product of disjoint cycles. Our results are summarized in Table 16.9.

For π_0, the identity of G, we write $\pi_0 = (1)(2)(3)(4)$, a product of four disjoint cycles. We shall represent this cycle structure algebraically by x_1^4, where x_1 indicates a cycle of length 1. The term x_1^4 is called the *cycle structure representation* of π_0. Here we interpret "disjoint" as "independent," in the sense that whatever color is used to paint the vertices in one cycle has no bearing on the choice of color for the vertices in another cycle. As long as all the vertices in a given cycle have the same color, we shall find configurations that are invariant under π_0^*. (Admittedly, this seems like mathematical overkill again, inasmuch as π_0^* fixes all 2-colorings of the square.) In addition, since we can paint the vertices in each cycle either red or white, we have 2^4 configurations, and we find that $(r + w)^4 =$

Table 16.9

Rigid Motions π (Elements of G)	Configurations in S that are Invariant under π^*	Cycle Structure Representation of π	Inventory of Configurations that are Invariant under π^*		
$\pi_0 = (1)(2)(3)(4)$	2^4: All configurations in S	x_1^4	$(r+w)^4$	$= r^4 + 4r^3w + 6r^2w^2 + 4rw^3 + w^4$	
$\pi_1 = (1234)$	2: C_1, C_{16}	x_4	$r^4 + w^4$	$= r^4$	$+ w^4$
$\pi_2 = (13)(24)$	2^2: $C_1, C_{10}, C_{11}, C_{16}$	x_2^2	$(r^2+w^2)^2$	$= r^4 \qquad + 2r^2w^2$	$+ w^4$
$\pi_3 = (1432)$	2: C_1, C_{16}	x_4	$r^4 + w^4$	$= r^4$	$+ w^4$
$r_1 = (14)(23)$	2^2: C_1, C_7, C_9, C_{16}	x_2^2	$(r^2+w^2)^2$	$= r^4 \qquad + 2r^2w^2$	$+ w^4$
$r_2 = (12)(34)$	2^2: C_1, C_6, C_8, C_{16}	x_2^2	$(r^2+w^2)^2$	$= r^4 \qquad + 2r^2w^2$	$+ w^4$
$r_3 = (13)(2)(4)$	2^3: $C_1, C_2, C_4, C_{10},$ $C_{11}, C_{13}, C_{15}, C_{16}$	$x_2 x_1^2$	$(r^2+w^2)(r+w)^2 = r^4 + 2r^3w + 2r^2w^2 + 2rw^3 + w^4$		
$r_4 = (1)(24)(3)$	2^3: $C_1, C_3, C_5, C_{10},$ $C_{11}, C_{12}, C_{14}, C_{16}$	$x_2 x_1^2$	$(r^2+w^2)(r+w)^2 = r^4 + 2r^3w + 2r^2w^2 + 2rw^3 + w^4$		
	$P_G(x_1, x_2, x_3, x_4) = \frac{1}{8}(x_1^4 + 2x_4 + 3x_2^2 + 2x_2x_1^2)$	Complete Inventory $\Big\}$	$= 8r^4 + 8r^3w + 16r^2w^2 + 8rw^3 + 8w^4$		

$r^4 + 4r^3w + 6r^2w^2 + 4rw^3 + w^4$ generates these 16 configurations. For example, from the term $6r^2w^2$ we find that there are six configurations with two red and two white vertices, as found in classes (3) and (4) of Fig. 16.5.

Turning to π_1, we find $\pi_1 = (1234)$, a cycle of length 4. This cycle structure is represented by x_4, and here there are only two invariant configurations. The fact that the cycle structure for π_1 has only one cycle tells us that for a configuration to be invariant under π_1^*, every vertex in this cycle must be painted the same color. With two colors to choose from, there are only two possible configurations, C_1 and C_{16}. In this case the term $(r^4 + w^4)$ generates these configurations.

Continuing with r_1, we have $r_1 = (14)(23)$, a product of two disjoint cycles of length 2; the term x_2^2 represents this cycle structure. For a configuration to be fixed by r_1^*, the vertices at 2 and 3 must be the same color; that is, we have two choices for coloring the vertices in (23). We also have two choices for coloring the vertices in (14). Consequently, we get 2^2 invariant configurations: $C_1(r^4)$, $C_7(r^2w^2)$, $C_9(r^2w^2)$, and $C_{16}(w^4)$. $[(r^2 + w^2)^2 = r^4 + 2r^2w^2 + w^4.]$

Finally, in the case of $r_3 = (13)(2)(4)$, $x_2 x_1^2$ indicates its decomposition into one cycle of length 2 and two of length 1. The vertices at 1 and 3 must be painted the same color if the configuration is to be fixed by r_3^*. With three cycles and two choices of color for each cycle, we find 2^3 invariant configurations. These are $C_1(r^4)$, $C_2(r^3w)$, $C_4(r^3w)$, $C_{10}(r^2w^2)$, $C_{11}(r^2w^2)$, $C_{13}(rw^3)$, $C_{15}(rw^3)$, and $C_{16}(w^4)$. The generating function for these configurations is $(r^2 + w^2)(r + w)^2$, for when we consider the cycle (13) we have two choices: both vertices red (r^2) or both vertices white (w^2). This gives us $(r^2 + w^2)$. For each single vertex in the two cycles of length 1, $r + w$ provides the choices for each cycle, $(r + w)^2$ the choices for the two. By the independence of choice of colors as we go from one cycle to another, $(r^2 + w^2)(r + w)^2$ generates the 2^3 configurations that are invariant under r_3^*.

Similar arguments provide the information in Table 16.9 for the permutations π_2, π_3, r_2, and r_4.

At this point we see that what determines the number of configurations that are invariant under π^*, for $\pi \in G$, depends on the cycle structure of π. Within each cycle the same color must be used, but that color can be selected from the two or more choices made available. For r_1, we had two cycles (of length 2) and 2^2 configurations. If three colors had been available, the number of invariant configurations would have been 3^2. For m colors, the number is m^2. Adding these terms for all the cycle structures that arise gives $\sum_{\pi \in G} \psi(\pi^*)$.

We now wish to place more emphasis on cycle structures, so we define the *cycle index*, P_G, for *group* G as

$$P_G(x_1, x_2, x_3, x_4) = \frac{1}{|G|} \sum_{\pi \in G} \text{(cycle structure representation of } \pi).$$

In this example,

$$P_G(x_1, x_2, x_3, x_4) = (1/8)(x_1^4 + 2x_4 + 3x_2^2 + 2x_2 x_1^2).$$

When each occurrence of x_1, x_2, x_3, x_4 is replaced by 2, we find that the number of nonequivalent 2-colorings is equal to

$$P_G(2, 2, 2, 2) = (1/8)(2^4 + 2(2) + 3(2^2) + 2(2)(2^2)) = 6.$$

We summarize our present findings in the following result.

THEOREM 16.19   Let S be a set of configurations that are acted on by a permutation group G. (G is a subgroup of S_n, the group of all permutations of $\{1, 2, 3, \ldots, n\}$, and the cycle index $P_G(x_1, x_2, x_3, \ldots, x_n)$ of G is

$$(1/|G|) \sum_{\pi \in G} \text{(cycle structure representation of } \pi).)$$

The number of nonequivalent m-colorings of S is then $P_G(m, m, m, \ldots, m)$.

We close this section with an example that uses this theorem.

Example 16.29   In how many distinct ways can we 4-color the vertices of a regular hexagon that is free to move in space?

For a regular hexagon there are twelve rigid motions: (a) the six clockwise rotations through $0°$, $60°$, $120°$, $180°$, $240°$, and $300°$; (b) the three reflections in diagonals through opposite vertices; and (c) the three reflections through lines passing through the midpoints of opposite edges.

In Fig. 16.10 we have listed each group element as a product of disjoint cycles, together with its cycle structure representation. Here

$$P_G(x_1, x_2, x_3, x_4, x_5, x_6) = (1/12)(x_1^6 + 2x_6 + 2x_3^2 + 4x_2^3 + 3x_1^2 x_2^2),$$

and there are

$$P_G(4,4,4,4,4,4) = (1/12)(4^6 + 2(4) + 2(4^2) + 4(4^3) + 3(4^2)(4^2)) = 430$$

nonequivalent 4-colorings of a regular hexagon. (*Note:* Even though neither x_4 nor x_5 occurs in a cycle structure representation, we may list these variables among the arguments of P_G.) □

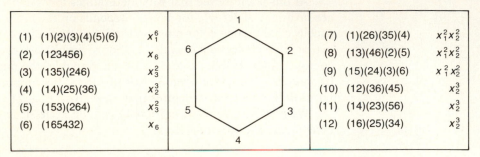

(1) (1)(2)(3)(4)(5)(6) x_1^6
(2) (123456) x_6
(3) (135)(246) x_3^2
(4) (14)(25)(36) x_2^3
(5) (153)(264) x_3^2
(6) (165432) x_6

(7) (1)(26)(35)(4) $x_1^2 x_2^2$
(8) (13)(46)(2)(5) $x_1^2 x_2^2$
(9) (15)(24)(3)(6) $x_1^2 x_2^2$
(10) (12)(36)(45) x_2^3
(11) (14)(23)(56) x_2^3
(12) (16)(25)(34) x_2^3

Figure 16.10

EXERCISES 16.10

1. In how many ways can we 5-color the vertices of a square that is free to move in (a) two dimensions? (b) three dimensions?

2. Answer Exercise 1 for a regular pentagon.

3. Find the number of nonequivalent 4-colorings of the vertices in the configurations shown in Fig. 16.11 when they are free to move in (a) two dimensions; (b) three dimensions.

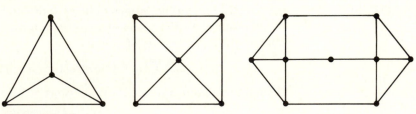

Figure 16.11

4. a) In how many ways can we 3-color the vertices of a regular hexagon that is free to move in space?

   b) Give a combinatorial argument to show that for all $m \in \mathbf{Z}^+$, $(m^6 + 2m + 2m^2 + 4m^3 + 3m^4)$ is divisible by 12.

5. a) In how many ways can we 5-color the vertices of a regular hexagon that is free to move in two dimensions?

   b) Answer part (a) if the hexagon is free to move in three dimensions.

   c) Find two 5-colorings that are equivalent for case (b) but distinct for case (a).

6. In how many distinct ways can we 3-color the *edges* in the configurations shown in Fig. 16.11 if they are free to move in (a) two dimensions; (b) three dimensions?

7. **a)** In how many distinct ways can we 3-color the *edges* of a square that is free to move in three dimensions?

   b) In how many distinct ways can we 3-color both the vertices and the edges of such a square?

   c) For a square that can move in three dimensions, let k, m, and n denote the number of distinct ways in which we can 3-color its vertices (alone), its edges (alone), and both its vertices and edges, respectively. Does $n = km$? (Give a geometric explanation.)

16.11
THE PATTERN INVENTORY: POLYA'S METHOD OF ENUMERATION

In this final section we return to Example 16.25, concentrating now on the pattern inventory and how it can be derived from the cycle index.

For $\pi_0 \in G$, every configuration in S is invariant. The cycle structure (representation) for π_0 is given by x_1^4, where for each cycle of length 1 we have a choice of coloring the vertex in that cycle red (r) or white (w). Using + to represent *or*, we write $(r + w)$ to denote the two choices for that vertex (cycle of length 1). With four such cycles, $(r + w)^4$ generates the patterns of the 16 configurations.

In the case of $\pi_1 = (1234)$, x_4 denotes the cycle structure, and here all four vertices must be the same color for the configuration to remain fixed under π_1^*. Consequently, we have all four vertices red or all four vertices white, and we express this algebraically by $r^4 + w^4$.

At this point we notice that for each of the permutations we have considered, the number of factors in the pattern inventory for that permutation equals the number of factors in its cycle structure (representation). Is this just a coincidence?

Continuing with $r_1 = (14)(23)$, we find the cycle structure x_2^2. For the cycle (14) we must color both of the vertices 1 and 4 either red or white. These choices are represented by $r^2 + w^2$. Since there are two such cycles of length 2, the generating function for the pattern inventory of configurations of S fixed by r_1^* is $(r^2 + w^2)^2$. Once again the number of factors in the cycle structure equals the number of factors in the corresponding term of the generating function.

Last, for $r_3 = (13)(2)(4)$, the cycle structure is $x_1^2 x_2$. For each of the cycles (2) and (4), $r + w$ represents the choices for each of these vertices, whereas $(r + w)^2$ accounts for all four colorings of the pair. The cycle (13) indicates that vertices 1 and 3 must have the same color; $r^2 + w^2$ accounts for the two possibilities. Hence the summand in the generating function for these configurations is $(r + w)^2 \cdot (r^2 + w^2)$, and we find three factors in both the cycle structure and the corresponding summand of the generating function. But even more comes to light here.

Looking at the terms in the cycle structures, we see that, for $1 \leq i \leq n$, the factor x_i in the cycle structure corresponds with the term $r^i + w^i$ in the associated summand of the generating function.

Continuing with the cycle structures for π_2, π_3, r_2, and r_4, we find that the pattern inventory can be obtained by replacing each x_i in $P_G(x_1, x_2, x_3, x_4)$ with $r^i + w^i$, for $1 \leq i \leq 4$. Consequently,

$$P_G(r + w, r^2 + w^2, r^3 + w^3, r^4 + w^4) = r^4 + r^3 w + 2r^2 w^2 + rw^3 + w^4.$$

(This result is $(1/8)$-th of the complete inventory listed in Table 16.9.)

If we had three colors (red, white, and blue), the replacement for x_i would be $r^i + w^i + b^i$, where $1 \leq i \leq 4$.

We generalize these observations in the following theorem.

THEOREM 16.20   (*Polya's Method of Enumeration*) Let S be a set of configurations that are acted on by a permutation group G, where G is a subgroup of S_n and has cycle index $P_G(x_1, x_2, \ldots, x_n)$. Then the generating function for the pattern inventory of nonequivalent m-colorings of S is given by

$$P_G\left(\sum_{i=1}^{m} c_i, \sum_{i=1}^{m} c_i^2, \ldots, \sum_{i=1}^{m} c_i^n\right),$$

where $c_1, c_2, \ldots, c_m$ denote the m colors that are available.

We apply this theorem in the following examples.

Example 16.30   A child's bracelet is formed by placing three beads—red, white, and blue—on a circular piece of wire. Bracelets are considered equivalent if one can be obtained from the other by a (planar) rotation. Find the pattern inventory for these bracelets.

Here G is the group of rotations of an equilateral triangle, so $G = \{(1)(2)(3), (123), (132)\}$, where 1, 2, 3 denote the vertices of the triangle. Then $P_G(x_1, x_2, x_3) = (1/3)(x_1^3 + 2x_3)$, and the pattern inventory is given by $(1/3)[(r + w + b)^3 + 2(r^3 + w^3 + b^3)] = (1/3)[3r^3 + 3r^2 w + 3r^2 b + 3rw^2 + 6rwb + 3rb^2 + 3w^3 + 3w^2 b + 3wb^2 + 3b^3] = r^3 + r^2 w + r^2 b + rw^2 + 2rwb + rb^2 + w^3 + w^2 b + wb^2 + b^3$.

If the bracelets can also be reflected, then G becomes $\{(1)(2)(3), (123), (132), (1)(23), (2)(13), (3)(12)\}$, and the pattern inventory here is the same as the one above, with one exception. Here we have rwb, instead of $2rwb$, because the nonequivalent (for rotations) patterns in Fig. 16.12 become equivalent when reflections are allowed. $\square$

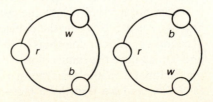

Figure 16.12

Example 16.31   Consider the 3-colorings of the configurations in Example 16.25. If the three colors are red, white, and blue, how many nonequivalent configurations have exactly two red vertices?

Given that $P_G(x_1, x_2, x_3, x_4) = (1/8)(x_1^4 + 2x_4 + 3x_2^2 + 2x_2 x_1^2)$, the answer is the sum of the coefficients of $r^2 w^2$, $r^2 b^2$, and $r^2 wb$ in $(1/8)[(r + w + b)^4 + 2(r^4 + w^4 + b^4) + 3(r^2 + w^2 + b^2)^2 + 2(r^2 + w^2 + b^2)(r + w + b)^2]$.

In $(r + w + b)^4$, we find the term $6r^2 w^2 + 6r^2 b^2 + 12r^2 wb$. For $3(r^2 + w^2 + b^2)^2$, we are interested in the term $6r^2 w^2 + 6r^2 b^2$, whereas $4r^2 w^2 + 4r^2 b^2 + 4r^2 bw$ arises in $2(r^2 + w^2 + b^2)(r + w + b)^2$.

Then $(1/8)[6r^2 w^2 + 6r^2 b^2 + 12r^2 wb + 6r^2 w^2 + 6r^2 b^2 + 4r^2 w^2 + 4r^2 b^2 + 4r^2 bw] = 2r^2 w^2 + 2r^2 b^2 + 2r^2 bw$, the inventory of the six nonequivalent configurations that contain exactly two red vertices.   □

Our next example deals with the pattern inventory for the 2-colorings of the vertices of a cube. (The colors are red and white.)

Example 16.32   For the cube in Fig. 16.13, we find that its group G of rigid motions consists of the following.

1. The identity transformation with cycle structure x_1^8.

2. Rotations through 90°, 180°, and 270° about an axis through the centers of two opposite faces: From Fig. 16.13(a) we have

90° rotation:	(1234)(5678)	Cycle structure: x_4^2
180° rotation:	(13)(24)(57)(68)	Cycle structure: x_2^4
270° rotation:	(1432)(5876)	Cycle structure: x_4^2

   Since there are two other pairs of opposite faces, these nine rotations account for the term $3x_2^4 + 6x_4^2$ in the cycle index.

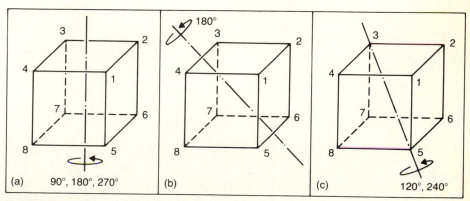

Figure 16.13

3. Rotations through 180° about an axis through the midpoints of two opposite edges: As in Fig. 16.13(b), we have the permutation (17)(28)(34)(56), whose

cycle structure is given by x_2^4. With six pairs of opposite edges, these rotations contribute the term $6x_2^4$ to the cycle index.

4. Rotations through 120° and 240° about an axis through two diagonally opposite vertices: From part (c) of the figure we have

120° rotation: (168)(274)(3)(5)   Cycle structure: $x_1^2 x_3^2$
240° rotation: (186)(247)(3)(5)   Cycle structure: $x_1^2 x_3^2$

Here there are four such pairs of vertices, and these result in the term $8x_1^2 x_3^2$ in the cycle index.

Therefore, $P_G(x_1, x_2, \ldots, x_8) = (1/24)(x_1^8 + 9x_2^4 + 6x_4^2 + 8x_1^2 x_3^2)$, and the pattern inventory for these configurations is

$$(1/24)[(r + w)^8 + 9(r^2 + w^2)^4 + 6(r^4 + w^4)^2 + 8(r + w)^2(r^3 + w^3)^2]$$
$$= r^8 + r^7 w + 3r^6 w^2 + 3r^5 w^3 + 7r^4 w^4 + 3r^3 w^5 + 3r^2 w^6 + rw^7 + w^8.$$

Replacing r and w by 1, we find 23 nonequivalent configurations here.   □

Since Polya's Method of Enumeration was first developed in order to count isomers of organic compounds, we close this section with an application that deals with a certain class of organic compounds. This is based on an example by C. L. Liu. (See pp. 152–154 of reference [7].)

Example 16.33   Here we are concerned with organic molecules of the form shown in Fig. 16.14, where C is a carbon atom and X denotes any of the following components: Br (bromine), H (hydrogen), CH_3 (methyl), or C_2H_5 (ethyl). For example, if each X is replaced by H the compound CH_4 (methane) results. Figure 16.14 should not be allowed to mislead us. The structure of these organic compounds is three-dimensional. Consequently we turn to the regular tetrahedron in order to model this structure. We place the carbon atom at the center of the tetrahedron and then place our selections for X at vertices 1, 2, 3, and 4 as shown in Fig. 16.15.

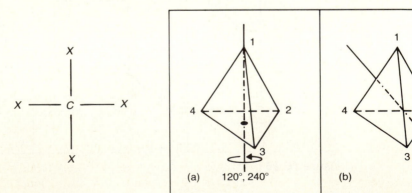

Figure 16.14   **Figure 16.15**

The group G acting on these configurations is given as follows:

1. The identity transformation $(1)(2)(3)(4)$ with cycle structure x_1^4.

2. Rotations through $120°$ or $240°$ about an axis through a vertex and the center of the opposite face: As Fig. 16.15(a) shows, we have

   $120°$ rotation: $(1)(243)$ with cycle structure $x_1 x_3$
   $240°$ rotation: $(1)(234)$ with cycle structure $x_1 x_3$

   By symmetry there are three other pairs of vertices and opposite faces, so these rigid motions account for the term $8x_1 x_3$ in $P_G(x_1, x_2, x_3, x_4)$.

3. Rotations of $180°$ about an axis through the midpoints of two opposite edges: The case shown in part (b) of the figure is the permutation $(14)(23)$ whose cycle structure is x_2^2. With three pairs of opposite edges, we get the term $3x_2^2$ in $P_G(x_1, x_2, x_3, x_4)$.

   Hence $P_G(x_1, x_2, x_3, x_4) = (1/12)[x_1^4 + 8x_1 x_3 + 3x_2^2]$ and $P_G(4, 4, 4, 4) = (1/12) \cdot [4^4 + 8(4^2) + 3(4^2)] = 36$, so there are 36 distinct organic compounds that can be formed in this way.

   Last, if we wish to know how many of these compounds have exactly two bromine atoms, we let w, x, y, and z represent the "colors" Br, H, CH_3, and C_2H_5, respectively, and find the sum of the coefficients of $w^2 x^2$, $w^2 y^2$, $w^2 z^2$, $w^2 xy$, $w^2 xz$, and $w^2 yz$ in the pattern inventory

   $$(1/12)[(w + x + y + z)^4 + 8(w + x + y + z)(w^3 + x^3 + y^3 + z^3) + 3(w^2 + x^2 + y^2 + z^2)^2].$$

   For $(w + x + y + z)^4$ the term is $6w^2 x^2 + 6w^2 y^2 + 6w^2 z^2 + 12w^2 xy + 12w^2 xz + 12w^2 yz$. The middle summand of the pattern inventory does not give rise to any of these situations, whereas in $3(w^2 + x^2 + y^2 + z^2)^2$ we find $6w^2 x^2 + 6w^2 y^2 + 6w^2 z^2$.

   Consequently the pattern inventory for the compounds containing exactly two bromine atoms is

   $$(1/12)[12w^2 x^2 + 12w^2 y^2 + 12w^2 z^2 + 12w^2 xy + 12w^2 xz + 12w^2 yz]$$

and there are six such organic compounds. $\square$

EXERCISES 16.11

1. **a)** Find the pattern inventory for the 2-colorings of the edges of a square that is free to move in (i) two dimensions; (ii) three dimensions. (Let the colors be red and white.)

   b) Answer part (a) for 3-colorings, where the colors are red, white, and blue.

2. If a regular pentagon is free to move in space and we can color its vertices with red, white, and blue paint, how many nonequivalent configurations have exactly three red vertices? How many have two red, one white, and two blue vertices?

3. Suppose that in Example 16.32 we 2-color the faces of the cube, which is free to move in space.

   a) How many distinct 2-colorings are there for this situation?

   b) If the available colors are red and white, determine the pattern inventory.

   c) How many nonequivalent colorings have three red and three white faces?

4. For the organic compounds in Example 16.33, how many have at least one bromine atom? How many have exactly three hydrogen atoms?

5. Find the pattern inventories for the 2-colorings of the vertices in the configurations in Fig. 16.11, when they are free to move in space. (Let the colors be green and gold.)

6. a) In how many ways can the seven (identical) horses on a carousel be painted with black, brown, and white paint in such a way that there are three black, two brown, and two white horses?

   b) In how many ways would there be equal numbers of black and brown horses?

   c) Give a combinatorial argument to verify that for all $n \in \mathbf{Z}^+$, $n^7 + 6n$ is divisible by 7.

7. a) In how many ways can we paint the eight squares of a 2×4 chessboard, using the colors red and white? (The back of the chessboard is black cardboard.)

   b) Find the pattern inventory for the colorings in part (a).

   c) How many of the colorings in part (a) have four red and four white squares? How many have six red and two white squares?

8. a) In how many ways can we 2-color the eight regions of the pinwheel shown in Fig. 16.16, using the colors black and gold, if the back of each region remains grey?

   b) Answer part (a) for the possible 3-colorings, using black, gold, and blue paints to color the regions.

   c) For the colorings in part (b), how many have four black, two gold, and two blue regions?

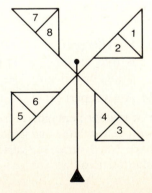

Figure 16.16

9. Let $m, n \in \mathbf{Z}^+$ with $n \geq 3$. How many distinct summands appear in the pattern inventory for the m-colorings of the vertices of a regular polygon of n sides?

16.12
SUMMARY AND HISTORICAL REVIEW

Although the notion of a group of transformations evolved slowly in the study of geometry, the major thrust in the development of the group concept came from the study of polynomial equations.

Methods for solving quadratic equations were known to the ancient Greeks. Then in the sixteenth century, advances were made toward solving cubic and quartic equations over $\mathbf{Q}$. Continuing with polynomials of fifth and higher degree, both Leonhard Euler (1707–1783) and Joseph-Louis Lagrange (1736–1813) attempted to solve the general quintic. Lagrange realized there had to be a connection between the degree n of a polynomial equation and the permutation group S_n. However, it was Niels Henrik Abel (1802–1829) who finally proved that it was not possible to find a formula for solving the general quintic using only addition, subtraction, multiplication, division, and root extraction. During this same period, the existence of a necessary and sufficient condition for when a polynomial of degree $n \geq 5$ with rational coefficients can be solved by radicals was investigated and solved by the illustrious French mathematician Evariste Galois (1811–1832). Since the work of Galois utilizes the structures of both groups and fields, we shall say more about him in the summary of Chapter 17.

Following the accomplishments of Galois, group theory established itself further in many areas of mathematics. During the late nineteenth century, for example, the German mathematician Felix Klein (1849–1929), in what has come to be known as the *Erlanger Programm*, attempted to codify all existing geometries according to the group of transformations under which the properties of the geometry were invariant.

Many other mathematicians, such as Augustin-Louis Cauchy (1789–1857), Arthur Cayley (1821–1895), Ludwig Sylow (1832–1918), Richard Dedekind (1831–1916), and Leopold Kronecker (1823–1891) contributed to the further development of certain types of groups. However, it was not until 1900 that lists of defining conditions were given for the general abstract group.

During the twentieth century a great deal of research has been carried on in the attempt to analyze the structure of finite groups. For finite abelian groups, it is known that any such group is isomorphic to a direct product of cyclic groups of prime power order. In the case of the finite nonabelian group, a great deal more was needed. Starting with the work of Galois, one finds particular attention paid to a special type of subgroup called a normal subgroup. For any group G, a subgroup H (of G) is called *normal* if, for all $g \in G$ and $h \in H$, we have $ghg^{-1} \in H$. In an abelian group every subgroup is normal, but this is not the case for nonabelian groups. In every group G, both $\{e\}$ and G are normal subgroups, but if *G* has no other normal subgroups it is called *simple*. During the past five decades mathematicians have sought to find all the finite simple groups and explain their role in the structure of all finite groups. Among the prime movers in this development are Professors Walter Feit, John Thompson, Daniel Gorenstein, Michael Aschbacher, and Robert Griess, Jr.

There are many texts one can turn to for further study in the theory of groups. At the introductory level, the text by V. Larney [6] provides further coverage beyond the introduction given in this chapter. The text by I. Herstein [5] is an excellent source and includes material on Galois theory.

The beginnings of algebraic coding theory can be traced to 1941, when Claude Elwood Shannon began his investigations of problems in communications. These problems were prompted by the needs of the war effort. His research resulted in many new ideas and principles that were later published in 1948 in the journal article [13]. As a result of this work, Shannon is acknowledged as the founder of information theory. After this publication, results by M. Golay [3] and R. Hamming [4] soon followed, giving further impetus to research in this area. The 1478 references listed in the bibliography at the end of Volume II of the texts by F. MacWilliams and N. Sloane [8] should convey some idea of the activity in this area after 1950.

Our coverage of coding theory followed the development in Chapter 5 of the text by L. Dornhoff and F. Hohn [2]. The text by V. Pless [9] provides a nice coverage of topics at a fairly intermediate level. More advanced work in coding can be found in the books by F. MacWilliams and N. Sloane [8] and by A. Street and W. Wallis [14]. An interesting application on the use of the pigeonhole principle in coding theory is given in Chapter XI of [14].

In Sections 9, 10, and 11 of the chapter, we came upon an enumeration technique whose development is attributed to the Hungarian mathematician George Polya (1887–1985). His article [10] provided the fundamental techniques for counting equivalence classes of chemical isomers, graphs, and trees. (To some extent, the ideas in this work were anticipated by J. Redfield [12].) Since then these techniques have been found invaluable for counting problems in such areas as the electronic realizations of Boolean functions. Polya's fundamental theorem was first generalized in the article by N. DeBruijn [1], and other extensions of these ideas can be found in the literature. The article by R. C. Read [11] relates the profound influence that Polya's Theorem has had on developments in combinatorial analysis. (The issue of the journal that contains this article also includes several other articles dealing with the life and work of George Polya.)

Our coverage of this topic follows the presentation given in the article by A. Tucker [15]. A more rigorous presentation of this method can be found in Chapter 5 of the text by C. L. Liu [7].

REFERENCES

1. DeBruijn, Nicolaas Govert. "Polya's Theory of Counting," Chapter 5 in *Applied Combinatorial Mathematics,* ed. by Edwin F. Beckenbach. New York: Wiley, 1964.
2. Dornhoff, Larry L., and Hohn, Franz E. *Applied Modern Algebra*. New York: Macmillan, 1978.

3. Golay, Marcel J. E. "Notes on Digital Coding." *Proceedings of the IRE 37,* 1949, p. 657.

4. Hamming, Richard W. "Error Detecting and Error Correcting Codes." *Bell System Technical Journal 29,* 1950, pp. 147–160.

5. Herstein, I. N. *Topics in Algebra,* 2nd ed. Lexington, Mass.: Xerox College Publishing, 1975.

6. Larney, Violet H. *Abstract Algebra: A First Course.* Boston: Prindle, Weber & Schmidt, 1975.

7. Liu, C. L. *Introduction to Combinatorial Mathematics.* New York: McGraw-Hill, 1968.

8. MacWilliams, F. Jessie, and Sloane, Neil J. A. *The Theory of Error-Correcting Codes,* Volumes I and II. Amsterdam: North-Holland, 1977.

9. Pless, Vera. *Introduction to the Theory of Error-Correcting Codes.* New York: Wiley, 1982.

10. Polya, George. "Kombinatorische Anzahlbestimmungen für Gruppen, Graphen und Chemishe Verbindungen." *Acta Mathematica 68,* 1937, pp. 145–254.

11. Read, R. C. "Polya's Theorem and Its Progeny." *Mathematics Magazine* 60, No. 5 (December 1987): pp. 275–282.

12. Redfield, J. Howard. "The Theory of Group Reduced Distributions." *American Journal of Mathematics 49,* 1927, pp. 433–455.

13. Shannon, Claude E. "The Mathematical Theory of Communication." *Bell System Technical Journal 27,* 1948, pp. 379–423, 623–656. Reprinted in C. E. Shannon and W. Weaver, *The Mathematical Theory of Communication,* (Urbana: University of Illinois Press, 1949).

14. Street, Anne Penfold, and Wallis, W. D. *Combinatorial Theory: An Introduction.* Winnipeg, Canada: The Charles Babbage Research Center, 1977.

15. Tucker, Alan. "Polya's Enumeration Formula by Example." *Mathematics Magazine 47,* 1974, pp. 248–256.

MISCELLANEOUS EXERCISES

1. Let $f: G \to H$ be a group homomorphism with e_H the identity in H. Prove that

   a) $K = \{x \in G \,|\, f(x) = e_H\}$ is a subgroup of G.

   b) If $g \in G$ and $x \in K$, then $gxg^{-1} \in K$.

2. If G, H, and K are groups and $G = H \times K$, prove that G contains subgroups that are isomorphic to H and K.

3. Let G be a group where $a^2 = e$ for all $a \in G$. Prove that G is abelian.

4. If G is a group of even order, prove that there is an element $a \in G$ with $a \neq e$ and $a = a^{-1}$.

5. Let $f: G \to H$ be a group homomorphism. If T is a subgroup of H, prove that $f^{-1}(T) = \{a \in G \,|\, f(a) \in T\}$ is a subgroup of G.

6. Considering the possible orders of elements, prove that the groups $\mathbf{Z}_9$ and $\mathbf{Z}_3 \times \mathbf{Z}_3$ are not isomorphic.

7. Let $f: G \to H$ be a group homomorphism *onto* H. If G is a cyclic group, prove that H is also cyclic.

8. Let $(G, \circ)$ be a group, X a set, and $f: G \to X$ a one-to-one function onto X. Define the

binary operation $*$ on X as follows: For $a, b \in X$, $a * b = f(f^{-1}(a) \circ f^{-1}(b))$. Prove that $(X, *)$ is a group.

9. Let G be a group. Define relation $\mathcal{R}$ on G by $x \,\mathcal{R}\, y$, if $x = gyg^{-1}$ for some $g \in G$. (Under these circumstances, x is called a *conjugate* of y.)

   a) Verify that $\mathcal{R}$ is an equivalence relation on G.

   b) Find the partition induced by this equivalence relation on the following groups: (i) S_3 (See Table 16.4); (ii) The group of rigid motions of a square (See Fig. 16.6); (iii) The Klein Four group of Example 16.13; (iv) Any finite abelian group.

   c) For any finite group G, prove that if $x, y \in G$ and x is a conjugate of y, then x and y have the same order. Is the converse of this result also true?

10. For $k, n \in \mathbf{Z}^+$ with $n \geq k \geq 1$, let $Q(n, k)$ count the number of permutations $\pi \in S_n$ where any representation of π, as a product of disjoint cycles, contains no cycle of length greater than k. Verify that

$$Q(n + 1, k) = \sum_{i=0}^{k-1} \binom{n}{i} (i!) Q(n - i, k).$$

11. For $k, n \in \mathbf{Z}^+$ where $n \geq 2$ and $1 \leq k \leq n$, let $P(n, k)$ denote the number of permutations $\pi \in S_n$ that have k cycles. (For example, $(1)(23) \in P(3, 2)$, $(12)(34) \in P(4, 2)$, and $(1)(23)(4) \in P(4, 3)$.)

   a) Verify that $P(n + 1, k) = P(n, k - 1) + nP(n, k)$.

   b) Determine $\sum_{k=1}^{n} P(n, k)$.

12. For $n \geq 1$, if $\sigma, \tau \in S_n$, define the distance $d(\sigma, \tau)$ between σ and τ by

$$d(\sigma, \tau) = \max\{|\sigma(i) - \tau(i)| \mid 1 \leq i \leq n\}.$$

   a) Prove that the following properties hold for d.

     (i) $d(\sigma, \tau) \geq 0$ for all $\sigma, \tau \in S_n$

     (ii) $d(\sigma, \tau) = 0$ if and only if $\sigma = \tau$

     (iii) $d(\sigma, \tau) = d(\tau, \sigma)$ for all $\sigma, \tau \in S_n$

     (iv) $d(\rho, \tau) \leq d(\rho, \sigma) + d(\sigma, \tau)$, for all $\rho, \sigma, \tau \in S_n$

   b) Let ϵ denote the identity element of S_n (that is, $\epsilon(i) = i$ for all $1 \leq i \leq n$). If $\pi \in S_n$ and $d(\pi, \epsilon) \leq 1$, what can we say about $\pi(n)$?

   c) For $n \geq 1$ let a_n be the number of permutations π in S_n, where $d(\pi, \epsilon) \leq 1$. Find and solve a recurrence relation for a_n.

13. a) Regard $\mathbf{Z}_2^n$ as a group code. Given that $a, b, c \in \mathbf{Z}_2^n$, prove that $d(a, b) = d(a + c, b + c)$.

   b) For the group code $C \subseteq \mathbf{Z}_2^n$, let a, b be distinct code words. With $x \in C$, define $d(x, C) = \min\{d(x, c) \mid c \in C, c \neq x\}$. Prove that $d(a, C) = d(b, C)$.

14. Let $C \subseteq \mathbf{Z}_2^n$ be a k–error-correcting group code, where $4k + 2 > n$.

   a) Show that if we list the code words (in a column), any column in this list contains all 0's or half 0's and half 1's.

   b) Verify that the total number of 1's in the list is at least $(2k + 1)[|C| - 1]$.

   c) Prove that $\dfrac{2(2k + 1)}{4k + 2 - n} \geq |C|$.

15. a) Define the relation $\mathcal{R}$ on $\mathbf{Z}_2^7$ by $a_1 a_2 \ldots a_7 \mathcal{R} b_1 b_2 \ldots b_7$ if $b_1 b_2 \ldots b_7$ is a cyclic shift of $a_1 a_2 \ldots a_7$. (For example, $a_1 a_2 \ldots a_7 = 0110011$ yields $1100110 = a_2 a_3 \ldots a_7 a_1$ after one cyclic shift, whereas 1001101 results from $a_1 a_2 \ldots a_7$ after two cyclic shifts. Each of these examples is referred to as *a cyclic shift*.) Prove that $\mathcal{R}$ is an equivalence relation on $\mathbf{Z}_2^7$.

b) How many equivalence classes result in the partition of $\mathbf{Z}_2^7$ induced by $\mathcal{R}$?

16. Let $C \subseteq \mathbf{Z}_2^7$, where C contains the elements 0000000, 0110100, 1001011, and 1111111 and all their resulting cyclic shifts.

a) Prove that C is a group code.

b) What are the error-correcting and error-detecting capabilities of this code?

17. A projectile undergoes testing by being fired at a target. Assume that the probability of a hit is 0.75 for any single test and that the results of successive firings are independent.

a) If four projectiles are fired, what is the probability of (i) exactly two hits? (ii) at least two hits?

b) How many projectiles must we fire in order for the probability of at least one hit to be at least 0.95?

18. a) In how many ways can Nicole paint the eight regions of the square shown in Fig. 16.17, if five colors are available?

b) In how many ways if she actually uses four of the colors?

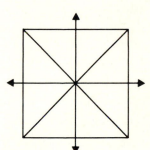

Figure 16.17

17
Finite Fields and Combinatorial Designs

It is time now to recall the ring structure of Chapter 14 as we examine rings of polynomials and their role in the construction of finite fields. We know that for any prime p, $(\mathbf{Z}_p, +, \cdot)$ is a finite field, but here we shall find other finite fields. Just as the order of a finite Boolean algebra is restricted to powers of 2, for finite fields the possible orders are p^n, where p is a prime and $n \in \mathbf{Z}^+$. Applications of these finite fields will include a discussion of such combinatorial designs as Latin squares. Finally, we shall investigate the structure of a finite geometry and discover how these geometries and combinatorial designs are interrelated.

17.1
POLYNOMIAL RINGS

We recall that a ring $(R, +, \cdot)$ consists of a nonempty set R, where $(R, +)$ is an abelian group, $(R, \cdot)$ is closed under the associative operation $\cdot$, and the two operations are related by the distributive laws: $a(b + c) = ab + ac$ and $(b + c)a = ba + ca$, for $a, b, c \in R$. (We write ab for $a \cdot b$.)

Let x denote an indeterminate—that is, a formal symbol that is not an element of the ring R. We use x to define the following.

DEFINITION 17.1 Given a ring $(R, +, \cdot)$, an expression of the form $f(x) = a_n x^n + a_{n-1} x^{n-1} + \cdots + a_1 x^1 + a_0 x^0$, where $a_i \in R$ for $0 \le i \le n$, is called a *polynomial in the indeterminate x over R*.

If a_n is not the zero element of R, then a_n is called *the leading coefficient of* $f(x)$ and we say that $f(x)$ has *degree n*. Hence the degree of a polynomial is the highest power of x that occurs in a summand of the polynomial. The term $a_0 x^0$ is called the *constant*, or *constant term*, of $f(x)$.

If $g(x) = b_m x^m + b_{m-1} x^{m-1} + \cdots + b_1 x^1 + b_0 x^0$ is also a polynomial in x over R, then $f(x) = g(x)$ if $m = n$ and $a_i = b_i$ for all $0 \le i \le n$.

Example 17.1

a) Over the ring $R = (\mathbf{Z}_6, +, \cdot)$, $5x^2 + 3x^1 - 2x^0$ is a polynomial of degree 2, with leading coefficient 5 and constant term $-2x^0$. Here we are using a to denote $[a]$. This polynomial can also be expressed as $5x^2 + 3x^1 + 4x^0$ since $[4] = [-2]$ in $\mathbf{Z}_6$.

b) If z is the zero element of ring R, then $zx^0 = z$ is said to have *no degree* and no leading coefficient. A polynomial over R that is the zero element or of degree 0 is called a *constant polynomial*. For example, the polynomial $5x^0$ over $\mathbf{Z}_7$ has degree 0 and is a constant polynomial. □

We introduce operations of addition and multiplication for these polynomials in order to obtain a new ring. For a ring $(R, +, \cdot)$, let

$$f(x) = a_n x^n + a_{n-1} x^{n-1} + \cdots + a_1 x^1 + a_0 x^0$$
$$g(x) = b_m x^m + b_{m-1} x^{m-1} + \cdots + b_1 x^1 + b_0 x^0,$$

where $a_i, b_j \in R$ for $0 \le i \le n, 0 \le j \le m$.
Assume that $n \ge m$. We define

$$f(x) + g(x) = \sum_{i=0}^{n} (a_i + b_i)x^i, \tag{1}$$

where $b_i = z$ for $i > m$, and

$$f(x)g(x) = (a_n b_m)x^{n+m} + (a_n b_{m-1} + a_{n-1} b_m)x^{n+m-1} + \cdots$$
$$+ (a_1 b_0 + a_0 b_1)x^1 + (a_0 b_0)x^0. \tag{2}$$

In the definition of $f(x) + g(x)$, each coefficient $(a_i + b_i)$, for $0 \le i \le n$, is obtained from the addition in R. For $f(x)g(x)$, the coefficient of x^t is $\sum_{k=0}^{t} a_{t-k} b_k$, where all additions and multiplications occur in R, and $0 \le t \le n + m$.

These operations are needed in the following result.

THEOREM 17.1

Given a ring R, let $R[x]$ denote the set of all polynomials in the indeterminate x with coefficients in R. Under the operations of addition and multiplication given above in Eqs. (1) and (2), $(R[x], +, \cdot)$ is a ring, called the *polynomial ring*, or *ring of polynomials*, over R.

Proof

The ring properties for $R[x]$ follow from those of R. Consequently, we shall prove the associative law of multiplication here and shall leave the proofs of the other properties to the reader. Let $h(x) = \sum_{k=0}^{p} c_k x^k$, with $f(x), g(x)$ as defined earlier. A typical summand in $(f(x)g(x))h(x)$ has the form Ax^t, where $0 \le t \le (m + n) + p$ and A is the sum of all products of the form $(a_i b_j)c_k$, with $0 \le i \le n$, $0 \le j \le m, 0 \le k \le p$, and $i + j + k = t$. In $f(x)(g(x)h(x))$ the coefficient of x^t is the sum of all products $a_i(b_j c_k)$, again with $0 \le i \le n$, $0 \le j \le m$, $0 \le k \le p$, and $i + j + k = t$. Since R is associative under multiplication, $(a_i b_j)c_k = a_i(b_j c_k)$ for each of these terms, and the coefficient of x^t in $(f(x)g(x))h(x)$ is the same as it is in $f(x)(g(x)h(x))$. Hence $(f(x)g(x))h(x) = f(x)(g(x)h(x))$. ■

COROLLARY 17.1   Let $R[x]$ be a polynomial ring.

        a) If R is commutative, then $R[x]$ is commutative.

        b) If R is a ring with unity, then $R[x]$ is a ring with unity.

        c) $R[x]$ is an integral domain if and only if R is an integral domain.

Proof   The proof of this corollary is left for the reader.   ■

From this point on, we shall write x instead of x^1. If R has unity u, we define $x^0 = u$, and for any $r \in R$ we write rx^0 as r.

Example 17.2   Let $f(x), g(x) \in \mathbf{Z}_8[x]$ with $f(x) = 4x^2 + 1$ and $g(x) = 2x + 3$. Then $f(x)$ has degree 2 and $g(x)$ has degree 1. From our past experiences with polynomials, we expect the degree of $f(x)g(x)$ to be 3, the sum of the degrees of $f(x)$ and $g(x)$. Here, however, $f(x)g(x) = (4x^2 + 1)(2x + 3) = 8x^3 + 12x^2 + 2x + 3 = 4x^2 + 2x + 3$, because $[8] = [0]$ in $\mathbf{Z}_8$. So the degree of $f(x)g(x) = 2 < 3 = $ degree $f(x) + $ degree $g(x)$.   □

What causes the result in Example 17.2 is the existence of zero divisors in the ring $\mathbf{Z}_8$. This observation leads us to the following theorem.

THEOREM 17.2   Let $(R, +, \cdot)$ be a commutative ring with unity u. Then R is an integral domain if and only if for all $f(x), g(x) \in R[x]$,

$$\text{degree } f(x)g(x) = \text{degree } f(x) + \text{degree } g(x).$$

Proof   Let $f(x) = \sum_{i=0}^{n} a_i x^i$, $g(x) = \sum_{j=0}^{m} b_j x^j$, with $a_n \neq z, b_m \neq z$. If R is an integral domain, then $a_n b_m \neq z$, so the degree of $f(x)g(x) = n + m = $ degree $f(x) + $ degree $g(x)$. Conversely, if R is not an integral domain, let $a, b \in R$ with $a \neq z, b \neq z$, but $ab = z$. The polynomials $f(x) = ax + u, g(x) = bx + u$ each have degree 1, but $f(x)g(x) = (a + b)x + u$ and degree $f(x)g(x) \leq 1 < 2 = $ degree $f(x) + $ degree $g(x)$.   ■

Let R be a ring with $f(x) = a_n x^n + \cdots + a_1 x + a_0 \in R[x]$. If $r \in R$, then $f(r) = a_n r^n + \cdots + a_1 r + a_0 \in R$. We are especially interested in those values of r for which $f(r) = z$.

DEFINITION 17.2   Let $f(x) \in R[x]$ and $r \in R$. If $f(r) = z$, then r is called a *root* of the polynomial $f(x)$.   ▬

Example 17.3   a) If $f(x) = x^2 - 2 \in \mathbf{R}[x]$, then $f(x)$ has $\sqrt{2}$ and $-\sqrt{2}$ as roots because $(\sqrt{2})^2 - 2 = 0 = (-\sqrt{2})^2 - 2$. In addition, we can write $f(x) = (x - \sqrt{2})(x + \sqrt{2})$, with $x - \sqrt{2}, x + \sqrt{2} \in \mathbf{R}[x]$. However, if we regard $f(x)$ as an element of $\mathbf{Q}[x]$, then $f(x)$ has no roots because $\sqrt{2}$ and $-\sqrt{2}$ are irrational numbers. Consequently, the existence of roots for a polynomial is dependent on the underlying ring of coefficients.

b) For $f(x) = x^2 + 3x + 2 \in \mathbf{Z}_6[x]$, we find that

$$f(0) = (0)^2 + 3(0) + 2 = 2$$
$$f(1) = (1)^2 + 3(1) + 2 = 6 = 0$$
$$f(2) = (2)^2 + 3(2) + 2 = 12 = 0$$
$$f(3) = (3)^2 + 3(3) + 2 = 20 = 2$$
$$f(4) = (4)^2 + 3(4) + 2 = 30 = 0$$
$$f(5) = (5)^2 + 3(5) + 2 = 42 = 0$$

Consequently, $f(x)$ has four roots: 1, 2, 4, and 5. This is more than we expected. In our prior experiences, a polynomial of degree 2 had at most two roots. $\square$

In this chapter we shall be concerned primarily with polynomial rings $F[x]$, where F is a field (and $F[x]$ is an integral domain). Consequently, we shall not dwell any further on situations where degree $f(x)g(x) <$ degree $f(x) +$ degree $g(x)$. In addition, unless it is stated otherwise, we shall denote the zero element of a field by 0 and use 1 to denote its unity.

As a result of Example 17.3(b), we shall now develop the concepts needed to find out when a polynomial of degree n has at most n roots.

DEFINITION 17.3   Let F be a field. For $f(x), g(x) \in F[x]$, $f(x)$ is called a *divisor* (or *factor*) of $g(x)$ if there exists $h(x) \in F[x]$ with $f(x)h(x) = g(x)$. In this situation we also say that $f(x)$ *divides* $g(x)$ and that $g(x)$ is a *multiple* of $f(x)$.   ▬

This leads to the *Division Algorithm* for polynomials. Before proving the general result, however, we shall examine two particular examples.

Example 17.4   Early in algebra we were taught how to perform the long division of polynomials. Given two polynomials $f(x), g(x)$ with degree $f(x) \leq$ degree $g(x)$, we organized our work in the form

$$
\begin{array}{r}
q(x) \\
f(x)\overline{)g(x)} \\
\underline{f(x)q(x)} \\
g(x) - f(x)q(x) \\
\cdots\cdots\cdots \\
\overline{ r(x)}
\end{array}
$$

where we continued to divide until we found either

$$r(x) = 0 \qquad \text{or} \qquad \text{degree } r(x) < \text{degree } f(x).$$

It then followed that $g(x) = q(x)f(x) + r(x)$.

For example, if $f(x) = x - 3$ and $g(x) = 7x^3 - 2x^2 + 5x - 2$, then $f(x), g(x) \in \mathbf{Q}[x]$ (or $\mathbf{R}[x]$, or $\mathbf{C}[x]$), and we find

$$
\begin{array}{r}
7x^2 + 19x + 62 \ (= q(x)) \\
x - 3 \overline{)\, 7x^3 - 2x^2 + 5x - 2\ } \\
7x^3 - 21x^2 \\
\hline
19x^2 + 5x - 2 \\
19x^2 - 57x \\
\hline
62x - 2 \\
62x - 186 \\
\hline
184 \ (= r(x))
\end{array}
$$

Checking these results, we have

$$q(x)f(x) + r(x) = (7x^2 + 19x + 62)(x - 3) + 184$$
$$= 7x^3 - 2x^2 + 5x - 2 = g(x). \quad \square$$

Example 17.5 The technique illustrated in Example 17.4 also applies when the coefficients of our polynomials are taken from a *finite field*.

If $f(x) = 3x^2 + 4x + 2$ and $g(x) = 6x^4 + 4x^3 + 5x^2 + 3x + 1$ are polynomials in $\mathbf{Z}_7[x]$, then the process of long division provides the following calculations:

$$
\begin{array}{r}
2x^2 + x + 6 \, (= q(x)) \\
3x^2 + 4x + 2 \overline{)\, 6x^4 + 4x^3 + 5x^2 + 3x + 1\ } \\
6x^4 + x^3 + 4x^2 \\
\hline
3x^3 + x^2 + 3x + 1 \\
3x^3 + 4x^2 + 2x \\
\hline
4x^2 + x + 1 \\
4x^2 + 3x + 5 \\
\hline
5x + 3 \ (= r(x))
\end{array}
$$

Performing (as above) all arithmetic in $\mathbf{Z}_7$, we find that

$$q(x)f(x) + r(x) = (2x^2 + x + 6)(3x^2 + 4x + 2) + (5x + 3)$$
$$= 6x^4 + 4x^3 + 5x^2 + 3x + 1 = g(x) \quad \square$$

We turn now to the general situation.

THEOREM 17.3 (*Division Algorithm*) Let $f(x), g(x) \in F[x]$ with $f(x) \neq 0$, the zero polynomial. There exist unique polynomials $q(x), r(x) \in F[x]$ such that $g(x) = q(x)f(x) + r(x)$, where $r(x) = 0$ or degree $r(x) <$ degree $f(x)$.

Proof   Let $S = \{g(x) - t(x)f(x) \,|\, t(x) \in F[x]\}$.

If $0 \in S$, then $0 = g(x) - t(x)f(x)$ for some $t(x) \in F[x]$. Then with $q(x) = t(x)$ and $r(x) = 0$, we have $g(x) = q(x)f(x) + r(x)$.

If $0 \notin S$, consider the degrees of the elements of S, and let $r(x) = g(x) - q(x)f(x)$ be an element in S of minimum degree. Since $r(x) \neq 0$, the result follows if degree $r(x) <$ degree $f(x)$. If not, let

$$r(x) = a_n x^n + a_{n-1} x^{n-1} + \cdots + a_2 x^2 + a_1 x + a_0, \qquad a_n \neq 0,$$
$$f(x) = b_m x^m + b_{m-1} x^{m-1} + \cdots + b_2 x^2 + b_1 x + b_0, \qquad b_m \neq 0,$$

with $n \geq m$. Define

$$h(x) = r(x) - [a_n b_m^{-1} x^{n-m}]f(x) = (a_n - a_n b_m^{-1} b_m)x^n + (a_{n-1} - a_n b_m^{-1} b_{m-1})x^{n-1}$$
$$+ \cdots + (a_{n-m} - a_n b_m^{-1} b_0)x^{n-m} + a_{n-m-1} x^{n-m-1} + \cdots + a_1 x + a_0.$$

Then $h(x)$ has degree less than n, the degree of $r(x)$. More important, $h(x) = [g(x) - q(x)f(x)] - [a_n b_m^{-1} x^{n-m}]f(x) = g(x) - [q(x) + a_n b_m^{-1} x^{n-m}]f(x)$, so $h(x) \in S$ and this contradicts the choice of $r(x)$ as having minimum degree. Consequently, degree $r(x) <$ degree $f(x)$ and we have the existence part of the theorem.

For uniqueness, let $g(x) = q_1(x)f(x) + r_1(x) = q_2(x)f(x) + r_2(x)$ where $r_1(x) = 0$ or degree $r_1(x) <$ degree $f(x)$, and $r_2(x) = 0$ or degree $r_2(x) <$ degree $f(x)$. Then $[q_2(x) - q_1(x)]f(x) = r_1(x) - r_2(x)$, and if $q_2(x) - q_1(x) \neq 0$, then degree $([q_2(x) - q_1(x)]f(x)) \geq$ degree $f(x)$ whereas $r_1(x) - r_2(x) = 0$ or degree $[r_1(x) - r_2(x)] <$ degree $f(x)$. Consequently, $q_1(x) = q_2(x)$, and $r_1(x) = r_2(x)$. ∎

The Division Algorithm provides the following results on roots and factors.

THEOREM 17.4   (*The Remainder Theorem*) For $f(x) \in F[x]$ and $a \in F$, the remainder in the division of $f(x)$ by $x - a$ is $f(a)$.

Proof   From the Division Algorithm, $f(x) = q(x)(x - a) + r(x)$, with $r(x) = 0$ or degree $r(x) <$ degree $(x - a) = 1$. Hence $r(x) = r$ is an element of F. Substituting a for x, we find $f(a) = q(a)(a - a) + r(a) = 0 + r = r$. ∎

THEOREM 17.5   (*The Factor Theorem*) If $f(x) \in F[x]$ and $a \in F$, then $x - a$ is a factor of $f(x)$ if and only if a is a root of $f(x)$.

Proof   If $x - a$ is a factor of $f(x)$, then $f(x) = q(x)(x - a)$. With $f(a) = q(a)(a - a) = 0$, a is a root of $f(x)$. Conversely, suppose that a is a root of $f(x)$. By the Division Algorithm, $f(x) = q(x)(x - a) + r$, where $r \in F$. Since $f(a) = 0$ it follows that $r = 0$, so $f(x) = q(x)(x - a)$, and $x - a$ is a factor of $f(x)$. ∎

Example 17.6   **a)** Let $f(x) = x^7 - 6x^5 + 4x^4 - x^2 + 3x - 7 \in \mathbf{Q}[x]$. From the Remainder Theorem, it follows that when $f(x)$ is divided by $x - 2$, the remainder is

$$f(2) = 2^7 - 6(2^5) + 4(2^4) - 2^2 + 3(2) - 7 = -5.$$

If we were to divide $f(x)$ by $x + 1$, then the remainder would be $f(-1) = -2$.

b) If $g(x) = x^5 + 3x^4 + x^3 + x^2 + 2x + 2 \in \mathbf{Z}_5[x]$ is divided by $x - 1$, then the remainder here is $g(1) = 1 + 3 + 1 + 1 + 2 + 2 = 0$ (in $\mathbf{Z}_5$). Consequently, $x - 1$ divides $g(x)$, and by the Factor Theorem,

$$g(x) = q(x)(x - 1) \qquad \text{(where degree } q(x) = 4).\ \square$$

Using the results of Theorems 17.4 and 17.5, we now establish the last major idea for this section.

THEOREM 17.6 If $f(x) \in F[x]$ has degree $n \geq 1$, then $f(x)$ has at most n distinct roots in F.

Proof The proof is by mathematical induction on the degree of $f(x)$. If $f(x)$ has degree 1, then $f(x) = ax + b$, for $a, b \in F$, $a \neq 0$. With $f(-a^{-1}b) = 0$, $f(x)$ has at least one root in F. If c_1 and c_2 are both roots, then $f(c_1) = ac_1 + b = 0 = ac_2 + b = f(c_2)$. By cancellation in a ring, $ac_1 + b = ac_2 + b \Rightarrow ac_1 = ac_2$. Since F is a field and $a \neq 0$, $ac_1 = ac_2 \Rightarrow c_1 = c_2$, so $f(x)$ has only one root in F.

Now assume the result of the theorem is true for all polynomials of degree k in $F[x]$. Consider a polynomial $f(x)$ of degree $k + 1$. If $f(x)$ has no roots in F, the theorem follows. Otherwise, let $r \in F$ with $f(r) = 0$. By the Factor Theorem, $f(x) = (x - r)g(x)$ where $g(x)$ has degree k. Consequently, by the induction hypothesis, $g(x)$ has at most k distinct roots in F, and $f(x)$, in turn, has at most $k + 1$ distinct roots. ∎

Example 17.7 **a)** Let $f(x) = x^2 - 6x + 9 \in \mathbf{R}[x]$. Then $f(x)$ has the roots 3, 3 and $f(x) = (x - 3)(x - 3)$, a factorization into first-degree, or *linear*, factors. Here we say that 3 is a root of *multiplicity* 2.

b) For $g(x) = x^2 + 4 \in \mathbf{R}[x]$, $g(x)$ has no real roots, but Theorem 17.6 is not contradicted. (Why?) In $\mathbf{C}[x]$, $g(x)$ has the roots $2i$, $-2i$ and can be factored as $g(x) = (x - 2i)(x + 2i)$.

c) If $h(x) = x^2 + 2x + 6 \in \mathbf{Z}_7[x]$, then $h(2) = 0$, $h(3) = 0$ and these are the only roots. Also, $h(x) = (x - 2)(x - 3) = x^2 - 5x + 6 = x^2 + 2x + 6$, because $[-5] = [2]$ in $\mathbf{Z}_7$.

d) As we saw in Example 17.3(b), the polynomial $x^2 + 3x + 2$ has four roots. This is not a contradiction to Theorem 17.6, because $\mathbf{Z}_6$ is not a field. Also, $x^2 + 3x + 2 = (x + 1)(x + 2) = (x + 4)(x + 5)$, two distinct factorizations. □

We close with one final remark, without proof, on the idea of factorization in $F[x]$. If $f(x) \in F[x]$ has degree n, and $r_1, r_2, \ldots, r_n$ are the roots of $f(x)$ in F (where it is possible for a root to be repeated), then $f(x) = a_n(x - r_1)(x - r_2) \cdots (x - r_n)$, where a_n is the leading coefficient of $f(x)$. This representation of $f(x)$ is unique up to the order of the first-degree factors.

EXERCISES 17.1

1. How many polynomials are there of degree 2 in $\mathbf{Z}_{11}[x]$? How many have degree 3? degree 4? degree n, $n \in \mathbf{N}$?

2. **a)** Find two nonzero polynomials $f(x)$, $g(x)$ in $\mathbf{Z}_{12}[x]$ where $f(x)g(x) = 0$.

   b) Find polynomials $h(x)$, $k(x) \in \mathbf{Z}_{12}[x]$ such that degree $h(x) = 5$, degree $k(x) = 2$, and degree $h(x)k(x) = 3$.

3. Complete the proofs of Theorem 17.1 and Corollary 17.1.

4. If $f(x) = ax^3 + bx^2 + cx + d$, $g(x) = 5x^2 + 3x - 7 \in \mathbf{Z}[x]$, and $f(x) = (3x + 1)g(x)$, find a, b, c, and d.

5. a) If $f(x) = x^4 - 16$, find its roots and factorization in $\mathbf{Q}[x]$.

   b) Answer part (a) for $f(x) \in \mathbf{R}[x]$.

   c) Answer part (a) for $f(x) \in \mathbf{C}[x]$.

   d) Answer parts (a), (b), and (c) for $f(x) = x^4 - 25$.

6. For each of the following pairs $f(x), g(x)$, find $q(x), r(x)$ so that $g(x) = q(x)f(x) + r(x)$, where $r(x) = 0$ or degree $r(x) <$ degree $f(x)$.

   a) $f(x), g(x) \in \mathbf{Q}[x]$, $f(x) = x^4 - 5x^3 + 7x$, $g(x) = x^5 - 2x^2 + 5x - 3$

   b) $f(x), g(x) \in \mathbf{Z}_2[x]$, $f(x) = x^2 + 1$, $g(x) = x^4 + x^3 + x^2 + x + 1$

   c) $f(x), g(x) \in \mathbf{Z}_5[x]$, $f(x) = x^2 + 3x + 1$, $g(x) = x^4 + 2x^3 + x + 4$

7. In each of the following, find the remainder when $f(x)$ is divided by $g(x)$.

   a) $f(x), g(x) \in \mathbf{Q}[x]$, $f(x) = x^8 + 7x^5 - 4x^4 + 3x^3 + 5x^2 - 4$, $g(x) = x - 3$.

   b) $f(x), g(x) \in \mathbf{Z}_2[x]$, $f(x) = x^{100} + x^{90} + x^{80} + x^{50} + 1$, $g(x) = x - 1$.

   c) $f(x), g(x) \in \mathbf{Z}_{11}[x]$, $f(x) = 3x^5 - 8x^4 + x^3 - x^2 + 4x - 7$, $g(x) = x + 9$.

8. a) Find all roots of $f(x) = x^2 + 4x$ if $f(x) \in \mathbf{Z}_{12}[x]$.

   b) Find four distinct linear polynomials $g(x), h(x), s(x), t(x) \in \mathbf{Z}_{12}[x]$ so that $f(x) = g(x)h(x) = s(x)t(x)$.

   c) Do the results in part (b) contradict the statements made in the paragraph following Example 17.7?

9. Does the Division Algorithm (for polynomials) hold in the integral domain $\mathbf{Z}[x]$? Explain.

10. For each of the following polynomials $f(x) \in \mathbf{Z}_7[x]$, determine all of the roots in $\mathbf{Z}_7$ and then write $f(x)$ as a product of first-degree polynomials.

    a) $f(x) = x^3 + 5x^2 + 2x + 6$.

    b) $f(x) = x^7 - x$.

11. How many units are there in the ring $\mathbf{Z}_5[x]$? How many in $\mathbf{Z}_7[x]$? How many in $\mathbf{Z}_p[x]$, p a prime?

12. If R is an integral domain, prove that if $f(x)$ is a unit in $R[x]$, then $f(x)$ is a constant and is a unit in R.

13. Verify that $f(x) = 2x + 1$ is a unit in $\mathbf{Z}_4[x]$. Does this contradict the result of Exercise 12?

14. For $n \in \mathbf{Z}^+$, $n \geq 2$, let $f(x) \in \mathbf{Z}_n[x]$. Prove that if $a, b \in \mathbf{Z}$ and $a \equiv b \pmod{n}$, then $f(a) \equiv f(b) \pmod{n}$.

15. Let R, S be rings, and let $g: R \to S$ be a ring homomorphism. Prove that the function $G: R[x] \to S[x]$ defined by

$$G\left(\sum_{i=0}^{n} r_i x^i\right) = \sum_{i=0}^{n} g(r_i)x^i$$

is a ring homomorphism.

17.2
IRREDUCIBLE POLYNOMIALS: FINITE FIELDS

We now wish to construct finite fields other than $(\mathbf{Z}_p, +, \cdot)$. The construction will use the following special polynomials.

DEFINITION 17.4   Let $f(x) \in F[x]$, with F a field and degree $f(x) \geq 2$. We call $f(x)$ *reducible* (over F) if there exist $g(x), h(x) \in F[x]$, where $f(x) = g(x)h(x)$ and degree $g(x), h(x) \geq 1$. If $f(x)$ is not reducible it is called *irreducible*, or *prime*. ▬

Theorem 17.7 contains some useful observations about irreducible polynomials.

THEOREM 17.7   For polynomials in $F[x]$,

a) any nonzero polynomial of degree ≤ 1 is irreducible.

b) if $f(x) \in F[x]$ with degree $f(x) = 2$ or 3, then $f(x)$ is reducible if and only if $f(x)$ has a root in F.

Proof   The proof is left for the reader.   ∎

Example 17.8   a) The polynomial $x^2 + 1$ is irreducible in $\mathbf{Q}[x]$ or $\mathbf{R}[x]$, but in $\mathbf{C}[x]$ we find $x^2 + 1 = (x + i)(x - i)$.

b) Let $f(x) = x^4 + 2x^2 + 1 \in \mathbf{R}[x]$. Although $f(x)$ has no real roots, it is reducible because $(x^2 + 1)^2 = x^4 + 2x^2 + 1$. Hence part (b) of Theorem 17.7 is not applicable for polynomials of degree > 3.

c) In $\mathbf{Z}_2[x]$, $f(x) = x^3 + x^2 + x + 1$ is reducible because $f(1) = 0$. But $g(x) = x^3 + x + 1$ is irreducible because $g(0) = g(1) = 1$.

d) Let $h(x) = x^4 + x^3 + x^2 + x + 1 \in \mathbf{Z}_2[x]$. Is $h(x)$ reducible in $\mathbf{Z}_2[x]$? Since $h(0) = h(1) = 1$, $h(x)$ has no first-degree factors, but perhaps we can find $a, b, c, d \in \mathbf{Z}_2$ such that $(x^2 + ax + b)(x^2 + cx + d) = x^4 + x^3 + x^2 + x + 1$.

By comparing coefficients of like powers of x, we find $a + c = 1$, $ac + b + d = 1$, $ad + bc = 1$, and $bd = 1$. With $bd = 1$, it follows that $b = 1$ and $d = 1$, so $ac + b + d = 1 \Rightarrow ac = 1 \Rightarrow a = c = 1 \Rightarrow a + c = 0$. This contradicts $a + c = 1$. Consequently $h(x)$ is irreducible in $\mathbf{Z}_2[x]$.   □

All of the polynomials in Example 17.8 share a common property, which we shall now define.

DEFINITION 17.5   A polynomial $f(x) \in F[x]$ is called *monic* if its leading coefficient is 1, the unity of F. ▬

The next results (up to and including the discussion in Example 17.10) awaken memories of Chapters 4 and 14.

DEFINITION 17.6   If $f(x), g(x) \in F[x]$, then $h(x) \in F[x]$ is a *greatest common divisor* of $f(x)$ and $g(x)$

a) if $h(x)$ divides each of $f(x)$ and $g(x)$, and

b) if $k(x) \in F[x]$ and $k(x)$ divides both $f(x), g(x)$, then $k(x)$ divides $h(x)$. ▬

We now state the following results on the existence and uniqueness of the greatest common divisor, which we write once again as g.c.d. Furthermore, there is a method for finding this g.c.d. that is called the Euclidean Algorithm for Polynomials. A proof for the first result is outlined in the section exercises.

THEOREM 17.8   Let $f(x), g(x) \in F[x]$, with at least one of $f(x), g(x)$ not the zero polynomial. Then any polynomial of minimum degree that can be written as a linear combination of $f(x), g(x)$—that is, $s(x)f(x) + t(x)g(x)$, for $s(x), t(x) \in F[x]$—will be a greatest common divisor of $f(x), g(x)$. If we require the g.c.d. to be monic, it will be unique.

THEOREM 17.9   (*Euclidean Algorithm for Polynomials*) Let $f(x), g(x) \in F[x]$ with degree $f(x) \le$ degree $g(x)$ and $f(x) \neq 0$. Applying the Division Algorithm, we write

$$
\begin{array}{ll}
g(x) = q(x)f(x) + r(x), & \text{degree } r(x) < \text{degree } f(x) \\
f(x) = q_1(x)r(x) + r_1(x), & \text{degree } r_1(x) < \text{degree } r(x) \\
r(x) = q_2(x)r_1(x) + r_2(x), & \text{degree } r_2(x) < \text{degree } r_1(x) \\
\quad\vdots & \quad\vdots \\
r_{k-2}(x) = q_k(x)r_{k-1}(x) + r_k(x), & \text{degree } r_k(x) < \text{degree } r_{k-1}(x) \\
r_{k-1}(x) = q_{k+1}(x)r_k(x) + r_{k+1}(x), & r_{k+1}(x) = 0.
\end{array}
$$

Then $r_k(x)$, the last nonzero remainder, is a constant multiple of the greatest common divisor of $f(x), g(x)$. (Multiplying $r_k(x)$ by the inverse of its leading coefficient allows us to obtain the unique monic polynomial we call *the* greatest common divisor.)

DEFINITION 17.7   If $f(x), g(x) \in F[x]$ and their g.c.d. is 1, then $f(x)$ and $g(x)$ are called *relatively prime*.

The last results we need in order to construct our new finite fields provide the analog of a construction we developed in Section 14.3.

THEOREM 17.10   Let $s(x) \in F[x], s(x) \neq 0$. Define relation $\mathcal{R}$ on $F[x]$ by $f(x) \mathcal{R} g(x)$ if $f(x) - g(x) = t(x)s(x)$, for some $t(x) \in F[x]$—that is, $s(x)$ divides $f(x) - g(x)$. Then $\mathcal{R}$ is an equivalence relation on $F[x]$.

Proof   The verification of the reflexive, symmetric, and transitive properties of $\mathcal{R}$ is left for the reader.   ■

When the situation in Theorem 17.10 occurs, we say that $f(x)$ is *congruent* to $g(x)$ *modulo* $s(x)$ and write $f(x) \equiv g(x) \,(\mathrm{mod}\, s(x))$. The relation $\mathcal{R}$ is referred to as *congruence modulo s(x)*.

Let us examine the equivalence classes for such a relation.

Example 17.9   Let $s(x) = x^2 + x + 1 \in \mathbf{Z}_2[x]$. Then

a) $[0] = [x^2 + x + 1]$
$= \{0, x^2 + x + 1, x^3 + x^2 + x, (x+1)(x^2 + x + 1), \ldots\}$
$= \{t(x)(x^2 + x + 1) \mid t(x) \in \mathbf{Z}_2[x]\}$

b) $[1] = \{1, x^2 + x, x(x^2 + x + 1) + 1, (x+1)(x^2 + x + 1) + 1, \ldots\}$
$= \{t(x)(x^2 + x + 1) + 1 \mid t(x) \in \mathbf{Z}_2[x]\}$

c) $[x] = \{x, x^2 + 1, x(x^2 + x + 1) + x, (x+1)(x^2 + x + 1) + x, \ldots\}$
$= \{t(x)(x^2 + x + 1) + x \mid t(x) \in \mathbf{Z}_2[x]\}$

d) $[x+1] = \{x+1, x^2, x(x^2 + x + 1) + (x+1), (x+1)(x^2 + x + 1)$
$+ (x+1), \ldots\} = \{t(x)(x^2 + x + 1) + (x+1) \mid t(x) \in \mathbf{Z}_2[x]\}$

Are these all of the equivalence classes? If $f(x) \in \mathbf{Z}_2[x]$, then by the Division Algorithm $f(x) = q(x)s(x) + r(x)$, where $r(x) = 0$ or degree $r(x) <$ degree $s(x)$. Since $f(x) - r(x) = q(x)s(x)$, $f(x) \equiv r(x) \, (\text{mod } s(x))$, so $f(x) \in [r(x)]$. Consequently, to determine all the equivalence classes, we consider the possibilities for $r(x)$. Here $r(x) = 0$ or degree $r(x) < 2$, so $r(x) = ax + b$, where $a, b \in \mathbf{Z}_2$. With only two choices for each of a, b, there are four possible choices for $r(x)$: $0, 1, x,$ $x + 1$.   □

We now place a ring structure on the equivalence classes of Example 17.9. Considering how this was accomplished in Chapter 14 for $\mathbf{Z}_n$, we define addition by $[f(x)] + [g(x)] = [f(x) + g(x)]$. Since degree $(f(x) + g(x)) \leq \max\{\text{degree } f(x),$ degree $g(x)\}$, we can find the equivalence class for $[f(x) + g(x)]$ without too much trouble. Here, for example, $[x] + [x+1] = [x + (x+1)] = [2x + 1] = [1]$ because $2 = 0$ in $\mathbf{Z}_2$.

In defining the multiplication of these equivalence classes, we run into a little more difficulty. For instance, what is $[x][x]$ in Example 17.9? If, in general, we define $[f(x)][g(x)] = [f(x)g(x)]$, it is possible for degree $f(x)g(x) \geq$ degree $s(x)$, so we may not readily find $[f(x)g(x)]$ in the list of equivalence classes. However, if degree $f(x)g(x) \geq$ degree $s(x)$, then using the Division Algorithm, we can write $f(x)g(x) = q(x)s(x) + r(x)$, where $r(x) = 0$ or degree $r(x) <$ degree $s(x)$. With $f(x)g(x) = q(x)s(x) + r(x)$, it follows that $f(x)g(x) \equiv r(x) \, (\text{mod } s(x))$, and we define $[f(x)g(x)] = [r(x)]$, where $[r(x)]$ does occur in the list of equivalence classes.

From these observations we construct Tables 17.1 and 17.2 for the addition and multiplication, respectively, of $\{[0], [1], [x], [x+1]\}$. (In these tables we write a for $[a]$).

Table 17.1

+	0	1	x	$x + 1$
0	0	1	x	$x + 1$
1	1	0	$x + 1$	x
x	x	$x + 1$	0	1
$x + 1$	$x + 1$	x	1	0

Table 17.2

$\cdot$	0	1	x	$x + 1$
0	0	0	0	0
1	0	1	x	$x + 1$
x	0	x	$x + 1$	1
$x + 1$	0	$x + 1$	1	x

From the multiplication table (Table 17.2), we find that these equivalence classes form not only a ring but also a field, where $[1]^{-1} = [1]$, $[x]^{-1} = [x + 1]$ and $[x + 1]^{-1} = [x]$. This field of order 4 is denoted by $\mathbf{Z}_2[x]/(x^2 + x + 1)$. In addition, for the nonzero elements of this field we find that $[x]^1 = [x]$, $[x]^2 = [x + 1]$, $[x]^3 = [1]$, so we have a cyclic group of order 3. But the nonzero elements of any field form a group under multiplication, and any group of order 3 is cyclic, so why bother with this observation? In general, the nonzero elements of *any* finite field form a cyclic group under multiplication. (A proof for this can be found in Chapter 12 of reference [9].)

The construction above is summarized in the following theorem. An outline of the proof is given in the section exercises.

THEOREM 17.11   Let $s(x)$ be a nonzero polynomial in $F[x]$.

a) The equivalence classes of $F[x]$ for the relation of congruence modulo $s(x)$ form a commutative ring with unity under the operations

$$[f(x)] + [g(x)] = [f(x) + g(x)], \qquad [f(x)][g(x)] = [f(x)g(x)] = [r(x)],$$

where $r(x)$ is the remainder obtained upon dividing $f(x)g(x)$ by $s(x)$. This ring is denoted by $F[x]/(s(x))$.

b) If $s(x)$ is irreducible in $F[x]$, then $F[x]/(s(x))$ is a field.

c) If $|F| = q$ and degree $s(x) = n$, then $F[x]/(s(x))$ contains q^n elements.

We now examine another example of a finite field that arises by virtue of Theorem 17.11.

Example 17.10   In $\mathbf{Z}_3[x]$ the polynomial $s(x) = x^2 + x + 2$ is irreducible because $s(0) = 2$, $s(1) = 1$, and $s(2) = 2$. Consequently, $\mathbf{Z}_3[x]/(s(x))$ is a field containing all equivalence classes of the form $[ax + b]$, where $a, b \in \mathbf{Z}_3$. These are the possible remainders when a polynomial $f(x) \in \mathbf{Z}_3[x]$ is divided by $s(x)$. The nine equivalence classes are $[0]$, $[1]$, $[2]$, $[x]$, $[x + 1]$, $[x + 2]$, $[2x]$, $[2x + 1]$, and $[2x + 2]$.

Instead of constructing a complete multiplication table, we examine four sample multiplications.

a) $[2x][x] = [2x^2] = [2x^2 + 0] = [2x^2 + (x^2 + x + 2)] = [3x^2 + x + 2] = [x + 2]$ because $3 = 0$ in $\mathbf{Z}_3$.

b) $[x + 1][x + 2] = [x^2 + 3x + 2] = [x^2 + 2] = [x^2 + 2 + 2(x^2 + x + 2)] = [2x]$.

c) $[2x + 2]^2 = [4x^2 + 8x + 4] = [x^2 + 2x + 1] = [(-x - 2) + (2x + 1)]$ since $x^2 \equiv (-x - 2) \pmod{s(x)}$. Consequently, $[2x + 2]^2 = [x - 1] = [x + 2]$.

d) Often we write the equivalence classes without brackets and concentrate on the coefficients of the powers of x. For example, 11 is written for $[x + 1]$ and 21 represents $[2x + 1]$. Consequently, $(21) \cdot (12) = [2x + 1][x + 2] = [2x^2 + 5x + 2] = [2x^2 + 2x + 2] = [2(-x - 2) + 2x + 2] = [-4 + 2] = [-2] = [1]$, so $(21)^{-1} = (12)$.

e) We make one more observation about this field.

$$[x]^1 = [x] \qquad [x]^3 = [2x + 2] \qquad [x]^5 = [2x] \qquad [x]^7 = [x + 1]$$
$$[x]^2 = [2x + 1] \qquad [x]^4 = [2] \qquad [x]^6 = [x + 2] \qquad [x]^8 = [1]$$

Therefore the nonzero elements of $\mathbf{Z}_3[x]/(s(x))$ form a cyclic group under multiplication. $\square$

In Example 17.9 (and in the discussion that follows it) and in Example 17.10, we constructed finite fields of orders $4 (= 2^2)$ and $9 (= 3^2)$, respectively. Now we shall close this section as we investigate other possibilities for the order of a finite field. To accomplish this we need the following idea.

DEFINITION 17.8   Let $(R, +, \cdot)$ be a ring. If there is a least positive integer n such that $nr = z$ (the zero of R) for all $r \in R$, we say that R has *characteristic n* and write $\text{char}(R) = n$. If no such integer exists, R is said to have *characteristic 0*.

Example 17.11   a) The ring $(\mathbf{Z}_3, +, \cdot)$ has characteristic 3; $(\mathbf{Z}_4, +, \cdot)$ has characteristic 4; in general, $(\mathbf{Z}_n, +, \cdot)$ has characteristic n.

b) The rings $(\mathbf{Z}, +, \cdot)$ and $(\mathbf{Q}, +, \cdot)$ both have characteristic 0.

c) A ring can be infinite and still have positive characteristic. For example, $\mathbf{Z}_3[x]$ is an infinite ring but it has characteristic 3.

d) The ring in Example 17.9 has characteristic 2. In Example 17.10 the characteristic of the ring is 3. Unlike the examples in (a), the order of a finite ring can be different from its characteristic.

   Examples 17.9 and 17.10, however, are more than just rings. They are fields with prime characteristic. Could this property be true for all finite fields? $\square$

THEOREM 17.12   Let $(F, +, \cdot)$ be a field. If $\text{char}(F) > 0$, then $\text{char}(F)$ must be prime.

Proof   In this proof we write the unity of F as u so that it is distinct from the positive integer 1. Let $\text{char}(F) = n > 0$. If n is not prime, we write $n = mk$, where $m, k \in \mathbf{Z}^+$ and $1 < m, k < n$. By the definition of characteristic, $nu = z$, the zero of F. Hence $(mk)u = z$. But

$$(mk)u = \underbrace{(u + u + \cdots + u)}_{mk \text{ summands}} = \underbrace{(u + u + \cdots + u)}_{m \text{ summands}}\underbrace{(u + u + \cdots + u)}_{k \text{ summands}} = (mu)(ku).$$

With F a field, $(mu)(ku) = z \Rightarrow (mu) = z$ or $(ku) = z$. Assume without loss of generality that $ku = z$. Then for any $r \in F$, $kr = k(ur) = (ku)r = zr = z$, contradicting the choice of n as the characteristic of F. Consequently, $\text{char}(F)$ is prime. ∎

(The proof of Theorem 17.12 actually requires only that F be an integral domain.)

If F is a finite field and $m = |F|$, then $ma = z$ for all $a \in F$, because $(F, +)$ is an additive group of order m. (See Exercise 7 of Section 16.3.) Consequently, F has positive characteristic and by Theorem 17.12 this characteristic is prime. This leads us to the following.

THEOREM 17.13   Any finite field F has order p^t, where p is a prime and $t \in \mathbf{Z}^+$.

Proof   Since F is a finite field, let $\text{char}(F) = p$, a prime, and let u denote the unity and z the zero element. Then $S_0 = \{u, 2u, 3u, \ldots, pu = z\}$ is a set of p distinct elements in F. If not, $mu = nu$ for $1 \le m < n \le p$ and $(n - m)u = z$, with $0 < n - m < p$. This contradicts $\text{char}(F) = p$. If $F = S_0$, then $|F| = p^1$ and the result follows. If not, let $a \in F - S_0$. Then $S_1 = \{ma + nu \mid 0 < m, n \le p\}$ is a subset of F with $|S_1| \le p^2$. If $|S_1| < p^2$, then $m_1 a + n_1 u = m_2 a + n_2 u$, with $0 < m_1, m_2, n_1, n_2 \le p$ and at least one of $m_1 - m_2, n_2 - n_1 \ne 0$. Should $m_1 - m_2 = 0$, then $(m_1 - m_2)a = z = (n_2 - n_1)u$, with $0 < |n_2 - n_1| < p$. Consequently, for all $x \in F$, $|n_2 - n_1|x = |n_2 - n_1|ux = zx$ with $0 < |n_2 - n_1| < p = \text{char}(F)$. Therefore, $(m_1 - m_2)a = (n_2 - n_1)u \ne z$. Choose $k \in \mathbf{Z}^+$ such that $0 < k < p$ and $k(m_1 - m_2) \equiv 1 \pmod{p}$. Then $a = k(m_1 - m_2)a = k(n_2 - n_1)u$, and $a \in S_0$, another contradiction. Hence $|S_1| = p^2$, and if $F = S_1$ the theorem is proved. If not, continue this process with an element $b \in F - S_1$. Then $S_2 = \{\ell b + ma + nu \mid 0 < \ell, m, n \le p\}$ will have order p^3. (Prove this.) Since F is finite, we reach a point where $F = S_{t-1}$ for $t \in \mathbf{Z}^+$, and $|F| = |S_{t-1}| = p^t$. ∎

As a result of this theorem there can be no finite fields with orders such as $6, 10, 12, \ldots$. In addition, for each prime p and each $t \in \mathbf{Z}^+$, there is really only one field of order p^t. Any two finite fields of the same order are isomorphic. These fields were discovered by the French mathematician Evariste Galois (1811–1832) in his work on the nonexistence of formulas for solving general polynomial equations of degree ≥ 5 over $\mathbf{Q}$. As a result, a finite field of order p^t is denoted by $GF(p^t)$, where the letters GF stand for *Galois field*.

EXERCISES 17.2

1. Determine whether or not each of the following polynomials is irreducible over the given fields. If it is reducible, provide a factorization into irreducible factors.

   a) $x^2 + 3x - 1$ over $\mathbf{Q}, \mathbf{R}, \mathbf{C}$ b) $x^4 - 2$ over $\mathbf{Q}, \mathbf{R}, \mathbf{C}$

   c) $x^2 + x + 1$ over $\mathbf{Z}_3, \mathbf{Z}_5, \mathbf{Z}_7$ d) $x^4 + x^3 + 1$ over $\mathbf{Z}_2$

   e) $x^3 + x + 1$ over $\mathbf{Z}_5$ f) $x^3 + 3x^2 - x + 1$ over $\mathbf{Z}_5$

   g) $x^4 + x^3 - x^2 + 3$ over $\mathbf{Z}_7$

2. Give an example of a polynomial $f(x) \in \mathbf{R}[x]$ where $f(x)$ has degree 6, is reducible, but has no real roots.

3. Determine all polynomials $f(x) \in \mathbf{Z}_2[x]$ such that $1 \le \text{degree } f(x) \le 3$ and $f(x)$ is irreducible (over $\mathbf{Z}_2$).

4. Let $f(x) = (2x^2 + 1)(5x^3 - 5x + 3)(4x - 3) \in Z_7[x]$. Write $f(x)$ as the product of a unit and three monic polynomials.

5. How many monic polynomials in $Z_7[x]$ have degree 5?

6. Prove Theorem 17.7.

7. Below is an outline for a proof of Theorem 17.8.

   a) Let $S = \{s(x)f(x) + t(x)g(x) \mid s(x), t(x) \in F[x]\}$. Select an element $m(x)$ of minimum degree in S. (Recall that the zero polynomial has no degree, so it is not selected.) Can we guarantee that $m(x)$ is monic?

   b) Show that if $h(x) \in F[x]$ and $h(x)$ divides both $f(x)$ and $g(x)$, then $h(x)$ divides $m(x)$.

   c) Show that $m(x)$ divides $f(x)$. If not, use the Division Algorithm and write $f(x) = q(x)m(x) + r(x)$, where $r(x) \neq 0$ and degree $r(x) <$ degree $m(x)$. Then show that $r(x) \in S$ and obtain a contradiction.

   d) Repeat the argument in part (c) to show that $m(x)$ divides $g(x)$.

8. Prove Theorems 17.9 and 17.10.

9. Use the Euclidean Algorithm for Polynomials to find the g.c.d. of each pair of polynomials, over the designated field F. Then write the g.c.d. as $s(x)f(x) + t(x)g(x)$, where $s(x), t(x) \in F[x]$.

   a) $f(x) = x^2 + x - 2$, $g(x) = x^5 - x^4 + x^3 + x^2 - x - 1$ in $Q[x]$

   b) $f(x) = x^4 + x^3 + 1$, $g(x) = x^2 + x + 1$ in $Z_2[x]$

   c) $f(x) = x^4 + 2x^2 + 2x + 2$, $g(x) = 2x^3 + 2x^2 + x + 1$ in $Z_3[x]$

10. If F is any field, let $f(x), g(x) \in F[x]$. If $f(x), g(x)$ are relatively prime, prove that there is no element $a \in F$ with $f(a) = 0$ and $g(a) = 0$.

11. Let $f(x), g(x) \in R[x]$ with $f(x) = x^3 + 2x^2 + ax - b, g(x) = x^3 + x^2 - bx + a$. Determine values for a, b so that the g.c.d. of $f(x), g(x)$ is a polynomial of degree 2.

12. For Example 17.9, determine which equivalence class contains each of the following:

   a) $x^4 + x^3 + x + 1$ **b)** $x^3 + x^2 + 1$ **c)** $x^4 + x^3 + x^2 + 1$

13. Below is an outline for the proof of Theorem 17.11.

   a) Prove that the operations defined in Theorem 17.11(a) are well-defined by showing that if $f(x) \equiv f_1(x) \pmod{s(x)}$ and $g(x) \equiv g_1(x) \pmod{s(x)}$, then $f(x) + g(x) \equiv f_1(x) + g_1(x) \pmod{s(x)}$ and $f(x)g(x) \equiv f_1(x)g_1(x) \pmod{s(x)}$.

   b) Verify the ring properties for the equivalence classes in $F[x]/(s(x))$.

   c) Let $f(x) \in F[x]$, with $f(x) \neq 0$ and degree $f(x) <$ degree $s(x)$. If $s(x)$ is irreducible in $F[x]$, why does it follow that 1 is the g.c.d. of $f(x)$ and $s(x)$?

   d) Use part (c) to prove that if $s(x)$ is irreducible in $F[x]$, then $F[x]/(s(x))$ is a field.

   e) If $|F| = q$ and degree $s(x) = n$, determine the order of $F[x]/(s(x))$.

14. a) Show that $s(x) = x^2 + 1$ is reducible in $\mathbf{Z}_2[x]$.

  b) Find the equivalence classes for the ring $\mathbf{Z}_2[x]/(s(x))$.

  c) Is $\mathbf{Z}_2[x]/(s(x))$ an integral domain?

15. For the field in Example 17.10, find each of the following:

  a) $[x + 2][2x + 2] + [x + 1]$ **b)** $[2x + 1] - [x + 1][2x + 1]$

  c) $[2x + 1]^2[x + 2]$ **d)** $(22)^{-1} = [2x + 2]^{-1}$

16. Let $s(x) = x^4 + x^3 + 1 \in \mathbf{Z}_2[x]$.

  a) Prove that $s(x)$ is irreducible.

  b) What is the order of the field $\mathbf{Z}_2[x]/(s(x))$?

  c) Find $[x^2 + x + 1]^{-1}$ in this field. (*Hint:* Find $a, b, c, d \in \mathbf{Z}_2$ so that $[x^2 + x + 1][ax^3 + bx^2 + cx + d] = [1]$.)

  d) Determine $[x^3 + x + 1][x^2 + 1]$ in this field.

17. For p a prime, let $s(x)$ be irreducible of degree n in $\mathbf{Z}_p[x]$.

  a) How many elements are there in the field $\mathbf{Z}_p[x]/(s(x))$?

  b) How many elements in $\mathbf{Z}_p[x]/(s(x))$ generate the multiplicative group of nonzero elements of this field?

18. Give the characteristic for each of the following rings:

  a) $\mathbf{Z}_{11}$ **b)** $\mathbf{Z}_{11}[x]$ **c)** $\mathbf{Q}[x]$

  d) $\mathbf{Z}[\sqrt{5}] = \{a + b\sqrt{5} \mid a, b \in \mathbf{Z}\}$, under the ordinary operations of addition and multiplication of real numbers

19. In each of the following rings, the operations are componentwise addition and multiplication, as in Exercise 18 of Section 14.2. Determine the characteristic in each case.

  a) $\mathbf{Z}_2 \times \mathbf{Z}_3$ **b)** $\mathbf{Z}_3 \times \mathbf{Z}_4$ **c)** $\mathbf{Z}_4 \times \mathbf{Z}_6$

  d) $\mathbf{Z}_m \times \mathbf{Z}_n$, for $m, n \in \mathbf{Z}^+, m, n \geq 2$

  e) $\mathbf{Z}_3 \times \mathbf{Z}$ **f)** $\mathbf{Z} \times \mathbf{Z}$ **g)** $\mathbf{Z} \times \mathbf{Q}$

20. For Theorem 17.13, prove that $|S_2| = p^3$.

21. Find the orders n for all fields $GF(n)$, where $100 \leq n \leq 150$.

17.3
LATIN SQUARES

Our first application for this chapter deals with the structure called a Latin square. Such configurations arise in the study of combinatorial designs and play a role in statistics—in the design of experiments. We introduce the structure in the following example.

Example 17.12   A petroleum corporation is interested in testing four types of gasoline additives to determine their effects on mileage. To do so, a research team designs an experiment wherein four different automobiles, denoted A, B, C, and D, are run on a fixed track in a laboratory. Each run uses the same prescribed amount of fuel with one of the additives present. In order to see how each additive affects each type of auto, the team follows the schedule in Table 17.3, where the additives are numbered 1, 2, 3, and 4. This schedule provides a way to test each additive thoroughly in each type of auto. If one additive produces the best results in all four types, the experiment will reveal its superior capability.

The same corporation is also interested in testing four other additives developed for cleaning engines. A similar schedule for these tests is shown in Table 17.4, where these engine-cleaning additives are also denoted as 1, 2, 3, and 4.

Furthermore, the research team is interested in the combined effect of both types of additives. It requires 16 days to test the 16 possible pairs of additives (one for improved mileage, the other for cleaning engines) in every automobile. If the results are needed in four days, the research team must design the schedules so that every pair is tested once by some auto. There are 16 ordered pairs in $\{1, 2, 3, 4\} \times \{1, 2, 3, 4\}$, so this can be done in the allotted time if the schedules in Tables 17.3 and 17.4 are superimposed to obtain the schedule in Table 17.5. Here, for example, the entry $(4, 3)$ indicates that on Tuesday, auto C is used to test the combined effect of the fourth additive for improved mileage and the third additive for maintaining a clean engine.

Table 17.3

Auto	Day Mon	Tues	Wed	Thurs
A	1	2	3	4
B	2	1	4	3
C	3	4	1	2
D	4	3	2	1

Table 17.4

Auto	Day Mon	Tues	Wed	Thurs
A	1	2	3	4
B	3	4	1	2
C	4	3	2	1
D	2	1	4	3

Table 17.5

Auto	Day Mon	Tues	Wed	Thurs
A	(1, 1)	(2, 2)	(3, 3)	(4, 4)
B	(2, 3)	(1, 4)	(4, 1)	(3, 2)
C	(3, 4)	(4, 3)	(1, 2)	(2, 1)
D	(4, 2)	(3, 1)	(2, 4)	(1, 3)

What has happened here leads us to the following concepts.

DEFINITION 17.9   An $n \times n$ *Latin square* is a square array of symbols, usually $1, 2, 3, \ldots, n$, where each symbol appears exactly once in each row and each column of the array.

Example 17.13

a) Tables 17.3 and 17.4 are examples of 4×4 Latin squares.

b) For any $n \geq 2$, using n instead of 0, we can obtain an $n \times n$ Latin square from the table of the group $(\mathbf{Z}_n, +)$.   □

From the two Latin squares in Example 17.12 we were able to produce all of the ordered pairs in $S \times S$, for $S = \{1, 2, 3, 4\}$. We now question whether or not we can do this for $n \times n$ Latin squares in general.

DEFINITION 17.10   Let $L_1 = (a_{ij})$, $L_2 = (b_{ij})$ be two $n \times n$ Latin squares, where $1 \leq i, j \leq n$ and each $a_{ij}, b_{ij} \in \{1, 2, 3, \ldots, n\}$. If the n^2 ordered pairs (a_{ij}, b_{ij}), $1 \leq i, j \leq n$, are distinct, then L_1, L_2 are a *pair of orthogonal Latin squares*.

Example 17.14

a) There is no pair of 2×2 orthogonal Latin squares, because the only possibilities are

$$L_1: \begin{matrix} 1 & 2 \\ 2 & 1 \end{matrix} \quad \text{and} \quad L_2: \begin{matrix} 2 & 1 \\ 1 & 2. \end{matrix}$$

b) In the 3×3 case, we find the orthogonal pair

$$L_1: \begin{matrix} 1 & 2 & 3 \\ 2 & 3 & 1 \\ 3 & 1 & 2 \end{matrix} \quad \text{and} \quad L_2: \begin{matrix} 1 & 2 & 3 \\ 3 & 1 & 2 \\ 2 & 3 & 1 \end{matrix}$$

c) The two 4×4 Latin squares in Example 17.12 form an orthogonal pair. The 4×4 Latin square shown in Table 17.6 is orthogonal to each of the Latin squares in that example.

Table 17.6

$$\begin{matrix} 1 & 2 & 3 & 4 \\ 4 & 3 & 2 & 1 \\ 2 & 1 & 4 & 3 \\ 3 & 4 & 1 & 2 \end{matrix}$$ □

We could continue listing some larger Latin squares, but we've seen enough of them at this point to ask the following questions:

1. What is the first value of $n > 2$ for which there is no pair of orthogonal $n \times n$ Latin squares?

2. For $n > 1$, what can we say about the number of $n \times n$ Latin squares that can be constructed so that each pair of them is orthogonal?

3. Is there a method to assist us in constructing a pair of orthogonal $n \times n$ Latin squares for certain values of $n > 2$?

Before we can examine these questions, we need to standardize some of our results.

DEFINITION 17.11   If L is an $n \times n$ Latin square, then L is said to be in *standard form* if its first row is 1 2 3 ... n.

Except for L_2 in Example 17.14(a), all the Latin squares we've seen in this section are in standard form. If a Latin square is not in standard form, it can be put in that form by interchanging some of the symbols.

Example 17.15   The 5×5 Latin square (a) is not in standard form. If, however, we replace each occurrence of 4 by 1, each occurrence of 5 by 4, and each occurrence of 1 by 5, then the result is the (standard) 5×5 Latin square (b).

$$
\begin{array}{ccccc}
4 & 2 & 3 & 5 & 1 \\
1 & 3 & 5 & 4 & 2 \\
3 & 4 & 2 & 1 & 5 \\
2 & 5 & 1 & 3 & 4 \\
5 & 1 & 4 & 2 & 3 \\
\end{array}
\qquad
\begin{array}{ccccc}
1 & 2 & 3 & 4 & 5 \\
5 & 3 & 4 & 1 & 2 \\
3 & 1 & 2 & 5 & 4 \\
2 & 4 & 5 & 3 & 1 \\
4 & 5 & 1 & 2 & 3 \\
\end{array}
$$
(a)   (b)

It is often convenient to deal with Latin squares in standard form. But will this affect our results on orthogonal pairs in any way?

THEOREM 17.14   Let L_1, L_2 be an orthogonal pair of $n \times n$ Latin squares. If L_1, L_2 are standardized as L_1^*, L_2^*, then L_1^*, L_2^* are orthogonal.

Proof   The proof of this result is left for the reader. ■

These ideas are needed for the main results of this section.

THEOREM 17.15   If $n \in \mathbf{Z}^+, n > 2$, then the largest number of $n \times n$ Latin squares that are orthogonal in pairs is $n - 1$.

Proof   Let $L_1, L_2, \ldots, L_k$ be k distinct $n \times n$ Latin squares that are in standard form and orthogonal in pairs. We write $a_{ij}^{(m)}$ to denote the entry in the ith row and jth column of L_m, where $1 \le i, j \le n$, $1 \le m \le k$. Since these Latin squares are in standard form, $a_{11}^{(m)} = 1, a_{12}^{(m)} = 2, \ldots$, and $a_{1n}^{(m)} = n$ for all $1 \le m \le k$. Now consider $a_{21}^{(m)}$, for $1 \le m \le k$. These entries in the second row and first column are below $a_{11}^{(m)} = 1$. Thus $a_{21}^{(m)} \ne 1$, for all $1 \le m \le k$, or the configuration is not a Latin square. Further, if there exists $1 \le \ell < m \le k$ with $a_{21}^{(\ell)} = a_{21}^{(m)}$, then the pair L_ℓ, L_m cannot be an orthogonal pair. (Why not?) Consequently, there are at best $n - 1$ choices for the a_{21} entries in any of our $n \times n$ Latin squares, and the result follows from this observation. ■

This theorem places an upper bound on the number of $n \times n$ Latin squares that are orthogonal in pairs. We shall find that for certain values of n, this upper bound can be attained. In addition, our next theorem provides a method for constructing these Latin squares, though initially not in standard form. The construction uses the structure of a finite field.

Before proving this theorem for the general situation, however, we shall examine one special case.

Example 17.16   Let $F = \{f_i \mid 1 \le i \le 5\} = \mathbf{Z}_5$ with $f_1 = 1$, $f_2 = 2$, $f_3 = 3$, $f_4 = 4$, and $f_5 = 5$, the zero of $\mathbf{Z}_5$.

For $1 \le k \le 4$, let L_k be the 5×5 array $(a_{ij}^{(k)})$, where $1 \le i, j \le 5$ and

$$a_{ij}^{(k)} = f_k f_i + f_j.$$

When $k = 1$, we construct $L_1 = (a_{ij}^{(1)})$ as follows. Here $a_{ij}^{(1)} = f_1 f_i + f_j = f_i + f_j$, for $1 \le i, j \le 5$. With $i = 1$, the first row of L_1 is calculated as follows:

$$a_{11}^{(1)} = f_1 + f_1 = 2 \qquad a_{12}^{(1)} = f_1 + f_2 = 3$$
$$a_{13}^{(1)} = f_1 + f_3 = 4 \qquad a_{14}^{(1)} = f_1 + f_4 = 5$$
$$a_{15}^{(1)} = f_1 + f_5 = 1$$

The entries in the second row of L_1 are computed when $i = 2$. Here we find

$$a_{21}^{(1)} = f_2 + f_1 = 3 \qquad a_{22}^{(1)} = f_2 + f_2 = 4$$
$$a_{23}^{(1)} = f_2 + f_3 = 5 \qquad a_{24}^{(1)} = f_2 + f_4 = 1$$
$$a_{25}^{(1)} = f_2 + f_5 = 2$$

Continuing these calculations, we obtain the Latin square L_1 as

$$
\begin{array}{ccccc}
2 & 3 & 4 & 5 & 1 \\
3 & 4 & 5 & 1 & 2 \\
4 & 5 & 1 & 2 & 3 \\
5 & 1 & 2 & 3 & 4 \\
1 & 2 & 3 & 4 & 5
\end{array}
$$

For $k = 2$, the entries of L_2 are given by the formula $a_{ij}^{(2)} = f_2 f_i + f_j = 2 f_i + f_j$. To obtain the first row of L_2, we set i equal to 1 and compute

$$a_{11}^{(2)} = 2f_1 + f_1 = 3 \qquad a_{12}^{(2)} = 2f_1 + f_2 = 4$$
$$a_{13}^{(2)} = 2f_1 + f_3 = 5 \qquad a_{14}^{(2)} = 2f_1 + f_4 = 1$$
$$a_{15}^{(2)} = 2f_1 + f_5 = 2$$

When i is set equal to 2, the entries in the second row of L_2 are calculated as follows:

$$a_{21}^{(2)} = 2f_2 + f_1 = 5 \qquad a_{22}^{(2)} = 2f_2 + f_2 = 1$$
$$a_{23}^{(2)} = 2f_2 + f_3 = 2 \qquad a_{24}^{(2)} = 2f_2 + f_4 = 3$$
$$a_{25}^{(2)} = 2f_2 + f_5 = 4$$

Similar calculations for $i = 3, 4$, and 5 result in the Latin square L_2 given by

$$
\begin{array}{ccccc}
3 & 4 & 5 & 1 & 2 \\
5 & 1 & 2 & 3 & 4 \\
2 & 3 & 4 & 5 & 1 \\
4 & 5 & 1 & 2 & 3 \\
1 & 2 & 3 & 4 & 5
\end{array}
$$

It is straightforward to check that the two Latin squares L_1 and L_2 are orthogonal. In Exercise 5 the reader will be asked to calculate L_3 and L_4. Our next result will verify that the four arrays L_1, L_2, L_3, and L_4 are Latin squares and that they are orthogonal in pairs. □

THEOREM 17.16 Let $n \in \mathbf{Z}^+$, $n > 2$. If p is a prime and $n = p^t$, $t \in \mathbf{Z}^+$, then there are $n - 1$ Latin squares that are $n \times n$ and orthogonal in pairs.

Proof Let $F = GF(p^t)$, the Galois field of order $p^t = n$. Consider $F = \{f_1, f_2, \ldots, f_n\}$, where f_1 is the unity and f_n is the zero element.

We construct $n - 1$ Latin squares as follows.

For any $1 \le k \le n - 1$, let L_k be the $n \times n$ array $(a_{ij}^{(k)})$, $1 \le i, j \le n$, where $a_{ij}^{(k)} = f_k f_i + f_j$.

First we show that each L_k is a Latin square. If not, there are two identical elements of F in the same row or column of L_k. Suppose that a repetition occurs in a column—that is, $a_{rj}^{(k)} = a_{sj}^{(k)}$ for $1 \le r, s \le n$. Then $a_{rj}^{(k)} = f_k f_r + f_j = f_k f_s + f_j = a_{sj}^{(k)}$. This implies that $f_k f_r = f_k f_s$, by the cancellation for addition in F. Since $k \ne n$, it follows that $f_k \ne f_n$, the zero of F. Consequently, f_k is invertible, so $f_r = f_s$ and $r = s$. A similar argument shows that there are no repetitions in any row of L_k. (This proof is left as an exercise.)

At this point we have $n - 1$ Latin squares, $L_1, L_2, \ldots, L_{n-1}$. Now we shall prove that they are orthogonal in pairs. If not, let $1 \le k < m \le n - 1$ with

$$
a_{ij}^{(k)} = a_{rs}^{(k)}, \quad a_{ij}^{(m)} = a_{rs}^{(m)}, \quad 1 \le i, j, r, s \le n, \quad \text{and} \quad (i, j) \ne (r, s).
$$

(Then the same ordered pair occurs twice when we superimpose L_k and L_m.) But

$$
a_{ij}^{(k)} = a_{rs}^{(k)} \Leftrightarrow f_k f_i + f_j = f_k f_r + f_s, \quad \text{and}
$$
$$
a_{ij}^{(m)} = a_{rs}^{(m)} \Leftrightarrow f_m f_i + f_j = f_m f_r + f_s.
$$

Subtracting these equations, we find that $(f_k - f_m)f_i = (f_k - f_m)f_r$. With $k \ne m$, $(f_k - f_m)$ is not the zero of F, so it is invertible and we have $f_i = f_r$. Putting this back into either of the prior equations, we find that $f_j = f_s$. Consequently, $i = r$ and $j = s$. Therefore for $k \ne m$, L_k and L_m form an orthogonal pair. ■

The first value of n that is not a power of a prime is 6. The existence of a pair of 6×6 orthogonal Latin squares was first investigated by Leonhard Euler (1707–1783) when he sought a solution to the "problem of the 36 officers." This problem deals with six different regiments wherein six officers, each with a different rank, are selected from each regiment. (There are only six possible ranks.) The objective is to arrange the 36 officers in a 6×6 array so that in any row or column of the array, every rank and every regiment is represented exactly once. Hence each officer in the square array corresponds to an ordered pair (i, j) where $1 \leq i, j \leq 6$, with i for his regiment and j for his rank. In 1782 Euler conjectured that the problem could not be solved—that there is no pair of 6×6 orthogonal Latin squares. He went further and conjectured that for any $n \in \mathbf{Z}^+$, if $n \equiv 2 \pmod{4}$, then there is no pair of $n \times n$ orthogonal Latin squares. In 1900 G. Tarry verified Euler's conjecture for $n = 6$ by a systematic enumeration of all possible 6×6 Latin squares. However, it was not until 1960, through the combined efforts of R. C. Bose, S. S. Shrikhande, and E. T. Parker, that the remainder of Euler's conjecture was proved false. They showed that if $n \in \mathbf{Z}^+$ with $n \equiv 2 \pmod{4}$ and $n > 6$, then there exists a pair of $n \times n$ orthogonal Latin squares.

For more on this result and Latin squares in general, the reader should consult the chapter references.

EXERCISES 17.3

1. **a)** Rewrite the following 4×4 Latin square in standard form.

$$\begin{array}{cccc}
1 & 3 & 4 & 2 \\
3 & 1 & 2 & 4 \\
2 & 4 & 3 & 1 \\
4 & 2 & 1 & 3
\end{array}$$

   b) Find a 4×4 Latin square in standard form that is orthogonal to the result in part (a).

   c) Apply the reverse of the process in part (a) to the result in part (b). Show that your answer is orthogonal to the given 4×4 Latin square.

2. Prove Theorem 17.14.

3. Complete the proof of the first part of Theorem 17.16.

4. The three 4×4 Latin squares in Tables 17.3, 17.4, and 17.6 are orthogonal in pairs. Can you find another 4×4 Latin square that is orthogonal to each of these three?

5. Complete the calculations in Example 17.16 to obtain the two 5×5 Latin squares L_3 and L_4. Rewrite each Latin square L_i, for $1 \leq i \leq 4$, in standard form.

6. Find three 7×7 Latin squares that are orthogonal in pairs. Rewrite these results in standard form.

7. Extend the experiment in Example 17.12 so that the research team needs three 4×4 Latin squares that are orthogonal in pairs.

8. A Latin square L is called *self-orthogonal* if L and its transpose L^{tr} form an orthogonal pair.

   a) Show that there is no 3×3 self-orthogonal Latin square.

b) Give an example of a 4×4 Latin square that is self-orthogonal.

c) If $L = (a_{ij})$ is an $n \times n$ self-orthogonal Latin square, prove that the elements a_{ii}, for $1 \le i \le n$, must all be distinct.

17.4
FINITE GEOMETRIES AND AFFINE PLANES

In the Euclidean geometry of the real plane, we find that (a) two distinct points determine a unique line and (b) if ℓ is a line in the plane, and P a point not on ℓ, then there is a unique line ℓ' that contains P and is parallel to ℓ. During the eighteenth and nineteenth centuries, non-Euclidean geometries were developed when alternatives to condition (b) were investigated. Yet all of these geometries contained infinitely many points and lines. The notion of a finite geometry did not appear until the end of the nineteenth century in the work of Gino Fano (*Giornale di Matematiche*, 1892).

How can we construct such a geometry? To do so, we return to the more familiar Euclidean geometry. In order to describe points and lines in this plane algebraically, we introduced a set of coordinate axes and identified each point P by an ordered pair (c, d) of real numbers. This description set up a one-to-one correspondence between the points in the plane and the set $\mathbf{R} \times \mathbf{R}$. By using the idea of slope, we could represent uniquely each line in this plane by either (1) $x = a$, where the slope is infinite, or (2) $y = mx + b$, where m is the slope; a, m, and b are any real numbers. We also found that two distinct lines are parallel if and only if they have the same slope. When their slopes are distinct, the lines intersect in a unique point.

Instead of using real numbers for the points (c, d) and the lines $x = a$, $y = mx + b$, we turn to a comparable *finite* structure, the finite field. Our objective is to construct what is called a (finite) affine plane.

DEFINITION 17.12   Let $\mathcal{P}$ be a finite set of points, and let $\mathcal{L}$ be a set of subsets of $\mathcal{P}$, called lines. A (*finite*) *affine plane* on the sets $\mathcal{P}$ and $\mathcal{L}$ is a finite structure satisfying the following conditions.

A1) Two distinct points of $\mathcal{P}$ are in only one element of $\mathcal{L}$; that is, they are on only one line.

A2) For any $\ell \in \mathcal{L}$, and any $P \in \mathcal{P}$ with $P \notin \ell$, there exists a unique element $\ell' \in \mathcal{L}$ where $P \in \ell'$ and ℓ, ℓ' have no point in common.

A3) There are four points in $\mathcal{P}$, no three of which are collinear.

The reason for condition (A3) is to avoid uninteresting situations like the one shown in Fig. 17.1. If only conditions (A1) and (A2) were considered, then this system would be an affine plane.

Figure 17.1

We return now to our construction. Let $F = GF(n)$, where $n = p^t$ for some prime p and $t \in \mathbf{Z}^+$. In constructing our affine plane, denoted by $AP(F)$, we let $\mathscr{P} = \{(c,d) \mid c, d \in F\}$. Thus we have n^2 points.

How many lines do we have for the set $\mathscr{L}$?

The lines fall into two categories. For a line of infinite slope the equation is $x = a$, where $a \in F$. Thus we have n such "vertical lines." The other lines are given algebraically by $y = mx + b$, where $m, b \in F$. With n choices for each of m and b, it follows that there are n^2 lines that are not "vertical." Hence $|\mathscr{L}| = n^2 + n$.

Before we verify that $AP(F)$, with $\mathscr{P}$ and $\mathscr{L}$ as constructed, is an affine plane, we make two other observations.

First, for any line $\ell \in \mathscr{L}$, if ℓ is given by $x = a$, then there are n choices for y on $\ell = \{(a,y) \mid y \in F\}$. Thus ℓ contains exactly n points. If ℓ is given by $y = mx + b$, for $m, b \in F$, then for each choice of x we have y uniquely determined, and again ℓ consists of n points.

Now consider any point $(c,d) \in \mathscr{P}$. This point is on the line $x = c$. Furthermore, on any line $y = mx + b$ of finite slope m, $d - mc$ uniquely determines b. With n choices for m, (c,d) is on the n lines of the form $y = mx + (d - mc)$. Overall, (c,d) is on $n + 1$ lines.

Thus far in our construction of $AP(F)$ we have a set $\mathscr{P}$ of points and a set $\mathscr{L}$ of lines where (a) $|\mathscr{P}| = n^2$; (b) $|\mathscr{L}| = n^2 + n$; (c) each $\ell \in \mathscr{L}$ contains n points; and (d) each point in $\mathscr{P}$ is on exactly $n + 1$ lines. We shall now prove that $AP(F)$ satisfies the three conditions to be an affine plane.

A1) Let $(c,d), (e,f) \in \mathscr{P}$. Using the two-point formula for the equation of a line, we have

$$(e - c)(y - d) = (f - d)(x - c) \tag{1}$$

as a line on which we find both (c,d) and (e,f). Each of these points is on $n + 1$ lines. Could there be a second line containing both of them?

The point (c,d) is on the line $x = c$. If (e,f) is also on that line, then $e = c$, but $f \neq d$ because the points are distinct. With $e = c$, Eq. (1) reduces to $0 = (f - d)(x - c)$, or $x = c$ because $f - d \neq 0$, and so we do not have a second line.

With $c \neq e$, if $(c,d), (e,f)$ are on a second line of the form $y = mx + b$, then $d = mc + b, f = me + b$, and $(f - d) = m(e - c)$. Our coefficients are taken from a field and $e \neq c$, so $m = (f - d)(e - c)^{-1}$ and $b = d - mc = d - (f - d)(e - c)^{-1}c$. Consequently this second line containing (c,d) and (e,f) is

$$y = (f - d)(e - c)^{-1}x + [d - (f - d)(e - c)^{-1}c],$$

or $(e - c)(y - d) = (f - d)(x - c)$, which is Eq. (1). Thus two points from $\mathscr{P}$ are on only one line, and condition (A1) is satisfied.

A2) To verify this condition, consider the point P and the line ℓ as shown in Fig. 17.2. Since there are n points on any line, let $P_1, P_2, \ldots, P_n$ be the points of ℓ. The point P is not on ℓ, so P and P_i determine a unique line ℓ_i, for each $1 \le i \le n$. We showed earlier that each point is on $n + 1$ lines, so now there is one additional line ℓ' with P on ℓ' and with ℓ' not intersecting ℓ.

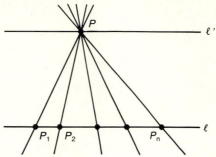

Figure 17.2

A3) The last condition uses the field F. Since $|F| \ge 2$, there is the unity 1 and the zero element 0 in F. Considering the points $(0, 0), (1, 0), (0, 1), (1, 1)$, if line ℓ contains any three of these points, then two of the points have the form $(c, c), (c, d)$. Consequently the equation for ℓ is given by $x = c$, which is not satisfied by either (d, c) or (d, d). Hence no three of these points are collinear.

We have now shown the following.

THEOREM 17.17   If F is a finite field, then the system based on the set $\mathscr{P}$ of points and the set $\mathscr{L}$ of lines, as described above, is an affine plane denoted by $AP(F)$.

Some particular examples will indicate a connection between these finite geometries, or affine planes, and the Latin squares of the previous section.

Example 17.17   For $F = (\mathbf{Z}_2, +, \cdot)$, we have $n = |F| = 2$. The affine plane in Fig. 17.3 has $n^2 = 4$ points and $n^2 + n = 6$ lines. For example, the line $\ell_4 = \{(1, 0), (1, 1)\}$, and ℓ_4 contains no other points that the figure might suggest. Furthermore, ℓ_5 and ℓ_6 are parallel lines in this finite geometry because they do not intersect. $\square$

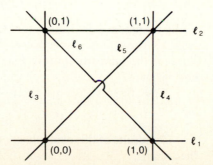

Figure 17.3

Example 17.18   Let $F = GF(2^2)$. Recall the notation of Example 17.10(d) and write $F = \{00, 01, 10, 11\}$, with addition and multiplication given by Table 17.7. We use this field to construct a finite geometry with $n^2 = 16$ points and $n^2 + n = 20$ lines. The 20 lines can be partitioned into five *parallel classes* of four lines each.

Table 17.7

+	00	01	10	11
00	00	01	10	11
01	01	00	11	10
10	10	11	00	01
11	11	10	01	00

·	00	01	10	11
00	00	00	00	00
01	00	01	10	11
10	00	10	11	01
11	00	11	01	10

Class 1: Here we have the lines of infinite slope. These four "vertical" lines are given by the equations $x = 00$, $x = 01$, $x = 10$, and $x = 11$.

Class 2: For the "horizontal" class, or class of slope 0, we have the four lines $y = 00$, $y = 01$, $y = 10$, and $y = 11$.

Class 3: The lines with slope 01 are those whose equations are $y = 01x + 00$, $y = 01x + 01$, $y = 01x + 10$, and $y = 01x + 11$.

Class 4: This class consists of the lines with equations $y = 10x + 00$, $y = 10x + 01$, $y = 10x + 10$, and $y = 10x + 11$.

Class 5: The last class contains the four lines given by $y = 11x + 00$, $y = 11x + 01$, $y = 11x + 10$, and $y = 11x + 11$.

Since each line in $AP(F)$ contains four points and each parallel class contains four lines, we shall see now how three of these parallel classes partition the 16 points of $AP(F)$.

For the class with $m = 01$, there are four lines: (1) $y = 01x + 00$; (2) $y = 01x + 01$; (3) $y = 01x + 10$; and (4) $y = 01x + 11$. Above each point in $AP(F)$ we write the number corresponding to the line it is on (see Fig. 17.4). This configuration can be given by the following Latin square:

$$
\begin{array}{cccc}
4 & 3 & 2 & 1 \\
3 & 4 & 1 & 2 \\
2 & 1 & 4 & 3 \\
1 & 2 & 3 & 4
\end{array}
$$

If we repeat this process for classes 4 and 5, we get the partitions shown in Figs. 17.5 and 17.6, respectively. In each class the lines are listed, for the given slope, in the same order as for Fig. 17.4. Within each figure is the corresponding Latin square.

These figures give us three 4×4 Latin squares that are orthogonal in pairs.   ☐

The results of this example are no accident, as demonstrated by the following theorem.

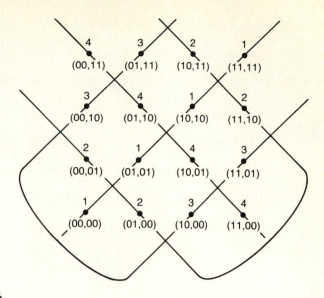

Figure 17.4

```
  4        2        1        3
  •        •        •        •
(00,11)  (01,11)  (10,11)  (11,11)


  3        1        2        4
  •        •        •        •
(00,10)  (01,10)  (10,10)  (11,10)


  2        4        3        1
  •        •        •        •
(00,01)  (01,01)  (10,01)  (11,01)
```

```
                                    4  2  1  3
  1        3        4        2      3  1  2  4
  •        •        •        •      2  4  3  1
(00,00)  (01,00)  (10,00)  (11,00)  1  3  4  2
```

Figure 17.5

```
  4        1        3        2
  •        •        •        •
(00,11)  (01,11)  (10,11)  (11,11)


  3        2        4        1
  •        •        •        •
(00,10)  (01,10)  (10,10)  (11,10)


  2        3        1        4
  •        •        •        •
(00,01)  (01,01)  (10,01)  (11,01)
```

```
                                    4  1  3  2
  1        4        2        3      3  2  4  1
  •        •        •        •      2  3  1  4
(00,00)  (01,00)  (10,00)  (11,00)  1  4  2  3
```

Figure 17.6

THEOREM 17.18   Let $F = GF(n)$, where $n \geq 3$ and $n = p^t$, p a prime. The Latin squares that arise from $AP(F)$ for the $n - 1$ parallel classes, where the slope is neither 0 nor infinite, are orthogonal in pairs.

Proof   A proof of this result is outlined in the section exercises.   ■

EXERCISES 17.4

1. Complete the following table dealing with affine planes.

Field	Number of points	Number of lines	Number of points on a line	Number of lines on a point
	25			
GF(3^2)				
		56		
				17
			31	

2. How many parallel classes do each of the affine planes in Exercise 1 determine? How many lines are in each class?

3. Construct the affine plane $AP(\mathbf{Z}_3)$. Determine its parallel classes and the corresponding Latin squares for the classes of finite nonzero slope.

4. Answer Exercise 3, replacing $\mathbf{Z}_3$ by $\mathbf{Z}_5$.

5. Determine each of the following lines.

   a) The line in $AP(\mathbf{Z}_7)$ that is parallel to $y = 4x + 2$ and contains $(3, 6)$.

   b) The line in $AP(\mathbf{Z}_{11})$ that is parallel to $2x + 3y + 4 = 0$ and contains $(10, 7)$.

   c) The line in $AP(F)$, where $F = GF(2^2)$, that is parallel to $10y = 11x + 01$ and contains $(11, 01)$. (See Table 17.7.)

6. Suppose we try to construct an affine plane $AP(\mathbf{Z}_6)$ as we did in this section.

   a) Determine which of the conditions (A1), (A2), and (A3) fail in this situation.

   b) Find how many lines contain a given point P and how many points are on a given line ℓ, for this "geometry."

7. The following provides an outline for a proof of Theorem 17.18.

   a) Consider a parallel class of lines given by $y = mx + b$, where $m \in F$, $m \neq 0$. Show that each line in this class intersects each "vertical" line and each "horizontal" line in exactly one point of $AP(F)$. Thus the configuration obtained by labeling the points of $AP(F)$, as in Figs. 17.4, 17.5, and 17.6, is a Latin square.

   b) To show that the Latin squares corresponding to two different classes, other than the classes of slope 0 or infinite slope, are orthogonal, assume that an ordered pair (i, j) appears more than once when one square is superimposed upon the other. How does this lead to a contradiction?

17.5
BLOCK DESIGNS AND PROJECTIVE PLANES

In this final section, we examine a type of combinatorial design and see how it is related to the structure of a finite geometry.

Example 17.19   Dick (d) and his wife Mary (m) go to New York City with their five children Richard (r), Peter (p), Christopher (c), Brian (b), and Julie (j). While staying in the city they receive three passes each day, for a week, to visit the Empire State Building. Can we make up a schedule for this family so that everyone gets to visit this attraction the same number of times?

The following schedule is one possibility.

(1) b, c, d	**(2)** b, j, r	**(3)** b, m, p
(4) c, j, m	**(5)** c, p, r	**(6)** d, j, p
(7) d, m, r		

Here the result was obtained by trial and error. For a problem of this size such a technique is feasible. However, in general a more effective strategy is needed. Furthermore, in asking for a certain schedule, we may be asking for something that doesn't exist. In the problem above, for example, each pair of family members is together on only one visit. If the family had received four passes each day, we would not be able to construct a schedule that maintained this property.   □

The situation in this example generalizes as follows.

DEFINITION 17.13   Let V be a set with v elements. A collection $\{B_1, B_2, \ldots, B_b\}$ of subsets of V is called a *balanced incomplete block design*, or (v, b, r, k, λ)-*design*, if the following conditions are satisfied:

a) For each $1 \le i \le b$, the subset B_i contains k elements, where k is a fixed constant and $k < v$.

b) Each element $x \in V$ is in $r \ (\le b)$ of the subsets B_i, $1 \le i \le b$.

c) Every pair x, y of elements of V appear together in $\lambda \ (\le b)$ of the subsets B_i, $1 \le i \le b$.   ▬

The elements of V are often called *varieties* because of the early applications in the design of experiments that dealt with tests on fertilizers and plants. The b subsets $B_1, B_2, \ldots, B_b$ of V, are called *blocks*, where each block contains k varieties. The number r is referred to as the *replication number* of the design. Finally, λ is termed the *covalency* for the design. This parameter makes the design balanced in the following sense. For general block designs we have a number λ_{xy} for each pair $x, y \in V$; if λ_{xy} is the same for all pairs of elements from V, then λ represents this common measure and the design is called balanced. In this text we deal only with balanced designs.

Example 17.20   **a)** The schedule in Example 17.19 is an example of a $(7, 7, 3, 3, 1)$-design.

b) For $V = \{1, 2, 3, 4, 5, 6\}$, the ten blocks

$$
\begin{array}{ccccc}
1\ 2\ 4 & 1\ 3\ 4 & 1\ 5\ 6 & 2\ 3\ 6 & 3\ 4\ 6 \\
1\ 2\ 6 & 1\ 3\ 5 & 2\ 3\ 5 & 2\ 4\ 5 & 4\ 5\ 6
\end{array}
$$

constitute a $(6, 10, 5, 3, 2)$-design.

c) If F is a finite field, with $|F| = n$, then the affine plane $AP(F)$ yields an $(n^2, n^2 + n, n + 1, n, 1)$-design. Here the varieties are the n^2 points in $AP(F)$; the $n^2 + n$ lines are the blocks of the design.   □

At this point there are five parameters determining our design. We now examine how these parameters are related.

THEOREM 17.19   For a (v, b, r, k, λ)-design, (1) $vr = bk$ and (2) $\lambda(v - 1) = r(k - 1)$.

Proof   **(1)** With b blocks in the design and k elements per block, listing all the elements of the blocks, we get bk symbols. This collection of symbols consists of the elements of V with each element appearing r times, for a total of vr symbols. Hence $vr = bk$.

(2) For this property we introduce the *pairwise incidence matrix* A for the design. With $|V| = v$, let $t = \binom{v}{2}$, the number of pairs of elements in V. We construct the $t \times b$ matrix $A = (a_{ij})$ by defining $a_{ij} = 1$ if the ith pair of elements from V is in the jth block of the design; if not, $a_{ij} = 0$.

$$
\begin{array}{c}
\\
x_1 x_2 \\
x_1 x_3 \\
\vdots \\
\\
x_1 x_v \\
x_2 x_3 \\
\vdots \\
\\
x_{v-1} x_v
\end{array}
\begin{array}{c}
\begin{array}{cccc}
B_1 & B_2 & \cdots & B_b
\end{array} \\
\left[
\begin{array}{cccc}
a_{11} & a_{12} & \cdots & a_{1b} \\
a_{21} & a_{22} & \cdots & a_{2b} \\
\vdots & \vdots & \vdots & \vdots \\
a_{v-1,1} & a_{v-1,2} & \cdots & a_{v-1,b} \\
a_{v,1} & a_{v,2} & \cdots & a_{v,b} \\
\vdots & \vdots & \vdots & \vdots \\
a_{t1} & a_{t2} & \cdots & a_{tb}
\end{array}
\right]
\end{array}
$$

We now count the number of 1's in matrix A in two ways.

a) Consider the rows. Since each pair x_i, x_j, for $1 \le i < j \le v$, appears in λ blocks, it follows that each row contains λ 1's. With t rows in the matrix, the number of 1's is then $\lambda t = \lambda v(v - 1)/2$.

b) Now consider the columns. As each block contains k elements, this determines $\binom{k}{2} = k(k-1)/2$ pairs, and this is the number of 1's in each column of matrix A. With b columns, the total number of 1's is $bk(k-1)/2$.

Then, $\lambda v(v-1)/2 = bk(k-1)/2 = vr(k-1)/2$, so $\lambda(v-1) = r(k-1)$. ∎

When n is a power of a prime, the $(n^2, n^2+n, n+1, n, 1)$-design can be obtained from the affine plane $AP(F)$, where $F = GF(n)$. Here the points are the varieties and the lines are the blocks. We shall now introduce a construction that enlarges $AP(F)$ to what is called a finite projective plane. From this projective plane we can construct the $(n^2+n+1, n^2+n+1, n+1, n+1, 1)$-design. First let us see how these two kinds of planes compare.

DEFINITION 17.14   If $\mathscr{P}'$ is a finite set of points and $\mathscr{L}'$ a set of lines formed from subsets of $\mathscr{P}'$, then the (finite) plane based on $\mathscr{P}'$ and $\mathscr{L}'$ is called a *projective plane* if the following conditions are satisfied.

P1) Two distinct points of $\mathscr{P}'$ are on only one line.

P2) Any two lines from $\mathscr{L}'$ intersect in a unique point.

P3) There are four points in $\mathscr{P}'$, no three of which are collinear. ▬

The difference between the affine and projective planes lies in the condition dealing with the existence of parallel lines. Here the parallel lines of the affine plane based on $\mathscr{P}$ and $\mathscr{L}$ will intersect when the given system is enlarged to the projective plane based on $\mathscr{P}'$ and $\mathscr{L}'$.

The construction proceeds as follows.

Example 17.21   Start with an affine plane $AP(F)$ where $F = GF(n)$. For each point $(x, y) \in \mathscr{P}$, rewrite the point as $(x, y, 1)$. We then think of the points as ordered triples (x, y, z) where $z = 1$. Rewrite the equations of the lines $x = c$ and $y = mx + b$ in $AP(F)$ as $x = cz$ and $y = mx + bz$, where $z = 1$. We still have our original $AP(F)$, but with a change of notation.

Add the set of points $\{(1, 0, 0)\} \cup \{(x, 1, 0) \mid x \in F\}$ to $\mathscr{P}$ to get the set $\mathscr{P}'$. Then $|\mathscr{P}'| = n^2 + n + 1$. Let ℓ_∞ be the subset of $\mathscr{P}'$ consisting of these new points. This new line can be given by the equation $z = 0$, with the stipulation that we never have $x = y = z = 0$. Hence $(0, 0, 0) \notin \mathscr{P}'$.

Let us examine these ideas for the affine plane $AP(\mathbf{Z}_2)$. Here $\mathscr{P} = \{(0, 0), (1, 0), (0, 1), (1, 1)\}$, so

$$\mathscr{P}' = \{(0, 0, 1), (1, 0, 1), (0, 1, 1), (1, 1, 1)\} \cup \{(1, 0, 0), (0, 1, 0), (1, 1, 0)\}.$$

The six lines in $\mathscr{L}$ were originally

$x = 0$: $\{(0, 0), (0, 1)\}$ $y = 0$: $\{(0, 0), (1, 0)\}$ $y = x$: $\{(0, 0), (1, 1)\}$
$x = 1$: $\{(1, 0), (1, 1)\}$ $y = 1$: $\{(0, 1), (1, 1)\}$ $y = x + 1$: $\{(0, 1), (1, 0)\}$

We now rewrite these as

$$\begin{array}{ccc} x = 0 & y = 0 & y = x \\ x = z & y = z & y = x + z \end{array}$$

and add a new line ℓ_∞ defined by $z = 0$. These constitute the set of lines $\mathscr{L}'$ for our projective plane. And now at this point we consider z as a *variable*. Consequently, the line $x = z$ consists of the points $(0, 1, 0)$, $(1, 0, 1)$, and $(1, 1, 1)$. In fact, each line of $\mathscr{L}$ that contained two points will now contain three points when considered in $\mathscr{L}'$. The set $\mathscr{L}'$ consists of the following seven lines.

$x = 0$: $\{(0, 0, 1), (0, 1, 0), (0, 1, 1)\}$ $y = z$: $\{(1, 0, 0), (0, 1, 1), (1, 1, 1)\}$

$y = 0$: $\{(0, 0, 1), (1, 0, 0), (1, 0, 1)\}$ $y = x$: $\{(0, 0, 1), (1, 1, 0), (1, 1, 1)\}$

$x = z$: $\{(0, 1, 0), (1, 0, 1), (1, 1, 1)\}$ $y = x + z$: $\{(0, 1, 1), (1, 1, 0), (1, 0, 1)\}$

$z = 0$ (ℓ_∞): $\{(1, 0, 0), (0, 1, 0), (1, 1, 0)\}$

In the original affine plane the lines $x = 0$ and $x = 1$ were parallel, because no point in this plane satisfied both equations simultaneously. Here in this new system $x = 0$ and $x = z$ intersect in the point $(0, 1, 0)$, so they are no longer parallel in the sense of $AP(\mathbf{Z}_2)$. Likewise, $y = x$ and $y = x + 1$ were parallel in $AP(\mathbf{Z}_2)$, whereas here the lines $y = x$ and $y = x + z$ intersect at $(1, 1, 0)$. We depict this projective plane based on $\mathscr{P}'$ and $\mathscr{L}'$ as shown in Fig. 17.7. Here the "circle" through $(1, 0, 1)$, $(1, 1, 0)$, and $(0, 1, 1)$ is the line $y = x + z$. Note that every line intersects ℓ_∞, which is often called the *line at infinity*. This line consists of the three *points at infinity*. We define two lines to be parallel in the projective plane when they intersect in a point at infinity, or on ℓ_∞.

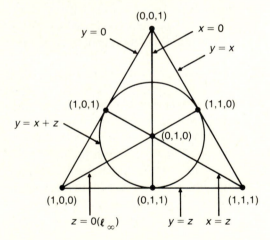

Figure 17.7

This projective plane provides us with a $(7, 7, 3, 3, 1)$-design like the one we developed by trial and error in Example 17.19.   □

We generalize the results of Example 17.21 as follows: Let n be a power of a prime. The affine plane $AP(F)$, for $F = GF(n)$, provides an example of an $(n^2, n^2 + n, n + 1, n, 1)$-design. In $AP(F)$ the $n^2 + n$ lines fall into $n + 1$ parallel classes. For each parallel class we add a point at infinity to $AP(F)$. The point $(0, 1, 0)$ is added for the class of lines $x = cz$, $c \in F$; the point $(1, 0, 0)$ for the class of lines $y = bz$, $b \in F$. When $m \in F$ and $m \neq 0$, then we add the point $(m^{-1}, 1, 0)$

for the class of lines $y = mx + bz, b \in F$. The line at infinity, ℓ_∞, is then defined as the set of $n + 1$ points at infinity. In this way we obtain the projective plane over $GF(n)$, which has $n^2 + n + 1$ points and $n^2 + n + 1$ lines. Here each point is on $n + 1$ lines, and each line contains $n + 1$ points. Furthermore, any two points in this plane are on only one line. Consequently, we have an example of an $(n^2 + n + 1, n^2 + n + 1, n + 1, n + 1, 1)$-design.

EXERCISES 17.5

1. Let $V = \{1, 2, \ldots, 9\}$. Determine the values of v, b, r, k, and λ for the design given by the blocks

$$
\begin{array}{cccccc}
1\ 2\ 6 & 1\ 4\ 7 & 2\ 3\ 4 & 2\ 7\ 9 & 3\ 7\ 8 & 4\ 6\ 8 \\
1\ 3\ 5 & 1\ 8\ 9 & 2\ 5\ 8 & 3\ 6\ 9 & 4\ 5\ 9 & 5\ 6\ 7
\end{array}
$$

2. Find an example of a $(4, 4, 3, 3, \lambda)$-design.

3. Find an example of a $(7, 7, 4, 4, \lambda)$-design.

4. Complete the following table so that the parameters v, b, r, k, λ in any row are possible in a balanced incomplete block design.

v	b	r	k	λ
4			3	2
9	12		3	
10		9		2
13		4	4	
	30	10		3

5. Is it possible to have a (v, b, r, k, λ)-design where (a) $b = 28$, $r = 4$, $k = 3$? (b) $v = 17$, $r = 8$, $k = 5$?

6. Given a (v, b, r, k, λ)-design with $b = v$, prove that if v is even, then λ is even.

7. A (v, b, r, k, λ)-design is called a *triple system* if $k = 3$. When $k = 3$ and $\lambda = 1$ we call the design a *Steiner triple system*.

   a) Prove that in every triple system, $\lambda(v - 1)$ is even and $\lambda v(v - 1)$ is divisible by 6.

   b) Prove that in every Steiner triple system, v is congruent to 1 or 3 modulo 6.

8. Verify that the following blocks constitute a Steiner triple system on nine varieties.

$$
\begin{array}{cccccc}
1\ 2\ 8 & 1\ 4\ 7 & 2\ 3\ 4 & 2\ 7\ 9 & 3\ 8\ 9 & 4\ 6\ 8 \\
1\ 3\ 5 & 1\ 6\ 9 & 2\ 5\ 6 & 3\ 6\ 7 & 4\ 5\ 9 & 5\ 7\ 8
\end{array}
$$

9. In a Steiner triple system with $b = 12$, find the values of v and r.

10. In each of the following, $\mathscr{P}'$ is a set of points and $\mathscr{L}'$ a set of lines formed from subsets of $\mathscr{P}'$. Which of the conditions (P1), (P2), and (P3) of Definition 17.14 hold for the given $\mathscr{P}'$ and $\mathscr{L}'$?

a) $\mathscr{P}' = \{a, b, c\}$
   $\mathscr{L}' = \{\{a, b\}, \{a, c\}, \{b, c\}\}$

b) $\mathscr{P}' = \{(x, y, z) \mid x, y, z \in \mathbf{R}\} = \mathbf{R}^3$
   $\mathscr{L}'$ is the set of all lines in $\mathbf{R}^3$.

c) $\mathscr{P}'$ is the set of all lines in $\mathbf{R}^3$ that pass through $(0, 0, 0)$.
   $\mathscr{L}'$ is the set of all planes in $\mathbf{R}^3$ that pass through $(0, 0, 0)$.

11. Bowling teams of five students each are formed from a class of 15 college freshmen. Each of the students bowls on the same number of teams; each pair of students bowls together on two teams. (a) How many teams are there in all? (b) On how many different teams does each student bowl?

12. In a programming class Professor Madge has n students, and she wants to assign m students to each of p computer projects. Each student must be assigned to the same number of projects. (a) In how many projects will each individual student be involved? (b) In how many projects will each pair of students be involved?

13. a) If a projective plane has six lines through every point, how many points does this projective plane have in all?

   b) If there are 57 points in a projective plane, how many points lie on each line of the plane?

14. In constructing the projective plane from $AP(\mathbf{Z}_2)$ in Example 17.21, why didn't we want to include the point $(0, 0, 0)$ in the set $\mathscr{P}'$?

15. Determine the values of v, b, r, k, and λ for the balanced incomplete block design associated with the projective plane that arises from $AP(F)$ for the following choices of F: (a) $\mathbf{Z}_5$; (b) $\mathbf{Z}_7$; (c) $GF(8)$.

16. a) List the points and lines in $AP(\mathbf{Z}_3)$. How many parallel classes are there for this finite geometry? What are the parameters for the associated balanced incomplete block design?

   b) List the points and lines for the projective plane that arises from $AP(\mathbf{Z}_3)$. Determine the points on ℓ_∞ and use them to determine the "parallel" classes for this geometry. What are the parameters for the associated balanced incomplete block design?

17.6
SUMMARY AND HISTORICAL REVIEW

The structure of a field was first developed in Chapter 14. In the present chapter, we examined polynomial rings and their role in the structure of finite fields, directing our attention to applications in finite geometries and combinatorial designs.

In Chapter 15 we saw that the order of a finite Boolean algebra could only be a power of 2. Now we find that for a finite field the order can only be a power of a prime and that for any prime p and any $n \in \mathbf{Z}^+$, there is only one field, up to isomorphism, of order p^n. This field is denoted by $GF(p^n)$, in honor of the French mathematician Evariste Galois (1811–1832).

The finite fields $(\mathbf{Z}_p, +, \cdot)$, for p a prime, were obtained in Chapter 14 by means of the equivalence relation, congruence modulo p, defined on $\mathbf{Z}$. Using these finite fields, we developed here the integral domains $\mathbf{Z}_p[x]$. Then, with $s(x)$ an irreducible polynomial of degree n in $\mathbf{Z}_p[x]$, a similar equivalence relation, namely congruence modulo $s(x)$, gave us a set of p^n equivalence classes, denoted $\mathbf{Z}_p[x]/(s(x))$. These p^n equivalence classes became the elements of the field $GF(p^n)$. (Although we did not prove every possible result in general, it can be shown that over any finite field $\mathbf{Z}_p$, there is an irreducible polynomial of degree n for any $n \in \mathbf{Z}^+$.)

The theory of finite fields was developed by Galois in his work on the solutions of polynomial equations. As we mentioned in the summary of Chapter 16, the study of polynomial equations was an area of mathematical research that challenged many mathematicians from the sixteenth to the nineteenth centuries. In the nineteenth century, Niels Henrik Abel (1802–1829) first showed that the solution of the general quintic could not be given by radicals. Galois showed that for any polynomial of degree n over a field F, there is a corresponding group G that is isomorphic to a subgroup of S_n, the group of permutations of $\{1, 2, 3, \ldots, n\}$. The essence of Galois's work is that such a polynomial equation can be solved by (addition, subtraction, multiplication, division, and) radicals if its corresponding group is *solvable*. Now what makes a group solvable? We say that a group G is solvable if it has a chain of subgroups $G = K_1 \supset K_2 \supset K_3 \supset \ldots \supset K_t = \{e\}$, where for all $2 \le i \le t$, K_i is a normal subgroup of K_{i-1} (that is, $xyx^{-1} \in K_i$ for all $y \in K_i$ and for all $x \in K_{i-1}$), and $|K_{i-1}| \div |K_i|$ is a prime. One finds that all subgroups of S_i, $1 \le i \le 4$, are solvable, but for $n \ge 5$ there are subgroups of S_n that are not solvable.

Though it seems that Galois theory is concerned predominantly with groups, there is a great deal more on the theory of fields that we have not mentioned. As a consequence of Galois's work, the areas of field theory and finite group theory became topics of great mathematical interest.

For more on *Galois theory*, the reader will find Chapter 6 of the text by V. Larney [7] and Chapter 12 in the book by N. H. McCoy and T. Berger [9] good places to start. Chapter 5 of I. N. Herstein [5] has more on the topic. A detailed presentation can be found in the text by O. Zariski and P. Samuel [13]. Appendix E in the text by V. Larney [7] includes an interesting short account of the life of Galois; more on his life can be found in the somewhat fictional account by L. Infeld [6]. The article by T. Rothman [10] provides a contemporary discussion of the inaccuracies and myths surrounding the life, and especially the death, of Galois.

The Latin squares, combinatorial designs, and finite geometries of the later sections of the chapter showed us how the finite field structure entered into problems of design. Dating back to the time of Leonhard Euler (1707–1783) and the problem of the "thirty-six officers," the study of orthogonal Latin squares has been developed considerably since 1900, and especially since 1960 with the work of R. C. Bose, S. S. Shrikhande, and E. T. Parker. Chapter 7 of the monograph by H. Ryser [11] provides the details of their accomplishments. The text by C. L. Liu [8] includes ideas from coding theory in its discussion of Latin squares.

The study of finite geometries can be traced back to the work of Gino Fano who, in 1892, considered a finite three-dimensional geometry consisting of 15 points, 35 lines, and 15 planes. However, it was not until 1906 that these geometries gained any notice, when O. Veblen and W. Bussey began their study of finite projective geometries. For more on this topic, the reader should find the texts by A. Albert and R. Sandler [1] and H. Dorwart [3] very interesting. The text by P. Dombowski [2] provides an extensive coverage for those seeking more advanced study.

Finally, the notion of designs was first studied by statisticians in the area called the design of experiments. Through the research of R. A. Fisher and his followers, this area has come to play an important role in the modern theory of statistical analysis. In our development, we examined conditions under which a (v, b, r, k, λ)-design could exist and how such designs were related to affine planes and finite projective planes. The text by M. Hall, Jr. [4] provides more on this topic, as does the work by A. Street and W. Wallis [12]. Chapter XIII of reference [12] includes material relating designs and coding theory.

REFERENCES

1. Albert, A. Adrian, and Sandler, R. *An Introduction to Finite Projective Planes.* New York: Holt, 1968.
2. Dombowski, Peter. *Finite Geometries.* New York: Springer-Verlag, 1968.
3. Dorwart, Harold L. *The Geometry of Incidence.* Englewood Cliffs, N.J.: Prentice-Hall, 1966.
4. Hall, Marshall, Jr. *Combinatorial Theory.* Waltham, Mass.: Blaisdell, 1967.
5. Herstein, I. N. *Topics in Algebra,* 2nd ed. Lexington, Mass.: Xerox College Publishing, 1975.
6. Infeld, Leopold. *Whom the Gods Love.* New York: McGraw-Hill, 1948.
7. Larney, Violet H. *Abstract Algebra: A First Course.* Boston: Prindle, Weber & Schmidt, 1975.
8. Liu, C. L. *Topics in Combinatorial Mathematics.* Mathematical Association of America, 1972.
9. McCoy, Neal H., and Berger, Thomas R. *Algebra: Groups, Rings, and Other Topics.* Boston: Allyn and Bacon, 1977.
10. Rothman, Tony. "Genius and Biographers: The Fictionalization of Evariste Galois." *The American Mathematical Monthly* 89, no. 2 (1982): pp. 84–106.
11. Ryser, Herbert J. *Combinatorial Mathematics.* Carus Mathematical Monographs, Number 14, Mathematical Association of America, 1963.
12. Street, Anne Penfold, and Wallis, W. D. *Combinatorial Theory: An Introduction.* Winnipeg, Canada: The Charles Babbage Research Center, 1977.
13. Zariski, Oscar, and Samuel, Pierre. *Commutative Algebra,* Vol. I. New York: Van Nostrand, 1958.

MISCELLANEOUS EXERCISES

1. Determine n if over $GF(n)$ there are 6561 monic polynomials of degree 5 with no constant term.

2. **a)** Let $f(x) = a_n x^n + \cdots + a_1 x + a_0 \in \mathbf{Z}[x]$. If $r/s \in \mathbf{Q}$, with $(r, s) = 1$ and $f(r/s) = 0$, prove that $s \mid a_n$ and $r \mid a_0$.

   b) Find the rational roots, if any exist, of the following polynomials over $\mathbf{Q}$. Factor $f(x)$ in $\mathbf{Q}[x]$.

   (i) $f(x) = 2x^3 + 3x^2 - 2x - 3$

   (ii) $f(x) = x^4 + x^3 - x^2 - 2x - 2$

   (iii) $f(x) = x^4 + x^3 - 3x^2 - 5x - 2$

   (iv) $f(x) = 3x^4 - 2x^3 - 5x^2 - 4x - 4$

   c) Show that the polynomial $f(x) = x^{100} - x^{50} + x^{20} + x^3 + 1$ has no rational root.

3. **a)** Verify that $s(x) = x^2 + 1$ is irreducible in $\mathbf{R}[x]$.

   b) Show that the field $\mathbf{R}[x]/(s(x))$ is isomorphic to the field $\mathbf{C}$ of complex numbers.

4. Verify that the polynomial $f(x) = x^4 + x^3 + x + 1$ is reducible over any field F (finite or infinite).

5. If p is a prime, prove that in $\mathbf{Z}_p[x]$,

$$x^p - x = \prod_{a \in \mathbf{Z}_p} (x - a).$$

6. For any field F, let $f(x) = x^n + a_{n-1}x^{n-1} + \cdots + a_1 x + a_0 \in F[x]$. If $r_1, r_2, \ldots, r_n$ are the roots of $f(x)$, and $r_i \in F$ for all $1 \le i \le n$, prove that

   a) $-a_{n-1} = r_1 + r_2 + \cdots + r_n$

   b) $(-1)^n a_0 = r_1 r_2 \cdots r_n$.

7. Let $f(x) \in \mathbf{R}[x]$. If $a + bi \in \mathbf{C}$ and $f(a + bi) = 0$, prove that $f(\overline{a + bi}) = f(a - bi) = 0$. (The properties noted in Miscellaneous Exercise 4(a) of Chapter 14 should prove helpful here.)

8. Let R be a commutative ring with unity. Under what condition(s) is each of the following statements true?

   a) For all $a, b \in R$, $(a + b)^2 = a^2 + b^2$.

   b) For all $a, b \in R$, $(a + b)^3 = a^3 + b^3$.

9. Four of the seven blocks in a $(7, 7, 3, 3, 1)$-design are $\{1, 3, 7\}$, $\{1, 5, 6\}$, $\{2, 6, 7\}$, and $\{3, 4, 6\}$. Determine the other three blocks.

10. Find the values of b and r for a Steiner triple system where $v = 63$.

11. **a)** If a projective plane has 73 points, how many points lie on each line?

    b) If each line in a projective plane passes through 10 points, how many lines are there in the projective plane?

12. A projective plane is coordinatized with the elements of a field F. If this plane contains 91 lines, what are $|F|$ and $\text{char}(F)$?

13. Let $V = \{x_1, x_2, \ldots, x_v\}$ be the set of varieties and $\{B_1, B_2, \ldots, B_b\}$ the collection of blocks for a (v, b, r, k, λ)-design. We define the *incidence matrix* A for the design by

$$A = (a_{ij})_{v \times b}, \qquad \text{where } a_{ij} = \begin{cases} 1, & \text{if } x_i \in B_j \\ 0, & \text{otherwise.} \end{cases}$$

a) How many 1's are there in each row and column of A?

b) Let $J_{m \times n}$ be the $m \times n$ matrix where every entry is 1. For $J_{n \times n}$ we write J_n. Prove that for the incidence matrix A, $A \cdot J_b = r \cdot J_{v \times b}$ and $J_v \cdot A = k \cdot J_{v \times b}$.

c) Show that

$$A \cdot A^{\text{tr}} = \begin{bmatrix} r & \lambda & \lambda & \ldots & \lambda \\ \lambda & r & \lambda & \ldots & \lambda \\ \lambda & \lambda & r & \ldots & \lambda \\ \ldots & \ldots & \ldots & \ldots & \ldots \\ \lambda & \lambda & \lambda & \ldots & r \end{bmatrix} = (r - \lambda)I_v + \lambda J_v,$$

where I_v is the $v \times v$ (multiplicative) identity.

d) Prove that $\det(A \cdot A^{\text{tr}}) = (r - \lambda)^{v-1}[r + (v - 1)\lambda] = (r - \lambda)^{v-1} rk$.

Answers

Sections 1.1 and 1.2 – p. 9

1. a) 13   **b)** 40   **c)** The rule of sum in part (a); the rule of product in part (b).

3. a) 288   **b)** 24   **c)** 8

5. 2^9 **7.** $4! = 24$

9. $26 + 26(36) + 26(36)^2 + \cdots + 26(36)^7 = \sum_{i=0}^{7} 26(36)^i$

11. a) $7! = 5040$ **b)** $(4!)(3!) = 144$ **c)** $(5!)(3!) = 720$ **d)** $2(3!)(4!) = 288$
    e) $(4!)(3!) = 144$

13. a) $12!/(3!2!2!2!)$ **b)** $2[11!/(3!2!2!2!)]$
    c) $[7!/(2!2!)][6!/(3!2!)]$ **d)** 1260 in part (c); 166,320 in part (a).

15. a) $n = 10$ **b)** $n = 5$ **c)** $n = 5$

17. a) $(10!)/(2!7!) = 360$ **b)** $(10!)/(2!7!) = 360$
    c) Let x, y, and z be any real numbers and let m, n, and p be any nonnegative
    integers. The number of paths from (x, y, z) to $(x + m, y + n, z + p)$, as described
    in part (a), is $(m + n + p)!/(m!n!p!)$.
    d) 210

19. a) $9 \times 9 \times 8 \times 7 \times 6 \times 5 = 136{,}080$ **b)** 9×10^5
      (i) (a) 68,880   (b) 450,000
      (ii) (a) 28,560   (b) 180,000
      (iii) (a) 33,600   (b) 225,000

21. a) 2^{10} **b)** 3^{10} **c)** $4^{10}; 5^{10}$

23. a) $6!$ **b)** $2(5!) = 240$

25. a) $10 + 128 + 128^2 = 16{,}522$ blocks; 8,459,264 bytes
    b) $10 + 128 + 128^2 + 128^3 = 2{,}113{,}674$ blocks; 1,082,201,088 bytes

Section 1.3 – p. 20

1. $\binom{6}{2} = 6!/(2!4!) = 15$. The selections of size 2 are $ab, ac, ad, ae, af, bc, bd, be, bf, cd, ce,$
$cf, de, df,$ and ef.

3. a) $\binom{20}{12} = 125{,}970$ **b)** $\binom{10}{6}\binom{10}{6} = 44{,}100$ **c)** $\sum_{i=1}^{5} \binom{10}{12-2i}\binom{10}{2i}$
    d) $\sum_{i=7}^{10} \binom{10}{i}\binom{10}{12-i}$ **e)** $\sum_{i=8}^{10} \binom{10}{i}\binom{10}{12-i}$

5. a) $\binom{8}{2}=28$ **b)** $\binom{8}{4}=70$ **c)** $\binom{8}{6}=28$ **d)** $\binom{8}{6}+\binom{8}{7}+\binom{8}{8}=37$

7. a) 120   **b)** 56   **c)** 110

9. a) $\binom{12}{3}\binom{9}{3}\binom{6}{3}\binom{3}{3}=12!/[(3!)^4]=369{,}600$ **b)** $\binom{12}{4}\binom{8}{4}\binom{4}{2}\binom{2}{2}=12!/[(4!)^2(2!)^2]=207{,}900$

11. $\binom{8}{4}\left(\dfrac{7!}{4!2!}\right)=7350$

13. a) $\binom{15}{2}=105$ **b)** $\binom{25}{3}=2300;\ \binom{25}{3};\ \binom{25}{4}=12{,}650$

15. a) $\binom{10}{3}+\binom{10}{1}\binom{9}{1}+\binom{10}{1}=220$ **b)** $\binom{10}{4}+\binom{10}{2}+\binom{10}{1}\binom{9}{2}+\binom{10}{1}\binom{9}{1}=705$
c) $2^{10}(\sum_{i=0}^{5}\binom{10}{2i}))$

17. a) $\binom{12}{9}$ **b)** $\binom{12}{9}(2^3)$ **c)** $\binom{12}{9}(2^9)(-3)^3$

19. a) $\binom{4}{1,1,2}=12$ **b)** $\binom{4}{0,1,1,2}=12$ **c)** $\binom{4}{1,1,2}(2)(-1)(-1)^2=-24$
d) $\binom{4}{1,1,2}(-2)(3^2)=-216$ **e)** $\binom{8}{3,2,1,2}(2^3)(-1)^2(3)(-2)^2=161{,}280$

21.
$$\binom{2n}{n}+\binom{2n}{n-1}=\frac{(2n)!}{n!n!}+\frac{(2n)!}{(n-1)!(n+1)!}$$
$$=\frac{(2n)!(n+1)^2}{(n+1)!(n+1)!}+\frac{(2n)!(n)(n+1)}{(n+1)!(n+1)!}$$
$$=\left(\frac{1}{2}\right)\left[\frac{(2n)!(2n+2)(n+1)+(2n)!(2n+2)n}{(n+1)!(n+1)!}\right]$$
$$=\left(\frac{1}{2}\right)\binom{2n+2}{n+1}$$

23.
$$n\binom{m+n}{m}=n\frac{(m+n)!}{m!n!}=\frac{(m+n)!}{m!(n-1)!}=(m+1)\frac{(m+n)!}{(m+1)(m!)(n-1)!}$$
$$=(m+1)\frac{(m+n)!}{(m+1)!(n-1)!}=(m+1)\binom{m+n}{m+1}$$

25. Consider the expansions of (a) $[(1+x)-x]^n$; (b) $[(2+x)-(x+1)]^n$; and (c) $[(2+x)-x]^n$.

Section 1.4 – p. 28

1. a) $\binom{14}{10}$ **b)** $\binom{9}{5}$ **c)** $\binom{12}{8}$ **3.** $\binom{23}{20}$ **5. a)** 2^5 **b)** 2^n

7. a) $\binom{35}{32}$ **b)** $\binom{31}{28}$ **c)** $\binom{11}{8}$ **d)** 1   **e)** $\binom{43}{40}$ **f)** $\binom{31}{28}-\binom{6}{3}$

9. a) $\binom{14}{5}$ **b)** $\binom{11}{5}+3\binom{10}{4}+3\binom{9}{3}+\binom{8}{2}$

11. a) $\binom{7}{4}$ **b)** $\sum_{i=0}^{3}\binom{9-2i}{7-2i}$ **13. a)** $\binom{16}{12}$ **b)** 5^{12}

15. a) $(5^3)\binom{29}{25}$ **b)** $[4\binom{23}{20}+4^2\binom{18}{15}+4^2\binom{13}{10}+4^3\binom{8}{5}]+[4^2\binom{28}{25}+4\binom{18}{15}+4^2\binom{13}{10}+4^2\binom{8}{5}+4^3]$

17. a) $\binom{8}{6}\binom{34}{31}$ **b)** $\binom{5}{3}\binom{34}{31}$ **19.** $\binom{23}{4}$ **21.** $24{,}310=\sum_{i=1}^{n}i$ (for $n=\binom{12}{3}$)

23. a) Consider the following Pascal program segment, where $i, j, k, m, n,$ and sum are integer variables.

```
sum:=0;
For i:=1 to n do
  For j:=1 to i do
    For k:=1 to j do
      For m:=1 to k do
        sum:=sum+1;
```

After the segment is executed, the value of sum is $\binom{n+3}{4}$, the number of ways to select i, j, k, m from $\{1, 2, 3, \ldots, n\}$, with repetitions allowed and $1 \le m \le k \le j \le i \le n$.

For each value of i, the *For* loops for j, k, and m result in $\binom{i+2}{3}$ executions of the statement   sum := sum + 1. Consequently, $\binom{n+3}{4} = \sum_{i=1}^{n} \binom{i+2}{3}$.

b) From Exercise 22 we know that $\sum_{i=1}^{n} i^2 = (1/6)(n)(n+1)(2n+1)$. We shall use this in step 4 of the following derivation.

(1) $\binom{n+3}{4} = \sum_{i=1}^{n} \binom{i+2}{3}$

(2) $(1/4!)(n+3)(n+2)(n+1)(n) = \sum_{i=1}^{n} (1/6)(i+2)(i+1)(i)$

(3) $(1/4)(n+3)(n+2)(n+1)(n) = \sum_{i=1}^{n} (i^3 + 3i^2 + 2i)$

(4) $(1/4)(n+3)(n+2)(n+1)(n) = \sum_{i=1}^{n} i^3 + (3)(1/6)(n)(n+1)(2n+1)$
$$+ (2)(1/2)(n+1)(n)$$

(5) $(1/4)(n+1)(n)[(n+3)(n+2) - 2(2n+1) - 4] = \sum_{i=1}^{n} i^3$

(6) $(1/4)(n+1)(n)[n^2 + 5n + 6 - 4n - 6] = \sum_{i=1}^{n} i^3$

(7) $(1/4)(n+1)^2 n^2 = \sum_{i=1}^{n} i^3$.

25. a) Place one of the m identical objects into each of the n distinct containers. This leaves $m - n$ identical objects to be placed into the n distinct containers. This results in $\binom{n+(m-n)-1}{m-n} = \binom{m-1}{m-n} = \binom{m-1}{n-1}$ distributions.

b) Place r of these (identical) objects into each of the n distinct containers. We can then put the remaining $m - rn$ identical objects into the n distinct containers in $\binom{n+(m-rn)-1}{m-rn} = \binom{m-1+(1-r)n}{m-rn} = \binom{m-1+(1-r)n}{n-1}$ ways.

Miscellaneous Exercises – p. 34

1. $\binom{4}{1}\binom{7}{7} + \binom{4}{2}\binom{7}{4} + \binom{4}{3}\binom{7}{6}$

3. a) $36 + 36^2 + \cdots + 36^8 = \sum_{i=1}^{8} 36^i$ **b)** $(36 + 36^2 + \cdots + 36^6)$

 c) $(51 + 51^2 + \cdots + 51^8)(36^3)$ **d)** $(51 + 51^2 + \cdots + 51^8)(36 \times 35 \times 34)$

5. a) 10^{25} **b)** $(10)(11)(12) \cdots (34) = 34!/9!$ **c)** $(25!)\binom{24}{9}$

7. a) $C(12, 8)$ **b)** $P(12, 8)$

9. $[(7!/2!)\binom{6}{1}](6)$

11. $(1/10)[10!/(4!3!3!)]$

13. a) (i) $\binom{5}{4} + \binom{5}{2}\binom{4}{2} + \binom{4}{4}$ (ii) $\binom{8}{4} + \binom{6}{2}\binom{5}{2} + \binom{7}{4}$ (iii) $\binom{8}{4} + \binom{6}{2}\binom{5}{2} + \binom{7}{4} - 9$

 b) (i) $\binom{5}{1}\binom{4}{3} + \binom{5}{3}\binom{4}{1}$ (ii) and (iii) $\binom{5}{1}\binom{6}{3} + \binom{7}{3}\binom{4}{1}$

15. a) $(5)(9!)$ **b)** $(3)(8!)$

17. $17 + (10)(2) + \binom{6}{2} + 5\binom{3}{2} + \binom{5}{4}$

19. $\binom{16}{12} - 5\binom{8}{4}$

21. $2\binom{n+r-k-1}{n-k} + (n + r - k - 1)\binom{n+r-k-2}{n-k}$

23. $0 = (1 + (-1))^n = \binom{n}{0} - \binom{n}{1} + \binom{n}{2} - \binom{n}{3} + \cdots + (-1)^n\binom{n}{n}$,

 so $\binom{n}{0} + \binom{n}{2} + \binom{n}{4} + \cdots = \binom{n}{1} + \binom{n}{3} + \binom{n}{5} + \cdots$

25. a) $\binom{9}{1} + \binom{10}{3} + \cdots + \binom{16}{15} = \sum_{k=0}^{8} \binom{9+k}{1+2k}$ **b)** $\sum_{k=0}^{9} \binom{9+k}{2k}$

 c) $n = 2k + 1, k \ge 0$: $\sum_{i=0}^{k} \binom{k+1+i}{1+2i}$

  $n = 2k, k \ge 1$: $\sum_{i=0}^{k} \binom{k+i}{2i}$

27. a) $\binom{r+(n-r)-1}{n-r} = \binom{n-1}{n-r} = \binom{n-1}{r-1}$

    b) $\sum_{r=1}^{n} \binom{n-1}{r-1} = \binom{n-1}{0} + \binom{n-1}{1} + \cdots + \binom{n-1}{n-1} = 2^{n-1}$

29. a) $11!/(7!4!)$ **b)** $[11!/(7!4!)] - [4!/(2!2!)][4!/(3!1!)]$

    c) $[11!/(7!4!)] + [10!/(6!3!1!)] + [9!/(5!2!2!)] + [8!/(4!1!3!)] + [7!/(3!4!)]$

        (in part (a))

        $\{[11!/(7!4!)] + [10!/(6!3!1!)] + [9!/(5!2!2!)] + [8!/(4!1!3!)] + [7!/(3!4!)]\}$

                 $- [\{[4!/(2!2!)] + [3!/(1!1!1!)] + [2!/2!]\} \times \{[4!/(3!1!)] + [3!/(2!1!)]\}]$

        (in part (b))

CHAPTER 2

Fundamentals of Logic

Section 2.1 – p. 47

1. Sentences (a), (c), (e), (f), (g), (h), and (j) are statements. The other four sentences are not.

3. a) If triangle ABC is equilateral, then it is isosceles.

    b) If triangle ABC is not isosceles, then it is not equilateral.

    c) Triangle ABC is equilateral if and only if it is equiangular.

    d) Triangle ABC is isosceles but it is not equilateral.

    e) If triangle ABC is equiangular, then it is isosceles.

5. Propositions (a), (e), (f), and (h) are tautologies.

7. a) $2^5 = 32$ **b)** 2^n

9. a) p: 0; r: 0; s: 0

    b) Statement s must have the truth value 1, but statement p can have the truth value 0 or 1, as can statement r.

11. 27

Section 2.2 – p. 58

1. a) (i)

p	q	r	$q \wedge r$	$p \rightarrow (q \wedge r)$	$p \rightarrow q$	$p \rightarrow r$	$(p \rightarrow q) \wedge (p \rightarrow r)$
0	0	0	0	1	1	1	1
0	0	1	0	1	1	1	1
0	1	0	0	1	1	1	1
0	1	1	1	1	1	1	1
1	0	0	0	0	0	0	0
1	0	1	0	0	0	1	0
1	1	0	0	0	1	0	0
1	1	1	1	1	1	1	1

(ii)

p	q	r	$p \vee q$	$(p \vee q) \rightarrow r$	$p \rightarrow r$	$q \rightarrow r$	$(p \rightarrow r) \wedge (q \rightarrow r)$
0	0	0	0	1	1	1	1
0	0	1	0	1	1	1	1
0	1	0	1	0	1	0	0
0	1	1	1	1	1	1	1
1	0	0	1	0	0	1	0
1	0	1	1	1	1	1	1
1	1	0	1	0	0	0	0
1	1	1	1	1	1	1	1

(iii)

p	q	r	$q \vee r$	$p \to (q \vee r)$	$p \to q$	$\bar{r} \to (p \to q)$
0	0	0	0	1	1	1
0	0	1	1	1	1	1
0	1	0	1	1	1	1
0	1	1	1	1	1	1
1	0	0	0	0	0	0
1	0	1	1	1	0	1
1	1	0	1	1	1	1
1	1	1	1	1	1	1

b)

$$[p \to (q \vee r)] \Leftrightarrow [\bar{r} \to (p \to q)] \qquad \text{From part (a)}$$

$$\Leftrightarrow [\bar{r} \to (\bar{p} \vee q)] \qquad \text{By the 2nd Substitution Rule,}$$
$$\text{and } (p \to q) \Leftrightarrow (\bar{p} \vee q)$$

$$\Leftrightarrow [\overline{(\bar{p} \vee q)} \to \bar{r}] \qquad \text{By the 1st Substitution Rule,}$$
$$\text{and } (s \to t) \Leftrightarrow (\bar{t} \to \bar{s})$$

$$\Leftrightarrow [(\bar{\bar{p}} \wedge \bar{q}) \to r] \qquad \text{By DeMorgan's Law, Double Negation}$$
$$\text{and the 2nd Substitution Rule}$$

$$\Leftrightarrow [(p \wedge \bar{q}) \to r] \qquad \text{By Double Negation and the}$$
$$\text{2nd Substitution Rule}$$

3. a) For any statement s, $s \vee \bar{s} \Leftrightarrow T_0$. Replace each occurrence of s by $p \vee (q \wedge r)$, and the result follows by the 1st Substitution Rule.

b) For any statements s, t, we have $(s \to t) \Leftrightarrow (\bar{t} \to \bar{s})$. Replace each occurrence of s by $p \vee q$ and each occurrence of t by r, and the result is a consequence of the 1st Substitution Rule.

c) For any statements a, b, and c, the Distributive Law of $\vee$ over $\wedge$ states that $a \vee (b \wedge c) \Leftrightarrow (a \vee b) \wedge (a \vee c)$. The result given here is a tautology when we apply the 1st Substitution Rule with the following replacements: a by $[(p \vee q) \to r]$; b by s; and, c by t.

5. a) Kelsey placed her studies before her interest in cheerleading, but she (still) did not get a good education.

b) Norma is not doing her mathematics homework, or Karen is not practicing her piano lesson.

c) Logan went on vacation and didn't worry about traveling by airplane, but he (still) did not enjoy himself.

d) Harold did not pass his Pascal course and he didn't finish his data structures project, but he (still) graduated at the end of the semester.

7. a)

p	q	$(\bar{p} \vee q) \wedge (p \wedge (p \wedge q))$	$p \wedge q$
0	0	0	0
0	1	0	0
1	0	0	0
1	1	1	1

b) $(\bar{p} \wedge q) \vee (p \vee (p \vee q)) \Leftrightarrow p \vee q$

9. a) Converse: If quadrilateral $ABCD$ is a rectangle, then quadrilateral $ABCD$ is a square.

Inverse: If quadrilateral $ABCD$ is not a square, then quadrilateral $ABCD$ is not a rectangle.

Contrapositive: If quadrilateral $ABCD$ is not a rectangle, then quadrilateral $ABCD$ is not a square.

b) Converse: If n^2 is an even integer, then n is an even integer.

Inverse: If n is not an even integer, then n^2 is not an even integer.

Contrapositive: If n^2 is not an even integer, then n is not an even integer.

c) Converse: If two distinct lines in the xy plane have the same slope, then they are parallel.

Inverse: If two distinct lines in the xy plane are not parallel, then they do not have the same slope.

Contrapositive: If two distinct lines in the xy plane do not have the same slope, then they are not parallel.

11. a) $(q \to r) \lor \bar{p}$ **b)** $(\bar{q} \lor r) \lor \bar{p}$

13.

z	Figure 2.1(a)	Figure 2.1(b)
a) 2	20	$10 + 1 = 11$
b) 6	20	$10 + 5 = 15$
c) 9	20	$10 + 8 = 18$
d) 10	20	$10 + 9 = 19$
e) 15	20	$10 + 10 = 20$
f) $n > 10$	20	$10 + 10 = 20$

15.

p	q	r	$[(p \leftrightarrow q) \land (q \leftrightarrow r) \land (r \leftrightarrow p)]$	$[(p \to q) \land (q \to r) \land (r \to p)]$
0	0	0	1	1
0	0	1	0	0
0	1	0	0	0
0	1	1	0	0
1	0	0	0	0
1	0	1	0	0
1	1	0	0	0
1	1	1	1	1

17. a) $(p \uparrow p)$ **b)** $(p \uparrow p) \uparrow (q \uparrow q)$ **c)** $(p \uparrow q) \uparrow (p \uparrow q)$ **d)** $p \uparrow (q \uparrow q)$
e) $(r \uparrow s) \uparrow (r \uparrow s)$, where r stands for $p \uparrow (q \uparrow q)$ and s for $q \uparrow (p \uparrow p)$

19.

p	q	$\overline{(p \downarrow q)}$	$(\bar{p} \uparrow \bar{q})$	$\overline{(\bar{p} \uparrow \bar{q})}$	$(\bar{p} \downarrow \bar{q})$
0	0	0	0	0	0
0	1	1	1	0	0
1	0	1	1	0	0
1	1	1	1	1	1

Section 2.3 – p. 73

1. a)

p	q	r	$p \to q$	$(p \lor q)$	$(p \lor q) \to r$
0	0	0	1	0	1
0	0	1	1	0	1
0	1	0	1	1	0
0	1	1	1	1	1
1	0	0	0	1	0
1	0	1	0	1	1
1	1	0	1	1	0
1	1	1	1	1	1

The validity of the argument follows from the results in the last row. (The first seven rows can be ignored.)

b)

p	q	r	$(p \wedge q) \to r$	$\overline{q}$	$p \to \overline{r}$	$\overline{p} \vee \overline{q}$
0	0	0	1	1	1	1
0	0	1	1	1	1	1
0	1	0	1	0	1	1
0	1	1	1	0	1	1
1	0	0	1	1	1	1
1	0	1	1	1	0	1
1	1	0	0	0	1	0
1	1	1	1	0	0	0

The validity of the argument follows from the results in rows 1, 2, and 5 of the table. The results in the other five rows may be ignored.

c)

p	q	r	$p \to q$	$p \vee r$	$(p \vee r) \to q$
0	0	0	1	0	1
0	0	1	1	1	0
0	1	0	1	0	1
0	1	1	1	1	1
1	0	0	0	1	0
1	0	1	0	1	0
1	1	0	1	1	1
1	1	1	1	1	1

The results in the last two rows of this table establish the validity of the given argument. The results in the first six rows of the table may be disregarded.

3. a) If p has the truth value 0, then so does $p \wedge q$.

b) When $p \vee q$ has the truth value 0, then the truth value of p (and that of q) is 0.

c) If q has the truth value 0, then the truth value of $[(p \vee q) \wedge \overline{p}]$ is 0, regardless of the truth value of p.

d) The statement $q \vee s$ has the truth value 0 only when each of q, s has the truth value 0. Then $(p \to q)$ has the truth value 1 when p has the truth value 0; $(r \to s)$ has the truth value 1 when r has the truth value 0. But then $(p \vee r)$ must have the truth value 0, not 1.

e) For $(\overline{p} \vee \overline{r})$ the truth value is 0 when both p and r have the truth value 1. This then forces q and s to have the truth value 1, in order for $(p \to q)$ and $(r \to s)$ to have the truth value 1. However, this results in the truth value 0 for $(\overline{q} \vee \overline{s})$.

f) The only time that $[(p \vee r) \to (q \vee r)]$ has the truth value 0 is for the assignment p: 1; r: 0; and q: 0. But under these conditions $(p \to q)$ has the truth value 0, not 1.

g) The statement $[(q \to r) \to (p \to r)]$ has the truth value 0 only for the assignment p: 1; q: 0; and r: 0. But then $(p \to q)$ will have the truth value 0.

5. (1) and (2)   Hypothesis

(3)   (1), (2), and the Rule of Detachment

(4)   Hypothesis

(5)   (4) and $(r \to \overline{q}) \Leftrightarrow (\overline{\overline{q}} \to \overline{r}) \Leftrightarrow (q \to \overline{r})$

(6)   (3), (5), and the Rule of Detachment

(7)   Hypothesis

(8)   (6), (7), and the Rule of Disjunctive Syllogism

(9)   (8) and the Rule of Disjunctive Amplification

7. a) (1)   Hypothesis (The Negation of the Conclusion)

(2)   (1) and $\overline{(\overline{q}\rightarrow s)}\Leftrightarrow\overline{(\overline{\overline{q}}\vee s)}\Leftrightarrow\overline{(q\vee s)}\Leftrightarrow\overline{q}\wedge\overline{s}$

(3)   (2) and the Rule of Conjunctive Simplification

(4)   Hypothesis

(5)   (3), (4), and the Rule of Disjunctive Syllogism

(6)   Hypothesis

(7)   (6) and $(p\rightarrow q)\Leftrightarrow(\overline{p}\vee q)$

(8)   (2) and the Rule of Conjunctive Simplification

(9)   (7), (8), and the Rule of Disjunctive Syllogism

(10)   Hypothesis

(11)   (9), (10), and the Rule of Disjunctive Syllogism

(12)   (5), (11), and the Rule of Conjunction

(13)   (1), (12), and the Rule for Proof by Contradiction

b) (1) $p\rightarrow q$ Hypothesis

(2) $\overline{q}\rightarrow\overline{p}$ (1) and $(p\rightarrow q)\Leftrightarrow(\overline{q}\rightarrow\overline{p})$

(3) $p\vee r$ Hypothesis

(4) $\overline{p}\rightarrow r$ (3) and $(p\vee r)\Leftrightarrow(\overline{p}\rightarrow r)$

(5) $\overline{q}\rightarrow r$ (2), (4) and the Law of the Syllogism

(6) $\overline{r}\vee s$ Hypothesis

(7) $r\rightarrow s$ (6) and $(\overline{r}\vee s)\Leftrightarrow(r\rightarrow s)$

(8) $\therefore\overline{q}\rightarrow s$ (5), (7) and the Law of the Syllogism

c) (1) $\overline{p}\leftrightarrow q$ Hypothesis

(2) $(\overline{p}\rightarrow q)\wedge(q\rightarrow\overline{p})$ (1) and $(\overline{p}\leftrightarrow q)\Leftrightarrow[(\overline{p}\rightarrow q)\wedge(q\rightarrow\overline{p})]$

(3) $\overline{p}\rightarrow q$ (2) and the Rule of Conjunctive Simplification

(4) $q\rightarrow r$ Hypothesis

(5) $\overline{p}\rightarrow r$ (3), (4) and the Law of the Syllogism

(6) $\overline{r}$ Hypothesis

(7) $\therefore p$ (5), (6) and Modus Tollens

9. If not, then $\sqrt[3]{2}=a/b$, where a and b are positive integers whose only common positive divisor is 1. But $\sqrt[3]{2}=a/b\Rightarrow 2=a^3/b^3\Rightarrow a^3=2b^3\Rightarrow a^3$ is even $\Rightarrow a$ is even, so $a=2c$ for some positive integer c. With $a=2c$, it follows that $2b^3=a^3=8c^3$ or $b^3=4c^3=2(2c^3)$. Consequently b^3 is even, so b is also even. But now that a and b are both even, we contradict the fact that their only common positive divisor is 1. This contradiction arises from the assumption that $\sqrt[3]{2}$ is irrational.

11. a) p: Rochelle gets the supervisor's position.

     q: Rochelle works hard.

     r: Rochelle gets a raise.

     s: Rochelle buys a new car.

$$(p\wedge q)\rightarrow r$$
$$r\rightarrow s$$
$$\overline{s}$$
$$\overline{}$$
$$\therefore\overline{p}\vee\overline{q}$$

(1) $\overline{s}$ Hypothesis

(2) $r\rightarrow s$ Hypothesis

(3) $\overline{r}$ (1), (2) and Modus Tollens

(4) $(p\wedge q)\rightarrow r$ Hypothesis

(5) $\overline{p\wedge q}$ (3), (4) and Modus Tollens

(6) $\therefore\overline{p}\vee\overline{q}$ (5) and $\overline{(p\wedge q)}\Leftrightarrow\overline{p}\vee\overline{q}$

b) p: Dominic goes to the racetrack.

     q: Helen gets mad.

r: Ralph plays cards all night.

s: Carmela gets mad.

t: Veronica is notified.

$$p \rightarrow q$$
$$r \rightarrow s$$
$$(q \vee s) \rightarrow t$$
$$\bar{t}$$
$$\overline{}$$
$$\therefore \bar{p} \wedge \bar{r}$$

(1)	$\bar{t}$	Hypothesis
(2)	$(q \vee s) \rightarrow t$	Hypothesis
(3)	$\overline{(q \vee s)}$	(1), (2) and Modus Tollens
(4)	$\bar{q} \wedge \bar{s}$	(3) and $\overline{(q \vee s)} \Leftrightarrow \bar{q} \wedge \bar{s}$
(5)	$\bar{q}$	(4) and the Rule of Conjunctive Simplification
(6)	$p \rightarrow q$	Hypothesis
(7)	$\bar{p}$	(5), (6) and Modus Tollens
(8)	$\bar{s}$	(4) and the Rule of Conjunctive Simplification
(9)	$r \rightarrow s$	Hypothesis
(10)	$\bar{r}$	(8), (9) and Modus Tollens
(11)	$\therefore \bar{p} \wedge \bar{r}$	(7), (10) and the Rule of Conjunction

c) *p*: Norma goes to her Tuesday morning meeting.

  q: Norma gets up early Tuesday morning.

  r: Norma goes to the rock concert Monday evening.

  s: Norma does not get home (Monday evening) until after 11:00 p.m.

  t: Norma goes to work after less than seven hours sleep.

$$p \rightarrow q$$
$$r \rightarrow s$$
$$(s \wedge q) \rightarrow t$$
$$\bar{t}$$
$$\overline{}$$
$$\therefore \bar{r} \vee \bar{p}$$

(1)	$(s \wedge q) \rightarrow t$	Hypothesis
(2)	$\bar{t}$	Hypothesis
(3)	$\overline{(s \wedge q)}$	(1), (2) and Modus Tollens
(4)	$\bar{s} \vee \bar{q}$	(3) and $\overline{(s \wedge q)} \Leftrightarrow \bar{s} \vee \bar{q}$
(5)	$p \rightarrow q$	Hypothesis
(6)	$r \rightarrow s$	Hypothesis
(7)	$\therefore \bar{r} \vee \bar{p}$	(5), (6), (4) and the Rule of the Destructive Dilemma

d) *p*: Margie teaches the Pascal class.

  q: Margie gets a raise.

  r: Margie vacations in Hawaii in the winter.

  s: Margie goes bird watching in Tennessee in the winter.

$$p \leftrightarrow q$$
$$q \rightarrow r$$
$$\bar{p} \rightarrow s$$
$$\bar{r} \vee s$$
$$\overline{\phantom{\bar{p} \rightarrow s}}$$
$$\therefore \bar{s}$$

The following truth value assignments provide one counterexample to the validity of this argument: *p*: 0; *q*: 0; *r*: 1; *s*: 1. A second possible assignment is *p*: 1; *q*: 1; *r*: 1; *s*: 1.

e) p: The integer variable N is not initialized in the program.
q: $N > 0$ (This is not actually a statement until we examine the value of N.)
r: The value of $N^2 + N$ is printed (later) during program execution.
s: The value of N is supplied by the user in a 'Read' statement.

$$
\begin{aligned}
p &\rightarrow q \\
q &\rightarrow r \\
s &\rightarrow p \\
\bar{r}& \\
\hline
\therefore \bar{s}&
\end{aligned}
$$

(1)	$\bar{r}$	Hypothesis
(2)	$q \rightarrow r$	Hypothesis
(3)	$\bar{q}$	(1), (2) and Modus Tollens
(4)	$s \rightarrow p$	Hypothesis
(5)	$p \rightarrow q$	Hypothesis
(6)	$s \rightarrow q$	(4), (5) and the Law of the Syllogism
(7)	$\therefore \bar{s}$	(6), (3) and Modus Tollens

f) p: There is a chance of rain.
q: Lois's red head scarf is missing.
r: Lois does not mow her lawn.
s: The temperature is over 80° F.

$$
\begin{aligned}
(p \vee q) &\rightarrow r \\
s &\rightarrow \bar{p} \\
s \wedge \bar{q}& \\
\hline
\therefore \bar{r}&
\end{aligned}
$$

The following truth value assignments provide a counterexample to the validity of this argument: p: 0; q: 0; r: 1; s: 1.

Section 2.4 – p. 88

1. a) (i) $\exists x\, q(x)$ (ii) $\forall x\, [q(x) \rightarrow \overline{t(x)}]$ (iii) $\forall x\, [q(x) \rightarrow \overline{t(x)}]$
(iv) $\exists x\, [q(x) \wedge t(x)]$ (v) $\forall x\, [(q(x) \wedge r(x)) \rightarrow s(x)]$
b) Statements (i), (iv) and (v) are true. Statements (ii) and (iii) are false: $x = 10$ provides a counterexample for either statement.
c) (i) If x is a perfect square, then $x > 0$.
(ii) If x is divisible by 4, then x is even.
(iii) If x is divisible by 4, then x is not divisible by 5.
(iv) There exists an integer that is divisible by 4 but it is not a perfect square.
(v) If x is an even perfect square, then x is divisible by 4.
d) (i) Let $x = 0$. (iii) Let $x = 20$.
3. a) (i) True (ii) False Consider $x = 3$.
(iii) True (iv) True
b) (i) True (ii) False Consider $x = 3$.
(iii) True (iv) True
c) (i) True (ii) True
(iii) True (iv) False For $x = 2$ or 5, the truth value of $p(x)$ is 1 while that of $r(x)$ is 0.

5. a) $\forall x\, [\overline{p(x)} \wedge \overline{q(x)}]$ **b)** $\exists x\, [\overline{p(x)} \vee q(x)]$ **c)** $\exists x\, [p(x) \wedge \overline{q(x)}]$
d) $\forall x\, [(p(x) \vee q(x)) \wedge \overline{r(x)}]$ **e)** $\forall x\, [p(x) \wedge (\overline{q(x)} \vee \overline{r(x)})]$

7. a) When the statement $\exists x\, [p(x) \vee q(x)]$ is true, there is at least one element c in the prescribed universe where $p(c) \vee q(c)$ is true. Hence at least one of the statements $p(c)$ and $q(c)$ has the truth value 1, so at least one of the statements $\exists x\, p(x)$ and $\exists x\, q(x)$ is true. Therefore, it follows that $\exists x\, p(x) \vee \exists x\, q(x)$ is true, and $\exists x\, [p(x) \vee q(x)] \Rightarrow \exists x\, p(x) \vee \exists x\, q(x)$. Conversely, if $\exists x\, p(x) \vee \exists x\, q(x)$ is true, then at least one of $p(a)$ and $q(b)$ has the truth value 1, for some a and b in the prescribed universe. Assume without loss of generality that it is $p(a)$. Then $p(a) \vee q(a)$ has the truth value 1, so $\exists x\, [p(x) \vee q(x)]$ is a true statement, and $\exists x\, p(x) \vee \exists x\, q(x) \Rightarrow \exists x\, [p(x) \vee q(x)]$.

b) First consider when the statement $\forall x\, [p(x) \wedge q(x)]$ is true. This occurs when $p(a) \wedge q(a)$ is true for each a in the prescribed universe. Then $p(a)$ is true (as is $q(a)$) for all a in the universe, so the statements $\forall x\, p(x)$ and $\forall x\, q(x)$ are true. Therefore, the statement $\forall x\, p(x) \wedge \forall x\, q(x)$ is true and $\forall x\, [p(x) \wedge q(x)] \Rightarrow \forall x\, p(x) \wedge \forall x\, q(x)$. Conversely, suppose that $\forall x\, p(x) \wedge \forall x\, q(x)$ is a true statement. Then $\forall x\, p(x)$ and $\forall x\, q(x)$ are both true. Now let c be any element in the prescribed universe. Then $p(c)$, $q(c)$, and $p(c) \wedge q(c)$ are all true. And since c was chosen arbitrarily, it follows that the statement $\forall x\, [p(x) \wedge q(x)]$ is true, and $\forall x\, p(x) \wedge \forall x\, q(x) \Rightarrow \forall x\, [p(x) \wedge q(x)]$.

9. a) True **b)** False **c)** True **d)** True **e)** True **f)** True
11. a) True **b)** False **c)** False **d)** True
13. a) (i) $\forall x\, [x \neq 0 \to \exists !y\, (xy = 1)]$ (ii) $\forall x\, \forall y\, \exists !z\, (z = x + y)$
(iii) $\forall x\, \exists !y\, (y = 3x + 7)$
b) (i) False (ii) True **c)** (i) True (ii) True
d) When the prescribed universe consists of the integers 0, 1, and 2, the statement p is true. For the universe consisting of 2, 3, and 4, the statement is false.

Miscellaneous Exercises – p. 93

1.

p	q	r	s	$q \wedge r$	$\overline{s \vee r}$	$[(q \wedge r) \to \overline{(s \vee r)}]$	$p \leftrightarrow t$
0	0	0	0	0	1	1	0
0	0	0	1	0	0	1	0
0	0	1	0	0	0	1	0
0	0	1	1	0	0	1	0
0	1	0	0	0	1	1	0
0	1	0	1	0	0	1	0
0	1	1	0	1	0	0	1
0	1	1	1	1	0	0	1
1	0	0	0	0	1	1	1
1	0	0	1	0	0	1	1
1	0	1	0	0	0	1	1
1	0	1	1	0	0	1	1
1	1	0	0	0	1	1	1
1	1	0	1	0	0	1	1
1	1	1	0	1	0	0	0
1	1	1	1	1	0	0	0

The column $[(q \wedge r) \to \overline{(s \vee r)}]$ is headed by t.

3. a)

p	q	r	$q \leftrightarrow r$	$p \leftrightarrow (q \leftrightarrow r)$	$(p \leftrightarrow q)$	$(p \leftrightarrow q) \leftrightarrow r$
0	0	0	1	0	1	0
0	0	1	0	1	1	1
0	1	0	0	1	0	1
0	1	1	1	0	0	0
1	0	0	1	1	0	1
1	0	1	0	0	0	0
1	1	0	0	0	1	0
1	1	1	1	1	1	1

b) The truth value assignments p: 0; q: 0; r: 0, result in the truth value 1 for $[p \rightarrow (q \rightarrow r)]$ and the truth value 0 for $[(p \rightarrow q) \rightarrow r]$. Consequently, these statements are not logically equivalent.

5. a) $(\bar{p} \vee \bar{q}) \wedge (F_0 \vee p) \wedge p$

b) $(\bar{p} \vee \bar{q}) \wedge (F_0 \vee p) \wedge p$

$\Leftrightarrow (\bar{p} \vee \bar{q}) \wedge (p \wedge p)$ $F_0 \vee p \Leftrightarrow p$

$\Leftrightarrow (\bar{p} \vee \bar{q}) \wedge p$ Idempotent Law of $\wedge$

$\Leftrightarrow p \wedge (\bar{p} \vee \bar{q})$ Commutative Law of $\wedge$

$\Leftrightarrow (p \wedge \bar{p}) \vee (p \wedge \bar{q})$ Distributive Law of $\wedge$ over $\vee$

$\Leftrightarrow F_0 \vee (p \wedge \bar{q})$ $p \wedge \bar{p} \Leftrightarrow F_0$

$\Leftrightarrow p \wedge \bar{q}$ F_0 is the identity for $\vee$.

7. a) contrapositive   **b)** inverse   **c)** contrapositive   **d)** inverse   **e)** inverse

f) contrapositive   **g)** converse   **h)** converse   **i)** converse   **j)** converse

9. a)

p	q	r	$p \veebar q$	$(p \veebar q) \veebar r$	$q \veebar r$	$p \veebar (q \veebar r)$
0	0	0	0	0	0	0
0	0	1	0	1	1	1
0	1	0	1	1	1	1
0	1	1	1	0	0	0
1	0	0	1	1	0	1
1	0	1	1	0	1	0
1	1	0	0	0	1	0
1	1	1	0	1	0	1

It follows from the results in columns 5 and 7 that $[(p \veebar q) \veebar r] \Leftrightarrow [p \veebar (q \veebar r)]$.

b) The given statements are not logically equivalent. The truth value assignments p: 1; q: 1; r: 1 provide a counterexample.

c) If p, q, and r all have the truth value 1, then $p \rightarrow (q \veebar r)$ is true while $(p \rightarrow q) \veebar (p \rightarrow r)$ is false, so the implication is *not* a logical implication.

d) There are only two cases where $p \rightarrow (q \veebar r)$ is false: (i) p: 1; q: 1; r: 1; and (ii) p: 1; q: 0; r: 0. In these cases, $(p \rightarrow q) \veebar (p \rightarrow r)$ has the truth value 0. Consequently the given result is a logical implication.

11. If not, there exist integers x and y such that $3x + \sqrt{2}y = 0$ and ($x \neq 0$ or $y \neq 0$). If $x = 0$, then $\sqrt{2}y = 0 \Rightarrow y = 0$; if $y = 0$, then $3x = 0 \Rightarrow x = 0$. Hence $x \neq 0$ and $y \neq 0$. But then $\sqrt{2} = -3x/y$, a quotient of integers, so $\sqrt{2}$ is rational.

13. This result is false. No such y and z exist when $x = 2$.

CHAPTER 3

Set Theory

Section 3.1 – p. 104

1. They are all the same set.

3. Parts (b) and (d) are false; the remaining parts are true.

5. a) $\forall x\, [x \in A \rightarrow x \in B] \land \exists x\, [x \in B \land x \notin A]$
   b) $\exists x\, [x \in A \land x \notin B] \lor \forall x\, [x \notin B \lor x \in A]$
   OR $\exists x\, [x \in A \land x \notin B] \lor \forall x\, [x \in B \rightarrow x \in A]$

7. a) $|A| = 6$ b) $|B| = 7$
   c) If B has 2^n subsets of odd cardinality, then $|B| = n + 1$.

9. a) 31   b) 30   c) 28

11. Let $W = \{1\}$, $X = \{\{1\}, 2\}$, and $Y = \{X, 3\}$

13. c) If $x \in A$, then $A \subseteq B \Rightarrow x \in B$, and $B \subset C \Rightarrow x \in C$. Hence $A \subseteq C$. Since $B \subset C$, there exists $y \in C$ with $y \notin B$. Also, $A \subseteq B$ and $y \notin B \Rightarrow y \notin A$. Consequently, $A \subseteq C$ and $y \in C$ with $y \notin A \Rightarrow A \subset C$.
    d) Since $A \subset B$, it follows that $A \subseteq B$. The result then follows from part (c).

15. $n = 20$

17. Let $A = \{x, y, a_1, a_2, \ldots, a_n\}$. The number of subsets of A of size r is $\binom{n+2}{r}$. These fall into four classes: (1) the $\binom{n}{r}$ subsets containing neither x nor y; (2) the $\binom{n}{r-1}$ subsets containing x but not y; (3) the $\binom{n}{r-1}$ subsets containing y but not x; (4) the $\binom{n}{r-2}$ subsets containing both x and y.

Section 3.2 – p. 115

1. a) $\{1, 2, 3, 5\}$ b) A c) and d) $\mathcal{U} - \{2\}$
   e) $\{4, 8\}$ f) $\{1, 2, 3, 4, 5, 8\}$ g) $\emptyset$
   h) $\{2, 4, 8\}$ i) $\{1, 3, 4, 5, 8\}$

3. $B = \{a, b, d\}$ or $B = \{a, b, e\}$

5. a) Let $\mathcal{U} = \{1, 2, 3\}$, $A = \{1\}$, $B = \{2\}$, and $C = \{3\}$. Then $A \cap C = B \cap C = \emptyset$ but $A \neq B$.
   b) For $\mathcal{U} = \{1, 2\}$, $A = \{1\}$, $B = \{2\}$, and $C = \mathcal{U}$, we have $A \cup C = B \cup C$ but $A \neq B$.
   c) $x \in A \Rightarrow x \in A \cup C \Rightarrow x \in B \cup C$. So $x \in B$ or $x \in C$. If $x \in B$, then $A \subseteq B$. If $x \in C$, then $x \in A \cap C = B \cap C$ and $x \in B$. In either case, $A \subseteq B$. Likewise, $y \in B \Rightarrow y \in B \cup C = A \cup C$, so $y \in A$ or $y \in C$. If $y \in C$, then $y \in B \cap C = A \cap C$. In either case, $y \in A$ and $B \subseteq A$. Hence $A = B$.
   d) Let $x \in A$. Consider two cases: (1) $x \in C \Rightarrow x \notin A \triangle C \Rightarrow x \notin B \triangle C \Rightarrow x \in B$. (2) $x \notin C \Rightarrow x \in A \triangle C \Rightarrow x \in B \triangle C \Rightarrow x \in B$ (because $x \notin C$). In either case, $x \in B$, so $A \subseteq B$. In a similar way it follows that $B \subseteq A$ and $A = B$.

7. 7; 1

9. a) $\emptyset = (A \cup B) \cap (A \cup \overline{B}) \cap (\overline{A} \cup B) \cap (\overline{A} \cup \overline{B})$
   b) $A = A \cup (A \cap B)$
   c) $A \cap B = (A \cup B) \cap (A \cup \overline{B}) \cap (\overline{A} \cup B)$
   d) $A = (A \cap B) \cup (A \cap \mathcal{U})$

11. a) Let $\mathcal{U} = \{1, 2, 3\}$, $A = \{1\}$, and $B = \{2\}$. Then $\{1, 2\} \in \mathcal{P}(A \cup B)$ but $\{1, 2\} \notin \mathcal{P}(A) \cup \mathcal{P}(B)$.
    b) $X \in \mathcal{P}(A \cap B) \Leftrightarrow X \subseteq A \cap B \Leftrightarrow X \subseteq A$ and $X \subseteq B \Leftrightarrow X \in \mathcal{P}(A)$ and $X \in \mathcal{P}(B) \Leftrightarrow X \in \mathcal{P}(A) \cap \mathcal{P}(B)$, so $\mathcal{P}(A \cap B) = \mathcal{P}(A) \cap \mathcal{P}(B)$.

13. a) 2^6 b) 2^n
    c) In the membership table, $A \subseteq B$ if the columns for A, B are such that whenever a 1 occurs in the column for A, there is a corresponding 1 in the column for B.

d)	A	B	C	$A \cup \overline{B}$	$(A \cap B) \cup (\overline{B \cap C})$
	0	0	0	1	1
	0	0	1	1	1
	0	1	0	0	1
	0	1	1	0	0
	1	0	0	1	1
	1	0	1	1	1
	1	1	0	1	1
	1	1	1	1	1

15. **a)** $[-6, 9]$　　**b)** $[-8, 12]$　　**c)** $\emptyset$　　**d)** $[-8, -6) \cup (9, 12]$　　**e)** $[-14, 21]$
f) $[-2, 3]$　　**g)** $\mathbf{R}$　　**h)** $[-2, 3]$

17. $x \in A \cup (\bigcap_{i \in I} B_i) \Leftrightarrow x \in A$ or $(x \in B_i$ for all $i \in I) \Leftrightarrow$ For all $i \in I$, $x \in A$ or
$x \in B_i \Leftrightarrow$ For all $i \in I$, $x \in A \cup B_i \Leftrightarrow x \in \bigcap_{i \in I} (A \cup B_i)$.
$[A \cap (\bigcup_{i \in I} B_i) = \bigcup_{i \in I} (A \cap B_i)$ now follows by duality]

Sections 3.3 and 3.4 – p. 121

1. **a)** $24! + 24! - 22!$　　**b)** $26! - [24! + 24! - 23!]$

3. $9! + 9! - 8!$　　　**5.** **a)** $2/9$　　**b)** $4/9$

7. **a)** $[13!/(2!)^3] - 3[12!/(2!)^2] + 3(11!/2!) - 10!$　　**b)** Divide the result in part (a) by
$[13!/(2!)^3]$.

9. $Pr(A) = 1/3$, $Pr(B) = 7/15$, $Pr(A \cap B) = 2/15$, $Pr(A \cup B) = 2/3$;
$Pr(A \cup B) = 2/3 = 1/3 + 7/15 - 2/15 = Pr(A) + Pr(B) - Pr(A \cap B)$

11. $3/28$　　　　**13.** $2/39$

Miscellaneous Exercises – p. 125

1. Suppose that $(A - B) \subseteq C$ and $x \in A - C$. Then $x \in A$ but $x \notin C$. If $x \notin B$, then
$[x \in A \wedge x \notin B] \Rightarrow x \in (A - B) \subseteq C$. So now we have $x \notin C$ and $x \in C$. This contradiction gives us $x \in B$, so $(A - C) \subseteq B$.
Conversely, if $(A - C) \subseteq B$, let $y \in A - B$. Then $y \in A$ but $y \notin B$. If $y \notin C$, then
$[y \in A \wedge y \notin C] \Rightarrow y \in (A - C) \subseteq B$. This contradiction—that is, $y \notin B$ and $y \in B$—
yields $y \in C$, so $(A - B) \subseteq C$.

3. **a)** The sets $\mathcal{U} = \{1, 2, 3\}$, $A = \{1, 2\}$, $B = \{1\}$, and $C = \{2\}$ provides a counterexample.

b) $A = A \cap \mathcal{U} = A \cap (C \cup \overline{C}) = (A \cap C) \cup (A \cap \overline{C}) = (A \cap C) \cup (A - C)$

$= (B \cap C) \cup (B - C) = (B \cap C) \cup (B \cap \overline{C}) = B \cap (C \cup \overline{C}) = B \cap \mathcal{U} = B$

c) The set assignments for part (a) also provide a counterexample for this situation.

d) Let $\mathcal{U} = \{1, 2, 3, 4\}$, $A = \{1, 2, 3\}$, and $B = \{1\}$. Then $|\mathcal{P}(A - B)| = 2^2 = 4$ while
$|\mathcal{P}(A)| - |\mathcal{P}(B)| = 6$, so the result is false, as demonstrated by this counter--
example.

5. **a)** 126 (if teams wear different uniforms); 63 (if teams are not distinguishable)
112 (if teams wear different uniforms); 56 (if teams are not distinguishable)

b) $2^n - 2$; $(1/2)(2^n - 2)$.　　$2^n - 2 - 2n$; $(1/2)(2^n - 2 - 2n)$.

7. **a)** 128　　**b)** $|A| = 8$

9. Suppose that $(A \cap B) \cup C = A \cap (B \cup C)$ and that $x \in C$. Then
$x \in C \Rightarrow x \in (A \cap B) \cup C \Rightarrow x \in A \cap (B \cup C) \subseteq A$, so $x \in A$, and $C \subseteq A$.
Conversely, suppose that $C \subseteq A$.

(1) If $y \in (A \cap B) \cup C$, then $y \in A \cap B$ or $y \in C$.

  (i) $y \in A \cap B \Rightarrow y \in (A \cap B) \cup (A \cap C) \Rightarrow y \in A \cap (B \cup C)$.

  (ii) $y \in C \Rightarrow y \in A$, because $C \subseteq A$. Also, $y \in C \Rightarrow y \in B \cup C$.

    So $y \in A \cap (B \cup C)$. In either case (i) or case (ii), we have $y \in A \cap (B \cup C)$, so $(A \cap B) \cup C \subseteq A \cap (B \cup C)$.

(2) Now let $z \in A \cap (B \cup C)$.

  Then $z \in A \cap (B \cup C) = (A \cap B) \cup (A \cap C) \subseteq (A \cap B) \cup C$, since $A \cap C \subseteq C$.

  From (1) and (2) it follows that $(A \cap B) \cup C = A \cap (B \cup C)$.

11. **a)** Assume that $A \cup B = \mathcal{U}$ and let $x \in \overline{A}$. Then $x \in \overline{A} \Rightarrow x \notin A \Rightarrow x \in B$, since $A \cup B = \mathcal{U}$. Consequently, $\overline{A} \subseteq B$.

  Conversely, let $y \in \mathcal{U}$. If $y \in A$, we are finished. If not, then $y \notin A$, so $y \in \overline{A} \subseteq B$. In either situation we have $y \in A \cup B$, so $\mathcal{U} \subseteq A \cup B$. But we always have $A \cup B \subseteq \mathcal{U}$, so it follows that $A \cup B = \mathcal{U}$.

b) If $x \in B$ and $A \cap B = \emptyset$, then $x \notin A$. But $x \notin A \Rightarrow x \in \overline{A}$, so $B \subseteq \overline{A}$.

  Conversely, if $B \subseteq \overline{A}$ and $y \in A \cap B$, then $y \in B$ and $y \in A$. However, $y \in B \Rightarrow y \in \overline{A}$, because $B \subseteq \overline{A}$. So $y \in A$ and $y \in \overline{A}$, or $y \in A \cap \overline{A} = \emptyset$ and $A \cap B \subseteq \emptyset$. We always have $\emptyset \subseteq A \cap B$, so it follows that $A \cap B = \emptyset$.

13. **a)** $[0, 14/3]$ **b)** $(0, 9/5]$ **c)** $\{0\}$ **d)** $\{0\} \cup (6, 12]$

    e) $[0, +\infty)$ **f)** $(0, +\infty)$ **g)** $\{0\}$ **h)** $\emptyset$

15. **a)**

A	B	$A \cap B$
0	0	0
0	1	0
1	0	0
1	1	1

Since $A \subseteq B$, consider only rows 1, 2, and 4. For these rows, $A \cap B = A$.

b)

A	B	C	$A \cap B$	$B \cup C$	$A \cup B \cup C$
0	0	0	0	0	0
0	0	1	0	1	1
0	1	0	0	1	1
0	1	1	0	1	1
1	0	0	0	0	1
1	0	1	0	1	1
1	1	0	1	1	1
1	1	1	1	1	1

Since $A \cap B = A$ and $B \cup C = C$, consider only rows 1, 2, 4, and 8. For these four rows, $A \cup B \cup C = C$.

c)

A	B	C	$(A \cap \overline{B}) \cup (B \cap \overline{C})$	$A \cap \overline{C}$
0	0	0	0	0
0	0	1	0	0
0	1	0	1	0
0	1	1	0	0
1	0	0	1	1
1	0	1	1	0
1	1	0	1	1
1	1	1	0	0

For $C \subseteq B \subseteq A$, consider only rows 1, 5, 7, and 8. Here $(A \cap \overline{B}) \cup (B \cap \overline{C}) = A \cap \overline{C}$.

d)

A	B	C	$A \triangle B$	$A \triangle C$	$B \triangle C$
0	0	0	0	0	0
0	0	1	0	1	1
0	1	0	1	0	1
0	1	1	1	1	0
1	0	0	1	1	0
1	0	1	1	0	1
1	1	0	0	1	1
1	1	1	0	0	0

When $A \triangle B = C$, we consider rows 1, 4, 6, and 7. In these cases, $A \triangle C = B$ and $B \triangle C = A$.

17. a) $\binom{r+1}{m}$ $(m \le r+1)$ b) $\binom{n-k+1}{k}$ $(2k \le n+1)$

19. 2/7

21. a) $\binom{15}{12}/\binom{19}{16}$ b) $\binom{11}{8}/\binom{19}{16}$ c) $1/\binom{19}{16}$

23. $7^{15} - 3(3^{15}) + 3$

25. $\binom{12}{4}\binom{10}{3}/\binom{22}{7} = 0.3483$

27. a) $\sum_{i=0}^{8} \binom{2i}{i}\binom{i+8}{8-i} = \sum_{i=0}^{8} \dfrac{(i+8)!}{i!i!(8-i)!}$

   b) (i) $\binom{12}{6}\binom{4}{2}/\left[\sum_{i=0}^{8} \binom{2i}{i}\binom{i+8}{8-i}\right]$ (ii) $\binom{12}{6}\binom{13}{1}/\left[\sum_{i=0}^{8} \binom{2i}{i}\binom{i+8}{8-i}\right]$

   (iii) $\left[\binom{16}{8} + \binom{12}{6}\binom{4}{2} + \binom{8}{4}\binom{12}{4} + \binom{4}{2}\binom{10}{6} + \binom{0}{0}\binom{8}{8}\right]/\left[\sum_{i=0}^{8}\binom{2i}{i}\binom{i+8}{8-i}\right]$

CHAPTER 4

Properties of the Integers: Mathematical Induction

Section 4.1 – p. 141

1. b) Since $1 \cdot 3 = (1)(2)(9)/6$, the result is true for $n = 1$. Assume the result for $n = k$: $1 \cdot 3 + 2 \cdot 4 + 3 \cdot 5 + \cdots + k(k+2) = k(k+1)(2k+7)/6$. Then consider the case for $n = k + 1$: $[1 \cdot 3 + 2 \cdot 4 + \cdots + k(k+2)] + (k+1)(k+3) = [k(k+1)(2k+7)/6] + (k+1)(k+3) = [(k+1)/6][k(2k+7) + 6(k+3)] = (k+1)(2k^2 + 13k + 18)/6 = (k+1)(k+2)(2k+9)/6$. Hence the result follows for all $n \in \mathbf{Z}^+$ by the principle of finite induction.

   c) $S(n)$: $\displaystyle\sum_{i=1}^{n} \frac{1}{i(i+1)} = \frac{n}{n+1}$

   $S(1)$: $\displaystyle\sum_{i=1}^{1} \frac{1}{i(i+1)} = \frac{1}{1(2)} = \frac{1}{1+1}$, so $S(1)$ is true.

   Assume $S(k)$: $\displaystyle\sum_{i=1}^{k} \frac{1}{i(i+1)} = \frac{k}{k+1}$. Consider $S(k+1)$.

   $\displaystyle\sum_{i=1}^{k+1} \frac{1}{i(i+1)} = \sum_{i=1}^{k} \frac{1}{i(i+1)} + \frac{1}{(k+1)(k+2)} = \frac{k}{(k+1)} + \frac{1}{(k+1)(k+2)}$

   $= [k(k+2) + 1]/[(k+1)(k+2)] = (k+1)/(k+2)$,

   so $S(k) \Rightarrow S(k+1)$ and the result follows for all $n \in \mathbf{Z}^+$ by the principle of finite induction.

3. a) 7626   b) 627,874

5. a) 506   b) 12,144

7. For $n = 11$, $11 - 2 = 9 < 9\frac{1}{6} = (11^2 - 11)/12$. Assume the result true for $n = k(\ge 11)$: $k - 2 < (k^2 - k)/12$. When $n = k + 1$, $k - 2 < (k^2 - k)/12 \Rightarrow (k-2) + 1 < ((k^2 - k)/12) + 1 \Rightarrow (k+1) - 2 < (k^2 - k + 12)/12$. For $k > 6$, $2k > 12$, and $k > 12 - k$, so

$(k + 1) - 2 < (k^2 + k)/12 = [(k + 1)^2 - (k + 1)]/12$. Consequently the result follows for all $n \geq 11$ by the principle of finite induction.

9. For $n = 5$, $2^5 = 32 > 25 = 5^2$. Assume the result for $n = k(\geq 5)$: $2^k > k^2$. For $k > 3$, $k(k - 2) > 1$, or $k^2 > 2k + 1$. $2^k > k^2 \Rightarrow 2^k + 2^k > k^2 + k^2 \Rightarrow 2^{k+1} > k^2 + k^2 > k^2 + (2k + 1) = (k + 1)^2$. Hence the result is true for $n \geq 5$ by the principle of mathematical induction.

11. Assume $S(k)$. For $S(k + 1)$, $\sum_{i=1}^{k+1} i = ([k + (1/2)]^2/2) + (k + 1) = (k^2 + k + (1/4) + 2k + 2)/2 = [(k + 1)^2 + (k + 1) + (1/4)]/2 = [(k + 1) + (1/2)]^2/2$. So $S(k) \Rightarrow S(k + 1)$. However, we have no first value of k where $S(k)$ is true: for any $k \geq 1$, $\sum_{i=1}^{k} i = (k)(k + 1)/2$ and $(k)(k + 1)/2 = [k + (1/2)]^2/2 \Rightarrow 0 = 1/4$.

13. **a)** For $n = 1$, $[\sin(2^n \theta)]/(2^n \sin \theta) = (\sin 2\theta)/(2 \sin \theta) = (2 \sin \theta \cos \theta)/(2 \sin \theta) = \cos \theta$, so the result is true for this case.

Assume the result for $n = k$:
$(\cos \theta)(\cos 2\theta) \cdots (\cos(2^{k-1})\theta) = [\sin(2^k \theta)]/[2^k \sin \theta]$. When $n = k + 1$, then

$(\cos \theta)(\cos 2\theta) \cdots (\cos(2^{k-1})\theta)(\cos(2^k \theta)) = [\sin(2^k \theta) \cos(2^k \theta)]/[2^k \sin \theta]$
$= [2 \sin(2^k \theta) \cos(2^k \theta)]/[2^{k+1} \sin \theta]$
$= [\sin 2(2^k \theta)]/[2^{k+1} \sin \theta]$
$= [\sin(2^{k+1} \theta)]/(2^{k+1} \sin \theta)$.

Consequently, the result is true for all $n \geq 1$, by the principle of mathematical induction.

b) When $n = 1$, the statement becomes $\cos \theta = (\sin 2\theta)/(2 \sin \theta)$, which is true because $(\sin 2\theta)/(2 \sin \theta) = (2 \sin \theta \cos \theta)/(2 \sin \theta) = \cos \theta$.

For $n = k$, we assume the truth of the statement:
$\cos \theta + \cos 3\theta + \cos 5\theta + \cdots + \cos(2k - 1)\theta = (\sin 2k\theta)/(2 \sin \theta)$.

When $n = k + 1$, we have $(\cos \theta + \cos 3\theta + \cdots + \cos(2k - 1)\theta) + \cos(2k + 1)\theta$
$= [(\sin 2k\theta)/(2 \sin \theta)] + \cos(2k + 1)\theta = [\sin 2k\theta + 2 \sin \theta \cos(2k + 1)\theta]/(2 \sin \theta)$.
To establish this case, we need to show that
$\sin(2k + 2)\theta = [\sin 2k\theta + 2 \sin \theta \cos(2k + 1)\theta]$. We find that

$\sin(2k + 2)\theta = \sin 2k\theta \cos 2\theta + \cos 2k\theta \sin 2\theta$
$= \sin 2k\theta[1 - 2 \sin^2 \theta] + 2 \sin \theta \cos \theta \cos 2k\theta$
$= \sin 2k\theta - 2 \sin \theta[\sin \theta \sin 2k\theta - \cos \theta \cos 2k\theta]$
$= \sin 2k\theta + 2 \sin \theta \cos(2k + 1)\theta$.

The result now follows for all $n \geq 1$, by the principle of finite induction.

15. Let $S(n)$ denote the following statement: For $x, n \in \mathbf{Z}^+$, if the program reaches the top of the While loop, then upon exiting the loop, the value of the integer variable Answer is $x(n!)$.

First consider $S(1)$, the statement for the case where $n = 1$. Here the program (if it reaches the top of the While loop) will result in one execution of the While loop: x will be assigned the value $x \cdot 1 = x(1!)$, and the value of n will be decreased to 0. With the value of n equal to 0, the loop is not processed again and the value of the variable Answer is $x(1!)$. Hence $S(1)$ is true.

Now assume the truth of $S(k)$: For $x, k \in \mathbf{Z}^+$, if the program reaches the top of the While loop, then upon exiting the loop, the value of the variable Answer is $x(k!)$. To establish the truth of $S(k + 1)$, if the program reaches the top of the While loop, then the following occur during the first execution:

The value assigned to the variable x is $x(k + 1)$.
The value of n is decreased to $(k + 1) - 1 = k$.

But then we can apply the induction hypothesis to the integers $x(k + 1)$ and k, and upon exiting the While loop, the value of the variable Answer is $(x(k + 1))(k!) = x(k + 1)!$

Consequently, $S(n)$ is true for all $n \geq 1$, and we have verified the correctness of this program segment by using the principle of mathematical induction.

17. **a)** (i) For $n = 2$, $x_1 + x_2$ denotes the ordinary sum of the real numbers x_1 and x_2.

(ii) For real numbers $x_1, x_2, \ldots, x_n, x_{n+1}$, we have $x_1 + x_2 + \cdots + x_n + x_{n+1} = (x_1 + x_2 + \cdots + x_n) + x_{n+1}$, the sum of the *two* real numbers $x_1 + x_2 + \cdots + x_n$ and x_{n+1}.

b) The truth of this result for $n = 3$ follows from the associative law of addition—since $x_1 + (x_2 + x_3) = (x_1 + x_2) + x_3$, there is no ambiguity in writing $x_1 + x_2 + x_3$.

Assuming the result true for any $k \geq 3$ and $1 \leq r < k$, let us examine the case for $k + 1$ real numbers. We find that

$$(x_1 + x_2 + \cdots + x_r) + (x_{r+1} + \cdots + x_k + x_{k+1})$$
$$= (x_1 + x_2 + \cdots + x_r) + [(x_{r+1} + \cdots + x_k) + x_{k+1}]$$
$$= [(x_1 + x_2 + \cdots + x_r) + (x_{r+1} + \cdots + x_k)] + x_{k+1}$$
$$= (x_1 + x_2 + \cdots + x_r + x_{r+1} + \cdots + x_k) + x_{k+1}$$
$$= x_1 + x_2 + \cdots + x_r + x_{r+1} + \cdots + x_k + x_{k+1},$$

so the result is true for all $n \geq 3$ and all $1 \leq r < n$, by the principle of mathematical induction.

19. Let $T(n)$ denote the following statement: For $n \in \mathbf{Z}^+$, $n \geq 2$, and the statements $p, q_1, q_2, \ldots, q_n$,

$$p \vee (q_1 \wedge q_2 \wedge \cdots \wedge q_n) \Leftrightarrow (p \vee q_1) \wedge (p \vee q_2) \wedge \cdots \wedge (p \vee q_n).$$

The statement $T(2)$ is true by virtue of the distributive law of $\vee$ over $\wedge$. Assuming $T(k)$, for $k \geq 2$, we now examine the situation for the statements $p, q_1, q_2, \ldots, q_k, q_{k+1}$. We find that $p \vee (q_1 \wedge q_2 \wedge \cdots \wedge q_k \wedge q_{k+1})$

$$\Leftrightarrow p \vee [(q_1 \wedge q_2 \wedge \cdots \wedge q_k) \wedge q_{k+1}]$$
$$\Leftrightarrow [p \vee (q_1 \wedge q_2 \wedge \cdots \wedge q_k)] \wedge (p \vee q_{k+1})$$
$$\Leftrightarrow [(p \vee q_1) \wedge (p \vee q_2) \wedge \cdots \wedge (p \vee q_k)] \wedge (p \vee q_{k+1})$$
$$\Leftrightarrow (p \vee q_1) \wedge (p \vee q_2) \wedge \cdots \wedge (p \vee q_k) \wedge (p \vee q_{k+1}).$$

It then follows by the principle of mathematical induction that the statement $T(n)$ is true for all $n \geq 2$.

21. **a)** For $n = 2$, the truth of the result $A \cap (B_1 \cup B_2) = (A \cap B_1) \cup (A \cap B_2)$ follows by virtue of the distributive law of $\cap$ over $\cup$.

Assuming the result for $n = k$, let us examine the case for the sets $A, B_1, B_2, \ldots, B_k, B_{k+1}$. We have

$$A \cap (B_1 \cup B_2 \cup \cdots \cup B_k \cup B_{k+1}) = A \cap [(B_1 \cup B_2 \cup \cdots \cup B_k) \cup B_{k+1}]$$
$$= [A \cap (B_1 \cup B_2 \cup \cdots \cup B_k)] \cup (A \cap B_{k+1})$$
$$= [(A \cap B_1) \cup (A \cap B_2) \cup \cdots \cup (A \cap B_k)] \cup (A \cap B_{k+1})$$
$$= (A \cap B_1) \cup (A \cap B_2) \cup \cdots \cup (A \cap B_k) \cup (A \cap B_{k+1}).$$

b) (i) The intersection of A_1, A_2 is $A_1 \cap A_2$.

(ii) The intersection of $A_1, A_2, \ldots, A_n, A_{n+1}$ is given by $A_1 \cap A_2 \cap \cdots \cap A_n \cap A_{n+1} = (A_1 \cap A_2 \cap \cdots \cap A_n) \cap A_{n+1}$, the intersection of the *two* sets $A_1 \cap A_2 \cap \cdots \cap A_n$ and A_{n+1}.

c) Let $S(n)$ denote the given statement. Then the truth of $S(3)$ follows from the associative law of $\cap$. Assuming $S(k)$ true for $k \geq 3$, consider the case for $k + 1$ sets.

Then

$$(A_1 \cap A_2 \cap \cdots \cap A_r) \cap (A_{r+1} \cap \cdots \cap A_k \cap A_{k+1})$$
$$= (A_1 \cap A_2 \cap \cdots \cap A_r) \cap [(A_{r+1} \cap \cdots \cap A_k) \cap A_{k+1}]$$
$$= [(A_1 \cap A_2 \cap \cdots \cap A_r) \cap (A_{r+1} \cap \cdots \cap A_k)] \cap A_{k+1}$$
$$= (A_1 \cap A_2 \cap \cdots \cap A_r \cap A_{r+1} \cap \cdots \cap A_k) \cap A_{k+1}$$
$$= A_1 \cap A_2 \cap \cdots \cap A_r \cap A_{r+1} \cap \cdots \cap A_k \cap A_{k+1},$$

and by the principle of mathematical induction, $S(n)$ is true for all $n \geq 3$ and all $1 \leq r < n$.

d) (i) For $n = 2$, the result follows from the DeMorgan Laws. Assuming the result for $n = k \geq 2$, consider the case for $k + 1$ sets $A_1, A_2, \ldots, A_k, A_{k+1}$. Then

$$\overline{A_1 \cap A_2 \cap \cdots \cap A_k \cap A_{k+1}}$$
$$= \overline{(A_1 \cap A_2 \cap \cdots \cap A_k) \cap A_{k+1}}$$
$$= \overline{(A_1 \cap A_2 \cap \cdots \cap A_k)} \cup \overline{A_{k+1}}$$
$$= [\overline{A_1} \cup \overline{A_2} \cup \cdots \cup \overline{A_k}] \cup \overline{A_{k+1}}$$
$$= \overline{A_1} \cup \overline{A_2} \cup \cdots \cup \overline{A_k} \cup \overline{A_{k+1}},$$

and the result is true for all $n \geq 2$, by the principle of mathematical induction.

(ii) The proof for this result is similar to the one in part (i). Simply replace each occurrence of $\cap$ by $\cup$, and vice versa. [We can also obtain (ii) from (i) by invoking the principle of duality—Theorem 3.5.]

e) Let $S(n)$ denote the given statement. We know that $S(2)$ is true by the distributive law of $\cup$ over $\cap$. So we assume the truth of $S(k)$ for $k \geq 2$ and consider $S(k + 1)$ for the sets $A, B_1, B_2, \ldots, B_k, B_{k+1}$. We find that

$$A \cup [B_1 \cap B_2 \cap \cdots \cap B_k \cap B_{k+1}]$$
$$= A \cup [(B_1 \cap B_2 \cap \cdots \cap B_k) \cap B_{k+1}]$$
$$= [A \cup (B_1 \cap B_2 \cap \cdots \cap B_k)] \cap (A \cup B_{k+1})$$
$$= [(A \cup B_1) \cap (A \cup B_2) \cap \cdots \cap (A \cup B_k)] \cap (A \cup B_{k+1})$$
$$= (A \cup B_1) \cap (A \cup B_2) \cap \cdots \cap (A \cup B_k) \cap (A \cup B_{k+1}),$$

and so $S(n)$ is true for all $n \geq 2$, by the principle of mathematical induction.

23. For $P(n) = (1000/\sqrt{5})[((1 + \sqrt{5})/2)^{n+1} - ((1 - \sqrt{5})/2)^{n+1}]$, we find (upon substitution of 1 and 2 for n) that $P(1) = 1000$ and $P(2) = 2000$, so $P(n)$ is valid for $n = 1, 2$. Assume the validity of $P(1), P(2), \ldots, P(k - 2), P(k - 1)$, for $k \geq 3$. Then

$$P(k) = P(k - 1) + P(k - 2) = (1000/\sqrt{5})[((1 + \sqrt{5})/2)^k - ((1 - \sqrt{5})/2)^k$$
$$+ ((1 + \sqrt{5})/2)^{k-1} - ((1 - \sqrt{5})/2)^{k-1}]$$
$$= (1000/\sqrt{5})[((1 + \sqrt{5})/2)^{k-1} \cdot [1 + ((1 + \sqrt{5})/2)]$$
$$- ((1 - \sqrt{5})/2)^{k-1} \cdot [1 + ((1 - \sqrt{5})/2)]]$$
$$= (1000/\sqrt{5})[((1 + \sqrt{5})/2)^{k-1} \cdot ((1 + \sqrt{5})/2)^2$$
$$- ((1 - \sqrt{5})/2)^{k-1} \cdot ((1 - \sqrt{5})/2)^2]$$
$$= (1000/\sqrt{5})[((1 + \sqrt{5})/2)^{k+1} - ((1 - \sqrt{5})/2)^{k+1}],$$

so $P(n)$ is valid for all $n \geq 1$, by mathematical induction (the alternative form).

25. Let $T = \{n \in \mathbf{Z}^+ \,|\, S(n) \text{ is false}\}$. $S(1)$ true $\Rightarrow 1 \notin T$. If $T \neq \emptyset$, then T has a least element r, since $T \subseteq \mathbf{Z}^+$. However, $S(1), S(2), \ldots, S(r - 1)$ true $\Rightarrow S(r)$ true. Hence $T = \emptyset$ and the result follows.

Section 4.2 – p. 152

1. e) If $a \,|\, x$ and $a \,|\, y$, then $x = ac$ and $y = ad$ for some $c, d \in \mathbf{Z}$. So $z = x - y = a(c - d)$, and $a \,|\, z$. The proofs for the other cases are similar.

g) Follows from part (f) by mathematical induction.

3. Since q is prime, its only positive divisors are 1 and q. With p a prime, $p > 1$. Hence $p \,|\, q \Rightarrow p = q$.

5. $[b\,|\,a$ and $b\,|\,(a+2)] \Rightarrow b\,|\,[ax+(a+2)y]$ for all $x,y \in \mathbf{Z}$. Let $x=-1, y=1$. Then $b>0$ and $b\,|\,2$, so $b=1$ or 2.

7. Let $a=2m+1$ and $b=2n+1$, for $m,n \geq 0$. Then $a^2+b^2 = 4(m^2+m+n^2+n)+2$, so $2\,|\,(a^2+b^2)$ but $4\nmid(a^2+b^2)$.

9.

	Base 10	Base 2	Base 16
a)	22	10110	16
b)	527	1000001111	20F
c)	1234	10011010010	4D2
d)	6923	1101100001011	1B0B

11.

	Base 2	Base 10	Base 16
a)	11001110	206	CE
b)	00110001	49	31
c)	11110000	240	F0
d)	01010111	87	57

13.

	Largest Integer	Smallest Integer
a)	$7=2^3-1$	$-8=-(2^3)$
b)	$127=2^7-1$	$-128=-(2^7)$
c)	$2^{15}-1$	$-(2^{15})$
d)	$2^{31}-1$	$-(2^{31})$
e)	$2^{n-1}-1$	$-(2^{n-1})$

15. $ax=ay \Rightarrow ax-ay=0 \Rightarrow a(x-y)=0$. In the system of integers, if $b,c \in \mathbf{Z}$ and $bc=0$, then $b=0$ or $c=0$. Since $a(x-y)=0$ and $a \neq 0$, it follows that $(x-y)=0$ and $x=y$.

21. a) Since $2\,|\,10^t$ for all $t \in \mathbf{Z}^+$, $2\,|\,n$ if and only if $2\,|\,r_0$.
b) Follows from the fact that $4\,|\,10^t$ for $t \geq 2$.
c) Follows from the fact that $8\,|\,10^t$ for $t \geq 3$. In general,
$$2^{t+1}\,|\,n \text{ if and only if } 2^{t+1}\,|\,(r_t \cdot 10^t + \cdots + r_1 \cdot 10 + r_0).$$

Section 4.3 – p. 158

1. a) $(1820,231)=7=1820(8)+231(-63)$
b) $(2597,1369)=1=2597(534)+1369(-1013)$
c) $(4001,2689)=1=4001(-1117)+2689(1662)$
d) $(7983,7982)=1=7983(1)+7982(-1)$.

3. $(n,n+1)=1; [n,n+1]=n(n+1)$.

5. $(a,b)=d \Rightarrow d=ax+by$, for some $x,y \in \mathbf{Z}$.
$(a,b)=d \Rightarrow a/d, b/d \in \mathbf{Z}$
$1=(a/d)x+(b/d)y \Rightarrow (a/d, b/d)=1$.

7. Let $(a,b)=h$ and $(b,d)=g$.
$(a,b)=h \Rightarrow [h\,|\,a$ and $h\,|\,b] \Rightarrow h\,|\,(a \cdot 1+bc) \Rightarrow h\,|\,d$.
$[h\,|\,b$ and $h\,|\,d] \Rightarrow h\,|\,g$.
$(b,d)=g \Rightarrow [g\,|\,b$ and $g\,|\,d] \Rightarrow g\,|\,(d \cdot 1+b(-c)) \Rightarrow g\,|\,a$.
$[g\,|\,b, g\,|\,a,$ and $h=(a,b)] \Rightarrow g\,|\,h. \; h\,|\,g, g\,|\,h$, with $g,h \in \mathbf{Z}^+ \Rightarrow g=h$.

9. There is no solution for $c \neq 12, 18$. For $c=12$, the solutions are $x=118-165k$, $y=-10+14k$, $k \in \mathbf{Z}$. For $c=18$, the solutions are $x=177-165k$, $y=-15+14k$, $k \in \mathbf{Z}$.

11. $(a,b)=1 \Rightarrow ax+by=1$, for some $x,y \in \mathbf{Z}$. Then $acx+bcy=c$. $a\,|\,acx, a\,|\,bcy$ (because $a\,|\,bc) \Rightarrow a\,|\,c$.

13. Let $a, b, c \in \mathbf{Z}^+$. If $ax + by = c$ has a solution $x_0, y_0 \in \mathbf{Z}$, then $ax_0 + by_0 = c$, and since (a, b) divides a and b, it follows that $(a, b) | c$. Conversely, suppose $(a, b) | c$. Then $c = (a, b)d$, for some $d \in \mathbf{Z}$. Since $(a, b) = as + bt$, for some $s, t \in \mathbf{Z}$, we have $a(sd) + b(td) = (a, b)d = c$ or $ax_0 + by_0 = c$, and $ax + by = c$ has a solution in $\mathbf{Z}$.

Section 4.4 – p. 160

1. a) $2^2 \cdot 3^3 \cdot 5^3 \cdot 11$　　**b)** $2^4 \cdot 3 \cdot 5^2 \cdot 7^2 \cdot 11^2$　　**c)** $3^2 \cdot 5^3 \cdot 7^2 \cdot 11 \cdot 13$

3. g.c.d. $= 3 \cdot 5^2 \cdot 11 = 825$

l.c.m. $= 2^4 \cdot 3^3 \cdot 5^3 \cdot 7^2 \cdot 11^2 \cdot 13 = 4162158000$

5. b) Let $\log_{10} p = a/b$, where $a, b \in \mathbf{Z}^+$, $(a, b) = 1$. Then $p = 10^{(a/b)}$ or $p^b = 10^a = 2^a \cdot 5^a$, contradicting the Fundamental Theorem of Arithmetic.

7. If $n \in \mathbf{Z}^+$ and n is a perfect square, then $n = p_1^{e_1} p_2^{e_2} \cdots p_k^{e_k}$ where p_i is a prime and e_i is a positive even integer for each $1 \le i \le k$. Hence the number of positive divisors of n is $(e_1 + 1)(e_2 + 1) \cdots (e_k + 1)$, a product of odd integers. Therefore the number of divisors of n is odd.

Conversely, if $n \in \mathbf{Z}^+$ and n is not a perfect square, then $n = p_1^{e_1} p_2^{e_2} \cdots p_k^{e_k}$ where e_i is odd for some $1 \le i \le k$ and $(e_i + 1)$ is even for that i. Since $(e_1 + 1)(e_2 + 1) \cdots (e_k + 1)$ has at least one $(e_i + 1)$ even, it follows that the number of divisors of n is even.

9. $n = 2 \cdot 3 \cdot 5^2 \cdot 7^2 = 7350$

Miscellaneous Exercises – p. 163

1. $a + (a + d) + (a + 2d) + \cdots + (a + (n - 1)d) = na + [(n - 1)nd]/2$. For $n = 1$, $a = a + 0$, and the result is true in this case. Assuming that

$$\sum_{i=1}^{k} [a + (i - 1)d] = ka + [(k - 1)kd]/2,$$

we have

$$\sum_{i=1}^{k+1} [a + (i - 1)d] = (ka + [(k - 1)kd]/2) + (a + kd) = (k + 1)a + [k(k + 1)d]/2,$$

so the result follows for all $n \in \mathbf{Z}^+$ by induction.

3. a)

n	$n^2 + n + 41$	n	$n^2 + n + 41$	n	$n^2 + n + 41$
1	43	4	61	7	97
2	47	5	71	8	113
3	53	6	83	9	131

b) For $n = 39$, $n^2 + n + 41 = 1601$, a prime. But for $n = 40$, $n^2 + n + 41 = (41)^2$, so $S(39) \not\Rightarrow S(40)$.

5. For $n = 1$, $\sum_{i=1}^{1} i^4 = 1 = [(1)(2)(3)(5)]/30$, so the formula is true in this first case. Now assume that the result is true for $n = k$—that is, $\sum_{i=1}^{k} i^4 = [k(k + 1)(2k + 1)(3k^2 + 3k - 1)]/30$. When $n = k + 1$, we have

$$\sum_{i=1}^{k+1} i^4 = [k(k + 1)(2k + 1)(3k^2 + 3k - 1)]/30 + (k + 1)^4$$

$$= [(k + 1)/30][k(2k + 1)(3k^2 + 3k - 1) + 30(k + 1)^3]$$

$$= [(k + 1)/30][6k^4 + 39k^3 + 91k^2 + 89k + 30]$$

$$= [(k + 1)/30][(k + 2)(2k + 3)(3k^2 + 9k + 5)]$$

$$= [(k + 1)/30][(k + 2)(2(k + 1) + 1)(3(k + 1)^2 + 3(k + 1) - 1)],$$

so the truth for $n = k + 1$ follows from that for $n = k$. Consequently the result follows for all $n \in \mathbf{Z}^+$ by the principle of mathematical induction.

7. **a)** For $n = 0$, $2^{2n+1} + 1 = 2 + 1 = 3$, so the result is true in this first case. Assuming that 3 divides $2^{2k+1} + 1$ for $n = k \in \mathbf{N}$, consider the case of $n = k + 1$. Since $2^{2(k+1)+1} + 1 = 2^{2k+3} + 1 = 4(2^{2k+1}) + 1 = 4(2^{2k+1} + 1) - 3$, and 3 divides both $2^{2k+1} + 1$ and 3, it follows that 3 divides $2^{2(k+1)+1} + 1$. Consequently, the result is true for $n = k + 1$ whenever it is true for $n = k$. So by the principle of finite induction, the result follows for all $n \in \mathbf{N}$.

   b) When $n = 0$, $0^3 + (0 + 1)^3 + (0 + 2)^3 = 9$, so the statement is true in this case. We assume the truth of the result when $n = k \geq 0$ and examine the result for $n = k + 1$. We find that $(k + 1)^3 + (k + 2)^3 + (k + 3)^3 = (k + 1)^3 + (k + 2)^3 + [k^3 + 9k^2 + 27k + 27] = [k^3 + (k + 1)^3 + (k + 2)^3] + [9(k^2 + 3k + 3)]$, where the first summand is divisible by 9 because of the induction hypothesis. Consequently, since the result is true for $n = 0$, and since the truth for $n = k (\geq 0)$ implies the truth for $n = k + 1$, it follows from the principle of mathematical induction that the statement is true for all integers $n \geq 0$.

   c) If $n = 0$, $(n^7/7) + (n^3/3) + (11n/21) = 0$, an integer. So the statement is true in this first case. Assuming the statement true for $n = k \geq 0$, we have $(k^7/7) + (k^3/3) + (11k/21) \in \mathbf{Z}$. And now we turn to the case of $n = k + 1$. Here we have $((k + 1)^7/7 + (k + 1)^3/3 + 11(k + 1)/21) = [(k^7/7) + (k^3/3) + (11k/21)] + [(7k^6 + 21k^5 + 35k^4 + 35k^3 + 21k^2 + 7k)/7] + [(3k^2 + 3k)/3] + [(1/7) + (1/3) + (11/21)]$, where the first summand is an integer by virtue of the induction hypothesis, and the second and third summands are integers since each has a numerator that is divisible by the given denominator. So the truth of the statement for $n = k + 1$ hinges on whether or not the last summand is an integer—but this is true because $(1/7) + (1/3) + (11/21) = (3 + 7 + 11)/21 = 1 \in \mathbf{Z}$. Hence the result is true for all $n \in \mathbf{N}$, by the principle of mathematical induction.

9. For $n = 2$ we find that $2^2 = 4 < 6 = \binom{4}{2} < 16 = 4^2$, so the statement is true in this first case. Assuming the result true for $n = k \geq 2$—that is, $2^k < \binom{2k}{k} < 4^k$, we now consider what happens for $n = k + 1$. Here we find that

$$\binom{2(k+1)}{k+1} = \binom{2k+2}{k+1} = \left[\frac{(2k+2)(2k+1)}{(k+1)(k+1)}\right]\binom{2k}{k} = 2[(2k+1)/(k+1)]\binom{2k}{k}$$
$$> 2[(2k+1)/(k+1)]2^k > 2^{k+1},$$

since $(2k+1)/(k+1) = [(k+1) + k]/(k+1) > 1$. In addition, $[(k+1) + k]/(k+1) < 2$, so $\binom{2k+2}{k+1} = 2[(2k+1)/(k+1)]\binom{2k}{k} < (2)(2)\binom{2k}{k} < 4^{k+1}$. Consequently, the result is true for all $n \geq 2$, by the principle of mathematical induction.

11. **a)** First we observe that the statement is true for all $n \in \mathbf{Z}^+$ where $64 \leq n \leq 68$. This follows from the calculations

   $64 = 2(17) + 6(5)$ $65 = 13(5)$ $66 = 3(17) + 3(5)$
   $67 = 1(17) + 10(5)$ $68 = 4(17)$

   Now assume the result is true for all n where $68 \leq n \leq k$, and consider the integer $k + 1$. Then $k + 1 = (k - 4) + 5$, and since $64 \leq k - 4 < k$, we can write $k - 4 = a(17) + b(5)$ for some $a, b \in \mathbf{N}$. Consequently, $k + 1 = a(17) + (b + 1)(5)$, and the result follows for all $n \geq 64$, by the alternative form of the principle of mathematical induction.

   b) The proof here is similar to that for part (a). The following calculations are needed to show that the result is true for $108 \leq n \leq 117$.

$$108 = 6(13) + 3(10) \qquad 109 = 3(13) + 7(10) \qquad 110 = 11(10)$$
$$111 = 7(13) + 2(10) \qquad 112 = 4(13) + 6(10) \qquad 113 = 1(13) + 10(10)$$
$$114 = 8(13) + 1(10) \qquad 115 = 5(13) + 5(10) \qquad 116 = 2(13) + 9(10)$$
$$117 = 9(13)$$

13. a) $r = r_0 + r_1 \cdot 10 + r_2 \cdot 10^2 + \cdots + r_n \cdot 10^n$
$$= r_0 + r_1(9) + r_1 + r_2(99) + r_2 + \cdots + r_n(\underbrace{99 \ldots 9}_{n\,9\text{'s}}) + r_n$$
$$= [9r_1 + 99r_2 + \cdots + (99 \ldots 9)r_n] + (r_0 + r_1 + r_2 + \cdots + r_n).$$
Hence $9\,|\,r$ if and only if $9\,|\,(r_0 + r_1 + r_2 + \cdots + r_n)$.

c) $3\,|\,t$ for $x = 1$ or 4 or 7; $9\,|\,t$ for $x = 7$.

15. a) $\binom{13}{9}$ **b)** $\binom{8}{4}$ **c)** $\binom{11}{7}$ for part (a); $\binom{7}{3}$ for part (b)

17. a) $1, 4, 9$ **b)** $1, 4, 9, 16, \ldots, k$, where k is the largest square less than or equal to n.

CHAPTER 5

Relations and Functions

Section 5.1 – p. 169

1. $A \times B = \{(1, 2), (2, 2), (3, 2), (4, 2), (1, 5), (2, 5), (3, 5), (4, 5)\}$
$B \times A = \{(2, 1), (2, 2), (2, 3), (2, 4), (5, 1), (5, 2), (5, 3), (5, 4)\}$
$A \cup (B \times C) = \{1, 2, 3, 4, (2, 3), (2, 4), (2, 7), (5, 3), (5, 4), (5, 7)\}$
$(A \cup B) \times C = \{(1, 3), (2, 3), (3, 3), (4, 3), (5, 3), (1, 4), (2, 4), (3, 4), (4, 4), (5, 4), (1, 7),$
$\qquad (2, 7), (3, 7), (4, 7), (5, 7)\} = (A \times C) \cup (B \times C)$

3. a) 9 **b)** 2^9 **c)** 2^9 **d)** 2^7 **e)** $\binom{9}{5}$ **f)** $\binom{9}{7} + \binom{9}{8} + \binom{9}{9}$

5. $\mathscr{R} = \{(2, 10), (2, 12), (2, 14), (3, 12), (4, 12), (5, 10), (6, 12), (7, 14)\}$

7.

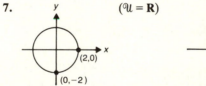 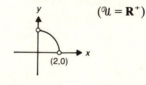

9. b) Follows from part (a) by duality.
c) $(x, y) \in (A \cap B) \times C \Leftrightarrow x \in A \cap B$ and $y \in C \Leftrightarrow (x \in A$ and $x \in B)$ and $y \in C \Leftrightarrow$
$(x \in A$ and $y \in C)$ and $(x \in B$ and $y \in C) \Leftrightarrow (x, y) \in A \times C$ and $(x, y) \in B \times C \Leftrightarrow$
$(x, y) \in (A \times C) \cap (B \times C)$
d) Follows from part (c) by duality.

11. $(x, y) \in A \times (B - C) \Leftrightarrow x \in A$ and $y \in B - C \Leftrightarrow x \in A$ and $(y \in B$ and $y \notin C) \Leftrightarrow$
$(x \in A$ and $y \in B)$ and $(x \in A$ and $y \notin C) \Leftrightarrow (x, y) \in A \times B$ and $(x, y) \notin A \times C \Leftrightarrow$
$(x, y) \in (A \times B) - (A \times C)$

Section 5.2 – p. 174

1. a) Function. Range $= \{7, 8, 11, 16, 23, \ldots\}$ **b)** Relation, not a function
c) Function. Range $= \mathbf{R}$. **d)** and **e)** Relation, not a function

3. a) (1) $\{(1, x), (2, x), (3, x), (4, x)\}$ (2) $\{(1, y), (2, y), (3, y), (4, y)\}$
(3) $\{(1, z), (2, z), (3, z), (4, z)\}$ (4) $\{(1, x), (2, y), (3, x), (4, y)\}$
(5) $\{(1, x), (2, y), (3, z), (4, x)\}$

b) 3^4 **c)** 0 **d)** 4^3 **e)** 24 **f)** 3^3 **g)** 3^2 **h)** 3^2

5. a) One-to-one; the range is the set of all odd integers.
b) One-to-one; the range is $\mathbf{Q}$.

   c) Not one-to-one; the range is $\{0, \pm 6, \pm 24, \pm 60, \dots\} = \{n^3 - n \mid n \in \mathbf{Z}\}$.

   d) One-to-one; the range is $(0, +\infty)$.

   e) One-to-one; the range is $[-1, 1]$.

   f) Not one-to-one; the range is $[0, 1]$.

7. 4^2

9. a) $f(A_1 \cup A_2) = \{y \in B \mid y = f(x), x \in A_1 \cup A_2\} = \{y \in B \mid y = f(x), x \in A_1 \text{ or } x \in A_2\} = \{y \in B \mid y = f(x), x \in A_1\} \cup \{y \in B \mid y = f(x), x \in A_2\} = f(A_1) \cup f(A_2)$

   c) From part (b), $f(A_1 \cap A_2) \subseteq f(A_1) \cap f(A_2)$. Conversely, $y \in f(A_1) \cap f(A_2) \Rightarrow y = f(x_1) = f(x_2)$, for $x_1 \in A_1, x_2 \in A_2 \Rightarrow y = f(x_1)$ and $x_1 = x_2$ (because f is injective) $\Rightarrow y \in f(A_1 \cap A_2)$. So f injective $\Rightarrow f(A_1 \cap A_2) = f(A_1) \cap f(A_2)$.

11. a) $f(a_{ij}) = 12(i-1) + j$ **b)** $f(a_{ij}) = 10(i-1) + j$

   c) $f(a_{ij}) = 7(i-1) + j$

13. a) (i) $f(a_{ij}) = n(i-1) + (k-1) + j$ (ii) $g(a_{ij}) = m(j-1) + (k-1) + i$

   b) $k + (mn - 1) \le r$

Section 5.3 – p. 179

1. a) $A = \{1, 2, 3, 4\}, B = \{v, w, x, y, z\}, f = \{(1, v), (2, v), (3, w), (4, x)\}$

   b) A, B as in (a), $f = \{(1, v), (2, x), (3, z), (4, y)\}$

   c) $A = \{1, 2, 3, 4, 5\}, B = \{w, x, y, z\}, f = \{(1, w), (2, w), (3, x), (4, y), (5, z)\}$

   d) $A = \{1, 2, 3, 4\}, B = \{w, x, y, z\}, f = \{(1, w), (2, x), (3, y), (4, z)\}$

3. (a), (b), (c), and (f) are one-to-one and onto.

   d) Neither one-to-one nor onto; range $= [0, +\infty)$.

   e) Neither one-to-one nor onto; range $= [-\frac{1}{4}, +\infty)$.

5. (For the case $n = 5, m = 3$):

$$\sum_{k=0}^{5} (-1)^k \binom{5}{5-k}(5-k)^3 = (-1)^0 \binom{5}{5}5^3 + (-1)^1 \binom{5}{4}4^3 + (-1)^2 \binom{5}{3}3^3 + (-1)^3 \binom{5}{2}2^3$$
$$+ (-1)^4 \binom{5}{1}1^3 + (-1)^5 \binom{5}{0}0^3$$
$$= 125 - 5(64) + 10(27) - 10(8) + 5 = 0$$

7. For any $r \in \mathbf{R}$ there is at least one $a \in \mathbf{R}$ such that $a^5 - 2a^2 + a - r = 0$, because the polynomial $x^5 - 2x^2 + x - r$ has odd degree and real coefficients. Consequently, f is onto. However, $f(0) = 0 = f(1)$, so f is not one-to-one.

9.

m \ n	1	2	3	4	5	6	7	8	9	10
9	1	255	3025	7770	6951	2646	462	36	1	
10	1	511	9330	34105	42525	22827	5880	750	45	1

Section 5.4 – p. 184

1. (a), (b), and (d) are commutative and associative; (c) is neither commutative nor associative.

3. a) $n^{(n^2-1)}$ **b)** $(n^n)(n^{(n^2-n-2)/2})$

5. $f(z_1, z_2) = z_2$ because z_1 is an identity for f.

7. a) $\pi_A(D) = [0, +\infty)$ $\pi_B(D) = \mathbf{R}$

   b) $\pi_A(D) = \mathbf{R}$ $\pi_B(D) = [-1, 1]$

   c) $\pi_A(D) = [-1, 1]$ $\pi_B(D) = [-1, 1]$

9. a) 5   **b)** A_3 A_4 A_5 **c)** A_1, A_2

25	25	6
25	2	4
60	40	20
25	40	10

Section 5.5 – p. 188

1. The pigeons are the socks; the pigeonholes are the colors.

3. a) 7   **b)** 13   **c)** $6(n-1)+1$

5. $26^2 + 1 = 677$

7. a) For each $x \in \{1, 2, 3, \dots, 300\}$ write $x = 2^n \cdot m$, where $n \geq 0$ and $(2, m) = 1$. There are 150 possibilities for m: $1, 3, 5, \dots, 299$. When we select 151 numbers from $\{1, 2, 3, \dots, 300\}$, there must be two numbers of the form $x = 2^s \cdot m$, $y = 2^t \cdot m$. If $x < y$, then $x \mid y$; otherwise $y < x$ and $y \mid x$.

b) If $n + 1$ integers are selected from the set $\{1, 2, 3, \dots, 2n\}$, then there must be two integers x, y in the selection where $x \mid y$ or $y \mid x$.

9. a) Here the pigeons are the integers $1, 2, 3, \dots, 25$ and the pigeonholes are the 13 sets $\{1, 25\}, \{2, 24\}, \dots, \{11, 15\}, \{12, 14\}, \{13\}$. In selecting 14 integers, we get the elements in at least one two-element subset, and these sum to 26.

b) If $S = \{1, 2, 3, \dots, 2n + 1\}$, for n a positive integer, then any subset of size $n + 2$ from S must contain two elements that sum to $2n + 2$.

11. a) For any $t \in \{1, 2, 3, \dots, 100\}$, $1 \leq \sqrt{t} \leq 10$. When we select 11 elements from $\{1, 2, 3, \dots, 100\}$ there must be two—say x and y—where $\lfloor \sqrt{x} \rfloor = \lfloor \sqrt{y} \rfloor$ so that $0 < |\sqrt{x} - \sqrt{y}| < 1$.

b) Let $n \in \mathbf{Z}^+$. If $n + 1$ elements are selected from $\{1, 2, 3, \dots, n^2\}$, then there exist two—say x and y—where $0 < |\sqrt{x} - \sqrt{y}| < 1$.

13.

Divide the interior of the square into four smaller congruent squares as shown in the figure. Each smaller square has diagonal length $1/\sqrt{2}$. Let region R_1 be the interior of square $AEKH$ together with the points on segment EK, excluding point E. Region R_2 is the interior of square $EBFK$ together with the points on segment FK, excluding points F and K. Regions R_3 and R_4 are defined in a similar way. Then if five points are chosen in the interior of square $ABCD$, at least two are in R_i for some $1 \leq i \leq 4$, and these points are within $1/\sqrt{2}$ (units) of each other.

15. Start by associating four students with each event (a total of 32 students). If we associate 6 more students with one of these events, we have used 38 students. Now do the same for a second event. At this point 44 students are assigned; two events have 10 students each, and the other six events each have 4 students. When we assign the last student, a third event will now have 5 or more students training for it.

17. a) Since $|S| \geq 3$ there exist $x, y \in S$ where x, y are both even or both odd. In either case, $x + y$ is even.

b) $5 = 2^2 + 1$ **c)** $|S| \geq 9 = 2^3 + 1$

d) For $n \in \mathbf{Z}^+$ let $S = \{(a_1, a_2, \ldots, a_n) \mid a_i \in \mathbf{Z}^+, 1 \le i \le n\}$. If $|S| \ge 2^n + 1$, then S contains two ordered n-tuples $(x_1, x_2, \ldots, x_n), (y_1, y_2, \ldots, y_n)$ such that $x_i + y_i$ is even for all $1 \le i \le n$.

e) 5, as in part (b)

Section 5.6 – p. 197

1. h is onto $\Leftrightarrow$ for all $b \in B, d \in D$, there exists $a \in A, c \in C$ with $h(a, c) = (b, d) \Leftrightarrow$ for all $b \in B, d \in D$, there exists $a \in A, c \in C$ with $f(a) = b, g(c) = d \Leftrightarrow f, g$ are onto. h is one-to-one $\Leftrightarrow$ [for all $a, a_1 \in A, c, c_1 \in C, h(a, c) = h(a_1, c_1) \Rightarrow a = a_1, c = c_1] \Leftrightarrow$ [for all $a, a_1 \in A, c, c_1 \in C, f(a) = f(a_1) \Rightarrow a = a_1$, and $g(c) = g(c_1) \Rightarrow c = c_1] \Leftrightarrow f, g$ are one-to-one.

3.
$$g^2(A) = g(T \cap (S \cup A)) = T \cap (S \cup [T \cap (S \cup A)])$$
$$= T \cap [(S \cup T) \cap (S \cup (S \cup A))] = T \cap [(S \cup T) \cap (S \cup A)]$$
$$= [T \cap (S \cup T)] \cap (S \cup A) = T \cap (S \cup A) = g(A)$$

5. $a = 3, b = -1; a = -3, b = 2$

7. a) $7! - 6! = 4320$

b) $n! - (n-1)! = (n-1)(n-1)!$

9. a) $(b, a) \in (\mathcal{R}_1 \cup \mathcal{R}_2)^c \Leftrightarrow (a, b) \in \mathcal{R}_1 \cup \mathcal{R}_2 \Leftrightarrow (a, b) \in \mathcal{R}_1$ or $(a, b) \in \mathcal{R}_2 \Leftrightarrow (b, a) \in \mathcal{R}_1^c$ or $(b, a) \in \mathcal{R}_2^c \Leftrightarrow (b, a) \in \mathcal{R}_1^c \cup \mathcal{R}_2^c$

b) $(b, a) \in (\mathcal{R}_1 \cap \mathcal{R}_2)^c \Leftrightarrow (a, b) \in \mathcal{R}_1 \cap \mathcal{R}_2 \Leftrightarrow (a, b) \in \mathcal{R}_1$ and $(a, b) \in \mathcal{R}_2 \Leftrightarrow (b, a) \in \mathcal{R}_1^c$ and $(b, a) \in \mathcal{R}_2^c \Leftrightarrow (b, a) \in \mathcal{R}_1^c \cap \mathcal{R}_2^c$

c) $(a, b) \in (\mathcal{R}_1^c)^c \Leftrightarrow (b, a) \in (\mathcal{R}_1^c) \Leftrightarrow (a, b) \in \mathcal{R}_1$

11. a) $f^{-1}(x) = (1/2)(\ln x - 5)$

b) For $x \in \mathbf{R}^+$,
$$(f \circ f^{-1})(x) = f((1/2)(\ln x - 5)) = e^{2((1/2)(\ln x - 5)) + 5} = e^{\ln x - 5 + 5} = e^{\ln x} = x.$$
For $x \in \mathbf{R}$,
$$(f^{-1} \circ f)(x) = f^{-1}(e^{2x+5}) = (1/2)[\ln(e^{2x+5}) - 5] = (1/2)[2x + 5 - 5] = x.$$

c)

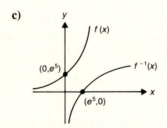

13. f, g invertible $\Rightarrow$ each of f, g is both one-to-one and onto $\Rightarrow g \circ f$ is one-to-one and onto $\Rightarrow g \circ f$ invertible. Since $(g \circ f) \circ (f^{-1} \circ g^{-1}) = 1_C$ and $(f^{-1} \circ g^{-1}) \circ (g \circ f) = 1_A, f^{-1} \circ g^{-1}$ is an inverse of $g \circ f$. By uniqueness of inverses, $f^{-1} \circ g^{-1} = (g \circ f)^{-1}$.

15. a) $[0, 2)$ **b)** $[-1, 2)$ **c)** $[0, 1)$ **d)** $[0, 2)$
    e) $[-1, 2)$ **f)** $[0, 2)$ **g)** $[-1, 3)$ **h)** $[-1, 0) \cup [2, 4)$

17. a) $a \in f^{-1}(B_1 \cap B_2) \Leftrightarrow f(a) \in B_1 \cap B_2 \Leftrightarrow f(a) \in B_1$ and $f(a) \in B_2 \Leftrightarrow a \in f^{-1}(B_1)$ and $a \in f^{-1}(B_2) \Leftrightarrow a \in f^{-1}(B_1) \cap f^{-1}(B_2)$

c) $a \in f^{-1}(\overline{B_1}) \Leftrightarrow f(a) \in \overline{B_1} \Leftrightarrow f(a) \notin B_1 \Leftrightarrow a \notin f^{-1}(B_1) \Leftrightarrow a \in \overline{f^{-1}(B_1)}$

19. a) (i) $f(x) = 2x$ (ii) $f(x) = \lfloor x/2 \rfloor$

   b) No. The set $\mathbf{Z}$ is not finite.

Section 5.7 – p. 202

1. a) $f \in O(n)$ **b)** $f \in O(1)$ **c)** $f \in O(n^3)$ **d)** $f \in O(n^2)$

   e) $f \in O(n^3)$ **f)** $f \in O(n^2)$ **g)** $f \in O(n^2)$

3. a) For all $n \in \mathbf{Z}^+$, $0 \le \log_2 n < n$. So let $k = 1$ and $m = 200$ in Definition 5.22. Then $|f(n)| = 100 \log_2 n = 200(\frac{1}{2} \log_2 n) < 200(\frac{1}{2}n) = 200|g(n)|$, so $f \in O(g)$.

   b) For $n = 6$, $2^n = 64 < 3096 = 4096 - 1000 = 2^{12} - 1000 = 2^{2n} - 1000$. Assuming that $2^k < 2^{2k} - 1000$ for $n = k \ge 6$, we find that $2 < 2^2 \Rightarrow 2(2^k) < 2^2(2^{2k} - 1000) < 2^2 2^{2k} - 1000$, or $2^{k+1} < 2^{2(k+1)} - 1000$, so $f(n) < g(n)$ for all $n \ge 6$. Therefore with $k = 6$ and $m = 1$ in Definition 5.22, we find that for $n \ge k$, $|f(n)| \le m|g(n)|$ and $f \in O(g)$.

   c) For all $n \ge 4$, $n^2 \le 2^n$. (A formal proof of this can be given by mathematical induction.) So let $k = 4$ and $m = 3$ in Definition 5.22. Then for $n \ge k$, $|f(n)| = 3n^2 \le 3(2^n) < 3(2^n + 2n) = m|g(n)|$ and $f \in O(g)$.

5. To show that $f \in O(g)$, let $k = 1$ and $m = 4$ in Definition 5.22. Then for any $n \ge k$, $|f(n)| = n^2 + n \le n^2 + n^2 = 2n^2 \le 2n^3 = 4((1/2)n^3) = 4|g(n)|$, and f is dominated by g. To show that $g \notin O(f)$, we follow the idea given in Example 5.51, namely that

$$\forall m \in \mathbf{R}^+ \, \forall k \in \mathbf{Z}^+ \, \exists n \in \mathbf{Z}^+ \, [(n \ge k) \wedge (|g(n)| > m|f(n)|)].$$

So no matter what the values of m and k are, choose $n > \max\{4m, k\}$. Then $|g(n)| = (\frac{1}{2})n^3 > (\frac{1}{2})(4m)n^2 = m(2n^2) \ge m(n^2 + n) = m|f(n)|$, so $g \notin O(f)$.

7. For all $n \ge 1$, $\log_2 n \le n$, so with $k = 1$ and $m = 1$ in Definition 5.22, we have $|g(n)| = \log_2 n \le n = m \cdot n = m|f(n)|$. Hence $g \in O(f)$. To show that $f \notin O(g)$, we first observe that $\lim\limits_{n \to \infty} \dfrac{n}{\log_2 n} = +\infty$. (This can be established by using L'Hospital's Rule from the calculus.) Since $\lim\limits_{n \to \infty} \dfrac{n}{\log_2 n} = +\infty$, we find that for every $m \in \mathbf{R}^+$ and $k \in \mathbf{Z}^+$, there is an $n \in \mathbf{Z}^+$ such that $\dfrac{n}{\log_2 n} > m$, or $|f(n)| = n > m \log_2 n = m|g(n)|$. Hence $f \notin O(g)$.

9. Since $f \in O(g)$, there exists $m \in \mathbf{R}^+, k \in \mathbf{Z}^+$ such that $|f(n)| \le m|g(n)|$ for all $n \ge k$. But then $|f(n)| \le [m/|c|]|cg(n)|$ for all $n \ge k$, so $f \in O(cg)$.

Section 5.8 – p. 211

1. a) $f \in O(n^2)$ **b)** $f \in O(n^3)$ **c)** $f \in O(n^2)$ **d)** $f \in O(\log_2 n)$

   e) $f \in O(n \log_2 n)$

3. a) $f_3 \in O(n^2)$ **b)** $f_3 \in O(\log_2 n)$ **c)** $f_3 \in O(n^2)$ **d)** $f_3 \in O(n^2)$

   e) $f_3 \in O(2^n)$

Miscellaneous Exercises – p. 215

1. a) $(a, b) \in A \times B \Rightarrow a \in A, b \in B \Rightarrow a \in C, b \in D$, since $A \subseteq C$ and $B \subseteq D \Rightarrow (a, b) \in C \times D$.

   b) If either A or B is $\emptyset$, the result need not follow. For example, let $A = \{1, 2\}$, $B = \emptyset$, $C = \{1\}$, and $D = \{1\}$; then $A \times B = \emptyset \subseteq C \times D$ but $A \not\subseteq C$. If neither A nor B is $\emptyset$, let $a \in A$, $b \in B$. Then $(a, b) \in A \times B \subseteq C \times D \Rightarrow a \in C, b \in D$. Hence $A \subseteq C$ and $B \subseteq D$.

3. a)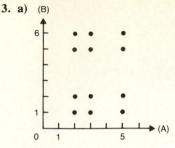

b)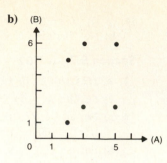

c) $2^{12} - 4^3$

5. a) $(x, y) \in (A \cap B) \times (C \cap D) \Leftrightarrow x \in A \cap B, y \in C \cap D \Leftrightarrow (x \in A, y \in C)$ and $(x \in B, y \in D) \Leftrightarrow (x, y) \in A \times C$ and $(x, y) \in B \times D \Leftrightarrow (x, y) \in (A \times C) \cap (B \times D)$

b) $(x, y) \in (A \cup B) \times (C \cup D) \Leftrightarrow x \in A \cup B, y \in C \cup D \Leftrightarrow (x \in A$ or $x \in B)$ and $(y \in C$ or $y \in D) \Leftrightarrow (x \in A$ and $y \in C)$ or $(x \in B$ and $y \in D)$ or $(x \in A$ and $y \in D)$ or $(x \in B$ and $y \in C) \Leftrightarrow ((x, y) \in A \times C)$ or $((x, y) \in B \times D)$ or $((x, y) \in A \times D)$ or $((x, y) \in B \times C) \Leftrightarrow (x, y) \in (A \times C) \cup (B \times D) \cup (A \times D) \cup (B \times C)$

7. $A(2, 3) = 9$ **9. a)** $(7!)/[2(7^5)]$

11. For $1 \le i \le 10$, let x_i be the number of letters typed on day i. Then $x_1 + x_2 + x_3 + \cdots + x_8 + x_9 + x_{10} = 84$, or $x_3 + \cdots + x_8 = 54$. Suppose that $x_1 + x_2 + x_3 < 25$, $x_2 + x_3 + x_4 < 25, \ldots, x_8 + x_9 + x_{10} < 25$. Then $x_1 + 2x_2 + 3(x_3 + \cdots + x_8) + 2x_9 + x_{10} < 8(25) = 200$, or $3(x_3 + \cdots + x_8) < 160$. Consequently, we obtain the contradiction $54 = x_3 + \cdots + x_8 < \frac{160}{3} = 53\frac{1}{3}$.

13. For $\prod_{k=1}^{n}(k - i_k)$ to be odd, $(k - i_k)$ must be odd for all $1 \le k \le n$, that is, one of k, i_k must be even and the other odd. Since n is odd, $n = 2m + 1$ and in the list $1, 2, \ldots, n$ there are m even integers and $m + 1$ odd integers. Let $1, 3, 5, \ldots, n$ be the pigeons and $i_1, i_3, i_5, \ldots, i_n$ the pigeonholes. At most m of the pigeonholes can be even integers, so $(k - i_k)$ must be even for at least one $k = 1, 3, 5, \ldots, n$. Consequently, $\prod_{k=1}^{n}(k - i_k)$ is even.

15. Let the n distinct objects be $x_1, x_2, \ldots, x_n$. Place x_n in a container. Now there are two *distinct* containers. For each of $x_1, x_2, \ldots, x_{n-1}$ there are two choices and this gives 2^{n-1} distributions. Among these there is one where $x_1, x_2, \ldots, x_{n-1}$ are in the container with x_n, so we remove this distribution and find $S(n, 2) = 2^{n-1} - 1$.

17. a) and **b)** $m! S(n, m)$

19. Fix $m = 1$. For $n = 1$ the result is true. Assume $f \circ f^k = f^k \circ f$ and consider $f \circ f^{k+1}$. $f \circ f^{k+1} = f \circ (f \circ f^k) = f \circ (f^k \circ f) = (f \circ f^k) \circ f = f^{k+1} \circ f$. Hence $f \circ f^n = f^n \circ f$ for all $n \in \mathbf{Z}^+$. Now assume that for $t \ge 1$, $f^t \circ f^n = f^n \circ f^t$. Then $f^{t+1} \circ f^n = (f \circ f^t) \circ f^n = f \circ (f^t \circ f^n) = f \circ (f^n \circ f^t) = (f \circ f^n) \circ f^t = (f^n \circ f) \circ f^t = f^n \circ (f \circ f^t) = f^n \circ f^{t+1}$, so $f^m \circ f^n = f^n \circ f^m$ for all $m, n \in \mathbf{Z}^+$.

21. Any $n \in \mathbf{Z}^+$ where n is odd.

23. $x < y \Rightarrow f(x) < f(y)$ (since f is increasing) $\Rightarrow g(f(x)) < g(f(y))$ (since g is increasing.) Hence $g \circ f$ is increasing.

25. $f \circ g = \{(x, z), (y, y), (z, x)\}$; $g \circ f = \{(x, x), (y, z), (z, y)\}$; $f^{-1} = \{(x, z), (y, x), (z, y)\}$; $g^{-1} = \{(x, y), (y, x), (z, z)\}$; $(g \circ f)^{-1} = \{(x, x), (y, z), (z, y)\} = f^{-1} \circ g^{-1}$; $g^{-1} \circ f^{-1} = \{(x, z), (y, y), (z, x)\}$.

27. a) 4   **b)** 10

29. a) $a \in A \Rightarrow f(a) \in f(A) \Rightarrow a \in f^{-1}(f(A))$. If f is one-to-one, $A = f^{-1}(f(A))$.

b) $b \in f(f^{-1}(B)) \Rightarrow b = f(a)$ for some $a \in f^{-1}(B) \Rightarrow b = f(a)$ where $f(a) \in B \Rightarrow b \in B$. If f is onto, $B = f(f^{-1}(B))$.

31. a) $(\pi \circ \sigma)(x) = (\sigma \circ \pi)(x) = x$

b) $\pi^n(x) = x - n; \sigma^n(x) = x + n \qquad (n \geq 2)$

c) $\pi^{-n}(x) = x + n; \sigma^{-n}(x) = x - n \qquad (n \geq 2)$

33. a) $\tau(n) = (e_1 + 1)(e_2 + 1) \cdots (e_k + 1)$

b) $k = 2: \tau(2) = \tau(3) = \tau(5) = 2 \qquad k = 3: \tau(2^2) = \tau(3^2) = \tau(5^2) = 3$

$k = 4: \tau(6) = \tau(8) = \tau(10) = 4 \qquad k = 5: \tau(2^4) = \tau(3^4) = \tau(5^4) = 5$

$k = 6: \tau(12) = \tau(18) = \tau(20) = 6$

c) For any $k > 1$ and any prime p, $\tau^{-1}(p^{k-1}) = k$.

35. a) $4! \, S(8, 4)$ **b)** $m! \, S(n, m)$

37. a) Let $m = 1$ and $k = 1$. Then for any $n \geq k$, $|f(n)| \leq 2 < 3 \leq |g(n)| = m|g(n)|$, so $f \in O(g)$.

b) Let $m = 4$ and $k = 1$. Then for any $n \geq k$, $|g(n)| \leq 4 = 4 \cdot 1 \leq 4|f(n)| = m|f(n)|$, so $g \in O(f)$.

39. First note that if $\log_a n = r$, then $n = a^r$ and $\log_b n = \log_b(a^r) = r \log_b a = (\log_b a)(\log_a n)$. Now let $m = (\log_b a)$ and $k = 1$. Then for any $n \geq k$, $|g(n)| = \log_b n = (\log_b a)(\log_a n) = m|f(n)|$, so $g \in O(f)$. Finally, with $m = (\log_b a)^{-1} = \log_a b$ and $k = 1$, we find that for any $n \geq k$, $|f(n)| = \log_a n = (\log_a b)(\log_b n) = m|g(n)|$. Hence $f \in O(g)$.

CHAPTER 6

Languages: Finite State Machines

Section 6.1 – p. 226

1. a) $25; 125$ **b)** 3906 **3.** 12 **5.** 780

7. Let Σ be an alphabet with $\emptyset \neq A \subseteq \Sigma^*$. If $|A| = 1$ and $x \in A$, then $xx = x$ since $A^2 = A$. But $\|xx\| = 2\|x\| = \|x\| \Rightarrow \|x\| = 0 \Rightarrow x = \lambda$. If $|A| > 1$, let $x \in A$ where $\|x\| > 0$ but $\|x\|$ is minimal. Then $x \in A^2 \Rightarrow x = yz$, for $y, z \in A$. Since $\|x\| = \|y\| + \|z\|$, if $\|y\|, \|z\| > 0$, then one of y, z is in A with length smaller than $\|x\|$. Consequently, one of $\|y\|$ or $\|z\|$ is 0, so $\lambda \in A$.

9. a) $x \in AC \Rightarrow x = ac$, for some $a \in A, c \in C \Rightarrow x \in BD$, since $A \subseteq B, C \subseteq D$.

b) If $A\emptyset \neq \emptyset$, let $x \in A\emptyset$. $x \in A\emptyset \Rightarrow x = yz$, for some $y \in A, z \in \emptyset$. But $z \in \emptyset$ is impossible. Hence $A\emptyset = \emptyset$. [In like manner, $\emptyset A = \emptyset$].

11. If $A = A^2$, then it follows by mathematical induction that $A = A^n$ for all $n \in \mathbf{Z}^+$. Hence $A = A^+$. By Exercise 7, $A = A^2 \Rightarrow \lambda \in A$. Hence $A = A^*$.

13. By Definition 6.11, $AB = \{ab \mid a \in A, b \in B\}$, and since it is possible to have $a_1 b_1 = a_2 b_2$ with $a_1, a_2 \in A$, $a_1 \neq a_2$, and $b_1, b_2 \in B$, $b_1 \neq b_2$, it follows that $|AB| \leq |A \times B| = |A||B|$.

15. a) $x \in A(\bigcup_{i \in I} B_i) \Leftrightarrow x = ab$, for some $a \in A, b \in \bigcup_{i \in I} B_i \Leftrightarrow x = ab$, where $a \in A$ and $b \in B_i$, for some $i \in I \Leftrightarrow x \in AB_i$, for some $i \in I \Leftrightarrow x \in \bigcup_{i \in I} AB_i$.

b) $y \in (\bigcup_{i \in I} B_i)A \Leftrightarrow y = dc$, for some $d \in \bigcup_{i \in I} B_i, c \in A \Leftrightarrow y = dc$, where $d \in B_i$ for some $i \in I$, and $c \in A \Leftrightarrow y \in B_i A$ for some $i \in I \Leftrightarrow y \in \bigcup_{i \in I} B_i A$.

Section 6.2 – p. 233

1. a) $0010101; s_1$

b) $0000000; s_1$

c) $001000000; s_0$

3. **a)** 010110   **b)**

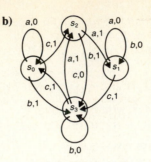

5. **a)** 010000; s_2 **b)** (s_1) 100000; s_2
(s_2) 000000; s_2
(s_3) 110010; s_2

c)

	ν		ω	
	0	1	0	1
s_0	s_0	s_1	0	0
s_1	s_1	s_2	1	1
s_2	s_2	s_2	0	0
s_3	s_0	s_3	0	1
s_4	s_2	s_3	0	1

d) s_1 **e)** $x = 101$ (unique)

7. **a)** (i) 15   (ii) 3^{15} (iii) 2^{15} **b)** 6^{15}

Section 6.3 – p. 242

1. **a)** 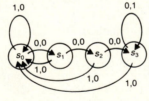   **b)**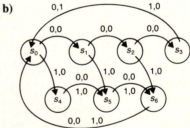

3. $(k = 1)$: $(k > 1)$:

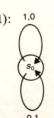 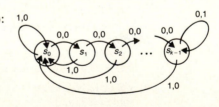

7. **a)** The transient states are s_0, s_1. State s_3 is a sink state. $\{s_1, s_2, s_3, s_4, s_5\}$, $\{s_4\}$, and $\{s_2, s_3, s_5\}$ (with the corresponding restrictions on the given function ν) constitute submachines. The strongly connected submachines are $\{s_4\}$ and $\{s_2, s_3, s_5\}$.

b) States s_2, s_3 are transient. The only sink state is s_4. The set $\{s_0, s_1, s_3, s_4\}$ provides the states for a submachine; $\{s_0, s_1\}$ and $\{s_4\}$ provide strongly connected submachines.

c) Here there are no transient states. State s_6 is a sink state. There are three sub-machines: $\{s_2, s_3, s_4, s_5, s_6\}$, $\{s_3, s_4, s_5, s_6\}$, and $\{s_6\}$. The only strongly connected sub-machine is $\{s_6\}$.

Miscellaneous Exercises – p. 244

1. a) True   **b)** True   **c)** True   **d)** False   **e)** True
   f) True   **g)** True   **h)** True   **i)** False   **j)** True

3. Let $x \in \Sigma$ and $A = \{x\}$. Then $A^2 = \{xx\}$ and $(A^2)^* = \{\lambda, x^2, x^4, \dots\}$.
   $A^* = \{\lambda, x, x^2, x^3, \dots\}$ and $(A^*)^2 = A^*$, so $(A^2)^* \neq (A^*)^2$.

5. $\mathcal{O}_{02} = \{1, 00\}^*\{0\}$
   $\mathcal{O}_{22} = \{0\}\{1, 00\}^*\{0\}$
   $\mathcal{O}_{11} = \emptyset$
   $\mathcal{O}_{00} = \{1, 00\}^* - \{\lambda\}$
   $\mathcal{O}_{10} = \{1\}\{1, 00\}^* \cup \{10\}\{1, 00\}^*$

9. Both 1110 and 1111 are transfer sequences from s_0 to s_5.

11.

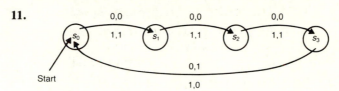

Start

13. Suppose that such a machine can be constructed and that it has n states, for $n \in \mathbf{Z}^+$. For the input string $0^n 1^{n+1}$, we want the output to be $0^{2n} 1$. However, as the 1's in this input string are processed, we obtain $n + 1$ states $s_1, s_2, \dots, s_n, s_{n+1}$ from the function ν. Consequently, by the pigeonhole principle, there are two states s_i, s_j where $i < j$ but $s_i = s_j$. As a result, if we remove the $j - i$ states $s_{i+1}, s_{i+2}, \dots, s_j$ and the 1's that were their corresponding inputs, we find that the machine recognizes the sequence $0^n 1^{n+1-(j-i)}$, where $n + 1 - (j - i) \leq n$. However, $0^n 1^{n+1-(j-i)} \notin A$.

15. a)

	ν		ω	
	0	1	0	1
(s_0, s_3)	(s_0, s_4)	(s_1, s_3)	1	1
(s_0, s_4)	(s_0, s_3)	(s_1, s_4)	0	1
(s_1, s_3)	(s_1, s_3)	(s_2, s_3)	1	1
(s_1, s_4)	(s_1, s_4)	(s_2, s_4)	1	1
(s_2, s_3)	(s_2, s_3)	(s_0, s_4)	1	1
(s_2, s_4)	(s_2, s_4)	(s_0, s_3)	1	0

b) $\omega(1101) = 1111$; M_1 is in state s_0, and M_2 is in state s_4.

CHAPTER 7

Relations: The Second Time Around

Section 7.1 – p. 253

1. a) $\{(1, 1), (2, 2), (3, 3), (4, 4), (1, 2), (2, 1), (2, 3), (3, 2)\}$
   b) $\{(1, 1), (2, 2), (3, 3), (4, 4), (1, 2)\}$ **c)** $\{(1, 1), (2, 2), (1, 2), (2, 1)\}$

3. a) Let $f_1, f_2, f_3 \in \mathcal{F}$ with $f_1(n) = n + 1$, $f_2(n) = 5n$, and $f_3(n) = 4n + 1/n$.

   b) Let $g_1, g_2, g_3 \in \mathcal{F}$ with $g_1(n) = 3$, $g_2(n) = 1/n$, and $g_3(n) = \sin n$.

5. a) reflexive, antisymmetric, transitive   **b)** transitive
    c) reflexive, symmetric, transitive   **d)** symmetric
    e) (even)-reflexive, symmetric, transitive; (odd)-symmetric
    f) (even)-reflexive, symmetric, transitive; (odd)-symmetric
    g) symmetric   **h)** reflexive, symmetric
    i) reflexive, transitive

7. a) For all $x \in A$, $(x,x) \in \mathcal{R}_1, \mathcal{R}_2$, so $(x,x) \in \mathcal{R}_1 \cap \mathcal{R}_2$ and $\mathcal{R}_1 \cap \mathcal{R}_2$ is reflexive.

    b) (i) $(x,y) \in \mathcal{R}_1 \cap \mathcal{R}_2 \Rightarrow (x,y) \in \mathcal{R}_1, \mathcal{R}_2 \Rightarrow (y,x) \in \mathcal{R}_1, \mathcal{R}_2 \Rightarrow (y,x) \in \mathcal{R}_1 \cap \mathcal{R}_2$ and $\mathcal{R}_1 \cap \mathcal{R}_2$ is symmetric.

      (ii) $(x,y), (y,x) \in \mathcal{R}_1 \cap \mathcal{R}_2 \Rightarrow (x,y), (y,x) \in \mathcal{R}_1, \mathcal{R}_2$. By the antisymmetry of $\mathcal{R}_1$ (or $\mathcal{R}_2$), $x = y$ and $\mathcal{R}_1 \cap \mathcal{R}_2$ is antisymmetric.

      (iii) $(x,y), (y,z) \in \mathcal{R}_1 \cap \mathcal{R}_2 \Rightarrow (x,y), (y,z) \in \mathcal{R}_1, \mathcal{R}_2 \Rightarrow (x,z) \in \mathcal{R}_1, \mathcal{R}_2$ (transitive property) $\Rightarrow (x,z) \in \mathcal{R}_1 \cap \mathcal{R}_2$, so $\mathcal{R}_1 \cap \mathcal{R}_2$ is transitive.

9. a) True   **b)** False: let $A = \{1,2\}$ and $\mathcal{R} = \{(1,2),(2,1)\}$.
    c) (i) Reflexive: true
      (ii) Symmetric: false. Let $A = \{1,2\}$, $\mathcal{R}_1 = \{(1,1)\}$, and $\mathcal{R}_2 = \{(1,1),(1,2)\}$.
      (iii) Antisymmetric and transitive: false. Let $A = \{1,2\}$, $\mathcal{R}_1 = \{(1,2)\}$, and $\mathcal{R}_2 = \{(1,2),(2,1)\}$.

    d) (i) Reflexive: false. Let $A = \{1,2\}$, $\mathcal{R}_1 = \{(1,1)\}$, and $\mathcal{R}_2 = \{(1,1),(2,2)\}$.
      (ii) Symmetric: false. Let $A = \{1,2\}$, $\mathcal{R}_1 = \{(1,2)\}$, and $\mathcal{R}_2 = \{(1,2),(2,1)\}$.
      (iii) Antisymmetric: true
      (iv) Transitive: false. Let $A = \{1,2\}$, $\mathcal{R}_1 = \{(1,2),(2,1)\}$, and $\mathcal{R}_2 = \{(1,1),(1,2),(2,1),(2,2)\}$.

    e) True.

11. There may exist an element $a \in A$ such that for all $b \in B$, neither (a,b) nor $(b,a) \in \mathcal{R}$.

13. $r - n$ counts the elements in $\mathcal{R}$ of the form (a,b), $a \neq b$. Since $\mathcal{R}$ is symmetric, $r - n$ is even.

Section 7.2 – p. 264

1. $\mathcal{R} \circ \mathcal{S} = \{(1,3),(1,4)\}$; $\mathcal{S} \circ \mathcal{R} = \{(1,2),(1,3),(1,4),(2,4)\}$;
$\mathcal{R}^2 = \mathcal{R}^3 = \{(1,4),(2,4),(4,4)\}$; $\mathcal{S}^2 = \mathcal{S}^3 = \{(1,1),(1,2),(1,3),(1,4)\}$

3. $(a,d) \in (\mathcal{R}_1 \circ \mathcal{R}_2) \circ \mathcal{R}_3 \Rightarrow (a,c) \in \mathcal{R}_1 \circ \mathcal{R}_2$, $(c,d) \in \mathcal{R}_3$ for some $c \in C \Rightarrow (a,b) \in \mathcal{R}_1$, $(b,c) \in \mathcal{R}_2$, $(c,d) \in \mathcal{R}_3$ for some $b \in B$, $c \in C \Rightarrow (a,b) \in \mathcal{R}_1$, $(b,d) \in \mathcal{R}_2 \circ \mathcal{R}_3 \Rightarrow (a,d) \in \mathcal{R}_1 \circ (\mathcal{R}_2 \circ \mathcal{R}_3)$, and $(\mathcal{R}_1 \circ \mathcal{R}_2) \circ \mathcal{R}_3 \subseteq \mathcal{R}_1 \circ (\mathcal{R}_2 \circ \mathcal{R}_3)$.

5. This follows by the pigeonhole principle. Here the pigeons are the $2^{n^2} + 1$ integers between 0 and 2^{n^2}, inclusive, and the pigeonholes are the 2^{n^2} relations on A.

7. 2^{21}

9. Consider the entry in the ith row and jth column of $M(\mathcal{R}_1 \circ \mathcal{R}_2)$. If this entry is a 1, then there exists $b_k \in B$ where $1 \leq k \leq n$ and $(a_i, b_k) \in \mathcal{R}_1$, $(b_k, c_j) \in \mathcal{R}_2$. Consequently, the entry in the ith row and kth column of $M(\mathcal{R}_1)$ is 1, and the entry in the kth row and jth column of $M(\mathcal{R}_2)$ is 1. This results in a 1 in the ith row and jth column in the product $M(\mathcal{R}_1) \cdot M(\mathcal{R}_2)$.

    Should the entry in row i and column j of $M(\mathcal{R}_1 \circ \mathcal{R}_2)$ be 0, then for each b_k, where $1 \leq k \leq n$, either $(a_i, b_k) \notin \mathcal{R}_1$ or $(b_k, c_j) \notin \mathcal{R}_2$. This means that in the matrices $M(\mathcal{R}_1)$, $M(\mathcal{R}_2)$, if the entry in the ith row and kth column of $M(\mathcal{R}_1)$ is 1, then the entry in the kth row and jth column of $M(\mathcal{R}_2)$ is 0. Hence the entry in the ith row and jth column of $M(\mathcal{R}_1) \cdot M(\mathcal{R}_2)$ is 0.

11. d) Let s_{xy} be the entry in row (x) and column (y) of M. Then s_{yx} appears in row (x) and column (y) of M^{tr}. $\mathcal{R}$ is antisymmetric $\Leftrightarrow (s_{xy} = s_{yx} = 1 \Rightarrow x = y) \Leftrightarrow M \cap M^{\text{tr}} \le I_n$.

13. $\mathcal{R}$ $\quad \mathcal{R}^2$ $\quad \mathcal{R}^3$ and $\mathcal{R}^4$

15. a) 2^{25} **b)** 2^{15}

17. a) $\mathcal{R}_1:$
$$\begin{bmatrix} 1 & 1 & 0 & 0 & 0 \\ 1 & 1 & 0 & 0 & 0 \\ 0 & 0 & 1 & 1 & 0 \\ 0 & 0 & 1 & 1 & 0 \\ 0 & 0 & 0 & 0 & 1 \end{bmatrix}$$
$\mathcal{R}_2:$
$$\begin{bmatrix} 1 & 1 & 1 & 0 & 0 \\ 1 & 1 & 1 & 0 & 0 \\ 1 & 1 & 1 & 0 & 0 \\ 0 & 0 & 0 & 1 & 1 \\ 0 & 0 & 0 & 1 & 1 \end{bmatrix}$$

b) Given an equivalence relation $\mathcal{R}$ on a finite set A, list the elements of A so that elements in the same cell of the partition (see Section 7.4) are adjacent. The resulting relation matrix will then have square blocks of 1's along the diagonal (from upper left to lower right).

Section 7.3 – p. 274

1.

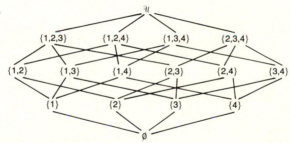

3. For all $a \in A$, $b \in B$, we have $a \mathcal{R}_1 a$ and $b \mathcal{R}_2 b$, so $(a, b) \mathcal{R} (a, b)$ and $\mathcal{R}$ is reflexive. $(a, b) \mathcal{R} (c, d)$, $(c, d) \mathcal{R} (a, b) \Rightarrow a \mathcal{R}_1 c$, $c \mathcal{R}_1 a$ and $b \mathcal{R}_2 d$, $d \mathcal{R}_2 b \Rightarrow a = c$, $b = d \Rightarrow (a, b) = (c, d)$, so $\mathcal{R}$ is antisymmetric. $(a, b) \mathcal{R} (c, d)$, $(c, d) \mathcal{R} (e, f) \Rightarrow a \mathcal{R}_1 c$, $c \mathcal{R}_1 e$ and $b \mathcal{R}_2 d$, $d \mathcal{R}_2 f \Rightarrow a \mathcal{R}_1 e$, $b \mathcal{R}_2 f \Rightarrow (a, b) \mathcal{R} (e, f)$, and this implies that $\mathcal{R}$ is transitive.

5. $\emptyset < \{1\} < \{2\} < \{3\} < \{1, 2\} < \{1, 3\} < \{2, 3\} < \{1, 2, 3\}$.
(There are other possibilities.)

7. a)

b) $3 < 2 < 1 < 4$ or $3 < 1 < 2 < 4$ **c)** 2

11. Let x, y both be least upper bounds. Then $x \mathcal{R} y$, since y is an upper bound and x is a least upper bound. Likewise, $y \mathcal{R} x$. $\mathcal{R}$ antisymmetric $\Rightarrow x = y$. (The proof for the glb is similar.)

13. Let $\mathcal{U} = \{1, 2\}$, $A = \mathcal{P}(\mathcal{U})$, and let $\mathcal{R}$ be the inclusion relation. Then $(A, \mathcal{R})$ is a poset but not a total order. Let $B = \{\emptyset, \{1\}\}$. Then $(B \times B) \cap \mathcal{R}$ is a total order.

15. $n + \binom{n}{2}$

17. a) The n elements of A are arranged along a vertical line. For if $A = \{a_1, a_2, \ldots, a_n\}$ where $a_1 \mathcal{R} a_2 \mathcal{R} a_3 \mathcal{R} \cdots \mathcal{R} a_n$, then the diagram can be drawn as follows:

b) $n!$

19.

	lub	glb
a)	$\{1, 2\}$	$\emptyset$
b)	$\{1, 2, 3\}$	$\emptyset$
c)	$\{1, 2\}$	$\emptyset$
d)	$\{1, 2, 3\}$	$\{1\}$
e)	$\{1, 2, 3\}$	$\emptyset$
f)	$\{1, 2, 3\}$	$\emptyset$

21. a) False. Let $\mathcal{U} = \{1, 2\}$, $A = \mathcal{P}(\mathcal{U})$, and let $\mathcal{R}$ be the inclusion relation. Then $(A, \mathcal{R})$ is a lattice where for any $S, T \in A$, $\mathrm{lub}\{S, T\} = S \cup T$ and $\mathrm{glb}\{S, T\} = S \cap T$. However, $\{1\}$ and $\{2\}$ are not related, so $(A, \mathcal{R})$ is not a total order.

b) If $(A, \mathcal{R})$ is a total order, then for any $x, y \in A$, $x \mathcal{R} y$ or $y \mathcal{R} x$. For $x \mathcal{R} y$, $\mathrm{lub}\{x, y\} = y$ and $\mathrm{glb}\{x, y\} = x$. Consequently, $(A, \mathcal{R})$ is a lattice.

23. a) a **b)** a **c)** c **d)** e **e)** z **f)** e **g)** v

$(A, \mathcal{R})$ is a lattice with z the greatest (and only maximal) element and a the least (and only minimal) element.

Section 7.4 – p. 279

1. $\mathcal{R} = \{(1, 1), (1, 2), (2, 1), (2, 2), (3, 3), (3, 4), (4, 3), (4, 4), (5, 5)\}$

3. $\mathcal{R}$ is not transitive since $1 \mathcal{R} 2$ and $2 \mathcal{R} 3$ but $1 \not\mathcal{R} 3$.

5. a) For all $(x, y) \in A$, $x + y = x + y \Rightarrow (x, y) \mathcal{R} (x, y)$.
$(x_1, y_1) \mathcal{R} (x_2, y_2) \Rightarrow x_1 + y_1 = x_2 + y_2 \Rightarrow x_2 + y_2 = x_1 + y_1 \Rightarrow (x_2, y_2) \mathcal{R} (x_1, y_1)$.
$(x_1, y_1) \mathcal{R} (x_2, y_2), (x_2, y_2) \mathcal{R} (x_3, y_3) \Rightarrow x_1 + y_1 = x_2 + y_2, x_2 + y_2 = x_3 + y_3$, so $x_1 + y_1 = x_3 + y_3$ and $(x_1, y_1) \mathcal{R} (x_3, y_3)$. Since $\mathcal{R}$ is reflexive, symmetric, and transitive, it is an equivalence relation.

b) $[(1, 3)] = \{(1, 3), (2, 2), (3, 1)\}$; $[(2, 4)] = \{(1, 5), (2, 4), (3, 3), (4, 2), (5, 1)\}$
$[(1, 1)] = \{(1, 1)\}$

c) $A = \{(1, 1)\} \cup \{(1, 2), (2, 1)\} \cup \{(1, 3), (2, 2), (3, 1)\} \cup \{(1, 4), (2, 3), (3, 2), (4, 1)\} \cup$
$\{(1, 5), (2, 4), (3, 3), (4, 2), (5, 1)\} \cup \{(2, 5), (3, 4), (4, 3), (5, 2)\} \cup$
$\{(3, 5), (4, 4), (5, 3)\} \cup \{(4, 5), (5, 4)\} \cup \{(5, 5)\}$

7. a) For any $X \subseteq A$, $B \cap X = B \cap X$, so $X \mathcal{R} X$ and $\mathcal{R}$ is reflexive. If $X, Y \subseteq A$, then $X \mathcal{R} Y \Rightarrow X \cap B = Y \cap B \Rightarrow Y \cap B = X \cap B \Rightarrow Y \mathcal{R} X$, so $\mathcal{R}$ is symmetric. And finally, if $W, X, Y \subseteq A$ with $W \mathcal{R} X$ and $X \mathcal{R} Y$, then $W \cap B = X \cap B$ and $X \cap B = Y \cap B$. Hence $W \cap B = Y \cap B$, so $W \mathcal{R} Y$ and $\mathcal{R}$ is transitive. Consequently $\mathcal{R}$ is an equivalence relation on $\mathcal{P}(A)$.

  b) $\{\emptyset, \{3\}\} \cup \{\{1\}, \{1,3\}\} \cup \{\{2\}, \{2,3\}\} \cup \{\{1,2\}, \{1,2,3\}\}$

  c) $[X] = \{\{1,3\}, \{1,3,4\}, \{1,3,5\}, \{1,3,4,5\}\}$

  d) 8—one for each subset of B

9. 300

11. Let $\{A_i\}_{i \in I}$ be a partition of a set A. Define $\mathcal{R}$ on A by $x \mathcal{R} y$ if for some $i \in I$, we have $x, y \in A_i$. For any $x \in A$, $x, x \in A_i$ for some $i \in I$, so $x \mathcal{R} x$ and $\mathcal{R}$ is reflexive. $x \mathcal{R} y \Rightarrow x, y \in A_i$, for some $i \in I \Rightarrow y, x \in A_i$, for some $i \in I \Rightarrow y \mathcal{R} x$, so $\mathcal{R}$ is symmetric. If $x \mathcal{R} y$ and $y \mathcal{R} z$, then $x, y \in A_i$ and $y, z \in A_j$ for some $i, j \in I$. Since $A_i \cap A_j$ contains y, by Theorem 7.6(c), it follows that $A_i = A_j$, so $i = j$. Hence $x, z \in A_i$, so $x \mathcal{R} z$ and $\mathcal{R}$ is transitive.

Section 7.5 – p. 286

1. a) s_2 and s_5 are equivalent.   **b)** s_2 and s_5 are equivalent.

  c) s_2 and s_7 are equivalent; s_3 and s_4 are equivalent.

3. a) s_1 and s_7 are equivalent; s_4 and s_5 are equivalent.

  b) (i) 0 0 0 0
     (ii) 0
     (iii) 0 0

	ν		ω	
M:	0	1	0	1
s_1	s_4	s_1	1	0
s_2	s_1	s_2	1	0
s_3	s_6	s_1	1	0
s_4	s_3	s_4	0	0
s_6	s_2	s_1	1	0

Miscellaneous Exercises – p. 289

1. a) False. Let $A = \{1, 2\}$, $I = \{1, 2\}$, $\mathcal{R}_1 = \{(1, 1)\}$, and $\mathcal{R}_2 = \{(2, 2)\}$. Then $\bigcup_{i \in I} \mathcal{R}_i$ is reflexive, but neither $\mathcal{R}_1$ nor $\mathcal{R}_2$ is reflexive. Conversely, however, if $\mathcal{R}_i$ is reflexive for all (actually, at least one) $i \in I$, then $\bigcup_{i \in I} \mathcal{R}_i$ is reflexive.

  b) True. $\bigcap_{i \in I} \mathcal{R}_i$ is reflexive $\Leftrightarrow (a, a) \in \bigcap_{i \in I} \mathcal{R}_i$ for all $a \in A \Leftrightarrow (a, a) \in \mathcal{R}_i$ for all $a \in A$, and $i \in I \Leftrightarrow \mathcal{R}_i$ is reflexive for all $i \in I$.

3. $(a, c) \in \mathcal{R}_2 \circ \mathcal{R}_1 \Rightarrow$ for some $b \in A$, $(a, b) \in \mathcal{R}_2$, $(b, c) \in \mathcal{R}_1$. With $\mathcal{R}_1, \mathcal{R}_2$ symmetric, $(b, a) \in \mathcal{R}_2$, $(c, b) \in \mathcal{R}_1$, so $(c, a) \in \mathcal{R}_1 \circ \mathcal{R}_2 \subseteq \mathcal{R}_2 \circ \mathcal{R}_1$. $(c, a) \in \mathcal{R}_2 \circ \mathcal{R}_1 \Rightarrow (c, d) \in \mathcal{R}_2$, $(d, a) \in \mathcal{R}_1$, for some $d \in A$. Then $(d, c) \in \mathcal{R}_2$, $(a, d) \in \mathcal{R}_1$ by symmetry, and $(a, c) \in \mathcal{R}_1 \circ \mathcal{R}_2$, so $\mathcal{R}_2 \circ \mathcal{R}_1 \subseteq \mathcal{R}_1 \circ \mathcal{R}_2$ and the result follows.

5. $(c, a) \in (\mathcal{R}_1 \circ \mathcal{R}_2)^c \Leftrightarrow (a, c) \in \mathcal{R}_1 \circ \mathcal{R}_2 \Leftrightarrow (a, b) \in \mathcal{R}_1$, $(b, c) \in \mathcal{R}_2$, for some $b \in B \Leftrightarrow (b, a) \in \mathcal{R}_1^c$, $(c, b) \in \mathcal{R}_2^c$, for some $b \in B \Leftrightarrow (c, a) \in \mathcal{R}_2^c \circ \mathcal{R}_1^c$.

7. Let $\mathcal{U} = \{1, 2, 3, 4, 5\}$, $A = \mathcal{P}(\mathcal{U}) - \{\mathcal{U}, \emptyset\}$. Under the inclusion relation, A is a poset with the five minimal elements $\{x\}$, $1 \leq x \leq 5$, but no least element. Also, A has five maximal elements—the five subsets of $\mathcal{U}$ of size 4—but no greatest element.

9. $n = 10$

11. a) For any $f \in \mathcal{F}$, $|f(n)| \leq 1|f(n)|$ for all $n \geq 1$, so $f \mathcal{R} f$ and $\mathcal{R}$ is reflexive. Second, if $f, g \in \mathcal{F}$, then $f \mathcal{R} g \Rightarrow (f \in O(g)$ and $g \in O(f)) \Rightarrow (g \in O(f)$ and $f \in O(g)) \Rightarrow g \mathcal{R} f$, so $\mathcal{R}$ is symmetric. Finally, let $f, g, h \in \mathcal{F}$ with $f \mathcal{R} g$, $g \mathcal{R} f$, $g \mathcal{R} h$, and $h \mathcal{R} g$. Then there exist $m_1, m_2 \in \mathbf{R}^+$, and $k_1, k_2 \in \mathbf{Z}^+$ such that $|f(n)| \leq m_1|g(n)|$ for all $n \geq k_1$, and $|g(n)| \leq m_2|h(n)|$ for all $n \geq k_2$. Consequently, for all $n \geq \max\{k_1, k_2\}$ we have $|f(n)| \le m_1|g(n)| \le m_1 m_2|h(n)|$, so $f \in O(h)$. And in a similar manner, $h \in O(f)$. So $f\mathcal{R}h$ and $\mathcal{R}$ is transitive.

b) For any $f \in \mathcal{F}$, f is dominated by itself, so $[f]\mathcal{S}[f]$ and $\mathcal{S}$ is reflexive. Second, if $[g],[h] \in \mathcal{F}'$ with $[g]\mathcal{S}[h]$ and $[h]\mathcal{S}[g]$, then $g\mathcal{R}h$, as in part (a), and $[g]=[h]$. Consequently, $\mathcal{S}$ is antisymmetric. Finally, if $[f],[g],[h] \in \mathcal{F}'$ with $[f]\mathcal{S}[g]$ and $[g]\mathcal{S}[h]$, then f is dominated by g and g is dominated by h. So, as in part (a), f is dominated by h and $[f]\mathcal{S}[h]$, making $\mathcal{S}$ transitive.

c) Let $f, f_1, f_2 \in \mathcal{F}$ with $f(n) = n$, $f_1(n) = n+3$, and $f_2(n) = 2-n$. Then $(f_1 + f_2)(n) = 5$ and $f_1 + f_2 \notin [f]$, because f is not dominated by $f_1 + f_2$.

13. No. Consider $S = \{q \in \mathbf{Q} \mid 0 < q < \sqrt{2}\}$.

15.

$$M(\mathcal{R}) = \begin{bmatrix} 1 & 1 & 1 & 1 & 1 & 1 & 1 & 1 & 1 \\ 0 & 1 & 1 & 0 & 1 & 1 & 1 & 0 & 1 \\ 0 & 0 & 1 & 0 & 0 & 0 & 0 & 0 & 0 \\ 0 & 1 & 1 & 1 & 1 & 1 & 1 & 0 & 1 \\ 0 & 0 & 1 & 0 & 1 & 0 & 0 & 0 & 0 \\ 0 & 0 & 1 & 0 & 1 & 1 & 0 & 0 & 0 \\ 0 & 0 & 1 & 0 & 1 & 1 & 1 & 0 & 1 \\ 1 & 1 & 1 & 1 & 1 & 1 & 1 & 1 & 1 \\ 0 & 0 & 1 & 0 & 1 & 1 & 1 & 0 & 1 \end{bmatrix}$$

17.

a)

Adjacency List		Index List	
1	2	1	1
2	3	2	2
3	1	3	3
4	4	4	5
5	5	5	6
6	3	6	8
7	5		

b)

Adjacency List		Index List	
1	2	1	1
2	3	2	2
3	1	3	3
4	5	4	4
5	4	5	5
		6	6

c)

Adjacency List		Index List	
1	2	1	1
2	3	2	2
3	1	3	3
4	4	4	6
5	5	5	7
6	1	6	8
7	4		

19. b) The cells of the partition are the connected components of G.

21. One possible order is 10, 3, 8, 6, 7, 9, 1, 4, 5, 2, where program 10 is run first and program 2 last.

23. a) $|A| = 2^5 = 32$

c) There are six equivalence classes—one for each of the weights 0, 1, 2, 3, 4, and 5. The number of elements in each equivalence class is given as follows: weight 0, $\binom{5}{0}$; weight 1, $\binom{5}{1}$; weight 2, $\binom{5}{2}$; weight 3, $\binom{5}{3}$; weight 4, $\binom{5}{4}$; and weight 5, $\binom{5}{5}$.

d) Replace 5 by n, where $n \in \mathbf{Z}^+$. Then there are $n+1$ equivalence classes—one for each of the weights $0, 1, 2, \ldots, n$. If $0 \le k \le n$, then there are $\binom{n}{k}$ elements of A in the equivalence class for weight k.

25. $4^n - 2(3^n) + 2^n$

27. a) (i) $B\mathcal{R}A\mathcal{R}C$;   (ii) $B\mathcal{R}C\mathcal{R}F$

$B\mathcal{R}A\mathcal{R}C\mathcal{R}F$ is a maximal chain. There are six such maximal chains.

b) Here $11\mathcal{R}385$ is a maximal chain of length 2, while $2\mathcal{R}6\mathcal{R}12$ is one of length 3. The length of a longest chain for this poset in 3.

c) (i) $\emptyset \subseteq \{1\} \subseteq \{1,2\} \subseteq \{1,2,3\} \subseteq \mathcal{U}$ (ii) $\emptyset \subseteq \{2\} \subseteq \{2,3\} \subseteq \{1,2,3\} \subseteq \mathcal{U}$

There are $4! = 24$ such maximal chains.

d) $n!$

29. Let $a_1 \mathcal{R} a_2 \mathcal{R} \ldots \mathcal{R} a_{n-1} \mathcal{R} a_n$ be a longest (maximal) chain in $(A, \mathcal{R})$. Then a_n is a maximal element in $(A, \mathcal{R})$ and $a_1 \mathcal{R} a_2 \mathcal{R} \ldots \mathcal{R} a_{n-1}$ is a maximal chain in $(B, \mathcal{R}')$. Hence the length of a longest chain in $(B, \mathcal{R}')$ is at least $n - 1$. If there is a chain $b_1 \mathcal{R}' b_2 \mathcal{R}' \ldots \mathcal{R}' b_n$ in $(B, \mathcal{R}')$ of length n, then this is also a chain of length n in $(A, \mathcal{R})$. But then b_n must be a maximal element of $(A, \mathcal{R})$, and this contradicts $b_n \in B$.

31. If $n = 1$, then for any $x, y \in A$, if $x \neq y$ then $x \not\mathcal{R} y$ and $y \not\mathcal{R} x$. Hence $(A, \mathcal{R})$ is an antichain, and the result follows. Now assume the result true for $n = k \geq 1$, and let $(A, \mathcal{R})$ be a poset where the length of a longest chain is $k + 1$. If M is the set of all maximal elements in $(A, \mathcal{R})$, then $M \neq \emptyset$ and M is an antichain in $(A, \mathcal{R})$. Also, by virtue of Exercise 29 above, $(A - M, \mathcal{R}')$, for $\mathcal{R}' = ((A - M) \times (A - M)) \cap \mathcal{R}$, is a poset with k the length of a longest chain. So by the induction hypothesis, $A - M = C_1 \cup C_2 \cup \cdots \cup C_k$, a partition into k antichains. Consequently, $A = C_1 \cup C_2 \cup \cdots \cup C_k \cup M$, a partition into $k + 1$ antichains.

CHAPTER 8
The Principle of Inclusion and Exclusion

Section 8.1 — p. 303

1. a) 534 **b)** 458 **c)** 76

3. a) $\binom{22}{19}$ **b)** $\binom{22}{19} - 4\binom{14}{11} + 6\binom{6}{3}$ **c)** $\binom{16}{13} - [2\binom{10}{7} + \binom{9}{6} + \binom{11}{8})] + [10 + 2\binom{5}{2})]$

5. $\binom{37}{31} - \binom{7}{1}\binom{27}{21} + \binom{7}{2}\binom{17}{11} - \binom{7}{3}\binom{7}{1}$

7. $26! - [3(23!) + 24!] + (20! + 21!)$

9. a) 2^{n-1} **b)** $2^{n-1}(p - 1)$

11. a) 1600 **b)** 4399

13. $[6^8 - \binom{6}{1}5^8 + \binom{6}{2}4^8 - \binom{6}{3}3^8 + \binom{6}{2}2^8 - \binom{6}{1}]/6^8$

15. $9!/[(3!)^3] - 3[7!/[(3!)^2]] + 3(5!/3!) - 3!$ **17. No**

19. If a number n is divisible by a prime p, the execution of the While loop divides out every occurrence of p. The first loop is for $p = 2$, the next loop for $p = 3$, and the third loop is for primes $p > 3$.

When we are working with primes, once we are past $p = 5$, there is no sense in examining even integers as possible candidates for primes.

Section 8.2 — p. 308

1. $E_0 = 768$; $E_1 = 205$; $E_2 = 40$; $E_3 = 10$; $E_4 = 0$; $E_5 = 1$. $\sum_{i=0}^{5} E_i = 1024 = N$.

3. a) $[14!/(2!)^5] - \binom{5}{1}[13!/(2!)^4] + \binom{5}{2}[12!/(2!)^3] - \binom{5}{3}[11!/(2!)^2] + \binom{5}{4}[10!/2!] - \binom{5}{5}[9!]$
   b) $E_2 = \binom{5}{2}[12!/(2!)^3] - \binom{5}{1}\binom{3}{3}[11!/(2!)^2] + \binom{5}{2}\binom{3}{4}[10!/2!] - \binom{5}{3}\binom{5}{5}[9!]$
   c) $L_3 = \binom{5}{3}[11!/(2!)^2] - \binom{3}{2}\binom{5}{4}[10!/2!] + \binom{4}{1}\binom{5}{5}[9!]$

5. a) $7! - d_7$ $(d_7 \doteq (7!)e^{-1})$; **b)** $d_{26} \doteq (26!)e^{-1}$

7. a) $[\sum_{i=0}^{3} (-1)^i \binom{4}{i} \binom{52 - 13i}{13}]/\binom{52}{13}$ **b)** $[\sum_{i=1}^{3} (-1)^{i+1}(i) \binom{4}{i} \binom{52 - 13i}{13}]/\binom{52}{13}$
   c) $[\binom{4}{2}\binom{26}{13} - 3\binom{4}{3}\binom{13}{13}]/\binom{52}{13}$

Section 8.3 — p. 310

1. $10! - \binom{5}{1}9! + \binom{5}{2}8! - \binom{5}{3}7! + \binom{5}{4}6! - \binom{5}{5}5!$

3. $(10!)d_{10} \doteq (10!)^2(e^{-1})$

5. a) $(d_{10})^2 \doteq (10!)^2 e^{-2}$ **b)** $\sum_{i=0}^{10} (-1)^i \binom{10}{i}[(10 - i)!]^2$

7. $\binom{n}{0}(n - 1)! - \binom{n}{1}(n - 2)! + \binom{n}{2}(n - 3)! - \cdots + (-1)^{n-1}\binom{n}{n-1}(0!) + (-1)^n\binom{n}{n}$

Sections 8.4 and 8.5 – p. 317

3. **a)** $\binom{8}{0} + \binom{8}{1}8x + \binom{8}{2}(8\cdot7)x^2 + \binom{8}{3}(8\cdot7\cdot6)x^3 + \binom{8}{4}(8\cdot7\cdot6\cdot5)x^4 + \cdots$

$\qquad + \binom{8}{8}(8!)x^8 = \sum_{i=0}^{8} \binom{8}{i}P(8,i)x^i$

  b) $\sum_{i=0}^{n} \binom{n}{i}P(n,i)x^i$

5. **a)** (i) $(1+2x)^3$ (ii) $1 + 8x + 14x^2 + 4x^3$

    (iii) $1 + 9x + 25x^2 + 21x^3$ (iv) $1 + 8x + 16x^2 + 7x^3$

  b) If the board C consists of n steps, and each step has k blocks, then $r(C,x) = (1+kx)^n$.

7. $5! - 8(4!) + 21(3!) - 20(2!) + 6(1!) = 20$

9. **a)** 20   **b)** 3/10   **c)** 3/5

11. $(6!/2!) - 9(5!/2!) + 27(4!/2!) - 31(3!/2!) + 12 = 63$

Miscellaneous Exercises – p. 320

1. 134   3. $\phi(17) = \phi(32) = \phi(48) = 16$

5. $\sum_{i=0}^{7}(-1)^i\binom{8}{i}(8-i)!$

7. $9! - \binom{5}{1}(2)(8!) + \binom{5}{2}(2^2)(7!) - \binom{5}{3}(2^3)(6!) + \binom{5}{4}(2^4)(5!) - \binom{5}{5}(2^5)(4!)$

9. Let $T = (13!)/(2!)^5$.

  a) $([\binom{5}{3}(10!)/(2!)^2] - [\binom{4}{1}\binom{5}{4}(9!)/(2!)] + [\binom{5}{2}\binom{5}{5}(8!)])/T$

  b) $[T - (E_4 + E_5)]/T,$

    where $E_4 = [\binom{5}{4}(9!)/(2!)] - [\binom{5}{1}\binom{5}{5}(8!)]$ and $E_5 = \binom{5}{5}(8!)$

11. $[16!/(4!)^4] - \binom{4}{1}[13!/(4!)^3] + \binom{4}{2}[10!/(4!)^2] - \binom{4}{3}(7!/4!) + (4!)$

13. **a)** $\binom{n-m}{r-m}$

CHAPTER 9

Generating Functions

Section 9.1 – p. 326

1. **a)** The coefficient of x^{20} in $(1 + x + x^2 + \cdots + x^7)^4$

  b) The coefficient of x^{20} in $(1 + x + x^2 + \cdots + x^{20})^2(1 + x^2 + x^4 + \cdots + x^{20})^2$ or $(1 + x + x^2 + \cdots)^2(1 + x^2 + x^4 + \cdots)^2$

  c) The coefficient of x^{30} in $(x^2 + x^3 + x^4)(x^3 + x^4 + \cdots + x^8)^4$

  d) The coefficient of x^{30} in $(1 + x + x^2 + \cdots + x^{30})^3(1 + x^2 + x^4 + \cdots + x^{30})\cdot$ $(x + x^3 + x^5 + \cdots + x^{29})$ or $(1 + x + x^2 + \cdots)^3(1 + x^2 + x^4 + \cdots)\cdot$ $(x + x^3 + x^5 + \cdots)$

3. **a)** The coefficient of x^{10} in $(1 + x + x^2 + x^3 + \cdots)^6$

  b) The coefficient of x^r in $(1 + x + x^2 + x^3 + \cdots)^n$

5. The answer is the coefficient of x^{31} in the generating function

$$(1 + x + x^2 + x^3 + \cdots)^3(1 + x + x^2 + \cdots + x^{10}).$$

Section 9.2 – p. 333

1. **a)** $(1+x)^8$ **b)** $8(1+x)^7$ **c)** $(1+x)^{-1}$ **d)** $x^3/(1-x)$

  e) $6x^3/(1+x)$ **f)** $(1-x^2)^{-1}$ **g)** $(1-2x)^{-1}$ **h)** $x^2/(1-ax)$

3. **a)** $g(x) = f(x) - a_3x^3 + 3x^3 = f(x) + (3 - a_3)x^3$

b) $g(x) = f(x) + (3 - a_3)x^3 + (7 - a_7)x^7$

c) $g(x) = 2f(x) + (1 - 2a_1)x + (3 - 2a_3)x^3$

d) $g(x) = 2f(x) + [5/(1-x)] + (1 - 2a_1 - 5)x + (3 - 2a_3 - 5)x^3 + (7 - 2a_7 - 5)x^7$

5. a) $\binom{21}{7}$ **b)** $\binom{n+6}{7}$

7. $\binom{14}{10} - 5\binom{9}{5} + \binom{5}{2}$

9. a) 0 **b)** $\binom{14}{12} - 5\binom{16}{14}$ **c)** $\binom{18}{15} + 4\binom{17}{14} + 6\binom{16}{13} + 4\binom{15}{12} + \binom{14}{11}$

11. $\binom{99}{96} - 4\binom{64}{61} + 6\binom{29}{26}$

13. $[\binom{29}{18} - \binom{12}{1}\binom{23}{12} + \binom{12}{2}\binom{17}{6} - \binom{12}{3}]/(6^{12})$

15. $(1/8)[1 + (-1)^n] + (1/4)\binom{n+1}{n} + (1/2)\binom{n+2}{n}$

17. $(1 - x - x^2 - x^3 - x^4 - x^5 - x^6)^{-1} = [1 - (x + x^2 + \cdots + x^6)]^{-1}$

$= 1 + \underbrace{(x + x^2 + \cdots + x^6)}_{\text{one roll}} + \underbrace{(x + x^2 + \cdots + x^6)^2}_{\text{two rolls}} + \underbrace{(x + x^2 + \cdots + x^6)^3}_{\text{three rolls}} + \cdots,$

where the 1 takes care of the case where the die is not rolled.

19. $a = 1, b = -2, k = -1/5$

21. a) The differences are 2, 3, 2, 7, and 0, and these sum to 14.

    b) $\{3, 5, 8, 15\}$ **c)** $\{1 + a, 1 + a + b, 1 + a + b + c, 1 + a + b + c + d\}$

23. $c_k = \sum_{i=0}^{k} i(k - i)^2 = k^2 \sum_{i=0}^{k} i - 2k \sum_{i=0}^{k} i^2 + \sum_{i=0}^{k} i^3$

$= (k^2)[k(k + 1)/2] - 2k[k(k + 1)(2k + 1)/6] + [k^2(k + 1)^2/4]$

$= (1/12)(k^2)(k^2 - 1)$

Section 9.3 – p. 338

1. $5; 4 + 1; 3 + 2; 3 + 1 + 1; 2 + 2 + 1; 2 + 1 + 1 + 1; 1 + 1 + 1 + 1 + 1.$
$7; 6 + 1; 5 + 2; 5 + 1 + 1; 4 + 3; 4 + 2 + 1; 4 + 1 + 1 + 1; 3 + 3 + 1;$
$3 + 2 + 2; 3 + 2 + 1 + 1; 3 + 1 + 1 + 1 + 1; 2 + 2 + 2 + 1;$
$2 + 2 + 1 + 1 + 1; 2 + 1 + 1 + 1 + 1 + 1; 1 + 1 + 1 + 1 + 1 + 1 + 1.$

3. The number of partitions of 6 into 1's, 2's, and 3's is 7.

5. a) and **b)**

$$(1 + x^2 + x^4 + x^6 + \cdots)(1 + x^4 + x^8 + \cdots)(1 + x^6 + x^{12} + \cdots) \cdots = \prod_{i=1}^{\infty} \frac{1}{1 - x^{2i}}$$

7. Let $f(x)$ be the generating function for the number of partitions of $n \in \mathbf{Z}^+$ where no summand appears more than twice. Then

$$f(x) = \prod_{i=1}^{\infty} (1 + x^i + x^{2i}).$$

Let $g(x)$ be the generating function for the number of partitions of n where no summand is divisible by 3. Here

$$g(x) = \frac{1}{1 - x} \cdot \frac{1}{1 - x^2} \cdot \frac{1}{1 - x^4} \cdot \frac{1}{1 - x^5} \cdot \frac{1}{1 - x^7} \cdots$$

But

$$f(x) = (1 + x + x^2)(1 + x^2 + x^4)(1 + x^3 + x^6)(1 + x^4 + x^8) \cdots$$

$$= \frac{1 - x^3}{1 - x} \cdot \frac{1 - x^6}{1 - x^2} \cdot \frac{1 - x^9}{1 - x^3} \cdot \frac{1 - x^{12}}{1 - x^4} \cdots$$

$$= \frac{1}{1 - x} \cdot \frac{1}{1 - x^2} \cdot \frac{1}{1 - x^4} \cdot \frac{1}{1 - x^5} \cdot \frac{1}{1 - x^7} \cdots$$

$$= g(x).$$

9. $(1 + x + x^2 + \cdots + x^{99})(1 + x^5 + x^{10} + \cdots + x^{95})(1 + x^{10} + \cdots + x^{90}) \cdot$
$(1 + x^{25} + x^{50} + x^{75})(1 + x^{50})$

11. This result follows from the one-to-one correspondence between the Ferrer's graphs with summands (rows) not exceeding m and the transpose graphs (also Ferrer's graphs) that have m summands (rows).

Section 9.4 – p. 342

1. a) e^{-x} **b)** e^{2x} **c)** e^{-ax} **d)** $e^{a^2 x}$ **e)** $ae^{a^2 x}$ **f)** xe^{2x}

3. a) $g(x) = f(x) + [(3 - a_3)/3!]x^3$

b) $g(x) = f(x) + [(-1 - a_3)/3!]x^3 = e^{5x} - [126x^3/(3!)]$

c) $g(x) = 2f(x) + [2 - 2a_1]x + [(4 - 2a_2)/2!]x^2$

d) $g(x) = 2f(x) + 3e^x + [2 - 2a_1 - 3]x + [(4 - 2a_2 - 3)/2!]x^2 + [(8 - 2a_3 - 3)/3!]x^3$

5. a) $(1 + x)^2 \left(1 + x + \dfrac{x^2}{2}\right)^2$

b) $(1 + x)\left(1 + x + \dfrac{x^2}{2}\right)\left(1 + x + \dfrac{x^2}{2} + \dfrac{x^3}{3!} + \dfrac{x^4}{4!}\right)^2$

c) $(1 + x)^3 \left(1 + x + \dfrac{x^2}{2}\right)^4$

7. The answer is the coefficient of $\dfrac{x^{25}}{25!}$ in $\left(\dfrac{x^3}{3!} + \dfrac{x^4}{4!} + \cdots + \dfrac{x^{10}}{10!}\right)^4$.

9. a) $(1/2)[3^{20} + 1]/(3^{20})$ **b)** $(1/4)[3^{20} + 3]/(3^{20})$ **c)** $(1/2)[3^{20} - 1]/(3^{20})$
d) $(1/2)[3^{20} - 1]/(3^{20})$ **e)** $(1/2)[3^{20} + 1]/(3^{20})$

11. a) $\left(\dfrac{x^5}{5!} + \dfrac{x^6}{6!} + \cdots\right)\left(\dfrac{x^2}{2!} + \dfrac{x^3}{3!} + \cdots\right)^4$

b) $\left(x + \dfrac{x^2}{2!} + \dfrac{x^3}{3!} + \cdots\right)^3 \left[(x)\left(\dfrac{x^2}{2!} + \dfrac{x^3}{3!} + \dfrac{x^4}{4!} + \dfrac{x^5}{5!}\right) + \left(\dfrac{x^2}{2!}\right)\left(\dfrac{x^3}{3!} + \dfrac{x^4}{4!} + \dfrac{x^5}{5!}\right)\right.$

$\left. + \left(\dfrac{x^3}{3!}\right)\left(\dfrac{x^4}{4!} + \dfrac{x^5}{5!}\right) + \left(\dfrac{x^4}{4!}\right)\left(\dfrac{x^5}{5!}\right)\right]$

Section 9.5 – p. 345

3. $a_0, a_1 - a_0, a_2 - a_1, a_3 - a_2, \ldots$

5. $f(x) = [e^x/(1 - x)]$

Miscellaneous Exercises – p. 348

1. a) $6/(1 - x) + 1/(1 - x)^2$ **b)** $1/(1 - ax)$ **c)** $1/[1 - (1 + a)x]$
d) $1/(1 - x) + 1/(1 - ax)$

5. $[\binom{15}{12} - \binom{4}{1}\binom{9}{6} + \binom{4}{2}]^2$

7. Let $f(x)$ be the generating function for the number of partitions of $n \in \mathbf{Z}^+$ in which no even summand is repeated (an odd summand may or may not be repeated). Then
$$f(x) = (1 + x + x^2 + x^3 + \cdots)(1 + x^2)(1 + x^3 + x^6 + x^9 + \cdots)(1 + x^4) \cdots$$

$$= \frac{1}{1 - x} \cdot (1 + x^2) \cdot \frac{1}{1 - x^3} \cdot (1 + x^4) \cdot \frac{1}{1 - x^5} \cdots$$

Let $g(x)$ be the generating function for the number of partitions of $n \in \mathbf{Z}^+$ where no summand occurs more than three times. Then

$$g(x) = (1 + x + x^2 + x^3)(1 + x^2 + x^4 + x^6)(1 + x^3 + x^6 + x^9) \cdots$$
$$= [(1+x)(1+x^2)][(1+x^2)(1+x^4)][(1+x^3)(1+x^6)] \cdots$$
$$= [(1-x^2)/(1-x)](1+x^2)[(1-x^4)/(1-x^2)](1+x^4) \cdot$$
$$[(1-x^6)/(1-x^3)](1+x^6) \cdots$$
$$= (1/(1-x))(1+x^2)(1/(1-x^3))(1+x^4)(1/(1-x^5))(1+x^6) \cdots = f(x).$$

9. $4^{10} - \binom{4}{1}(10)(3^9) + \binom{4}{2}(10)(9)(2^8) - \binom{4}{3}(10)(9)(8)$

11. a) $1, 5, (5)(7), (5)(7)(9), (5)(7)(9)(11), \ldots$ **b)** $a = 4, b = -\frac{7}{4}$

13. a) $\binom{19}{8}$ **b)** $\binom{9}{4}^2 / \binom{19}{8}$

CHAPTER 10

Recurrence Relations

Section 10.1 – p. 360

1. a) $a_n = 5a_{n-1}, n \geq 1, a_0 = 2$ **b)** $a_n = -3a_{n-1}, n \geq 1, a_0 = 6$

    c) $a_n = (1/3)a_{n-1}, n \geq 1, a_0 = 1$ **d)** $a_n = (2/5)a_{n-1}, n \geq 1, a_0 = 7$

3. $c = \pm(3/7)$ **5.** 141 months

7. $3(7)^{10/3}$

9. a) 145   **b)** 45

11. a) 21345   **b)** 52143, 52134   **c)** 21534, 21354, 21345

    d) 21543 is 113th, 35421 is 67th, 31524 is 43rd.

13. b) Assuming that a "bubble sort" type of procedure is used in part (a), the worst-case complexity is $O(n^2)$.

Section 10.2 – p. 368

1. a) $a_n = (3/7)(-1)^n + (4/7)(6)^n, n \geq 0$ **b)** $a_n = 4(1/2)^n - 2(5)^n, n \geq 0$

    c) $a_n = 4 + 3(-1/3)^n, n \geq 0$ **d)** $a_n = 3\sin(n\pi/2), n \geq 0$

    e) $a_n = 2^n[\cos(n\pi/2) + (1/2)\sin(n\pi/2)], n \geq 0$ **f)** $a_n = (5-n)3^n, n \geq 0$

    g) $a_n = (\sqrt{2})^n[\cos(3\pi n/4) + 4\sin(3\pi n/4)], n \geq 0$ **h)** $a_n = 3(5)^n, n \geq 0$

    i) $a_n = (-2/3)^n(1 - 7n), n \geq 0$ **j)** $a_n = [(-5 + 6n)/9](-1)^n + (14/9)(2^n), n \geq 0$

3. $a_n = (1/10)[7^n - (-3)^n], n \geq 0$

5. a)

$$F_1 = F_2 - F_0$$
$$F_3 = F_4 - F_2$$
$$F_5 = F_6 - F_4$$
$$\vdots$$
$$\underline{F_{2n-1} = F_{2n} - F_{2n-2}}$$
$$F_1 + F_3 + F_5 + \cdots + F_{2n-1} = F_{2n} - F_0 = F_{2n}$$

b)

$$F_2 = F_3 - F_1$$
$$F_4 = F_5 - F_3$$
$$F_6 = F_7 - F_5$$
$$\vdots$$
$$\underline{F_{2n} = F_{2n+1} - F_{2n-1}}$$
$$F_2 + F_4 + \cdots + F_{2n} = F_{2n+1} - F_1$$
$$F_0 + F_2 + F_4 + \cdots + F_{2n} = F_{2n+1} - 1 \ (F_0 = 0; F_1 = 1).$$

  c)
$$F_1 + F_3 + \cdots + F_{2n-1} = F_{2n}$$
$$F_2 + F_4 + \cdots + F_{2n} = F_{2n+1} - 1$$
$$\overline{F_1 - F_2 + F_3 - F_4 + \cdots + F_{2n-1} - F_{2n}} = (F_{2n} - F_{2n+1}) + 1 = -F_{2n-1} + 1$$

7. $a_n = (1/\sqrt{5})[((1 + \sqrt{5})/2)^{n+1} - ((1 - \sqrt{5})/2)^{n+1}], n \geq 0$

9. $a_n = [(5 + \sqrt{21})/(2\sqrt{21})][(3 + \sqrt{21})/2]^n - [(5 - \sqrt{21})/(2\sqrt{21})][(3 - \sqrt{21})/2]^n$

11. $x_n = 4(2^n) - 3, n \geq 0$

13. $a_n = \sqrt{51(4^n) - 35}, n \geq 0$

15. Since $(F_1, F_0) = 1 = (F_2, F_1)$, consider $n \geq 2$. Then
$$F_3 = F_2 + F_1(= 1)$$
$$F_4 = F_3 + F_2$$
$$F_5 = F_4 + F_3$$
$$\vdots$$
$$F_{n+1} = F_n + F_{n-1}$$

Reversing the order of these equations, we have the steps in the Euclidean Algorithm for computing (F_{n+1}, F_n), the g.c.d. of F_{n+1} and F_n, $n \geq 2$. Since the last nonzero remainder is $F_1 = 1$, it follows that $(F_{n+1}, F_n) = 1$ for all $n \geq 2$.

Section 10.3 – p. 375

1. a) $a_n = (n + 1)^2, \ n \geq 0$ **b)** $a_n = 3 + n(n - 1)^2, \ n \geq 0$

 c) $a_n = 6(2^n) - 5, \ n \geq 0$ **d)** $a_n = 2^n + n(2^{n-1}), \ n \geq 0$

3. a) $a_n = a_{n-1} + n, \ n \geq 1, \ a_0 = 1$ $a_n = 1 + [n(n + 1)]/2, \ n \geq 0$

 b) $b_n = b_{n-1} + 2, \ n \geq 2, \ b_1 = 2$ $b_n = 2n, \ n \geq 1, \ b_0 = 1$

5. a) $a_n = (3/4)(-1)^n - (4/5)(-2)^n + (1/20)(3)^n, \ n \geq 0$

 b) $a_n = (2/9)(-2)^n - (5/6)(n)(-2)^n + (7/9), \ n \geq 0$

 c) $a_n = (-55/54)n(-2)^n + (n^2/9) - (4n/27), \ n \geq 0$

 d) $a_n = (5/4) - (1/4)(-1)^n - (1/2) \sin(n\pi/2)$

7. $a_n = A + Bn + Cn^2 - (3/4)n^3 + (5/24)n^4$

9. $P = \$117.68$

11. a) $a_n = [(3/4)(3)^n - 5(2)^n + (7n/2) + (21/4)]^{1/2}, \ n \geq 0$

 b) $a_n = (1/2)[(-1)^n + 1]n!, \ n \geq 0$

 c) $a_n = 2, \ n \geq 0$

Section 10.4 – p. 381

1. a) $a_n = (1/2)[1 + 3^n], \ n \geq 0$

 b) $a_n = 1 + [n(n - 1)(2n - 1)]/6, \ n \geq 0$

 c) $a_n = 2^n, \ n \geq 0$

 d) $a_n = [n^3 - 3n^2 + 8n + 6]/6$

3. a) $a_n = 2^n(1 - 2n), \ b_n = n(2^{n+1}), \ n \geq 0$

 b) $a_n = (-3/4) + (1/2)(n + 1) + (1/4)(3^n)$,

  $b_n = (3/4) + (1/2)(n + 1) - (1/4)(3^n), \ n \geq 0$

Section 10.5 – p. 387

1. $b_4 = (8!)/[(5!)(4!)] = 14$

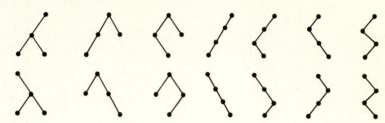

3. $\dbinom{2n-1}{n} - \dbinom{2n-1}{n-2} = \left[\dfrac{2n-1)!}{n!(n-1)!}\right] - \left[\dfrac{(2n-1)!}{(n-2)!(n+1)!}\right]$

$$= \left[\dfrac{(2n-1)!(n+1)}{(n+1)!(n-1)!}\right] - \left[\dfrac{(2n-1)!(n-1)}{(n-1)!(n+1)!}\right]$$

$$= \left[\dfrac{(2n-1)!}{(n+1)!(n-1)!}\right][(n+1)-(n-1)]$$

$$= \dfrac{(2n-1)!(2)}{(n+1)!(n-1)!} = \dfrac{(2n-1)!(2n)}{(n+1)!n!} = \dfrac{(2n)!}{(n+1)(n!)(n!)}$$

$$= \dfrac{1}{(n+1)}\dbinom{2n}{n}$$

5. a) $b_{n+1} = \left(\dfrac{1}{n+2}\right)\dbinom{2n+2}{n+1} = \left(\dfrac{1}{n+2}\right)\left[\dfrac{(2n+2)!}{(n+1)!(n+1)!}\right]$

$$= \dfrac{(2n+2)(2n+1)(2n)!}{(n+2)(n+1)^2(n!)^2} = \dfrac{2(2n+1)}{(n+2)} \cdot \dfrac{(2n)!}{(n+1)(n!)^2} = \dfrac{2(2n+1)}{(n+2)}\left[\dfrac{1}{n+1}\dbinom{2n}{n}\right]$$

$$= \dfrac{2(2n+1)}{(n+2)}b_n$$

Section 10.6 – p. 396

1. a) $f(n) = (5/3)(4n^{\log_3 4} - 1)$ and $f \in O(n^{\log_3 4})$ for $n \in \{3^i \,|\, i \in \mathbf{N}\}$

b) $f(n) = 7(\log_5 n + 1)$ and $f \in O(\log_5 n)$ for $n \in \{5^i \,|\, i \in \mathbf{N}\}$

c) $f(n) = (3/7)(8n^3 - 1)$ and $f \in O(n^3)$ for $n \in \{2^i \,|\, i \in \mathbf{N}\}$

3. a) $f \in O(\log_b n)$ on $\{b^k \,|\, k \in \mathbf{N}\}$ **b)** $f \in O(n^{\log_b a})$ on $\{b^k \,|\, k \in \mathbf{N}\}$

5. a) $f(1) = 0$ $f(n) = 2f(n/2) + 1$

From Exercise 2(b), $f(n) = n - 1$.

b) The equation $f(n) = f(n/2) + (n/2)$ arises as follows: There are $n/2$ matches played in the first round. Then there are $n/2$ players remaining, so we need $f(n/2)$ additional matches to determine the winner.

7. $O(1)$

9. a)

$$f(n) \le af(n/b) + cn$$
$$af(n/b) \le a^2 f(n/b^2) + ac(n/b)$$
$$a^2 f(n/b^2) \le a^3 f(n/b^3) + a^2 c(n/b^2)$$
$$a^3 f(n/b^3) \le a^4 f(n/b^4) + a^3 c(n/b^3)$$
$$\vdots \qquad\qquad \vdots \qquad\qquad \vdots$$
$$a^{k-1} f(n/b^{k-1}) \le a^k f(n/b^k) + a^{k-1} c(n/b^{k-1})$$

Hence $f(n) \le a^k f(n/b^k) + cn[1 + (a/b) + (a/b)^2 + \cdots + (a/b)^{k-1}] = a^k f(1) + cn[1 + (a/b) + (a/b)^2 + \cdots + (a/b)^{k-1}]$, since $n = b^k$. Since $f(1) \le c$ and $(n/b^k) = 1$, we have $f(n) \le cn[1 + (a/b) + (a/b)^2 + \cdots + (a/b)^{k-1} + (a/b)^k] = (cn) \sum_{i=0}^{k} (a/b)^i$.

b) When $a = b$, $f(n) \le (cn) \sum_{i=0}^{k} 1^i = (cn)(k + 1)$, where $n = b^k$, or $k = \log_b n$. Hence $f(n) \le (cn)(\log_b n + 1)$ so $f \in O(n \log_b n) = O(n \log n)$, for any base greater than 1.

c) For $a \ne b$,

$$cn \sum_{i=0}^{k} (a/b)^i = cn \left[\frac{1 - (a/b)^{k+1}}{1 - (a/b)}\right]$$

$$= (c)(b^k)\left[\frac{1 - (a/b)^{k+1}}{1 - (a/b)}\right] = c\left[\frac{b^k - (a^{k+1}/b)}{1 - (a/b)}\right]$$

$$= c\left[\frac{b^{k+1} - a^{k+1}}{b - a}\right] = c\left[\frac{a^{k+1} - b^{k+1}}{a - b}\right].$$

d) From part (c), $f(n) \le (c/(a - b))[a^{k+1} - b^{k+1}] = (ca/(a - b))a^k - (cb/(a - b))b^k$. But $a^k = a^{\log_b n} = n^{\log_b a}$ and $b^k = n$, so $f(n) \le (ca/(a - b))n^{\log_b a} - (cb/(a - b))n$.

(i) When $a < b$, then $\log_b a < 1$, and $f \in O(n)$ on $\mathbf{Z}^+$.

(ii) When $a > b$, then $\log_b a > 1$, and $f \in O(n^{\log_b a})$ on $\mathbf{Z}^+$.

Miscellaneous Exercises – p. 401

1. $\binom{n}{k+1} = \frac{n!}{(k+1)!(n-k-1)!} = \frac{(n-k)}{(k+1)} \cdot \frac{n!}{k!(n-k)!} = \binom{n-k}{k+1}\binom{n}{k}$

3. There are two cases to consider. Case 1 (1 is a summand): Here there are $p(n-1, k-1)$ ways to partition $n-1$ into exactly $k-1$ summands. Case 2 (1 is not a summand): Here each summand $s_1, s_2, \ldots, s_k > 1$. For $1 \le i \le k$, let $t_i = s_i - 1 \ge 1$. Then $t_1, t_2, \ldots, t_k$ provide a partition of $n - k$ into exactly k summands. These cases are exhaustive and disjoint, so by the rule of sum,
$p(n, k) = p(n-1, k-1) + p(n-k, k)$.

5. $a_n = a_{n-1} + a_{n-2}$, for $n \ge 3$, $a_1 = a_2 = 1$

$a_n = F_n$, the nth Fibonacci number

7. a)
$$A^2 = \begin{bmatrix} 2 & 1 \\ 1 & 1 \end{bmatrix} = \begin{bmatrix} F_3 & F_2 \\ F_2 & F_1 \end{bmatrix}, \quad A^3 = \begin{bmatrix} 3 & 2 \\ 2 & 1 \end{bmatrix} = \begin{bmatrix} F_4 & F_3 \\ F_3 & F_2 \end{bmatrix},$$

$$A^4 = \begin{bmatrix} 5 & 3 \\ 3 & 2 \end{bmatrix} = \begin{bmatrix} F_5 & F_4 \\ F_4 & F_3 \end{bmatrix}$$

b) Conjecture: For $n \in \mathbf{Z}^+$, $A^n = \begin{bmatrix} F_{n+1} & F_n \\ F_n & F_{n-1} \end{bmatrix}$, where F_n denotes the nth Fibonacci number.

Proof: For $n = 1$, $A = A^1 = \begin{bmatrix} 1 & 1 \\ 1 & 0 \end{bmatrix} = \begin{bmatrix} F_2 & F_1 \\ F_1 & F_0 \end{bmatrix}$, so the result is true in this case.

Assume the result true for $n = k \ge 1$. That is, $A^k = \begin{bmatrix} F_{k+1} & F_k \\ F_k & F_{k-1} \end{bmatrix}$. For $n = k + 1$,

$$A^n = A^{k+1} = A^k \cdot A = \begin{bmatrix} F_{k+1} & F_k \\ F_k & F_{k-1} \end{bmatrix}\begin{bmatrix} 1 & 1 \\ 1 & 0 \end{bmatrix}$$

$$= \begin{bmatrix} F_{k+1} + F_k & F_{k+1} \\ F_k + F_{k-1} & F_k \end{bmatrix} = \begin{bmatrix} F_{k+2} & F_{k+1} \\ F_{k+1} & F_k \end{bmatrix}.$$

Consequently, the result is true for all $n \in \mathbf{Z}^+$, by the principle of mathematical induction.

9. **a)** For any derangement, 1 is placed in position i, where $2 \leq i \leq n$. Two things then occur. Case 1 (i is in position 1): Here the other $n - 2$ integers are deranged in d_{n-2} ways. With $n - 1$ choices for i, this results in $(n - 1)d_{n-2}$ such derangements. Case 2 (i is not in position 1 (or position i)): Here we consider 1 as the new natural position for i, so there are $n - 1$ elements to derange. With $n - 1$ choices for i, we have $(n - 1)d_{n-1}$ derangements. Since the two cases are exhaustive and disjoint, the result follows from the rule of sum.

   b) $d_0 = 1$ **c)** $d_n - nd_{n-1} = d_{n-2} - (n - 2)d_{n-3}$

11. $a_n = n^2 - n + 2$, $n \geq 1$; $a_0 = 1$

13. **a)** $p_n = (0.6)^n p_0$, $n \geq 0$ **b)** $p_0 = 2/5$

15. **a)** $a_n = \binom{2n}{n}$, $n \geq 0$ **b)** $r = 1$, $s = -4$, $t = -1/2$

   d) $b_n = (1/(2n - 1))\binom{2n}{n}$, $n \geq 1$; $b_0 = 0$

17. When $|B| = n = 1$ and $|A| = m$, $f\colon A \rightarrow B$ where $f(a) = b$ for all $a \in A$ and $\{b\} = B$ is the only onto function from A to B. Hence $a(m, 1) = 1$.

   For $m \geq n > 1$, $n^m =$ the total number of functions $f\colon A \rightarrow B$. If $1 \leq i \leq n - 1$, there are $\binom{n}{i}a(m, i)$ onto functions g with domain A and range a subset of B of size i. Furthermore, any function $h\colon A \rightarrow B$ that is not onto is found among these functions g. Consequently, $a(m, n) = n^m - \sum_{i=1}^{n-1} \binom{n}{i}a(m, i)$.

CHAPTER 11

An Introduction to Graph Theory

Section 11.1 – p. 410

1. **a)** To represent the air routes traveled among a certain set of cities by a particular airline.

   b) To represent an electrical network. Here the vertices can represent switches, transistors, etc., and an edge (x, y) indicates the existence of a wire connecting x to y.

   c) Let the vertices represent a set of job applicants and a set of open positions in a corporation. Draw an edge (A, b) to denote that applicant A is qualified for position b. Then all open positions can be filled if the resulting graph provides a matching between the applicants and open positions.

3. 6

5. **a)**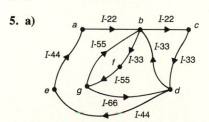

   b) $\{(g, d), (d, e), (e, a)\}$;
      $\{(g, b), (b, c), (c, d), (d, e), (e, a)\}$.

   c) Two: one of $\{(b, c), (c, d)\}$ and one of $\{(b, f), (f, g), (g, d)\}$.

   d) No

   e) Yes: Travel the path
      $\{(c, d), (d, e), (e, a), (a, b), (b, f), (f, g)\}$.

   f) Yes. Travel the trail
      $\{(g, b), (b, f), (f, g), (g, d), (d, b), (b, c), (c, d), (d, e), (e, a), (a, b)\}$.

7. If $\{a, b\}$ is not part of a cycle, then its removal disconnects a and b (and G). If not, there is a path P from a to b, and P together with $\{a, b\}$ provides a cycle containing $\{a, b\}$. Conversely, if the removal of $\{a, b\}$ from G disconnects G, then there exist $x, y \in V$ such that the only path P from x to y contains $e = \{a, b\}$. If e were part of a cycle C, then the edges in $(P - \{e\}) \cup (C - \{e\})$ would contain a second path connecting x to y.

9. a) Yes **b)** No **c)** $n-1$

11. The partition of V induced by $\mathcal{R}$ yields the (connected) components of G.

13. $\kappa(G) = 3$

Section 11.2 — p. 421

1. a) 3 **b)** $G_1 = \langle U \rangle$, where $U = \{a, b, d, f, g, h, i, j\}$; $G_1 = G - \{c\}$

  c) $G_2 = \langle W \rangle$, where $W = \{b, c, d, f, g, i, j\}$; $G_2 = G - \{a, h\}$

  d)

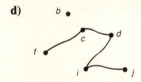

  e)

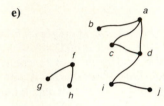

  f) (i, ii, and iii)

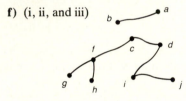

  g) Four: (i) G (ii) $G - e$, for $e = \{a, c\}$

    (iii) $G - e'$, for $e' = \{a, d\}$ (iv) $G - e''$, for $e'' = \{c, d\}$

3. G is (or is isomorphic to) K_n, where $n = |V|$.

5. (i)

	R		Y		W		B	
B	1	Y	R	2	B	Y	3	R	W	4	W
	W			B			Y			R	

(ii) No solution.

(iii)

	W		B		Y		R	
R	1	W	W	2	B	Y	3	R	B	4	Y
	Y			R			B			W	

7. (1) No **(2)** No **(3)** Yes. Correspond a with u, b with w, c with x, d with y, e with v, and f with z.

9. $\binom{v}{2} - e$

11. a) If $G_1 = (V_1, E_1)$ and $G_2 = (V_2, E_2)$ are isomorphic, then there is a function $f: V_1 \to V_2$ that is one-to-one and onto and preserves adjacencies. If $x, y \in V_1$ and $\{x, y\} \notin E_1$, then $\{f(x), f(y)\} \notin E_2$. Hence the same function f preserves adjacencies for $\overline{G_1}, \overline{G_2}$ and can be used to define an isomorphism for $\overline{G_1}, \overline{G_2}$. The converse follows in a similar way.

  b) They are not isomorphic. The complement of the graph containing vertex a is a cycle of length 8. The complement of the other graph is the disjoint union of two cycles of length 4.

13.

If G is the cycle with edges $\{a, b\}, \{b, c\}, \{c, d\}, \{d, e\}$, and $\{e, a\}$, then $\bar{G}$ is the cycle with edges $\{a, c\}, \{c, e\}, \{e, b\}, \{b, d\}$, and $\{d, a\}$. Hence G and $\bar{G}$ are isomorphic. Conversely, if G is a cycle on n vertices and $G, \bar{G}$ are isomorphic, then $n = \frac{1}{2}\binom{n}{2}$, or $n = \frac{1}{4}(n)(n-1)$, and $n = 5$.

15. a) Here f must also maintain directions. So $(a, b) \in E_1$ if and only if $(f(a), f(b)) \in E_2$.

b)

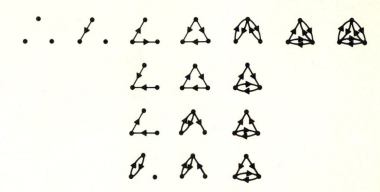

c) They are not isomorphic. Consider vertex a in the first graph. It is incident to one vertex and incident from two other vertices. No vertex in the other graph has this property.

Section 11.3 – p. 430

1. a) $|V| = 6$ **b)** $|V| = 1$ or 2 or 3 or 5 or 6 or 10 or 15 or 30
(In the first four cases, G is a multigraph; when $|V| = 30$, G is disconnected.)
c) $|V| = 6$

3. $\delta|V| \le \sum_{v \in V} \deg(v) \le \Delta|V|$. Since $2|E| = \sum_{v \in V} \deg(v)$, it follows that $\delta|V| \le 2|E| \le \Delta|V|$, so $\delta \le 2(e/n) \le \Delta$.

5. (Corollary 11.1). Let $V = V_1 \cup V_2$, where $V_1(V_2)$ contains all vertices of odd (even) degree. Then $2|E| - \sum_{v \in V_2} \deg(v) = \sum_{v \in V_1} \deg(v)$ is an even integer. For $|V_1|$ odd, $\sum_{v \in V_1} \deg(v)$ is odd.

(Corollary 11.2). For the converse let $G = (V, E)$ have an Euler trail with a, b as the starting and terminating vertices. Add the edge $\{a, b\}$ to G to form the larger graph $G_1 = (V, E_1)$ where G_1 has an Euler circuit. Hence G_1 is connected and each vertex in G_1 has even degree. When we remove edge $\{a, b\}$ from G_1, the vertices in G will have the same even degree except for a, b; $\deg_G(a) = \deg_{G_1}(a) - 1$, $\deg_G(b) = \deg_{G_1}(b) - 1$, so the vertices a, b have odd degree in G. Also, since the edges in G form an Euler trail, G is connected.

7. a) Let $a, b, c, x, y \in V$ with $\deg(a) = \deg(b) = \deg(c) = 1$, $\deg(x) = 5$, and $\deg(y) = 7$. Since $\deg(y) = 7$, y is adjacent to all of the other (seven) vertices in V. Therefore vertex x is not adjacent to any of the vertices a, b, and c. Since x cannot be adjacent to itself, unless we have loops, it follows that $\deg(x) \le 4$, and we cannot draw a graph for the given conditions.

b)

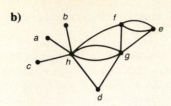

9. n odd; $n = 2$ **11.** Yes

13. $\sum_{v\in V} \deg^+(v) = |E| = \sum_{v\in V} \deg^-(v)$

15. From Exercise 13, $\sum_{v\in V} [\deg^+(v) - \deg^-(v)] = 0$. For each $v\in V$, $\deg^+(v) + \deg^-(v) = n - 1$, so

$$0 = (n-1)\cdot 0 = \sum_{v\in V} (n-1)[\deg^+(v) - \deg^-(v)]$$

$$= \sum_{v\in V} [\deg^+(v) + \deg^-(v)][\deg^+(v) - \deg^-(v)]$$

$$= \sum_{v\in V} [(\deg^+(v))^2 - (\deg^-(v))^2],$$

and the result follows.

17. a) and b)

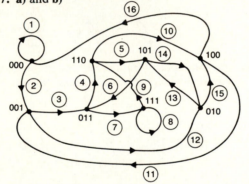

c)

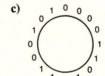

19. Let $|V| = n \geq 2$. Since G is loop-free and connected, for all $x \in V$ we have $1 \leq \deg(x) \leq n - 1$. Apply the pigeonhole principle with the n vertices as the pigeons and the $n - 1$ possible degrees as the pigeonholes.

21. a)

$$A = \begin{array}{c} \\ v_1 \\ v_2 \\ v_3 \\ v_4 \\ v_5 \end{array}
\begin{array}{c} v_1\ v_2\ v_3\ v_4\ v_5 \\
\begin{bmatrix} 0 & 1 & 1 & 0 & 1 \\ 1 & 0 & 1 & 1 & 1 \\ 1 & 1 & 1 & 1 & 1 \\ 0 & 1 & 1 & 0 & 1 \\ 1 & 1 & 1 & 1 & 1 \end{bmatrix}
\end{array}$$

$$I = \begin{array}{c} \\ v_1 \\ v_2 \\ v_3 \\ v_4 \\ v_5 \end{array}
\begin{array}{c} e_1\ e_2\ e_3\ e_4\ e_5\ e_6\ e_7\ e_8\ e_9\ e_{10}\ e_{11} \\
\begin{bmatrix} 1 & 1 & 1 & 0 & 0 & 0 & 0 & 0 & 0 & 0 & 0 \\ 0 & 0 & 1 & 1 & 1 & 1 & 0 & 0 & 0 & 0 & 0 \\ 1 & 0 & 0 & 0 & 1 & 0 & 0 & 1 & 0 & 1 & 1 \\ 0 & 0 & 0 & 0 & 0 & 1 & 0 & 1 & 1 & 0 & 0 \\ 0 & 1 & 0 & 1 & 0 & 0 & 1 & 0 & 1 & 1 & 0 \end{bmatrix}
\end{array}$$

b) If there is a walk of length two between v_i and v_j, denote this by $\{v_i, v_k\}, \{v_k, v_j\}$. Then $a_{ik} = a_{kj} = 1$ in A, and the i, j entry in A^2 is 1. Conversely, if the i, j entry of A^2 is 1, then there is at least one value of k, for $1 \le k \le n$, such that $a_{ik} = a_{kj} = 1$, and this indicates the existence of a walk $\{v_i, v_k\}, \{v_k, v_j\}$ between the ith and jth vertices of V.

c) For any $1 \le i, j \le n$, the i, j entry of A^2 counts the number of distinct walks of length two between the ith and jth vertices of V.

d) For v at the top of the column, the column sum is the degree of v if there is no loop at v. Otherwise, $\deg(v) = [(\text{column sum for } v) - 1] + 2(\text{number of loops at } v)$.

e) For each column of I the column sum is 1 for a loop and 2 for an edge that is not a loop.

Section 11.4 – p. 445

1. 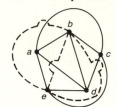 In this situation vertex b is in the region formed by the edges $\{a, d\}, \{d, c\}, \{c, a\}$, and vertex e is outside of this region. Hence the edge $\{b, e\}$ will cross one of the edges $\{a, d\}, \{d, c\}$, or $\{a, c\}$ (as shown).

3. a)

Graph	Number of vertices	Number of edges
$K_{4,7}$	11	28
$K_{7,11}$	18	77
$K_{m,n}$	$m + n$	mn

b) $m = 6$

c) Let the vertices of K_{m+n} be listed as $\{x_1, x_2, \ldots, x_m, y_1, y_2, \ldots, y_n\}$, and let $V_1 = \{x_1, x_2, \ldots, x_m\}$ and $V_2 = \{y_1, y_2, \ldots, y_n\}$. The number of edges in K_{m+n} is $\binom{m+n}{2}$. Remove all edges of the form $\{x_i, y_j\}$, where $1 \le i \le m$, $1 \le j \le n$. There are mn such edges. This leaves the subgraph $\overline{K_{m,n}}$ which has K_m and K_n as its two components. The number of edges in $\overline{K_{m,n}}$ is $\binom{m}{2} + \binom{n}{2}$.

5. Partition V as $V_1 \cup V_2$ with $|V_1| = m$, $|V_2| = v - m$. Since G is bipartite, the maximum number of edges that G can have is $m(v - m) = -[m - (v/2)]^2 + v/2)^2$, a function of m. For a given value of v, when v is even, $m = v/2$ maximizes $m(v - m) = (v/2)[v - (v/2)] = (v/2)^2$. For v odd, $m = (v - 1)/2$ or $m = (v + 1)/2$ maximizes $m(v - m) = [(v - 1)/2][v - ((v - 1)/2)] = [(v - 1)/2][(v + 1)/2] = [(v + 1)/2][v - ((v + 1)/2)] = (v^2 - 1)/4 = \lfloor (v/2)^2 \rfloor < (v/2)^2$. Hence if $|E| > (v/2)^2$, then G cannot be bipartite.

7. a)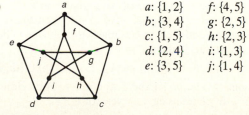

a: $\{1, 2\}$	f: $\{4, 5\}$
b: $\{3, 4\}$	g: $\{2, 5\}$
c: $\{1, 5\}$	h: $\{2, 3\}$
d: $\{2, 4\}$	i: $\{1, 3\}$
e: $\{3, 5\}$	j: $\{1, 4\}$

b) G is (isomorphic to) the Petersen graph. [See Fig. 11.42(a).]

9. 10

11. If not, $\deg(v) \ge 6$ for all $v \in V$. Then $2e = \sum_{v \in V} \deg(v) \ge 6|V|$ so $e \ge 3|V|$, contradicting $e \le 3|V| - 6$ (Corollary 11.3).

13. **a)** $2e \geq kr = k(2 + e - v) \Rightarrow (2 - k)e \geq k(2 - v) \Rightarrow e \leq [k/(k - 2)](v - 2)$

   b) 4

   c) In $K_{3,3}$, we have $e = 9$ and $v = 6$. $[k/(k - 2)](v - 2) = (4/2)(4) = 8 < 9 = e$. Since $K_{3,3}$ is connected, it must be nonplanar.

   d) Here $k = 5$, $v = 10$, $e = 15$, and $[k/(k - 2)](v - 2) = (5/3)(8) = (40/3) < 15 = e$. The Petersen graph is connected, so it must be nonplanar.

15. The dual for the tetrahedron [Fig. 11.45(b)] is the graph itself. For the graph (cube) in Fig. 11.45(d) the dual is the octahedron, and vice versa. Likewise, the dual of the dodecahedron is the icosahedron, and vice versa.

17. (i) $\{\{k, m\}, \{m, s\}\}$ (ii) $\{\{j, k\}, \{k, s\}, \{k, m\}\}$

19.

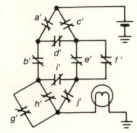

Section 11.5 – p. 454

1. **a)**   **b)**   **c)**   **d)**

3. **a)** Hamilton cycle: $a \to g \to k \to i \to h \to b \to c \to d \to j \to f \to e \to a$

   b) Hamilton cycle: $a \to d \to b \to e \to g \to j \to i \to f \to h \to c \to a$

   c) Hamilton cycle: $a \to h \to e \to f \to g \to i \to d \to c \to b \to a$

5. **a)** $(1/2)(n - 1)!$ **b)** 10   **c)** 9

7. Let $G = (V, E)$ be a loop-free undirected graph with no odd cycles. We assume that G is connected—otherwise, we work with the components of G. Select any vertex x in V, and let $V_1 = \{v \in V \mid d(x, v)$, the length of a shortest path between x and v, is odd$\}$ and $V_2 = \{w \in V \mid d(x, w)$, the length of a shortest path between x and w, is even$\}$. Note that (i) $x \in V_2$, (ii) $V = V_1 \cup V_2$, and (iii) $V_1 \cap V_2 = \emptyset$. We claim that each edge $\{a, b\}$ in E has one vertex in V_1 and the other vertex in V_2. For suppose that $e = \{a, b\} \in E$ with $a, b \in V_1$. (The proof for $a, b \in V_2$ is similar.) Let $E_a = \{\{a, v_1\}, \{v_1, v_2\}, \ldots, \{v_{m-1}, x\}\}$ be the m edges in a shortest path from a to x, and let $E_b = \{\{b, v_1'\}, \{v_1', v_2'\}, \ldots, \{v_{n-1}', x\}\}$ be the n edges in a shortest path from b to x. Note that m and n are both odd. If $\{v_1, v_2, \ldots, v_{m-1}\} \cap \{v_1', v_2', \ldots, v_{n-1}'\} = \emptyset$, then the set of edges $E' = \{\{a, b\}\} \cup E_a \cup E_b$ provides an odd cycle in G. Otherwise, let $w (\neq x)$ be the first vertex where the paths come together, and let $E'' =$

$$\{\{a, b\}\} \cup \{\{a, v_1\}, \{v_1, v_2\}, \ldots, \{v_i, w\}\} \cup \{\{b, v_1'\}, \{v_1', v_2'\}, \ldots, \{v_j', w\}\},$$

for some $1 \leq i \leq m - 1$ and $1 \leq j \leq n - 1$. Then either E'' provides an odd cycle for G or $E' - E''$ contains an odd cycle for G.

9. **a)**

b)

 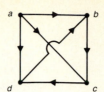

$$\begin{array}{ll} \deg^+(a) = 3 & \deg^-(a) = 0 \\ \deg^+(b) = 2 & \deg^-(b) = 1 \\ \deg^+(c) = 0 & \deg^-(c) = 3 \\ \deg^+(d) = 1 & \deg^-(d) = 2 \end{array} \qquad \begin{array}{ll} \deg^+(a) = 3 & \deg^-(a) = 0 \\ \deg^+(b) = 1 & \deg^-(b) = 2 \\ \deg^+(c) = 1 & \deg^-(c) = 2 \\ \deg^+(d) = 1 & \deg^-(d) = 2 \end{array}$$

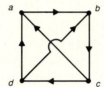 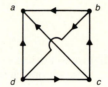

$$\begin{array}{ll} \deg^+(a) = 1 & \deg^-(a) = 2 \\ \deg^+(b) = 1 & \deg^-(b) = 2 \\ \deg^+(c) = 2 & \deg^-(c) = 1 \\ \deg^+(d) = 2 & \deg^-(d) = 1 \end{array} \qquad \begin{array}{ll} \deg^+(a) = 0 & \deg^-(a) = 3 \\ \deg^+(b) = 2 & \deg^-(b) = 1 \\ \deg^+(c) = 2 & \deg^-(c) = 1 \\ \deg^+(d) = 2 & \deg^-(d) = 1 \end{array}$$

11. For all $x, y \in V$, $\deg(x) + \deg(y) \geq 2[(n-1)/2] = n-1$, so the result follows from Theorem 11.8.

13. This follows from Theorem 11.9, since for all (nonadjacent) $x, y \in V$, $\deg(x) + \deg(y) = 12 > 11 = |V|$.

15. a) (i) $\{a, c, f, h\}, \{a, g\}$ (ii) $\{z\}, \{u, w, y\}$

    b) (i) $\beta(G) = 4$ (ii) $\beta(G) = 3$

    c) (i) 3 (ii) 3 (iii) 3 (iv) 4 (v) 6 (vi) The maximum of m and n

    d) The complete graph on $|I|$ vertices

    e) $\binom{n}{2} - m$

Section 11.6 – p. 463

1. Draw a vertex for each species of fish. If two species x, y must be kept in separate aquaria, draw the edge $\{x, y\}$. The smallest number of aquaria needed is then the chromatic number of the resulting graph.

3. a) 2   **b)** 2 (n even); 3 (n odd)

    c) Figure 11.45(d): 2; Fig. 11.48: 3; Fig. 11.68(a): 2; Fig. 11.68(b): 3.

5. Let G be a cycle on n vertices, where n is odd and $n \geq 5$.

7. a) (1) $\lambda(\lambda - 1)^2(\lambda - 2)^2$ (2) $\lambda(\lambda - 1)(\lambda - 2)(\lambda^2 - 2\lambda + 2)$

        (3) $\lambda(\lambda - 1)(\lambda - 2)(\lambda^2 - 5\lambda + 7)$

    b) (1) 3 (2) 3 (3) 3

    c) (1) 720 (2) 1020 (3) 420

9. The result is false. If G is the cycle on four vertices, then $P(G, \lambda) = \lambda(\lambda - 1)^3 - \lambda(\lambda - 1)(\lambda - 2) = \lambda(\lambda - 1)(\lambda^2 - 3\lambda + 3)$, where $\lambda^2 - 3\lambda + 3$ has no real roots.

11. a) $\lambda(\lambda - 1)(\lambda - 2)$

    b) Follows from Theorem 11.10.

    c) Follows by the rule of product.

d)
$$P(C_n, \lambda) = P(L_n, \lambda) - P(C_{n-1}, \lambda) = \lambda(\lambda - 1)^{n-1} - P(C_{n-1}, \lambda)$$
$$= [(\lambda - 1) + 1](\lambda - 1)^{n-1} - P(C_{n-1}, \lambda)$$
$$= (\lambda - 1)^n + (\lambda - 1)^{n-1} - P(C_{n-1}, \lambda), \text{ so}$$
$$P(C_n, \lambda) - (\lambda - 1)^n = (\lambda - 1)^{n-1} - P(C_{n-1}, \lambda).$$

Replacing n by $n - 1$ yields
$$P(C_{n-1}, \lambda) - (\lambda - 1)^{n-1} = (\lambda - 1)^{n-2} - P(C_{n-2}, \lambda).$$

Hence
$$P(C_n, \lambda) - (\lambda - 1)^n = P(C_{n-2}, \lambda) - (\lambda - 1)^{n-2}$$

e) Continuing from part (d),
$$P(C_n, \lambda) = (\lambda - 1)^n + (-1)^{n-3}[P(C_3, \lambda) - (\lambda - 1)^3]$$
$$= (\lambda - 1)^n + (-1)^{n-1}[\lambda(\lambda - 1)(\lambda - 2) - (\lambda - 1)^3]$$
$$= (\lambda - 1)^n + (-1)^n(\lambda - 1).$$

13. From Theorem 11.13, the expansion for $P(G, \lambda)$ will contain exactly one occurrence of the chromatic polynomial of K_n. Since no larger graph occurs, this term determines the degree as n and the leading coefficient as 1.

15. $\lambda(\lambda - 1)(\lambda - 2)^2(\lambda^2 - 3\lambda + 3)$

Miscellaneous Exercises – p. 469

1. $n = 17$

3. a) For $v, w \in V$, if v, w differ in k components, then there is a path of length k from v to w obtained by starting with v and changing one component at a time until vertex w is reached.

    b) $|V| = 2^n$; $|E| = (n)(2^{n-1})$

5. a) Label the vertices of K_6 with $a, b, \ldots, f$. Of the five edges on a, at least three have the same color, say red. Let these edges be $\{a, b\}, \{a, c\}, \{a, d\}$. If the edges $\{b, c\}$, $\{c, d\}, \{b, d\}$ are all blue, the result follows. If not, one of these edges, say $\{c, d\}$, is red. Then the edges $\{a, c\}, \{a, d\}, \{c, d\}$ yield a red triangle.

    b) Consider the six people as vertices. If two people are friends (strangers) draw a red (blue) edge connecting their respective vertices. The result then follows from part (a).

7. a) We can redraw G_2 as

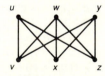

    b) 72

9. a) Let I be independent and $\{a, b\} \in E$. If neither a nor b is in $V - I$, then $a, b \in I$, and since they are adjacent, I is not independent. Conversely, if $I \subseteq V$ with $V - I$ a covering of G, then if I is not independent there are vertices $x, y \in I$ with $\{x, y\} \in E$. But $\{x, y\} \in E \Rightarrow$ either x or y is in $V - I$.

    b) Let I be a largest maximal independent set in G and K a minimum covering. From part (a), $|K| \leq |V - I| = |V| - |I|$ and $|I| \geq |V - K| = |V| - |K|$, or $|K| + |I| \geq |V| \geq |K| + |I|$.

11. a) (i) 3^n (ii) 2^n (iii) 4^n

b) Let $G = (V, E)$ be an undirected loop-free graph with labeled vertices and components $C_1, C_2, \ldots, C_n$. If each component C_i contains k_i maximal independent sets, for $1 \le i \le n$, then G contains $k_1 \cdot k_2 \cdots k_n$ maximal independent sets.

13. a) $|V| = 2n$; $|E| = 3n - 2$ $(n \ge 1)$

b) $a_n = a_{n-1} + a_{n-2}$, $a_0 = a_1 = 1$

$a_n = F_{n+1}$, the $(n + 1)$st Fibonacci number

15. a) $\gamma(G) = 2$; $\beta(G) = 3$; $\chi(G) = 4$

b) G has neither an Euler trail nor an Euler circuit; G does have a Hamiltonian cycle.

c) G is not bipartite but it is planar.

17. a) $\chi(G) \ge \omega(G)$.   **b)** $\omega(G) = \max\{\omega(C_i) \mid 1 \le i \le k\}$.

c) No, if $\omega(G) = 8$, then G has at least $\binom{8}{2} = 28$ edges.   **d)** They are equal.

19. a) The constant term is 3, not 0. This contradicts Theorem 11.11.

b) The leading coefficient is 3, not 1. This contradicts the result in Exercise 13 of Section 11.6.

c) The sum of the coefficients is -1, not 0. This contradicts Theorem 11.12.

d) Here there is no λ^3 term. This violates the result of Exercise 14 in Section 11.6.

21. a) K_n **b)** $|V| = \sum_{i=1}^{n} p_i$, $|E| = \sum_{1 \le i < j \le n} p_i p_j$ **c)** $\kappa(\overline{G}) = n$; $\overline{G}$ has $\sum_{i=1}^{n} \binom{p_i}{2}$ edges.

d) $n = 2$: $p_1 = 1$, $p_2 \ge 1$

$p_1 = 2$, $p_2 \ge 2$

$n = 3$: $p_1 = 1$, $p_2 = 1$, $p_3 \ge 1$

$p_1 = 1$, $p_2 = 2$, $p_3 = 2$

$p_1 = 2$, $p_2 = 2$, $p_3 = 2$

$n = 4$: $p_1 = 1$, $p_2 = 1$, $p_3 = 1$, $p_4 = 1$

$p_1 = 1$, $p_2 = 1$, $p_3 = 1$, $p_4 = 2$

CHAPTER 12

Trees

Section 12.1 – p. 479

1. a)

b) 5

3. a) $e = \upsilon - \kappa$ **b)** $\kappa - 1$

5.

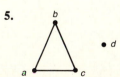

7. If there is a unique path between any pair of vertices in G, then G is connected. If G contains a cycle, then there is a pair of vertices x, y with two distinct paths connecting x and y. Hence, G is a loop-free connected undirected graph with no cycles, so G is a tree.

9. $\binom{n}{2}$

A-54

ANSWERS

11. n

13. a) If the complement of T contains a cut set, then the removal of these edges disconnects G, and there are vertices x, y with no path connecting them. Hence T is not a spanning tree for G.

b) If the complement of C contains a spanning tree, then every pair of vertices in G has a path connecting them, and this path includes no edges of C. Hence the removal of the edges in C from G does not disconnect G, so C is not a cut set for G.

15. a) (i) 3, 4, 6, 3, 8, 4   (ii) 3, 4, 6, 6, 8, 4

b) No pendant vertex of the given tree appears in the sequence, so the result is true for these vertices. When an edge $\{x, y\}$ is removed and y is a pendant vertex (of the tree or one of the resulting subtrees), the $\deg(x)$ is decreased by 1 and x is placed in the sequence. As the process continues, either (i) this vertex x becomes a pendant vertex in a subtree and is removed but not recorded again in the sequence, or (ii) the vertex x is left as one of the last two vertices of an edge. In either case x has been listed in the sequence $(\deg(x) - 1)$ times.

c)

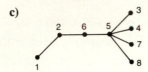

d) From the given sequence the degree of each vertex in the tree is known.

Step 1: Set the counter i to 1.

Step 2: From among the vertices of degree 1, select the vertex v with the smallest label. This determines the edge $\{v, x_i\}$. Remove v from the set of labels and reduce the degree of x_i by 1.

Step 3: If $i < n - 2$, increase i by 1 and return to step 2.

Step 4: If $i = n - 2$, the vertices (labels) x_{n-3}, x_{n-2} are connected by an edge if $x_{n-3} \neq x_{n-2}$. (The tree is then complete.)

Section 12.2 – p. 497

1. a) f, h, k, p, q, s, t **b)** a **c)** d
d) e, f, j, q, s, t **e)** q, t **f)** 2
g) k, p, q, s, t **h)** 4   **i)** No

3. a) $/ + w - xy * \pi \uparrow z3$ **b)** 0.4

c) (1)   (2)

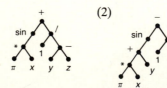

5. Preorder: $r, j, h, g, e, d, b, a, c, f, i, k, m, p, s, n, q, t, v, w, u$
  Inorder: $h, e, a, b, d, c, g, f, j, i, r, m, s, p, k, n, v, t, w, q, u$
  Postorder: $a, b, c, d, e, f, g, h, i, j, s, p, m, v, w, t, u, q, n, k, r$

7. a) 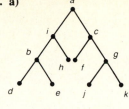   **b)**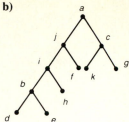

9. a) (i) and (iii) 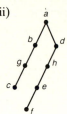   (ii)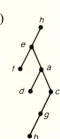

b) (i)    (ii)    (iii)

11. G is connected.

13. Theorem 12.6
  a) Each internal vertex has m children, so there are mi vertices that are the children of some other vertex. This accounts for all vertices in the tree except the root. Hence $n = mi + 1$.
  b) $\ell + i = n = mi + 1 \Rightarrow \ell = (m - 1)i + 1$
  c) $\ell = (m - 1)i + 1 \Rightarrow i = (\ell - 1)/(m - 1)$
    $n = mi + 1 \Rightarrow i = (n - 1)/m$.
  Corollary 12.1
  Since the tree is balanced, $m^{h-1} < \ell \leq m^h$ by Theorem 12.7.
  $$m^{h-1} < \ell \leq m^h \Rightarrow \log_m(m^{h-1}) < \log_m(\ell) \leq \log_m(m^h)$$
  $$\Rightarrow (h - 1) < \log_m \ell \leq h \Rightarrow h = \lceil \log_m \ell \rceil$$

15. a) 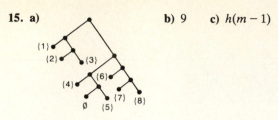 **b)** 9 **c)** $h(m-1)$

17. $21845; 1 + m + m^2 + \cdots + m^{h-1} = (m^h - 1)/(m - 1)$

19.

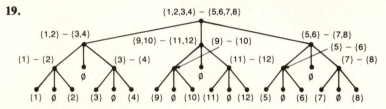

Section 12.3 − p. 505

1. a) L_1: $1, 3, 5, 7, 9$ L_2: $2, 4, 6, 8, 10$

   b) Assume that $m < n$.

   L_1: $1, 3, 5, 7, \ldots, 2m - 3, m + n$
   L_2: $2, 4, 6, 8, \ldots, 2m - 2, 2m - 1, 2m, 2m + 1, \ldots, m + n - 1$

3. a)

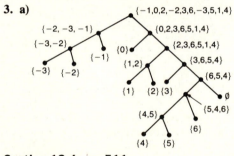

Section 12.4 − p. 511

1. a) tear **b)** tatener **c)** rant

3. a: 001   c: 1000   e: 01   g: 00001   i: 11
   b: 000001   d: 0001   f: 1001   h: 101   j: 000000

5. 55,987

Section 12.5 − p. 515

1. The articulation points are b, e, f, h, j, k. The biconnected components are B_1: $\{\{a, b\}\}$;
B_2: $\{\{d, e\}\}$; B_3: $\{\{b, c\}, \{c, f\}, \{f, e\}, \{e, b\}\}$; B_4: $\{\{f, g\}, \{g, h\}, \{h, f\}\}$;
B_5: $\{\{h, i\}, \{i, j\}, \{j, h\}\}$; B_6: $\{\{j, k\}\}$; B_7: $\{\{k, p\}, \{p, n\}, \{n, m\}, \{m, k\}, \{p, m\}\}$.

3. a) T can have as few as one or as many as $n - 2$ articulation points. If T contains a
   vertex of degree $(n - 1)$, then this vertex is the only articulation point. If T is a path
   with n vertices and $n - 1$ edges, then the $n - 2$ vertices of degree 2 are all articulation points.

   b) In all cases, a tree on n vertices has $n - 1$ biconnected components. Each edge is a
   biconnected component.

5. $\chi(G) = \max\{\chi(B_i) \mid 1 \le i \le k\}$.

7. No $G:$ B_1 $G':$ B'_1 B'_2 B'_3

B_3 B_2

9. We always have low $(y_2) = $ low $(y_1) = 1$. (Note: Vertices y_2 and y_1 are always in the same biconnected component.)

Miscellaneous Exercises — p. 519

1. a) If G is a tree, consider G a rooted tree. Then there are λ choices for coloring the root of G and $(\lambda - 1)$ choices for coloring each of its descendants. The result then follows by the rule of product.

Conversely, if $P(G, \lambda) = \lambda(\lambda - 1)^{n-1}$, then since the factor λ occurs only once, the graph G is connected.

$$P(G, \lambda) = \lambda(\lambda - 1)^{n-1}$$
$$= \lambda^n - (n - 1)\lambda^{n-1} + \cdots + (-1)^{n-1}\lambda \Rightarrow G \text{ has } n \text{ vertices}$$

and $(n - 1)$ edges. Hence G is a tree [by part (d) of Theorem 12.5].

b) From part (a), $P(G, 1) = 0$ and $P(G, 2) = 2 > 0$, so $\chi(G) = 2$.

c) For any graph $G = (V, E)$, if $e \in E$, then $P(G, \lambda) = P(G_e, \lambda) - P(G'_e, \lambda)$, so $P(G, \lambda) \le P(G_e, \lambda)$. If G is connected but not a tree, delete an edge from each cycle of G until the subgraph obtained is a spanning tree T of G. Then $P(G, \lambda) \le P(T, \lambda) = \lambda(\lambda - 1)^{n-1}$. [Recall that $P(G, \lambda)$ is the number of ways to properly color the vertices in G with λ colors. Hence $P(G, \lambda) \le P(T, \lambda)$ is interpreted here as an inequality for two numbers, not two polynomials.]

3. a) 10   **b)** $0, a_1, a_1.a_2, a_1.a_2.a_3, \ldots, a_1.a_2.a_3.\ldots.a_8.a_9.a_{10}$

5. If H is a Hamilton path in G, then H is connected, contains every vertex of G, and contains no cycles. Hence H is a spanning tree for G. Since H is a path, deg(v) (in H) is 2 for all vertices on the path except the endpoints which have (in H) degree 1.

Conversely, if H is a spanning tree for G, then if each vertex of H has degree (in H) at most 2, H is a path. Consequently H is a path containing all vertices of G, so H is a Hamilton path for G.

7. a) In going from level h to level $h + 1$, we add on the m children of exactly one of the leaves at level h.

b) $x_h = (m/2)(h^2 + h)$, $h \ge 0$

9. a) 1011001010100

b) (i)   (ii)

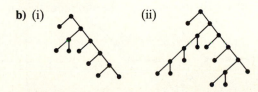

c) Since the last two vertices visited in a preorder traversal are leaves, the last two symbols in the characteristic sequence of any complete binary tree are 00.

11. a) 32; 31; 62   **b)** 2^h; $2^h - 1$; $2(2^h - 1)$

13. We assume that $G = (V, E)$ is connected—otherwise we work with a component of G. Since G is connected, and deg(v) ≥ 2 for all $v \in V$, it follows from Theorem 12.4 that

G is not a tree. But any loop-free connected undirected graph that is not a tree must contain a cycle.

15. For $1 \le i (<n)$, let $x_i =$ the number of vertices v where $\deg(v) = i$. Then
$x_1 + x_2 + \cdots + x_{n-1} = |V| = |E| + 1$, so $2|E| = 2(-1 + x_1 + x_2 + \cdots + x_{n-1})$. But
$2|E| = \sum_{v \in V} \deg(v) = (x_1 + 2x_2 + 3x_3 + \cdots + (n-1)x_{n-1})$. Solving
$2(-1 + x_1 + x_2 + \cdots + x_{n-1}) = x_1 + 2x_2 + \cdots + (n-1)x_{n-1}$ for x_1, we find
that $x_1 = 2 + x_3 + 2x_4 + 3x_5 + \cdots + (n-3)x_{n-1} = 2 + \sum_{\deg(v_i) \ge 3}[\deg(v_i) - 2]$.

17. If the edge e is *not* part of a cycle, then it must be in every spanning tree of G.

19. **a)** G^2 is isomorphic to K_5.   **b)** G^2 is isomorphic to K_4.

c) G^2 is isomorphic to K_{n+1}, so the number of new edges is $\binom{n+1}{2} - n = \binom{n}{2}$.

d) If G^2 has an articulation point x, then there exists $u, v \in V$ such that every path (in G^2) from u to v passes through x. (This follows from Exercise 2 of Section 12.5) Since G is connected, there exists a path P (in G) from u to v. If x is not on this path (which is also a path in G^2), then we contradict x's being an articulation point in G^2. Hence the path P (in G) passes through x, and we can write $P: u \to u_1 \to \cdots \to u_{n-1} \to u_n \to x \to v_m \to v_{m-1} \to \cdots \to v_1 \to v$. But then in G^2 we add the edge $\{u_n, v_m\}$, and the path P' (in G^2) given by $P': u \to u_1 \to \cdots \to u_{n-1} \to u_n \to v_m \to v_{m-1} \to \cdots \to v_1 \to v$ does not pass through x. So x is not an articulation point of G^2, and G^2 has no articulation points.

21. **a)**

For n even, $\ell_1 = n/2$ and $\ell_2 = \ell_1 + 1$.

For n odd, $\ell_1 = \ell_2 + 1$ and $\ell_2 = \lceil n/2 \rceil$.

b) Label the vertex of degree n with the label 1. Label the other n vertices (one vertex per label) with the labels $2, 3, \ldots, n, n+1$.

c) For $|V| = 4$ the only trees are a path of length 3 and $K_{1,3}$. These are handled by parts (a) and (b), respectively. For $|V| = 5$ there are three trees: (1) A path of length 4; (2) $K_{1,4}$; and (3) the tree with a vertex of degree 3. Trees (1) and (2) are handled by parts (a) and (b), respectively. The third tree may be labeled as follows:

For $|V| = 6$, there are six trees. The paths of length 5 and $K_{1,5}$ are dealt with by parts (a) and (b), respectively. The other four trees may be labeled as follows:

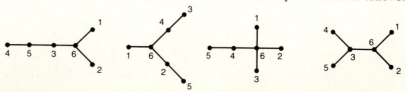

CHAPTER 13

Optimization and Matching

Section 13.1 – p. 530

1. **a)** If not, let $v_i \in \bar{S}$, where $1 \le i \le m$ and i is the smallest such subscript. Then $d(v_0, v_i) < d(v_0, v_{m+1})$, and we contradict the choice of v_{m+1} as a vertex v in $\bar{S}$ for which $d(v_0, v)$ is a minimum.

b) Suppose there is a shorter directed path (in G) from v_0 to v_k. If this path passes through a vertex in $\bar{S}$, then from part (a) we have a contradiction. Otherwise, we have a shorter directed path P'' from v_0 to v_k, and P'' only passes through vertices in S. But then $P'' \cup \{(v_k, v_{k+1}), (v_{k+1}, v_{k+2}), \ldots, (v_{m-1}, v_m), (v_m, v_{m+1})\}$ is a directed path (in G) from v_0 to v_{m+1}, and it is shorter than path P.

3. a) $d(a, b) = 5$; $d(a, c) = 6$; $d(a, f) = 12$; $d(a, g) = 16$; $d(a, h) = 12$
  b) $f: (a, c), (c, f)$ $g: (a, b), (b, h), (h, g)$ $h: (a, b), (b, h)$

5. False. Consider the following weighted graph.

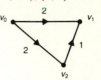

Section 13.2 — p. 536

1. Kruskal's algorithm generates the following sequence (of forests), which terminates in a minimal spanning tree T of weight 18.

  (1) $F_1 = \{\{e, h\}\}$ (2) $F_2 = F_1 \cup \{\{a, b\}\}$ (3) $F_3 = F_2 \cup \{\{b, c\}\}$
  (4) $F_4 = F_3 \cup \{\{d, e\}\}$ (5) $F_5 = F_4 \cup \{\{e, f\}\}$ (6) $F_6 = F_5 \cup \{\{a, e\}\}$
  (7) $F_7 = F_6 \cup \{\{d, g\}\}$ (8) $F_8 = T = F_7 \cup \{\{f, i\}\}$

3. a) Evansville–Indianapolis (168); Bloomington–Indianapolis (51): South Bend–Gary (58); Terre Haute–Bloomington (58); South Bend–Fort Wayne (79); Indianapolis–Fort Wayne (121).
  b) Fort Wayne–Gary (132); Evansville–Indianapolis (168); Bloomington–Indianapolis (51); Gary–South Bend (58); Terre Haute–Bloomington (58); Indianapolis–Fort Wayne (121).

5. a) To determine an optimal tree of maximal weight, replace the two occurrences of "small" in Kruskal's algorithm by "large."
  b) Use the edges: South Bend–Evansville (303); Fort Wayne–Evansville (290); Gary–Evansville (277); Fort Wayne–Terre Haute (201); Gary–Bloomington (198); Indianapolis–Evansville (168).

7. When the weights of the edges are all distinct, in each step of Kruskal's algorithm a unique edge is selected.

Section 13.3 — p. 547

1. a) $s = 2$; $t = 4$; $w = 5$; $x = 9$; $y = 4$ **b)** 18
  c) (i) $P = \{a, b, h, d, g, i\}$; $\bar{P} = \{z\}$ (ii) $P = \{a, b, h, d, g\}$; $\bar{P} = \{i, z\}$
    (iii) $P = \{a, h\}$; $\bar{P} = \{b, d, g, i, z\}$

3. (1)

b 15,14 d
8,8   8,6 6,6   20,20
a   g 5   z
    15,15   8,3
10,9   6,6   15,12
h 6,6 i

The maximal flow is 32,
which is $c(P, \bar{P})$ for
$P = \{a, b, d, g, h\}$ and $\bar{P} = \{i, z\}$.

(2)

b 6,6 d
8,8   9,9 g 7,2 7,4   10,10
a   7,3 h   z
6,6   7   6,6   6,1   14,13
i 8,6 j 12,12 k

The maximal flow is 23,
which is $c(P, \bar{P})$ for
$P = \{a\}$ and $\bar{P} = \{b, g, i, j, d, h, k, z\}$.

5. Here $c(e)$ is a positive integer for each $e \in E$, and the initial flow is defined as $f(e) = 0$ for all $e \in E$. The result follows because in each application of the labeling procedure, increases in flow come about from the second component of a label, and this component is always a nonnegative integer.

7.

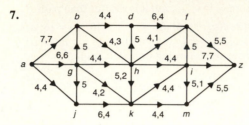

Section 13.4 – p. 556

1. $5/\binom{8}{4} = 1/14$

3. Let the committees be represented as $c_1, c_2, \ldots, c_6$, according to the way they are listed in the exercise.

    a) Select the members as follows: $c_1 - A$; $c_2 - G$; $c_3 - M$; $c_4 - N$; $c_5 - K$; $c_6 - R$.

    b) Select the nonmembers as follows: $c_1 - K$; $c_2 - A$; $c_3 - G$; $c_4 - J$; $c_5 - M$; $c_6 - P$.

5. Yes, such an assignment can be made by Brenda. Let X be the set of student applicants and Y the set of part-time jobs. Then for all $x \in X$, $y \in Y$, draw the edge (x, y) if applicant x is qualified for part-time job y. Then $\deg(x) \geq 4 \geq \deg(y)$ for all $x \in X$, $y \in Y$, and the result follows from Corollary 13.3.

7. a) (i) Select i from A_i for $1 \leq i \leq 4$.

       (ii) Select $i + 1$ from A_i for $1 \leq i \leq 3$, and 1 from A_4.

    b) 2

9. Each university can send one of its six graduates to one of the six different corporations, thereby maintaining the hiring policy. Since we have seven universities, we may assign seven graduates (one from each university) to each corporation and still remain within the limits set forth by these corporations. Consequently, all 42 graduates may be hired under these circumstances. [However, if any of the last three corporations (each seeking only seven graduates) decides to hire fewer than seven graduates, then at least one of these 42 graduates will not be hired by these six corporations.]

11. a) $\delta(G) = 1$. A maximal matching of X into Y is given by
$\{\{x_1, y_4\}, \{x_2, y_2\}, \{x_3, y_1\}, \{x_5, y_3\}\}$.

    b) If $\delta(G) = 0$, there is a complete matching of X into Y, and $\beta(G) = |Y|$, or $|Y| = \beta(G) - \delta(G)$. If $\delta(G) = k > 0$, let $A \subseteq X$ where $|A| - |R(A)| = k$. Then $A \cup (Y - R(A))$ is a largest maximal independent set in G and $\beta(G) = |A| + |Y - R(A)| = |Y| + (|A| - |R(A)|) = |Y| + \delta(G)$, so $|Y| = \beta(G) - \delta(G)$.

    c) Fig. 13.27(a): $\{x_1, x_2, x_3, y_2, y_4, y_5\}$; Fig. 13.28: $\{x_3, x_4, y_2, y_3, y_4\}$.

Miscellaneous Exercises – p. 561

1. $d(a, b) = 5$ $d(a, c) = 11$ $d(a, d) = 7$ $d(a, e) = 8$
$d(a, f) = 19$ $d(a, g) = 9$ $d(a, h) = 14$

    (Note that the loop at vertex g and the edges (c, a) of weight 9 and (f, e) of weight 5 are of no significance.)

3. a) The edge e_1 will always be selected in the first step of Kruskal's algorithm.

    b) Again using Kruskal's algorithm, edge e_2 will be selected in the first application of step 2 unless each of the edges e_1, e_2 is incident with the same two vertices—that is, the edges e_1, e_2 form a circuit and G is a multigraph.

5.

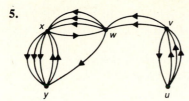

7. There are d_n, the number of derangements of $\{1, 2, 3, \ldots, n\}$.

9. The vertices [in the line graph $L(G)$] determined by E' form a maximal independent set.

CHAPTER 14

Rings and Modular Arithmetic

Section 14.1 – p. 567

1. (Example 14.4): $-a = a$, $-b = e$, $-c = d$, $-d = c$, $-e = b$
(Example 14.5): $-s = s$, $-t = y$, $-v = x$, $-w = w$, $-x = v$, $-y = t$

3. a) x **b)** $-s = t$, $-t = s$, $-x = x$, $-y = y$ **c)** y **d)** Yes   **e)** No   **f)** s, y

5. b) $1, -1, i, -i$

7. $\begin{bmatrix} a & b \\ c & d \end{bmatrix}^{-1} = (1/(ad - bc)) \begin{bmatrix} d & -b \\ -c & a \end{bmatrix}$, $ad - bc \neq 0$

9. a) For example, $\oplus$ is associative since $(a \oplus b) \oplus c = (a + b - 1) \oplus c = (a + b - 1 + c) - 1 = (a + b + c) - 2$, and $a \oplus (b \oplus c) = a \oplus (b + c - 1) = (a + b + c - 1) - 1 = (a + b + c) - 2$. Also, 1 is the additive identity, and the additive inverse of any $a \in \mathbf{Z}$ is $2 - a$.
b) Yes   **c)** The unity is 0. The units are 0 and 2.
d) The ring is an integral domain, but not a field.

Section 14.2 – p. 572

1. Theorem 14.10(a). If $(S, +, \cdot)$ is a subring of R, then $a - b$, $ab \in S$ for any $a, b \in S$. Conversely, let $a \in S$. Then $a - a = z \in S$ and $z - a = -a \in S$. Also, if $b \in S$, then $-b \in S$, so $a - (-b) = a + b \in S$, and S is a subring by Theorem 14.9.

3. a) $(ab)(b^{-1}a^{-1}) = aua^{-1} = aa^{-1} = u$ and $(b^{-1}a^{-1})(ab) = b^{-1}ub = b^{-1}b = u$, so ab is a unit. Since the multiplicative inverse of a unit is unique, $(ab)^{-1} = b^{-1}a^{-1}$.

b) $A^{-1} = \begin{bmatrix} 2 & -7 \\ -1 & 4 \end{bmatrix}$ $B^{-1} = \begin{bmatrix} 1 & -2 \\ -2 & 5 \end{bmatrix}$ $(AB)^{-1} = \begin{bmatrix} 4 & -15 \\ -9 & 34 \end{bmatrix}$

$(BA)^{-1} = \begin{bmatrix} 16 & -39 \\ -9 & 22 \end{bmatrix}$ $B^{-1}A^{-1} = \begin{bmatrix} 4 & -15 \\ -9 & 34 \end{bmatrix}$

5. $(-a)^{-1} = -(a^{-1})$

7. a) The subrings are $\{x\}$, $\{x, y\}$, and $\{x, y, s, t\}$.
b) All of the subrings in part (a) are also ideals.

9. $z \in S, T \Rightarrow z \in S \cap T \Rightarrow S \cap T \neq \emptyset$. $a, b \in S \cap T \Rightarrow a, b \in S$ and $a, b \in T \Rightarrow a + b, ab \in S$ and $a + b$, $ab \in T \Rightarrow a + b$, $ab \in S \cap T$. $a \in S \cap T \Rightarrow a \in S$ and $a \in T \Rightarrow -a \in S$ and $-a \in T \Rightarrow -a \in S \cap T$. So $S \cap T$ is a subring of R.

11. b) $\begin{bmatrix} 1 & 0 \\ 0 & 1 \end{bmatrix}$ **c)** $\begin{bmatrix} 1 & 0 \\ 0 & 0 \end{bmatrix}$

d) S is an integral domain, while R is a noncommutative ring with unity.

13. Since $za = z$, $z \in N(a)$ and $N(a) \neq \emptyset$. If $r_1, r_2 \in N(a)$, then $(r_1 - r_2)a = r_1 a - r_2 a =$

$z - z = z$, so $r_1 - r_2 \in N(a)$. Finally, if $r \in N(a)$ and $s \in R$, then $(rs)a = (sr)a = s(ra) = sz = z$, so $rs, sr \in N(a)$. Hence $N(a)$ is an ideal, by Definition 14.6.

15. 2

17. a) $a = au \in aR$ since $u \in R$, so $aR \neq \emptyset$. If $ar_1, ar_2 \in aR$, then $ar_1 - ar_2 = a(r_1 - r_2) \in aR$. Also, for $ar_1 \in aR$ and $r \in R$, we have $r(ar_1) = (ar_1)r = a(r_1 r) \in aR$. Hence aR is an ideal of R.

b) Let $a \in R$, $a \neq z$. Then $a = au \in aR$ so $aR = R$. Since $u \in R = aR$, $u = ar$ for some $r \in R$, and $r = a^{-1}$. Hence R is a field.

19. a) $\binom{4}{2}(49)$ **b)** 7^4

c) Yes, the element (u, u, u, u) **d)** 4^4

21. b) If R has a unity u, define $a^0 = u$, for $a \in R$, $a \neq z$. If a is a unit of R, define a^{-n} as $(a^{-1})^n$, for $n \in \mathbf{Z}^+$.

Section 14.3 – p. 579

1. a) $-6, 1, 8, 15$ **b)** $-9, 2, 13, 24$ **c)** $-7, 10, 27, 44$

3. b) No, $2\mathcal{R}3$ and $3\mathcal{R}5$, but $5\not\mathcal{R}8$. Also, $2\mathcal{R}3$ and $2\mathcal{R}5$, but $4\not\mathcal{R}15$.

5. a) $[17]^{-1} = [831]$ **b)** $[100]^{-1} = [111]$ **c)** $[777]^{-1} = [735]$

7. (a) 16 units, 0 proper zero divisors; (b) 72 units, 44 proper zero divisors; (c) 1116 units, 0 proper zero divisors.

9. $[\binom{334}{3} + 2\binom{333}{3} + \binom{334}{1}\binom{333}{1}^2]/\binom{1000}{3}$

11. a) For $n = 0$ we have $10^0 = 1 = 1(-1)^0$, so $10^0 \equiv (-1)^0$ (mod 11). [Since $10 - (-1) = 11$, $10 \equiv (-1)$ (mod 11), or $10^1 \equiv (-1)^1$ (mod 11). Hence the result is also true for $n = 1$.] Assume the result true for $n = k \geq 1$ and consider the case for $k + 1$. Then, since $10^k \equiv (-1)^k$ (mod 11) and $10 \equiv (-1)$(mod 11), we have $10^{k+1} = 10^k \cdot 10 \equiv (-1)^k(-1) = (-1)^{k+1}$ (mod 11). The result now follows for all $n \in \mathbf{N}$, by the principle of mathematical induction.

b) If $x_n x_{n-1} \ldots x_2 x_1 x_0 = x_n \cdot 10^n + x_{n-1} \cdot 10^{n-1} + \cdots + x_2 \cdot 10^2 + x_1 \cdot 10 + x_0$ denotes an $(n + 1)$-digit integer, then

$$x_n x_{n-1} \ldots x_2 x_1 x_0 \equiv (-1)^n x_n + (-1)^{n-1} x_{n-1} + \cdots + x_2 - x_1 + x_0 \text{ (mod 11)}.$$

Proof:

$$x_n x_{n-1} \ldots x_2 x_1 x_0 = x_n \cdot 10^n + x_{n-1} \cdot 10^{n-1} + \cdots + x_2 \cdot 10^2 + x_1 \cdot 10 + x_0$$
$$\equiv x_n(-1)^n + x_{n-1}(-1)^{n-1} + \cdots + x_2(-1)^2 + x_1(-1) + x_0$$
$$= (-1)^n x_n + (-1)^{n-1} x_{n-1} + \cdots + x_2 - x_1 + x_0 \text{ (mod 11)}.$$

c) (i) $d = 1$ (ii) $d = 4$

13. Let $g = (a, n)$, $h = (b, n)$. $[a \equiv b \pmod{n}] \Rightarrow [a = b + kn$, for some $k \in \mathbf{Z}] \Rightarrow [g \mid b$ and $h \mid a]$. $[g \mid b$ and $g \mid n] \Rightarrow g \mid h$; $[h \mid a$ and $h \mid n] \Rightarrow h \mid g$. Since $g, h > 0$, it follows that $g = h$.

15. a) 112   **b)** $031 - 43 - 3464$

Section 14.4 – p. 586

1. $s \to 0, t \to 1, v \to 2, w \to 3, x \to 4, y \to 5$

3. Let $(R, +, \cdot), (S, \oplus, \odot)$, and $(T, +', \cdot')$ be the rings. For any $a, b \in R$, $(g \circ f)(a + b) = g(f(a + b)) = g(f(a) \oplus f(b)) = g(f(a)) +' g(f(b)) = (g \circ f)(a) +' (g \circ f)(b)$. Also, $(g \circ f)(a \cdot b) = g(f(a \cdot b)) = g(f(a) \odot f(b)) = g(f(a)) \cdot' g(f(b)) = (g \circ f)(a) \cdot' (g \circ f)(b)$. Hence $g \circ f$ is a ring homomorphism.

5. a) Since $f(z_R) = z_S$, it follows that $z_R \in K$ and $K \neq \emptyset$. If $x, y \in K$, then $f(x - y) = f(x + (-y)) = f(x) \oplus f(-y) = f(x) \ominus f(y) = z_S \ominus z_S = z_S$, so $x - y \in K$. Finally,

if $x \in K$ and $r \in R$, then $f(rx) = f(r) \odot f(x) = f(r) \odot z_s = z_s$, and
$f(xr) = f(x) \odot f(r) = z_s \odot f(r) = z_s$, so $rx, xr \in K$. Consequently, K is an ideal of R.

b) The kernel is $\{6n \mid n \in \mathbf{Z}\}$.

c) If f is one-to-one, then for any $x \in K$, $[f(x) = z_s = f(z_R)] \Rightarrow [x = z_R]$, so $K = \{z_R\}$. Conversely, if $K = \{z_R\}$, let $x, y \in R$ with $f(x) = f(y)$. Then
$z_s = f(x) \ominus f(y) = f(x - y)$, so $x - y \in K = \{z_R\}$. Consequently, $x - y = z_R \Rightarrow x = y$, and f is one-to-one.

7. a)

x (in $\mathbf{Z}_{20}$)	$f(x)$ (in $\mathbf{Z}_4 \times \mathbf{Z}_5$)	x (in $\mathbf{Z}_{20}$)	$f(x)$ (in $\mathbf{Z}_4 \times \mathbf{Z}_5$)
0	(0, 0)	10	(2, 0)
1	(1, 1)	11	(3, 1)
2	(2, 2)	12	(0, 2)
3	(3, 3)	13	(1, 3)
4	(0, 4)	14	(2, 4)
5	(1, 0)	15	(3, 0)
6	(2, 1)	16	(0, 1)
7	(3, 2)	17	(1, 2)
8	(0, 3)	18	(2, 3)
9	(1, 4)	19	(3, 4)

b) (i) $f((17)(19) + (12)(14)) = (1, 2)(3, 4) + (0, 2)(2, 4) = (3, 3) + (0, 3) = (3, 1)$, and $f^{-1}(3, 1) = 11$

(ii) $f((18)(11) - (9)(15)) = (2, 3)(3, 1) - (1, 4)(3, 0) = (2, 3) - (3, 0) = (3, 3)$, and $f^{-1}(3, 3) = 3$

9. a) 4   **b)** 1   **c)** No

11. No. $\mathbf{Z}_4$ has two units, while the ring in Example 14.3 has only one unit.

13. It is better to use the form in the program, since matrix multiplication is more accurate than matrix inversion. If the inverse is computed first, then round-off error may be compounded as the inverse is raised to the fourth power.

Miscellaneous Exercises – p. 589

1. a) False. Let $R = \mathbf{Z}$ and $S = \mathbf{Z}^+$.

b) False. Let $R = \mathbf{Z}$ and $S = \{2x \mid x \in \mathbf{Z}\}$.

c) False. Let $R = M_2(\mathbf{Z})$ and $S = \left\{ \begin{bmatrix} a & 0 \\ 0 & 0 \end{bmatrix} \,\middle|\, a \in \mathbf{Z} \right\}$.

d) and e) True.

f) False. $(\mathbf{Z}, +, \cdot)$ is a subring (but not a field) in $(\mathbf{Q}, +, \cdot)$.

g) False. For any prime p, $\{a/(p^n) \mid a, n \in \mathbf{Z}, n \geq 0\}$ is a subring in $(\mathbf{Q}, +, \cdot)$.

h) False. $f(6) = 12 = f(2 \cdot 3)$, but $f(2) \cdot f(3) = (4)(6) = 24$.

i) False. Consider the field in Table 14.6.   **j)** True.

k) False. Let $R = (\mathbf{Z}, +, \cdot)$, $S = \{2x \mid x \in \mathbf{Z}\}$, $T = \{3x \mid x \in \mathbf{Z}\}$.

l) False. In $\mathbf{Z}_5$, $1^2 = 4^2$ but $1 \neq 4$.

3. a) $[a + a = (a + a)^2 = a^2 + a^2 + a^2 + a^2 = (a + a) + (a + a)] \Rightarrow [a + a = 2a = z]$. Hence $-a = a$.

b) For any $a \in R$, $a + a = z \Rightarrow a = -a$. For $a, b \in R$, $(a + b) = (a + b)^2 = a^2 + ab + ba + b^2 = a + ab + ba + b \Rightarrow ab + ba = z \Rightarrow ab = -ba = ba$, so R is commutative.

5. Since $az = z = za$ for all $a \in R$, we have $z \in C$ and $C \neq \emptyset$. If $x, y \in C$, then $(x + y)a =$

$xa + ya = ax + ay = a(x + y)$, $(xy)a = x(ya) = x(ay) = (xa)y = (ax)y = a(xy)$, and
$(-x)a = -(xa) = -(ax) = a(-x)$, for all $a \in R$, so $x + y, xy, -x \in C$. Consequently,
C is a subring of R.

7. **b)** (i) $\mathbf{Z} = \{0\} \cup \{1, -1\} \cup \{2, -2\} \cup \{3, -3\} \cup \cdots$
   (ii) $\{a + bi \mid a, b \in \mathbf{Z}, i^2 = -1\} = \bigcup_{x,y \in \mathbf{Z}} \{x + yi, -x - yi, -y + xi, y - xi\}$
   (iii) $\mathbf{Z}_7 = \{0\} \cup \{1, 2, 3, 4, 5, 6\}$

9. Let $x = a_1 + b_1$, $y = a_2 + b_2$, for $a_1, a_2 \in A$ and $b_1, b_2 \in B$. Then
   $x - y = (a_1 - a_2) + (b_1 - b_2) \in A + B$. If $r \in R$ and $a + b \in A + B$, with $a \in A$ and
   $b \in B$, then $ra \in A$, $rb \in B$, and $r(a + b) \in A + B$. Similarly, $(a + b)r \in A + B$, and
   $A + B$ is an ideal of R.

11. Consider the numbers $x_1, x_1 + x_2, x_1 + x_2 + x_3, \ldots, x_1 + x_2 + x_3 + \cdots + x_n$. If one of
    these numbers is congruent to 0 modulo n, the result follows. If not, there exist
    $1 \le i < j \le n$ with $(x_1 + x_2 + \cdots + x_i) \equiv (x_1 + \cdots + x_i + x_{i+1} + \cdots + x_j)$ (mod n). Hence
    n divides $(x_{i+1} + \cdots + x_j)$.

CHAPTER 15

**Boolean Algebra
and Switching
Functions**

Section 15.1 – p. 598

1. **a)** 1   **b)** 1   **c)** 1   **d)** 1   **e)** 1

3. **a)** 2^n **b)** $2^{(2^n)}$

5. **a)** 2^{64} **b)** 2^6 **c)** 2^6 **d)** 2^{49} **e)** 2^7 **f)** 2^{32}; 2^8

7. $m + k = 2^n$

9. **a)** $y + x\bar{z}$ **b)** $x + y$ **c)** $wx + z$ **d)** $x_1 + x_2 + x_3 + x_4 + \cdots$

11. **a)** (i)

f	g	h	fg	$\bar{f}h$	gh	$fg + \bar{f}h + gh$	$fg + \bar{f}h$
0	0	0	0	0	0	0	0
0	0	1	0	1	0	1	1
0	1	0	0	0	0	0	0
0	1	1	0	1	1	1	1
1	0	0	0	0	0	0	0
1	0	1	0	0	0	0	0
1	1	0	1	0	0	1	1
1	1	1	1	0	1	1	1

Alternately, $fg + \bar{f}h = (fg + \bar{f})(fg + h) = (f + \bar{f})(g + \bar{f})(fg + h) =$
$1(g + \bar{f})(fg + h) = fgg + gh + \bar{f}fg + \bar{f}h = fg + gh + 0g + \bar{f}h = fg + gh + \bar{f}h$.
   (ii) $fg + f\bar{g} + \bar{f}g + \bar{f}\bar{g} = f(g + \bar{g}) + \bar{f}(g + \bar{g}) = f \cdot 1 + \bar{f} \cdot 1 = f + \bar{f} = 1$
   b) (i) $(f + g)(\bar{f} + h)(g + h) = (f + g)(\bar{f} + h)$
   (ii) $(f + g)(f + \bar{g})(\bar{f} + g)(\bar{f} + \bar{g}) = 0$

13. **a)** $f \oplus f = 0$; $f \oplus \bar{f} = 1$; $f \oplus 1 = \bar{f}$; $f \oplus 0 = f$
    b) (i) $f \oplus g = 0 \Leftrightarrow f\bar{g} + \bar{f}g = 0 \Rightarrow f\bar{g} = \bar{f}g = 0$. $[f = 1$ and $f\bar{g} = 0] \Rightarrow g = 1$. $[f = 0$ and
    $\bar{f}g = 0] \Rightarrow g = 0$. Hence $f = g$.
    (iii) $\bar{f} \oplus \bar{g} = \bar{f}\bar{\bar{g}} + \bar{\bar{f}}\bar{g} = \bar{f}g + f\bar{g} = f\bar{g} + \bar{f}g = f \oplus g$
    (iv) This is the only result that is not true. When f has value 1, g has value 0 and
    h value 1 (or g has value 1 and h value 0), then $f \oplus gh$ has value 1 but
    $(f \oplus g)(f \oplus h)$ has value 0.
    (v) $fg \oplus fh = \overline{fg}fh + fg\overline{fh} = (\bar{f} + \bar{g})fh + fg(\bar{f} + \bar{h}) = \bar{f}fh + \bar{f}gh + f\bar{f}g + fg\bar{h} =$
    $f\bar{g}h + fg\bar{h} = f(\bar{g}h + g\bar{h}) = f(g \oplus h)$

(vi) $\bar{f} \oplus g = \bar{\bar{f}} \bar{g} + fg = fg + \bar{f}\bar{g} = f \oplus \bar{g}$

$\overline{f \oplus g} = \overline{f}\overline{g} + \overline{f}\overline{g} = (\bar{f} + g)(f + \bar{g}) = \bar{f}\bar{g} + fg = \bar{f} \oplus g$

(vii) $[f \oplus g = f \oplus h] \Rightarrow [f \oplus (f \oplus g) = f \oplus (f \oplus h)] \Rightarrow [(f \oplus f) \oplus g =$
$(f \oplus f) \oplus h] \Rightarrow [\mathbf{0} \oplus g = \mathbf{0} \oplus h] \Rightarrow [g = h]$

Section 15.2 – p. 608

1. a) $x \oplus y = (x + y)(\overline{xy})$

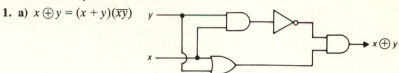

b) $\overline{xy}$

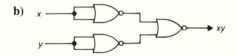

c) $\overline{x + y}$

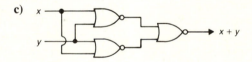

3. a)

b)

c)

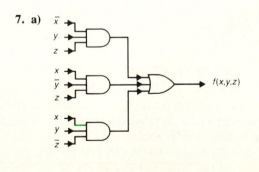

5. $f(w, x, y, z) = \overline{w}\,\overline{x}y\overline{z} + (w + x + \overline{y})z$

7. a)

b)

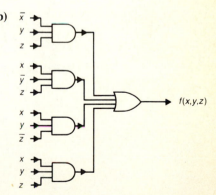

9. $f(w, x, y, z) = (w + y)(\overline{x} + y)(\overline{w} + \overline{y} + z)(x + \overline{y} + z)$

11. a) 64   **b)** 32   **c)** 16   **d)** 8

Section 15.3 – p. 613

1. $uv + wvy + uxz + uyz + wz$

3. Parts (a) and (c) are true.

   b) In Fig. 15.11, let $D_1 = \{c, d\}$, $D_2 = \{a, g\}$.

   d) Let $D_1 = \{a, d\}$, $D_2 = \{a\}$ in Fig. 15.11.

   e) Let $D_1 = \{a\}$, $D_2 = \{d\}$ in Fig. 15.11.

5. **a)** $f(w, x, y, z) = z$

   b) $f(w, x, y, z) = \bar{x}\,\bar{y}\,\bar{z} + x\bar{y}z + xy\bar{z}$

   c) $f(v, w, x, y, z) = v\bar{y}\,\bar{z} + \overline{w}\,\overline{x}yz + \bar{v}\,\overline{w}\,\bar{z} + \bar{u}x\bar{y}$

Section 15.4 – p. 623

3. **a)** 30   **b)** 30   **c)** 1   **d)** 21   **e)** 30   **f)** 70

5. $y \le x$

7. From Theorem 15.5(a), with x_1, x_2 distinct atoms, if $x_1 x_2 \ne 0$, then $x_1 = x_1 x_2 = x_2 x_1 = x_2$, a contradiction.

9. No. Let $\mathcal{U} = \{a, b, c\}$, and $\mathcal{B} = (\mathcal{P}(\mathcal{U}), \cup, \cap, \bar{\ }, \emptyset, \mathcal{U})$. If $\mathcal{B}_1 = \{\emptyset, \mathcal{U}, \{a\}, \{b\}, \{a, b\}\}$, then $\mathcal{B}_1$ satisfies the stated conditions, but it is not a subalgebra of $\mathcal{B}$. For example, $\overline{\{a\}} \notin \mathcal{B}_1$.

11. **a)** $f(0) = f(x\bar{x})$ for any $x \in \mathcal{B}_1$. $f(x\bar{x}) = f(x)f(\bar{x}) = f(x)\overline{f(x)} = 0$.

   b) Follows from (a) by duality.

   c) $x \le y \Leftrightarrow xy = x \Rightarrow f(xy) = f(x) \Rightarrow f(x)f(y) = f(x) \Leftrightarrow f(x) \le f(y)$

   e) $S_1 \ne \emptyset \Rightarrow f(S_1) \ne \emptyset$. Let $x_2, y_2 \in f(S_1)$ with $x_1, y_1 \in S_1$ and $f(x_1) = x_2, f(y_1) = y_2$. Then $f(x_1 + y_1) = f(x_1) + f(y_1) = x_2 + y_2$, and $x_1 + y_1 \in S_1$. Hence $x_2 + y_2 \in f(S_1)$. Also, $\overline{x_2} = \overline{f(x_1)} = f(\overline{x_1})$, and since $\overline{x_1} \in S_1$, it follows that $\overline{x_2} \in f(S_1)$. By Exercise 8(d), $f(S_1)$ is a subalgebra of $\mathcal{B}_2$.

13. **a)** $f(xy) = f(\overline{\overline{x} + \overline{y}}) = \overline{f(\overline{x} + \overline{y})} = \overline{f(\overline{x}) + f(\overline{y})} = \overline{f(\overline{x})} \cdot \overline{f(\overline{y})} = f(\overline{\overline{x}}) \cdot f(\overline{\overline{y}}) = f(x) \cdot f(y)$

   b) Let $\mathcal{B}_1$, $\mathcal{B}_2$ be Boolean algebras with $f: \mathcal{B}_1 \to \mathcal{B}_2$ one-to-one and onto. Then f is an isomorphism if $f(\bar{x}) = \overline{f(x)}$ and $f(xy) = f(x)f(y)$ for all $x, y \in \mathcal{B}_1$. (Follows from (a) by duality.)

15. For any $1 \le i \le n$, $(x_1 + x_2 + \cdots + x_n)x_i = x_1 x_i + x_2 x_i + \cdots + x_{i-1}x_i + x_i x_i + x_{i+1}x_i + \cdots + x_n x_i = 0 + 0 + \cdots + 0 + x_i + 0 + \cdots + 0 = x_i$, by part (b) of Theorem 15.5. Consequently, it follows from Theorem 15.7 that $(x_1 + x_2 + \cdots + x_n)x = x$ for all $x \in \mathcal{B}$. Since the one element is unique (from Exercise 10), we conclude that $1 = x_1 + x_2 + \cdots + x_n$.

Miscellaneous Exercises – p. 625

1. **a)** (i) When $n = 2$, $x_1 + x_2$ denotes the Boolean sum of x_1 and x_2. For $n \ge 2$, we define $x_1 + x_2 + \cdots + x_n + x_{n+1}$ recursively by $(x_1 + x_2 + \cdots + x_n) + x_{n+1}$. [A similar definition can be given for the Boolean product.] For $n = 2$, $\overline{x_1 + x_2} = \overline{x_1}\,\overline{x_2}$ is true; this is one of the DeMorgan Laws. Assume the result for $n = k \, (\ge 2)$ and consider the case of $n = k + 1$.

$$\overline{(x_1 + x_2 + \cdots + x_k + x_{k+1})} = \overline{(x_1 + x_2 + \cdots + x_k) + x_{k+1}}$$
$$= \overline{(x_1 + x_2 + \cdots + x_k)}\,\overline{x_{k+1}}$$
$$= \overline{x_1}\,\overline{x_2}\cdots\overline{x_k}\,\overline{x_{k+1}}$$

Consequently, the result follows for all $n \ge 2$, by the principle of finite induction.

(ii) Follows from (i) by duality.

(iii) $(\sum_{i=1}^{n} x_i)(\overline{\prod_{i=1}^{n} x_i}) = (x_1 + x_2 + \cdots + x_n)(\overline{x_1} + \overline{x_2} + \cdots + \overline{x_n})$. This has value 0 if $x_i = 0$ for all $1 \le i \le n$, or if $\overline{x_i} = 0$ for all $1 \le i \le n$. In either case, $x_1 \overline{x_2} + x_2 \overline{x_3} + \cdots + x_n \overline{x_1}$ has value 0. In every other case, the expression has value 1 and there exist $x_i, x_{i+1}, 1 \le i \le n-1$, where one has the value 0 and the other 1, or x_n has value 0(1) while x_1 has value 1(0). In these cases, $x_1 \overline{x_2} + x_2 \overline{x_3} + \cdots + x_n \overline{x_1}$ also has value 1.

b) $(\prod_{i=1}^{n} x_i)(\overline{\sum_{i=1}^{n} x_i}) = (x_1 + \overline{x_2})(x_2 + \overline{x_3}) \cdots (x_{n-1} + \overline{x_n})(x_n + \overline{x_1})$

3. She can invite only Nettie and Cathy.

5. a) $f(w, x, y, z) = \overline{w}\,\overline{x} + xy$

   b) $g(v, w, x, y, z) = \overline{v}\,\overline{w}\,yz + xz + w\overline{y}\,\overline{z} + \overline{x}\,\overline{y}\,\overline{z}$

7. a) $2^{(2^{n-1})}$ **b)** $2^4; 2^{n+1}$

9. a) If $n = 60$, there are 12 divisors, and no Boolean algebra contains 12 elements since 12 is not a power of 2.

   b) If $n = 120$, there are 16 divisors. However, if $x = 4$, then $\overline{x} = 30$ and $x \cdot \overline{x} = (x, \overline{x}) = (4, 30) = 2$, which is not the zero element. So the Inverse Laws are not satisfied.

11. If $c \le a$, then $ac = c$, so $ab + c = ab + ac = a(b + c)$. Conversely, if $ab + c = a(b + c) = ab + ac$, then $ac = ac + 0 = ac + (ab + \overline{ab}) = (ab + ac) + \overline{ab} = (ab + c) + \overline{ab} = c + (ab + \overline{ab}) = c$, and $ac = c \Rightarrow c \le a$.

CHAPTER 16

Groups, Coding Theory, and Polya's Method of Enumeration

Section 16.1 – p. 634

1. a) Yes. The identity is 1 and each element is its own inverse.

   b) No. The set is not closed under addition and there is no identity.

   c) No. The set is not closed under addition.

   d) Yes. The identity is 0; the inverse of $10n$ is $10(-n)$ or $-10n$.

   e) No. The function $f: A \to A$ where $f(x) = 1$ for all $x \in A$ is not invertible.

   f) Yes. The identity is 1_A and the inverse of $g: A \to A$ is $g^{-1}: A \to A$.

   g) Yes. The identity is 0; the inverse of $a/(2^n)$ is $(-a)/(2^n)$.

3. Subtraction is not an associative binary operation. For example, $(3 - 2) - 4 = -3 \ne 5 = 3 - (2 - 4)$.

5. Since $x, y \in \mathbf{Z} \Rightarrow x + y + 1 \in \mathbf{Z}$, the operation is a binary operation (or $\mathbf{Z}$ is closed under $\circ$). For any $w, x, y \in \mathbf{Z}$, $w \circ (x \circ y) = w \circ (x + y + 1) = w + (x + y + 1) + 1 = (w + x + 1) + y + 1 = (w \circ x) \circ y$, so the binary operation is associative. Furthermore, $x \circ y = x + y + 1 = y + x + 1 = y \circ x$, for all $x, y \in \mathbf{Z}$, so $\circ$ is also commutative. If $x \in \mathbf{Z}$ then $x \circ (-1) = x + (-1) + 1 = x [= (-1) \circ x]$, so -1 is the identity element for $\circ$. And finally, for any $x \in \mathbf{Z}$, we have $-x - 2 \in \mathbf{Z}$ and $x \circ (-x - 2) = x + (-x - 2) + 1 = -1 [= (-x - 2) + x]$, so $-x - 2$ is the inverse for x under $\circ$. Consequently, $(\mathbf{Z}, \circ)$ is an abelian group.

7. a) The result follows from Theorem 16.1(b) because both $(a^{-1})^{-1}$ and a are inverses of a^{-1}.

   b) $\quad (b^{-1}a^{-1})(ab) = b^{-1}(a^{-1}a)b = b^{-1}(e)b = b^{-1}b = e$ and $(ab)(b^{-1}a^{-1})$

$$= a(bb^{-1})a^{-1} = a(e)a^{-1} = aa^{-1} = e$$

   So $b^{-1}a^{-1}$ is an inverse of ab, and by Theorem 16.1(b), $(ab)^{-1} = b^{-1}a^{-1}$

9. a) $\{0\}; \{0, 6\}; \{0, 4, 8\}; \{0, 3, 6, 9\}; \{0, 2, 4, 6, 8, 10\}; \mathbf{Z}_{12}$

   b) $\{1\}; \{1, 10\}; \{1, 3, 4, 5, 9\}; \mathbf{Z}_{11}^{*}$

   c) $\{\pi_0\}; \{\pi_0, \pi_1, \pi_2\}; \{\pi_0, r_1\}; \{\pi_0, r_2\}; \{\pi_0, r_3\}; S_3$

11. a) There are 10: five rotations through $i(72°)$, $0 \le i \le 4$, and five reflections about lines containing a vertex and the midpoint of the opposite side.

b) For a regular n-gon ($n \ge 3$) there are $2n$ rigid motions. There are the n rotations through $i(360°/n)$, $0 \le i \le n-1$. There are n reflections. For n odd, each reflection is about a line through a vertex and the midpoint of the opposite side. For n even, there are $n/2$ reflections about lines through opposite vertices and $n/2$ reflections about lines through the midpoints of opposite sides.

13. Since $eg = ge$ for all $g \in G$, it follows that $e \in H$ and $H \ne \emptyset$. If $x, y \in H$, then $xg = gx$ and $yg = gy$ for all $g \in G$. Consequently, $(xy)g = x(yg) = x(gy) = (xg)y = (gx)y = g(xy)$ for all $g \in G$, and we have $xy \in H$. Finally, for any $x \in H$, $g \in G$, $xg^{-1} = g^{-1}x$. So $(xg^{-1})^{-1} = (g^{-1}x)^{-1}$, or $gx^{-1} = x^{-1}g$, and $x^{-1} \in H$. Therefore H is a subgroup of G.

15. b) (i) 216

(ii) $H_1 = \{(x,0,0) \mid x \in \mathbf{Z}_6\}$ is a subgroup of order 6

$H_2 = \{(x,y,0) \mid x,y \in \mathbf{Z}_6, y = 0, 3\}$ is a subgroup of order 12

$H_3 = \{(x,y,0) \mid x,y \in \mathbf{Z}_6\}$ has order 36

(iii) $-(2,3,4) = (4,3,2)$; $-(4,0,2) = (2,0,4)$; $-(5,1,2) = (1,5,4)$

Section 16.2 – p. 638

1. b) $f(a^{-1}) \cdot f(a) = f(a^{-1} \cdot a) = f(e_G) = e_H$ and $f(a) \cdot f(a^{-1}) = f(a \cdot a^{-1}) = f(e_G) = e_H$, so $f(a^{-1})$ is an inverse of $f(a)$. By the uniqueness of inverses [Theorem 16.1(b)], it follows that $f(a^{-1}) = [f(a)]^{-1}$.

3. $f(0) = (0,0) \qquad f(1) = (1,1) \qquad f(2) = (2,0)$
$f(3) = (0,1) \qquad f(4) = (1,0) \qquad f(5) = (2,1)$

5. a) $o(\pi_0) = 1$, $o(\pi_1) = o(\pi_2) = 3$, $o(r_1) = o(r_2) = o(r_3) = 2$

b) (See Fig. 16.6) $o(\pi_0) = 1$, $o(\pi_1) = o(\pi_3) = 4$, $o(\pi_2) = o(r_1) = o(r_2) = o(r_3) = o(r_4) = 2$

7. $\mathbf{Z}_5^* = \langle 2 \rangle = \langle 3 \rangle$; $\mathbf{Z}_7^* = \langle 3 \rangle = \langle 5 \rangle$; $\mathbf{Z}_{11}^* = \langle 2 \rangle = \langle 6 \rangle = \langle 7 \rangle = \langle 8 \rangle$

9. This result is false. The Klein Four group of Example 16.13 provides a counterexample.

11. a) $(\mathbf{Z}_{12}, +) = \langle 1 \rangle = \langle 5 \rangle = \langle 7 \rangle = \langle 11 \rangle$

$(\mathbf{Z}_{16}, +) = \langle 1 \rangle = \langle 3 \rangle = \langle 5 \rangle = \langle 7 \rangle = \langle 9 \rangle = \langle 11 \rangle = \langle 13 \rangle = \langle 15 \rangle$

$(\mathbf{Z}_{24}, +) = \langle 1 \rangle = \langle 5 \rangle = \langle 7 \rangle = \langle 11 \rangle = \langle 13 \rangle = \langle 17 \rangle = \langle 19 \rangle = \langle 23 \rangle$

b) Let $G = \langle a^k \rangle$. Since $G = \langle a \rangle$, $a = (a^k)^s$ for some $s \in \mathbf{Z}$. Then $a^{1-ks} = e$, so $1 - ks = tn$ since $o(a) = n$. $1 - ks = tn \Rightarrow 1 = ks + tn \Rightarrow (k,n) = 1$. Conversely, let $G = \langle a \rangle$ where $a^k \in G$ and $(k,n) = 1$. Then $\langle a^k \rangle \subseteq G$. $(k,n) = 1 \Rightarrow 1 = ks + tn$, for some $s, t \in \mathbf{Z} \Rightarrow a = a^1 = a^{ks+nt} = (a^k)^s (a^n)^t = (a^k)^s (e)^t = (a^k)^s \in \langle a^k \rangle$. Hence $G \subseteq \langle a^k \rangle$. So $G = \langle a^k \rangle$, or a^k generates G.

c) $\phi(n)$.

Section 16.3 – p. 641

1. a) $\{ \begin{pmatrix}1&2&3&4\\2&3&4&1\end{pmatrix}, \begin{pmatrix}1&2&3&4\\3&4&1&2\end{pmatrix}, \begin{pmatrix}1&2&3&4\\4&1&2&3\end{pmatrix}, \begin{pmatrix}1&2&3&4\\1&2&3&4\end{pmatrix} \}$

b) $\begin{pmatrix}1&2&3&4\\2&1&4&3\end{pmatrix}H = \{ \begin{pmatrix}1&2&3&4\\3&2&1&4\end{pmatrix}, \begin{pmatrix}1&2&3&4\\4&3&2&1\end{pmatrix}, \begin{pmatrix}1&2&3&4\\1&4&3&2\end{pmatrix}, \begin{pmatrix}1&2&3&4\\2&1&4&3\end{pmatrix} \}$

$\begin{pmatrix}1&2&3&4\\4&2&3&1\end{pmatrix}H = \{ \begin{pmatrix}1&2&3&4\\1&3&4&2\end{pmatrix}, \begin{pmatrix}1&2&3&4\\2&4&1&3\end{pmatrix}, \begin{pmatrix}1&2&3&4\\3&1&2&4\end{pmatrix}, \begin{pmatrix}1&2&3&4\\4&2&3&1\end{pmatrix} \}$

$\begin{pmatrix}1&2&3&4\\1&2&4&3\end{pmatrix}H = \{ \begin{pmatrix}1&2&3&4\\2&3&1&4\end{pmatrix}, \begin{pmatrix}1&2&3&4\\3&4&2&1\end{pmatrix}, \begin{pmatrix}1&2&3&4\\4&1&3&2\end{pmatrix}, \begin{pmatrix}1&2&3&4\\1&2&4&3\end{pmatrix} \}$

$\begin{pmatrix}1&2&3&4\\1&3&2&4\end{pmatrix}H = \{ \begin{pmatrix}1&2&3&4\\2&4&3&1\end{pmatrix}, \begin{pmatrix}1&2&3&4\\3&1&4&2\end{pmatrix}, \begin{pmatrix}1&2&3&4\\4&2&1&3\end{pmatrix}, \begin{pmatrix}1&2&3&4\\1&3&2&4\end{pmatrix} \}$

$\begin{pmatrix}1&2&3&4\\1&4&2&3\end{pmatrix}H = \{ \begin{pmatrix}1&2&3&4\\2&1&3&4\end{pmatrix}, \begin{pmatrix}1&2&3&4\\3&2&4&1\end{pmatrix}, \begin{pmatrix}1&2&3&4\\4&3&1&2\end{pmatrix}, \begin{pmatrix}1&2&3&4\\1&4&2&3\end{pmatrix} \}$

$\begin{pmatrix}1&2&3&4\\1&2&3&4\end{pmatrix}H = H$

3. 12

5. a) Since $(0,0) \in H$, it follows that $H \neq \emptyset$. Let $(a,0), (b,0) \in H$. Then $(a,0) + (b,0) = (a+b,0) \in H$. Also $(-a,0) \in H$ for any $(a,0) \in H$. Hence H is a subgroup.

    b) Each coset consists of the points on a horizontal line.

7. Let $o(a) = k$. Then $|\langle a \rangle| = k$, so by Lagrange's Theorem, k divides n. Hence $a^n = a^{km} = (a^k)^m = e^m = e$.

9. a) If H is a proper subgroup of G, then by Lagrange's Theorem, $|H|$ is 2 or p. If $|H| = 2$, then $H = \{e, x\}$ where $x^2 = e$, so $H = \langle x \rangle$. If $|H| = p$, let $y \in H$, $y \neq e$. Then $o(y) = p$, so $H = \langle y \rangle$.

    b) Let $x \in G$, $x \neq e$. Then $o(x) = p$ or $o(x) = p^2$. If $o(x) = p$, then $|\langle x \rangle| = p$. If $o(x) = p^2$, then $G = \langle x \rangle$ and $\langle x^p \rangle$ is a subgroup of G of order p.

11. b) Let $x \in H \cap K$. If the order of x is r, then r must divide both m and n. Since $(m,n) = 1$, it follows that $r = 1$, so $x = e$ and $H \cap K = \{e\}$.

Section 16.4 — p. 646

1. a) $e = 0001001$ **b)** $r = 1111011$ **c)** $c = 0101000$

3. a) (i) $D(111101100) = 101$ (ii) $D(000100011) = 000$
    (iii) $D(010011111) = 011$

    b) 000000000, 000000001, 100000000

    c) 64

Sections 16.5 and 16.6 — p. 653

1. $S(101010, 1) = \{101010, 001010, 111010, 100010, 101110, 101000, 101011\}$
$S(111111, 1) = \{111111, 011111, 101111, 110111, 111011, 111101, 111110\}$

3. a) $|S(x,1)| = 11$; $|S(x,2)| = 56$; $|S(x,3)| = 176$

    b) $|S(x,k)| = 1 + \binom{n}{1} + \binom{n}{2} + \cdots + \binom{n}{k} = \sum_{i=0}^{k} \binom{n}{i}$

5. a) The minimum distance between code words is 3. The code can detect all errors of weight ≤ 2 and can correct all single errors.

    b) The minimum distance between code words is 5. The code can detect all errors of weight ≤ 4 and can correct all errors of weight ≤ 2.

    c) The minimum distance is 2. The code detects all single errors but has no correction capability.

    d) The minimum distance is 3. The code detects all errors of weight ≤ 2 and corrects all single errors.

7. a) $C = \{00000, 10110, 01011, 11101\}$. The minimum distance between code words is 3, so the code can detect all errors of weight ≤ 2 and can correct all single errors.

    b) $H = \begin{bmatrix} 1 & 0 & 1 & 0 & 0 \\ 1 & 1 & 0 & 1 & 0 \\ 0 & 1 & 0 & 0 & 1 \end{bmatrix}$

    c) (i) 01   (ii) 11   (v) 11   (vi) 10

        For (iii) and (iv) the syndrome is $(111)^{tr}$ which is not a column of H. Assuming a double error, if $(111)^{tr} = (110)^{tr} + (001)^{tr}$, then the decoded received word is 01 [for (iii)] and 10 [for (iv)]. If $(111)^{tr} = (011)^{tr} + (100)^{tr}$, we get 10 [for (iii)] and 01 [for (iv)].

9. $G = [I_8 | A]$ where I_8 is the 8×8 multiplicative identity matrix and A is a column of eight 1's. $H = [A^{tr} | 1] = [11111111 | 1]$.

11. Compare the generator (parity-check) matrix in Exercise 9 with the parity-check (generator) matrix in Exercise 10.

Sections 16.7 and 16.8 – p. 660

1. $\binom{256}{2}$; 255

3. a)

Syndrome	Coset Leader			
000	00000	10110	01011	11101
110	10000	00110	11011	01101
011	01000	11110	00011	10101
100	00100	10010	01111	11001
010	00010	10100	01001	11111
001	00001	10111	01010	11100
101	11000	01110	10011	00101
111	01100	11010	00111	10001

(The last two rows are not unique.)

b)

Received Word	Code Word	Decoded Message
11110	10110	10
11101	11101	11
11011	01011	01
10100	10110	10
10011	01011	01
10101	11101	11
11111	11101	11
01100	00000	00

5. a) G is 57×63; H is 6×63 **b)** The rate is $\frac{57}{63}$.

7. a) $(0.99)^7 + \binom{7}{1}(0.99)^6(0.01)$ **b)** $[(0.99)^7 + \binom{7}{1}(0.99)^6(0.01)]^5$

Section 16.9 – p. 667

1. a) $\pi_2^* =$

$$\begin{pmatrix} C_1 & C_2 & C_3 & C_4 & C_5 & C_6 & C_7 & C_8 & C_9 & C_{10} & C_{11} & C_{12} & C_{13} & C_{14} & C_{15} & C_{16} \\ C_1 & C_4 & C_5 & C_2 & C_3 & C_8 & C_9 & C_6 & C_7 & C_{10} & C_{11} & C_{14} & C_{15} & C_{12} & C_{13} & C_{16} \end{pmatrix}$$

$r_4^* =$

$$\begin{pmatrix} C_1 & C_2 & C_3 & C_4 & C_5 & C_6 & C_7 & C_8 & C_9 & C_{10} & C_{11} & C_{12} & C_{13} & C_{14} & C_{15} & C_{16} \\ C_1 & C_4 & C_3 & C_2 & C_5 & C_9 & C_8 & C_7 & C_6 & C_{10} & C_{11} & C_{12} & C_{15} & C_{14} & C_{13} & C_{16} \end{pmatrix}$$

b) $(\pi_1^{-1})^* =$

$$\begin{pmatrix} C_1 & C_2 & C_3 & C_4 & C_5 & C_6 & C_7 & C_8 & C_9 & C_{10} & C_{11} & C_{12} & C_{13} & C_{14} & C_{15} & C_{16} \\ C_1 & C_5 & C_2 & C_3 & C_4 & C_9 & C_6 & C_7 & C_8 & C_{11} & C_{10} & C_{15} & C_{12} & C_{13} & C_{14} & C_{16} \end{pmatrix}$$

$= (\pi_1^*)^{-1}$

c) $\pi_3^* r_4^* =$

$$\begin{pmatrix} C_1 & C_2 & C_3 & C_4 & C_5 & C_6 & C_7 & C_8 & C_9 & C_{10} & C_{11} & C_{12} & C_{13} & C_{14} & C_{15} & C_{16} \\ C_1 & C_5 & C_4 & C_3 & C_2 & C_6 & C_9 & C_8 & C_7 & C_{11} & C_{10} & C_{13} & C_{12} & C_{15} & C_{14} & C_{16} \end{pmatrix}$$

$= (\pi_3 r_4)^*$

3. a) $o(\alpha) = 7$; $o(\beta) = 12$; $o(\gamma) = 3$; $o(\delta) = 6$

b) Let $\alpha \in S_n$, with $\alpha = c_1 c_2 \dots c_k$, a product of disjoint cycles. Then $o(\alpha)$ is the l.c.m. of $\ell(c_1), \ell(c_2), \dots, \ell(c_k)$, where $\ell(c_i) =$ length of c_i, $1 \le i \le k$.

5. a) 8   **b)** 39

7. a) 70   **b)** 55

9. Triangular Figure: **a)** 8   **b)** 8
Square Figure: **a)** 12   **b)** 12

11. a) 140   **b)** 102
13. 315

Section 16.10 – p. 672

1. a) 165   **b)** 120
3. Triangular Figure: **a)** 96   **b)** 80
     Square Figure: **a)** 280   **b)** 220
     Hexagonal Figure: **a)** 131,584   **c)** 70,144
5. a) 2635   **b)** 1505

   c)

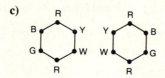

7. a) 21   **b)** 954
   c) No: $k = 21$ and $m = 21$, so $km = 441 \neq 954 = n$. Here the location of a certain edge
     must be considered relative to the location of the vertices. For example,

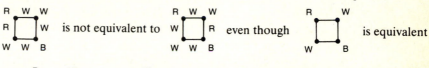

Section 16.11 – p. 677

1. a) (i) and (ii) $r^4 + w^4 + r^3 w + 2r^2 w^2 + rw^3$
   b) (i) $(1/4)[(r + b + w)^4 + 2(r^4 + b^4 + w^4) + (r^2 + b^2 + w^2)^2]$
     (ii) $(1/8)[(r + b + w)^4 + 2(r^4 + b^4 + w^4) + 3(r^2 + b^2 + w^2)^2$
               $+ 2(r + b + w)^2(r^2 + b^2 + w^2)]$
3. a) 10
   b) $(1/24)[(r + w)^6 + 6(r + w)^2(r^4 + w^4) + 3(r + w)^2(r^2 + w^2)^2 + 6(r^2 + w^2)^3$
               $+ 8(r^3 + w^3)^2]$
   c) 2
5. Let g = green and y = gold.
   Triangular Figure: $(1/6)[(g + y)^4 + 2(g + y)(g^3 + y^3) + 3(g + y)^2(g^2 + y^2)]$
   Square Figure: $(1/8)[(g + y)^5 + 2(g + y)(g^4 + y^4) + 3(g + y)(g^2 + y^2)^2$
                   $+ 2(g + y)^3(g^2 + y^2)]$
   Hexagonal Figure: $(1/4)[(g + y)^9 + 2(g + y)(g^2 + y^2)^4 + (g + y)^5(g^2 + y^2)^2]$
7. a) 136
   b) $(1/2)[(r + w)^8 + (r^2 + w^2)^4]$
   c) 38; 16
9. $\binom{m+n-1}{n}$

Miscellaneous Exercises – p. 681

1. a) Since $f(e_G) = e_H$, it follows that $e_G \in K$ and $K \neq \emptyset$. If $x, y \in K$, then $f(x) = f(y) = e_H$ and $f(xy) = f(x)f(y) = e_H e_H = e_H$, so $xy \in K$. Also, for $x \in K$, $f(x^{-1}) = [f(x)]^{-1} = e_H^{-1} = e_H$, so $x^{-1} \in K$. Hence K is a subgroup of G.

b) If $x \in K$, then $f(x) = e_H$. For all $g \in G$,
$$f(gxg^{-1}) = f(g)f(x)f(g^{-1}) = f(g)e_H f(g^{-1}) = f(g)f(g^{-1}) = f(gg^{-1}) = f(e_G) = e_H.$$
Hence, for all $x \in K$, $g \in G$, we find that $gxg^{-1} \in K$.

3. Let $a, b \in G$. Then $a^2 b^2 = ee = e = (ab)^2 = abab$. But $a^2 b^2 = abab \Rightarrow aabb = abab \Rightarrow ab = ba$, so G is abelian.

5. If T is a subgroup of H, then $e_H \in T$ and $f(e_G) = e_H$, so $e_G \in f^{-1}(T)$. Hence $f^{-1}(T) \neq \emptyset$. If $x, y \in f^{-1}(T)$, then $f(x), f(y) \in T$. T a subgroup of $H \Rightarrow f(x)f(y) = f(xy) \in T \Rightarrow xy \in f^{-1}(T)$. Also, $x \in f^{-1}(T) \Rightarrow f(x) \in T \Rightarrow [f(x)]^{-1} \in T \Rightarrow f(x^{-1}) \in T \Rightarrow x^{-1} \in f^{-1}(T)$. Consequently, $f^{-1}(T)$ is a subgroup of G.

7. Let $G = \langle g \rangle$ and let $h = f(g)$. If $h_1 \in H$, then $h_1 = f(g^n)$ for some $n \in \mathbf{Z}$, since f is onto. Therefore, $h_1 = f(g^n) = [f(g)]^n = h^n$, and $H = \langle h \rangle$.

9. a) For any $x \in G$, $x = exe^{-1} \Rightarrow x \mathcal{R} x$, so $\mathcal{R}$ is reflexive. If $x, y \in G$ and $x \mathcal{R} y$, we find that $x \mathcal{R} y \Rightarrow x = gyg^{-1} \Rightarrow g^{-1}x(g^{-1})^{-1} = y \Rightarrow y \mathcal{R} x$, so $\mathcal{R}$ is symmetric. Finally, let $w, x, y \in G$ with $w \mathcal{R} x$ and $x \mathcal{R} y$. Then $w = hxh^{-1}$ and $x = gyg^{-1}$ for some $h, g \in G$, so $w = h(gyg^{-1})h^{-1} = (hg)y(g^{-1}h^{-1}) = (hg)y(hg)^{-1}$ and $w \mathcal{R} y$. Consequently $\mathcal{R}$ is transitive. Since $\mathcal{R}$ is reflexive, symmetric, and transitive, it is an equivalence relation.

b) (i) $S_3 = \{\pi_0\} \cup \{\pi_1, \pi_2 = r_1 \pi_1 r_1^{-1}\} \cup \{r_1, r_2 = \pi_1 r_1 \pi_1^{-1}, r_3 = \pi_2 r_1 \pi_2^{-1}\}$

(ii) $G = \{\pi_0\} \cup \{\pi_1, \pi_3\} \cup \{\pi_2\} \cup \{r_1, r_2\} \cup \{r_3, r_4\}$ (iii). $G = \{e\} \cup \{a\} \cup \{b\} \cup \{c\}$

(iv) Let $G = \{x_1, x_2, \ldots, x_n\}$, $n \geq 1$. The partition induced by $\mathcal{R}$ is $G = \{x_1\} \cup \{x_2\} \cup \cdots \cup \{x_n\}$.

c) Let $o(x) = m$, $o(y) = n$, with $x = gyg^{-1}$. Then $x^n = (gyg^{-1})^n = (gyg^{-1})(gyg^{-1}) \cdots (gyg^{-1}) = (gy)(g^{-1}g)y(g^{-1}g) \cdots (g^{-1}g)(yg^{-1}) = gy^n g^{-1} = geg^{-1} = gg^{-1} = e$, so $m \mid n$. But $x = gyg^{-1} \Rightarrow y = g^{-1}xg$, and $y^m = (g^{-1}xg)^m = g^{-1}x^m g = e$, so $n \mid m$. Hence $o(x) = m = n = o(y)$.

The converse of this result is false. The partition in subsection (ii) of part (b) provides a counterexample.

11. a) Consider a permutation σ that is counted in $P(n+1, k)$. If $(n+1)$ is a cycle (of length 1) in σ, then σ is counted in $P(n, k-1)$. Otherwise, consider any permutation τ that is counted in $P(n, k)$. For any cycle of τ, say $(a_1 a_2 \ldots a_r)$, there are r locations in which to place $n+1$—(1) Between a_1 and a_2; (2) Between a_2 and a_3; $\ldots$; $(r-1)$ Between a_{r-1} and a_r; and (r) Between a_r and a_1. Hence there are n locations, in total, to locate $n+1$ in τ. Consequently, $P(n+1, k) = P(n, k-1) + nP(n, k)$.

b) $\sum_{k=1}^{n} P(n, k)$ counts all of the permutations in S_n, which has $n!$ elements.

13. a) $d(a+c, b+c) = wt(a+c+b+c) = wt(a+b+(c+c))$
$$= wt(a+b+\mathbf{0}) = wt(a+b) = d(a, b)$$

b) Let $c_1, c_2 \in C$ with $d(a, c_1), d(b, c_2)$ minimal.
$$d(a, c_1) = wt(a+c_1) = wt(b+a+b+c_1) = d(b, a+b+c_1) \geq d(b, c_2)$$
$$d(b, c_2) = wt(b+c_2) = wt(a+b+a+c_2) = d(a, b+a+c_2) \geq d(a, c_1)$$
Hence $d(a, C) = d(a, c_1) = d(b, c_2) = d(b, C)$.

15. b) 20

17. a) (i) $\binom{4}{2}(0.75)^2(0.25)^2$

(ii) $\binom{4}{2}(0.75)^2(0.25)^2 + \binom{4}{3}(0.75)^3(0.25) + \binom{4}{4}(0.75)^4$

b) At least 3

CHAPTER 17

Finite Fields and Combinatorial Designs

Section 17.1 – p. 691

1. $(10)(11)^2$; $(10)(11)^3$; $(10)(11)^4$; $(10)(11)^n$

5. a) and **b)** $f(x) = (x^2 + 4)(x - 2)(x + 2)$; the roots are ± 2.
c) $f(x) = (x + 2i)(x - 2i)(x - 2)(x + 2)$; the roots are $\pm 2, \pm 2i$.

d) (a) $f(x) = (x^2 - 5)(x^2 + 5)$; there are no rational roots.
(b) $f(x) = (x - \sqrt{5})(x + \sqrt{5})(x^2 + 5)$; the roots are $\pm\sqrt{5}$.
(c) $f(x) = (x - \sqrt{5})(x + \sqrt{5})(x - \sqrt{5}i)(x + \sqrt{5}i)$; the roots are $\pm\sqrt{5}, \pm i\sqrt{5}$.

7. a) $f(3) = 8060$ **b)** $f(1) = 1$ **c)** $f(-9) = f(2) = 6$

9. Since $(\mathbf{Z}, +, \cdot)$ is an integral domain but *not* a field, the Division Algorithm for Polynomials does not hold in $\mathbf{Z}[x]$.

11. $4; 6; p - 1$

13. In $\mathbf{Z}_4[x]$, $(2x + 1))(2x + 1) = 1$, so $(2x + 1)$ is a unit. This does not contradict Exercise 12 because $(\mathbf{Z}_4, +, \cdot)$ is not an integral domain.

15. Let $f(x) = \sum_{i=0}^{m} a_i x^i$ and $h(x) = \sum_{i=0}^{k} b_i x^i$, where $a_i \in R$ for $0 \le i \le m$, $b_i \in R$ for $0 \le i \le k$, and $m \le k$. Then $f(x) + h(x) = \sum_{i=0}^{k}(a_i + b_i)x^i$, where $a_{m+1} = a_{m+2} = \cdots = a_k = z$, the zero of R, so $G(f(x) + h(x)) = G(\sum_{i=0}^{k}(a_i + b_i)x^i) = \sum_{i=0}^{k} g(a_i + b_i)x^i = \sum_{i=0}^{k}[g(a_i) + g(b_i)]x^i = \sum_{i=0}^{k} g(a_i)x^i + \sum_{i=0}^{k} g(b_i)x^i = G(f(x)) + G(h(x))$. Also, $f(x)h(x) = \sum_{i=0}^{m+k} c_i x^i$, where $c_i = a_i b_0 + a_{i-1} b_1 + \cdots + a_1 b_{i-1} + a_0 b_i$, and

$$G(f(x)h(x)) = G(\sum_{i=0}^{m+k} c_i x^i) = \sum_{i=0}^{m+k} g(c_i)x^i.$$

Since $g(c_i) = g(a_i)g(b_0) + g(a_{i-1})g(b_1) + \cdots + g(a_1)g(b_{i-1}) + g(a_0)g(b_i)$, it follows that $\sum_{i=0}^{m+k} g(c_i)x^i = (\sum_{i=0}^{m} g(a_i)x^i)(\sum_{i=0}^{k} g(b_i)x^i) = G(f(x)) \cdot G(h(x))$.

Consequently, $G: R[x] \to S[x]$ is a ring homomorphism.

Section 17.2 – p. 698

1. a) $x^2 + 3x - 1$ is irreducible over $\mathbf{Q}$. Over $\mathbf{R}, \mathbf{C}$,

$$x^2 + 3x - 1 = [x - ((-3 + \sqrt{13})/2)][x - ((-3 - \sqrt{13})/2)].$$

b) $x^4 - 2$ is irreducible over $\mathbf{Q}$.
Over $\mathbf{R}$, $x^4 - 2 = (x - \sqrt[4]{2})(x + \sqrt[4]{2})(x^2 + \sqrt{2})$;
$x^4 - 2 = (x - \sqrt[4]{2})(x + \sqrt[4]{2})(x - \sqrt[4]{2}i)(x + \sqrt[4]{2}i)$ over $\mathbf{C}$.

c) $x^2 + x + 1 = (x + 2)(x + 2)$ over $\mathbf{Z}_3$. Over $\mathbf{Z}_5, x^2 + x + 1$ is irreducible;
$x^2 + x + 1 = (x + 5)(x + 3)$ over $\mathbf{Z}_7$.

d) $x^4 + x^3 + 1$ is irreducible over $\mathbf{Z}_2$.

e) $x^3 + x + 1$ is irreducible over $\mathbf{Z}_5$.

f) $x^3 + 3x^2 - x + 1$ is irreducible over $\mathbf{Z}_5$.

g) $x^4 + x^3 - x^2 + 3 = (x + 2)(x^3 + 6x^2 + x + 5)$ over $\mathbf{Z}_7$.

3. Degree 1: $x; x + 1$ Degree 2: $x^2 + x + 1$ Degree 3: $x^3 + x^2 + 1; x^3 + x + 1$

5. 7^5

7. a) Yes, since the coefficients of the polynomials are from a field.

b) $h(x)|f(x), g(x) \Rightarrow f(x) = h(x)u(x), g(x) = h(x)v(x)$, for some $u(x), v(x) \in F[x]$. $m(x) = s(x)f(x) + t(x)g(x)$ for some $s(x), t(x) \in F[x]$, so $m(x) = h(x)[s(x)u(x) + t(x)v(x)]$ and $h(x)|m(x)$.

c) If $m(x) \nmid f(x)$, then $f(x) = q(x)m(x) + r(x)$, where $0 < \deg r(x) < \deg m(x)$. $m(x) = s(x)f(x) + t(x)g(x)$ so $r(x) = f(x) - q(x)[s(x)f(x) + t(x)g(x)]$
$= (1 - q(x))s(x)f(x) - q(x)t(x)g(x)$, so $r(x) \in S$.
With $\deg r(x) < \deg m(x)$ we contradict the choice of $m(x)$. Hence $r(x) = 0$ and $m(x)|f(x)$.

9. a) The g.c.d. is $(x - 1) = (1/17)(x^5 - x^4 + x^3 + x^2 - x - 1)$
$- (1/17)(x^2 + x - 2)(x^3 - 2x^2 + 5x - 8)$.

b) The g.c.d. is $1 = (x + 1)(x^4 + x^3 + 1) + (x^3 + x^2 + x)(x^2 + x + 1)$.

c) The g.c.d. is $x^2 + 2x + 1 = (x^4 + 2x^2 + 2x + 2) + (x + 2)(2x^3 + 2x^2 + x + 1)$.

11. $a = 0, b = 0; a = 0, b = 1$

13. a) $f(x) \equiv f_1(x)(\bmod s(x)) \Rightarrow f(x) = f_1(x) + h(x)s(x)$, and $g(x) \equiv g_1(x)(\bmod s(x)) \Rightarrow g(x) = g_1(x) + k(x)s(x)$.
Hence $f(x) + g(x) = f_1(x) + g_1(x) + (h(x) + k(x))s(x)$, so $f(x) + g(x) \equiv f_1(x) + g_1(x)(\bmod s(x))$, and $f(x)g(x) = f_1(x)g_1(x) + (f_1(x)k(x) + g_1(x)h(x) + h(x)k(x)s(x))s(x)$, so $f(x)g(x) \equiv f_1(x)g_1(x)(\bmod s(x))$.

b) These properties follow from the corresponding properties for $F[x]$. For example, for the distributive law,

$$[f(x)]([g(x)] + [h(x)]) = [f(x)][g(x) + h(x)] = [f(x)(g(x) + h(x))]$$
$$= [f(x)g(x) + f(x)h(x)] = [f(x)g(x)] + [f(x)h(x)]$$
$$= [f(x)][g(x)] + [f(x)][h(x)].$$

c) If not, there exists $g(x) \in F[x]$, where $\deg g(x) > 0$ and $g(x)|f(x), s(x)$. But then $s(x)$ would be reducible.

d) A nonzero element of $F[x]/(s(x))$ has the form $[f(x)]$, where $f(x) \neq 0$ and $\deg f(x) < \deg s(x)$. With $f(x), s(x)$ relatively prime, there exist $r(x), t(x)$ with $1 = f(x)r(x) + s(x)t(x)$, so $1 \equiv f(x)r(x)(\bmod s(x))$ or $[1] = [f(x)][r(x)]$. Hence $[r(x)] = [f(x)]^{-1}$.

e) q^n

15. a) $[2x + 1]$ **b)** $[x + 1]$ **c)** $[2x + 1$ **d)** $[2x]$

17. a) p^n **b)** $\phi(p^n - 1)$

19. a) 6   **b)** 12   **c)** 12   **d)** $[m, n]$, the l.c.m. of m, n
e) 0   **f)** 0   **g)** 0

21. 101, 103, 107, 109, 113, 121, 125, 127, 128, 131, 137, 139, 149

Section 17.3 – p. 706

1. a)

1	2	3	4
2	1	4	3
4	3	2	1
3	4	1	2

b)

1	2	3	4
3	4	1	2
2	1	4	3
4	3	2	1

c)

1	3	4	2
4	2	1	3
3	1	2	4
2	4	3	1

3. $a_{ri}^{(k)} = a_{rj}^{(k)} \Rightarrow f_k f_r + f_i = f_k f_r + f_j \Rightarrow f_i = f_j \Rightarrow i = j$

5. L_3:

4	5	1	2	3
2	3	4	5	1
5	1	2	3	4
3	4	5	1	2
1	2	3	4	5

L_4:

5	1	2	3	4
4	5	1	2	3
3	4	5	1	2
2	3	4	5	1
1	2	3	4	5

In standard form the Latin squares L_i, $1 \le i \le 4$, become

L_1': 1 2 3 4 5 L_2': 1 2 3 4 5
         2 3 4 5 1   3 4 5 1 2
         3 4 5 1 2   5 1 2 3 4
         4 5 1 2 3   2 3 4 5 1
         5 1 2 3 4   4 5 1 2 3

L_3': 1 2 3 4 5 L_4': 1 2 3 4 5
         4 5 1 2 3   5 1 2 3 4
         2 3 4 5 1   4 5 1 2 3
         5 1 2 3 4   3 4 5 1 2
         3 4 5 1 2   2 3 4 5 1

7. Introduce a third factor, such as four types of transmission fluid or four types of tires.

Section 17.4 – p. 712

1.

Field	Number of points	Number of lines	Number of points on a line	Number of lines on a point
$GF(5)$	25	30	5	6
$GF(3^2)$	81	90	9	10
$GF(7)$	49	56	7	8
$GF(2^4)$	256	272	16	17
$GF(31)$	961	992	31	32

3. There are nine points and twelve lines. These lines fall into four parallel classes.
  (i) Slope of 0: $y = 0$; $y = 1$; $y = 2$
  (ii) Infinite slope: $x = 0$; $x = 1$; $x = 2$
  (iii) Slope 1: $y = x$; $y = x + 1$; $y = x + 2$
  (iv) Slope 2 (as shown in the figure): (1) $y = 2x$ (2) $y = 2x + 1$ (3) $y = 2x + 2$

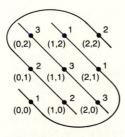

The Latin square corresponding to the fourth parallel class is

$$3 \ 1 \ 2$$
$$2 \ 3 \ 1$$
$$1 \ 2 \ 3$$

5. a) $y = 4x + 1$ **b)** $y = 3x + 10$ or $2x + 3y + 3 = 0$
  c) $y = 10x$ or $10y = 11x$
7. a) Vertical line: $x = c$. The line $y = mx + b$ intersects this vertical line at the unique point $(c, mc + b)$. As b takes on the values of F, there are no two column entries (on the line $x = c$) that are the same.

Horizontal line: $y = c$. The line $y = mx + b$ intersects this horizontal line at the unique point $(m^{-1}(c - b), c)$. As b takes on the values of F, no two row entries (on the line $y = c$) are the same.

b) Let L_i be the Latin square for the parallel class of slope m_i, $i = 1, 2$, $m_i \neq 0$, m_i finite. If an ordered pair (j, k) appears more than once when L_1, L_2 are superimposed, then there are two pairs of lines: (1) $y = m_1 x + b_1$, $y = m_2 x + b_2$ and (2) $y = m_1 x + b'_1$, $y = m_2 x + b'_2$ that both intersect at (j, k). But then $b_1 = k - m_1 j = b'_1$ and $b_2 = k - m_2 j = b'_2$.

Section 17.5 – p. 717

1. $v = 9$, $b = 12$, $r = 4$, $k = 3$, $\lambda = 1$

3. $\lambda = 2$

$$
\begin{array}{llll}
1\ 2\ 3\ 4 & 1\ 3\ 5\ 7 & 2\ 3\ 6\ 7 \\
1\ 2\ 5\ 6 & 1\ 4\ 6\ 7 & 2\ 5\ 4\ 7 & 3\ 4\ 5\ 6
\end{array}
$$

5. a) No   **b)** No

7. a) $\lambda(v - 1) = r(k - 1) = 2r \Rightarrow \lambda(v - 1)$ is even.
$\lambda v(v - 1) = vr(k - 1) = bk(k - 1) = b(3)(2) \Rightarrow 6 \mid \lambda v(v - 1)$

b) $(\lambda = 1)$ $6 \mid \lambda v(v - 1) \Rightarrow 6 \mid v(v - 1) \Rightarrow 3 \mid v(v - 1) \Rightarrow 3 \mid v$ or $3 \mid (v - 1)$
$\lambda(v - 1)$ even $\Rightarrow (v - 1)$ even $\Rightarrow v$ odd
$3 \mid v \Rightarrow v = 3t$, t odd $\Rightarrow v = 3(2s + 1) = 6s + 3$ and $v \equiv 3 \pmod 6$
$3 \mid (v - 1) \Rightarrow v - 1 = 3t$, t even $\Rightarrow v - 1 = 6x \Rightarrow v = 6x + 1$ and $v \equiv 1 \pmod 6$

9. $v = 9$, $r = 4$ **11. a)** $b = 21$ **b)** $r = 7$ **13. a)** 31   **b)** 8

15. a) $v = b = 31$; $r = k = 6$; $\lambda = 1$ **b)** $v = b = 57$; $r = k = 8$; $\lambda = 1$
c) $v = b = 73$; $r = k = 9$; $\lambda = 1$

Miscellaneous Exercises – p. 721

1. $n = 9$

3. a) There is no real number r such that $r^2 + 1 = 0$, so $x^2 + 1$ is irreducible in $\mathbf{R}[x]$.

b) Define f: $\mathbf{R}[x]/s(x) \to \mathbf{C}$ by $f(a + bx) = a + bi$. $a + bx = c + dx \Leftrightarrow a = c$, $b = d \Leftrightarrow a + bi = c + di$, so f is a one-to-one function. For any $a + bi \in \mathbf{C}$, with $a, b \in \mathbf{R}$, $a + bx \in \mathbf{R}[x]/(s(x))$ and $f(a + bx) = a + bi$, so f is onto. $f((a + bx) + (c + dx)) = f((a + c) + (b + d)x) = (a + c) + (b + d)i = (a + bi) + (c + di) = f(a + bx) + f(c + dx)$; $f((a + bx)(c + dx)) = f((ac + bdx^2) + (ad + bc)x) = f((ac - bd) + (ad + bc)x)$ (because $x^2 \equiv -1 \pmod{s(x)}$) $= (ac - bd) + (ad + bc)i = (a + bi)(c + di) = f(a + bx) \cdot f(c + dx)$. Hence f preserves addition and multiplication and is an isomorphism.

5. For all $a \in \mathbf{Z}_p$, $a^p = a$ [See part (a) of Exercise 8 at the end of Section 16.3.], so a is a root of $x^p - x$, and $x - a$ is a factor of $x^p - x$. Since $(\mathbf{Z}_p, +, \cdot)$ is a field, the polynomial $x^p - x$ can have at most p roots. Therefore $x^p - x = \prod_{a \in \mathbf{Z}_p} (x - a)$.

7. From part (a) of Miscellaneous Exercise 4 of Chapter 14, we know that for $z_1, z_2 \in \mathbf{C}$, $r \in \mathbf{R}$, and $n \in \mathbf{N}$, we have (1) $\overline{z_1 + z_2} = \overline{z_1} + \overline{z_2}$; (2) $\overline{r z_1} = \overline{r}\,\overline{z_1} = r \overline{z_1}$; and (3) $(\overline{z_1^n}) = (\overline{z_1})^n$. Since $f(x) \in \mathbf{R}[x]$, we may write $f(x) = a_m x^m + a_{m-1} x^{m-1} + \cdots + a_2 x^2 + a_1 x + a_0$, where $a_j \in \mathbf{R}$ for $0 \leq j \leq m$. Since $f(a + bi) = 0$ for $a + bi \in \mathbf{C}$, we find that $\sum_{j=0}^{m} a_j (a + bi)^j = 0$. Therefore, $0 = \overline{0} = \overline{\sum_{j=0}^{m} a_j (a + bi)^j} = \sum_{j=0}^{m} \overline{a_j (a + bi)^j} = \sum_{j=0}^{m} \overline{a_j}\,\overline{(a + bi)^j} = \sum_{j=0}^{m} a_j \overline{(a + bi)}^j = \sum_{j=0}^{m} a_j (a - bi)^j$ and $a - bi$ is a root of $f(x)$.

9. $\{1, 2, 4\}, \{2, 3, 5\}, \{4, 5, 7\}$

11. a) 9   **b)** 91

13. **a)** r 1's in each row; k 1's in each column

b) $A \cdot J_b$ is a $v \times b$ matrix whose (i,j)th entry is r, since there are r 1's in each row of A and every entry in J_b is 1. Hence $A \cdot J_b = rJ_{v \times b}$. Likewise, $J_v \cdot A$ is a $v \times b$ matrix whose (i,j)th entry is k, because there are k 1's in each column of A and every entry in J_v is 1. Hence $J_v \cdot A = k \cdot J_{v \times b}$.

c) The (i,j) entry in $A \cdot A^{tr}$ is obtained from the componentwise multiplication of rows i and j of A. If $i = j$, this results in the number of 1's in row i, which is r. For $i \neq j$, the number of 1's is the number of times x_i and x_j appear in the same block—which is given by λ. Hence $A \cdot A^{tr} = (r - \lambda)I_v + \lambda J_v$.

d)

$$
\begin{vmatrix}
r & \lambda & \lambda & \lambda & \cdots & \lambda \\
\lambda & r & \lambda & \lambda & \cdots & \lambda \\
\lambda & \lambda & r & \lambda & \cdots & \lambda \\
\lambda & \lambda & \lambda & r & \cdots & \lambda \\
\cdots & \cdots & \cdots & \cdots & \cdots & \cdots \\
\lambda & \lambda & \lambda & \lambda & \cdots & r
\end{vmatrix}
$$

$$
\overset{(1)}{=}
\begin{vmatrix}
r & \lambda - r & \lambda - r & \lambda - r & \cdots & \lambda - r \\
\lambda & r - \lambda & 0 & 0 & \cdots & 0 \\
\lambda & 0 & r - \lambda & 0 & \cdots & 0 \\
\lambda & 0 & 0 & r - \lambda & \cdots & 0 \\
\cdots & \cdots & \cdots & \cdots & \cdots & \cdots \\
\lambda & 0 & 0 & 0 & \cdots & r - \lambda
\end{vmatrix}
$$

$$
\overset{(2)}{=}
\begin{vmatrix}
r + (v - 1)\lambda & 0 & 0 & 0 & \cdots & 0 \\
\lambda & r - \lambda & 0 & 0 & \cdots & 0 \\
\lambda & 0 & r - \lambda & 0 & \cdots & 0 \\
\lambda & 0 & 0 & r - \lambda & \cdots & 0 \\
\cdots & \cdots & \cdots & \cdots & \cdots & \cdots \\
\lambda & 0 & 0 & 0 & \cdots & r - \lambda
\end{vmatrix}
$$

$$
= [r + (v - 1)\lambda](r - \lambda)^{v-1} = (r - \lambda)^{v-1}[r + r(k - 1)] = rk(r - \lambda)^{v-1}
$$

Key: (1) Multiply column 1 by -1 and add it to the other $v - 1$ columns.

   (2) Add rows 2 through v to row 1.

Index

NOTATION

SPECIAL SETS OF NUMBERS	$\mathbf{Z}$	the set of integers: $\{0, 1, -1, 2, -2, 3, -3, \ldots\}$
	$\mathbf{N}$	the set of nonnegative integers or natural numbers: $\{0, 1, 2, 3, \ldots\}$
	$\mathbf{Z}^+$	the set of positive integers: $\{1, 2, 3, \ldots\} = \{x \in \mathbf{Z} \mid x > 0\}$
	$\mathbf{Q}$	the set of rational numbers: $\{a/b \mid a, b \in \mathbf{Z}, b \neq 0\}$
	$\mathbf{Q}^+$	the set of positive rational numbers
	$\mathbf{Q}^*$	the set of nonzero rational numbers
	$\mathbf{R}$	the set of real numbers
	$\mathbf{R}^+$	the set of positive real numbers
	$\mathbf{R}^*$	the set of nonzero real numbers
	$\mathbf{C}$	the set of complex numbers: $\{x + yi \mid x, y \in \mathbf{R}, i^2 = -1\}$
	$\mathbf{C}^*$	the set of nonzero complex numbers
	$\mathbf{Z}_n$	$\{0, 1, 2, \ldots, n - 1\}$, for $n \in \mathbf{Z}^+$
	$[a, b]$	the closed interval from a to b: $\{x \in \mathbf{R} \mid a \leq x \leq b\}$
	(a, b)	the open interval from a to b: $\{x \in \mathbf{R} \mid a < x < b\}$
	$[a, b)$	a half-open interval from a to b: $\{x \in \mathbf{R} \mid a \leq x < b\}$
	$(a, b]$	a half-open interval from a to b: $\{x \in \mathbf{R} \mid a < x \leq b\}$
ALGEBRAIC STRUCTURES	$(R, +, \cdot)$	R is a ring with binary operations $+$ and $\cdot$
	$R[x]$	the ring of polynomials over ring R
	$(G, \circ)$	G is a group under the binary operation $\circ$
	S_n	the symmetric group on n symbols
	aH	a left coset of subgroup H (in group G): $\{ah \mid h \in H\}$
	$(\mathcal{B}, +, \cdot, {}^-, 0, 1)$	the Boolean algebra $\mathcal{B}$ with binary operations $+$ and $\cdot$, the unary operation $^-$, and identity elements 0 (for $+$) and 1 (for $\cdot$)
GRAPH THEORY	$G = (V, E)$	G is a graph with vertex set V and edge set E
	K_n	the complete graph on n vertices
	$\overline{G}$	the complement of graph G
	$\deg(v)$	the degree of vertex v (in an undirected graph G)
	$\deg^+(v)$	the out degree of vertex v (in a directed graph G)
	$\deg^-(v)$	the in degree of vertex v (in a directed graph G)
	$\kappa(G)$	the number of connected components of graph G
	$K_{m,n}$	the complete bipartite graph on $V = V_1 \cup V_2$ where $V_1 \cap V_2 = \emptyset$, $|V_1| = m$, $|V_2| = n$
	$\beta(G)$	the independence number of G
	$\chi(G)$	the chromatic number of G
	$P(G, \lambda)$	the chromatic polynomial of G
	$\gamma(G)$	the domination number of G
	$L(G)$	the line graph of G
	$T = (V, E)$	T is a tree with vertex set V and edge set E
	$N = (V, E)$	N is a (transport) network with vertex set V and edge set E